AF786834

Natural Bioactive Compounds and Human Health

Natural Bioactive Compounds and Human Health

Editors

Arunaksharan Narayanankutty
Ademola C. Famurewa
Eliza Oprea

MDPI

Basel • Beijing • Wuhan • Barcelona • Belgrade • Novi Sad • Cluj • Manchester

Editors
Arunaksharan Narayanankutty
PG and Research Department of Zoology
St. Joseph's College (Autonomous)
Calicut
India

Ademola C. Famurewa
Medical Biochemistry Department
Alex-Ekwueme Federal University
Abakaliki
Nigeria

Eliza Oprea
Department of Microbiology
University of Bucharest
Bucharest
Romania

Editorial Office
MDPI AG
Grosspeteranlage 5
4052 Basel, Switzerland

This is a reprint of articles from the Special Issue published online in the open access journal *Molecules* (ISSN 1420-3049) (available at: www.mdpi.com/journal/molecules/special_issues/Natural_Bioactive_Health).

For citation purposes, cite each article independently as indicated on the article page online and as indicated below:

Lastname, A.A.; Lastname, B.B. Article Title. *Journal Name* **Year**, *Volume Number*, Page Range.

ISBN 978-3-7258-1664-4 (Hbk)
ISBN 978-3-7258-1663-7 (PDF)
doi.org/10.3390/books978-3-7258-1663-7

© 2024 by the authors. Articles in this book are Open Access and distributed under the Creative Commons Attribution (CC BY) license. The book as a whole is distributed by MDPI under the terms and conditions of the Creative Commons Attribution-NonCommercial-NoDerivs (CC BY-NC-ND) license.

Contents

Preface

The intricate relationship between natural bioactive compounds and human health has long fascinated researchers and practitioners alike. In this comprehensive volume, "Natural Bioactive Compounds and Human Health", we delve into the multifaceted world of phytochemicals, antioxidants, polyphenols, and other biologically active substances derived from nature. These compounds, abundant in plants, fungi, marine organisms, and even certain animals, have been celebrated for their potential to promote well-being and prevent disease.

This reprint represents a collective effort to explore the diverse roles that these bioactive compounds play in maintaining human health across various cultures and lifestyles. From traditional remedies to cutting-edge research, each chapter uncovers new insights into the mechanisms through which these compounds exert their beneficial effects on the human body. We examine their antioxidant properties, anti-inflammatory actions, antimicrobial potential, and their impact on metabolic pathways and cellular function.

Furthermore, "Natural Bioactive Compounds and Human Health" highlights the growing body of evidence supporting their therapeutic applications in combating chronic diseases, such as cancer, cardiovascular disorders, neurodegenerative conditions, and metabolic syndrome. We also discuss their role in enhancing immune function, promoting gastrointestinal health, and even influencing mental well-being.

As editors of this volume, we have strived to gather contributions from experts across disciplines, fostering a dialogue that bridges the gap between traditional wisdom and modern scientific inquiry. Whether you are a researcher, healthcare professional, student, or simply curious about the profound intersection of nature and health, this reprint aims to serve as a valuable resource.

Ultimately, our hope is that this compilation not only informs but also inspires further exploration into the potential of natural bioactive compounds in order to optimize human health and well-being. By understanding and harnessing the power of these compounds, we move closer to unlocking nature's potential in order to enhance the quality of life for individuals worldwide.

Arunaksharan Narayanankutty, Ademola C. Famurewa, and Eliza Oprea
Editors

molecules

MDPI

Review

Flavonoids with Anti-Angiogenesis Function in Cancer

Qiang Wei * and Yi-han Zhang

School of Medicine, Anhui Xinhua University, 555 Wangjiang West Road, Hefei 230088, China; zhangyihan@axhu.edu.cn
* Correspondence: weiqiang@axhu.edu.cn; Tel.: +86-0551-6587-2736

Abstract: The formation of new blood vessels, known as angiogenesis, significantly impacts the development of multiple types of cancer. Consequently, researchers have focused on targeting this process to prevent and treat numerous disorders. However, most existing anti-angiogenic treatments rely on synthetic compounds and humanized monoclonal antibodies, often expensive or toxic, restricting patient access to these therapies. Hence, the pursuit of discovering new, affordable, less toxic, and efficient anti-angiogenic compounds is imperative. Numerous studies propose that natural plant-derived products exhibit these sought-after characteristics. The objective of this review is to delve into the anti-angiogenic properties exhibited by naturally derived flavonoids from plants, along with their underlying molecular mechanisms of action. Additionally, we summarize the structure, classification, and the relationship between flavonoids with their signaling pathways in plants as anti-angiogenic agents, including main HIF-1α/VEGF/VEGFR2/PI3K/AKT, Wnt/β-catenin, JNK1/STAT3, and MAPK/AP-1 pathways. Nonetheless, further research and innovative approaches are required to enhance their bioavailability for clinical application.

Keywords: flavonoids; plants; anti-angiogenesis; cancer; mechanism; signaling pathway

Citation: Wei, Q.; Zhang, Y.-h. Flavonoids with Anti-Angiogenesis Function in Cancer. *Molecules* **2024**, *29*, 1570. https://doi.org/10.3390/molecules29071570

Academic Editors: Arunaksharan Narayanankutty, Ademola C. Famurewa and Eliza Oprea

Received: 10 March 2024
Revised: 23 March 2024
Accepted: 29 March 2024
Published: 31 March 2024

Copyright: © 2024 by the authors. Licensee MDPI, Basel, Switzerland. This article is an open access article distributed under the terms and conditions of the Creative Commons Attribution (CC BY) license (https://creativecommons.org/licenses/by/4.0/).

1. Introduction

The term "flavonoid" finds its roots in the Latin word "flavus", which signifies the color yellow, as these compounds often manifest as secondary metabolites in a diverse range of plants [1]. Since the human body cannot synthesize flavonoids on its own, it is necessary to consume fruits, vegetables, or extracts from medicinal plants as a nutritional supplement. Many scientific researchers have showcased the pivotal role played by flavonoids in reducing the risk of cancer development viavarious mechanisms [2–4]. These mechanisms include safeguarding against DNA damage [5], inducing autophagy and apoptosis in tumor cells [6], inhibiting tumor cell invasion and metastasis [7], and suppressing angiogenesis [8], as well as modulating xenobiotic enzymes and antioxidant status [9].

The establishment of new blood vessels heavily accelerates the growth and metastasis of tumor cells, which provide essential blood support [10]. Consequently, angiogenesis inhibitors can impede tumor vascularization by inhibiting, regressing, or normalizing existing blood vessels. These inhibitors offer novel therapeutic approaches to hinder tumor growth, including the use of anti-angiogenic drugs comprised predominantly of antibodies (such as bevacizumab) and small molecule drugs (like sunitinib, sorafenib, and vandetanib) [11]. In general, tyrosine kinase (TK) receptors and their associated growth factors play a crucial role in stimulating endothelial cells to form new blood vessels. Thus, targeting these growth factors, their receptors, and related downstream signaling pathways has emerged as a promising strategy for drug discovery in angiogenesis inhibition [12]. For instance, anti-angiogenic antibodies selectively bind to VEGF and prevent its interaction with corresponding receptors [13]. On the other hand, small molecule agents act as tyrosine kinase inhibitors (TKIs) and directly target the VEGF receptor [14]. Additionally, cysteine proteases, like cathepsins L and B, have been linked to tumor growth and metastasis, playing roles in both intracellular proteolysis and extracellular matrix remodeling. Inhibition of cathepsins B or L also impedes the formation of new vascular structures [15].

In the quest for future cancer therapy drugs that are more potent, less toxic, and safer the exploration of angiogenesis inhibitors becomes imperative. This review article delves into the crucial structural classifications and molecular mechanisms of 47 flavonoids from natural products, elucidating their antiangiogenic properties. It provides a comprehensive analysis of the distinct signal pathways through which these substances exert their antiangiogenic effects, offering a fresh perspective in this domain.

2. Flavonoids

Figure 1 illustrates flavonoids, a compound class distinguished by the attachment of two benzene rings to a central oxygenated ring C [16]. As depicted in Figure 1, research has identified six subclasses of flavonoids with antiangiogenic potential: flavones, flavanones, flavonols, flavanols, isoflavones, and chalcones. Flavones are characterized by a C-2 connected to ring B, highlighting a double bond of C-2 to C-3 [17]. Flavanones, also known as dihydroflavones, possess only a saturated, oxidized ring C. Much like flavones, flavonols possess a hydroxyl group linked to C-3 [18]. Flavanols share characteristics with flavones but have an unoxidized, saturated ring C with a hydroxyl group at C-3. Isoflavones are distinguished by a ring B attached to C-3 [16]. Finally, chalcones are the ring C-opening isomers of dihydroflavones and lack an oxygen-containing ring [16].

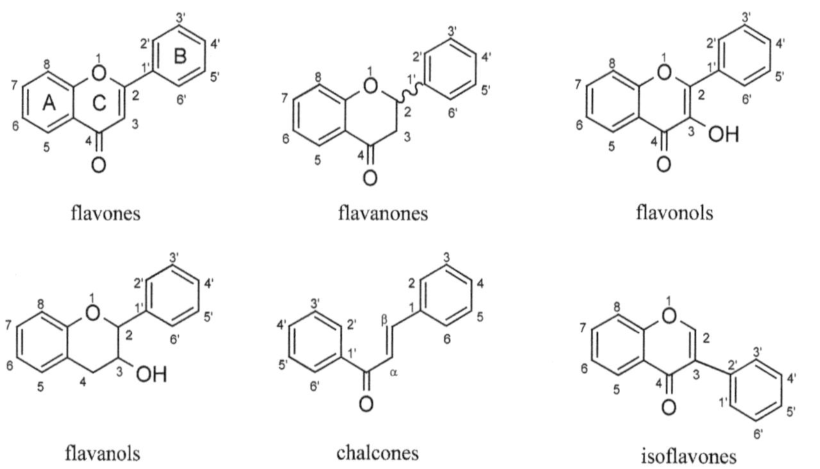

Figure 1. The partial structure of flavonoids.

In order to explore the relationship between flavonoids and their anti-tumor vascular effects, we searched for the literature recorded in Google Scholar, Web of Science, and Pub Med to classify, summarize, and analyze the relation contents, including chemical structure of every flavonoid compound, its anti-angiogenesis mechanism in cancer, relational cancer models or cell lines, and the rise or fall of related indicators. Based on the study of pathological mechanism of cancer angiogenesis and flavonoids' pharmacological action, we try to give flavonoids' structure, classification, and signaling pathways as anti-angiogenic agents.

3. Results

3.1. Chemical Components

There are a total of 47 flavonoids related to their antiangiogenic effects, and their classification and composition are shown in Figure 2 and Table 1, and their structure is shown in Figure 3. A more detailed table of statistics on antiangiogenic molecular mechanisms of different flavonoids can be found in Supplementary Materials.

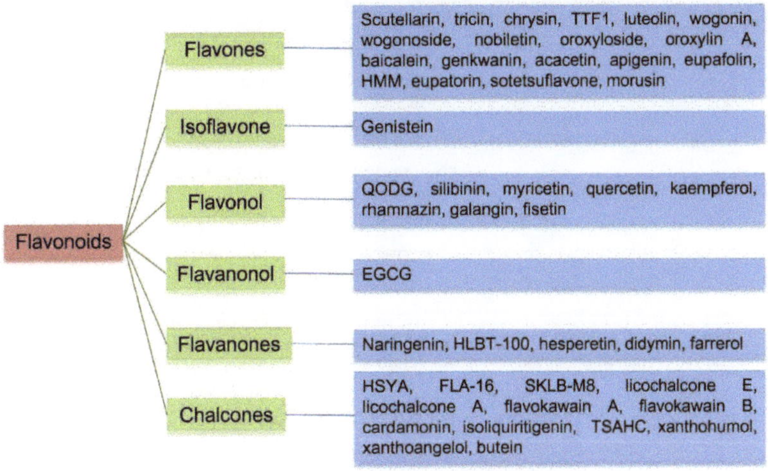

Figure 2. Flow diagram showing detailed classification of flavonoids.

3.2. *Anti-Angiogenic Flavonoids in Cancer*

3.2.1. Flavones

Scutellarin (4′,5,6-trihydroxy-7-[(β-D-glucopyranuronosyl)oxy]flavone) is a compound found in Erigeron breviscapus. The VEGF signaling pathway regulates angiogenesis in endothelial cells viaephrinb2 and ephb4 [19]. In the context of colorectal cancer, scutellarin inhibits angiogenesis induced by cancer cells, migration of HUVECs, tube formation of HUVECs, and microvessel formation in chick embryo CAM. This inhibition is achieved by suppressing the expression of ephrinb2 [20]. Additionally, scutellarin may control the level of the transcription factor AP-1 to inhibit tumor angiogenesis in oral squamous cell carcinoma(OSCC), which is accomplished by decreasing the MMP-9 and MMP-2 levels, alongside decreasing integrin αvβ6 levels regarding human tongue cancer SAS cells. Notably, overexpression of integrin αvβ6 activates TGF-β, causing the generation of MMPs, cell migration, and survival [21]. Considering the crucial function of EMT activation in carcinoma metastasis, cell migration, invasion, and adhesion, scutellarin's anti-angiogenic effects have been investigated in relation to its suppression of EMT, achieved by suppressing the PI3K/Akt/mTOR pathway [22], as seen in Table 1.

Table 1. Antiangiogenic molecular mechanisms of different flavonoids.

Type	Compounds (No.)	Mechanism	References
Flavone	Scutellarin (**1**)	Targeting ephrinb2 signaling; possibly regulating transcription factor AP-1; inhibiting PI3K/Akt/mTOR pathway	[20–22]
	Tricin (**2**)	Reducing ROS; inhibiting HIF-1αaccumulation	[23]
	Chrysin (**3**)	Regulating PI3K/Akt signaling; downregulating JAK1/STAT3 pathway and VEGF/VEGFR2 expression	[24–26]
	TTF1 (**4**)	Downregulating VEGF, KDR, bFGF, HIF-1α, and COX-2	[27]
	Luteonin (**5**)	Downregulating AEG-1, MMP-2, MMP-9, HIF-1α, and STAT3; stimulating immune response; inhibiting AKT/ERK/mTOR/P70S6K/MMPs pathway or PI3K/Akt/mTOR pathway; elevating JNK phosphorylation; inhibiting NF-κB-DNA binding activity; modulating IL-6/STAT3 pathway	[28–36]

Table 1. *Cont.*

Type	Compounds (No.)	Mechanism	References
Flavone	Wogonin (**6**)	Degrading HIF-1α protein; modulating c-Myc/HIF-1α/VEGF signaling axis; inhibiting VEGFR2phosphorylation or PI3K/Akt/NF-κB signaling	[37–40]
	Wogonoside (**7**)	Suppressing Wnt/β-catenin pathway	[41]
	Nobiletin (**8**)	Inhibiting VEGF- and bFGF-induced signaling; activating caspase pathway; inhibiting Akt phosphorylation; mediating Src/FAK/STAT3 signaling	[42–44]
	Oroxylin A (**9**)	Blocking KDR/Flk-1 phosphorylation	[45]
	Oroxyloside (**10**)	Inhibiting Akt/MAPK/NF-κB pathway	[46]
	Baicalein (**11**)	Partly mediating VEGF and FGFR-2 signalling; regulating p53/Rb signaling and TRAF6-mediated TLR4 pathway;inhibiting VEGF, HIF-1α, cMyc, NFκB, MMP-2, ROS, and PI 3K/Akt pathway, as well as ERK1/2 and p38 MAPK phospho-activation	[47–52]
	Genkwanin (**12**)	Inhibiting invasion and tube formation	[53]
	Acacetin (**13**)	Inhibiting AKT/HIF-1α pathway or STAT-VEGF axis	[54,55]
	Apigenin (**14**)	Blocking the ERK and ERK 1/2 survival signaling or IGF-I/IGFBP-3 signaling; regulating PI3K/AKT/p70S6K1 and HDM2/p53 pathways; downregulating HIF-1α, GLUT-1, and VEGF	[56–59]
	Eupafolin (**15**)	Blocking VEGFR2activation, ERK1/2, and Akt phosphorylation	[60]
	HMM (**16**)	Inhibiting cathepsins B and L	[61]
	Eupatorin (**17**)	Blocking the phospho-Akt pathway and cell cycle	[62]
	Sotetsuflavone (**18**)	Modulating PI3K/AKT and TNF-α/NF-κB pathways; inhibiting TGF-β, STAT3,and β-catenin; increasing endostatin and ZO-1	[63,64]
	Morusin (**19**)	Attenuating IL-6/STAT3 signaling; inhibiting VEGF and COX-2genes	[65,66]
Isoflavone	Genistein (**20**)	Suppressing autocrine and paracrine signalings, hypoxic activation of HIF-1, MMP-1, VEGF, PDGF-A, TF, uPA, MMP-2, and MMP-9; upregulating PAI-1, endostatin, angiostatin, TSP-1, CTGF, and CTAP; downregulating type IV collagenase, uPAR, protease M, PAR-2, VEGF, VEGFR, TGF-b, BPGF, LPA, TSP, JNK, and p38 activation; modulating TIMP-1 and -2 and PAI-1	[67–72]
Flavonol	Quercetin (**21**)	Regulating AKT/mTOR/P70S6K signaling; inhibiting NF-κB and MMP-2/MMP-9 signalings, the H-ras protein synthesis, VEGF and bFGF, STAT3 tyrosine phosphorylation, and Akt phosphorylation, NF-κB activity, eNOS, and early M-phase cell cyclearrest, p300 signaling and the binding of multiple transactivators to COX-2 promoter; upregulating TSP-1	[73–80]
	QODG (**22**)	Suppressing VEGFR2-mediated signaling	[81]
	Silibinin (**23**)	Downregulating survivin, VEGF, VEGFR-2, bFGF, NOS, COX, HIF-1α; increasing p53; inhibiting Akt and NF-kB signaling, MMP-2 secretion, PI3K/Akt signaling or Raf/MEK/ERK pathway, VEGF and endothelial cell growth, or NF-κB signaling; inducingapoptosis; upregulating VEGFR-1	[82–90]

Table 1. *Cont.*

Type	Compounds (No.)	Mechanism	References
Flavonol	Myricetin (**24**)	Suppressing PI-3 kinase activity or PI3K/Akt/mTOR signaling; attenuating Akt/p70S6K phosphorylation; modulating Akt/p70S6K/HIF-1α/VEGF and p21/HIF-1α/VEGF pathways	[91–93]
	Kaempferol (**25**)	Regulating ERK-NF-κB-cMyc-p21-VEGF and VEGFR2 pathways, ERK/p38 MAPK and PI3K/Akt/mTOR pathways, Akt/HIF and ESRRA pathways; inhibiting VEGFR2 expression, VEGF and FGF pathways or PI3K/AKT, MEK, and ERK pathways	[94–98]
	Rhamnazin (**26**)	Regulating VEGF and PEDF; downregulating the VEGFR2/STAT3/MAPK/Akt pathway	[99,100]
	Galangin (**27**)	Downregulating CD44 and VEGF; modulating Akt/p70S6K/HIF-1α/VEGF pathway	[92,101]
	Fisetin (**28**)	Inhibiting MMPs, MMP-8, and MMP-13, p38 MAPK-dependent NF-κB pathway, NF-κB, MAPK, Wnt, Akt, and mTOR; G1 phrase-G2/M arrest; downregulating cyclin D1, survivin, VEGF, eNOS, iNOS, Bcl-2; inducing p53 and p21, Bax expression and cleavage of caspases-3 and -7, and PARP; regulated by HO-1 viatranscription factor Nrf2; inactivating PI3K/Akt and JNK pathways; diminishing NF-κB and AP-1 DNA-binding activities	[102–112]
Flavanonol	EGCG (**29**)	Inhibiting VEGF-induced VEGFR2 signaling or NF-κB and ERK1/2 signalings, endoglin/pSmad1 signaling, DNA synthesis, cell proliferation, and signal transduction pathway, and PI3K/AKT/mTOR/HIF1α pathway; downregulating VEGF, uPA, angiopoietin 1 and 2, VEGFR-1 and -2, ERK-1 and -2, MMP-2 and -9, HIF-1α, and CXCL12; suppressing HIF-1 and VEGF/VEGFR axis activation, VEGF, IL-8, and CD31 and Akt activation, NF-κB, and MT1-MMP; increasing endostatin and TIMP1; modulating the genes transcription	[113–133]
Flavanones	Naringenin (**30**)	Mediating ERRα/VEGF/KDR signaling	[134]
	HLBT-001 (**31**)	Not mentioned	[135]
	Hesperetin (**32**)	Modulating PI3K/AKT, ERK and p38 MAPK signalings	[136]
		Inhibiting angiogenic growth factors and COX-2 mRNA expression	[137]
	Didymin (**33**)	Preventing NF-κB and expression of adhesion molecules	[138]
	Farrerol (**34**)	Downregulating Akt/mTOR, Erk and Jak2/Stat3 signalings	[139]
Chalcone	HSYA (**35**)	Inhibiting tumor vascularization; blocking ERK/MAPK and NF-κB signaling or p38 MAPK phosphorylation; downregulating VEGF, bFGF and MMP-9	[140–144]
	FLA-16 (**36**)	Modulating PI3K/Akt signaling through the inhibition of CYP4A	[145]
	SKLB-M8 (**37**)	Decreasing ERK1/2 phosphorylation	[146]
	LicA (**38**)	Blocking VEGF/VEGFR-2 signaling	[147]
	LicE (**39**)	Decreasing VEGFR2, VEGF-A, HIF-1α, COX-2 and iNOS	[148]
	FKB (**40**)	Reducing angiogenin, F3, SDF-1, serpin F1, and TSP-2; suppressing the formation of vessels	[149,150]
	FKA (**41**)	Inhibiting new blood vessels; downregulating the androgen receptor	[151,152]

Table 1. *Cont.*

Type	Compounds (No.)	Mechanism	References
Chalcone	Cardamonin (**42**)	Inhibiting HIF-α and VEGF; regulating ERK1/2 and AKT signaling; downregulating miR-21	[153–155]
	Isoliquiritigenin (**43**)	Hampering MAPK signaling of JNK and p38, VEGF/VEGFR2 pathway, ERK1/2 and VEGF; promoting PEDF expression or JNK	[156–160]
	TSAHC (**44**)	Disturbing protein–protein interaction between TM4SF5 and other membrane receptors	[161]
	Xanthohumol (**45**)	Mitigating NF-κB activity, AMPK and AKT/mTOR pathways, and Akt/NF-kB signaling; modulating NF-κB signalling; inhibiting ICAM-1, MMP-9, VEGF, and NF-κB activity	[113–116,162–164]
	Xanthoangelol (**46**)	Inhibiting tube formationand the binding of VEGF to vascular endothelial cells	[117]
	Butein (**47**)	Targeting the AKT/mTOR translation-dependent signaling; inhibiting NF-κB signaling	[118,119]

Tricin, namely 4′,5,7-trihydroxy-3′,5′-dimethoxyflavone, is a substance found in some foods such as rice and wheat [120]. Inhibitory effects of tricin on the proliferation and invasion triggered by VEGF, as well as HUVEC tube assembly and angiogenesis of CAM. These effects are achieved by downregulating the signal transduction of VEGFR2, partially through diminishing the generation of ROS in endothelial cells. Additionally, tricin inhibits the VEGF expression by preventing the accumulation of HIF-1α in tumor cells [23].

Chrysin (5,7-dihydroxyflavone) is a bioactive component available in diverse fruits, vegetables, and mushrooms [165]. Dampened activities of chrysin on angiogenesis and the HIF-1α level in DU145 prostatic carcinoma cells is achieved viathe PI3K/Akt pathway. Boosting the oxygen-controlled degradation domain prolyl hydroxylation, chrysin triggers increased ubiquitin tagging and degradation of HIF-1a, simultaneously interfering with a HIF-1α interacting a heat shock protein 90 (HSP90) [24]. Another study reveals that the anti-angiogenic effects of chrysin are associated with the inhibition of sIL-6R, gp130, phosphorylated JAK1, STAT3, and VEGF expression in HUVECs and CAM assay [25]. Flavonoids are known to inhibit inflammation-induced angiogenesis, and chrysin specifically inhibits lipopolysaccharide (LPS)-induced CAM neovascular density, downregulating VEGF/VEGFR2 expression, and disrupting an IL-6/IL-6R self-regulatory loop using HUVECs [26].

Isolated from Sorbariasorbifolia, TTF1, also known as 5,2′,4′-Trihydroxy-6,7,5′-trimethoxyflavone, has showcased its capacity to inhibit tumor angiogenesis elicited by HepG-2 cells via reducing the VEGF, KDR, bFGF, COX-2, and HIF-1α levels of RNA and protein in key factors regulating angiogenesis [27].

Luteolin (3′,4′,5,7-tetrahydroxyflavone), a compound commonly existing in medicative plants or veggies [121], been demonstrated to exhibit potent anti-angiogenic effects on the chick chorioallantoic membrane and to possess anti-invasive activity against breast cancer cells. It achieves this by downregulating the expression of various angiogenesis-related factors, including dampening the MMP-2 and astrocyte elevated gene 1 (AEG-1) expression [28]. Furthermore, it suppresses angiogenesis through the decrease expression of VEGF-A and MMP-9 [29]. Similar effects on tumor angiogenesis have been observed with luteolin, as it reduces phosphorylated VEGFR2 induced by VEGF-A and inhibits subsequent proteins including mTOR, AKT, P70S6K, ERK, MMP-2, and MMP-9 [30]. In addition, luteolin exerts its action of anti-angiogenesis by diminishing VEGF secretion through mitigating VEGF mRNA production, which is regulated by inhibiting the NF-κB transcription activity [31]. Gas6 initiates the activation of Axl receptor tyrosine kinase (Axl), which, in turn, drives the Gas6/Axl pathway. This pathway not only fosters growth, movement, infiltration, and tubulogenesis but also further governs the PI3K/Akt/mTOR pathway [32].

Luteolin also inhibits the Gas6/Axl pathway and its downstream PI3K/Akt/mTOR pathway in HMECs-1 in vitro, thus exerting an anti-angiogenic effect [33]. The activated STAT3 protein contributes to cell growth, differentiation, and the upregulation of VEGF expression, leading to induction of tumor angiogenesis [34]. Luteolin exhibits anti-angiogenic effects by inhibiting the HIF-1α and phosphorylated STAT3 (p-STAT3) pathways, particularly within the alternatively activated TAMs [35]. Additionally, luteolin exerts inhibitory effects viathe IL-6/STAT3 pathway, which modulates the level of IL-6 receptor (IL-6Rα), resulting in decreased MMP-2 levels and increased the protein expression of suppressor of cytokine signaling (SOCS3) [36]. SOCS3, regulated by STAT3, negatively regulates JAK activation and IL-6-mediated signaling, thereby impacting the JAK/STAT signaling [122].

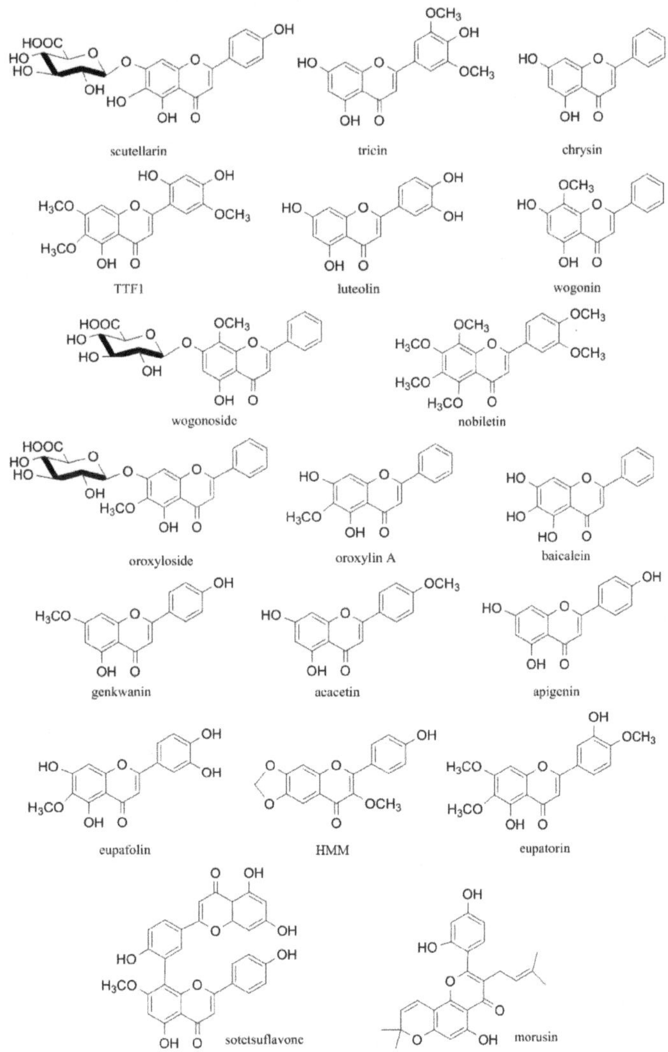

(**A**): Flavones

Figure 3. *Cont.*

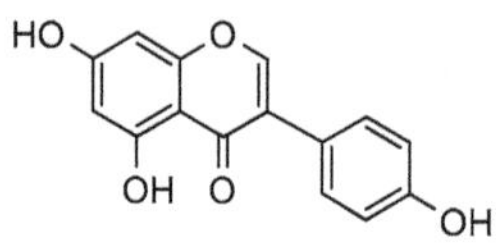

genistein

(**B**): Isoflavones

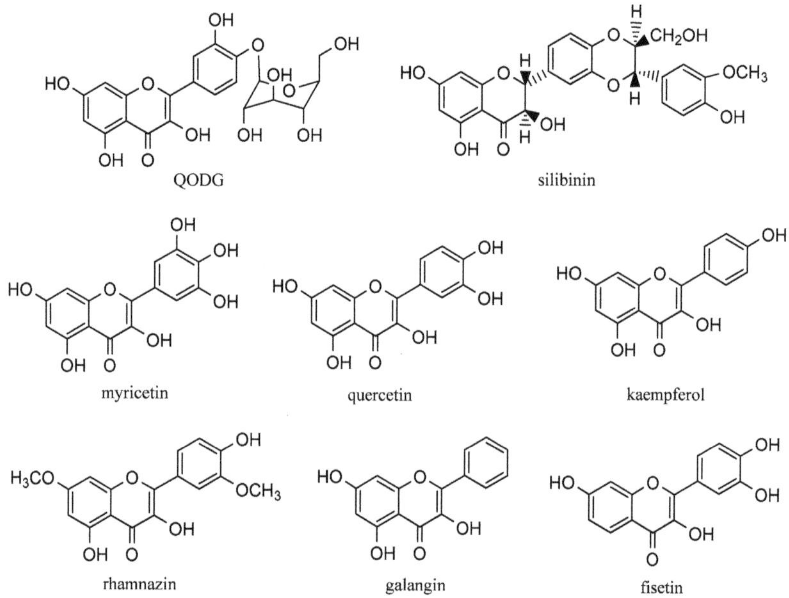

(**C**): Flavonols

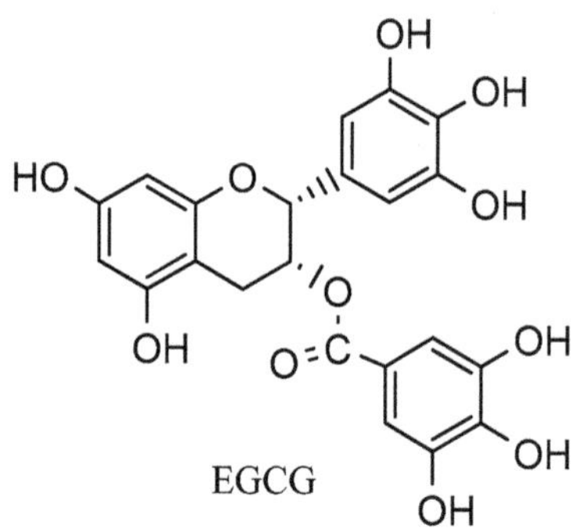

(**D**): Flavanonols

Figure 3. *Cont.*

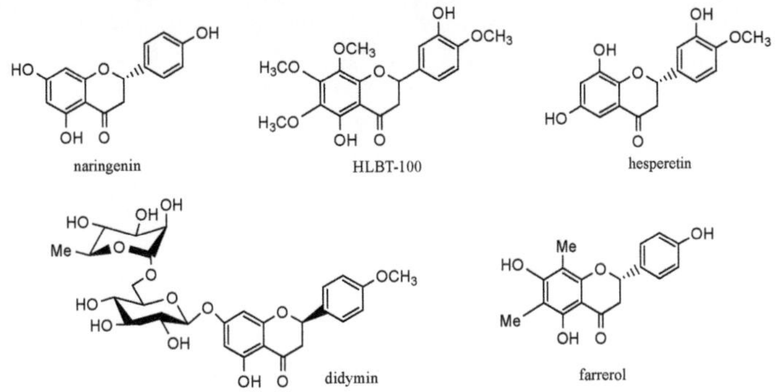

(E): Flavanones

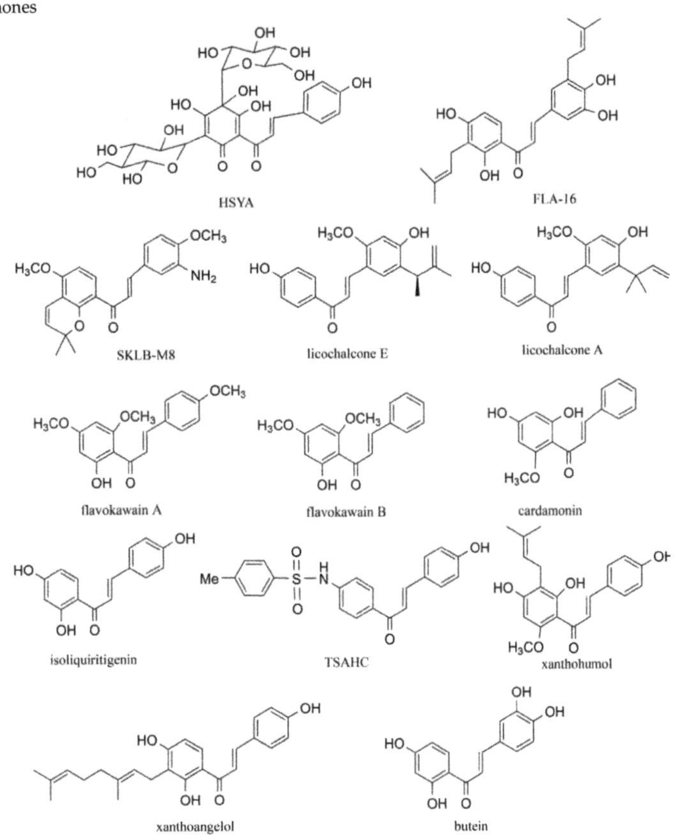

(F): Chalcones

Figure 3. Structure of anti-angiongenic flavonoids in cancer.

Obtained from Scutellariabaicalensis, wogonin (5,7-dihydroxy-8-methoxyflavone) has demonstrated an ability to suppress angiogenesis. It achieves this by decreasing the HIF-1α level and increasing PHD1, 2, 3 expression, as well as VHL E3 ubiquitin ligase. This, in turn, reduces the heatshock protein 90 (Hsp90), client proteins such as EGFR, Cdk4, and survivin, leading to their degradation in the proteasome. Additionally, wogonin hinders the binding between Hsp90 and HIF-1α, resulting in the downstream reduction inVEGF secretion [37]. Angiogenesis is closely linked to the progression of multiple myeloma

(MM). Wogonin is shown to suppress MM-stimulated angiogenesis and reduce levels of secreted VEGF, platelet-derived growth factor (PDGF), and bFGF by way of the c-Myc/HIF-1α/VEGF pathway [38]. Further examination of the signaling pathway reveals that wogonin hinders HUVEC migration and tube assembly triggered by VEGF and blocks VEGF-induced tyrosine phosphorylation of VEGFR2. Moreover, significant decreases inAKT, ERK, and p38 phosphorylation induced by VEGF are observed [39]. In addition, wogonin manifests inhibitory action on H_2O_2-induced angiogenesis in HUVECs based on the suppression of the signaling pathway of PI3K/Akt and NF-κB [40]. Another flavone glycoside derived from Scutellariabaicalensis Georgi, called wogonoside (wogonin-7-*O*-glucuronide), has also exhibited anti-angiogenic activity by suppressing the Wnt/β-catenin signaling in MCF-7 cells. This results in reduced intracellular levels of Wnt3a, an increase in GSK-3β and AXIN expression, and enhanced β-catenin phosphorylation, facilitating its degradation by the proteasome [41].

Nobiletin (5,6,7,8,3′,4′-hexamethoxyflavone), a polymethoxyflavonoid present in specific citrus fruits [123], has demonstrated the ability to hinder various endothelial cell functions and angiogenesis by thwarting the phosphorylated ERK1/2 and JNK induced by FGF in HUVECs [42]. Moreover, nobiletin has shown the capability to hinder cancerous growth and angiogenic processes by reducing AKT activation, consequently inhibiting VEGF, NF-κB, and HIF-1α. By downregulating AKT, HIF-1α secretion is inhibited, leading to the suppression of VEGF in ovarian cancer cells [43]. Additionally, nobiletin has been discovered to impede cancer angiogenesis in mammary carcinoma showing estrogen receptor positivity by restraining the signaling mediated by Src, FAK, and STAT3 while fostering PXN gene expression [44].

Oroxylin A (5,7-dihydroxy-6-methoxyflavone) is a prominent bioactive flavonone found in the roots of Scutellariabaicalensis. It effectively exerts antiangiogenic activity by significantly inhibiting the phosphorylated VEGFR2 induced by VEGF as well as its subsequent signaling factors such as Akt, p38 MAPK, and ERK1/2 [45]. Oroxyloside, known as Oroxylin A 7-*O*-glucuronide, one of the primary metabolites of oroxylin A, also exhibits anti-angiogenic properties. It suppresses the autophosphorylation of VEGFR2/Flk-1 and upregulates the expression of R-Ras and E-cadherin by inhibiting the downstream Akt/MAPK/NF-κB pathways, leading to a decrease inthe process that proteins translocate into the nucleus and NF-κB's capacity to bind DNA [46].

Derived from Scutellariabaicalensis Georgi, baicalein, also called 5,6,7-trihydroxyflavone [124]. Its anti-angiogenesis activity is attributed to the VEGF suppression and FGFR-2, and the downregulation of corresponding factors such as FGF2, VEGF, MMP1, TEK, and ANGPT1 [47]. Baicalein also effectively suppresses MMP-2 activity associated with cells, thereby inhibiting migration, proliferation, and in vitro capillary formation in bFGF-treated HUVECs [48]. Furthermore, baicalein has exhibited the ability to hinder angiogenesis via multiple pathways, reducing expression levels of HIF-α, VEGF, NF-κB, and c-Myc in ovarian cancer cells [49], decreases the hypoxia-induced genes expression of hypoxia-responsive COX-2, VEGF, and iNOS by hindering the PI3K/Akt and ROS signaling in BV2 microglia [50], and attenuates the phosphorylation of VEGF2 and ERK in baicalein-treated HUVECs by lowering G1-related proteins expression, leading to stalling the cell cycle at the G1/S boundary, and affecting the p53/Rb signaling [51]. Additionally, this leads to a notable reduction in TRAF6 triggered by LPS and the phosphorylation of ERK, AKT, and p38, further inhibiting HUVECs proliferation, migration, and tube-like structure generation [52]. Overall, baicalein demonstrates diverse anti-angiogenesis properties and regulates multiple molecular pathways to impede angiogenesis in various contexts.

Genkwanin (4′,5-dihydroxy-7-methoxyflavone) is a compound found in various plants, including Alnus Glutinosa (Betulaceae) [125], Leonurus turkestanicus [126], and Thymus taxa [127]. Studies have shown that genkwanin exhibits a stronger inhibitory effect on angiogenesis stimulated by VEGF in HUVECs compared to apigenin, showing efficiently the inhibition of invasion and tube formation without affecting endothelial cell viability [53].

Acacetin (5,7-dihydroxy-4′-methoxyflavone), isolated from various plants, seeds, and flowers, has demonstrated the ability to suppress VEGF level via activating AKT and degrading subsequent its downstream target, such as HIF-1α protein [54]. This inhibitory effect is also mediated by blocking the STAT/VEGF pathway in both tumor cells and endothelial cells, leading to a decrease in phosphorylated STAT-1 and phosphorylated STAT-3, along with a downregulation of angiogenesis-promoting factors like VEGF, bFGF, iNOS, eNOS, and MMP-2 in HUVECs [55].

Apigenin (5,7,4′-trihydroxyflavone) is a widely encountered flavone from various fruits and veggies such as parsley, grapes, and apples [128]. It has demonstrated significant anti-angiogenic effects, potentially achieved by inhibiting vascular tube assembly and inducing apoptotic process in endothelial cells assembling tubes. This mechanism entails the suppression of ERK 1/2-mediated cell survival pathway and accompanied by the downregulation of cell adhesion molecules associated with angiogenesis, involving vascular PECAM-1and E-cadherin. Apigenin additionally triggers caspase-3 activation, leading to the cleavage of PARP and lamin A/C [56]. In addition, apigenin exerts its anti-angiogenic activity by suppressing VEGF and HIF-1α levels viathe PI3K/AKT/p70S6K1 and human murine double minute 2 (HDM2)/p53 pathways [57]. It downregulates the levels of HIF-1α, GLUT-1, and VEGF in pancreatic carcinoma [58] and inhibits the level of MMP-2, MMP-9, uPA, and VEGF in TRAMP mouse by downregulating IGF-I/IGFBP-3 signaling [59].

Eupafolin (5,7,3′,4′-tetrahydroxy-6-methoxy-flavone) is a flavone compound that can be found in various plants including Artemisia princeps Pampanini [129], Eupatorium litoralle [130], and Phyla nodiflora [131]. Studies have shown that eupafolin hampers tumor cell growth, movement, and tubular structure development induced by VEGF bycurbing the VEGFR2 activation, leading to reduced phosphorylation of ERK1/2 and Akt signaling pathways [60].

Cathepsin B is an enzyme that is predominantly found in microvascular endothelial cells surrounding human glioblastoma and prostate carcinomas [132]. It significantly influences the MMPs and TIMPs balance. By inactivating TIMP-2 and TIMP-3, cathepsin B shifts this balance, thereby promoting an angiogenic environment [133]. Another cathepsin enzyme, cathepsin L, is upregulated in various human cancers. It interacts with the glycosaminoglycan components of proteoglycans present on cell surfaces and extracellular matrices, leading to enhanced tumor metastasis and angiogenesis [166]. A novel flavone called 4′-Hydroxy-6,7-methylenedioxy-3-methoxyflavone(HMM), isolated from Dulaciaegleri, has demonstrated anti-angiogenic effects by specifically inhibiting the cysteine proteases cathepsins B and L [61].

Eupatorin (3′-hydroxy-5,6,7,4′-tetramethoxyflavone), which is obtained from Orthosiphon stamineus [167], has been found to inhibit the emergence of neovascularization and the process of angiogenesis in mouse aorta. This effect is achieved by blocking the phospho-Akt pathway in human breast cancer cells. Additionally, eupatorin demonstrates a correlation with the downregulation of Bcl2L11, VEGFA, and HIF1A genes, further supporting its anti-angiogenic properties [62].

Sotetsuflavone (7-*O*-methylamentoflavone), a biflavone obtained from Selaginella denticulata [168] or Torreyayunnanensis [169], has demonstrated the ability to dwindle VEGF level in A549 cells due to an increase in angiostatin expression by reducing HIF-1α levels and downregulating TNF-α, thereby inhibiting NF-κB expression [63]. Moreover, sotetsuflavone exhibits an effective anti-angiogenic result using the inhibition of TGF-β, STAT3, and β-catenin expression, while simultaneously upregulating the levels of endostatin and ZO-1 [64].

Morusin, an isoprenylated flavone derived from Morus alba Linn. [170], demonstrates significant constricting effects on migration, invasion, and angiogenesis in hepatocellular carcinoma cells (HepG2 and Hep3B) as well as HUVECs. This inhibition correlates with a decrease inkey factors, including VEGFR2, VEGF, MMP9, and MMP2, by effectively suppressing the IL-6-mediated STAT3 pathway [65]. Furthermore, it effectively reduces the expression of VEGF and COX-2 genes, both of which are implicated in angiogenesis of A549 cells [66].

3.2.2. Isoflavone

Genistein (4′,5,7-trihydroxyisoflavone), a soy phytoestrogen, has been the subject of multiple studies revealing its significant ability to suppress the baseline and hypoxia-triggered VEGF level in HUVECs and prostate cancer. This suppression is associated-withareduction inHIF-1α nuclear gathering [67]. In pancreatic carcinoma cells, genistein inhibits the activated procedure of HIF-1 induced by hypoxia, resulting in the inhibition of hypoxia-mediated upregulation of VEGF gene expression [68]. Furthermore, genistein demonstrates anti-angiogenic properties by partly blocking the uPA and MMP-1 expression together with the activated pro-MMP-2 stimulated by VEGF/bFGF, via modulating the secretion of their inhibitors of PAI-1, TIMP-1, and TIMP-2 [69]. In human bladder cancer cells, genistein suppresses production or release of angiogenic proteins of PDGF-A, tissue factor (TF), and VEGF165, as well as enzymes involved in matrix degradation for MMP-2, MMP-9, and uPA. Additionally, it boosts the level of PAI-1, angiostatin, TSP-1, and endostatin as anti-angiogenic agents [70]. The anti-angiogenic activity of genistein is also investigated in prostate cancer, where it downregulates the genes level of uPAR, MMP-9, VEGF, neuropilin, TSP-1, TSP, TGF-β2, BPGF, PAR-2, protease M, and LPA. Conversely, it upregulates the expression of genes such as CTAP and CTGF [71]. Moreover, genistein inhibits the production and effect of p38, MMP-9, JNK, and MMP-2 viathe inhibition of the activities of PTK and MAPK in VEGF-stimulated endothelial cells [72].

3.2.3. Flavonol

Quercetin (3,3′,4′,5,7-pentahydroxyflavone), a flavonol derived from various vegetables and fruits [171], has exhibited anti-angiogenesis properties viamodulation of the AKT/mTOR/P70S6K signaling cascade regulated by VEGFR2 [73]. It also hinders MMP-2 and MMP-9 signaling through curbing the MAPK and PI3K/AKT signaling cascades [74]. Moreover, it inhibits VEGF and bFGF expression, leading to MVD reduction, and suppresses the synthesis of H-ras protein, consequently halting cancerous cell proliferation and angiogenesis in a mammary carcinoma model induced by DMBA [75]. Quercetin also upregulates the TSP-1 level as a factor hindering angiogenesis [76], inhibits hypoxia-induced VEGF expression by suppressing STAT3 tyrosine phosphorylation viaan alternative mechanism independent of nuclear HIF levels, inducing HIF-1a expression under hypoxic conditions [77], substantially suppresses VEGF expression by inhibiting NF-κB activity [78], and hampers eNOS phosphorylation and early M-phase cell cycle arrest [79]. Additionally, the inhibition of COX-2-mediated angiogenesis by quercetin links to impeding COX-2 generation mainly regulated by hindering p300 HAT function [80]. Alternatively, quercetin-4′-*O*-β-D-glucopyranoside (QODG) found in Hypericum attenuatum, has been proved to inhibit angiogenesis in HUVECs by suppressing the phosphorylated VEGFR2 triggered by VEGF, leading to the inhibition of subsequent kinases of p70S6K, c-Src, AKT, FAK, mTOR, and ERK [81].

Silybum marianum contains a flavonol called silibinin, also known as silibinin, namely,5,7-trihydroxy-2-[3-(S)-(4-hydroxy-3-methoxyphenyl)-2-(S)-(hydroxy-methyl)-2,3-dihydro-1,4-benzodioxin-6-yl]chroman-4-one) [172]. It has been found to inhibit tumor angiogenesis in RT4 xenografts by downregulating survivin and increasing p53 expression [82]. Silibinin's anti-angiogenic effects are connected with a decline in VEGF secretion and an enhancement in VEGFR1 gene expression [83]. Similarly, the downregulation of VEGF and improvement of IGFBP-3 lead to an inhibitory effect on the mitogenic action of IGF-I [84]. Silibinin also demonstrates a defense activity against sunlight-induced skin cancer by downregulating angiogenic response, involving gene-regulating proteins of NF-κB, COX-2, STAT3, HIF-1α, iNOS, and their potential upstream regulators, p-STAT3 and phospho-p65 in skin cancer triggered by ultraviolet [85]. Additionally, it decreases the levels of VEGF and VEGFR2 expression while attenuating circulating levels of bFGF [86].

Silibinin also reduces the expression of VEGF, COX-2, COX-1, NOS, HIF-1α, and NOS3 [87]. It further inhibits the secretion of MMP-2, possibly entailing the downregulation of survivin and hindrance of the NF-κB together with Akt pathways, in which NF-kB activation is not dependent on the Akt pathway and has been found to be associated with LY294002 targets other than PI3K in HUVECs [88]. Moreover, silibinin blocks the accumulation of HIF-1α and hypoxia-induced VEGF secretion by downregulating the mTOR/p70S6K/4E-BP1 signaling [89]. Another study indicated that silibininprevents the levels of VEGF and MMP-9 stimulated by TPA viathe restraint of the Raf/MEK/ERK signaling [90].

Myricetin (3,5,7,3′,4′,5′-hexahydroxyflavone) exists in various berries, vegetables, walnuts, tea, onions, grapes, and medicinal herbs [173–175]. It has been shown to inhibit ultraviolet B-induced angiogenesis by blocking the expression of HIF-1α viadirect inhibition of PI3K activity, which is a critical target. This inhibition leads to a decrease in the level of MMP-9, VEGF, and MMP-13 [91]. The anti-angiogenic effects of myricetin are attributed, at least in part, to its effects on the VEGF, HIF-1α, p70S6K, and Akt signaling, resulting in a reduction in VEGF secretion and proteins expression of HIF-1α, p-Akt, and p-70S6K. Additionally, it has been discovered that myricetin hinders angiogenesis via an unconventional signaling that encompasses HIF-1α, VEGF, and p21 in OVCAR-3 cells [92]. Myricetin also exerts antiangiogenic effects by inhibiting cell migration and tube formation, triggering apoptosis induced by ROS while suppressing the signaling factors of Akt, PI3K, mTOR [93].

Kaempferol (3,5,7,4′-tetrahydroxyflavone), which is abundant in vegetative foods [176], shows the inhibitory effect on VEGF and angiogenesis by regulating the signaling factors of VEGF, cMyc, NF-κB, ERK, and p21. It downregulates the phosphorylated ERK, and the levels of cMyc and NF-κB while promoting p21 expression [94]. Another study also reported that kaempferol's anti-angiogenic effect and inhibition of VEGF level is performed by reducing HIF-1α expression via two different pathways, Akt/HIF and ESRRA, in which ESRRA, an orphan nuclear receptor, shares significant sequence similarity and engages in intense crosstalk with estrogen receptors [95]. Additionally, kaempferol was found to significantly suppress both VEGF and FGF signaling pathways, leading to a direct reduction in VEGFR2 expression [96]. Moreover, kaempferol lowers VEGF/VEGFR2 levels and diminishes its subsequent signaling, such as AKT, MEK1/2, p-mTOR, ERK1/2,and PI3K [97]. It additionally blocks the activated process of mTOR, Akt, and the subsequent p70S6K effector. Furthermore, kaempferol decreases the activated VEGFR2 and HIF-1α in endothelium and phosphorylated p38, Akt, mTOR, and ERK, thereby reducing VEGFR2 and HIF-1αphosphorylation viathe restriction of ERK/p38 MAPK and PI3K/Akt/mTOR signaling pathways [98].

Rhamnazin (7,3′-dimethoxy-3,5,4′-trihydroxyflavone) in therapeutic botanicals such as Ginkgo biloba [99], has shown potent anti-angiogenic activities in neovascularization after alkaline burn. It directly reduces the VEGF-stimulated phosphorylation of VEGFR2, leading to the restraint of downstream signaling of MAPK, STAT3, and AKTwithout affecting their total levels [99,100].

Extensive evidence has compellingly demonstrated that CD44, known for its vigorous angiogenic properties, is critical for angiogenesis through its modification of VEGF. This makes CD44 an important therapeutic target in the context of glioblastoma [101]. Galangin (3,5,7-trihydroxyflavone), which is found in galagal root, india root, and Alpinis officinarum [177], exhibits inhibitory effects on angiogenesis by downregulating VEGF expression in HUVECs [101]. Similar to myricetin's anti-angiogenic properties, galangin suppresses VEGF expression and downregulates p-Akt, p-p70S6K, and HIF-1α levels, contributing to its inhibitory effect on angiogenesis [92].

Fisetin (3,7,3′,4′-tetrahydroxyflavone) in strawberries, grapes, apples, onions, and cucumbers [178]. It exhibits anti-angiogenic activity by suppressing the activity of MMPs [102]. Specifically, it inhibits MMP-1, MMP-3, MMP-7, MMP-9, and MMP-14, which hinders the expansion of HUVECs and the activated process of proMMP-2 regulated by MMP-14 in human fibrosarcoma cells [102]. Based on a molecular docking method, it also inhibits post-translational forms of MMP-8 and MMP-13 in colorectal cancer progression and significantly palliates colorectal cancer invasion and metastasis [103]. Furthermore, fisetin suppresses HUVEC cell migration and VEGF-induced conditions by inducing G1 phase arrest and mild G2/M arrest. It also inhibits the level of cyclin D1, survivin, VEGF, Bcl-2, eNOS, iNOS, and MMPs and elevates the level of caspases-3, caspases-7, PARP, p53, p21, and Bax in prostate carcinoma and lung cancer [104,105]. The HO-1 elevation has been associated with tumor angiogenesis [179]. Fisetin inhibits cell migration in mammary carcinoma by silencing the gene regulatory protein Nrf2 in the nuclear fraction, leading to reduced activity of MMP-9 and MMP-2 [106]. Similar downregulation of MMP-2 and MMP-9 levels has been observed in prostatic carcinoma by inhibiting the signaling of JNK, PI3K, Akt to reduce NF-κB and AP-1 DNA-binding activities [107]. uPA is involved in matrix degradation, migration, invasion, metastasis, and tumor angiogenesis [180,181]. Fisetin partially suppresses the uPA-dependent increase in cell migration and invasion by inhibiting p38 MAPK activation. This inhibition is achieved byreducing the translocation of p38 MAPK to the nucleus and decreasing NF-κB's DNA-binding activities in adenocarcinoma of the cervix. Consequently, uPA expression is downregulated [108]. Additionally, Death receptor 3 (DR3), a TNF family member acting as a receptor for TNF, is a cell surface protein that has been revealed to induce NF-κB activation when overexpressed in animal cells. NF-κB is known to provoke angiogenesis and metastasis viathe regulation of VEGF and MMPs. Inhibition of NF-κB signaling in pancreatic carcinoma is attributed to the inhibition of DR3-mediated NF-κB activation [109,110]. Fisetin's inhibition of uPA enzyme in the capillary vessels surrounding the tumor could be involved in areduction inangiogenesis and subsequently impede neoplastic expansion [111,112].

3.2.4. Flavanonol

Epigallocatechin-3-gallate (EGCG), from tea, demonstrates anti-angiogenic activity by inhibiting VEGF-induced VEGFR2 signaling in HUVECs. This effect is attributed to the direct interaction between EGCG and the VEGF peptide [182,183]. EGCG also suppresses VEGF expression, hinders the attachment between growth factor with VEGFR2, or impedes the receptor's phosphorylation process, showing the suppressing the internal cellular signaling and stimulation of mitosis triggered by VEGF [184–186]. These additional actions result in increased endostatin expression and inhibition of VEGF mediated by EGCG [187]. Furthermore, stimulating the VEGF/VEGFR pathway leads to lowered protein levels such as HIF-1α protein, heregulin mRNAs, etc., as well as subsequent expressions of VEGFR2, p-VEGFR2, in SW837 colorectal malignancy cells [188]. EGCG exhibits comparable effects on angiogenesis in lung cancer cells. In both cases, it suppressed the secretion of HIF-1α protein, and CD31, VEGF, and IL-8 were triggered by HIF-1α while also activating Akt [189].

EGCG also exerts inhibitory effects on VEGF level and secretion triggered by IL-6, as well as gastric carcinoma cells' angiogenesis by suppressing STAT3 activity. This ultimately targets the STAT3/VEGF signaling pathway, which leads to the declined VEGF level induced by IL-6. This is attained byinhibiting the process of STAT3 translocation, access to nucleus and combination with VEGF promoter in gastric carcinoma [190,191], as well as the inhibition of VEGF protein level, secretion, and mRNA expression due to reduced activation of STAT3 in gastric carcinoma [192].

In A549 cells, nicotine enhances related proteins levels, such as VEGF, COX-2, HIF-1α, p-Akt, etc. However, the compound EGCG downregulates these expressions, leading to the suppression of HIF-1α-induced angiogenesis [193]. In human prostate carcinoma LNCaP cells, EGCG induces anti-angiogenesis by suppressing the expression of VEGF, angiopoietin 1 and 2, etc., and increasing the level of TIMP1, in which angiopoietins (specifically, angiopoietin 1 and 2) are signaling proteins that exert a pivotal function in promoting angiogenesis to form fully developed blood vessels, while TIMP1 is seen as the tissue inhibitor of MMP-9 [194]. Additionally, endoglin, a TGF-β co-receptor, is instrumental in maintaining the equilibrium of signaling cascades between TGF-β/ALK1/Smad1/5 and ALK5/Smad2/3, which effectively inhibits cell motility and division in physiological conditions viaSmad2/3. In semaxanib-treated HUVECs, EGCG demonstrates a marked inhibitory activity on the upregulation of endoglin and decreases the levels of phosphorylated Smad1, thereby suppressing angiogenic capacity [195]. It also regulates the expression of multiple genes or proteins participating in angiogenesis, including TNFAIP2, EFNA1, PDGFA, CXCL6, IFN-β1, ID1, THBS-1, ANGPTL4, IL-1β, TGF-β2, and CCL2. These mediate proliferative, adhesion, invasion, and migratory actions of endothelial cells in human cervical cancer cells (HeLa) [196]. EGCG prevents cell migration toward VEGF in EPCs and TECs, but not in NEC. This corresponds to the downregulation of MMP-9 expression, inhibition of Akt phosphorylation in TEC, and suppression of VEGF-triggered migration of CD133/VEGFR2 cells into the bloodstream [197]. Moreover, EGCG blocks DNA replication, cellular proliferation, phosphorylation of ERK1/2, VEGFR-1 and -2, and EGR1 mRNA level triggered by VEGF in HUVECs [198].

EGCG has also been shown to directly suppress VEGF expression, inhibiting VEGF-induced tumor growth, proliferation, migration, and angiogenesis in breast carcinoma [199]. In A549 cells, EGCG suppresses pulmonary carcinoma angiogenesis stimulated by IGF-I by reducing the level of HIF-1 and VEGF [200]. EGCG inhibits tumor angiogenesis in HUVECs by inhibiting MT1-MMP, which degrades collagen type I and subsequent MMP-2 activation [201]. It has also been shown to inhibit MMP-2 and MMP-9 in SK-N-BE human neuroblastoma and HT1080 human fibrosarcoma cells [202]. Considering that CXCL12 attracts infiltration of tumor-associated macrophages (TAMs), which are a significant source of VEGFA and vital chemoattractant for macrophages. A modified form of EGCG called peracetate-protected EGCG has been developed as a precursor drug, which reduces VEGFA secreted by cancer cells and by TAM in endometrial carcinoma by inhibiting the PI3K/AKT/mTOR/HIF1 pathway and infiltration of VEGFA-expressing TAMs viaCXCL12 mediation in stromal cells [203].

3.2.5. Flavanones

Research has demonstrated that the flavonoid naringenin, which is abundant in tomatoes and oranges, can inhibit angiogenesis. ERRα serves as a pivotal regulator in preserving energy balance and fostering generation of mitochondria by engaging with PPAR γ coactivator-1α and 1β. According to the literature, one of the mechanisms through which naringenin exerts its antiangiogenic effect is relevant to inducing cell cycle stasis at the G0-G1 phase and promoting apoptosis, and suppressing the release of cytokines MCP-1, IL-6, and ICAM-1involved in inflammation, and directly inhibiting the tyrosine phosphorylation function of KDR and blocking phosphorylated procedure of Akt, FAK, and paxillin induced by VEGF, and downregulating ERRα transactivation and expression, consequently leading to the inhibition of VEGF production [134].

In a laboratory experiment conducted with rat aortic rings, the flavanone 5,3′-dihydroxy-6,7,8,4′-tetramethoxyflavanone (HLBT-001) sourced from *Tillandsia recurvata* (L.) L. exhibited dose-dependent antiangiogenic potential by suppressing capillary sprout and tube development [135].

Hesperetin (3′,5,7-trihydroxy-4′-methoxyflavanone) derived from citrus fruits exerted an anti-angiogenic effect by inhibiting the creation of tubular structures, cell movement, and the proliferation of endothelial cells in VEGF-stimulated HUVECs [204]. This effect was found to be linked to the limitation of the VEGFR2-mediated signalings of AKT, p38 MAPK, PI3K, and ERK [136]. Additionally, Hesperetin demonstrated an inhibition of neovascularization factors like bFGF, EGF, and VEGF, along with a downregulation of mRNA COX-2 expression in rat colon carcinogenesis [137].

Monocytes exert a notable influence in angiogenesis and immune response regulation. Their adherence is influenced by the enhanced level of cell adhesion proteins like ICAM-1 and VCAM-1, as well as elevated secretion of chemokines [205]. Didymin (5,7-dihydroxy-7-rutinoside-4′-methoxy-flavone), a flavone glycoside found in oranges, lemons, and mandarins, has been shown to inhibit the VCAM-1 and ICAM-1 levels, effectively preventing high-glucose-induced monocyte adhesion to endothelial cells [206]. Moreover, Didymin reduces ROS production, NF-κB activation, and E-selectin, ICAM-1, and VCAM-1production in HUVECs induced by VEGF [138].

Farrerol (6,8-dimethyl-4′,5,7-trihydroxy-flavone), extracted from Rhododendron dauricum L., exhibits anti-vascular growth characteristics via the restraint of endothelial cell growth, migration, and generation of tubules. This effect is achieved by downregulating the Akt/mTOR, ERK, and JAK2/STAT3 signaling pathways in both HUVECs and HMECs-1 [139].

3.2.6. Chalcones

A chalcone glycoside, hydroxysafflower yellow A (HSYA) (2′,3′,4′,4-tetrahydroxy-3′,5′-di-*O*-β-D-glucopyranosyl-chalcone), was discovered in Carthamus tinctorius L. and has been shown to inhibit tumor angiogenesis. In a study using transplanted human gastric adenocarcinoma BGC-823 cells, HSYAreduced MVD, integrated optical density (IOD) and downregulated the mRNA expression of VEGF and bFGF, which are crucial growth factors involved in cancer angiogenesis [140,141]. The inhibitory effects of HSYA on tumor angiogenesis were further demonstrated by decreasing the expression of MMP-9 and bFGF, as well as MMP-9 mRNA, regulating the degradation of the blood vessel basilar membrane, blood vessel migration, and tumor vascularization in BGC-823 cells [142]. CD105 has been assessed as a marker for neovascularization in hepatocellular carcinoma (HCC) [207]. HSYA displayed an inhibitory effect on angiogenesis factor secretion by suppressing the expression of CD105, VEGFA, bFGF, and VEGFR1, in addition to the phosphorylated ERK and c-raf. It also inhibited the phosphorylated process of IκB and prevented degradation within the cytoplasm of IκB-α, causing reduced NF-κB transcriptional activity by suppressing the signaling of MAPK, NF-κB, and ERK. HSYA downregulated the transcription of genes like cyclinD1, c-myc, and c-Fos, known to be phosphorylated by activated ERK and translocated into the nucleus. This was achieved by reducing the expression of nuclear p65, increasing cytoplasmic levels of p65, blocking IκB phosphorylation, and preventing cytoplasmic degradation of IκB-α [143,208,209]. Furthermore, activation of the transcription factor 2 (ATF-2), an AP1 transcription factor family member, is responsible for regulating the transcriptional activation of target genes. Several studies signify that ATF2 mediates VEGF-stimulated angiogenic processes while also exerting an inhibitory effect on angiogenesis by negatively regulating Notch-related genes such as DLL4, HEY1, and NRARP [210]. HSYA reduced the COX-2, MMP-9, and MMP-2 levels in HepG2 cell lysates by inhibiting p38 MAPK phosphorylation [144].

Tumor-infiltrating EPCs contributes to resistance against anti-VEGF therapy by indirectly secreting TGF-β, HIF-1α, and VEGF as angiogenesis-promoting growth factors [211]. The enzymes of the CYP 4 family are responsible for catalyzing the conversion of diverse fatty acids by hydroxylation, including ARA. Among these hydroxylation products, 20-HETE as a principal product results from the ARA hydroxylation by CYP4A and acts as a key facilitator in angiogenesis mediated by VEGF [212]. FLA-16 (2,3′,4,4′-tetrahydroxy-3,5′-diprenylchalcone) is a compound derived from Glycyrrhiza glabra [213]. It has been observed that CYP450 4A-derived 20-hydroxyeicosatetraenoic acid, in conjunction with the induction of pro-angiogenic growth factors by tumor-associated macrophages (TAMs), promotes angiogenesis during anti-VEGF treatment. Additionally, by specifically targeting TAMs and EPCs viathe PI3K/Akt signaling pathway, FLA-16 effectively normalizes the vasculature in glioma, resulting in a considerable drop in both VEGF and TGF-β expression [145].

SKLB-M8, namely (*E*)-3-(3-amino-4-methoxyphenyl)-1-(5-methoxy-2,2-dimethyl-2H-chromen-8-yl)prop-2-en-1-one hydrochloride, which is derived from Millettiapachycarpa Benth, has demonstrated inhibitory effects on the proliferative and migratory activities of HUVECs by suppressing ERK1/2 activation. Moreover, treatment with SKLB-M8 has been shown to effectively reduce angiogenesis within alginate beads implanted in mice and significantly decrease the uptake of fluorescein isothiocyanate (FITC)-dextran in mouse models bearing melanoma B16F10 tumors [146].

Licochalcone A (LicA) has been detected in Glycyrrhiza inflata and shown to possess inhibitory effects on endothelial cell movement and tube-like structure creation, along with blood vessel formation. It achieves this by downregulating the activation of VEGFR2, VEGF-stimulated phosphorylation of carbon store regulator C (cSrc) and inhibiting angiogenic growth factors of IL-8 and IL-6 [147]. The proteolytic cleavage of extracellular matrix proteins is a crucial process in cancer cell migration. Licochalcone E (LicE) demonstrates effective inhibition of VEGF-A secretion, a crucial factor in tumor angiogenesis, and suppresses VEGF-R2 activation. Additionally, it induces a reduction in tumor angiogenesis and brings about alterations in the local tumor tissue environment. This includes inhibiting the infiltration of leukocytes into tumor tissues, reducing HIF-1α expression, as well as downregulating the proinflammatory enzymes COX-2 and iNOS in tumor tissues [148].

Flavokawain A (FKA, 2′-hydroxy,4,4′,6′-trimethoxy chalcone) and flavokawain B (FKB, 2′-hydroxy,4′,6′-dimethoxy chalcone), two chalcones, can be extracted from the Piper methysticum or Alpinia pricei [214]. Angiogenin is a ribonuclease enzyme known for its crucial role in neovascularization, particularly within embryo, post-birth, and uterine lining tissues. Continual expression of F3 can contribute to the onset of thrombosis associated with cancer. FKB significantly decreases the expression of numerous angiogenic-associated factors, such as VEGF, F3, TSP-2, SDF-1, angiogenin, serpin F1, and pentraxin 3 [149]. FOXM1 acts as a transcriptional regulator of promoting cell cycle advancement, while GLUT1 serves as a carrier for glucose uptake viaglycolysis, particularly in tumors and often connected to cancer phenotypes. FKB effectively impedes the production of vascular-like structures by downregulating several key factors, including MMP9, VEGF, GLUT1, and FOXM1, with a specific emphasis on angiogenesis [150]. ICAM-1 is a glycoprotein known for its involvement in immune response and tumorigenesis. FKA demonstrates effective inhibition of VEGF, GLUT1, and ICAM-1 expression in the mammary carcinoma, resulting in a significant suppression of new vascular formation. In an ex vivo model employing rat aortic rings, increasing doses of FKA effectively impede vessel outgrowth from the fragmented aorta [151]. Additionally, FKA likely exerts an anti-angiogenic effect by downregulating the androgen receptor in bladder cancer to some extent, resulting in a more pronounced reduction in tumor growth in male transgenic mice compared to females [152].

Isolated from Alpinia katsumadai or Alpinia conchigera, cardamonin, namely 2′,4′-dihydroxy-6′-methoxychalcone, shows inhibitory activity on the protein levels of VEGF, HIF-1α, HIF-2α, and ribosomal S6 kinase 1 (S6K1) in CoCl2-induced hypoxic SKOV3 cells. These effects were partially attributed to mTOR inhibition [153]. Furthermore, cardamonin demonstrates anti-angiogenic properties by suppressing VEGF-induced angiogenesis and attenuating the VEGF-stimulated ERK and AKT phosphorylation in both HUVECs and mouse aortic ring assay [154]. MicroRNAs (miRNAs) represent short non-coding RNAs, regulating genetic transcription by interacting with mRNA targets, notably within the 3′ untranslated region. They impact various endothelial cell functions like apoptosis, differentiation, and angiogenesis. In a recent study, the blocking effect of VEGF-triggered angiogenesis by cardamonin is performed by diminishing the miR-21 expression, resulting in reduced cell growth, movement, and blood vessel formation stimulated by VEGF [155].

Isoliquiritigenin (4,2′,4′-trihydroxychalcone), derived from shallot, bean sprouts, or Glycyrrhiza glabra L. [215,216], exhibits inhibitory effects on the level and functionality of PMA-induced MMP-2 and MT1-MMP. Furthermore, it enhances the production of TIMP, upregulated by PMA, in a biphasic manner, while suppressing the MAPK-dependent signaling cascades involving p38 and JNK [156]. Isoliquiritigenin significantly inhibitsHUVEC proliferation stimulated by VEGF. This inhibition encompasses various anti-angiogenic processes, including the suppression of the capacity of HUVECs to form tubes, invade, and migrate, as well as prevention of VEGF-induced sprouting in aortic rings. Molecular mechanism studies revealed that isoliquiritigenin promotes HIF-1α proteasome degradation, resulting in a significant inhibition of VEGF expression in mammary carcinoma Moreover, isoliquiritigenin directly interacts with VEGFR2 to inhibit its kinase function Another report demonstrates that isoliquiritigenin effectively suppresses the proliferation of breast cancer cells and the formation of new blood vessels, accompanied by a suppressed VEGF/VEGFR2 pathway [157]. Isoliquiritigenin inhibits tumor angiogenesis in adenoid cystic carcinoma (ACC) cells through the obstruction of mTOR-dependent VEGF expression This effect is achieved by activating JNK and inactivating ERK, resulting in a remarkable reduction inmicrovascular density in grafted tumors [158]. Moreover, isoliquiritigenin consistently exhibits anti-angiogenic activity by inhibiting the phosphorylation of ERK1/2, suppressing FGF-induced in vivo angiogenesis in mice, as well as cell proliferation, migration, and angiogenesis [159]. The maintenance of avascularity is delicately regulated by maintaining a delicate equilibrium between various anti-angiogenic factors like PEDF, and pro-angiogenic factors like FGF-2 and HGF, in which PEDF, a constituent of the serpin family, regulates cell multiplication and enhances neuronal survival. Isoliquiritigenin downregulates VEGF expression while upregulating PEDF level [160].

The overexpression of TM4SF5, a protein, is a crucial factor in encouraging the proliferation of tumor cells and is consistently found in hepatocarcinoma patients, contributing to hightened expression of TM4SF5 connected to VEGF level and angiogenesis [217]. During TM4SF5-induced multilayer growth, TM4SF5 activates FAK by increasing phosphorylation at Tyr577 (pY577FAK) and enhances the expression and stability of p27Kip1 in the cytosol. These processes lead to the inactivation of RhoA and contribute to the development of elongated morphology and epithelial–mesenchymal transition (EMT), which are responsible for promoting multilayer growth. A synthetic chalcone compound called TSAHC has been shown to effectively suppresses TM4SF5-mediated EMT, impedes multilayer growth, inhibits migration and invasion, and prevents the onset of tumors, showing the decrease inpY577FAK and p27Kip1 levels [161].

Xanthohumol(XN), namely 2′,4′,4-trihydroxy-6′-methoxy-3′-prenylchalcone, present in beer or Humulus lupulus, exhibit antiangiogenic effects. This is believed to occur viaareduction inNF-κB activity in HUVECs, leading to the inhibition of viability, invasion, and formation of capillary-like structures in endothelial cells [162,218]. XN demonstrates a dual anti-cancer effect by effectively suppressing NF-κB activity and IL1β expression in MCF7 cells and endothelial cells, impacting both tumor cells and angiogenesis [163]. Furthermore, XN displays a more potent anti-angiogenic effect compared to EGCG, which is linked to the upregulation of AMPK phosphorylation and activity mediated by CAMMKβ, as well as a reduction in eNOS phosphorylation partly induced by AKT signaling inhibition [164]. The angiogenesis process in leukemias relies heavily on the activation of Akt/NF-κB [219]. Therefore, XN suppresses angiogenesis by reducing VEGF secretion in leukemic cells, interfering with the cascades inducing NF-κB activation, and inhibiting Akt phosphorylation via suppression of Akt/NF-κB signaling in U937 chronic myelogenous leukemia [113], similar to the observed inhibitory effects in HUVECs [114]. Additionally, XN regulates the anti-angiogenic effects of NF-κB by suppressing the level of MMP-9, VEGF, and ICAM-1 stimulated by TNF in chronic myelogenous leukemia [115]. In the case of pancreatic cancer, XN blocks the induction of NF-κB to suppress angiogenesis, along with the subsequent expressions of VEGF and interleukin-8 (IL-8) [116].

Xanthoangelol, a natural chalcone also known as 2′,4,4′-trihydroxy-3′-[(*E*)-3,7-dimethyl-2,6-octadienyl]chalcone, was derived from Angelica keiskei [220]. It showed potent inhibitory effects on Lewis lung carcinoma-induced angiogenesis by blocking capillary-like tube formation and the interaction of HUVECs with VEGF [117].

Butein (3,4,2′,4′-tetrahydroxychalcone), a derivative from plants like Dalbergia odorifera, has revealed effective inhibition effects on cell proliferation, cell movement, and tubular structure induced by serum and VEGF in EPCs viathe downregulation of Akt and mTOR phosphorylation, as well as their major downstream effectors p70S6K, 4E-BP1, and eIF4E [118]. Additionally, butein exhibits anti-angiogenic properties by inhibiting NF-κB activity, as demonstrated in other studies. It specifically targets NF-κB, leading to the suppression of PMA- and TNF-α-stimulated MMP-9 and VEGF expression in prostatic carcinoma [119].

Tumor angiogenesis comprises a multifaceted process encompassing multiple factors and steps, including cellular expansion, migration, and degradation of the extracellular matrix. Many malignant solid tumors heavily rely on vascular supply and are typically associated with cascades of vascular dysplasia. The balance between angiogenesis promotion and inhibition factors plays a crucial role in this dependence [221].

As shown in Figure 4, there is a main pathway involving the proteins of HIF-1α, VEGF, VEGFR2, PI3K, and AKT, relating to the tumor anti-angiogenesis, including the most flavonoids in our findings. HIF-1α, an oxygen-dependent transcription factor, is highly expressed in tumor cells under hypoxic conditions, contributing significantly to tumorigenesis, development, invasion, metastasis, and apoptosis [222]. As shown in Figure 4, compounds **13**, **17**, **18**, **23**, and **24** exert anti-tumor angiogenic effects by inhibiting HIF-1. VEGF, the most powerful and specific factor for promoting tumor angiogenesis known to date, enhances endothelial cell proliferation, migration, and growth upon combination with VEGF [223]. VEGF is a crucial downstream target of HIF-1α, as its expression is upregulated by HIF-1α, ultimately leading to the formation of tumor angiogenesis, which can be inhibited by compounds **12**, **27**, **30**, **41**, **42** and **46** in Figure 4 [224]. Notably, HIF-1α also participates in cancer tissue growth, invasion, and metastasis, further emphasizing the importance of blocking the HIF-1α/VEGF pathway to effectively inhibit tumor angiogenesis [225].

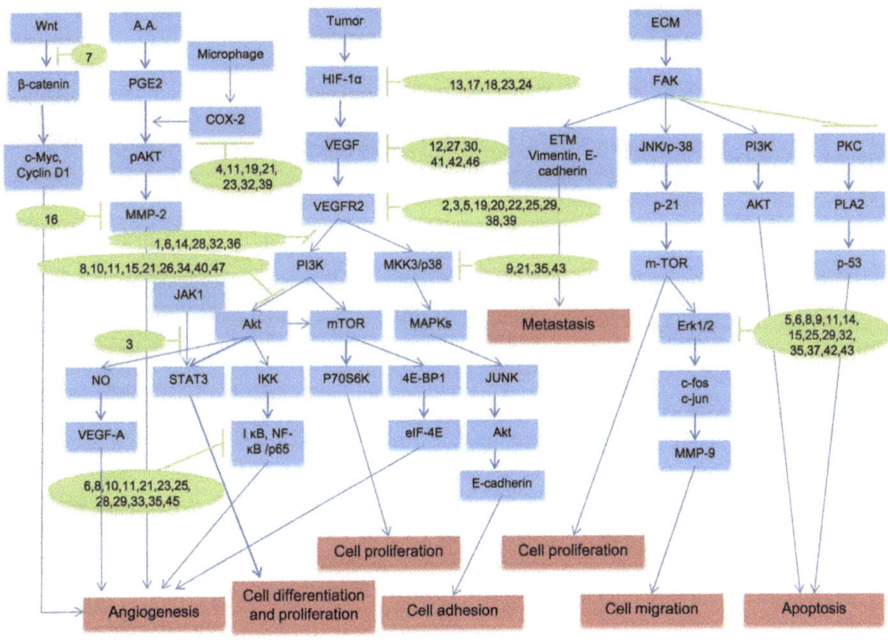

Figure 4. Partial mechanism of anti-angiongenic flavonoids in cancer (The blue represents the relative targets or pathways; the green represents the relative chemical compounds; the red represents the effects triggered by the targets or pathways).

VEGFR2 serves as a dominant mediator of both physiological and pathological angiogenesis [226]. HIF-1α enables VEGFR2 activation by binding to the hypoxia response element situated within the VEGF gene's promoter region [227]. Therefore, compounds **2**, **3**, **5**, **19**, **20**, **22**, **25**, **29**, **38**, **39** in Figure 4 play a role of VEGFR2 inhibitor. The PH domain of AKT interacts with and binds to PI3K, leading to AKT's structural alterations followed by phosphorylation. After activation, AKT relocates from the cytoplasm to the cellular membrane. This cascade facilitates the direct or indirect activation of downstream molecular proteins, including NF-κB and mTOR [228]. Akt is involved in controlling normal blood vessel formation and pathological angiogenesis, and its activation alone can stimulate VEGF expression in human cancer cells. mTOR, an essential protein kinase within the PI3K/Akt pathway, governs various critical cellular functions; its dysregulation has been implicated in tumorigenesis [229]. Once PI3K/Akt is activated, it can trigger mTOR activation, consequently elevating HIF-1α and VEGF expression as a pivotal player in the growth and formation of blood vessels by endothelial cells [230]. To some degree, compounds **1**, **6**,**14**, **28**, **32**, **36** and **8**, **10**, **11**, **15**, **21**, **26**, **34**, **40**, **47** in Figure 4 are related to the signal suppression PI3K/Akt pathway, whereas compounds **6**, **8**, **10**, **11**, **21**, **23**, **25**, **28**, **29**, **33**, **35**, and **45** entail the inhibition of NF-κB signal pathway.

In addition, many other important signal pathways are highly relevant to anti-angiogenic effects for future cancer therapy in Figure 4. COX-2, an inducible enzyme, is specifically generated in response to stimulation from associated cytokines, tumor inducers, and tumor genes [231]. The main pro-angiogenic activity of COX-2 occursviathe action of three ARA metabolites: PGI2, TXA2, and PGE2 [232]. Three eicosanoid products have a pivotal significance in downstream proangiogenic activities such as VEGF production, facilitation of blood vessel sprouting, migration, and tube development [233], improved endothelial cell survival via activating Akt production and increasing Bcl-2 level [234], stimulation of MMPs, initiation of EGFR-driven angiogenesis, and inhibition of interleukin-12 generation [235]. Compounds **4**, **11**, **19**, **21**, **23**, **32**, **39** in Figure 4 mitigate COX-2 in different ways and exerting the effect of inhibiting tumor neovascularization.

Wnt proteins are a family of signaling molecules that initiate a cascade of events within cells, influencing various developmental processes and maintaining tissue homeostasis. β-catenin transduces signals to the nucleus, activating Wnt-specific genes that regulate cell fate across various tissues [236]. The Wnt/β-catenin pathway activation is pivotal in driving angiogenesis by promoting growth of endothelial cells and facilitating neovascularization, interacting with various angiogenic proteins, like FGF and VEGF to influence their expression and activity [237]. Wnt signaling contributes to the establishment and maintenance of stable blood vessels and regulates the recruitment and function of pericytes, impacting vessel stability [238], which is achieved partly by compound **7** as an anti-angiogenic influence bysuppressing the Wnt/β-catenin pathway, as shown in in Figure 4.

JNK1 is a member of the MAPK family. In the context of angiogenesis, JNK1 signaling has been found to influence endothelial cell behaviors like cellular expansion, migration, and tubule creation for neovascularization [239]. STAT3 is a JNK1 downstream transcription factor controlling gene expression linked to cell survival and proliferation, and inflammation. In angiogenesis, STAT3 activation in endothelial cells regulates VEGF level, endothelial cell survival and migration, and blood vessel formation [240]. In some cancers, aberrant activation of JNK1/STAT3 pathways can promote tumor angiogenesis by initiating angiogenesis to facilitate the delivery of oxygen and nutrients to the tumor [241], which can be prevented by compound **3** as an anti-angiogenic agent in Figure 4.

The MAPK/AP-1 pathway holds significant importance in governing the expression of genes linked to metastasis. AP-1, a transcriptional regulator, is influenced by MAPKs including p38 MAPK, JNK1/2, and ERK1/2. MAPK phosphorylation regulation dictates the nuclear concentration of AP-1 subunits, which, in turn, determines the transcriptional activity of AP-1 [242]. ERK1/2 is pivotal in inducing immediate early genes (such as c-fos and c-jun) and orchestrates genetic transcriptional activation induced by growth factors, hormones, and cytokines [243]. The decline in genetic activity of COX-2, MMP-9, and MMP-3 corresponds to reduced phosphorylation levels of JNK1/2and ERK1/2, consequently diminishing nuclear c-Jun and c-Fos [242]. Furthermore, ERK1 and ERK2 collaborate in orchestrating endothelial cell proliferation and migration during angiogenesis [244].

However, compounds **5**, **6**, **8**, **9**, **11**, **14**, **15**, **26**, **29**, **32**, **35**, **37**, and **42** exhibit the capability to inhibit these processes in Figure 4.

4. Conclusions

The "tumor-starvation therapy" is a treatment method that cuts off the blood supply to the tumor, causing it to enter a state of starvation and thus limiting its growth. The main result of this research is to deeply explore the role and underlying molecular mechanisms of natural flavonoids in anti-angiogenesis. Additionally, this article systematically summarizes the structural characteristics and classification system of flavonoids, as well as the major signaling pathways involved in their function as plant-derived anti-angiogenic agents. These pathways include but are not limited to HIF-1α/VEGF/VEGFR2/PI3K/AKT, Wnt/β-catenin, JNK1/STAT3, and MAPK/AP-1, which may provide some new active ingredients or targets for future cancer treatment. At present, it is highly necessary to find non-toxic and cost-effective molecules based on or derived from the flavonoids' structure for the advancement among cancer angiogenesis therapy, including luteonin, baicalein, apigenin, genistein, quercetin, silibinin, myricetin, kaempferol, fisetin, and EGCG, as deeply researched flavonoids in pharmacology. In addition, the validation of in vivo experiments also should receive high attention, not just cell experiments. It requires not only a significant amount of funding for investment but also enough experimental data through animal experiments and clinical trials, with multiple animals, models, and levels to determine the accuracy of targets and their treatment effectiveness.

Supplementary Materials: The following supporting information can be downloaded at: https://www.mdpi.com/article/10.3390/molecules29071570/s1, Table S1. Antiangiogenic molecular mechanisms of different flavonoids.

Author Contributions: Conceptualization, Y.-h.Z. and writing—review and editing, Q.W. All authors have read and agreed to the published version of the manuscript.

Funding: This research was funded by the Natural Science Foundation Project of the Anhui Educational Committee (grant number: KJ2020A0791) and the Key Subject of Anhui Xinhua University (grant number: zdxk202104).

Conflicts of Interest: The authors declare no conflicts of interest.

References

1. Arora, V.; Sharma, N.; Tarique, M.; Vyas, G.; Sharma, R.B. An overview of flavonoids: A diverse group of bioactive phytoconstituents. *Curr. Tradit. Med.* **2023**, *9*, 1–12. [CrossRef]
2. Hazafa, A.; Rehman, K.U.; Jahan, N.; Jabeen, Z. The role of polyphenol (flavonoids) compounds in the treatment of cancer cells. *Nutr. Cancer* **2020**, *72*, 386–397. [CrossRef] [PubMed]
3. Khan, A.U.; Dagur, H.S.; Khan, M.; Malik, N.; Alam, M.; Mushtaque, M. Therapeutic role of flavonoids and flavones in cancer prevention: Current trends and future perspectives. *Eur. J. Med. Chem. Rep.* **2021**, *3*, 100010. [CrossRef]
4. Slika, H.; Mansour, H.; Wehbe, N.; Nasser, S.A.; Iratni, R.; Nasrallah, G.; Shaito, A.; Ghaddar, T.; Kobeissy, F.; Eid, A.H. Therapeutic potential of flavonoids in cancer: ROS-mediated mechanisms. *Biomed. Pharmacother.* **2022**, *146*, 112442. [CrossRef] [PubMed]
5. Scagliarini, A.; Mathey, A.; Aires, V.; Delmas, D. Xanthohumol, a prenylated flavonoid from hops, induces DNA damages in colorectal cancer cells and sensitizes SW480 cells to the SN38 chemotherapeutic agent. *Cells* **2020**, *9*, 932. [CrossRef]
6. Pang, X.; Zhang, X.; Jiang, Y.; Su, Q.; Li, Q.; Li, Z. Autophagy: Mechanisms and therapeutic potential of flavonoids in cancer. *Biomolecules* **2021**, *11*, 135. [CrossRef]
7. Zhang, Z.; Wang, M.; Xing, S.; Zhang, C. Flavonoids of *Rosa rugosa* Thunb. inhibit tumor proliferation and metastasis in human hepatocellular carcinoma HepG2 cells. *Food Sci. Hum. Wellness* **2022**, *11*, 374–382. [CrossRef]
8. Subbaraj, G.K.; Kumar, Y.S.; Kulanthaivel, L. Antiangiogenic role of natural flavonoids and their molecular mechanism: An update. *Egypt. J. Intern. Med.* **2021**, *33*, 29. [CrossRef]
9. Vaiyapuri, M.; Natesan, K.; Vasamsetti, B.M.; Mekapogu, M.; Swamy, M.K.; Thangaraj, K. Orientin: A C-glycosyl flavonoid that mitigates colorectal cancer. In *Plant-Derived Bioactives: Chemistry and Mode of Action*; Springer: Singapore, 2020; pp. 1–9. [CrossRef]
10. Fares, J.; Fares, M.Y.; Khachfe, H.H.; Salhab, H.A.; Fares, Y. Molecular principles of metastasis: A hallmark of cancer revisited. *Signal Transduct. Target. Ther.* **2020**, *5*, 28. [CrossRef]
11. Hyytiäinen, A.; Wahbi, W.; Väyrynen, O.; Saarilahti, K.; Karihtala, P.; Salo, T.; Al-Samadi, A. Angiogenesis inhibitors for head and neck squamous cell carcinoma treatment: Is there still hope? *Front. Oncol.* **2021**, *2123*, 683570. [CrossRef]

12. Qin, S.; Li, A.; Yi, M.; Yu, S.; Zhang, M.; Wu, K. Recent advances on anti-angiogenesis receptor tyrosine kinase inhibitors in cancer therapy. *J. Hematol. Oncol.* **2019**, *12*, 27. [CrossRef] [PubMed]
13. Abdulkadir, S.; Li, C.; Jiang, W.; Zhao, X.; Sang, P.; Wei, L.; Hu, Y.; Li, Q.; Cai, J. Modulating angiogenesis by proteomimetics of vascular endothelial growth factor. *J. Am. Chem. Soc.* **2021**, *144*, 270–281. [CrossRef]
14. Fallahi, P.; Ferrari, S.M.; Galdiero, M.R.; Varricchi, G.; Elia, G.; Ragusa, F.; Paparo, S.R.; Benvenga, S.; Antonelli, A. Molecular targets of tyrosine kinase inhibitors in thyroid cancer. *Semin. Cancer Biol.* **2022**, *79*, 180–196. [CrossRef] [PubMed]
15. Breznik, B.; Mitrović, A.; Lah, T.T.; Kos, J. Cystatins in cancer progression: More than just cathepsin inhibitors. *Biochimie* **2019**, *166*, 233–250. [CrossRef] [PubMed]
16. Wei, Q.; Li, Q.Z.; Wang, R.L. Flavonoid components, distribution, and biological activities in *Taxus*: A review. *Molecules* **2023**, *28*, 1713. [CrossRef] [PubMed]
17. Wang, X.; Cao, Y.; Chen, S.; Lin, J.; Bian, J.; Huang, D. Anti-inflammation activity of flavones and their structure–activity relationship. *J. Agric. Food Chem.* **2021**, *69*, 7285–7302. [CrossRef] [PubMed]
18. Kota, S.; Senthil, R.R. Flavonoids-apoptotic inducers in cancer therapy. *NeuroQuantology* **2022**, *20*, 733–741.
19. Du, E.; Li, X.; He, S.; Li, X.; He, S. The critical role of the interplays of EphrinB2/EphB4 and VEGF in the induction of angiogenesis. *Mol. Biol. Rep.* **2020**, *47*, 4681–4690. [CrossRef] [PubMed]
20. Zhu, P.T.; Mao, M.; Liu, Z.G.; Tao, L.; Yan, B.C. Scutellarin suppresses human colorectal cancer metastasis and angiogenesis by targeting ephrinb2. *Am. J. Transl. Res.* **2017**, *9*, 5094.
21. Li, H.X.; Huang, D.Y.; Gao, Z.X.; Chen, Y.; Zhang, L.; Zheng, J.-H. Scutellarin inhibits the growth and invasion of human tongue squamous carcinoma through the inhibition of matrix metalloproteinase-2 and-9 and αvβ6 integrin. *Int. J. Oncol.* **2013**, *42*, 1674–1681. [CrossRef]
22. Li, C.Y.; Wang, Q.; Wang, X.M.; Li, G.X.; Wei, X.L. Scutellarin inhibits the invasive potential of malignant melanoma cells through the suppression epithelial-mesenchymal transition and angiogenesis via the PI3K/Akt/mTOR signaling pathway. *Eur. J. Pharmacol.* **2019**, *858*, 172463. [CrossRef] [PubMed]
23. Han, J.M.; Kwon, H.J.; Jung, H.J. Tricin, 4′,5,7-trihydroxy-3′,5′-dimethoxyflavone, exhibits potent antiangiogenic activity in vitro. *Int. J. Oncol.* **2016**, *49*, 1497–1504. [CrossRef] [PubMed]
24. Fu, B.; Xue, J.; Li, Z.; Shi, X.; Jiang, B.-H.; Fang, J. Chrysin inhibits expression of hypoxia-inducible factor-1alpha through reducing hypoxia-inducible factor-1alpha stability and inhibiting its protein synthesis. *Mol. Cancer Ther.* **2007**, *6*, 220–226. [CrossRef] [PubMed]
25. Lin, C.M.; Shyu, K.G.; Wang, B.W.; Chang, H.; Chen, Y.H.; Chiu, J.H. Chrysin suppresses IL-6-induced angiogenesis via down-regulation of JAK1/STAT3 and VEGF: An in vitro and in ovo approach. *J. Agric. Food Chem.* **2010**, *58*, 7082–7087. [CrossRef] [PubMed]
26. Lin, C.M.; Chang, H.; Li, S.Y.; Wu, I.H.; Chiu, J.H. Chrysin inhibits lipopolysaccharide-induced angiogenesis via down-regulation of VEGF/VEGFR-2 (KDR) and IL-6/IL-6R pathways. *Planta Med.* **2006**, *72*, 708–714. [CrossRef] [PubMed]
27. Liu, C.; Li, X.W.; Cui, L.M.; Li, L.C.; Chen, L.Y.; Zhang, X.-W. Inhibition of tumor angiogenesis by TTF1 from extract of herbal medicine. *World J. Gastroenterol.* **2011**, *17*, 4875. [CrossRef] [PubMed]
28. Jiang, Y.; Xie, K.P.; Huo, H.N.; Wang, L.M.; Zou, W.; Xie, M.J. Inhibitory effect of luteolin on the angiogenesis of chick chorioallantoic membrane and invasion of breast cancer cells via downregulation of AEG-1 and MMP-2. *Acta Physiol. Sin.* **2013**, *65*, 513–518.
29. Lu, X.Y.; Li, Y.H.; Xiao, X.W.; Li, X.B. Inhibitory effects of luteolin on human gastric carcinoma xenografts in nude mice and its mechanism. *Nat. Med. J. China* **2013**, *93*, 142–146.
30. Pratheeshkumar, P.; Son, Y.O.; Budhraja, A.; Wang, X.; Ding, S.-Z.; Wang, L.; Hitron, A.; Lee, J.C.; Kim, D.; Divya, S.P.; et al. Luteolin inhibits human prostate tumor growth by suppressing vascular endothelial growth factor receptor 2-mediated angiogenesis. *PLoS ONE* **2012**, *7*, e52279. [CrossRef]
31. Cai, X.; Lu, W.; Ye, T.; Lu, M.; Wang, J.; Huo, J.; Qian, S.-H.; Wang, X.-N.; Cao, P. The molecular mechanism of luteolin-induced apoptosis is potentially related to inhibition of angiogenesis in human pancreatic carcinoma cells. *Oncol. Rep.* **2012**, *28*, 1353–1361. [CrossRef]
32. Gui, S.; Zhou, S.; Liu, M.; Zhang, Y.; Gao, L.; Wang, T.; Zhou, R. Elevated levels of soluble Axl (sAxl) regulates key angiogenic molecules to induce placental endothelial dysfunction and a preeclampsia-like phenotype. *Front. Physiol.* **2021**, *12*, 619137. [CrossRef] [PubMed]
33. Li, X.; Chen, M.; Lei, X.; Huang, M.; Ye, W.; Zhang, R.; Zhang, D. Luteolin inhibits angiogenesis by blocking Gas6/Axl signaling pathway. *Int. J. Oncol.* **2017**, *51*, 677–685. [CrossRef] [PubMed]
34. Wang, L.; Astone, M.; Alam, S.K.; Zhu, Z.; Pei, W.; Frank, D.A.; Burgess, S.M.; Hoeppner, L.H. Suppressing STAT3 activity protects the endothelial barrier from VEGF-mediated vascular permeability. *Dis. Models Mech.* **2021**, *14*, dmm049029. [CrossRef] [PubMed]
35. Fang, B.; Chen, X.; Wu, M.; Kong, H.; Chu, G.; Zhou, Z.; Zhang, C.; Chen, B. Luteolin inhibits angiogenesis of the M2-like TAMs via the downregulation of hypoxia inducible factor-1α and the STAT3 signalling pathway under hypoxia. *Mol. Med. Rep.* **2018**, *18*, 2914–2922. [CrossRef] [PubMed]
36. Lamy, S.; Akla, N.; Ouanouki, A.; Lord-Dufour, S.; Beliveau, R. Diet-derived polyphenols inhibit angiogenesis by modulating the interleukin-6/STAT3 pathway. *Exp. Cell Res.* **2012**, *318*, 1586–1596. [CrossRef] [PubMed]

37. Song, X.M.; Yao, J.Y.; Wang, F.; Zhou, M.; Zhou, Y.X.; Wang, H.; Wei, L.B.; Zhao, L.; Li, Z.Y.; Lu, N.; et al. Wogonin inhibits tumor angiogenesis via degradation of HIF-1α protein. *Toxicol. Appl. Pharmacol.* **2013**, *271*, 144–155. [CrossRef] [PubMed]
38. Fu, R.; Chen, Y.; Wang, X.P.; An, T.; Tao, L.; Zhou, Y.X.; Huang, Y.-J.; Chen, B.-A.; Li, Z.Y.; You, Q.D.; et al. Wogonin inhibits multiple myeloma-stimulated angiogenesis via the c-Myc/VHL/HIF-1α signaling axis. *Oncotarget* **2016**, *7*, 5715–5727. [CrossRef]
39. Lu, N.; Gao, Y.; Ling, Y.; Chen, Y.; Gu, H.Y.; Qi, Q.; Liu, W.; Wang, X.T.; You, Q.D.; Guo, Q.L. Wogonin suppresses tumor growth in vivo and VEGF-induced angiogenesis through inhibiting tyrosine phosphorylation of VEGFR2. *Life Sci.* **2008**, *82*, 956–963. [CrossRef] [PubMed]
40. Zhou, M.; Song, X.; Huang, Y.; Wei, L.; Li, Z.; You, Q.; Guo, Q.; Lu, N. Wogonin inhibits H_2O_2-induced angiogenesis via suppressing the PI3K/Akt/NF-κB signaling pathway. *Vasc. Pharmacol.* **2014**, *60*, 110–119. [CrossRef]
41. Huang, Y.; Zhao, K.; Hu, Y.; Zhou, Y.; Luo, X.; Li, X.; Wei, L.; Li, Z.; You, Q.; Guo, Q.; et al. Wogonoside inhibits angiogenesis in breast cancer via suppressing the Wnt/β-catenin pathway. *Mol. Carcinog.* **2016**, *55*, 1598–1612. [CrossRef]
42. Kunimasa, K.; Ikekita, M.; Sato, M.; Ohta, T.; Yamori, Y.; Ikeda, M.; Kuranuki, S.; Oikawa, T. Nobiletin, a citrus polymethoxyflavonoid, suppresses multiple angiogenesis-related endothelial cell functions and angiogenesis in vivo. *Cancer Sci.* **2010**, *101*, 2462–2469. [CrossRef] [PubMed]
43. Chen, J.; Chen, A.Y.; Huang, H.; Ye, X.; Rollyson, W.D.; Perry, H.E.; Brown, K.C.; Rojanasakul, Y.; Rankin, G.O.; Dasgupta, P.; et al. The flavonoid nobiletin inhibits tumor growth and angiogenesis of ovarian cancers via the Akt pathway. *Int. J. Oncol.* **2015**, *46*, 2629–2638. [CrossRef] [PubMed]
44. Sp, N.; Kang, D.Y.; Joung, Y.H.; Park, J.H.; Kim, W.S.; Lee, H.K.; Song, K.D.; Park, Y.M.; Yang, Y.M. Nobiletin inhibits angiogenesis by regulating Src/FAK/STAT3-mediated signaling through PXN in ER(+) breast cancer cells. *Int. J. Mol. Sci.* **2017**, *18*, 935. [CrossRef] [PubMed]
45. Gao, Y.; Lu, N.; Ling, Y.; Chen, Y.; Wang, L.; Zhao, Q.; Qi, Q.; Liu, W.; Zhang, H.; You, Q.; et al. Oroxylin A inhibits angiogenesis through blocking vascular endothelial growth factor-induced KDR/Flk-1 phosphorylation. *J. Cancer Res. Clin.* **2010**, *136*, 667–675. [CrossRef]
46. Zhao, K.; Li, X.; Lin, B.; Yang, D.; Zhou, Y.; Li, Z.; Guo, Q.; Lu, N. Oroxyloside inhibits angiogenesis through suppressing internalization of VEGFR2/Flk-1 in endothelial cells. *J. Cell Physiol.* **2018**, *233*, 3454–3464. [CrossRef] [PubMed]
47. Cathcart, M.C.; Useckaite, Z.; Drakeford, C.; Semik, V.; Lysaght, J.; Gately, K.; O'Byrne, K.J.; Pidgeon, G.P. Anti-cancer effects of baicalein in non-small cell lung cancer in vitro and in vivo. *BMC Cancer* **2016**, *16*, 707. [CrossRef] [PubMed]
48. Liu, J.J.; Huang, T.S.; Cheng, W.F.; Lu, F.J. Baicalein and baicalin are potent inhibitors of angiogenesis: Inhibition of endothelial cell proliferation, migration and differentiation. *Int. J. Cancer* **2003**, *106*, 559–565. [CrossRef] [PubMed]
49. Chen, J.; Li, Z.; Chen, A.Y.; Ye, X.; Luo, H.; Rankin, G.O.; Chen, Y.C. Inhibitory effect of baicalin and baicalein on ovarian cancer cells. *Int. J. Mol. Sci.* **2013**, *14*, 6012–6025. [CrossRef] [PubMed]
50. Hwang, K.Y.; Oh, Y.T.; Yoon, H.; Lee, J.; Kim, H.; Choe, W.; Kang, I. Baicalein suppresses hypoxia-induced HIF-1alpha protein accumulation and activation through inhibition of reactive oxygen species and PI3-kinase/Akt pathway in BV2 murine microglial cells. *Neurosci. Lett.* **2008**, *444*, 264–269. [CrossRef]
51. Ling, Y.; Chen, Y.; Chen, P.; Hui, H.; Song, X.; Lu, Z.; Li, C.; Lu, N.; Guo, Q. Baicalein potently suppresses angiogenesis induced by vascular endothelial growth factor through the p53/Rb signaling pathway leading to G1/S cell cycle arrest. *Exp. Biol. Med.* **2011**, *236*, 851–858. [CrossRef]
52. Ling, Y.; Wang, L.; Chen, Y.; Feng, F.; You, Q.; Lu, N.; Guo, Q. Baicalein inhibits angiogenesis induced by lipopolysaccharide through TRAF6 mediated toll-like receptor 4 pathway. *Biomed. Prev. Nutr.* **2011**, *1*, 172–179. [CrossRef]
53. Koirala, N. Metabolic Engineering of *Eescherichia coli* BL21 (DE3) for the Production of Methylated/Glycosylated Flavonoids and Their Biological Activities. Ph.D. Thesis, Sun Moon University, Asan-si, Republic of Korea, 2017. Available online: http://archive.nnl.gov.np:8080/handle/123456789/245 (accessed on 9 March 2024).
54. Liu, L.Z.; Jing, Y.; Jiang, L.L.; Jiang, X.E.; Jiang, Y.; Rojanasakul, Y.; Jiang, B.H. Acacetin inhibits VEGF expression, tumor angiogenesis and growth through AKT/HIF-1α pathway. *Biochem. Biophys. Res. Commun.* **2011**, *413*, 299–305. [CrossRef]
55. Bhat, T.A.; Nambiar, D.; Tailor, D.; Pal, A.; Agarwal, R.; Singh, R.P. Acacetin inhibits in vitro and in vivo angiogenesis and downregulates STAT signaling and VEGF expression. *Cancer Prev. Res.* **2013**, *6*, 1128–1139. [CrossRef] [PubMed]
56. Kim, H.J.; Ahn, M.-R. Apigenin suppresses angiogenesis by inhibiting tube formation and inducing apoptosis. *Nat. Prod. Commun.* **2016**, *11*, 1433–1436. [CrossRef]
57. Fang, J.; Xia, C.; Cao, Z.; Zheng, J.Z.; Reed, E.; Jiang, B.H. Apigenin inhibits VEGF and HIF-1 expression via PI3K/AKT/p70S6K1 and HDM2/p53 pathways. *FASEB J.* **2005**, *19*, 342–353. [CrossRef]
58. Melstrom, L.G.; Salabat, M.R.; Ding, X.Z.; Strouch, M.J.; Grippo, P.J.; Mirzoeva, S.; Pelling, J.C.; Bentrem, D.J. Apigenin downregulates the hypoxia response genes: HIF-1α, GLUT-1, and VEGF in human pancreatic cancer cells. *J. Surg. Res.* **2011**, *167*, 173–181. [CrossRef]
59. Shukla, S.; MacLennan, G.T.; Fu, P.; Gupta, S. Apigenin attenuates insulin-like growth Factor-I signaling in an autochthonous mouse prostate cancer model. *Pharm. Res.* **2012**, *29*, 1506–1517. [CrossRef] [PubMed]
60. Jiang, H.; Wu, D.; Xu, D.; Yu, H.; Zhao, Z.; Ma, D.; Jin, J. Eupafolin exhibits potent anti-angiogenic and antitumor activity in hepatocellular carcinoma. *Int. J. Biol. Sci.* **2017**, *13*, 701–711. [CrossRef] [PubMed]

61. de Novais, L.M.; de Arueira, C.C.; Ferreira, L.F.; Ribeiro, T.A.; Sousa, P.T., Jr.; Jacinto, M.J.; de Carvalho, M.G.; Judice, W.A.; Jesus, L.O.; de Souza, A.A.; et al. 4′-Hydroxy-6,7-methylenedioxy-3-methoxyflavone: A novel flavonoid from *Dulacia egleri* with potential inhibitory activity against cathepsins B and L. *Fitoterapia* **2019**, *132*, 26–29. [CrossRef]
62. Razak, N.A.; Abu, N.; Ho, W.Y.; Zamberi, N.R.; Tan, S.W.; Alitheen, N.B.; Long, K.; Yeap, S.K. Cytotoxicity of eupatorin in MCF-7 and MDA-MB-231 human breast cancer cells via cell cycle arrest, anti-angiogenesis and induction of apoptosis. *Sci. Rep.* **2019**, *9*, 1514. [CrossRef]
63. Wang, S.; Yan, Y.; Cheng, Z.; Hu, Y.; Liu, T. Sotetsuflavone suppresses invasion and metastasis in non-small-cell lung cancer A549 cells by reversing EMT via the TNF-α/NF-κB and PI3K/AKT signaling pathway. *Cell Death Discov.* **2018**, *4*, 26. [CrossRef] [PubMed]
64. Cheng, Z.K.; Wang, S.H.; Hu, Y.L.; Yan, Y.; Liu, T.X. Effects of sotetsuflavone on expression of endostatin, TGF-β, STAT3, β-catenin and ZO-1 in non-small cell lung cancer A549 cells. *TMR Cancer* **2018**, *1*, 58–65. [CrossRef]
65. Gao, L.; Wang, L.; Sun, Z.; Li, H.; Wang, Q.; Yi, C.; Wang, X. Morusin shows potent antitumor activity for human hepatocellular carcinoma in vitro and in vivo through apoptosis induction and angiogenesis inhibition. *Drug Des. Dev. Ther.* **2017**, *11*, 1789–1802. [CrossRef]
66. Yin, X.L.; Lv, Y.; Wang, S.; Zhang, Y.Q. Morusin suppresses A549 cell migration and induces cell apoptosis by downregulating the expression of COX 2 and VEGF genes. *Oncol. Rep.* **2018**, *40*, 504–510. [CrossRef] [PubMed]
67. Guo, Y.; Wang, S.; Hoot, D.R.; Clinton, S.K. Suppression of VEGF-mediated autocrine and paracrine interactions between prostate cancer cells and vascular endothelial cells by soy isoflavones. *J. Nutr. Biochem.* **2007**, *18*, 408–417. [CrossRef] [PubMed]
68. Büchler, P.; Reber, H.A.; Büchler, M.W.; Friess, H.; Lavey, R.S.; Hines, O.J. Antiangiogenic activity of genistein in pancreatic carcinoma cells is mediated by the inhibition of hypoxia-inducible factor-1 and the down-regulation of VEGF gene expression. *Cancer* **2004**, *100*, 201–210. [CrossRef]
69. Kim, M.H. Flavonoids inhibit VEGF/bFGF-induced angiogenesis in vitro by inhibiting the matrix-degrading proteases. *J. Cell Biochem.* **2003**, *89*, 529–538. [CrossRef]
70. Su, S.J.; Yeh, T.M.; Chuang, W.J.; Ho, C.L.; Chang, K.L.; Cheng, H.L.; Liu, H.S.; Cheng, H.L.; Hsu, P.Y.; Chow, N.H. The novel targets for anti-angiogenesis of genistein on human cancer cells. *Biochem. Pharmacol.* **2005**, *69*, 307–318. [CrossRef] [PubMed]
71. Li, Y.; Sarkar, F.H. Down-regulation of invasion and angiogenesis-related genes identified by cDNA microarray analysis of PC3 prostate cancer cells treated with genistein. *Cancer Lett.* **2002**, *186*, 157–164. [CrossRef]
72. Yu, X.; Zhu, J.; Mi, M.; Chen, W.; Pan, Q.; Wei, M. Anti-angiogenic genistein inhibits VEGF induced endothelial cell activation by decreasing PTK activity and MAPK activation. *Med. Oncol.* **2012**, *29*, 349–357. [CrossRef]
73. Pratheeshkumar, P.; Budhraja, A.; Son, Y.O.; Wang, X.; Zhang, Z.; Ding, S.-Z.; Wang, L.; Hitron, A.; Lee, J.C.; Xu, M.; et al. Quercetin inhibits angiogenesis mediated human prostate tumor growth by targeting VEGFR 2 regulated AKT/mTOR/P70S6K signaling pathways. *PLoS ONE* **2012**, *7*, e47516. [CrossRef] [PubMed]
74. Lai, W.W.; Hsu, S.C.; Chueh, F.S.; Chen, Y.Y.; Yang, J.S.; Lin, J.P.; Lien, J.C.; Tsai, C.H.; Chung, J.G. Quercetin inhibits migration and invasion of SAS human oral cancer cells through inhibition of NF-κB and matrix metalloproteinase-2/-9 signaling pathways. *Anticancer Res.* **2013**, *33*, 1941–1950. [PubMed]
75. Kong, L.; Wu, K.; Lin, H. Inhibitory effects of quercetin on angiogenesis of experimental mammary carcinoma. *Chin. J. Clin. Oncol.* **2005**, *2*, 631–636. [CrossRef]
76. Yang, F.; Jiang, X.; Song, L.; Wang, H.; Mei, Z.; Xu, Z.; Xing, N. Quercetin inhibits angiogenesis through thrombospondin-1 upregulation to antagonize human prostate cancer PC-3 cell growth in vitro and in vivo. *Oncol. Rep.* **2016**, *35*, 1602–1610. [CrossRef] [PubMed]
77. Ansó, E.; Zuazo, A.; Irigoyen, M.; Urdaci, M.C.; Rouzaut, A.; Martínez-Irujo, J.J. Flavonoids inhibit hypoxia-induced vascular endothelial growth factor expression by a HIF-1 independent mechanism. *Biochem. Pharmacol.* **2010**, *79*, 1600–1609. [CrossRef] [PubMed]
78. Huang, D.Y.; Dai, Z.R.; Li, W.-M.; Wang, R.G.; Yang, S.M. Inhibition of EGF expression and NF-κB activity by treatment with quercetin leads to suppression of angiogenesis in nasopharyngeal carcinoma. *Saudi. J. Biol. Sci.* **2018**, *25*, 826–831. [CrossRef] [PubMed]
79. Jackson, S.J.; Venema, R.C. Quercetin inhibits eNOS, microtubule polymerization, and mitotic progression in bovine aortic endothelial cells. *J. Nutr.* **2006**, *136*, 1178–1184. [CrossRef]
80. Xiao, X.; Shi, D.; Liu, L.; Wang, J.; Xie, X.; Kang, T.; Deng, W. Quercetin suppresses cyclooxygenase-2 expression and angiogenesis through inactivation of P300 signaling. *PLoS ONE* **2011**, *6*, e22934. [CrossRef]
81. Lin, C.; Wu, M.; Dong, J. Quercetin-4′-*O*-β-D-glucopyranoside (QODG) inhibits angiogenesis by suppressing VEGFR2-mediated signaling in zebrafish and endothelial cells. *PLoS ONE* **2012**, *7*, e31708. [CrossRef]
82. Singh, R.P.; Tyagi, A.; Sharma, G.; Mohan, S.; Agarwal, R. Oral silibinin inhibits in vivo human bladder tumor xenograft growth involving down-regulation of survivin. *Clin. Cancer Res.* **2008**, *14*, 300–308. [CrossRef]
83. Yang, S.-H.; Lin, J.K.; Huang, C.J.; Chen, W.S.; Li, S.Y.; Chiu, J.H. Silibinin inhibits angiogenesis via Flt-1, but not KDR, receptor up-regulation. *J. Surg. Res.* **2005**, *128*, 140–146. [CrossRef] [PubMed]
84. Singh, R.P.; Sharma, G.; Dhanalakshmi, S.; Agarwal, C.; Agarwal, R. Suppression of advanced human prostate tumor growth in athymic mice by silibinin feeding is associated with reduced cell proliferation, increased apoptosis, and inhibition of angiogenesis. *Cancer Epidemiol. Biomark. Prev.* **2003**, *12*, 933–939.

85. Gu, M.; Singh, R.P.; Dhanalakshmi, S.; Agarwal, C.; Agarwal, R. Silibinin inhibits inflammatory and angiogenic attributes in photocarcinogenesis in SKH-1 hairless mice. *Cancer Res.* **2007**, *67*, 3483–3491. [CrossRef] [PubMed]
86. Singh, R.P.; Raina, K.; Sharma, G.; Agarwal, R. Silibinin inhibits established prostate tumor growth, progression, invasion, and metastasis and suppresses tumor angiogenesis and epithelial-mesenchymal transition in transgenic adenocarcinoma of the mouse prostate model mice. *Clin. Cancer Res.* **2008**, *14*, 7773–7780. [CrossRef]
87. Singh, R.P.; Gu, M.; Agarwal, R. Silibinin inhibits colorectal cancer growth by inhibiting tumor cell proliferation and angiogenesis. *Cancer Res.* **2008**, *68*, 2043–2050. [CrossRef]
88. Singh, R.P.; Dhanalakshmi, S.; Agarwal, C.; Agarwal, R. Silibinin strongly inhibits growth and survival of human endothelial cells via cell cycle arrest and downregulation of survivin, Akt and NF-κB: Implications for angioprevention and antiangiogenic therapy. *Oncogene* **2005**, *24*, 1188–1202. [CrossRef] [PubMed]
89. García-Maceira, P.; Mateo, J. Silibinin inhibits hypoxia-inducible factor-1alpha and mTOR/p70S6K/4E-BP1 signaling pathway in human cervical and hepatoma cancer cells: Implications for anticancer therapy. *Oncogene* **2009**, *28*, 313. [CrossRef] [PubMed]
90. Kim, S.; Choi, J.H.; Lim, H.I.; Lee, S.K.; Kim, W.W.; Kim, J.S.; Kim, J.H.; Choe, J.H.; Yang, J.H.; Nam, S.J.; et al. Silibinin prevents TPA-induced MMP-9 expression and VEGF secretion by inactivation of the Raf/MEK/ERK pathway in MCF-7 human breast cancer cells. *Phytomedicine* **2009**, *16*, 573–580. [CrossRef]
91. Jung, S.K.; Lee, K.W.; Byun, S.; Lee, E.J.; Kim, J.E.; Bode, A.M.; Dong, Z.; Lee, H.J. Myricetin inhibits UVB-induced angiogenesis by regulating PI-3 kinase in vivo. *Carcinogenesis* **2010**, *31*, 911–917. [CrossRef]
92. Huang, H.; Chen, A.Y.; Rojanasakul, Y.; Ye, X.; Rankin, G.O.; Chen, Y.C. Dietary compounds galangin and myricetin suppress ovarian cancer cell angiogenesis. *J. Funct. Foods* **2015**, *15*, 464–475. [CrossRef]
93. Kim, G.D. Myricetin inhibits angiogenesis by inducing apoptosis and suppressing PI3K/Akt/mTOR signaling in endothelial cells. *J. Cancer Prev.* **2017**, *22*, 219–227. [CrossRef]
94. Luo, H.; Rankin, G.O.; Juliano, N.; Jiang, B.H.; Chen, Y.C. Kaempferol inhibits VEGF expression and in vitro angiogenesis through a novel ERK-NFkappaB-cMyc-p21 pathway. *Food Chem.* **2012**, *130*, 321–328. [CrossRef] [PubMed]
95. Luo, H.; Rankin, G.O.; Liu, L.; Daddysman, M.K.; Jiang, B.H.; Chen, Y.C. Kaempferol inhibits angiogenesis and VEGF expression through both HIF dependent and independent pathways in human ovarian cancer cells. *Nutr. Cancer* **2009**, *61*, 554–563. [CrossRef]
96. Liang, F.; Han, Y.; Gao, H.; Xin, S.; Chen, S.; Wang, N.; Qin, W.; Zhong, H.; Lin, S.; Yao, X.; et al. Kaempferol identified by zebrafish assay and fine fractionations strategy from *Dysosma versipellis* inhibits angiogenesis through VEGF and FGF pathways. *Sci. Rep.* **2015**, *5*, 14468. [CrossRef]
97. Chin, H.K.; Horng, C.T.; Liu, Y.S.; Lu, C.C.; Su, C.Y.; Chen, P.S.; Chiu, H.Y.; Tsai, F.J.; Shieh, P.C.; Yang, J.S. Kaempferol inhibits angiogenic ability by targeting VEGF receptor-2 and downregulating the PI3K/AKT, MEK and ERK pathways in VEGF-stimulated human umbilical vein endothelial cells. *Oncol. Rep.* **2018**, *39*, 2351–2357. [CrossRef] [PubMed]
98. Kim, G.D. Kaempferol Inhibits Angiogenesis by Suppressing HIF-1α and VEGFR2 Activation via ERK/p38 MAPK and PI3K/Akt/mTOR Signaling Pathways in Endothelial Cells. *Prev. Nutr. Food Sci.* **2017**, *22*, 320–326. [CrossRef] [PubMed]
99. Yu, Y.; Zhou, X.Z.; Ye, L.; Yuan, Q.; Freeberg, S.; Shi, C.; Zhu, P.W.; Bao, J.; Jiang, N.; Shao, Y. Rhamnazin attenuates inflammation and inhibits alkali burn-induced corneal neovascularization in rats. *RSC Adv.* **2018**, *8*, 26696–26706. [CrossRef] [PubMed]
100. Yu, Y.; Cai, W.; Pei, C.G.; Shao, Y. Rhamnazin, a novel inhibitor of VEGFR2 signaling with potent antiangiogenic activity and antitumor efficacy. *Biochem. Biophys. Res. Commun.* **2015**, *458*, 913–919. [CrossRef]
101. Chen, D.; Li, D.; Xu, X.B.; Qiu, S.; Luo, S.; Qiu, E.; Rong, Z.; Zhang, J.; Zheng, D. Galangin inhibits epithelial-mesenchymal transition and angiogenesis by downregulating CD44 in glioma. *J. Cancer* **2019**, *10*, 4499. [CrossRef]
102. Khodadadi, N.; Abbasi, B. The effect of Fisetin on the colorectal cancer: A review. *Food Health* **2020**, *3*, 35–44.
103. Arowosegbe, M.A.; Ogunleye, A.J.; Eniafe, G.O.; Omotuyi, O.I.; Ehima, V.O.; Metibemu, D.S.; Ogungbe, B.; Kanmodi, R.I.; Ogunmola, O.J. Applications of in silico methodologies in exploring the inhibitory potentials of fisetin on MMP-8 and MMP-13 in colorectal cancer progression. *Int. J. Drug Dev. Res.* **2017**, *9*, 9–15.
104. Bhat, T.A.; Nambiar, D.; Pal, A.; Agarwal, R.; Singh, R.P. Fisetin inhibits various attributes of angiogenesis in vitro and in vivo—Implications for angioprevention. *Carcinogenesis* **2012**, *33*, 385–393. [CrossRef] [PubMed]
105. Park, J.H.; Jang, Y.J.; Choi, Y.J.; Jang, J.W.; Kim, J.H.; Rho, Y.K.; Kim, I.J.; Kim, H.J.; Leem, M.J.; Lee, S.T. Fisetin inhibits matrix metalloproteinases and reduces tumor cell invasiveness and endothelial cell tube formation. *Nutr. Cancer* **2013**, *65*, 1192–1199. [CrossRef] [PubMed]
106. Tsai, C.F.; Chen, J.H.; Chang, C.N.; Lu, D.Y.; Chang, P.C.; Wang, S.L.; Yeh, W.L. Fisetin inhibits cell migration via inducing HO-1 and reducing MMPs expression in breast cancer cell lines. *Food Chem. Toxicol.* **2018**, *120*, 528–535. [CrossRef] [PubMed]
107. Chien, C.-S.; Shen, K.-H.; Huang, J.-S.; Ko, S.-C.; Shih, Y.-W. Antimetastatic potential of fisetin involves inactivation of the PI3K/Akt and JNK signaling pathways with downregulation of MMP-2/9 expressions in prostate cancer PC-3 cells. *Mol. Cell. Biochem.* **2009**, *333*, 169–180. [CrossRef]
108. Chou, R.H.; Hsieh, S.C.; Yu, Y.L.; Huang, M.H.; Huang, Y.C.; Hsieh, Y.H. Fisetin inhibits migration and invasion of human cervical cancer cells by down-regulating urokinase plasminogen activator expression through suppressing the p38 MAPK-dependent NF-κB signaling pathway. *PLoS ONE* **2013**, *8*, e71983. [CrossRef] [PubMed]
109. Murtaza, I.; Adhami, V.M.; Hafeez, B.B.; Saleem, M.; Mukhtar, H. Fisetin, a natural flavonoid, targets chemoresistant human pancreatic cancer AsPC-1 cells through DR3-mediated inhibition of NF-κB. *Int. J. Cancer* **2009**, *125*, 2465–2473. [CrossRef] [PubMed]

110. Bassères, D.S.; Baldwin, A.S. Nuclear factor-κB and inhibitor of κB kinase pathways in oncogenic initiation and progression *Oncogene* **2006**, *25*, 6817–6830. [CrossRef]
111. Syed, D.N.; Adhami, V.M.; Khan, M.I.; Mukhtar, H. Inhibition of Akt/mTOR signaling by the dietary flavonoid fisetin. *Anti-Cancer Agents Med. Chem.* **2013**, *13*, 995–1001. [CrossRef]
112. Aksamitiene, E.; Kiyatkin, A.; Kholodenko, B.N. Cross-talk between mitogenic Ras/MAPK and survival PI3K/Akt pathways A fine balance. *Biochem. Soc. Trans.* **2012**, *40*, 139–146. [CrossRef]
113. Dell'Eva, R.; Ambrosini, C.; Vannini, N.; Piaggio, G.; Albini, A.; Ferrari, N. AKT/NF-κB inhibitor xanthohumol targets cell growth and angiogenesis in hematologic malignancies. *Cancer* **2007**, *110*, 2007–2011. [CrossRef] [PubMed]
114. Albini, A.; Dell'Eva, R.; Vené, R.; Ferrari, N.; Buhler, D.R.; Noonan, D.M.; Fassina, G. Mechanisms of the antiangiogenic activity by the hop flavonoid xanthohumol: NF-κB and Akt as targets. *FASEB J.* **2006**, *20*, 527–529. [CrossRef] [PubMed]
115. Harikumar, K.B.; Kunnumakkara, A.B.; Ahn, K.S.; Anand, P.; Krishnan, S.; Guha, S.; Aggarwal, B.B. Modification of the cysteine residues in IκBα kinase and NF-κB (p65) by xanthohumol leads to suppression of NF-κB–regulated gene products and potentiation of apoptosis in leukemia cells. Blood. *J. Am. Soc. Hematol.* **2009**, *113*, 2003–2013. [CrossRef] [PubMed]
116. Saito, K.; Matsuo, Y.; Imafuji, H.; Okubo, T.; Maeda, Y.; Sato, T.; Shamoto, T.; Tsuboi, K.; Morimoto, M.; Takahashi, H.; et al Xanthohumol inhibits angiogenesis by suppressing nuclear factor-κB activation in pancreatic cancer. *Cancer Sci.* **2018**, *109*, 132–140. [CrossRef] [PubMed]
117. Kimura, Y.; Baba, K. Antitumor and antimetastatic activities of *Angelica keiskei* roots, part 1: Isolation of an active substance, xanthoangelol. *Int. J. Cancer* **2003**, *106*, 429–437. [CrossRef] [PubMed]
118. Chung, C.H.; Chang, C.H.; Chen, S.S.; Wang, H.H.; Yen, J.Y.; Hsiao, C.J.; Wu, N.L.; Chen, Y.L.; Huang, T.F.; Wang, P.C. Butein inhibits angiogenesis of human endothelial progenitor cells via the translation dependent signaling pathway. *Evid. Based Complement. Altern. Med.* **2013**, *2013*, 943187. [CrossRef] [PubMed]
119. Moon, D.O.; Choi, Y.H.; Moon, S.K.; Kim, W.J.; Kim, G.Y. Butein suppresses the expression of nuclear factor-kappa B-mediated matrix metalloproteinase-9 and vascular endothelial growth factor in prostate cancer cells. *Toxicol. In Vitro* **2010**, *24*, 1927–1934. [CrossRef] [PubMed]
120. Jiang, B.; Song, J.; Jin, Y. A flavonoid monomer tricin in Gramineous plants: Metabolism, bio/chemosynthesis, biological properties, and toxicology. *Food Chem.* **2020**, *320*, 126617. [CrossRef]
121. Saleem, H.; Anwar, S.; Alafnan, A.; Ahemad, N. Luteolin in a centum of valuable plant bioactives. In *A Centum of Valuable Plant Bioactives*; Academic Press: Cambridge, MA, USA, 2021; pp. 509–523. [CrossRef]
122. Yin, Y.; Qu, L.; Zhu, D.; Wu, Y.; Zhou, X. Effect of SOCS3 on apoptosis of human trophoblasts via adjustment of the JAK2/STAT3 signaling pathway in preterm birth. *Transl. Pediatr.* **2021**, *10*, 1637. [CrossRef]
123. Murata, T.; Ishiwa, S.; Lin, X.; Nakazawa, Y.; Tago, K.; Funakoshi-Tago, M. The citrus flavonoid, nobiletin inhibits neuronal inflammation by preventing the activation of NF-κB. *Neurochem. Int.* **2023**, *171*, 105613. [CrossRef]
124. Wang, C.; Gao, M.Q. Research progress on the antidepressant effects of baicalin and its aglycone baicalein: A systematic review of the biological mechanisms. *Neurochem. Res.* **2023**, *49*, 14–28. [CrossRef] [PubMed]
125. Kim, S.H.; Lee, Y.C. Plant-derived nanoscale-encapsulated antioxidants for oral and topical uses: A brief review. *Int. J. Mol. Sci.* **2022**, *23*, 3638. [CrossRef]
126. Bozorova, M.I.; Maulyanov, S.A.; Abdumalikov, I.I.; Amanova, Z.K.; Sharipova, F.S.; Kungirotova, A.I.; Toshmatov, Z.O. Flavonoids from *Leonurus turkestanicus*. *Chem. Nat. Compd.* **2021**, *57*, 152–153. [CrossRef]
127. Adzet, T.; Martinez, F. Flavonoids in the leaves of Thymus: A chemotaxonomic survey. *Biochem. Syst. Ecol.* **1981**, *9*, 293–295. [CrossRef]
128. Daneshvar, S.; Zamanian, M.Y.; Ivraghi, M.S.; Golmohammadi, M.; Modanloo, M.; Kamiab, Z.; Pourhosseini, S.M.E.; Heidari, M.; Bazmandegan, G. A comprehensive view on the apigenin impact on colorectal cancer: Focusing on cellular and molecular mechanisms. *Food Sci. Nutr.* **2023**, *11*, 6789–6801. [CrossRef] [PubMed]
129. Thakur, M.; Kumar, I.; Dhiman, S. A review on artemisia princeps: Pharmacology and anti-Inflammatory potential of phytoconstituents. *Asian J. Pharm. Technol.* **2023**, *13*, 285–292. [CrossRef]
130. Patel, D.K. Protective role of eupafolin against tumor, inflammation, melanogenesis, viral disease and renal injury: Pharmacological and analytical aspects through scientific data analysis. *Curr. Chin. Sci.* **2022**, *2*, 143–151. [CrossRef]
131. Alafnan, A.; Nadeem, M.F.; Ahmad, S.F.; Sarfraz, M.; Aalamri, A.; Khalifa, N.E.; Ahemad, N.; Saleem, H. A comprehensive assessment of phytochemicals from *Phyla nodiflora* (L.) Greene as a potential enzyme inhibitor, and their biological potential: An in-silico, in-vivo, and in-vitro approach. *Arab. J. Chem.* **2023**, *16*, 105233. [CrossRef]
132. Mustafa, A.; Elkhamisy, F.; Arghiani, N.; Pranjol, M.Z.I. Potential crosstalk between pericytes and cathepsins in the tumour microenvironment. *Biomed. Pharmacother.* **2023**, *164*, 114932. [CrossRef] [PubMed]
133. Zakiyanov, O.; Kalousová, M.; Zima, T.; Tesař, V. Matrix metalloproteinases and tissue inhibitors of matrix metalloproteinases in kidney disease. *Adv. Clin. Chem.* **2021**, *105*, 141–212. [CrossRef]
134. Li, Q.; Wang, Y.; Zhang, L.; Chen, L.; Du, Y.; Ye, T.; Shi, X. Naringenin exerts anti-angiogenic effects in human endothelial cells: Involvement of ERRα/VEGF/KDR signaling pathway. *Fitoterapia* **2016**, *111*, 78–86. [CrossRef] [PubMed]
135. Lowe, H.I.; Toyang, N.J.; Watson, C.T.; Ayeah, K.N.; Bryant, J. HLBT-100: A highly potent anti-cancer flavanone from *Tillandsia recurvata* (L.) L. *Cancer Cell Int.* **2017**, *17*, 38. [CrossRef]

136. Kim, G.D. Hesperetin inhibits vascular formation by suppressing of the PI3K/AKT, ERK, and p38 MAPK signaling pathways. *Prev. Nutr. Food Sci.* **2014**, *19*, 299–306. [CrossRef] [PubMed]
137. Nalini, N.; Aranganathan, S.; Kabalimurthy, J. Chemopreventive efficacy of hesperetin (citrus flavonone) against 1,2-dimethylhydrazine-induced rat colon carcinogenesis. *Toxicol. Mech. Methods* **2012**, *22*, 397–408. [CrossRef] [PubMed]
138. Shukla, K.; Sonowal, H.; Saxena, A.; Ramana, K.V. Didymin by suppressing NF-κB activation prevents VEGF-induced angiogenesis in vitro and in vivo. *Vascul. Pharmacol.* **2019**, *115*, 18–25. [CrossRef] [PubMed]
139. Dai, F.; Gao, L.; Zhao, Y.; Wang, C.; Xie, S. Farrerol inhibited angiogenesis through Akt/mTOR, Erk and Jak2/Stat3 signal pathway. *Phytomedicine* **2016**, *23*, 686–693. [CrossRef] [PubMed]
140. Xi, S.Y.; Zhang, Q.; Liu, C.Y.; Xie, H.; Yue, L.F.; Gao, X.M. Effects of hydroxy safflower yellow-A on tumor capillary angiogenesis in transplanted human gastric adenocarcinoma BGC-823 tumors in nude mice. *J. Tradit. Chin. Med.* **2012**, *32*, 243–248. [CrossRef] [PubMed]
141. Xi, S.Y.; Zhang, Q.; Xie, H.; Liu, L.; Liu, C.; Gao, X.; Zhang, J.; Wu, L.; Qian, L.; Deng, X. Effects of hydroxy safflor yellow A on blood vessel and mRNA expression with VEGF and bFGF of transplantation tumor with gastric adenocarcinoma cell line BGC-823 in nude mice. *China J. Chin. Mater. Med.* **2009**, *34*, 605–610.
142. Xi, S.; Zhang, Q.; Liu, C.; Xie, H.; Yue, L.; Zhao, Y.; Zang, B.; Gao, X. Effects of HSYA on expression of bFGF protein and MMP-9 in BGC-823 transplantation tumor of nude mice. *China J. Chin. Mater. Med.* **2010**, *35*, 2877–2881. [CrossRef]
143. Yang, F.; Li, J.; Zhu, J.; Wang, D.; Chen, S.; Bai, X. Hydroxysafflor yellow A inhibits angiogenesis of hepatocellular carcinoma via blocking ERK/MAPK and NF-κB signaling pathway in H22 tumor-bearing mice. *Eur. J. Pharmacol.* **2015**, *754*, 105–114. [CrossRef]
144. Zhang, J.; Li, J.; Song, H.; Xiong, Y.; Liu, D.; Bai, X. Hydroxysafflor yellow A suppresses angiogenesis of hepatocellular carcinoma through inhibition of p38 MAPK phosphorylation. *Biomed. Pharmacother.* **2019**, *109*, 806–814. [CrossRef]
145. Wang, C.; Li, Y.; Chen, H.; Zhang, J.; Zhang, J.; Qin, T.; Duan, C.; Chen, X.; Liu, Y.; Zhou, X.; et al. Inhibition of CYP4A by a novel flavonoid FLA-16 prolongs survival and normalizes tumor vasculature in glioma. *Cancer Lett.* **2017**, *402*, 131–141. [CrossRef] [PubMed]
146. Wang, J.; Yang, Z.; Wen, J.; Ma, F.; Wang, F.; Yu, K.; Tang, M.; Wu, W.; Dong, Y.; Cheng, X.; et al. SKLB-M8 induces apoptosis through the AKT/mTOR signaling pathway in melanoma models and inhibits angiogenesis with a decrease of ERK1/2 phosphorylation. *J. Pharmacol. Sci.* **2014**, *126*, 198–207. [CrossRef]
147. Kim, Y.H.; Shin, E.K.; Kim, D.H.; Lee, H.H.; Park, J.H.Y.; Kim, J.K. Antiangiogenic effect of licochalcone A. *Biochem. Pharmacol.* **2010**, *80*, 1152–1159. [CrossRef]
148. Kwon, S.J.; Park, S.Y.; Kwon, G.T.; Lee, K.W.; Kang, Y.H.; Choi, M.S.; Yun, J.W.; Jeon, J.H.; Jun, J.G.; Park, J.H.Y. Licochalcone E present in licorice suppresses lung metastasis in the 4T1 mammary orthotopic cancer model. *Cancer Prev. Res.* **2013**, *6*, 603–613. [CrossRef]
149. Abu, N.; Mohamed, N.E.; Yeap, S.K.; Lim, K.L.; Akhtar, M.N.; Zulfadli, A.J.; Kee, B.B.; Abdullah, M.P.; Omar, A.R.; Alitheen, N.B. In vivo antitumor and antimetastatic effects of flavokawain B in 4T1 breast cancer cell-challenged mice. *Drug Des. Devel. Ther.* **2015**, *9*, 1401–1417. [CrossRef] [PubMed]
150. Abu, N.; Akhtar, M.N.; Yeap, S.K.; Lim, K.L.; Ho, W.Y.; Abdullah, M.P.; Ho, C.L.; Omar, A.R.; Ismail, J.; Alitheen, N.B. Flavokawain B induced cytotoxicity in two breast cancer cell lines, MCF-7 and MDA-MB231 and inhibited the metastatic potential of MDA-MB231 via the regulation of several tyrosine kinases in vitro. *BMC Complement. Altern. Med.* **2016**, *16*, 86. [CrossRef]
151. Abu, N.; Akhtar, M.N.; Yeap, S.K.; Lim, K.L.; Ho, W.Y.; Zulfadli, A.J.; Omar, A.R.; Sulaiman, M.R.; Abdullah, M.P.; Alitheen, N.B. Flavokawain A induces apoptosis in MCF-7 and MDA-MB231 and inhibits the metastatic process in vitro. *PLoS ONE* **2014**, *9*, e105244. [CrossRef]
152. Liu, Z.; Xu, X.; Li, X.; Liu, S.; Simoneau, A.R.; He, F.; Wu, X.R.; Zi, X. Kava chalcone, flavokawain A, inhibits urothelial tumorigenesis in the UPII-SV40T transgenic mouse model. *Cancer Prev. Res.* **2013**, *6*, 1365–1375. [CrossRef]
153. Xue, Z.G.; Niu, P.G.; Shi, D.H.; Liu, Y.; Deng, J.; Chen, Y.Y. Cardamonin inhibits angiogenesis by mTOR downregulation in SKOV3 cells. *Planta Med.* **2016**, *82*, 70–75. [CrossRef]
154. Tian, S.S.; Jiang, F.S.; Zhang, K.; Zhu, X.X.; Jin, B.; Lu, J.J.; Ding, Z.S. Flavonoids from the leaves of *Carya cathayensis* Sarg. inhibit vascular endothelial growth factor-induced angiogenesis. *Fitoterapia* **2014**, *92*, 34–40. [CrossRef] [PubMed]
155. Jiang, F.S.; Tian, S.S.; Lu, J.J.; Ding, X.H.; Qian, C.D.; Ding, B.; Ding, Z.S.; Jin, B. Cardamonin regulates miR-21 expression and suppresses angiogenesis induced by vascular endothelial growth factor. *BioMed Res. Int.* **2015**, *2015*, 501581. [CrossRef] [PubMed]
156. Kang, S.W.; Choi, J.S.; Choi, Y.J.; Bae, J.Y.; Li, J.; Kim, D.S.; Kim, J.L.; Shin, S.Y.; Lee, Y.J.; Kwun, I.S.; et al. Licorice isoliquiritigenin dampens angiogenic activity via inhibition of MAPK-responsive signaling pathways leading to induction of matrix metalloproteinases. *J. Nutr. Biochem.* **2010**, *21*, 55–65. [CrossRef] [PubMed]
157. Wang, Z.; Wang, N.; Han, S.; Wang, D.; Mo, S.; Yu, L. Dietary compound isoliquiritigenin inhibits breast cancer neoangiogenesis via VEGF/VEGFR-2 signaling pathway. *PLoS ONE* **2013**, *8*, e68566. [CrossRef]
158. Sun, Z.J.; Chen, G.; Zhang, W.; Hu, X.; Huang, C.F.; Wang, Y.F.; Jia, J.; Zhao, Y.F. Mammalian target of rapamycin pathway promotes tumor-induced angiogenesis in adenoid cystic carcinoma: Its suppression by isoliquiritigenin through dual activation of c-Jun NH2-terminal kinase and inhibition of extracellular signal-regulated kinase. *J. Pharmacol. Exp. Ther.* **2010**, *334*, 500–512. [CrossRef] [PubMed]
159. He, W.; Li, Y. Phytochemical isoliquiritigenin inhibits angiogenesis ex vivo and corneal neovascularization in mice. *Altern. Integr. Med.* **2014**, *3*, 176. [CrossRef]

160. Jhanji, V.; Liu, H.; Law, K.; Lee, V.Y.-W.; Huang, S.-F.; Pang, C.-P.; Yam, G.H.-F. Isoliquiritigenin from licorice root suppressed neovascularization in experimental ocular angiogenesis models. *Brit. J. Ophthalmol.* **2011**, *95*, 1309–1315. [CrossRef] [PubMed]
161. Lee, S.A.; Lee, M.; Ryu, H.; Kwak, T.; Kim, H.; Kang, M.; Jung, O.; Kim, H.; Park, K.; Lee, J. Differential inhibition of transmembrane 4 L six family member 5 (TM4SF5)-mediated tumorigenesis by TSAHC and sorafenib. *Cancer Biol. Ther.* **2011**, *11*, 330–336. [CrossRef]
162. Negrão, R.; Incio, J.; Lopes, R.; Azevedo, I.; Soares, R. Evidence for the effects of xanthohumol in disrupting angiogenic, but not stable vessels. *Int. J. Biomed. Sci.* **2007**, *3*, 279. [CrossRef]
163. Monteiro, R.; Calhau, C.; Silva, A.O.E.; Pinheiro-Silva, S.; Guerreiro, S.; Gärtner, F.; Azevedo, I.; Soares, R. Xanthohumol inhibits inflammatory factor production and angiogenesis in breast cancer xenografts. *J. Cell. Biochem.* **2008**, *104*, 1699–1707. [CrossRef]
164. Gallo, C.; Dallaglio, K.; Bassani, B.; Rossi, T.; Rossello, A.; Noonan, D.M.; D'Uva, G.; Bruno, A.; Albini, A. Hop derived flavonoid xanthohumol inhibits endothelial cell functions via AMPK activation. *Oncotarget* **2016**, *7*, 59917. [CrossRef] [PubMed]
165. Gao, S.; Siddiqui, N.; Etim, I.; Du, T.; Zhang, Y.; Liang, D. Developing nutritional component chrysin as a therapeutic agent: Bioavailability and pharmacokinetics consideration, and ADME mechanisms. *Biomed. Pharmacother.* **2021**, *142*, 112080. [CrossRef] [PubMed]
166. Karamanos, N.K.; Theocharis, A.D.; Piperigkou, Z.; Manou, D.; Passi, A.; Skandalis, S.S.; Vynios, D.H.; Orian-Rousseau, V.; Ricard-Blum, S.; Schemlzer, C.E.H.; et al. A guide to the composition and functions of the extracellular matrix. *FEBS J.* **2021**, *288*, 6850–6912. [CrossRef] [PubMed]
167. Chriscensia, E.; Aqila Arham, A.; Chrestella Wibowo, E.; Gracius, L.; Nathanael, J.; Hartrianti, P. Eupatorin from *Orthosiphon aristatus*: A review of the botanical origin, pharmacological effects and isolation methods. *Curr. Bioact. Compd.* **2023**, *19*, 45–60. [CrossRef]
168. López-Sáez, J.A.; Pérez-Alonso, M.J.; Negueruela, A.V. Biflavonoids of *Selaginella denticulata* growing in Spain. *Z. Naturforschung C* **1994**, *49*, 267–270. [CrossRef]
169. Li, S.H.; Zhang, H.J.; Niu, X.M.; Yao, P.; Sun, H.D.; Fong, H.H. Chemical constituents from *Amentotaxus yunnanensis* and *Torreya yunnanensis*. *J. Nat. Prod.* **2003**, *66*, 1002–1005. [CrossRef] [PubMed]
170. Chan, E.W.C.; Wong, S.K.; Tangah, J.; Inoue, T.; Chan, H.T. Phenolic constituents and anticancer properties of *Morus alba* (white mulberry) leaves. *J. Integr. Med.* **2020**, *18*, 189–195. [CrossRef] [PubMed]
171. Mansour, F.R.; Abdallah, I.A.; Bedair, A.; Hamed, M. Analytical methods for the determination of quercetin and quercetin glycosides in pharmaceuticals and biological Samples. *Crit. Rev. Anal. Chem.* **2023**, 1–26. [CrossRef]
172. Wang, X.; Zhang, Z.; Wu, S.C. Health benefits of Silybum marianum: Phytochemistry, pharmacology, and applications. *J. Agric. Food. Chem.* **2020**, *68*, 11644–11664. [CrossRef]
173. González-Sarrías, A.; Tomás-Barberán, F.A.; García-Villalba, R. Structural diversity of polyphenols and distribution in foods. In *Dietary Polyphenols: Their Metabolism and Health Effects*; John Wiley & Sons, Inc.: Hoboken, NJ, USA, 2020; pp. 1–29. [CrossRef]
174. Simón, J.; Casado-Andrés, M.; Goikoetxea-Usandizaga, N.; Serrano-Maciá, M.; Martínez-Chantar, M.L. Nutraceutical properties of polyphenols against liver diseases. *Nutrients* **2020**, *12*, 3517. [CrossRef]
175. Prabhu, S.; Molath, A.; Choksi, H.; Kumar, S.; Mehra, R. Classifications of polyphenols and their potential application in human health and diseases. *Int. J. Physiol. Nutr. Phys. Educ.* **2021**, *6*, 293–301. [CrossRef]
176. Kumar, N.; Jain, V. Kaempferol: A key emphasis on its counter-wired potential. *Int. J. Innov. Sci. Res. Technol.* **2023**, *8*, 2534–2540.
177. Abbaszadeh, H.; Keikhaei, B.; Mottaghi, S. A review of molecular mechanisms involved in anticancer and antiangiogenic effects of natural polyphenolic compounds. *Phytother. Res.* **2019**, *33*, 2002–2014. [CrossRef] [PubMed]
178. Khan, N.; Syed, D.N.; Ahmad, N.; Mukhtar, H. Fisetin: A dietary antioxidant for health promotion. *Antioxid. Redox Signal.* **2013**, *19*, 151–162. [CrossRef] [PubMed]
179. Nitti, M.; Ivaldo, C.; Traverso, N.; Furfaro, A.L. Clinical significance of heme oxygenase 1 in tumor progression. *Antioxidants* **2021**, *10*, 789. [CrossRef] [PubMed]
180. Masucci, M.T.; Minopoli, M.; Di Carluccio, G.; Motti, M.L.; Carriero, M.V. Therapeutic strategies targeting urokinase and its receptor in cancer. *Cancers* **2022**, *14*, 498. [CrossRef]
181. Kretschmer, M.; Rüdiger, D.; Zahler, S. Mechanical aspects of angiogenesis. *Cancers* **2021**, *13*, 4987. [CrossRef] [PubMed]
182. Meng, F.; Abedini, A.; Plesner, A.; Verchere, C.B.; Raleigh, D.P. The flavanol (−)-epigallocatechin 3-gallate inhibits amyloid formation by islet amyloid polypeptide, disaggregates amyloid fibrils, and protects cultured cells against IAPP-induced toxicity. *Biochemistry.* **2010**, *49*, 8127–8133. [CrossRef]
183. Moyle, C.W.; Cerezo, A.B.; Winterbone, M.S.; Hollands, W.J.; Alexeev, Y.; Needs, P.W.; Kroon, P.A. Potent inhibition of VEGFR-2 activation by tight binding of green tea epigallocatechin gallate and apple procyanidins to VEGF: Relevance to angiogenesis. *Mol. Nutr. Food Res.* **2015**, *59*, 401–412. [CrossRef]
184. Munir, S.; Shah, A.A.; Shahid, M.; Ahmed, M.S.; Shahid, A.; Rajoka, M.S.; Akash, M.S.; Akram, M.; Khurshid, M. Anti-angiogenesis potential of phytochemicals for the therapeutic management of tumors. *Curr. Pharm. Des.* **2020**, *26*, 265–278. [CrossRef]
185. Fassina, G.; Vene, R.; Morini, M.; Minghelli, S.; Benelli, R.; Noonan, D.M.; Albini, A. Mechanisms of inhibition of tumor angiogenesis and vascular tumor growth by epigallocatechin-3-gallate. *Clin. Cancer Res.* **2004**, *10*, 4865–4873. [CrossRef]
186. Zhou, F.; Zhou, H.; Wang, T.; Mu, Y.; Wu, B.; Guo, D.L.; Zhang, X.M.; Wu, Y. Epigallocatechin-3-gallate inhibits proliferation and migration of human colon cancer SW620 cells in vitro. *Acta Pharmacol. Sin.* **2012**, *33*, 120–126. [CrossRef]

187. Sakamoto, Y.; Terashita, N.; Muraguchi, T.; Fukusato, T.; Kubota, S. Effects of epigallocatechin-3-gallate (EGCG) on A549 lung cancer tumor growth and angiogenesis. *Biosci. Biotechnol. Biochem.* **2013**, *77*, 1799–1803. [CrossRef]
188. Shimizu, M.; Shirakami, Y.; Sakai, H.; Yasuda, Y.; Kubota, M.; Adachi, S.; Tsurumi, H.; Hara, Y.; Moriwaki, H. (−)-Epigallocatechin gallate inhibits growth and activation of the VEGF/VEGFR axis in human colorectal cancer cells. *Chem. Biol. Interact.* **2010**, *185*, 247–252. [CrossRef]
189. He, L.; Zhang, E.; Shi, J.; Li, X.; Zhou, K.; Zhang, Q.; Le, A.D.; Tang, X. (−)-Epigallocatechin-3-gallate inhibits human papillomavirus (HPV)-16 oncoprotein-induced angiogenesis in non-small cell lung cancer cells by targeting HIF-1α. *Cancer Chemother. Pharmacol.* **2013**, *71*, 713–725. [CrossRef]
190. Rashidi, B.; Malekzadeh, M.; Goodarzi, M.; Masoudifar, A.; Mirzaei, H. Green tea and its anti-angiogenesis effects. *Biomed. Pharmacother.* **2017**, *89*, 949–956. [CrossRef]
191. Zhu, B.H.; Chen, H.Y.; Zhan, W.H.; Wang, C.Y.; Cai, S.R.; Wang, Z.; Zhang, C.H.; He, Y.L. (−)-Epigallocatechin-3-gallate inhibits VEGF expression induced by IL-6 via Stat3 in gastric cancer. *World J. Gastroenterl.* **2011**, *17*, 2315–2325. [CrossRef]
192. (154) Zhu, B.H.; Zhan, W.H.; Li, Z.R.; Wang, Z.; He, Y.L.; Peng, J.S.; Cai, S.R.; Ma, J.P.; Zhang, C.H. (−)-Epigallocatechin-3-gallate inhibits growth of gastric cancer by reducing VEGF production and angiogenesis. *World J. Gastroenterol.* **2007**, *13*, 1162. [CrossRef]
193. Shi, J.L.; Liu, F.; Zhang, W.Z.; Liu, X.; Lin, B.H.; Tang, X.D. Epigallocatechin-3-gallate inhibits nicotine-induced migration and invasion by the suppression of angiogenesis and epithelial-mesenchymal transition in non-small cell lung cancer cells. *Oncol. Rep.* **2015**, *33*, 2972–2980. [CrossRef]
194. Siddiqui, I.A.; Malik, A.; Adhami, V.M.; Asim, M.; Hafeez, B.B.; Sarfaraz, S.; Mukhtar, H. Green tea polyphenol EGCG sensitizes human prostate carcinoma LNCaP cells to TRAIL-mediated apoptosis and synergistically inhibits biomarkers associated with angiogenesis and metastasis. *Oncogene* **2008**, *27*, 2055–2063. [CrossRef]
195. Chen, C.Y.; Lin, Y.J.; Wang, C.C.N.; Lan, Y.H.; Lan, S.J.; Sheu, M.J. Epigallocatechin-3-gallate inhibits tumor angiogenesis: Involvement of endoglin/Smad1 signaling in human umbilical vein endothelium cells. *Biomed. Pharmacother.* **2019**, *120*, 109491. [CrossRef]
196. Tudoran, O.; Soritau, O.; Balacescu, O.; Balacescu, L.; Braicu, C.; Rus, M.; Gherman, C.; Virag, P.; Irimie, F.; Berindan-Neagoe, I. Early transcriptional pattern of angiogenesis induced by EGCG treatment in cervical tumour cells. *J. Cell. Mol. Med.* **2012**, *16*, 520–530. [CrossRef]
197. Ohga, N.; Hida, K.; Hida, Y.; Muraki, C.; Tsuchiya, K.; Matsuda, K.; Ohiro, Y.; Totsuka, Y.; Shindoh, M. Inhibitory effects of epigallocatechin-3 gallate, a polyphenol in green tea, on tumor-associated endothelial cells and endothelial progenitor cells. *Cancer Sci.* **2009**, *100*, 1963–1970. [CrossRef]
198. Neuhaus, T.; Pabst, S.; Stier, S.; Weber, A.A.; Schrör, K.; Sachinidis, A.; Vetter, H.; Ko, Y.D. Inhibition of the vascular-endothelial growth factor-induced intracellular signaling and mitogenesis of human endothelial cells by epigallocatechin-3 gallate. *Eur. J. Pharmacol.* **2004**, *483*, 223–227. [CrossRef]
199. Gu, J.W.; Makey, K.L.; Tucker, K.B.; Chinchar, E.; Mao, X.; Pei, I.; Thomas, E.Y.; Miele, L. EGCG, a major green tea catechin suppresses breast tumor angiogenesis and growth via inhibiting the activation of HIF-1α and NFκB, and VEGF expression. *Vasc. Cell.* **2013**, *5*, 9. [CrossRef]
200. Li, X.; Feng, Y.; Liu, J.; Feng, X.; Zhou, K.; Tang, X. Epigallocatechin-3-gallate inhibits IGF-I-stimulated lung cancer angiogenesis through downregulation of HIF-1α and VEGF expression. *J. Nutr. Nutr.* **2013**, *6*, 169–178. [CrossRef] [PubMed]
201. Yamakawa, S.; Asai, T.; Uchida, T.; Matsukawa, M.; Akizawa, T.; Oku, N. (−)-Epigallocatechin gallate inhibits membrane-type 1 matrix metalloproteinase, MT1-MMP, and tumor angiogenesis. *Cancer Lett.* **2004**, *210*, 47–55. [CrossRef] [PubMed]
202. Garbisa, S.; Sartor, L.; Biggin, S.; Salvato, B.; Benelli, R.; Albini, A. Tumor gelatinases and invasion inhibited by the green tea flavanol epigallocatechin-3-gallate. *Cancer* **2001**, *91*, 822–832. [CrossRef] [PubMed]
203. Wang, J.; Man, G.C.W.; Chan, T.H.; Kwong, J.; Wang, C.C. A prodrug of green tea polyphenol (–)-epigallocatechin-3-gallate (Pro-EGCG) serves as a novel angiogenesis inhibitor in endometrial cancer. *Cancer Lett.* **2018**, *412*, 10–20. [CrossRef]
204. Khan, A.; Ikram, M.; Hahm, J.R.; Kim, M.O. Antioxidant and anti-inflammatory effects of citrus flavonoid hesperetin: Special focus on neurological disorders. *Antioxidants* **2020**, *9*, 609. [CrossRef] [PubMed]
205. Abu El-Asrar, A.M.; Nawaz, M.I.; Ahmad, A.; Dillemans, L.; Siddiquei, M.; Allegaert, E.; Gikandi, P.W.; Hertogh, G.D.; Opdenakker, G.; Struyf, S. CD40 Ligand–CD40 Interaction is an intermediary between inflammation and angiogenesis in proliferative diabetic retinopathy. *Int. J. Mol. Sci.* **2023**, *24*, 15582. [CrossRef]
206. Shukla, K.; Sonowal, H.; Saxena, A.; Ramana, K.V. Didymin prevents hyperglycemia-induced human umbilical endothelial cells dysfunction and death. *Biochem. Pharmacol.* **2018**, *152*, 1–10. [CrossRef] [PubMed]
207. Wang, X.; Lai, Z.; Pang, Y.; Sun, Q.; Yang, W.; Wang, W. PD-1 blocking strategy for enhancing the anti-tumor effect of CAR T cells targeted CD105. *Heliyon* **2023**, *9*, e12688. [CrossRef] [PubMed]
208. Ao, H.; Feng, W.; Peng, C. Hydroxysafflor yellow A: A promising therapeutic agent for a broad spectrum of diseases. *Evid.-Based Complement. Altern. Med.* **2018**, *2018*, 8259280. [CrossRef] [PubMed]
209. Cheng, B.F.; Gao, Y.X.; Lian, J.J.; Guo, D.D.; Wang, L.; Wang, M.; Yang, H.J.; Feng, Z.W. Hydroxysafflor yellow A inhibits IL-1β-induced release of IL-6, IL-8, and MMP-1 via suppression of ERK, NF-κB and AP-1 signaling in SW982 human synovial cells. *Food Funct.* **2016**, *7*, 4516–4522. [CrossRef] [PubMed]

210. Kalyanakrishnan, K. Effect of ATF2 Transcription Factor on DLL4 Gene Expression in Angiogenesis. Master's Thesis, University of Wolverhampton, Wolverhampton, UK, 2021. Available online: https://wlv.openrepository.com/handle/2436/624703 (accessed on 1 November 2021).
211. Haibe, Y.; Kreidieh, M.; El Hajj, H.; Khalifeh, I.; Mukherji, D.; Temraz, S.; Shamseddine, A. Resistance mechanisms to anti-angiogenic therapies in cancer. *Front. Oncol.* **2020**, *10*, 221. [CrossRef] [PubMed]
212. Colombero, C.; Cárdenas, S.; Venara, M.; Martin, A.; Pennisi, P.; Barontini, M.; Nowicki, S. Cytochrome 450 metabolites of arachidonic acid (20-HETE, 11, 12-EET and 14, 15-EET) promote pheochromocytoma cell growth and tumor-associated angiogenesis. *Biochimie* **2020**, *171*, 147–157. [CrossRef] [PubMed]
213. Lafta, H.A.; AbdulHussein, A.H.; Al-Shalah, S.A.; Alnassar, Y.S.; Mohammed, N.M.; Akram, S.M.; Maytham, T.; Najafi, M. Tumor-associated macrophages (TAMs) in cancer resistance; modulation by natural products. *Curr. Top. Med. Chem.* **2023**, *23*, 1104–1122. [CrossRef] [PubMed]
214. Celentano, A.; Yiannis, C.; Paolini, R.; Zhang, P.; Farah, C.S.; Cirillo, N.; Yap, T.; McCullough, M. Kava constituents exert selective anticancer effects in oral squamous cell carcinoma cells in vitro. *Sci. Rep.* **2020**, *10*, 15904. [CrossRef] [PubMed]
215. Angeline, P.; Thomas, A.; Sankaranarayanan, S.A.; Rengan, A.K. Effect of pH on Isoliquiritigenin (ISL) fluorescence in lipo-polymeric system and metallic nanosystem. *Spectrochim. Acta A* **2021**, *252*, 119545. [CrossRef] [PubMed]
216. Plyusnin, S.N.; Babak, T.V.; Orlovskaya, N.V.; Ulyasheva, N.S.; Golubev, D.A.; Alekseev, A.; Shaposhnikov, M.V.; Moskalev, A. Sources of potential geroprotectors in the flora of the European Northeast. *J. Herb. Med.* **2023**, *41*, 100717. [CrossRef]
217. Rahim, N.S.; Wu, Y.S.; Sim, M.S.; Velaga, A.; Bonam, S.R.; Gopinath, S.; Subramaniyan, V.; Choy, K.W.; Teow, S.-Y.; Fareez, I.M.; et al. Three members of transmembrane-4-superfamily, TM4SF1, TM4SF4, and TM4SF5, as emerging anticancer molecular targets against cancer phenotypes and chemoresistance. *Pharmaceuticals* **2023**, *16*, 110. [CrossRef] [PubMed]
218. Yao-Lei, S.; Tian-Shuang, X.; Yi, J.; Na-Ni, W.; Ling-Chuan, X.; Ting, H.; Hai-Liang, X. Humulus lupulus L. extract and its active constituent xanthohumol attenuate oxidative stress and nerve injury induced by iron overload via activating AKT/GSK3 β and Nrf2/NQO1 pathways. *J. Nat. Med.* **2023**, *77*, 12–27. [CrossRef]
219. Javid, H.; Afshari, A.R.; Zahedi Avval, F.; Asadi, J.; Hashemy, S.I. Aprepitant promotes caspase-dependent apoptotic cell death and G2/M arrest through PI3K/Akt/NF-κB axis in cancer stem-like esophageal squamous cell carcinoma spheres. *BioMed Res. Int.* **2021**, *2021*, 8808214. [CrossRef] [PubMed]
220. Mottin, M.; Caesar, L.K.; Brodsky, D.; Mesquita, N.C.; de Oliveira, K.Z.; Noske, G.D.; Sousa, B.; Ramos, P.; Jarmer, H.; Loh, B.; et al. Chalcones from *Angelica keiskei* (ashitaba) inhibit key Zika virus replication proteins. *Bioorgan. Chem.* **2022**, *120*, 105649. [CrossRef] [PubMed]
221. Yetkin-Arik, B.; Kastelein, A.W.; Klaassen, I.; Jansen, C.H.; Latul, Y.P.; Vittori, M.; Biri, A.; Kahraman, K.; Griggioen, A.W.; Amant, F.; et al. Angiogenesis in gynecological cancers and the options for anti-angiogenesis therapy. *Biochim. Biophys. Acta (BBA)-Rev. Cancer* **2021**, *1875*, 188446. [CrossRef] [PubMed]
222. Jin, X.; Dai, L.; Ma, Y.; Wang, J.; Liu, Z. Implications of HIF-1α in the tumorigenesis and progression of pancreatic cancer. *Cancer Cell Int.* **2020**, *20*, 273. [CrossRef]
223. Vimalraj, S. A concise review of VEGF, PDGF, FGF, Notch, angiopoietin, and HGF signalling in tumor angiogenesis with a focus on alternative approaches and future directions. *Int. J. Biol. Macromol.* **2022**, *221*, 1428–1438. [CrossRef] [PubMed]
224. Liu, Q.; Guan, C.; Liu, C.; Li, H.; Wu, J.; Sun, C. Targeting hypoxia-inducible factor-1alpha: A new strategy for triple-negative breast cancer therapy. *Biomed. Pharmacother.* **2022**, *156*, 113861. [CrossRef] [PubMed]
225. Han, J.M.; Choi, Y.S.; Dhakal, D.; Sohng, J.K.; Jung, H.J. Novel nargenicin A1 analog inhibits angiogenesis by downregulating the endothelial VEGF/VEGFR2 signaling and tumoral HIF-1α/VEGF pathway. *Biomedicines* **2020**, *8*, 252. [CrossRef] [PubMed]
226. Eguchi, R.; Kawabe, J.I.; Wakabayashi, I. VEGF-independent angiogenic factors: Beyond VEGF/VEGFR2 signaling. *J. Vasc. Res.* **2022**, *59*, 78–89. [CrossRef] [PubMed]
227. You, L.; Wu, W.; Wang, X.; Fang, L.; Adam, V.; Nepovimova, E.; Wu, Q.; Kuca, K. The role of hypoxia-inducible factor 1 in tumor immune evasion. *Med. Res. Rev.* **2021**, *41*, 1622–1643. [CrossRef] [PubMed]
228. Liu, R.; Chen, Y.; Liu, G.; Li, C.; Song, Y.; Cao, Z.; Li, W.; Hu, J.; Lu, C.; Liu, Y. PI3K/AKT pathway as a key link modulates the multidrug resistance of cancers. *Cell Death Dis.* **2020**, *11*, 797. [CrossRef] [PubMed]
229. Miricescu, D.; Balan, D.G.; Tulin, A.; Stiru, O.; Vacaroiu, I.A.; Mihai, D.A.; Popa, C.C.; Papacocea, R.I.; Enyedi, M.; Sorin, N.A.; et al. PI3K/AKT/mTOR signalling pathway involvement in renal cell carcinoma pathogenesis. *Exp. Ther. Med.* **2021**, *21*, 540. [CrossRef] [PubMed]
230. Chen, J.; Zhang, X.; Liu, X.; Zhang, C.; Shang, W.; Xue, J.; Chen, R.; Xing, Y.; Song, D.; Xu, R. Ginsenoside Rg1 promotes cerebral angiogenesis via the PI3K/Akt/mTOR signaling pathway in ischemic mice. *Eur. J. Pharmacol.* **2019**, *856*, 172418. [CrossRef] [PubMed]
231. Ji, X.K.; Madhurapantula, S.V.; He, G.; Wang, K.Y.; Song, C.H.; Zhang, J.Y.; Wang, K.J. Genetic variant of cyclooxygenase-2 in gastric cancer: More inflammation and susceptibility. *World J. Gastroenterol.* **2021**, *27*, 4653. [CrossRef] [PubMed]
232. Xu, Y.; Wang, J.; He, Z.; Rao, Z.; Zhang, Z.; Zhou, J.; Zhou, T.; Wang, H. A review on the effect of COX-2-mediated mechanisms on development and progression of gastric cancer induced by nicotine. *Biochem. Pharmacol.* **2024**, *220*, 115980. [CrossRef] [PubMed]
233. Wu, Y.; Liu, M.; Zhou, H.; He, X.; Shi, W.; Yuan, Q.; Zuo, Y.; Li, B.; Hu, Q.; Xie, Y. COX-2/PGE2/VEGF signaling promotes ERK-mediated BMSCs osteogenic differentiation under hypoxia by the paracrine action of ECs. *Cytokine* **2023**, *161*, 156058. [CrossRef] [PubMed]

234. Szweda, M.; Rychlik, A.; Babińska, I.; Pomianowski, A. Significance of cyclooxygenase-2 in oncogenesis. *J. Vet. Res.* **2019**, *63*, 215. [CrossRef] [PubMed]
235. Suryanti, S.; Agustina, H.; Aziz, A.; Yulianti, H.; Suryawathy, B.; Putri, L. High immunoexpression of COX-2 as a metastatic risk factor in ccRCC without PD-L1 involvement. *Res. Rep. Urol.* **2021**, *13*, 623–630. [CrossRef] [PubMed]
236. Pond, K.W.; Doubrovinski, K.; Thorne, C.A. Wnt/β-catenin signaling in tissue self-organization. *Genes* **2020**, *11*, 939. [CrossRef] [PubMed]
237. Mankuzhy, P.; Dharmarajan, A.; Perumalsamy, L.R.; Sharun, K.; Samji, P.; Dilley, R.J. The role of Wnt signaling in mesenchymal stromal cell-driven angiogenesis. *Tissue Cell* **2023**, *85*, 102240. [CrossRef] [PubMed]
238. Bats, M.L.; Peghaire, C.; Delobel, V.; Dufourcq, P.; Couffinhal, T.; Duplàa, C. Wnt/frizzled signaling in endothelium: A major player in blood-retinal- and blood-brain-barrier integrity. *CSH Perspect. Med.* **2022**, *12*, a041219. [CrossRef] [PubMed]
239. Wang, Z.; Yang, J.; Qi, J.; Jin, Y.; Tong, L. Activation of NADPH/ROS pathway contributes to angiogenesis through JNK signaling in brain endothelial cells. *Microvasc. Res.* **2020**, *131*, 104012. [CrossRef]
240. Zhou, W.; Yang, L.; Nie, L.; Lin, H. Unraveling the molecular mechanisms between inflammation and tumor angiogenesis. *Am. J. Cancer Res.* **2021**, *11*, 301. [PubMed]
241. Wan, M.L.; Wang, Y.; Zeng, Z.; Deng, B.; Zhu, B.S.; Cao, T.; Li, Y.K.; Xiao, J.; Han, Q.; Wu, Q. Colorectal cancer (CRC) as a multifactorial disease and its causal correlations with multiple signaling pathways. *Biosci. Rep.* **2020**, *40*, BSR20200265. [CrossRef] [PubMed]
242. Lee, M.G.; Lee, K.S.; Nam, K.S. Anti-metastatic effects of arctigenin are regulated by MAPK/AP-1 signaling in 4T-1 mouse breast cancer cells. *Mol. Med. Rep.* **2020**, *21*, 1374–1382. [CrossRef] [PubMed]
243. Kim, M.; Kim, J.G.; Kim, K.Y. Trichosanthes kirilowii extract promotes wound healing through the phosphorylation of ERK1/2 in Keratinocytes. *Biomimetics* **2022**, *7*, 154. [CrossRef] [PubMed]
244. Ricard, N.; Zhang, J.; Zhuang, Z.W.; Simons, M. Isoform-specific roles of ERK1 and ERK2 in arteriogenesis. *Cells* **2019**, *9*, 38. [CrossRef] [PubMed]

Disclaimer/Publisher's Note: The statements, opinions and data contained in all publications are solely those of the individual author(s) and contributor(s) and not of MDPI and/or the editor(s). MDPI and/or the editor(s) disclaim responsibility for any injury to people or property resulting from any ideas, methods, instructions or products referred to in the content.

molecules

MDPI

Systematic Review

Systematic Review of Chemical Compounds with Immunomodulatory Action Isolated from African Medicinal Plants

Wendwaoga Arsène Nikiema [1,2], Moussa Ouédraogo [1,2,3,*], Windbedma Prisca Ouédraogo [1,2,3], Souleymane Fofana [2,4], Boris Honoré Amadou Ouédraogo [1,2], Talwendpanga Edwige Delma [1,2], Belem Amadé [1,2], Gambo Moustapha Abdoulaye [1,2], Aimé Serge Sawadogo [3], Raogo Ouédraogo [2] and Rasmané Semde [1,2,3]

1 Laboratoire de Développement du Médicament, Ecole Doctorale Sciences et Santé, Université Joseph KI—ZERBO, 03 BP 7021 Ouagadougou 03, Burkina Faso; nikiemarsn@ujkz.bf (W.A.N.); windbedema.ouedraogo@ujkz.bf (W.P.O.); bigboris16.cforem@ujkz.bf (B.H.A.O.); edwige.delma@ujkz.bf (T.E.D.); hamade_belem@ujkz.bf (B.A.); moustapha_abdoulaye@ujkz.bf (G.M.A.); rasmane.semde@ujkz.bf (R.S.)

2 Centre d'Excellence Africain, Centre de Formation, de Recherche et d'Expertises en sciences du Médicament (CEA-CFOREM), Université Joseph KI—ZERBO, 03 BP 7021 Ouagadougou 03, Burkina Faso; fof.soul@ujkz.bf (S.F.); raogo.ouedraogo@ujkz.bf (R.O.)

3 Unité de Formation et de Recherche, Sciences de la Santé, Université Joseph KI—ZERBO, 03 BP 7021 Ouagadougou 03, Burkina Faso; serge.sawadogo@ujkz.bf

4 Institut des Sciences de la Santé, Université NAZI Boni, 01 BP 1091 Bobo-Dioulasso 01, Burkina Faso

* Correspondence: moussa.ouedraogo@ujkz.bf

Citation: Nikiema, W.A.; Ouédraogo, M.; Ouédraogo, W.P.; Fofana, S.; Ouédraogo, B.H.A.; Delma, T.E.; Amadé, B.; Abdoulaye, G.M.; Sawadogo, A.S.; Ouédraogo, R.; et al. Systematic Review of Chemical Compounds with Immunomodulatory Action Isolated from African Medicinal Plants. *Molecules* **2024**, *29*, 2010. https://doi.org/10.3390/molecules29092010

Academic Editors: Arunaksharan Narayanankutty, Ademola C. Famurewa and Eliza Oprea

Received: 8 January 2024
Revised: 25 March 2024
Accepted: 29 March 2024
Published: 26 April 2024

Copyright: © 2024 by the authors. Licensee MDPI, Basel, Switzerland. This article is an open access article distributed under the terms and conditions of the Creative Commons Attribution (CC BY) license (https://creativecommons.org/licenses/by/4.0/).

Abstract: A robust, well-functioning immune system is the cornerstone of good health. Various factors may influence the immune system's effectiveness, potentially leading to immune system failure. This review aims to provide an overview of the structure and action of immunomodulators isolated from African medicinal plants. The research was conducted according to PRISMA guidelines. Full-text access research articles published in English up to December 2023, including plant characteristics, isolated phytochemicals, and immuno-modulatory activities, were screened. The chemical structures of the isolated compounds were generated using ChemDraw® (version 12.0.1076), and convergent and distinctive signaling pathways were highlighted. These phytochemicals with demonstrated immunostimulatory activity include alkaloids (berberine, piperine, magnoflorine), polysaccharides (pectin, glucan, acemannan, CALB-4, GMP90-1), glycosides (syringin, cordifolioside, tinocordiside, aucubin), phenolic compounds (ferulic acid, vanillic acid, eupalitin), flavonoids (curcumin, centaurein, kaempferin, luteolin, guajaverin, etc.), terpenoids (oleanolic acid, ursolic acid, betulinic acid, boswellic acids, corosolic acid, nimbidin, andrographolides). These discussed compounds exert their effects through various mechanisms, targeting the modulation of MAPKs, PI3K-Akt, and NF-kB. These mechanisms can support the traditional use of medicinal plants to treat immune-related diseases. The outcomes of this overview are to provoke structural action optimization, to orient research on particular natural chemicals for managing inflammatory, infectious diseases and cancers, or to boost vaccine immunogenicity.

Keywords: medicinal plants; phytochemicals; immunomodulators; transduction mechanisms

1. Introduction

The immune system comprises a complex network of cells and biological mediators that safeguard the body against harm from foreign invaders like microbes and malignant cell infiltration while preventing excessive immune activation. It is distinguished by innate and adaptive immunities, working synergistically to protect the body [1].

Several endogenous and exogenous factors may influence the immune system's effectiveness, potentially leading to malfunction. In conditions such as infectious diseases, e.g.,

COVID-19, dengue fever, and autoimmune diseases, e.g., celiac disease, type 1 diabetes, Addison's disease, Graves' disease, and Rheumatoid polyarthritis, there is an inappropriate immune response [2–4]. For instance, the interaction of the dengue virus with immune cells triggers a cytokine storm (involving IL1β, IL6, and tumor necrosis factor α), exacerbating the disease [5]. Also, in autoimmune diseases, self-reactive T cells and the exaggerated production of antibodies against the body's tissues result in persistent inflammation [6]. In both categories of diseases, controlling the immune response is crucial.

Immunomodulation refers to any modification of the immune response and may involve the induction, expression, amplification, or inhibition of a part or phase of the immune response [7,8]. The concept of immunomodulation has gained significant attention, particularly with the resurgence of infectious diseases in recent years. Immunomodulators are categorized into immunostimulants, immunoadjuvants, and immunosuppressants.

Immunostimulants are pharmacological agents capable of strengthening the body's resistance to infection. In healthy individuals, they serve as preventive measures and potentiators by enhancing the immune response. They can be used in immunotherapy for individuals with compromised immune systems. Notably, immunostimulants show promise in cancer treatment [9]. Immunosuppressants are critical in preventing organ transplant rejection and managing autoimmune diseases and immune-related disorders linked to infections. Immunoadjuvants stimulate the immune system by enhancing the antigenicity of vaccines without exerting a specific antigenic effect. They serve three main functions: aiding in antigen-targeting immune cells, acting as depots for the gradual release of the antigen, and modulating and reinforcing the type of immune response induced. They can influence cellular and humoral response choices, Th1 and Th2, immune protection versus immune destruction, and regeneration [10,11]. These constitute a new and promising application for immunoadjuvants.

Although synthetic immunomodulatory drugs offer many advantages, their undesirable side-effect profile and broad impact on the entire immune system are significant limitations to the extended use of these drugs, justifying the search for more effective and safer agents with targeted immunomodulatory activity. Using natural substances as immunoadjuvants during vaccine development to enhance immunogenicity is very promising [12]. Previous studies have shown that natural products with immunomodulatory activity have already been used to treat autoimmune diseases, inflammatory disorders, and cancer [13]. These substances constitute a valuable source of biologically active secondary metabolites, including alkaloids, polysaccharides, terpenoids, flavonoids, coumarins, glycosides, and proteins. This review aims to compile a comprehensive database of molecules from African medicinal plants capable of modulating immunity and improve our knowledge of the potential signaling pathways.

2. Results

2.1. Database Search Results

In the course of our database search, a total of 610 articles were initially identified (Figure 1). Upon removing duplicate entries (n = 587), 495 studies were subsequently excluded due to their classification as review articles or because the studies focused solely on crude extracts without evaluating isolated bioactive substances. Consequently, we scrutinized 92 full-text articles for eligibility, ultimately excluding 43. Our final dataset comprised 49 studies investigating the immunomodulatory activity of African medicinal plants.

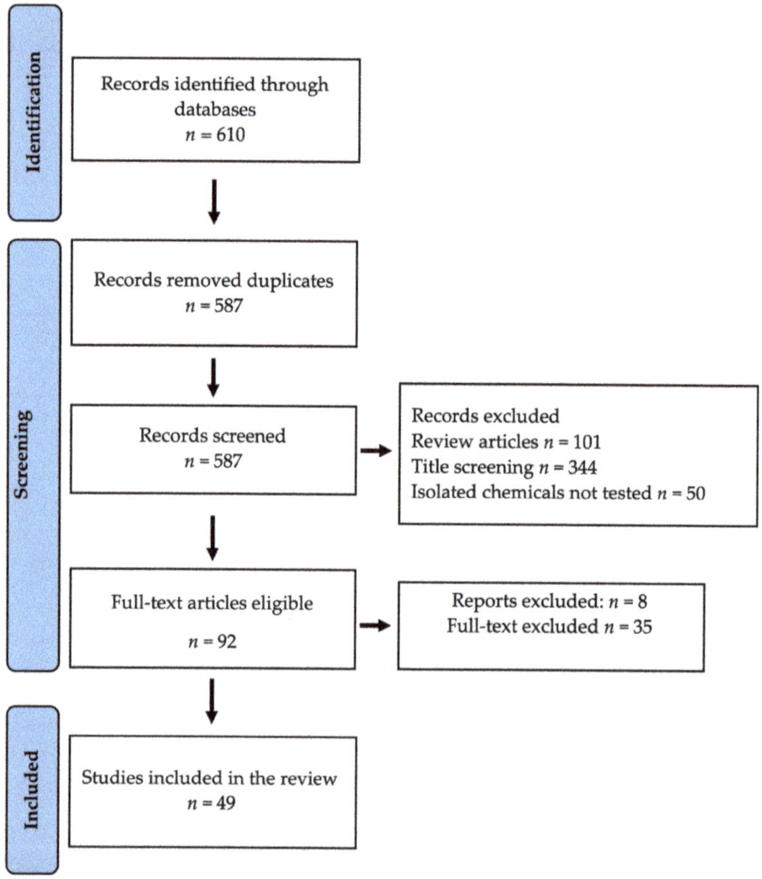

Figure 1. PRISMA flow diagram of the study selection.

The selected articles were published from 1994 to 2022. A total of 35 medicinal plant species belonging to 25 families were identified. The Liliaceae and Menispermaceae families were most represented with four and three plants species, respectively.

From medicinal plants, 86 molecules were isolated, 80 molecules were evaluated for their immunomodulatory proprieties (Table 1), and 56 had their structures represented (Figures 3–7).

Table 1. African medicinal plants-based isolated immunomodulators.

Species and Families of Plants	Parts of Plant Used	Solvent	Chemical Groups	Isolated Molecules	Other Biological Activity	Reference
Cissampelos pareira L., Menispermaceae	Roots	Methanol	Alkaloids	Berberine (**1**), Tetrandrine.	Antioxidant, antibacterial	[14,15]
Tinospora crispa, Menispermaceae	Stem	Ethanol	Alkaloids, glycosides, terpenoids	N-formyllannonaine (**2**), N-formylnornuceferine (**3**), Lysicamine (**4**) Magnoflorine (**5**) Syringin (**50**) 1-Octacosanol.	Anti-inflammatory, antioxidant	[16,17]
Tinospora cordifolia (Wild) Hook. F. & Thomson, Menispermaceae	Stem	Methanol, n-hexane, chloroform, ethyl acetate and n-butanol	Alkaloids, glycosides, proteins	11-hydroxymustakone (**51**), N-methyl-2-pyrrolidone (**52**), N-formylannonaine (**2**) Cordifolioside (**53**), Tinocordiside (**54**) Syringin (**50**)		[18–20]
Piper longum Linn. Piperaceae	Fruits	Methanol	Alkaloids	Piperine (**6**)	Anti-inflammatory, anti-infectious, antitumor, analgesic	[21]
Echinacea purpura, Echinaceae	Whole plant, Root	Methanol, ethanol, aqueous	Polysaccharides, flavonoids	Polysaccharides, alkyl amides, Arabinogalactans, Caffeic acid(**34**)	Antioxidant, anti-inflammatory	[22–24]
Fructus aurantii, Rutaceae	Fruit		Polysaccharides	Pectic polysaccharide: CALB-4	Anti-carcinogenic, antimicrobial	[25]
Garcinia mangostana L., Guttiferae	Bark	Methanol	Polysaccharides	Arabinofuran (GMP90-1)	Antioxidant, anti-inflammatory, antimicrobial	[26]
Siraitia grosvenorii, Cucurbitaceae	Whole plant	Aqueous	Polysaccharides	Polysaccharides	Antioxidant, anti-inflammatory	[27]
Aesculus hippocastanum, Hippocastanaceae	Seed	Alcoholic	Saponins triterpenoides	β-aescin (**18**)	Antiviral	[28]
Andrographis paniculata, Acanthaceae	Whole plant	Methanol-water	Terpenoids	Andrographolide (**10**) 14-deoxyandrographolide (**11**): 14-deoxy-11,12-didehydroandrographolide (**12**).	Anticancer, Anti-inflammatory	[29,30]

Table 1. *Cont.*

Species and Families of Plants	Parts of Plant Used	Solvent	Chemical Groups	Isolated Molecules	Other Biological Activity	Reference
Azadirachta *indica*, Meliaceae	Oil		Terpenoids	Nimbidin (**14**)	Anti-inflammatory, anti-arthritic	[31]
Ocimum *sanctum* Lamiaceae	Whole plant	Alcoholic, aqueous	Terpenoids, essential oils, phenols, flavonoids	Eugenol, Carvacrol, Oleanolic acid (**7**), Ursolic acid (**8**),	Anti-inflammatory, antiallergic	[32]
Boswellia serrata Roxb. Burseraceae	Oleogum resin		Terpenoids	Boswellic acids (**13**)	Anti-inflammatory	[33]
Pogostemon cablin Benth. Lamiaceae	Aerial parts	Ethanol aqueous	Terpenoids	PA: Patchouli alcoholic (**19**)	Antioxidant, Antimicrobial	[34]
Biden Pilosa, Asteraceae	Whole plant	n-butanol	Flavonoids	Polyacetylene 2-O-β-D-glucosyltrideca-11^{E}-en-3,5,7,9-tetrayn-1,2-diol (PA-1), Centaurein (**24**), Centaureidin (**25**)	Anti-inflammatory, antihyperglycemic	[35,36]
Callistenom viridiflorus, Myrtaceae	Leaves	Ethanol	Phenols, flavonoids	Apigenin 4′-O-β-d-glucopyranosyl-(1‴ → 4″)-O-β-d-glucopyranoside, Kaempferide (**28**), Isoquercetin (**27**), Hyperin (**29**)	Anti-inflammatory, analgesic, antibacterial, antifungal.	[37]
Curcuma longa, Zingiberaceae	Rhizome		Flavonoids	Curcumin (**23**)	Anti-inflammatory, antimutagenic	[38,39]
Justicia spicigera Schltdl. Acanthaceae	Leaves	Ethanol	Flavonoids	Kaempferitrin (**26**)	Antioxidant, antitumor	[40,41]
Phyllantus amarus, Euphorbiaceae	Leaves	Ethanol, fractions: ethyl acetate, dichloromethane	Flavonoids, lignan	Corosolic acid (**9**), Oleanolic acid (**7**), Phyllanthin, Hypophyllanthin (**30**)	Anti-inflammatory, antiviral, antimutagenic.	[42–44]
Psidium guajava, Myrtaceae	Leaves	Ethanol	Flavonoids, glycosides, phenolic compounds, terpenoids	Ellagic acid, Hyperin (**29**), Isoquercitin (**27**), Guajaverin (**31**), Avicularin (**32**), Asiatic acid (**21**), Maslinic acid (**20**), Corosolic acid (**9**), Oleanolic acid (**7**), Ursolic acid (**8**)	Antiallergic, antitumoral, anti-inflammatory, analgesic, antimicrobial	[45,46]

Table 1. *Cont.*

Species and Families of Plants	Parts of Plant Used	Solvent	Chemical Groups	Isolated Molecules	Other Biological Activity	Reference
Teucrium ramosissimum Desf., Lamiaceae	Aerials parts	Chloroform	Flavonoids	Apigenin-7-glucoside (**44**), Genkwanin (**43**) Naringenin (**41**)	Antioxidant, anti-inflammatory	[47,48]
Ferula szowitsiana, Apiaceae	Roots	Methanol	Coumarins terpenoids	Methyl galbanate (**49**), Umbelliprenin (**48**)	Anti-inflammatory, antioxidant	[49,50]
Aloe vera, Liliaceae	Whole roots	Chloroform	Coumarins, flavonoids, phenolics, carbohydrates, lignans	Esculetin (6,7-dihydrocoumarin) (**47**) Acemannann	Anti-inflammatory, antioxidant	[51–53]
Allium sativum, Alliaceae	Bulbs	PBS	Proteins	Proteins (QR-1, QR-2, QR-3), Fructans, proteins (QA-1, QA-2, QA-3)	Anti-inflammatory, antioxidant, antimicrobial, antitumor	[54–56]
Allium cepa, *Alliaceae*	Bulbs	Ethanol	Proteins, polysaccharides, lectins	Pectin, FOS (fructo-oligosaccharides), Agglutinin	Antimicrobial	[57–59]
Astragalus membranaceus, Fabaceae	Waste	Alkali solvent	Proteins, saponins, alkaloids, polysaccharides, glucosides	Proteins: AMWPDG2, AMWPDG4, AMWPDG6, Astragaloside IV (**15**), Astragaloside VII (**16**), Macrophyllosaponin B (**17**)	Immunoadjuvants	[60–62]
Plantago sp. (*P. major*, *P. asiatica*) Plantiginaceae.	Leaves	Aqueous	Flavonoids, phenols, terpenoids, iridoids,	Aucubin (**55**), Chlorogenic acid (**35**), Ferulic acid (**36**), p-Ccumaric acid (**37**), Vanillic acid (**38**), Luteolin (**42**), Ursolic acid (**8**), Oleanolic acid (**7**), Baicalein (**33**), Baicalin (**33′**).	Anticancer, antimicrobial, anti-inflammatory, antioxidant	[63,64]
Mangifera indica L. Anacardiaceae	Leaves	-	Xanthone glucoside	Mangiferin (**56**)	Antioxidant, antitumoral	[65,66]
Nigella sativa L. Ranunculaceae	Seeds	Ethanolic	Volatile oil	Thymoquinone (**46**)	Anti-inflammatory, antioxidant.	[67]
Zingiber officinale Zingiberaceae	Dried ginger	Distilled water	Volatile oil, polyphenols	6-Gingerol (**45**)	Antibacterial, anti-inflammatory, antitumoral	[68–70]

Table 1. *Cont.*

Species and Families of Plants	Parts of Plant Used	Solvent	Chemical Groups	Isolated Molecules	Other Biological Activity	Reference
Boerhavia diffusa Nyctaginaceae	Leaves	Hexane, chloroform, ethanol	Flavonoids	Eupalitin (BdI) (**39**), Eupalitin-3-O-β-D-galactopyranoside (BdII) (**40**)	Anti-inflammatory	[71]
Tamarindus indica Leguminoseae	Seeds	Water	Polysaccharides	Polysaccharides	Antitumoral	[72]
Salvia officinalis L. Lamiaceae	Arial parts	Methanol-chloroform	Polysaccharides, proteins	Arabinogalactans (A), Pectins (B), Glucurunoxylan polymers (D).	Anti-inflammatory	[73]
Moringa oleifera, Moringaceae	Mature pods	Aqueous	Polysaccharides	(1→4)-α-D glucan	Anti-inflammatory	[74]

The chemical compounds or groups (n = 86) were isolated from 35 medicinal plants species belonging to 25 different families. The numbers in parentheses are the appearance order numbers in the text of the 57 isolated chemicals whose structures were drawn (Figures 3–7).

2.2. *African Medicinal Plants Used for Immunomodulation*

Numerous medicinal plants used in traditional medicine systems have attracted the attention of scientists worldwide (Table 1). As discussed below, these medicinal plants exhibit immunomodulatory activity due to various chemical groups (Figure 2) and others medicinal properties, including antioxidant, anti-inflammatory, analgesic, and anti-arthritic activity.

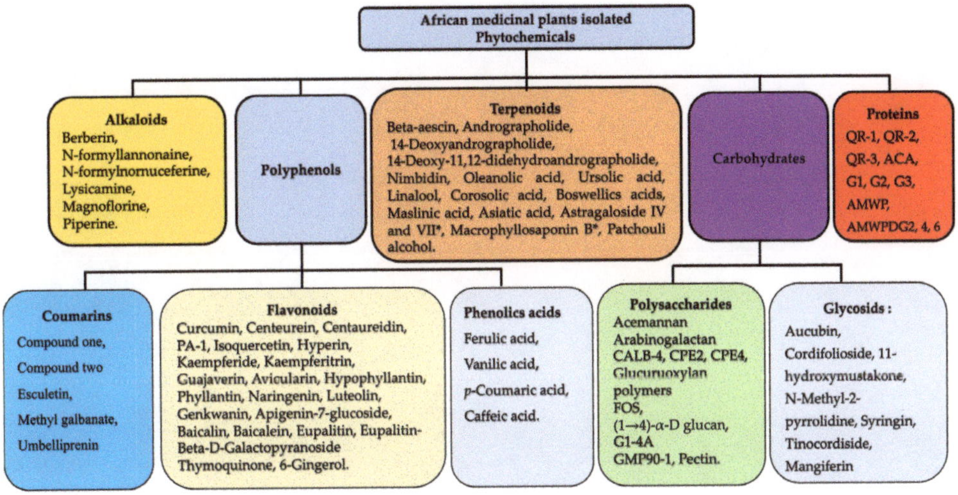

Figure 2. Plant- derived immunomodulators. The isolated phytochemicals with immunomodulatory activities include polyphenols (carbohydrates, alkaloids, and proteins). * NB: Astragalus VII and Macrophyllosaponin B were isolated from the Astragalus genius in particular *Astragalus trojanus Stev.* and *Astragalus oleifolius DC*, which are not distributed in Africa.

2.3. *Chemistry of Plant-Derived Immunomodulators*

2.3.1. Alkaloids

Alkaloids isolated (Figure 3) from *C. pareira*, *T. crispa*, *T. cordifolia*, and *P. longum* exhibited immunomodulatory activities (Table 2).

Humoral immunity produces antigen-specific antibodies and is primarily driven by B cells. On the other hand, cell-mediated immunity does not depend on antibodies for its adaptive immune functions [75]. Mature T cells, macrophages, and the release of cytokines in response to an antigen primarily drive it. The alkaloid fraction of *C. pareira* tested on humoral and cell-mediated immunity by measuring the hemagglutination antibody titer and the delayed-type hypersensitivity (DTH) response demonstrated significant immunosuppressive effects at 25 to 100 mg/kg doses. It significantly ($p < 0.01$) reduced the humoral antibody titer and suppressed the DTH response ($p < 0.01$) at 75 mg/kg [76]. Berberine (compound **1**), an isoquinoline alkaloid isolated from *C. pareira*, displayed no effect on splenocyte proliferation. However, it downregulated the Th1/Th2 cytokines' expression (TNF-α, IL-2, IL-4, IL-10) in a mouse primary splenocytes model assay at a range of concentrations (0.8, 1.6, and 3.3 μM, or 0.5 mL/well) [14].

Phagocytosis is an essential cell-defense mechanism against foreign, non-self organisms, and has been used as a critical non-specific immunological parameter to evaluate immune functions. Phagocytes also kill microbes via an oxygen-independent mechanism, although not as effectively as oxygen-dependent mechanisms. Macrophages are essential for the phagocytosis mechanism.

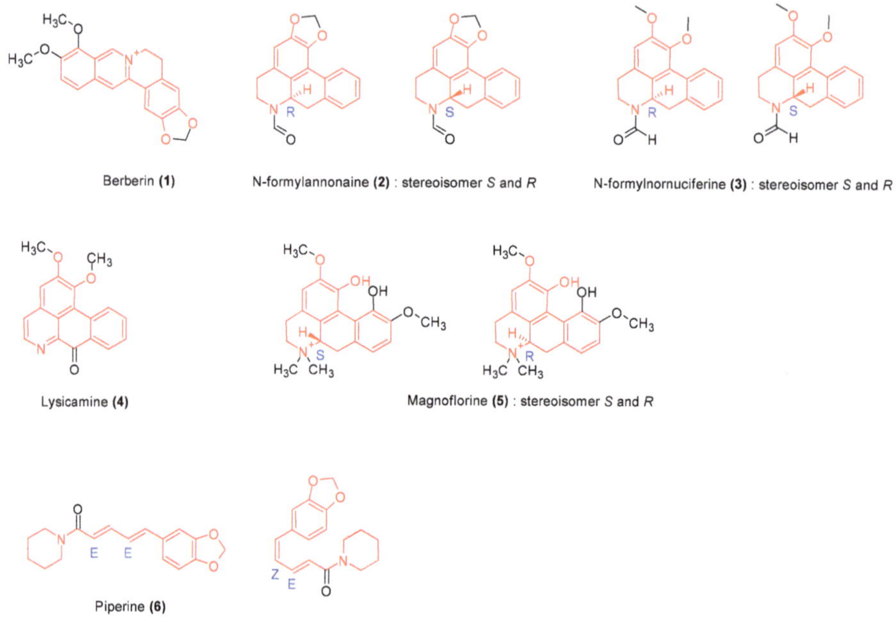

Figure 3. Structures of the identified alkaloids with immunomodulatory activities. The isolated alkaloids comprise a protoberberine isoquinoleic (**1**), aporphine isoquinoleine alkaloids (**2–5**), and a piperidine alkaloid (**6**). They have common scaffold highlighted in red. All have been tested and showed immunomodulatory activities.

Table 2. Isolated alkaloids with immunomodulatory activities.

Isolated Molecules (n°)	Models	Pharmacodynamic Parameters ED_{50}/IC_{50}	Biological Effects	Cellular Effect	References
1	In vitro mouse primary splenocyte assay	6.6 µM	No significant effect on cell viability at 0.8, 1.6, and 3.3 µM	Downregulates splenocytes cytokines (IL2, 4, 10, TNFα) expression.	[14]
2	Mouse macrophage RAW 264.7 viability, chemotactic, phagocytic assay, ROS, NO, PGE2, and cytokine production, monocyte chemoattractant Protein-1 (MCP-1) production.	nd	Isolated compounds **2**, **3**, **4**, and **5** at concentrations above 25 µg/mL showed toxic effects on macrophages' viability (<90%). Stimulation of cell migration. Increase in macrophage migration. Stimulation of cell migration, strong enhancement of macrophage phagocytic activity (81.01% compound **5**). Augmentation of ROS and NO generation.	Significantly stimulates PGE2 production, enhances the MCP-1 level. Significantly increases IL-1β, IL6, and TNFα production.	[17,18]
3		nd			
4		nd			
5		23.8 mM			
6	In vivo hematological assay, in vitro Dalton's lymphoma ascites (DLA), Ehrlich ascites carcinoma (EAC) cells assay, L929 cells	nd	Increase in white cell count (138.9%), stimulation of stem cell proliferation, enhancement of the number of plaque-forming cells (71.4%) Cytotoxicity on DLA, EAC at 200 µg/mL, and L929 at 50 µg/mL.	Enhancement of the antibody production.	[21]

nd = non-determined.

In a murine RAW macrophages in vitro assay, the crude extract of *T. crispa* (concentrations of 25–200 μg/mL) and the isolated alkaloids (compounds **2**, **3**, **4**, and **5**) at various concentrations (1.56, 3.12, 6.25, 12.5, and 25 μg/mL) increased chemotactic activity and enhanced macrophage phagocytic activity. Furthermore, they significantly increased cytokine levels (TNFα, IL1β, and IL6) [17]. Cytokines like TNFα contribute to antitumoral effects.

The treatment of Balb/c mice with the alcoholic extract (10 mg/dose/animal) of fruits of *Piper longum* and piperine (compound **6**), a purified alkaloid, at 1.14 mg/dose/animal intraperitoneally for five consecutive days, yielded an increase in the white blood cell count by 142.8% and 138.9%, respectively. In addition, a cytotoxic effect was observed against L929 cancer cells at 100 μg/mL for the crude extract and 50 μg/mL for piperine [21].

2.3.2. Polysaccharides

Numerous studies have shown that plant polysaccharides can regulate the immune system in multiple ways and levels (Table 3). They not only activate immune cells, including T cells, B lymphocytes, macrophages, and dendritic cells, but they also activate and promote the production of cytokines (NO, TNFα, and IL6), thus showing regulatory effects on the immune system in various ways.

Dendritic cells (DCs) act as initiators of the initial immune response and play an essential part in regulating the immune system [77]. DCs recognize, capture, process, and present antigens to naive T cells, which stimulate the activation and proliferation of naive T cells for adaptive immune responses. Assessing immunomodulatory activity on DCs has been performed. Polysaccharides isolated from *E. purpurea* and *Plantago asiatica* could upregulate the maturation of DCs [22,23,64]. They act on cell maturation markers by enhancing the expression of surface molecules, including CD80, CD86, and MHC class II.

The thymus is an organ of the immune system, and is the site of production and maturation of T lymphocytes. The spleen is the body's largest secondary lymphoid organ and, as such, hosts a wide range of immunological functions in addition to its hematopoietic function [78]. In addition to B and T cells, a small amount of macrophages and other cells, such as dendritic cells, are included in splenocytes. The activity of plant extracts has been studied on spleen and thymus cells.

The immunomodulatory effects of plant polysaccharides on macrophages are mainly achieved through the generation of reactive oxygen species (ROS), the secretion of cytokines, cell proliferation, and the phagocytic activity of macrophages. A water-soluble pectic extract from *Allium cepa* exhibited the capacity to enhance NO production in murine macrophages and stimulate the proliferation of splenocytes and thymocytes. An optimal concentration for proliferation was observed at 50 μg/mL [57]. In vitro, fructo-oligosaccharides (FOS) provoked a significant increase in the mitogenic activity of murine splenocytes and thymocytes after 24 h incubation at 5 and 50 μg/mL concentrations. Macrophage activation is involved in the first phase of the immune response, and interestingly, onion FOS significantly induced macrophage phagocytosis and NO release [58]. A water-soluble glucan isolated from *M. oleifera* exhibited significant macrophage activation and phagocytic activity along with the induction of monocyte NO release at 0.1 μg/mL [74]. The effect of GMP90-1 polysaccharides isolated from *G. mangostana* on the viability of RAW 264.7 macrophage was studied using an MTT assay. These studies showed that GMP90-1 polysaccharides inhibited cell growth at 400 μg/mL, whereas 50–200 μg/mL concentrations had no inhibitory effect. Additionally, GMP90-1 polysaccharides increased macrophage phagocytosis and induced NO production and cytokine expression (TNF-α, IL-6, IL-1β) at concentrations of 50, 100, and 200 μg/mL [26].

Table 3. Isolated polysaccharides with immunomodulator activities.

Sources	Extraction Method	Molecular Weight (kDa)	Monosaccharide Composition	Active Substance	Biological Activity	References
Allium cepa	Hot water	1.8×10^2	D-galactose: 6-O-Me-D-galactose: 3-O-acetyl-D-methyl galacturonate: D-methyl galacturonate1:1:1:1	Pectin	Enhancement of NO production in macrophage, stimulation of splenocyte and thymocyte proliferation.	[57]
	Hot ethanol			FOS: monosaccharide to hexasaccharide	Increase in splenocytes/thymocytes proliferation (~3-fold), macrophage phagocytic activity, NO production (~2.5-fold).	[58]
Moringa oleifera	Distilled water	70	Gluc.	(1→4)-α-D glucan	Increase in macrophage phagocytic activity, and in the number and percentage of globulin.	[74]
Garcinia mangostana L.	Water extraction	5.3	Ara., Gal., Rham.	GMP90-1 = arabinofurane	Enhancement of phagocytic activity (28.0%; 40.3% at 100 and 200 μg/mL, respectively), increase in NO secretion (2.2, 3.9, and 10.3 times at the concentrations of 50, 100, and 200, respectively), IL1β (38.42% at 200 μg/mL), IL6 (4.6, 5.1, and 8.5 times at 50, 100, and 200, respectively), TNFα (5.6, 41.7, and 200.1% at 50, 100, and 200 μg/mL, respectively).	[26]
Aloe vera	Distilled water	-	Man, Gluc, Gal. 62.9:13.1:0.6	Heteroglycan or acemannan	Increase in splenocyte proliferation (5.7 and 7.1% after 24 and 48 h, respectively). Increase in IL-1 and TNFα secretion in irradiated mice (2.34 and 1.32~fold, respectively).	[52]
Echinaneae purpurea L.	Water		Diploid, tetraploid	CPE2, CPE4	Stimulation of lymphocyte proliferation and cytokine secretion.	[24]
			Gal, Ara	Arabinogalactane		[79,80]
Fructus aurantii	Cold water, hot water	3.14×10^2	Man, Rha, GlcUA, GalUA, Gal, Ara 16.3:4.0:2.9:3.4:21.7:41.7	Pectic polysaccharide CALB-4	Promotion of PBMC proliferation. Upregulation of NO production. Affects TNFα, IL1β, IL6, and IL8 secretion. Increases of proIL-1 expression.	[25]
Siraitia grosvenorii	Hot water		Gluc, Gal. Ara. Rham 5.8:0.77:0.38:0.12		Promotion of B and T lymphocyte proliferation. Increase in thymus index. Increase in IL-2 and decrease in IL-1.	[27]
T. cordifolia	Acetone extract			G1-4A	Upregulation of TNFα, IL1β, IL6, IL10, IL12, and IFNγ expression. Enhancement of NO level.	[20]
Tamarindus indica	Fresh water		Gal., Man., Gluc.		Increase in phagocytic activity. Inhibition of PHA-induced lymphocyte proliferation and leukocyte migration by 63–70.%	[72]
Salvia officinalis L.	Ethanol-water	10,000 < Mw > 50,000	Rham. Ara., Xyl., Man., Gluc., Gal., UA.	Arabinogalactans (A), Pectins (B), Glucurunoxylan polymers (D).	Polysaccharides-induced thymocyte proliferation.	[73]

Ara: arabinose; Gal: galactose; GalUA: galacturonic acid; Gluc: glucose; GlcUA: glucuronic acid; Man: manose; Rham: rhamnose; FOS: fructo-oligosaccharides; UA.: uronic acid.

Acemannan, a bioactive compound isolated from *A. vera*, displayed immunomodulatory activity. Studies in immunosuppressed mice indicated that acemannan treatment increased animal survivability and reduced mortality. It was established that acemannan upregulated cytokines' (TNFα, IL1β, and IL6) production and improved peripheral lymphocyte counts, spleen cellularity, and the spleen index. Moreover, acemannan stimulated the macrophage nitric oxide release, surface molecule expression, and cell morphology in RAW 264.7 cells, a mouse macrophage cell line [52].

In vitro experiments demonstrated that tetraploid and diploid *E. purpura* enhanced the stimulation of mouse spleen lymphocytes by Concanavalin A. Tetraploid forms exhibited higher activity at lower concentrations and strongly promoted the release of IL-2 and IFNγ secretion [24].

Zampeng Shu et al. found that pectic polysaccharides (CALB-4) extracted from *F. aurantii* stimulated NO, TNF-α, IL-1β, IL-6, and IL-8 production depending on the concentration between 24 and 48 h. These polysaccharides also stimulated splenocyte proliferation and increased cyclophosphamide-induced carbon clearance [25]. Lymphocytes, which are essential contributors to the humoral immune response, were stimulated by polysaccharides extracted from *S. grosvenonii* [27]. These polysaccharides promoted splenocyte and thymocyte proliferation in an in vitro MTT model at 12.5 and 200 μg/mL concentrations. The effect on cytokine secretion was marked by a significant increase in IL-2 at 50 mg/kg and a notable decrease in IL-1β production at 400 mg/kg [27].

Polysaccharides isolated and purified from *T. indica* showed immunomodulatory activity by blocking mitotic activity induced by PHA on lymphocytes; they enhanced macrophages' phagocytic activity, and inhibited leukocyte migration [72]. The immunomodulatory activity of water (A), ammonium oxalate (B), and potassium extractable (D) polysaccharide extracted from *S. officinalis* was evaluated via an in vitro co-mitogenic thymocyte test. Fraction A had an inhibitory effect at 300 μg/mL, and fractions B and D were at 1000 μg/mL. The inhibition was significant with fraction D. The optimum dose of this fraction was 100 μg/mL. Moreover, fraction D had a more marked SI_{comit}/SI_{mito} rate (3–4) than fractions A and B (≈2) [73].

2.3.3. Triterpenoids

Terpenoids, sometimes called isoprenoids, are a large and diverse class of naturally occurring organic chemicals that are similar to terpenes and are derived from assembled five-carbon isoprene units. Triterpenoids possess a rich chemistry and pharmacology with several pentacyclic motifs. They are used in inflammatory diseases and cancer therapeutics [81,82]. Numerous compounds (Figure 4) falling under the class of triterpenoids isolated from diverse medicinal plant species showed immunomodulatory properties.

Oleanolic acid (compound **7**) and ursolic acid (compound **8**): these pentacyclic terpenoids, extracted from various species of the Plantago genus, such as *P. major*, *Ocimum sanctum*, *Psydium guajava*, and *Phyllantus amarus*, have exhibited a range of pharmacological activities, including antioxidant and anti-inflammatory effects, functioning as immunoinhibitors. Compounds **7** and **8** inhibited the peripheral blood proliferation of mononuclear cells (PMBCs) at 1.25 and 20 μg/mL, respectively. Ursolic acid displayed high activity at 40 μg/mL [63]. These compounds enhanced interferon-gamma (IFN-γ) secretion. In a model of human keratinocytes (HKLs), ursolic acid exhibited inhibitory effects on cell viability, while compounds **7** (15 μM) and **8** (30 μM) significantly increased respiratory burst levels. Both oleanolic acid (30 μM) and ursolic acid (30 μM) increased lysosomal enzyme activity, but ursolic acid (7.5 μM) inhibited lysosomal enzyme activity [46]. Corosolic acid (compound **9**) isolated from *P. guajava* reduced HKL cell viability and significantly enhanced the cell respiratory burst level after 24 h of incubation. It also increased lysozyme activity without affecting NO production.

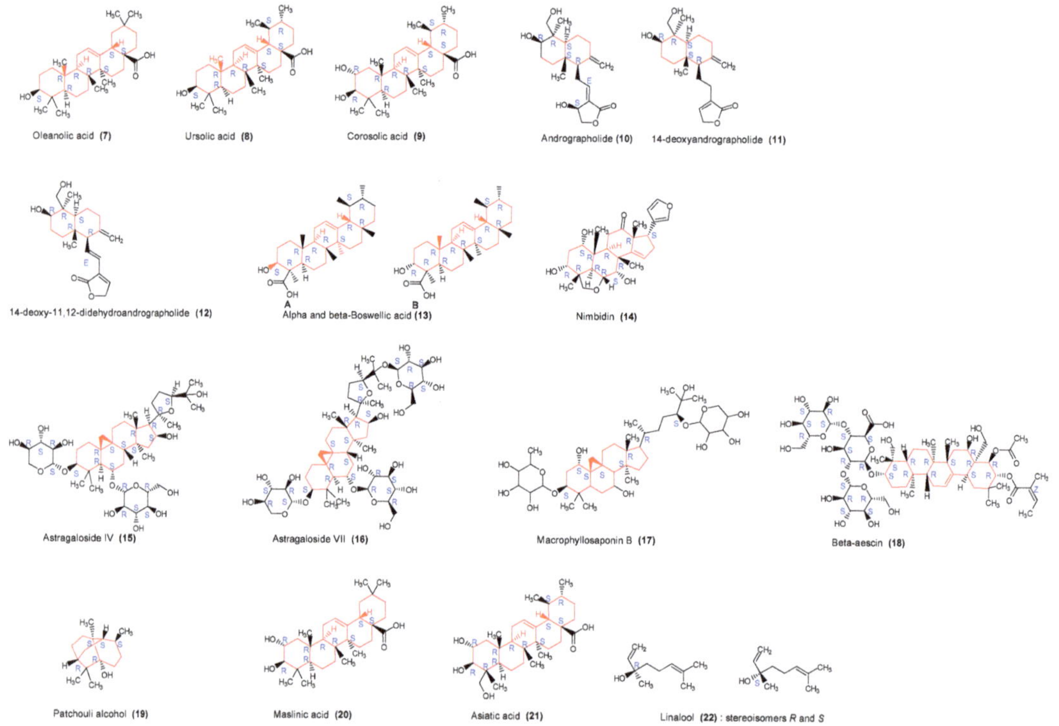

Figure 4. Structures of identified terpenoids with immunomodulatory activities. These terpenoids are distinguished in triterpenoid pentacyclic saponin (**7**, **8**, **9**, **13**, **20**, **21**), lactone sesquiterpenoid (**10**, **11**, **12**), tetracyclic saponin heterosid (**14**, **15**, **16**, **17**), pentacyclic saponin heterosid (**18**), diterpenoid (**19**) and Acyclic monoterpenoid (**22**). They have in common adjacent 2–5 rings, except for linalool (**22**). It is an advantage to have simple and complex structures like in linalool and beta-aescin (**18**), respectively, exhibiting immunomodulatory activities, which will lengthen the list of analogues.

Andrographolides: in vivo and in vitro animal models were used to evaluate the immunomodulatory activity of these diterpenoids (compounds **10**–**12**) extracted from *A. paniculata*. Mice treated with different doses displayed a significant rise in the hemagglutination (HA) titer and in plaque-forming cells (PFC) in the spleen of sheep red blood cell (SRBC)-sensitized mice. Phagocytic activity assessed through carbon clearance exhibited a dose-dependent increase, and white blood cell counts were significantly increased [29].

The effect of compound **13** on the SRBC-induced delayed-type hypersensitivity (DTH) response indicated that oral administration inhibited the expression of the DTH response in mice. Splenocyte proliferation was inhibited at concentrations greater than 3.9 μg/mL, and the macrophage phagocytic activity function was enhanced [33].

Nimbidin (compound **14**), isolated from *A. indica*, possesses immunomodulatory activity by inhibiting macrophage cell migration, phagocytosis, and phorbol myristate acetate (PMA)-stimulated respiratory bursts. Nimbidin also exhibited inhibitory effects on IL-1β release and NO and PGE2 production [31].

Astragalosides (AST IV compound **15**; AST VII compound **16**): these cycloartane triterpenes with saponin-like structures are mainly extracted from the Astragalus genus species. AST VII and Macrophyllosaponin B (compound **17**) displayed low hemolytic activity at 500 μg/mL and increased splenocyte proliferation induced by Concanavalin A, lipopolysaccharide (LPS), and bovine serum albumin (BSA) in immunized mice. Immunoglobulin G1 and G2 antibody titers were increased by AST VII (120 μg) and Macrophyllosaponin B

(90 μg), which stimulated IFNγ [83]. According to Nalbantsoy et al., Macrophyllosaponin B (156 μg/mL) exerts a suppressive effect on Th2 lymphocytes and a positive effect on Th1 lymphocytes by stimulating the release of specific cytokines (IL-2, IFNγ). It also inhibits the activity of inducible nitric oxide synthase (iNOS). In a murine model of a lymphoproliferation assay using MTT and a hemolysin spectrophotometry assay, AST IV increased T and B lymphocyte proliferation at 50–200 mg/kg. The activity of IL-1β at 1 nmol/L was increased. Additionally, TNF-α activity was inhibited with or without LPS stimulation [84].

β-aescin (compound **18**), isolated from the roots of *A. hippocastanum*, has demonstrated important antiviral and virucidal activity against the dengue and VSH viruses by targeting their envelope. The crude extract increased the secretion of pro-inflammatory cytokines (TNF-α and IL-6), while the extracted compound β-aescin from *A. hippocastanum* displayed a synergistic effect with glucocorticoids, enhancing anti-inflammatory activity. β-aescin, in an in vitro model of RAW264.7 cells, decreased the concentration of TNF-α, IL-1β, and NO in a concentration-dependent manner [28,85].

Compound **19**, a sesquiterpene tricyclic isolated from *P. cablin*, was studied for immunomodulatory activity in a mouse model. Oral administration significantly increased macrophage phagocytosis and boosted circulating immunoglobulin (IgM and IgG) while significantly decreasing the DTH response [34]. In the human peripheral blood mononuclear cells (PBMC) assay, compound **22** showed weak activity stimulating cell proliferation and a moderate stimulation of IFNγ secretion.

2.3.4. Polyphenols

The immune system plays a vital role in human well-being by increasing the immune response and providing protection. Polyphenols have well-demonstrated immunomodulatory effects, as they regulate the immune cells, macrophages, cytokines, and signaling pathways, and influence dendritic cells and lymphocytes (B and T), suppress T cell activation and natural killer cells, and suppress tumor-associated macrophages (Table 4).

Polyphenols are a heterogeneous group of phenolic compounds with two major classes: flavonoids and phenolic acids (Figure 5). They show immunomodulatory activity.

Curcumin (compound **23**) has demonstrated an in vitro immunomodulatory effect. On human PBMCs, compound **23** isolated from *C. longa* exhibited no effect on cell viability but significantly inhibited PHA-stimulated lymphocyte proliferation and enhanced natural killer cytotoxicity. Curcumin inhibited PHA-stimulated IL-2 production and had a weak effect on TNFα release [38]. Another model evaluating curcumin (2.5 μg/mL) activity on splenocytes revealed an inhibition of mitogen-induced splenocyte proliferation and IL-2 synthesis [39]. Centaurein flavonoids: centaurein (compound **24**) and its aglycone centaureidin (compound **25**), isolated from n-butanol fractions of *B. pilosa*, were studied for immunomodulatory activity through a cell transfection model with plasmids. They demonstrated increased IFNγ and induced the nuclear factor of activated T-cells (NFAT) and NFκB activity [35]. Another flavonoid, PA-1 (IC_{50} = 1.25–2.5 μg/mL), exhibited inhibitory activity on lymphocyte proliferation in an in vitro model of murine lymphocyte stimulation [36].

Table 4. Polyphenols isolated from African medicinal plants acting as immunomodulators.

Isolated Molecules (n°)	Models	Pharmacodynamic Parameters		Biological Effects	Cellular Effect	References
		ED_{50}	IC_{50}			
23	Mouse macrophage, lymphocytes (PBMCs) proliferation assay, natural killer cytotoxicity assay	nd	nd	No effect on cell viability. Inhibition of lymphocyte proliferation. Enhancement of NK cytotoxicity	Inhibition of PHA-induced IL2 release, weak inhibition of TNFα production in PBMC	[38]
	Splenocytes assay	3.5 μg/mL		Inhibition of splenocyte proliferation	Inhibition IL-2 synthesis	[39]
18	PBMC cell proliferation assay		1.5 mg/mL	Suppression of lymphocyte proliferation		[36]
24, **25**	IFNγ promoter-driven luciferase reporter and T cells assay.	75 0.9 mg/mL		Modulation of IFNγ transcription		[35]
27, **28**, **29**	in vitro RAW 264.7 macrophage proliferation assay	nd	nd	Increase in macrophages' proliferation (by 1.53-fold for compound **27** and 1.43-fold for compound **28**). No significant increase was observed for compound **29**.		[37]
26	Proliferation of murine splenocytes, macrophages, and human PBMCs. NO production, lysosomal enzyme activity, and neutral red uptake assay		nd	Increase in cell viability in the absence of LPS (macrophages 23%, splenocytes 17%, human PBMCs 24%). Increase in lysosome activity (57%) in a concentration-dependent manner. Lack of effect on neutral red uptake. Stimulation of NK cell activity (11%)	No effect on NO release.	[41]
30, **31**	HKLs	nd	nd	Modulation affects the viability of HKLs. No effect on cell viability Increase in lysozyme activity		[46]
41, **43**, **44**	In vitro mouse splenocyte proliferation assay, NK cell activity, cytotoxicity T lymphocyte activity, lysosomal enzyme activity	nd	nd	Induction of splenocyte proliferation in the presence or absence of mitogen. Enhancement of NK activity. Inhibition of lysosomal function in a dose-dependent manner	Reduction of NO production (from 53.37 μM to 22.33 μM for compound **44**; 20.66 μM for compound **43**; and 28.64 μM compound **41**)	[50]
32–38	Human PBMCs assay	nd nd	nd nd	Stimulation of PBMC proliferation	Stimulation of IFNγ secretion	[46]
42	Human PBMCs assay	nd	nd	Inhibition of cell proliferation	Inhibition of IL2 secretion. Inhibition of NO release	[71]
39, **40**	Human PBMCs assay, RAW cells assay	nd	nd	Inhibition of cell proliferation (n°39). Inhibition of lymphocyte proliferation. No effect on NK cytotoxicity		

nd = non-determined. The isolated polyphenol molecules with immunomodulatory activities reduced IL2 and NO secretion while increased IFNγ secretion.

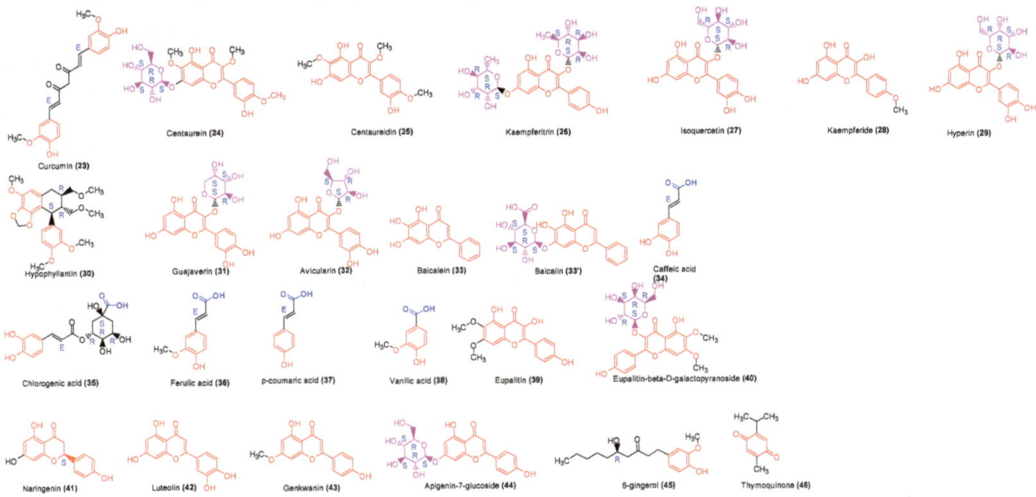

Figure 5. The structures of identified polyphenols acting as immunomodulators. The isolated polyphenols include flavonoids with the scaffold highlighted in red, polyphenolic acids with the acid group highlighted in blue, and ceto-phenolics. The thymoquinone is classed among polyphenols because in vivo metabolism gives phenolic compounds. The sugar group (highlighted in pink) is critical in the structure–activity relationship of flavonoids. Nevertheless, the active molecules have shown a similar way of modulating the immune system. They comprise phenol acid (**38**), cinnamic acid (**23**, **34–37**), lignan (**30**), flavonols (**25**, **28**, **39**), flavones (**33**, **42**, **43**), flavanone (**41**), heterosid flavonols (**24**, **26**, **27**, **29**, **31**, **32**, **40**), heterosid flavones (**33′**, **44**), and other phenolic compounds (**45**, **46**).

The ethanolic extract of *B. diffusa* roots exhibited antiproliferative activity on various human and murine cell lines and human PBMCs. Two flavonoids (compounds **39** and **40**) were isolated from the ethanolic and chloroform extracts of *B. diffusa* roots. These compounds inhibited PBMC proliferation induced by PHA and the mixed lymphocyte reaction (MLR), natural killer cytotoxicity, and LPS-induced NO production. In RAW264.7 cells, compound **40** inhibited PHA-induced IL-2 release and LPS-induced TNFα [71,86].

Compound **26** at concentrations of 25 mM showed no toxic effect on murine splenocytes, macrophages, and human PBMCs. It induced cell proliferation, stimulated lysosomal activity, and increased the neutral red uptake in macrophages and the natural killer cell activity [41].

Flavonoids isolated from the ethanolic extract of *C. viridiflorus* leaves exhibited varying effects on RAW264.7 cells. Compounds **27** and **28** significantly increased cell proliferation, while compound **29** had no effect [87]. Certain flavonoids isolated from different extracts of *P. guajava* showed immunomodulatory activity in head-kidney leucocyte assays [46]. Hypophyllantin (compound **30**) and guajaverin (compound **31**) significantly affected cell viability, whereas avicularin (compound **32**) did not. Compound **30** significantly increased the production of RBA, whereas compound **31** did not. Compound **30** also significantly increased the NOS production and lysozyme activity in HKL after 24 h of contact.

Compounds **32** and **33**, isolated from *P. major*, exhibited significant ($p < 0.05$) human PBMC proliferation and IFNγ secretion stimulation. Compound **37** possesses intense activity stimulating human PBMC proliferation and IFNγ secretion (181 pg/mL). The stimulation index was 4.59. Compound **38** enhanced human PBMC proliferation at the range of 5 and 40 mg/mL concentrations. The activity of compound **36** was lesser than that of compound **37**; however, this compound possesses higher activity than compounds **34** and **36** **[63]**. Compound **42** possesses inhibitory activity on human PBMC proliferation. The volatile oil extracted from *Z. officinale* (0.001–10 ng/mL) inhibited IL1α release in mice peritoneal macrophages. The DTH induced by DNFB was inhibited in a dose-dependent manner, and the inhibition rates were 31.6% ($p < 0.01$), 34.4% ($p < 0.01$), and 35.0 ($p < 0.01$), respectively. The thymus and spleen index decreased at 0.125, 0.25, and 0.5 g/kg bw doses. 6-gingerol (compound **45**) is a main pharmacologic substance isolated from *Z. officinale*. The combination of LPS and 6-gingerol had no significant cytotoxicity on RAW264.7 cells. Nitrite production was also significantly ($p < 0.05$) inhibited dose-dependently: 6-gingerol significantly suppressed iNOS proteins as well as mRNA levels, TNFα, and IL10 release. The molecule had a protective effect by preventing the calcium overload induced by H_2O_2. It blocked PKC-α translocation and suppressed LPS-induced cytoplasmic I-κBα phosphorylation [68,70]. The effect of thymoquinone (compound **46**), isolated from *N. sativa*, was evaluated on a rat *Wistar* cell proliferation model. A low concentration (1 µg/mL) did not significantly affect splenocyte viability and cell proliferation. At 5 µg/mL, a significant ($p < 0.05$) reduction of cell viability and proliferation was observed, whereas there was no effect on cytokines (IL4, IFNγ) production [67].

2.3.5. Coumarins

Coumarins isolated from various plant species (Figure 6) have demonstrated immunomodulatory effects (Table 5). Two coumarins isolated from *A. vera*, named compounds one and two were studied in an in vitro model. The compound one increased the macrophage phagocytic function in a concentration-dependent manner (50 to 200 µg/mL), with a maximum effect observed at 200 µg/mL, whereas compound two had no effect [51].

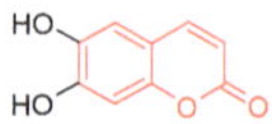

6,7-dihydrocoumarin **(47)**

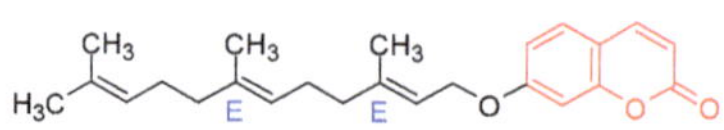

Umbelliprenin **(48)**

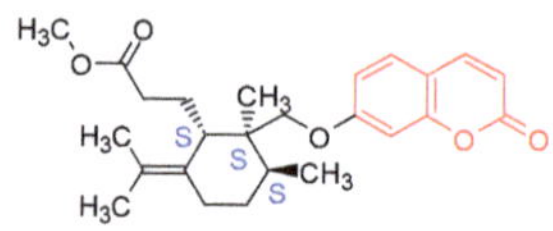

Methyl galbanate **(49)**

Figure 6. Structures of African medicinal isolated coumarins acting as immunomodulators. Sesquiterpenyl coumarins (**48–49**) with a long carbon chain show an identical mechanism of action while simple coumarin (**47**) has a different mechanism of action. The coumarin basic core is in red.

Esculetin (6,7-dihydrocoumarin, compound **47**), another isolated coumarin, from *A. vera* enhanced the mitogenic effect of splenocytes stimulated with LPS and concanavalin A. It induced lymphokine-activated killer (LAK) activity in lymphocytes [53].

Two coumarin terpenoids isolated from *F. szowitsiana*, compounds **48** and **49**, reduced PHA-induced splenocyte proliferation and preferentially induced IL-4 while suppressing IFNγ secretion [50].

Table 5. Isolated coumarins' immunomodulator activities.

Isolated Molecules (n°)	Models	Pharmacodynamic Parameters		Biological Effects	Cellular Effect	References
		ED_{50}	IC_{50}			
47	Murine macrophages and lymphocytes assay	nd	nd	No effect on macrophage viability. Enhancement of endocytic activity induced by LPS on macrophages at concentrations of 80 and 120 mM. Increase in mutagenic-induced cell proliferation. Induction of LAK activity of splenic lymphocytes.	Enhances NO production and iNOS gene expression	[53]

Table 5. *Cont.*

Isolated Molecules (n°)	Models	Pharmacodynamic Parameters		Biological Effects	Cellular Effect	References
		ED50	IC50			
48, **49**	Murine splenocytes assay	nd	nd	No effect on cell viability for tested concentrations (0.5–15 μM). Compound **48** at concentration >0.5 μM decreased splenocytes stimulation index. Compound **49** decreased cell proliferation at lowest dose. Suppression of PHA-induced cell proliferation.	Significantly augments IL4 secretion. Inhibits IFNγ production. Inhibits NO production by stimulated macrophages. Compound **48** increases PGE2 release; however, compound **49** inhibits it.	[50]

nd = non-determined.

2.3.6. Other Molecules: Glycosides

Syringin (compound **50**), a phenolic glucoside isolated from *T. crispa* and *T. cordifolia* (Figure 7), was investigated in an in vitro study using RAW 264.7 mouse macrophage cultures. It significantly reduced macrophage phagocytic activity and cell chemotaxis. The impact on cytokine production included reducing TNFα, IL1β, and IL6 production [17,18].

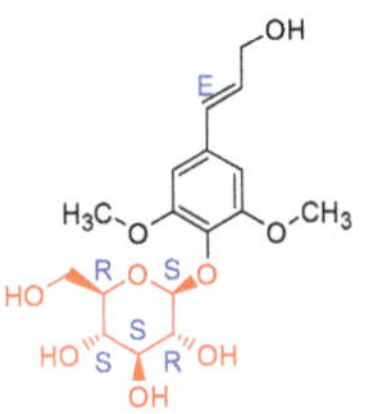

Syringin **(50)**

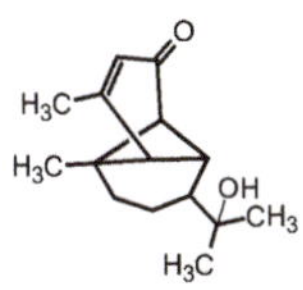

11-hydroxymustakone **(51)**

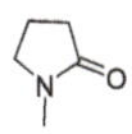

N-methyl-2-pyrrolidine **(52)**

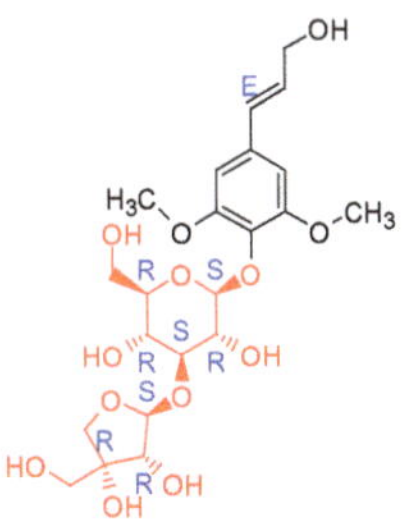

Cordifolioside **(53)**

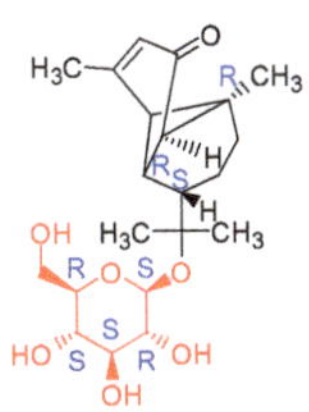

Tinocordiside **(54)**

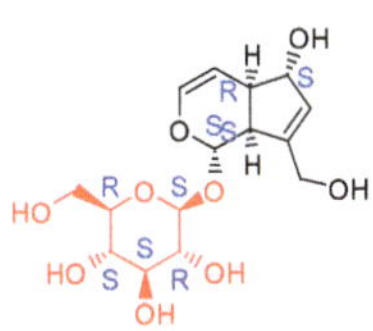

Aucubin **(55)**

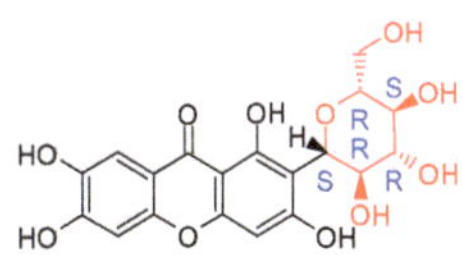

Mangiferin **(56)**

Figure 7. Structures of identified glycosides acting as immunomodulators. They comprise monosaccharides (**50**, **54**, **55**, **56**), disaccharides (**53**) and aglycans (**51**, **52**). The activity of glycosides is not linked only to the structure of the sugar unit (represented in red), but the aglycan part plays an important role. For example, the 11-hydroxymustakone molecule, without a sugar unit, has comparable activity to that of tinocordiside and cordifolioside.

Compounds **51**, **52**, **53**, and **54** (Figure 7 and Table 6) increased cell phagocytic activity and NO production in cells [18]. Compound **55** tested on PBMC showed significant stimulation of cell proliferation and enhanced IFNγ secretion [63].

Mangiferin (compound **56**), an isolated natural xanthone glucoside, demonstrated immunomodulatory activity. The 100 mg/kg dose significantly increased IgG and IgM levels, whereas IgA levels decreased in in vivo mice model [66].

Table 6. Isolated glycosides' immunomodulator activity.

Isolated Molecules (n°)	Models	Pharmacodynamic Parameters		Biological Effects	Cellular Effect	References
		ED_{50}	IC_{50}			
50	Murine RAW 264.7 cell viability assay Chemotaxis assay Phagocytosis assay NO, ROS, PGE2 production Monocyte chemoattractant Protein-1 production Cytokine production	nd	nd	Toxicity effect above 25 μg/mL. Reduction of cell chemotactic and phagocytosis activities. Diminution of MCP-1 production (IC_{50} = 48.3)	Reduction of NO production. Inhibition of PGE2 production (IC50 = 12.08 μM). Decrease in IL1β, IL6, and TNFα production.	[17]
51, 52, 53, 54	PMN cells viability assay Phagocytosis assay ROS, NO production assay	nd	nd	Increase in phagocytosis activity	Dose-dependent increase in NO and superoxide production	[18]
55	Human mononuclear cells assay Lymphocytes transformations test	nd	nd	Stimulation of PBMC proliferation	Enhancement of IFN-γ production	[63]

nd = non-determined.

2.3.7. Proteins

The analysis of raw garlic extract revealed the presence of several proteins within the 10–75 kD range (Table 7). The mitogenic activity on peripheral blood lymphocytes (PBL) demonstrated a significant increase in cell proliferation at 10 mg/mL concentrations. Notably, protein QR-2 exhibited the highest mitogenic activity. Additionally, the modulatory effect on splenocytes and thymocytes displayed a stimulatory effect on cell proliferation, while QR-1 and -2 showed agglutination in rabbit erythrocytes [54].

The thymus is the primary immune organ that produces functional T cells. The effect of onion agglutinin (ACA) on thymocyte proliferation showed an 4- and 3.5-fold increase in cell proliferation at 0.01 μg/well and 0.1 μg/well, respectively, at 24 h. On the other hand, ACA showed a weak increase in LPS-induced B-lymphocyte proliferation (1.3-fold). It significantly elevated the expression of IL-2 and IFN-γ. The macrophages are the first line of defense of the body against infections. At 0.1 μg/well, ACA induces a significant increase (6–8-fold) in NO production by RAW264.7 at 24 h. The release of cytokines (TNFα and IL-12) was significantly stimulated [59].

Proteins isolated from *A. membranaceus* displayed immunomodulatory activities. These proteins significantly affected the proliferation of splenocytes, murine peritoneal macrophages, and bone marrow-derived dendritic cells (BMDCs) at 10–90 μg/mL, except at 10 μg/mL. The optimal activity was observed at a concentration of 50 μg/mL. These proteins promoted the phagocytosis effect of murine peritoneal macrophages, with the compound AMWPDG2 displaying the highest activity. Furthermore, these proteins significantly promoted the secretion of various cytokines and chemokines, including TNFα, IL-6, IL12p40, IL-1β, IL-1α, nitric oxide, hydrogen peroxide, and CXCL1 and CXCL3 secretion [61].

Table 7. Isolated proteins with immunomodulator activity.

Sources	Extraction Method	Isolated Proteins	Molecular Weight (kDa)	Biological Effects	References
Allium sativum		QR-1, QR-2, QR3 (7:28:1)	13	Mitogenic activity on human PBMC, murine splenocytes and thymocytes. QR-1 and QR-2 showed hemagglutination and mannose-binding activities.	[54]
Allium cepa	Dialysis-D-mannose chromatography	ACA: *Allium cepa* Agglutinin	12	ACA at 0.1 μg/well and 0.01 μg/well enhance thymocyte proliferation by ~4- and 3.5-fold, respectively, with a marginal effect on B cells proliferation (~1.3-fold at 0.01 μg/well), significantly increased cytokine production (TNFα, IL12), and IFN-γ and IL2 expression. ACA induced an ~8-fold increase in NO production by rat peritoneal cells at 12 and 24 h. ACA (0.01–10 μg/well) significantly enhanced IL12 (~3-fold) and TNFα (~2–3-fold) release. The phagocytosis activity is enhanced by 2-fold by ACA (0.1; 1; 10 μg).	[59]
Tinospora cordifolia	Chromatography	G1, G2, G3	10–80	The proteins at a concentration range of 1–10 μg/mL showed mitogenic activity (3-fold) in murine splenocytes at 1–10 μg/mL and ~5–7-fold in thymocytes. They induced NO release by macrophages and enhanced macrophage phagocytosis activity.	[19]
Astragalus membranaceus,	Alkali extraction	AMWP (16 aa) AMWPDG2 (16 aa), AMWPDG4 (15 aa), AMWPDG6 (15 aa)	- 406.115 268.795 342.281	All proteins contain seven essential amino acids: Thr, Val., Met., Ile., Leu., Phe., and Lys. Proteins at 50 μg/mL significantly promoted in murine peritoneal macrophage phagocytosis activity, secretion of immunomodulatory factors like NO (AMWPDG2 > AMWPDG4 = AMWPDG6) and H_2O_2 (AMWPDG2 > AMWPDG6 > AMWPDG4) and inflammatory cytokines (TNFa and IL6)	[61]

aa: amino acid; Thr.: threonine; Val.: valine; Ile: Isoleucine; Leu: leucine; Phe.: phenylalanine; Lys.: lysine; Met.: methionine.

2.4. Mechanism of Action of Plant-Derived Immunomodulators

Numerous studies have elucidated the molecular mechanism underlying the immunomodulatory effects of phytochemicals. These compounds activate macrophages and other cells, such as dendritic and lymphocyte cells, through Toll-like Receptors (TLR) [88]. Proinflammatory cytokines and other immune system mediators are closely associated with the induction of transcription factors, including NF-kB, the nuclear factor of activated T lymphocytes, signal transduction, and transcription activator (STAT).

Cell signaling is initiated by receptor stimulation, with most receptors belonging to the TLR family. TLRs are pivotal receptors for many natural substances, including lipopolysaccharides (LPS), natural polysaccharides, alkaloids, and terpenoids. The activation of TLRs triggers the recruitment of MyD88 and subsequently activates specific intracellular pathways. All TLR signaling pathways ultimately lead to the activation of the transcription factor NF-kappa B, which regulates the expression of numerous inflammatory cytokine genes. Three main cell-signaling pathways, including phosphoinositide (PI3K-Akt), Mitogen-Activated Protein Kinases (MAPKs), and nuclear factor kappa B (NF-kB), can activate and transcribe NF-kB.

The MAPK pathway consists of a three-tier kinase cascade in which MAP3K activates MAP2K, which activates MAPKs. MAPK signaling pathways include the activation of the extracellular-related kinase (ERK1/2), p38 isoforms (p38), and c-Jun NH2-terminal kinase (JNK1/2). Once activated, MAPKs can be phosphorylated and translocated into the nucleus, leading to the expression of related genes and cellular responses, such as the secretion of signaling molecules (NO and ROS) and cytokines (IL-1β, IL6, TNFα).

In resting cells, NF-κB in the cytoplasm tightly associates with the inhibitory protein IκB, forming the IκB kinase (IKK) complex. Activation of the NF-κB signaling pathway provokes the disintegration of the IKK complex, leading to the phosphorylation and degradation of IκB-α. Consequently, NF-κB is released, translocated into the nucleus, and bound to DNA, initiating the transcription of proinflammatory-related genes.

The isolated phytochemicals target many proteins in these signaling ways (Table 8). Oleanolic and ursolic acid inhibit TLR4. Oleanolic acid blocks TLR3 activation and inhibits mRNA expression while suppressing the activation of IKKα/β proteins [89]. Magnoflorine and G1-4A activate MAPKs while being inhibited by flavonoids, such as immunomodulator curcumin.

Ursolic acid and betulinic acid inhibit the degradation of Iκ-Bα, the phosphorylation of IkBα and p64 protein, the activation of Iκ-Bα kinase, and the translocation of p65 [90]. Astragaloside IV activates the phosphorylation of the p65, p38, ERK, and JNK proteins [91]. Magnoflorine, an alkaloid extracted from *T. crispa*, activates the phosphorylation of p65 and increases the phosphorylation and degradation of IκB. It also increases the phosphorylation of the JNK, ERK, and p38 proteins [92]. Curcumin inhibits LPS-induced NF-kB activation by suppressing the MAPK pathway [93]. The possible molecular mechanisms of isolated bioactive-induced immunomodulation are shown in Figure 8 and Table 8.

Table 8. Convergent mechanism of action of isolated compounds of African medicinal plants with immunomodulator activities.

Compounds N°	Phytochemical Group	Cellular Model	Receptor	Transduction Pathway	Mechanism of Action	Cellular Actions	References
5	Alkaloids	Macrophages (U937)	TLR4	MAPKs, PI3K-Akt	Augmentation of Akt phosphorylation, induction of JNK, ERK, and p38 phosphorylation	Enhancement of upregulation of TNFα, IL1β, PGE2, COX-2	[92]
G1-4A	Polysaccharides	Macrophages	TLR4/MyD88	MAPKs	Activation of JNK, ERK, and p38 phosphorylation	Upregulation of the expression of TNFα, IL6, IL12, IL10	[20]
8	Triterpenoids	Macrophages	TLR4	TLR4-MyD88	Blocking TLR4/MyD88	Decrease in TNF-α, IL-1β et IL-6 release	[94]
7	Triterpenoids	THP1 cells	TLR3	MAPKs	Inhibition of IκB phosphorylation and NF-κB translocation		[89]

Table 8. *Cont.*

Compounds N°	Phytochemical Group	Cellular Model	Receptor	Transduction Pathway	Mechanism of Action	Cellular Actions	References
15	Terpenoids saponins	Macrophages		MAPKs/NFκB	Increase in the phosphorylation of p65, p38, JNK, and ERK, and a decrease in their protein expression	Increase in IL1β, IL6, TNFα, and inducible nitric oxide synthase	[91]
23	Flavonoids	Dendritic cells		MAPKs/NFκB	Suppression of MAPKs and p65 activation	Reduction of inducible NO synthase and IL-12	[93]
27	Flavonoids	Macrophages	TLR4	MAPKs	Suppression of phosphorylation of proteins p50/p65	Increase in TNFα, IL1β, iNOS	[95]
55	Glycosides	3T3-L1 adipocytes		NF-κB	Suppression of ERK phosphorylation and IκBα degradation	Inhibiting TNFα production	[96]
56	Glycosides	Mouse primary hepatocytes		MAPKs	Inhibiting the activation of c-JNK and ERK $\frac{1}{2}$		[97]

The proinflammatory mediators such TNFα, IL1β, IL6, IL10, COX-2, PGE2, and NO release are inversely modulated by inhibitors (**7**, **8**, **23**, **55**, **56**) and activators (**5**, **15**, **27**, **G1-4A**).

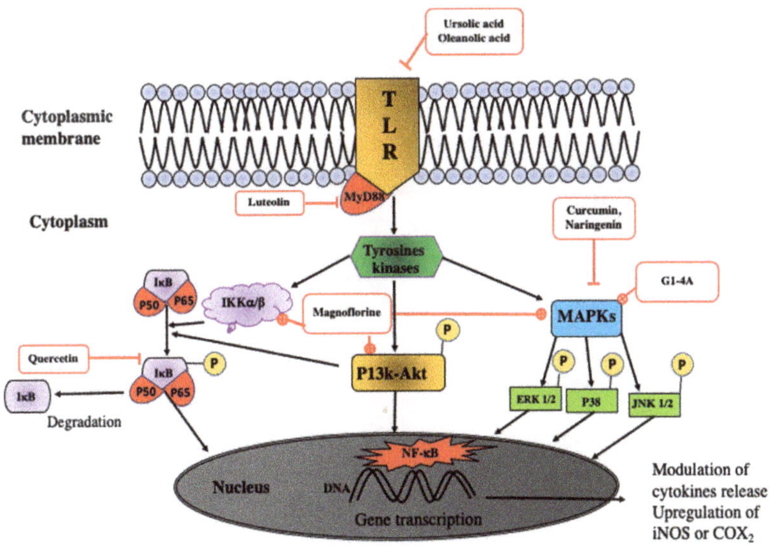

Figure 8. Immunomodulatory signal-transduction pathways of some molecules. TLR: toll-like receptor; MyD88: myeloid differentiation primary response gene 88; MAPKs: mitogen-activated protein kinases; P13k-Akt: phosphatidylinositol-3-kinase-kinase B protein; Ik-Bk: inhibitory kappa B kinase; IKK: I-kB kinase; p38, p50, p65: proteins 38, 50, 65; P: phosphate; NF-kB: nuclear factor kappa B; ERK: extracellular signal-regulated kinase; JNK: c-Jun N-terminal kinase; —●: activation or stimulation by phytochemical; —|: inhibition or reduction by phytochemical; →: transduction way or activation.

3. Discussion and Perspectives

This systematic review reported the immunomodulatory properties of 86 isolated phytochemicals or groups from 35 medicinal plant species belonging to 25 different families of African flora. Fifty-seven molecules had their structures identified and tested for immunomodulatory properties.

The review covered data from accessible full-text publications and did not take into account grey literature or others' protected data. Nevertheless, based on the obtained data, an analysis of the chemical-based immunomodulatory properties of isolated molecules in various immune cell models, including macrophages, splenocytes, thymocytes, T and B lymphocytes, dendritic cells, and human PBMCs, was carried out.

The chemical structures of the isolated immunomodulators are organized according to phytochemical groups. These molecules offer a wide range of therapeutic options, such as treating immune-related inflammation, cancer, and oxidant stress-related diseases.

3.1. Alkaloids

Among the identified alkaloids, berberine, piperine, and magnoflorine showed relevant prospects.

Studies have shown that berberine, an iso-quinoline isolated from various medicinal plants, has a variety of pharmacological actions, including antiarthritic, antibacterial, and anticancer effects [98,99]. Berberine is active on several cancer cell lines and has remarkable antiviral activity on several viruses, justified by its immunomodulatory mechanism.

Berberine has been reported to attenuate the radio-resistance of colon cancer cells by repressing P-gp expression [100] and to sensitize breast cancer cells to different chemotherapeutic drugs [101]. Berberine also attenuates ovarian cancer cell resistance to cisplatin (DDP) by targeting the miRNA-21 to regulate the post-transcriptional expression of the tumor suppressor programmed cell death 4 [102]. Pretreatment with berberine promoted the antitumor effects of DDP on laryngeal cancer cells [103]. Pandey et al. showed the potential actions of berberine in attenuating resistance to 5-fluorouracil in gastric cancer cells [104].

The isoquinoline alkaloids have multi-target potential in multifactorial chronic diseases, including immune, metabolic, and neurological disorders, and present a high perspective in therapy [105].

The alkaloids also had antiviral activities on both RNA and DNA viruses. Their broad spectrum of activity is of interest for identifying new applications, like for berberine, an isoquinoline with clinical promise. Berberine effectively disrupts the replication process of various DNA and RNA viruses, including the human immunodeficiency virus and the new severe acute respiratory syndrome linked to coronavirus-2 (SARCOV-2) [106–108].

Berberine targets antiviral activity by inhibiting RTase [109] with an $IC_{50} = 0.13$ mM on HIV-1 NL4.3, activating protein kinase signaling pathways in several viruses (see the figure describing the mechanism of action: the kinase cascade) or the ERK and JNK signaling cascades to prevent the generation of virions. By activating the MAPK pathway, berberine reduces the virion titer during CHIKV infection [110] and suppresses RSV replication [111].

Piperine, another alkaloid isolated from *P. longa*, showed antitumoral, anti-inflammatory activities correlated with its immunomodulatory activity. It inhibited the translocation of NF-κB subunits like p50, p65, and c-Rel, as well as CREB, ATF-2, and c-Fos [112]. In an in vitro LPS-induced osteoarthritis model, piperine treatment showed anti-inflammatory activity by downregulating miR-127 and MyD88 expression. Piperine triggers apoptosis in ovarian cancer cells by increasing the JNK and p38 MAPK phosphorylation [113].

Based on these results, isolated alkaloids could be candidates for therapeutic agents for diseases-based NF-κB signaling pathways' activation.

3.2. Polysaccharides and Proteins

Peritoneal macrophages possess important immunological functions, such as immune defense, surveillance, regulation, and antigen presentation [114]. Phagocytosis is the first step in the macrophage response to invading microorganisms, and the activation of phagocytosis elevates the innate immune response. Activated peritoneal macrophages induce NO production, and various immunostimulatory factors, such as IL-6 and TNF-α, play important roles in the phagocytosis, antigen presentation, and inflammatory regulation of macrophages [115]. It is important to measure the activity of immunostimulatory factors to highlight the immune-based mechanism of protection or pathophysiology genesis of immune-related diseases to guide medication. Numerous plant-isolated molecules act on macrophages, enhancing phagocytic activity and stimulating NO or TNFα release. The spleen is the largest immune organ reflecting the systemic immune status in the human body, and it plays an important role in anti-infection and anticancer activities [116]. The proliferation of spleen lymphocytes is key to activating cellular and humoral immunomodulatory responses. Some proteins (QR-1, QR-2, QR-3, AMWP, AMWPDG2, AMWPDG4, and AMWPDG6) and polysaccharides (Pectin, FOS, Glucan, CPE2, CPE4, and CALB-4) stimulate splenocytes' and thymocytes' proliferation. Polysaccharides help boost the immune system while also possessing anti-tumor properties.

3.3. *Terpenoids*

Among the terpenoids studied, oleanolic acid, ursolic acid, and betulinic acid have divergent effects depending on the concentration. Oleanolic acid stimulates T lymphocyte proliferation at 0.5 μg/mL, while ursolic and betulinic acid have an inhibitory effect with IC_{50} of 3.01 and 50 μg/mL, respectively [117]. Depending on their concentration, molecules belonging to the same phytochemical family can have opposite effects.

The chemical structure plays a significant role in the biological effect observed. Studying the structure–activity relationship will help elucidate the mechanism. Ursolic and oleanolic acids correlate with different biological activities, such as anti-inflammatory, anticancer, and antidiabetic [118]. Different mechanisms, including immunomodulation, could justify these proprieties.

Immunomodulators act by stimulating or suppressing the immune response. Therefore, they can be used in several therapeutic areas.

In vaccine development, new and improved adjuvants are needed to mitigate adverse effects and increase immunogenicity. AST VII and Mac B isolated from Astragalus have shown an immunostimulant effect by increasing IgG and IgG1 antibody titers with a smaller hemolytic effect. They also act by activating T and B lymphocytes during immunization with BSA [83]. The glucuronoxylan-D polymer isolated from *S. officinalis*, which has a significant comitogenic effect, has potential adjuvant properties [73].

Immunomodulators offer exciting prospects in the treatment of cancer. Those that strengthen the body's ability to identify and eliminate cancer cells are revolutionizing treatment and can also help to strengthen the immune system. Immunomodulators have great potential in the treatment of autoimmune diseases. They, therefore, help to restore immunological balance.

They also offer excellent prospects for treating infectious diseases and combating bacterial resistance. B-aescin isolated from the seeds of *A. hippocastanum* has virucidal and antiviral activity in addition to its immunomodulatory effect [119]. Combining these effects would provide comprehensive treatment while strengthening the immune system.

3.4. *Polyphenols*

The negative impacts of synthetic drugs and the quest for natural alternatives for therapy have led to an increased demand for the multi-target action of phenols for enhancing immunity [120].

Combining quercetin with piperine, flavonoid, and no-flavonoid compounds presented an effective and potent anti-inflammatory strategy for treating acute colitis in mice. This anti-inflammatory effect was mediated by impaired DC immune responses [121]. Several flavonoids also display antitumor effects directly or indirectly. Luteolin enhanced the sensitivity of lapatinib in human breast cancer cells, and the combination of baicalein and cisplatin increased the apoptosis of gastric cancer and A549 lung adenocarcinoma cells in vitro [122,123].

The immunomodulatory effect is one of the most important mechanisms for the anti-tumor effect of flavonoids. They enhanced the cytotoxicity of NK and CTL cells to tumor cells via the upregulation of their activating receptor [124,125].

Flavonoids inhibit the production of various pro-inflammatory cytokines (IL6 and IL1β). Undeniably, the inflammatory tumor microenvironment is essential in the progression of malignant [126]. Baicalein (**33**) and baicalin (**33′**) are directly cytotoxic to some tumors. In addition to direct cytotoxicity, these two flavonoids stimulate the T cell-mediated immune response against tumors by reducing PD-L1 expression in cancer cells [127]. Kaempferol, curcumin, and quercetin inhibit the proliferation of numerous cell line [128].

NF-κB is essential in human cancer initiation, development, metastasis, and treatment resistance [129–132]. Many human cancers exhibit constitutive NF-κB activity due to the inflammatory microenvironment and various oncogenic mutations. NF-κB activity is associated with tumor cell proliferation [132]. In addition to suppressing apoptosis and promoting angiogenesis, it also induces an epithelial–mesenchymal transition, facilitating distant metastasis formation. NF-κB activation can also remodel the local metabolism and stimulate the immune system to promote tumor growth. The suppression of NF-κB in myeloid or tumor cells generally results in tumor regression, making the NF-κB pathway a promising therapeutic target. Howeve r, due to its vital role in various biological activities, selective targeting of components of the NF-κB pathway must be achieved for therapeutic purposes.

Since NF-κB plays an essential role in both tumor cells and the tumor microenvironment, targeting NF-κB as an anticancer therapy has been explored extensively over the last few decades. Hundreds of natural and synthetic compounds have been reported to inhibit NF-κB. However, their clinical application has shown little efficacy, except for certain types of lymphoma and leukemia [132].

Flavonoids such as curcumin, naringenin, and quercetin inhibit NF-κB through various signaling pathways, which could justify their anticancer activity.

3.5. Glycosides

The immunomodulatory properties of mangiferin have been demonstrated in several studies. It inhibits NF-KB by reducing the translocation of the p65 protein subunit. It also inhibits activation of the AGE-RAGE/mitogen-activated protein kinase (MAPK), the c-Jun N-terminal kinase (JNK), and the p38 pathways. The expression of extracellular regulated kinase 1/2 (ERK1/2) in the myocardium is also increased. Mangiferin could have beneficial effects in diabetic cardiomyopathy [133,134]. Mangiferin has antioxidant and anti-apoptotic properties via the MAPK/NF-κB/mitochondria-dependent pathways. It has an immunoprotective effect during cancer therapy [65,135]. Obesity is closely associated with a state of chronic inflammation, characterized by the abnormal production of cytokines and the activation of inflammatory signaling pathways in adipose tissue. The signaling pathways involve MAPKs.

Aucubin (**55**) suppressed the activation of extracellular signal-regulated kinase (ERK), the degradation of inhibitory kappa Ba (IκBα), and the subsequent activation of nuclear factor kappa B (NF-κB) [96]. Aucubin could improve obesity-induced atherosclerosis by attenuating TNF-α-induced inflammatory responses.

4. Methods

4.1. Search Strategy

The research was conducted according to PRISMA guidelines to identify studies about African medicinal plants used for their immunomodulatory properties. This exploration involved using the scholarly search engine Google Scholar and a comprehensive screening of prominent international databases, including PubMed, ScienceDirect, African

Journal Online, and Embase. Our search queries incorporated specific keywords such as "immunomodulator" OR "immunity" AND "medicinal plant" OR "herbal plant" AND "phytochemicals".

We exclusively considered scientific research articles published in English until December 2023, ensuring that they were accessible without restrictions. Article titles and abstracts were screened according to the research objectives, which focused on the immunomodulatory effects of African medicinal plant chemicals. This examination encompassed instances involving either full or partial isolation, coupled with the subsequent determination of the chemical structure of active compounds. Articles deemed relevant following independent evaluation were included.

4.2. Data Extraction

Eligible records were extracted into Microsoft Excel, 2013 by NWA and double-checked by OM, OWP, and RO. We systematically gathered data from each publication, including scientific names, botanical families, used parts of plants, extraction solvents, phytochemical groups, isolated compounds, and experimental models for objectifying immunomodulatory activities. These data were meticulously documented using a standardized Excel sheet form.

The risk of bias and the quality of each article were assessed by two independent reviewers.

Following this, we analyzed proposed mechanisms underpinning immunomodulatory activities to elucidate convergent and distinctive signaling pathways associated with the active compounds. The chemical structures of the isolated compounds were generated using ChemDraw® (version 12.0.1076). The specific scaffold (alkaloids, terpenoid, phenol, polyphenol, coumarin, or glycoside) was drawn in red, the sugar group and acid group were respectively drawn in pink and blue. In purpose to demonstrate stereochemistry, R and S chirality centers configuration were determined using the Cahn Ingold Prelog priority rules, and the E and Z absolute configurations for double bounds were indicated.

This systematic review protocol is registered on inplasy.com at number INPLASY202410116 (Accessed on 29 January 2024).

5. Conclusions

The present review is a contribution to highlighting African medicinal plants with isolated immunomodulatory molecules. The isolated chemicals, such oleanolic acid, ursolic acid, boswellic acid, betulinic acid, astragalosides, magnoflorine, luteolin, curcumin, andrographolides, centeurein, centaureidin, quercetin, guaverin, corosolic acid, naringenin, pectin, acemannan, nimbidin, syringin, esculetin, umbeliprenin, methyl galbanate, hypophyllantin, and baicalein, possess significant immunomodulatory properties. These molecules belong to various phytochemical groups: alkaloids, polyphenols, terpenoids, carbohydrates, glycosides, and proteins.

The chemical structures' stereoisomers' demonstration of the isolated compounds allows an analytical approach to the structure–activity relationship, as their main immunomodulatory transduction pathways are proposed. The pharmacological properties of the chemicals can be improved by chemical group optimizing to yield analogs with improved pharmacokinetic and pharmacodynamic properties.

The chemical actions comprise modulating the expression or phosphorylation status of various accessory proteins associated with the TLRs, STAT3, NF-κB, MAPKs, and PI3K/Akt pathways. The therapeutic perspectives of such immunomodulators are infectious, cancer, and chronic inflammatory diseases' care, and the development of immunoadjuvants for vaccines.

The review data can also contribute to the registration of immunomodulatory plant-based traditional medicines, to orient researchers towards the screening of new immunomodulatory chemicals from the species or genera of other medicinal plants, or to allow the comparison with immunomodulatory medicinal plants and phytochemicals of other continents.

Author Contributions: Conceptualization, M.O. and W.A.N.; writing—original draft preparation, W.A.N.; writing—review and editing, M.O., W.P.O. and R.O.; Read and corrections: T.E.D., B.H.A.O., B.A., A.S.S., G.M.A., S.F. and R.S. All authors have read and agreed to the published version of the manuscript.

Funding: This research received no external funding.

Institutional Review Board Statement: Not applicable.

Informed Consent Statement: Not applicable.

Data Availability Statement: Any supplementary information can be obtained on request addressed to the corresponding author.

Acknowledgments: We would like to thank the ACE-CFOREM, a project of the Word Bank giving to N.W.A a scholarship and facilities to perform his studies.

Conflicts of Interest: The authors declare no conflicts of interest. The ACE-CFOREM, which gave the scholarship, had no role in the design of the study, in the collection, analyses, or interpretation of data, in the writing of the manuscript, or in the decision to publish the results.

References

1. Nicholson, L.B. The Immune System. *Essays Biochem.* **2016**, *60*, 275–301. [CrossRef]
2. Mohandas, S.; Jagannathan, P.; Henrich, T.J.; Sherif, Z.A.; Bime, C.; Quinlan, E.; Portman, M.A.; Gennaro, M.; Rehman, J. RECOVER Mechanistic Pathways Task Force Immune Mechanisms Underlying COVID-19 Pathology and Post-Acute Sequelae of SARS-CoV-2 Infection (PASC). *Elife* **2023**, *12*, e86014. [CrossRef]
3. Nielsen, D.G. The Relationship of Interacting Immunological Components in Dengue Pathogenesis. *Virol. J.* **2009**, *6*, 211. [CrossRef]
4. Pisetsky, D.S. Pathogenesis of Autoimmune Disease. *Nat. Rev. Nephrol.* **2023**, *19*, 509–524. [CrossRef]
5. Khan, M.A.; Khan, Z.A.; Charles, M.; Pratap, P.; Naeem, A.; Siddiqui, Z.; Naqvi, N.; Srivastava, S. Cytokine Storm and Mucus Hypersecretion in COVID-19: Review of Mechanisms. *J. Inflamm. Res.* **2021**, *14*, 175–189. [CrossRef]
6. Castro, C.; Gourley, M. Diagnostic Testing and Interpretation of Tests for Autoimmunity. *J. Allergy Clin. Immunol.* **2010**, *125*, S238–S247. [CrossRef]
7. Mohamed, S.I.A.; Jantan, I.; Haque, M.A. Naturally Occurring Immunomodulators with Antitumor Activity: An Insight on Their Mechanisms of Action. *Int. Immunopharmacol.* **2017**, *50*, 291–304. [CrossRef]
8. Shantilal, S.; Vaghela, J.S.; Sisodia, S.S. Review on Immunomodulation and Immunomodulatory Activity of Some Medicinal Plant. *Eur. J. Biomed.* **2018**, *5*, 163–174.
9. Zhao, Z.; Zheng, L.; Chen, W.; Weng, W.; Song, J.; Ji, J. Delivery Strategies of Cancer Immunotherapy: Recent Advances and Future Perspectives. *J. Hematol. Oncol.* **2019**, *12*, 126. [CrossRef]
10. Cox, E.; Verdonck, F.; Vanrompay, D.; Goddeeris, B. Adjuvants Modulating Mucosal Immune Responses or Directing Systemic Responses towards the Mucosa. *Vet. Res.* **2006**, *37*, 511–539. [CrossRef]
11. Lutsiak, M.E.C.; Kwon, G.S.; Samuel, J. Biodegradable Nanoparticle Delivery of a Th2-Biased Peptide for Induction of Th1 Immune Responses. *J. Pharm. Pharmacol.* **2006**, *58*, 739–747. [CrossRef]
12. Zhang, A.; Yang, X.; Li, Q.; Yang, Y.; Zhao, G.; Wang, B.; Wu, D. Immunostimulatory Activity of Water-Extractable Polysaccharides from Cistanche Deserticola as a Plant Adjuvant In Vitro and In Vivo. *PLoS ONE* **2018**, *13*, e0191356. [CrossRef]
13. Bdallah, E.; Koko, W. Medicinal Plants of Antimicrobial and Immunomodulating Properties. *Antimicrob. Res. Nov. Bioknowl. Educ. Programs* **2017**, *6*, 127–139.
14. Lin, W.-C.; Lin, J.-Y. Berberine Down-Regulates the Th1/Th2 Cytokine Gene Expression Ratio in Mouse Primary Splenocytes in the Absence or Presence of Lipopolysaccharide in a Preventive Manner. *Int. Immunopharmacol.* **2011**, *11*, 1984–1990. [CrossRef]
15. Lee, C.-H.; Chen, J.-C.; Hsiang, C.-Y.; Wu, S.-L.; Wu, H.-C.; Ho, T.-Y. Berberine Suppresses Inflammatory Agents-Induced Interleukin-1β and Tumor Necrosis Factor-α Productions via the Inhibition of IκB Degradation in Human Lung Cells. *Pharmacol. Res.* **2007**, *56*, 193–201. [CrossRef]
16. Okon, E.; Kukula-Koch, W.; Halasa, M.; Jarzab, A.; Baran, M.; Dmoszynska-Graniczka, M.; Angelis, A.; Kalpoutzakis, E.; Guz, M.; Stepulak, A.; et al. Magnoflorine—Isolation and the Anticancer Potential against NCI-H1299 Lung, MDA-MB-468 Breast, T98G Glioma, and TE671 Rhabdomyosarcoma Cancer Cells. *Biomolecules* **2020**, *10*, 1532. [CrossRef]
17. Ahmad, W.; Jantan, I.; Kumolosasi, E.; Haque, M.A.; Bukhari, S.N.A. Immunomodulatory Effects of *Tinospora crispa* Extract and Its Major Compounds on the Immune Functions of RAW 264.7 Macrophages. *Int. Immunopharmacol.* **2018**, *60*, 141–151. [CrossRef]
18. Sharma, U.; Bala, M.; Kumar, N.; Singh, B.; Munshi, R.K.; Bhalerao, S. Immunomodulatory Active Compounds from *Tinospora cordifolia*. *J. Ethnopharmacol.* **2012**, *141*, 918–926. [CrossRef]
19. Aranha, I.; Clement, F.; Venkatesh, Y.P. Immunostimulatory Properties of the Major Protein from the Stem of the Ayurvedic Medicinal Herb, Guduchi (*Tinospora cordifolia*). *J. Ethnopharmacol.* **2012**, *139*, 366–372. [CrossRef]

20. Gupta, P.K.; Rajan, M.G.R.; Kulkarni, S. Activation of Murine Macrophages by G1-4A, a Polysaccharide from *Tinospora cordifolia*, in TLR4/MyD88 Dependent Manner. *Int. Immunopharmacol.* **2017**, *50*, 168–177. [CrossRef]
21. Sunila, E.S.; Kuttan, G. Immunomodulatory and Antitumor Activity of *Piper longum* Linn. and Piperine. *J. Ethnopharmacol.* **2004**, *90*, 339–346. [CrossRef] [PubMed]
22. El-Ashmawy, N.E.; El-Zamarany, E.A.; Salem, M.L.; El-Bahrawy, H.A.; Al-Ashmawy, G.M. In Vitro and In Vivo Studies of the Immunomodulatory Effect of Echinacea Purpurea on Dendritic Cells. *J. Genet. Eng. Biotechnol.* **2015**, *13*, 185–192. [CrossRef] [PubMed]
23. Benson, J.M.; Pokorny, A.J.; Rhule, A.; Wenner, C.A.; Kandhi, V.; Cech, N.B.; Shepherd, D.M. Echinacea Purpurea Extracts Modulate Murine Dendritic Cell Fate and Function. *Food Chem. Toxicol.* **2010**, *48*, 1170–1177. [CrossRef] [PubMed]
24. Yang, G.; Li, K.; Liu, C.; Peng, P.; Bai, M.; Sun, J.; Li, Q.; Yang, Z.; Yang, Y.; Wu, H. A Comparison of the Immunostimulatory Effects of Polysaccharides from Tetraploid and Diploid Echinacea Purpurea. *BioMed Res. Int.* **2018**, *2018*, e8628531. [CrossRef] [PubMed]
25. Shu, Z.; Yang, Y.; Xing, N.; Wang, Y.; Wang, Q.; Kuang, H. Structural Characterization and Immunomodulatory Activity of a Pectic Polysaccharide (CALB-4) from Fructus Aurantii. *Int. J. Biol. Macromol.* **2018**, *116*, 831–839. [CrossRef] [PubMed]
26. Zhang, S.; Li, Z.; Wang, X.; An, L.; Bao, J.; Zhang, J.; Cui, J.; Li, Y.; Jin, D.-Q.; Tuerhong, M.; et al. Isolation, Structural Elucidation, and Immunoregulation Properties of an Arabinofuranan from the Rinds of *Garcinia mangostana*. *Carbohydr. Polym.* **2020**, *246*, 116567. [CrossRef] [PubMed]
27. Ling, Z.; Ting Jun, H.; Chun Ni, L. Immunomodulatory and Antioxidant Activity of a Siraitia Grosvenorii Polysaccharide in Mice. *Afr. J. Biotechnol.* **2011**, *10*, 10045–10053. [CrossRef]
28. Michelini, F.M.; Alché, L.E.; Bueno, C.A. Virucidal, Antiviral and Immunomodulatory Activities of β-Escin and Aesculus Hippocastanum Extract. *J. Pharm. Pharmacol.* **2018**, *70*, 1561–1571. [CrossRef]
29. Low, M.; Khoo, C.S.; Münch, G.; Govindaraghavan, S.; Sucher, N.J. An In Vitro Study of Anti-Inflammatory Activity of Standardised *Andrographis paniculata* Extracts and Pure Andrographolide. *BMC Complement. Altern. Med.* **2015**, *15*, 18. [CrossRef]
30. Naik, S.R.; Hule, A. Evaluation of Immunomodulatory Activity of an Extract of Andrographolides from *Andographis paniculata*. *Planta Med.* **2009**, *75*, 785–791. [CrossRef]
31. Kaur, G.; Sarwar Alam, M.; Athar, M. Nimbidin Suppresses Functions of Macrophages and Neutrophils: Relevance to Its Antiinflammatory Mechanisms. *Phytother. Res.* **2004**, *18*, 419–424. [CrossRef]
32. Vaghasiya, J.; Datani, M.; Nandkumar, K.; Malaviya, S.; Jivani, N. Comparative Evaluation of Alcoholic and Aqueous Extracts of Ocimum Sanctum for Immunomodulatory Activity. *Int. J. Pharm. Biol. Res.* **2010**, *1*, 25e9.
33. Sharma, M.L.; Kaul, A.; Khajuria, A.; Singh, S.; Singh, G.B. Immunomodulatory Activity of Boswellic Acids (Pentacyclic Triterpene Acids) from *Boswellia serrata*. *Phytother. Res.* **1996**, *10*, 107–112. [CrossRef]
34. Liao, J.; Wu, D.; Peng, S.; Xie, J.; Li, Y.; Su, J.; Chen, J.; Su, Z. Immunomodulatory Potential of Patchouli Alcohol Isolated from Pogostemon Cablin (Blanco) Benth (Lamiaceae) in Mice. *Trop. J. Pharm. Res* **2013**, *12*, 559–565. [CrossRef]
35. Chang, S.-L.; Chiang, Y.-M.; Chang, C.L.-T.; Yeh, H.-H.; Shyur, L.-F.; Kuo, Y.-H.; Wu, T.-K.; Yang, W.-C. Flavonoids, Centaurein and Centaureidin, from *Bidens pilosa*, Stimulate IFN-γ Expression. *J. Ethnopharmacol.* **2007**, *112*, 232–236. [CrossRef] [PubMed]
36. Pereira, R.L.; Ibrahim, T.; Lucchetti, L.; da Silva, A.J.; Gonçalves de Moraes, V.L. Immunosuppressive and Anti-Inflammatory Effects of Methanolic Extract and the Polyacetylene Isolated from *Bidens pilosa* L. *Immunopharmacology* **1999**, *43*, 31–37. [CrossRef] [PubMed]
37. Abdelhady, M.I.S.; Kamal, A.M.; Rauf, A.; Mubarak, M.S.; Hadda, T.B. Bioassay-Guided Isolation and POM Analyses of a New Immunomodulatory Polyphenolic Constituent from *Callistemon viridiflorus*. *Nat. Prod. Res.* **2016**, *30*, 1131–1135. [CrossRef] [PubMed]
38. Yadav, V.S.; Mishra, K.P.; Singh, D.P.; Mehrotra, S.; Singh, V.K. Immunomodulatory Effects of Curcumin. *Immunopharmacol. Immunotoxicol.* **2005**, *27*, 485–497. [CrossRef]
39. Ranjan, D.; Chen, C.; Johnston, T.D.; Jeon, H.; Nagabhushan, M. Curcumin Inhibits Mitogen Stimulated Lymphocyte Proliferation, NFkappaB Activation, and IL-2 Signaling. *J. Surg. Res.* **2004**, *121*, 171–177. [CrossRef]
40. Alonso-Castro, A.J.; Ortiz-Sánchez, E.; Domínguez, F.; Arana-Argáez, V.; Juárez-Vázquez, M.d.C.; Chávez, M.; Carranza-Álvarez, C.; Gaspar-Ramírez, O.; Espinosa-Reyes, G.; López-Toledo, G.; et al. Antitumor and Immunomodulatory Effects of *Justicia spicigera* Schltdl (Acanthaceae). *J. Ethnopharmacol.* **2012**, *141*, 888–894. [CrossRef]
41. Del Carmen Juárez-Vázquez, M.; Josabad Alonso-Castro, A.; García-Carrancá, A. Kaempferitrin Induces Immunostimulatory Effects In Vitro. *J. Ethnopharmacol.* **2013**, *148*, 337–340. [CrossRef] [PubMed]
42. Harikrishnan, H.; Jantan, I.; Haque, M.A.; Kumolosasi, E. Anti-Inflammatory Effects of *Phyllanthus amarus* Schum. & Thonn. through Inhibition of NF-κB, MAPK, and PI3K-Akt Signaling Pathways in LPS-Induced Human Macrophages. *BMC Complement. Altern. Med.* **2018**, *18*, 224. [CrossRef] [PubMed]
43. Kassuya, C.A.L.; Leite, D.F.P.; de Melo, L.V.; Rehder, V.L.G.; Calixto, J.B. Anti-Inflammatory Properties of Extracts, Fractions and Lignans Isolated from *Phyllanthus amarus*. *Planta Med.* **2005**, *71*, 721–726. [CrossRef] [PubMed]
44. Jantan, I.; Ilangkovan, M.; Yuandani; Mohamad, H.F. Correlation between the Major Components of *Phyllanthus amarus* and *Phyllanthus urinaria* and Their Inhibitory Effects on Phagocytic Activity of Human Neutrophils. *BMC Complement. Altern. Med.* **2014**, *14*, 429. [CrossRef]

45. Shabbir, A.; Butt, H.; Shahzad, M.; Arshad, H.; Waheed, I. Immunostimulatory Effect of Methanolic Leaves Extract of *Psidium guajava* (Guava) on Humoral and Cell-Mediated Immunity in Mice. *J. Anim. Plant Sci.* **2016**, *26*, 1492–1500.
46. Nhu, T.Q.; Dam, N.P.; Bich Hang, B.T.; Bach, L.T.; Thanh Huong, D.T.; Buu Hue, B.T.; Scippo, M.-L.; Phuong, N.T.; Quetin-Leclercq, J.; Kestemont, P. Immunomodulatory Potential of Extracts, Fractions and Pure Compounds from *Phyllanthus amarus* and *Psidium guajava* on Striped Catfish (*Pangasianodon hypophthalmus*) Head Kidney Leukocytes. *Fish. Shellfish. Immunol.* **2020**, *104*, 289–303. [CrossRef] [PubMed]
47. Ben Sghaier, M.; Krifa, M.; Mensi, R.; Bhouri, W.; Ghedira, K.; Chekir-Ghedira, L. In Vitro and In Vivo Immunomodulatory and Anti-Ulcerogenic Activities of *Teucrium ramosissimum* Extracts. *J. Immunotoxicol.* **2011**, *8*, 288–297. [CrossRef]
48. Nasr-Bouzaiene, N.; Sassi, A.; Bedoui, A.; Krifa, M.; Chekir-Ghedira, L.; Ghedira, K. Immunomodulatory and Cellular Antioxidant Activities of Pure Compounds from *Teucrium ramosissimum* Desf. *Tumour Biol.* **2016**, *37*, 7703–7712. [CrossRef] [PubMed]
49. Askari, V.R.; Alavinezhad, A.; Rahimi, V.B.; Rezaee, S.A.; Boskabady, M.H. Immuno-Modulatory Effects of Methanolic Extract of *Ferula szowitsiana* on Isolated Human Th1/Th2/Treg Cytokines Levels, and Their Genes Expression and Nitric Oxide Production. *Cytokine* **2021**, *138*, 155387. [CrossRef]
50. Zamani Taghizadeh Rabe, S.; Iranshahi, M.; Mahmoudi, M. In Vitro Anti-Inflammatory and Immunomodulatory Properties of Umbelliprenin and Methyl Galbanate. *J. Immunotoxicol.* **2016**, *13*, 209–216. [CrossRef]
51. Zhang, X.; Wang, H.; Song, Y.; Nie, L.; Wang, L.; Liu, B.; Shen, P.; Liu, Y. Isolation, Structure Elucidation, Antioxidative and Immunomodulatory Properties of Two Novel Dihydrocoumarins from Aloe Vera. *Bioorg. Med. Chem. Lett.* **2006**, *16*, 949–953. [CrossRef] [PubMed]
52. Kumar, S.; Tiku, A.B. Immunomodulatory Potential of Acemannan (Polysaccharide from Aloe Vera) against Radiation Induced Mortality in Swiss Albino Mice. *Food Agric. Immunol.* **2016**, *27*, 72–86. [CrossRef]
53. Leung, K.; Leung, P.; Kong, L.; Leung, P. Immunomodulatory Effects of Esculetin (6,7-Dihydroxycoumarin) on Murine Lymphocytes and Peritoneal Macrophages. *Cell Mol. Immunol.* **2005**, *2*, 181–188. [PubMed]
54. Clement, F.; Pramod, S.N.; Venkatesh, Y.P. Identity of the Immunomodulatory Proteins from Garlic (*Allium sativum*) with the Major Garlic Lectins or Agglutinins. *Int. Immunopharmacol.* **2010**, *10*, 316–324. [CrossRef] [PubMed]
55. Chandrashekar, P.M.; Prashanth, K.V.H.; Venkatesh, Y.P. Isolation, Structural Elucidation and Immunomodulatory Activity of Fructans from Aged Garlic Extract. *Phytochemistry* **2011**, *72*, 255–264. [CrossRef] [PubMed]
56. Chandrashekar, P.M.; Venkatesh, Y.P. Identification of the Protein Components Displaying Immunomodulatory Activity in Aged Garlic Extract. *J. Ethnopharmacol.* **2009**, *124*, 384–390. [CrossRef] [PubMed]
57. Patra, P.; Sen, I.K.; Bhanja, S.K.; Nandi, A.K.; Samanta, S.; Das, D.; Devi, K.S.P.; Maiti, T.K.; Islam, S.S. Pectic Polysaccharide from Immature Onion Stick (*Allium cepa*): Structural and Immunological Investigation. *Carbohydr. Polym.* **2013**, *92*, 345–352. [CrossRef] [PubMed]
58. Kumar, V.P.; Prashanth, K.V.H.; Venkatesh, Y.P. Structural Analyses and Immunomodulatory Properties of Fructo-Oligosaccharides from Onion (*Allium cepa*). *Carbohydr. Polym.* **2015**, *117*, 115–122. [CrossRef] [PubMed]
59. Prasanna, V.K.; Venkatesh, Y.P. Characterization of Onion Lectin (*Allium cepa* Agglutinin) as an Immunomodulatory Protein Inducing Th1-Type Immune Response In Vitro. *Int. Immunopharmacol.* **2015**, *26*, 304–313. [CrossRef]
60. Farag, M.R.; Alagawany, M. The Role of *Astragalus membranaceus* as Immunomodulator in Poultry. *World's Poult. Sci. J.* **2019**, *75*, 43–54. [CrossRef]
61. Huang, H.; Luo, S.-H.; Huang, D.-C.; Cheng, S.-J.; Cao, C.-J.; Chen, G.-T. Immunomodulatory Activities of Proteins from *Astragalus membranaceus* Waste. *J. Sci. Food Agric.* **2019**, *99*, 4174–4181. [CrossRef] [PubMed]
62. Liang, Y.; Zhang, Q.; Zhang, L.; Wang, R.; Xu, X.; Hu, X. *Astragalus membranaceus* Treatment Protects Raw264.7 Cells from Influenza Virus by Regulating G1 Phase and the TLR3-Mediated Signaling Pathway. *Evid.-Based Complement. Altern. Med.* **2019**, *2019*, 2971604. [CrossRef] [PubMed]
63. Chiang, L.-C.; Ng, L.T.; Chiang, W.; Chang, M.-Y.; Lin, C.-C. Immunomodulatory Activities of Flavonoids, Monoterpenoids, Triterpenoids, Iridoid Glycosides and Phenolic Compounds of Plantago Species. *Planta Med.* **2003**, *69*, 600–604. [CrossRef] [PubMed]
64. Huang, D.; Nie, S.; Jiang, L.; Xie, M. A Novel Polysaccharide from the Seeds of Plantago Asiatica L. Induces Dendritic Cells Maturation through Toll-like Receptor 4. *Int. Immunopharmacol.* **2014**, *18*, 236–243. [CrossRef] [PubMed]
65. García, D.; Leiro, J.; Delgado, R.; Sanmartín, M.L.; Ubeira, F.M. *Mangifera Indica* L. Extract (Vimang) and Mangiferin Modulate Mouse Humoral Immune Responses. *Phytother. Res.* **2003**, *17*, 1182–1187. [CrossRef] [PubMed]
66. Rajendran, P.; Jayakumar, T.; Nishigaki, I.; Ekambaram, G.; Nishigaki, Y.; Vetriselvi, J.; Sakthisekaran, D. Immunomodulatory Effect of Mangiferin in Experimental Animals with Benzo(a)Pyrene-Induced Lung Carcinogenesis. *Int. J. Biomed. Sci.* **2013**, *9*, 68–74. [CrossRef] [PubMed]
67. Gholamnezhad, Z.; Rafatpanah, H.; Sadeghnia, H.R.; Boskabady, M.H. Immunomodulatory and Cytotoxic Effects of *Nigella sativa* and Thymoquinone on Rat Splenocytes. *Food Chem. Toxicol.* **2015**, *86*, 72–80. [CrossRef] [PubMed]
68. Zhou, H.; Deng, Y.; Xie, Q. The Modulatory Effects of the Volatile Oil of Ginger on the Cellular Immune Response In Vitro and In Vivo in Mice. *J. Ethnopharmacol.* **2006**, *105*, 301–305. [CrossRef] [PubMed]
69. Carrasco, F.; Schmidt, G.; Lopes Romero, A.; Sartoretto, J.; Caparroz-Assef, S.; Bersani-Amado, C.; Cuman, R. Immunomodulatory Activity of *Zingiber officinale* Roscoe, *Salvia officinalis* L. and *Syzygium aromaticum* L. Essential Oils: Evidence for Humor- and Cell-Mediated Responses. *J. Pharm. Pharmacol.* **2009**, *61*, 961–967. [CrossRef]

70. Lee, T.-Y.; Lee, K.-C.; Chen, S.-Y.; Chang, H.-H. 6-Gingerol Inhibits ROS and iNOS through the Suppression of PKC-α and NF-κB Pathways in Lipopolysaccharide-Stimulated Mouse Macrophages. *Bioch. Biophys. Res. Commun.* **2009**, *382*, 134–139. [CrossRef]
71. Pandey, R.; Maurya, R.; Singh, G.; Sathiamoorthy, B.; Naik, S. Immunosuppressive Properties of Flavonoids Isolated from *Boerhaavia diffusa* Linn. *Int. Immunopharmacol.* **2005**, *5*, 541–553. [CrossRef]
72. Sreelekha, T.T.; Vijayakumar, T.; Ankanthil, R.; Vijayan, K.K.; Nair, M.K. Immunomodulatory Effects of a Polysaccharide from Tamarindus Indica. *Anti-Cancer Drugs* **1993**, *4*, 209–212. [CrossRef] [PubMed]
73. Capek, P.; Hrıbalová, V.; Švandová, E.; Ebringerová, A.; Sasinková, V.; Masarová, J. Characterization of Immunomodulatory Polysaccharides from *Salvia officinalis* L. *Int. J. Biol. Macromol.* **2003**, *33*, 113–119. [CrossRef] [PubMed]
74. Mondal, S.; Chakraborty, I.; Pramanik, M.; Rout, D.; Islam, S. Structural Studies of an Immunoenhancing Polysaccharide Isolated from Mature Pods (Fruits) of *Moringa oleifera* (Sajina). *Med. Chem. Res.* **2004**, *13*, 390–400. [CrossRef]
75. Sadeghalvad, M.; Mohammadi-Motlagh, H.-R.; Rezaei, N. Structure and Function of the Immune System. In *Encyclopedia of Infection and Immunity*; Rezaei, N., Ed.; Elsevier: Oxford, UK, 2022; pp. 24–38. ISBN 978-0-323-90303-5.
76. Bafna, A.; Mishra, S. Antioxidant and Immunomodulatory Activity of the Alkaloidal Fraction of *Cissampelos pareira* Linn. *Sci. Pharm.* **2010**, *78*, 21–31. [CrossRef] [PubMed]
77. Liu, K. Dendritic Cells. *Encycl. Cell Biol.* **2016**, 741–749. [CrossRef]
78. Parker, G.A.; Papenfuss, T.L. Immune System. In *Atlas of Histology of the Juvenile Rat*; Parker, G.A., Picut, C.A., Eds.; Academic Press: Boston, MA, USA, 2016; pp. 293–347. ISBN 978-0-12-802682-3.
79. Luettig, B.; Steinmuller, C.; Gifford, G.E.; Wagner, H.; Lohmann-Matthes, M.-L. Macrophage Activation by the Polysaccharide Arabinogalactan Isolated From Plant Cell Cultures of Echinacea Purpurea. *J. Natl. Cancer Inst.* **1989**, *81*, 669–675. [CrossRef] [PubMed]
80. Steinmüller, C.; Roesler, J.; Gröttrup, E.; Franke, G.; Wagner, H.; Lohmann-Matthes, M.L. Polysaccharides Isolated from Plant Cell Cultures of Echinacea Purpurea Enhance the Resistance of Immunosuppressed Mice against Systemic Infections with Candida Albicans and Listeria Monocytogenes. *Int. J. Immunopharmacol.* **1993**, *15*, 605–614. [CrossRef]
81. Miranda, R.d.S.; Jesus, B.d.S.M.; Silva Luiz, S.R.; Viana, C.B.; Adão Malafaia, C.R.; Figueiredo, F.d.S.; Carvalho, T.d.S.C.; Silva, M.L.; Londero, V.S.; Costa-Silva, T.A.; et al. Antiinflammatory Activity of Natural Triterpenes—An Overview from 2006 to 2021. *Phytother. Res.* **2022**, *36*, 1459–1506. [CrossRef]
82. Renda, G.; Gökkaya, İ.; Şöhretoğlu, D. Immunomodulatory Properties of Triterpenes. *Phytochem. Rev.* **2022**, *21*, 537–563. [CrossRef]
83. Nalbantsoy, A.; Nesil, T.; Erden, S.; Calış, I.; Bedir, E. Adjuvant Effects of *Astragalus saponins* Macrophyllosaponin B and Astragaloside VII. *J. Ethnopharmacol.* **2011**, *134*, 897–903. [CrossRef] [PubMed]
84. Wang, Y.-P.; Li, X.-Y.; Song, C.-Q.; Hu, Z.-B. Effect of Astragaloside IV on T, B Lymphocyte Proliferation and Peritoneal Macrophage Function in Mice. *Acta Pharmacol. Sin.* **2002**, *23*, 263–266. [PubMed]
85. Xin, W.; Zhang, L.; Sun, F.; Jiang, N.; Fan, H.; Wang, T.; Li, Z.; He, J.; Fu, F. Escin Exerts Synergistic Anti-Inflammatory Effects with Low Doses of Glucocorticoids In Vivo and In Vitro. *Phytomedicine* **2011**, *18*, 272–277. [CrossRef] [PubMed]
86. Mehrotra, S.; Singh, V.K.; Agarwal, S.S.; Maurya, R.; Srimal, R.C. Antilymphoproliferative Activity of Ethanolic Extract of *Boerhaavia diffusa* Roots. *Exp. Mol. Pathol.* **2002**, *72*, 236–242. [CrossRef] [PubMed]
87. Kawasaki, T.; Kawai, T. Toll-like Receptor Signaling Pathways. *Front. Immunol.* **2014**, *5*, 461. [CrossRef] [PubMed]
88. Lim, H.J.; Jang, H.-J.; Kim, M.H.; Lee, S.; Lee, S.W.; Lee, S.-J.; Rho, M.-C. Oleanolic Acid Acetate Exerts Anti-Inflammatory Activity via IKKα/β Suppression in TLR3-Mediated NF-κB Activation. *Molecules* **2019**, *24*, 4002. [CrossRef] [PubMed]
89. Zhao, J.; Zheng, H.; Sui, Z.; Jing, F.; Quan, X.; Zhao, W.; Liu, G. Ursolic Acid Exhibits Anti-Inflammatory Effects through Blocking TLR4-MyD88 Pathway Mediated by Autophagy. *Cytokine* **2019**, *123*, 154726. [CrossRef] [PubMed]
90. Kim, K.-S.; Lee, D.-S.; Kim, D.-C.; Yoon, C.-S.; Ko, W.; Oh, H.; Kim, Y.-C. Anti-Inflammatory Effects and Mechanisms of Action of Coussaric and Betulinic Acids Isolated from Diospyros Kaki in Lipopolysaccharide-Stimulated RAW 264.7 Macrophages. *Molecules* **2016**, *21*, 1206. [CrossRef] [PubMed]
91. Li, Y.; Meng, T.; Hao, N.; Tao, H.; Zou, S.; Li, M.; Ming, P.; Ding, H.; Dong, J.; Feng, S.; et al. Immune Regulation Mechanism of Astragaloside IV on RAW264.7 Cells through Activating the NF-κB/MAPK Signaling Pathway. *Int. Immunopharmacol.* **2017**, *49*, 38–49. [CrossRef]
92. Haque, M.A.; Jantan, I.; Harikrishnan, H.; Abdul Wahab, S.M. Magnoflorine Enhances LPS-Activated Pro-Inflammatory Responses via MyD88-Dependent Pathways in U937 Macrophages. *Planta Med.* **2018**, *84*, 1255–1264. [CrossRef]
93. Kim, G.-Y.; Kim, K.-H.; Lee, S.-H.; Yoon, M.-S.; Lee, H.-J.; Moon, D.-O.; Lee, C.-M.; Ahn, S.-C.; Park, Y.C.; Park, Y.-M. Curcumin Inhibits Immunostimulatory Function of Dendritic Cells: MAPKs and Translocation of NF-Kappa B as Potential Targets. *J. Immunol.* **2005**, *174*, 8116–8124. [CrossRef] [PubMed]
94. Hwang, Y.-J.; Song, J.; Kim, H.-R.; Hwang, K.-A. Oleanolic Acid Regulates NF-κB Signaling by Suppressing MafK Expression in RAW 264.7 Cells. *BMB Rep* **2014**, *47*, 524–529. [CrossRef] [PubMed]
95. Zhong, R.; Miao, L.; Zhang, H.; Tan, L.; Zhao, Y.; Tu, Y.; Prieto, M.A.; Simal-Gandara, J.; Chen, L.; He, C.; et al. Anti-Inflammatory Activity of Flavonols via Inhibiting MAPK and NF-κB Signaling Pathways in RAW264.7 Macrophages. *Curr. Res. Food Sci.* **2022**, *5*, 1176–1184. [CrossRef] [PubMed]
96. Park, K.S. Aucubin, a Naturally Occurring Iridoid Glycoside Inhibits TNF-α-Induced Inflammatory Responses through Suppression of NF-κB Activation in 3T3-L1 Adipocytes. *Cytokine* **2013**, *62*, 407–412. [CrossRef] [PubMed]

97. Pal, P.B.; Sinha, K.; Sil, P.C. Mangiferin, a Natural Xanthone, Protects Murine Liver in Pb(II) Induced Hepatic Damage and Cell Death via MAP Kinase, NF-κB and Mitochondria Dependent Pathways. *PLoS One* **2013**, *8*, e56894. [CrossRef] [PubMed]
98. Suchal, K.; Malik, S.; Khan, S.I.; Malhotra, R.K.; Goyal, S.N.; Bhatia, J.; Kumari, S.; Ojha, S.; Arya, D.S. Protective Effect of Mangiferin on Myocardial Ischemia-Reperfusion Injury in Streptozotocin-Induced Diabetic Rats: Role of AGE-RAGE/MAPK Pathways. *Sci. Rep.* **2017**, *7*, 42027. [CrossRef]
99. Imenshahidi, M.; Hosseinzadeh, H. Berberis Vulgaris and Berberine: An Update Review. *Phytother. Res.* **2016**, *30*, 1745–1764. [CrossRef] [PubMed]
100. Wang, K.; Feng, X.; Chai, L.; Cao, S.; Qiu, F. The Metabolism of Berberine and Its Contribution to the Pharmacological Effects. *Drug Metab. Rev.* **2017**, *49*, 139–157. [CrossRef] [PubMed]
101. Guan, X.; Zheng, X.; Vong, C.T.; Zhao, J.; Xiao, J.; Wang, Y.; Zhong, Z. Combined Effects of Berberine and Evodiamine on Colorectal Cancer Cells and Cardiomyocytes In Vitro. *Eur. J. Pharmacol.* **2020**, *875*, 173031. [CrossRef]
102. Gao, X.; Wang, J.; Li, M.; Wang, J.; Lv, J.; Zhang, L.; Sun, C.; Ji, J.; Yang, W.; Zhao, Z.; et al. Berberine Attenuates XRCC1-mediated Base Excision Repair and Sensitizes Breast Cancer Cells to the Chemotherapeutic Drugs. *J. Cell. Mol. Med.* **2019**, *23*, 6797–6804. [CrossRef]
103. Liu, S.; Fang, Y.; Shen, H.; Xu, W.; Li, H. Berberine Sensitizes Ovarian Cancer Cells to Cisplatin through miR-21/PDCD4 Axis. *Acta Biochim. Biophys. Sin. (Shanghai)* **2013**, *45*, 756–762. [CrossRef] [PubMed]
104. Palmieri, A.; Iapichino, A.; Cura, F.; Scapoli, L.; Carinci, F.; Mandrone, M.; Martinelli, M. Pre-Treatment with Berberine Enhances Effect of 5-Fluorouracil and Cisplatin in HEP2 Laryngeal Cancer Cell Line. *J. Biol. Regul. Homeost. Agents* **2018**, *32*, 167–177.
105. Pandey, A.; Vishnoi, K.; Mahata, S.; Tripathi, S.C.; Misra, S.P.; Misra, V.; Mehrotra, R.; Dwivedi, M.; Bharti, A.C. Berberine and Curcumin Target Survivin and STAT3 in Gastric Cancer Cells and Synergize Actions of Standard Chemotherapeutic 5-Fluorouracil. *Nutr. Cancer* **2015**, *67*, 1293–1304. [CrossRef] [PubMed]
106. Plazas, E.; Avila M, M.C.; Muñoz, D.R.; Cuca, S.L.E. Natural Isoquinoline Alkaloids: Pharmacological Features and Multi-Target Potential for Complex Diseases. *Pharmacol. Res.* **2022**, *177*, 106126. [CrossRef]
107. Cecil, C.E.; Davis, J.M.; Cech, N.B.; Laster, S.M. Inhibition of H1N1 Influenza A Virus Growth and Induction of Inflammatory Mediators by the Isoquinoline Alkaloid Berberine and Extracts of Goldenseal (*Hydrastis canadensis*). *Int. Immunopharmacol.* **2011**, *11*, 1706–1714. [CrossRef] [PubMed]
108. Chin, L.W.; Cheng, Y.-W.; Lin, S.-S.; Lai, Y.-Y.; Lin, L.-Y.; Chou, M.-Y.; Chou, M.-C.; Yang, C.-C. Anti-Herpes Simplex Virus Effects of Berberine from *Coptidis rhizoma*, a Major Component of a Chinese Herbal Medicine, Ching-Wei-San. *Arch. Virol.* **2010**, *155*, 1933–1941. [CrossRef]
109. Varghese, F.S.; van Woudenbergh, E.; Overheul, G.J.; Eleveld, M.J.; Kurver, L.; van Heerbeek, N.; van Laarhoven, A.; Miesen, P.; den Hartog, G.; de Jonge, M.I.; et al. Berberine and Obatoclax Inhibit SARS-CoV-2 Replication in Primary Human Nasal Epithelial Cells In Vitro. *Viruses* **2021**, *13*, 282. [CrossRef]
110. Bodiwala, H.S.; Sabde, S.; Mitra, D.; Bhutani, K.K.; Singh, I.P. Synthesis of 9-Substituted Derivatives of Berberine as Anti-HIV Agents. *Eur. J. Med. Chem.* **2011**, *46*, 1045–1049. [CrossRef]
111. Luganini, A.; Mercorelli, B.; Messa, L.; Palù, G.; Gribaudo, G.; Loregian, A. The Isoquinoline Alkaloid Berberine Inhibits Human Cytomegalovirus Replication by Interfering with the Viral Immediate Early-2 (IE2) Protein Transactivating Activity. *Antiviral Res.* **2019**, *164*, 52–60. [CrossRef]
112. Shin, H.-B.; Choi, M.-S.; Yi, C.-M.; Lee, J.; Kim, N.-J.; Inn, K.-S. Inhibition of Respiratory Syncytial Virus Replication and Virus-Induced P38 Kinase Activity by Berberine. *Int. Immunopharmacol.* **2015**, *27*, 65–68. [CrossRef]
113. Pradeep, C.R.; Kuttan, G. Piperine Is a Potent Inhibitor of Nuclear Factor-κB (NF-κB), c-Fos, CREB, ATF-2 and Proinflammatory Cytokine Gene Expression in B16F-10 Melanoma Cells. *Int. Immunopharmacol.* **2004**, *4*, 1795–1803. [CrossRef] [PubMed]
114. Si, L.; Yang, R.; Lin, R.; Yang, S. Piperine Functions as a Tumor Suppressor for Human Ovarian Tumor Growth via Activation of JNK/P38 MAPK-Mediated Intrinsic Apoptotic Pathway. *Biosci. Rep.* **2018**, *38*, BSR20180503. [CrossRef] [PubMed]
115. Hirayama, D.; Iida, T.; Nakase, H. The Phagocytic Function of Macrophage-Enforcing Innate Immunity and Tissue Homeostasis. *Int. J. Mol. Sci.* **2017**, *19*, 92. [CrossRef]
116. Zhang, C.; Yang, M.; Ericsson, A.C. Function of Macrophages in Disease: Current Understanding on Molecular Mechanisms. *Front. Immunol.* **2021**, *12*, 620510. [CrossRef] [PubMed]
117. Aliyu, M.; Zohora, F.; Saboor-Yaraghi, A.A.; Aliyu, M.; Zohora, F.; Saboor-Yaraghi, A.A. Spleen in Innate and Adaptive Immunity Regulation. *AIMS Allergy Immunol.* **2021**, *5*, 1–17. [CrossRef]
118. Ayatollahi, A.M.; Ghanadian, M.; Afsharypour, S.; Abdella, O.M.; Mirzai, M.; Askari, G. Pentacyclic Triterpenes in *Euphorbia microsciadia* with Their T-Cell Proliferation Activity. *Iran. J. Pharm. Res.* **2011**, *10*, 287–294. [PubMed]
119. Mioc, M.; Milan, A.; Malița, D.; Mioc, A.; Prodea, A.; Racoviceanu, R.; Ghiulai, R.; Cristea, A.; Căruntu, F.; Șoica, C. Recent Advances Regarding the Molecular Mechanisms of Triterpenic Acids: A Review (Part I). *Int. J. Mol. Sci.* **2022**, *23*, 7740. [CrossRef] [PubMed]
120. Salinas, F.M.; Vázquez, L.; Gentilini, M.V.; O´Donohoe, A.; Regueira, E.; Nabaes Jodar, M.S.; Viegas, M.; Michelini, F.M.; Hermida, G.; Alché, L.E.; et al. Aesculus Hippocastanum L. Seed Extract Shows Virucidal and Antiviral Activities against Respiratory Syncytial Virus (RSV) and Reduces Lung Inflammation In Vivo. *Antiviral Res.* **2019**, *164*, 1–11. [CrossRef] [PubMed]
121. Rathod, N.B.; Elabed, N.; Punia, S.; Ozogul, F.; Kim, S.-K.; Rocha, J.M. Recent Developments in Polyphenol Applications on Human Health: A Review with Current Knowledge. *Plants* **2023**, *12*, 1217. [CrossRef]

122. Cavalcanti, E.; Vadrucci, E.; Delvecchio, F.R.; Addabbo, F.; Bettini, S.; Liou, R.; Monsurrò, V.; Huang, A.Y.-C.; Pizarro, T.T.; Santino, A.; et al. Administration of Reconstituted Polyphenol Oil Bodies Efficiently Suppresses Dendritic Cell Inflammatory Pathways and Acute Intestinal Inflammation. *PLOS ONE* **2014**, *9*, e88898. [CrossRef]
123. Li, P.; Hu, J.; Shi, B.; Tie, J. Baicalein Enhanced Cisplatin Sensitivity of Gastric Cancer Cells by Inducing Cell Apoptosis and Autophagy via Akt/mTOR and Nrf2/Keap 1 Pathway. *Biochem. Biophys. Res. Commun.* **2020**, *531*, 320–327. [CrossRef]
124. Zhang, L.; Liu, Q.; Huang, L.; Yang, F.; Liu, A.; Zhang, J. Combination of Lapatinib and Luteolin Enhances the Therapeutic Efficacy of Lapatinib on Human Breast Cancer through the FOXO3a/NQO1 Pathway. *Biochem. Biophys. Res. Commun.* **2020**, *531*, 364–371. [CrossRef]
125. Bahrami, A.; Fereidouni, M.; Pirro, M.; Bianconi, V.; Sahebkar, A. Modulation of Regulatory T Cells by Natural Products in Cancer. *Cancer Lett.* **2019**, *459*, 72–85. [CrossRef] [PubMed]
126. Chen, X.; Du, Y.; Lin, X.; Qian, Y.; Zhou, T.; Huang, Z. CD4+CD25+ Regulatory T Cells in Tumor Immunity. *Int. Immunopharmacol.* **2016**, *34*, 244–249. [CrossRef] [PubMed]
127. Casey, S.C.; Amedei, A.; Aquilano, K.; Azmi, A.S.; Benencia, F.; Bhakta, D.; Bilsland, A.E.; Boosani, C.S.; Chen, S.; Ciriolo, M.R.; et al. Cancer Prevention and Therapy through the Modulation of the Tumor Microenvironment. *Semin. Cancer Biol.* **2015**, *35*, S199–S223. [CrossRef] [PubMed]
128. Ke, M.; Zhang, Z.; Xu, B.; Zhao, S.; Ding, Y.; Wu, X.; Wu, R.; Lv, Y.; Dong, J. Baicalein and Baicalin Promote Antitumor Immunity by Suppressing PD-L1 Expression in Hepatocellular Carcinoma Cells. *Int. Immunopharmacol.* **2019**, *75*, 105824. [CrossRef] [PubMed]
129. Bhosale, P.B.; Ha, S.E.; Vetrivel, P.; Kim, H.H.; Kim, S.M.; Kim, G.S. Functions of Polyphenols and Its Anticancer Properties in Biomedical Research: A Narrative Review. *Transl. Cancer Res.* **2020**, *9*, 7619–7631. [CrossRef] [PubMed]
130. Meylan, E.; Dooley, A.L.; Feldser, D.M.; Shen, L.; Turk, E.; Ouyang, C.; Jacks, T. Requirement for NF-κB Signalling in a Mouse Model of Lung Adenocarcinoma. *Nature* **2009**, *462*, 104–107. [CrossRef]
131. Xia, Y.; Yeddula, N.; Leblanc, M.; Ke, E.; Zhang, Y.; Oldfield, E.; Shaw, R.J.; Verma, I.M. Reduced Cell Proliferation by IKK2 Depletion in a Mouse Lung-Cancer Model. *Nat. Cell Biol.* **2012**, *14*, 257–265. [CrossRef]
132. Basséres, D.S.; Ebbs, A.; Levantini, E.; Baldwin, A.S. Requirement of the NF-kappaB Subunit P65/RelA for K-Ras-Induced Lung Tumorigenesis. *Cancer Res.* **2010**, *70*, 3537–3546. [CrossRef]
133. Keutgens, A.; Robert, I.; Viatour, P.; Chariot, A. Deregulated NF-kappaB Activity in Haematological Malignancies. *Biochem. Pharmacol.* **2006**, *72*, 1069–1080. [CrossRef] [PubMed]
134. Hou, J.; Zheng, D.; Fung, G.; Deng, H.; Chen, L.; Liang, J.; Jiang, Y.; Hu, Y. Mangiferin Suppressed Advanced Glycation End Products (AGEs) through NF-κB Deactivation and Displayed Anti-Inflammatory Effects in Streptozotocin and High Fat Diet-Diabetic Cardiomyopathy Rats. *Can. J. Physiol. Pharmacol.* **2016**, *94*, 332–340. [CrossRef] [PubMed]
135. Muruganandan, S.; Lal, J.; Gupta, P.K. Immunotherapeutic Effects of Mangiferin Mediated by the Inhibition of Oxidative Stress to Activated Lymphocytes, Neutrophils and Macrophages. *Toxicology* **2005**, *215*, 57–68. [CrossRef]

Disclaimer/Publisher's Note: The statements, opinions and data contained in all publications are solely those of the individual author(s) and contributor(s) and not of MDPI and/or the editor(s). MDPI and/or the editor(s) disclaim responsibility for any injury to people or property resulting from any ideas, methods, instructions or products referred to in the content.

molecules MDPI

Article

Vitamin D_3 Bioaccessibility from Supplements and Foods—Gastric pH Effect Using a Static In Vitro Gastrointestinal Model

Evangelia Pasidi and Patroklos Vareltzis *

Department of Chemical Engineering, Faculty of Engineering, Aristotle University of Thessaloniki, 54124 Thessaloniki, Greece; egpasidi@cheng.auth.gr
* Correspondence: pkvareltzis@cheng.auth.gr; Tel.: +30-231-099-6162

Abstract: Vitamin D_3 deficiency is a global phenomenon, which can be managed with supplementation and food fortification. However, vitamin D_3 bioaccessibility may depend on factors such as matrix composition and interactions throughout the gastrointestinal (GI) tract. This research focused on the effect of different matrices on vitamin D_3 content during digestion, as well as the effect of pH on its bioaccessibility. The INFOGEST protocol was employed to simulate digestion. Three different types of commercial supplements, two foods naturally rich in vitamin D_3, and three fortified foods were investigated. High-Performance Liquid Chromatography was used to determine the initial vitamin D_3 content in the supplements and foods, as well as after each digestion stage. The results indicate that the foods exhibited higher bioaccessibility indices compared to the supplements and a higher percentage retention at the end of the gastric phase. The pH study revealed a positive correlation between an increased gastric pH and the corresponding content of vitamin D_3. Interestingly, exposing the matrix to a low pH during the gastric phase resulted in an increased intestinal content of D_3. Vitamin D_3 is more bioaccessible from foods than supplements, and its bioaccessibility is susceptible to changes in gastric pH. Fasting conditions (i.e., gastric pH = 1) enhance the vitamin's bioaccessibility.

Keywords: vitamin D_3 bioaccessibility; in vitro digestion; gastric pH

Citation: Pasidi, E.; Vareltzis, P. Vitamin D_3 Bioaccessibility from Supplements and Foods—Gastric pH Effect Using a Static In Vitro Gastrointestinal Model. *Molecules* **2024**, *29*, 1153. https://doi.org/10.3390/molecules29051153

Academic Editors: Eliza Oprea, Arunaksharan Narayanankutty and Ademola C. Famurewa

Received: 29 January 2024
Revised: 28 February 2024
Accepted: 4 March 2024
Published: 5 March 2024

Copyright: © 2024 by the authors. Licensee MDPI, Basel, Switzerland. This article is an open access article distributed under the terms and conditions of the Creative Commons Attribution (CC BY) license (https://creativecommons.org/licenses/by/4.0/).

1. Introduction

Vitamin D is a biologically active compound found mainly in the forms of ergocalciferol (vitamin D_2), cholecalciferol (vitamin D_3), and 25-hydroxycholecalciferol (25(OH)D_3) (Scheme 1). Vitamin D_3 is a micronutrient, essential for maintaining the overall health and wellness of humans, as it is associated with bone health and immune system boosting [1]. It plays a key role in calcium absorption [2] and has been proposed to act against cancer cell growth [3]. In addition, vitamin D_3 has been linked with a lower risk of developing multiple sclerosis and rheumatoid arthritis, as well as type 1 and type 2 diabetes mellitus [4]. The protective effect of vitamin D_3 against SARS-CoV-2 has also been examined and showed that the vitamin can potentially prevent severe illness [5]. Vitamin D_3 can be photosynthesised through skin exposure to ultraviolet radiation [6] or ingested through foods and supplements. However, vitamin D_3 deficiency is a global concern [7]. Supplementation, as well as the consumption of foods rich in vitamin D_3, can aid in coping with this phenomenon.

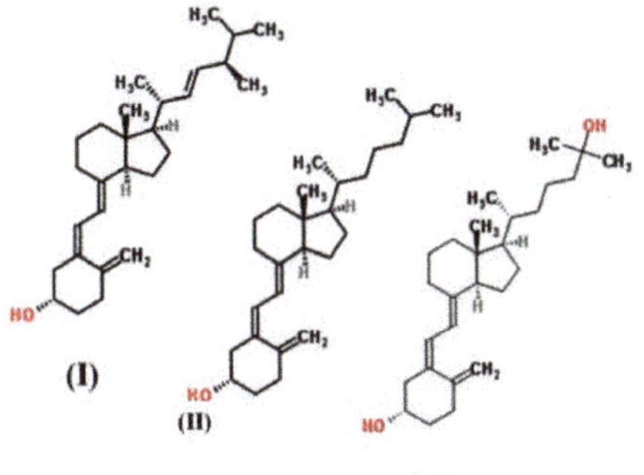

Scheme 1. Chemical structures of main dietary forms of vitamin D: **(I)** ergocalciferol (vitamin D_2), **(II)** cholecalciferol (vitamin D_3), **(III)** 25-hydroxy-cholecalciferol (25(OH)D_3) [8].

Supplements are available in different forms, such as tablets, capsules, or oil-emulsified drops [9]. Vitamin D_3 in supplements may be encapsulated in microcapsules, micelles, or liposomes to increase its bioavailability [10–12]. A meta-analysis of several clinical studies concluded that vitamin D_3 bioavailability is better in oil vehicles (capsules or liquid) than in powder tablets (cellulose or lactose) or ethanol [13]. However, another study testing oil and tablets showed that they were equally efficient in raising serum 25-hydroxyvitamin D, though the authors speculated that these results may be due to the timing of measuring serum concentrations [14].

Foods can either naturally contain vitamin D_3, such as fish and eggs, or be fortified with the vitamin, such as milk, orange juice, plant oils, flour, bread, and cereals. Different food matrices can result in different bioaccessibility and bioavailability levels of the vitamin. The structure of the food matrix, the amount and type of dietary lipids (chain length and degree of saturation), and the dietary fibres can affect the final bioavailability [8,15]. Foods fortified with vitamin D_3 have demonstrated comparable effectiveness to supplements in increasing serum 25(OH)D_3 levels [16,17]. It has also been suggested that vitamin D_3 absorption is protein-mediated at low concentrations, such as that found in dietary sources, while at high pharmacological concentrations, the absorption mechanism shifts to passive diffusion [18]. The differences observed between different foods and supplements indicate that research on various matrices is necessary.

When a vitamin D_3-containing matrix is ingested, it undergoes physiological conditions encountered during digestion, including enzyme activity and pH fluctuations. The digestive process is initiated in the mouth with amylase catalysing starch hydrolysis, followed by the stomach, where proteins and lipids are hydrolysed by pepsin and gastric lipase, forming gastric chyme [19]. As the gastric chyme moves to the intestine, pancreatin further breaks down the food with assistance from intestinal peristaltic movements. Pepsin and trypsin may play a role in releasing vitamin D_3 from its food matrix by disrupting the binding of proteins to vitamin D_3. Digestive enzymes in the duodenum, including amylase, lipase, and protease, continue to liberate vitamin D_3 from its food matrix [15]. The released vitamin D_3 integrates into the mixed micelles formed during digestion, consisting of phospholipids, cholesterol, lipid digestion products, and bile salts [8,15,20]. The composition of mixed micelles is influenced by the types of lipids present during digestion [21,22].

pH variation is another critical factor that might impact the final bioaccessibility of vitamin D_3. A lower pH has been shown to lead to a decreasing stability of vitamin D_3 [23]. Vitamin D_3 is isomerised to isotachysterol at acidic pH [24]. Encapsulation of the vitamin has been proposed to protect it from degradation at different pH values [25]. Many encapsulation techniques and materials, such as β-lactoglobulin [26], ovalbumin–pectin nanocomplexes [27], gum arabic, maltodextrin, whey protein concentrate, and soy isolate protein [28], have been used to produce systems that are stable under different pH conditions [25]. Food intake alters the basal gastric pH. Different food compositions result in varying gastric pH values, which may take up to 3 h to return to basal levels [29]. Consequently, supplements taken after different foods or during fasting may encounter different gastric pH conditions. The timing of supplement intake following food consumption can also lead to variations in the encountered pH values [30].

The aim of this research was to investigate the influence of diverse matrices (including natural foods with vitamin D_3 with or without heat treatment, fortified foods, and supplements) on the fate of vitamin D_3 at different stages along the gastrointestinal (GI) tract. Using the INFOGEST protocol, these matrices underwent in vitro digestion, and the vitamin content was determined at different stages of the protocol to evaluate its bioaccessibility. Additionally, this study investigated the effect of gastric pH by testing four different pH values to simulate conditions during fasting and the consumption of various foods. The findings from this research contribute to a deeper understanding of how each digestion stage influences vitamin D_3 and the impact of gastric pH variations on its bioaccessibility.

2. Results

Foods naturally containing vitamin D_3 (eggs and salmon), fortified foods (milk, cereals, and sour cherry juice), and supplements (tablets, capsules containing an oil-based emulsion, and oil-based liquid drops) were subjected to in vitro digestion (INFOGEST protocol) [31]. High-Performance Liquid Chromatography (HPLC) was used to determine the vitamin D_3 content and losses at each stage. The eggs and salmon were thermally processed until their core temperature reached 70 °C [32] before being subjected to digestion. The effect of gastric pH variation was also examined by subjecting the vitamin D_3 liquid oil-based supplement to four different gastric pH values.

2.1. Vitamin D_3 Content of Foods and Supplements

The detected content of the vitamin in the tablet, capsule, and liquid supplement was 20.99 ± 1.17 µg/tablet, 20.24 ± 0.78 µg/capsule, and 95.93 ± 0.64 µg/mL, respectively (Figure 1, t = 0 min). The liquid supplement had the highest content, followed by the tablet and capsule, which had similar contents.

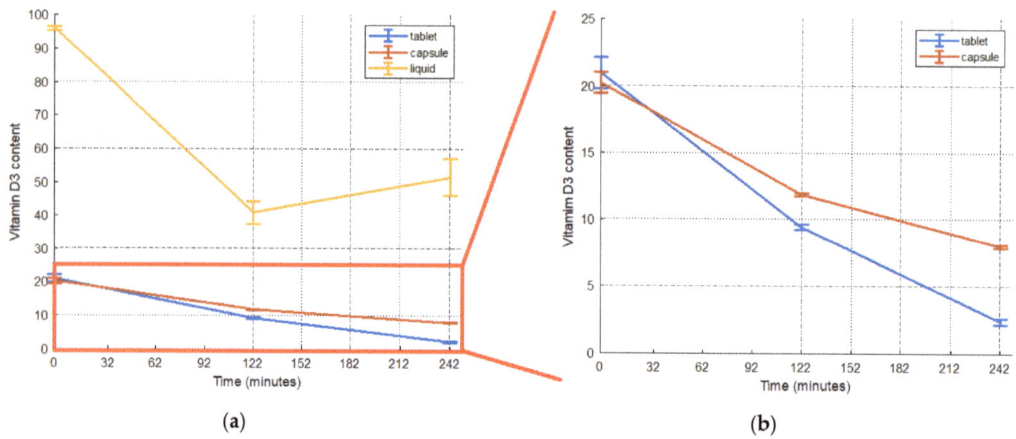

Figure 1. Detected vitamin D_3 content during in vitro digestion of (**a**) commercial supplements: tablet; capsule; liquid. (**b**) Close view of tablet and capsule (tablet in µg/tablet, capsule in µg/capsule, and liquid in µg/mL). On *x*-axis, 0, 122, and 242 minutes refer respectively to the initial content, content after gastric stage, and content after intestinal stage.

The fortified foods had a higher vitamin content than the natural foods, as expected (Table 1). Between the two natural foods examined, the salmon had a higher vitamin D_3 content than the eggs, as seen in other studies [33]. An HPLC analysis of the egg and salmon showed a second peak, before vitamin D_3's peak (Figures S3 and S4—Supplementary file), which may correspond to the hydroxylated form 25(OH)D_3 [34]. This form is naturally present in these foods [35,36].

Table 1. Detected vitamin D_3 content and bioaccessibility indices (BIs) of foods after INFOGEST protocol.

Food Sample		Detected Vitamin D_3 Content (μg/g [1])				Bioaccessibility Index (BI)
		Initial	Thermally Processed	Stomach	Intestine	
Natural	Egg	0.06 ± 0.004^{b}	0.03 ± 0.005^{c}	0.08 ± 0.007^{a}	0.06 ± 0.008^{b}	1.06 ± 0.153
	Salmon	0.50 ± 0.021^{c}	0.38 ± 0.020^{d}	0.74 ± 0.015^{a}	0.55 ± 0.019^{b}	1.10 ± 0.060
Fortified	Milk	1.53 ± 0.056^{a}	N/A	0.62 ± 0.007^{b}	0.61 ± 0.004^{b}	0.40 ± 0.015
	Cereals	$0.89 \pm 0.040^{a,b}$	N/A	0.84 ± 0.005^{b}	0.92 ± 0.006^{a}	1.04 ± 0.046
	Sour cherry juice	1.15 ± 0.005^{c}	N/A	1.20 ± 0.008^{b}	1.24 ± 0.003^{a}	1.08 ± 0.054

Table values are means ± standard deviations. Different superscript letters (a, b, c, d) represent statistical differences in the same row ($p \leq 0.05$). [1] g in vitamin D3 content refers to the initial food sample.

2.2. Vitamin D_3 Bioaccessibility

2.2.1. Vitamin D_3 Bioaccessibility from Supplements

The bioaccessibility index (BI) shows the amount of vitamin D_3 remaining after digestion processes and available for absorption, and it was calculated according to Equation (2). In Figure 1, the remaining detected content of vitamin D_3 is presented. Among the supplements, the liquid one had the highest bioaccessibility, followed by the capsule and the tablet (Figure 2).

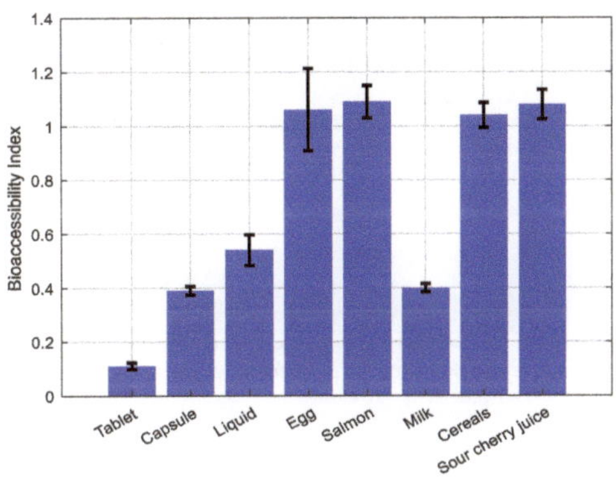

Figure 2. Bioaccessibility indices of supplements: tablet; capsule; liquid; and foods: egg, salmon, milk, cereals, sour cherry juice.

At the end of the gastric phase, the vitamin D_3 content in the tablet, capsule, and liquid supplements was reduced by 55%, 41%, and 43%, respectively. Further losses at the end of the intestinal phase were recorded (75% and 20% for the tablet and capsule). On the contrary, in the case of the liquid supplement, there appeared to be a 25% increase in the vitamin D_3 content in the intestinal stage compared to the gastric stage. Greater losses were observed for the tablet at each stage. The reduction in the vitamin content during the intestinal phase was more significant for the tablet compared to that during the gastric phase. In contrast, for the capsule, the reduction was more pronounced for the gastric content compared to the initial content.

2.2.2. Vitamin D_3 Bioaccessibility from Foods

The detected vitamin contents in each digestion step of the foods, as well as the corresponding bioaccessibility, are presented in Table 1. The sour cherry juice, egg, salmon, and cereals had BIs around 1. The milk had the lowest BI at 0.40, which is rather low in comparison with the other samples. The foods, except for the milk, exhibited higher BIs than the supplements, as shown in Figure 2.

The thermal processing of the egg and salmon seemed to decrease the vitamin D_3 content by 43% and 25%, respectively (Table 1).

In the natural food samples, the vitamin D_3 content seemed to increase after the gastric step. The vitamin D_3 content in the gastric chyme of the egg and salmon samples was increased by 33% and 48%, respectively, compared to the initial concentration. The intestinal content compared to the gastric content was decreased by 29% for the eggs and by 26% for the salmon.

The results for the fortified food samples showed a 60% and 5% decrease in the vitamin content in the gastric phase for the milk and cereals, respectively. For the sour cherry juice, there was a slight increase (4%) in the gastric content compared to the initial content. For the milk samples, there was no significant difference between the gastric and intestinal contents. The cereals and sour cherry juice showed an increase in the intestinal content (10% and 3%, respectively).

The gastric step seemed to have a greater impact on vitamin D_3 for all food samples, either by increasing or decreasing the content.

2.3. *Gastric pH Effect on Vitamin D_3 Bioaccessibility*

Four different pH values were simulated to investigate the effect of the gastric stage pH on vitamin D_3 bioaccessibility. The sample tested was the liquid supplement, as it was the most bioaccessible among the supplements. The gastric and intestinal contents of the vitamin, as well as the calculated BIs, are presented in Table 2.

Table 2. Detected vitamin D_3 content and bioaccessibility indices (BIs) at four different gastric pH values after INFOGEST application.

Gastric pH Value	Detected Vitamin D_3 Content (μg/mL)			BI
	Initial	Stomach	Intestine	
1		$39.87 \pm 8.97^{b,B}$	$70.86 \pm 4.58^{c,A}$	0.74 ± 0.05
3	95.93 ± 0.64^{a}	$40.95 \pm 2.69^{b,B}$	$51.71 \pm 5.46^{c,B}$	0.54 ± 0.06
5		$47.14 \pm 3.71^{b,A,B}$	$51.62 \pm 2.08^{b,B}$	0.54 ± 0.02
7		$53.65 \pm 6.55^{b,A}$	$41.28 \pm 2.89^{c,C}$	0.43 ± 0.03

Table values are means ± standard deviations. Different superscript lowercase letters (a, b, c) represent statistical differences ($p \leq 0.05$) in the same row. Different superscript uppercase (A, B, C) letters represent statistical differences ($p \leq 0.05$) in the same column.

There was a profound effect of the gastric digestion step on the vitamin content. The vitamin's decrease during this stage ranged from 44 to 58%. D_3's gastric content was the highest at pH 7 and the lowest at pH 1 ($p < 0.05$). Even at pH 7, there was a 44% decrease in the vitamin D_3 content. This suggests that vitamin D_3 stability might be affected not only by pH but also by the presence of other components of gastric fluids. A low pH has been shown to negatively affect vitamin D_3 [23]. Different pH values may have caused the degradation of vitamin D_3 to isomers [34,37].

At every pH level, there was an increase in the vitamin D_3 content at the end of the intestinal digestion phase, except for at pH 7. The percentage increases were 78%, 26%, and 10% at pH 1, 3, and 5, respectively. The lower the pH of the gastric phase, the higher the increase in the vitamin D_3 content in the intestinal phase. On the contrary, when the sample was exposed to gastric pH 7, a notable 23% reduction in the vitamin D_3 content was observed, from 54 μg/mL after the stomach phase to 41 μg/mL after the intestinal phase. Exposure to the lowest pH value of 1 resulted in the highest BI, while pH 7 led to the lowest BI. pH values 3 and 5 had similar BIs.

To determine the possible effect of the carrier's oxidation (sunflower oil) on the vitamin D_3 content in each digestion stage, sunflower seed oil oxidation was investigated at two different pH values (Figure 3). Primary oxidation was more profound at pH 3 than at pH 7. The concentration of primary oxidation products peaked during the gastric phase at 75 min at pH 3 and at 135 min at pH 7. Even though oxidation at pH 7 was significantly delayed during the gastric phase, it reached the same peak concentrations of oxidation products at pH 3 ($p \geq 0.05$). Secondary oxidation peaked during the intestinal phase of digestion, when primary oxidation products had the lowest concentrations (195 min). The concentration of secondary oxidation products was greater at pH 7; however, the difference was not statistically significant ($p \geq 0.05$).

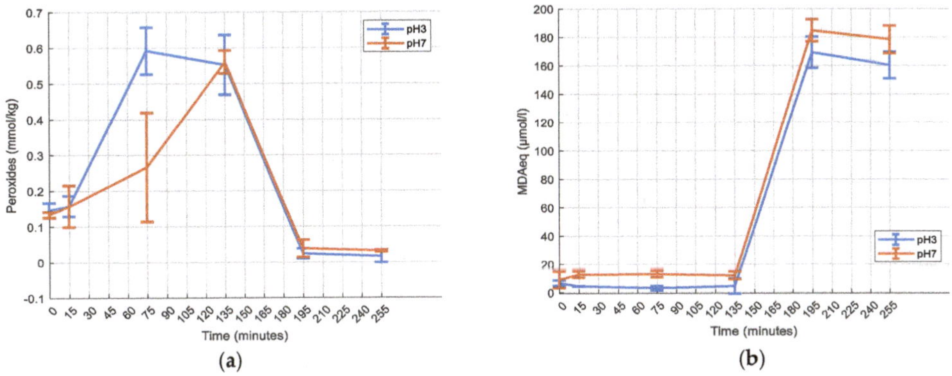

Figure 3. Effect of gastric pH on (**a**) primary oxidation and (**b**) secondary oxidation of sunflower seed oil. On x-axis, 0 min refers to the initial concentration, 15 min refers to the concentration after oral phase, 75 and 135 min refer to the concentration after the first and second hours of gastric phase, and 195 and 225 min refer to the concentration after the first and second hours of intestinal phase.

3. Discussion

Supplements and foods containing vitamin D_3, either naturally or from fortification, can be used to battle vitamin D_3 deficiency [38,39]. When ingested, vitamin D_3 is exposed to GI tract conditions, which can affect the stability of the vitamin and its final bioaccessibility.

Of the three commercial supplements, the oil-based liquid drops had the highest vitamin content. The in vitro digestion of the supplements showed the highest BI for the oil-based liquid drops, followed by the capsule and the tablet, which is in accordance with previous studies testing vitamin D_3 bioavailability [13]. Vitamin D_3 is a lipophilic vitamin and is more stable in oil vehicles [23].

The tablets exhibited higher gastric losses compared to the capsules and liquid supplements, with the intestinal stage exerting a more significant impact on the vitamin content of the tablets. Conversely, for the capsules, the gastric stage had a more pronounced effect. In the case of the liquid supplement, there was a decrease in content from the initial to the gastric stage, followed by an increase from the gastric to the intestinal stage. This phenomenon may be attributed to the enhanced release of the vitamin from its matrix during this stage, potentially facilitated by the action of pancreatin on the oil matrix (sunflower oil). A similar behaviour was noted for carotenoids, as they were undetected in the gastric stage but present in measurable concentrations during the intestinal stage. The authors attributed this outcome partly to the presence of pancreatin in the intestinal stage [40]. Additionally, the antioxidant capacity of the α-tocopherol present as an additive in the liquid supplement may have protected vitamin D_3 from degradation during in vitro digestion [41]. Differences in the initial concentrations among the supplements might also have contributed to the different behaviours during digestion. Previous research has shown that the BI of omega-3 supplements can be dependent on the initial concentration of the lipophilic components [42].

Heat treatment can adversely affect the vitamin D_3 content of foods by decreasing it, depending on the method of heating [43,44]. In our study, the thermal processing of egg and salmon decreased the vitamin D_3 content, with the egg being more affected than the salmon (42% vs. 25% decrease, respectively). Vitamin D_3, as a lipophilic vitamin, may be better protected in salmon than eggs, as salmon has a greater lipid content. This can result in a greater retention of vitamin D_3 in salmon after thermal processing. Vitamin D_3 converts to pre-vitamin D_3 reversibly when heated, especially at higher temperatures [37]. The reversibility of this conversion may be the explanation for the increase observed in the gastric step of both the eggs and salmon after the heat treatment.

The eggs and salmon had a lower vitamin D_3 content than the fortified foods, as expected. The salmon had a higher content than the eggs, as shown in other studies [33]. From the fortified foods, the milk had the highest content, followed by the cereals and sour cherry juice. The foods exhibited higher BIs than the supplements, apart from the milk, which had a rather low BI, closer to that of the supplements. Previous research has shown that naturally formulated vitamin D_3 extracted from agricultural products had a higher bioaccessibility than synthetic vitamin D_3 [45]. An investigation on vitamin E bioaccessibility revealed that the incorporation of vitamin E-loaded Pickering emulsions into foods led to an increased bioaccessibility of the vitamin, surpassing the bioaccessibility observed when the emulsion was digested alone. This observation was attributed to the natural presence of macronutrients in foods [46]. These findings are in accordance with our results concerning the better bioaccessibility of vitamin D_3 from foods.

In the natural foods, the gastric phase showed a beneficial impact, leading to an increase in the vitamin D_3 content, while the intestinal phase adversely affected the vitamin's content. The observed increase during the gastric step may be due to the release of the vitamin from the food matrix, which made it available for detection. The percentage increase in the gastric step was higher than the percentage decrease in the intestinal step, which indicates that the gastric step had a greater effect on the vitamin D_3 content. The intrinsic antioxidant mechanisms of fish tissue may have acted as a protective agent for vitamin D_3. Greater lipid oxidation may cause the degradation of the vitamin [47,48]. The enzymatic antioxidants in fish, such as glutathione peroxidase (GPx), can reduce lipid peroxides [49], thus protecting vitamin D_3 by decreasing lipid oxidation. Vitamin C and vitamin E, which act as antioxidants, are also present in fish tissue [50]. These vitamins may also have functioned as protective agents against vitamin D_3 degradation. Regarding eggs, their digestion causes the release of amino acids and antioxidant peptides, which raise their antioxidative capacity while preserving the bioaccessibility of their naturally occurring antioxidants, zeaxanthin and lutein [51–53]. This phenomenon may have aided in protecting the vitamin D_3 present in the eggs during digestion.

Among the fortified foods, the milk exhibited a notable reduction in the vitamin D_3 content from the initial to the gastric step. However, the decrease from the gastric to the intestinal step was comparatively minimal and lacked statistical significance. This suggests that, like the natural foods, the gastric step had a more pronounced impact on the milk. Previous studies have shown a low bioaccessibility of vitamin D_3 in milk [54,55]. The bioaccessibility of vitamin D_3 in milk has been found to vary in different types of milk (skim, partially defatted, whole, and infant formula milk) [54], indicating the possible role of not only the fat content but also the type of fats present in the matrix. The low bioaccessibility may also be attributed to the interference of calcium with vitamin D_3 absorption. Previous research on fortified plant-based milk has shown that calcium forms insoluble calcium soaps that trap the vitamin [56]. Similar results were obtained for water-in-oil-in-water emulsions, where vitamin D_3 bioaccessibility was reduced in the presence of calcium [57]. Furthermore, vitamin D_3 can bind to milk proteins, such as β-lactoglobulin and β-casein, under both acidic and alkaline conditions with different binding affinities [58]. This may also have resulted in decreased bioaccessibility, as vitamin D_3 may not be able to be separated from milk proteins during saponification and extraction.

The vitamin D_3 content in the cereals decreased in the gastric stage compared to the initial content, while for the sour cherry juice, a slight increase after the gastric stage was observed. After the intestinal stage, there was an increase in the content compared to the one in the gastric stage for both foods. A study on vitamin D_3 bioaccessibility from test meals showed that semolina meal had the highest bioaccessibility [59], though not as high as in our study. The cereals used in this study contained whole wheat flour and corn semolina, which are high in antioxidants [60,61]. A study on the in vitro digestion of juice extracts found that the content of some phenolic acids and flavonoids increased either during the gastric stage or the intestinal stage, as well as that of some monosaccharides and oligosaccharides, which was attributed to the increased release during digestion [62]. The antioxidant capacity of these compounds during digestion was maintained at elevated levels. Antioxidants have been shown to protect vitamin D_3 against degradation [41,63]. The behaviour of vitamin D_3 during the cereal and juice digestion can be attributed to its increased release during digestion, as well as the antioxidant capacity of the phenolic acids and flavonoids present in the cereals and juice, which may have acted as protective agents against vitamin D_3 degradation.

Regarding the effect of pH on the vitamin D_3 liquid supplement, there were two main observations: On the one hand, in the gastric phase, the lower the pH, the higher the decrease in D_3. On the other hand, exposure to a lower pH during the gastric phase led to a higher content of vitamin D_3 in the intestinal phase; i.e., the content of vitamin D_3 was higher when the matrix was exposed to pH 1 and lower when exposed to pH 7. A study on vitamin D_3 stability in aqueous solutions found that a lower pH had a negative effect on its stability [23]. The stability and content of vitamin D_3 in the GI tract may be affected by lipid oxidation, hydrolysis, and enzyme action. Metal ions, present in the gastric chyme, can also destabilise vitamin D_3, as its degradation may be catalysed by them [23]. In this case, the matrix of the supplement consists of sunflower seed oil, which is not affected by the pepsin present in the gastric phase, as pepsin is a proteolytic enzyme [64]. The decreased content at low pH values can also be attributed to the faster primary oxidation of sunflower seed oil at lower pH values, as lipid oxidation can affect vitamin D_3 by promoting its degradation [47,48]. The intestinal content is affected more by gastric pH changes. A lower pH leads to greater lipid hydrolysis and the release of free fatty acids, which are mixed micelles' structural components [20]. More free fatty acids can form more mixed micelles available to incorporate vitamin D_3, which may lead to better bioaccessibility. A study found that sunflower oil hydrophilicity increases as the pH decreases [65]. Decreased hydrophobicity may affect the formation of mixed micelles regarding their size, shape, and stability, which, by extension, can affect the vitamin's bioaccessibility. The increased content of vitamin D_3 in the intestine could also be attributed to the isomerisation processes that take place at different pH values. Vitamin D_3 is isomerised to isotachysterol under acidic conditions [37], as well as lumisterol and tachysterol [34]. The isomerisation to tachysterol and lumisterol can be reversed, and pre-vitamin D_3 is formed [66], which is then converted to vitamin D_3. The lower pH in the gastric stage may have caused the vitamin's isomerisation (Figure S5—Supplementary File). Based on the elution order of vitamin D_3 and its isomers from similar published HPLC analysis results, it is suggested that the three peaks in Figure S5 (A, B, and C—Supplementary File) may correspond to isotachysterol, lumisterol, and pre-vitamin D_3 [34,37]. As the gastric pH increases, the isomerisation processes can be of a smaller magnitude. This phenomenon, in combination with lipid oxidation, may explain the decrease in the D_3 content in the gastric phase, as well as the corresponding increase observed in the intestinal stage. However, it is important to exercise caution when interpreting these findings, as vitamin D_3 is prone to isomerisation and degradation under diverse conditions. This makes its stability in food products potentially uncertain and its analysis challenging. Early studies suggest that factors like the substrate/reactant ratio, solvents, and time can have varying impacts on the generation pathway of vitamin D isomers [44].

This research highlights that vitamin D_3 is more bioaccessible from foods than supplements, and its bioaccessibility is susceptible to changes in gastric pH. Even though exposure to low gastric pH values, i.e., pH = 1, led to a lower detected vitamin D_3 content, the corresponding intestinal content significantly increased. The mechanism(s) behind this phenomenon should be further explored. It is crucial to understand the behaviour and stability of vitamin D_3 during digestion, as its effectiveness when consumed through foods or supplements relies on its bioaccessibility. Understanding how vitamin D_3 interacts with other components in the digestive system and under GI conditions is essential for developing supplements and foods that optimise its stability and absorption.

4. Materials and Methods

4.1. Chemicals and Reagents

Bile bovine dried, potassium chloride (KCl), calcium chloride ($CaCl_2(H_2O)_2$), and magnesium chloride ($MgCl_2(H_2O)_6$) were purchased from Merck & Co. (Rahway, NJ, USA). Sodium chloride (NaCl), sodium sulphate (Na_2SO_4), potassium dihydrogen phosphate (KH_2PO_4), potassium hydroxide (KOH), hydrochloric acid 37% (HCl), methanol (CH_3OH), and ethanol (C_2H_5OH) were purchased from Chem-Lab NV (Zedelgem, Belgium). Diastase (α-amylase, malt diastase), porcine pepsin, pancreatin, and ammonium carbonate $(NH_4)_2CO_3$ were purchased from Central Drug House Ltd. (New Delhi, India). Sodium hydroxide (NaOH) was purchased from Lach-Ner Ltd. (Neratovice, Czech Republic). Ascorbic acid (vitamin C, $C_6H_8O_6$) was purchased from Sigma-Aldrich (Buchs, Switzerland). Hexane ($H_3C(CH_2)_4CH_3$) was purchased from Avantor Performance Materials (Radnor, PA, USA). Vitamin D_3 standard was purchased from Carl Roth GmbH + Co. KG (Karlsruhe, Germany). All the chemicals and reagents used in this study were of analytical or HPLC-grade. The food samples tested were purchased from local vendors, while supplements were purchased from local pharmacies.

4.2. Digestion Procedure

Digestion was simulated in vitro using the INFOGEST protocol [31]. Enzyme activity must be determined for each enzyme used. In this study, amylase (mouth), pepsin (stomach), and pancreatin (intestine), as well as bile bovine, were used. The activity of enzymes not declared by the manufacturer was calculated according to the protocol. Simulated digestion fluids were prepared according to the protocol, containing KCl, KH_2PO_4, NaCl, $MgCl_2(H_2O)_6$, $(NH_4)_2CO_3$, HCl, and $Cacl_2(H_2O)_2$. $CaCl_2(H_2O)_2$ was added immediately before use at each step due to precipitation issues.

The samples used for the digestion experiments were 3 different types of supplements (tablets, capsules containing an oil-based emulsion, oil-based liquid drops), naturally containing vitamin D_3 foods (eggs, salmon), and fortified foods (milk, cereals, sour cherry juice).

For each food, 5 g was used in the first step. For the tablet and capsule, an amount corresponding to 1200 IU was used, while for the liquid supplement, 5000 IU was used (diluted with water to 2 mL final volume). Solid foods were diluted and minced to achieve a paste-like consistency. Thermal processing (70 °C core temperature for 15 s) of eggs and salmon was conducted by heating the samples in a water bath [32]. Gastric pH effect experiments were conducted using the supplement with the highest bioaccessibility, as determined from the first round of experiments.

4.2.1. Oral Phase

Firstly, the amount of sample was weighted, and simulated salivary fluid (SSF) 1.25× was added. Distilled water was added to reach 1× concentration of SSF. If the sample contained starch, amylase (75 U/mL) was also added. The sample was stirred for 2 min at 37 °C (ONE 14-SV 1422, Memmert, Schwabach, Germany).

4.2.2. Gastric Phase

Simulated gastric fluid (SGF) 1.25×, pepsin (2000 U/mL), and distilled water were added to the mixture at the end of the oral phase. pH adjustment to 3 (protocol value), 1, 4, or 7 (for the pH study) was achieved by adding HCl 1 M (pH 211, HANNA instruments, Woonsocket, RI, USA). The sample was gently shaken for 2 h at 37 °C (ONE 14-SV 1422, Memmert, Germany).

4.2.3. Intestinal Phase

Simulated intestinal fluid (SIF) 1.25× and bile salts (10 mM) were added to the gastric chyme. The mixture was stirred for 30 min at 37 °C until complete bile solubilisation. Afterwards, pancreatin (100 U/mL trypsin activity) was added, the pH was adjusted to 7 (NaOH 1 M) (pH 211, HANNA instruments, USA), and distilled water was added. The sample was stirred for 2 h at 37 °C (ONE 14-SV 1422, Memmert, Germany).

All samples were stored at −20 °C until further analysis.

4.3. Vitamin D_3 Isolation

4.3.1. Samples with Saponification

The isolation method used was based on Yanhai et al. [67], with some modifications. Raw and thermally processed food were diluted in an appropriate amount of water. The samples from the digestion steps were not diluted. To each sample, 15 g/L solution of vitamin C in ethanol in a 1:2 ratio (*v*/*v*) and 1.25 g/mL solution of KOH in water in a 2:1 ratio (*v*/*v*) were added. The sample was heated at 60 °C for 45 min with continuous stirring to achieve lipid saponification. Afterwards, the sample was cooled at room temperature and underwent 2 subsequent extractions with hexane in a 1:2 ratio (*v*/*v*). For each extraction, hexane was added to the sample and vortexed for 5 min. Then, the mixture was placed in a separating funnel until complete phase separation. The water phase was removed. The hexane phases from the two extraction steps were collected and combined. Na_2SO_4 was added to remove any residual water. To remove Na_2SO_4, the sample was filtered through filter paper (retention 10–15 µm). Subsequently, the sample was placed in a rotary evaporator at 40 °C (Laborota 4003, Heidolph, Schwabach, Germany) and evaporated to dryness. Solids were redissolved with 2 mL methanol and filtered through a 0.22 µm filter (PTFE).

4.3.2. Samples without Saponification

For the supplements, the isolation of vitamin D_3 was conducted as follows: The capsule and tablet were diluted with 5 mL of water. The liquid supplement was used undiluted. Methanol was added to the samples in a 1:2 (*v*/*v*) ratio, vortexed for 2–3 min, and placed in an ultrasonic bath (LBS1 10Lt, FALC instruments, Treviglio, Italy) for 10 min. Then, the mixture was vortexed again for 2–3 min and centrifuged at 2.938× *g* (unicen 21, Ortoalresa, Madrid, Spain) for 15 min to achieve complete phase separation. The organic phase was collected and evaporated to dryness (40 °C, Laborota 4003, Heidolph, Germany). Solids were redissolved in 2 mL methanol and filtered through a 0.22 µm filter (PTFE).

The juice and the digestion fractions of the juice and supplements were extracted twice with hexane. The procedure followed was as described in the previous Section 4.3.1.

4.4. High-Performance Liquid Chromatography (HPLC)

The vitamin D_3 concentration was determined using HPLC with a UV detector (KNAUER 1200 system, Burladingen, Germany). The column used for separation was Eurospher II 100-5 C18A (250 × 4 mm). The mobile phase was HPLC-grade methanol and 0.1% formic acid with a constant flow rate at 1 mL/min and 25 °C. The injection volume was 20 µL. The UV detector was set to a 265 nm wavelength. Vitamin D was eluted at 4.8–4.9 min.

The vitamin D_3 concentration in each sample was determined based on a standard curve. The standard curve was constructed using the vitamin D_3 standard (Figures S1 and S2—Supplementary File). Different concentrations of the standard in the range of 5 to 30 ppm were analysed in the HPLC system to determine the corresponding peak areas The limit of detection (LOD) was 0.05 ppm, and the limit of quantification (LOQ) was 0.17 ppm (Tables S1 and S2—Supplementary File).

Recovery was determined by spiking raw foods and supplements with a known amount of the vitamin D_3 standard and analysing the sample. Recovery was calculated according to the following formula:

$$\%recovery = \left[\left(A_{spiked} - A_{unspiked}\right) / A_{standard}\right] \times 100\% \quad (1)$$

where A_{spiked} is the peak area of the spiked sample; $A_{uspiked}$ is the peak area of the unspiked sample; and $A_{standard}$ is the peak area of the vitamin D_3 standard, which was used for spiking the sample.

All values in the tables and figures were corrected based on the recovery of each sample.

4.5. Bioaccessibility Index

The vitamin D_3 bioaccessibility index (BI) was calculated using the following formula:

$$BI = C_b / C_a \quad (2)$$

where C_a and C_b are the amounts of vitamin D_3 before and after digestion [68].

4.6. Oxidation Measurement

4.6.1. Peroxide Value

Peroxide value measurement was performed as described by Richards et al. [69], with some modifications. In each sample, 500 ppm BHT was added to stop the oxidation process and vortexed to achieve homogenisation. Next, 10 mL of $CHCl_3$-CH_3OH (2:1 v/v) was added to 1 g of the sample. Then, 1.5 mL of NaCl (0.5%) was added, and the samples were vortexed and centrifuged at 2.798× g (unicen 21, Ortoalresa, Spain) for 10 min at ambient temperature. The lower phase of $CHCl_3$ was collected, and $CHCl_3$-CH_3OH (2:1 v/v) was added until 10 mL final volume was reached. Following this, 25 μL of NH_4SCN solution (30% w/v) and 25 μL of freshly prepared $FeCl_2$ solution (0.66% w/v) were added, and the mixture was vortexed for 2-4 s. A proper amount of the sample was transferred to a Quartz cell, and the absorbance was measured in a spectrometer (uniSPEC 2 UV/VIS-Spectrometer, LLG, Meckenheim, Germany) at 500 nm. Furthermore, 10 mL of $CHCl_3$-CH_3OH (2:1 v/v) was used as blind. The oxidation products are expressed as mmol/kg of the lipid phase using a standard curve formed with cumene hydroperoxide solutions [70–72].

4.6.2. Thiobarbituric Acid Method (TBARS)

The TBARS method was performed according to Lemon [73], with some modifications. First, 1.5 g of the sample was transferred to a test tube containing 5 mL of TCA (7.5% w/v) and vortexed. The mixture was centrifuged for 30 min at 2.798× g (unicen 21, Ortoalresa, Spain). A 2 mL aliquot was mixed with 2 mL of TBA solution (0.02 M). The mixture was heated in a water bath for 40 min at a constant temperature of 100 °C. The samples were then cooled down to room temperature under running tap water and transferred to a Quartz cell to measure the absorbance (uniSPEC 2 UV/VIS-Spectrometer, LLG, Germany) at 532 nm. TBA:TCA solution (1:1 v/v) was used as blind. The oxidation products are expressed as MDAeq (μmol/L) with the help of the standard curve constructed using TEP solutions.

4.7. Statistical Analysis

Three independent digestion experiments (n = 3) were conducted, and the experimental results are expressed as means ± standard deviations. Minitab 21 Statistical Software (Minitab LLC, State College, PA, USA) was used to statistically process the data by carrying out a one-way ANOVA with Fisher's test for means comparison. Differences were considered significant at $p < 0.05$.

Supplementary Materials: The following supporting information can be downloaded at: https://www.mdpi.com/article/10.3390/molecules29051153/s1, Table S1. Average peak areas (n = 3) of low concentrations of vitamin D_3 standard; Table S2. LINEST function parameters; Figure S1: Vitamin D_3 standard curve; Figure S2: Vitamin D_3 standards for standard curve; Figure S3: Raw salmon sample spiked with vitamin D_3; Figure S4: Raw egg sample spiked with vitamin D_3; Figure S5: Chromatograph of liquid supplement after gastric digestion at pH 1. (A), (B), and (C) may be isomers of vitamin D_3, produced during digestion due to acidic degradation.

Author Contributions: Conceptualisation, E.P. and P.V.; methodology, E.P. and P.V; software, E.P. and P.V.; validation, E.P. and P.V.; formal analysis, E.P. and P.V.; investigation, E.P. and P.V.; resources, E.P. and P.V.; data curation, E.P. and P.V.; writing—original draft preparation, E.P.; writing—review and editing, E.P. and P.V.; visualisation, E.P. and P.V.; supervision, P.V.; project administration, P.V. All authors have read and agreed to the published version of the manuscript.

Funding: This research was funded by the Ministry of Agriculture, Program for Agricultural Development, ESPA 2014-2020, Submeasures 16.1 and 16.2, grant number M16ΣΥΝ-00807.

Institutional Review Board Statement: Not applicable.

Informed Consent Statement: Not applicable.

Data Availability Statement: Data are available upon request to the corresponding author: pkvareltzis @cheng.auth.gr.

Conflicts of Interest: The authors declare no conflicts of interest.

References

1. Combs, G.F.; McClung, J.P. Sources of the Vitamins. In *The Vitamins*; Elsevier: Amsterdam, The Netherlands, 2017; pp. 501–530.
2. Janoušek, J.; Pilařová, V.; Macáková, K.; Nomura, A.; Veiga-Matos, J.; Silva, D.D.d.; Remião, F.; Saso, L.; Malá-Ládová, K.; Malý, J.; et al. Vitamin D: Sources, Physiological Role, Biokinetics, Deficiency, Therapeutic Use, Toxicity, and Overview of Analytical Methods for Detection of Vitamin D and Its Metabolites. *Crit. Rev. Clin. Lab. Sci.* **2022**, *59*, 517–554. [CrossRef]
3. Welsh, J. Cellular and Molecular Effects of Vitamin D on Carcinogenesis. *Arch. Biochem. Biophys.* **2012**, *523*, 107–114. [CrossRef]
4. Khazai, N.; Judd, S.E.; Tangpricha, V. Calcium and Vitamin D: Skeletal and Extraskeletal Health. *Curr. Rheumatol. Rep.* **2008**, *10*, 110. [CrossRef]
5. Meng, J.; Li, X.; Liu, W.; Xiao, Y.; Tang, H.; Wu, Y.; Xiong, Y.; Gao, S. The Role of Vitamin D in the Prevention and Treatment of SARS-CoV-2 Infection: A Meta-Analysis of Randomized Controlled Trials. *Clin. Nutr.* **2023**, *42*, 2198–2206. [CrossRef]
6. Bendik, I.; Friedel, A.; Roos, F.F.; Weber, P.; Eggersdorfer, M. Vitamin D: A Critical and Essential Micronutrient for Human Health. *Front. Physiol.* **2014**, *5*, 248. [CrossRef]
7. Cui, A.; Zhang, T.; Xiao, P.; Fan, Z.; Wang, H.; Zhuang, Y. Global and Regional Prevalence of Vitamin D Deficiency in Population-Based Studies from 2000 to 2022: A Pooled Analysis of 7.9 Million Participants. *Front. Nutr.* **2023**, *10*, 1070808. [CrossRef]
8. Borel, P.; Caillaud, D.; Cano, N.J. Vitamin D Bioavailability: State of the Art. *Crit. Rev. Food Sci. Nutr.* **2015**, *55*, 1193–1205. [CrossRef]
9. Traub, M.L.; Finnell, J.S.; Bhandiwad, A.; Oberg, E.; Suhaila, L.; Bradley, R. Impact of Vitamin D3 Dietary Supplement Matrix on Clinical Response. *J. Clin. Endocrinol. Metab.* **2014**, *99*, 2720–2728. [CrossRef] [PubMed]
10. Fox, C.B.; Kim, J.; Le, L.V.; Nemeth, C.L.; Chirra, H.D.; Desai, T.A. Micro/Nanofabricated Platforms for Oral Drug Delivery. *J. Control. Release* **2015**, *219*, 431–444. [CrossRef] [PubMed]
11. Joye, I.J.; Davidov-Pardo, G.; McClements, D.J. Nanotechnology for Increased Micronutrient Bioavailability. *Trends Food Sci. Technol.* **2014**, *40*, 168–182. [CrossRef]
12. Šimoliūnas, E.; Rinkūnaitė, I.; Bukelskienė, Ž.; Bukelskienė, V. Bioavailability of Different Vitamin D Oral Supplements in Laboratory Animal Model. *Medicina* **2019**, *55*, 265. [CrossRef]
13. Grossmann, R.E.; Tangpricha, V. Evaluation of Vehicle Substances on Vitamin D Bioavailability: A Systematic Review. *Mol. Nutr. Food Res.* **2010**, *8*, 1055–1061. [CrossRef]
14. Helde Frankling, M.; Norlin, A.C.; Hansen, S.; Wahren Borgström, E.; Bergman, P.; Björkhem-Bergman, L. Are Vitamin D3 Tablets and Oil Drops Equally Effective in Raising S-25-Hydroxyvitamin D Concentrations? A Post-Hoc Analysis of an Observational Study on Immunodeficient Patients. *Nutrients* **2020**, *12*, 1230. [CrossRef]
15. Maurya, V.K.; Aggarwal, M. Factors Influencing the Absorption of Vitamin D in GIT: An Overview. *J. Food Sci. Technol.* **2017**, *54*, 3753–3765. [CrossRef]

16. Natri, A.-M.; Salo, P.; Vikstedt, T.; Palssa, A.; Huttunen, M.; Kärkkäinen, M.U.M.; Salovaara, H.; Piironen, V.; Jakobsen, J.; Lamberg-Allardt, C.J. Bread Fortified with Cholecalciferol Increases the Serum 25-Hydroxyvitamin D Concentration in Women as Effectively as a Cholecalciferol Supplement. *J. Nutr.* **2006**, *136*, 123–127. [CrossRef]
17. Biancuzzo, R.M.; Young, A.; Bibuld, D.; Cai, M.H.; Winter, M.R.; Klein, E.K.; Ameri, A.; Reitz, R.; Salameh, W.; Chen, T.C.; et al. Fortification of Orange Juice with Vitamin D2 or Vitamin D 3 Is as Effective as an Oral Supplement in Maintaining Vitamin D Status in Adults. *Am. J. Clin. Nutr.* **2010**, *91*, 1621–1626. [CrossRef] [PubMed]
18. Reboul, E.; Goncalves, A.; Comera, C.; Bott, R.; Nowicki, M.; Landrier, J.F.; Jourdheuil-Rahmani, D.; Dufour, C.; Collet, X.; Borel, P. Vitamin D Intestinal Absorption Is Not a Simple Passive Diffusion: Evidences for Involvement of Cholesterol Transporters. *Mol. Nutr. Food Res.* **2011**, *55*, 691–702. [CrossRef] [PubMed]
19. Hornbuckle, W.E.; Simpson, K.W.; Tennant, B.C. Gastrointestinal Function. In *Clinical Biochemistry of Domestic Animals*; Academic Press: Cambridge, MA, USA, 2008; pp. 413–457.
20. Reboul, E. Intestinal Absorption of Vitamin D: From the Meal to the Enterocyte. *Food Function* **2015**, *6*, 356–362. [CrossRef] [PubMed]
21. Ozturk, B.; Argin, S.; Ozilgen, M.; McClements, D.J. Nanoemulsion Delivery Systems for Oil-Soluble Vitamins: Influence of Carrier Oil Type on Lipid Digestion and Vitamin D3 Bioaccessibility. *Food Chem.* **2015**, *187*, 499–506. [CrossRef] [PubMed]
22. Goncalves, A.; Gleize, B.; Roi, S.; Nowicki, M.; Dhaussy, A.; Huertas, A.; Amiot, M.J.; Reboul, E. Fatty Acids Affect Micellar Properties and Modulate Vitamin D Uptake and Basolateral Efflux in Caco-2 Cells. *J. Nutr. Biochem.* **2013**, *24*, 1751–1757. [CrossRef] [PubMed]
23. Temova Rakuša, Ž.; Pišlar, M.; Kristl, A.; Roškar, R. Comprehensive Stability Study of Vitamin D3 in Aqueous Solutions and Liquid Commercial Products. *Pharmaceutics* **2021**, *13*, 617. [CrossRef]
24. Jin, X.; Yang, X.; Yang, L.; Liu, Z.L.; Zhang, F. Autoxidation of Isotachysterol. *Tetrahedron* **2004**, *60*, 2881–2888. [CrossRef]
25. Esmaeili, M.; Yekta, R.; Abedi, A.S.; Ghanati, K.; Derav, R.Z.; Houshyarrad, A.; Dehkordi, Z.S.; Ajami, M.; Mahmoudzadeh, M. Encapsulating Vitamin D: A Feasible and Promising Approach to Combat Its Deficiency. In *Pharmaceutical Sciences*; Tabriz University of Medical Sciences: Tabriz, Iran, 2022; pp. 194–207.
26. Diarrassouba, F.; Remondetto, G.; Liang, L.; Garrait, G.; Beyssac, E.; Subirade, M. Effects of Gastrointestinal PH Conditions on the Stability of the β-Lactoglobulin/Vitamin D3 Complex and on the Solubility of Vitamin D3. *Food Res. Int.* **2013**, *52*, 515–521. [CrossRef]
27. Xiang, C.; Gao, J.; Ye, H.; Ren, G.; Ma, X.; Xie, H.; Fang, S.; Lei, Q.; Fang, W. Development of Ovalbumin-Pectin Nanocomplexes for Vitamin D3 Encapsulation: Enhanced Storage Stability and Sustained Release in Simulated Gastrointestinal Digestion. *Food Hydrocoll.* **2020**, *106*, 105926. [CrossRef]
28. Sharifi, F.; Jahangiri, M. Investigation of the Stability of Vitamin D in Emulsion-Based Delivery Systems. *Chem. Ind. Chem. Eng. Q.* **2017**, *24*, 157–167. [CrossRef]
29. Sams, L.; Paume, J.; Giallo, J.; Carrière, F. Relevant PH and Lipase for in Vitro Models of Gastric Digestion. *Food Funct.* **2016**, *7*, 30–45. [CrossRef]
30. Wang, X.; Ye, A.; Lin, Q.; Han, J.; Singh, H. Gastric Digestion of Milk Protein Ingredients: Study Using an in Vitro Dynamic Model. *J. Dairy. Sci.* **2018**, *101*, 6842–6852. [CrossRef]
31. Brodkorb, A.; Egger, L.; Alminger, M.; Alvito, P.; Assunção, R.; Ballance, S.; Bohn, T.; Bourlieu-Lacanal, C.; Boutrou, R.; Carrière, F.; et al. INFOGEST Static in Vitro Simulation of Gastrointestinal Food Digestion. *Nat. Protoc.* **2019**, *14*, 991–1014. [CrossRef]
32. Food Code 2022 | FDA. Available online: https://www.fda.gov/food/fda-food-code/food-code-2022 (accessed on 25 July 2023).
33. Schmid, A.; Walther, B. Natural Vitamin D Content in Animal Products. *Adv. Nutr.* **2013**, *4*, 453–462. [CrossRef]
34. Temova, Ž.; Roškar, R. Stability-Indicating HPLC–UV Method for Vitamin D3 Determination in Solutions, Nutritional Supplements and Pharmaceuticals. *J. Chromatogr. Sci.* **2016**, *54*, 1180–1186. [CrossRef]
35. Dunlop, E.; Cunningham, J.; Sherriff, J.; Lucas, R.; Greenfield, H.; Arcot, J.; Strobel, N.; Black, L. Vitamin D3 and 25-Hydroxyvitamin D3 Content of Retail White Fish and Eggs in Australia. *Nutrients* **2017**, *9*, 647. [CrossRef] [PubMed]
36. Jakobsen, J.; Smith, C.; Bysted, A.; Cashman, K.D. Vitamin D in Wild and Farmed Atlantic Salmon (Salmo Salar)—What Do We Know? *Nutrients* **2019**, *11*, 982. [CrossRef] [PubMed]
37. Mahmoodani, F.; Perera, C.O.; Fedrizzi, B.; Abernethy, G.; Chen, H. Degradation Studies of Cholecalciferol (Vitamin D3) Using HPLC-DAD, UHPLC-MS/MS and Chemical Derivatization. *Food Chem.* **2017**, *219*, 373–381. [CrossRef]
38. Flores-Aldana, M.; Rivera-Pasquel, M.; García-Guerra, A.; Pérez-Cortés, J.G.; Bárcena-Echegollén, J.E. Effect of Vitamin D Supplementation on (25(OH)D) Status in Children 12–30 Months of Age: A Randomized Clinical Trial. *Nutrients* **2023**, *15*, 2756. [CrossRef]
39. Villamor, E.; Oliveros, H.; Marín, C.; López-Arana, S.; Agudelo-Cañas, S. Increased Serum Total and Free 25-Hydroxyvitamin D with Daily Intake of Cholecalciferol-Fortified Skim Milk: A Randomized Controlled Trial in Colombian Adolescents. *J. Nutr.* **2023**, *153*, 1189–1198. [CrossRef]
40. Sollano-mendieta, X.C.; Meza-márquez, O.G.; Osorio-revilla, G.; Téllez-medina, D.I. Effect of In Vitro Digestion on the Antioxidant Compounds and Antioxidant Capacity of 12 Plum (*Spondias purpurea* L.) Ecotypes. *Foods* **2021**, *10*, 1995. [CrossRef]
41. Hemery, Y.M.; Fontan, L.; Moench-Pfanner, R.; Laillou, A.; Berger, J.; Renaud, C.; Avallone, S. Influence of Light Exposure and Oxidative Status on the Stability of Vitamins A and D3 during the Storage of Fortified Soybean Oil. *Food Chem.* **2015**, *184*, 90–98. [CrossRef] [PubMed]

42. Floros, S.; Toskas, A.; Pasidi, E.; Vareltzis, P. Bioaccessibility and Oxidative Stability of Omega-3 Fatty Acids in Supplements, Sardines and Enriched Eggs Studied Using a Static In Vitro Gastrointestinal Model. *Molecules* **2022**, *27*, 415. [CrossRef]
43. Jakobsen, J.; Knuthsen, P. Stability of Vitamin D in Foodstuffs during Cooking. *Food Chem.* **2014**, *148*, 170–175. [CrossRef]
44. Szlinder-Richert, J.; Malesa-Ciećwierz, M. Effect of Household Cooking Methods on Nutritional Value of Cod and Salmon-Twin Fillet Approach. *Carpathian J. Food Sci. Technol.* **2018**, *10*, 142–157.
45. Lee, H.J.; Shin, C.; Chun, Y.S.; Kim, J.; Jung, H.; Choung, J.; Shim, S.M. Physicochemical Properties and Bioavailability of Naturally Formulated Fat-Soluble Vitamins Extracted from Agricultural Products for Complementary Use for Natural Vitamin Supplements. *Food Sci. Nutr.* **2020**, *8*, 5660–5672. [CrossRef]
46. Ribeiro, A.; Gonçalves, R.F.S.; Pinheiro, A.C.; Manrique, Y.A.; Barreiro, M.F.; Lopes, J.C.B.; Dias, M.M. In Vitro Digestion and Bioaccessibility Studies of Vitamin E-Loaded Nanohydroxyapatite Pickering Emulsions and Derived Fortified Foods. *LWT* **2022**, *154*, 112706. [CrossRef]
47. Mahmoodani, F.; Perera, C.O.; Abernethy, G.; Fedrizzi, B.; Chen, H. Lipid Oxidation and Vitamin D3 Degradation in Simulated Whole Milk Powder as Influenced by Processing and Storage. *Food Chem.* **2018**, *261*, 149–156. [CrossRef] [PubMed]
48. Mahmoodani, F.; Perera, C.O.; Abernethy, G.; Fedrizzi, B.; Greenwood, D.; Chen, H. Identification of Vitamin D3 Oxidation Products Using High-Resolution and Tandem Mass Spectrometry. *J. Am. Soc. Mass. Spectrom.* **2018**, *29*, 1442–1455. [CrossRef]
49. Bochkov, V.N.; Oskolkova, O.V.; Birukov, K.G.; Levonen, A.L.; Binder, C.J.; Stöckl, J. Generation and Biological Activities of Oxidized Phospholipids. *Antioxid. Redox Signal* **2010**, *12*, 1009. [CrossRef]
50. Vélez-Alavez, M.; Méndez-Rodriguez, L.C.; De Anda Montañez, J.A.; Mejía, C.H.; Galván-Magaña, F.; Zenteno-Savín, T. Vitamins C and E Concentrations in Muscle of Elasmobranch and Teleost Fishes. *Comp. Biochem. Physiol. A Mol. Integr. Physiol.* **2014**, *170*, 26–30. [CrossRef]
51. Rao, S.; Sun, J.; Liu, Y.; Zeng, H.; Su, Y.; Yang, Y. ACE Inhibitory Peptides and Antioxidant Peptides Derived from in Vitro Digestion Hydrolysate of Hen Egg White Lysozyme. *Food Chem.* **2012**, *135*, 1245–1252. [CrossRef] [PubMed]
52. Young, D.; Nau, F.; Pasco, M.; Mine, Y. Identification of Hen Egg Yolk-Derived Phosvitin Phosphopeptides and Their Effects on Gene Expression Profiling against Oxidative Stress-Induced Caco-2 Cells. *J. Agric. Food Chem.* **2011**, *59*, 9207–9218. [CrossRef]
53. Remanan, M.K.; Wu, J. Antioxidant Activity in Cooked and Simulated Digested Eggs. *Food Funct.* **2014**, *5*, 1464–1474. [CrossRef]
54. Lipkie, T.E.; Ferruzzi, M.G.; Weaver, C.M. Low Bioaccessibility of Vitamin D2 from Yeast-Fortified Bread Compared to Crystalline D2 Bread and D3 from Fluid Milks. *Food Funct.* **2016**, *7*, 4589–4596. [CrossRef]
55. Hernández-Olivas, E.; Muñoz-Pina, S.; Sánchez-García, J.; Andrés, A.; Heredia, A. Understanding the Role of Food Matrix on the Digestibility of Dairy Products under Elderly Gastrointestinal Conditions. *Food Res. Int.* **2020**, *137*, 109454. [CrossRef]
56. Zhou, H.; Zheng, B.; Zhang, Z.; Zhang, R.; He, L.; McClements, D.J. Fortification of Plant-Based Milk with Calcium May Reduce Vitamin D Bioaccessibility: An in Vitro Digestion Study. *J. Agric. Food Chem.* **2021**, *69*, 4223–4233. [CrossRef]
57. Dima, C.; Dima, S. Bioaccessibility Study of Calcium and Vitamin D3 Co-Microencapsulated in Water-in-Oil-in-Water Double Emulsions. *Food Chem.* **2020**, *303*, 125416. [CrossRef] [PubMed]
58. Forrest, S.A.; Yada, R.Y.; Rousseau, D. Interactions of Vitamin D3 with Bovine β-Lactoglobulin A and β-Casein. *J. Agric. Food Chem.* **2005**, *53*, 8003–8009. [CrossRef]
59. Antoine, T.; Icard-Vernière, C.; Scorrano, G.; Salhi, A.; Halimi, C.; Georgé, S.; Carrière, F.; Mouquet-Rivier, C.; Reboul, E. Evaluation of Vitamin D Bioaccessibility and Mineral Solubility from Test Meals Containing Meat and/or Cereals and/or Pulses Using in Vitro Digestion. *Food Chem.* **2021**, *347*, 128621. [CrossRef]
60. Li, Y.; Ma, D.; Sun, D.; Wang, C.; Zhang, J.; Xie, Y.; Guo, T. Total Phenolic, Flavonoid Content, and Antioxidant Activity of Flour, Noodles, and Steamed Bread Made from Different Colored Wheat Grains by Three Milling Methods. *Crop J.* **2015**, *3*, 328–334. [CrossRef]
61. Siyuan, S.; Tong, L.; Liu, R.H. Corn Phytochemicals and Their Health Benefits. *Food Sci. Hum. Wellness* **2018**, *7*, 185–195. [CrossRef]
62. Aguillón-Osma, J.; Luzardo-Ocampo, I.; Cuellar-Nuñez, M.L.; Maldonado-Celis, M.E.; Loango-Chamorro, N.; Campos-Vega, R. Impact of in Vitro Gastrointestinal Digestion on the Bioaccessibility and Antioxidant Capacity of Bioactive Compounds from Passion Fruit (Passiflora Edulis) Leaves and Juice Extracts. *J. Food Biochem.* **2019**, *43*, e12879. [CrossRef]
63. Goebel, S.; Avallone, S.; Detchewa, P.; Prasajak, P.; Sriwichai, W. Natural and Synthetic Antioxidants Prevent the Degradation of Vitamin D3fortification in Canola Oil during Baking and in Vitro Digestion. *Appl. Sci. Eng. Prog.* **2021**, *14*, 247–258. [CrossRef]
64. Blanco, A.; Blanco, G. Digestion—Absorption. *Med. Biochem.* **2017**, 251–273. [CrossRef]
65. Wang, W.; Cui, C.; Wang, Q.; Sun, C.; Jiang, L.; Hou, J. Effect of PH on Physicochemical Properties of Oil Bodies from Different Oil Crops. *J. Food Sci. Technol.* **2019**, *56*, 49. [CrossRef] [PubMed]
66. Bikle, D. Nonclassic Actions of Vitamin D. *J. Clin. Endocrinol. Metab.* **2009**, *94*, 26–34. [CrossRef] [PubMed]
67. Yanhai, Z.; Yan, J.; Qun, X.; Rohrer, J. Simultaneous Determination of Vitamins A, E, and D3 in Milk-Based Nutritionals by On-Line Two-Dimensional HPLC. Available online: https://www.thermofisher.cn/document-connect/document-connect.html?url=https://assets.thermofisher.cn/TFS-Assets/CMD/Application-Notes/AN-1117-HPLC-Vitamins-Milk-AN71511-EN.pdf (accessed on 10 December 2022).
68. Zhu, Y.; Yang, S.; Huang, Y.; Huang, J.; Li, Y. Effect of in Vitro Gastrointestinal Digestion on Phenolic Compounds and Antioxidant Properties of Soluble and Insoluble Dietary Fibers Derived from Hulless Barley. *J. Food Sci.* **2021**, *86*, 628–634. [CrossRef] [PubMed]
69. Richards, M.P.; Hultin, H.O. Effect of PH on Lipid Oxidation Using Trout Hemolysate as a Catalyst: A Possible Role for Deoxyhemoglobin. *J. Agric. Food Chem.* **2000**, *48*, 3141–3147. [CrossRef]

70. Peroxide Value Method. Available online: https://www.protocols.io/view/Peroxide-Value-Method-4rm7vz12lx1w/v1 (accessed on 1 August 2023).
71. Christie, W.; Han, X. *Lipid Analysis*, 5th ed.; Woodhead: Oxford, UK, 2012; pp. 181–211.
72. Shantha, N.C.; Decker, E.A. Rapid, Sensitive, Iron-Based Spectrophotometric Methods for Determination of Peroxide Values of Food Lipids. *J. AOAC Int.* **1994**, *77*, 421–424. [CrossRef]
73. Lemons, D.W. Fisheries and Marine Service. 1975. Available online: http://icnaf.nafo.int/docs/1974/res-07.pdf (accessed on 10 March 2021).

Disclaimer/Publisher's Note: The statements, opinions and data contained in all publications are solely those of the individual author(s) and contributor(s) and not of MDPI and/or the editor(s). MDPI and/or the editor(s) disclaim responsibility for any injury to people or property resulting from any ideas, methods, instructions or products referred to in the content.

molecules

Article

Phytochemical Composition and Antioxidant and Anti-Inflammatory Activities of *Humboldtia sanjappae* Sasidh. & Sujanapal, an Endemic Medicinal Plant to the Western Ghats

Jameema Sidhic [1], Satheesh George [1,*], Ahmed Alfarhan [2], Rajakrishnan Rajagopal [2], Opeyemi Joshua Olatunji [3] and Arunaksharan Narayanankutty [4,*]

1 Phytochemistry and Pharmacology Division, PG & Research Department of Botany, St. Joseph's College (Autonomous), Calicut 673008, India
2 Department of Botany and Microbiology, College of Science, King Saud University, P.O. Box 2455, Riyadh 11451, Saudi Arabia; alfarhan@ksu.edu.sa (A.A.); rrajagopal@ksu.edu.sa (R.R.)
3 African Genome Center, Mohammed VI Polytechnic University, Ben Guerir 43150, Morocco
4 Division of Cell and Molecular Biology, PG & Research Department of Zoology, St. Joseph's College (Autonomous), Calicut 673008, India
* Correspondence: george.satheesh@gmail.com (S.G.); arunaksharann@devagiricollege.org (A.N.)

Abstract: Ethnomedicinal plants are important sources of drug candidates, and many of these plants, especially in the Western Ghats, are underexplored. *Humboldtia*, a genus within the Fabaceae family, thrives in the biodiversity of the Western Ghats, Kerala, India, and holds significant ethnobotanical importance. However, many *Humboldtia* species remain understudied in terms of their biological efficacy, while some lack scientific validation for their traditional uses. However, *Humboldtia sanjappae*, an underexplored plant, was investigated for the phytochemical composition of the plant, and its antioxidant, enzyme-inhibitory, anti-inflammatory, and antibacterial activities were assessed. The LC-MS analysis indicated the presence of several bioactive substances, such as Naringenin, Luteolin, and Pomiferin. The results revealed that the ethanol extract of *H. sanjappae* exhibited significant in vitro DPPH scavenging activity (6.53 ± 1.49 μg/mL). Additionally, it demonstrated noteworthy FRAP (Ferric Reducing Antioxidant Power) activity (8.46 ± 1.38 μg/mL). Moreover, the ethanol extract of *H. sanjappae* exhibited notable efficacy in inhibiting the activities of α-amylase (47.60 ± 0.19μg/mL) and β-glucosidase (32.09 ± 0.54 μg/mL). The pre-treatment with the extract decreased the LPS-stimulated release of cytokines in the Raw 264.7 macrophages, demonstrating the anti-inflammatory potential. Further, the antibacterial properties were also evident in both Gram-positive and Gram-negative bacteria. The observed high zone of inhibition in the disc diffusion assay and MIC values were also promising. *H. sanjappae* displays significant anti-inflammatory, antioxidant, antidiabetic, and antibacterial properties, likely attributable to its rich composition of various biological compounds such as Naringenin, Luteolin, Epicatechin, Maritemin, and Pomiferin. Serving as a promising reservoir of these beneficial molecules, the potential of *H. sanjappae* as a valuable source for bioactive ingredients within the realms of nutraceutical and pharmaceutical industries is underscored, showcasing its potential for diverse applications.

Keywords: *Humboldtia sanjappae*; LC-MS analysis; radical scavenging; anti-inflammatory activity; cytokine release

Citation: Sidhic, J.; George, S.; Alfarhan, A.; Rajagopal, R.; Olatunji, O.J.; Narayanankutty, A. Phytochemical Composition and Antioxidant and Anti-Inflammatory Activities of *Humboldtia sanjappae* Sasidh. & Sujanapal, an Endemic Medicinal Plant to the Western Ghats. *Molecules* **2023**, *28*, 6875. https://doi.org/10.3390/molecules28196875

Academic Editors: Satyajit D Sarker, Artur M. S. Silva and De-Xing Hou

Received: 31 July 2023
Revised: 11 September 2023
Accepted: 27 September 2023
Published: 29 September 2023

Copyright: © 2023 by the authors. Licensee MDPI, Basel, Switzerland. This article is an open access article distributed under the terms and conditions of the Creative Commons Attribution (CC BY) license (https://creativecommons.org/licenses/by/4.0/).

1. Introduction

Humans have always battled with various infections. In addition to these, recent decades have witnessed a significant increase in the occurrence of various non-communicable diseases. These diseases have been associated with increased mortality globally. The changes in lifestyle comprising dietary changes and reduced physical activity have resulted in a sudden increase in the number of patients. The role of oxidative stress and inflammation is impeccable in the onset of these diseases. Oxidative stress, the imbalance between the generation of reactive molecules and its scavenging, plays significant roles in non-communicable diseases [1]. Together with this, numerous studies have established that inflammation has a role in the progression of different diseases. The link between inflammatory response and the onset of different cancers such as ovarian cancer and pancreatic cancer has been studied well [2,3]. Recent studies suggest that neuroinflammation in Alzheimer's will escalate the disease progression Recently, studies on *Gynostemma pentaphyllum* (Thunb.) Makino demonstrated a protective effect against the inflammation [4,5]. New therapies indicate that the anti-inflammatory treatments associated with cardiovascular disease are a promising strategy to bring down the succession of the disease [6]. Inflammatory processes in the host defense system should be highly regulated, and the loss of control is problematic. So, the inflammatory molecules have become the primary target for the prevention of various diseases, in which the main signaling pathways such as the nuclear factor-κB (NF-κB) signaling pathway, Janus kinase/signal transducer and activator of transcription (JAK/STAT) signaling pathway, and mitogen-activated protein kinase (MAPK) signaling pathway might be brought under control to prevent the diseases [7–9]; the JAK/STAT is focused on more because this pathway is associated with the pathogenesis of different inflammatory diseases such as Rheumatoid Arthritis (RA) and Inflammatory Bowel Disease (IBD) [10,11]. Even though inflammation is an evolutionarily conserved mechanism for the organism's survival [12], it is essential to control the prolonged release of anti-inflammatory mediators to prevent the development of various diseases [13]. Different natural products have recently been investigated and have given satisfactory results in this respect [14,15]. Excess inflammation in the body will lead to the development of Reactive Oxygen Species (ROS); if the concentration of the same exceeds a limit, the body will not be able to neutralize the same [16]. Due to this reason, pharmaceutical and food industries have considerably used natural products with antioxidant capabilities, and herbal products that promote health have also become highly popular in recent years [17,18].

It has been discovered that several plant medicines have a variety of pharmacological properties, including antioxidant, anti-inflammatory, anticancer, neuroprotective, and hypolipidemic properties [8,19,20]. These activities are attributed to the non-nutritive chemicals present in the plants [10]. Findings from studies conducted in mice suggest that the leaves of *Hemigraphis alternata* exhibit anti-inflammatory, anti-nociceptive, and anti-diarrheal activities [21]. These kinds of phytochemicals are reported in several plant families such as Lamiaceae, Zingiberaceae, Malvaceae, Acanthaceae, and Apocyanacea [22,23]. Fabaceae is one such family with an abundance of different phytochemicals which are good in curing various diseases [24].

The genus *Humboldtia* belongs to the family Fabaceae. The plants of the particular genus are well known for their traditional uses and pharmacological properties [25]. Research has been conducted on certain species of *Humboldtia*, revealing their rich phytochemical composition. These plants have been found to contain valuable phytochemicals, including phenols, flavonoids, saponins, tannins, terpenoids, cardiac glycosides, apigenin, steroids, phlobatannins, and more [26–28]. In the realm of traditional medicine, the bark of *Humboldtia* species held curative significance, being employed to address conditions such as biliousness, leprosy, ulcers, and epilepsy and acting as an anticonvulsant [29]. The remediation of biliousness, impurities in the blood, ulcers, and epilepsy all involved the preparation of a decoction from the bark powder [30].

Humboldtia brunonis Wall, *Humboldtia unijuga* Bedd., and *Humboldtia vahliana* Wight are well known for their pharmacological efficiency and their antioxidant, anti-inflammatory, anticancer, antimicrobial, and anti-depressant effects [28,29,31–33]. *H. brunonis* fulfilled roles as a styptic, demulcent, anthelmintic, ulcer remedy, stomachic, astringent, and treatment for menstrual and urinary issues [34]. Furthermore, the local populations residing in Karnataka's Shiradi and Bisle Ghats harnessed the leaves and bark of *H. brunonis* for arthritis and diabetes treatments, a practice detailed by Prasad and Kumar [26]. It was documented that the *H. brunonis* bark and leaves were utilized for addressing wounds, menstruation disorders, and overgrowth issues [35]. *Humboldtia unijuga*, referred to as 'palakan' by the Kani tribes in Agasthyamala, was employed to treat ailments such as headaches, chickenpox, and snake bites [36]. It has been discovered that the plant possesses Erythrodiol-3-acetate and 2,4-di-tert-butylphenol, which have been demonstrated to exhibit anti-inflammatory and anticancer properties [25].

The plant species *Humboldtia sanjappae*, belonging to the Fabaceae family, and its related species exhibit a diverse range of pharmacological properties [28,29,31–33]. However, the phytochemical constituents and pharmacological effects of this plant remain largely unexplored. Currently, there are no existing reports on the antioxidant and anti-inflammatory activities of this plant, and data regarding its phytochemical composition are also lacking. Consequently, this study represents a pioneering effort to investigate the phytochemical profile of the plant and evaluate its potential antioxidant, enzyme-inhibitory, anti-inflammatory, and antibacterial properties. To identify the bioactive phytochemicals, LC-MS analysis is employed as a key analytical technique.

2. Results and Discussion

2.1. *Quantitative and Qualitative Estimation of Phytochemicals in H. sanjappae*

The *H. sanjappae* extract was analyzed using LC-MS, which indicated the presence of flavonoids, including Naringenin, Luteolin, and Pomiferin, as well as phenols such as Epicatechin and Maritemin (Figure 1, Table 1). Flavonoids and phenolic substances not only have antioxidant properties, but they also work well as anti-inflammatory agents [37]. Various studies provide support for the immune-modulating effects of polyphenols and flavonoids. Seed polyphenols extracted from *Nigella sativa* L. were evaluated for their analgesic and anti-inflammatory properties. The study findings demonstrated that these polyphenols effectively reduced paw edema induced by carrageenan [38]. Concentrated extract derived from *Dendrobium loddigesii* Rolfe, rich in polyphenols, was administered to treat diabetic mice. The outcomes indicated that this extract exhibited the capability to lower blood glucose levels, reduce body weight, decrease levels of low-density lipoprotein cholesterol, and elevate insulin levels within the mice [39]. Flavonoids are part of the category of polyphenolic natural compounds, encompassing over 4000 identified variations. The advantageous biological effects of flavonoids are unquestionably intertwined with their structural composition and properties, rendering them prime contenders for pharmaceutical development. Numerous inflammatory molecules such as TNF-α, IL-1, IL-6, IL-17, and IFN-γ, released via the activation of various signaling pathways, primarily the NF-κB pathway, have been demonstrated to be inhibited upon administration of flavonoid [40]. Scientists detected that supplementation of Epicatechin potentially contributed to reducing inflammation and enhancing insulin sensitivity in visceral adipose tissue of high-fat fed mice [41]. The presence of phytochemicals such as Epicatechin in the plant may be responsible for its observed antidiabetic activity. However, additional experiments and studies are necessary to validate this hypothesis and confirm the specific compounds and mechanisms involved in the plant's potential benefits for diabetes management. Certainly, the antioxidant properties of the extract can play a crucial role in managing oxidative stress in individuals with diabetes. As diabetes can cause substantial cellular damage, including in the brain, combating oxidative stress with antioxidant compounds becomes important [42]. By reducing oxidative stress, the extract's antioxidant properties may help protect cells, mitigate damage, and contribute to improved overall health in diabetic patients.

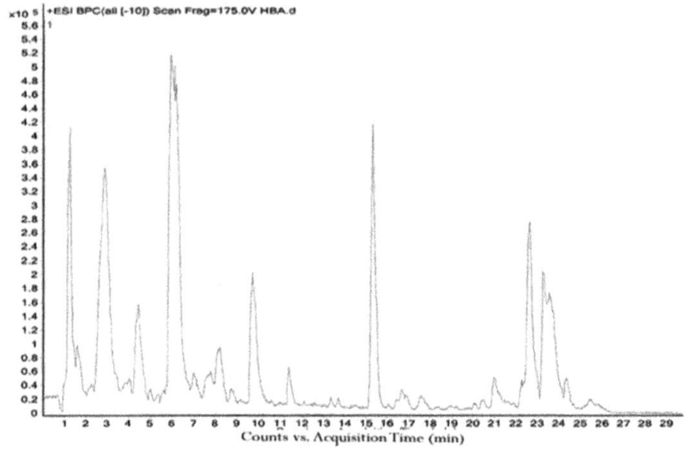

Figure 1. The LC−MS total ion chromatogram of *H. sanjappae* ethanol extract.

Table 1. LC−MS analysis and chemical composition of *H. sanjappae*.

Sl. No.	RT	*m/z* [a]	*m/z* [b]	Name of the Compound	Fragments	Mol. Wt.	Chemical Formula	Structure
1	3.575	577.14	577.14	Richotomine	483.13, 315.11, 197.80	532.14	$C_{30}H_{20}N_4O_6$	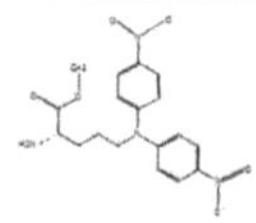
2	3.773	579.15	579.15	Procyanidin B7	579.14, 443.15, 383.12, 227.17	578.14	$C_{30}H_{26}O_{12}$	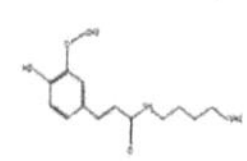
3	4.166	289.07	289.07	(−)−Epicatechin	289.07, 226.07	290.08	$C_{15}H_{14}O_6$	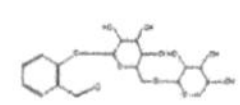
4	4.633	513.14	513.14	2′′,6′′−Di-O-acetylononin	513.14, 289.07	514.15	$C_{26}H_{26}O_{11}$	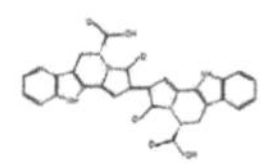
5	5.908	494.24	494.24	Ryanodine	494.23, 189.07	493.23	$C_{25}H_{35}NO_9$	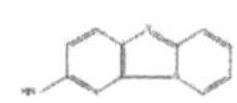
6	5.919	465.16	465.16	Pomiferin	421.16, 213.07	420.16	$C_{25}H_{24}O_6$	
7	6.312	549.22	549.22	Cymorcin diglucoside	431.10, 253.03	490.21	$C_{22}H_{34}O_{12}$	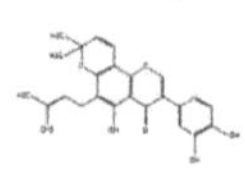
8	8.763	287.05	287.06	Maritimetin	287.05, 283.15, 267.15	286.05	$C_{15}H_{10}O_6$	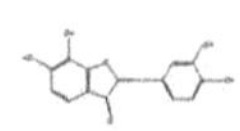
9	9.101	285.04	285.04	Luteolin	285.04, 215.09	286.05	$C_{15}H_{10}O_6$	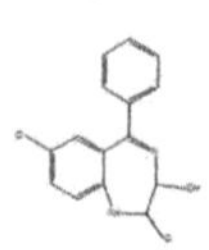

Table 1. *Cont.*

Sl. No.	RT	*m/z* [a]	*m/z* [b]	Name of the Compound	Fragments	Mol. Wt.	Chemical Formula	Structure
10	9.475	271.06	271.06	Naringenin	271.06, 259.12, 248.97	272.07	$C_{15}H_{12}O_5$	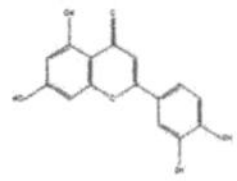
11	9.681	271.06	271.06	Coriandrone E	251.16, 179.10	248.07	$C_{13}H_{12}O_5$	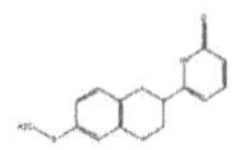
12	10.067	271.06	271.06	Morindone	271.05, 267.15, 253.14	270.05	$C_{15}H_{10}O_5$	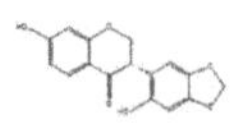
13	10.197	301.07	301.07	(+)-Sophorol	271.05, 301.06, 295.18, 277.17	300.06	$C_{16}H_{12}O_6$	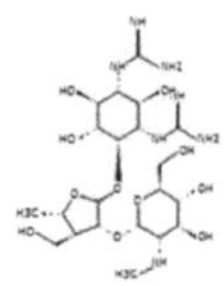
14	10.485	269.08	269.08	Formononetin	258.04, 179.10, 139.15	268.07	$C_{16}H_{12}O_4$	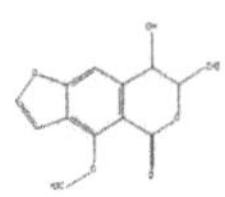
15	11.846	283.19	283.19	Lactapiperanol C	279.09, 265.17	282.18	$C1_6H_{26}O_4$	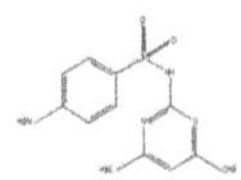
16	15.229	507.23	507.22	Limonoate	507.22, 351.25, 238.12	506.22	$C_{26}H_{34}O_{10}$	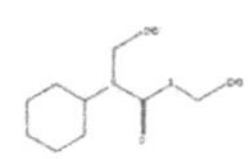
17	16.573	295.23	295.23	17−Hydroxylinolenic acid	295.22, 284.32, 277.21	294.22	$C_{18}H_{30}O_3$	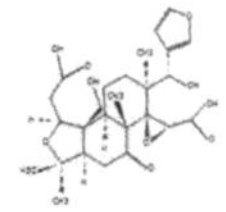
18	19.676	645.42	645.42	Capsanthin 5,6−epoxide	529.30, 403.26, 238.89	600.42	$C_{40}H_{56}O_4$	
19	21.644	593.27	593.27	Ganoderic acid F	415.35, 227.17, 570.41	570.28	$C_{32}H_{42}O_9$	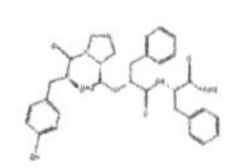
20	22.179	471.35	471.35	delta−Maslinic acid	471.85, 311.17, 248.97	472.36	$C_{30}H_{48}O_4$	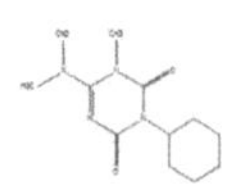
21	23.291	413.26	413.27	D8′−Merulinic acid A	391.28, 279.15, 149.02	390.28	$C_{24}H_{38}O_4$	

[a]: calculated m/Z ratio, [b]: Reference m/Z ratio.

Several investigators have noted natural substances' anti-inflammatory properties, including multiple preclinical experiments [43–46]. As a result, we infer that the antioxidant and anti-inflammatory properties exhibited by the ethanol extract of *H. sanjappae* must be due to the presence of various phytochemical components such as polyphenols, flavonoids, isocoumarins, etc., and also that these many different phytochemicals in plants offer a valuable source of antimicrobial compounds with immense therapeutic potential [47].

Antibiotic-resistant bacteria have emerged as a significant global health concern [48,49]. Plants have long been renowned for their antibacterial powers as nature's medicine [50]. Polyphenols and essential oils, among other bioactive substances, have powerful antibacterial properties [30,51–53]. Adopting plant phenomena might open the way for innovative and long-lasting antibacterial treatments because traditional antibiotics are becoming less effective, leading to a pressing need for the development of new and effective antimicrobial agents. In this context, the rich diversity of plant species provides a vast array of bioactive compounds that can be explored for their antimicrobial properties. Polyphenols and flavonoids possess well-documented antimicrobial properties [54,55], exhibiting inhibitory effects against a broad spectrum of bacteria [55,56].

Our research has supplied substantiating proof of the elevated levels of these compounds in the ethanol extract of *H. sanjappae*. Specifically, the total phenolic content measured at 378.77 ± 6.62 µg equivalent per milligram and the total flavonoid content recorded at 204.76 ± 6.10 µg equivalent per milligram both emphasize the concentration within the extract (Table 2). Given its rich content of polyphenols and flavonoids, the extract from *H. sanjappae* shows promise as a potential source for the development of novel antibiotics. Considering its antimicrobial potential, the polyphenol- and flavonoid-rich extract of *H. sanjappae* warrants further exploration in the search for new antibiotic compounds.

Table 2. The total polyphenol and flavonoid contents of *H. sanjappae* ethanol extract.

Assay	µg Equivalent/mg of Extract
Total phenolic content	378.77 ± 6.62
Total flavonoid content	204.76 ± 6.10

2.2. *In Vitro Antioxidant Activities of* H. sanjappae *Extract*

Species of the Fabaceae family consist of phytochemicals responsible for the plant's antioxidant potential [57]. The different genera of the family are established as having antioxidant potential [58]. In vivo studies of *Tamarindus*, a related genus of *Humbodtia*, showed that it exhibits potent antioxidant activity [59], and the antioxidant potential of species within the genus *Humboldtia* has been previously explored, and their effectiveness has been reported [27,31]. The IC_{50} value of *H. sanjappae* bark extract in the anti-DPPH radical scavenging assay was shown to be 6.53 ± 1.49 µg/mL. Likewise, Table 3 shows other antioxidant activity in Ferric Reducing Antioxidant Power, represented by its value of 8.46 ± 1.38 µg/mL. The antioxidant properties of the plant must be assigned to the different phytochemicals present in the extract; those bioactive compounds identified from LC-MS analysis are listed in the Table 1. For example, previous studies demonstrated that anticancer properties of Epicatechin are linked to its antioxidative potential [60]. Another component present in the extract is Luteolin, which is found in glycosylated forms in a variety of vegetables and fruits and is classified within the flavone subclass of flavonoids. Its documented effects include in vivo anti-inflammatory [61], antioxidative [62–64], antidiabetic [61], antimicrobial [65], and anticancer [66,67] activities. The antioxidant properties of Naringenin [68,69], Morindone [70,71], Capsanthin 5,6-epoxide [72,73], and Ganoderic acid F [74] have been previously established. Therefore, these compounds could potentially account for the robust antioxidant activity observed in the extract. Oxidative stress plays a critical role as an independent factor in the development of numerous chronic diseases, including cancer, diabetes, and cardiovascular diseases [75–78]. Therefore, the antioxidant properties found in the plant extract can be beneficial in the management of diseases that are linked to oxidative stress.

Table 3. In vitro antioxidant and antidiabetic activities of *H. sanjappae* expressed as IC_{50} values (μg/mL).

Activity	IC_{50} Value(μg/mL)		
	HSE	Ascorbic Acid	Acarbose
DPPH scavenging	6.53 ± 1.49	2.11 ± 0.25	>200
FRAP value	8.46 ± 1.38	4.15 ± 0.47	>200
α-amylase	47.60 ± 0.19	33.92 ± 2.54	122.18 ± 3.08
α-glucosidase	32.09 ± 0.54	29.85 ± 2.01	103.45 ± 2.68

2.3. *Enzyme Inhibitory Properties of H. sanjappae Ethanol Extract*

The extract was examined for its enzyme-inhibitory properties against key enzymes associated with type 2 diabetes mellitus, namely α-amylase and α-glucosidase. The IC_{50} value for the inhibition of α-amylase and α-glucosidase by the extract was determined to be 47.60 ± 0.19 μg/mL and for the inhibition of β-glucosidase, 32.09 ± 0.54 μg/mL (Table 3). The standard antidiabetic drug acarbox exibited an IC_{50} value of 122.18 ± 3.08 and 103.45 ± 2.68 μg/mL against α-amylase and α-glucosidase enzymes, respectively. Hence, the plant extract contains stronger antidiabetic compounds compared to the acarbose. The α-amylase and α-glucosidase are enzymes that play crucial roles in carbohydrate metabolism and are frequently targeted by antidiabetic medications [79]. Indeed, the inhibition of α-amylase and α-glucosidase by the *H. sanjappae* extract may contribute to its potent antidiabetic activity. By inhibiting these enzymes involved in carbohydrate metabolism, the extract can potentially help regulate blood glucose levels and manage diabetes effectively. The enzyme-inhibitory properties are well corroborated by the major bioactive substances observed in the plant using LC-MS analysis. Epicatechin, Luteolin, and Naringenin were reported to inhibit the α-amylase and α-glucosidase in in vitro and animal model studies [80,81]. In addition, the reports clearly indicated the antidiabetic properties of these bioactive compounds in independent studies.

2.4. *Anti-Inflammatory Activity of H. sanjappae*

The anti-inflammatory activity of the extract was evaluated using Raw 264.7 macrophages as the model. Raw 264.7 cells stimulated with lipopolysacchride (LPS) are a widely utilized cellular model of inflammation [82]. The lipopolysaccharide is the cell wall component of most of the Gram-negative bacteria; the LPS stimulates the macrophage in a toll-like-receptor-dependent manner [83]. In the present study, the normal macrophage was estimated for the level of IL-1β, and it was estimated to be 64.6 ± 1.9 pg/mg protein. However, there was observed a significant elevation in the IL-1β levels upon stimulation with the lipopolysaccharide to 573.4 ± 4.5 pg/mg protein. The increased level of IL-1β is an indicator of inflammation in the cellular conditions [84–86]. However, the pre-treatment of macrophages with the different doses of HSE resulted in a significant reduction in IL-1β levels (Table 4). The pre-treatment of Raw 264.7 cells with 5 μg/mL of HSE resulted in cellular IL-1β levels of 403.7 ± 6.2 pg/mg protein ($p < 0.01$). Similarly, the pre-treatment with 10 μg/mL (298.5 ± 8.4 pg/mg protein) and 20 μg/mL of HSE (156.2 ± 3.4 pg/mg protein) resulted in lower IL-1β levels ($p < 0.001$). The reduction in IL-1β levels is indicative of the anti-inflammatory potential of the HSE at the respective treatment doses.

Together with IL-1β, IL-6 was also found to significantly influence the inflammation in macrophages [87,88]. IL-6 has a major role in the innate immune defense systems [89]; however, the same molecule is associated with the progression of various diseases [90,91]. The level of IL-6 in the untreated macrophages without LPS stimulation was estimated to be 133.4 ± 5.8 pg/mg protein; however, the exposure of LPS elevated the cellular IL-6 levels to 628.5 ± 8.2 pg/mg protein. On the contrary, the level was brought down by the treatment with 5 and 10 μg/mL of *H. sanjappae* extract, which reduced the cellular IL-6 levels to 507.1 ± 8.1 ($p < 0.05$) and 388.4 ± 2.8 pg/mg protein ($p < 0.001$). In the highest concentration of *H. sanjappae* extract treatment, the IL-6 was estimated to be 291.3 ± 6.6 pg/mg protein ($p < 0.001$).

Table 4. Anti-inflammatory activity of *H. sanjappae* extract against lipopolysaccharide-induced activation of Raw 264.7 cells and comparison with standard aspirin (1 mM).

	IL-1β (pg/mg Protein)	IL-6 (pg/mg Protein)	TNF-α (pg/mg Protein)	NO (μM/mg Protein)
Untreated	64.6 ± 1.9	133.4 ± 5.8	115.2 ± 3.1	10.7 ± 0.64
LPS Control	573.4 ± 4.5	628.5 ± 8.2	856.0 ± 11.2	75.2 ± 2.1
Aspirin (1 mM)	147.5 ± 5.1 ***	209.5 ± 9.1 ***	247.5 ± 6.3 ***	22.7 ± 1.7 ***
HSE 5 μg/mL	403.7 ± 6.2 **	507.1 ± 8.1 *	715.2 ± 8.8 *	58.8 ± 3.4 *
HSE 10 μg/mL	298.5 ± 8.4 ***	388.4 ± 2.8 ***	602.8 ± 5.2 ***	42.3 ± 1.9 ***
HSE 20 μg/mL	156.2 ± 3.4 ***	291.3 ± 6.6 ***	493.7 ± 6.4 ***	30.7 ± 2.5 ***

* indicates significant variation with respect to LPS control ($p < 0.05$); ** indicates higher significant variation with respect to that of LPS control ($p < 0.01$), and *** indicates highest significant variation with respect to that of LPS control ($p < 0.001$). All the results are indicated as mean ± standard deviation of six independent experiments.

The TNF-α levels are crucial for the survival and proliferation of various cancer cells [92]. The cytokine is also important in the progression events of cancers including metastasis and stemness [93]. The level of TNF-α in the untreated and unstimulated Raw 264.7 macrophages was estimated to be 115.2 ± 3.1 pg/mg protein. However, the level was elevated to 856.0 ± 11.2 pg/mg protein upon stimulation by the LPS. This clearly indicated the induction of acute inflammation in the experimental condition. In 5 μg/mL *H. sanjappae* treated macrophages, the level of TNF-α was reduced to 715.2 ± 8.8 pg/mg protein. A similar decrease in the TNF-α level was also noted in the 10 and 20 μg/mL *H. sanjappae* treatment, which brought down the TNF-α level to 602.8 ± 5.2 and 493.7 ± 6.4 pg/mg protein.

The nitric oxide level is also an important inflammatory indicator in cells; the inducible nitric oxide synthase is an enzyme responsible for the overwhelming load of nitric oxide in the body [94]. Despite its physiological and immunological importance, nitric oxide is often associated with chronic inflammation and is thereby involved in many of the degenerative diseases [95]. The level of nitric oxide in the untreated macrophage cells was estimated to be 10.7 ± 0.64 μM/mg protein. The level was increased to 75.2 ± 2.1 μM/mg protein in the macrophages exposed to LPS. Interestingly, the level was brought down by the pre-treatment with the 5 μg/mL of HSE (58.8 ± 3.4 μM/mg protein). In the 10 μg/mL of *H. sanjappae* extract treatment, the NO level was estimated to be 42.3 ± 1.9 μM/mg protein, and, in the 20 μg/mL HSE treatment, it was further reduced to 30.7 ± 2.5 μM/mg protein. Hence, it is clearly indicated that the pre-treatment with different doses of HSE dose-dependently reduced the inflammatory insults in cultured macrophages.

Pathogen-associated molecular pattern molecules (PAMPs) or damage-associated molecular pattern molecules (DAMPs) are the two types of molecules that trigger the production and release of IL-1β. LPS is the main outer surface membrane component and is a highly potent PAMP which stimulates innate or natural immunity in various eukaryotic cells [96]. LPS-induced inflammatory responses are linked to the production of ROS in cells [97]. *H. sanjappae* extract was found to possess anti-inflammatory potential in a dose-dependent manner (Table 4). In vitro analysis revealed that it inhibits nitric oxide (NO) radicals. The inflammatory insults caused by LPS are prevented in cells pre-treated with *H. sanjappae* extract, and cytokine level is also reduced as a result. Interleukin-1β (IL- 1β) is a potent proinflammatory cytokine vital in the host cell defense reaction to infection [98]. After LPS stimulation, macrophages were shown to have a considerably higher level of IL-1β; however, pre-treatment with *H. sanjappae* extract at various dosages dramatically reduced the IL-1β levels in the macrophages. Like IL-β, interleukin-6 (IL-6) and tumor necrosis factor-α (TNF-α) are significant mediators of innate immunity [99]. LPS also elevated the level of these cytokines. Application of the *H. sanjappae* extract reduced the elevated levels of TNF-α and IL-6.

The bioactive flavonoids present in the *H. sanjappae* extract such as Epicatechin, Luteolin, and Naringenin might play an important role in the anti-inflammatory potential of the plant. Independent studies have reported the anti-inflammatory potential of these bioactive flavonoid molecules in cultured macrophages and animals [37].

To further explain the mechanism of anti-inflammatory activity, the expression of genes NF-KB and COX2 was assessed. Compared to the untreated LPS control, the *H. sanjappae* extract treatment significantly brought down the expression of NF-KB and COX2. The expression of NF-KB is a crucial event in inflammation, and it is associated with various diseases including cancers. Likewise, the COX2 is a well-known inflammatory enzyme associated with the production of prostaglandins. The expression of COX2 is also evident in different forms of cancers (Figure 2).

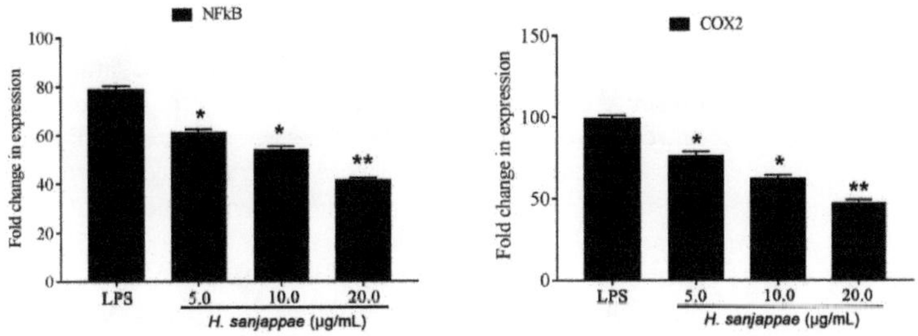

Figure 2. The expression of NF-KB and COX2 genes in the Raw 264.7 macrophages treated with LPS and different doses of *H. sanjappae* leaf methanol extract. * indicates significant variation with respect to LPS control ($p < 0.05$); ** indicates higher significant variation with respect to that of LPS control ($p < 0.01$).

2.5. Antibacterial Activity of *H. sanjappae*

The antibacterial activity of *H. sanjappae* is being documented for the first time, while antimicrobial properties of *H. brunonis* have previously been reported [100]. Previous studies have investigated the antimicrobial potential of different extracts from *H. brunonis*, where the methanolic extract of the leaves [101] and aqueous extract of stem and leaf [33] exhibited significant antibacterial activity. It is worth noting that many species within the genus have not been extensively studied in this regard. Therefore, there remains a considerable knowledge gap regarding the antimicrobial potential of the majority of species within the genus. The present study observed a significant antibacterial activity of the plant against different pathogenic microbes (Table 5). The highest activity was observed against *Pseudomonas aeruginosa* (24.1 ± 0.3 mm) followed by *Salmonella enterica* (22.1 ± 0.1 mm). The lowest activity was observed against *E. coli* (18.5 ± 0.2 mm). The standard antibiotic gentamicin (20 µg) had growth inhibition zones of 21.7 ± 0.5, 22.1 ± 0.1, 19.7 ± 0.2, and 20.5 ± 0.2 mm against *E. coli, P. aeuginosa, S. aureus,* and *S. enterica*, respectively. Likewise, the minimum inhibitory concentration was found to be highly effective against *P. aeruginosa* (0.625 ± 0.02 mg/mL) followed by *Salmonella enterica* (1.00 ± 0.01 mg/mL). The lowest activity was observed against *E. coli* (1.50 ± 0.01 mg/mL). The MIC values of gentamicin were found to range between 0.325 and 0.625 mg/mL (Table 5).

Table 5. Efficacy of *H. sanjappae* (HSE) as an antimicrobial agent estimated using disc diffusion method and minimum inhibitory concentrations and comparison with gentamicin (GM).

Bacteria	Zone of Inhibition (mm)		MIC Concentration (mg/mL)	
	HSE	GM (20 µg)	HSE	GM
Escherichia coli	18.5 ± 0.2	21.7 ± 0.5	1.50 ± 0.01	0.325
Pseudomonas aeruginosa	24.1 ± 0.3	22.1 ± 0.1	0.625 ± 0.02	0.325
Staphylococcus aureus	20.6 ± 0.3	19.7 ± 0.2	1.25 ± 0.04	0.625
Salmonella enterica	22.1 ± 0.1	20.5 ± 0.2	1.00 ± 0.01	0.625

All the results are indicated as mean ± standard deviation of six independent experiments.

3. Materials and Methods

3.1. *Humboldtia sanjappae* Collection and Extraction

The *Humboldtia sanjappae* plant samples were collected on 12 December 2022 from Ernakulam District, Western Ghats of Kerala (10.04829° N, 76.8399° E). The mature leaves and the bark of the plants collected were carefully cleaned of all kinds of dust via washing. These materials were dried under shade for 21 days and powdered using a mixer grinder; the powder was extracted with 100% ethanol. About 5 g of each material was subjected to 6 h of extraction. All the extracts were evaporated to dryness using a rotary evaporator, and extract yield was calculated for the same.

3.2. *Phytochemical Analysis of Humboldtia sanjappae*

A preliminary phytochemical analysis of all samples was performed to determine the presence of biologically important secondary metabolites such as alkaloids, flavonoids, terpenoids, steroids, carbohydrates, saponins, etc. All tests were conducted following the conventional procedures described by Yadav, Agarwala, and Harborne [102,103]. The total phenolic content (TPC) and the total flavonoid content (TFC) were determined spectrophotometrically. TPC was found using the Folin–Ciocalteu reagent assay [104]. A standard curve was constructed using Gallic Acid standards. TPC was measured in Gallic Acid Equivalents (GAE). TFC was determined using an aluminium chloride colorimetric assay [105]. The flavonoid content was estimated using the standard quercetin calibration curve. TFC was measured in terms of quercetin equivalents.

The LC-MS analysis was carried out according to the previous methods of House et al. [106]. Briefly, the HR-LCMS-Q-TOF analysis was carried out using Agilent 1290 UHPLC system (Agilent Technologies, Santa Clara, CA, USA). Accurately, 10 µL of the extract was injected into the system, and the run was carried out using water (0.1% formic acid *v*/*v*) (A) and methanol (B) as solvents. The gradient elution mode was used as follows: 1–10 min 95% A, 10–20 min 75% A, 20–25 min 50% A, 25–30 min 30% A, 30–40 min 95% B. The flow rate was set at 0.3 mL/min, and pressure was 1200 bar.

3.3. *Analysis of the Antioxidant Activity of H. sanjappae Ethanol Extract*

The antioxidant activities were assessed by evaluating the scavenging potentials of various radicals, such as diphenyl picryl hydrazyl (DPPH) and FRAP (Ferric Reducing Antioxidant Power). These methods allow for the measurement of the ability of the tested samples to neutralize or reduce these radicals, providing insights into their antioxidant properties [96,97]. A solution of DPPH was prepared by dissolving it in methanol at a concentration of 0.1 mM. Varying concentrations of the extract were mixed with the DPPH solution. The resulting mixture was then incubated in a dark environment at a temperature of 30 degrees Celsius for 20 min. The change in absorbance of the solution was measured and used to estimate the percentage inhibition [96]. The stock solutions consisted of a 300 mM acetate buffer (prepared by dissolving 3.1 g of sodium acetate and 16 mL of acetic acid) with a pH of 3.6, a 10 mM TPTZ (2, 4, 6-tripyridyl-s-triazine) solution (3.12 mg/mL) in 40 mM HCl, and a 20 mM FeCl3 solution (3.25 mg/mL). To prepare the fresh working solution, 25 mL of acetate buffer, 2.5 mL of TPTZ solution, and 2.5 mL of FeCl3 solution were mixed together. The resulting solution was warmed at 37 °C before use. For the FRAP assay, a test sample of varying concentrations was mixed with 2.80 mL of the prepared FRAP solution. The mixture was allowed to react for 30 min under dark conditions. After the incubation period, the absorbance of the colored product, known as the ferrous tripyridyltriazine complex, was measured at a wavelength of 593 nm [98]. The control in the experiment refers to the reaction mixture in which the test sample was not added.

3.4. Analysis of the H. sanjappae Ethanol Extract on Activities of Enzymes

To assess the enzyme-inhibitory properties of the test samples, specific enzymes related to diabetes and secondary diabetic complications were targeted. The inhibitory effects on α-amylase [45] and α-glucosidase [46] were examined using standard procedures.

3.5. Effect of H. sanjappae Ethanol Extract on Lipopolysaccharide-Induced Anti-Inflammatory Activity in Macrophages

The murine Raw 264.7 cells were seeded at 1×10^7 cells/mL in a 24-well plate containing complete growth media. The *H. sanjappae* extract was diluted in RPMI-1640 media at different concentrations (5, 10, and 20 μg/mL). A standard anti-inflammatory compound, aspirin, was also used as positive control at a concentration of 1 mM. The cells were then treated with lipopolysaccharide (LPS) at a concentration of 1 μg/mL for 24 h. The levels of inflammatory cytokines such as interleukin-1β, interleukin-6, and tumor necrosis factor-α were measured using PeproTech ELISA kits. The production of nitric oxide in the media by the macrophages was quantified biochemically using the Griess method [106]. The gene expression of NF-KB and Cyclooxygenase 2 was determined by real-time PCR according to the ΔΔCT method. Briefly, the total RNA was isolated, and cDNA was synthesized using standard kits from Takara (Banglore, India). The real-time PCR analysis was carried out using the temperature cycle of 95 °C (melting temperature) for 15 s, 60 °C (annealing temperature) for 45 s, and 73 °C (extension temperature) for 30 s. The cycle was repeated 40 times, and CT values were calculated using the Applied Biosystems 7300 software. The primer sequences used in the study are listed in Supplementary Table S5.

3.6. Antibacterial Activity of H. sanjappae Ethanol Extract

The antibacterial activity of *H. sanjappae* was estimated in terms of the disc diffusion method according to the methods of Webber et al. [107]. The extracts were placed in circular discs and kept in the bacterial culture plate at 80 mm distance to one another. The growth inhibition zone in each of the bacterial cultures was determined and expressed as zone of inhibition in mm. The MIC value was determined according to the previous methods of Morgan et al. [108]. The gentamicin was used as a standard antibacterial agent at a concentration of 20 μg.

3.7. Statistical Analysis

The data were represented as mean of three independent experiments with triplicate analysis. The statistical operations were carried out using GraphPad Prism 7.0.

4. Conclusions

Most of the species in the genus *Humboldtia* have not been evaluated for their pharmacological potential despite their relevance in ethnomedicine. The present study for the first time reports the phytochemical composition and antioxidant, antimicrobial, and anti-inflammatory activities of *H. sanjappae*, a native of the Western Ghats of India. The study concludes that the plant has strong antioxidant properties in terms of radical scavenging and reducing potentials, and it is also effective as an antibacterial agent. Further, the extract inhibited the cytokine levels in Raw 264.7 macrophages, which is indicative of its anti-inflammatory properties. Enzymes such as α-amylase and α-glucosidase are important in controlling how our bodies absorb carbs and are frequently targeted by diabetic drugs [79]. Indeed, the potent antidiabetic actions of HSE may be connected to its capacity to inhibit α-amylase and α-glucosidase. The extract may help regulate blood sugar levels and successfully manage diabetes by inhibiting certain carbohydrate-processing enzymes.

By carefully studying and testing, we have clearly shown that bark extract of *H. sanjappae* made with ethanol is really good at reducing inflammation, controlling diabetes, fighting bacteria, and acting as an antioxidant. These different benefits not only highlight how valuable *H. sanjappae* is, but also remind us that using plants for medicine has always been a great way to create a variety of medicines. There are many examples from history where medicinal plants have led to big changes in medicine. For example, aspirin, which comes from the bark of the willow tree (*Salix alba* L.), changed how we manage pain and helped develop other drugs such as NSAIDs (non-steroidal anti-inflammatory drugs) [109]. Additionally, the Madagascar periwinkle plant (*Catharanthus roseus* L.) gave us vinblastine and vincristine, powerful compounds that have really changed how we treat cancer [110].

Significantly, more than half of the drugs utilized worldwide in modern pharmaceuticals have their origins in natural sources [111,112]. The worldwide commercial success of established and effective pharmaceuticals taken from many plant kinds demonstrates the importance of medicinal plants as potential drug reservoirs. Quinine, an anti-malarial alkaloid derived from the bark of *Cinchona officinalis* L., is one example. Furthermore, chloroquine, derived from quinine, not only modulates inflammatory autoimmune responses but has recently shown promise in anticancer therapy [112,113].

The polypharmacological potential of *H. sanjappae*, as evidenced by its diverse array of beneficial properties, aligns perfectly with this lineage of discovery. Its capacity to simultaneously tackle a range of health factors—spanning from inflammation and diabetes to bacterial infections and oxidative stress—resonates with the holistic approach of medicinal plants. These qualities offer the potential for more complete and refined therapeutic treatments, acknowledging the complexities of human health.

Supplementary Materials: The following supporting information can be downloaded at: https://www.mdpi.com/article/10.3390/molecules28196875/s1. Table S1: Qualitative analysis of phytochemicals present in different extracts of *Humboldtia sanjappae;* Table S2: Percentage yield of different extracts of *H. sanjappae;* Table S3: Antioxidant activity of different extracts of *H. sanjappae;* Table S4: Total phenol and total flavonoid contents of different extracts of *H. sanjappae;* Table S5: The forward and reverse primer sequences of different genes used for real-time PCR analysis.

Author Contributions: J.S.: analysis, manuscript preparation, experimentation. S.G.: study design, methodology, experimentation, analysis, funding acquisition, manuscript editing. R.R.: analysis, manuscript preparation, experimentation. A.A.: study design, methodology, experimentation, analysis, funding acquisition, manuscript editing. O.J.O.: study design, methodology, experimentation, analysis, funding acquisition, manuscript editing. A.N.: study design, methodology, experimentation, analysis, funding acquisition, manuscript editing. All authors have read and agreed to the published version of the manuscript.

Funding: The authors acknowledge the funding support from Researchers Supporting Project Number (RSP2023R11), King Saud University, Riyadh, Saudi Arabia. Authors acknowledge the financial support from Mohammed VI Polytechnic University, Morocco.

Institutional Review Board Statement: Not applicable.

Informed Consent Statement: Not applicable.

Data Availability Statement: The data may be shared upon valid request.

Acknowledgments: The authors acknowledge the funding support from Researchers Supporting Project Number (RSP2023R11), King Saud University, Riyadh, Saudi Arabia. The DBT-STAR (project number: BT/HRD/11/09/2020) scheme supported infrastructural development in St. Joseph's College (Autonomous), Devagiri, Calicut.

Conflicts of Interest: The authors declare no conflict of interest.

Sample Availability: Samples of the compounds are available from the authors.

References

1. Kumar, V.; Bishayee, K.; Park, S.; Lee, U.; Kim, J. Oxidative stress in cerebrovascular disease and associated diseases. *Front. Endocrinol.* **2023**, *14*, 1124419. [CrossRef]
2. Jia, D.; Nagaoka, Y.; Katsumata, M.; Orsulic, S. Inflammation is a key contributor to ovarian cancer cell seeding. *Sci. Rep.* **2018**, *8*, 12394. [CrossRef]
3. Hausmann, S.; Kong, B.; Michalski, C.; Erkan, M.; Friess, H. The role of inflammation in pancreatic cancer. *Inflamm. Cancer* **2014**, *816*, 129–151.

4. Mastinu, A.; Bonini, S.A.; Premoli, M.; Maccarinelli, G.; Mac Sweeney, E.; Zhang, L.; Lucini, L.; Memo, M. Protective Effects of *Gynostemma pentaphyllum* (var. Ginpent) against Lipopolysaccharide-Induced Inflammation and Motor Alteration in Mice. *Molecules* **2021**, *26*, 570. [CrossRef] [PubMed]
5. Rahman, S.; Atikullah, M.; Islam, M.N.; Mohaimenul, M.; Ahammad, F.; Islam, M.S.; Saha, B.; Rahman, H. Anti-inflammatory, antinociceptive and antidiarrhoeal activities of methanol and ethyl acetate extract of *Hemigraphis alternata* leaves in mice. *Clin. Phytoscience* **2019**, *5*, 16. [CrossRef]
6. Golia, E.; Limongelli, G.; Natale, F.; Fimiani, F.; Maddaloni, V.; Pariggiano, I.; Bianchi, R.; Crisci, M.; D'Acierno, L.; Giordano, R. Inflammation and cardiovascular disease: From pathogenesis to therapeutic target. *Curr. Atheroscler. Rep.* **2014**, *16*, 435. [CrossRef]
7. Iqbal, J.; Abbasi, B.A.; Mahmood, T.; Kanwal, S.; Ali, B.; Shah, S.A.; Khalil, A.T. Plant-derived anticancer agents: A green anticancer approach. *Asian Pac. J. Trop. Biomed.* **2017**, *7*, 1129–1150. [CrossRef]
8. Kasote, D.M.; Katyare, S.S.; Hegde, M.V.; Bae, H. Significance of antioxidant potential of plants and its relevance to therapeutic applications. *Int. J. Biol. Sci.* **2015**, *11*, 982. [CrossRef]
9. Oguntibeju, O.O. Medicinal plants with anti-inflammatory activities from selected countries and regions of Africa. *J. Inflamm. Res.* **2018**, *11*, 307. [CrossRef] [PubMed]
10. Coulibaly, A.Y.; Hashim, R.; Sulaiman, S.F.; Sulaiman, O.; Ang, L.Z.P.; Ooi, K.L. Bioprospecting medicinal plants for antioxidant components. *Asian Pac. J. Trop. Med.* **2014**, *7*, S553–S559. [CrossRef] [PubMed]
11. Okach, D.; Nyunja, A.; Opande, G. Phytochemical screening of some wild plants from Lamiaceae and their role in traditional medicine in Uriri District-Kenya. *Int. J. Herb. Med.* **2013**, *1*, 135–143.
12. Liu, C.H.; Abrams, N.D.; Carrick, D.M.; Chander, P.; Dwyer, J.; Hamlet, M.R.; Macchiarini, F.; PrabhuDas, M.; Shen, G.L.; Tandon, P. Biomarkers of chronic inflammation in disease development and prevention: Challenges and opportunities. *Nat. Immunol.* **2017**, *18*, 1175–1180. [CrossRef]
13. Nunes, C.d.R.; Barreto Arantes, M.; Menezes de Faria Pereira, S.; Leandro da Cruz, L.; de Souza Passos, M.; Pereira de Moraes, L.; Vieira, I.J.C.; Barros de Oliveira, D. Plants as sources of anti-inflammatory agents. *Molecules* **2020**, *25*, 3726. [CrossRef] [PubMed]
14. Zammel, N.; Saeed, M.; Bouali, N.; Elkahoui, S.; Alam, J.M.; Rebai, T.; Kausar, M.A.; Adnan, M.; Siddiqui, A.J.; Badraoui, R. Antioxidant and anti-Inflammatory effects of *Zingiber officinale* roscoe and *Allium subhirsutum*: In silico, biochemical and histological Study. *Foods* **2021**, *10*, 1383. [CrossRef] [PubMed]
15. Apaza Ticona, L.; Pérez-Uz, B.; García Esteban, M.T.; Aguilar Rico, F.; Slowing, K. Anti-melanogenic and Anti-inflammatory Activities of *Hibiscus sabdariffa*. *Rev. Bras. De Farmacogn.* **2022**, *32*, 127–132. [CrossRef]
16. Ďuračková, Z. Some current insights into oxidative stress. *Physiol. Res.* **2010**, *59*, 459–469. [CrossRef]
17. Rodriguez-Garcia, I.; Silva-Espinoza, B.A.; Ortega-Ramirez, L.A.; Leyva, J.M.; Siddiqui, M.W.; Cruz-Valenzuela, M.R.; Gonzalez-Aguilar, G.A.; Ayala-Zavala, J.F. Oregano Essential Oil as an Antimicrobial and Antioxidant Additive in Food Products. *Crit. Rev. Food Sci. Nutr.* **2016**, *56*, 1717–1727. [CrossRef]
18. Mhatre, S.; Srivastava, T.; Naik, S.; Patravale, V. Antiviral activity of green tea and black tea polyphenols in prophylaxis and treatment of COVID-19: A review. *Phytomedicine* **2021**, *85*, 153286. [CrossRef]
19. Alagumanivasagam, G.; Veeramani, P. A review on medicinal plants with hypolipidemic activity. *Int. J. Pharm. Anal. Res.* **2015**, *4*, 129–134.
20. Fu, W.; Zhuang, W.; Zhou, S.; Wang, X. Plant-derived neuroprotective agents in Parkinson's disease. *Am. J. Transl. Res.* **2015**, *7*, 1189.
21. Shamsudin, N.F.; Ahmed, Q.U.; Mahmood, S.; Shah, S.A.A.; Sarian, M.N.; Khattak, M.M.A.K.; Khatib, A.; Sabere, A.S.M.; Yusoff, Y.M.; Latip, J. Flavonoids as Antidiabetic and Anti-Inflammatory Agents: A Review on Structural Activity Relationship-Based Studies and Meta-Analysis. *Int. J. Mol. Sci.* **2022**, *23*, 12605. [CrossRef]
22. Gangaram, S.; Naidoo, Y.; Dewir, Y.H.; El-Hendawy, S. Phytochemicals and Biological Activities of *Barleria* (Acanthaceae). *Plants* **2021**, *11*, 82. [CrossRef]
23. Olivia, N.U.; Goodness, U.C.; Obinna, O.M. Phytochemical profiling and GC-MS analysis of aqueous methanol fraction of *Hibiscus asper* leaves. *Future J. Pharm. Sci.* **2021**, *7*, 59. [CrossRef]
24. Njamen, D.; Djiogue, S.; Zingue, S.; Mvondo, M.A.; Nkeh-Chungag, B.N. In vivo and in vitro estrogenic activity of extracts from *Erythrina poeppigiana* (Fabaceae). *J. Complement. Integr. Med.* **2013**, *10*, 63–73. [CrossRef] [PubMed]
25. Nair, R.V.; Jayasree, D.V.; Biju, P.G.; Baby, S. Anti-inflammatory and anticancer activities of erythrodiol-3-acetate and 2, 4-di-tert-butylphenol isolated from *Humboldtia unijuga*. *Nat. Prod. Res.* **2018**, *34*, 2319–2322. [CrossRef]
26. Kumar, J.K.; Prasad, A.D.; Chaturvedi, V. Phytochemical screening of five medicinal legumes and their evaluation for in vitro anti-tubercular activity. *Ayu* **2014**, *35*, 98.
27. Pavithra, G.; Naik, A.S.; Siddiqua, S.; Vinayaka, K.; TR, P.K.; Mukunda, S. Antioxidant and antibacterial activity of flowers of *Calycopteris floribunda* (Roxb.) Poiret, *Humboldtia brunonis* Wall and *Kydia calycina* Roxb. *Int. J. Drug Dev. Res.* **2013**, *5*, 301–310.
28. Sindhu, S.; Manorama, S.; Sumathi, P.; Adira, S. Antimicrobial studies on the endemic medicinal plant *Humboldtia brunonis* wall. (Caesalpiniaceae). *Int. J. Pharmaceut. Sci. Health Care* **2014**, 4.
29. Asirvatham, R.; Yesudanam, S. Neuropharmacological study of *Humboldtia vahliana* Wight. *Sch. Acad. J. Pharm.* **2018**, *7*, 171–183.
30. Sanjappa, M. A revision of the genus Humboldtia Vahl (Leguminosae-Caesalpinioideae). *Blumea Biodivers. Evol. Biogeogr. Plants* **1986**, *31*, 329–339.

31. Asirvatham, R.; Yesudanam, S. Evaluation of antioxidant potential of Humboldtia Vahliana Wight in Neuropharmacological screening on mice. *J. Int. Res. Med. Pharm. Sci.* **2017**, *11*, 41–53.
32. John, B.; Sulaiman, C.; George, S.; Reddy, V. Total phenolics and flavonoids in selected medicinal plants from Kerala. *Int. J. Pharm. Pharm. Sci.* **2014**, *6*, 406–408.
33. Sheik, S.; Chandrashekar, K. Antimicrobial and antioxidant activities of *Kingiodendron pinnatum* (DC.) Harms and *Humboldtia brunonis* Wallich: Endemic plants of the Western Ghats of India. *J. Natl. Sci. Found. Sri Lanka* **2014**, *42*, 307. [CrossRef]
34. Nisbet, L.J.; Moore, M. Will natural products remain an important source of drug research for the future? *Curr. Opin. Biotechnol.* **1997**, *8*, 708–712. [CrossRef] [PubMed]
35. Nagabhushan, R.K.; Raveesha, A. Ethnobotanical survey and scientific validation of medicinal plants used in the treatment of fungal infections in Agumbe region of Western Ghats, India. *Int. J. Pharm. Pharm. Sci.* **2015**, *7*, 273–277.
36. Vijayan, A.; Liju, V.B.; John, R.J.V.; Parthipan, B.; Renuka, C. *Traditional Remedies of Kani Tribes of Kottoor Reserve Forest, Agasthyavanam, Thiruvananthapuram, Kerala*; CSIR: New Delhi, India, 2007.
37. Al-Khayri, J.M.; Sahana, G.R.; Nagella, P.; Joseph, B.V.; Alessa, F.M.; Al-Mssallem, M.Q. Flavonoids as potential anti-inflammatory molecules: A review. *Molecules* **2022**, *27*, 2901. [CrossRef]
38. Ghannadi, A.; Hajhashemi, V.; Jafarabadi, H. An investigation of the analgesic and anti-inflammatory effects of *Nigella sativa* seed polyphenols. *J. Med. Food* **2005**, *8*, 488–493. [CrossRef]
39. Li, X.-W.; Chen, H.-P.; He, Y.-Y.; Chen, W.-L.; Chen, J.-W.; Gao, L.; Hu, H.-Y.; Wang, J. Effects of Rich-Polyphenols Extract of Dendrobium loddigesii on Anti-Diabetic, Anti-Inflammatory, Anti-Oxidant, and Gut Microbiota Modulation in db/db Mice. *Molecules* **2018**, *23*, 3245. [CrossRef]
40. Ginwala, R.; Bhavsar, R.; Chigbu, D.G.I.; Jain, P.; Khan, Z.K. Potential Role of Flavonoids in Treating Chronic Inflammatory Diseases with a Special Focus on the Anti-Inflammatory Activity of Apigenin. *Antioxidants* **2019**, *8*, 35. [CrossRef]
41. Bettaieb, A.; Cremonini, E.; Kang, H.; Kang, J.; Haj, F.G.; Oteiza, P.I. Anti-inflammatory actions of (−)-epicatechin in the adipose tissue of obese mice. *Int. J. Biochem. Cell Biol.* **2016**, *81*, 383–392. [CrossRef]
42. Reagan, L.P.; Magarinos, A.M.; McEWEN, B.S. Neurological changes induced by stress in streptozotocin diabetic rats. *Ann. N. Y. Acad. Sci.* **1999**, *893*, 126–137. [CrossRef] [PubMed]
43. Freitas, L.M.; Antunes, F.T.T.; Obach, E.S.; Correa, A.P.; Wiiland, E.; de Mello Feliciano, L.; Reinicke, A.; Amado, G.J.V.; Grivicich, I.; Fialho, M.F.P. Anti-inflammatory effects of a topical emulsion containing *Helianthus annuus* oil, glycerin, and vitamin B3 in mice. *J. Pharm. Investig.* **2021**, *51*, 223–232. [CrossRef]
44. Ou, Z.; Zhao, J.; Zhu, L.; Huang, L.; Ma, Y.; Ma, C.; Luo, C.; Zhu, Z.; Yuan, Z.; Wu, J. Anti-inflammatory effect and potential mechanism of betulinic acid on λ-carrageenan-induced paw edema in mice. *Biomed. Pharmacother.* **2019**, *118*, 109347. [CrossRef] [PubMed]
45. Kwon, Y.-I.I.; Vattem, D.A.; Shetty, K. Evaluation of clonal herbs of *Lamiaceae* species for management of diabetes and hypertension. *Asia Pac. J. Clin. Nutr.* **2006**, *15*, 107. [PubMed]
46. Shai, L.; Magano, S.; Lebelo, S.; Mogale, A. Inhibitory effects of five medicinal plants on rat alpha-glucosidase: Comparison with their effects on yeast alpha-glucosidase. *J. Med. Plants Res.* **2011**, *5*, 2863–2867.
47. Matu, E.N.; Van Staden, J. Antibacterial and anti-inflammatory activities of some plants used for medicinal purposes in Kenya. *J. Ethnopharmacol.* **2003**, *87*, 35–41. [CrossRef]
48. Frieri, M.; Kumar, K.; Boutin, A. Antibiotic resistance. *J. Infect. Public Health* **2017**, *10*, 369–378. [CrossRef]
49. MacGowan, A.; Macnaughton, E. Antibiotic resistance. *Medicine* **2017**, *45*, 622–628. [CrossRef]
50. Alibi, S.; Crespo, D.; Navas, J. Plant-derivatives small molecules with antibacterial activity. *Antibiotics* **2021**, *10*, 231. [CrossRef]
51. Guimarães, A.C.; Meireles, L.M.; Lemos, M.F.; Guimarães, M.C.C.; Endringer, D.C.; Fronza, M.; Scherer, R. Antibacterial activity of terpenes and terpenoids present in essential oils. *Molecules* **2019**, *24*, 2471. [CrossRef]
52. Simirgiotis, M.J.; Burton, D.; Parra, F.; López, J.; Muñoz, P.; Escobar, H.; Parra, C. Antioxidant and antibacterial capacities of *Origanum vulgare* L. essential oil from the arid Andean Region of Chile and its chemical characterization by GC-MS. *Metabolites* **2020**, *10*, 414. [CrossRef]
53. El Moussaoui, A.; Jawhari, F.Z.; Almehdi, A.M.; Elmsellem, H.; Benbrahim, K.F.; Bousta, D.; Bari, A. Antibacterial, antifungal and antioxidant activity of total polyphenols of *Withania frutescens* L. *Bioorganic Chem.* **2019**, *93*, 103337. [CrossRef]
54. Reglodi, D.; Renaud, J.; Tamas, A.; Tizabi, Y.; Socías, S.B.; Del-Bel, E.; Raisman-Vozari, R. Novel tactics for neuroprotection in Parkinson's disease: Role of antibiotics, polyphenols and neuropeptides. *Prog. Neurobiol.* **2017**, *155*, 120–148. [CrossRef]
55. Ramata-Stunda, A.; Petriņa, Z.; Valkovska, V.; Borodušķis, M.; Gibnere, L.; Gurkovska, E.; Nikolajeva, V. Synergistic effect of polyphenol-rich complex of plant and green propolis extracts with antibiotics against respiratory infections causing bacteria. *Antibiotics* **2022**, *11*, 160. [CrossRef]
56. Haghjoo, B.; Lee, L.H.; Habiba, U.; Tahir, H.; Olabi, M.; Chu, T.-C. The synergistic effects of green tea polyphenols and antibiotics against potential pathogens. *Adv. Biosci. Biotechnol.* **2013**, *4*, 959. [CrossRef]
57. Usman, M.; Khan, W.R.; Yousaf, N.; Akram, S.; Murtaza, G.; Kudus, K.A.; Ditta, A.; Rosli, Z.; Rajpar, M.N.; Nazre, M. Exploring the phytochemicals and anti-cancer potential of the members of Fabaceae family: A comprehensive review. *Molecules* **2022**, *27*, 3863. [CrossRef]

58. Zonyane, S.; Fawole, O.A.; La Grange, C.; Stander, M.A.; Opara, U.L.; Makunga, N.P. The implication of chemotypic variation on the anti-oxidant and anti-cancer activities of *Sutherlandia frutescens* (L.) R.Br.(Fabaceae) from different geographic locations. *Antioxidants* **2020**, *9*, 152. [CrossRef]
59. Borquaye, L.S.; Doetse, M.S.; Baah, S.O.; Mensah, J.A. Anti-inflammatory and anti-oxidant activities of ethanolic extracts of *Tamarindus indica* L. (Fabaceae). *Cogent Chem.* **2020**, *6*, 1743403. [CrossRef]
60. Abdulkhaleq, L.A.; Assi, M.A.; Noor, M.H.M.; Abdullah, R.; Saad, M.Z.; Taufiq-Yap, Y.H. Therapeutic uses of epicatechin in diabetes and cancer. *Vet. World* **2017**, *10*, 869–872. [CrossRef] [PubMed]
61. Dong, H.; Yang, X.; He, J.; Cai, S.; Xiao, K.; Zhu, L. Enhanced antioxidant activity, antibacterial activity and hypoglycemic effect of luteolin by complexation with manganese (II) and its inhibition kinetics on xanthine oxidase. *RSC Adv.* **2017**, *7*, 53385–53395. [CrossRef]
62. Alshehri, S.; Imam, S.S.; Altamimi, M.A.; Hussain, A.; Shakeel, F.; Elzayat, E.; Mohsin, K.; Ibrahim, M.; Alanazi, F. Enhanced dissolution of luteolin by solid dispersion prepared by different methods: Physicochemical characterization and antioxidant activity. *ACS Omega* **2020**, *5*, 6461–6471. [CrossRef] [PubMed]
63. Kim, N.M.; Kim, J.; Chung, H.Y.; Choi, J.S. Isolation of luteolin 7-O-rutinoside and esculetin with potential antioxidant activity from the aerial parts of *Artemisia montana*. *Arch. Pharmacal Res.* **2000**, *23*, 237–239. [CrossRef] [PubMed]
64. Xu, H.; Linn, B.S.; Zhang, Y.; Ren, J. A review on the antioxidative and prooxidative properties of luteolin. *React. Oxyg. Species* **2019**, *7*, 136–147. [CrossRef]
65. Guo, Y.; Liu, Y.; Zhang, Z.; Chen, M.; Zhang, D.; Tian, C.; Liu, M.; Jiang, G. The antibacterial activity and mechanism of action of luteolin against *Trueperella pyogenes*. *Infect. Drug Resist.* **2020**, *13*, 1697–1711. [CrossRef]
66. Çetinkaya, M.; Baran, Y. Therapeutic Potential of Luteolin on Cancer. *Vaccines* **2023**, *11*, 554. [CrossRef]
67. Potočnjak, I.; Šimić, L.; Gobin, I.; Vukelić, I.; Domitrović, R. Antitumor activity of luteolin in human colon cancer SW620 cells is mediated by the ERK/FOXO3a signaling pathway. *Toxicol. Vitr.* **2020**, *66*, 104852. [CrossRef]
68. Cavia-Saiz, M.; Busto, M.D.; Pilar-Izquierdo, M.C.; Ortega, N.; Perez-Mateos, M.; Muñiz, P. Antioxidant properties, radical scavenging activity and biomolecule protection capacity of flavonoid naringenin and its glycoside naringin: A comparative study. *J. Sci. Food Agric.* **2010**, *90*, 1238–1244. [CrossRef]
69. Patel, K.; Singh, G.K.; Patel, D.K. A review on pharmacological and analytical aspects of naringenin. *Chin. J. Integr. Med.* **2018**, *24*, 551–560. [CrossRef]
70. Ismail, N.H.; Mohamad, H.; Mohidin, A.; Lajis, N.H. Antioxidant activity of anthraquinones from *Morinda elliptica*. *Nat. Prod. Sci.* **2002**, *8*, 48–51.
71. Chee, C.W.; Zamakshshari, N.H.; Lee, V.S.; Abdullah, I.; Othman, R.; Lee, Y.K.; Hashim, N.M.; Rashid, N.N. Morindone from Morinda citrifolia as a potential antiproliferative agent against colorectal cancer cell lines. *PLoS ONE* **2022**, *17*, e0270970. [CrossRef]
72. Guil-Guerrero, J.; Martínez-Guirado, C.; del Mar Rebolloso-Fuentes, M.; Carrique-Pérez, A. Nutrient composition and antioxidant activity of 10 pepper (*Capsicum annuun*) varieties. *Eur. Food Res. Technol.* **2006**, *224*, 1–9. [CrossRef]
73. Sun, T.; Xu, Z.; Wu, C.T.; Janes, M.; Prinyawiwatkul, W.; No, H. Antioxidant activities of different colored sweet bell peppers (*Capsicum annuum* L.). *J. Food Sci.* **2007**, *72*, S98–S102. [CrossRef] [PubMed]
74. Ahmad, M.F.; Wahab, S.; Ahmad, F.A.; Ashraf, S.A.; Abullais, S.S.; Saad, H.H. *Ganoderma lucidum*: A potential pleiotropic approach of ganoderic acids in health reinforcement and factors influencing their production. *Fungal Biol. Rev.* **2022**, *39*, 100–125. [CrossRef]
75. Dubois-Deruy, E.; Peugnet, V.; Turkieh, A.; Pinet, F. Oxidative stress in cardiovascular diseases. *Antioxidants* **2020**, *9*, 864. [CrossRef] [PubMed]
76. Ndrepepa, G. Myeloperoxidase–A bridge linking inflammation and oxidative stress with cardiovascular disease. *Clin. Chim. Acta* **2019**, *493*, 36–51. [CrossRef] [PubMed]
77. Hayes, J.D.; Dinkova-Kostova, A.T.; Tew, K.D. Oxidative stress in cancer. *Cancer Cell* **2020**, *38*, 167–197. [CrossRef]
78. Yaribeygi, H.; Sathyapalan, T.; Atkin, S.L.; Sahebkar, A. Molecular mechanisms linking oxidative stress and diabetes mellitus. *Oxidative Med. Cell. Longev.* **2020**, *2020*, 8609213. [CrossRef]
79. Alqahtani, A.S.; Hidayathulla, S.; Rehman, M.T.; ElGamal, A.A.; Al-Massarani, S.; Razmovski-Naumovski, V.; Alqahtani, M.S.; El Dib, R.A.; AlAjmi, M.F. Alpha-Amylase and Alpha-Glucosidase Enzyme Inhibition and Antioxidant Potential of 3-Oxolupenal and Katononic Acid Isolated from *Nuxia oppositifolia*. *Biomolecules* **2019**, *10*, 61. [CrossRef]
80. Semaan, D.G.; Igoli, J.O.; Young, L.; Marrero, E.; Gray, A.I.; Rowan, E.G. In vitro anti-diabetic activity of flavonoids and pheophytins from *Allophylus cominia* Sw. on PTP1B, DPPIV, alpha-glucosidase and alpha-amylase enzymes. *J. Ethnopharmacol.* **2017**, *203*, 39–46. [CrossRef]
81. Feunaing, R.T.; Tamfu, A.N.; Gbaweng, A.J.Y.; Mekontso Magnibou, L.; Ntchapda, F.; Henoumont, C.; Laurent, S.; Talla, E.; Dinica, R.M. In vitro Evaluation of α-amylase and α-glucosidase Inhibition of 2,3-Epoxyprocyanidin C1 and Other Constituents from *Pterocarpus erinaceus* Poir. *Molecules* **2022**, *28*, 126. [CrossRef]
82. Facchin, B.M.; dos Reis, G.O.; Vieira, G.N.; Mohr, E.T.B.; da Rosa, J.S.; Kretzer, I.F.; Demarchi, I.G.; Dalmarco, E.M. Inflammatory biomarkers on an LPS-induced RAW 264.7 cell model: A systematic review and meta-analysis. *Inflamm. Res.* **2022**, *71*, 741–758. [CrossRef] [PubMed]

83. Hoppstädter, J.; Dembek, A.; Linnenberger, R.; Dahlem, C.; Barghash, A.; Fecher-Trost, C.; Fuhrmann, G.; Koch, M.; Kraegeloh, A.; Huwer, H.; et al. Toll-Like Receptor 2 Release by Macrophages: An Anti-inflammatory Program Induced by Glucocorticoids and Lipopolysaccharide. *Front. Immunol.* **2019**, *10*, 1634. [CrossRef] [PubMed]
84. Kaneko, N.; Kurata, M.; Yamamoto, T.; Morikawa, S.; Masumoto, J. The role of interleukin-1 in general pathology. *Inflamm. Regen.* **2019**, *39*, 12. [CrossRef]
85. Pyrillou, K.; Burzynski, L.C.; Clarke, M.C.H. Alternative Pathways of IL-1 Activation, and Its Role in Health and Disease. *Front. Immunol.* **2020**, *11*, 613170. [CrossRef] [PubMed]
86. Bent, R.; Moll, L.; Grabbe, S.; Bros, M. Interleukin-1 Beta—A Friend or Foe in Malignancies? *Int. J. Mol. Sci.* **2018**, *19*, 2155. [CrossRef]
87. Chen, J.; Wang, W.; Ni, Q.; Zhang, L.; Guo, X. Interleukin 6-regulated macrophage polarization controls atherosclerosis-associated vascular intimal hyperplasia. *Front. Immunol.* **2022**, *13*, 952164. [CrossRef]
88. Hirani, D.; Alvira, C.M.; Danopoulos, S.; Milla, C.; Donato, M.; Tian, L.; Mohr, J.; Dinger, K.; Vohlen, C.; Selle, J.; et al. Macrophage-derived IL-6 trans-signalling as a novel target in the pathogenesis of bronchopulmonary dysplasia. *Eur. Respir. J.* **2022**, *59*, 2002248. [CrossRef] [PubMed]
89. Rose-John, S.; Jenkins, B.J.; Garbers, C.; Moll, J.M.; Scheller, J. Targeting IL-6 trans-signalling: Past, present and future prospects. *Nat. Rev. Immunol.* **2023**, *23*, 666–681. [CrossRef]
90. Chen, J.; Wei, Y.; Yang, W.; Huang, Q.; Chen, Y.; Zeng, K.; Chen, J. IL-6: The Link Between Inflammation, Immunity and Breast Cancer. *Front. Oncol.* **2022**, *12*, 903800. [CrossRef]
91. Rašková, M.; Lacina, L.; Kejík, Z.; Venhauerová, A.; Skaličková, M.; Kolář, M.; Jakubek, M.; Rosel, D.; Smetana, K., Jr.; Brábek, J. The Role of IL-6 in Cancer Cell Invasiveness and Metastasis-Overview and Therapeutic Opportunities. *Cells* **2022**, *11*, 3698. [CrossRef]
92. Al Obeed, O.A.; Alkhayal, K.A.; Al Sheikh, A.; Zubaidi, A.M.; Vaali-Mohammed, M.A.; Boushey, R.; McKerrow, J.H.; Abdulla, M.H. Increased expression of tumor necrosis factor-α is associated with advanced colorectal cancer stages. *World J. Gastroenterol.* **2014**, *20*, 18390–18396. [CrossRef]
93. Zhao, P.; Zhang, Z. TNF-α promotes colon cancer cell migration and invasion by upregulating TROP-2. *Oncol. Lett.* **2018**, *15*, 3820–3827. [CrossRef] [PubMed]
94. Król, M.; Kepinska, M. Human Nitric Oxide Synthase—Its Functions, Polymorphisms, and Inhibitors in the Context of Inflammation, Diabetes and Cardiovascular Diseases. *Int. J. Mol. Sci.* **2021**, *22*, 56. [CrossRef] [PubMed]
95. Iwata, M.; Inoue, T.; Asai, Y.; Hori, K.; Fujiwara, M.; Matsuo, S.; Tsuchida, W.; Suzuki, S. The protective role of localized nitric oxide production during inflammation may be mediated by the heme oxygenase-1/carbon monoxide pathway. *Biochem. Biophys. Rep.* **2020**, *23*, 100790. [CrossRef] [PubMed]
96. Liu, D.; Guo, Y.; Wu, P.; Wang, Y.; Golly, M.K.; Ma, H. The necessity of walnut proteolysis based on evaluation after in vitro simulated digestion: ACE inhibition and DPPH radical-scavenging activities. *Food Chem.* **2020**, *311*, 125960. [CrossRef]
97. Lai, S.-C.; Ho, Y.-L.; Huang, S.-C.; Huang, T.-H.; Lai, Z.-R.; Wu, C.-R.; Lian, K.-Y.; Chang, Y.-S. Antioxidant and antiproliferative activities of *Desmodium triflorum* (L.) DC. *Am. J. Chin. Med.* **2010**, *38*, 329–342. [CrossRef]
98. Konaté, K.; Souza, A.; Coulibaly, A.; Meda, N.; Kiendrebeogo, M.; Lamien-Meda, A.; Millogo-Rasolodimby, J.; Lamidi, M.; Nacoulma, O. In vitro antioxidant, lipoxygenase and xanthine oxidase inhibitory activities of fractions from *Cienfuegosia digitata* Cav. *Sida alba L. and Sida acuta Burn f.(Malvaceae). Pak. J. Biol. Sci. PJBS* **2010**, *13*, 1092–1098.
99. Tanaka, T.; Narazaki, M.; Kishimoto, T. IL-6 in inflammation, immunity, and disease. *Cold Spring Harb. Perspect. Biol.* **2014**, *6*, a016295. [CrossRef]
100. Samaraweera, U.; Sotheeswaran, S.; Uvais, M.; Sultanbawa, S. 3,5,7,3′, 5′-Pentahydroxyflavan and 3α-methoxyfriedelan from *Humboldtia laurifolia. Phytochemistry* **1983**, *22*, 565–567. [CrossRef]
101. Dyamavvanahalli, S.L.; Raveesha, K.A.; Nagabhushan, S. Bioprospecting of selected medicinal plants for antibacterial activity against some pathogenic bacteria. *J. Med. Plants Res.* **2011**, *5*, 4087–4093.
102. Yadav, R.; Agarwala, M. Phytochemical analysis of some medicinal plants. *J. Phytol.* **2011**, *3*, 10–14.
103. Harborne, A. *Phytochemical Methods a Guide to Modern Techniques of Plant Analysis*; Springer Science & Business Media: London, UK, 1998.
104. Singleton, V.L.; Rossi, J.A. Colorimetry of total phenolics with phosphomolybdic-phosphotungstic acid reagents. *Am. J. Enol. Vitic.* **1965**, *16*, 144–158. [CrossRef]
105. Zhishen, J.; Mengcheng, T.; Jianming, W. The determination of flavonoid contents in mulberry and their scavenging effects on superoxide radicals. *Food Chem.* **1999**, *64*, 555–559. [CrossRef]
106. House, N.C.; Puthenparampil, D.; Malayil, D.; Narayanankutty, A. Variation in the polyphenol composition, antioxidant, and anticancer activity among different *Amaranthus* species. *S. Afr. J. Bot.* **2020**, *135*, 408–412. [CrossRef]
107. Webber, D.M.; Wallace, M.A.; Burnham, C.D. Stop Waiting for Tomorrow: Disk Diffusion Performed on Early Growth Is an Accurate Method for Antimicrobial Susceptibility Testing with Reduced Turnaround Time. *J. Clin. Microbiol.* **2022**, *60*, e0300720. [CrossRef]
108. Morgan, B.L.; Depenbrock, S.; Martínez-López, B. Identifying Associations in Minimum Inhibitory Concentration Values of *Escherichia coli* Samples Obtained From Weaned Dairy Heifers in California Using Bayesian Network Analysis. *Front. Vet. Sci.* **2022**, *9*, 771841. [CrossRef] [PubMed]

109. Di Simone, S.C.; Acquaviva, A.; Libero, M.L.; Chiavaroli, A.; Recinella, L.; Leone, S.; Brunetti, L.; Politi, M.; Giannone, C.; Campana, C.; et al. The association of *Tanacetum parthenium* and *Salix alba* extracts reduces cortex serotonin turnover, in an ex vivo experimental model of migraine. *Processes* **2022**, *10*, 280. [CrossRef]
110. Iskandar, N.N.; Iriawati, I. Vinblastine and Vincristine production on Madagascar Periwinkle (*Catharanthus roseus* (L.) G. Don) callus culture treated with polethylene glycol. *Makara J. Sci.* **2016**, *20*, 7–16. [CrossRef]
111. Rao, P.; Knaus, E.E. Evolution of nonsteroidal anti-inflammatory drugs (NSAIDs): Cyclooxygenase (COX) inhibition and beyond. *J. Pharm. Pharm. Sci.* **2008**, *11*, 81s–110s. [CrossRef]
112. Khumalo, G.P.; Van Wyk, B.E.; Feng, Y.; Cock, I.E. A review of the traditional use of southern African medicinal plants for the treatment of inflammation and inflammatory pain. *J. Ethnopharmacol.* **2022**, *283*, 114436. [CrossRef] [PubMed]
113. Bharadwaj, K.C.; Gupta, T.; Singh, R.M. Alkaloid group of *Cinchona officinalis*: Structural, synthetic, and medicinal aspects. In *Synthesis of Medicinal Agents from Plants*; Elsevier: Amsterdam, The Netherlands, 2018; pp. 205–227.

Disclaimer/Publisher's Note: The statements, opinions and data contained in all publications are solely those of the individual author(s) and contributor(s) and not of MDPI and/or the editor(s). MDPI and/or the editor(s) disclaim responsibility for any injury to people or property resulting from any ideas, methods, instructions or products referred to in the content.

molecules

MDPI

Review

Phytocompounds from Amazonian Plant Species against Acute Kidney Injury: Potential Nephroprotective Effects

Alberto Souza Paes [1,2], Rosemary de Carvalho Rocha Koga [1,2], Priscila Faimann Sales [1,2], Hellen Karine Santos Almeida [2,3], Thiago Afonso Carvalho Celestino Teixeira [1,2,3] and José Carlos Tavares Carvalho [1,2,3,*]

1 Pharmaceutical Innovation Program, Department of Biological and Health Sciences, Federal University of Amapá, Rodovia Juscelino Kubitschek, km 02, Macapá CEP 68903-419, Amapá, Brazil; dralbertopaes343ap@hotmail.com (A.S.P.); rosemarykoga@unifap.br (R.d.C.R.K.); pfaimann@gmail.com (P.F.S.); thafonsoteixeira@gmail.com (T.A.C.C.T.)

2 Research Laboratory of Drugs, Department of Biological and Health Sciences, Federal University of Amapá, Rodovia Juscelino Kubitschek, km 02, Macapá CEP 68903-419, Amapá, Brazil; hellenkarine1612@gmail.com

3 University Hospital, Federal University of Amapá, Rodovia Josmar Chaves Pinto, km 02, Macapá CEP 68903-419, Amapá, Brazil

* Correspondence: farmacos@unifap.br

Abstract: There are several Amazonian plant species with potential pharmacological validation for the treatment of acute kidney injury, a condition in which the kidneys are unable to adequately filter the blood, resulting in the accumulation of toxins and waste in the body. Scientific production on plant compounds capable of preventing or attenuating acute kidney injury—caused by several factors, including ischemia, toxins, and inflammation—has shown promising results in animal models of acute kidney injury and some preliminary studies in humans. Despite the popular use of Amazonian plant species for kidney disorders, further pharmacological studies are needed to identify active compounds and subsequently conduct more complex preclinical trials. This article is a brief review of phytocompounds with potential nephroprotective effects against acute kidney injury (AKI). The classes of Amazonian plant compounds with significant biological activity most evident in the consulted literature were alkaloids, flavonoids, tannins, steroids, and terpenoids. An expressive phytochemical and pharmacological relevance of the studied species was identified, although with insufficiently explored potential, mainly in the face of AKI, a clinical condition with high morbidity and mortality.

Keywords: Amazonian traditional medicine; phytotherapy; hypoxia; oxidative stress; nephroprotection; antioxidant; anti-inflammatory; diuretic

Citation: Paes, A.S.; Koga, R.d.C.R.; Sales, P.F.; Santos Almeida, H.K.; Teixeira, T.A.C.C.; Carvalho, J.C.T. Phytocompounds from Amazonian Plant Species against Acute Kidney Injury: Potential Nephroprotective Effects. *Molecules* **2023**, *28*, 6411. https://doi.org/10.3390/molecules28176411

Academic Editors: Arunaksharan Narayanankutty, Ademola C. Famurewa and Eliza Oprea

Received: 15 June 2023
Revised: 25 August 2023
Accepted: 28 August 2023
Published: 2 September 2023

Copyright: © 2023 by the authors. Licensee MDPI, Basel, Switzerland. This article is an open access article distributed under the terms and conditions of the Creative Commons Attribution (CC BY) license (https://creativecommons.org/licenses/by/4.0/).

1. Introduction

Natural products have been a source of important biologically active substances, and this is due to the diversity of chemical compounds that can be found in plants, fungi, and bacteria, among other organisms [1]. These compounds have a wide variety of structures and biological activity, which makes them potential candidates for the development of new drugs [2]. Products derived from plant species, in particular, those from the Amazonian biodiversity, are a rich source of compounds with promising pharmacological properties, including nephroprotective, anti-inflammatory, and antioxidant activity [3].

Some of the classes of Amazonian plant compounds have been investigated for their potential activity in kidney protection, among them, plant species belonging to the alkaloid classes, for their nephroprotective effects in experimental models of acute kidney injury (AKI); flavonoids, due to antioxidant and anti-inflammatory properties with renal protective activity; tannins, due to their antioxidant and anti-inflammatory potential, with probable prevention or reduction of kidney damage induced by toxins and other substances; steroids, with anti-inflammatory and antioxidant activity, which may help to prevent acute kidney injury induced by oxidative stress; and terpenoids, for their potent anti-inflammatory, antioxidant, and immunomodulatory activity [4–6].

Therefore, knowing this Amazonian biodiversity allows us to exploit it in the best way and protect it. This is mainly due to great expectations regarding the environmentally correct exploitation of natural resources for production and processing and a fair return for the traditional population [7].

These natural compounds can be isolated and used as models for the development of new drugs, or they can be chemically modified to improve their pharmacokinetic and pharmacodynamic properties, resulting in new compounds with potential therapeutic activity against AKI [8–10]. AKI is a medical condition that can lead to acute renal failure, with a high risk of morbidity and mortality [11]. AKI is characterized by an abrupt reduction in renal function, with accumulation of toxic metabolites and electrolytes in the body, triggering serious complications such as hemodynamic disorders, pulmonary edema, metabolic disorders, and even death [12].

AKI can affect up to 7% of hospitalized patients and up to 50% of critically ill patients, being one of the main causes of mortality in this group of patients. The dysfunction comes from a variety of factors, including the use of nephrotoxic drugs, renal ischemia, heart failure, hypovolemia, and sepsis, among others [13]. Advanced age, the presence of comorbidities such as diabetes and hypertension, and the use of invasive procedures such as cardiac surgery are also associated with a higher risk of developing AKI [12].

This clinical condition generates several social problems, especially in countries with precarious health systems, resulting in (a) increased health costs, due to the need for intensive and prolonged treatment, including intensive care, and, in some cases, dialysis; (b) reduced quality of life, especially in the most serious cases, where recovery can be slow and complicated, resulting in loss of productivity, inability to work, and the need for special care; (c) socioeconomic inequalities, disproportionately affecting more vulnerable populations, including elderly patients, people with low socioeconomic status, or those with limited access to adequate health care; (d) overload of healthcare systems, especially in countries with limited resources, due to the requirement for intensive and long-term care [14].

AKI involves a series of complex mechanisms, including hemodynamic disturbances, in which there is a reduction in renal perfusion, with consequent hypoxia and ischemia, resulting in cell damage and inflammation [12]. AKI is often accompanied by kidney inflammation, which can contribute to cell damage and tubular dysfunction. This reduction in kidney function with damage to the epithelial cells of the renal tubules, releases toxic metabolites and electrolytes into the body, resulting in serious complications. In addition to tubular dysfunction, it can reduce the ability to reabsorb water and electrolytes as well as the ability to excrete toxic metabolites [13].

Based on its pathophysiology, it is classified into three types: pre-renal, due to a reduction in renal perfusion due to hypovolemia, heart failure, or hypotension, resulting in hypoxia and cell damage; intrinsic, due to direct damage to the kidneys, including nephrotoxicity from medications, exposure to toxic chemicals, infections, or autoimmune diseases; postrenal, caused by kidney stones, tumors, or other obstructions to the flow of urine, resulting in accumulation of urine in the kidneys and subsequent damage [15].

Therefore, AKI can be potentially fatal if not treated properly, and conventional treatments often have significant side effects [12]. In this context, the Amazonian population uses several plant species with bioactive compounds to treat diseases of the renal and urinary system [16]. Traditional Amazonian medicine has contributed to the discovery of new bioactive products [1]. Therefore, developing drugs from plant species from the Amazon against AKI can help combat resistance to existing drugs. This can also minimize significant side effects, including damage to the liver and other organs; improve effectiveness; and provide new treatment options for AKI.

2. Method

This study constitutes, methodologically, an analytical bibliographic review related to the mapping of secondary metabolites found in Amazonian plant species with potential to treat AKI, belonging to the classes of alkaloid compounds, flavonoids, tannins, terpenoids, and steroids. Among the species highlighted are *Banisteriopsis caapi* (Spruce ex Griseb.) Morton, *Peganum harmala* L., *Passiflora edulis* Sims, *Annona muricata* L., *Uncaria tomentosa* (Willd.) DC., *Hymenaea courbaril* L., *Echinodorus macrophyllus* (Kunth) Micheli, *Acmella oleracea* (L.) R. K. Jansen, and *Rosmarinus officinalis* L., in addition to studies on potential nephroprotective effects. Data collection was carried out from September 2022 to February 2023, using the following databases: CAPES journals, PubMed, Science Direct from Elsevier, Wiley Online Library, Springer-Nature, Taylor and Francis, BMC, Hindawi, Scielo, ACS—American Chemical Society, and Google Scholar, as well as databases of scientific articles and patents "The LENS" and "ORBIT Intelligence".

The inclusion criteria for this work included original articles exclusive to the genus and species studied, with full text available in Portuguese, English, and other languages. Exclusion criteria included abstracts, online sites without scientific sources, incomplete texts, and unrelated and repeated articles.

As for the search strategy, the descriptive words used in this work were as follows: species *Banisteriopsis caapi* (Spruce ex Griseb.) Morton, *Peganum harmala* L., *Passiflora edulis* Sims, *Annona muricata* L., *Uncaria tomentosa* (Willd.) DC., *Hymenaea courbaril* L., *Echinodorus macrophyllus* (Kunth) Micheli, *Acmella oleracea* (L.) R. K. Jansen, and *Rosmarinus officinalis* L., correlated with secondary metabolites and their nephroprotective potential. The articles were selected by reading the titles and abstracts of the publications, associated with the Boolean descriptor "AND", in order to refine the samples.

The review is based primarily on articles published after 2010. However, some older articles were also mentioned to provide relevant background or when providing well-documented information. The study shows the expressive phytochemical and pharmacological relevance of the studied species, although many of them with insufficiently explored potential, mainly in the face of AKI.

3. Secondary Metabolites and Nephroprotective Potential in Amazonian Plant Species

3.1. Classes of Compounds Present in Amazonian Plant Species

3.1.1. Alkaloids

Alkaloids are a class of nitrogenous organic compounds that occur naturally in plant species [17]. Such constituents are a class of chemical compounds with alkaline properties that contain at least one nitrogen atom in their structure, being produced by plants as a form of defense against herbivores and pathogens [18].

Alkaloids are characterized by having a heterocyclic ring structure with at least one nitrogen atom, unlike aliphatic nitrogen compounds, which are non-cyclic [19]. These substances can occur as homoligomeric or heteroligomeric monomers, dimers, trimers, or tetramers. There are two main groups of alkaloids: those with a heterocyclic or non-heterocyclic chemical structure, and those of biological or natural origin, which come from specific sources [20].

Alkaloids can be divided into different classes or groups based on their chemical structures and properties: indole (serotonin, melatonin, and tryptamine), isoquinoline (morphine, codeine, and papaverine), terpenic (atropine, scopolamine, and ephedrine), and pyrrolizidine (senecionin and retronecin) [21].

Many alkaloids have pharmacological properties such as analgesics [22], hallucinogens [23], anesthetics [24], antidiabetic [25], and anticancer [18]. Some studies suggest that plants that produce harmine, harmaline, and tetrahydroharmine alkaloids may have nephroprotective biological activity; that is, they may protect the kidneys against damage [26].

However, it is worth mentioning the toxicity of certain metabolites belonging to the class of alkaloids, for example, saponins. According to Fang et al. [27], in an acute toxicity test, with crude extracts of *Albizia coriaria* (used by the traditional population of Uganda), with a significant percentage of saponins in phytochemical screening, extracts of the species possibly produced acute renal toxicity, based on histopathological identification of tissue with severe acute multifocal nephritis, characterized by infiltration of inflammatory cells at various sites in the renal interstitium. In addition, the animal models used clinically had excessive urination, with a probable link to toxicity.

3.1.2. Flavonoids

Flavonoids are a class of organic compounds widely distributed in nature, characterized by their flavone chemical structure, which includes two aromatic rings joined by a three-carbon bridge. They have been the subject of studies due to the wide range of biological properties derived from their bioactive compounds [28]. Based on their chemical structures and biological properties, flavonoids are divided into several subclasses, such as flavones (a hydroxyl group at the 4 position of the B ring: luteolin and apigenin), flavonols (a hydroxyl group at the 3 position of the C ring: quercetin and kaempferol), flavanones (without a hydroxyl group at position 3 of the C ring: hesperidin and naringin), flavanols (a hydroxyl group at position 3 and a hydroxyl group at position 4 of the C ring: catechin and epicatechin), anthocyanins (water-soluble pigments responsible for the red, purple, and blue colors of many fruits and vegetables), and isoflavones (found primarily in legumes) [29].

Several preclinical and clinical studies have documented the pharmacological activity of flavonoids, mainly their antioxidant properties [30,31], antidiabetics [32], antiobesity [33], antihyperlipidemic [34], anti-inflammatory [35], anti-osteoporotic effects [36], antiallergic, and antithrombotic [37], in addition to being hepatoprotective [38], neuroprotective [39], nephroprotectors [40–42], chemopreventives, and anticancers [43], as well as having antibacterial, antifungal, and antiviral activity [44]. Flavonoids can inhibit in vitro proliferation of several cancer cell lines and reduce tumor growth in animal models [43]. They are recognized as antioxidants and have properties that eliminate free radicals. Thus, they act as divalent cation chelators and have free radical scavenging properties, inhibiting lipid peroxidation, capillary permeability, and platelet aggregation and fragility [45].

These compounds are able to increase the activity of endogenous free radical metabolizing enzymes, including catalase (CAT), superoxide dismutase (SOD), glutathione peroxidase, and glutathione (GSH), which are crucial for the elimination of ROS and consequently the increase of antioxidant activity. Several studies have shown that pretreatment with flavonoids can play a key role in ischemia and reperfusion injury [46].

In a study carried out in a renal model of AKI in Wistar rats, pretreatment with rutin significantly reduced renal failure, in addition to inhibiting the production of malondialdehyde (MDA) and restoring depleted levels of GSH and superoxide dis-mutase activity [47]. The compound apigenin increased SOD and glutathione peroxidase activity, as well as being able to reduce MDA in a rat model of AKI through activation of the JAK2/STAT3 signaling pathway [48].

Flavonoids modulate oxidative stress through the pathway of nuclear factor 2 related to erythroid 2 (Nrf2), a transcription factor that regulates the expression of several cytoprotective and antioxidant genes [49]; inhibit the production of peroxynitrite, suppressing the iNOS activity and NO production; and are essential mediators of the pathological and physiological processes of AKI [50].

In addition to promoting an inflammatory response in the biological processes of AKI, they inhibit the tumor necrosis factor-α (TNF-α), which initiates the inflammatory cascade and the positive regulation of chemokines and cytokines such as IL-6 and IL-1β, which can damage renal cells directly [51]. In addition, in toll-like receptor 4 (TLR4) signaling, they are an important modulator of chemokines and pro-inflammatory cytokines, migration and infiltration of leukocytes, and apoptosis of renal tubular epithelial cells [52].

In addition, flavonoids regulate biological systems through the inhibition of several enzymes, including hydrolase, lipase, α-glucosidase, aldose reductase, cyclooxygenase, xanthine oxidase, hyaluronidase, alkaline phosphatase, arylsulfatase, lipoxygenase, Ca^{+2}-ATPase, cAMP phosphodiesterase, and various kinases [53].

3.1.3. Tannins

Tannins are the most abundant secondary metabolites produced by plants [54]. Tannins are widespread in the plant kingdom and occur in different concentrations in all parts of plant material, be it bark, fruit, wood, or roots [55]. Vegetable tannins are generally classified into two groups: pyrogallol tannins or hydrolysable tannins and catechol tannins or condensable tannins. Those of the hydrolysable type, in turn, are subdivided into two groups: gallotannins, which produce gallic acid and glucose, and ellagitannins, which provide ellagic acid and glucose. Condensable tannins are not prone to hydrolysis but are amenable to oxidation and polymerization to form insoluble products known as red tannins/phlobaphenes [56].

Tannins have been the subject of several studies due to their potential pharmacological effects. They are known to have antioxidant activity, protecting the body's cells against oxidative damage caused by free radicals [57]. They promote anti-inflammatory action [58] and are also able to kill or inhibit the growth of bacteria, fungi, and viruses, thus exhibiting antimicrobial activity [59]. Tannins have been studied for their potential antitumor effect by inhibiting the growth and proliferation of tumor cells [60]. They also have hypoglycemic effects in vivo [61] and are excellent promoters of hepatoprotection, against damage caused by toxins and other harmful substances [62]. Among the cardiovascular effects, some studies suggest that tannins can help reduce the risk of cardiovascular diseases, including atherosclerosis and hypertension [63].

In terms of nephroprotection, the tannins obtained from the methanolic extract of the *Jatropha tanjorensis* leaf improved the serum levels of the renal metabolites urea, uric acid, and creatinine in animals exposed to renal damage by sodium benzoate; treatment with the extract reversed significantly changes these important markers of kidney damage in a dose-dependent manner [64].

In Brazil, tea from the plant *Phyllanthus niruri* known as 'stone breaker' is commonly used in cases of kidney stones. The species does not present acute or chronic toxicity; it has a uricosuric effect and increases glomerular filtration, which suggests its potential use not only as a lytic and/or preventive effect on the formation of urinary calculi but also as a possible use in hyperuricemic patients (by the uricosuric effect) and patients with renal failure [65]. According to the Brazilian Pharmacopoeia [66], the plant drug is derived from the dried aerial parts of *Phyllanthus niruri* L. [syn. *Phyllanthus niruri* ssp. niruri L. and *Phyllanthus niruri* ssp. lathyroides (Kunth) G.L.Webster] containing at least 6.5% of total tannins and 0.15% of gallic acid ($C_7H_6O_5$, 170.12) for herbal marketing.

3.1.4. Steroids

Steroids are organic compounds that occur naturally in plants; they have a characteristic sterol ring molecular structure [67]. Each type of steroid has a specific function, playing important roles in the physiology and biochemistry of those plant species in which they are found. For example, phytosterols play an important role in regulating cell membrane permeability while brassinosteroids act as growth and development hormones [68]. Although natural products are often associated with deleterious health effects, plant steroids have many medicinal applications, and research continues to explore these secondary metabolites as potential leaders in drug design and discovery [69].

Several natural steroids have been extensively studied for pharmacological efficacy as antihormones [70], contraceptive drugs [71], anticancer agents [72], cardiovascular agents [73], osteoporosis medications [74], antibiotics, anesthetics, anti-inflammatories and antiasthmatics [75].

Important roles for steroids have been evaluated, in diuresis and renal protection, from secondary metabolites of plant species traditionally used by the Amazonian population [76]. Studies in animal models of nephrotoxicity have shown nephroprotective activity [77], reduction of lipid peroxidation and renal fibrosis, and improvement of renal function in models of acute kidney injury [78].

3.1.5. Terpenoids

Terpenoids constitute the largest class of secondary metabolites and generally do not contain nitrogen or sulfur in their structures. As a consequence, many terpenoids have pronounced pharmacological activity and are therefore interesting for medicine and biotechnology [79].

Terpenoids are classified by the number of five-carbon (isoprene) units they contain. Thus, the smallest terpenes contain a single isoprene, called hemiterpene, monoterpenoids (with two isoprene units), sesquiterpenoids (three units), diterpenoids (four units), triterpenoids (six units), tetraterpenoids (eight units), and polyterpenoids (more than eight units). Each group of terpenoids has distinct physical and chemical properties, which influence their biological and pharmacological activity [80].

Several in vitro, preclinical, and clinical studies have confirmed that this class of compounds exhibits a wide range of very important pharmacological properties: analgesic, anti-inflammatory, anticancer, anticonvulsant, antibacterial, antiparasitic, and nutraceutical activity [81,82].

Hence, the diverse collection of terpenoid structures and functions triggers increased interest in their commercial use, resulting in some with well-established medical applications being registered as drugs on the market [81].

3.2. *Nefroprotective Potential of Compounds from Amazonian Plant Species*

Ischemic injury is a complex process of severe vasoconstriction and hypoxia, mainly in the renal cortex, with impairment of cellular integrity. The pathogenesis of ischemia and reperfusion injury involves multiple cellular and extracellular mechanisms [83].

Certain classes of plant compounds present in the Amazon, such as alkaloids, flavonoids, tannins, steroids, and terpenoids, have been investigated for their promising nephroprotective activity, including the modulation of different cellular mechanisms, thus promoting antioxidant activity, as well as anti-inflammatory activity [3].

The renal inflammatory process is characterized by an increase in chemotactic factors, including the chemokine protein chemotaxic protein-1 for monocytes (MCP-1) [84] and granulocyte and macrophage colony-stimulating factors (GM-CSF) [85]. Endothelial damage favors the formation of intercellular adhesion molecules (ICAM-1), adhesion molecules (VCAM), and P and E selectins, which promote leukocyte-endothelium interaction, platelet adhesion, and mechanical obstruction of the renal microvasculature. Monocytes cross the vascular endothelium and migrate to the damaged tissue, generating macrophages that produce inflammatory mediators, including transforming growth factor beta (TGF-β), tumor necrosis factor alpha (TNF-α), and interleukins 1, 6, and 12, Figure 1 [86].

CXC motif chemokine ligand 1 (CXCL1) is a cytokine belonging to the CXC subfamily of chemokines whose main receptor is CXC motif chemokine 2 (CXCR2), causing the migration and infiltration of neutrophils to sites of high expression [87]. CXCL1 plays a role in many adverse conditions associated with inflammation and neutrophil accumulation. Neutrophils phagocytose using reactive oxygen species (ROS), reactive nitrogen species (RNS), and other reactive small molecule compounds in affected tissues [88].

Factors such as pro-inflammatory cytokines, including interleukin-1β (IL-1β) and tumor necrosis factor α (TNF-α), act to increase the expression of CXCL1; these cytokines activate the nuclear factor κB (NF-κB), which also attenuated the increase of its expression [89]. The most significant CXCL1 receptor is CXCR2, a G protein-coupled receptor that acts in signal transduction by activating several signaling pathways such as extracellular signal-regulated kinase (ERK), mitogen-activated protein kinase (MAPK), and focal adhesion kinase (FAK) [90].

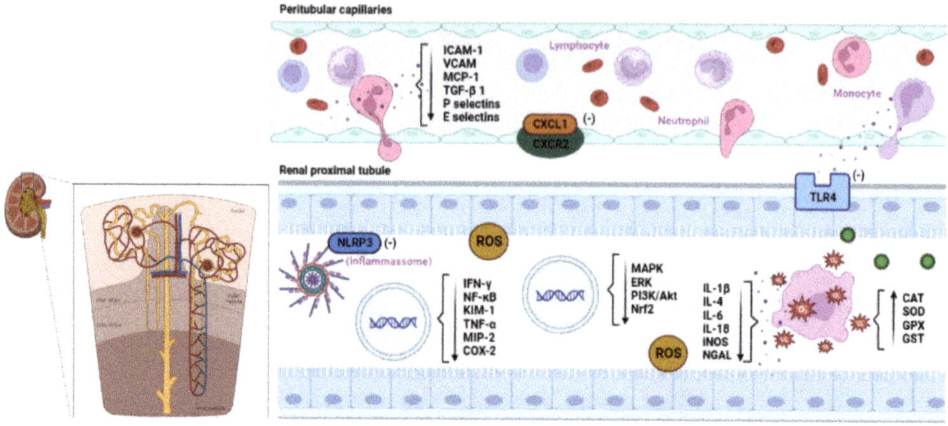

Figure 1. Amazonian phytocompounds against acute kidney injury. The proposed mechanism applies mainly to the kidneys. Classes of Amazonian plant compounds such as alkaloids, flavonoids, tannins, steroids, and terpenoids have significant biological activity especially for the treatment of acute kidney injury (AKI). Endothelial injury in induced AKI favors the formation of intercellular adhesion molecules (ICAM-1), adhesion molecules (VCAM), chemokine chemotaxic protein-1 for monocytes (MCP-1), transforming growth factor β-1 (TGF-β 1), and P and E selectins, which promote leukocyte–endothelium interaction, platelet adhesion, and mechanical obstruction of the renal microvasculature; however, active biological substances can act in the expressive reduction of these chemokines. CXC motif chemokine ligand 1 (CXCL1) is a cytokine belonging to the CXC subfamily of chemokines, whose main receptor is CXC motif chemokine 2 (CXCR2), a G protein-coupled receptor that causes neutrophil migration and infiltration. Activation of CXCR2 results in signal transduction through multiple pathways. Flavonoids and tannins act to reduce this expression; pro-inflammatory cytokines, such as interleukin-1β (IL-1β) and tumor necrosis factor α (TNF-α), activate the nuclear factor κB (NF-κB), which increases the expression of the CXCL1 gene. CXCL1/CXCR2 mediates the activation of phosphatidylinositol-4,5-bisphosphate 3-kinase (PI3K), and C-β phospholipase (PLC-β). PI3K induces activation of protein kinase B (PKB)/Akt, and signals pathways including extracellular signal-regulated kinase (ERK), mitogen-activated protein kinase (MAPK), and focal adhesion kinase (FAK). Toll-like receptor 4 (TLR4) is responsible for initiating the production of inflammatory cytokines; its inhibition results in decreased inflammation and renal dysfunction during nephrotoxicity. The NLRP3 receptor regulates post-transcriptional processes, which lead to the formation of inflammasome, responsible for the maturation of the inactive forms pro-IL-1β and pro-IL-18; the decrease of its expression leads to the inactivation of macrophages and lymphocytes. The activation of NF-κB promotes the transcription of specific genes that encode inflammatory mediators; however, the exogenous induction of active compounds decreases the expression of inflammatory mediators such as interferon-γ (IFN-γ), which leads to a reduction in the expression of renal injury molecule 1 (KIM-1), tumor necrosis factor-α (TNF-α) in renal tubular cells, and consequently the inactivation of a large network of pro-inflammatory cytokines, such as interleukin-1, 4, 6 (IL-1β, IL-4, IL-6, IL-18), macrophage inflammatory protein 2 (MIP-2) chemokines, among others. It also induces a decrease in the expression of the inflammatory enzyme cyclooxygenase (COX-2) and oxide nitric synthase (iNOS), in the renal medulla, in the glomerular mesangial cells, and in the endothelial cells of the renal vasculature, thereby reducing the tubular stress marker lipocalin associated with neutrophil gelatinase (NGAL). The signaling system of mitogen-activated protein kinases (MAPKs) is induced by cellular stress, by inflammatory responses. MAPK activation leads to the degradation of Iκβ (NF-κβ inhibitor), consequently promoting NF-κβ activation and migration to the nucleus. Flavonoids, tannins, steroids, and terpenoids also regulate many pathways such as MAPK, extracellular signal-regulated kinase (ERK), phosphoinositide 3 kinase (PI3K)/Akt, and related protein kinase pathways to reduce oxidative stress and inflammation, and they may have nephroprotective effects.

These compounds also induce transcription factors such as Nrf2, an antioxidant responsive element (ARE), which mediates the expression of antioxidant proteins. Nrf2 suppresses MCP-1 and VCAM-1 expression and thus decreases monocyte adhesion and transmigration to endothelial cells, which reduces MAPK expression. These mechanisms enhance the activity of endogenous antioxidants such as catalase (CAT), superoxide dismutase (SOD), glutathione peroxidase (GPX), and glutathione S-transferase (GST), which act together to provide a line of defense against oxidative damage.

Toll-like receptors (TLR)—especially subtype 4 (TLR4) and the protein cluster of differentiation 14 (CD14) existing on the surface of monocytes, macrophages, dendritic cells, and neutrophils—act in the immunomodulatory activity [91]. TLR4 is responsible for initiating the production of inflammatory cytokines, increasing renal oxidative stress and macrophage-mediated inflammation, as well as activating the nuclear factor κB (NF-κB), which is essential in initiating the intrarenal inflammatory response in the event of nephrotoxicity [92,93].

NLR family pyrin domain containing 3 (NLRP3) are investigated for their association with chronic metabolic and inflammatory diseases [94]. This receptor regulates post-transcriptional processes, with formation of the inflammasome, composed of active caspase-1 and ASC adapter protein (an apoptosis-associated speck-like protein containing a CARD domain), which promote the cleavage of inactive pro-IL-1β interleukins and pro-IL-18 in their active forms [95,96].

NF-κB activation requires the transcription of specific genes that intervene in the encoding of inflammatory mediators, promoting immune, proliferative, anti-apoptotic, and anti-inflammatory responses [97]. This causes increased expression of tumor necrosis factor-α (TNF-α) in renal tubular cells, an important cytokine involved in systemic inflammation that coordinates the activation of a large network of pro-inflammatory cytokines, such as interleukin-1, 4, 6 (IL-1β, IL-4, IL-6), transforming growth factor β-1 (TGF-β 1), and the chemokine chemotaxic protein-1 for monocytes (MCP-1) [98,99].

The increased expression of renal injury molecule 1 (KIM-1), a tubular transmembrane protein, has a signaling function since it is associated with the activation of T cells and the immune response. When chronically expressed, it results in progressive renal fibrosis and chronic renal failure [100].

The inflammatory response with release of TNF-α, interferon-γ (IFN-γ), and IL-1 induces the expression of the enzyme inducible nitric oxide synthase (iNOS) in the renal medulla, in the glomerular mesangial cells, and in the endothelium cells of the renal vasculature [101]. With the release of inflammatory enzymes such as COX-2, whose transcription is dependent on NK-κB [102], as well as nitric oxide (NO), free radicals have the ability to interact with ROS and form toxic molecules such as peroxidonitrite, which oxidize and damage cell membrane proteins with even greater toxicity [101]. Therefore, phytochemical constituents found in Amazonian species may act to inhibit or attenuate these pathways.

The mitogen-activated protein kinases (MAPKs) signaling system consists of protein pathways with serine/threonine kinase activity. MAPKs are induced by cellular stress, inflammatory responses, and apoptotic pathways initiated by a variety of biological stressors [103]. MAPKs lead to the degradation of Iκβ (NF-κβ inhibitor); consequently, they act in the activation and migration of NF-κβ to the nucleus, producing pro-inflammatory cytokines including TNF-α [104].

Flavonoids and other classes of compounds also regulate many pathways such as MAPK, extracellular signal-regulated kinase (ERK), phosphoinositide 3 kinase (PI3K)/Akt, and protein kinase-related pathways to reduce oxidative stress and inflammation, potentially having nephroprotector effects [105].

Additionally, these natural compounds can reduce inflammation by acting on many regulatory substances. These include inhibition of NF-κB, activator protein-1 (AP-1), interleukin-1beta (IL-1β), tumor necrosis factor alpha (TNF-α), IL-6, IL-8, and COX2 [105,106].

Likewise, they act on promising and potent antioxidant molecules that confer anti-inflammatory activity, inducing transcription factors such as Nrf2, an antioxidant responsive element (ARE), which mediates the expression of antioxidant proteins. Nrf2 acts by suppressing the expression of MCP-1 and VCAM-1 and, thus, decreasing monocyte adhesion and transmigration to endothelial cells, which consequently reduces MAPK expression [107,108].

Therefore, mechanisms involving the activity of compounds such as alkaloids, flavonoids, tannins, steroids, and terpenoids may promote the ability to increase antioxidant enzymes, such as superoxide dismutase (SOD), chloramphenicol acetyltransferase (CAT), and plasma glutathione peroxidase (GSH-Px), as well as reduce the expression of inducible NO synthase (iNOS) and nitrites in the cell, thus protecting the renal cells [3,109].

3.3. *Nefroprotective Potential of Amazonian Plant Species*

There is a growing interest in the nephroprotective potential of Amazonian plant species. Several studies have investigated the potential of these species to prevent or treat kidney disease [110], Figure 2.

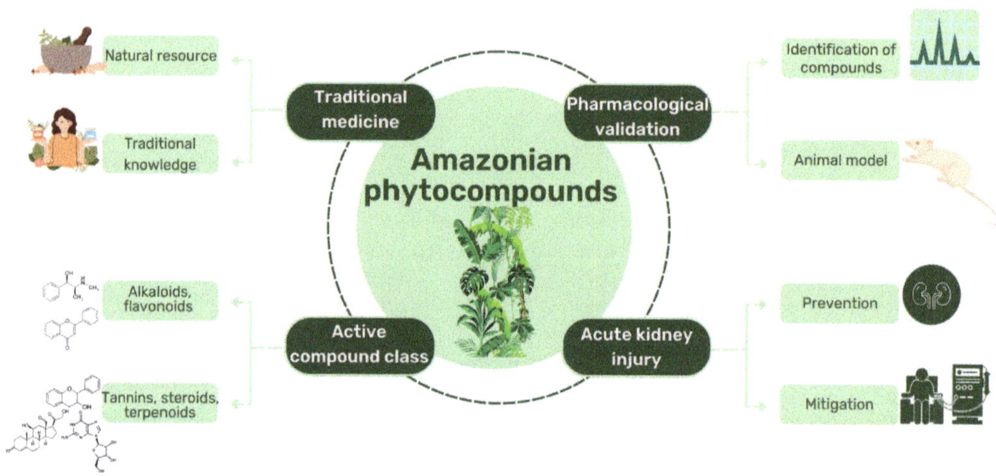

Figure 2. Potential of Amazonian phytocompounds against acute kidney injury.

3.3.1. *Banisteriopsis caapi* (Spruce ex Griseb.) Morton

Banisteriopsis caapi (Spruce ex Griseb.) Morton is a woody liana, common around the Amazon Basin, belonging to the kingdom Plantae, class Equisetopsida C. Agardh, order Malpighiales Juss. Former Bercht. & J. Presl, family Malpighiaceae Juss., genus *Banisteriopsis* CB Rob. and species *Banisteriopsis caapi* (Spruce ex Griseb.) Morton, which has broad ethnopharmacological use by the Amazonian people [111]. It is a species used as the main ingredient of the hallucinogenic drink, called ayahuasca, consumed by religious groups in Brazil to treat various ailments [112]. It is consumed for its hallucinogenic properties, which have been known by many of the indigenous people of the Amazon for centuries [113].

Some species of the Malpighiaceae family are known to produce alkaloids, among them, *B. caapi* (Spruce ex Griseb.) Morton. Harmine, one of the main alkaloids found in *B. caapi* (Spruce ex Griseb.) Morton, a plant widely consumed in the ayahuasca drink It is a β-carboline alkaloid widely disseminated due to its monoaminoxidase (MAO) inhibitory activity. As seen in the studies by Samoylenko et al. [114], harmine and harmaline (obtained from aqueous extracts of fresh and dried branches of *B. caapi* (Spruce ex Griseb.) Morton), have potent effects on MAO inhibitory and antioxidant activity. In addition, strong antioxidant activity for inhibition of cellular reactive oxygen species (ROS) generation by phorbol-12-myristate-13-acetate (PMA) has also been observed.

The effects of harmine against nicotine-induced damage in mouse kidneys were detailed by Salahshoor et al. [26]. In the study, administration of harmine to nicotine-treated animals significantly improved renal malondialdehyde (MDA), blood urea nitrogen (BUN), creatinine, and nitrite oxide levels. It increased the number of glomeruli and the level of power tissue ferric reducer/antioxidant (FRAP) compared to the nicotine group ($p < 0.05$), Table 1.

Table 1. Phytocompounds from Amazonian plant species and their pharmacological activity.

Species	Parts Used	Isolated or Characterized Constituents	Pharmacological Activity
Banisteriopsis caapi (Spruce ex Griseb.) Morton	Stem	Harmine **(1)**, harmaline **(2)** [114], tetrahydroharmine **(3)**, and harmalinic acid **(4)** [115]	Analgesic [22], hallucinogen [23], anesthetic [24], antidiabetic [25], anticancerogenic [18], nephroprotective, diuretic [26]
Peganum harmala L.	Seeds	Harmol **(5)**, harmalol **(6)**, harmine **(1)**, and harmaline **(2)** [116]	Antioxidant, nephroprotective, anti-inflammatory, anti-apoptotic [116]
Passiflora edulis Sims	Fruit peel, leaves, flowers, seeds	Orientin **(7)** and isoorientin **(8)** [117]	Anxiolytic, sedative, neuropathic pain [118], anticonvulsant [119], cognitive function and degenerative diseases [120], antioxidant action, antitumor action, hypoglycemic action, obesity, insomnia, nephroprotector [121]
Annona muricata L.	Leaves	Acetogenin **(9)** [122], δ-Cadinene **(10)**, and α-Muurolene **(11)** [123]	Anticancerogenic, hepatoprotective, neurotoxic, antinociceptive, antiulcerative, chemopreventive, nephroprotective [124]
Uncaria tomentosa (Willd.) DC.	Stem	Uncarine F **(12)**, speciophylline **(13)**, and mitraphylline **(14)** [125]	Antioxidant and immunomodulator, anti-inflammatory, analgesic, anticancer, and diuretic [126]
Hymenaea courbaril L.	Stem and leaves	Fisetin **(15)**, cyclosativene **(16)**, caryophyllene **(17)**, and α-himachalene **(18)** [127]	Antioxidant, antiulcerogenic, anti-inflammatory, antitumor, and diuretic [128]
Echinodorus macrophyllus (Kunth) Micheli	Leaves	Linalool **(19)**, α-caryophyllene **(20)**, β-caryophyllene **(21)** [129], isovitexin **(22)**, and isoorientin **(8)** [130]	Diuretic, anti-inflammatory, treatment of kidney and liver disorders [131]
Acmella oleracea (L.) R. K. Jansen	Flowers and leaves	Spilanthol **(23)**, spermidine **(24)**, spermine **(25)**, and 3-acetylaleuritolic acid **(26)** [132–134]	Aphrodisiac, treatment of male sexual dysfunctions, diuretic, and anti-inflammatory [135,136]
Rosmarinus officinalis L.	Leaves	Camphene **(27)**, limonene **(28)**, camphor **(29)**, borneol **(30)**, cineol **(31)**, and linalool **(19)** [137]	Analgesic, anti-inflammatory, anticarcinogenic, antirheumatic, nephroprotective, spasmolytic, antihepatotoxic, atherosclerotic [138]

Table 1. *Cont.*

Species	Parts Used	Isolated or Characterized Constituents		Pharmacological Activity
Harmine 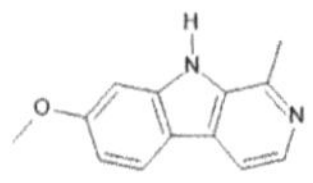 **(1)**	**Harmaline** 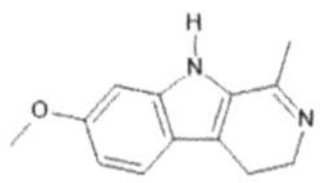 **(2)**	**Tetrahydroharmine** 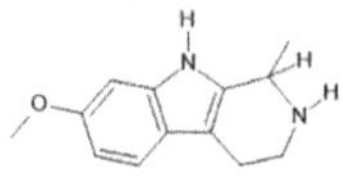 **(3)**	**Harmalinic acid** 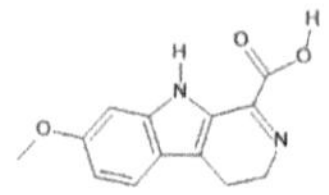 **(4)**	**Harmol** 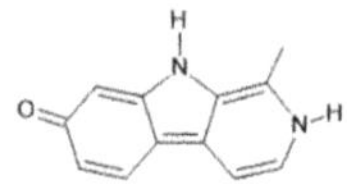 **(5)**
Harmalol 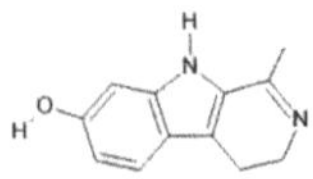 **(6)**	**Orientin** 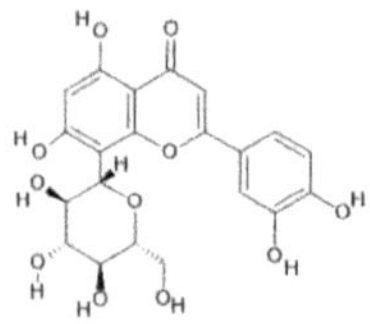 **(7)**	**Isoorientin** 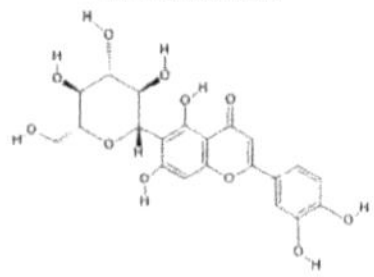 **(8)**	**Acetogenin** 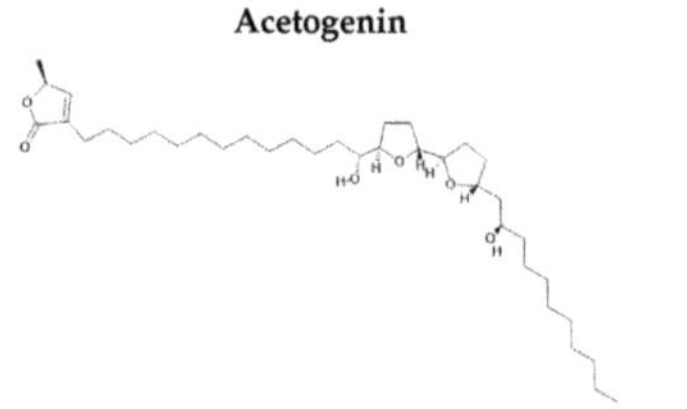 **(9)**	
δ-Cadinene 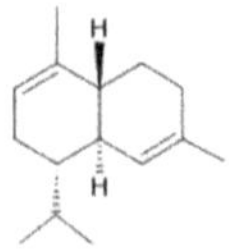 **(10)**	**α-Muurolene** 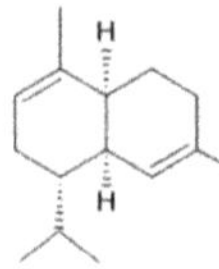 **(11)**	**Uncarine F** 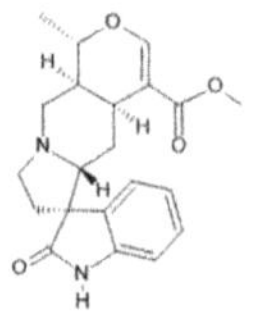 **(12)**	**Speciophylline** 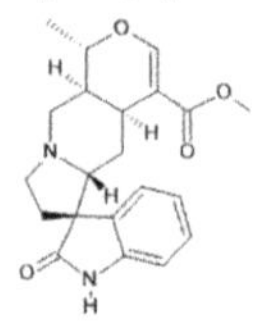 **(13)**	**Mitraphylline** 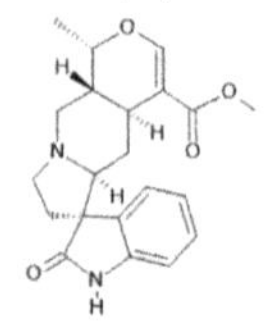 **(14)**
Fisetin 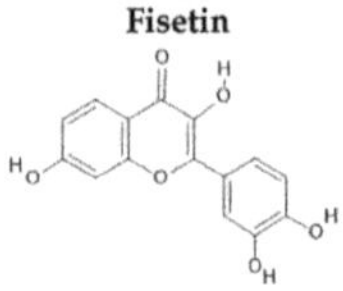 **(15)**	**Cyclosativene** 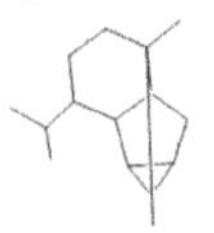 **(16)**	**Caryophyllene** 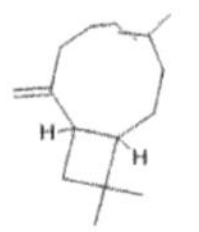 **(17)**	**α-himachalene** 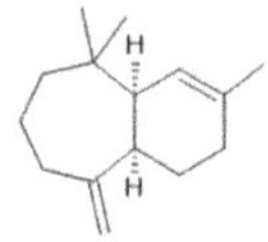 **(18)**	**Linalool** 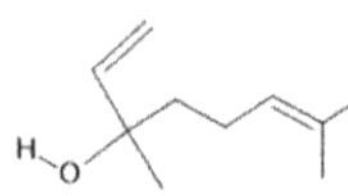 **(19)**
α-caryophyllene 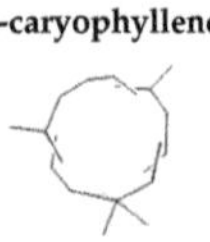 **(20)**	**β-caryophyllene** 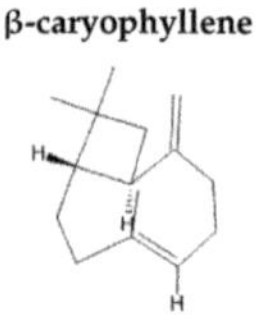 **(21)**	**Isovitexin** 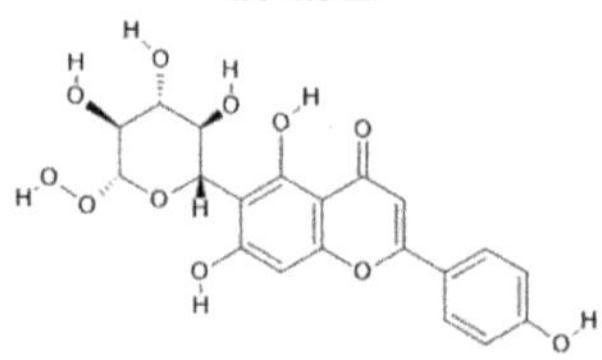 **(22)**		**Spilanthol** 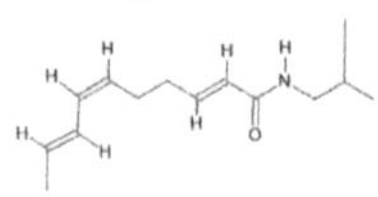 **(23)**
Spermidine 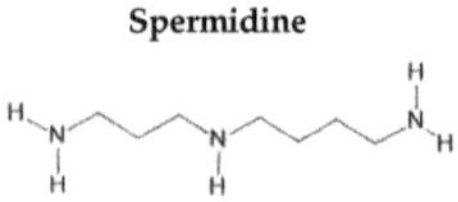 **(24)**	**Spermine** 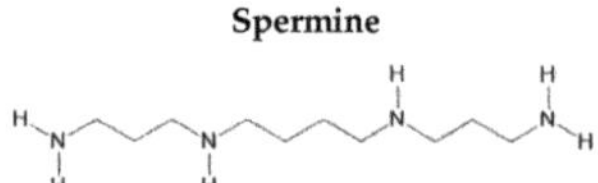 **(25)**		**3-acetylaleuritolic acid** 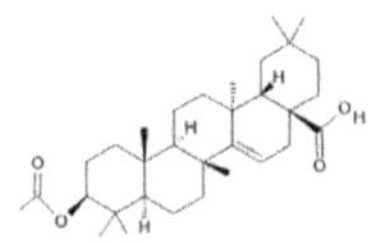 **(26)**	**Camphene** 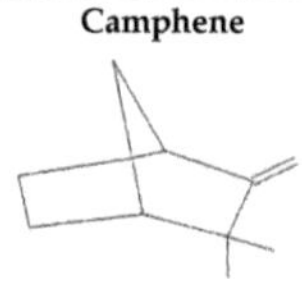 **(27)**
Limonene 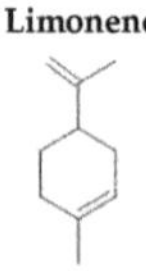 **(28)**	**Camphor** 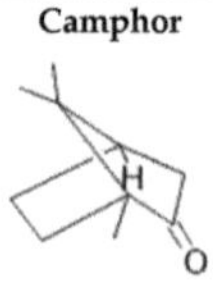 **(29)**	**Borneol** 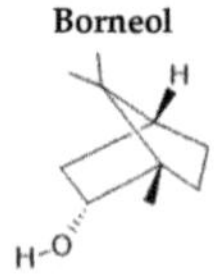 **(30)**	**Cineol** 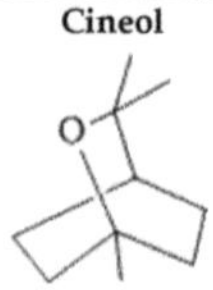 **(31)**	

The numbers in bold correspond to the molecular structures shown below.

3.3.2. *Peganum harmala* L.

The studied group belongs to the kingdom Plantae, class Equisetopsida C. Agardh, order Sapindales Juss. ex Bercht. & J. Presl, family Nitrariaceae Lindl., genus *Peganum* L. and species *Peganum harmala* L. [139]. The herbaceous *P. harmala* L. is perennial and branched, with leaves sectioned into three to five linear lobes. It produces whitish-yellow flowers and fruits in globular capsules with three chambers, containing black angular seeds [140]. It is commonly called wild rue, Syrian rue, or African rue [141].

Most species of the Nitrariaceae family contain alkaloids, which have been the subject of studies for their possible biological and pharmacological activity. For example, the studies by Niu et al. [142], when investigating the protective effect of harmine—the major compound isolated from *P. harmala* L.—in renal inflammation induced by lipopolysaccharide (LPS), as well as the respective molecular mechanisms involved, showed that pretreatment with harmine markedly alleviated the lesion kidney, reducing the release of renal biomarkers, inflammatory mediators, and the formation of malondialdehyde (MDA) and myeloperoxidase (MPO), while increasing superoxide dismutase (SOD) and glutathione (GSH) and reducing renal histopathological changes. Furthermore, in immunohistochemical staining and western blot analysis, the study indicated that the treatment with harmine suppressed the expression of the toll-like receptor 4 (TLR4), phosphorylation of nuclear factor kappa B (NF-κB) p65, and κBα inhibitor (IκBα), while the treatment also inhibited the expression of NLRP3, caspase-1, and interleukin-1β (IL-1β). In summary, pretreatment with harmine extracted from *P. harmala* L. can protect against LPS-induced acute kidney injury by attenuating oxidative stress and inflammatory responses and increasing antioxidant activity. The underlying mechanisms of harmine in mice with LPS-induced acute kidney injury may be related to the inhibition of the TLR4-NF-κB and NLRP3 pathways of the inflammasome.

Another study observed the effects of harmine on the renal activity of mice after cisplatin administration. The researchers demonstrated that there was a significant decrease in the total antioxidant capacity of the renal tissue, in the diameter of the renal corpuscles, and in the level of IL-10 expression in the group treated with cisplatin in relation to the control group, while the values of these parameters were significantly similar to those of the control group in the moderate or high dose groups treated with harmine + cisplatin. In addition, they noted significant increases in serum levels of urea and creatinine, Bowman's space, amount of malondialdehyde, apoptosis rate, and gene expressions of TNF-α, NF-κB, IL-1β, and caspase-3 in the renal tissue of the cisplatin group compared to the control group, while these criteria did not differ in the harmine + cisplatin moderate or high dose groups. Thus, the study considered that harmine protected the kidneys against damage induced by cisplatin, and the antioxidant, anti-inflammatory, and anti-apoptotic properties of this compound were involved in the observed curative effect [118].

3.3.3. *Passiflora edulis* Sims

The species under study belongs to the kingdom Plantae, class Equisetopsida C. Agardh, order Malpighiales Juss. ex Bercht. & J. Presl, family Passifloraceae Juss. ex Roussel, genus *Passiflora* L. and species *Passiflora edulis* Sims, with wide ethnopharmacological use by the people of the Amazon. *Passiflora edulis* Sims is a vine, supported by axillary tendrils [143]. It consists of palmate leaves, usually three-lobed with serrated margins; large flowers, with long peduncles, whitish, with a purple and pink triple crown; fruits, oval-shaped berries, containing abundant flat ovoid seeds, covered by a yellowish or brownish aril [144]. It has a vast geographic distribution: Brazil, Paraguay, Argentina, Antilles (West Indies islands), Central America, Venezuela, and Ecuador. It is commonly called passion fruit [121].

Pharmacological trials have shown numerous activity from compounds obtained from *P. edulis* Sims, including anxiolytic, sedative, neuropathic pain [118], activity linked to alcoholism and narcotics use [145], anticonvulsant and anxiolytic activity [119], cognitive function and degenerative diseases [120], antioxidant, antitumor action, hypoglycemic action, obesity, and insomnia [121].

The species is rich in natural bioactive compounds, among them, a significant content of flavonoids. For example, orientin and isoorientin are compounds with potential hypoglycemic effects, pointed out in the study by Galdino et al. [117] when evaluating the therapeutic effect of the aqueous extract of the fruit peel of *P. edulis* Sims as an adjuvant to insulin, to confer nephroprotection against diabetes induced by streptozotocin in Wistar rats. In the study, those animals treated with *P. edulis* extract showed superior glycemic control, which resulted in a reduction in the urinary albumin/creatinine ratio; maintenance of basal levels of mRNA expression of Nphs1, Nphs2, and Wt1n in the renal tissue; expression of mRNA Lrp2; prevention of protein loss from the renal tissue to the urinary space; and maintenance of glomerular basement membrane thickness, hyalinization, and glomerular and tubulointerstitial fibrosis with values close to those of the control group and significantly lower than those in the diabetic group. Therefore, the extract of *P. edulis* revealed potential therapeutic action of nephroprotection due to the reduction and prevention of the development of diabetic kidney disease.

The protective effect of flavonoids from *P. edulis* Sims was evaluated in alloxan-induced diabetic *Rattus norvegicus*, in which researchers observed renal dysfunction in uncontrolled diabetic groups, given the increased production of free radicals, with probable cellular damage and tubular damage, resulting in renal inflammation. In the study, the biomarkers urea and creatinine were measured in the animals' bloodstream. Diabetic animals that received the flavonoid fraction of *P. edulis* Sims had lower urea and creatinine values when compared to the control group [146].

3.3.4. *Annona muricata* L.

This plant species belongs to the kingdom Plantae, class Equisetopsida C. Agardh, order Magnoliales Bromhead, family Annonaceae Juss., genus *Annona* L. and species *Annona muricata* L. It has various uses in traditional indigenous medicine [139]. It is medium to large in size, reaching up to 10 m in height. Its leaves are green and shiny, with an oval shape and smooth texture, while its flowers are large and solitary, with thick, yellowish petals [147]. The species is widely distributed geographically, being found in several tropical regions of the world, including Central and South America, Africa, and Asia. It is known by several popular names, such as soursop, fruit of the count, and heart of queen, and is cultivated for its edible fruits, which have a sweet and slightly acidic taste [148].

In addition to its gastronomic uses, *A. muricata* L. is also consumed due to its medicinal properties. Many of these properties come from bioactive compounds present in its leaves, seeds, and fruits, with antioxidant, anti-inflammatory, antiparasitic, and anticancer activity [124].

A well-described example in the scientific literature is the species *A. muricata* L. This species has been studied for its bioactive metabolites, including acetogenins, and its constituents may have anticancer, hepatoprotective, neurotoxic, antinociceptive, antiulcerative, and chemopreventive activity [122].

A study of veterinary pharmacology and toxicology [149] demonstrated that *A. muricata* L. attenuates glycerol-induced nephrotoxicity in male albino rats through angiotensin-converting enzyme (ACE) signaling pathways. The methanolic extract of the leaves of *A. muricata* L., in that study, caused a significant decrease in the expression of the kidney injury molecule 1 (KIM-1) and exhibited antioxidant properties. This nephroprotective effect of the extract was observed by improving the levels of enzymatic and non-enzymatic antioxidants, suppressing inflammatory processes and inhibiting lipid peroxidation, thus revealing such antioxidant and anti-inflammatory properties.

3.3.5. *Uncaria tomentosa* (Willd.) DC.

The group studied has wide ethnopharmacological use in the Amazon. It belongs to the kingdom Plantae, class Equisetopsida C. Agardh, order Gentianales Juss. ex Bercht. & J. Presl, family Rubiaceae Juss., genus *Uncaria* Schreb. and species *Uncaria tomentosa* (Willd.) DC. [150]. Preliminary phytochemical screenings demonstrated the marked presence of tannins in species belonging to the Rubiaceae family. These plant species are widely disseminated in the culture of traditional people and communities, due to their richness in the production of bioactive compounds [151,152].

Other phytochemical studies have found tetracyclic and pentacyclic oxindole alkaloids, indole and β-carbonyl alkaloids, flavonoids [153], coumarins [154], proanthocyanidins, steroids, ursan-derived triterpenoids, and quinovic acid glycosides [125].

The species *U. tomentosa* (Willd.) DC. has been associated with several health benefits, such as antioxidant and immunomodulatory, anti-inflammatory, analgesic and anticancer action, in addition to other medicinal properties. In traditional medicine, the plant is used to treat a variety of conditions, including infections, arthritis, diabetes, gastrointestinal problems, and "kidney cleansing" [126].

The renal benefits of herbal medicines such as *U. tomentosa* (Willd.) DC. have been demonstrated in the studies by Vattimo and Silva [155], when performing a pretreatment with *U. tomentosa* (Willd.) DC. in experimental models of ischemia/reperfusion, in which there was functional protection assessed by increased creatinine clearance, reduced peroxidation, and urinary thiobarbituric acid reactive substances (TBARS), probably related to the antioxidant activity of the herbal medicine.

3.3.6. *Hymenaea courbaril* L.

The studied group belongs to the kingdom Plantae, class Equisetopsida C. Agardh, order Fabales Bromhead, family Fabaceae Lindl., genus *Hymenaea* L. and species *Hymenaea courbaril* L. [156]. The Fabaceae family is known to produce a wide variety of bioactive compounds, including tannins. The tannins present in *Hymenaea courbaril* L. have been the object of research for their potential biological effects, such as antioxidant, antiulcerogenic, anti-inflammatory, and antitumor properties [128]. Some studies also report the antiviral and antibacterial activity of these compounds [157].

Hymenaea courbaril L. has a wide distribution in South America and Central America; it is a large tree, reaching 15 to 20 m in height, and the trunk can be up to 1 m in diameter [158]. The flowers are pollinated by bats. Ripe fruits are eaten by rodents, birds, and monkeys, which, when breaking the fruits, release the seeds [159]. Its wood is considered valuable due to its high density and resistance to attack by xylophagous organisms. In the Amazon region, the species is known as jassaí, jataí, jataíba, jataíba stone, jataúba, jatel, jati, jatobá de anta, jutaí, jutaí açu, jutaí white, jutaí grande, and jutaí catinga [160].

The main phytochemical constituents found In the species are flavonoids, such as quercetin, kaempferol, and isorhamnetin, which are present in the leaves and fruits; tannins, such as catechins and proanthocyanidins, most commonly found in bark and seeds; fatty acids, including oleic and linoleic acid (seeds); and stilbenes, such as trans-resveratrol, found in the bark and fruit [161,162].

The tea produced from the bark of *H. courbaril* L. is indicated to treat kidney problems [163]. Pereira et al. [164] demonstrated the oxidizing activity of the methanolic fraction of *H. courbaril* L. seeds in mice treated with acetaminophen; the study showed the probable restoration of renal glutathione (GSH) levels in animals treated with the extract, in addition to reversing the increase in carbonylated proteins. Another study, using aqueous extracts of seed or bark of *H. courbaril* L., observed a reduction in renal levels of reactive substances to thiobarbituric acid 7 days after treatment [165].

3.3.7. *Echinodorus macrophyllus* (Kunth) Micheli

This species belongs to the kingdom Plantae, class Equisetopsida C. Agardh, order Alismatales R. Br. ex Bercht. & J. Presl, family Alismataceae Vent., genus *Echinodorus*

Rich. and the species *Echinodorus macrophyllus* (Kunth) Micheli, widely used by traditional medicine in Brazil [166]. The Alismataceae family is known for the presence of several bioactive compounds. Many Alismataceae species have traditionally been used in folk medicine for the treatment of a vast range of diseases, due to their diuretic and anti-inflammatory effects, as well as in kidney and liver disorders [131]. Some of the bioactive compounds present in plants of this family, such as *Echinodorus macrophyllus*, have been the object of scientific investigation for their pharmacological properties and possible therapeutic applications [167].

Echinodorus macrophyllus is a perennial plant, herbaceous or subshrub, of aquatic origin, emerging from the water. It has rhizomes and can reach between 1 and 2 m in height. Its leaves are petiolate and oval, with a heart-shaped base and a sharp tip [168]. Popularly, the species is known as leather hat, water hyacinth, campanha tea, brejo tea, poor man's tea, mineiro tea, congonha do brejo, brejo herb, and swamp herb [169].

According to Silva et al. [129], among the constituents produced by the species are the terpenic profile containing linalool, α- and β-caryophyllene, *E*-nerolidol, and phytol as predominant, as well as a variety of diterpenoids belonging to the same classes, such as the chapecoderines of the group labdanos. Furthermore, a (+)-3-carene derivative was detected, along with a significant proportion of carotenoids. Gasparotto et al. [130] demonstrated the presence of the flavonoids vitexin and isovitexin. While Garcia et al. [170] found the presence of phenylpropanoids in the species, such as ferulic and *E*-caffeoyl tartronic acid (2-*E*-caffeoyloxymalonic acid).

Traditionally, the Amazonian population uses extracts from the leaves of *E. macrophyllus*, from infusion, decoction, or maceration methods, in water or alcohol, to treat urinary system disorders, as they are known to be powerful diuretic agents. In view of popular usage, Nascimento et al. [171] demonstrated that preconditioning with *E. macrophyllus* attenuated cyclophosphamide-induced acute kidney injury in rats as evidenced by increased creatinine clearance and reduced oxidative metabolites in urine and increased reserve of antioxidant enzymes in renal tissue.

Studies carried out in a model of acute kidney injury induced by gentamicin found similar results of antioxidant protection of *E. macrophyllus*, when administering crude ethanolic extracts of leaves and fractions of *E. macrophyllus* by endogastric route, in normal rats or with acute tubular necrosis induced by gentamicin-cine. Thus, it was demonstrated that it produced a dose-dependent reduction in urine output. The extracts in question were effective in reversing all changes induced by gentamicin, such as polyuria and reduction in the glomerular filtration rate; in addition, the morphological changes induced by gentamicin were not observed in animals that were treated with extracts of *E. macrophyllus* concomitantly with gentamycin [167].

3.3.8. *Acmella oleracea* (L.) R. K. Jansen

The studied group belongs to the kingdom Plantae, class Equisetopsida C. Agardh, order Asterales Link, family Asteraceae Bercht. & J. Presl, genus *Acmella* Rich. ex Pers. and the species *Acmella oleracea* (L.) R. K. Jansen, which is widely used in medicine and cooking by traditional Amazonian people [172]. The Asteraceae family is known for the presence of bioactive compounds, such as sesquiterpene lactones and flavonoids [173]. Primarily for their medicinal properties, species in this family have traditionally been used to treat a wide range of ailments, including respiratory problems, inflammation, headaches, gastrointestinal problems, and infectious diseases [24]. Scientific research has focused on some of the bioactive compounds present in plants of this family, such as *Acmella oleracea*, in search of possible therapeutic applications, such as analgesic, anti-inflammatory, antimicrobial, and antioxidant activity [136].

Acmella oleracea is an important medicinal herb, which occurs in tropical and subtropical regions of the planet. It is an annual, perennial herbaceous, 30–40 cm high, semi-straight, or creeping, with cylindrical, a fleshy stem and decumbent branches, usually without roots at the nodes. The main root is pivotal, with abundant lateral branches. The leaves are opposite, membranous, and petiolate [174]. The species is popularly known as jambu, cress from pará, abecedária, cress bravo, cress from Brazil, cress from the north, buttercup, crazy herb, jabuaçu, and nhambu [136].

Acmella oleracea is used in northern Brazil for the treatment of various diseases, such as tuberculosis, flu, cough, and rheumatism, and as an anti-inflammatory; in addition, hydroethanolic formulations with this species are popularly used as a female aphrodisiac, for treatment of male sexual dysfunctions, and as a diuretic [135,136].

Regarding the production of metabolites, *A. oleracea* is a rich source of secondary metabolites, and its phytochemistry has been widely investigated [175]. Borges et al. [176] observed an increase of 31.6% in the content of spilanthol and 16.8% of flavonoids in the inflorescences and higher contents of total phenols, carotenoids, spermidine, and spermine in the leaves and flowers of jambú. The work by Abeysiri et al. [132] revealed that alkaloids, flavonoids, saponins, steroid glycosides, and tannins are distributed in all parts of the plant. Going into more detail about the phytochemical composition of *A. oleracea*, several triterpenoids were found, such as 3-acetylaleuritolic acid, β-sitostenone, and stigmasterol. Furthermore, steroidal glycosides, namely, stigmasteryl-3-O-β-D-glucopyranoside and β-sitosteryl-3-O-β-D-glucopyranoside, have been identified. Several phenolic compounds were also detected, such as vanillic, trans-ferulic, and trans-isoferulic acids; scopoletin; and fatty acids such as n-hexadecanoic and n-tetradecanoic acids [132–134].

Some studies have observed a marked diuretic action of aqueous extract of *A. oleracea* inflorescences in rats; the authors have described an increase in Na^+ and K^+ levels and a reduction in osmolarity in the urine of animals treated with the extract [177]. Yadav and collaborators [178] showed that the ethanolic extract of *A. oleracea* in rats provided diuresis similar to that produced by the action of furosemide. Gerbino et al. [179] consider that the inhibition of cyclic AMP induced by spilanthol negatively modulates the mechanisms of urine concentration. Furthermore, the mechanisms of action on the kidney show that *A. oleracea* is a promising source of compounds with diuretic activity.

3.3.9. *Rosmarinus officinalis* L.

The species belongs to the kingdom Plantae, class Equisetopsida C. Agardh, order Lamiales Bromhead, family Lamiaceae Martinov, genus *Rosmarinus* L. and species *Rosmarinus officinalis* L.; it has wide ethnopharmacological use [180]. The Lamiaceae family is one of the most important herbaceous families; it is composed of an immense variety of plant species with biological and medicinal applications [138]. This family includes numerous aromatic spices, including *Rosmarinus officinalis* L., a plant species commonly known as rosemary, which is useful in cooking due to its characteristic aroma; it is widely used by indigenous populations where it grows spontaneously [181].

Rosmarinus officinalis is a shrubby herb, widely used in culinary, medicinal, and commercial applications, including the fragrance and food industries [182]. The leaves (fresh or dried) are consumed due to the characteristic odor that they offer to the dish. They are also consumed in small amounts in the form of tea, while extracts of *R. officinalis* are regularly used for their active natural antioxidant properties to improve shelf life of perishable foods [183].

Phytochemical screenings carried out on the species revealed 0.5% to 2.5% volatile oil in the leaves. Among the phytocompounds, the species exhibits the presence of monoterpene hydrocarbons (alpha and beta-pinene), camphene, limonene, camphor (10% to 20%), borneol, cineol, linalool, and verbinol [137]. In addition to numerous volatile and aromatic components, the species has flavonoids, such as diosmetin, diosmin, genkwanin, luteolin, hispidulin, and apigenin, as well as terpenoid compounds such as triterpenes (oleanolic and ursolic acid) and diterpene carnosol. Among the phenols found in the species are caffeic, chlorogenic, labiatic, neochlorogenic, and rosmarinic acids, as well as a considerable number of salicylates [182–184].

Among the ethnomedicinal applications for *R. officinalis* are analgesic, anti-inflammatory, anticarcinogenic, antirheumatic, nephroprotective, spasmolytic, antihepatotoxic, atherosclerotic, carminative, and choleretic action. It also offers protection against UV and gamma radiation and improvement of stress [138].

Zohrabi et al. [185] investigated the effect of an oral extract of *R. officinalis* on acute renal failure (ARF) disorders induced by ischemia/reperfusion in rats. The authors showed that the aqueous extract of *R. officinalis* suffered the oxidative stress marker malondialdehyde (MDA), increased the ferric antioxidant reducing power (FRAP) compared to the vehicle groups and, regarding the histopathological analyses, observed a significant reduction in vessel management, disturbance of the tubules, and Bowman's Capsule space compared to the vehicle groups.

Another study evaluated the effectiveness of *R. officinalis* essential oil (REO) against changes induced by potassium dichromate in the kidneys of male rats, in which they injected hexavalent chromium to induce renal dysfunction (oxidative damage and alterations in the antioxidant defense system, and histopathological and immunohistochemical alterations). The animals were treated with REO before or after the induction of renal dysfunction, resulting in an improvement in the toxic effect by extinguishing, chelating, and detoxifying free radicals and enhancing the state of antioxidant defense [186].

4. Conclusions

Knowing the phytocompounds with potential nephroprotective effects against AKI based on the traditional Amazonian knowledge of treating different ailments that disturb/affect the health of the kidneys is generally passed on over generations by healers, housewives, and elderly people from riverside communities, who, due to limited access to health services, use this precious information about the natural resources of the Amazon as their only resource. The pharmacotoxicological validation of this information is highly necessary, considering that it subsidizes the knowledge of the medicinal potential of the Amazonian flora, substantially improving the phytochemical and pharmacological relevance of these species, especially in the face of AKI, a clinical condition with high morbidity and mortality. Although much of the research on the nephroprotective potential of Amazonian plant species is still in the preclinical stage, these plants show promise as a potential source of new therapies for kidney disease. However, more research is needed to fully understand its mechanisms of action and possible side effects, as well as to develop safe and effective dosages for human use.

Author Contributions: Conceptualization, A.S.P. and R.d.C.R.K.; writing—original draft preparation, A.S.P., P.F.S., R.d.C.R.K. and H.K.S.A.; writing—review and editing, R.d.C.R.K., T.A.C.C.T. and J.C.T.C.; supervision, J.C.T.C. All authors have read and agreed to the published version of the manuscript.

Funding: National Council for Scientific and Technological Development—CNPq No. 12/2020—Proc.: 403587/2020-4, Master's and Doctoral Program for Innovation—MAI/DAI.

Acknowledgments: This work was supported in part by the National Council for Scientific and Technological Development—CNPq No. 12/2020—Proc.: 403587/2020-4, Master's and Doctoral Program for Innovation—MAI/DAI.

Conflicts of Interest: The authors declare no conflict of interest.

References

1. De Carvalho Koga, R.; de Lima Teixeira dos Santos, A.V.T.; do Socorro Ferreira Rodrigues Sarquis, R.; Carvalho, J.C.T. *Bauhinia guianensis* Aubl., a Plant from Amazon Biome with Promising Biologically Active Properties: A Systematic Review. *Pharmacogn. Rev.* **2021**, *15*, 76–81. [CrossRef]
2. Sales, P.F.; de Carvalho Rocha Koga, R.; de Souza, A.A.; do Nascimento, A.L.; Pinheiro, F.C.; Alberto, A.K.M.; da Costa, M.J.; Carvalho, J.C.T. Pharmacological Potential and Mechanisms of Action Involved in Oil Species from the Brazilian Amazon: The Case of *Abelmoschus esculentus* L. Moench, Euterpe Oleracea Martius and Bixa Orellana Linné. *Pharmacogn. Rev* **2023**, *17*, 24–42. [CrossRef]
3. Méril-Mamert, V.; Ponce-Mora, A.; Sylvestre, M.; Lawrence, G.; Bejarano, E.; Cebrián-Torrejón, G. Antidiabetic Potential of Plants from the Caribbean Basin. *Plants* **2022**, *11*, 1360. [CrossRef]

4. Rates, S.M.K. Plants as Source of Drugs. *Toxicon* **2001**, *39*, 603–613. [CrossRef]
5. Da Silva, H.R.; de Assis, D.D.C.; Ariadna, L.P.; Hady, K.; Jesus, R.R.A.; José, C.T.C. Euterpe Oleracea Mart. (Aai): An Old Known Plant with a New Perspective. *Afr. J. Pharm. Pharmacol.* **2016**, *10*, 995–1006. [CrossRef]
6. Carvalho, J. *Fitoterápicos Anti-Inflamatórios*, 2nd ed.; Pharmabooks: São Paulo, Brazil, 2017; ISBN 139788589731805.
7. Carvalho, J.C.T.; Perazzo, F.F.; Machado, L.; Bereau, D. Biologic Activity and Biotechnological Development of Natural Products. *BioMed Res. Int.* **2013**, *2013*, 971745. [CrossRef]
8. Isgut, M.; Rao, M.; Yang, C.; Subrahmanyam, V.; Rida, P.C.G.; Aneja, R. Application of Combination High-Throughput Phenotypic Screening and Target Identification Methods for the Discovery of Natural Product-Based Combination Drugs. *Med. Res. Rev.* **2018**, *38*, 504–524. [CrossRef]
9. Najmi, A.; Javed, S.A.; al Bratty, M.; Alhazmi, H.A. Modern Approaches in the Discovery and Development of Plant-Based Natural Products and Their Analogues as Potential Therapeutic Agents. *Molecules* **2022**, *27*, 349. [CrossRef]
10. Kang, H.G.; Lee, H.K.; Cho, K.B.; Park, S.I. A Review of Natural Products for Prevention of Acute Kidney Injury. *Medicina* **2021**, *57*, 1266. [CrossRef]
11. Neyra, J.A.; Chawla, L.S. Acute Kidney Disease to Chronic Kidney Disease. *Crit. Care Clin.* **2021**, *37*, 453–474. [CrossRef]
12. Kellum, J.A.; Romagnani, P.; Ashuntantang, G.; Ronco, C.; Zarbock, A.; Anders, H.-J. Acute Kidney Injury. *Nat. Rev. Dis. Primers* **2021**, *7*, 52. [CrossRef]
13. Macedo, E.; Bouchard, J.; Soroko, S.H.; Chertow, G.M.; Himmelfarb, J.; Ikizker, T.A.; Paganini, E.P.; Mehta, R.L. Fluid Accumulation, Recognition and Staging of Acute Kidney Injury in Critically-Ill Patients. *Crit. Care* **2010**, *14*, R82. [CrossRef]
14. Lewington, A.J.P.; Cerdá, J.; Mehta, R.L. Raising Awareness of Acute Kidney Injury: A Global Perspective of a Silent Killer. *Kidney Int.* **2013**, *84*, 457–467. [CrossRef]
15. Makris, K.; Spanou, L. Acute Kidney Injury: Definition, Pathophysiology and Clinical Phenotypes. *Clin. Biochem. Rev.* **2016**, *37*, 85–98. [PubMed]
16. Pamunuwa, G.; Karunaratne, D.N.; Waisundara, V.Y. Antidiabetic Properties, Bioactive Constituents, and Other Therapeutic Effects of *Scoparia dulcis*. *Evid.-Based Complement. Altern. Med.* **2016**, *2016*, 8243215. [CrossRef]
17. Yang, L.; Stöckigt, J. Trends for Diverse Production Strategies of Plant Medicinal Alkaloids. *Nat. Prod. Rep.* **2010**, *27*, 1469. [CrossRef]
18. Zenk, M.H.; Juenger, M. Evolution and Current Status of the Phytochemistry of Nitrogenous Compounds. *Phytochemistry* **2007**, *68*, 2757–2772. [CrossRef] [PubMed]
19. Adhikari, B. Roles of Alkaloids from Medicinal Plants in the Management of Diabetes Mellitus. *J. Chem.* **2021**, *2021*, 2691525. [CrossRef]
20. Putra, I.M.W.A.; Fakhrudin, N.; Nurrochmad, A.; Wahyuono, S. A Review of Medicinal Plants with Renoprotective Activity in Diabetic Nephropathy Animal Models. *Life* **2023**, *13*, 560. [CrossRef]
21. Gangasani, J.K.; Pemmaraju, D.B.; Murthy, U.S.N.; Rengan, A.K.; Naidu, V.G.M. Chemistry of Herbal Biomolecules. In *Herbal Biomolecules in Healthcare Applications*; Elsevier: Amsterdam, The Netherlands, 2022; pp. 63–79.
22. Sawant, M.; Isaac, J.C.; Narayanan, S. Analgesic Studies on Total Alkaloids and Alcohol Extracts of *Eclipta Alba* (Linn.) Hassk. *Phytother. Res.* **2004**, *18*, 111–113. [CrossRef]
23. Morales-García, J.A.; de la Fuente Revenga, M.; Alonso-Gil, S.; Rodríguez-Franco, M.I.; Feilding, A.; Perez-Castillo, A.; Riba, J. The Alkaloids of Banisteriopsis Caapi, the Plant Source of the Amazonian Hallucinogen Ayahuasca, Stimulate Adult Neurogenesis in Vitro. *Sci. Rep.* **2017**, *7*, 5309. [CrossRef]
24. Batista, L.L.; do Nascimento, L.C.; Guimarães, G.F.; Matias Pereira, A.C.; Koga, R.d.C.R.; Teixeira dos Santos, A.V.T.d.L.; Fernandes, C.P.; Teixeira, T.A.; Hu, Y.; Hu, X.; et al. A Review of Medicinal Plants Traditionally Used to Treat Male Sexual Dysfunctions—The Overlooked Potential of *Acmella oleracea* (L.) R.K. Jansen. *Pharmacogn. Rev.* **2021**, *15*, 1–11. [CrossRef]
25. Rasouli, H.; Yarani, R.; Pociot, F.; Popović-Djordjević, J. Anti-Diabetic Potential of Plant Alkaloids: Revisiting Current Findings and Future Perspectives. *Pharmacol. Res.* **2020**, *155*, 104723. [CrossRef] [PubMed]
26. Salahshoor, M.; Roshankhah, S.; Motavalian, V.; Jalili, C. Effect of Harmine on Nicotine-Induced Kidney Dysfunction in Male Mice. *Int. J. Prev. Med.* **2019**, *10*, 97. [CrossRef] [PubMed]
27. Fang, C.; Lou, D.; Zhou, L.; Wang, J.; Yang, B.; He, Q.; Wang, J.; Weng, Q. Natural Products: Potential Treatments for Cisplatin-Induced Nephrotoxicity. *Acta Pharmacol. Sin.* **2021**, *42*, 1951–1969. [CrossRef] [PubMed]
28. Sharma, V.; Boonen, J.; Chauhan, N.S.; Thakur, M.; de Spiegeleer, B.; Dixit, V.K. *Spilanthes acmella* Ethanolic Flower Extract: LC–MS Alkylamide Profiling and Its Effects on Sexual Behavior in Male Rats. *Phytomedicine* **2011**, *18*, 1161–1169. [CrossRef]
29. Beecher, G.R. Overview of Dietary Flavonoids: Nomenclature, Occurrence and Intake. *J. Nutr.* **2003**, *133*, S3248–S3254. [CrossRef] [PubMed]
30. Banjarnahor, S.D.S.; Artanti, N. Antioxidant Properties of Flavonoids. *Med. J. Indones.* **2015**, *23*, 239–244. [CrossRef]
31. Van Acker, S.A.B.E.; Van Den Berg, D.; Tromp, M.N.J.L.; Griffioen, D.H.; Van Bennekom, W.P.; Van Der Vijgh, W.J.F.; Bast, A. Structural Aspects of Antioxidant Activity of Flavonoids. *Free Radic. Biol. Med.* **1996**, *20*, 331–342. [CrossRef]
32. Vinayagam, R.; Xu, B. Antidiabetic Properties of Dietary Flavonoids: A Cellular Mechanism Review. *Nutr. Metab.* **2015**, *12*, 60. [CrossRef]
33. Rufino, A.T.; Costa, V.M.; Carvalho, F.; Fernandes, E. Flavonoids as Antiobesity Agents: A Review. *Med. Res. Rev.* **2021**, *41*, 556–585. [CrossRef] [PubMed]

34. Unnikrishnan, M.K.; Veerapur, V.; Nayak, Y.; Mudgal, P.P.; Mathew, G. Antidiabetic, Antihyperlipidemic and Antioxidant Effects of the Flavonoids. In *Polyphenols in Human Health and Disease*; Elsevier: Amsterdam, The Netherlands, 2014; pp. 143–161.
35. Maleki, S.J.; Crespo, J.F.; Cabanillas, B. Anti-Inflammatory Effects of Flavonoids. *Food Chem.* **2019**, *299*, 125124. [CrossRef] [PubMed]
36. Di Carlo, G.; Mascolo, N.; Izzo, A.A.; Capasso, F. Flavonoids: Old and New Aspects of a Class of Natural Therapeutic Drugs. *Life Sci.* **1999**, *65*, 337–353. [CrossRef] [PubMed]
37. Al Aboody, M.S.; Mickymaray, S. Anti-Fungal Efficacy and Mechanisms of Flavonoids. *Antibiotics* **2020**, *9*, 45. [CrossRef]
38. Lu, X.; Liu, T.; Chen, K.; Xia, Y.; Dai, W.; Xu, S.; Xu, L.; Wang, F.; Wu, L.; Li, J.; et al. Isorhamnetin: A Hepatoprotective Flavonoid Inhibits Apoptosis and Autophagy via P38/PPAR-α Pathway in Mice. *Biomed. Pharmacother.* **2018**, *103*, 800–811. [CrossRef]
39. Vauzour, D.; Vafeiadou, K.; Rodriguez-Mateos, A.; Rendeiro, C.; Spencer, J.P.E. The Neuroprotective Potential of Flavonoids: A Multiplicity of Effects. *Genes Nutr.* **2008**, *3*, 115–126. [CrossRef] [PubMed]
40. Al-Khayri, J.M.; Sahana, G.R.; Nagella, P.; Joseph, B.V.; Alessa, F.M.; Al-Mssallem, M.Q. Flavonoids as Potential Anti-Inflammatory Molecules: A Review. *Molecules* **2022**, *27*, 2901. [CrossRef]
41. Vargas, F.; Romecín, P.; García-Guillén, A.I.; Wangesteen, R.; Vargas-Tendero, P.; Paredes, M.D.; Atucha, N.M.; García-Estañ, J. Flavonoids in Kidney Health and Disease. *Front. Physiol.* **2018**, *9*, 394. [CrossRef]
42. Silva, W.T.; dos Santos, J.G.; Watanabe, M.; Vattimo, M.d.F.F. Efeito Renoprotetor Dos Flavonoides Do Vinho Na Nefrotoxicidade Do Imunossupressor Tacrolimus. *Acta Paul. Enferm.* **2011**, *24*, 388–392. [CrossRef]
43. Galati, G.; O'Brien, P.J. Potential Toxicity of Flavonoids and Other Dietary Phenolics: Significance for Their Chemopreventive and Anticancer Properties. *Free Radic. Biol. Med.* **2004**, *37*, 287–303. [CrossRef] [PubMed]
44. Orhan, D.D.; Özçelik, B.; Özgen, S.; Ergun, F. Antibacterial, Antifungal, and Antiviral Activities of Some Flavonoids. *Microbiol. Res.* **2010**, *165*, 496–504. [CrossRef]
45. Kumar, G.; Banu, G.S.; Pappa, P.V.; Sundararajan, M.; Pandian, M.R. Hepatoprotective Activity of *Trianthema portulacastrum* L. against Paracetamol and Thioacetamide Intoxication in Albino Rats. *J. Ethnopharmacol.* **2004**, *92*, 37–40. [CrossRef]
46. Peng, P.; Zou, J.; Zhong, B.; Zhang, G.; Zou, X.; Xie, T. Protective Effects and Mechanisms of Flavonoids in Renal Ischemia-Reperfusion Injury. *Pharmacology* **2023**, *108*, 27–36. [CrossRef]
47. Korkmaz, A.; Kolankaya, D. Protective Effect of Rutin on the Ischemia/Reperfusion Induced Damage in Rat Kidney. *J. Surg. Res.* **2010**, *164*, 309–315. [CrossRef]
48. Liu, Y.; Wang, L.; Du, Y.; Chen, Z.; Guo, J.; Weng, X.; Wang, X.; Wang, M.; Chen, D.; Liu, X. Effects of Apigenin Pretreatment against Renal Ischemia/Reperfusion Injury via Activation of the JAK2/STAT3 Pathway. *Biomed. Pharmacother.* **2017**, *95*, 1799–1808. [CrossRef]
49. Liu, M.; Grigoryev, D.N.; Crow, M.T.; Haas, M.; Yamamoto, M.; Reddy, S.P.; Rabb, H. Transcription Factor Nrf2 Is Protective during Ischemic and Nephrotoxic Acute Kidney Injury in Mice. *Kidney Int.* **2009**, *76*, 277–285. [CrossRef]
50. Viñas, J.L.; Sola, A.; Genescà, M.; Alfaro, V.; Pí, F.; Hotter, G. NO and NOS Isoforms in the Development of Apoptosis in Renal Ischemia/Reperfusion. *Free Radic Biol. Med.* **2006**, *40*, 992–1003. [CrossRef] [PubMed]
51. Okusa, M.D. The Inflammatory Cascade in Acute Ischemic Renal Failure. *Nephron* **2002**, *90*, 133–138. [CrossRef]
52. Wu, H.; Chen, G.; Wyburn, K.R.; Yin, J.; Bertolino, P.; Eris, J.M.; Alexander, S.I.; Sharland, A.F.; Chadban, S.J. TLR4 Activation Mediates Kidney Ischemia/Reperfusion Injury. *J. Clin. Investig.* **2007**, *117*, 2847–2859. [CrossRef] [PubMed]
53. Kumar, G.; Murugesan, A.G. Hypolipidaemic Activity of *Helicteres isora* L. Bark Extracts in Streptozotocin Induced Diabetic Rats. *J. Ethnopharmacol.* **2008**, *116*, 161–166. [CrossRef] [PubMed]
54. Barbehenn, R.V.; Constabel, C.P. Tannins in Plant–Herbivore Interactions. *Phytochemistry* **2011**, *72*, 1551–1565. [CrossRef] [PubMed]
55. White, T. Tannins—Their Occurrence and Significance. *J. Sci. Food Agric.* **1957**, *8*, 377–385. [CrossRef]
56. Bacelo, H.A.M.; Santos, S.C.R.; Botelho, C.M.S. Tannin-Based Biosorbents for Environmental Applications—A Review. *Chem. Eng. J.* **2016**, *303*, 575–587. [CrossRef]
57. Zhang, L.; Lin, Y. Tannins from Canarium Album with Potent Antioxidant Activity. *J. Zhejiang Univ. Sci. B* **2008**, *9*, 407–415. [CrossRef]
58. Liu, J.-B.; Ding, Y.-S.; Zhang, Y.; Chen, J.-B.; Cui, B.-S.; Bai, J.-Y.; Lin, M.-B.; Hou, Q.; Zhang, P.-C.; Li, S. Anti-Inflammatory Hydrolyzable Tannins from *Myricaria bracteata*. *J. Nat. Prod.* **2015**, *78*, 1015–1025. [CrossRef]
59. Tamokou, J.D.D.; Mbaveng, A.T.; Kuete, V. Antimicrobial Activities of African Medicinal Spices and Vegetables. In *Medicinal Spices and Vegetables from Africa*; Elsevier: Amsterdam, The Netherlands, 2017; pp. 207–237.
60. Okuda, T. Systematics and Health Effects of Chemically Distinct Tannins in Medicinal Plants. *Phytochemistry* **2005**, *66*, 2012–2031. [CrossRef] [PubMed]
61. Ajebli, M.; Eddouks, M. The Promising Role of Plant Tannins as Bioactive Antidiabetic Agents. *Curr. Med. Chem.* **2019**, *26*, 4852–4884. [CrossRef]
62. Wei, X.; Luo, C.; He, Y.; Huang, H.; Ran, F.; Liao, W.; Tan, P.; Fan, S.; Cheng, Y.; Zhang, D.; et al. Hepatoprotective Effects of Different Extracts From Triphala Against CCl4-Induced Acute Liver Injury in Mice. *Front. Pharmacol.* **2021**, *12*, 664607. [CrossRef]
63. Smeriglio, A.; Barreca, D.; Bellocco, E.; Trombetta, D. Proanthocyanidins and Hydrolysable Tannins: Occurrence, Dietary Intake and Pharmacological Effects. *Br. J. Pharmacol.* **2017**, *174*, 1244–1262. [CrossRef] [PubMed]

64. Oladele, J.O.; Oladele, O.T.; Ademiluyi, A.O.; Oyeleke, O.M.; Awosanya, O.O.; Oyewole, O.I. Chaya (*Jatropha tanjorensis*) Leafs Protect against Sodium Benzoate Mediated Renal Dysfunction and Hepatic Damage in Rats. *Clin. Phytoscience* **2020**, *6*, 13. [CrossRef]
65. Barros, M.E.; Schor, N.; Boim, M.A. Effects of an Aqueous Extract from Phyllantus Niruri on Calcium Oxalate Crystallization in Vitro. *Urol. Res.* **2003**, *30*, 374–379. [CrossRef]
66. Coordenação da Farmacopeia Brasileira. *Farmacopeia Brasileira*, 6th ed.; Distrito Federal—ANVISA: Brasilia, Brazil, 2019.
67. Kolekar, S.M.; Jain, B.U.; Kondawarkar, M.S. A Review on Steroids and Terpenoids (Stereochemistry, Structural Elucidation, Isolation of Steroids and Terpenoids). *Res. J. Pharm. Dos. Forms Technol.* **2019**, *11*, 126. [CrossRef]
68. Moreau, R.A.; Nyström, L.; Whitaker, B.D.; Winkler-Moser, J.K.; Baer, D.J.; Gebauer, S.K.; Hicks, K.B. Phytosterols and Their Derivatives: Structural Diversity, Distribution, Metabolism, Analysis, and Health-Promoting Uses. *Prog. Lipid Res.* **2018**, *70*, 35–61. [CrossRef]
69. Asami, T. Brassinosteroid Biosynthesis Inhibitors. *Trends Plant Sci.* **1999**, *4*, 348–353. [CrossRef] [PubMed]
70. Jovanović-Šanta, S.S.; Petri, E.T.; Klisurić, O.R.; Szécsi, M.; Kovačević, R.; Petrović, J.A. Antihormonal Potential of Selected D-Homo and D-Seco Estratriene Derivatives. *Steroids* **2015**, *97*, 45–53. [CrossRef] [PubMed]
71. Singh, A.R.; Bajaj, V.K.; Shekhawat, P.S.; Singh, K. Screening of Potential Male Contraceptive Drugs from Natural Resources: An Overview. *Int. J. Pharm. Sci. Res.* **2013**, *4*, 1654–1668.
72. Thao, N.P.; Luyen, B.T.T.; Kim, E.J.; Kang, J., II; Kang, H.K.; Cuong, N.X.; Nam, N.H.; Van Kiem, P.; Van Minh, C.; Kim, Y.H. Steroidal Constituents from the Edible Sea Urchin *Diadema Savignyi* Michelin Induce Apoptosis in Human Cancer Cells. *J. Med. Food* **2015**, *18*, 45–53. [CrossRef]
73. Rattanasopa, C.; Phungphong, S.; Wattanapermpool, J.; Bupha-Intr, T. Significant Role of Estrogen in Maintaining Cardiac Mitochondrial Functions. *J. Steroid. Biochem. Mol. Biol.* **2015**, *147*, 1–9. [CrossRef]
74. Sultan, A. Steroids: A Diverse Class of Secondary Metabolites. *Med. Chem.* **2015**, *5*, 7. [CrossRef]
75. Aav, R.; Kanger, T.; Pehk, T.; Lopp, M. Unexpected Reactivity of Ethyl 2-(Diethylphosphono)Propionate Toward 2,2-Disubstituted-1,3-Cyclopentanediones. *Phosphorus Sulfur Silicon Relat. Elem.* **2005**, *180*, 1739–1748. [CrossRef]
76. Santo, B.L.S.d.E.; Santana, L.F.; Kato Junior, W.H.; Araújo, F.d.O.d.; Tatara, M.B.; Croda, J.; Bogo, D.; Freitas, K.d.C.; Guimarães, R.d.C.A.; Hiane, P.A.; et al. Medicinal Potential of Garcinia Species and Their Compounds. *Molecules* **2020**, *25*, 4513. [CrossRef] [PubMed]
77. Mahipal, P.; Pawar, R.S. Nephroprotective Effect of Murraya Koenigii on Cyclophosphamide Induced Nephrotoxicity in Rats. *Asian Pac. J. Trop. Med.* **2017**, *10*, 808–812. [CrossRef]
78. Dennis, J.; Witting, P. Protective Role for Antioxidants in Acute Kidney Disease. *Nutrients* **2017**, *9*, 718. [CrossRef] [PubMed]
79. Ashour, M.; Wink, M.; Gershenzon, J. Biochemistry of Terpenoids: Monoterpenes, Sesquiterpenes and Diterpenes. In *Biochemistry of Plant Secondary Metabolism*; Wiley-Blackwell: Oxford, UK, 2010; pp. 258–303.
80. Croteau, R.; Kutchan, T.M.; Lewis, N.G. Natural Products (Secondary Metabolites). In *Biochemistry and Molecular Biology of Plants*; Buchanan, B.B., Gruissem, W., Jones, R.L., Eds.; American Society of Plant Physiologists: Rockville, MA, USA, 2000; Volume 24, pp. 1250–1319.
81. Ludwiczuk, A.; Skalicka-Woźniak, K.; Georgiev, M.I. Terpenoids. In *Pharmacognosy*; Elsevier: Amsterdam, The Netherlands, 2017; pp. 233–266.
82. Jahangeer, M.; Fatima, R.; Ashiq, M.; Basharat, A.; Qamar, S.A.; Bilal, M.; Iqbal, H.M.N. Therapeutic and Biomedical Potentialities of Terpenoids—A Review. *J. Pure Appl. Microbiol.* **2021**, *15*, 471–483. [CrossRef]
83. Orban, J.-C.; Quintard, H.; Cassuto, E.; Jambou, P.; Samat-Long, C.; Ichai, C. Effect of N-Acetylcysteine Pretreatment of Deceased Organ Donors on Renal Allograft Function. *Transplantation* **2015**, *99*, 746–753. [CrossRef] [PubMed]
84. Jo, S.-K.; Sung, S.-A.; Cho, W.-Y.; Go, K.-J.; Kim, H.-K. Macrophages Contribute to the Initiation of Ischaemic Acute Renal Failure in Rats. *Nephrol. Dial. Transplant.* **2006**, *21*, 1231–1239. [CrossRef] [PubMed]
85. Ricardo, S.D.; van Goor, H.; Eddy, A.A. Macrophage Diversity in Renal Injury and Repair. *J. Clin. Investig.* **2008**, *118*, 3522–3530. [CrossRef]
86. Bonventre, J.V.; Zuk, A. Ischemic Acute Renal Failure: An Inflammatory Disease? *Kidney Int.* **2004**, *66*, 480–485. [CrossRef]
87. Jablonska, J.; Wu, C.-F.; Andzinski, L.; Leschner, S.; Weiss, S. CXCR2-Mediated Tumor-Associated Neutrophil Recruitment Is Regulated by IFN-β. *Int. J. Cancer* **2014**, *134*, 1346–1358. [CrossRef]
88. Glennon-Alty, L.; Hackett, A.P.; Chapman, E.A.; Wright, H.L. Neutrophils and Redox Stress in the Pathogenesis of Autoimmune Disease. *Free Radic. Biol. Med.* **2018**, *125*, 25–35. [CrossRef]
89. Issa, R.; Xie, S.; Lee, K.-Y.; Stanbridge, R.D.; Bhavsar, P.; Sukkar, M.B.; Chung, K.F. GRO-α Regulation in Airway Smooth Muscle by IL-1β and TNF-α: Role of NF-KB and MAP Kinases. *Am. J. Physiol.-Lung Cell. Mol. Physiol.* **2006**, *291*, L66–L74. [CrossRef]
90. Korbecki, J.; Kupnicka, P.; Chlubek, M.; Gorący, J.; Gutowska, I.; Baranowska-Bosiacka, I. CXCR2 Receptor: Regulation of Expression, Signal Transduction, and Involvement in Cancer. *Int. J. Mol. Sci.* **2022**, *23*, 2168. [CrossRef]
91. Gay, N.J.; Symmons, M.F.; Gangloff, M.; Bryant, C.E. Assembly and Localization of Toll-like Receptor Signalling Complexes. *Nat. Rev. Immunol.* **2014**, *14*, 546–558. [CrossRef]
92. Sollinger, D.; Eißler, R.; Lorenz, S.; Strand, S.; Chmielewski, S.; Aoqui, C.; Schmaderer, C.; Bluyssen, H.; Zicha, J.; Witzke, O.; et al. Damage-Associated Molecular Pattern Activated Toll-like Receptor 4 Signalling Modulates Blood Pressure in l-NAME-Induced Hypertension. *Cardiovasc. Res.* **2014**, *101*, 464–472. [CrossRef] [PubMed]

93. Bomfim, G.F.; Dos Santos, R.A.; Oliveira, M.A.; Giachini, F.R.; Akamine, E.H.; Tostes, R.C.; Fortes, Z.B.; Webb, R.C.; Carvalho, M.H.C. Toll-like Receptor 4 Contributes to Blood Pressure Regulation and Vascular Contraction in Spontaneously Hypertensive Rats. *Clin. Sci.* **2012**, *122*, 535–543. [CrossRef]
94. Kim, S.-M.; Lee, S.-H.; Kim, Y.-G.; Kim, S.-Y.; Seo, J.-W.; Choi, Y.-W.; Kim, D.-J.; Jeong, K.-H.; Lee, T.-W.; Ihm, C.-G.; et al. Hyperuricemia-Induced NLRP3 Activation of Macrophages Contributes to the Progression of Diabetic Nephropathy. *Am. J. Physiol.-Ren. Physiol.* **2015**, *308*, F993–F1003. [CrossRef]
95. Anders, H.-J.; Schaefer, L. Beyond Tissue Injury—Damage-Associated Molecular Patterns, Toll-Like Receptors, and Inflammasomes Also Drive Regeneration and Fibrosis. *J. Am. Soc. Nephrol.* **2014**, *25*, 1387–1400. [CrossRef] [PubMed]
96. Bauernfeind, F.G.; Horvath, G.; Stutz, A.; Alnemri, E.S.; MacDonald, K.; Speert, D.; Fernandes-Alnemri, T.; Wu, J.; Monks, B.G.; Fitzgerald, K.A.; et al. Cutting Edge: NF-KB Activating Pattern Recognition and Cytokine Receptors License NLRP3 Inflammasome Activation by Regulating NLRP3 Expression. *J. Immunol.* **2009**, *183*, 787–791. [CrossRef] [PubMed]
97. Miller, R.P.; Tadagavadi, R.K.; Ramesh, G.; Reeves, W.B. Mechanisms of Cisplatin Nephrotoxicity. *Toxins* **2010**, *2*, 2490–2518. [CrossRef]
98. Zhang, B.; Ramesh, G.; Norbury, C.C.; Reeves, W.B. Cisplatin-Induced Nephrotoxicity Is Mediated by Tumor Necrosis Factor-α Produced by Renal Parenchymal Cells. *Kidney Int.* **2007**, *72*, 37–44. [CrossRef]
99. Peres, L.A.B.; da Cunha, A.D., Jr. Acute Nephrotoxicity of Cisplatin: Molecular Mechanisms. *J. Bras. Nefrol.* **2013**, *35*, 332–340. [CrossRef] [PubMed]
100. Han, W.K.; Bailly, V.; Abichandani, R.; Thadhani, R.; Bonventre, J.V. Kidney Injury Molecule-1 (KIM-1): A Novel Biomarker for Human Renal Proximal Tubule Injury. *Kidney Int.* **2002**, *62*, 237–244. [CrossRef] [PubMed]
101. Regueira, T.; Andresen, M.; Mercado, M.; Downey, P. Fisiopatología de La Insuficiencia Renal Aguda Durante La Sepsis. *Med. Intensiva* **2011**, *35*, 424–432. [CrossRef] [PubMed]
102. Bastos, V.P.D.; Gomes, A.S.; Lima, F.J.B.; Brito, T.S.; Soares, P.M.G.; Pinho, J.P.M.; Silva, C.S.; Santos, A.A.; Souza, M.H.L.P.; Magalhães, P.J.C. Inhaled 1,8-Cineole Reduces Inflammatory Parameters in Airways of Ovalbumin-Challenged Guinea Pigs. *Basic Clin. Pharmacol. Toxicol.* **2011**, *108*, 34–39. [CrossRef] [PubMed]
103. dos Santos, N.A.G.; Carvalho Rodrigues, M.A.; Martins, N.M.; dos Santos, A.C. Cisplatin-Induced Nephrotoxicity and Targets of Nephroprotection: An Update. *Arch. Toxicol.* **2012**, *86*, 1233–1250. [CrossRef]
104. Ramesh, G.; Reeves, W.B. P38 MAP Kinase Inhibition Ameliorates Cisplatin Nephrotoxicity in Mice. *Am. J. Physiol. Renal Physiol.* **2005**, *289*, F166–F174. [CrossRef]
105. Ullah, A.; Munir, S.; Badshah, S.L.; Khan, N.; Ghani, L.; Poulson, B.G.; Emwas, A.-H.; Jaremko, M. Important Flavonoids and Their Role as a Therapeutic Agent. *Molecules* **2020**, *25*, 5243. [CrossRef]
106. Leyva-López, N.; Gutierrez-Grijalva, E.; Ambriz-Perez, D.; Heredia, J. Flavonoids as Cytokine Modulators: A Possible Therapy for Inflammation-Related Diseases. *Int. J. Mol. Sci.* **2016**, *17*, 921. [CrossRef]
107. Ye, M.; Wang, Q.; Zhang, W.; Li, Z.; Wang, Y.; Hu, R. Oroxylin A Exerts Anti-Inflammatory Activity on Lipopolysaccharide-Induced Mouse Macrophage via Nrf2/ARE Activation. *Biochem. Cell Biol.* **2014**, *92*, 337–348. [CrossRef]
108. Chen, X.-L.; Dodd, G.; Thomas, S.; Zhang, X.; Wasserman, M.A.; Rovin, B.H.; Kunsch, C. Activation of Nrf2/ARE Pathway Protects Endothelial Cells from Oxidant Injury and Inhibits Inflammatory Gene Expression. *Am. J. Physiol. -Heart Circ. Physiol.* **2006**, *290*, H1862–H1870. [CrossRef]
109. Yu, J.S.; Lim, S.H.; Lee, S.R.; Choi, C.-I.; Kim, K.H. Antioxidant and Anti-Inflammatory Effects of White Mulberry (*Morus alba* L.) Fruits on Lipopolysaccharide-Stimulated RAW 264.7 Macrophages. *Molecules* **2021**, *26*, 920. [CrossRef] [PubMed]
110. Vattimo, M.d.F.F.; da Silva, N.O. Uncária Tomentosa e a Lesão Renal Aguda Isquêmica Em Ratos. *Rev. Esc. Enferm. USP* **2011**, *45*, 194–198. [CrossRef] [PubMed]
111. Morton, C.V. Tropicos.Org. Missouri Botanical Garden. Available online: http://www.tropicos.org (accessed on 14 June 2023).
112. Schwarz, M.J.; Houghton, P.J.; Rose, S.; Jenner, P.; Lees, A.D. Activities of Extract and Constituents of Banisteriopsis Caapi Relevant to Parkinsonism. *Pharmacol. Biochem. Behav.* **2003**, *75*, 627–633. [CrossRef]
113. Pic-Taylor, A.; da Motta, L.G.; de Morais, J.A.; Junior, W.M.; Santos, A.d.F.A.; Campos, L.A.; Mortari, M.R.; von Zuben, M.V.; Caldas, E.D. Behavioural and Neurotoxic Effects of Ayahuasca Infusion (*Banisteriopsis Caapi* and *Psychotria Viridis*) in Female Wistar Rat. *Behav. Process.* **2015**, *118*, 102–110. [CrossRef] [PubMed]
114. Samoylenko, V.; Rahman, M.d.M.; Tekwani, B.L.; Tripathi, L.M.; Wang, Y.-H.; Khan, S.I.; Khan, I.A.; Miller, L.S.; Joshi, V.C.; Muhammad, I. Banisteriopsis Caapi, a Unique Combination of MAO Inhibitory and Antioxidative Constituents for the Activities Relevant to Neurodegenerative Disorders and Parkinson's Disease. *J. Ethnopharmacol.* **2010**, *127*, 357–367. [CrossRef]
115. Santos, B.W.L.; Moreira, D.C.; Borges, T.K.d.S.; Caldas, E.D. Components of Banisteriopsis Caapi, a Plant Used in the Preparation of the Psychoactive Ayahuasca, Induce Anti-Inflammatory Effects in Microglial Cells. *Molecules* **2022**, *27*, 2500. [CrossRef]
116. Ghanbari, A.; Jalili, C.; Salahshoor, M.; Javanmardy, S.; Ravankhah, S.; Akhshi, N. Harmine Mitigates Cisplatin-Induced Renal Injury in Male Mice through Antioxidant, Anti-Inflammatory, and Anti-Apoptosis Effects. *Res. Pharm. Sci.* **2022**, *17*, 417. [CrossRef] [PubMed]
117. Araújo Galdino, O.; de Souza Gomes, I.; Ferreira de Almeida Júnior, R.; Conceição Ferreira de Carvalho, M.I.; Abreu, B.J.; Abbott Galvão Ururahy, M.; Cabral, B.; Zucolotto Langassner, S.M.; Costa de Souza, K.S.; Augusto de Rezende, A. The Nephroprotective Action of Passiflora edulis in Streptozotocin-Induced Diabetes. *Sci. Rep.* **2022**, *12*, 17546. [CrossRef]

118. Deng, J.; Zhou, Y.; Bai, M.; Li, H.; Li, L. Anxiolytic and Sedative Activities of *Passiflora edulis* f. *Flavicarpa*. *J. Ethnopharmacol.* **2010**, *128*, 148–153. [CrossRef]
119. Sena, L.M.; Zucolotto, S.M.; Reginatto, F.H.; Schenkel, E.P.; De Lima, T.C.M. Neuropharmacological Activity of the Pericarp of *Passiflora edulis Flavicarpa* Degener: Putative Involvement of C -Glycosylflavonoids. *Exp. Biol. Med.* **2009**, *234*, 967–975. [CrossRef]
120. Doungue, H.T.; Kengne, A.P.N.; Kuate, D. Neuroprotective Effect and Antioxidant Activity of Passiflora edulis Fruit Flavonoid Fraction, Aqueous Extract, and Juice in Aluminum Chloride-Induced Alzheimer's Disease Rats. *Nutrire* **2018**, *43*, 23. [CrossRef]
121. Taïwe, G.S.; Kuete, V. Passiflora edulis. In *Medicinal Spices and Vegetables from Africa*; Elsevier: Amsterdam, The Netherlands, 2017; pp. 513–526.
122. Chan, W.-J.J.; McLachlan, A.J.; Hanrahan, J.R.; Harnett, J.E. The Safety and Tolerability of *Annona muricata* Leaf Extract: A Systematic Review. *J. Pharm. Pharmacol.* **2019**, *72*, 1–16. [CrossRef] [PubMed]
123. Gyesi, J.N.; Opoku, R.; Borquaye, L.S. Chemical Composition, Total Phenolic Content, and Antioxidant Activities of the Essential Oils of the Leaves and Fruit Pulp of *Annona muricata* L. (Soursop) from Ghana. *Biochem. Res. Int.* **2019**, *2019*, 4164576. [CrossRef]
124. Adewole, S.; Caxton-Martins, E. Morphological Changes and Hypoglycemic Effects of *Annona muricata* Linn. (Annonaceae) Leaf Aqueous Extract on Pancreatic β-Cells of Streptozotocin-Treated Diabetic Rats. *Afr. J. Biomed. Res.* **2009**, *9*. [CrossRef]
125. Pilarski, R.; Zieliński, H.; Ciesiołka, D.; Gulewicz, K. Antioxidant Activity of Ethanolic and Aqueous Extracts of Uncaria tomentosa (Willd.) DC. *J. Ethnopharmacol.* **2006**, *104*, 18–23. [CrossRef]
126. Batiha, G.E.-S.; Magdy Beshbishy, A.; Wasef, L.; Elewa, Y.H.A.; Abd El-Hack, M.E.; Taha, A.E.; Al-Sagheer, A.A.; Devkota, H.P.; Tufarelli, V. *Uncaria tomentosa* (Willd. Ex Schult.) DC.: A Review on Chemical Constituents and Biological Activities. *Appl. Sci.* **2020**, *10*, 2668. [CrossRef]
127. Khoo, S.F.; Oehlschlager, A.C.; Ourisson, G. Structure and Stereochemistry of the Diterpenes of *Hymenaea courbaril* (Caesalpinioideae) Seed Pod Resin. *Tetrahedron* **1973**, *29*, 3379–3388. [CrossRef]
128. Spera, K.D.; Figueiredo, P.A.; Santos, P.C.; Barbosa, F.C.; Alves, C.P.; Dokkedal, A.L.; Saldanha, L.L.; Silva, L.P.; Figueiredo, C.R.; Ferreira, P.C.; et al. Genotoxicity, Anti-Melanoma and Antioxidant Activities of *Hymenaea courbaril* L. Seed Extract. *An. Acad. Bras. Cienc.* **2019**, *91*, e20180446. [CrossRef] [PubMed]
129. Silva, T.M.; Miranda, R.R.S.; Ferraz, V.P.; Pereira, M.T.; de Siqueira, E.P.; Alcântara, A.F.C. Changes in the Essential Oil Composition of Leaves of Echinodorus macrophyllus Exposed to γ-Radiation. *Rev. Bras. Farmacogn.* **2013**, *23*, 600–607. [CrossRef]
130. Gasparotto, F.; Lívero, F.; Palozi, R.; Ames, M.; Nunes, B.; Donadel, G.; Ribeiro, R.; Lourenço, E.; Kassuya, C.; Junior, A. Heart-Protective Effects of *Echinodorus grandiflorus* in Rabbits That Are Fed a High-Cholesterol Diet. *Planta Med.* **2018**, *84*, 1271–1279. [CrossRef]
131. Fernandes, D.C.; Martins, B.P.M.P.; Medeiros, D.L.; Santos, S.V.; Gayer, C.R.; Velozo, L.S.; Coelho, M.G. Antinociceptive and Anti-Inflammatory Activities of the Hexanic Extract of "*Echinodorus macrophyllus*" (Kunth) Micheli in Mice. *Braz. J. Health Biomed. Sci.* **2019**, *18*, 25–32.
132. Spinozzi, E.; Ferrati, M.; Baldassarri, C.; Cappellacci, L.; Marmugi, M.; Caselli, A.; Benelli, G.; Maggi, F.; Petrelli, R. A Review of the Chemistry and Biological Activities of *Acmella oleracea* ("Jambù", Asteraceae), with a View to the Development of Bioinsecticides and Acaricides. *Plants* **2022**, *11*, 2721. [CrossRef] [PubMed]
133. Abeysiri, G.R.P.I.; Dharmadasa, R.M.; Abeysinghe, D.C.; Samarasinghe, K. Screening of Phytochemical, Physico-Chemical and Bioactivity of Different Parts of *Acmella oleraceae* Murr. (Asteraceae), a Natural Remedy for Toothache. *Ind. Crops Prod.* **2013**, *50*, 852–856. [CrossRef]
134. Prachayasittikul, S.; Suphapong, S.; Worachartcheewan, A.; Lawung, R.; Ruchirawat, S.; Prachayasittikul, V. Bioactive Metabolites from *Spilanthes acmella* Murr. *Molecules* **2009**, *14*, 850–867. [CrossRef]
135. Batista, L.L.; Koga, R.d.C.R.; Teixeira, A.V.T.d.L.; Teixeira, T.A.; de Melo, E.L.; Carvalho, J.C.T. Clinical Safety of a Pharmaceutical Formulation Containing an Extract of *Acmella oleracea* (L.) in Patients With Premature Ejaculation: A Pilot Study. *Am. J. Men's Health* **2023**, *17*, 155798832311678. [CrossRef] [PubMed]
136. Souza, G.; Dias Ribeiro da Silva, I.; Duarte Viana, M.; Costa de Melo, N.; Sánchez-Ortiz, B.; Maia Rebelo de Oliveira, M.; Ramos Barbosa, W.; Maciel Ferreira, I.; Tavares Carvalho, J. Acute Toxicity of the Hydroethanolic Extract of the Flowers of *Acmella oleracea* L. in Zebrafish (*Danio rerio*): Behavioral and Histopathological Studies. *Pharmaceuticals* **2019**, *12*, 173. [CrossRef] [PubMed]
137. Borges, R.S.; Ortiz, B.L.S.; Pereira, A.C.M.; Keita, H.; Carvalho, J.C.T. *Rosmarinus officinalis* Essential Oil: A Review of Its Phytochemistry, Anti-Inflammatory Activity, and Mechanisms of Action Involved. *J. Ethnopharmacol.* **2019**, *229*, 29–45. [CrossRef]
138. Uritu, C.M.; Mihai, C.T.; Stanciu, G.-D.; Dodi, G.; Alexa-Stratulat, T.; Luca, A.; Leon-Constantin, M.-M.; Stefanescu, R.; Bild, V.; Melnic, S.; et al. Medicinal Plants of the Family Lamiaceae in Pain Therapy: A Review. *Pain Res. Manag.* **2018**, *2018*, 7801543. [CrossRef]
139. Linnaeus, C.V. Tropicos.Org. Missouri Botanical Garden. Available online: http://www.tropicos.org (accessed on 14 June 2023).
140. Shahrajabian, M.H.; Sun, W.; Cheng, Q. Improving Health Benefits with Considering Traditional and Modern Health Benefits of Peganum Harmala. *Clin. Phytosci.* **2021**, *7*, 18. [CrossRef]
141. Niroumand, M.C.; Farzaei, M.H.; Amin, G. Medicinal Properties of Peganum Harmala L. in Traditional Iranian Medicine and Modern Phytotherapy: A Review. *J. Tradit. Chin. Med.* **2015**, *35*, 104–109. [CrossRef]
142. Niu, X.; Yao, Q.; Li, W.; Zang, L.; Li, W.; Zhao, J.; Liu, F.; Zhi, W. Harmine Mitigates LPS-Induced Acute Kidney Injury through Inhibition of the TLR4-NF-KB/NLRP3 Inflammasome Signalling Pathway in Mice. *Eur. J. Pharmacol.* **2019**, *849*, 160–169. [CrossRef] [PubMed]

143. Sims, J. Tropicos.Org. Missouri Botanical Garden. Available online: http://www.tropicos.org (accessed on 14 June 2023).
144. De Melo, N.F.; Cervi, A.C.; Guerra, M. Karyology and Cytotaxonomy of the Genus *Passiflora* L. (Passifloraceae). *Plant Syst. Evol.* **2001**, *226*, 69–84. [CrossRef]
145. Zhang, Y.-J.; Zhou, T.; Wang, F.; Zhou, Y.; Li, Y.; Zhang, J.-J.; Zheng, J.; Xu, D.-P.; Li, H.-B. The Effects of *Syzygium samarangense*, *Passiflora edulis* and *Solanum muricatum* on Alcohol-Induced Liver Injury. *Int. J. Mol. Sci.* **2016**, *17*, 1616. [CrossRef] [PubMed]
146. Salles, B.C.C.; Leme, K.C.; da Silva, M.A.; da Rocha, C.Q.; Tangerina, M.M.P.; Vilegas, W.; Figueiredo, S.A.; da Silveria Duarte, S.M.; Rodrigues, M.R.; de Araújo Paula, F.B. Protective Effect of Flavonoids from *Passiflora edulis* Sims on Diabetic Complications in Rats. *J. Pharm. Pharmacol.* **2021**, *73*, 1361–1368. [CrossRef] [PubMed]
147. Leboeuf, M.; Cavé, A.; Bhaumik, P.K.; Mukherjee, B.; Mukherjee, R. The Phytochemistry of the Annonaceae. *Phytochemistry* **1980**, *21*, 2783–2813. [CrossRef]
148. Moghadamtousi, S.; Fadaeinasab, M.; Nikzad, S.; Mohan, G.; Ali, H.; Kadir, H. *Annona muricata* (Annonaceae): A Review of Its Traditional Uses, Isolated Acetogenins and Biological Activities. *Int. J. Mol. Sci.* **2015**, *16*, 15625–15658. [CrossRef]
149. Adedapo, A.A.; Oni, O.A.; Falayi, O.O.; Ogunmiluyi, I.O.; Ogunpolu, B.S.; Omobowale, T.O.; Oyagbemi, A.A.; Oguntibeju, O.O.; Yakubu, M.A. Annona muricata Mitigates Glycerol-Induced Nephrotoxicities in Male Albino Rats through Signaling Pathways of Angiotensin Conversion Enzyme, Kidney Injury Molecule-1, and Antioxidant Properties. *Sci. Afr.* **2022**, *16*, e01225. [CrossRef]
150. Bremekamp, C.E.B. Tropicos.Org. Missouri Botanical Garden. Available online: http://www.tropicos.org (accessed on 14 June 2023).
151. Conserva, L.; Ferreira, J. *Borreria* and *Spermacoce* Species (Rubiaceae): A Review of Their Ethnomedicinal Properties, Chemical Constituents, and Biological Activities. *Pharmacogn. Rev.* **2012**, *6*, 46. [CrossRef]
152. Kala, S.C. Medicinal Attributes of Family Rubiaceae. *Int. J. Pharm. Biol. Sci.* **2015**, *5*, 179–181.
153. Heitzman, M.E.; Neto, C.C.; Winiarz, E.; Vaisberg, A.J.; Hammond, G.B. Ethnobotany, Phytochemistry and Pharmacology of Uncaria (Rubiaceae). *Phytochemistry* **2005**, *66*, 5–29. [CrossRef]
154. Valente, L.M.M.; Bizarri, C.H.B.; Liechocki, S.; Barboza, R.S.; da Paixão, D.; Almeida, M.B.S.; Benevides, P.J.C.; Magalhães, A.; Siani, A.C. Kaempferitrin from Uncaria Guianensis (Rubiaceae) and Its Potential as a Chemical Marker for the Species. *J. Braz. Chem. Soc.* **2009**, *20*, 1041–1045. [CrossRef]
155. Machado, D.I.; de Oliveira Silva, E.; Ventura, S.; Vattimo, M.d.F.F. The Effect of Curcumin on Renal Ischemia/Reperfusion Injury in Diabetic Rats. *Nutrients* **2022**, *14*, 2798. [CrossRef] [PubMed]
156. Mostacedo, C.B.; Uslar, Y. Plantas Silvestres Con Frutos y Semillas Comestibles Del Departamento de Santa Cruz, Bolivia: Un Inventario Preliminar. *Rev. Soc. Boliv. Botánica* **1999**, *2*, 203–226.
157. Sales, G.W.P.; Batista, A.H.M.; Rocha, L.Q.; Nogueira, N.A.P. Efeito Antimicrobiano e Modulador Do Óleo Essencial Extraído Da Casca de Frutos Da *Hymenaea courbaril* L. *Rev. Ciências Farm. Básica E Apl.* **2014**, *35*, 709–715.
158. Tiago, P.V.; Larocca, D.; da Silva, I.V.; Carpejani, A.A.; Tiago, A.V.; Dardengo, J.d.F.E.; Rossi, A.A.B. Caracterização Morfoanatômica, Fitoquímica e Histoquímica de *Hymenaea courbaril* (Leguminosae), Ocorrente Na *Amazônia meridional*. *Rodriguésia* **2020**, *71*, e02182018. [CrossRef]
159. Duarte, M.M.; de Paula, S.R.P.; Ferreira, F.R.d.L.; Nogueira, A.C. Morphological Characterization of Fruit, Seed and Seedling and Germination of *Hymenaea courbaril* L. (Fabaceae) ('Jatobá'). *J. Seed Sci.* **2016**, *38*, 204–211. [CrossRef]
160. Gorchov, D.L.; Palmeirim, J.M.; Ascorra, C.F. Dispersal of Seeds of *Hymenaea courbaril* (Fabaceae) in a Logged Rain Forest in the Peruvian Amazonian. *Acta Amazon.* **2004**, *34*, 251–259. [CrossRef]
161. Dos Santos, A.G.; Sivieri, K.; Miglioli da Mata, B.P.; Salgaço, M.K.; Silva do Sacramento, L.V. Jatobá (*Hymenaea courbaril* L.). In *Handbook of Phytonutrients in Indigenous Fruits and Vegetables*; CABI: Wallingford, UK, 2022; pp. 266–280.
162. Delgado, C.; Mendez-Callejas, G.; Celis, C. Caryophyllene Oxide, the Active Compound Isolated from Leaves of *Hymenaea courbaril* L. (Fabaceae) with Antiproliferative and Apoptotic Effects on PC-3 Androgen-Independent Prostate Cancer Cell Line. *Molecules* **2021**, *26*, 6142. [CrossRef]
163. Campelo, D.S.; Campelo, T.P.T.; Ferraz, A.B.F. *Avaliação das Características Químicas e Biológicas da Garrafada de Carobinha*; Digital Editora: Canoas, Brazil, 2021.
164. Gindri-Sinhorin, V. Avaliação Antioxidante Do Extrato Da Semente de *Hymenaea courbaril* l. (Jatobá) Em Camundongos Tratados Com Acetaminofeno. *Rev. Cuba. Plantas Med.* **2020**, *25*, 1–13.
165. de Jesus Lisboa, E.M.; Albiero, L.R.; Melchiors, N.; de Paula Borges, W.S.; da Silva Lima, V.; Rodrigues, F.D.; Sinhorin, V.D.G.; Castoldi, L. Evaluation of the Antitumor and Antioxidant Effects of Jatobá (*Hymenaea courbaril*) Extracts/Avaliação Do Efeito Antitumoral e Antioxidante de Extratos Do Jatobá (*Hymenaea courbaril*). *Braz. J. Dev.* **2021**, *7*, 116001–116018. [CrossRef]
166. Micheli, M. Tropicos.org. Missouri Botanical Garden. Available online: http://www.tropicos.org (accessed on 14 June 2023).
167. Portella, V.G.; Cosenza, G.P.; Diniz, L.R.L.; Pacheco, L.F.; Cassali, G.D.; Caliari, M.V.; Brandão, M.d.G.L.; Vieira, M.A.R. Nephroprotective Effect of Echinodorus macrophyllus Micheli on Gentamicin-Induced Nephrotoxicity in Rats. *Nephron Extra* **2012**, *2*, 177–183. [CrossRef]
168. Dutra, R.C.; Tavares, C.Z.; Ferraz, S.O.; Sousa, O.V.; Pimenta, D.S. Investigação Das Atividades Analgésica e Antiinflamatória Do Extrato Metanólico Dos Rizomas de *Echinodorus grandiflorus*. *Rev. Bras. Farmacogn.* **2006**, *16*, 469–474. [CrossRef]
169. Marques, A.M.; Provance, D.W.; Kaplan, M.A.C.; Figueiredo, M.R. *Echinodorus grandiflorus*: Ethnobotanical, Phytochemical and Pharmacological Overview of a Medicinal Plant Used in Brazil. *Food Chem. Toxicol.* **2017**, *109*, 1032–1047. [CrossRef]

170. De Fafia Garcia, E.; de Oliveira, M.A.; Godin, A.M.; Ferreira, W.C.; Bastos, L.F.S.; de Matos Coelho, M.; Braga, F.C. Antiedematogenic Activity and Phytochemical Composition of Preparations from *Echinodorus grandiflorus* Leaves. *Phytomedicine* **2010**, *18*, 80–86. [CrossRef] [PubMed]
171. Do Nascimento, E.L.; Watanabe, M.; da Fonseca, C.D.; Schlottfeldt, F.d.S.; Vattimo, M.d.F.F. Renoprotective Effect of the Echinodorus macrophyllus in Induced Renal Injury. *Acta Paul. Enferm.* **2014**, *27*, 12–17. [CrossRef]
172. Jansen, R.K.; Tropicos.org. Missouri Botanical Garden. Available online: http://www.tropicos.org (accessed on 14 June 2023).
173. Kostić, A.Ž.; Janaćković, P.; Kolašinac, S.M.; Dajić Stevanović, Z.P. Balkans' Asteraceae Species as a Source of Biologically Active Compounds for the Pharmaceutical and Food Industry. *Chem. Biodivers.* **2020**, *17*, e2000097. [CrossRef] [PubMed]
174. De Oliveira, P.R.; Anholeto, L.; Ferreira Rodrigues, R.; Arnosti, A.; Bechara, G.; de Carvalho Castro, K.; Camargo-Mathias, M. Cytotoxic Effects of Extract of *Acmella oleracea* in the Ovaries and Midgut of *Rhipicephalus sanguineus* Latreille, 1806 (Acari: Ixodidae) Female Ticks. *J. Microsc. Ultrastruct.* **2019**, *7*, 28. [CrossRef] [PubMed]
175. Borges, L.d.S.; Vieir, M.A.; Marqu, M.O.; Vianello, F.; Lima, G.P. Influence of Organic and Mineral Soil Fertilization on Essential Oil of Spilanthes Oleracea Cv. Jambuarana. *Am. J. Plant Physiol.* **2012**, *7*, 135–142. [CrossRef]
176. Da Silva Borges, L.; de Souza Vieira, M.C.; Vianello, F.; Goto, R.; Lima, G.P.P. Antioxidant Compounds of Organically and Conventionally Fertilized Jambu (*Acmella oleracea*). *Biol. Agric. Hortic.* **2016**, *32*, 149–158. [CrossRef]
177. Ratnasooriya, W.D.; Pieris, K.P.P.; Samaratunga, U.; Jayakody, J.R.A.C. Diuretic Activity of *Spilanthes acmella* Flowers in Rats. *J. Ethnopharmacol.* **2004**, *91*, 317–320. [CrossRef]
178. Yadav, R.; Kharya, M.D.; Yadav, N.; Savadi, R. Diuretic Activity of *Spilanthes acmella* Murr. Leaves Extract in Rats. *Int. J. Res. Pharm. Chem.* **2011**, *1*, 57–61.
179. Gerbino, A.; Schena, G.; Milano, S.; Milella, L.; Barbosa, A.F.; Armentano, F.; Procino, G.; Svelto, M.; Carmosino, M. Spilanthol from *Acmella oleracea* Lowers the Intracellular Levels of CAMP Impairing NKCC2 Phosphorylation and Water Channel AQP2 Membrane Expression in Mouse Kidney. *PLoS ONE* **2016**, *11*, e0156021. [CrossRef] [PubMed]
180. Zappi, D.C.; Filardi, F.L.R.; Leitman, P.; Souza, V.C.; Walter, B.M.T.; Pirani, J.R.; Morim, M.P.; Queiroz, L.P.; Cavalcanti, T.B.; Mansano, V.F.; et al. Growing Knowledge: An Overview of Seed Plant Diversity in Brazil. *Rodriguésia* **2015**, *66*, 1085–1113. [CrossRef]
181. Fernandez, L.; Duque, S.; Sanchez, I.; Quiñones, D.; Rodriguez, F.; Garcia-Abujeta, J.L. Allergic Contact Dermatitis from Rosemary (*Rosmarinus officinalis* L.). *Contact Dermat.* **1997**, *37*, 248–249. [CrossRef]
182. Emami, F.; Ali-Beig, H.; Farahbakhs, S.; Mojabi, N.; Rastegar-M, B.; Arbabian, S.; Kazemi, M.; Tekieh, E.; Golmanesh, L.; Ranjbaran, M.; et al. Hydroalcoholic Extract of Rosemary (*Rosmarinus officinalis* L.) and Its Constituent Carnosol Inhibit Formalin-Induced Pain and Inflammation in Mice. *Pak. J. Biol. Sci.* **2013**, *16*, 309–316. [CrossRef]
183. Sotelo-Félix, J.I.; Martinez-Fong, D.; Muriel, P.; Santillán, R.L.; Castillo, D.; Yahuaca, P. Evaluation of the Effectiveness of *Rosmarinus officinalis* (Lamiaceae) in the Alleviation of Carbon Tetrachloride-Induced Acute Hepatotoxicity in the Rat. *J Ethnopharmacol.* **2002**, *81*, 145–154. [CrossRef]
184. Almela, L.; Sánchez-Muñoz, B.; Fernández-López, J.A.; Roca, M.J.; Rabe, V. Liquid Chromatograpic–Mass Spectrometric Analysis of Phenolics and Free Radical Scavenging Activity of Rosemary Extract from Different Raw Material. *J. Chromatogr. A* **2006**, *1120*, 221–229. [CrossRef] [PubMed]
185. Zohrabi, M. The Study of 24 h Post Treatment Effects of the Aqueous Extract of *Rosmarinus officinalis* after Renal Ischemia/Reperfusion in Rat. *J. Physiol. Pathophysiol.* **2012**, *3*, 12–19. [CrossRef]
186. El-Demerdash, F.M.; El-Sayed, R.A.; Abdel-Daim, M.M. *Rosmarinus officinalis* Essential Oil Modulates Renal Toxicity and Oxidative Stress Induced by Potassium Dichromate in Rats. *J. Trace Elem. Med. Biol.* **2021**, *67*, 126791. [CrossRef]

Disclaimer/Publisher's Note: The statements, opinions and data contained in all publications are solely those of the individual author(s) and contributor(s) and not of MDPI and/or the editor(s). MDPI and/or the editor(s) disclaim responsibility for any injury to people or property resulting from any ideas, methods, instructions or products referred to in the content.

molecules

MDPI

Article

Synthesis of Bis-Chalcones and Evaluation of Its Effect on Peroxide-Induced Cell Death and Lipopolysaccharide-Induced Cytokine Production

Alby Tom [1], Jisha Jacob [2], Manoj Mathews [3], Rajakrishnan Rajagopal [4], Ahmed Alfarhan [4], Damia Barcelo [5] and Arunaksharan Narayanankutty [1,*]

1 Division of Cell and Molecular Biology, PG and Research Department of Zoology, St. Joseph's College Devagiri (Autonomous), Calicut 673008, Kerala, India; albytom333@gmail.com
2 Molecular Microbial Ecology Lab, PG and Research Department of Zoology, St. Joseph's College Devagiri (Autonomous), Calicut 680555, Kerala, India; jishajacob@devagiricollege.org
3 PG and Research Department of Chemistry, St. Joseph's College Devagiri (Autonomous), Calicut 680555, Kerala, India; manojmathews@devagiricollege.org
4 Department of Botany and Microbiology, College of Science, King Saud University, P.O. Box 2455, Riyadh 11451, Saudi Arabia; rrajagopal@ksu.edu.sa (R.R.); alfarhan@ksu.edu.sa (A.A.)
5 Water and Soil Research Group, Department of Environmental Chemistry, Idaea-Csic, Jordi Girona 18-26, 08034 Barcelona, Spain; damia.barcelo@idaea.csic.es
* Correspondence: arunaksharann@devagiricollege.org

Abstract: Plant secondary metabolites are important sources of biologically active compounds with wide pharmacological potentials. Among the different classes, the chalcones form integral pharmacologically active agents. Natural chalcones and bis-chalcones exhibit high antioxidant and anti-inflammatory properties in various experiments. Studies are also underway to explore more biologically active bis-chalcones by chemical synthesis of these compounds. In this study, the effects of six synthetic bis-chalcones were evaluated in intestinal epithelial cells (IEC-6); further, the anti-inflammatory potentials were studied in lipopolysaccharide-induced cytokine production in macrophages. The synthesized bis-chalcones differ from each other first of all by the nature of the aromatic cores (functional group substitution, and their position) and by the size of a central alicycle. The exposure of IEC-6 cells to peroxide radicals reduced the cell viability; however, pre-treatment with the bis-chalcones improved the cell viability in these cells. The mechanism of action was observed to be the increased levels of glutathione and antioxidant enzyme activities. Further, these bis-chalcones also inhibited the LPS-stimulation-induced inflammatory cytokine production in RAW 264.7 macrophages. Overall, the present study indicated the cytoprotective and anti-inflammatory abilities of synthetic bis-chalcones.

Keywords: bis-chalcones; antioxidant activity; cytoprotection; peroxide toxicity; anti-inflammatory activity

Citation: Tom, A.; Jacob, J.; Mathews, M.; Rajagopal, R.; Alfarhan, A.; Barcelo, D.; Narayanankutty, A. Synthesis of Bis-Chalcones and Evaluation of Its Effect on Peroxide-Induced Cell Death and Lipopolysaccharide-Induced Cytokine Production. *Molecules* **2023**, *28*, 6354. https://doi.org/10.3390/molecules28176354

Academic Editor: Kyoko Nakagawa-Goto

Received: 31 July 2023
Revised: 19 August 2023
Accepted: 23 August 2023
Published: 30 August 2023

Copyright: © 2023 by the authors. Licensee MDPI, Basel, Switzerland. This article is an open access article distributed under the terms and conditions of the Creative Commons Attribution (CC BY) license (https://creativecommons.org/licenses/by/4.0/).

1. Introduction

Colorectal cancers (CRCs) are one among the major concerns of healthcare systems in developed and developing countries. Colorectal cancer is the second most common cancer in the United States and it also accounts for approximately 52,550 deaths in the country [1]. In addition, the report emphasized the increase in the occurrence of CRCs among those younger than 50 years of age [1–3]. In the Indian population, the incidence of CRC is much lower with an age-standardized ratio of 7.2 per 100,000; however, it should be noted that the country accounts for a significantly lower 5-year survival rate in CRC with respect to global standards [4,5]. Oxidative stress is a key event in the onset and progression of intestinal disorders including colorectal cancers [6,7]. The initial events of oxidative stress begin with the imbalance in the production of reactive radical moieties in the cells and also with the scavenging of these reactive molecules [8]. These free radicals include the oxygen, nitrogen and lipid-derived radicals and among these, the predominant ones are the peroxides [9]. The peroxide radicals are commonly formed in the intestinal epithelial cells and are known to induce genomic instability and cell death [10]. The increased levels of peroxides are also shown to increase the expression of angiogenic factors in CRCs [11]. Increased oxidative damage is also associated with the metastatic phenotype in CRCs [12]. The intestinal peroxide radicals are also contributors to inducing ulcerative colitis by driving inflammatory events in the colon [13]. The inflammatory cascades are also associated with the promotion of various cancers including colorectal cancer [14,15]. The inflammatory cytokines and other inflammatory mediators are reported to be elevated in the advanced stages of colorectal cancers. Similarly, increased levels of inflammatory NF-κB are associated with the metastatic phenotype of colorectal cancer cells [16]. Tumor-associated macrophages are also involved in the induction of the inflammatory tumor microenvironment and subsequent progression of colorectal cancers [17]. Hence, the prevention of oxidative stress and inflammation forms a strategic response in the prevention and onset of colorectal cancers.

Natural products, especially those of plant origin, are important radical scavengers and anti-inflammatory molecules [18]. Among the different classes, the polyphenols, flavonoids, anthocyanins, terpenoids, anthraquinones, etc. are important compounds. Among these, chalcones and bis-chalcones are important compounds which belong to the flavonoid family. Chalcones are the unsaturated α, β-ketones with terminal aromatic fragments [19]. The Asteraceae, Leguminosae and Moraceae members of plants contain a large quantity of natural chalcones [20,21]. Several biological and pharmacological properties are attributed to these chalcones. The compounds are reported to inhibit the oxidative radical formation in different in vitro and in vivo models [22,23]. Together with this, the compounds are reported to enhance antioxidant defense in animals [24,25]. It has also been reported that chalcones and bis-chalcones reduce cytokine production and nitric oxide synthase activity in macrophages [26,27]. It has further been reported that chalcones effectively regulate prostaglandin synthesis and NF-κB signaling and thereby reduce inflammation in the cellular milieu [28]. The yield of chalcones in plants is low and the purification process is difficult. However, synthesis of chalcones is possible and is an alternative for yielding high-purity chalcones within a short period of time [29,30].

Hence, the present study aimed to analyze the cytoprotective and antioxidant abilities of synthetic bis-chalcones (Figure 1) that were proven to have antioxidant properties in our previous study [31]. The present work focused on three compounds from the previous study, compounds **1**, **2** and **4**. In addition, we evaluated the anti-inflammatory properties of these compounds in LPS-stimulated macrophages.

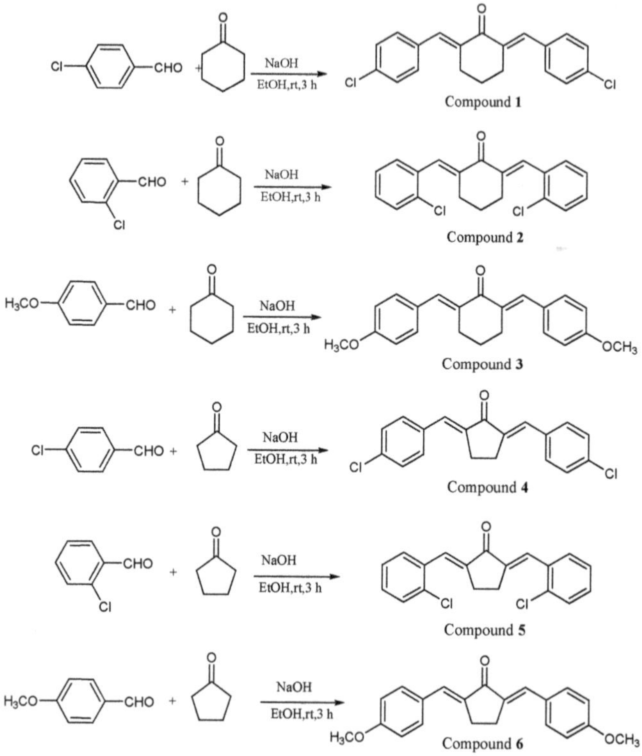

Figure 1. The synthesis and structure of different synthetic bis-chalcones.

2. Results and Discussion

2.1. Characterization of Compounds

Chalcones and bis-chalcones are important bio-organic compounds that are naturally occurring in plants [32]. The chalcones are synthesized in plants using chalcones synthase enzymes and they act as intermediates in the production of flavonoids [33]. They also perform integral roles as floral pigments, insect repellents, and antimicrobial and UV protective molecules in plants [34]. However, due to the difficulty in extraction and limited availability and yield from plants, the synthetic preparations are now considered to replace the natural chalcones and bis-chalcones. Compared to chalcones, bis-chalcones are less explored for their biological efficacies. In the present study, we synthesized six bis-chalcones and evaluated them for their cytoprotective and anti-inflammatory effects (the data of only three are shown, since compounds **3**, **4** and **5** had no activity).

Compound **1** was obtained as yellow crystals at 66% yield by mixed solvent recrystallization. The physico-chemical properties were as follows; MP: 220–225 °C. $R_f = 0.66$ in 20% EtOAc—Hexane. The FTIR spectra represented peaks (cm^{-1}) as 3058, 2963, 1670, 1605, 1575, 1488, 1401, 1319, 1264, 1091, 977, 929, 835, 821, 798, and 524. The result of the NMR was: ^{1}H NMR (500 MHz, $CDCl_3$): δ ppm 7.7(s, 2H, Olefinic-H), 7.35 (t, 8H, Ar-H), 2.86–2.84 (m, 4H, 2xCH_2), and 1.81–1.74 (m, 2H, 1xCH_2).

Compound **2** was obtained as yellow crystals by mixed solvent recrystallization at 90.6% yield. The physico-chemical properties were as follows; MP: 220–225 °C. $R_f = 0.4883$ in 20% EtOAc—Hexane. The FTIR spectra represented peaks (cm^{-1}) as 3070, 2970, 1663, 1601, 1575, 1467, 1432, 1260, 1169, 1145, 1119.5, 977, 929, 835, 821, 766, and 735. The result of the NMR was: ^{1}H NMR (500 MHz, $CDCl_3$): δ ppm 7.90 (s, 2H, Olefinic-H), 7.43–7.41 (m,

2H, Ar-H), 7.33–7.31 (m, 2H, Ar-H), 7.28–7.25 (m, 4H, Ar-H), 2.75 (t, 4H, J = 5 Hz, $2xCH_2$), and 1.34 (m, 2H, $1xCH_2$).

Compound **3** was obtained as yellow crystals by mixed solvent recrystallization at 48.8% yield. R_f = 0.73 in 20% EtOAc—Hexane. The FTIR spectra represented peaks (cm^{-1}) as 2963, 2916, 2843, 2051, 1915, 1696, 1616, 1597, 1508, 1438, 1418, 1365, 1305, 1253, 1171, 1119, 1029, 987, 929, 835, 819, 689, 607, 536, and 517. The result of the NMR was: ^{1}H NMR (500 MHz, $CDCl_3$): δ ppm 7.76 (s, 2H,Olefinic-H), 7.46–7.44 (m,4H, Ar-H), 6.94–6.92 (m, 4H, Ar-H), 3.83 (s, 6H, 2 × $-OCH_3$), 1.8–1.6 (m, 2H, $1xCH_2$), and 2.92–2.89 (m, 4H, $2xCH_2$).

Compound **4** was obtained as yellow crystals by mixed solvent recrystallization at 49.4% yield. The physico-chemical properties were as follows: R_f = 0.58 in 20%. EtOAc—Hexane. The FTIR spectra represented peaks (cm^{-1}) as 3378, 2919, 2366, 2342, 1913, 1694, 1621, 1607, 1584, 1556, 1489, 1404, 1306, 1278, 1253, 1178, 1106, 1092, 1009, 985, 929, 833, 820, 729, 685, 611, and 520. The result of the NMR was: ^{1}H NMR (500 MHz, $CDCl_3$): δ ppm 7.53–7.40 (m, 6 H, J = 8.5Hz, 2H (Olefinic-H), 4H (Ar-H), 7.34 (d, 4H, J = 8.5 Hz, Ar-H), and 3.01 (s, 4H, $2xCH_2$).

Compound **5** was obtained as yellow crystals by mixed solvent recrystallization at 69.5% yield. The physico-chemical properties were as follows: R_f = 0.51 in 20% EtOAc—Hexane. The FTIR spectra represented peaks (cm^{-1}) as 3069, 2912, 2361, 1687, 1620, 1599, 1587, 1560, 1465, 1449, 1431, 1277, 1239, 1182, 1155, 1124, 1038, 987, 941, 759, 751, 726, 688, 622, 544, and 532. ^{1}H NMR (500 MHz, $CDCl_3$): δ ppm 7.90 (s, 2H, Olefinic-H), 7.53–7.42 (m, 4H, Ar-H), 7.30–7.27 (d, 4H, J = 1.5 Hz, Ar-H), and 2.97 (s, 4H, 2 × CH_2). The result of the NMR was: ^{13}C NMR (500 MHz, $CDCl_3$): δ ppm 195.57, 139. 36, 136.13, 133.91, 130.30, 130.19, 130.13, 126.71, and 26.61.

Compound **6** was obtained as yellow crystals by mixed solvent recrystallization at 49.7% yield. The physico-chemical properties were as follows: R_f = 0.64 in 20% EtOAc—Hexane. The FTIR spectra represented peaks (cm^{-1}) as 3069, 2963, 2937, 2913, 2846, 2057, 2007, 1696, 1616, 1597, 1508, 1439, 1365, 1306, 1252, 1171, 1029, 929, 835, 819, 694, 608, 536, and 517. The result of NMR was: ^{1}H NMR (500 MHz, $CDCl_3$): δ ppm 7.57–7.55 (m, 6H, Ar-H), 6.96–6.95 (m, 4H, Ar-H), 3.85 (s, 6H, 2 × OCH_3), and 3.06 (s, 4H, $2xCH_2$).

The characterization details of these chalcones using FT-IR and NMR spectroscopy are included in Supplementary Figures S1–S9. In Supplementary, the synthesis protocol of all six bis-chalcones has been included. (Characterization details of compounds **1**, **2** and **4** were included in the Supplementary Files because other compounds did not exhibit any activity.)

2.2. Cytotoxicity of Peroxide Radicals and Bio-Safe Concentration of Bis-Chalcones

The cytotoxicity analysis by MTT assay revealed that the IC50 value of hydrogen peroxide was 391.8 ± 4.6 μM. Hence, the study opted for 400 μM as the test concentration of peroxide radicals to induce cytotoxicity in the IEC-6 cells (Figure 2).

The cytotoxicity evaluation of synthetic bis-chalcones (data not included) revealed that the IC50 values of the compounds were above 250 μg/mL (Figure 2). Hence, the biological safety analysis was limited to a concentration of 2.5 μg/mL; among these, the concentrations 0.5 and 1.0 μg/mL had no significant toxic effects on the cells and therefore those doses were opted for further cytoprotective studies.

2.3. Cytoprotective Effect of Bis-Chalcones

The cytoprotective effect was analyzed against peroxide radicals by pre-incubating the cells with different doses of bis-chalcones for 2 h and exposing them to the radicals for 24 h. The untreated normal IEC-6 cells were considered to be 100% viable and the exposure of these cells to peroxide radicals reduced the net cell viability to 48.2 ± 2.3% (Figure 3). Increased peroxide radicals in cells are also shown to reduce the cell viability by inducing apoptotic cell death mediated through lipid peroxidation and DNA damage [35,36]. It is therefore possible that the increased cell death in peroxide exposure may be due to the apoptotic cell death in colon cells. However, pre-treatment with compound **1** increased the cell viability to 78.5 ± 3.4 and 84.3 ± 1.9% at the respective doses of 0.5 and 1.0 μg/mL. Likewise, the treatment with compound **2** was found to reduce the toxic insults of peroxide radicals and thereby increase cell viability to 62.7 ± 2.4 and 79.3 ± 2.7% at the respective doses of 0.5 and 1.0 μg/mL. Compound **3** also improved the reduction in cell viability induced by the radical to 69.4 ± 3.1 and 80.6 ± 2.0% at 0.5 and 1.0 μg/mL pre-treatment. However, compounds **3**, **5** and **6** had no protective effect and therefore are not shown in the present results. It is thus possible that compounds **1**, **2** and **4** may have been involved in the free radical scavenging and improvement in the cellular antioxidant defense and subsequently increased the cell viability.

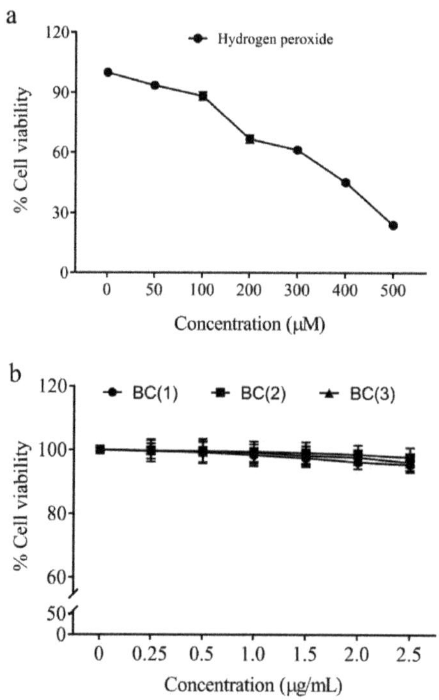

Figure 2. Cytotoxicity of hydrogen peroxide (**a**) and biologically safer concentration of synthetic bis-chalcones (**b**).

The level of reduced glutathione content in the normal IEC-6 cells was estimated to be 4.68 ± 0.22 μmoles/mg protein. Upon exposure to the hydrogen peroxide radicals at a concentration of (400 μM) for 24 h, the level of reduced glutathione was significantly brought down to 2.31 ± 0.24 μmoles/mg protein ($p < 0.05$). It has been previously reported that increased peroxide radicals often tend to reduce the cellular glutathione pool [37]. Glutathione is the central antioxidant and its depletion is essential for the induction of apoptotic death in cells [38,39]; hence, the increased cell death in peroxide-exposed cells may be attributed to the glutathione depletion events.

Compared to the control cells, the treatment with compound **1** [BC(**1**)] significantly increased the cellular-reduced glutathione pool to 3.07 ± 0.21 ($p < 0.05$) and 3.78 ± 0.35 μmoles/mg protein ($p < 0.01$) at their respective doses of 0.5 μg/mL and 1.0 μg/mL. In compound **2** [BC(**2**)], the level of glutathione was increased to 2.94 ± 0.23 and 3.15 ± 0.18 μmoles/mg protein. In BC(**3**) treated cells, the glutathione content was elevated to 3.13 ± 0.30 and 3.85 ± 0.41 μmoles/mg protein. Hence, it is clear that treatment with bis-chalcones increased the cellular pool of reduced glutathione. A previous study by Kachadourian et al. [40] indicated the possible potential of chalcones as inducers of glutathione biosynthesis; hence, it is also possible that the treated chalcones may have induced glutathione biosynthesis in the cell and subsequently increased the cellular GSH levels (Table 1).

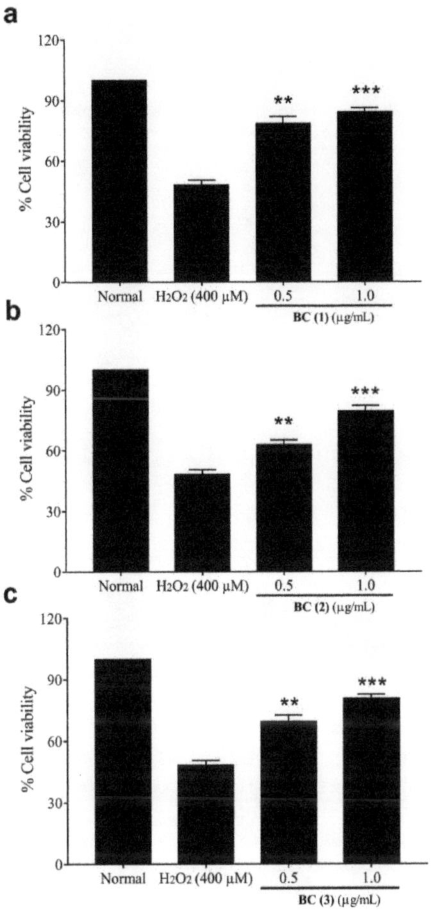

Figure 3. Protective effect of synthetic bischalcones BC (**1**) (**a**), BC (**2**) (**b**), BC (**3**) (**c**) against hydrogen peroxide-induced cell death in intestinal epithelial cells. (** indicate $p < 0.01$ and *** indicate $p < 0.001$).

Table 1 represents the results of various treatment regimens on the catalase activity in IEC-6 cells. The normal IEC-6 cells exhibited a catalase activity of 47.64 ± 3.7 U/mg protein. However, exposure to the peroxide radicals over 24 h resulted in a shoot in catalase activity (88.19 ± 4.3 U/mg protein). The catalase enzyme plays an important role in the degradation of peroxide radicals to yield water and therefore detoxification [41]. In addition, the increased level of peroxide exposure has also been reported to increase the activity of catalase in cells [42]. Therefore, it may be possible that the exposure of cells to peroxide radicals resulted in an increase in catalase activity. Interestingly, the cell pre-treated with compound **1** [BC(**1**)] (at 0.5 µg/mL and 1.0 µg/mL) showed a significant reduction in the activity of the catalase enzyme (65.62 ± 3.4 and 50.04 ± 4.2 U/mg protein). Furthermore, the other compounds also indicated a similar reduction in the catalase activities to 72.11 ± 2.3 and 59.15 ± 6.1 U/mg protein ((BC(**2**)) as well as 70.55 ± 4.8, and 61.82 ± 6.4 U/mg protein (BC(**3**)). The restoration of catalase activity to near normal possibly indicates the protective effect of synthetic bis-chalcones on peroxide toxicity.

Table 1. Changes in the cellular redox status in cells exposed to peroxide radicals and the effect of synthetic bis-chalcones.

Treatment	Catalase (U/mg Protein)	GSH (μmoles/mg Protein)	GPx (U/mg Protein)	TBARS (nmoles/mg Protein)
Normal	47.64 ± 3.7	4.68 ± 0.22	65.65 ± 3.94	1.75 ± 0.34
H_2O_2 (400 μM)	88.19 ± 4.3	2.31 ± 0.24	103.10 ± 4.82	6.55 ± 0.45
BC(**1**) 0.5 μg/mL	65.62 ± 3.4 **	3.07 ± 0.21 *	85.64 ± 4.12 **	4.59 ± 0.21 *
BC(**1**) 1.0 μg/mL	50.04 ± 4.2 ***	3.78 ± 0.35 **	75.05 ± 3.14 ***	3.25 ± 0.42 **
BC(**2**) 0.5 μg/mL	72.11 ± 2.3 *	2.94 ± 0.23 **	90.04 ± 2.15 *	4.76 ± 0.40 *
BC(**2**) 1.0 μg/mL	59.15 ± 6.1 ***	3.15 ± 0.18	84.11 ± 3.45 **	4.05 ± 0.51 **
BC(**3**) 0.5 μg/mL	70.55 ± 4.8 *	3.13 ± 0.30 *	80.17 ± 4.03 *	5.01 ± 0.17 *
BC(**3**) 1.0 μg/mL	61.82 ± 6.4 **	3.85 ± 0.41 **	70.52 ± 4.16 **	4.11 ± 0.38 **

BC(**1**)—Bis-chalcone **1** (compound **1**); BC(**2**)—Bis-chalcone **2** (compound **2**); BC(**3**)—Bis-chalcone **3** (compound **3**). The * indicates significant variation at $p < 0.05$, ** indicate $p < 0.01$ and *** indicate $p < 0.001$.

Glutathione peroxidase is another cellular enzyme associated with peroxide neutralization [43]; the glutathione-derived hydrogen moieties are utilized by this enzyme for the detoxification of the peroxide radicals [44]. The GPx is also important during excessive peroxide-induced inactivation of the catalase; under such conditions, the glutathione peroxide takes the central responsibility of the redox balance in cells [45]. In the present study, the GPx activity of normal cells was 65.65 ± 3.94 U/mg protein. Peroxide radical exposure resulted in an elevation in the GPx activity to 103.10 ± 4.82 U/mg protein ($p < 0.001$). On the contrary, the pre-treatment with BC(**1**), BC(**2**), and BC(**3**) brought down the peroxide-induced elevation in the GPx activity (Table 1). Hence, it is possible that the acute exposure to peroxide may have resulted in the elevation in GPx activity; however, pre-treatment with chalcones may have partially scavenged the radicals and thereby resulted in the reduction in GPx activity.

The altered antioxidant defense is often expressed as increased lipid peroxidation products [46]; TBARS is considered to be an important marker of the extent of lipid peroxidation [47]. In the present study, the level of TBARS in normal cells was 1.75 ± 0.34 nmoles/mg protein. The peroxide exposure resulted in a significant increase in the TBARS level (6.55 ± 0.45 nmoles/mg protein). However, pre-treatment with BC(**1**) reduced the levels to 4.59 ± 0.21 and 3.25 ± 0.42 nmoles/mg protein; the level in BC(**2**) treatment was found to be 4.76 ± 0.40 and 4.05 ± 0.51 nmoles/mg protein. However, the BC(**3**) was less effective in preventing TBARS formation and the level was estimated to be 5.01 ± 0.17 and 4.11 ± 0.38 nmoles/mg protein (Table 1).

2.4. *Anti-Inflammatory Effects of Bis-Chalcones*

Inflammation plays a crucial role in the oncogenic transformation of colorectal epithelial cells [48,49]. Hence, the management of inflammatory insults also becomes important to prevent the colon carcinogenesis process. Being an integral ingredient of functional foods, the chalcones are one among the promising anticancer dietary agents [50]. In the present study, LPS-mediated inflammation was used as a model; LPS induces inflammatory cytokine release from macrophages by stimulating toll-like receptor mediated signaling [51,52].

Likewise, the level of IL-1β in untreated cells was 54.5 ± 2.9 pg/mg protein. Exposure of RAW 264.7 cells to LPS resulted in the increased secretion of IL-1β (503.2 ± 12.3 pg/mg protein). On the contrary, the pre-treatment with BC(**1**) at 0.5 and 1.0 μg/mL resulted in reduced IL-1β levels to 407.8 ± 15.6 and 298.4 ± 12.4 pg/mg protein. Furthermore, pre-treatment with BC(**2**) at the same concentrations reduced the level of IL-1β to 421.8 ± 14.5 and 365.7 ± 15.5 pg/mg protein. Likewise, pre-treatment with BC(**3**) reduced IL-1β levels to 434.5 ± 10.9 and 389.4 ± 16.2 pg/mg protein at 0.5 and 1.0 μg/mL. The role of IL-1β in colon cancer is evident as it promoted the proliferation potential of the cells and also triggered epithelial to mesenchymal transition in these cells [53,54]. Hence, inhibition of IL-1β secretion is therefore beneficial in intestinal conditions.

IL-6 is another pro-inflammatory cytokine that triggers transformation in colon epithelial cells by triggering STAT3 signaling cascades [55,56]. In untreated cells, the level of IL-6 was estimated to be 103.4 ± 10.2 pg/mg protein. However, LPS stimulation increased the secretion of IL-6 and thereby the level was elevated to 1185.2 ± 24.6 pg/mg protein. Treatment with BC(**1**) reduced the IL-6 levels to 851.1 ± 20.6 and 756.1 ± 22.4 pg/mg protein at its low and high doses. Likewise, pre-treatment with BC(**2**) (927.5 ± 27.3 and 835.1 ± 17.2 pg/mg protein) and BC(**3**) (964.7 ± 19.5 and 876.1 ± 27.4 pg/mg protein) effectively inhibited the IL-6 production in macrophages.

Untreated cells had a TNF-α level of 259.4 ± 10.9 pg/mg protein, which was increased upon exposure to LPS (1635.0 ± 22.5 pg/mg protein). On the contrary, pre-treatment with BC(**1**), BC(**2**) and BC(**3**) at their low (0.5 µg/mL) and high (1.0 µg/mL) doses inhibited the production of TNF-α levels (Table 2). The TNF-α levels are crucial for the survival and proliferation of various cancer cells including colon [57]. The cytokine is also important in the progression events of colon cancer such as metastasis and stemness [58]. Numerous studies have indicated the potential of various chalcones against the production of inflammatory cytokines [59,60].

Table 2. Effect of synthetic bis-chalcones on LPS-stimulated cytokine release in RAW 264.7 cells.

	IL-1β (pg/mg Protein)	**IL-6 (pg/mg Protein)**	**TNF-α (pg/mg Protein)**	**NO (µM/mg Protein)**
Untreated	54.5 ± 2.9	103.4 ± 10.2	259.4 ± 10.9	8.5 ± 0.7
LPS	503.2 ± 12.3	1185.2 ± 24.6	1635.0 ± 22.5	67.8 ± 2.7
BC(**1**) 0.5 µg/mL	407.8 ± 15.6 *	851.1 ± 20.6 *	1367.0 ± 18.3 **	39.5 ± 1.2 *
BC(**1**) 1.0 µg/mL	298.4 ± 12.4 ***	756.1 ± 22.4 ***	1015.1 ± 18.6 ***	26.5 ± 1.4 ***
BC(**2**) 0.5 µg/mL	421.8 ± 14.5 *	927.5 ± 27.3 *	1475.8 ± 10.4 *	51.0 ± 0.5 ns
BC(**2**) 1.0 µg/mL	365.7 ± 15.5 **	835.1 ± 17.2 **	1300.7 ± 33.4 **	39.2 ± 1.1 *
BC(**3**) 0.5 µg/mL	434.5 ± 10.9 *	964.7 ± 19.5 *	1404.1 ± 28.2 *	46.7 ± 0.8 *
BC(**3**) 1.0 µg/mL	389.4 ± 16.2 **	876.1 ± 27.4 **	1288.5 ± 16.3 **	31.2 ± 1.8 **

The values are represented as mean ± SD of three independent experiments, each carried in triplicate. (ns indicates not significant; * indicates significant difference $p < 0.05$; ** indicate significant difference $p < 0.01$ and *** indicate significant difference $p < 0.001$).

The untreated RAW 264.7 cells had a nitric oxide level of 8.5 ± 0.7 µM/mg protein. However, exposure to LPS stimulated the production of NO and the level was elevated to 67.8 ± 2.7 µM/mg protein. Treatment with BC(**1**) reduced the nitric oxide level in the cells to 39.5 ± 1.2 and 26.5 ± 1.4 µM/mg protein at the low and high doses, respectively. Likewise, pre-treatment with BC(**2**) brought down the cellular NO levels to 51.0 ± 0.5 and 39.2 ± 1.1 µM/mg protein. The nitric oxide level in the BC(**3**) pre-treated cells (0.5 µg/mL and 1.0 µg/mL) was found to be 46.7 ± 0.8 and 31.2 ± 1.8 µM/mg protein.

Overall, the results of the present study confirm the cytoprotective and anti-inflammatory potentials of the three synthetic bis-chalcones. However, further studies are necessary to study the molecular mechanism of action of these bis-chalcones, especially on the Nrf2/ARE axis of antioxidant defense.

3. Materials and Methods

3.1. Chemicals, Cells and Media

The chemicals were sourced from Sigma-Aldrich (St. Louis, MO, USA) and all the materials obtained were of analytical grade. The cells were procured from NCCS, Pune, India, and maintained under standard conditions in RPMI-1640 media (containing 10% FBS).

3.2. Synthesis and Characterization of Bis-Chalcones

The bis-chalcones were synthesized using cyclohexanone and cyclopentanone as core compounds. The detailed synthesis methods are illustrated in Supplementary Figures S1–S9. The compounds were previously synthesized as per our own publication [31].

3.3. Cytotoxicity Analysis of Peroxides and Biologically Safe Concentration of Bis-Chalcones

The non-cancer cell line IEC-6 (intestinal epithelial cells) and RAW 264.7 cells were procured from the National Centre for Cell Science and maintained in RPMI-1640 and DMEM media. The IEC-6 cells and RAW 264.7 were then plated in 96-well plates at a seeding density of 1×10^6 cells/mL. The cells were treated with different doses of various bis-chalcones (0–250 μg/mL) for 48 h. At the end of incubation, the media were replaced with fresh media containing 5 mg/mL MTT. The cell viability was determined according to the previous descriptions of Khanapure et al. [61]. Similarly, the IC50 value of the peroxide radical was also determined using hydrogen peroxide as a radical source.

3.4. Analysis of the Effect of Bis-Chalcones against Peroxide-Induced Damage in Cells

In order to assess the protective effect against the peroxide radicals, the cells were plated as described in Section 3.3. The cells were pre-treated for 2 h with two biologically safe concentrations of individual bis-chalcones. The cells were further exposed to peroxide radicals after incubation for another 24 h. A peroxide control and untreated normal cells were also maintained to determine the extent of protection as indicated by the improvement in cell viability. The cell viability of each treatment was determined using MTT assay as described previously in Section 3.3 and the percentage of protection was observed.

Furthermore, to assess the mechanism of protection, the cells were again exposed to similar experimental conditions in Petri dishes. The cells were collected by mechanical scrapers and lysed in Tris buffer. The lysate was centrifuged at $8000\times g$ for 15 min in a refrigerated centrifuge to yield clear supernatant. The activity of enzymes such as catalase, glutathione peroxidase and levels of glutathione as well as thiobarbituric acid reactive substances were also estimated according to our previous protocols [62].

3.5. Effect of Bis-Chalcones on LPS-Stimulated Macrophages

The anti-inflammatory activity was estimated in LPS-primer macrophages. Briefly, the macrophages were cultured in DMEM media for 48 h. The actively dividing cells were again plated onto a 6-well-plate. After 24 h, the cell was pre-treated with different biosafe concentrations of various bis-chalcones. Subsequently, the cells were exposed to 1 mg/mL concentration of lipopolysaccharide. At the end of the incubation, the cell and media were collected and stored at -80 °C for the analysis of inflammatory cytokine and biochemical detection of the nitric oxide formed [63]. The cytokine levels were determined using ELISA kits from Peprotech.

3.6. Statistical Analysis

The obtained results were processed in MS Office Excel and the statistical comparison was carried out using two-way ANOVA in a GraphPad Prism version 10.0 (La Jolla, CA, USA).

4. Conclusions

The study analyzed the cytoprotective and anti-inflammatory activity of six different synthetic bis-chalcones. Among those tested, three were found to be active as anti-inflammatory and cytoprotective agents; they include compounds **1**, **2** and **4**. The cytoprotective effect was efficiently modulated through the restoration of antioxidant enzyme activities and reduction of subsequent lipid peroxidation. The inhibition of inflammatory cytokine production is found to be the mechanism of anti-inflammatory activity. Overall, the synthetic bis-chalcones **1**, **2** and **4** are promising as possible therapeutic candidates against oxidative insult and inflammation. To be more specific, compound **2** was found to be more effective and seems to be promising as a pharmacological agent for future use. However, further studies are necessary on the toxicity aspects of these compounds in animals as well as confirmation of their protective effect in rat/mice models.

Supplementary Materials: The following supporting information can be downloaded at: https://www.mdpi.com/article/10.3390/molecules28176354/s1. Figure S1: Synthesis of Compound **1**; Figure S2: FT-IR spectrum of compound **1**; Supplementary Figure S3: ^{1}H NMR Spectrum of compound **1** in $CDCl_3$; Figure S4: Synthesis of Compound **2**; Figure S5: FT-IR spectrum of compound **2**; Figure S6: ^{1}H NMR Spectrum of compound **2** in $CDCl_3$; Figure S7: Synthesis of Compound **4**; Figure S8: FT-IR spectrum of compound **4**; Figure S9: ^{1}H NMR Spectrum of compound **4** in $CDCl_3$.

Author Contributions: A.T.: Analysis, manuscript preparation, experimentation; R.R.: Analysis, manuscript preparation, experimentation; A.N.: Study design, methodology, experimentation, analysis, funding acquisition, manuscript editing; M.M.: Study design, methodology, experimentation, analysis, funding acquisition, manuscript editing; J.J.: Experimentation, supervision, manuscript editing; A.A.: Study design, methodology, experimentation, analysis, funding acquisition, manuscript editing; D.B.: Study design, analysis, manuscript editing. All authors have read and agreed to the published version of the manuscript.

Funding: This research was funded by Researchers Supporting Project Number (RSP2023R11), King Saud University, Riyadh, Saudi Arabia.

Institutional Review Board Statement: Not applicable.

Informed Consent Statement: Not applicable.

Data Availability Statement: The data may be shared upon valid request.

Acknowledgments: The authors acknowledge the funding support from Researchers Supporting Project Number (RSP2023R11), King Saud University, Riyadh, Saudi Arabia. DBT-STAR (Project number: BT/HRD/11/09/2020) scheme supported infrastructural development at St. Joseph's College (Autonomous), Devagiri, Calicut.

Conflicts of Interest: The authors declare no conflict of interest.

Sample Availability: Samples of the compounds are available from the authors.

References

1. Siegel, R.L.; Wagle, N.S.; Cercek, A.; Smith, R.A.; Jemal, A. Colorectal cancer statistics, 2023. *CA Cancer J. Clin.* **2023**, *73*, 233–254. [CrossRef] [PubMed]
2. Wang, Y.; Huang, X.; Cheryala, M.; Aloysius, M.; Zheng, B.; Yang, K.; Chen, B.; Fang, Q.; Chowdary, S.B.; Abougergi, M.S.; et al. Global increase of colorectal cancer in young adults over the last 30 years: An analysis of the Global Burden of Disease Study 2019. *J. Gastroenterol. Hepatol.* **2023**. [CrossRef] [PubMed]
3. Sifaki-Pistolla, D.; Poimenaki, V.; Fotopoulou, I.; Saloustros, E.; Mavroudis, D.; Vamvakas, L.; Lionis, C. Significant Rise of Colorectal Cancer Incidence in Younger Adults and Strong Determinants: 30 Years Longitudinal Differences between under and over 50s. *Cancers* **2022**, *14*, 4799. [CrossRef] [PubMed]
4. Allemani, C.; Weir, H.K.; Carreira, H.; Harewood, R.; Spika, D.; Wang, X.S.; Bannon, F.; Ahn, J.V.; Johnson, C.J.; Bonaventure, A.; et al. Global surveillance of cancer survival 1995–2009: Analysis of individual data for 25,676,887 patients from 279 population-based registries in 67 countries (CONCORD-2). *Lancet* **2015**, *385*, 977–1010. [CrossRef]
5. Patil, P.S.; Saklani, A.; Gambhire, P.; Mehta, S.; Engineer, R.; De'Souza, A.; Chopra, S.; Bal, M. Colorectal Cancer in India: An Audit from a Tertiary Center in a Low Prevalence Area. *Indian J. Surg. Oncol.* **2017**, *8*, 484–490. [CrossRef]
6. Cao, S.; Chen, C.; Gu, D.; Wang, Z.; Xu, G. Establishment and external verification of an oxidative stress-related gene signature to predict clinical outcomes and therapeutic responses of colorectal cancer. *Front. Pharmacol.* **2023**, *13*, 991881. [CrossRef]
7. Basak, D.; Uddin, M.N.; Hancock, J. The Role of Oxidative Stress and Its Counteractive Utility in Colorectal Cancer (CRC). *Cancers* **2020**, *12*, 3336. [CrossRef]
8. Ahmadinejad, F.; Geir Møller, S.; Hashemzadeh-Chaleshtori, M.; Bidkhori, G.; Jami, M.-S. Molecular Mechanisms behind Free Radical Scavengers Function against Oxidative Stress. *Antioxidants* **2017**, *6*, 51. [CrossRef]
9. Sharifi-Rad, M.; Anil Kumar, N.V.; Zucca, P.; Varoni, E.M.; Dini, L.; Panzarini, E.; Rajkovic, J.; Tsouh Fokou, P.V.; Azzini, E.; Peluso, I.; et al. Lifestyle, Oxidative Stress, and Antioxidants: Back and Forth in the Pathophysiology of Chronic Diseases. *Front. Physiol.* **2020**, *11*, 694. [CrossRef]

10. Azzolin, V.F.; Cadoná, F.C.; Machado, A.K.; Berto, M.D.; Barbisan, F.; Dornelles, E.B.; Glanzner, W.G.; Gonçalves, P.B.; Bica, C.G.; da Cruz, I.B. Superoxide-hydrogen peroxide imbalance interferes with colorectal cancer cells viability, proliferation and oxaliplatin response. *Toxicol. Vitr.* **2016**, *32*, 8–15. [CrossRef]
11. Zhu, J.W.; Yu, B.M.; Ji, Y.B.; Zheng, M.H.; Li, D.H. Upregulation of vascular endothelial growth factor by hydrogen peroxide in human colon cancer. *World J. Gastroenterol. WJG* **2002**, *8*, 153–157. [CrossRef] [PubMed]
12. van der Waals, L.M.; Jongen, J.M.J.; Elias, S.G.; Veremiyenko, K.; Trumpi, K.; Trinh, A.; Laoukili, J.; Ubink, I.; Schenning-van Schelven, S.J.; van Diest, P.J.; et al. Increased Levels of Oxidative Damage in Liver Metastases Compared with Corresponding Primary Colorectal Tumors: Association with Molecular Subtype and Prior Treatment. *Am. J. Pathol.* **2018**, *188*, 2369–2377. [CrossRef] [PubMed]
13. Pravda, J. Evidence-based pathogenesis and treatment of ulcerative colitis: A causal role for colonic epithelial hydrogen peroxide. *World J. Gastroenterol. WJG* **2022**, *28*, 4263–4298. [CrossRef] [PubMed]
14. Tuomisto, A.E.; Mäkinen, M.J.; Väyrynen, J.P. Systemic inflammation in colorectal cancer: Underlying factors, effects, and prognostic significance. *World J. Gastroenterol. WJG* **2019**, *25*, 4383–4404. [CrossRef]
15. Schmitt, M.; Greten, F.R. The inflammatory pathogenesis of colorectal cancer. *Nat. Rev. Immunol.* **2021**, *21*, 653–667. [CrossRef]
16. Soleimani, A.; Rahmani, F.; Ferns, G.A.; Ryzhikov, M.; Avan, A.; Hassanian, S.M. Role of the NF-κB signaling pathway in the pathogenesis of colorectal cancer. *Gene* **2020**, *726*, 144132. [CrossRef]
17. Wang, H.; Tian, T.; Zhang, J. Tumor-Associated Macrophages (TAMs) in Colorectal Cancer (CRC): From Mechanism to Therapy and Prognosis. *Int. J. Mol. Sci.* **2021**, *22*, 8470. [CrossRef]
18. Cory, H.; Passarelli, S.; Szeto, J.; Tamez, M.; Mattei, J. The Role of Polyphenols in Human Health and Food Systems: A Mini-Review. *Front. Nutr.* **2018**, *5*, 87. [CrossRef]
19. Marotta, L.; Rossi, S.; Ibba, R.; Brogi, S.; Calderone, V.; Butini, S.; Campiani, G.; Gemma, S. The green chemistry of chalcones: Valuable sources of privileged core structures for drug discovery. *Front. Chem.* **2022**, *10*, 988376. [CrossRef]
20. Orlikova, B.; Tasdemir, D.; Golais, F.; Dicato, M.; Diederich, M. Dietary chalcones with chemopreventive and chemotherapeutic potential. *Genes Nutr.* **2011**, *6*, 125–147. [CrossRef]
21. Wu, X.; Zhang, S.; Liu, X.; Shang, J.; Zhang, A.; Zhu, Z.; Zha, D. Chalcone synthase (CHS) family members analysis from eggplant (*Solanum melongena* L.) in the flavonoid biosynthetic pathway and expression patterns in response to heat stress. *PLoS ONE* **2020**, *15*, e0226537. [CrossRef]
22. Oldoni, T.L.C.; Cabral, I.S.R.; d'Arce, M.A.B.R.; Rosalen, P.L.; Ikegaki, M.; Nascimento, A.M.; Alencar, S.M. Isolation and analysis of bioactive isoflavonoids and chalcone from a new type of Brazilian propolis. *Sep. Purif. Technol.* **2011**, *77*, 208–213. [CrossRef]
23. Escobar-Ramos, A.; Lobato-García, C.E.; Zamilpa, A.; Gómez-Rivera, A.; Tortoriello, J.; González-Cortazar, M. Homoisoflavonoids and Chalcones Isolated from *Haematoxylum campechianum* L., with Spasmolytic Activity. *Molecules* **2017**, *22*, 1405. [CrossRef] [PubMed]
24. Akhtar, M.S.; Rehman, A.U.; Arshad, H.; Malik, A.; Fatima, M.; Tabassum, T.; Raza, A.R.; Bukhsh, M.; Murtaza, M.A.; Mehmood, M.H.; et al. In Vitro Antioxidant Activities and the Therapeutic Potential of Some Newly Synthesized Chalcones against 4-Acetaminophenol Induced Hepatotoxicity in Rats. *Dose-Response* **2021**, *19*, 1559325821996955. [CrossRef]
25. Karimi-Sales, E.; Mohaddes, G.; Alipour, M.R. Chalcones as putative hepatoprotective agents: Preclinical evidence and molecular mechanisms. *Pharmacol. Res.* **2018**, *129*, 177–187. [CrossRef] [PubMed]
26. Wang, L.; Yang, X.; Zhang, Y.; Chen, R.; Cui, Y.; Wang, Q. Anti-inflammatory Chalcone–Isoflavone Dimers and Chalcone Dimers from Caragana jubata. *J. Nat. Prod.* **2019**, *82*, 2761–2767. [CrossRef] [PubMed]
27. Cai, X.; Sha, F.; Zhao, C.; Zheng, Z.; Zhao, S.; Zhu, Z.; Zhu, H.; Chen, J.; Chen, Y. Synthesis and anti-inflammatory activity of novel steroidal chalcones with 3β-pregnenolone ester derivatives in RAW 264.7 cells in vitro. *Steroids* **2021**, *171*, 108830. [CrossRef]
28. ur Rashid, H.; Xu, Y.; Ahmad, N.; Muhammad, Y.; Wang, L. Promising anti-inflammatory effects of chalcones via inhibition of cyclooxygenase, prostaglandin E2, inducible NO synthase and nuclear factor κb activities. *Bioorg. Chem.* **2019**, *87*, 335–365. [CrossRef]
29. Pozzetti, L.; Ibba, R.; Rossi, S.; Taglialatela-Scafati, O.; Taramelli, D.; Basilico, N.; D'Alessandro, S.; Parapini, S.; Butini, S.; Campiani, G.; et al. Total Synthesis of the Natural Chalcone Lophirone E, Synthetic Studies toward Benzofuran and Indole-Based Analogues, and Investigation of Anti-Leishmanial Activity. *Molecules* **2022**, *27*, 463. [CrossRef]
30. Jasim, H.A.; Nahar, L.; Jasim, M.A.; Moore, S.A.; Ritchie, K.J.; Sarker, S.D. Chalcones: Synthetic Chemistry Follows Where Nature Leads. *Biomolecules* **2021**, *11*, 1203. [CrossRef]
31. Kuttithodi, A.M.; Nikhitha, D.; Jacob, J.; Narayanankutty, A.; Mathews, M.; Olatunji, O.J.; Rajagopal, R.; Alfarhan, A.; Barcelo, D. Antioxidant, Antimicrobial, Cytotoxicity, and Larvicidal Activities of Selected Synthetic Bis-Chalcones. *Molecules* **2022**, *27*, 8209. [CrossRef] [PubMed]
32. Salehi, B.; Quispe, C.; Chamkhi, I.; El Omari, N.; Balahbib, A.; Sharifi-Rad, J.; Bouyahya, A.; Akram, M.; Iqbal, M.; Docea, A.O.; et al. Pharmacological Properties of Chalcones: A Review of Preclinical Including Molecular Mechanisms and Clinical Evidence. *Front. Pharmacol.* **2021**, *11*, 592654. [CrossRef] [PubMed]
33. Rudrapal, M.; Khan, J.; Dukhyil, A.A.B.; Alarousy, R.M.I.I.; Attah, E.I.; Sharma, T.; Khairnar, S.J.; Bendale, A.R. Chalcone Scaffolds, Bioprecursors of Flavonoids: Chemistry, Bioactivities, and Pharmacokinetics. *Molecules* **2021**, *26*, 7177. [CrossRef]
34. Dao, T.T.; Linthorst, H.J.; Verpoorte, R. Chalcone synthase and its functions in plant resistance. *Phytochem. Rev.* **2011**, *10*, 397–412. [CrossRef] [PubMed]

35. Vilema-Enríquez, G.; Arroyo, A.; Grijalva, M.; Amador-Zafra, R.I.; Camacho, J. Molecular and Cellular Effects of Hydrogen Peroxide on Human Lung Cancer Cells: Potential Therapeutic Implications. *Oxidative Med. Cell. Longev.* **2016**, *2016*, 1908164. [CrossRef]
36. Xiang, J.; Wan, C.; Guo, R.; Guo, D. Is Hydrogen Peroxide a Suitable Apoptosis Inducer for All Cell Types? *BioMed Res. Int.* **2016**, *2016*, 7343965. [CrossRef]
37. Lubos, E.; Loscalzo, J.; Handy, D.E. Glutathione peroxidase-1 in health and disease: From molecular mechanisms to therapeutic opportunities. *Antioxid. Redox Signal.* **2011**, *15*, 1957–1997. [CrossRef]
38. Franco, R.; Panayiotidis, M.I.; Cidlowski, J.A. Glutathione depletion is necessary for apoptosis in lymphoid cells independent of reactive oxygen species formation. *J. Biol. Chem.* **2007**, *282*, 30452–30465. [CrossRef]
39. Liu, N.; Ma, X.; Luo, X.; Zhang, Y.; He, Y.; Dai, Z.; Yang, Y.; Wu, G.; Wu, Z. L-Glutamine Attenuates Apoptosis in Porcine Enterocytes by Regulating Glutathione-Related Redox Homeostasis. *J. Nutr.* **2018**, *148*, 526–534. [CrossRef]
40. Kachadourian, R.; Day, B.J.; Pugazhenti, S.; Franklin, C.C.; Genoux-Bastide, E.; Mahaffey, G.; Gauthier, C.; Di Pietro, A.; Boumendjel, A. A synthetic chalcone as a potent inducer of glutathione biosynthesis. *J. Med. Chem.* **2012**, *55*, 1382–1388. [CrossRef]
41. Andrés, C.M.; Pérez de la Lastra, J.M.; Juan, C.A.; Plou, F.J.; Pérez-Lebeña, E. Chemistry of Hydrogen Peroxide Formation and Elimination in Mammalian Cells, and Its Role in Various Pathologies. *Stresses* **2022**, *2*, 256–274. [CrossRef]
42. Martins, D.; English, A.M. Catalase activity is stimulated by H_2O_2 in rich culture medium and is required for H_2O_2 resistance and adaptation in yeast. *Redox Biol.* **2014**, *2*, 308–313. [CrossRef] [PubMed]
43. Ighodaro, O.M.; Akinloye, O.A. First line defence antioxidants-superoxide dismutase (SOD), catalase (CAT) and glutathione peroxidase (GPX): Their fundamental role in the entire antioxidant defence grid. *Alex. J. Med.* **2018**, *54*, 287–293. [CrossRef]
44. Pei, J.; Pan, X.; Wei, G.; Hua, Y. Research progress of glutathione peroxidase family (GPX) in redoxidation. *Front. Pharmacol.* **2023**, *14*, 1147414. [CrossRef] [PubMed]
45. Baud, O.; Greene, A.E.; Li, J.; Wang, H.; Volpe, J.J.; Rosenberg, P.A. Glutathione peroxidase-catalase cooperativity is required for resistance to hydrogen peroxide by mature rat oligodendrocytes. *J. Neurosci.* **2004**, *24*, 1531–1540. [CrossRef] [PubMed]
46. Feng, J.; Wang, J.; Wang, Y.; Huang, X.; Shao, T.; Deng, X.; Cao, Y.; Zhou, M.; Zhao, C. Oxidative Stress and Lipid Peroxidation: Prospective Associations Between Ferroptosis and Delayed Wound Healing in Diabetic Ulcers. *Front. Cell Dev. Biol.* **2022**, *10*, 898657. [CrossRef]
47. Aguilar Diaz De Leon, J.; Borges, C.R. Evaluation of Oxidative Stress in Biological Samples Using the Thiobarbituric Acid Reactive Substances Assay. *J. Vis. Exp.* **2020**, *12*, 61122.
48. Hirano, T.; Hirayama, D.; Wagatsuma, K.; Yamakawa, T.; Yokoyama, Y.; Nakase, H. Immunological Mechanisms in Inflammation-Associated Colon Carcinogenesis. *Int. J. Mol. Sci.* **2020**, *21*, 3062. [CrossRef]
49. Means, A.L.; Freeman, T.J.; Zhu, J.; Woodbury, L.G.; Marincola-Smith, P.; Wu, C.; Meyer, A.R.; Weaver, C.J.; Padmanabhan, C.; An, H.; et al. Epithelial Smad4 Deletion Up-Regulates Inflammation and Promotes Inflammation-Associated Cancer. *Cell. Mol. Gastroenterol. Hepatol.* **2018**, *6*, 257–276. [CrossRef]
50. Ouyang, Y.; Li, J.; Chen, X.; Fu, X.; Sun, S.; Wu, Q. Chalcone Derivatives: Role in Anticancer Therapy. *Biomolecules* **2021**, *11*, 894. [CrossRef]
51. Buccini, D.F.; Roriz, B.C.; Rodrigues, J.M.; Franco, O.L. Antimicrobial peptides could antagonize uncontrolled inflammation via Toll-like 4 receptor. *Front. Bioeng. Biotechnol.* **2022**, *10*, 1037147. [CrossRef] [PubMed]
52. Li, W.; Cai, Z.; Schindler, F.; Bahiraii, S.; Brenner, M.; Heiss, E.H.; Weckwerth, W. Norbergenin prevents LPS-induced inflammatory responses in macrophages through inhibiting NFκB, MAPK and STAT3 activation and blocking metabolic reprogramming. *Front. Immunol.* **2023**, *14*, 1117638. [CrossRef] [PubMed]
53. Voronov, E.; Apte, R.N. IL-1 in Colon Inflammation, Colon Carcinogenesis and Invasiveness of Colon Cancer. *Cancer Microenviron.* **2015**, *8*, 187–200. [CrossRef] [PubMed]
54. Li, Y.; Wang, L.; Pappan, L.; Galliher-Beckley, A.; Shi, J. IL-1β promotes stemness and invasiveness of colon cancer cells through Zeb1 activation. *Mol. Cancer* **2012**, *11*, 1476–4598. [CrossRef]
55. Lin, Y.; He, Z.; Ye, J.; Liu, Z.; She, X.; Gao, X.; Liang, R. Progress in understanding the IL-6/STAT3 pathway in colorectal cancer. *OncoTargets Ther.* **2020**, *13*, 13023–13032. [CrossRef]
56. Waldner, M.J.; Foersch, S.; Neurath, M.F. Interleukin-6—A key regulator of colorectal cancer development. *Int J. Biol. Sci.* **2012**, *8*, 1248–1253. [CrossRef]
57. Al Obeed, O.A.; Alkhayal, K.A.; Al Sheikh, A.; Zubaidi, A.M.; Vaali-Mohammed, M.A.; Boushey, R.; McKerrow, J.H.; Abdulla, M.H. Increased expression of tumor necrosis factor-α is associated with advanced colorectal cancer stages. *World J. Gastroenterol.* **2014**, *20*, 18390–18396. [CrossRef]
58. Zhao, P.; Zhang, Z. TNF-α promotes colon cancer cell migration and invasion by upregulating TROP-2. *Oncol. Lett.* **2018**, *15*, 3820–3827. [CrossRef]
59. Martinez, R.M.; Pinho-Ribeiro, F.A.; Steffen, V.S.; Caviglione, C.V.; Fattori, V.; Bussmann, A.J.C.; Bottura, C.; Fonseca, M.J.V.; Vignoli, J.A.; Baracat, M.M.; et al. trans-Chalcone, a flavonoid precursor, inhibits UV-induced skin inflammation and oxidative stress in mice by targeting NADPH oxidase and cytokine production. *Photochem. Photobiol. Sci.* **2017**, *16*, 1162–1173. [CrossRef]
60. Lee, J.S.; Bukhari, S.N.; Fauzi, N.M. Effects of chalcone derivatives on players of the immune system. *Drug Des. Devel. Ther.* **2015**, *9*, 4761–4778.

61. Khanapure, S.; Jagadale, M.; Bansode, P.; Choudhari, P.; Rashinkar, G. Anticancer activity of ruthenocenyl chalcones and their molecular docking studies. *J. Mol. Struct.* **2018**, *1173*, 142–147. [CrossRef]
62. Narayanankutty, A.; Kunnath, K.; Famurewa, A.C.; Ramesh, V.; Rajagopal, R.; Alfarhan, A. Variations in the composition, cytoprotective and anti-inflammatory effects of natural polyphenols of edible oils extracted from fresh and dried coconut testa. *Physiol. Mol. Plant Pathol.* **2022**, *117*, 101742. [CrossRef]
63. Xia, G.; Zhou, L.; Ma, J.; Wang, Y.; Ding, L.; Zhao, F.; Chen, L.; Qiu, F. Sesquiterpenes from the essential oil of Curcuma wenyujin and their inhibitory effects on nitric oxide production. *Fitoterapia* **2015**, *103*, 143–148. [CrossRef] [PubMed]

Disclaimer/Publisher's Note: The statements, opinions and data contained in all publications are solely those of the individual author(s) and contributor(s) and not of MDPI and/or the editor(s). MDPI and/or the editor(s) disclaim responsibility for any injury to people or property resulting from any ideas, methods, instructions or products referred to in the content.

molecules

MDPI

Review

A Comprehensive Review on Bioactive Compounds Found in *Caesalpinia sappan*

Twinkle Vij [1,†], Pawase Prashant Anil [2,†], Rafeeya Shams [1], Kshirod Kumar Dash [3,*], Rhythm Kalsi [1], Vinay Kumar Pandey [4,5], Endre Harsányi [6,*], Béla Kovács [7,*] and Ayaz Mukarram Shaikh [7]

1 Department of Food Technology and Nutrition, Lovely Professional University, Phagwara 144411, Punjab, India
2 MIT School of Food Technology, MIT ADT University, Pune 412201, Maharashtra, India
3 Department of Food Processing Technology, Ghani Khan Choudhury Institute of Engineering and Technology (GKCIET), Malda 732141, West Bengal, India
4 Division of Research & Innovation (DRI), School of Applied & Life Sciences, Uttaranchal University, Dehradun 248007, Uttarakhand, India
5 Department of Bioengineering, Integral University, Lucknow 226026, Uttar Pradesh, India
6 Agricultural Research Institutes and Academic Farming (AKIT), Faculty of Agriculture, Food Science and Environmental Management, University of Debrecen, 4032 Debrecen, Hungary
7 Faculty of Agriculture, Food Science and Environmental Management, Institute of Food Science, University of Debrecen, 4032 Debrecen, Hungary
* Correspondence: kshirod@tezu.ernet.in (K.K.D.); harsanyie@agr.unideb.hu (E.H.); kovacsb@agr.unideb.hu (B.K.)
† These authors contributed equally to this work.

Citation: Vij, T.; Anil, P.P.; Shams, R.; Dash, K.K.; Kalsi, R.; Pandey, V.K.; Harsányi, E.; Kovács, B.; Shaikh, A.M. A Comprehensive Review on Bioactive Compounds Found in *Caesalpinia sappan*. *Molecules* **2023**, *28*, 6247. https://doi.org/10.3390/molecules28176247

Academic Editors: Adele Papetti and Hyun-Ock Pae

Received: 24 June 2023
Revised: 22 July 2023
Accepted: 16 August 2023
Published: 25 August 2023

Copyright: © 2023 by the authors. Licensee MDPI, Basel, Switzerland. This article is an open access article distributed under the terms and conditions of the Creative Commons Attribution (CC BY) license (https://creativecommons.org/licenses/by/4.0/).

Abstract: Sappan wood (*Caesalpinia sappan*) is a tropical hardwood tree found in Southeast Asia. Sappan wood contains a water-soluble compound, which imparts a red color named brazilin. Sappan wood is utilized to produce dye for fabric and coloring agents for food and beverages, such as wine and meat. As a valuable medicinal plant, the tree is also known for its antioxidant, anti-inflammatory, and anticancer properties. It has been observed that sappan wood contains various bioactive compounds, including brazilin, brazilein, sappan chalcone, and protosappanin A. It has also been discovered that these substances have various health advantages; they lower inflammation, enhance blood circulation, and are anti-oxidative in nature. Sappan wood has been used as a medicine to address a range of illnesses, such as gastrointestinal problems, respiratory infections, and skin conditions. Studies have also suggested that sappan wood may have anticarcinogenic potential as it possesses cytotoxic activity against cancer cells. Based on this, the present review emphasized the different medicinal properties, the role of phytochemicals, their health benefits, and several food and nonfood applications of sappan wood. Overall, sappan wood has demonstrated promising medicinal properties and is an important resource in traditional medicine. The present review has explored the potential role of sappan wood as an essential source of bioactive compounds for drug development.

Keywords: sappan wood; Brazilwood; Suou; Indian Redwood; heartwood; chemical constituents; coloring agent

1. Introduction

Sappan wood is typically found in Southeast Asia and the Pacific Islands. The scientific name of sappan wood is *Caesalpinia sappan* L. (genus Fabaceae), and it is also known as Sappan, Brazilwood, and Suou in diverse parts of the world. Southeast Asia, Malaysia, India, and Indonesia are home to the native variety of the tropical heartwood sappan tree, also known as the Indian Redwood [1,2]. Due to its commercial and cultural significance, sappan wood has been extensively harvested in many parts of Southeast Asia, leading to concerns about its sustainability. The tree is harvested every 6–8 years and when the trunk has attained 5–6 cm diameter. The tree is cut about 1 m above the ground to allow sprouts to grow from the stump. The main branches along with the stump are harvested. The average yield of inner pulp is about 80 kg/tree. Seeds can be harvested right from the second year of planting, but the heartwood is ready only after 6–12 years. A yield of 2000–2500 kg of pods may be obtained, which in turn may yield 200–250 kg of seeds per hectare. The harvested wood is chipped into pieces, and the dye is extracted by boiling them in water. While extracting, a few paddy grains are thrown into the boiling liquid to check the completion of the extraction process. If the husk scales are off, boiling is considered sufficient. The wood dye yield varies with varietal and cultural factors. To assure the long-term survival of this tree, campaigns are being launched to support ethical harvesting methods and protect its natural habitat [2]. This plant thrives at an elevation of 1000 m above sea level and is well suited for mountainous regions with moderate temperatures. It is a small tree, typically reaching a height of 5–10 m with a diameter of 15–25 cm. The trunk and branches are spiny and covered in reddish-brown hairs, while the stem is round and brownish-green in color. The flowers are arranged in panicles, which are terminal (at the ends of branches) and located in the axils of the upper leaves. The panicles measure about 30–40 cm in length, and the pedicles (flower stalks) are approximately 1.3–1.5 cm long. The stamens (male reproductive parts) are delicate, waxy-white, and have filaments that are densely woolly at the base. The pods, measuring 7–10 cm in length and 3.8–5 cm in width, are woody, slanted, and elongated [3].

Bottom of Form

Sappan wood has been used for various purposes for centuries, including as a natural dye, medicinal herb, and in traditional medicine practices [4,5]. The heartwood contains a natural dye called brazilin, which produces a range of red hues. Sappan wood has also been used for dyeing textiles made from using silk and cotton, producing high-quality furniture and decorative items, and coloring foods, like rice and noodles, in Southeast Asia. Additionally, the dye is employed in the food business to tend items like cheese, smoked fish, and meat [6,7]. Sappan wood has been used in traditional medicine practices for its anti-inflammatory, antioxidant, and antimicrobial properties. Many diseases, including cough, fever, gastrointestinal issues, and skin conditions, have all been treated with it [8]. Additionally, it has been employed as a blood cleanser and astringent [3,9].

The United States Department of Agriculture states that the *Caesalpinia sappan* L. classification is as follows: Kingdom (Plantae), Subkingdom (Tracheobionta), Division (Magnoliophyta), Class (Magnoliopsida), Order (Fabales), Family (Fabaceae—pea family), Genus (*Caesalpinia* L.), Species (*Caesalpinia sappan* L.—Sappan wood) [10]. *Caesalpinia sappan* L. has different vernacular names in different regions varying by language, including English (sappan wood), Tamil (Patungam), Hindi (Bakam), Telugu (Vakama), Malayalam (Sappanam, Pathimukham), Sanskrit (Patrangah, Patangah), Kannada (Chappanga), and Gujarati (Patang) [11]. The sappan wood plant is a small tree with a spherical and brownish-green stem, pubescent rufous limbs, and a height of 5 to 10 m. This wood is excellent for turning and also yields a crimson dye. It is hard, weighty, prickly, hefty, and orange-red. After a year of growth, fruit production can begin, typically during the rainy season. Flowering can then occur after about six months. Due to its cultural and economic significance, sappan wood has been designated as a state tree in some Indian states, including Kerala and Tamil Nadu [11]. Due to overharvesting and habitat loss, sappan wood is regarded as a threatened species in many areas of Southeast Asia. Efforts are underway to promote sustainable harvesting practices and conserve this tree's natural habitats. Some countries have also established protected areas to preserve sappan wood populations. Using sappan wood for dyeing fabric has a long history in Southeast Asia. In some countries, such as Thailand, the traditional dyeing process using sappan wood is still practiced today [12]. The purpose of this review is to discuss the various bioactive components found in sappan wood, as well as the therapeutic properties of sappan wood. The conventional and novel extraction techniques of sappan wood's bioactive components are addressed. Furthermore, the commercial uses of sappan wood are highlighted.

The conservatively accepted *Caesalpinia sappan* L. has medicinal properties based on the literature basis. Various studies have proved the uncountable benefits of sappan wood, but the exact mechanism behind the therapeutic benefits is still unknown. The use of sappan wood has begun in the cosmetic industry, but its impact should not be limited to it. It has potential to flourish in the pharmaceutical and nutraceutical industries as well. Future research will be required to determine the mechanism of action and isolation of active ingredients from *Caesalpinia sappan* L., which has extraordinarily stimulated biological effects and a significant body of traditional myths based on natural resources [13].

2. Bioactive Compounds of Sappan Wood

2.1. Bioactive Compounds of Sappan Wood

The phytochemical composition of sappan wood has been studied extensively, and it has been found that it contains various bioactive compounds. The major constituents of sappan wood are flavonoids, phenolic acids, and anthraquinones. The major metabolites recognized from *Caesalpinia sappan* L. with identified chemical structures are presented in Figure 1.

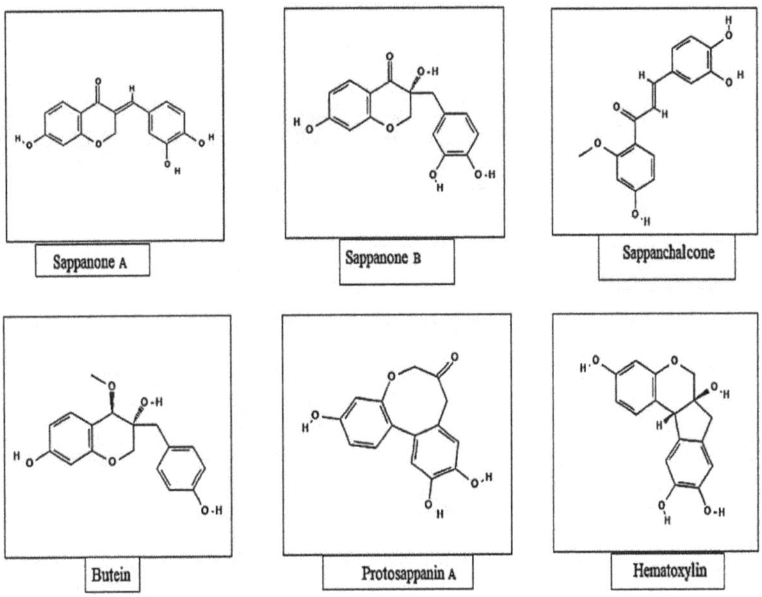

Figure 1. Major metabolites recognized from *Caesalpinia sappan* with identified chemical structures.

Other substances found in sappan wood, in addition to flavonoids and phenolic ones, are triterpenoids, steroids, alkaloids, saponins, and tannins. Many different biological actions, such as anti-inflammation, anticancer, and antioxidant characteristics, are exhibited by the class of polyphenolic compounds and have been connected to several health benefits [14]. Several structurally unique phenolic components, such as brazilin, xanthone, one coumarin, chalcones, flavones, and homo isoflavonoids, are present in sappan wood [15,16]. The different bioactive compounds present in sappan wood are illustrated in Table 1.

Table 1. Different bioactive compounds present in sappan wood.

Bioactive Compound	Properties	References
Xanthone	The xanthones existing in the pericarp, whole fruit, heartwood, and leaf of mangosteen have been found to possess a broad variety of pharmacologic properties, including antioxidant, antitumor, antiallergic, anti-inflammatory, antibacterial, antifungal, and antiviral activities (Garcinia mangostana Linn., GML).	[15]
Coumarin	Coumarin is a substance that smells like vanilla and is present in many plants. It was historically used to flavor meals.	[14]
Chalcones	The golden crystalline ketone $C_6H_5CHCHCOC_6H_5$, as well as many of its many derivatives, some of which are flavone-related plant colors, are created by combining benzaldehyde and acetophenone. Currently, a wide range of chalcones are used as dietary additives, cosmetic ingredients, and for the treatment of gastritis, stomach cancer, viral illnesses, cardiovascular diseases, pain, and for the treatment of pain.	[16]
Flavones	A colorless, crystalline compound that serves as the building block for several yellow or whitish plant pigments. The anti-inflammatory properties of phytonutrients, like flavonoids, are advantageous, and they shield our cells from oxidative harm that can cause diseases.	[17]
Homo isoflavonoids	Within a select few plant families, the rare compound is spread. These natural compounds can be found in large quantities in the genus Caesalpinia. Numerous bioactivities of homoisoflavonoids have been noted, including antimicrobial, antimutagenic, antidiabetic, and vasorelaxant properties.	[18]
Brazilin	$C_{16}H_{14}O_5$, a white or pale phenolic substance, is primarily used for dyeing and is derived from the Caesalpinia Brazilwood species.	[19]

2.1.1. Flavonoids

The flavonoids found in sappan wood are brazilin, haematoxylin, and protosappanin. Sappan wood includes conjugated aromatic benzene groups as mentioned above; flavonoid compounds are hypothesized to be able to block UV (ultraviolet) rays, protecting skin from exposure to the sun. The flavonoids in sappan wood have the potential to be used as sunscreens [20]. The content of flavonoid and anthocyanin components in sappan wood extract (*Caesalpinia sappan* L.) was studied. The five concentration levels used in this experiment were 20%, 40%, 60%, 80%, and 100%. Fifteen experimental units were produced after three iterations of the study. The information was shown using tables, illustrations, and descriptive statistics. The analysis revealed that the sappan wood extract had a flavonoid content of 6.02% and an anthocyanin content of 2.43% [21].

2.1.2. Phenolic Acids

Phenolic acids are found in a variety of plant species and have been connected to several health benefits, including antioxidant and anti-inflammatory properties. Sappan wood contains numerous phenolic acids, including chlorogenic acid, caffeic acid, and gallic acid [22]. During research, sappan wood extract from Bone Regency, South Sulawesi Province, was extracted using ultrasonic-assisted solvent extraction (ultrasonic extraction), with water serving as the solvent. The content of polyphenols in this extract was then determined. The Follin–Ciocalteu visible spectrophotometer was employed, with pH conditions of 6, 7, and 8. The results show that the polyphenol content of sappan wood at three pH values was 34.33% (pH 6), 13.70% (pH 7), and 12.66% (pH 8). The analysis findings show that pH 6 has the highest polyphenol level and that as pH increases, polyphenol content decreases [23].

2.1.3. Anthraquinones

Various pharmacological actions, such as antimicrobial, anticancer, and anti-inflammatory properties, have been linked to a family of organic compounds known as anthraquinones. Sappan wood contains several anthraquinones, including brazilin, brazilein, and sappanone A [24]. Anthroquinones, also known as simple anthrones or bianthrones, are chemical compounds. The free anthraquinone aglycones do not have much of a therapeutic effect. The sugar residue makes it easier for the aglycone to be absorbed and transported to the region of action. Anthraquinones and associated glycosides are stimulant cathartics that work by making the smooth muscle of the large intestinal wall more toned. The large intestine is where the glycosides are expelled after being reabsorbed from the small intestine, where they induce irritation of the colon mucosa and promote motility to have a laxative effect. In conclusion, rats responded to dosages of *Caesalpinia sappan* L. wood extracts by becoming laxative. All extracts except for ethanolic extract were shown to be more powerful and to only display a dose-dependent laxative effect in a drug-induced constipation paradigm. In models with low-fiber diets, all extracts showed equivalent effectiveness [25].

2.1.4. Triterpenoids and Steroids

The class of organic compounds known as terpenoids is found in plants and has been shown to have a variety of pharmacological effects, such as anticancer, anti-inflammatory, and antimicrobial characteristics. Sappan wood contains several triterpenoids, including lupeol, β-amyrin, and cycloartenol [26]. Steroids are a different family of organic compounds widely distributed in the plant kingdom and exhibit various biological activities, such as anticancer and anti-inflammatory characteristics. Sappan wood contains several steroids, including stigmasterol and β-sitosterol [27].

2.1.5. Alkaloids and Tannins

Alkaloids are biological complexes that contain nitrogen and are present in a variety of plant species. They demonstrate a variety of pharmacological effects, such as analgesic, anti-inflammatory, and anticancer characteristics. Sappan wood contains several alkaloids, including sappan chalcone and sappanone B [28]. Sappan chalcone has been demonstrated to have effects on the growth of human prostate cancer cells, and sappanone B has been demonstrated to have effects on inflammation by suppressing the production of proinflammatory cytokines [27]. Tannins are polyphenolic substances that are frequently present in plants and are responsible for giving them an astringent flavor. Sappan wood contains condensed tannins, sometimes proanthocyanidins, and hydrolyzable tannins, such as ellagic acid and gallic acid. Syamsunarno et al. [29] stated that the diverse range of phytochemical compounds found in sappan wood may have an extensive range of potential well-being advantages. More research is necessary to completely comprehend the compounds' pharmacological actions and their potential therapeutic uses. During a study, the bark of the tree was successively extracted with n-hexane, ethyl acetate, methanol, and hot water before being analyzed with GC-MS to determine its total phenolic content (TPC), total flavonoid content (TFC), and total tannin content (TTC) as well as its antioxidant activity (DPPH scavenging activity). The methanol extract of the bark had the greatest concentrations of TPC (824.1662 $\pm$ 28 mg GAE/g), TFC (185.031 $\pm$ 91 mg QE/g), and TTC (987.0730 $\pm$ 98 mg TAE/g) [30].

2.2. Extraction of Bioactive Compounds from Sappan L.

Due to the high cost of synthetic pharmaceuticals and the negative side effects of synthetic molecules, researchers are more interested in discovering bioactive substances from natural sources, such as drugs. As a result, plant resources are continually assessed in an effort to find bioactive molecules that can treat diseases. One of the most significant bioactive natural substances found in *Caesalpinia sappan* L. heartwood is brazilin, which has a wide range of industrial applications in the textile, cosmetics, and pharmaceutical industries [31,32]. *Caesalpinia sappan* L. bioactive metabolites obtained from various parts of the plant are presented in Table 2.

Table 2. *Caesalpinia Sappan* L. bioactive metabolites from various parts of plant.

Part	Principal Compound	Reference
Stem	Flavonoids, tannins, alkaloids, sterols, and terpenoids	[20]
Seeds	Caesalpinia R and S Caesalsappanins A–L Caesalsappanins M–N	[33–35]
Bark	Alkaloids, flavonoids, tannins, terpenoids, and steroids	[27]
Wood	Brazilin, sappanone B, and protosappanin A	[36]
Leaves	Glycosides, phenols, tannins, saponins, flavonoids, and steroids	[11]
Heartwood	Ceasalpiniaphenols A–D Sappanchalcone, ceasalpiniaphenol G, and quercetin Brazilin, protosappanin A, protosappanin B, protosappanin C, protosappanin D, and protosappanin E 3′-Deoxy-4-O-methylepisappanol, (+)-(8S,8′S)-bisdihydrosiringenin Brazilein palmitic acid Protosappanins E-1 and E-2 Lupeol, vanillin, β-sitosterol, linoleic acid, stigmasterol, and friedelin	[37–39]

Caesalpinia sappan L. heartwood is a dark red color that is commonly used for furniture, musical instruments, and decorative items made from wood, which is valued for its deep red hue. In addition to its esthetic properties, the heartwood of sappan wood is also known for its medicinal properties. It has been used for many years in traditional medicine to address a wide range of illnesses, such as inflammation, digestive issues, and respiratory issues [40]. The heartwood is used as a decoction in the Namya-utai solution, which possesses antithirst and cardiotonic properties. In northern Thailand, a decoction of *Caesalpinia sappan* L. heartwood, especially in the provinces of Chiang Mai, Nan, and Lampang, uses this as an anti-inflammatory drug to treat arthritis and traumatic disease. It is mostly utilized in Thailand as a coloring agent in clothing, cosmetics, beverages, and foods [41]. Traditional Chinese medicine primarily used sappan wood as an emmenagogue, hemostatic, analgesic, anti-inflammatory, and blood flow-promoting drug for traumatic diseases. Furthermore, *Caesalpinia sappan* L. heartwood decoction is used to treat a variety of conditions, including high blood pressure, cataracts, digestion, dysmenorrhea, burning sensations, ear infections, gonorrhea, heart problems, jaundice, nervous disorders, obesity, ophthalmic illnesses, stomach aches, syphilis, spermatorrhoea, urinary diseases, and vascular diseases [42]. Heartwood contains a variety of compounds, including brazilin and brazilein, which are known to have anti-inflammatory and antioxidant properties. Overall, the heartwood of sappan wood is a valuable natural resource with both esthetic and medicinal properties. Its sustainable use is important for preserving both its cultural and ecological significance [15]. The three main water-soluble flavonoids in heartwood are brazilin, protosappanin, and hematoxylin.

2.2.1. Extraction of Bioactive Compounds

Extraction of Anticonvulsant Compounds

Using EtOAc, n-BuOH, and water, 80% aqueous methanolic extracts from *Caesalpinia sappan* wood were fractionated. This wood exhibited exceptional anticonvulsant efficacy. One of them, succinic semialdehyde dehydrogenase (SSADH), and succinic semialdehyde reductase (SSAR), were strongly inhibited by the ethyl acetate fraction. The separation of the two major active components was accomplished using many column chromatographies for the percentage determined by an activity test. Based on spectrum data, their chemical structures were identified as sappanchalcone and brazilin. The SSAR activities were inactivated by the pure compounds sappanchalcone and brazilin in a dose-dependent manner, whereas SSADH was only partly inhibited by sappanchalcone and not by brazilin [43].

Brazilin

Sappan wood (*Caesalpinia sappan* L.) contains several active compounds, but the most notable one is brazilin. The structures of brazilin and brazilein are presented in Figure 2. Brazilin is a red pigment that gives sappan wood its characteristic red color. It is a flavonoid that is part of a class of substances called chalcones. Brazilin has been found to have various biological activities, including anticancer, antioxidant properties, and anti-inflammatory ones [44]. Its medicinal advantages have also been utilized in traditional medicine. Brazilin is a white phenolic molecule with a 5-membered ring that contains two aromatic rings and one pyron. But when the hydroxyl group of brazilin is oxidized, it changes to the carbonyl group resulting in its structural change and the production of brazilein is formed (a colorful substance). Brazilein has a long history of use as a natural colorant [45]. The Caesalpinia wood-based red dye has been rendered attainable by brazilein. Currently, brazilein is regarded to be yellow. Both substances are tetracyclic because they have two aromatic rings, one pyrone ring, and five carbons. The presence of the carbonyl group causes a rise in the delocalization of electrons, transforming the yellow brazilin into the red brazilein [15]. In 95% ethanol for two hours, wood extraction yield is highest. The analysis of sappan wood's chemical constituents led to the identification of many structural kinds of phenolic components, including brazilin, one xanthone, one coumarin, three chalcones, two flavones, and three homoisoflavonoids.

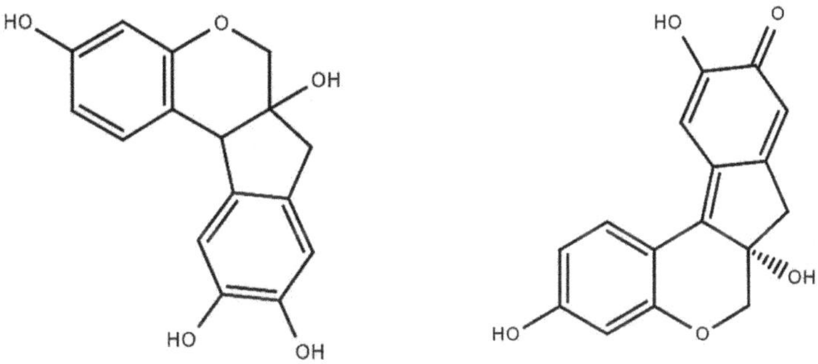

(**a**) (**b**)

Figure 2. Structure of (**a**) brazilin and (**b**) brazilein.

2.3. *Medicinal Characteristics of Sappan Wood*

Sappan wood exhibits various pharmacological properties. These characteristics are thought to be a result of the flavonoids and phenolic compounds found in sappan wood. According to researchers, sappan wood extract demonstrated potent antioxidant activity in vitro, as judged by its capacity to scavenge free radicals and prevent lipid for oxidation. Another study reported that sappan wood extract had a protective effect against oxidative damage in mice [46]. Sappan wood also has an antimicrobial activity against a variety of pathogenic bacteria and fungi. In one study, for instance, it was discovered that sappan wood extract had antibacterial activity against both Gram-positive and Gram-negative bacteria, such as Staphylococcus aureus and Escherichia coli. Another study found that sappan wood extract had antifungal activity against several strains of Candida [26]. Furthermore, sappan wood has been traditionally used to treat various diseases, including diabetes, cancer, and cardiovascular diseases. The studies have reported on the potential health benefits of sappan wood extracts in animal models. For instance, the study found that sappan wood extract had a hypoglycaemic effect in diabetic rats. A different study found that sappan wood extract had a cardioprotective effect in mice with an induced myocardial infarction [47]. Mode of action of various pharmacological properties of *Caesalpinia sappan* L. is presented in Table 3. Overall, the diverse range of bioactive compounds found in sappan wood and their potential health benefits suggest that it may have a role in traditional medicine and as a potential source of new drugs.

Table 3. Mode of action of various pharmacological properties of *Caesalpinia sappan* L.

Characteristics	Mode of Action	References
Anti-inflammatory	Inhibition of iNOS gene expression Nuclear factor kappa B (NF-κB)	[38,42,48–50]
	Tumor necrosis factor-α production	[38,42,50,51]
Antioxidant	DPPH radical scavenging assay	[49]
	Ferric reduction assay	[46,51]
Antiacne	Zone of inhibition	[52–55]
Antibacterial	MIC (minimum inhibitory concentration)/MIB (minimum bactericidal concentration)	[56–58]
Hepatoprotective	Inhibition of CCl_4 intoxication	[59,60]

2.3.1. Anti-Inflammatory Properties

The heartwood of sappan wood has an extended history of use in medicine due to its anti-inflammatory properties. Additionally, sappan wood extract has been demonstrated to have analgesic (pain-relieving) properties, which can also aid in reducing discomfort brought on by inflammation [51]. Studies have shown that sappan wood contains various bioactive compounds that have anti-inflammatory effects. Brazilin is one of these compounds, and studies have shown that it can stop cells from releasing inflammatory molecules, like interleukin-1 beta (IL-1 β) and tumor necrosis factor-alpha (TNF-α). Nuclear factor kappa B (NF-κB), both crucial transcription factors that regulate the expression of genes linked to inflammation, has also been shown to be inhibited by brazilin [42]. Another component found in sappan wood called protosappanin A has been shown to stop the production of proinflammatory molecules, like cyclooxygenase-2 (COX2), nitric oxide (NO), and prostaglandin E2 (PGE2). By blocking the simulation of NF-B, protosappanin A has also been discovered to decrease the expression of proinflammatory genes, like COX-2 and inducible nitrogen oxide (iNOS) [38,50]. Moreover, sappan wood extract has been demonstrated to reduce inflammation in a variety of animal models. For instance, in a study on colitis-infected rats, sappan wood extract was shown to lessen the severity of inflammation and mucosal damage in the colon by inhibiting the activation of NF-B and the production of proinflammatory cytokines [39]. The bioactive components in sappan wood, like brazilin and protosappanin A, have been shown to inhibit the production of proinflammatory molecules and decrease the expression of proinflammatory genes by inhibiting NF-B activation [61]. These bioactive compounds are largely responsible for sappan wood's anti-inflammatory properties [62]. The diagrammatic representation of the extraction process that affects sappan lignum biological activity is shown in Figure 3, which indicates that anti-inflammatory actions on 70% ethanol microwave extraction is more effective than 70% ethanol heat extraction.

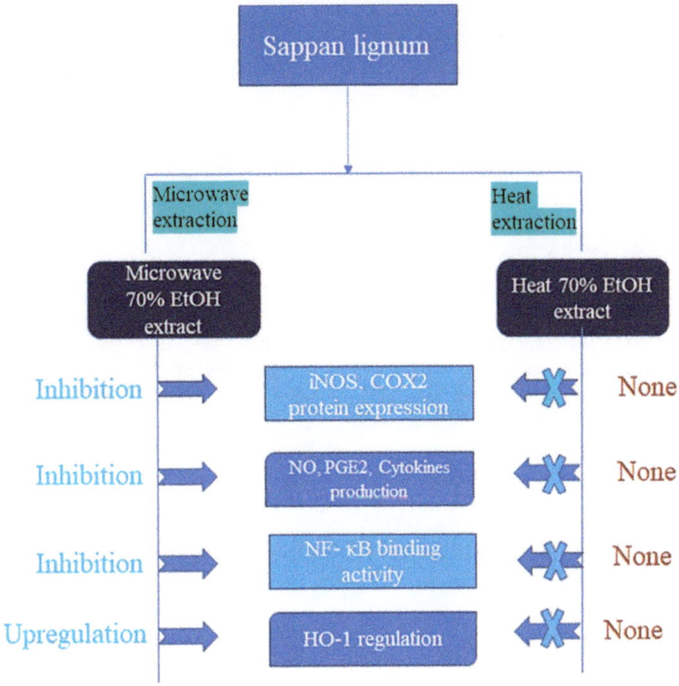

Figure 3. Diagrammatic representation of the extraction process that affects the *Sappan lignum* biological activity.

2.3.2. Antioxidant Properties

One of the properties of sappan wood is its antioxidant activity. This refers to the wood's capacity to combat hazardous molecules known as free radicals that can hurt cells and decrease the number of diseases. Several studies have examined sappan wood's antioxidant activity and found that it contains various compounds that can act as antioxidants. For example, sappan wood contains compounds, such as brazilin, protosappanin, and sappanone, which have antioxidant effects [51]. Sasaki et al. [49] discovered that the methanolic extract of sappan wood demonstrated strong antioxidant activity in a number of in vitro assays, including the DPPH radical scavenging assay, the ABTS radical scavenging assay, and the reducing power assay. Settharaksa et al. [46] estimated the sappan wood extract antioxidant activity by using DPPH (2,2-diphenyl-1-picrylhydrazyl) radical scavenging assay and FRAP (ferric-reducing antioxidant power) assay. The results showed that the extract possessed significant antioxidant activity in both assays.

Another study investigated the antioxidant activity of sappan wood extract. In cooked ground beef, researchers found that the extract was able to significantly reduce lipid oxidation and improve the sensory quality of the beef [62]. Another investigation evaluated the antioxidant activity of sappan wood extract in human colon cancer cells. The extract may have promise as a natural anticancer agent because the researchers discovered that it was able to cause cell death in cancer cells via an oxidative stress pathway [63]. Similarly, in human skin cells, they found that the extract could protect cells from the oxidative harm caused by UV rays, suggesting that it might be useful as a natural sunscreen ingredient [64]. Sarumathy et al. [65] also investigated the antioxidant activity of sappan wood extract in rats with liver damage caused by acetaminophen. According to their research, they found that the extract was able to significantly lessen oxidative stress and liver damage in rats. The extract may have promise as a natural treatment for liver diseases.

2.3.3. Antiacne Properties

The heartwood of the tree is used to treat skin disorders, such as acne. Numerous studies have investigated the antiacne activity of sappan wood extract. Mitani et al. [54] observed that sappan wood extracts have anti-inflammatory and antibacterial effects on propionibacterium acnes, the bacteria that cause acne. According to the study, the extract has significant anti-inflammatory and antibacterial capabilities, as well as the capacity to inhibit the development of *P. acnes*. They concluded that sappan wood extract could be a promising natural substitute for the treatment of acne. Madhubala et al. [55] discovered that the compounds had powerful antibacterial and anti-inflammatory properties, suggesting that they could be used as acne treatment options. Several compounds were isolated and tested for their antiacne activity from sappan wood extract. Pattananandecha et al. [66] discovered that the extract significantly reduced inflammation and antibacterial effects on microorganisms that cause acne by using sappan wood extract. It has been discovered that the extract greatly decreased the production of proinflammatory cytokines and the severity of acne [52]. It has been suggested that the sappan wood extract significantly reduced the sebum quantity produced by human sebocytes [53]. Overall, these studies suggest that sappan wood extract has the potential as a natural treatment for acne due to its anti-inflammatory and antibacterial properties. To ascertain its efficiency and safety in long-term use, more studies are required.

2.3.4. Antibacterial Properties

Srinivasan et al. [56] found that sappan wood extract has significant antibacterial activity against both Gram-positive and Gram-negative bacteria. The extract was found to inhibit the growth of various bacterial strains, including Salmonella typhimurium. *E. coli* and *S. aureus*. They suggested that the presence of different phytochemicals, such as flavonoids and tannins, in sappan wood extract serve as the main reason for its antibacterial action. Rina et al. [57] examined that sappan wood extract suppresses the antibacterial activity against Helicobacter pylori, a bacterium that can cause stomach cancer and peptic ulcers. The extract of sappan wood exhibits antibacterial activity against H. pylori, which was dose-dependent. The researchers suggested that the antibacterial activity of sappan wood extract may be due to its ability to inhibit the activity of enzymes involved in bacterial cell wall synthesis.

Puttipan et al. [58] examined the antibacterial action of sappan wood extract against dental caries bacteria. According to the investigators, the sappan wood extract inhibits a number of oral microbes, such as Lactobacillus Acidophilus and *S. mutans*. They claimed that the antibacterial action of sappan wood extract might be due to its ability to disrupt bacterial cell membranes and inhibit bacterial DNA synthesis. According to these studies, sappan wood extract may have broad-spectrum antibacterial activity and be successful against a variety of bacterial strains, including those that are resistant to traditional antibiotics. However, more investigation is required to establish the ideal dosage and administration strategy as well as any possible toxicity or adverse effects.

2.3.5. Hepatoprotective Properties

One of its purported benefits is its hepatoprotective activity, meaning its ability to protect the liver from damage. Srilakshmi et al. [59] examined the hepatoprotective activity of sappan wood extract against liver damage induced by carbon tetrachloride (CCl_4) in rats. The extract, as indicated by reduced liver enzyme levels and more effective histological characteristics, can significantly reduce liver damage. Researchers hypothesized that the sappan wood extract lowered lipid peroxidation and increased antioxidant enzyme levels. It may exercise its hepatoprotective effects by reducing oxidative stress in the liver. Gupta et al. [60] investigated the hepatoprotective activity of sappan wood extract against liver damage induced by paracetamol (acetaminophen) in rats. Kadir et al. [67] provide further evidence of the hepatoprotective activity of sappan wood extract and suggest that it may be a useful natural remedy for protecting the liver against various toxins and stressors.

2.3.6. Other Medicinal Properties

In addition to their previous activities, from *Caesalpinia sappan* L., isolated compounds and crude extracts have been reported that exhibit antiviral, cytotoxic, anticancer, anticonvulsant, hypolipidemic, cognitive-enhancing activity, and analgesic activity [68]. The various pharmacological properties of *Caesalpinia sappan* are presented in Table 3.

2.4. Methods of Extraction of Bioactive Compounds of Sappan Wood

The term "bioactive molecules" refers to substances that interact with living beings and change them in some way. The compounds can be created synthetically or obtained from natural sources, like plants and food. To discover substitutes for synthetic materials generated from natural resources, many studies have been conducted. The different procedures for the extraction of bioactive compounds from *Ceasalpinia sappan* L. are presented in Table 4. Due to the abundance of agricultural by-products that could act as a low-cost source of raw materials to extract plant bioactive compounds, this substitution may have economic advantages [69]. The antibacterial, antioxidant, and anticancer potential of bioactive substances produced by plants is well documented. Plant extracts are commonly employed in the pharmaceutical, food, and cosmetic industries [70]. The substances that are found in wood log extractives are a fusion of phenolic compounds, such as terpenoids, alkaloids, terpenes, and saponins. The extraction process is the initial step in separating and using the bioactive compounds that plants contain. In order to enhance the extraction yield of these compounds, it is crucial to use an appropriate extraction process [71]. The process yield is greatly influenced by the pretreatment techniques, the physical and chemical properties of the plant and the target complex, the choice of the extraction process, and the extraction process operating parameters [72]. Various methods of extracting bioactive compounds are presented in Figure 4.

Table 4. Different procedures for extraction of bioactive compounds from *Ceasalpinia sappan* L.

	Conventional Extraction			
Raw Material	**Technique**	**Experimental Data**	**Results**	**References**
Conventional extraction		Solvent: Ethanol–water (80:20) SLR:1:20 (*w*:*v*) 72 h	Brazilin produces 3.12 mg/g of extract.	[73]
	Maceration	Solvent: Ethanol–water (96:4) SLR:1:20 (*w*:*v*) 72 h	Brazilin produces 4.58 mg/g of extract.	
	Soxhlet extraction	Solvent: Ethanol–water (96:4) SLR:1:20 (*w*:*v*) 3 h	Brazilin produces 5.43 mg/g of extract.	[22]
Novel extraction	Deep eutectic solvents	Betain:lactic acid with 60% water SLR: 1:20 (*w*:*v*)	Brazilin produces 4.49 mg/g of extract.	[22,73]

Abbreviations: SLR: Solid/liquid ratio.

2.4.1. Conventional Extraction Technique

In this technique, procedures are found for the removal of the complexes in accordance with their varying polarities while using suitable diluters, considering that the primary component in this procedure is the solvent employed. These time-tested techniques use solid/liquid extraction techniques that have been widely used in both the laboratory and the industrial setting to isolate solid matrix chemicals. The added factors that have the most prominent effects on conventional techniques are the environmental characteristics, liquid–solid ratio, pressure, temperature, and abstraction time [71].

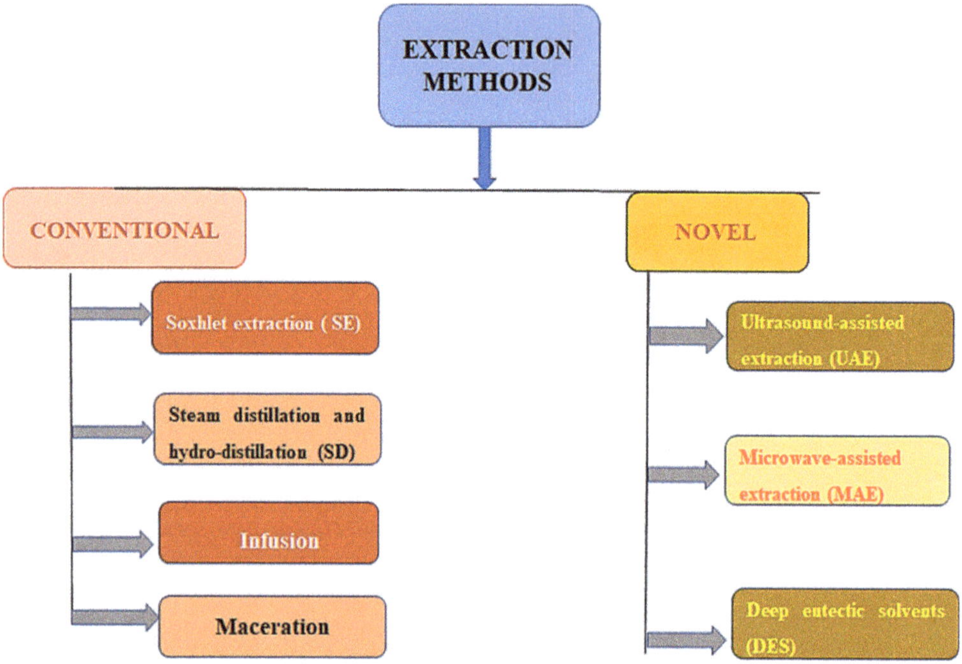

Figure 4. Various methods of extracting bioactive compounds.

Soxhlet Extraction (SE)

To perform a Soxhlet extraction, a heat source is used to vaporize the solvent, which is then compressed in a condenser that produces reflux and discharged over the specimen's container. Baron Von Soxhlet, who developed this method in the middle of the nineteenth century, is the source of the name for this technique [72]. The solvent containing the substance that was extracted is drawn back down to the bottom by a siphon once it reaches the top of the specimen's container. This procedure can be carried out for a predetermined number of Soxhlet cycles or repeatedly over a predetermined number of times over a predetermined amount of time [74]. The temperature of the extraction changes depending on the solvent that is being used because the extraction is performed at the solvent's boiling point [75]. Soxhlet extraction has a number of benefits, including the maintenance of relatively high extraction temperature due to the heat of the distillation flask, the preservation of the sample in contact with fresh solvent, the lack of filtration required after lixiviation, the effective simplicity, and the low cost. Santos et al. [75] produced 5.10 weight percent for ethyl acetate, 1.57 weight percent for n-hexane, and 7.23 weight percent for ethanol when studying the Soxhlet extraction of wood, i.e., Eremanthus erythropappus using solvents that varied in polarity. This most likely happened as the yield of Soxhlet extraction typically increases with solvent polarity, which is understandable given the poor selectivity of polar solvents. Due to this poor selectivity, there is an increase in the extraction of nonvolatile substances, like steroids, coumarins, flavonoids, tannins, saponins, and triterpenes. Similarly, Bukhanko et al. [74] studied from a biorefinery point of view the branches, cones, needles, and bark of the *Piceaabies* L. species with the goal of extracting different complexes.

Steam Distillation and Hydrodistillation

The most common method to extract steam from water is by boiling it all at once. The plant material, which is positioned in a separate location, is passed through by the generated vapor. Direct steam can be used as a stand-in for this process to remove the substance. But in hydrodistillation, plant materials, and water are combined in one vessel and processed simultaneously. Both of these techniques involve dissolving extracts in steam to extract the liquid [76]. Despite the techniques' similarities, steam distillation produces more essential oils and a better mixture of volatile chemicals. Because there is no direct contact between the plant material and the water during the steam distillation process, it can prevent degradation and produce essential oils of higher quality [77]. The study of Meullemiestre et al. [69] on the hydrodistillation on Eremanthus erythropappus (DC) Macle-like wood using the particles that Tyler 28 and 32 mesh preserved (30% and 70%, respectively) attained a particle diameter of 520 m on average and weighed the sample precisely. Sawdust is used as the raw material in this process because the wood is typically pulverized before being distilled, which maximizes the extraction yield. Santos et al. [75] discovered that extraction time, which varies on an hourly basis, is another crucial element for extraction yield.

Infusion

A common extraction method used frequently in conventional medicines is infusion. The infusion is made by mixing a small quantity of plant matter with a solvent at a high temperature, which prepares the mixture quickly [78,79]. The plant matter must be chopped up into little bits to make the extraction of the chemicals easier. Utilizing this technique results in extracts that are rich in glycosides and essential oils [80]. The study into Prunus avium stems can provide a variety of flavonoids when heated in 200 mL of water with 1 g of the sample, then cooled for 5 min at room temperature; i.e., quercetin-3-O-rutinoside, quercetin-3-*O*-glucoside, kaempferol-3-*O*-glucoside, kaempferol-3-O-rutinoside, and catechin were among the flavonoids found [81]. Mu'nisa et al. [82] reported that sappan wood extracted using distilled solvent was conducted using infusion for 15 min at a temperature of 90 °C. Infusion was performed of sappan wood making by introducing 20 g simplicia of sappan wood into the pot infusion and adding 100 mL of distilled water into a measuring cup. The solution was heated at 90 °C for 15 min, after which it was filtered using a flannel cloth, hereinafter called the filtered water infused wooden cup. Sappan wood infusion can be stored at a temperature of 4 °C. Based on the research results, infusing sappan wood (*Caesalpinia sappan* L.) using distilled water solvent was analyzed by using a phenolic compound i.e., gallic acid standard solution of 551.663 mgGAE/mL infusion, while the analysis test flavonoid quercetin used a standard solution of 1.103 mgQAE/mL infusion. Infused sappan wood can reduce free radical DPPH. The capacity to reduce free radical DPPH sappan wood is greater than the ability of quercetin. The IC50 value of the sappan wood is 0.047 mg/mL higher than quercetin is 5054 mg/mL.

Maceration

It is easy to extract materials using this procedure. Although not entirely successful in removing bioactive compounds from plants, this approach has the advantage of being straightforward. The bioactive components are extracted using this approach, which involves submerging a plant sample in a solvent for a long time. In order to recover as many of the scattered compounds as possible, the liquid is drained, and the solid material that remains at the bottom is pressed [76,79]. Variable amounts of time are used during the maceration process.

Using (70:30 v/v) water–ethanol solution and maceration with n-hexane as solvents, tannins, flavonoids, and polyphenols were extracted from *Populus nigra* L. wood [83]. Bostyn et al. [84] reported that macerated sawdust could be used to create extractives, a source of various useful compounds that can be utilized in industries. Two of the most important flavonoids of *Robiniapseudoacacia* L. wood, robinetin and dihydrorobinetin, were obtained through the maceration of the wood using a water–ethanol solution as the solvent, yielding concentrations of 670 mg/L and 3000 mg/L, respectively. Extracts produced by maceration have strong antioxidant properties. Several researchers have verified that using the maceration technique, the extract yields have virtual total phenolic content (TPC) values [83,85]. Batubara et al. [86] macerated CS heartwood (500 g) powder in 5 L of MeOH for 12 h. This process was repeated twice, and the extract was concentrated by using a rotary evaporator. The crude extract (10 g) was separated on a silica gel column and eluted with hexane, ethyl acetate, and methanol. Brazilin was purified from the ethyl acetate fraction by preparative HPLC using a gradient solvent system from 5% to 100% methanol in 0.05% trifluoroacetic acid at a flow rate of 10 mL/min for 45 min.

2.4.2. Novel Extraction Technique

Novel extraction methods typically operate at low temperatures. In this technique, the stability of the extracted chemicals can be preserved by reducing or eliminating the usage of organic solvents through the abstraction procedure. Furthermore, because of the high solvent temperatures and prolonged extraction times, it is possible that the bioactive components will be thermally degraded. To avoid the problems with traditional methods, different extraction techniques have been created [87]. In addition, novel techniques typically generate higher yields and higher-quality products because they use less energy and time during extraction [75].

Ultrasound-Assisted Extraction (UAE)

Ultrasonic wave utilization during cavitation facilitates the liberation of the required compounds by cellular rupture of plant cells through the preparation stage or during both liquid and solid extraction. This process is known as ultrasound-assisted extraction. In this process, ultrasonic waves produce cycles of extension and compress the molecules that make up the medium as they travel through the so-called cavitation phenomenon. This results in the expansion, bursting of froths, and formation in a liquid standard because of the consequence of these alternating fluctuations in pressure [88]. It has been demonstrated that ultrasound-assisted extraction is effective at obtaining a variety of products, including proteins, vital oil polysaccharides, pigments, dyes, and bioactive complexes. The benefits of utilizing UAE for liquid and solid extraction include improved yield of extraction and decreased time period, solvent intake, energy, and condensed operative temperature, which enables the abstraction of complexes that are sensitive to heat. UAE is a relatively low-cost and easy-to-use method that has the ability to use a variety of solvents. In order to evaluate the outcomes, extraction by maceration is carried out in a similar way. It was discovered that this method requires roughly 24 h to obtain yield values [89]. Kurniasari et al. [90] reported the extraction of the phenolic compound of sappan wood using ultrasound-assisted extraction (UAE). It has been considered a green technology that gives better quality products with higher extraction rates in a shorter time and with less energy. The extraction of sappan wood using a UAE probe with varying ethanol concentrations (50%, 60%, 70%, 80%, and 90% v/v) revealed that the extract yield increased with higher ethanol concentrations up to 80%. However, beyond this point, the yield decreased. When comparing the UAE and Soxhlet methods, it was observed that UAE provided a higher yield (10.33%) in a shorter duration of time (20 min) compared to the Soxhlet method, which yielded 9.67% over a longer period (180 min). Yuniati et al. [91] carried out a study using the ultrasound-assisted extraction method to optimize and characterize the process of extracting colorants from sappan wood. The research identified the optimal operating conditions for sappan wood extraction, including a frequency of 40 kHz, a temperature of 60 °C, a ratio of 0.0050 g mL^{-1}, an extraction time of 20 min, and the use of a 60% ethanol solvent. The extracted sappan wood yielded a color ranging from yellow to reddish-orange under acidic pH levels (2–6), turned red under neutral pH (7), and shifted toward a purplish-red hue with increasing pH. A qualitative analysis confirmed the presence of quinone, flavonoid, quinone, and tannin compounds, as well as several phenolic compounds detected through GCMS.

Microwave-Assisted Extraction (MAE)

The basic devices of ionic polarization and molecular redirection are the foundation of microwave-assisted extraction. The polar molecules start to vibrate when exposed to an electrical pitch produced by microwave radiation, keeping their dipoles unceasingly allied with the electric field [92]. Microwaves are electromagnetic radiation with frequencies ranging from 0.3 to 300 GHz, with 2.45 GHz being the typical frequency utilized in commercial systems [77,93]. Moreira et al. [94] utilized water and ethanol as the solvent for the extraction of phenols from the Malus domestica Borkh heartwood using MAE with an output of 23 mg of GAE/g of dry wood extracted. According to Meullemiestre et al. [69], a 43 min MAE extraction produced an output of 0.43 percent (*w*/*w*) and 74.62 mg of GAE/g obtained in the TPC. However, after an 8 h hydrodistillation, the yield was just 0.28 percent (*w*/*w*), and the GAE TPC was 54.14 mg/g extract. So, using MAE produced an extraction yield that was comparable to or greater than hydrodistillation in a shorter time. Ahmad et al. [95] reported brazilin levels from sappan wood (*Caesalpinia sappan* L.) using the optimized ionic liquid-based microwave-assisted extraction (IL-MAE) method. According to his results, the IL-MAE (ionic liquid-based microwave-assisted extraction) method condition's optimization was carried out using response surface methodology with the Box–Behnken design. Brazilin levels were determined by the high-performance liquid chromatography (HPLC) gradient method (0.3% acetic acid in water and acetonitrile). HPLC analysis of sappan extract showed brazilin levels of 807.56–948.12 mg/g extract. In his present study, imidazolium basic IL-MAE was optimized and first applied to elevate the brazilin levels from sappan wood. Thus, this is an optimum extraction condition for elevating sappan wood's brazilin levels rapidly, efficiently, quickly, and in an environmentally friendly manner.

Extraction by Using Deep Eutectic Solvents (DES)

In order to maximize extraction efficiency, DES compositions can vary depending on the characteristics of the matrix of plants from which they will be isolated. DES is made up of nonionic types (molecular components or salts) that are connected via hydrogen bonds. This effectiveness is related to the chemical and physical characteristics of DES, including hydrogen bonding, miscibility, viscosity, density, and polarity [77]. DES is composed of multiple solid components that combine to generate a eutectic combination substance with a melting point less than the combined melting points of a separate component [96]. It has been discovered that natural-based ingredients are used to make naturally deep eutectic solvents (NADESs), improving the solvents' environmental safety. As a result, NADES is regarded as the primary green solvent [22]. The benefits of DES include biodegradability, low cost, low volatility, ease of use, low toxicity, high stability, sustainability, recyclability, and very little vapor pressure [73]. DES based on glycerol and choline chloride was used to extract the brazilin from the heartwood of *Caesalpinia sappan* L., and ultrasound was used to monitor the process. Brazilin was produced at a rate of 368.67 g/mL, 6.4 times more than the obtained value when it was macerated with water and ethanol (57.38 g/mL), despite solvent ingestion and the duration being higher [22]. Similar methods were used in the investigation to extract brazilin from sappan wood using betain:lactic acid. They utilized betain:lactic acid for 30 min, yielding 4.49 mg/g of brazilin. They carried out a SE via water–ethanol used for 3 h, yielding 5.43 mg/g, and extraction by maceration using water–ethanol used for 72 h, yielding 4.58 mg/g. Due to the shorter extraction time and competitive extraction yield achieved, DES extraction has thus far shown to be quite appealing [73].

2.5. Commercial Applications

2.5.1. Natural Colorant

Sappan wood is an excellent source of natural red dye. The wood contains a pigment called brazilin, which is used to dye fabrics, leather, papers, and other materials. The dye obtained from sappan wood is ecofriendly and has excellent color fastness properties, which is a glycoside hematoxylin [32,97]. Using sappan wood as a natural colorant has gained renewed interest, especially in the food industry. SWE (sappan wood extract) has been considered a safe and effective natural colorant for various food products, including confectioneries, beverages, and meat products. The utilization of SWE as a natural colorant in food products has been approved by several regulatory agencies, such as the European Union (EU) and the US Food and Drug Administration (FDA). The heartwood of the sappan tree is rich in a red pigment called brazilin, which is responsible for its vivid color. For example, in the use of SWE as a natural colorant in strawberry-flavored milk, the results depicted that SWE was effective in producing a stable red color in the milk without affecting its sensory properties. Also, SWE can be efficiently used as a natural colorant in various applications, including food, cosmetics, and textiles [7,98].

2.5.2. Textile Industry

SWE has been shown to be used as a natural colorant in the textile industry. The extract can be used to dye cotton, silk, and wool fabrics, producing shades of pink, red, and purple. In addition to its coloring properties, sappan wood extract has also been found to have antibacterial and antifungal activities, making it a potential candidate for use in textile preservation [99,100].

2.5.3. Beverages

SWE can be used as a natural colorant in beverages, like tea and fruit juices, without affecting their taste or aroma. The extract has also been found to have antioxidant properties, which can potentially enhance the numerous health benefits of these beverages [9,12].

2.5.4. Confectionery Products

SWE has been evaluated for its potential as a natural food colorant in various applications, including candies, baked goods, confectioneries, and meat products. The extract has been found to be stable under different processing conditions, and its coloring properties have been comparable to those of synthetic food colorants [101]. Thus, sappan wood is a promising natural colorant with various biological activities. Several regulatory agencies have approved its use as a natural colorant in food products, and several studies have reported its effectiveness in producing a stable red color in various food products. Different commercial applications of sappan wood are presented in Figure 5.

2.5.5. Nutraceuticals

Sappan wood contains several bioactive complexes, such as terpenoids, phenolic acids, and flavonoids, with potential health benefits. Extract from sappan wood could be developed into nutraceutical products, such as dietary supplements. Sappan wood compounds have been shown to possess anti-inflammatory, antioxidant, and anticancer properties [102].

2.5.6. Natural Remedies

Sappan wood is utilized in traditional medicine to treat numerous ailments, including diarrhea, inflammation, and fever. It could be developed into herbal remedies for these conditions [103].

2.5.7. Biofuel

Sappan wood could be used as a source of biofuel. A study conducted in Brazil found that sappan wood has a high energy content and is suitable for use as a solid biofuel for cooking and heating [104].

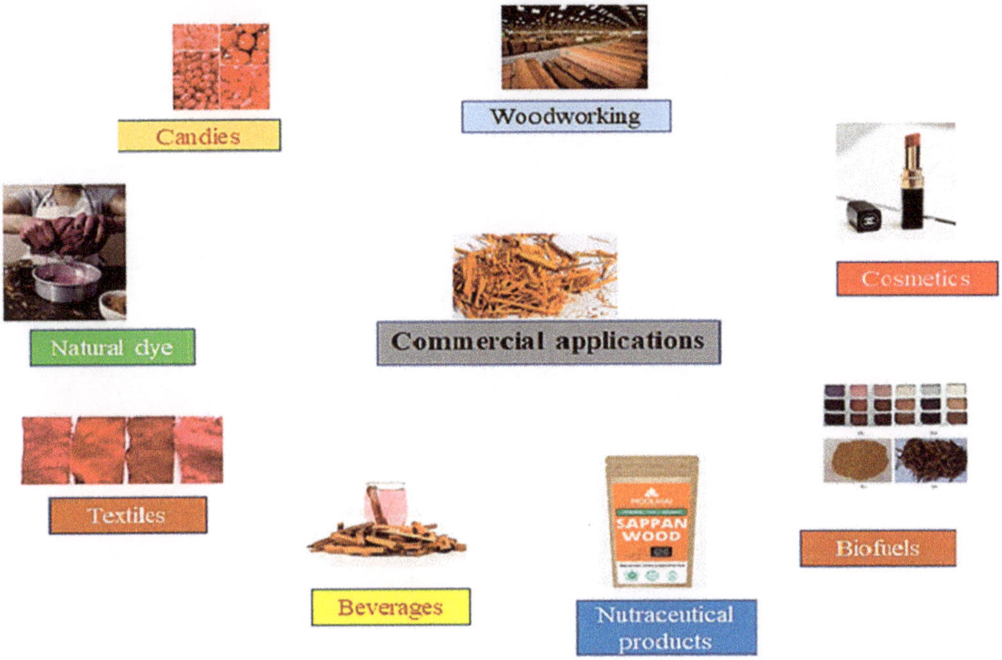

Figure 5. Different commercial applications of sappan wood.

2.5.8. Cosmetics

Sappan wood extract has been found to possess antiaging and skin-lightening properties. It could be used as an active ingredient in cosmetic products, such as an antiaging creams, lotions, and face masks, and it might be used as an active ingredient [105].

2.5.9. Woodworking

Sappan wood is a durable hardwood with a rich reddish-brown color. It is used in woodworking to make furniture, flooring, and decorative items. It is also used in traditional Japanese lacquerware, which is called Suou. The heartwood of sappan wood is dense and durable. It has high strength and stiffness and is suitable for use in structural applications [106]. Authors should discuss the results and how they can be interpreted from the perspective of previous studies and of the working hypotheses. The findings and their implications should be discussed in the broadest context possible. Future research directions may also be highlighted.

3. Conclusions

The conservatively accepted *Caesalpinia sappan* has medicinal properties based on the literature basis. It can be said that *Caesalpinia sappan* heartwood has significant potential for medicinal and coloring uses. It has a long history of use as a colorant in wines, meat, and textiles. In Kerala, Caesalpinia is traditionally used as herbal mineral water, which shows high therapeutic value. Medicinally, it is advised to use wood instead of logwood. It is extensively used in Ayurveda and Unani medicine. Brazilin has high latent pharmacological activities, such as antitumor, antidiabetic, anti-inflammatory, and immunostimulant properties and blood purifying action and healing properties. Various research has proved the uncountable benefits of sappan wood, but the exact mechanism behind the therapeutic benefits is still unknown. The use of sappan wood has begun in the cosmetic industry, but its impact should not be limited to it. It has a potential to flourish in the pharmaceutical and nutraceutical industries as well. Future research will be required to determine the mechanism of action and isolation of active ingredients from *Caesalpinia sappan* L., which has extraordinarily stimulated biological effects and a significant body of traditional myths based on natural resources.

Author Contributions: T.V., P.P.A., R.S. and K.K.D. wrote the main manuscript. T.V. and V.K.P. prepared the figures and tables. P.P.A., K.K.D., E.H., A.M.S., R.K. and B.K. reviewed the manuscript and corrected whenever required. Data curation and methodology: K.K.D., V.K.P., R.S., E.H., B.K. and A.M.S. Funding and support: E.H. and B.K. All authors reviewed the manuscript. All authors have read and agreed to the published version of the manuscript.

Funding: Project No. TKP2021-NKTA-32 has been implemented with support from the National Research, Development, and Innovation Fund of Hungary, financed under the TKP2021-NKTA funding scheme.

Institutional Review Board Statement: Not applicable.

Informed Consent Statement: Not applicable.

Data Availability Statement: Data will be made available on request.

Conflicts of Interest: The authors declare no conflict of interest.

Sample Availability: Not applicable.

References

1. Badami, S.; Moorkoth, S. *Caesalpinia sappan*–A medicinal and dye yielding plant. *Nat. Prod. Radiance* **2004**, *3*.
2. Thanayutsiri, T.; Patrojanasophon, P.; Opanasopit, P.; Ngawhirunpat, T.; Laiwattanapaisal, W.; Rojanarata, T. Rapid and efficient microwave-assisted extraction of *Caesalpinia sappan* Linn. heartwood and subsequent synthesis of gold nanoparticles. *Green. Process. Synth.* **2023**, *12*, 20228109. [CrossRef]
3. Mekala, K.; Radha, R. A Review on Sappan Wood-A Therapeutic Dye Yielding Tree. *Res. J. Pharmacogn. Phytochem.* **2015**, *7*, 227. [CrossRef]
4. Mathew, G.; Skaria, B.P.; Mathew, S.; Joy, P.P. Caesalpinia Sappan-an Economic Medicinal Tree for the Tropics. In Proceedings of the National Symposium on Medicinal and Aromatic Plants for the Economic Benefit of Rural People (MAPER), Thiruvananthapuram, India, 16–18 February 2007; pp. 1–7.
5. Rosniawaty, S.; Anjarsari, I.R.D.; Sudirja, R.; Mubarok, S.; Fatmawati, D. Effect of growth regulators and organic matter on the growth of Sappan wood seedlings (*Caesalpinia sappan* L.). *Res. Crops* **2023**, *24*, 198–209. [CrossRef]
6. Lee, D.K.; Cho, D.H.; Lee, J.H.; Shin, H.Y. Fabrication of nontoxic natural dye from sappan wood. *Korean J. Chem. Eng.* **2008**, *25*, 354–358. [CrossRef]
7. Nathan, V.K.; Rani, M.E. Natural dye from *Caesalpinia sappan* L. heartwood for eco-friendly coloring of recycled paper-based packing material and its in silico toxicity analysis. *Environ. Sci. Pollut. Res.* **2021**, *28*, 28713–28719. [CrossRef]
8. Saini, R.; Dhiman, N.K. Natural Anti-inflammatory and Anti-allergy Agents: Herbs and Botanical Ingredients. *Anti-Inflamm. Anti-Allergy Agents Med.* **2022**, *21*, 90–114. [CrossRef]
9. Sarofa, U.; Doko, F.M. Characterizing Instant Powder Drink Mixed of Kersen Leaves (Muntingiacalabura) and Secang Wood (*Caesalpinia sappan* L.) with Maltodextrin Addition. *MATEC Web Conf.* **2022**, *372*, 02007. [CrossRef]
10. Vardhani, A.K. *Caesalpinia sappan* L. In Proceedings of the International Conference on Applied Science and Health, Paris, France, 15–16 April 2019; Volume 4, pp. 302–308.
11. Thangal, A.H.; Prasanth, C.S.; Prasanth, M.L.; Anu, V. Pharmacognostic, phytochemical and pharmacological aspects of *Caesalpinia sappan* plant. *Int. J. Pharm. Sci. Res.* **2022**, *10*.
12. Abu, A. Natural Dyes from Secang (Biancaeasappan) Wood in Sutera. *J. Phys. Conf. Ser.* **2019**, *1387*, 012001.
13. Shad, A.A.; Ahmad, S.; Ullah, R.; AbdEl-Salam, N.M.; Fouad, H.; Ur Rehman, N.; Hussain, H.; Saeed, W. Phytochemical and biological activities of four wild medicinal plants. *Sci. World J.* **2014**, *2014*, 857363. [CrossRef] [PubMed]
14. Shan, T.; Ma, Q.; Guo, K.; Liu, J.; Li, W.; Wang, F.; Wu, E. Xanthones from mangosteen extracts as natural chemopreventive agents: Potential anticancer drugs. *Curr. Mol. Med.* **2011**, *11*, 666–677. [CrossRef] [PubMed]
15. Nirmal, N.P.; Rajput, M.S.; Prasad, R.G.; Ahmad, M. Brazilin from *Caesalpinia sappan* heartwood and its pharmacological activities: A review. *Asian Pac. J. Trop. Med.* **2015**, *8*, 421–430. [CrossRef] [PubMed]
16. Niu, Y.; Wang, S.; Li, C.; Wang, J.; Liu, Z.; Kang, W. Effective compounds from *Caesalpinia sappan* L. on the tyrosinase in vitro and in vivo. *Nat. Prod. Commun.* **2020**, *15*, 1934578X20920055. [CrossRef]
17. Wu, S.Q.; Otero, M.; Unger, F.M.; Goldring, M.B.; Phrutivorapongkul, A.; Ciiari, C.; Toegel, S. Anti-inflammatory activity of an ethanolic *Caesalpinia sappan* extract in human chondrocytes and macrophages. *J. Ethnopharmacol.* **2011**, *138*, 364–372. [CrossRef]

18. Du Toit, K.; Elgorashi, E.E.; Malan, S.F.; Drewes, S.E.; van Staden, J.; Crouch, N.R.; Mulholland, D.A. Anti-inflammatory activity and QSAR studies of compounds isolated from Hyacinthaceae species and TachiadenuslongiflorusGriseb. (Gentianaceae). *Bioorg. Med. Chem.* **2005**, *13*, 2561–2568. [CrossRef]
19. Dapson, R.W.; Bain, C.L. Brazilwood, sappanwood, brazilin and the red dye brazilein: From textile dyeing and folk medicine to biological staining and musical instruments. *Biotech. Histochem.* **2015**, *90*, 401–423. [CrossRef]
20. Hafshah, M.; Rohmah, A.; Mardliyah, A. Potential of Secang Wood (*Caesalpinia sappan* L.) Ethanol Extractas Anti-Oxidant and Sun-Protection. *Al-Kimia* **2022**, *10*. [CrossRef]
21. Nomer, N.M.G.R.; Duniaji, A.S.; Nocianitri, K.A. kandungansenyawa flavonoid dan antosianinekstrakkayusecang (*Caesalpinia sappan* L.) sertaaktivitasantibakteriterhadap Vibrio cholerae. *J. Ilmu Dan Teknol. Pangan* **2019**, *8*, 216–225. [CrossRef]
22. Sakti, A.S.; Saputri, F.C.; Mun'im, A. Optimization of choline chloride-glycerol based natural deep eutectic solvent for extraction bioactive substances from Cinnamomum burmannii barks and *Caesalpinia sappan* heartwoods. *Heliyon* **2019**, *5*, e02915. [CrossRef]
23. Asfar, A.M.I.A.; Asfar, A.M.I.T. Polyphenol in Sappan wood (*Caesalpinia sappan* L.) extract results of ultrasonic-assisted solvent extraction. *AIP Conf. Proc.* **2023**, *2719*, 1.
24. Ahmad, I.; Arifianti, A.E.; Sakti, A.S.; Saputri, F.C.; Mun'im, A. Simultaneous natural deep eutectic solvent-based ultrasonic-assisted extraction of bioactive compounds of cinnamon bark and sappan wood as a dipeptidyl peptidase IV inhibitor. *Molecules* **2020**, *25*, 3832. [CrossRef]
25. Bharathi, K.N.; Nair, D. Evaluation of Laxative Activity of *Caesalpinia sappan* Wood Extracts in Rats. *Int. J. Pharm. Res. Appl.* **2022**, *7*, 64–69.
26. Rajput, M.S.; Nirmal, N.P.; Nirmal, S.J.; Santivarangkna, C. Bio-actives from *Caesalpinia sappan* L.: Recent advancements in phytochemistry and pharmacology. *S. Afr. J. Bot.* **2022**, *151*, 60–74. [CrossRef]
27. Tu, W.C.; Ding, L.F.; Peng, L.Y.; Song, L.D.; Wu, X.D.; Zhao, Q.S. Cassane diterpenoids from the seeds of Caesalpinia bonduc and their nitric oxide production and α-glucosidase inhibitory activities. *Phytochemistry* **2022**, *193*, 112973. [CrossRef]
28. Tamburini, D. Investigating Asian colourants in Chinese textiles from Dunhuang (7th-10th century AD) by high performance liquid chromatography tandem mass spectrometry–Towards the creation of a mass spectra database. *Dyes Pigm.* **2019**, *163*, 454–474. [CrossRef]
29. Syamsunarno, M.R.A.; Safitri, R.; Kamisah, Y. Protective effects of *Caesalpinia sappan* Linn. and its bioactive compounds on cardiovascular organs. *Front. Pharmacol.* **2021**, *12*, 725745. [CrossRef]
30. Purba, B.A.V.; Pujiarti, R.; Masendra, M.; Lukmandaru, G. Total Phenolic, Flavonoid, Tannin Content and DPPH Scavenging Activity of *Caesalpinia sappan* Linn. Bark. *Wood Res. J.* **2022**, *13*, 63–68. [CrossRef]
31. Surup, G.R.; Hunt, A.J.; Attard, T.; Budarin, V.L.; Forsberg, F.; Arshadi, M.; Trubetskaya, A. The effect of wood composition and supercritical CO_2 extraction on charcoal production in ferroalloy industries. *Energy* **2020**, *193*, 116696. [CrossRef]
32. Mandal, S.; Venkatramani, J. A review of plant-based natural dyes in leather application with a special focus on color fastness characteristics. *Environ. Sci. Pollut. Res.* **2023**, *30*, 48769–48777. [CrossRef]
33. Ma, G.; Wu, H.; Chen, D.; Zhu, N.; Zhu, Y.; Sun, Z.; Xu, X. Antimalarial and antiproliferative cassane diterpenes of *Caesalpinia sappan*. *J. Nat. Prod.* **2015**, *78*, 2364–2371. [CrossRef] [PubMed]
34. Bao, H.; Zhang, L.L.; Liu, Q.Y.; Feng, L.; Ye, Y.; Lu, J.J.; Lin, L.G. Cytotoxic and pro-apoptotic effects of cassane diterpenoids from the seeds of *Caesalpinia sappan* in cancer cells. *Molecules* **2016**, *21*, 791. [CrossRef] [PubMed]
35. Zhu, N.L.; Sun, Z.H.; Hu, M.G.; Wu, T.Y.; Yuan, J.Q.; Wu, H.F.; Xu, X.D. New cassane diterpenoids from *Caesalpinia sappan* and their antiplasmodial activity. *Molecules* **2017**, *22*, 1751. [CrossRef] [PubMed]
36. Ngernnak, C.; Panyajai, P.A.W.A.R.E.T.; Anuchapreeda, S.O.N.G.Y.O.T.; Wongkham, W.E.E.R.A.H.; Saiai, A. Phytochemical and cytotoxic investigations of the heartwood of caesalpiniasappanlinn. *Asian J. Pharm. Clin. Res.* **2018**, *11*, 336–339. [CrossRef]
37. Wang, Z.; Sun, J.B.; Qu, W.; Guan, F.Q.; Li, L.Z.; Liang, J.Y. Caesappin A and B, two novel protosappanins from *Caesalpinia sappan* L. *Fitoterapia* **2014**, *92*, 280–284. [CrossRef]
38. Mueller, M.; Weinmann, D.; Toegel, S.; Holzer, W.; Unger, F.M.; Viernstein, H. Compounds from *Caesalpinia sappan* with anti-inflammatory properties in macrophages and chondrocytes. *Food Funct.* **2016**, *7*, 1671–1679. [CrossRef]
39. Zhao, J.; Zhu, A.; Sun, Y.; Zhang, W.; Zhang, T.; Gao, Y.; Shan, D.; Wang, S.; Li, G.; Wang, Q.; et al. Beneficial effects of sappanone A on lifespan and thermotolerance in Caenorhabditis elegans. *Eur. J. Pharm.* **2020**, *888*, 173558. [CrossRef]
40. Min, B.S.; Cuong, T.D.; Hung, T.M.; Min, B.K.; Shin, B.S.; Woo, M.H. Compounds from the heartwood of *Caesalpinia sappan* and their anti-inflammatory activity. *Bioorg. Med. Chem. Lett.* **2012**, *22*, 7436–7439. [CrossRef]
41. Saenjum, C.; Chaiyasut, C.; Kadchumsang, S.; Chansakaow, S.; Suttajit, M. Antioxidant activity and protective effects on DNA damage of *Caesalpinia sappan* L. extract. *J. Med. Plants Res.* **2010**, *4*, 1594–1600.
42. Wang, Y.Z.; Sun, S.Q.; Zhou, Y.B. Extract of the dried heartwood of *Caesalpinia sappan* L. attenuates collagen-induced arthritis. *J. Ethnopharmacol.* **2011**, *136*, 271–278. [CrossRef]
43. Baek, N.I.; Jeon, S.G.; Ahn, E.M.; Hahn, J.T.; Bahn, J.H.; Jang, J.S.; Choi, S.Y. Anticonvulsant compounds from the wood of *Caesalpinia sappan* L. *Arch. Pharmacal Res.* **2000**, *23*, 344–348. [CrossRef] [PubMed]
44. Han, D.; Ma, T.; Sun, S.; Zhang, Y.; Song, L. *Brazilin Inhibits the Inflammatory Immune Response Induced by LPS in THP-1 Cells*; Research Square Platform LLC: Durham, NC, USA, 2023.
45. Batubara, I.; Mitsunaga, T.; Ohashi, H. Brazilin from *Caesalpinia sappan* wood as an antiacne agent. *J. Wood Sci.* **2010**, *56*, 77–81. [CrossRef]

46. Settharaksa, S.; Songsangsirisak, C.; Monton, C. Computer-based estimation of antioxidant activity of *Caesalpinia sappan* L. *Thai J. Pharm. Sci.* **2018**, *42*.
47. Nagaraju, S.; Thippeswamy, T.G.; Ravi Kumar, H. Evaluating the Lipid Profile and Mineral Composition of the Seed Oil of *Caesalpinia sappan* L.(Caesalpiniaceae). In *Natural Product Experiments in Drug Discovery*; Springer: New York, NY, USA, 2022; pp. 75–86.
48. Bae, I.-K.; Min, H.-Y.; Han, A.-R.; Seo, E.-K.; Lee, S.K. Suppression of lipopolysaccharide-induced expression of inducible nitric oxide synthase by brazilin in RAW 264.7 macrophage cells. *Eur. J. Pharm.* **2005**, *513*, 237–242. [CrossRef]
49. Sasaki, Y.; Hosokawa, T.; Nagai, M.; Nagumo, S. In vitro study for inhibition of NO production about constituents of Sappan Lignum. *Biol. Pharm. Bull.* **2007**, *30*, 193–196. [CrossRef]
50. Hu, C.M.; Liu, Y.H.; Cheah, K.P.; Li, J.S.; Lam, C.S.K.; Yu, W.Y.; Choy, C.S. Heme oxygenase-1 mediates the inhibitory actions of brazilin in RAW264. 7 macrophages stimulated with lipopolysaccharide. *J. Ethnopharmacol.* **2009**, *121*, 79–85. [CrossRef]
51. Yohana, W.; Wihardja, R.; Sufiawati, I.; Khaerunnisa, D.F. Decreasing Respiratory rate, blood pressure and pulse after drinking secang in prehypertension patients. *J. Adv. Med. Dental Sci. Res.* **2019**, *7*, 34–36.
52. Jung, E.G.; Han, K.I.; Hwang, S.G.; Kwon, H.J.; Patnaik, B.B.; Kim, Y.H.; Han, M.D. Brazilin isolated from *Caesalpinia sappan* L. inhibits rheumatoid arthritis activity in a type-II collagen induced arthritis mouse model. *BMC Complement. Altern. Med.* **2015**, *15*, 1–11. [CrossRef]
53. Działo, M.; Mierziak, J.; Korzun, U.; Preisner, M.; Szopa, J.; Kulma, A. The potential of plant phenolics in prevention and therapy of skin disorders. *Int. J. Mol. Sci.* **2016**, *17*, 160. [CrossRef]
54. Mitani, K.; Takano, F.; Kawabata, T.; Allam, A.E.; Ota, M.; Takahashi, T.; Ohta, T. Suppression of melanin synthesis by the phenolic constituents of sappanwood (*Caesalpinia sappan*). *Planta Medica* **2013**, *79*, 37–44. [CrossRef]
55. Madhubala, S.; Poongothai, M.; Mahesh, K.E. Antibacterial and anti acne activity of *Caesalpinia sappan* L. and Cinnamomum verum J. Presl—A comparison. *Int. J. Adv. Res. Biol. Sci.* **2018**, *5*, 118–122.
56. Srinivasan, R.; Karthik, S.; Mathivanan, K.; Baskaran, R.; Karthikeyan, M.; Gopi, M.; Govindasamy, C. In vitro antimicrobial activity of *Caesalpinia sappan* L. *Asian Pac. J. Trop. Biomed.* **2012**, *2*, S136–S139. [CrossRef]
57. Rina, O.; Ibrahim, S.; Dharma, A.; Afrizal, I.; Wirawati, C.U. Screening for active agent to anti-diarrhea by an evaluation of antimicrobial activities from three fractions of sappan wood (*Caesalpinia sappan* L.). *Der Pharma Chem.* **2016**, *8*, 114–117.
58. Puttipan, R.; Wanachantararak, P.; Khongkhunthian, S.; Okonogi, S. Effects of *Caesalpinia sappan* on pathogenic bacteria causing dental caries and gingivitis. *Drug Discov. Ther.* **2017**, *11*, 316–322. [CrossRef]
59. Srilakshmi, V.S.; Vijayan, P.; Raj, P.V.; Dhanaraj, S.A.; Chandrashekhar, H.R. Hepatoprotective properties of *Caesalpinia sappan* Linn. heartwood on carbon tetrachloride induced toxicity. *Indian J. Exp. Biol.* **2010**, *48*, 905–910. [PubMed]
60. Gupta, N.; Chauhan, P.; Nayeem, M.; Safhi, M.M.; Agarwal, M. Hepatoprotective effect of Caesalpinia crista Linn. against CCl4 and paracetamol induced hepatotoxicity in albino rats. *Afr. J. Pharm. Pharmacol.* **2014**, *8*, 485–491.
61. Vij, T. A narrative review on Sappan wood (*Caesalpinia sappan* L.). *J. Pharm. Innov.* **2023**, *12*, 2861–2865.
62. Mashau, M.E.; Ramatsetse, K.E.; Ramashia, S.E. Effects of adding Moringa oleifera leaves powder on the nutritional properties, lipid oxidation and microbial growth in ground beef during cold storage. *Appl. Sci.* **2021**, *11*, 2944. [CrossRef]
63. Lee, Y.M.; Kim, Y.C.; Choi, B.J.; Lee, D.W.; Yoon, J.H.; Kim, E.C. Mechanism of sappanchalcone-induced growth inhibition and apoptosis in human oral cancer cells. *Toxicol. Vitr.* **2011**, *25*, 1782–1788. [CrossRef]
64. Gavamukulya, Y.; Wamunyokoli, F.; El-Shemy, H.A. Annona muricata: Is the natural therapy to most disease conditions including cancer growing in our backyard? A systematic review of its research history and future prospects. *Asian Pac. J. Trop. Med.* **2017**, *10*, 835–848. [CrossRef]
65. Sarumathy, K.; Vijay, T.; Palani, S.; Sakthivel, K.; Rajan, M.D. Antioxidant and hepatoprotective effects of *Caesalpinia sappan* against acetaminophen-induced hepatotoxicity in rats. *Int. J. Clin. Pharmacol. Ther.* **2011**, *1*, 19–31.
66. Pattananandecha, T.; Apichai, S.; Julsrigival, J.; Ogata, F.; Kawasaki, N.; Saenjum, C. Antibacterial Activity against Foodborne Pathogens and Inhibitory Effect on Anti-Inflammatory Mediators' Production of Brazilin-Enriched Extract from *Caesalpinia sappan* Linn. *Plants* **2022**, *11*, 1698. [CrossRef] [PubMed]
67. Kadir, F.A.; Kassim, N.M.; Abdulla, M.A.; Kamalidehghan, B.; Ahmadipour, F.; Yehye, W.A. PASS-predicted hepatoprotective activity of *Caesalpinia sappan* in thioacetamide-induced liver fibrosis in rats. *Sci. World J.* **2014**, *2014*, 301879. [CrossRef] [PubMed]
68. Prashith, T.R.; Vinayaka, K.S.; Raghavendra, H.S. *Caesalpinia sappan* L. (Caesalpiniaceae): A Review on its Phytochemistry and Pharmacological Activities. In: *Medicinal and aromatic plants:* Traditional Uses. *Phytochem. Pharmacol. Potential* **2021**.
69. Meullemiestre, A.; Petitcolas, E.; Maache-Rezzoug, Z.; Chemat, F.; Rezzoug, S.A. Impact of ultrasound on solid–liquid extraction of phenolic compounds from maritime pine sawdust waste. Kinetics, optimization and large scale experiments. *Ultrason. Sonochem.* **2016**, *28*, 230–239. [CrossRef] [PubMed]
70. Essien, S.O.; Young, B.; Baroutian, S. Recent advances in subcritical water and supercritical carbon dioxide extraction of bioactive compounds from plant materials. *Trends Food Sci. Technol.* **2020**, *97*, 156–169. [CrossRef]
71. Zhang, J.; Wen, C.; Zhang, H.; Duan, Y.; Ma, H. Recent advances in the extraction of bioactive compounds with subcritical water: A review. *Trends Food Sci. Technol.* **2020**, *95*, 183–195. [CrossRef]
72. Kou, D.; Mitra, S. Extraction of semivolatile organic compounds from solid matrices. *Sample Prep. Technol. Anal. Chem.* **2003**, *162*, 139–182.

73. Setiawan, H.; Angela, I.L.; Wijaya, O.; Mun'Im, A. Application of Natural Deep Eutectic Solvents (NADES) for Sappan Wood (*Caesalpinia sappan* L.) extraction to test for inhibition of DPP IV activity. *J. Res. Pharm.* **2020**, *24*, 380–388. [CrossRef]
74. Bukhanko, N.; Attard, T.; Arshadi, M.; Eriksson, D.; Budarin, V.; Hunt, A.J.; Clark, J. Extraction of cones, branches, needles and bark from Norway spruce (Piceaabies) by supercritical carbon dioxide and soxhlet extractions techniques. *Ind. Crops Prod.* **2020**, *145*, 112096. [CrossRef]
75. Santos, K.A.; Gonçalves, J.E.; Cardozo-Filho, L.; da Silva, E.A. Pressurized liquid and ultrasound-assisted extraction of α-bisabolol from candeia (Eremanthuserythropappus) wood. *Ind. Crops Prod.* **2019**, *130*, 428–435. [CrossRef]
76. Selvamuthukumaran, M.; Shi, J. Recent advances in extraction of antioxidants from plant by-products processing industries. *Food Qual. Saf.* **2017**, *1*, 61–81. [CrossRef]
77. Ali, A.; Chua, B.L.; Chow, Y.H. An insight into the extraction and fractionation technologies of the essential oils and bioactive compounds in *Rosmarinus officinalis* L.: Past, present and future. *TrAC-Trends Anal. Chem.* **2019**, *118*, 338–351. [CrossRef]
78. Gamboa-Gómez, C.I.; Simental-Mendía, L.E.; González-Laredo, R.F.; Alcantar-Orozco, E.J.; Monserrat-Juarez, V.H.; Ramírez-España, J.C.; Rocha-Guzmán, N.E. In vitro and in vivo assessment of anti-hyperglycemic and antioxidant effects of Oak leaves (Quercus convallata and Quercus arizonica) infusions and fermented beverages. *Food Res. Int.* **2017**, *102*, 690–699. [CrossRef] [PubMed]
79. Patra, J.K.; Das, G.; Lee, S.; Kang, S.S.; Shin, H.S. Selected commercial plants: A review of extraction and isolation of bioactive compounds and their pharmacological market value. *Trends Food Sci. Technol.* **2018**, *82*, 89–109. [CrossRef]
80. Rakotoniaina, E.N.; Donno, D.; Randriamampionona, D.; Harinarivo, H.L.; Andriamaniraka, H.; Solo, N.R.; Beccaro, G.L. Insights into an endemic medicinal plant species of Madagascar and Comoros: The case of Famelona (Chrysophyllumboivinianum (Pierre) Baehni, Sapotaceae family). *S. Afr. J. Bot.* **2018**, *117*, 110–118. [CrossRef]
81. Bastos, C.; Barros, L.; Duenas, M.; Calhelha, R.C.; Queiroz, M.J.R.P.; Santos-Buelga, C.; Ferreira, I.C.F.R. Chemical characterisation and bioactive properties of Prunus avium L.: The widely studied fruits and the unexplored stems. *Food Chem.* **2015**, *173*, 1045–1053. [CrossRef]
82. Mu'nisa, A.; Hala, Y.; Muflihunna, A. Analysis of Phenols and Antioxidants Infused Sappan Wood (*Caesalpinia sappan* L.). *Int. J. Sci. Res.* **2017**, *2*, 89–93.
83. Todaro, L.; Russo, D.; Cetera, P.; Milella, L. Effects of thermo-vacuum treatment on secondary metabolite content and antioxidant activity of poplar (*Populus nigra* L.) wood extracts. *Ind. Crops Prod.* **2017**, *109*, 384–390. [CrossRef]
84. Bostyn, S.; Destandau, E.; Charpentier, J.P.; Serrano, V.; Seigneuret, J.M.; Breton, C. Optimization and kinetic modelling of robinetin and dihydrorobinetin extraction from Robiniapseudoacacia wood. *Ind. Crops Prod.* **2018**, *126*, 22–30. [CrossRef]
85. Cetera, P.; Russo, D.; Milella, L.; Todaro, L. Thermo-treatment affects Quercus cerris L. wood properties and the antioxidant activity and chemical composition of its by-product extracts. *Ind. Crops Prod.* **2018**, 380–388. [CrossRef]
86. Batubara, I.; Mitsunaga, T.; Ohashi, H. Screening antiacne potency of Indonesian medicinal plants: Antibacterial, lipase inhibition, and antioxidant activities. *J. Wood Sci.* **2009**, *55*, 230–235. [CrossRef]
87. Tiwari, B.K. Ultrasound: A clean, green extraction technology. *Trends Anal. Chem.* **2015**, *71*, 100–109. [CrossRef]
88. Patil, H.; Mudaliar, S.; Athalye, A. Ultrasound-assisted enzymatic scouring of jute optimised by response surface methodology and its natural dyeing. *Color Tech.* **2023**, *139*, 97–108. [CrossRef]
89. St-Pierre, F.; Achim, A.; Stevanovic, T. Composition of ethanolic extracts of wood and bark from Acer saccharum and Betula alleghaniensis trees of different vigor classes. *Ind. Crops Prod.* **2013**, *41*, 179–187. [CrossRef]
90. Kurniasari, L.; Djaeni, M.; Kumoro, A.C. Ultrasound-Assisted Extraction (UAE) of sappan wood (*Caesalpinia sappan* L.): Effect of solvent concentration and kinetic studies. *Braz. J. Food Technol.* **2023**, *26*, e2022140. [CrossRef]
91. Yuniati, Y.; Azmi, D.D.; Nurandriea, E.; Qadariyah, L.; Mahfud, M. Parametric Study and Characterization of Sappan Wood (*Caesalpinia sappan* Linn) Natural Red Colorant Extract with Ultrasonic Assisted Extraction Method. *ASEAN J. Chem. Eng.* **2023**, *23*, 103–112. [CrossRef]
92. Chartarrayawadee, W.; Too, C.O.; Ross, S.; Ross, G.M.; Jumpatong, K.; Noimou, A.; Settha, A. Green Synthesis and Stabilization of Earthworm-Like Gold Nanostructure and Quasi-Spherical Shape Using Caesalpinia Sappan Linn. Extract. *Green Process. Synth.* **2018**, *7*, 424–432. [CrossRef]
93. Zwingelstein, M.; Draye, M.; Besombes, J.L.; Piot, C.; Chatel, G. Viticultural wood waste as a source of polyphenols of interest: Opportunities and perspectives through conventional and emerging extraction methods. *Waste Manag.* **2020**, *102*, 782–794. [CrossRef]
94. Moreira, M.M.; Barroso, M.F.; Boeykens, A.; Withouck, H.; Morais, S.; Delerue-Matos, C. Valorization of apple tree wood residues by polyphenols extraction: Comparison between conventional and microwave-assisted extraction. *Ind. Crops Prod.* **2017**, *104*, 210–220. [CrossRef]
95. Ahmad, I.; Setyaningsih, E.P.; Arifianti, A.E.; Saputri, F.C.; Munim, A. Optimization of ionic liquid-based microwave-assisted extraction on brazilin levels from sappan wood and its dipeptidyl peptidase IV inhibition activity. *J. Appl. Pharm Sci.* **2021**, *11*, 072–079.
96. Choi, Y.H.; Verpoorte, R. Green solvents for the extraction of bioactive compounds from natural products using ionic liquids and deep eutectic solvents. *Curr. Opin. Food Sci.* **2019**, *26*, 87–93. [CrossRef]
97. Ghorpade, B.; Darvekar, M.; Vankar, P.S. Ecofriendly cotton dyeing with Sappan wood dye using ultrasound energy. *Colourage* **2000**, *47*, 27–30.

98. Rosanti, I.; Sadikin, A.; Prasetia, R. Determination of the absorbility of natural dyes of secang wood (*Caesalpinia sappan*) and teak leaves (*Tectona grandis* L.) in organic kenaf fiber industry. *J. Ris. Ind. Has. Hutan* **2023**, *14*, 55–66.
99. Ohama, P.; Tumpat, N. Textile dyeing with natural dye from sappan tree (*Caesalpinia sappan* Linn.) extract. *Int. J. Mat. Text. Eng.* **2014**, *8*, 432–434.
100. Vankar, P.S.; Gangwar, A. Natural dyeing mediated by atmospheric air pressure plasma treatment of polyester. *Pigment. Resin. Technol.* **2023**. [CrossRef]
101. Ngamwonglumlert, L.; Devahastin, S.; Chiewchan, N.; Raghavan, G.V. Color and molecular structure alterations of brazilein extracted from *Caesalpinia sappan* L. under different pH and heating conditions. *Sci. Rep.* **2020**, *10*, 12386. [CrossRef]
102. Bukke, A.N.; Hadi, F.N.; Babu, K.S. In vitro studies data on anticancer activity of *Caesalpinia sappan* L. heartwood and leaf extracts on MCF7 and A549 cell lines. *Data Bri.* **2018**, *19*, 868–877. [CrossRef]
103. Januariyatun, A.; Wahyuningsih, M.S.H. Effect of Secang Drink (*Caesalpinia sappan* L.) on Plasma Nitric Oxide Level and Blood Pressure in Prehypertension Peoples. *KnE Life Sci.* **2018**, 193–202. [CrossRef]
104. Du, H.; Cheng, J.; Wang, M.; Tian, M.; Yang, X.; Wang, Q. Red dye extracted sappan wood waste derived activated carbons characterization and dye adsorption properties. *Diam. Rel. Mat.* **2020**, *102*, 107646. [CrossRef]
105. Herawati, E.; Yulastri, L. Formulation and Evaluation of Secang Stem Extract (*Caesalpinia sappan* L) in Decorative Cosmetics. *IOP Conf. Ser. Earth Environ. Sci.* **2021**, *810*, 012057. [CrossRef]
106. Lakshmi, C.; Gahlot, M. The Quintessential Naqqashi Nirmal Painting Art of Telangana: Source of Inspiration for Innovative Textile Design Ideas. *Int. J. Curr. Microbiol. Appl. Sci.* **2020**, *9*, 803–814. [CrossRef]

Disclaimer/Publisher's Note: The statements, opinions and data contained in all publications are solely those of the individual author(s) and contributor(s) and not of MDPI and/or the editor(s). MDPI and/or the editor(s) disclaim responsibility for any injury to people or property resulting from any ideas, methods, instructions or products referred to in the content.

molecules

MDPI

Article

Ethanolic Extract Propolis-Loaded Niosomes Diminish Phospholipase B1, Biofilm Formation, and Intracellular Replication of *Cryptococcus neoformans* in Macrophages

Kritapat Kietrungruang [1], Sanonthinee Sookkree [1], Sirikwan Sangboonruang [1], Natthawat Semakul [2], Worrapan Poomanee [3], Kuntida Kitidee [4], Yingmanee Tragoolpua [5,6] and Khajornsak Tragoolpua [1,2,*]

1 Division of Clinical Microbiology, Department of Medical Technology, Faculty of Associated Medical Sciences, Chiang Mai University, Chiang Mai 50200, Thailand; kritapat_k@cmu.ac.th (K.K.); sanonthinee_sookkree@cmu.ac.th (S.S.); sirikwan.sang@cmu.ac.th (S.S.)
2 Department of Chemistry, Faculty of Science, Chiang Mai University, Chiang Mai 50200, Thailand; natthawat.semakul@cmu.ac.th
3 Department of Pharmaceutical Sciences, Faculty of Pharmacy, Chiang Mai University, Chiang Mai 50200, Thailand; worrapan.p@cmu.ac.th
4 Center for Research Innovation and Biomedical Informatics, Faculty of Medical Technology, Mahidol University, Salaya, Nakhon Pathom 73170, Thailand; kuntida.kit@mahidol.ac.th
5 Natural Extracts and Innovative Products for Alternative Healthcare Research Group, Faculty of Science, Chiang Mai University, Chiang Mai 50200, Thailand; yingmanee.t@cmu.ac.th
6 Department of Biology, Faculty of Science, Chiang Mai University, Chiang Mai 50200, Thailand
* Correspondence: khajornsak.tr@cmu.ac.th

Abstract: Secretory phospholipase B1 (PLB1) and biofilms act as microbial virulence factors and play an important role in pulmonary cryptococcosis. This study aims to formulate the ethanolic extract of propolis-loaded niosomes (Nio-EEP) and evaluate the biological activities occurring during PLB1 production and biofilm formation of *Cryptococcus neoformans*. Some physicochemical characterizations of niosomes include a mean diameter of 270 nm in a spherical shape, a zeta-potential of -10.54 ± 1.37 mV, and $88.13 \pm 0.01\%$ entrapment efficiency. Nio-EEP can release EEP in a sustained manner and retains consistent physicochemical properties for a month. Nio-EEP has the capability to permeate the cellular membranes of *C. neoformans*, causing a significant decrease in the mRNA expression level of *PLB1*. Interestingly, biofilm formation, biofilm thickness, and the expression level of biofilm-related genes (*UGD1* and *UXS1*) were also significantly reduced. Pre-treating with Nio-EEP prior to yeast infection reduced the intracellular replication of *C. neoformans* in alveolar macrophages by 47%. In conclusion, Nio-EEP mediates as an anti-virulence agent to inhibit PLB1 and biofilm production for preventing fungal colonization on lung epithelial cells and also decreases the intracellular replication of phagocytosed cryptococci. This nano-based EEP delivery might be a potential therapeutic strategy in the prophylaxis and treatment of pulmonary cryptococcosis in the future.

Keywords: propolis; niosomes; pulmonary cryptococcosis; *Cryptococcus neoformans*; phospholipase B1; biofilm formation; phagocytosis

Citation: Kietrungruang, K.; Sookkree, S.; Sangboonruang, S.; Semakul, N.; Poomanee, W.; Kitidee, K.; Tragoolpua, Y.; Tragoolpua, K. Ethanolic Extract Propolis-Loaded Niosomes Diminish Phospholipase B1, Biofilm Formation, and Intracellular Replication of *Cryptococcus neoformans* in Macrophages. *Molecules* **2023**, *28*, 6224. https://doi.org/10.3390/molecules28176224

Academic Editors: Eliza Oprea, Arunaksharan Narayanankutty and Ademola C. Famurewa

Received: 24 July 2023
Revised: 21 August 2023
Accepted: 22 August 2023
Published: 24 August 2023

Copyright: © 2023 by the authors. Licensee MDPI, Basel, Switzerland. This article is an open access article distributed under the terms and conditions of the Creative Commons Attribution (CC BY) license (https://creativecommons.org/licenses/by/4.0/).

1. Introduction

Pulmonary cryptococcosis is an opportunistic and invasive mycosis usually found in immunocompromised patients [1]. The pathogenesis usually originates from *Cryptococcus neoformans* through inhaling spores and small infective particles, ultimately resulting in respiratory infection [2]. Several virulence factors, such as polysaccharide capsules and degrading enzymes, are produced to allow yeast pathogen adhesion, invasion, and damage to the host cells. Among the virulence-associated enzymes, secretory phospholipase B1 (PLB1) plays a crucial role in facilitating the adhesion and destabilization of the host cell membrane and the phospholipid lung surfactant [3,4]. PLB1 is also associated with the escape of yeast cells from the pulmonary macrophages through nonlytic exocytosis or vomocytosis [5,6]. Additionally, the extracellular polymeric matrix (EPM), or biofilm, constitutes the dynamic communities of microorganisms via adhesion/matrix proteins signaling and directional proliferation of the original-adhered yeast cells [7]. The biofilm formation begins with yeast cell adhesion, releasing enzyme, and biofilm maturation [8]. It promotes the survival of yeast cells and protects them from host immunity as well as antifungal drugs [9]. Although the treatment of pulmonary cryptococcosis is currently based on the first-line antifungal drug amphotericin B (AMB), the adverse effects, especially nephrotoxicity, remain a serious concern [10]. Hence, the application of a natural antimicrobial agent has become increasingly attractive as an alternative treatment.

Propolis is a complex mixture of natural and resinous substances that contain various active ingredients such as gallic acid, quercetin, pinocembrin, chrysin, and galangin [11]. Notably, it is a rich source of therapeutic properties and has antioxidant, anti-inflammatory, antimicrobial, and immunomodulatory capabilities [12–14]. Several reports have revealed potent in vitro anti-fungal activity of the ethanolic extract propolis, or EEP, against *Candida albicans* [15] and *C. neoformans* [16]. Moreover, our previous work found the anti-cryptococcal activity of EEP to reduce the growth rate and major virulence factors of *C. neoformans*, including the polysaccharide capsule, melanin, and urease [17]. Subsequently, the encapsulation of EEP based on the poly (n-butyl cyanoacrylate) (PBCA) nanosystem was further explored for therapeutic application in cryptococcal meningoencephalitis through blood–brain barrier (BBB) drug delivery [18]. However, administration via the pulmonary route is restricted due to mucus and lung surfactant [19]. To achieve therapeutic use in the lungs, non-ionic surfactant vesicles, or niosomes, were considered as an effective nanocarrier of EEP.

Niosomes are composed of self-assembled non-ionic surfactants and cholesterol, containing a hydrophilic head and hydrophobic tail, to form a vesicle. The niosomes are modified with a liposomal vesicle usually made of phospholipids. Due to the fragile phospholipid membranes, the liposome has low physical stability compared to the niosome and can cause drug leakage [20]. Moreover, the niosomal structure contains lipids, a similar component of lung surfactant, which improves nanoparticle penetration in the mucus layer and provides sustainable drug release [21]. These advantages of niosomes are beneficial to the encapsulation of both hydrophobic and lipophilic molecules [22]. Previously, the formulation of AMB–niosomes was established by significantly reducing the fungal burden, in an animal model, with invasive pulmonary aspergillosis [23]. In addition, the encapsulation of EEP was also successfully achieved in niosomes against *Staphylococcus aureus*, *C. albicans* [18], and *Mycobacterium tuberculosis* [24,25]. Therefore, this study aims to investigate the efficacy of EEP through the niosome system against *C. neoformans* for an in vitro model of pulmonary cryptococcosis.

In this work, we fabricated Nio-EEP and evaluated its anti-virulence factors, including PLB1, and the biofilm formation in vitro. Furthermore, the influence of Nio-EEP-induced phagocytosis and the killing of *C. neoformans* by macrophages were also investigated.

2. Results

2.1. Physicochemical Characterization of Nio-EEP

2.1.1. Particle Size, Polydispersity Index (PDI), Zeta Potential (ZP), Entrapment Efficiency (EE), Loading Capacity (LC), and Morphology

Niosomes were successfully fabricated with different proportions of non-ionic surfactant and CHOL. As shown in Table 1, all niosomal formulations had an average particle size of approximately 100–280 nm with PDI 0.32–0.37. The ZP measurements exhibited a negative surface charge of niosomal formulations, ranging from −10 to −12 mV. The EE of each formulation was greater than 85% while the LC showed a variation. The highest LC was approximately 83%, observed in the F1 formulation; in contrast, F2 and F3 had a lower capacity of approximately 46–50%. Due to the highest LC, F1 was chosen as the suitable formulation to characterize the physicochemical structure further. The particle number of F1 was calculated according to the previous study [26]. The number of particles was 6.5×10^{11} vesicles/mL corresponding to 3.52 ± 0.01 mg/mL of EEP. The scanning transmission electron microscope (STEM) images displayed the morphological particles in nanometer scales and presented the spherical vesicles as shown in Figure 1a,b.

Table 1. Particle size, polydispersity index (PDI), zeta potential (ZP), entrapment efficiency (EE), and loading capacity (LC) of the niosomes.

Formulations	Nio			Nio-EEP				
	Size (nm)	PDI	ZP (mV)	Size (nm)	PDI	ZP (mV)	EE (%)	LC (%)
F1	255.53 ± 25.36	0.37 ± 0.06	−9.38 ± 1.58	268.53 ± 10.89	0.32 ± 0.01	−10.54 ± 1.37	88.13 ± 0.01	82.95 * ± 0.01
F2	108.30 ± 6.53	0.34 ± 0.03	−8.62 ± 1.41	152.15 ± 34.00	0.35 ± 0.05	−10.05 ± 0.20	88.45 ± 0.00	45.76 # ± 0.00
F3	253.59 ± 20.49	0.32 ± 0.04	−10.51 ± 0.88	168.07 ± 23.91	0.32 ± 0.05	−9.92 ± 0.30	86.75 ± 0.05	50.49 ± 0.03

All data are represented as the mean ± SEM of three independent trials. * $p < 0.05$, significant for F1 compared to F2 and F3, and # $p < 0.05$, significant for F2 compared to F3.

2.1.2. Chemical Composition

To verify the existence of EEP in niosomes, nuclear magnetic resonance (NMR) spectroscopy might be a reliable tool, especially for studying the encapsulation of EEP into niosomal vesicles. The NMR spectra of Tween 80 (TW80), Span 60 (SP60), cholesterol (CHOL), and niosomes were recorded using DMSO-d_6 as a solvent, as shown in Figure 1c. The NMR spectra of TW80 and SP60 showed similar chemical shifts around 1–4 ppm corresponding to aliphatic protons [27]. It should be noted that TW80 featured chemical shifts around 4–5 ppm which ascribes to the olefinic protons [28]. The NMR spectrum of CHOL showed chemical shifts in two regions, i.e., chemical shifts around 1–2.5 ppm which are ascribed to aliphatic protons and chemical shifts around 4–5.5 ppm which are ascribed to olefinic protons [29]. Nio also showed a profile of absorption peaks similar to the surfactants and CHOL, indicating the successful formulation of niosomes. Next, the NMR analysis of Nio, EEP, and Nio-EEP was individually recorded, as shown in Figure 1d. The ^{1}H NMR spectrum of EEP revealed all phytochemical and other chemical constituents that can be interpreted based on chemical shift fingerprints. Examples of molecules with chemical shifts of aliphatic protons in the 0.5–3.0 ppm range include terpenoids, steroids, and linear fatty acid side chains for fats, oils, and waxes. Additionally, peaks in the chemical shift range of 3.5–5.5 ppm are due to sugar components. It is important to note that HPLC typically does not detect these compounds, thereby rendering NMR a useful alternative. Interestingly, chemical shifts around 6.0–8.1 ppm are also observed which correspond to the protons belonging to aromatic phenolic compounds [30,31]. It was found that the ^{1}H NMR spectrum of the Nio-EEP sample displayed chemical shifts that resembled the niosome components as well as the EEP components at a chemical shift between 7.25 and 7.75 ppm. In the HPLC analysis, our EEP sample consists of several phenolic compounds [11]. This result implies that phenolic compounds were successfully encapsulated into the niosomal formulation.

2.1.3. In Vitro Release Study

The study of EEP released from niosomal vesicles was carried out in modified stimulated lung fluid (mSLF) at pH 6.6 which mimics the pathological conditions of pneumonia. As shown in Figure 1e, the initial burst release of EEP was 11.2% in the first 3 h followed by a gradual release for up to 24 h, indicating that Nio-EEP acts as a sustained-release formulation. According to these physicochemical characteristics, F1 provided preferable properties for further investigation of its bioactivity against *C. neoformans*.

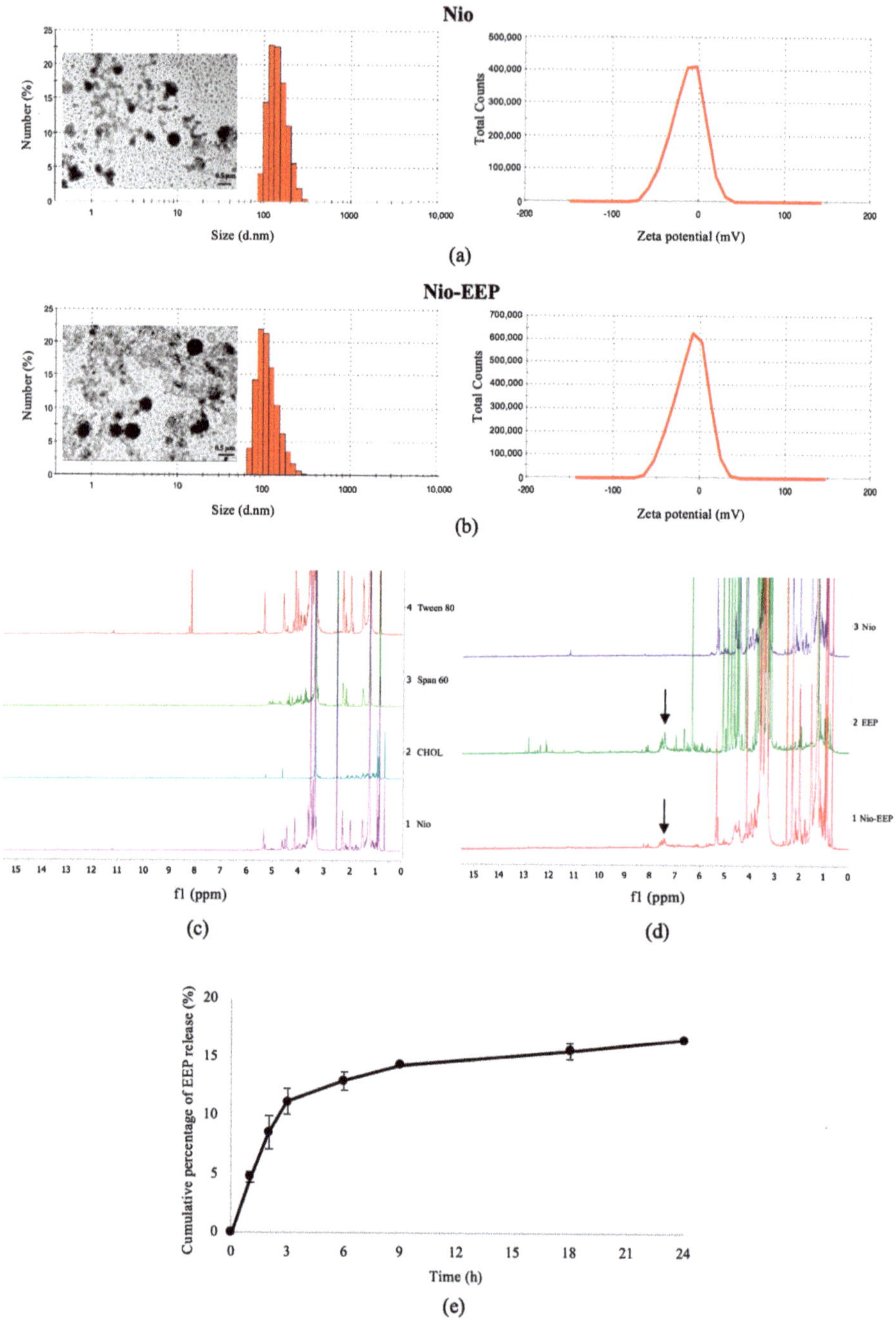

Figure 1. Characteristics of F1 formulation. Nanostructure, size distribution, and ZP distribution curves of (**a**) Nio and (**b**) Nio-EEP based on STEM and dynamic light scattering (DLS) analysis, respectively. The scale bar represents 0.5 µm. NMR spectra of (**c**) Nio composition and (**d**) Nio, EEP, and Nio-EEP are presented. (**e**) In vitro cumulative release of Nio-EEP in mSLF, pH 6.6, at 37 °C for 24 h. The data are represented as the mean ± SEM of three independent trials.

2.1.4. Stability Testing

The stability of the formulations was determined after storage at 4 °C for 1 month. As shown in Figure 2, the average size, PDI, and ZP values showed slight alterations at different times. Notably, EEP remained the same in the formulation; approximately 88% of EE was not different from the initial time point. This result indicates that the Nio-EEP and Nio were stable for 1 month of storage at 4 °C.

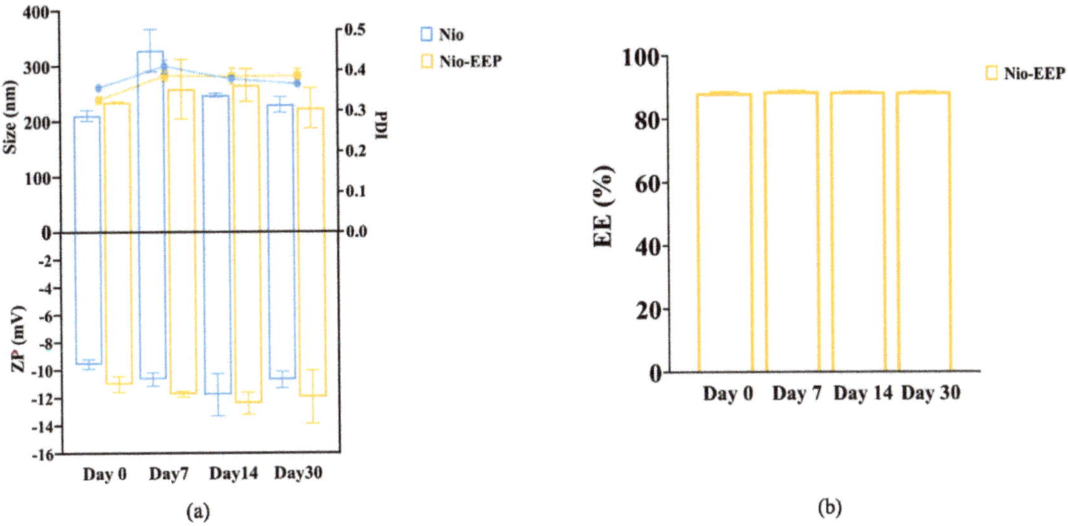

Figure 2. Stability of niosomes during storage at 4 °C for 1 month. (**a**) • Blue dot and ■ Yellow square on the PDI line represent Nio and Nio-EEP, respectively. (**b**) Entrapment efficiency (EE) of Nio-EEP.

2.2. *In Vitro Biological Activity of Nio-EEP*

2.2.1. Cytotoxicity Assay of Niosomes

The cytotoxicity of Nio-EEP was evaluated on A549 and NR8383 cells after treatment with nanoparticles between 0.325 and 6.5 × 10^{11} vesicles/mL. The results show a significant reduction in metabolic activity in both cell lines when treated with nanoparticles between 3.25 and 6.5 × 10^{11} vesicles/mL (Figure 3a,b). Niosomes with a number of particles below 3.25 × 10^{11} vesicles/mL are considered non-cytotoxic. Therefore, these concentrations of Nio-EEP were selected for investigation in further experiments.

2.2.2. Anti-Fungal Susceptibility Testing

The inhibitory effects of Nio-EEP on the growth of yeast cells was evaluated by a colony forming unit (CFU) assay. The results established that none of the concentrations of Nio-EEP and Nio reduced the growth of *C. neoformans* (Figure 3c). On the other hand, the metabolic activity was reduced based on a 3-[4,5-dimethylthiazol-2-yl]-2,5 diphenyltetrazolium bromide (MTT) assay. As shown in Figure 3d, there was no statistically significant difference in the metabolic activity of yeast cells between the control and Nio groups. Remarkably, Nio-EEP significantly reduced the metabolic activity of the yeast cells by approximately 25% and 40% at 1.0 and 2.0 × 10^{11} vesicles/mL, respectively, in contrast to Nio. Based on the results, Nio-EEP has the efficacy to inhibit the metabolic activity of *C. neoformans*.

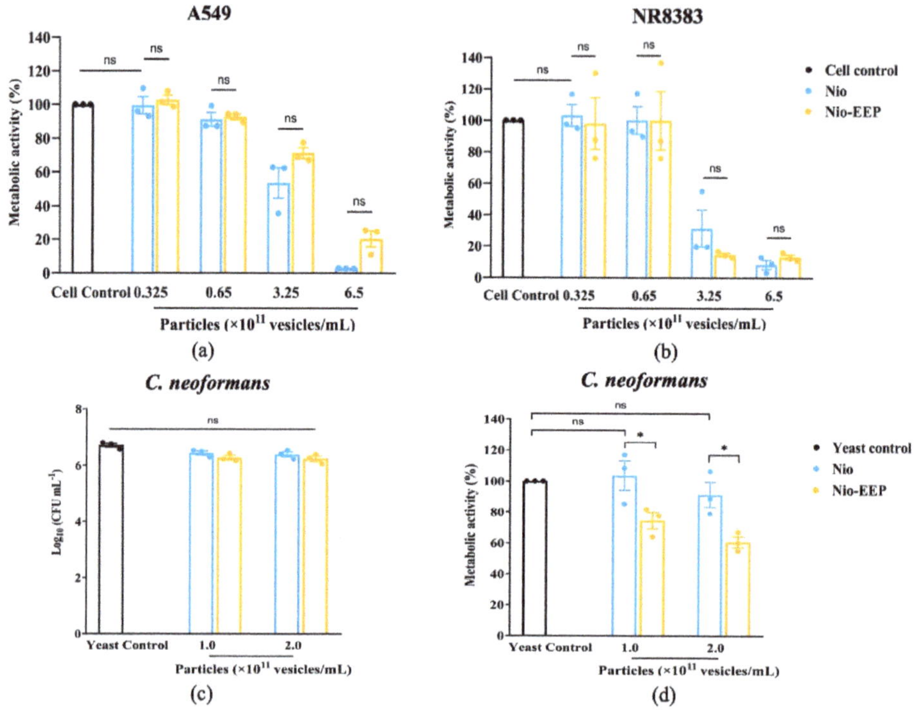

Figure 3. Determination of cytotoxicity and anti-fungal activity of Nio-EEP. Cytotoxicity of the niosomal formulations on (**a**) A549 and (**b**) NR8383 cell lines was evaluated by MTT assay. (**c**) Viability of treated yeast cells was assessed by CFU assay. (**d**) Reduction in metabolic activity in yeast cells from Nio-EEP. All values are expressed as mean ± SEM of three independent experiments performed in triplicate. Note: ns—not significant when compared to each group; * $p < 0.05$ when compared to each group. Black, blue, and yellow bars represent cell control or yeast control, Nio, and Nio-EEP, respectively.

2.2.3. Localization of Nio-EEP

To ensure the uptake of niosomes by the yeast cells, an assessment of niosome localization was performed. As shown in Figure 4a, the Nio-EEP was tracked by Nile red (NR) labeling (red) while the yeast cells were stained with calcofluor white (CFW) (blue). After the incubation period, the accumulation of Nio-EEP was observed inside the yeast cytoplasm. The orthogonal imaging analysis confirmed that Nio-EEP was located within the yeast cells (Figure 4b) and therefore could be up-taken by the yeast cells.

2.3. Anti-Virulence Factors of Nio-EEP

2.3.1. Phospholipase Production

Enzymatic phospholipase activity has been found to promote the binding of *C. neoformans* during lung infection; therefore, the effect of Nio-EEP on yeast phospholipase activity was preliminarily assessed. It was found that the phenotypic phospholipase activity of yeast was not reduced by niosomes, as determined by the EYA assay (Figure 5a). While the genotypic determination of the phospholipase-related gene, *PLB1*, showed a significant reduction in expression level after treatment with niosomes, both particle concentrations of Nio did not significantly affect *PLB1* expression. Interestingly, at 2×10^{11} vesicles/mL, Nio-EEP exhibited a significant down-regulation of *PLB1* levels by 0.54-fold changes in contrast to Nio (Figure 5b). These results imply that EEP might contribute to the interference of phospholipase production at the transcriptional level, leading to an attenuation of the virulence factor.

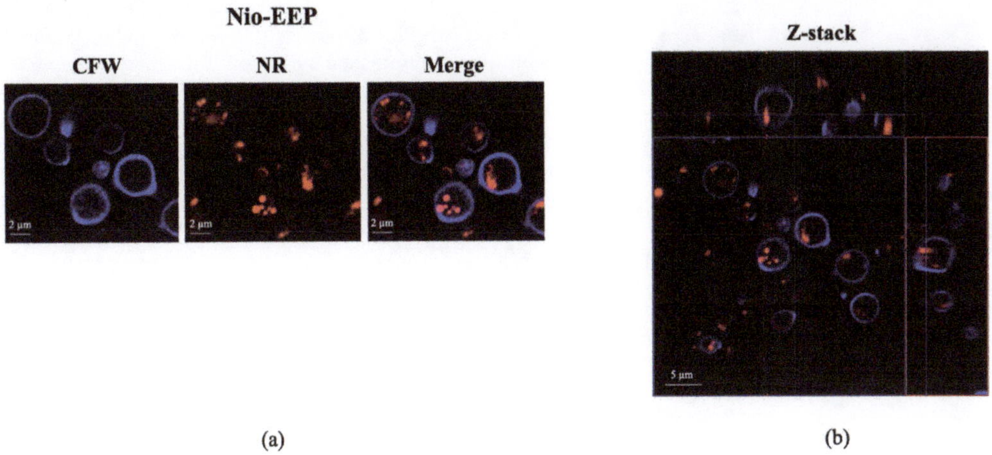

Figure 4. Intracellular uptake of niosomes by *C. neoformans*. (**a**) Localization of Nio-EEP inside the yeast cells. The Nio-EEP was pre-stained with NR (red) and subsequently incubated with CFW-labeled yeast cells (blue). The overlay of fluorescent images demonstrates the accumulation of Nio-EEP (red) within the yeast cells (blue). Scale bars represent 2 μm. (**b**) Orthogonal imaging analysis was performed to confirm the localization of Nio-EEP in the yeast cells. Scale bars represent 5 μm.

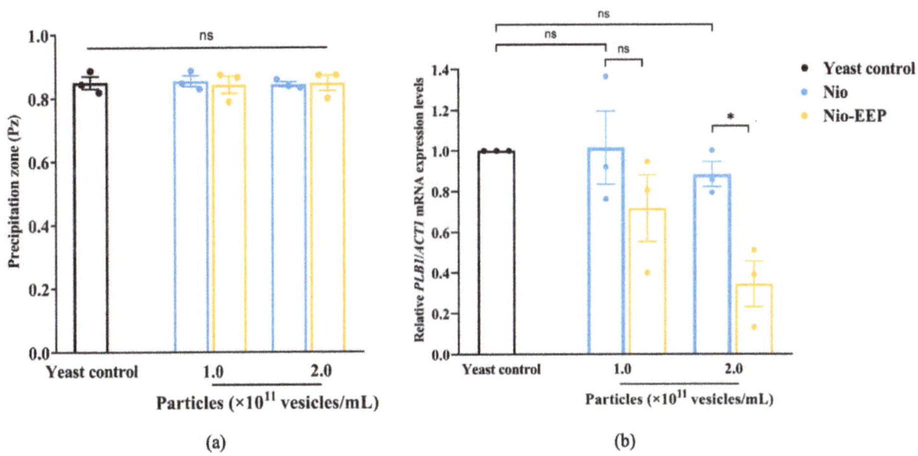

Figure 5. Effects of Nio-EEP on phospholipase production. (**a**) The production of phospholipase (Pz) was examined on EYA. (**b**) Down-regulation of *PLB1* on mRNA levels induced by Nio-EEP. Relative mRNA expression was normalized to *ACT1* and represented as a fold change in $2^{-\Delta\Delta CT}$ compared to the yeast control. Note: ns—not significant when compared to each group; * $p < 0.05$, significant compared to Nio. Black, blue, and yellow bars represent yeast control, Nio, and Nio-EEP, respectively.

2.3.2. Biofilm Formation

Following adhesion on lung epithelial cells, a biofilm of *C. neoformans* is formed and consequently self-produces an extracellular polymeric matrix (EPM) as a defense mechanism. To investigate the effects of Nio-EEP on biofilm formation, an examination of the formation of biofilm was conducted by MTT assay and confocal laser scanning microscope (CLSM) imaging analysis. There was no statistically significant difference in the biofilm formation of yeast cells between the control and Nio groups. Interestingly, the production of yeast biofilm was significantly reduced by 54 and 57% after treatment with 1.0 and 2.0 × 10^{11} vesicles/mL of Nio-EEP, respectively, in contrast to Nio (Figure 6a). To ensure that the formed biofilm was diminished by Nio-EEP, the three-dimensional (3D) structure of biofilm was evaluated by CLSM imaging analysis. The fluorescent images presented the metabolically active yeast cells with FUN-1 (red) and EPM with Concanavalin A (Con A) Alexa Flour 488 conjugate (green). The biofilm thickness of the yeast control, Nio, and Nio-EEP was approximately 25, 23, and 16 μm, respectively. The results exhibit, interestingly, that the biofilm thickness of Nio-EEP-treated yeasts was clearly reduced by about 30% in contrast to Nio, as shown in Figure 6b. These findings suggest that Nio-EEP might interrupt the mitochondrial activity and lead to a decrease in biofilm formation.

Aside from the phenotypic change in biofilm formation, the molecular expression levels of biofilm-related genes, including the *UGD1*, *UXS1*, and *MAN1* genes, were further assessed. As shown in Figure 6c, Nio did not statistically affect the expression of these three genes compared with the yeast control. Nio-EEP remarkably suppressed the mRNA expression levels of *UGD1* and *UXS1* but did not change the *MAN1* gene expression. The yeast cells treated with 1.0 × 10^{11} and 2.0 × 10^{11} vesicles/mL of Nio-EEP showed a significant down-regulation of mRNA levels by approximately 0.35 and 0.24-fold changes for *UGD1* and 0.86 and 0.40-fold changes for *UXS1*, respectively, contrasting the results with Nio. These findings suggest that Nio-EEP might influence the production of biofilm through the down-regulation of the *UGD1* and *UXS1* genes.

2.3.3. Nio-EEP-Induced Intracellular Killing

To evaluate the function of macrophages to phagocytose the treated yeasts, a phagocytosis assay was conducted. The NR8383 cells were challenged with the Nio-EEP-treated yeasts and stained with Wright-Giemsa. The phagocytosed yeast cells were presented (Figure 7a) and the percentages of phagocytosis were approximately 27–41% in all groups (Figure 7b) while the phagocytosis index was 0.3–0.4 cells/macrophage, as shown in Figure 7c. These results indicate that treatment with Nio-EEP and Nio did not affect the phagocytosis activity of macrophages. From these findings, we then hypothesized whether treating *C. neoformans* with Nio-EEP would decrease the survival rate of yeast cells in macrophages and the survival of intracellular yeasts was then carried out by a CFU assay. As shown in Figure 7d, the survival rate of Nio-EEP-treated yeasts was reduced by 20% compared to Nio. Noticeably, the survival rate of Nio-EEP-treated yeasts was reduced by 47% when compared to the yeast control. Based on the results, Nio-EEP could induce the killing of intracellular *C. neoformans* by alveolar macrophages.

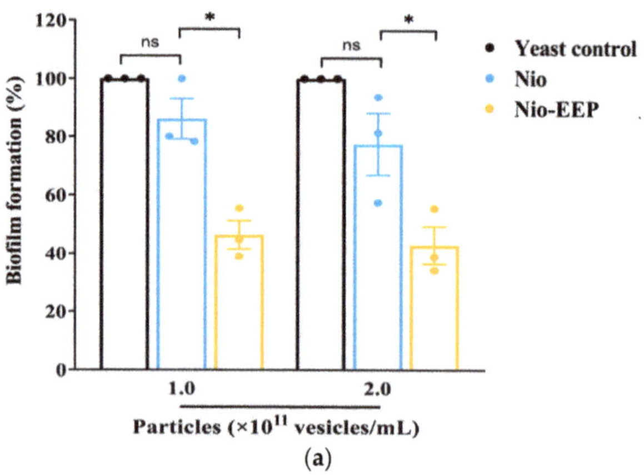

(a)

Figure 6. *Cont.*

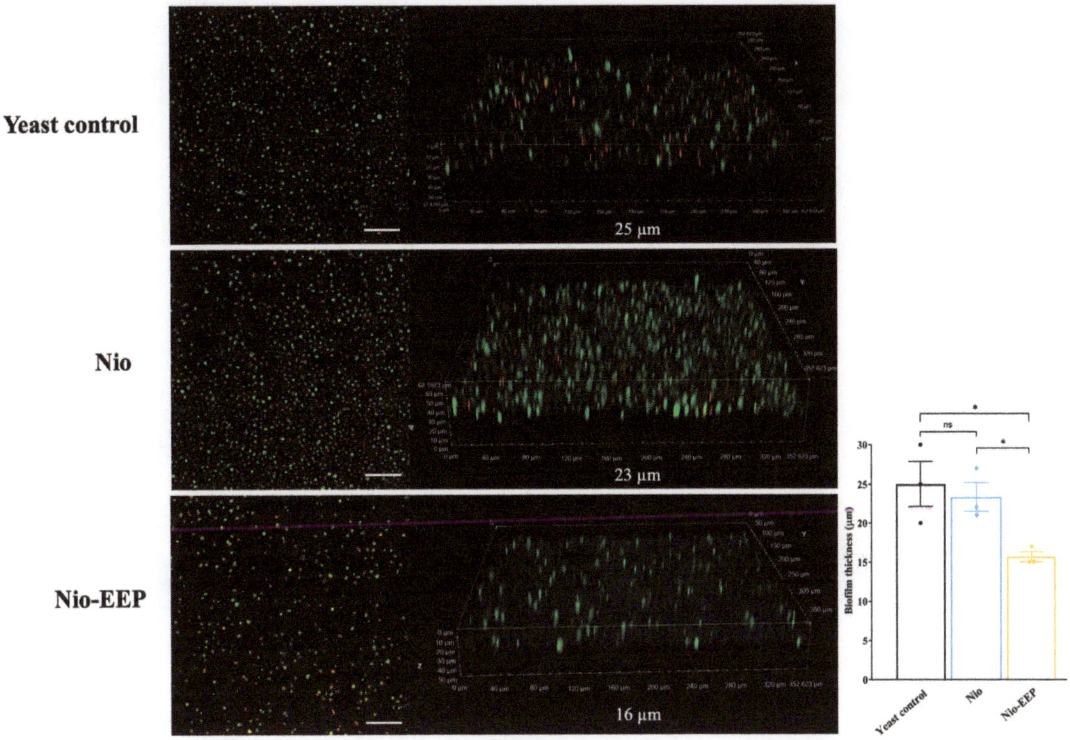

(**b**)

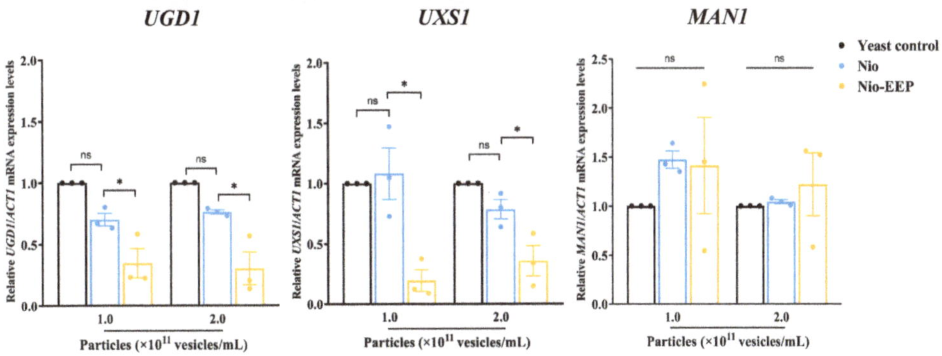

(**c**)

Figure 6. Reduction in biofilm formation by Nio-EEP. The surface-adhered yeast cells were treated with Nio-EEP for 48 h for mature biofilm. (**a**) Biofilm production was evaluated by an MTT assay. (**b**) Fluorescent and 3D images of biofilm thickness at 2×10^{11} vesicles/mL were taken by CLSM. The scale bar presents 50 μm. (**c**) Expression levels of biofilm-related genes, including *UGD1*, *UXS1*, and *MAN1* were assessed after Nio-EEP treatment. Relative mRNA expression was normalized to *ACT1* and expressed as a fold change. The error bars show mean ± SEM from three independent experiments performed in triplicate. Note: ns—not significant when compared to each group; * $p < 0.05$, significant compared to yeast control or Nio. Black, blue, and yellow bars represent yeast control, Nio, and Nio-EEP, respectively.

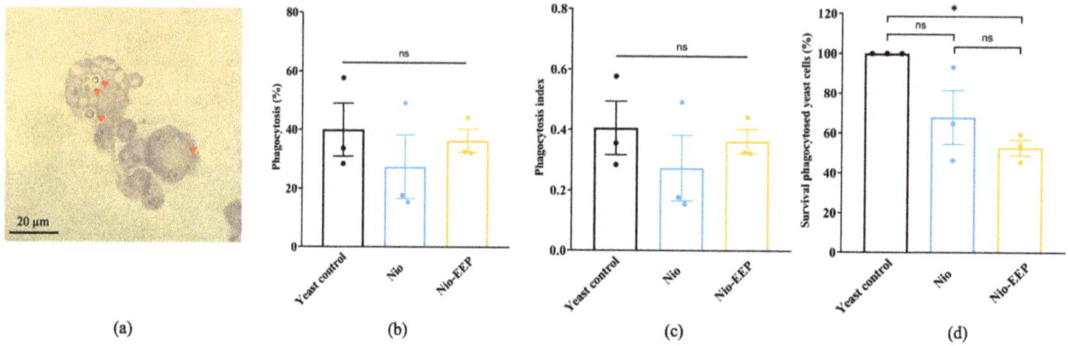

Figure 7. The Nio-EEP-induced killing of *C. neoformans* by alveolar macrophages. The treated yeast cells were opsonized with anti-GXM mAb (Clone 18b7) prior to infection in macrophages. (**a**) The macrophages infected with *C. neoformans* (red arrowhead) were stained by Wright-Giemsa, magnification 100×. (**b**) Phagocytosis (%). (**c**) Phagocytosis index determination. (**d**) Survival of treated yeast cells in the macrophages (%) after 24 h of incubation. The survival of yeast cells was assessed based on a CFU assay. The results show the mean ± SEM from three independent experiments performed in triplicate. Note: ns—not significant when compared to each group; * $p < 0.05$, significant compared to yeast control. Black, blue, and yellow bars represent yeast control, Nio, and Nio-EEP, respectively.

3. Discussion

The World Health Organization (WHO) has reported *C. neoformans* as one of the most critical fungal pathogens causing the greatest threat to human health [32]. Pulmonary cryptococcosis remains a significant concern, especially in an immunocompromised patient. Management of this invasive cryptococcal infection currently relies on the first-line drug, AMB. However, AMB-induced nephrotoxicity is a principal issue that limits effective treatment [1]. Thus, alternative treatments derived from natural substances have gained more attention as therapeutic agents. EEP, a natural product from bees, is a source of several effective molecules that exhibit antimicrobial activity [11]. Our previous study reported that EEP was found to have anti-fungal properties against major cryptococcal virulence factors, such as polysaccharide capsules, melanin pigment, and urease [17]. However, the direct application in the pulmonary system is still restricted due to the water solubility of the EEP [33] and lung surfactant permeability [34].

Improving the therapeutic properties of EEP was achieved through a nanocarrier-based drug delivery system. Niosomes are lipid-based nanocarriers produced from non-ionic surfactants and CHOL, which might be useful for delivery to the pulmonary system. This study investigated niosomal nanocarriers using different ratios of non-ionic surfactants (SP60 and TW80) and the additive CHOL. The non-ionic surfactant structure with a single alkyl tail generally forms a niosomal vesicle in aqueous solutions [28] and the insertion of CHOL into the niosomal membrane requires the membrane rigidity to increase, stabilizing the vesicular structure [35–37]. It was reported that these non-ionic surfactants were greater for high encapsulation of natural products [38,39] and that the EEP was successfully encapsulated by SP60 or TW80 with approximately 70% EE [24,40]. In this study, the optimization of the different SP60 and TW80 concentrations was initiated according to Sangboonruang et al. [41]. Moreover, the ratio of surfactants was further optimized by reducing the concentrations of both SP60 and TW80 (F1) as well as SP60 (F3) only. In the case of reducing only TW80, we found the undesirable characteristic [41].

As a result, the niosomal dispersion exhibited vesicular particle sizes of Nio ranging from 108 to 255 nm while Nio-EEP was 152 to 268 nm. The mean particle size of Nio did not change between the formulations F1 and F3. This finding means that a double concentration of TW80 in F3 did not affect the particle size. However, the particle size of

F2 was reduced due to high surfactant concentrations, possibly inducing micelle, rather than vesicle formations [42,43]. In addition, Nio-EEP particle size was markedly larger than Nio in F1 and F2. This might be due to EEP being encapsulated into the hydrophobic layer, leading to increased particle size [41]. Conversely, F3 exhibited a mean particle size of Nio-EEP smaller than Nio, likely due to the interaction between the surfactant and the extract, enhancing niosomal cohesion and resulting in a decreased vesicle size [44,45].

The distribution of vesicles in the solution was determined with PDI values by DLS analysis. The PDI values of all three formulations ranged from 0.32 to 0.37, indicating a relatively homogeneous vesicle population [41]. Furthermore, the presence of a surface charge, or ZP, can produce a repulsive force between the vesicles, causing a distributed suspension. In theory, ZP values outside of −30 mV to +30 mV are generally considered to have sufficient repulsive force, attaining better physical colloidal stability [46]. However, our results showed the ZP values of all formulations with a negative charge, ranging from −9.38 to −10.54 mV. To further improve the ZP values, some modifications with charge-inducing agents, such as diacetyl phosphate (DCP) for negative charges or stearyl amine (STR) for positive charges, [47] may be needed.

Regarding EE, our formulations of TW80, SP60, and CHOL exhibited high encapsulation rates of more than 85% in all formulations. In other works, it was reported that similar components produced from TW80 and CHOL were at 70% EE [40] and from SP60 and CHOL were at 71.29% EE [24]. Differences in the efficiency of drug encapsulation may depend on several factors, such as 3D chemical structure, hydrophilicity, the ratio of surfactant, and the structure of the surfactant [48,49]. In addition, the CHOL distributed between the lipid bilayer enhances encapsulation due to its membrane-stabilizing effect and the prevention of drug leakage [47,50]. The capacity of the nanovesicle to load EEP is determined by %LC. F2 and F3 contained LC below 50%, even after increasing the lipid phase composition, indicating the maximum capacity of the EEP–lipid interaction. The appropriate niosomal composition of F1 could occupy the EEP and result in the highest % LC. Therefore, F1 was chosen for further investigation.

The physicochemical characteristics of niosomes were confirmed by STEM and NMR spectroscopy, as shown in Figure 2. The F1 formulation showed spherical morphology with the approximate particle size correlating to the DLS results. In addition, the NMR spectra of Nio, EEP, and Nio-EEP exhibited minor constituents of EEP at a region between 7.25 and 7.75 ppm, corresponding to EEP derived from phenolic compound regions, as reported by Ilhan-Ayisigi et al. [40]. Hence, these results support the encapsulation of EEP in the niosomal system. The stability of the obtained formulation was also tested under storage conditions of 4 °C for 1 month. The particles' size, PDI, and ZP did not change and the loaded EEP was retained in the nanovesicles by more than 85% throughout the study period. This indicates a stable property of this nanoformulation. A drug release profile is one of the most important characteristics describing the process of payload migration from the niosome to the outer system [51]. In this study, the in vitro release profile of Nio-EEP revealed an initial burst-release in the mSLF at 3 h and a sustained release during the experiment period of 24 h. Additionally, the in vitro toxicity of the niosomes was not found in the A549 and NR8383 cells at a number of particles below 3.25×10^{11} vesicles/mL. The overloaded number of NPs with induced cellular cytotoxicity had been described elsewhere [52].

For anti-fungal activity, Nio-EEP vesicles did not affect the growth of *C. neoformans* in terms of the colony count. However, Nio-EEP exhibited an inhibitory effect on the metabolic activity of *C. neoformans* at approximately 40%. Based on the MTT assay reflecting the metabolic activity rather than direct cell viability, we suggest that Nio-EEP had the anti-fungal ability through the interference of yeast mitochondrial function. In concordance with previous reports, the released EEP from Nio-EEP might potentially interfere with the mitochondrial enzyme activity [11,53] and inhibit the electron transport chain (ETC) complexes (complex I to V), eventually resulting in mitochondrial dysfunction [54]. The Nio-EEP intracellular uptake in the yeast cells was visualized by CLSM. According to a previous study, this evidence can be described by non-phagocytic eukaryotic cells having an uptake nanoparticle size ranging from 200–500 nm via endocytosis [55]. One of the endocytosis processes in the yeasts might be clathrin-mediated endocytosis, initiated by cytosolic proteins assembling to promote plasma membrane blending and transforming the flat plasma membrane to clathrin-coated vesicles [56]. Consequently, it might be implied that the reduction of metabolic activity is capable of releasing intracellular EEP.

Nio-EEP presented properties against important virulence factors related to the adhesion and biofilm production of *C. neoformans* and PLB1 is one of the virulence-associated enzymes that play a crucial role in promoting yeast cell adhesion on the pulmonary epithelial cell surface [57]. Using a screening method with an EYA assay, the PLB1 production of niosome-treated *C. neoformans* was not found. The EYA method is based on precipitation zone production which has a low sensitivity [58,59]. Even though the radiolabeling method is specific to detect PLB1 activity, there are more practical difficulties. Thus, gene expression analysis is recommended and considered to be an accurate evaluation method [57,58]. As a result, *PLB1* expression at the transcriptional level was significantly disrupted by Nio-EEP. Therefore, the regulation of phospholipase synthesis might be defective, and eventually, the PLB1 enzyme activity was reduced.

Once the yeast cells adhere to and colonize the epithelial host cells, cell-to-cell communication can lead to the formation of EPM or biofilm. Yeast biofilm is another virulence factor that provides defensive activity from anti-fungal drug penetration and the pulmonary immune response, thus promoting yeast survival. In this work, the biofilm production indicating the yeast community was significantly decreased by approximately 50%. Also, the physical structure of yeast biofilm was reduced from 25 μm to 16 μm in thickness while Nio did not have this effect. These observations could imply that the biofilm decreased as a result of releasing EEP. In agreement with previous work by Kumari et al., it was found that a phenolic compound had reduced the biofilm formation of *C. neoformans*. It can be explained that the phenolic compounds trigger reactive oxygen species (ROS) generation and oxidative stress, sequentially reducing EPM biosynthesis [60]. Likewise, Iadnut et al. reported that the biofilm mass and gene-related expression of biofilms in *C. albicans* were reduced by the EEP-loaded PLGA-NPs [11]. Major components related to EPM in the biofilm are glycosyl compositions, such as mannose, glucuronic acid, and xylose, and these sugar molecules are processed through glycan synthetic pathways. The nucleotide sugars are the donor molecules for structural polysaccharide capsule synthesis. Guanosine diphosphate-mannose (GDP-Man) is made through the sequential action of phosphomannose isomerase (MAN1), which is encoded by the *MAN1* gene [61]. Uridine diphosphate-glucuronic acid (UDP-GlcA) is produced through the dehydrogenase uridine diphosphate-glucose (UDP-Glc) pathway by uridine diphosphate-glucose dehydrogenase (UGD1). UDP-GlcA is sequentially decarboxylated by uridine diphosphate-xylose decarboxylase (UXS1) to produce the uridine diphosphate-xylose (UDP-Xyl) [62,63]. For a clearer understanding, the mRNA expression of genes associated with the formation of *C. neoformans* biofilm was further assessed. The expression of *UGD1* and *UXS1* mRNA was suppressed in Nio-EEP-treated biofilm while *MAN1* had no change. These findings can be explained by the fact that the phenolic compounds of the released EEP may be interacting with glycosyltransferase, leading to the lack of a specific sugar donor to supply the downstream product of the EPM biosynthesis [60,64,65]. This suggests that Nio-EEP has the ability to disrupt the yeast community structure and inhibit the formation of biofilm via the glycosyl component interruption. To fill this gap, the proteomic profiles in mature biofilm, metabolic product accumulation, and a quorum sensing mechanism should be further studied. The integration of multidisciplinary fields will promote more understanding and development of strategies to reduce the virulence of fungal pathogenesis and prevent the extrapulmonary dissemination of *C. neoformans*.

Additionally, Nio-EEP was further investigated for its ability to enhance phagocytosis or kill yeast via alveolar macrophages. The results demonstrated no differences in the phagocytosis rate and phagocytosis index, suggesting Nio-EEP did not affect yeast recognition and phagocytosis via Fc–FcγR interactions [66]. In consideration of the survival of phagocytosed yeast cells, we found that Nio-EEP significantly decreased the survival rate of *C. neoformans*. We also suspect that the lower growth and survival rate of intracellular *C. neoformans* is due to the decrease in PLB1 activity. Taken together, the reduction of urease and melanin induced by EEP has been reported [17,67]. *C. neoformans* inhibited the acidification of the phagolysosome by the urease enzyme which degrades urea into CO_2 and ammonia. Melanin also plays a protective role against free radicals [68]. Additionally, phospholipase serves as a phospholipid hydrolysis enzyme in the phagolysosome [69]. These virulence factors are involved in cryptococcal survival, yeast escape, and macrophage killing. Therefore, the reduction of virulence factors not only detriments yeast survival but also increases sensitivity to killing via hydrolytic enzymes, reactive oxygen species, and reactive nitrogen species (ROS/RNS). To fill this knowledge gap, the investigation of PLB1 activity and other virulence factors, as well as the host immune response to the intracellular pathogen with Nio-EEP, should be further performed.

4. Materials and Methods

4.1. Materials

Propolis powder was kindly provided by Bee Product Industry Co., Ltd., Lamphun, Thailand. Sorbitan monostearate (Span 60; SP60), polysorbate 80 (Tween 80; TW80), and cholesterol (CHOL) were purchased from Sigma-Aldrich (St. Louis, MO, USA). All other chemicals and reagents used in this study were of analytical grade, including Sabouraud Dextrose Agar (SDA) (HiMedia, Mumbai, India), Sabouraud Dextrose Broth (SDB) (HiMedia, Mumbai, India), Dulbecco's modified Eagle's medium (DMEM) (Gibco, Carlsbad, CA, USA), Kaighn's modification of Ham's F12 medium (F-12K) (Caisson laboratories Inc., Smithfield, UT, USA), Chloroform (RCI Labscan, Taipei, Taiwan), Potassium phosphotungstic acid (TED PELLA Inc., Redding, CA, USA), 3-[4,5-dimethylthiazol-2-yl]-2,5 diphenyltetrazolium bromide (MTT) (Bio Basic Inc., Markham, ON, Canada), Calcofluor white (CFW) (Sigma-Aldrich, St. Louis, MO, USA), Nile-red dye (Sigma-Aldrich, St. Louis, MO, USA), ProLong Gold anti-fade reagent (Thermo Fisher Scientific, Waltham, MA, USA), Egg Yolk Tellurite Emulsion (HiMedia, Mumbai, India), FUN-1 (Molecular Probe, Waltham, MA, USA), Concanavalin A (Con A)-Alexa Flour 488 conjugate (Thermo Fisher Scientific, CA, USA), lipopolysaccharide (LPS) (Sigma-Aldrich, MO, USA), and interferon-γ (IFN-γ) (Biolegend, San Diego, CA, USA).

4.2. Yeast and Cell Lines

C. neoformans H99 was kindly provided by Assoc. Prof. Pojana Sriburee (Department of Microbiology, Faculty of Medicine, Chiang Mai University, Chiang Mai, Thailand). Yeast cells were maintained on SDA and incubated at 37 °C for 72 h. A few isolated colonies were selected, cultured in SDB, incubated at 37 °C for 16–18 h, and then shaken before experimentation.

The human lung epithelial cancer cell line (A549) was kindly obtained from Asst. Prof. Dr. Khanittha Punturee (Department of Medical Technology, Faculty of Associated Medical Sciences, Chiang Mai University, Chiang Mai, Thailand) and was cultured in DMEM supplemented with 10% (*v*/*v*) fetal bovine serum (FBS), 100 units/mL of penicillin, and 100 μg/mL of streptomycin. Alveolar macrophage cell line (NR8383) (ATCC, Manassas, VA, USA) was cultured in F-12K supplemented with 15% (*v*/*v*) FBS, 100 units/mL of penicillin, and 100 μg/mL of streptomycin. The cells were maintained in a humidified atmosphere of 5% CO_2 at 37 °C.

4.3. Formulation of Nio-EEP

Niosomal formulations were prepared by the thin–film hydration (TFH) technique according to our previous study with some modifications [41]. Briefly, different molar ratios of SP60, TW80, and CHOL were dissolved in 9 mL chloroform and supplemented with 1 mL ethanol solution of EEP (20 mg/mL) for Nio-EEP or without EEP for empty niosomes (Nio) in a round-bottom flask. The organic solvent was evaporated using a rotary evaporator under a vacuum at 60 °C and 100 rpm rotation to obtain a thin lipid film on the inner flask wall. The lipid thin film was hydrated with 10 mL PBS, pH 7.4, under mechanical stirring at 60 °C for 30 min. The obtained niosomal suspension was then subjected to an ultrasonic probe sonicator (Hielscher UP50H, Wanaque, NJ, USA) at 80% amplitude for 30 min in an ice bath to achieve size reduction. The constituents of the different niosomal formulations are indicated in Table 2. A schematic representation of a Nio-EEP is shown in Figure 8.

Table 2. Composition of the Nio-EEP formulations.

Formulations	SP60: TW80: CHOL (mM Ratio)	SP60 (mg)	TW80 (mg)	CHOL (mg)	EEP (mg/mL)
F1	1:1:1	4.3	13.1	3.8	2
F2	2:2:1	8.6	26.2	3.8	2
F3	1:2:1	4.3	26.2	3.8	2

Abbreviations: SP60, Span 60; TW80, Tween 80; CHOL, Cholesterol; EEP, Ethanolic extract propolis.

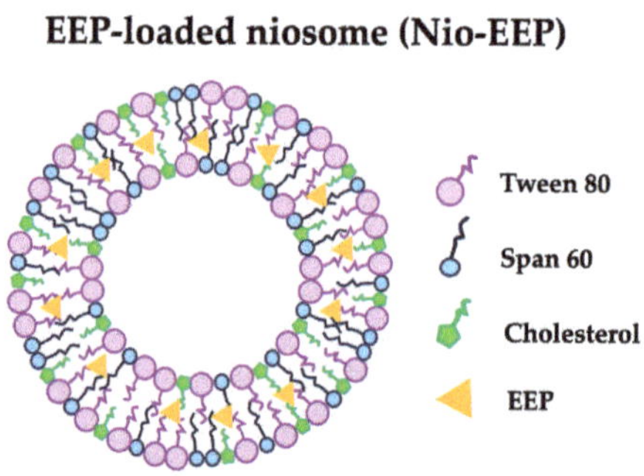

Figure 8. Schematic representation of a Nio-EEP.

4.4. *Physicochemical Characterization of Niosomes*

4.4.1. Particle Size, Polydispersity Index (PDI), Zeta Potential (ZP), Morphological Analysis, and Stability Testing

The particle size, PDI, and ZP of niosomes were measured using a Malvern Zetasizer Nano ZSP system (Malvern Instruments, Worcestershire, UK). Niosomal samples were suspended in PBS at a 1:100 dilution. The analyses were performed based on triplicates in three individual runs.

The morphological characteristics of niosomal vesicles were examined by scanning transmission electron microscope (STEM). A drop of the niosomal sample was placed onto a carbon-coated copper grid and stained with 1% (*w*/*v*) phosphotungstic potassium acid aqueous solution. The morphology was observed by JSM-IT800 Ultrahigh Resolution Field Emission SEM (JEOL, Peabody, MA, USA).

To investigate the stability of the formulations, the niosomal vesicles were kept at 4 °C, and measured in size, PDI, and ZP on days 7, 14, and 30.

4.4.2. Nio-EEP Chemical Structure

To further verify that the Nio-EEP was successfully performed, the chemical structure of niosomal vesicles was investigated. Nio-EEP, Nio, and EEP were lyophilized before the analysis comparison to the samples with chemical standards. All samples were dissolved in 700 μL deuterated dimethyl sulfoxide (DMSO-d_6) then filtered into a nuclear magnetic resonance spectroscopy (NMR) tube. ^{1}H NMR spectra were recorded by 500 MHz NMR spectroscopy (Bruker AV-500 NEO^{TM}, Berlin, Germany) and are internally referenced to residual proton signals in DMSO-d_6 (2.50 ppm).

4.4.3. Entrapment Efficiency (EE) and Loading Capacity (LC)

Free EEP was determined after separation from Nio-EEP by the centrifugation method with a membrane molecular weight cut-off (MWCO) filter, 10 kDa at 8000× g, and 4 °C for 2 h 30 min [41]. Then, the niosomal residues were re-suspended in 1 mL of sterile PBS, pH 7.4. Finally, the un-entrapped filtrate solution was determined at 290 nm by a UV/Vis spectrophotometer (Specord Plus, Jena, Germany) and the EEP concentration was calculated using the EEP standard calibration curve.

The percentages of entrapment efficiency (%EE) and loading capacity (%LC) were calculated by the following equation [70]:

$$\%EE = \left[\frac{(C_t - C_f)}{C_t}\right] \times 100 \quad (1)$$

where C_t is the concentration of total EEP and C_f is the concentration of free EEP in filtrate.

The amount of EEP-loaded per weight unit of lipid phase was calculated as shown on the %LC [71]:

$$\%LC = \frac{A_t}{L_t} \times 100 \quad (2)$$

where A_t is the total amount of Nio-EEP and L_t is the total weight of lipid phase.

EEP retained in the formulation corresponding to % EE was determined at 4 °C on days 7, 14, and 30.

4.4.4. In Vitro Release Study

The release of EEP from the niosomes was investigated using the modified dissolution method [11]. Briefly, the Nio-EEP was dissolved in 3 mL modified and stimulated lung fluid (mSLF) solution [72] and adjusted to pH 6.6 to mimic acidic pathological conditions [73]. The samples were rotated at 37 °C and the supernatant was collected by centrifugation at different time points and then replaced with the same volume of fresh mSLF solution. The amount of released EEP was analyzed using a UV/Vis spectrophotometer at 290 nm and compared with the EEP standard calibration curve.

4.5. In Vitro Bioactivity of Nio-EEP

4.5.1. Cytotoxicity Assay

The cytotoxicity of the niosomal formulation was examined on A549 and NR8383 cells by MTT assay [74]. Briefly, A549 (1×10^4 cells/well) or NR8383 (1×10^5 cells/well) was seeded in a 96-well tissue culture plate and cultured for 24 h. In addition, NR8383 (1×10^5 cells/well) was cultured in a 96-well tissue culture plate for 48 h. Then, various numbers of nanoparticles were added to the cells. After another 24 h incubation, 20 µL of MTT solution (5 mg/mL) were added to the treated cells and incubated at 37 °C for 4 h. Then, the supernatant was removed and 200 µL of dimethyl sulfoxide (DMSO) were added to solubilize the MTT-formazan produced by living cells. The optical density (OD) was measured at 540 and 630 nm. The metabolic activity was further evaluated and calculated by the equation shown below.

$$\%\text{Metabolic activity} = \frac{(OD_{540} - OD_{630})}{(OD_{540} - OD_{630})} \times 100 \quad (3)$$

4.5.2. Anti-Fungal Susceptibility Testing

The antifungal activity of Nio-EEP was determined by modified the. Clinical Laboratory Standard Institute (CLSI) broth microdilution method (M27-A3) [17]. In brief, the Nio-EEP (or Nio) was serially diluted with various numbers of particles in a 96-well microtiter plate. Then, 100 µL of yeast suspension (1×10^3 CFU/mL) were seeded and incubated at 37 °C for 72 h. After incubation, the yeast proliferation was determined by a colony forming unit (CFU) counted on an SDA plate and the metabolic activity was further examined by a MTT assay as previously described.

4.5.3. Yeast Cell Uptake of Nio-EEP

Nio-EEP uptake by yeast cells was examined using a confocal laser scanning electron microscope (CLSM). The yeast cells (1×10^8 CFU/mL) were stained with 2 mg/mL CFW and incubated for 30 min in the dark at room temperature. Then, the CFW-labeled yeast cells were washed and re-suspended in PBS. Meanwhile, 250 µL of Nio-EEP were stained with 40 µL of Nile Red (NR) solution (0.25 mg/mL) in PBS and incubated for 30 min in the dark at room temperature. Following washing with PBS, the NR-labeled Nio-EEP was re-suspended in PBS and further incubated with the CFW-labeled yeast cells. After 3 h incubation at 37 °C, the excess niosomes were removed, re-suspended with ProLong Gold anti-fade reagent and analyzed by CLSM (LSM900 Airyscan 2; Zeiss, Oberkochen, Germany).

4.6. *Effect of Nio-EEP on Virulence Factors of C. neoformans*

4.6.1. Phospholipase Production

The phenotypic phospholipase enzyme activity was examined using the egg yolk agar (EYA) method [75]. The yeast cell suspension (1×10^8 CFU/mL) was pre-treated with Nio-EEP (or Nio) and incubated at 37 °C for 4 h with rotation. Following washing with PBS, the concentration of treated yeast cells was adjusted. Five microliters (1×10^6 cells) were dropped on EYA and incubated at 37 °C for 4 days. The precipitation zone and the diameter of the colony was measured and the phospholipase production (Pz) value was determined using the following equation below:

$$Pz = \frac{\text{Colony diameter}}{(\text{Precipitation zone} + \text{Colony diameter})} \quad (4)$$

4.6.2. Biofilm Formation

The biofilm formation was evaluated by CLSM [76]. Briefly, one hundred microliters of yeast suspension (1×10^6 cells) were seeded into a 96-well plate and incubated at 37 °C in a 5% CO_2 humidified atmosphere for 4 h. Following the adhesion stage, the non-adherent yeast cells were removed and washed thrice with PBS. Then, 100 µL of niosomes were added and incubated continuously for 48 h. The medium was removed and the biofilm formation activity was determined by an MTT assay, as previously described. Next, the biofilms were stained with 10 µM of FUN-1 and 20 µg/mL of Con A-Alexa Flour 488 conjugate, incubated at 37 °C for 30 min, photographed, and analyzed by CLSM (Nikon AX; Nikon Instruments Inc., Melville, NY, USA).

4.6.3. Virulence-Related mRNA Expression

To observe the genotypic expression levels of virulence factor-related genes, a quantitative reverse-transcription polymerase chain reaction (qRT-PCR) assay was performed. Niosome-treated *C. neoformans* was harvested and the total RNA was extracted using TRIZOL® reagent (Invitrogen, Carlsbad, CA, USA) according to the manufacturer's instructions. Total RNA was reverse transcribed into cDNA according to the RevertAid First Strand cDNA Synthesis Kit (Thermo Fisher Scientific, Waltham, MA, USA). The amplification was carried out by SYBR Green qPCR Master Mix (Thermo Fisher Scientific, MA, USA) and specific primers. The PCR primer sequences were designed according to *PLB1* (accession number CNAG_06085), *MAN1* (accession number CNAG_04312), *UGD1* (accession number CNAG_04969), *UXS1* (accession number CNAG_03322), and actin (*ACT1*) (accession number CNAG_00483). The sequences of the primers are listed in Table 3. The PCR reactions were performed in 35 cycles: initial denaturation at 94 °C for 30 s, annealing at 58 °C for 30 s, and extension at 70 °C for 60 s followed by cooling at 37 °C for 30 s. The mRNA expression levels were analyzed by the $2^{-\Delta\Delta CT}$ method and are expressed as the relative fold change when normalized with *ACT1* as a housekeeping gene [77].

Table 3. The specific primer sequences.

Primers	Primer Sequences (5′-3′)	References
PLB1	TGATGAATGAGAGCACGGAAGC CTCAGACCAGCCCAGTAGCT	[78]
MAN1	GGCCTACGCTGAATTATGGA GTAAAGAGCCGTCCTTGCAG	This study
UGD1	GAGGAGGCTTGTGCTAATGC GACGACCTTGAAACCGATGT	This study
UXS1	AGCTGCATTTTACTCATCCCT TCCTTGATGTAGGCGGGAGA	This study
ACT1	CCTTGCTCCTTCTTCTAT CTCGTCGTATTCGCTCTT	[67]

4.7. Phagocytosis Assay

The alveolar macrophages were activated by adding 0.6 μg/mL of LPS and 100 ng/mL of IFN-γ and incubated at 37 °C in a 5% CO_2 humidified incubator for 24 h [79]. Nio-EEP (or Nio)-treated yeast was opsonized with a 1:10 dilution of anti-glucuronoxylomannan (GXM) monoclonal antibody (Clone 18b7) at 37 °C in a 5% CO_2 humidified incubator for 1 h 30 min. The cells were then infected with 5 MOI of opsonized and niosome-treated yeast and incubated at 37 °C in a 5% CO_2 humidified incubator for 2 h. After washing with PBS, the cells were subjected to Wright-Giemsa staining and the percentage of phagocytosis and phagocytosis index were determined using the following equations below [67]:

$$\%\text{Phagocytosis} = \frac{\text{Phagocytosed cryptococci}}{\text{Five-hundred macrophages}} \times 100 \quad (5)$$

$$\text{Phagocytosis index} = \frac{\text{Phagocytosed cryptococci}}{\text{Five-hundred macrophages}} \quad (6)$$

Alternatively, after 2 h of the phagocytosis process, the supernatant was removed, washed, and continuously incubated at 37 °C in a 5% CO_2 humidified incubator for 24 h in fresh medium. Subsequently, the cells were lysed using sterile deionized (DI) water for 30 min and then the intracellular yeasts were counted and expressed in CFU.

4.8. Statistical Analysis

All data were presented as a mean ± standard error of the mean (SEM) in triplicate following three independent experiments. The Shapiro–Wilk test was used to check for a normal distribution and was followed by one-way analysis of variance (ANOVA) and Tukey's post hoc test. Data without a normal distribution (% metabolic activity of A549 and NR8383) were analyzed using the Krustal–Wallis test and Dunn's post hoc test. Significant differences (* $p < 0.05$) for all analyses were considered. All graphics were generated using Graph Pad Prism version 9.0 (GraphPad Software Inc., San Diego, CA, USA).

5. Conclusions

In this study, Nio-EEP vesicles were successfully formulated on a nanometer scale with favorable physicochemical properties and were up-taken by *C. neoformans*. Moreover, the biological properties of Nio-EEP were introduced as anti-virulence factors, including the *PLB1* gene, biofilm formation, glycosyl components, and synthesis related genes, such as *UGD1* and *UXS1*. Furthermore, the intracellular replication of *C. neoformans* within alveolar macrophages was reduced after treatment with Nio-EEP. Regarding current studies, Nio-

EEP could be a potential anti-virulence agent and be applied with multimodal treatments for pulmonary cryptococcosis.

Author Contributions: Conceptualization, K.T.; methodology, K.K. (Kritapat Kietrungruang), S.S. (Sanonthinee Sookkree), S.S. (Sirikwan Sangboonruang), N.S. and K.K. (Kuntida Kitidee); validation, K.T., S.S. (Sirikwan Sangboonruang) and N.S.; formal analysis, K.K. (Kritapat Kietrungruang); investigation, K.K. (Kritapat Kietrungruang); resources, K.T., Y.T., W.P. and N.S.; data curation, K.T.; writing—original draft preparation, K.K. (Kritapat Kietrungruang); writing—review and editing, S.S. (Sirikwan Sangboonruang), N.S. and K.T.; supervision, K.T.; project administration, K.T.; funding acquisition, K.T. All authors have read and agreed to the published version of the manuscript.

Funding: This research work was supported by scholarships from the Department of Medical Technology, Faculty of Associated Medical Sciences, Chiang Mai University, and also partially supported by Chiang Mai University.

Institutional Review Board Statement: Not applicable.

Informed Consent Statement: Not applicable.

Data Availability Statement: Not applicable.

Acknowledgments: We would like to thank Ratchada Cressey, Piyawan Bunpo, and Kanyamas Choocheep from the Division of Clinical Chemistry, Department of Medical Technology, Faculty of Associated Medical Sciences, Chiang Mai University, for laboratory equipment. We would like to thank the application specialist from Rushmore Precision Co., Ltd., for supporting CLSM (LSM 900 with Airyscan 2, Zeiss, Germany) and the product specialist from the Scientific Instrument Division, Hollywood International Co., Ltd., for supporting CLSM (Nikon AX, Nikon, USA), respectively. We also thank Julia Marie Akins for the professional English proofreading and manuscript editing.

Conflicts of Interest: The authors declare no conflict of interest.

Sample Availability: Not available.

References

1. Setianingrum, F.; Rautemaa-Richardson, R.; Denning, D.W. Pulmonary cryptococcosis: A review of pathobiology and clinical aspects. *Med. Mycol.* **2019**, *57*, 133–150. [CrossRef]
2. Sabiiti, W.; May, R.C. Mechanisms of infection by the human fungal pathogen *Cryptococcus neoformans*. *Future Microbiol.* **2012**, *7*, 1297–1313. [CrossRef]
3. Chen, S.C.; Wright, L.C.; Golding, J.C.; Sorrell, T.C. Purification and characterization of secretory phospholipase B, lysophospholipase and lysophospholipase/transacylase from a virulent strain of the pathogenic fungus *Cryptococcus neoformans*. *Biochem. J.* **2000**, *347*, 431–439. [CrossRef]
4. Taylor-Smith, L.M. *Cryptococcus*-epithelial interactions. *J. Fungi* **2017**, *3*, 53. [CrossRef]
5. Alvarez, M.; Casadevall, A. Phagosome extrusion and host-cell survival after *Cryptococcus neoformans* phagocytosis by macrophages. *Curr. Biol.* **2006**, *16*, 2161–2165. [CrossRef] [PubMed]
6. Ma, H.; Croudace, J.E.; Lammas, D.A.; May, R.C. Expulsion of live pathogenic yeast by macrophages. *Curr. Biol.* **2006**, *16*, 2156–2160. [CrossRef] [PubMed]
7. Lopes, W.; Vainstein, M.H.; De Sousa Araujo, G.R.; Frases, S.; Staats, C.C.; de Almeida, R.M.C.; Schrank, A.; Kmetzsch, L.; Vainstein, M.H. Geometrical distribution of *Cryptococcus neoformans* mediates flower-like biofilm development. *Front. Microbiol.* **2017**, *8*, 2534. [CrossRef] [PubMed]
8. Martinez, L.R.; Casadevall, A. Biofilm formation by *Cryptococcus neoformans*. *Microbiol. Spectr.* **2015**, *3*, 3. [CrossRef]
9. Martinez, L.R.; Casadevall, A. *Cryptococcus neoformans* cells in biofilms are less susceptible than planktonic cells to antimicrobial molecules produced by the innate immune system. *Infect. Immun.* **2006**, *74*, 6118–6123. [CrossRef] [PubMed]
10. Perfect, J.R.; Dismukes, W.E.; Dromer, F.; Goldman, D.L.; Graybill, J.R.; Hamill, R.J.; Harrison, T.S.; Larsen, R.A.; Lortholary, O.; Nguyen, M.H.; et al. Clinical practice guidelines for the management of cryptococcal disease: 2010 update by the infectious diseases society of america. *Clin. Infect. Dis.* **2010**, *50*, 291–322. [CrossRef] [PubMed]
11. Iadnut, A.; Mamoon, K.; Thammasit, P.; Pawichai, S.; Tima, S.; Preechasuth, K.; Kaewkod, T.; Tragoolpua, Y.; Tragoolpua, K. In Vitro antifungal and antivirulence activities of biologically synthesized ethanolic extract of propolis-loaded PLGA nanoparticles against *Candida albicans*. *Evid. Based Complement. Alternat Med.* **2019**, *2019*, 3715481. [CrossRef]
12. Gheflati, A.; Dehnavi, Z.; Ghannadzadeh, Y.A.; Khorasanchi, Z.; Raeisi-Dehkordi, H.; Ranjbar, G. The effects of propolis supplementation on metabolic parameters: A systematic review and meta-analysis of randomized controlled clinical trials. *Avicenna J. Phytomed* **2021**, *11*, 551–565. [CrossRef]

13. Hallajzadeh, J.; Milajerdi, A.; Amirani, E.; Attari, V.E.; Maghsoudi, H.; Mirhashemi, S.M. Effects of propolis supplementation on glycemic status, lipid profiles, inflammation and oxidative stress, liver enzymes, and body weight: A systematic review and meta-analysis of randomized controlled clinical trials. *J. Diabetes Metab. Disord.* **2021**, *20*, 831–843. [CrossRef] [PubMed]
14. Zulhendri, F.; Lesmana, R.; Tandean, S.; Christoper, A.; Chandrasekaran, K.; Irsyam, I.; Suwantika, A.A.; Abdulah, R.; Wathoni, N. Recent update on the anti-inflammatory activities of propolis. *Molecules* **2022**, *27*, 8473. [CrossRef]
15. Corrêa, J.L.; Veiga, F.F.; Jarros, I.C.; Costa, M.I.; Castilho, P.F.; de Oliveira, K.M.P.; Rosseto, H.C.; Bruschi, M.L.; Svidzinski, T.I.E.; Negri, M. Propolis extract has bioactivity on the wall and cell membrane of *Candida albicans*. *J. Ethnopharmacol.* **2020**, *256*, 112791. [CrossRef]
16. Fernandes, F.F.; Dias, A.L.; Ramos, C.L.; Ikegaki, M.; de Siqueira, A.M.; Franco, M.C. The "In Vitro" antifungal activity evaluation of propolis g12 ethanol extract on *Cryptococcus neoformans*. *Rev. Inst. Med. Trop. Sao Paulo* **2007**, *49*, 93–95. [CrossRef]
17. Thammasit, P.; Iadnut, A.; Mamoon, K.; Khacha-Ananda, S.; Chupradit, K.; Tayapiwatana, C.; Kasinrerk, W.; Tragoolpua, Y.; Tragoolpua, K. A potential of propolis on major virulence factors of *Cryptococcus neoformans*. *Microb. Pathog.* **2018**, *123*, 296–303. [CrossRef]
18. Thammasit, P.; Tharinjaroen, C.S.; Tragoolpua, Y.; Rickerts, V.; Georgieva, R.; Bäumler, H.; Tragoolpua, K. Targeted propolis-loaded poly (butyl) cyanoacrylate nanoparticles: An alternative drug delivery tool for the treatment of cryptococcal meningitis. *Front. Pharmacol.* **2021**, *12*, 723727. [CrossRef]
19. Garcia-Verdugo, I.; Descamps, D.; Chignard, M.; Touqui, L.; Sallenave, J.M. Lung protease/anti-protease network and modulation of mucus production and surfactant activity. *Biochimie* **2010**, *92*, 1608–1617. [CrossRef] [PubMed]
20. Pasarin, D.; Ghizdareanu, A.I.; Enascuta, C.E.; Matei, C.B.; Bilbie, C.; Paraschiv-Palada, L.; Veres, P.A. Coating materials to increase the stability of liposomes. *Polymers* **2023**, *15*, 782. [CrossRef] [PubMed]
21. Terzano, C.; Allegra, L.; Alhaique, F.; Marianecci, C.; Carafa, M. Non-phospholipid vesicles for pulmonary glucocorticoid delivery. *Eur. J. Pharm. Biopharm.* **2005**, *59*, 57–62. [CrossRef]
22. Milan, A.; Mioc, A.; Prodea, A.; Mioc, M.; Buzatu, R.; Ghiulai, R.; Racoviceanu, R.; Caruntu, F.; Şoica, C. The optimized delivery of triterpenes by liposomal nanoformulations: Overcoming the challenges. *Int. J. Mol. Sci.* **2022**, *23*, 1140. [CrossRef] [PubMed]
23. Alsaadi, M.; Italia, J.L.; Mullen, A.B.; Ravi Kumar, M.N.; Candlish, A.A.; Williams, R.A.; Shaw, C.D.; Al Gawhari, F.; Coombs, G.H.; Wiese, M.; et al. The efficacy of aerosol treatment with non-ionic surfactant vesicles containing amphotericin B in rodent models of leishmaniasis and pulmonary aspergillosis infection. *J. Control Release* **2012**, *160*, 685–691. [CrossRef] [PubMed]
24. Patel, J.; Ketkar, S.; Patil, S.; Fearnley, J.; Mahadik, K.R.; Paradkar, A.R. Potentiating antimicrobial efficacy of propolis through niosomal-based system for administration. *Integr. Med. Res.* **2015**, *4*, 94–101. [CrossRef]
25. Sangboonruang, S.; Semakul, N.; Suriyaprom, S.; Kitidee, K.; Khantipongse, J.; Intorasoot, S.; Tharinjaroen, C.; Wattananandkul, U.; Butr-Indr, B.; Phunpae, P.; et al. Nano-delivery system of ethanolic extract of propolis targeting mycobacterium tuberculosis via aptamer-modified-niosomes. *Nanomaterials* **2023**, *13*, 269. [CrossRef] [PubMed]
26. Mozafari, M.; Mazaheri, E.; Dormiani, K. Simple equations pertaining to the particle number and surface area of metallic, polymeric, lipidic and vesicular nanocarriers. *Sci. Pharm.* **2021**, *89*, 15. [CrossRef]
27. Wenfei, L.; Ruixiang, Y.; Yang, L.; Zhanhai, Y.; Shumei, L. Preparation and characterization of reactive type dripping agent containing α-double bond. *Asian J. Chem.* **2014**, *26*, 8577–8580. [CrossRef]
28. Rustandi, R.R. Polysorbate 80 and histidine quantitative analysis by nmr in the presence of virus-like particles. *Electrophor.* **2022**, *43*, 1408–1414. [CrossRef]
29. Khatun, R.; Hunter, H.; Magcalas, W.; Sheng, Y.; Carpick, B.; Kirkitadze, M. Nuclear magnetic resonance (NMR) study for the detection and quantitation of cholesterol in HSV529 therapeutic vaccine candidate. *Comput. Struct. Biotechnol. J.* **2017**, *15*, 14–20. [CrossRef]
30. Kasote, D.M.; Pawar, M.V.; Bhatia, R.S.; Nandre, V.S.; Gundu, S.S.; Jagtap, S.D.; Kulkarni, M.V. HPLC, NMR based chemical profiling and biological characterisation of indian propolis. *Fitoterapia* **2017**, *122*, 52–60. [CrossRef]
31. Tran, C.T.N.; Brooks, P.R.; Bryen, T.J.; Williams, S.; Berry, J.; Tavian, F.; McKee, B.; Tran, T.D. Quality assessment and chemical diversity of australian propolis from *Apis mellifera* bees. *Sci. Rep.* **2022**, *12*, 13574. [CrossRef] [PubMed]
32. World Health Organization (WHO) Report. WHO Fungal Priority Pathogens List to Guide Research, Development and Public Health Action. Available online: https://www.who.int/publications/i/item/9789240060241 (accessed on 16 December 2022).
33. Bonifácio, B.V.; Silva, P.B.; Ramos, M.A.; Negri, K.M.; Bauab, T.M.; Chorilli, M. Nanotechnology-based drug delivery systems and herbal medicines: A review. *Int. J. Nanomed.* **2014**, *9*, 1–15. [CrossRef]
34. Hidalgo, A.; Garcia-Mouton, C.; Autilio, C.; Carravilla, P.; Orellana, G.; Islam, M.N.; Bhattacharya, J.; Bhattacharya, S.; Cruz, A.; Pérez-Gil, J. Pulmonary surfactant and drug delivery: Vehiculization, release and targeting of surfactant/tacrolimus formulations. *J. Control Release* **2021**, *329*, 205–222. [CrossRef] [PubMed]
35. Junyaprasert, V.B.; Singhsa, P.; Suksiriworapong, J.; Chantasart, D. Physicochemical properties and skin permeation of Span 60/Tween 60 niosomes of ellagic acid. *Int. J. Pharm.* **2012**, *423*, 303–311. [CrossRef]
36. Nematollahi, M.H.; Pardakhty, A.; Torkzadeh-Mahani, M.; Mehrabani, M.; Asadikaram, G. Changes in physical and chemical properties of niosome membrane induced by cholesterol: A promising approach for niosome bilayer intervention. *RSC Adv.* **2017**, *7*, 49463–49472. [CrossRef]

37. Vyas, S.P.; Singh, R.P.; Jain, S.; Mishra, V.; Mahor, S.; Singh, P.; Gupta, P.N.; Rawat, A.; Dubey, P. Non-ionic surfactant based vesicles (niosomes) for non-invasive topical genetic immunization against hepatitis B. *Int. J. Pharm.* **2005**, *296*, 80–86. [CrossRef] [PubMed]
38. Soliman, M.S.; Abd-Allah, F.I.; Hussain, T.; Saeed, N.M.; El-Sawy, H.S. Date seed oil loaded niosomes: Development, optimization and anti-inflammatory effect evaluation on rats. *Drug Dev. Ind. Pharm.* **2018**, *44*, 1185–1197. [CrossRef]
39. Chinembiri, T.N.; Gerber, M.; du Plessis, L.H.; du Preez, J.L.; Hamman, J.H.; du Plessis, J. Topical delivery of *Withania somnifera* crude extracts in niosomes and solid lipid nanoparticles. *Pharmacogn. Mag.* **2017**, *13*, 663–671. [CrossRef]
40. Ilhan-Ayisigi, E.; Ulucan, F.; Saygili, E.; Saglam-Metiner, P.; Gulce-Iz, S.; Yesil-Celiktas, O. Nano-vesicular formulation of propolis and cytotoxic effects in a 3D spheroid model of lung cancer. *J. Sci. Food Agric.* **2020**, *100*, 3525–3535. [CrossRef]
41. Sangboonruang, S.; Semakul, N.; Obeid, M.A.; Ruano, M.; Kitidee, K.; Anukool, U.; Pringproa, K.; Chantawannakul, P.; Ferro, V.A.; Tragoolpua, Y.; et al. Potentiality of melittin-loaded niosomal vesicles against vancomycin-intermediate *Staphylococcus aureus* and staphylococcal skin infection. *Int. J. Nanomed.* **2021**, *16*, 7639–7661. [CrossRef]
42. Gupta, A.; Aggarwal, G.; Singla, S.; Arora, R. Transfersomes: A novel vesicular carrier for enhanced transdermal delivery of sertraline: Development, characterization, and performance evaluation. *Sci. Pharm.* **2012**, *80*, 1061–1080. [CrossRef] [PubMed]
43. Opatha, S.A.T.; Titapiwatanakun, V.; Chutoprapat, R. Transfersomes: A promising nanoencapsulation technique for transdermal drug delivery. *Pharmaceutics* **2020**, *12*, 855. [CrossRef] [PubMed]
44. Obeid, M.A.; Gany, S.A.S.; Gray, A.I.; Young, L.; Igoli, J.O.; Ferro, V.A. Niosome-encapsulated balanocarpol: Compound isolation, characterisation, and cytotoxicity evaluation against human breast and ovarian cancer cell lines. *Nanotechnology* **2020**, *31*, 195101. [CrossRef] [PubMed]
45. Un, R.; Barlas, F.; Yavuz, M.; Ag Seleci, D.; Seleci, M.; Gümüş, Z.; Guler Celik, E.; Demir, B.; Can, M.; Coşkunol, H.; et al. Phyto-niosomes: In vitro assessment of the novel nanovesicles containing marigold extract. *Int. J. Polymer Mater.* **2015**, *64*, 927–937. [CrossRef]
46. Joseph, E.; Singhvi, G. Chapter 4—Multifunctional nanocrystals for cancer therapy: A potential nanocarrier. In *Nanomaterials for Drug Delivery and Therapy*, 1st ed.; Grumezescu, A.M., Ed.; William Andrew: Norwich, CT, USA; Elsevier: Amsterdam, The Netherlands, 2019; pp. 91–116. ISBN 978-0-12-816505-8.
47. Chen, S.; Hanning, S.; Falconer, J.; Locke, M.; Wen, J. Recent advances in non-ionic surfactant vesicles (niosomes): Fabrication, characterization, pharmaceutical and cosmetic applications. *Eur. J. Pharm. Biopharm.* **2019**, *144*, 18–39. [CrossRef]
48. Hao, Y.; Zhao, F.; Li, N.; Yang, Y.; Li, K. Studies on a high encapsulation of colchicine by a niosome system. *Int. J. Pharm.* **2002**, *244*, 73–80. [CrossRef]
49. Hichmah, H.N. Effect of surfactant concentration on the entrapment efficiency niosomes aqueous extract of cassava leaves (*Manihot esculenta* Crantz). *Asian J. Pharm.* **2019**, *13*, 276–281. [CrossRef]
50. Shirsand, S.; Para, M.; Nagendrakumar, D.; Kanani, K.; Keerthy, D. Formulation and evaluation of ketoconazole niosomal gel drug delivery system. *Int. J. Pharm. Investig.* **2012**, *2*, 201–207. [CrossRef]
51. Fu, Y.; Kao, W.J. Drug release kinetics and transport mechanisms of non-degradable and degradable polymeric delivery systems. *Expert. Opin. Drug Deliv.* **2010**, *7*, 429–444. [CrossRef]
52. Sukhanova, A.; Bozrova, S.; Sokolov, P.; Berestovoy, M.; Karaulov, A.; Nabiev, I. Dependence of nanoparticle toxicity on their physical and chemical properties. *Nanoscale Res. Lett.* **2018**, *13*, 44. [CrossRef]
53. Kyselova, Z. Toxicological aspects of the use of phenolic compounds in disease prevention. *Interdiscip. Toxicol.* **2011**, *4*, 173–183. [CrossRef] [PubMed]
54. Yang, Y.; He, P.Y.; Zhang, Y.; Li, N. Natural products targeting the mitochondria in cancers. *Molecules* **2020**, *26*, 92. [CrossRef]
55. Rejman, J.; Oberle, V.; Zuhorn, I.S.; Hoekstra, D. Size-dependent internalization of particles via the pathways of clathrin- and caveolae-mediated endocytosis. *Biochem. J.* **2004**, *377*, 159–169. [CrossRef] [PubMed]
56. Figueiredo Borgognoni, C.; Kim, J.H.; Zucolotto, V.; Fuchs, H.; Riehemann, K. Human macrophage responses to metal-oxide nanoparticles: A review. *Artif. Cells Nanomed. Biotechnol.* **2018**, *46*, 694–703. [CrossRef] [PubMed]
57. Ganendren, R.; Carter, E.; Sorrell, T.; Widmer, F.; Wright, L. Phospholipase B activity enhances adhesion of *Cryptococcus neoformans* to a human lung epithelial cell line. *Microbes Infect.* **2006**, *8*, 1006–1015. [CrossRef] [PubMed]
58. Cox, G.M.; McDade, H.C.; Chen, S.C.; Tucker, S.C.; Gottfredsson, M.; Wright, L.C.; Sorrell, T.C.; Leidich, S.D.; Casadevall, A.; Ghannoum, M.A.; et al. Extracellular phospholipase activity is a virulence factor for *Cryptococcus neoformans*. *Mol. Microbiol.* **2001**, *39*, 166–175. [CrossRef]
59. Ghannoum, M.A. Potential role of phospholipases in virulence and fungal pathogenesis. *Clin. Microbiol. Rev.* **2000**, *13*, 122–143. [CrossRef]
60. Kumari, P.; Arora, N.; Chatrath, A.; Gangwar, R.; Pruthi, V.; Poluri, K.M.; Prasad, R. Delineating the biofilm inhibition mechanisms of phenolic and aldehydic terpenes against *Cryptococcus neoformans*. *ACS Omega* **2019**, *4*, 17634–17648. [CrossRef]
61. Wills, E.A.; Roberts, I.S.; Del Poeta, M.; Rivera, J.; Casadevall, A.; Cox, G.M.; Perfect, J.R. Identification and characterization of the *Cryptococcus neoformans* phosphomannose isomerase-encoding gene, *Man1*, and its impact on pathogenicity. *Mol. Microbiol.* **2001**, *40*, 610–620. [CrossRef]
62. Bar-Peled, M.; Griffith, C.L.; Doering, T.L. Functional cloning and characterization of a UDP- glucuronic acid decarboxylase: The pathogenic fungus *Cryptococcus neoformans* elucidates UDP-xylose synthesis. *Proc. Natl. Acad. Sci. USA* **2001**, *98*, 12003–12009. [CrossRef]

63. Bar-Peled, M.; Griffith, C.L.; Ory, J.J.; Doering, T.L. Biosynthesis of UDP-GlcA, a key metabolite for capsular polysaccharide synthesis in the pathogenic fungus *Cryptococcus neoformans*. *Biochem. J.* **2004**, *381*, 131–136. [CrossRef] [PubMed]
64. Griffith, C.L.; Klutts, J.S.; Zhang, L.; Levery, S.B.; Doering, T.L. UDP-glucose dehydrogenase plays multiple roles in the biology of the pathogenic fungus *Cryptococcus neoformans*. *J. Biol. Chem.* **2004**, *279*, 51669–51676. [CrossRef] [PubMed]
65. Ozdal, T.; Capanoglu, E.; Altay, F. A review on protein–phenolic interactions and associated changes. *Food Res. Int.* **2013**, *51*, 954–970. [CrossRef]
66. Casadevall, A.; Pirofski, L.A. A new synthesis for antibody-mediated immunity. *Nat. Immunol.* **2011**, *13*, 21–28. [CrossRef] [PubMed]
67. Mamoon, K.; Thammasit, P.; Iadnut, A.; Kitidee, K.; Anukool, U.; Tragoolpua, Y.; Tragoolpua, K. Unveiling the properties of thai stingless bee propolis via diminishing cell wall-associated cryptococcal melanin and enhancing the fungicidal activity of macrophages. *Antibiotics* **2020**, *9*, 420. [CrossRef] [PubMed]
68. Zaragoza, O. Basic principles of the virulence of cryptococcus. *Virulence* **2019**, *10*, 490–501. [CrossRef]
69. Chayakulkeeree, M.; Johnston, S.A.; Oei, J.B.; Lev, S.; Williamson, P.R.; Wilson, C.F.; Zuo, X.; Leal, A.L.; Vainstein, M.H.; Meyer, W.; et al. *Sec14* is a specific requirement for secretion of phospholipase B1 and pathogenicity of *Cryptococcus neoformans*. *Mol. Microbiol.* **2011**, *80*, 1088–1101. [CrossRef]
70. Shukr, M.H. Novel In Situ gelling ocular inserts for voriconazole-loaded niosomes: Design, In Vitro characterisation and In Vivo evaluation of the ocular irritation and drug pharmacokinetics. *J. Microencapsul.* **2016**, *33*, 71–79. [CrossRef]
71. Cao, L.B.; Zeng, S.; Zhao, W. Highly stable pegylated poly(lactic-co-glycolic acid) (PLGA) nanoparticles for the effective delivery of docetaxel in prostate cancers. *Nanoscale Res. Lett.* **2016**, *11*, 305. [CrossRef]
72. Son, Y.J.; McConville, J.T. Development of a standardized dissolution test method for inhaled pharmaceutical formulations. *Int. J. Pharm.* **2009**, *382*, 15–22. [CrossRef]
73. Zajac, M.; Dreano, E.; Edwards, A.; Planelles, G.; Sermet-Gaudelus, I. Airway surface liquid pH regulation in airway epithelium current understandings and gaps in knowledge. *Int. J. Mol. Sci.* **2021**, *22*, 3384. [CrossRef] [PubMed]
74. Honarvari, B.; Karimifard, S.; Akhtari, N.; Mehrarya, M.; Moghaddam, Z.S.; Ansari, M.J.; Jalil, A.T.; Matencio, A.; Trotta, F.; Yeganeh, F.E.; et al. Folate-targeted curcumin-loaded niosomes for site-specific delivery in breast cancer treatment: In silico and In Vitro study. *Molecules* **2022**, *27*, 4634. [CrossRef] [PubMed]
75. Chen, S.C.; Muller, M.; Zhou, J.Z.; Wright, L.C.; Sorrell, T.C. Phospholipase activity in *Cryptococcus neoformans*: A new virulence factor? *J. Infect. Dis.* **1997**, *175*, 414–420. [CrossRef]
76. Benaducci, T.; Sardi, J.; Lourencetti, N.; Scorzoni, L.; Gullo, L.F.; Rossi, S.; Derissi, J.; Prata, M.; Almeida, A.; Mendes, G.M.J. Virulence of *Cryptococcus* sp. biofilms In Vitro and In Vivo using *Galleria mellonella* as an alternative model. *Front. Microbiol.* **2016**, *7*, 290. [CrossRef] [PubMed]
77. Fonseca, F.L.; Guimarães, A.J.; Kmetzsch, L.; Dutra, F.F.; Silva, F.D.; Taborda, C.P.; Araujo, G.D.S.; Frases, S.; Staats, C.C.; Bozza, M.T.; et al. Binding of the wheat germ lectin to *Cryptococcus neoformans* chitooligomers affects multiple mechanisms required for fungal pathogenesis. *Fungal Genet. Biol.* **2013**, *60*, 64–73. [CrossRef] [PubMed]
78. Latouche, G.N.; Sorrell, T.C.; Meyer, W. Isolation and characterisation of the phospholipase B gene of *Cryptococcus neoformans* var. *gattii*. *FEMS Yeast Res.* **2002**, *2*, 551–561. [CrossRef] [PubMed]
79. Nicola, A.M.; Casadevall, A. In Vitro measurement of phagocytosis and killing of *Cryptococcus neoformans* by macrophages. *Methods Mol. Biol.* **2012**, *844*, 189–197. [CrossRef]

Disclaimer/Publisher's Note: The statements, opinions and data contained in all publications are solely those of the individual author(s) and contributor(s) and not of MDPI and/or the editor(s). MDPI and/or the editor(s) disclaim responsibility for any injury to people or property resulting from any ideas, methods, instructions or products referred to in the content.

molecules

MDPI

Article

Therapeutic Applications of Essential Oils from Native and Cultivated Ecuadorian Plants: Cutaneous Candidiasis and Dermal Anti-Inflammatory Activity

Lilian Sosa [1,2], Lupe Carolina Espinoza [3,4], Eduardo Valarezo [3], Núria Bozal [5], Ana Calpena [4,6], María-José Fábrega [7], Laura Baldomà [8], María Rincón [9] and Mireia Mallandrich [4,6,*]

1 Microbiological Research Institute (IIM), National Autonomous University of Honduras (UNAH), Tegucigalpa 11101, Honduras; lilian.sosa@unah.edu.hn
2 Research Institute of Applied Sciences and Technology, National Autonomous University of Honduras (UNAH), Tegucigalpa 11101, Honduras
3 Departamento de Química, Universidad Técnica Particular de Loja, Loja 1101608, Ecuador; lcespinoza@utpl.edu.ec (L.C.E.); bevalarezo@utpl.edu.ec (E.V.)
4 Institut de Nanociència i Nanotecnologia (IN2UB), Universitat de Barcelona (UB), 08028 Barcelona, Spain; anacalpena@ub.edu
5 Departament de Biologia, Sanitat i Medi Ambient, Facultat de Farmàcia i Ciències de l'Alimentació, Universitat de Barcelona (UB), 08028 Barcelona, Spain; nuriabozaldefebrer@ub.edu
6 Departament Farmàcia, Tecnologia Farmacèutica, i Físicoquímica, Facultat de Farmàcia i Ciències de l'Alimentació, Universitat de Barcelona (UB), 08028 Barcelona, Spain
7 Department of Experimental and Health Sciences, Parc of Biomedical Research of Barcelona, Pompeu Fabra University, 08003 Barcelona, Spain; mjfabrega.f@gmail.com
8 Departament de Bioquímica i Fisiologia, Facultat de Farmàcia i Ciències de l'Alimentació, Universitat de Barcelona (UB), 08028 Barcelona, Spain; lbaldoma@ub.edu
9 Departament de Ciència de Materials i Química Física, Facultat de Química, Universitat de Barcelona (UB), 08028 Barcelona, Spain; m.rincon@ub.edu
* Correspondence: mireia.mallandrich@ub.edu

Citation: Sosa, L.; Espinoza, L.C.; Valarezo, E.; Bozal, N.; Calpena, A.; Fábrega, M.-J.; Baldomà, L.; Rincón, M.; Mallandrich, M. Therapeutic Applications of Essential Oils from Native and Cultivated Ecuadorian Plants: Cutaneous Candidiasis and Dermal Anti-Inflammatory Activity. *Molecules* **2023**, *28*, 5903. https://doi.org/10.3390/molecules28155903

Academic Editors: Arunaksharan Narayanankutty, Ademola C. Famurewa and Eliza Oprea

Received: 15 July 2023
Revised: 28 July 2023
Accepted: 29 July 2023
Published: 5 August 2023

Copyright: © 2023 by the authors. Licensee MDPI, Basel, Switzerland. This article is an open access article distributed under the terms and conditions of the Creative Commons Attribution (CC BY) license (https://creativecommons.org/licenses/by/4.0/).

Abstract: Essential oils are a complex mixture of aromatic substances whose pharmacological actions, including antimicrobial, antioxidant, anticancer, and anti-inflammatory activities, have been widely reported. This study aimed to evaluate the anti-*Candida* and dermal anti-inflammatory activity of essential oils from native and cultivated Ecuadorian plants. Essential oils from *Bursera graveolens*, *Dacryodes peruviana*, *Mespilodaphne quixos*, and *Melaleuca armillaris* were isolated by hydrodistillation and were characterized physically and chemically. Its tolerance was analyzed by in vitro and in vivo studies. The antifungal activity was studied against *Candida albicans*, *Candida glabrata*, and *Candida parapsilosis*, whereas the anti-inflammatory effect was evaluated by a mouse ear edema model. The main compounds were limonene, α-phellandrene, (*E*)-methyl cinnamate, and 1,8-cineole, respectively. All essential oils showed high tolerability for skin application, antifungal activity against the three *Candida* strains, and anti-inflammatory efficacy by decreasing edema and overexpression of pro-inflammatory cytokines. *Dacryodes peruviana* essential oil showed the highest antifungal activity. On the other hand, *Dacryodes peruviana* and *Melaleuca armillaris* showed the greatest anti-inflammatory potential, decreasing edema by 53.3% and 65.25%, respectively, and inhibiting the overexpression of TNF-α, IL-8, IL-17A, and IL-23. The results suggest that these essential oils could be used as alternative therapies in the treatment of both cutaneous candidiasis and dermal inflammation.

Keywords: *Candida albicans*; *Candida glabrata*; *Candida parapsilosis*; skin inflammation; essential oil; *Bursera graveolens*; *Dacryodes peruviana*; *Mespilodaphne quixos*; *Melaleuca armillaris*

1. Introduction

Fungal infections represent a public health problem affecting almost one billion people worldwide. These infections appear in different anatomical sites such as the skin, nails, scalp, and vagina [1]. The skin is the largest organ in the body and can suffer from fungal infections such as candidiasis. Cutaneous candidiasis is a superficial form of mycosis caused by the proliferation of *Candida* spp. fungi in the skin, especially in intertriginous areas causing itching, erosion, inflammation, and pustules [2,3].

Candida albicans is the species that produces most of the cases of candidiasis. However, in America and Europe (except for Spain), the cases have increased due to the appearance of *Candida glabrata*, while in South America, Japan, and Spain, *Candida parapsilosis* has been the main cause of candidiasis [4]. The clinical importance of non-albicans strains lies in infections caused by strains such as *Candida glabrata*. The incidence of this strain is high in patients with acquired immunodeficiency syndrome (AIDS), cancer, and diabetes and in those who use medical devices. Likewise, it has been reported as a fungus resistant to azoles (first-line treatment), with a faster propagation than *Candida albicans* and a higher mortality rate [5,6]. In the case of infections caused by *Candida parapsilosis*, this strain is responsible for a third of neonatal *Candida* infections reaching a mortality rate of approximately 10%. In addition, patients who require prolonged use of a central venous catheter are also at increased risk of infection due to the innate ability of this fungus to adhere to prosthetic surfaces and implanted medical devices due to the biofilm they produce. The structure of this biofilm shows high variability in different clinical isolates of *Candida parapsilosis* [7].

The treatments used for dermal candidiasis consist of the use of topical azoles that inhibit the biosynthesis of the ergosterol component of the cell membrane and some macrolides such as nystatin and amphotericin B, which bind to the ergosterol of the membrane to induce porosity and cause cell death through leakage of cytosolic components [8]. Likewise, drugs such as caspofungin have been reported in the literature as possible therapeutic alternatives. This drug inhibits fungal cell membrane formation by disrupting the synthesis of the structural component 1,3-β-d-glucan [9]. Despite advances in antifungal treatments, many patients frequently experience complications related to hepatoxicity and increased fungal resistance to conventional drugs [10].

Inflammation is the response of an organism to tissue damage caused by a foreign agent that can be physical, chemical, or biological [11]. When an inflammatory process is generated, macrophages are activated, which produce pro-inflammatory mediators such as prostaglandins, cyclooxygenases, reactive oxygen species, and cytokines [12]. Tumor necrosis factor alpha (TNF-α), interleukin (IL)-8, IL-17A, and IL-23 are cytokines secreted rapidly in response to an inflammatory process. An inappropriate or excessive inflammatory response in the skin can trigger chronic inflammation causing skin disorders such as dermatitis, psoriasis, and rosacea [13]. TNF-α is produced during acute inflammation by macrophages/monocytes, and it induces the expression of pro-inflammatory mediators such as IL-6, IL-8, and IL-1β [14]. IL-8 is a member of the chemokine family that attracts neutrophils, basophils, and T-cells to sites of inflammation and is released from different cells, including monocytes, macrophages, neutrophils, fibroblasts, endothelial cells, and keratinocytes [15]. IL-17A is produced by a subset of activated CD4 T cells, and it is associated with allergic processes and autoimmune diseases including psoriasis, and turn, increases the levels of IL-6 and IL-8 [16]. IL-23 is a dimeric cytokine that promotes the development and differentiation of effector Th17 cells, and thus, it has been implicated in chronic immune-inflammatory disorders such as arthritis, colitis, gastritis, and psoriasis [17].

Nonsteroidal anti-inflammatory drugs (NSAIDs), including diclofenac, fenoprofen, ibuprofen, ketoprofen, ketorolac, and naproxen, are one of the most widely used groups of drugs for treating inflammation, pain, and fever. However, these drugs have been associated with adverse gastrointestinal, renal, hepatic, and skin reactions [18,19].

In the search for effective compounds and alternative therapies to treat cutaneous candidiasis and dermal inflammation, natural products, especially essential oils, stand out as promising candidates in antifungal and anti-inflammatory treatments [20].

An essential oil (EO) is a complex mixture of aromatic substances isolated from diverse types of aromatic plants. Various scientific studies have reported numerous pharmacological actions for essential oils, including antimicrobial, antioxidant, anticancer, and anti-inflammatory properties [21]. Essential oils are used in the pharmaceutical, cosmetic, and food industries [22,23].

Burseraceae is a family of 18 genera and about 650 species [24]. A total of 40 species of this family have been reported in the megadiverse country of Ecuador [25]. Some studies show that the fruits of Burseraceae species contain a high yield (2–5%) of essential oil [26]. *Bursera graveolens* (Kunth) Triana and Planch is a shrub or tree commonly known as "palo santo" that grows in dry forests of tropical America and on the Pacific coast of South America, especially in Colombia, Costa Rica, El Salvador, Ecuador, Guatemala, Honduras, Mexico, Nicaragua, and Peru. In Ecuador, *B. graveolens* can mainly be found in the provinces of Galapagos, Guayas, Imbabura, Loja, and Manabí [25]. Essential oil of *B. graveolens* can be found in its leaves, trunk, and fruits and has shown acaricidal, antimicrobial, antioxidant, and repellent activities [26,27].

Dacryodes peruviana (Loes.) H.J. Lam (Burseraceae) is a tree of the Burseraceae family known as "copal", which grows in the humid forests of Colombia, Peru, Ecuador, and Venezuela [28]. In Ecuador, *D. peruviana* has the status of native and is widely distributed in the Amazonian and Andean Ecuadorian regions between 0 and 2500 m above sea level [25]. The essential oil of copal fruits has shown antimicrobial and repellent activity [29].

Lauraceae comprises approximately 2978 accepted species in nearly 68 genera worldwide [30]. Within this family, the *Mespilodaphne* is a small genus with 15 individuals [31]. *Mespilodaphne quixos* (Lam.) Rohwer (Syn. *Ocotea quixos* (Lam.,) Kosterm, accepted name *Mespilodaphne quixos* and validated name *Ocotea quixos*) [32] is an Ecuadorian native and cultivated aromatic species, widely distributed in the Andean and Amazonian regions between 0 and 1000 m above sea level [33]. The species is commonly known as "ishpingo" or "canela" because it smells similar to cinnamon ("canela" in Spanish). Essential oil of this species has been isolated from its leaves, trunk, fruits, and fruit calices and has been shown to have antimicrobial, anti-inflammatory, antioxidant, antithrombotic, and larvicidal activities [34–38].

Myrtaceae is a family of 129 genera and 6233 species distributed in tropical regions worldwide. *Myrtaceae* is a family of evergreen trees or shrubs rich in essential oils [39]. *Melaleuca armillaris* (Sol. ex Gaertn.) Sm. is a species of the *Myrtaceae* family cultivated in Ecuador known by the common name of bracelet honey myrtle or "árbol de papel". The essential oil of the leaves of *M. armillaris* has shown antimalarial, antioxidant, anticancer, antimicrobial, and phytotoxic activities [40,41].

Based on these remarkable findings, the purpose of this study was to evaluate the therapeutic applications of essential oils isolated from native and cultivated plants of Ecuador in the treatment of cutaneous candidiasis and dermal inflammation, as well as analyze their skin tolerance, to provide effective and safe alternatives in the treatment of these two pathologies.

2. Results

2.1. Isolation and Physical Properties of Essential Oil

The yield in essential oil and physical properties density, refractive index, and optical activity for the four essential oils used are shown in Table 1. The physical properties of an essential oil are used as a quality parameter. The yield of the fruits, in general, is higher than that of the leaves of a species. According to the classification made by Molares et al. in 2009, the yield of *Melaleuca armillaris* is intermediate (values between 5 mL/kg and 10 mL/kg), and the yields of the other species are high (values greater than 10 mL/kg) [42]. The yield, together with the biomass availability of a species, provides an idea of the amount of essential oil available to scale a project, which makes the essential oil of these species suitable for industrial applications.

Table 1. Yield and physical properties of the essential oils from *Bursera graveolens* (BGR), *Dacryodes peruviana* (DPE), *Mespilodaphne quixos* (MQU), and *Melaleuca armillaris* (MAR).

	BGR		DPE		MQU		MAR	
	Mean	SD	Mean	SD	Mean	SD	Mean	SD
Yield (mL/kg)	31	2	45	3	15	1	9	1
Density, ρ (g/cm^3)	0.8385	0.0010	0.8456	0.0023	0.8758	0.0013	0.9053	0.0045
Refractive index, n20	1.4760	0.0011	1.4751	0.0002	1.4790	0.0012	1.4641	0.0023
Specific rotation, [α] (°)	47.7	1.1	12.2	0.7	11.2	0.2	4.3	1.1

2.2. Essential Oil Compounds Identification

The identification of volatile compounds contained in *Bursera graveolens, Dacryodes peruviana, Ocotea quixos* (Accepted name *Mespilodaphne quixos*), and *Melaleuca armillaris* essential oils was carried out using gas chromatography equipped with a flame ionization detector and gas chromatography coupled to a mass spectrometer detector using capillary nonpolar column DB-5Ms. The compounds were classified into six groups: monoterpene hydrocarbons (MH), oxygenated monoterpene (OM), sesquiterpene hydrocarbons (SH), oxygenated sesquiterpene (OS), phenylpropanoids, and other compounds (OC) (Table 2). All essential oil compounds from *Dacryodes peruviana* and most essential oil compounds from *Bursera graveolens* belong to the MH group. The main compounds were limonene, α-phellandrene, (*E*)-methyl cinnamate, and 1,8-cineole in the essential oils of *Bursera graveolens*, *Dacryodes peruviana*, *Mespilodaphne quixos*, and *Melaleuca armillaris*, respectively.

Table 2. Majority compounds (>1%) of the essential oils from *Bursera graveolens* (BGR), *Dacryodes peruviana* (DPE), *Mespilodaphne quixos* (MQU), and *Melaleuca armillaris* (MAR).

Compounds	RIC	RIR	BGR		DPE		MQU		MAR		Type
			%	SD	%	SD	%	SD	%	SD	
α-Thujene	926	924	-		1.9	0.1	-		-		MH
α-Pinene	932	932	-		8.3	0.3	1.6	0.1	3.6	0.1	MH
Sabinene	969	969	-		1.4	0.4	-		-		MH
β-Pinene	973	974	-		2.6	0.9	1.3	0.1	1.4	0.1	MH
Myrcene	986	988	1.1	0.4	-		-		1.3	0.1	MH
α-Phellandrene	1005	1002	35.9	1.3	50.3	3.3	-		-		MH
p-Cymene	1021	1020	-		3.1	0.8	-		-		MH
Limonene	1025	1024	49.4	2.2	23.0	1.5	-		-		MH
1,8-Cineole	1025	1026	-		-		1.3	0.1	83.4	2.5	OM
Terpinolene	1082	1086	-		5.2	0.9	-		-		MH
Menthofuran	1159	1159	6.6	1.2	-		-		-		OM
α-Terpineol	1187	1186	-		-		-		8.0	0.1	OM
(*E*)-Cinnamaldehyde	1268	1267	-		-		10.0	1.4	-		PP
α-Copaene	1372	1374	-		-		4.5	0.9	-		SH
(*E*)-Methyl cinnamate	1376	1376	-		-		19.3	1.3	-		PP
(*E*)-Caryophyllene	1415	1417	-		-		15.8	1.8	-		SH
6,9-Guaiadiene	1442	1442	-		-		4.5	0.5	-		SH
(*E*)-Cinnamyl acetate	1445	1443	-		-		12.5	0.7	-		PP
Germacrene D	1476	1480	1.5	0.1	-		-		-		SH
β-Selinene	1489	1489	-		-		5.8	0.5	-		SH
Bicyclogermacrene	1496	1500	-		-		4.2	0.4	-		SH
Anisyl propanoate	1510	1511	-		-		2.7	0.4	-		OC
7-*epi*-α-Selinene	1520	1520	-		-		2.5	0.3	-		SH
(*E*)-γ-Bisabolene	1527	1529	-		-		3.1	0.3	-		SH
Caryophyllene oxide	1580	1582	-		-		2.5	0.4	-		OS

Table 2. *Cont.*

Compounds	RIC	RIR	BGR		DPE		MQU		MAR		Type
			%	SD	%	SD	%	SD	%	SD	
Monoterpene hydrocarbons (MH)			86.4		95.8		3.0		6.2		
Oxygenated monoterpene (OM)			6.6		-		1.3		91.4		
Sesquiterpene hydrocarbons (SH)			1.5		-		40.4		-		
Oxygenated sesquiterpene (OS)			-		-		2.5		-		
Phenylpropanoids (PP)			-		-		41.8				
Other compounds (OC)			-		-		2.7		-		
Total identified			94.6		95.8		91.7		97.6		

RIC: calculated retention rates; RIR: reference retention indices, Adams, R.P. (2007) [43]; %: mean percentage obtained from 9 replicates; SD: standard deviation.

2.3. *Cytotoxicity Studies by Methylthiazolyldiphenyl-Tetrazolium Bromide (MTT) Method*

Cell viability of human keratinocytes HaCaT exposed at the different concentrations of the essential oils studied was evaluated by MTT cytotoxicity assay. After 24 h of incubation, it was observed that the essential oils from *Bursera graveolens*, *Dacryodes peruviana*, and *Melaleuca armillaris* at the assayed concentrations from 500 to 5000 µg/mL did not affect cell viability, which was greater to 80% whereas the essential oil from *Mespilodaphne quixos* (MQU) showed cell viability greater to 80% from 500 to 1000 µg/mL (Figure 1). Therefore, these results suggest that the essential oils do not generate toxicity in the keratinocytes at these concentrations.

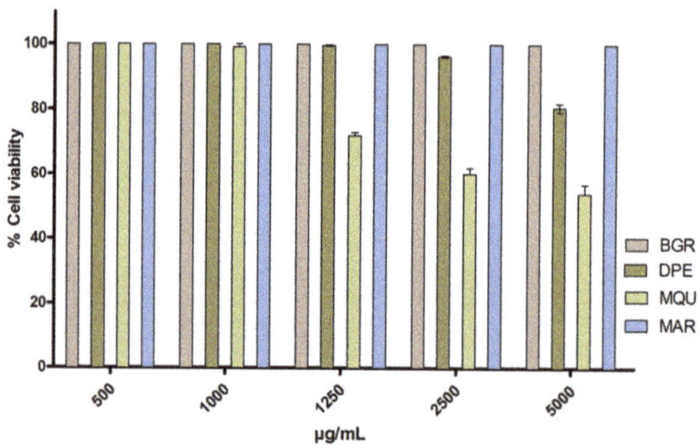

Figure 1. Cytotoxicity of HaCaT cell line exposed to essential oils from *Bursera graveolens* (BGR), *Dacryodes peruviana* (DPE), *Mespilodaphne quixos* (MQU), and *Melaleuca armillaris* (MAR).

2.4. *In Vivo Tolerance Studies by Evaluation of Transepidermal Water Loss (TEWL)*

Figure 2A shows that *Bursera graveolens* essential oil did not cause significant differences in TEWL value after topical application on the ventral area of the forearm of the volunteers compared with the basal state, whereas *Dacryodes peruviana*, *Mespilodaphne quixos*, and *Melaleuca armillaris* essential oils significantly reduced the TEWL values after 10 min of topical application (Figure 2B–D). However, a tendency to return to the basal states was observed after 2 h of the experiment.

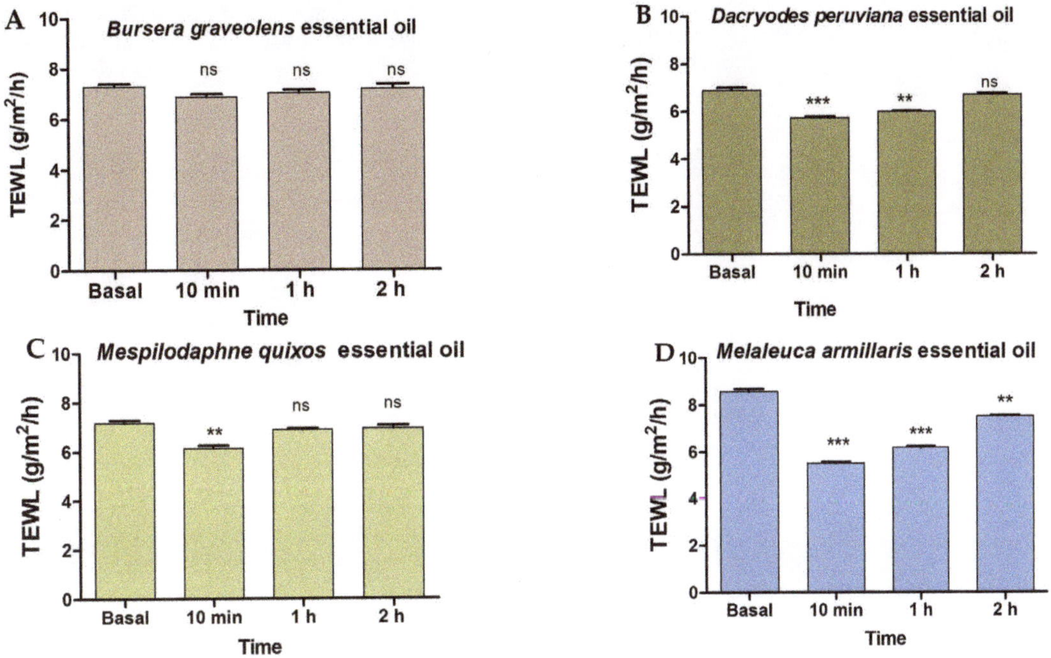

Figure 2. Tolerance studies by transepidermal water loss (TEWL): (**A**) *Bursera graveolens* essential oil; (**B**) *Dacryodes peruviana* essential oil; (**C**) *Mespilodaphne quixos essential oil*; (**D**) *Melaleuca armillaris* essential oil. Statistically significant differences: ** = $p < 0.01$; *** = $p < 0.001$; ns = not significant.

2.5. Efficacy Studies: Antifungal Activity

The Minimum Inhibitory Concentrations (MICs) of essential oil from *Bursera graveolens, Dacryodes peruviana, Mespilodaphne quixos*, and *Melaleuca armillaris* are shown in Table 3. All the essential oils presented antifungal activity against the three *Candida* strains. *Dacryodes peruviana* essential oil showed the lowest MIC value, and therefore it has greater antifungal activity. On the contrary, the essential oil with minor efficacy was *Melaleuca armillaris* essential oil.

Table 3. Antifungal activity of essential oils from *Bursera graveolens* (BGR), *Dacryodes peruviana* (DPE), *Mespilodaphne quixos* (MQU), and *Melaleuca armillaris* (MAR).

	MIC (μg/mL)		
	Candida albicans	*Candida glabrata*	*Candida parapsilosis*
Amphotericin B	0.15	0.60	0.30
BGR	524.06	262.03	524.06
DPE	132.13	32.98	65.96
MQU	273.69	273.69	136.84
MAR	565.81	565.81	565.81
Blank	-	-	-

2.6. In Vivo Anti-Inflammatory Activity

2.6.1. Arachidonic Acid (AA)-Induced Mouse Ear Edema

Topical application of AA on the mice's ears caused immediate erythema followed by the development of edema represented by the increase in the ear thickness as a result of the inflammatory process. However, this parameter decreased notably after topical treatment with the different essential oils tested mainly with *Dacryodes peruviana* (DPE) and *Melaleuca armillaris* (MAR), which decreased edema by 53.3% and 65.25%, respectively, although without reaching the efficacy of the reference drug (ibuprofen) that inhibits edema by 81.25%. The blank solution did not cause changes in the inflammatory process induced by AA (Figure 3).

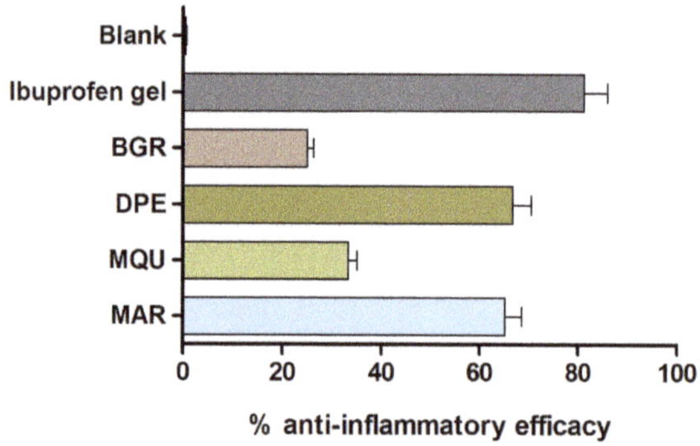

Figure 3. Anti-inflammatory efficacy after AA-induced mouse ear edema. Mean ± SD (n = 3). Ibuprofen gel = treatment with reference drug; Blank = Transcutol P:water; treatment with *Bursera graveolens* (BGR), *Dacryodes peruviana* (DPE), *Mespilodaphne quixos* (MQU), and *Melaleuca armillaris* (MAR).

2.6.2. Pro-Inflammatory Cytokines Determination

The positive control showed a significant increase in the expression of pro-inflammatory cytokines TNF-α, IL-8, IL-17A, and IL-23 by the inflammation triggered by skin application of AA. When compared with the positive control, topical treatment with *Bursera graveolens* essential oil significantly decreased the expression of IL-8, IL-17A, and IL-23, whereas *Mespilodaphne quixos* essential oil reduced the expression of IL-17A and IL-23. Essential oils from *Dacryodes peruviana* (DPE) and *Melaleuca armillaris* (MAR) significantly reduced the expression of all pro-inflammatory cytokines studied TNF-α, IL-8, IL-17A, and IL-23. These results suggest that these essential oils could regulate the inflammatory processes and serve as a treatment or adjuvant in anti-inflammatory therapies (Figure 4).

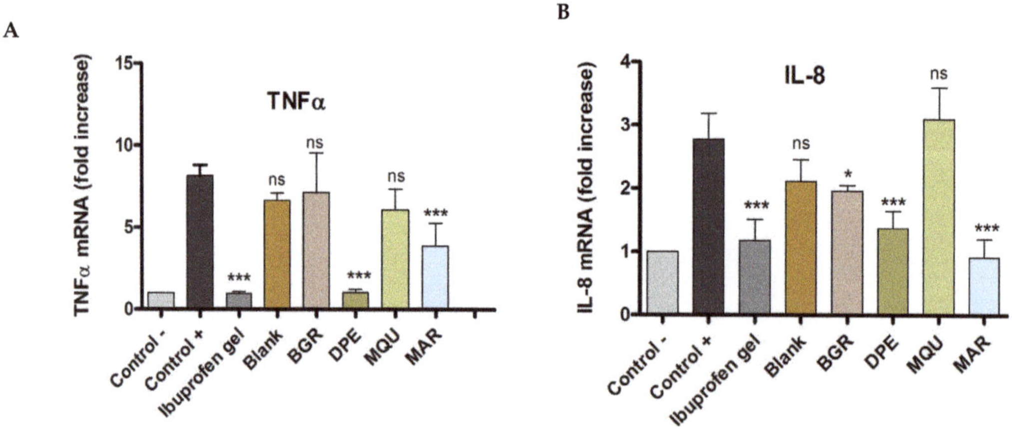

Figure 4. *Cont.*

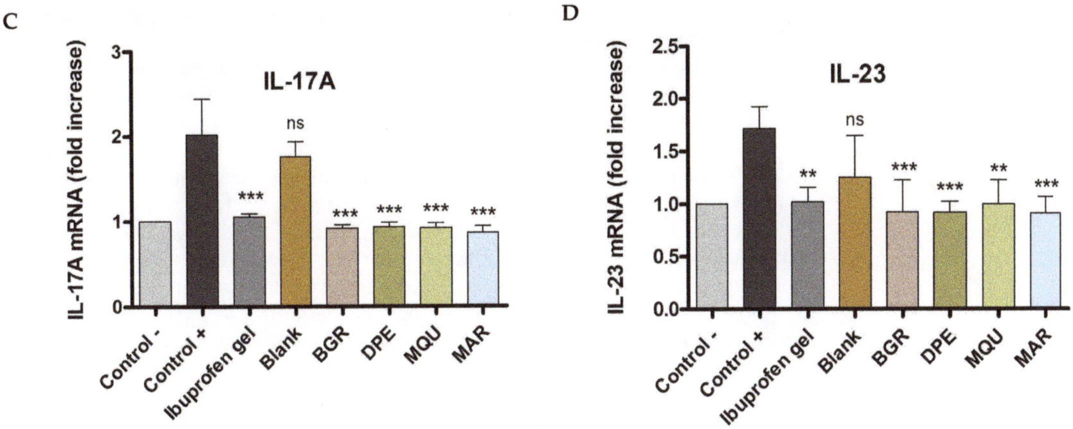

Figure 4. Relative expression of pro-inflammatory cytokines measured by quantitative reverse transcription polymerase chain reaction: (**A**) TNFα = Tumor necrosis factor-alpha; (**B**) IL–8 = interleukin–8; (**C**) IL–17A = interleukin–17A; (**D**) IL-23 = interleukin–23. Control − = negative control; Control + = positive control; Ibuprofen gel = treatment with reference drug; Blank = Transcutol P:water; treatment with *Bursera graveolens* (BGR), *Dacryodes peruviana* (DPE), *Mespilodaphne quixos* (MQU) and *Melaleuca armillaris* (MAR). Statistically significant differences: * = $p < 0.05$; ** = $p < 0.01$; *** = $p < 0.001$; ns = not significant.

3. Discussion

Essential oils are a complex mixture of aromatic substances, which have shown several pharmacological properties thanks to their various components [44,45]. The uses attributed to essential oils include antiseptic, irritant, spasmolytic, sedative, anti-inflammatory, and antimicrobial potential [2,44,46]. The antimicrobial activity is of great interest because of the phenomenon of resistance to conventional antimicrobial drugs [47]. Specifically, in the present study, the essential oils obtained from four different plants: *Bursera graveolens*, *Dacryodes peruviana*, *Mespilodaphne quixos*, and *Melaleuca armillaris* were studied physically, chemically, and microbiologically to establish their possible pharmacological use in the treatment of both cutaneous candidiasis and dermal inflammation. The yields of essential oils obtained from these species are suitable for various industrial applications, including pharmacological use. High yields were observed for the essential oils of *Bursera graveolens*, *Dacryodes peruviana*, and *Mespilodaphne quixos* with values greater than 10 mL/kg. In comparison, the yield for the essential oil of *Melaleuca armillaris* was intermediate, with values between 5 mL/kg and 10 mL/kg [42].

Chemical characterization showed that the main compounds for the essential oils of *Bursera graveolens*, *Dacryodes peruviana*, *Mespilodaphne quixos*, and *Melaleuca armillaris* are limonene, α-phellandrene, (*E*)-methyl cinnamate, and 1,8-cineole, respectively. In previous studies carried out on the essential oils of these species, similar compounds were found, although with different percentages. For example, limonene was the main compound in essential oil isolated from the fruits of *Bursera graveolens*, with percentages from 43.6% [26] to 49.89% [27]. In the essential oil of fruits *of Dacryodes peruviana*, α-phellandrene (52.4%) and limonene (22.5%) were determined as the major compounds [46]. In essential oil extracted from the leaves of *Mespilodaphne quixos*, the main components were β-caryophyllene (15.1%) and cinnamyl acetate (11.4%) [48]; however, in a study on the variability of the chemical composition of this essential oil with the amount of shade, type of soil and altitude, it was determined that there is a great variation in the percentages of the major compounds (*E*)-cinnamyl acetate (5.96–41.65%), (*E*)-methyl cinnamate (0.38–37.91%), and trans-caryophyllene (8.77–37.02%) [49]. In the essential oil extracted from leaves of *Melaleuca armillaris*, the 1,8-cineole was determined as the main component with percentages from 68.9% [50] to 85.8% [40].

To evaluate the tolerability of these essential oils for topical application on the skin, in vitro and in vivo were performed. In vitro models allowed for screening the toxicity of substances prior to in vivo assessment. Cell lines are highly used for toxicity screening in a living system because they are generally easy to cultivate, fast to grow, and sensitive to toxic irritation [51]. In this study, the results confirmed that the essential oils at the tested dilutions do not induce relevant cytotoxic effects, and thus they have high biocompatibility with human keratinocytes. This result was confirmed by further studies in volunteers by evaluation of the transepidermal water loss (TEWL) parameter, which allows for the evaluation of the skin barrier's integrity after exposure to physical or chemical agents. TEWL is a biomechanical parameter of the skin that indicates water loss across the stratum corneum. Its determination is a noninvasive in vivo valuable method in dermatology to detect disturbances of the skin barrier in early stages, even before they become visible. In damaged skin, TEWL increases, and subsequently, the hydration declines, causing cracks and fissures in the stratum corneum [51]. As a general result, the essential oils studied did not cause skin irritation in the volunteers indicating biocompatibility and suitability for human use. *Bursera graveolens* essential oil did not cause significant differences in TEWL value after skin application compared with the basal state, whereas a significant reduction in TEWL after 10 min of skin application of *Dacryodes peruviana*, *Mespilodaphne quixos*, and *Melaleuca armillaris* essential oils was observed but with a tendency to return to the basal states after 2 h of the experiment. These results suggest an enhancement of the hydration level of stratum corneum, probably due to the occlusive effect of these essential oils.

Regarding antifungal activity, all the essential oils were effective against *Candida albicans*, *Candida glabrata*, and *Candida parapsilosis*. Essential oils from *Bursera morelensis* belonging to the same family as *Bursera graveolens*, have shown an antifungal effect, specifically on *Candida albicans* [52]. In addition, previous studies have reported the antifungal activity of *Bursera graveolens* essential oil against *Candida albicans* [53]. Likewise, it has been shown that this essential oil could enhance the antifungal effect of drugs such as amphotericin B, improving its activity against various strains of *Candida* [2]. *Dacryodes peruviana* essential oil showed the lowest MIC value, and therefore it has greater antifungal activity. The antifungal activity of *Bursera graveolens* and *Dacryodes peruviana* essential oil could be due to its rich content in limonene and α-phellandrene. Specifically, *Bursera graveolens* essential oil showed the highest content in limonene ($49.4 \pm 2.2\%$) and *Dacryodes peruviana* essential oil showed the highest content of α-phellandrene ($50.3 \pm 3.3\%$). Studies have shown that limonene exhibits various biological effects, including excellent anti-*Candida* activity [54]. It is considered that the antifungal activity of this compound against *Candida* could be due to failure of ion transshipment and ATP generation in the membrane; disruption of intracellular ion homeostasis; and extensive leakage of intracellular substance caused by potential, permeability, and integrity of the cell membrane. In addition, limonene also leads to suppression or intracellular metabolic disorders in *Candida* [55,56]. On the other hand, the activity exerted by α-phellandrene could be because this molecule disrupts the integrity of the fungal cell membrane, causing leakage of potassium ions and cell constituents, which would trigger an increase in total lipid content, extracellular pH, and membrane permeability, causing the death of the fungus [57]. Additionally, it has been shown that the biotransformation of α-phellandrene to 5-p-menthene-1,2-diol would generate antifungal activity against *Candida* [58]. The effect of *Dacryodes peruviana* essential oil against candida is interesting since its antifungal effect has been demonstrated, but only against *Trichophyton rubrum* and *Trichophyton mentagrophytes* [29].

The antifungal effect of *Mespilodaphne quixos* essential oil has been reported in previous studies against *Candida albicans* and *Saccharomyces cerevisiae* [59]. This efficacy could be due to the synergy of the components of this essential oil, as well as its high content of (*E*)-methyl cinnamate ($19.3 \pm 1.3\%$) and trans-caryophyllene ($15.8 \pm 1.8\%$). Both components have shown an effect on *Candida* species [35,60]. However, their mechanism of action is not fully elucidated, and more studies are necessary to establish the antifungal mechanism exerted against these pathogenic fungi [61]. Finally, *Melaleuca armillaris* essential oil showed higher MIC values than those found for *Dacryodes peruviana* and *Mespilodaphne quixos* essential oil and similar to those found for *Bursera graveolens* essential oil. However, an antifungal effect was observed against three *Candida* strains tested. Previous studies have reported the anti-*Candida* effect of essential oils obtained from various species of *Melaleuca* spp. [62]. The main component found in *Melaleuca armillaris* essential oil was 1,8-cineole (83.4 $\pm$ 2.5%). This component has demonstrated its antifungal effect against *Candida albicans*, either alone or in combination with other drugs [63–65]. More in-depth studies are necessary to determine the antifungal mechanism of action of this essential oil against *Candida*. Together, these results suggest that the four essential oils studied could be used as active antifungal components or enhancers of conventional drugs in treating cutaneous candidiasis.

The anti-inflammatory potential of the studied essential oils was evaluated by an arachidonic acid (AA)-induced edema model in mouse ear. Previous studies have documented the inflammatory response of mouse ears to topical AA, and research indicates that this reaction occurs as a result of AA metabolites being produced through both the cyclooxygenase and lipoxygenase pathways. Arachidonic acid can trigger various events of an inflammatory response, such as an increase in skin thickness and the release of several inflammatory mediators [66]. An increase in skin thickness is among the initial events observed during inflammation, suggesting the presence of edema, epidermal hyperplasia, and increased vascular permeability [46]. In this study, topical applications of AA markedly increased the thickness of the ear skin of mice with respect to the basal state. Conversely, topical treatment with essential oils from *Bursera graveolens*, *Dacryodes peruviana*, *Mespilodaphne quixos*, and *Melaleuca armillaris* reduced this parameter by 25%, 53.3%, 33.42%, and 65.25%, respectively (Figure 3). These results were consistent with the biochemical studies, in which several pro-inflammatory cytokines, including TNF-α, IL-8, IL-17A, and IL-23, were evaluated by Quantitative reverse transcription polymerase chain reaction (RT-qPCR). An increase in the expression of cytokines TNF-α, IL-8, IL-17A, and IL-23 was observed in the positive control. However, this inflammatory response was counteracted by topical treatment of the studied essential oils, mainly *Dacryodes peruviana* and *Melaleuca armillaris*, which caused a decrease in the production of all studied cytokines (Figure 4A–D). On the other hand, *Bursera graveolens* essential oil reduced the expression of IL-8, IL-17A, and IL-23 (Figure 4B–D), whereas *Mespilodaphne quixos* essential oil reduced the expression of IL-17A and IL-23 (Figure 4C,D).

The anti-inflammatory activity exhibited that these essential oils may be due to the major compounds present in them. Specifically, limonene, α-phellandrene, (*E*)-methyl cinnamate, and 1,8-cineole have been reported in previous studies for their role in regulating the inflammatory process. Limonene has been suggested as a potential inductor for the production of the anti-inflammatory cytokine IL-10 and reduction in the expression of TNF-α, IL-1β, and IL-6 [67,68]. Studies have demonstrated the anti-inflammatory effects of limonene in colitis, dermatitis, and lung inflammation caused by allergies [69,70]. α-phellandrene inhibits neutrophil migration toward the inflamed area and reduces cytokines expression induced by TNF-α. Some studies suggest the benefit of α-Phellandrene in inflammatory diseases, such as rheumatoid arthritis, osteoarthritis, and allergic conditions [71]. Regarding methyl cinnamate, reports of its anti-inflammatory activity are scarce, although potential anti-inflammatory applications against periodontal disease and related systemic conditions have been suggested [72]. Finally, published research has documented that the monoterpene 1,8-Cineol reduces the expression of cytokines such as IL-4 and IL-6 in bronchial epithelial cells, as well as influences immune cells, where decreased expression in TNF-α, IL-1β, and IL-6 from monocytes, and also decreased the expression of IL-4 and IL-5 from lymphocytes. Even in acne, 1,8-Cineol-containing leaf extracts suppress the expression of IL-1β and IL-6 [73]. The large number of scientific publications that support the anti-inflammatory potential of α-phellandrene and 1,8-Cineol in the regulation of the inflammatory process and their promising application in the treatment of different inflammatory diseases agree with the results obtained in this investigation, in which the two essential oils that showed greater efficacy correspond to the species *Dacryodes peruviana* and *Melaleuca armillaris*, which are mainly composed of these two components. Therefore, these results suggest that the studied essential oils could be used as alternative therapy or adjuvant to increase the efficacy of conventional therapies in the treatment of dermal inflammation.

4. Materials and Methods

4.1. Materials

Helium was purchased from INDURA (Quito, Ecuador). The standard aliphatic hydrocarbons (C_9–C_{25}) were purchased from ChemService (West Chester, PA, USA). Anhydrous sodium sulfate and dichloromethane (DCM) were purchased from Sigma-Aldrich (San Luis, MO, USA). All chemicals were of analytical grade and used without further purification. Amphotericin B was obtained from Acofarma (Barcelona, Spain), RPMI-1640 medium, MOPS, glucose, and chloramphenicol were obtained from Sigma-Aldrish (Darmstadt, Germany). Three yeast species were used for the assay: *Candida albicans* ATCC 10231, *Candida glabrata* ATCC 66032, and *Candida parapsilosis* ATCC 22019.

4.2. Plant Material

The information on the collection of the four species is shown in Table 4. The species *Mespilodaphne quixos* and *Melaleuca armillaris* were collected in the vegetative stage. Three samples were collected. After being collected, the plant material was stored and transferred in airtight plastic containers (without refrigeration). The botanical specimens were identified by Dr. Nixón Cumbicus at the herbarium of the Universidad Técnica Particular de Loja (HUTPL). A voucher specimen is preserved in the HUTPL.

Table 4. Location and collection conditions of plant species.

Species	Plant Part	Ambient Conditions		Parish	Canton	Province	Coordinates		Altitude (m a.s.l.)
		T (°C)	P (atm)				Latitude	Longitude	
Bursera graveolens	Fruits	26	0.98	Garza Real	Zapotillo	Loja	4°19′13″ S	80°17′58″ W	150
Dacryodes peruviana	Fruits	25	0.89	La Paz	Yacuambi	Zamora Chinchipe	3°40′13″ S	78°54′21″ W	1025
Mespilodaphne quixos	Leaves	24	0.91	Pano	Tena	Napo	1°01′12″ S	77°51′57″ W	650
Melaleuca armillaris	Leaves	29	0.79	Guayllabamba	Quito	Pichincha	0°04′43″ S	78°20′59″ W	2171

T: temperature; P: pressure; a.s.l: above sea level.

4.3. Essential Oil Isolation

The plant material of Bursera graveolens (10 Kg), Dacryodes peruviana (9 kg), Mespilodaphne quixos (15 Kg), and Melaleuca armillaris (18 Kg) were processed fresh immediately after arriving at the laboratory, between 2 and 6 h after being collected. Extraction of the essential oil was carried out by hydrodistillation in Clevenger-type apparatus (80 L distiller). Initially, 16 L of water was placed in the distiller, then the plant material and the extraction process began [74]. The process was maintained for 3 h. The essential oil was separated from the water by decantation, dried using anhydrous sodium sulfate, and stored at 4 °C until being used in analysis. Each plant species was distilled three times.

4.4. Determination of Physical Properties of Essential Oil

The density of the essential oils was determined using the ISO 279:1998 standard [75] (equivalent to the AFNOR NF T 75-111 standard) using an analytical balance (Mettler AC 100,

Mettler Toledo, Columbus, OH, USA). The refractive index was determined using the standard ISO 280:1998 [76] (similarly to AFNOR NF T 75-112) using a refractometer (model ABBE, BOECO, Hamburg, Germany). The optical rotation of essential oils was determined according to the standard ISO 592:1998 [77] using an automatic polarimeter (Mrc-P810, MRC, Holon, Israel). The procedures were carried out in triplicate, and all measurements were performed at 20 °C.

4.5. Essential Oil Compounds Identification

4.5.1. Quantitative Analysis

The quantitative analysis was performed using gas chromatography coupled to flame ionization detector (GC-FID) for which an Agilent gas chromatograph (GC) (6890N series), a flame ionization detector (FID), a nonpolar GC column (DB-5ms, stationary phase 5%-phenyl-methylpolyxilosane, 30 m of length, 0.25 mm of diameter, and 0.25 µm of stationary layer thickness), and an automatic injector (7683 automatic liquid sampler) were used (all equipment was from Agilent Technologies, Santa Clara, CA, USA). The procedures were carried out as described by Valarezo et al. [49]. Briefly, 1 µL of solution (1/100, *v*/*v*, EO/DCM) was injected with a split ratio of 1:50. Helium was used as a carrier gas at 1 mL/min in constant flow mode and an average velocity of 25 cm/s. The injector and detector temperatures were 250 °C. The oven temperature program included an initial isotherm of 50 °C for 3 min, followed by a temperature ramp to 260 °C at 3 °C/min, and a final isotherm for 3 min. The relative amounts of the compounds were calculated based on the GC peak area (FID response) without using a correction factor.

4.5.2. Qualitative Analysis

The quantitative analysis was performed using gas chromatography coupled to mass spectrometry (GC-MS) for which the same equipment as in the quantitative analysis was used, except for the detector that was replaced by a mass spectrometer (MS) (quadrupole) detector (series 5973 inert, Agilent Technologies, Santa Clara, CA, USA). The procedures were carried out as described by Valarezo et al. [49]. The sample concentration and temperatures (ramp, injector, and detector) were the same as for qualitative analyses. Helium was used as a carrier gas at 0.9 mL/min in constant flow mode and an average velocity of 34 cm/s. For the identification of the compounds, the retention index IR and the mass spectra were compared with published data [43,78].

4.6. Cytotoxicity Studies by Methylthiazolyldiphenyl-Tetrazolium Bromide (MTT) Method

An in vitro MTT cytotoxicity assay was conducted using the HaCaT human keratinocyte cell line. The cells were adjusted to a concentration of 2×10^5 cells/mL and seeded in a 96-well plate for 24 h. They were cultured using Dubelcco's Modified Eagle's Medium (DMEM), which contained a high glucose level and buffered with 25 mM HEPES (4-(2-hydroxyethyl)-1-piperazineethanesulfonic acid). The medium was supplemented with 1% non-essential amino acids, 100 U/mL penicillin, 100 µg/mL streptomycin, and 10% heat-inactivated Fetal Bovine Serum (FBS). Following the incubation period, the cells were exposed to various dilutions of the essential oils from 500 to 5000 µg/mL using the culture medium as a dilution agent for 24 h. Subsequently, the HaCaT cells underwent a gentle washing process using 1% sterile PBS, followed by incubation with a solution of MTT (Sigma-Aldrich Chemical Co, St. Louis, MO, USA) at a concentration of 2.5 mg/mL for 2 h at 37 °C. After carefully removing the medium, a solubilization reagent (99% DMSO) was added in a volume of 0.1 mL to facilitate cell lysis and dissolve the purple MTT crystals. Cell viability was assessed at a wavelength of 570 nm using a microplate photometer, Varioskan TM LUX (Thermo Scientific, Waltham, MA, USA). In order to make a comparative analysis, a negative control (untreated HaCaT cells) was concurrently processed using the same method. This experiment was carried out in triplicate. The results were expressed as a percentage of cell survival relative to the negative control (100% viability) using Equation (1):

$$\%\text{Cell viability} = \frac{\text{Abs sample}}{\text{Abs control}} \times 100 \quad (1)$$

4.7. In Vivo Tolerance Studies

Transepidermal water loss (TEWL) measurements were conducted on the ventral area of the forearm in 10 volunteers (aged between 20 and 40 years) who had healthy skin. The participants provided their written informed consent prior to the study. The research protocol received approval from the Ethics Committee of the University of Barcelona (IRB00003099), adhering to the principles of the Declaration of Helsinki. Each essential oil from the four species was diluted to 5% using 55% Transcutol-P and 40% water. TEWL values were determined at basal state and 10 min, 1 h, and 2 h after skin application of essential oils using a Tewameter TM HEX (Courage & Khazaka Electronics GmbH, Cologne, Germany). The probe was pressed on the skin for 20 s, and the results are presented in units of $g/cm^2/h$.

4.8. Antifungal Efficacy Studies

4.8.1. Preparation of Culture Medium

A total of 5.20 g of RPMI 1640 medium powder was dissolved in 150 mL of distilled water. Subsequently, 17.26 g of MOPS buffer and 9 g of glucose were added until 250 mL was obtained. The culture medium was adjusted to pH 7.0 with 0.1 M NaOH, and 500 µg/mL chloramphenicol was added and passed through a 0.22-micron filter.

4.8.2. Inoculum Preparation

The Candida strains were plated on Sabouraud agar for 48 h at 30 °C. A second replanting of 24 h was then carried out. Isolated colonies (approximately four) were resuspended in Ringer's solution by vortexing to about 0.5 McFarland turbidity. A standard was prepared separately by adding 0.5 mL of solution A (1.175% *w*/*v* $BaCl_2.2H_2O$) and 99.50 mL of solution B (1% H_2SO_4), which were taken to a spectrophotometer and measured at 530 nm. The absorbance should be between 0.120 and 0.150. The colony suspension was adjusted by performing a 1/10 dilution with Ringer's medium until obtaining an absorbance similar to our standard. The final suspension would be close to 0.5–2.5 $\times$ 10^5 CFU/mL. Assays were as in *EUCAST Assays Definitive Document EDef 7.1: A Method for Determination of Broth Dilution MICs of Antifungal Agents for Fermentative Yeasts* [79].

4.8.3. Antifungal Activity

In a 96-well plate, 100 µL of the culture medium (mentioned in Section 4.8.1) was added (in each well). Subsequently, in the first well of the first two rows (vertically), a previously dissolved solution of Amphotericin B was added to a final concentration of 150 µg/mL (this drug was used as standard). In the following wells (first well vertically), the samples to be tested were added: *Bursera graveolens, Dacryodes peruviana, Mespilodaphne quixos*, and *Melaleuca armillaris* essential oils (dissolved at 5% with 55% Transcutol P and 40% water), and serial doubling dilutions were made (horizontally). Finally, 100 µL of the inoculum was added to all the wells. The culture medium without inoculum and without drug was used as blank, and the inoculum was used as the negative control. MIC was determined by spectrophotometry using a model 680 microplate reader spectrophotometer (Bio-Rad, Madrid, Spain) at a wavelength of 620 nm. The plates were read at t0 and then at 24 h and 48 h after incubation at 30 °C [79].

4.9. In Vivo Anti-Inflammatory Activity: Arachidonic Acid (AA)-Induced Edema

4.9.1. Study Protocol

In vivo experiments were conducted in accordance with the ethical protocol approved by the Bioethics Committee of the University of Barcelona (CEEA/UB ref. 4/16 and

Generalitat ref. 8756. Date: 28 January 2016). Male BALB/c mice aged 4–5 months old were used for the study. A solution of 60 μL of AA at 5 mg/mL diluted in PBS was topically applied to the right ears to induce inflammation. Ear thickness was measured before and 15 after the induction of inflammation using a micrometer. A group of inflamed animals (n = 3) served as a positive control, whereas a group of untreated mice (n = 3) was used as a negative group to compare the results. After 20 min of inducing inflammation, four groups of animals (n = 3) were topically treated with 60 μL of *Bursera graveolens*, *Dacryodes peruviana*, *Mespilodaphne quixos*, and *Melaleuca armillaris* essential oils (dissolved at 5% with 55% Transcutol P and 40% water), respectively. In parallel, a group (n = 3) was treated with a commercial anti-inflammatory product (60 mg of ibuprofen gel 50 mg/g; reference: 886,192.7), and a last group (referred to as the Blank group) was treated with Transcutol-P:water (60:40, v/v). After 20 min of these treatments, ear thickness was again measured, and the animals were euthanized by cervical dislocation. Oedema was expressed as the difference between the basal ear thickness and the ear thickness after AA application. The inhibition of inflammation percentage or anti-inflammatory efficacy was calculated based on the decrease in ear thickness after each treatment. Finally, the right ears were cut to evaluate the expression of pro-inflammatory cytokines.

4.9.2. Time Quantitative PCR to Assay Inflammatory Biomarkers

Small portions of mouse ear tissue were disrupted using a Bead Beater homogenizer (MP Biomedicals, Madrid, Spain) and the TRI-Reagent® solution (Sigma Aldrich, St. Louis, MO, USA), a phenol-based reagent (named as TRIzol). Total RNA was extracted following the manufacturer's protocol. Integrity and total amount of RNA were tested by NanoDrop One spectrophotometer (Thermo Scientific, Waltham, MA, USA). Then, 500 ng of RNA was retrotranscribed into complementary DNA using the High-Capacity cDNA Reverse Transcription Kit (Applied Biosystems, Foster City, CA, USA). Finally, a dilution of 1/10 of cDNA was used to perform the RT-qPCR. Therefore, the SYBR Green PCR Master Mix (Applied Biosystems, Foster City, CA, USA) and specific pair of primers for inflammatory genes (IL-8, IL-17A, IL-23, and TNFα) were added to the reaction. As a housekeeping gene, the GAPDH was selected and used for normalization. Relative changes in gene expression were analyzed by applying the $2^{-\Delta\Delta Ct}$ method. Table 5 shows the primers used for the Real-Time qPCR.

Table 5. Gene-specific primers used in the Real-Time qPCR.

Gene	Primer Sequence (5′ to 3′)
GAPDH	FW: AGCTTGTCATCAACGGGAAG RV: TTTGATGTTAGTGGGGTCTCG
IL-8	FW: GCTGTGACCCTCTCTGTGAAG RV: CAAACTCCATCTTGTTGTGTC
IL-23	FW: GAGCCTTCTCTGCTCCCTGATA RV: GACTGAGGCTTGGAATCTGCTG
IL-17A	FW: TTTTCAGCAAGGAATGTGGA RV: TTCATTGTGGAGGGCAGAC
TNFα	FW: AACTAGTGGTGCCAGCCGAT RV: CTTCACAGAGCAATGACTCC

GAPDH = Glyceraldehyde-3-Phosphate Dehydrogenase; IL-8 = interleukin-8; IL-23 = interleukin-23; IL-17A = interleukin-17A; TNFα = Tumor necrosis factor alpha; FW = forward primer; and RV = reverse primer.

4.10. Statistical Analysis

The results were presented as mean ± SD. Statistical analysis was carried out using GraphPad Prism, version 6.0 software (GraphPad Software Inc., San Diego, CA, USA). One-way ANOVA, followed by Tukey's test, was used to compare the mean values. Statistical significance was set at a p value less than 0.05.

5. Conclusions

In conclusion, this study demonstrates through its different experiments that the essential oils obtained from *Bursera graveolens*, *Dacryodes peruviana*, *Mespilodaphne quixos*, and *Melaleuca armillaris* can be topically used in dermal treatments as they exhibited high biocompatibility with human keratinocytes as well as an occlusive effect on the skin of healthy volunteers, improving hydration without causing irritation or damage to the integrity of the stratum corneum. Additionally, these essential oils showed antifungal activity against *Candida albicans*, *Candida glabrata*, and *Candida parapsilosis*, possibly thanks to their rich content in limonene, α-phellandrene, (*E*)-methyl cinnamate, and 1,8-cineole, respectively. Similarly, these essential oils effectively counteracted and alleviated two critical events in dermal inflammatory processes, such as edema and the overexpression of pro-inflammatory cytokines, including TNF-α, IL-8, IL-17A, and IL-23. The pharmacological potential of these essential oils suggests that they could be used both as active antifungal and anti-inflammatory components or adjuvants to increase the efficacy of conventional therapies in the treatment of both cutaneous candidiasis and dermal inflammation.

Author Contributions: Conceptualization, L.S., L.C.E. and M.M.; methodology, N.B., M.R. and L.B.; formal analysis, A.C.; investigation, E.V., M.-J.F. and L.C.E.; data curation, L.C.E.; writing—original draft preparation, E.V., L.S. and L.C.E.; writing—review and editing, L.C.E. and M.M.; supervision, A.C.; project administration, M.M. All authors have read and agreed to the published version of the manuscript.

Funding: This research received no external funding.

Institutional Review Board Statement: The study was conducted according to the guidelines of the Declaration of Helsinki and approved by the Institutional Review Board (or Ethics Committee) of the University of Barcelona (IRB00003099, dated 30 January 2019) and Bioethics Committee of the University of Barcelona (CEEA/UB ref. 4/16 and Generalitat ref. 8756. Date: 28 January 2016).

Informed Consent Statement: Informed consent was obtained from all subjects involved in the study.

Data Availability Statement: Data are contained within the article.

Acknowledgments: L.C.E. acknowledges the support of the Secretaría de Educación Superior, Ciencia, Tecnología e Innovación (SENESCYT—Ecuador).

Conflicts of Interest: The authors declare no conflict of interest.

References

1. Sanchez, D.A.; Schairer, D.; Tuckman-Vernon, C.; Chouake, J.; Kutner, A.; Makdisi, J.; Friedman, J.M.; Nosanchuk, J.D.; Friedman, A.J. Amphotericin B releasing nanoparticle topical treatment of *Candida* spp. in the setting of a burn wound. *Nanomed. Nanotechnol. Biol. Med.* **2014**, *10*, 269–277. [CrossRef]
2. Espinoza, L.C.; Sosa, L.; Granda, P.C.; Bozal, N.; Díaz-Garrido, N.; Chulca-Torres, B.; Calpena, A.C. Development of a Topical Amphotericin B and *Bursera graveolens* Essential Oil-Loaded Gel for the Treatment of Dermal Candidiasis. *Pharmaceuticals* **2021**, *14*, 1033. [PubMed]
3. Hussein, M.; Wong, L.J.M.; Zhao, J.; Rees, V.E.; Allobawi, R.; Sharma, R.; Rao, G.G.; Baker, M.; Li, J.; Velkov, T. Unique mechanistic insights into pathways associated with the synergistic activity of polymyxin B and caspofungin against multidrug-resistant Klebsiella pneumoniae. *Comput. Struct. Biotechnol. J.* **2022**, *20*, 1077–1087. [CrossRef] [PubMed]
4. Ben-Ami, R. Treatment of Invasive Candidiasis: A Narrative Review. *J. Fungi* **2018**, *4*, 97.
5. Hassan, Y.; Chew, S.Y.; Than, L.T.L. Candida glabrata: Pathogenicity and Resistance Mechanisms for Adaptation and Survival. *J. Fungi* **2021**, *7*, 667.
6. Kawai, A.; Yamagishi, Y.; Mikamo, H. In vitro efficacy of liposomal amphotericin B, micafungin and fluconazole against non-albicans Candida species biofilms. *J. Infect. Chemother.* **2015**, *21*, 647–653. [CrossRef]
7. Branco, J.; Miranda, I.M.; Rodrigues, A.G. Candida parapsilosis Virulence and Antifungal Resistance Mechanisms: A Comprehensive Review of Key Determinants. *J. Fungi* **2023**, *9*, 80.
8. Pérez-González, N.; Espinoza, L.C.; Rincón, M.; Sosa, L.; Mallandrich, M.; Suñer-Carbó, J.; Bozal-de Febrer, N.; Calpena, A.C.; Clares-Naveros, B. Gel Formulations with an Echinocandin for Cutaneous Candidiasis: The Influence of Azone and Transcutol on Biopharmaceutical Features. *Gels* **2023**, *9*, 308.
9. Sosa, L.; Calpena, A.C.; Silva-Abreu, M.; Espinoza, L.C.; Rincón, M.; Bozal, N.; Domenech, O.; Rodríguez-Lagunas, M.J.; Clares, B. Thermoreversible Gel-Loaded Amphotericin B for the Treatment of Dermal and Vaginal Candidiasis. *Pharmaceutics* **2019**, *11*, 312.

10. Gündel, S.d.S.; de Godoi, S.N.; Santos, R.C.V.; da Silva, J.T.; Leite, L.B.d.M.; Amaral, A.C.; Ourique, A.F. In vivo antifungal activity of nanoemulsions containing eucalyptus or lemongrass essential oils in murine model of vulvovaginal candidiasis. *J. Drug Deliv. Sci. Technol.* **2020**, *57*, 101762. [CrossRef]
11. Margraf, A.; Perretti, M. Immune Cell Plasticity in Inflammation: Insights into Description and Regulation of Immune Cell Phenotypes. *Cells* **2022**, *11*, 1824.
12. Mosser, D.M.; Edwards, J.P. Exploring the full spectrum of macrophage activation. *Nat. Rev. Immunol.* **2008**, *8*, 958–969. [CrossRef] [PubMed]
13. Espinoza, L.C.; Vera-García, R.; Silva-Abreu, M.; Domènech, Ò.; Badia, J.; Rodríguez-Lagunas, M.J.; Clares, B.; Calpena, A.C. Topical Pioglitazone Nanoformulation for the Treatment of Atopic Dermatitis: Design, Characterization and Efficacy in Hairless Mouse Model. *Pharmaceutics* **2020**, *12*, 255. [PubMed]
14. Parameswaran, N.; Patial, S. Tumor necrosis factor-α signaling in macrophages. *Crit. Rev. Eukaryot. Gene Expr.* **2010**, *20*, 87–103. [CrossRef] [PubMed]
15. Hoover, L.; Bochicchio, G.V.; Napolitano, L.M.; Joshi, M.; Bochicchio, K.; Meyer, W.; Scalea, T.M. Systemic inflammatory response syndrome and nosocomial infection in trauma. *J. Trauma* **2006**, *61*, 310–316; discussion 316–317. [CrossRef] [PubMed]
16. Fossiez, F.; Djossou, O.; Chomarat, P.; Flores-Romo, L.; Ait-Yahia, S.; Maat, C.; Pin, J.J.; Garrone, P.; Garcia, E.; Saeland, S.; et al. T cell interleukin-17 induces stromal cells to produce proinflammatory and hematopoietic cytokines. *J. Exp. Med.* **1996**, *183*, 2593–2603. [CrossRef]
17. Chan, T.C.; Hawkes, J.E.; Krueger, J.G. Interleukin 23 in the skin: Role in psoriasis pathogenesis and selective interleukin 23 blockade as treatment. *Ther. Adv. Chronic Dis.* **2018**, *9*, 111–119. [CrossRef]
18. Whelton, A. Nephrotoxicity of nonsteroidal anti-inflammatory drugs: Physiologic foundations and clinical implications. *Am. J. Med.* **1999**, *106*, 13s–24s. [CrossRef]
19. Sriuttha, P.; Sirichanchuen, B.; Permsuwan, U. Hepatotoxicity of Nonsteroidal Anti-Inflammatory Drugs: A Systematic Review of Randomized Controlled Trials. *Int. J. Hepatol.* **2018**, *2018*, 5253623. [CrossRef]
20. Zuo, X.; Gu, Y.; Wang, C.; Zhang, J.; Zhang, J.; Wang, G.; Wang, F. A Systematic Review of the Anti-Inflammatory and Immunomodulatory Properties of 16 Essential Oils of Herbs. *Evid.-Based Complement. Altern. Med. Ecam* **2020**, *2020*, 8878927. [CrossRef]
21. Nakatsu, T.; Lupo, A.T.; Chinn, J.W.; Kang, R.K.L. Biological activity of essential oils and their constituents. In *Studies in Natural Products Chemistry*; Atta-ur, R., Ed.; Elsevier: Amsterdam, The Netherlands, 2000; Volume 21, pp. 571–631.
22. Sharmeen, J.B.; Mahomoodally, F.M.; Zengin, G.; Maggi, F. Essential Oils as Natural Sources of Fragrance Compounds for Cosmetics and Cosmeceuticals. *Molecules* **2021**, *26*, 666. [CrossRef] [PubMed]
23. Pandey, A.K.; Kumar, P.; Saxena, M.J.; Maurya, P. Distribution of aromatic plants in the world and their properties. In *Feed Additives*; Florou-Paneri, P., Christaki, E., Giannenas, I., Eds.; Academic Press: Cambridge, MA, USA, 2020; Chapter 6; pp. 89–114.
24. The Plant List. Burseraceae. Available online: http://www.theplantlist.org (accessed on 3 July 2020).
25. Jørgesen, P.M.; León-Yáñez, S. Catalogue of the Vascular Plants of Ecuador. Available online: http://legacy.tropicos.org/ProjectAdvSearch.aspx?projectid=2 (accessed on 11 July 2020).
26. Viteri Jumbo, L.O.; Corrêa, M.J.M.; Gomes, J.M.; González Armijos, M.J.; Valarezo, E.; Mantilla-Afanador, J.G.; Machado, F.P.; Rocha, L.; Aguiar, R.W.S.; Oliveira, E.E. Potential of *Bursera graveolens* essential oil for controlling bean weevil infestations: Toxicity, repellence, and action targets. *Ind. Crops Prod.* **2022**, *178*, 114611. [CrossRef]
27. Rey-Valeirón, C.; Guzmán, L.; Saa, L.R.; López-Vargas, J.; Valarezo, E. Acaricidal activity of essential oils of *Bursera graveolens* (Kunth) Triana & Planch and Schinus molle L. on unengorged larvae of cattle tick Rhipicephalus (Boophilus) microplus (Acari:Ixodidae). *J. Essent. Oil Res.* **2017**, *29*, 344–350. [CrossRef]
28. Torre, L.d.l.; Navarrete, H.; Muriel, M.P.; Macía Barco, M.J.; Balslev, H. *Enciclopedia de las Plantas Útiles del Ecuador*; Herbario QCA de la Escuela de Ciencias Biológicas de la Pontificia Universidad Católica del Ecuador and Herbario AAU del Departamento de Ciencias Biológicas de la Universidad de Aarhus: Quito, Ecuador; Aarhus, Danmark, 2008.
29. Valarezo, E.; Ojeda-Riascos, S.; Cartuche, L.; Andrade-González, N.; González-Sánchez, I.; Meneses, M.A. Extraction and Study of the Essential Oil of Copal (Dacryodes peruviana), an Amazonian Fruit with the Highest Yield Worldwide. *Plants* **2020**, *9*, 1658.
30. The Plant List. Lauraceae. Available online: http://www.theplantlist.org/1.1/browse/A/Piperaceae/ (accessed on 4 July 2022).
31. Encyclopædia Britannica. Lauraceae. Available online: https://www.britannica.com/plant/Meliaceae (accessed on 3 July 2021).
32. Trofimov, D.; de Moraes, P.L.R.; Rohwer, J.G. Towards a phylogenetic classification of the Ocotea complex (Lauraceae): Classification principles and reinstatement of Mespilodaphne. *Bot. J. Linn. Soc.* **2019**, *190*, 25–50. [CrossRef]
33. Palacios, W. *Árboles del Ecuador: Familias y Géneros*; Ministerio del Ambiente del Ecuador-MAE: Quito, Ecuador, 2011.
34. Ballabeni, V.; Tognolini, M.; Giorgio, C.; Bertoni, S.; Bruni, R.; Barocelli, E. *Ocotea quixos* Lam. essential oil: In vitro and in vivo investigation on its anti-inflammatory properties. *Fitoterapia* **2010**, *81*, 289–295. [CrossRef]
35. Bruni, R.; Medici, A.; Andreotti, E.; Fantin, C.; Muzzoli, M.; Dehesa, M.; Romagnoli, C.; Sacchetti, G. Chemical composition and biological activities of Ishpingo essential oil, a traditional Ecuadorian spice from *Ocotea quixos* (Lam.) Kosterm. (Lauraceae) flower calices. *Food Chem.* **2004**, *85*, 415–421. [CrossRef]
36. Ballabeni, V.; Tognolini, M.; Bertoni, S.; Bruni, R.; Guerrini, A.; Rueda, G.M.; Barocelli, E. Antiplatelet and antithrombotic activities of essential oil from wild *Ocotea quixos* (Lam.) Kosterm. (Lauraceae) calices from Amazonian Ecuador. *Pharmacol. Res.* **2007**, *55*, 23–30. [CrossRef]

37. Scalvenzi, L.; Radice, M.; Toma, L.; Severini, F.; Boccolini, D.; Bella, A.; Guerrini, A.; Tacchini, M.; Sacchetti, G.; Chiurato, M.; et al. Larvicidal activity of Ocimum campechianum, *Ocotea quixos* and Piper aduncum essential oils against Aedes aegypti. *Parasite* **2019**, *26*, 23. [CrossRef]
38. Passos, B.G.; de Albuquerque, R.D.D.G.; Muñoz-Acevedo, A.; Echeverria, J.; Llaure-Mora, A.M.; Ganoza-Yupanqui, M.L.; Rocha, L. Essential oils from Ocotea species: Chemical variety, biological activities and geographic availability. *Fitoterapia* **2022**, *156*, 105065. [CrossRef]
39. WFO Plant List. Myrtaceae Juss. Available online: https://wfoplantlist.org/plant-list (accessed on 20 February 2023).
40. Chabir, N.; Romdhane, M.; Valentin, A.; Moukarzel, B.; Marzoug, H.N.B.; Brahim, N.B.; Mars, M.; Bouajila, J. Chemical Study and Antimalarial, Antioxidant, and Anticancer Activities of *Melaleuca armillaris* (Sol Ex Gateau) Sm Essential Oil. *J. Med. Food* **2011**, *14*, 1383–1388. [CrossRef]
41. Amri, I.; Mancini, E.; De Martino, L.; Marandino, A.; Lamia, H.; Mohsen, H.; Bassem, J.; Scognamiglio, M.; Reverchon, E.; De Feo, V. Chemical Composition and Biological Activities of the Essential Oils from Three Melaleuca Species Grown in Tunisia. *Int. J. Mol. Sci.* **2012**, *13*, 16580–16591. [CrossRef] [PubMed]
42. Molares, S.; González, S.B.; Ladio, A.; Agueda Castro, M. Etnobotánica, anatomía y caracterización físico-química del aceite esencial de Baccharis obovata Hook. et Arn. (Asteraceae: Astereae). *Acta Bot. Bras.* **2009**, *23*, 578–589.
43. Adams, R.P. *Identification of Essential Oil Components by Gas Chromatography/Mass Spectrometry*, 4th ed.; Allured Publishing Corporation: Carol Stream, IL, USA, 2007.
44. Soares, G.A.B.; Bhattacharya, T.; Chakrabarti, T.; Tagde, P.; Cavalu, S. Exploring Pharmacological Mechanisms of Essential Oils on the Central Nervous System. *Plants* **2022**, *11*, 21.
45. Mekonnen, A.; Tesfaye, S.; Christos, S.G.; Dires, K.; Zenebe, T.; Zegeye, N.; Shiferaw, Y.; Lulekal, E. Evaluation of Skin Irritation and Acute and Subacute Oral Toxicity of Lavandula angustifolia Essential Oils in Rabbit and Mice. *J. Toxicol.* **2019**, *2019*, 5979546. [CrossRef]
46. Espinoza, L.C.; Valarezo, E.; Fábrega, M.J.; Rodríguez-Lagunas, M.J.; Sosa, L.; Calpena, A.C.; Mallandrich, M. Characterization and In Vivo Anti-Inflammatory Efficacy of Copal (*Dacryodes peruviana* (Loes.) H.J. Lam) Essential Oil. *Plants* **2022**, *11*, 3104.
47. Cowen, L.E.; Sanglard, D.; Howard, S.J.; Rogers, P.D.; Perlin, D.S. Mechanisms of Antifungal Drug Resistance. *Cold Spring Harb. Perspect. Med.* **2014**, *5*, a019752. [CrossRef]
48. Sacchetti, G.; Guerrini, A.; Noriega, P.; Bianchi, A.; Bruni, R. Essential oil of wild *Ocotea quixos* (Lam.) Kosterm. (Lauraceae) leaves from Amazonian Ecuador. *Flavour Fragr. J.* **2006**, *21*, 674–676. [CrossRef]
49. Valarezo, E.; Vullien, A.; Conde-Rojas, D. Variability of the Chemical Composition of the Essential Oil from the Amazonian Ishpingo Species (*Ocotea quixos*). *Molecules* **2021**, *26*, 3961. [CrossRef]
50. Hayouni, E.A.; Bouix, M.; Abedrabba, M.; Leveau, J.-Y.; Hamdi, M. Mechanism of action of *Melaleuca armillaris* (Sol. Ex Gaertu) Sm. essential oil on six LAB strains as assessed by multiparametric flow cytometry and automated microtiter-based assay. *Food Chem.* **2008**, *111*, 707–718. [CrossRef]
51. Espinoza, L.C.; Silva-Abreu, M.; Calpena, A.C.; Rodriguez-Lagunas, M.J.; Fabrega, M.J.; Garduno-Ramirez, M.L.; Clares, B. Nanoemulsion strategy of pioglitazone for the treatment of skin inflammatory diseases. *Nanomedicine* **2019**, *19*, 115–125. [CrossRef]
52. Rivera-Yañez, C.R.; Terrazas, L.I.; Jimenez-Estrada, M.; Campos, J.E.; Flores-Ortiz, C.M.; Hernandez, L.B.; Cruz-Sanchez, T.; Garrido-Fariña, G.I.; Rodriguez-Monroy, M.A.; Canales-Martinez, M.M. Anti-Candida Activity of Bursera morelensis Ramirez Essential Oil and Two Compounds, α-Pinene and γ-Terpinene—An In Vitro Study. *Molecules* **2017**, *22*, 2095. [CrossRef]
53. Mendez, A.H.S.; Cornejo, C.G.F.; Coral, M.F.C.; Arnedo, M.C.A. Chemical Composition, Antimicrobial and Antioxidant Activities of the Essential Oil of *Bursera graveolens* (*Burseraceae*) From Perú. *Indian J. Pharm. Educ. Res.* **2017**, *51*, s429–s436. [CrossRef]
54. Yu, H.; Lin, Z.-X.; Xiang, W.-L.; Huang, M.; Tang, J.; Lu, Y.; Zhao, Q.-H.; Zhang, Q.; Rao, Y.; Liu, L. Antifungal activity and mechanism of d-limonene against foodborne opportunistic pathogen Candida tropicalis. *Lwt* **2022**, *159*, 113144. [CrossRef]
55. Thakre, A.; Zore, G.; Kodgire, S.; Kazi, R.; Mulange, S.; Patil, R.; Shelar, A.; Santhakumari, B.; Kulkarni, M.; Kharat, K.; et al. Limonene inhibits Candida albicans growth by inducing apoptosis. *Med. Mycol.* **2018**, *56*, 565–578. [CrossRef]
56. Zore, G.; Thakre, A.D.; Jadhav, S.; Karuppayil, S.M. Terpenoids inhibit Candida albicans growth by affecting membrane integrity and arrest of cell cycle. *Phytomedicine Int. J. Phytother. Phytopharm.* **2011**, *18*, 1181–1190. [CrossRef]
57. Zhang, J.H.; Sun, H.L.; Chen, S.Y.; Zeng, L.; Wang, T.T. Anti-fungal activity, mechanism studies on alpha-Phellandrene and Nonanal against Penicillium cyclopium. *Bot. Stud.* **2017**, *58*, 13. [CrossRef]
58. Thangaleela, S.; Sivamaruthi, B.S.; Kesika, P.; Tiyajamorn, T.; Bharathi, M.; Chaiyasut, C. A Narrative Review on the Bioactivity and Health Benefits of Alpha-Phellandrene. *Sci. Pharm.* **2022**, *90*, 57. [CrossRef]
59. Belletti, N.; Ndagijimana, M.; Sisto, C.; Guerzoni, M.E.; Lanciotti, R.; Gardini, F. Evaluation of the Antimicrobial Activity of Citrus Essences on Saccharomyces cerevisiae. *J. Agric. Food Chem.* **2004**, *52*, 6932–6938. [CrossRef]
60. Huang, Q.-S.; Zhu, Y.-J.; Li, H.-L.; Zhuang, J.-X.; Zhang, C.-L.; Zhou, J.-J.; Li, W.-G.; Chen, Q.-X. Inhibitory Effects of Methyl trans-Cinnamate on Mushroom Tyrosinase and Its Antimicrobial Activities. *J. Agric. Food Chem.* **2009**, *57*, 2565–2569. [CrossRef]
61. Ali, N.A.M.; Rahmani, M.; Shaari, K.; Ali, A.M.; Lian, G.E.C. Antimicrobial Activity of Cinnamomum impressicostatum and *C. pubescens* and Bioassay-Guided Isolation of Bioactive (E)-Methyl Cinnamate. *J. Biol. Sci.* **2010**, *10*, 101–106. [CrossRef]
62. Ruiz-Duran, J.; Torres, R.; Stashenko, E.E.; Ortiz, C. Antifungal and Antibiofilm Activity of Colombian Essential Oils against Different Candida Strains. *Antibiotics* **2023**, *12*, 668. [CrossRef]

63. Silva, R.A.d.; Antonieti, F.M.P.M.; Röder, D.V.D.d.B.; Pedroso, R.d.S. Essential Oils of Melaleuca, Citrus, Cupressus, and Litsea for the Management of Infections Caused by Candida Species: A Systematic Review. *Pharmaceutics* **2021**, *13*, 1700. [CrossRef]
64. de Macêdo, D.G.; de Almeida Souza, M.M.; Morais-Braga, M.F.B.; Coutinho, H.D.M.; dos Santos, A.T.L.; Machado, A.J.T.; Rodrigues, F.F.G.; da Costa, J.G.M.; de Menezes, I.R.A. Seasonality influence on the chemical composition and antifungal activity of Psidium myrtoides O. Berg. *S. Afr. J. Bot.* **2020**, *128*, 9–17. [CrossRef]
65. Hendry, E.R.; Worthington, T.; Conway, B.R.; Lambert, P.A. Antimicrobial efficacy of eucalyptus oil and 1,8-cineole alone and in combination with chlorhexidine digluconate against microorganisms grown in planktonic and biofilm cultures. *J. Antimicrob. Chemother.* **2009**, *64*, 1219–1225. [CrossRef]
66. Young, J.M.; Spires, D.A.; Bedord, C.J.; Wagner, B.; Ballaron, S.J.; de Young, L.M. The Mouse Ear Inflammatory Response to Topical Arachidonic Acid. *J. Investig. Dermatol.* **1984**, *82*, 367–371. [CrossRef]
67. Ku, C.M.; Lin, J.Y. Anti-inflammatory effects of 27 selected terpenoid compounds tested through modulating Th1/Th2 cytokine secretion profiles using murine primary splenocytes. *Food Chem.* **2013**, *141*, 1104–1113. [CrossRef]
68. Lappas, C.M.; Lappas, N.T. D-Limonene modulates T lymphocyte activity and viability. *Cell Immunol.* **2012**, *279*, 30–41. [CrossRef]
69. d'Alessio, P.A.; Mirshahi, M.; Bisson, J.-F.; Bene, M.C. Skin repair properties of dLimonene and perillyl alcohol in murine models. Anti-Inflammatory Anti-Allergy Agents. *Med. Chem.* **2014**, *13*, 29–35.
70. Hansen, J.S.; Norgaard, A.W.; Koponen, I.K.; Sorli, J.B.; Paidi, M.D.; Hansen, S.W.; Clausen, P.A.; Nielsen, G.D.; Wolkoff, P.; Larsen, S.T. Limonene and its ozone-initiated reaction products attenuate allergic lung inflammation in mice. *J. Immunotoxicol.* **2016**, *13*, 793–803. [CrossRef]
71. Valsalan Soba, S.; Babu, M.; Panonnummal, R. Ethosomal Gel Formulation of Alpha Phellandrene for the Transdermal Delivery in Gout. *Adv. Pharm. Bull.* **2021**, *11*, 137–149. [CrossRef] [PubMed]
72. Murakami, Y.; Kawata, A.; Suzuki, S.; Fujisawa, S. Cytotoxicity and Pro-/Anti-inflammatory Properties of Cinnamates, Acrylates and Methacrylates Against RAW264.7 Cells. *In Vivo* **2018**, *32*, 1309–1322. [CrossRef]
73. Pries, R.; Jeschke, S.; Leichtle, A.; Bruchhage, K.-L. Modes of Action of 1,8-Cineol in Infections and Inflammation. *Metabolites* **2023**, *13*, 751. [CrossRef] [PubMed]
74. de Elguea-Culebras, G.O.; Panamá-Tapia, L.A.; Melero-Bravo, E.; Cerro-Ibáñez, N.; Calvo-Martínez, A.; Sánchez-Vioque, R. Comparison of the phenolic composition and biological capacities of wastewater from Origanum vulgare L., Rosmarinus officinalis L., Salvia lavandulifolia Vahl. and Thymus mastichina L. resulting from two hydrodistillation systems: Clevenger and MAE. *J. Appl. Res. Med. Aromat. Plants.* **2023**, *34*, 100480. [CrossRef]
75. *ISO 279:1998*; Essential Oils-Determination of Relative Density at 20 °C—Reference Method. International Organization for Standardization (ISO): Geneva, Switzerland, 2020; pp. 1–4.
76. *ISO 280:1998*; Essential Oils-Determination of Refractive Index. International Organization for Standardization (ISO): Geneva, Switzerland, 2020; pp. 1–3.
77. *ISO 592:1998*; Essential Oils-Determination of Optical Rotation. International Organization for Standardization (ISO): Geneva, Switzerland, 2020; pp. 1–4.
78. van Den Dool, H.; Dec. Kratz, P. A generalization of the retention index system including linear temperature programmed gas—Liquid partition chromatography. *J. Chromatogr. A* **1963**, *11*, 463–471. [CrossRef]
79. Rodriguez-Tudela, J.L.; Arendrup, M.C.; Barchiesi, F.; Bille, J.; Chryssanthou, E.; Cuenca-Estrella, M.; Dannaoui, E.; Denning, D.W.; Donnelly, J.P.; Dromer, F.; et al. EUCAST Definitive Document EDef 7.1: Method for the determination of broth dilution MICs of antifungal agents for fermentative yeasts: Subcommittee on Antifungal Susceptibility Testing (AFST) of the ESCMID European Committee for Antimicrobial Susceptibility Testing (EUCAST)∗. *Clin. Microbiol. Infect.* **2008**, *14*, 398–405. [CrossRef]

Disclaimer/Publisher's Note: The statements, opinions and data contained in all publications are solely those of the individual author(s) and contributor(s) and not of MDPI and/or the editor(s). MDPI and/or the editor(s) disclaim responsibility for any injury to people or property resulting from any ideas, methods, instructions or products referred to in the content.

molecules

MDPI

Review

Medicinal Mushrooms: Their Bioactive Components, Nutritional Value and Application in Functional Food Production—A Review

Paulina Łysakowska, Aldona Sobota * and Anna Wirkijowska

Department of Plant Food Technology and Gastronomy, University of Life Sciences in Lublin, Skromna 8 Street, 20-704 Lublin, Poland; paulina.lysakowska@up.lublin.pl (P.Ł.); anna.wirkijowska@up.lublin.pl (A.W.)
* Correspondence: aldona.sobota@up.lublin.pl

Abstract: Medicinal mushrooms, e.g., Lion's Mane (*Hericium erinaceus* (Bull.) Pers.), Reishi (*Ganoderma lucidum* (Curtis) P. Karst.), Chaga (*Inonotus obliquus* (Ach. ex Pers.) Pilát), Cordyceps (*Ophiocordyceps sinensis* (Berk.) G.H. Sung, J.M. Sung, Hywel-Jones and Spatafora), Shiitake (*Lentinula edodes* (Berk.) Pegler), and Turkey Tail (*Trametes versicolor* (L.) Lloyd), are considered new-generation foods and are of growing interest to consumers. They are characterised by a high content of biologically active compounds, including (1,3)(1,6)-β-D-glucans, which are classified as dietary fibre, triterpenes, phenolic compounds, and sterols. Thanks to their low-fat content, they are a low-calorie product and are classified as a functional food. They have a beneficial effect on the organism through the improvement of its overall health and nutritional level. The biologically active constituents contained in medicinal mushrooms exhibit anticancer, antioxidant, antidiabetic, and immunomodulatory effects. In addition, these mushrooms accelerate metabolism, help fight obesity, and slow down the ageing processes thanks to their high antioxidant activity. The vast therapeutic properties of mushrooms are still not fully understood. Detailed mechanisms of the effects of medicinal mushrooms on the human organism still require long-term clinical studies to confirm their nutraceutical effects, their safety of use, and their dosage. Medicinal mushrooms have great potential to be used in the design of innovative functional foods. There is a need for further research on the possibility of incorporating mushrooms into food products to assess the interactions of their bioactive substances with ingredients in the food matrix. This review focuses on the properties of selected medicinal mushrooms and their effects on the human organism and presents current knowledge on the possibilities of their use in the production of functional foods.

Keywords: medicinal mushrooms; bioactive compounds; polysaccharides; functional foods; nutraceuticals; new generation foods; (1,3)(1,6)-β-D-glucans; polyphenols; antioxidant activity

Citation: Łysakowska, P.; Sobota, A.; Wirkijowska, A. Medicinal Mushrooms: Their Bioactive Components, Nutritional Value and Application in Functional Food Production—A Review. *Molecules* **2023**, *28*, 5393. https://doi.org/10.3390/molecules28145393

Academic Editors: Arunaksharan Narayanankutty, Ademola C. Famurewa, Eliza Oprea and José Sousa Câmara

Received: 11 April 2023
Revised: 11 July 2023
Accepted: 11 July 2023
Published: 14 July 2023

Copyright: © 2023 by the authors. Licensee MDPI, Basel, Switzerland. This article is an open access article distributed under the terms and conditions of the Creative Commons Attribution (CC BY) license (https://creativecommons.org/licenses/by/4.0/).

1. Introduction

For many years, mushrooms have accompanied humans both as food and medicine. Data from the literature indicate that, with the onset of hunting, mushrooms began to play an important role in the human diet [1]. Fruiting bodies, i.n., the visible part above the substrate commonly referred to as the mushroom, are the edible elements of some filamentous fungi [2]. Fungi form a separate kingdom alongside the kingdoms of prokaryotes, eukaryotes, plants, and animals [3]. About 2.2–3.8 million species of fungi in the world have been identified, of which 150,000 species have been described, 2000 species are considered edible, and over 200 species of wild mushrooms are considered medicinal [4,5]. Edible mushrooms, unlike medicinal mushrooms, are mainly consumed as fresh mushrooms with fruiting bodies or dried products. They can also be consumed as boiled, fried, roasted, soups, tinctures, teas, and many different dishes, while medicinal mushrooms are mostly used in biopharmaceutical applications in powdered, loose, or liquid extract forms [6]. In culinary terms, mushrooms are wrongly classified as vegetables and are informally categorised as 'white vegetables' [7]. According to the USDA (United States Department of Agriculture), they can be used as a substitute for vegetables in the diet at a ratio of 1:1 (USDA, 2022). Due to their content of biologically active compounds with beneficial health effects, medicinal mushrooms have been used worldwide in folk medicine for centuries. They are particularly popular in Asian countries, e.g., China, Japan, Taiwan, and Korea. Due to the presence of numerous biologically active compounds, including polysaccharides, proteins, peptides, terpenoids, polyphenols, vitamins, and mineral elements, they are ascribed, e.g., anti-cancer, anti-inflammatory, antioxidant, hypocholesterolemic, hypoglycaemic, and immunomodulatory effects [8,9] However, it should be remembered that the consumption of medicinal mushrooms is not always advisable. The safety of their use during pregnancy, lactation, and in children is still poorly reported. The selected bioactive compounds found in mushrooms may potentially limit the absorption of nutrients, trace elements, and vitamins. As a result, it is recommended that the elderly and children avoid the excessive consumption of mushrooms. Additionally, individuals taking medications or herbs should exercise caution when using mushrooms due to the potential for interactions with their bioactive compounds.

The chemical profile of medicinal mushrooms varies according to species, strain, cultivation conditions (cultured or growing wild) [10], the degree of maturity [11], and the proportion of individual anatomical parts in the total mass of the mushroom [12]. This is largely determined by environmental (access to water, light, UV radiation) [9,13] and biological (type of substrate/host, presence of competing fungi) factors. Song et al. [8] compared the chemical composition and functional properties of wood-cultured and sack-cultured Shiitake (*Lentinula edodes* (Berk.) Pegler) and proved that the wood-cultured fungus had a higher content of terpenoids and phenolic components and concurrently exhibited higher antioxidant and hypoglycaemic potential compared to the sack-cultured Shiitake (*Lentinula edodes* (Berk.) Pegler). In the case of Chaga (*Inonotus obliquus* (Ach. ex Pers.) Pilát), which is a parasite of various deciduous trees, only sclerotia derived from birch trunks have contained tree-specific compounds (betulin and betulinic acid) showing anticarcinogenic activity. Equally great importance for the chemical composition and health-promoting potential of medicinal mushrooms is ascribed to the world region from which they originate [14,15]. Chaga (*Inonotus obliquus* (Ach. ex Pers.) Pilát) sclerotia collected in France, Ukraine, and Canada were characterised by their different contents of betulin, betulinic acid, and inotodiol and showed differential biological activity in different cancer cells [15]. The bioactive substances present in fungi are primary and secondary metabolites that can be synthesised in response to specific environmental stimuli [9,13]. Their content depends on the species of fungus and their growing conditions [10,16]. However, Peng and Shahidi [17] emphasise that the cultivation of medicinal mushrooms in standard conditions offers the possibility to stimulate the synthesis of selected biologically active substances and yields raw materials with a reproducible chemical composition, comparable biological effects, and greater health safety (with a lower content of heavy metals, which are often found in excess in wild mushrooms growing in polluted environments).

The existence of a huge number of medicinal mushroom species with their diverse chemical composition and content of biologically active compounds and thus multidirectional effects on the human organism could make mushrooms objects of growing consumer interest. In 2020, the size of the global mushroom market was 14.35 million tonnes; it is estimated to grow to 24.05 million tonnes in 2028. The most popular mushrooms among consumers include Reishi (*Ganoderma lucidum*), Lion's Mane (*Hericium erinaceus*), Chaga (*Inonotus obliquus* (Ach. ex Pers.) Pilát), Turkey Tail (*Trametes versicolor* (L.) Lloyd)), Shiitake (*Lentinula edodes* (Berk.) Pegler), and Cordyceps (*Ophiocordyceps sinensis* (Berk.) G.H. Sung, J.M. Sung, Hywel-Jones and Spataforaprior name *Cordyceps sinensis*). It is, therefore, expedient to compile and systematise existing knowledge on the most popular medicinal mushrooms, compare their functional potential, and discuss the possibilities of their use in the food industry.

2. Nutritional Value and Bioactive Components

Due to their high-water content (about 80–90%), the fruiting bodies of medicinal mushrooms are low in calories (50–70 kcal/100 g) [18]. After drying, their moisture content is at the level of about 3–13% [12,19,20]. The chemical composition of medicinal mushrooms is shown in Table 1. These mushrooms are a source of such nutrients as carbohydrates (65.6–87.13% d.b.), dietary fibre (16–53% d.b.), protein (3.87–37.4% d.b.), minerals (6.2–9.7% d.b), and fats (1–5.62% d.b) [21].

Table 1. The chemical composition of medicinal mushrooms (g/100 g dried mushrooms).

Common Name	Latin Name	Moisture	Protein	Carbohydrates	Lipids	Dietary Fibre	Ash	The Literature Source
Reishi	*Ganoderma lucidum* (Curtis) P. Karst.	7.5–12.99	13.3–23.6	42.8–82.3	3–5.8	14.81	4	[19,22,23]
Lion's Mane	*Hericium erinaceus* (Bull.) Pers.	7.03 *	22.3	57.0	3.5	3.3–7.8	7.1	[19,23]
Chaga	*Inonotus obliquus* (Ach. ex Pers.) Pilát	3.5	2.4	10.3	1.7	67.5	n.d.	[20]
Cordyceps	*Ophiocordyceps sinensis* (Berk.) G.H. Sung, J.M. Sung, Hywel-Jones and Spatafora*prior name Cordyceps sinensis*	3.5 *	21.9–23.1	24.2–49.3	5.5–8.2	7.7	13.13	[19,23]
Shiitake	*Lentinula edodes* (Berk.) Pegler	7.14	17.2–27.09	38.1–66.0	1.26–2.95	46.19–49.09 (IDF: 40.7–44.2 and SDF: 1.95–8.4)	6.05–6.73	[24–27]
Turkey Tail It also known as:, Cloud mushroom, Yun Zhi, Kawaritake	*Trametes versicolor* (L.) Lloyd	-	11.07	-	1.35	-	-	[28]

(-)—no data; (*)—unpublished own research, IDF—water-insoluble dietary fibre, SDF—water-soluble dietary fibre.

2.1. Polysaccharides

Carbohydrates present in fungi are represented by monosaccharides (glucose, fructose, galactose), alcohol sugars (mannitol), oligosaccharides (trehalose, malezitose), and polysaccharides, among which homopolysaccharides (glucans, chitin, glycogen) and heteropolysaccharides (xylomannan, α-(1→4)-D-glucopyranosyl and β-(1→6)-D-galactopyranosyl with branches at *O*-6 of glucose and *O*-2 of galactose, 6-*O*-galactopyranoses substituted at *O*-2 by 3-*O*-D-mannopyranosyl-L-fucopyranosyl, α-D-mannopyranosyl, and α-L-fucopyranosyl, α-(1→3)-linked galactose, with β-(1→4),(1→6)-glucose and fucose branches, mucilage composed of glucose and galactose can be distinguished). Carbohydrates can also occur in complexes with other compounds (e.g., proteins) and may include various sugar subunits [29–31]. Depending on their structure, bond type, and molecular weight, carbohydrates have different functional properties. The main indigestible polysaccharides present in fungi are chitin and β-D-glucans. They are composed of sugar units that are linked by β-glycosidic bonds. The monomer in chitin is β-glucosamine and is linked by 1-4-β-glycosidic bonds, while β-glucans are made up of glucopyranose molecules. The molecules linked by β-(1,3) and β-(1,4) glycosidic bonds form linear segments to which side chains are attached via β-(1,6) glycosidic bonds [29]. These compounds are classified as dietary fibres. They are found in fungal fruiting bodies and in fungal cells at both the vegetative and generative stages of ontogenesis and play a structural role in co-forming cell walls. A special physiological role is attributed to β-D-glucans and complexes of these compounds with proteins [32,33].

Their structure takes the form of a single helix, a triple helix, or a random helix. Depending on their molecular weight, the type of β-glycosidic bonds present in the molecule, and the chain conformation, these compounds exhibit different functional properties [34]. Beta-glucans with a triple helix structure show a greater ability to inhibit tumour growth than β-glucans in a single helix form [35]. As reported by Sletmoen and Stokke [36] and Brown and Gordon [37], compounds with a higher molecular weight and lower water solubility are more potent immunostimulators, while β-glucans with a low MW and a short side chain are considered less active. On the contrary, Rop et al. [34] found that water-soluble β-glucans had stronger immunomodulatory properties than water-insoluble β-glucans. Macrophages mainly act as antigen-presenting immune cells, which contribute to the immunomodulatory effect of β-glucans by stimulating the fight against bacteria and viruses. High molecular weight molecules stimulate the action of specific lymphocyte—NK cells, which show cytotoxic effects against tumour cells.

In addition, they upregulate the expression of cytokines that are associated with immune response, including interferon-γ, TNF-α, IFN-g, IL-1, and IL-12, which inhibit tumour cell proliferation and induce their apoptosis, thereby exerting anti-tumour, antibacterial, and antiviral effects [23,37,38]. These compounds are often used as adjuvants in traditional cancer chemotherapy [39–44].

The content of β-glucans in mushrooms varies between 3.79% d.b. for Cordyceps (*Ophiocordyceps sinensis* (Berk.) G.H. Sung, J.M. Sung, Hywel-Jones and Spatafora) and 60.79% d.b. for Turkey Tail (*Trametes versicolor* (L.) Lloyd) (Table 2) [44,45]. In general, edodes are a better source of these compounds than caps. The Shiitake (*Lentinula edodes* (Berk.) Pegler) mushroom is a rich source of β-glucans [34]. It takes its specific name from its β-glucan lentinan, which stimulates immune cells to attack cancer cells. Lentinan enhances the production of T lymphocytes and can potentiate the effect of AZT (3′-Azido-3′-deoxythymidine) in the anti-viral treatment of AIDS [27]. Its positive effects have been proved in the treatment of, e.g., glioma (human astrocytoma U251 cells) [46], breast cancer [47] and liver cancer [48]. In turn, the Turkey Tail (*Trametes versicolor* (L.) Lloyd) contains characteristic proteoglucans. One of these is crestin, also known as polysaccharide-K (PSK), which contains about 25–38% of the protein in the molecule. This proteoglucan is effective in the treatment of, e.g., gastric, oesophageal, colon, rectal, and lung cancer [29].

Table 2. Beta-glucan content of different medicinal mushrooms [31,44,48].

Common Name	Latin Name	Content of β-Glucans (g/100 g d.b.)
Reishi	*Ganoderma lucidum* (Curtis) P. Karst.	4.3–23.6
Lion's Mane	*Hericium erinaceus* (Bull.) Pers.	35.3
Chaga	*Inonotus obliquus* (Ach. ex Pers.) Pilát	8.5
Shiitake cap/steam	*Lentinula edodes* (Berk.) Pegler	20.0/25.3
Turkey tail	*Trametes versicolor* (L.) Lloyd	60.79
Cordyceps	*Ophiocordyceps sinensis* (Berk.) G.H. Sung, J.M. Sung, Hywel-Jones and Spatafora *prior name Cordyceps sinensis*	3.79

Another type of glucan that is complexed with protein present in Turkey Tail (*Trametes versicolor* (L.) Lloyd) mushrooms is called PSP (Poly Saccharo Peptide) and activates immune cells by increasing the production of cytokines, chemokines, histamine, and prostagladin E. It reduces the detrimental effects of chemotherapy by alleviating fatigue, loss of appetite, vomiting, a dry mouth, and other related discomforts [49]. In addition to β-glucans, biological activity has also been attributed to poly- and monosaccharides occurring in complexes with other compounds. An example is the cordycepin present in cordyceps (*Ophiocordyceps sinensis* (Berk.) G.H. Sung, J.M. Sung, Hywel-Jones, and Spatafora). Its chemical structure resembles that of the nucleoside adenosine (ribose + adenine sugar); however, it lacks one hydroxyl group at position three of the five-membered ring of the ribose moiety. Adenosine is involved in DNA and/or RNA synthesis in cells. Thanks to its analogy to adenosine, cordycepin can build into the RNA and DNA structures of bacteria and viruses and interfere with the biosynthesis and modification of nucleic acids, thereby limiting the growth of these microorganisms. It increases the proliferation and secretion of T and B lymphocytes and has anti-inflammatory effects through a reduction in the expression of pro-inflammatory cytokines and chemokines. Additionally, it inhibits platelet aggregation and shows suppressive properties against tumour cells [50]. The positive effects of polysaccharides as well as other phytochemicals present in mushrooms, are shown in Table 3.

Table 3. Bioactive components in medicinal mushrooms and their health-promoting effects.

Common Name	Latin Name	Compounds with Bioactive Potential	Health-Promoting Effects	References
Reishi	*Ganoderma lucidum* (Curtis) P. Karst.	Polysaccharides Glycoproteins (lectins) Phenols Steroids Triterpenoids Nucleotides Fatty acids Vitamins Minerals	Anti-inflammatory Anticancer Antiviral (including HIV) Antimicrobial Hypotensive effect Cardiotonic Immunomodelling Nephrotonic Hepatoprotective Neurotonic Anti-asthmatic	[21] [a], [51], [52] [a,b], [53]
Lion's Mane	*Hericium erinaceus* (Bull.) Pers.	Hericerins, Erinacins, Glycoprotein, Polysaccharides Beta-glucans, Sterols, Lactone, Fatty acids Volatile compounds (e.g., hexadecanoic acid, linoleic acid, phenylacetaldehyde, benzaldehyde)	Anticancer, Antioxidant, Anti-ageing, Imunomodelling, Neurotonic, Anti-asmatic, Hypoglycemic effects Hypocholesterolemic effects	[46] [a,b], [53], [54] [a,b]
Chaga	*Inonotus obliquus* (Ach. ex Pers.) Pilát	Polysaccharides Fatty acids Hydroxy acids Poliphenols (phenolic acids, flavonoids, coumarins, quinones, and styrylpyrones) Triterpenoids (lanosterol) Steroids (ergosterol and ergosterol peroxide)	Antioxidant, Anti-ageing, Antimicrobial activity, Antitumor activity, Anti-inflammatory hypoglycemic effect, Antilipidemic effect, Antiglication effect, Immunoregulatory Cardioprotective effects	[14] [b], [17,54,55]
Cordyceps	*Ophiocordyceps sinensis* (Berk.) G.H. Sung, J.M. Sung, Hywel-Jones & Spatafora *prior name Cordyceps sinensis*	Cordycepin (purine alkaloid) Cordymin (peptide) Adenosine Cordycepic acid (D-mannitol) Trehalose Polysaccharide Beta-glucans Saponins Polyunsaturated fatty acids, Ergosterol δ-tocopherol Hydroxybenzoic acid	Antitumor, Hypoglycemic effect Hypocholesterolemic effect, Anti-inflammatory, Antioxidant, Antiaging activity, Antimicrobial activity, Anticonvulsant activity, Cardiovascular protection (reduces cardiac arrhythmia and chronic heart failure)	[49], [56] [ab], [57] [a], [58] [a], [59] [a]
Shiitake	*Lentinula edodes* (Berk.) Pegler	Polysaccharides, Beta-glucans (lentinan) Glycoproteins, Phenols, Steroids, Terpenoids, Nucleotides	Immune-enhancing effects, Antitumor, Antioxidant, Antiaging activity, Antimicrobial activity, Hypocholesterolemic effect, Reduction in blood pressure	[26], [27] [a], [60]
Turkey Tail It also known as: Cloud mushroom, Yun Zhi, Kawaritake	*Trametes versicolor* (L.) Lloyd	Polysaccharopeptide (PSP) and polysaccharide K (PSK) (1,3)(1,6)-β-D-glucans, Poliphenols (phenolic acids: p-hydroxy benzoic, protocatechuic, vanillic, and homogentisic), Vitamin B, Fatty acids (linoleic, oleic, stearic, linolenic)	Antitumor Immunoregulatory, Antioxidant activity Prevent obesity, Antimicrobial, Antidiabetic AChE inhibitorY	[28,61], [62] [b]

[a]—in vivo studies. [b]—in vitro studies.

2.2. Proteins

In addition to polysaccharides, proteins, and peptides are important bioactive components that are present in mushrooms. Their content ranges widely from 4.6 to 56.3 g/100 g and is mainly determined by the mushroom species. Of the mushrooms discussed, Lion's Mane (*Hericium erinaceus* (Bull.) Pers.), Cordyceps (*Ophiocordyceps sinensis* (Berk.) G.H. Sung, J.M. Sung, Hywel-Jones and Spatafora), and Shiitake (*Lentinula edodes* (Berk.) Pegler) have the highest protein content (more than 20%) [23]. The amino acid composition and sequence and the length of the polypeptide chain can determine the specific biological activity of these compounds. They are most commonly ascribed to hypotensive, angiotensin-converting enzyme (ACE) inhibition, antioxidant, anticancer, antiviral, and antibacterial activities [63]. The most important bioactive fungal proteins include lectins (glycoproteins), immunomodulatory proteins, and proteins with enzymatic activity, e.g., nucleases, ribonucleases, laccase, and ergotionein [64]. Lectins increase insulin secretion and contribute to lowering blood sugar levels. In addition, they activate the immune system and show chemo-preventive effects against various types of cancer, e.g., hepatocellular carcinoma [65,66]. These compounds are present, e.g., in Reishi (*Ganoderma lucidum*). This type of protein, named TVC, was also isolated by Li et al. [67] from the fungus *Trametes versicolor*. As demonstrated by the authors, TVC increases the proliferation of human peripheral blood lymphocytes and is responsible for the increased necrosis of alpha tumour cells that are induced by mouse macrophages [67]. A characteristic low molecular weight peptide (cordymin) is present in cordyceps (*Ophiocordyceps sinensis* (Berk.) G.H. Sung, J.M. Sung, Hywel-Jones and Spatafora). Studies have demonstrated a protective role for this compound in lowering blood glucose levels in alloxan-induced hyperglycaemic rats. A dose of 50–100 mg/kg of the body weight of the animals also resulted in a reduction in aglycated haemoglobin (HbA11C) levels 5 weeks after the study. The oxidative stress induced by high sugar levels and the animal body weight decreased [68]. Numerous studies have shown that mushroom-derived protein has a complete amino acid profile. As highlighted by Thatoi and Singdevsachan [69], its nutritional value is even greater than that of milk, meat, or egg proteins. The protein present in mushrooms can be characterised by a high content of essential amino acids and glutamic acid, aspartic acid, or arginine. Pop et al. [70] reported that the tramates versicolor contained as many as 18 types of amino acids like aspartic acid, threonine, serine, glutamic acid, glycine, alanine, valine, and leucine, which were identified. Furthermore, studies have confirmed the presence of ornithine, which is known for its particular physiological activity, and the non-protein neurotransmitter γ-aminobutyric acid (GABA) [71].

2.3. Lipids

The fat content in mushrooms varies depending on the species but can range from 0.1 to 5.9 g/100 g [19]. Among medicinal mushrooms, Cordyceps (*Ophiocordyceps sinensis* (Berk.) G.H. Sung, J.M. Sung, Hywel-Jones, and Spatafora) and Shiitake (*Lentinula edodes* (Berk.) Pegler) are the most abundant in fat. About 52–87% of the fat is made up of unsaturated fatty acids (UFAs) such as oleic (C18:1) and linoleic (C18:2) acids [1,25,71,72]. These acids predominate, for example, in Cordyceps (*Ophiocordyceps sinensis* (Berk.) G.H. Sung, J.M. Sung, Hywel-Jones and Spatafora). The minor fatty acid in this mushroom is saturated fatty acids, e.g., palmitic (C16:0) and stearic (C18:0) acids. Guo et al. [72] found that the fatty acid profile could vary depending on geographical origin in the example of cordyceps (*Ophiocordyceps sinensis* (Berk.) G.H. Sung, J.M. Sung, Hywel-Jones and Spatafora).

Comparative examinations between indoor-cultivated and wild *Ophiocordyceps sinensis* (Berk.) G.H. Sung, J.M. Sung, Hywel-Jones, and Spatafora, conducted by Guo et al. [72], demonstrated that the wild mushrooms were characterized by a higher PUFAs (Polyunsaturated Fatty Acids) content with indoor-cultivated mushrooms. Such fatty acids as oleic acid, hydroxydocosanoic acid, hydroxytricosanoic acid, hydroxytetracosanoic acid, and hydroxypentacosanoic acid are predominate in Shiitake (*Lentinula edodes* (Berk.) Pegler) mushrooms. Stearic acid, hydroxyhexacosanoic acid, linoleic acid, palmitic acid, hydroxyarachidic acid, hydroxyheneicosanoic acid, and hydroxy-tricosenoic acid are present in smaller amounts [17].

Linoleic acid is known to have anticancer effects on breast, colon, and prostate cancer; thus, as a natural source of this acid, medicinal mushrooms also exhibit such properties [73]. Furthermore, unsaturated fatty acids can be used for the production of tissue hormones and are useful in preventing excessive blood clotting.

2.4. Sterols

Mushrooms are also a source of sterols classified as bioactive compounds. The most common of these is ergosterol. This compound undergoes photolysis to vitamin D2 when exposed to UV radiation [74]. A study conducted by Zheng et al. [75] showed that ergosterol exhibited cytotoxicity towards acute promyelocytic leukaemia cancer cells and liver cancer cells. At the same time, the authors noted moderate antimicrobial activity against selected bacteria and fungi. A characteristic sterol named H1-A, which resembles testosterone and dehydroepiandrosterone in its structure, was isolated from Cordyceps. In vivo studies in mice have shown that this compound could be effective for the treatment of selected autoimmune diseases [76].

2.5. Polyphenols

Thanks to the presence of polyphenols, including mainly phenolic acids represented by benzoic acid and cinnamic acid derivatives, medicinal mushrooms can be attributed to antioxidant activity. As reported by Ahmed et al. [77], gallic, caffeic, and p-coumaric acids are the predominant phenolic compounds in mushrooms. Phenolic compounds that are present in mushrooms exhibit strong antioxidant properties [78]. They inhibit free radicals and limit peroxide decomposition, scavenge reactive oxygen species, and block the action of metals when catalysing oxidation reactions [79–81]. Thus, they prevent mutations of cellular DNA and reduce the processes of carcinogenesis [82]. Peng and Shahidi [17] analysed Chaga ethanol extracts and detected 111 different phenolic compounds, including phenolic acids, flavonoids, coumarins, quinones, and styrylpyrones. Flavonoids in medicinal mushrooms are represented by myricetin, rutin, naringenin, quercetin, morin, and hesperetin [83]. Research conducted by Sharpe et al. [78] showed that, among many medicinal mushrooms, Chaga (*Inonotus obliquus* (Ach. Ex Pers.) Pilát) had the highest polyphenolic content and the highest antioxidant activity. The total phenolic content in this mushroom was at 97 μmol GAE/mg, while the content in reishi (*Ganoderma lucidum* (Curtis) P. Karst.), shiitake, and turkey tail (*Trametes versicolor* (L.) Lloyd) was 21, 13, and 0.1 μmol GAE/mg, respectively. The water-ethanol extract from Chaga (*Inonotus obliquus* (Ach. Ex Pers.) Pilát) exhibited approximately five times higher antioxidant activity against DPPH than other mushrooms. Mushroom polyphenols exhibit multidirectional beneficial effects on the human body: anticancer, antioxidant, hypoglycemic, slowing down the aging process, and preventing the degenerative diseases of the nervous system and cardiovascular diseases. When used as a food additive, they reduce fat oxidation processes and extend the shelf life of products [79].

2.6. Terpenes and Terpenoids

Another group of compounds includes terpenes, with the general formula $(C_5H_8)_n$, and terpenoids containing additional functional groups (-OH, -CHO, =CO, -COOH, -O-O-). Triterpenes are the main biologically active metabolites of terpenoid nature and are synthesized by *Ganoderma lucidum* (Curtis) P. Karst. and *Inonotus obliquus* (Ach. ex Pers.) Pilát. Data in the literature have reported that large amounts of these compounds, e.g., in reishi (*Ganoderma lucidum* (Curtis) P. Karst.) and chaga (*Inonotus obliquus* (Ach. ex Pers.) Pilát). Terpenes exert primarily anti-inflammatory effects. Triterpenes isolated from *Ganoderma lucidum* (Curtis) P. Karst. and *Inonotus obliquus* (Ach. ex Pers.) Pilát reduced the secretion of pro-inflammatory cytokines in macrophages (such as TNF-α, IL-1β, and IL-6) and the inflammatory mediators of nitric oxide (NO) and prostaglandin E2 (PGE2) [84,85]. Similarly, anti-inflammatory properties were exhibited by lanostane-type triterpene acids present in *Ganoderma lucidum* (Curtis) P. Karst., which, as shown by Akihisa et al. [85], inhibited the inflammatory process induced in mouse macrophages. In addition to anticholinesterase activity, the beneficial effects of mushroom terpenes have been reported in anticancer, antiviral, antimalarial, and antimalarial treatments [86,87]. The pharmacological effect of triterpenoids has been employed in the treatment of neurodegenerative diseases, including Alzheimer's disease [88].

2.7. Vitamins and Minerals

The nutritional value of medicinal mushrooms is also related to their high vitamin and micronutrient content. The vitamins present in mushrooms are mainly fat-soluble vitamins, including A and E. as well as vitamin D2 (ergocalciferol) and provitamin D2 (ergosterol). Interestingly, medicinal mushrooms are considered to be the only non-animal raw material that contains vitamin D [27,89]. Thanks to their tocopherol content, medicinal mushrooms exhibit antioxidant properties [89]. In addition, medicinal mushrooms are a very good source of water-soluble B vitamins (B1, B2, B3, B6, B9, B12) and vitamin C [19]. The vitamin B12 found in medicinal mushrooms was an analogue of that found in beef, fish, and liver, indicating its highly bioavailable. Therefore, mushrooms can be a valuable addition to vegetarian and vegan diets [90,91]. Shitake is rich in vitamins that exhibit antioxidant properties such as A, E, and C [27]. Medicinal mushrooms are rich in valuable mineral elements, including K, P, Na, Ca, and Mg, and, in smaller amounts, Cu, Zn, Fe, Mo, and Cd [27,79]. Given the ability of fungi to accumulate such heavy metals as Cd, Pb, Ar, Cu, Ni, Ag, Cr, and Hg, it is important that they grow in the least contaminated environment possible [92].

3. Possibilities of Using Medicinal Mushrooms for Functional Food Production

Medicinal mushrooms and mushroom-derived preparations containing bioactive compounds are classified as nutraceuticals. According to the European Food Safety Agency, they can be used as supplements due to their health-promoting and disease-preventing activity [93]. The production of nutraceuticals requires a great deal of knowledge of the functional properties of individual mushroom species. Due to the possible presence of substances that are harmful to health, it is necessary to control the origin, cultivation conditions, and raw material processing in order to ensure the health and safety of nutraceutical products on the one hand and an adequate level of biologically active compounds on the other to guarantee the beneficial effects of its preparation on health [94]. Clinical studies have shown that the recommended dose of nutraceutical preparations varies depending on the diagnosis and the patient [95]. Currently, a variety of fungal preparations are commercially available, most commonly in the dry extract form. There is growing interest in exploring the possibility of using various medicinal mushroom preparations to develop functional foods. An example of a popular product with medicinal mushrooms such as Chaga (*Inonotus obliquus* (Ach. ex Pers.) Pilát), cordyceps (*Ophiocordyceps sinensis* (Berk.) G.H. Sung, J.M. Sung, Hywel-Jones & Spatafora), shiitake, lion's mane (*Hericium erinaceus* (Bull.) Pers.) or reishi (*Ganoderma lucidum* (Curtis) P. Karst.) is coffee. Its consumption regulates blood pressure, prevents heartburn, stimulates mental performance, boosts energy, and strengthens the immune system and performance of the organism [96]. Some medicinal mushrooms have also been used to enrich cereals, meat, fish, and beverage products (Table 4). Of the mushrooms discussed so far, reishi (*Ganoderma lucidum* (Curtis) P. Karst.) and shiitake (*Lentinula edodes* (Berk.) Pegler) have been used most commonly. In all food products, the addition of dried and powdered mushrooms resulted in an increase in protein and the total and insoluble dietary fibre and significantly increased the micronutrient content [66]. The introduction of mushroom powder at 5% in such bakery products as bread and biscuits did not have adverse effects of their quality [97]. In the case of additions above 5%, a deterioration in texture was often noted not only in bread but also in pasta, yoghurt, and cured meats (Table 4). There is no information in the literature on the possibility of using turkey tail (*Trametes versicolor* (L.) Lloyd), cordyceps (*Ophiocordyceps sinensis* (Berk.) G.H. Sung, J.M. Sung, Hywel-Jones and Spatafora), lion's mane (*Hericium erinaceus* (Bull.) Pers.), or Chaga (*Inonotus obliquus* (Ach. ex Pers.) Pilát) preparations for food enrichment. Given the high health-promoting potential of these mushrooms, further research into the possibility of developing new functional foods with the above-mentioned mushrooms is advisable. The effect on the addition of selected medicinal mushrooms on quality parameters and the chemical composition of food products is presented in Table 4. An important issue in the design and implementation of new food products is sensory quality. Scientific studies have shown that the addition of medicinal mushrooms to foods, especially in a crushed or powdered form, can have a negative effect on the taste, texture, flavour, colour, and appearance of products. The addition of alcoholic or aqueous mushroom extracts has a less negative impact on the sensory quality and, with a small amount (up to 4%), can even improve the selected sensory characteristics of products.

Table 4. Use of medicinal mushrooms for food enrichment.

Common Name	Latin Name	Product/ Size of Additive	Impact on Chemical Composition (~) Lack of Impact (↓) Decrease (↑) Increase	Impact on Quality Parameters	References
Reishi	*Ganoderma lucidum* (Curtis) P. Karst.	Smoked fish sausage 1% of crushed mushroom	(↑) Antioxidant properties (↑) Total phenol content: + (↓)Moisture: − (↑) Ash: + (↑) Protein: + (↓)Fat: − Fiber: +	(↑) Shelf life (↓) Texture Sensory evaluation: (↓) flavour, (↓) colour, (↓) taste, (↓) texture, (↓) appearance, (↓) overall	[98]
		1% of water extract	(↑) Antioxidant properties (↑) Total phenol content (↓) Moisture: − (↑) Ash (↓) Protein: − (~) Fat (↑) Fiber: +	(↑) Shelf life (↑) Texture - Sensory evaluation: (↑) flavour, (↑) colour, (↑) taste, (↑) texture, (~) appearance, (↑) overall	
		0.25% of spore	(↑) Antioxidant properties (↑) Total phenol content (↓) Moisture: − (↑) Ash (↑) Protein (↑) Fat (↓) Fiber	(↑) Shelf life (~) Texture Sensory evaluation: (↓) flavour, (↓) colour, (↓) taste, (↓) texture, (~) appearance, (↓) overall	
Reishi	*Ganoderma lucidum* (Curtis) P. Karst.	Pilzner beer 0.1–1.5 mL/L of alcohol extract		Sensory evaluation: (~) aroma (↑) taste (↑) body (↑) bitterness (↑) liveliness (↑) overall impression	[99]
Reishi	*Ganoderma lucidum* (Curtis) P. Karst.	Emulsion Type Sausage 1% of dried fruiting bodies	(↑) Antioxidant properties	Sensory evaluation: (~) texture − (↓) taste (↓) Colour − (↓) Smell (↓) Acceptability (~) Peroxide value (↑) Microbiological analysis +	[100]

Table 4. *Cont.*

Common Name	Latin Name	Product/ Size of Additive	Impact on Chemical Composition (~) Lack of Impact (↓) Decrease (↑) Increase	Impact on Quality Parameters	References
Reishi	*Ganoderma lucidum* (Curtis) P. Karst.	Bread 2/4/6/8% water extract		(↑) Baking loss (↓) Bitterness Sensory evaluation: (↑) 2–4%, (↓) 6–8% Texture: (~) 2–4% (↓) 6–8% –	[101]
Reishi	*Ganoderma lucidum* (Curtis) P. Karst.	Yoghurt 2% Industrial waste (residues from aqueous extraction)	(↑) anti-coli effect, (↑) against *E. coli*	(↓) Texture (↓) Taste	[102]
Reishi	*Ganoderma lucidum* (Curtis) P. Karst.	Semolina pasta enriched with 2.5 and 5% of mushroom powder	(~) Phenolic compounds (↑) ABTS antiradical properties (↑) Syringic acid (~) β-glucan content (~) Anticancer properties	Not analyzed	[103]
Lion's Mane	*Hericium erinaceus* (Bull.) Pers.	Semolina pasta enriched with 2.5 and 5% of mushroom powder	(~) Phenolic compounds (~) Antioxidant properties (~) ABTS antiradical properties (↑) Vanilin (~) β-glucan content (~) Anticancer properties	Not analyzed	[103]
Shiitake	*Lentinula edodes* (Berk.) Pegler	Biscuits with mushroom powder 10%	(↑) Protein (↑) Mineral (Fe, P, Zn, Ca) (↑) Total and insoluble dietary fibre	Sensory evaluation: (~) aroma, (~) colour, (~) texture, (~) shelf life	[104]
Shiitake	*Lentinula edodes* (Berk.) Pegler	Bread enriched with 5–15% addition of mushroom powder	(↑) Dietary fiber	Bread dough: (↑) water absorption; (↓) development time; (↓) stability; >5% decreased the dough strength Bread quality physical: (↓) loaf height; (↑) moisture content; (↓) specific volume; >5% (↑) bread's gumminess; >5% bread's (↑) hardness; (↓) porosity	[97]
		Pork patties 0–6% addition to mushroom powder	Not analyzed	(↑) texture +; (↑) juiciness +; (↑) moisture +	[105]
		Semolina pasta enriched with 5–15% addition of mushroom powder	Not analyzed	(↑) Cooking loss (~) Water absorption; (~) Moisture content; (~) Tensile strength; (↑) Firmness	[106]

4. Conclusions or Concluding Remarks

To date, a great deal of research has already been conducted into medicinal mushrooms; however, given the diversity of species and the amount of bioactive substances contained therein, this area still appears to be incompletely explored. It seems advisable to conduct research not only to isolate and identify the bioactive substances present in mushrooms but also to conduct clinical experiments to confirm the therapeutic effect of these substances. Such studies could facilitate a determination of the dose and duration of use for mushroom nutraceuticals. Toxicological studies confirming the safety of medicinal mushrooms are also needed. In the context of using medicinal mushrooms for the development of functional foods, it is important to study the interactions between the biologically active compounds present in mushrooms and food ingredients. It is important to bear in mind that the components present in the food matrix may act synergistically or antagonistically with mycochemicals, increasing or reducing their beneficial physiological effects, respectively. Based on the analysis of available information and scientific research, it can be concluded that the addition of medicinal mushrooms to foods, especially cereal products, can make their chemical composition more attractive due to their great health-promoting properties and the presence of biologically active compounds. Medicinal mushrooms are known for their potential to improve immunity, regulate metabolism, and prevent many diseases. The abundance of polysaccharides, polyphenols, amino acids, and vitamins in medicinal mushrooms is a valuable source of biologically active compounds that can contribute to maintaining the health and well-being of the body. At the same time, further scientific research is needed to confirm these benefits and develop optimal methods for the addition of medicinal mushrooms to foods, taking into account technological, sensory, and food safety aspects.

Funding: This research was supported by project no. SD/54/TŻ/2022 provided by the University of Life Sciences in Lublin, Poland.

Institutional Review Board Statement: Not applicable.

Informed Consent Statement: Not applicable.

Data Availability Statement: Data sharing not applicable.

Conflicts of Interest: The authors declare no conflict of interest.

References

1. Wani, A.; Bodha, R.H.; Wani, A.H. Nutritional and medicinal importance of mushrooms. *J. Med. Plants Res.* **2010**, *4*, 2598–2604. [CrossRef]
2. Feeney, M.J.; Miller, A.M.; Roupas, P. Mushrooms-Biologically Distinct and Nutritionally Unique: Exploring a "Third Food Kingdom". *Nutr. Today* **2014**, *49*, 301–307. [CrossRef] [PubMed]
3. Verma, A.K.; Prakash, S. Status of Animal Phyla in Different Kingdom Systems of Biological Classification. *Int. J. Biol. Innov.* **2020**, *2*, 149–154. [CrossRef]
4. Beulah, H.; Margret, A.A.; Nelson, J. Marvelous Medicinal Mushrooms. *Int. J. Pharma Bio Sci.* **2013**, *3*, 611–615.
5. Hyde, K.D. The numbers of fungi. *Fungal Divers.* **2022**, *114*, 1. [CrossRef]
6. Elkhateeb, W.A.; Daba, G.M.; Thomas, P.W.; Wen, T.-C. Medicinal Mushrooms as a Source of Natural Therapeutic Bioactive Compounds. *Egypt. Pharm. J.* **2019**, *18*, 145–155. [CrossRef]
7. Weaver, C.; Marr, E.T. White vegetables: A forgotten source of nutrients: Purdue roundtable executive summary. *Adv. Nutr.* **2013**, *4*, 318–326. [CrossRef]
8. Song, T.; Zhang, Z.; Liu, S.; Chen, J.; Cai, W. Effect of Cultured Substrates on the Chemical Composition and Biological Activities of Lingzhi or Reishi Medicinal Mushroom, Ganoderma lucidum (*Agaricomycetes*). *Int. J. Med. Mushrooms* **2020**, *22*, 1183–1190. [CrossRef] [PubMed]
9. Elkhateeb, W.A.; Daba, G.M. Medicinal mushroom: What should we know? *Int. J. Pharm. Chem. Anal.* **2022**, *9*, 1–9. [CrossRef]
10. Cateni, F.; Gargano, M.L.; Procida, G.; Venturella, G.; Cirlincione, F.; Ferraro, V. Mycochemicals in wild and cultivated mushrooms: Nutrition and health. *Phytochem. Rev.* **2022**, *21*, 339–383. [CrossRef]
11. Barros, L.; Baptista, P.; Estevinho, L.M.; Ferreira, I.C.F.R. Effect of Fruiting Body Maturity Stage on Chemical Composition and Antimicrobial Activity of Lactarius sp. Mushrooms. *J. Agric. Food Chem.* **2007**, *55*, 8766–8771. [CrossRef]
12. Safin, R.R.; Gainullin, R.H.; Safina, A.V.; Gainullin, R.H. Methods for evaluating chaga extraction effectiveness based on its porosity change. *J. Phys. Conf. Ser.* **2022**, *2373*, 042007. [CrossRef]
13. Huang, G.; Cai, W.; Xu, B. Vitamin D2, Ergosterol, and Vitamin B2 Content in Commercially Dried Mushrooms Marketed in China and Increased Vitamin D2 Content Following UV-C Irradiation. *Int. J. Vitam. Nutr. Res.* **2017**, *87*, 237–246. [CrossRef] [PubMed]

14. Glamočlija, J.; Ćirić, A.; Nikolić, M.; Fernandes, Â.; Barros, L.; Calhelha, R.C.; Ferreira, I.C.F.R.; Soković, M.; van Griensven, L.J.L.D. Chemical characterization and biological activity of Chaga (*Inonotus obliquus*), a medicinal "mushroom". *J. Ethnopharmacol.* **2015**, *162*, 323–332. [CrossRef]
15. Géry, A.; Dubreule, C.; André, V.; Rioult, J.P.; Bouchart, V.; Heutte, N.; Eldin de Pécoulas, P.; Krivomaz, T.; Garon, D. Chaga (*Inonotus obliquus*), a Future Potential Medicinal Fungus in Oncology? A Chemical Study and a Comparison of the Cytotoxicity Against Human Lung Adenocarcinoma Cells (A549) and Human Bronchial Epithelial Cells (BEAS-2B). *Integr Cancer Ther.* **2018**, *17*, 832–843. [CrossRef]
16. Yang, F.Q.; Li, D.Q.; Feng, K.; Hu, D.J.; Li, S.P. Determination of nucleotides, nucleosides and their transformation products in Cordyceps by ion-pairing reversed-phase liquid chromatography-mass spectrometry. *J. Chromatogr. A.* **2010**, *1217*, 5501–5510. [CrossRef]
17. Peng, H.; Shahidi, F. Qualitative Analysis of Secondary Metabolites of Chaga Mushroom (*Inonotus obliquus*): Phenolics, Fatty Acids, and Terpenoids. *J. Food Bioact.* **2022**, *17*, 56–57. [CrossRef]
18. Golianek, A.; Mazurkiewicz-Zapałowicz, K. Mushrooms in the human diet—Nutritional and pro-health value. *Kosmos* **2016**, *65*, 513–522.
19. Dimopoulou, M.; Kolonas, A.; Mourtakos, S.; Androutsos, O.; Gortzi, O. Nutritional Composition and Biological Properties of Sixteen Edible Mushroom Species. *Appl. Sci.* **2022**, *12*, 8074. [CrossRef]
20. Lu, Y.; Jia, Y.; Xue, Z.; Li, N.; Liu, J.; Chen, H. Recent Developments in *Inonotus obliquus* (*Chaga mushroom*) Polysaccharides: Isolation, Structural Characteristics, Biological Activities and Application. *Polymers* **2021**, *13*, 1441. [CrossRef] [PubMed]
21. Kyanko, M.V.; Canel, R.S.; Ludemann, V.; Pose, G.; Wagner, J.R. β-Glucan Content and Hydration Properties of Filamentous Fungi. *Prikl Biokhim Mikrobiol.* **2013**, *49*, 48–52. [CrossRef]
22. Parepalli, Y.; Chavali, M.; Sami, R.; Khojah, E.; Elhakem, A.; El Askary, A.; Singh, M.; Sinha, S.; El-Chaghaby, G. Evaluation of Some Active Nutrients, Biological Compounds and Health Benefits of Reishi Mushroom (*Ganoderma lucidum*). *Int. J. Pharmacol.* **2021**, *17*, 243–250. [CrossRef]
23. Chaturvedi, V.K.; Agarwal, S.; Gupta, K.K.; Ramteke, P.W.; Singh, M.P. Medicinal Mushroom: Boon for Therapeutic Applications. *3 Biotech* **2018**, *8*, 334. [CrossRef] [PubMed]
24. Martínez-Flores, H.E.; Maya-Cortés, D.C.; Figueroa-Cárdenas, J.D.; Garnica-Romo, M.G.; Ponce-Saavedra, J. Chemical composition and physicochemical properties of shiitake mushroom and high fiber products. *J. Food* **2009**, *7*, 7–14. [CrossRef]
25. Riaz, S.; Ahmad, A.; Farooq, R.; Ahmed, M.; Shaheryar, M.; Hussain, M. Edible Mushrooms, a Sustainable Source of Nutrition, Biochemically Active Compounds and Its Effect on Human Health. In *Current Topics in Functional Food*; IntechOpen: Rijeka, Croatia, 2022. [CrossRef]
26. Antunes, P.S.; Erpen-Dalla Corte, L.; Bueno, J.C.; Spinosa, W.A.; Resende, J.T.V.; Hata, F.T.; Cabrera, L.C.; Zeffa, D.M.; Gonçalves, L.S.; Constantino, L.V. Firmness and biochemical composition of Shitake and Shimeji commercialized in natura and consumers' opinion survey. *Hortic. Bras.* **2021**, *39*, 425–431. [CrossRef]
27. Reguła, J.; Siwulski, M. Dried Shiitake (*Lentinula edodes*) and Oyster (*Pleurotus ostreatus*) Mushrooms as a Good Source of Nutrient. *Acta Sci. Pol. Technol. Aliment.* **2007**, *6*, 135–142.
28. Kıvrak, I.; Kivrak, S.; Karababa, E. Assessment of Bioactive Compounds and Antioxidant Activity of Turkey Tail Medicinal Mushroom Trametes versicolor (*Agaricomycetes*). *Int. J. Med. Mushrooms* **2020**, *22*, 559–571. [CrossRef]
29. Vetter, J. The Mushroom Glucans: Molecules of High Biological and Medicinal Importance. *Foods* **2023**, *12*, 1009. [CrossRef]
30. Valverde, M.E.; Hernandez-Perez, T.; Paredes-Lopez, O. Edible mushroom: Improving human health and promoting quality life. *Int. J. Microbiol.* **2015**, *2015*, 376387. [CrossRef] [PubMed]
31. Villares, A.; Mateo-Vivaracho, L.; Guillamón, E. Structural Features and Healthy Properties of Polysaccharides Occurring in Mushrooms. *Agriculture* **2012**, *2*, 452–471. [CrossRef]
32. Cerletti, C.; Esposito, S.; Iacoviello, L. Edible Mushrooms and Beta-Glucans: Impact on Human Health. *Nutrients* **2021**, *13*, 2195. [CrossRef] [PubMed]
33. Meng, Y.; Lyu, F.; Xu, X.; Zhang, L. Recent advances in chain conformation and bioactivities of triple-helix polysaccharides. *Biomacromolecules* **2020**, *21*, 1653–1677. [CrossRef]
34. Rop, O.; Mlcek, J.; Jurikova, T. Beta-glucans in higher fungi and their health effects. *Nutr. Rev.* **2009**, *67*, 624–631. [CrossRef]
35. Falch, B.H.; Espevik, T.; Ryan, L.; Stokke, B.T. The cytokine stimulating activity of (1→3)-beta-D-glucans is dependent on the triple helix conformation. *Carbohydr. Res.* **2000**, *329*, 587–596. [CrossRef] [PubMed]
36. Sletmoen, M.; Stokke, B.T. Higher order structure of (1,3)-beta-D-glucans and its influence on their biological activities and complexation abilities. *Biopolymers* **2008**, *89*, 310–321. [CrossRef]
37. Brown, G.D.; Gordon, S. Fungal beta-glucans and mammalian immunity. *Immunity* **2003**, *19*, 311–315. [CrossRef] [PubMed]
38. Nitschke, J.; Modick, H.; Busch, E.; von Rekowski, R.W.; Altenbach, H.J.; Mölleken, H. A New Colorimetric Method to Quantify β-1,3–1,6-Glucans in Comparison with Total β-1,3-Glucans in Edible Mushrooms. *Food Chem.* **2011**, *127*, 791–796. [CrossRef]
39. Rasmy, G.E.; Botros, W.A.; Kabeil, S.; Daba, A.S. Preparation of Glucan from Lentinula edodes Edible Mushroom and Elucidation of Its Medicinal Value. *Aust. J. Basic Appl. Sci.* **2010**, *4*, 5717–5726.
40. Mirończuk-Chodakowska, I.; Witkowska, A.M. Evaluation of Polish Wild Mushrooms as Beta-Glucan Sources. *Int. J. Environ. Res. Public Health* **2020**, *17*, 7299. [CrossRef]

41. Trivedi, S.; Patel, K.; Belgamwar, V.; Wadher, K. Functional polysaccharide lentinan: Role in anti-cancer therapies and management of carcinomas. *Pharmacol. Res. Mod. Chin. Med.* **2022**, *2*, 100045. [CrossRef]
42. Wu, J.-Y.; Siu, K.-C.; Geng, P. Bioactive Ingredients and Medicinal Values of Grifola frondosa (Maitake). *Foods* **2021**, *10*, 95. [CrossRef] [PubMed]
43. Del Cornò, M.; Gessani, S.; Conti, L. Shaping the Innate Immune Response by Dietary Glucans: Any Role in the Control of Cancer? *Cancers* **2020**, *12*, 155. [CrossRef]
44. Song, H.-N. Functional Cordyceps Coffee Containing Cordycepin and β-Glucan Hyo-Nam Song. *Prev. Nutr. Food Sci.* **2020**, *25*, 184–193. [CrossRef] [PubMed]
45. Sari, M.; Prange, A.; Lelley, J.I.; Hambitzer, R. Screening of beta-glucan contents in commercially cultivated and wild growing mushrooms. *Food Chem.* **2017**, *216*, 45–51. [CrossRef]
46. Yuan, M.; Li, C.; Xiao, X.; Wan, D.; Xi, B.; Jiang, X.; Zhang, J. Effect of lentinan on proliferation and apoptosis of human astrocytoma U251 cells. *Pol J Pathol.* **2023**, *3*, 47758. [CrossRef] [PubMed]
47. Ataollahi, H.; Larypoor, M. Fabrication and investigation potential effect of lentinan and docetaxel nanofibers for synergistic treatment of breast cancer in vitro. *Polym. Adv. Technol.* **2022**, *33*, 1468–1480. [CrossRef]
48. Wang, Z.; Qu, K.; Zhou, L.; Ren, L.; Ren, B.; Meng, F.; Yu, W.; Wang, H.; Fan, H. Apaf1 NanoLuc biosensors identified lentinan as a potent synergizer of cisplatin in targeting hepatocellular carcinoma cells. *Biochem. Biophys. Res. Commun.* **2021**, *577*, 45–51. [CrossRef]
49. Abascal, K.Y.; Yarnell, E. A turkey tail polysaccharide as an immunochemotherapy agent in cancer. *Altern. Complement. Ther.* **2007**, *13*, 178–182. [CrossRef]
50. Thuy, D.T.P.; Anh, T.T.N.; Thuy, N.T.T.; Intaparn, P.; Tapingkae, T.; Mai, N.T. Simple and Efficient Method for the Detection and Quantification of Cordycepin Content in Cordyceps. *Chiang Mai J. Sci.* **2021**, *48*, 420–428.
51. Karishma, R.; Rachana, M. Potential Secondary Bioactive Compounds of Ganoderma lucidum (Reishi Mushroom) against Various Pathogenic Activity. *Pharmacologyonline* **2021**, *3*, 1923–1944.
52. Ahmad, M.F. Ganoderma lucidum: Persuasive biologically active constituents and their health endorsement. *Biomed. Pharmacother.* **2018**, *107*, 507–519. [CrossRef] [PubMed]
53. Turło, J. Large-flowered mushrooms—An underestimated source of medicinal substances. *Stud. I Mater. CEPL* **2015**, *17*, 138–151.
54. Thongbai, B.; Rapior, S.; Hyde, K.D.; Wittstein, K.; Stadler, M. *Hericium erinaceus*, an Amazing Medicinal Mushroom. *Mycol. Prog.* **2015**, *14*, 91. [CrossRef]
55. Doi, N.; Araki, K.; Fukuta, Y.; Kuwagaito, Y.; Yamauchi, Y.; Sasai, Y.; Kondo, S.; Kuzuya, M. Anti-glycation and antioxidant effects of Chaga mushroom decoction extracted with a fermentation medium. *Food Sci. Technol. Res.* **2023**, *29*, 155–161. [CrossRef]
56. Liu, Y.; Wang, J.; Wang, W.; Zhang, H.; Zhang, X.; Han, C. The Chemical Constituents and Pharmacological Actions of Cordyceps sinensis. Evid. Based Complement. *Altern. Med.* **2015**, *2015*, 575063. [CrossRef]
57. Wang, M.; Meng, X.Y.; Yang, R.L.; Qin, T.; Wang, X.Y.; Zhang, K.Y.; Fei, C.Z.; Li, Y.; Hu, Y.L.; Xue, F.Q. Cordyceps militaris polysaccharides can enhance the immunity and antioxidation activity in immunosuppressed mice. *Carbohydr. Polym.* **2012**, *89*, 461–466. [CrossRef]
58. Zhao, C.S.; Yin, W.T.; Wang, J.Y.; Zhang, Y.; Yu, H.; Cooper, R.; Smidt, C.; Zhu, J.S. CordyMax Cs-4 improves glucose metabolism and increases insulin sensitivity in normal rats. *J. Altern. Complement. Med.* **2002**, *8*, 309–314. [CrossRef]
59. Yan, X.-F.; Zhang, Z.-M.; Yao, H.-Y.; Guan, Y.; Zhu, J.-P.; Zhang, L.-H.; Jia, Y.-L.; Wang, R.-W. Cardiovascular protection and antioxidant activity of the extracts from the mycelia of Cordyceps sinensis act partially via adenosine receptors. *Phytother. Res.* **2013**, *27*, 1597–1604. [CrossRef]
60. Vetvicka, V.; Vetvickova, J. Immune-Enhancing Effects of Maitake (*Grifola frondosa*) and Shiitake (*Lentinula edodes*) Extracts. *Ann. Transl. Med.* **2014**, *2*, 14. [CrossRef]
61. Miletić, D.; Turło, J.; Podsadni, P.; Sknepnek, A.; Szczepańska, A.; Lević, S.; Nedović, V.; Nikšić, M. Turkey Tail Medicinal Mushroom, Trametes versicolor (*Agaricomycetes*), Crude Exopolysaccharides with Antioxidative Activity. *Int. J. Med. Mushrooms* **2020**, *22*, 885–895. [CrossRef]
62. Benson, K.F.; Stamets, P.; Davis, R.; Nally, R.; Taylor, A.; Slater, S.; Jensen, G.S. The mycelium of the *Trametes versicolor* (Turkey tail) mushroom and its fermented substrate each show potent and complementary immune activating properties in vitro. *BMC Complement. Altern. Med.* **2019**, *19*, 342. [CrossRef] [PubMed]
63. Landi, N.; Clemente, A.; Pedone, P.V.; Ragucci, S.; Di Maro, A. An Updated Review of Bioactive Peptides from Mushrooms in a Well-Defined Molecular Weight Range. *Toxins* **2022**, *14*, 84. [CrossRef] [PubMed]
64. Sousa, A.S.; Araújo-Rodrigues, H.; Pintado, M.E. The health-promoting potential of edible mushroom proteins. *Current Pharm. Des.* **2023**, *29*, 804–823. [CrossRef]
65. El-Maradny, Y.A.; El-Fakharany, E.M.; Abu-Serie, M.M.; Hashish, M.H.; Selim, H.S. Lectins purified from medicinal and edible mushrooms: Insights into their antiviral activity against pathogenic viruses. *Int. J. Biol. Macromol.* **2021**, *179*, 239–258. [CrossRef]
66. Singh, R.S.; Kaur Preet, H.; Kanwar, J.R. Mushroom lectins as promising anticancer substances. *Curr. Protein Pept. Sci.* **2016**, *17*, 797–807. [CrossRef]
67. Li, F.; Wen, H.; Zhang, Y.; Aa, M.; Liu, X. Purification and characterization of a novel immunomodulatory protein from the medicinal mushroom Trametes versicolor. *Sci. China Life Sci.* **2011**, *54*, 379–385. [CrossRef]

68. Qi, W.; Zhang, Y.; Yan, Y.B.; Lei, W.; Wu, Z.X.; Liu, N.; Liu, S.; Shi, L.; Fan, Y. The Protective Effect of Cordymin, a Peptide Purified from the Medicinal Mushroom Cordyceps sinensis, on Diabetic Osteopenia in Alloxan-Induced Diabetic Rats. *Evid. Based Complement. Alternat. Med.* **2013**, *2013*, 985636. [CrossRef]
69. Thatoi, H.; Singdevsachan, S.K. Diversity, Nutritional Composition and Medicinal Potential of Indian Mushrooms: A Review. *Afr. J. Biotechnol.* **2014**, *13*, 523–545. [CrossRef]
70. Pop, R.M.; Puia, I.C.; Puia, A.; Chedea, V.S.; Leopold, N.; Bocsan, I.C.; Buzoianu, A.D. Characterization of Trametes versicolor: Medicinal Mushroom with Important Health Benefits. *Not. Bot. Horti Agrobo.* **2018**, *46*, 343–349. [CrossRef]
71. Tagkouli, D.; Kaliora, A.; Bekiaris, G.; Koutrotsios, G.; Christea, M.; Zervakis, G.I.; Kalogeropoulos, N. Free Amino Acids in Three Pleurotus Species Cultivated on Agricultural and Agro-Industrial By-Products. *Molecules* **2020**, *25*, 4015. [CrossRef]
72. Guo, L.-X.; Xu, X.-M.; Wu, C.-F.; Lin, L.; Zou, S.-C.; Luan, T.-G.; Yuan, J.-P.; Wang, J.-H. Fatty acid composition of lipids in wild Cordyceps sinensis from major habitats in China. *Biomed. Prev. Nutr.* **2012**, *2*, 42–50. [CrossRef]
73. Kim, J.-H.; Hubbard, N.E.; Ziboh, V.; Kelly, L. Erickson. Conjugated Linoleic Acid Reduction of Murine Mammary Tumor Cell Growth through 5-Hydroxyeicosatetraenoic. *Acid. Biochim. Biophys. Acta* **2005**, *1687*, 103–109. [CrossRef]
74. Urbain, P.; Singler, F.; Ihorst, G.; Biesalski, H.K.; Bertz, H. Bioavailability of vitamin D2 from UV-B-irradiated button mushrooms in healthy adults deficient in serum 25-hydroxyvitamin D: A randomized controlled trial. *Eur. J. Clin. Nutr.* **2011**, *65*, 965–971. [CrossRef] [PubMed]
75. Zheng, J.; Wang, Y.; Wang, J.; Liu, P.; Li, J.; Zhu, W. Antimicrobial Ergosteroids and Pyrrole Derivatives from Halotolerant Aspergillus flocculosus PT05-1 Cultured in a Hypersaline Medium. *Extremophiles* **2013**, *17*, 963–971. [CrossRef] [PubMed]
76. Yang, L.-Y.; Huang, W.-J.; Hsieh, H.-G.; Lin, C.-Y. H1-A Extracted from Cordyceps sinensis, Suppresses the Proliferation of Human Mesangial Cells and Promotes Apoptosis, Probably by Inhibiting the Tyrosine Phosphorylation of Bcl-2 and Bcl-XL. *J. Lab. Clin. Med.* **2003**, *141*, 74–83. [CrossRef]
77. Ahmed, A.F.; Mahmoud, G.A.-E.; Hefzy, M.; Liu, Z.; Ma, C. Overview on the edible mushrooms in Egypt. *J. Future Foods* **2023**, *3*, 8–15. [CrossRef]
78. Sharpe, E.; Farragher-Gnadt, A.; Igbanugo, M.; Huber, T.; Michelotti, J.C.; Milenkowic, A.; Ludlam, S.; Walker, M.; Hanes, D.; Bradley, R.; et al. Comparison of Antioxidant Activity and Extraction Techniques for Commercially and Laboratory Prepared Extracts from Six Mushroom Species. *J. Agric. Food Res.* **2021**, *4*, 100130. [CrossRef]
79. Das, A.K.; Nanda, P.K.; Dandapat, P.; Bandyopadhyay, S.; Gullón, P.; Sivaraman, G.K.; McClements, D.J.; Gullón, B.; Lorenzo, J.M. Edible Mushrooms as Functional Ingredients for Development of Healthier and More Sustainable Muscle Foods: A Flexitarian Approach. *Molecules* **2021**, *26*, 2463. [CrossRef]
80. Podkowa, A.; Kryczyk-Poprawa, A.; Opoka, W.; Kozarski, M.; Wróbel, M.S. Culinary–Medicinal Mushrooms: A Review of Organic Compounds and Bioelements with Antioxidant Activity. *Eur. Food Res. Technol.* **2021**, *247*, 513–533. [CrossRef]
81. Ma, G.; Yang, W.; Zhao, L.; Pei, F.; Fang, D.; Hu, Q. A critical review on the health promoting effects of mushrooms nutraceuticals. *Food Sci. Hum. Wellness* **2018**, *7*, 125–133. [CrossRef]
82. Taşkın, H.; Süfer, Ö.; Attar, S.H.; Kılıç, Ö.; Güzel, M.; Atakol, O. Total Phenolics, Antioxidant Activities and Fatty Acid Profiles of Six Morchella Species. *J. Food Sci. Technol.* **2021**, *58*, 692–700. [CrossRef] [PubMed]
83. Saltarelli, R.; Palma, F.; Gioacchini, A.M.; Bucchini, A.; Chiarini, A.; Pellegrini, A.; Rocchi, M.B.L.; Stocchi, V. Phytochemical Composition, Antioxidant and Antiproliferative Activities and Effects on Nuclear DNA of Ethanolic Extract from an Italian Mycelial Isolate of Ganoderma lucidum. *J. Ethnopharmacol.* **2019**, *231*, 464–473. [CrossRef] [PubMed]
84. Van, Q.; Nayak, B.N.; Reimer, M.; Jones, P.J.H.; Fulcher, R.G.; Rempel, C.B. Anti-inflammatory effect of Inonotus obliquus, Polygala senega L.; and Viburnum trilobum in a cell screening assay. *J. Ethnopharmacol.* **2009**, *125*, 487–493. [CrossRef] [PubMed]
85. Akihisa, T.; Nakamura, Y.; Tagata, M.; Tokuda, H.; Yasukawa, K.; Uchiyama, E.; Suzuki, T.; Kimura, Y. Anti-Inflammatory and Anti-Tumor-Promoting Effects of Triterpene Acids and Sterols from the Fungus Ganoderma lucidum. *Chem. Biodivers.* **2007**, *4*, 105–255. [CrossRef]
86. Wang, S.; Bao, L.; Zhao, F.; Wang, Q.; Li, S.; Ren, J.; Li, L.; Wen, H.; Guo, L. Isolation, Identification, and Bioactivity of Monoterpenoids and Sesquiterpenoids from the Mycelia of Edible Mushroom Pleurotus cornucopiae. *J. Agric. Food Chem.* **2013**, *61*, 5122–5129. [CrossRef]
87. Dasgupta, A.; Acharya, K. Mushrooms: An Emerging Resource for Therapeutic Terpenoids. *3 Biotech.* **2019**, *9*, 369. [CrossRef]
88. Zhang, X.; Zhang, S.; Yang, Y.; Wang, D.; Gao, H. Natural barrigenol-like triterpenoids: A comprehensive review of their contributions to medicinal chemistry. *Phytochemistry* **2019**, *161*, 41–74. [CrossRef]
89. Teichmann, A.; Dutta, P.C.; Staffas, A.; Jägerstad, M. Sterol and vitamin D2 concentrations in cultivated and wild grown mushrooms: Effects of UV irradiation. *LWT Food Sci. Technol.* **2007**, *40*, 815–822. [CrossRef]
90. Feeney, M.J.; Dwyer, J.; Hasler-Lewis, C.M.; Milner, J.A.; Noakes, M.; Rowe, S.; Wach, M.; Beelman, R.B.; Caldwell, J.; Cantorna, M.T.; et al. Mushrooms and Health Summit Proceedings. *J. Nutr.* **2014**, *144*, 1128S–1136S. [CrossRef]
91. Gründemann, C.; Reinhardt, J.K.; Lindequist, U. European medicinal mushrooms: Do they have potential for modern medicine?–An update. *Phytomedicine* **2020**, *66*, 153131. [CrossRef]
92. Waktola, G.; Temesgen, T. Application of Mushroom as Food and Medicine. *Adv. Biotechnol. Microbiol.* **2018**, *11*, 555817. [CrossRef]
93. Sachdeva, V.; Roy, A.; Bharadvaja, N. Current Prospects of Nutraceuticals: A Review. *Curr. Pharm. Biotechnol.* **2020**, *21*, 884–896. [CrossRef] [PubMed]
94. Benkeblia, N. *Polysaccharides Natural Fibres in Food and Nutrition*; CRC Press: Boca Raton, FL, USA, 2014. [CrossRef]

95. Rathore, H.; Prasad, S.; Sharma, S. Mushroom Nutraceuticals for Improved Nutrition and Better Human Health: A Review. *PharmaNutrition* **2017**, *5*, 35–46. [CrossRef]
96. Krzystyniak, K.L.; Klonowska, J. *New Trends in Dietetics*; University of Engineering and Health: Warsaw, Poland, 2019; p. 27. ISBN 978-83-942432-4-1.
97. Lu, X.; Brennan, M.A.; Serventi, L.; Brennan, C.S. Incorporation of Mushroom Powder into Bread Dough—Effects on Dough Rheology and Bread Properties. *Cereal Chem.* **2018**, *95*, 418–427. [CrossRef]
98. Wannasupchue, W.; Siriamornpun, S.; Huaisan, K.; Huaisan, J.; Meeso, N. Effect of Adding Ling-zhi (*Ganoderma lucidum*) on Oxidative Stability, Textural and Sensory Properties of Smoked Fish Sausage. *Thai J. Agric. Sci.* **2011**, *44*, 505–512.
99. Leskosek-Cukalovic, I.; Despotovic, S.; Lakic, N.; Niksic, M.; Nedovic, V.; Tesevic, V. Ganoderma lucidum—Medical Mushroom as a Raw Material for Beer with Enhanced Functional Properties. *Food Res. Int.* **2010**, *43*, 2262–2269. [CrossRef]
100. Ghobadi, R.; Mohammadi, R.; Chabavizade, J.; Sami, M. Effect of Ganoderma lucidum Powder on Oxidative Stability, Microbial and Sensory Properties of Emulsion Type Sausage. *Adv. Biomed. Res.* **2018**, *7*, 135. [CrossRef]
101. Chung, H.C.; Lee, J.T.; Kwon, O.J. Bread Properties Utilizing Extracts of Ganoderma lucidum (GL). *J. Korean Soc. Food Sci. Nutr.* **2004**, *33*, 1201–1205. [CrossRef]
102. Jovanović, M.; Vojvodić, P.; Petrović, M.; Radić, D.; Mitić-Ćulafić, D.; Kostić, M.; Veljović, S. Yogurt Fortified with GABA-Producing Strain and Ganoderma lucidum Industrial Waste. *Czech J. Food Sci.* **2022**, *40*, 456–464. [CrossRef]
103. Szydłowska-Tutaj, M.; Szymanowska, U.; Tutaj, K.; Domagała, D.; Złotek, U. The Addition of Reishi and Lion's Mane Mushroom Powder to Pasta Influences the Content of Bioactive Compounds and the Antioxidant, Potential Anti-Inflammatory, and Anticancer Properties of Pasta. *Antioxidants* **2023**, *12*, 738. [CrossRef]
104. Singh, J.; Sindhu, S.C.; Sindhu, A.; Yadav, A. Development and Evaluation of Value Added Biscuits from Dehydrated Shiitake (*Lentinus edodes*) Mushroom. *Int. J. Curr. Res.* **2016**, *8*, 27155–27159.
105. Chun, S.; Chambers, E., IV.; Chambers, D. Perception of Pork Patties with Shiitake (*Lentinus edodes*) Mushroom Powder and Sodium Tripolyphosphate as Measured by Korean and United States Consumers. *J. Sens. Stud.* **2005**, *20*, 156–166. [CrossRef]
106. Lu, X.; Brennan, M.A.; Serventi, L.; Mason, S.; Brennan, C.S. How the Inclusion of Mushroom Powder Can Affect the Physicochemical Characteristics of Pasta. *Int. J. Food Sci. Technol.* **2016**, *51*, 2433–2439. [CrossRef]

Disclaimer/Publisher's Note: The statements, opinions and data contained in all publications are solely those of the individual author(s) and contributor(s) and not of MDPI and/or the editor(s). MDPI and/or the editor(s) disclaim responsibility for any injury to people or property resulting from any ideas, methods, instructions or products referred to in the content.

molecules

MDPI

Article

Comprehensive Quality Evaluation of *Polygonatum cyrtonema* and Its Processed Product: Chemical Fingerprinting, Determination and Bioactivity

Jianguang Zhang [1,2], Junjun Wang [1], Li Yang [1], Yue Wang [1], Wenfang Jin [1], Jing Li [1] and Zhifeng Zhang [1,*]

1 Tibetan Plateau Ethnic Medicinal Resources Protection and Utilization Key Laboratory of National Ethnic Affairs Commission of the People's Republic of China, Southwest Minzu University, Chengdu 610041, China; zhangjg0777@sina.com (J.Z.); 18308211183@163.com (L.Y.); yuewang724@163.com (Y.W.); jingli3485@163.com (J.L.)

2 Qin Zhou Provincial Health School, Qinzhou 535009, China

* Correspondence: zfzhang@swun.edu.cn; Tel.: +86-028-89165778; Fax: +86-028-85524382

Abstract: Processing of Chinese herbal medicines (CHMs) is a traditional pharmaceutical technology in Chinese medicine. Traditionally, proper processing of CHMs is necessary to meet the specific clinical requirements of different syndromes. Processing with black bean juice is considered one of the most important techniques in traditional Chinese pharmaceutical technology. Despite the long-standing practice of processing *Polygonatum cyrtonema* Hua (PCH), there is little research on the changes in chemical constituents and bioactivity before and after processing. This study investigated the influence of black bean juice processing on the chemical composition and bioactivity of PCH. The results revealed significant changes in both composition and contents during processing. Saccharide and saponin content significantly increased after processing. Moreover, the processed samples exhibited considerably stronger DPPH and ABTS radical scavenging capacity, as well as FRAP-reducing capacity, compared to the raw samples. The IC_{50} values for DPPH were 1.0 ± 0.12 mg/mL and 0.65 ± 0.10 mg/mL for the raw and processed samples, respectively. For ABTS, the IC_{50} values were 0.65 ± 0.07 mg/mL and 0.25 ± 0.04 mg/mL, respectively. Additionally, the processed sample demonstrated significantly higher inhibitory activity against α-glucosidase and α-amylase ($IC_{50} = 1.29 \pm 0.12$ mg/mL and 0.48 ± 0.04 mg/mL) compared to the raw sample ($IC_{50} = 5.58 \pm 0.22$ mg/mL and 0.80 ± 0.09 mg/mL). These findings underscore the significance of black bean processing in enhancing the properties of PCH and lay the foundation for its further development as a functional food. The study elucidates the role of black bean processing in PCH and offers valuable insights for its application.

Keywords: *Polygonatum cyrtonema* Hua; UPLC-Q-Exactive-MS/MS; GC-MS; processing; bioactivities

Citation: Zhang, J.; Wang, J.; Yang, L.; Wang, Y.; Jin, W.; Li, J.; Zhang, Z. Comprehensive Quality Evaluation of *Polygonatum cyrtonema* and Its Processed Product: Chemical Fingerprinting, Determination and Bioactivity. *Molecules* **2023**, *28*, 4341. https://doi.org/10.3390/molecules28114341

Academic Editors: Eliza Oprea, Arunaksharan Narayanankutty and Ademola C. Famurewa

Received: 5 May 2023
Revised: 22 May 2023
Accepted: 23 May 2023
Published: 25 May 2023

Copyright: © 2023 by the authors. Licensee MDPI, Basel, Switzerland. This article is an open access article distributed under the terms and conditions of the Creative Commons Attribution (CC BY) license (https://creativecommons.org/licenses/by/4.0/).

1. Introduction

Polygonatum cyrtonema Hua (PCH) belongs to the genus Polygonatum (Asparagaceae), which is mainly distributed in the North Temperate zones of Asia, such as India, Afghanistan, Pakistan, Korea, China and Japan [1]. The plant is also widely planted in Sichuan, Guizhou, Hunan, Jiangxi, and other regions in China. The rhizomes of PCH, called "Huangjing" in China, have been widely used as a functional food and herbal medicine [2]. With the function of invigorating Qi, nourishing Yin and moistening the lungs, it is often used to replenish energy, strengthen immunity and treat fatigue, weakness, diabetes and lung disorders [1]. Modern pharmacological studies have shown that PCH exhibits antioxidant [3], antidiabetic [4], antitumour [5], anti-inflammatory [6] and anti-osteoporotic effects [1]. Multiple components of PCH can interact in a synergistic fashion to potentiate its pharmacological activity. The open innovation platform such as International Natural Product

Sciences Taskforce could help researchers to easily identify the major constituent(s) of PCH's exerting activity [7].

In China, PCH is used after processing, because processed PCH exhibits better effects than the raw [1]. Generally, the raw rhizomes of PCH contain mucilaginous ingredients, and they are rarely taken orally directly due to its stimulating tongue. Therefore, the raw must be processed to remove these ingredients and enhance its tonic function. Depending on the theories of traditional Chinese medicine, PCH is processed by steaming using wine, black soybean, ginger and honey before clinical applications [8]. The most common traditional processing method is to steam and sun-dry the rhizomes of PCH nine times until the rhizomes turn black, soft and sweet. Numerous studies have indicated that some chemical compositions are increased or decreased during processing, and these variations might have a significant effect on biological activities [3,9,10]. Processing with black beans is a specific pre-treatment for many Chinese herbs before drying, which is usually applied to Chinese herbs, including toxic or irritating components, which is also beneficial in enhancing the efficacy of nourishing the liver and kidney and to reduce the toxicity and side effects of herbs [11,12]. Processing of PCH with black beans is a method recorded in the Processing Standard of Chinese Herbal Pieces in Sichuan Province 2015 Edition [13], while fewer studies have reported processing with black beans in rhizomes of PCH. Moreover, no studies have reported the differences in chemical composition and bioactivities among the raw and processed black beans of PCH. There is a lack of credible information to clarify the variations during PCH processing. Thus, it is meaningful to study the variations in chemical composition and bioactivities from processing with black beans in PCH.

Studies on phytochemicals have indicated that PCH contains many types of bioactive components, including polysaccharides [14], oligosaccharides [15], steroidal saponins [16], triterpenoid saponins [17], flavonoids [18] and so on. Polysaccharides are regarded as the main chemical constituents with various bioactivities, such as antioxidant protection and antidiabetic action. Polysaccharide content was regarded as an index for the quality control of PCH in the Chinese Pharmacopoeia 2020 Edition (Volume I) [19]. Concerning the analysis of small biological molecules, ultrahigh-performance liquid chromatography coupled with time-of-flight mass spectrometry (UHPLC-Q-TOF-MS/MS), with the advantages of rapid analysis time and high selectivity and sensitivity, is a powerful tool for detecting and analysing small biomolecules [20–22]. It is also a good analytical technique to screen differential compounds during processing [23,24]. Polysaccharides are macromolecules and saccharides with complex structures. Many saccharide test methods have been developed, including gas chromatography-mass spectrometry (GC-MS), high-performance anion exchange chromatography (HPAEC), liquid chromatography-mass spectrometry (LC-MS), gas chromatography (GC) and high-performance liquid chromatography-evaporative light scattering detection (ELSD). GC-MS has the advantages of high sensitivity, strong separation ability and accurate qualitative identification, and it has been widely used in the analysis of saccharides [25,26]. Chemometrics analysis technology, combining mathematics and statistics, can objectively analyse complicated data, and it has been applied to the quality control of traditional Chinese medicine [27–29].

The objectives of this study were to investigate of the differences in chemical components and biological activity between a raw sample and a black bean processed sample of PCH; UHPLC-Q-Exactive-MS/MS was used to identify the steroidal saponins, triterpenoid saponins, flavonoids and so on. The polysaccharides and saponin contents were determined by UV. GC-QQQ-MS/MS was used to quantify the monosaccharide composition of PCH. The bioactivities of the raw and the processed PCH were evaluated in vitro by antioxidant and antidiabetic assays. This study aims to elucidate the physicochemical properties and activity regulation of polysaccharides and saponins present in the active ingredients of black bean-processed PCH and investigate the scientific implications of antioxidant and hypoglycaemic activities through the modulation of these polysaccharides and saponins. Furthermore, the study's results may provide valuable insights into the development of novel foods and functional products utilizing polysaccharides and saponins from black bean processed PCH.

2. Results and Discussion

2.1. Method Optimization

To obtain satisfactory separation and more peak pattern, the influence of mobile phase (water-acetonitrile, water-methanol, 0.1% aqueous formic acid-acetonitrile and 0.1% aqueous formic acid-methanol), column type, flow rate (0.2, 0.3 and 0.5 mL/min), detection wavelength (210, 254 and 280 nm) and column temperature (25, 30 and 35 °C) were evaluated. As a result, a Hypersil GOLD C18 column (2.1 × 100 mm, 1.8 μm) was chosen, with the best chromatographic separation of samples. The mixture of 0.1% aqueous formic acid (A)-acetonitrile (B) was used for the mobile phase, with its better peak shapes, better resolution and higher response values. The column temperature was set at 30 °C, and the flow rate was 0.3 mL/min. As shown in Figure 1, 39 peaks of the samples were separated and detected within 25 min.

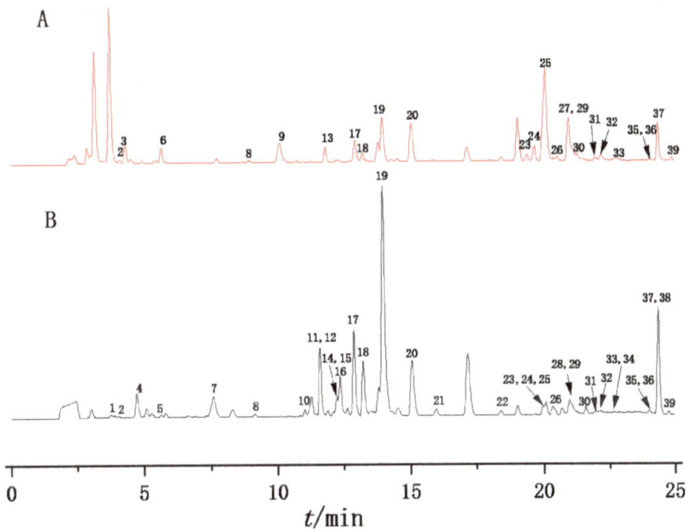

Figure 1. The TIC of processed PCH (**A**) and raw PCH (**B**).

2.2. Identification of Chemical Compounds by UPLC-Q-Exactive-MS/MS

In this work, the chemical components of raw and processed PCH samples were detected and identified by UPLC-Q-Exactive-MS/MS in both positive and negative ion modes. The positive ion showed a preferable detection effect with the peak purity and number of detected peaks. The total ion currents (TICs) of the raw and processed PCH samples are shown in Figure 1. In total, 39 compounds were detected, and 33 of them were identified, including flavonoids, alkaloids, coumarins, fatty acids, benzene, substituted derivatives and other compounds. Of these, compounds **9** and **19** were unambiguously identified by comparison with reference standards. The detailed information of these compounds is displayed in Table 1, including fragment ions, retention time, molecular formula and so on.

Table 1. The results of UPLC-Q-Exactive-MS/MS identification of chemical constituents from raw PCH and processed PCH.

No.	*t*/min	Precursor Ion (*m*/*z*)	Error /ppm	Fragment Ions (*m*/*z*)	Molecular Formula	Identification	Raw PCH	Processed PCH
1	3.668	144.08156 $[M + H]^+$	4.93	128.05013, 115.05492, 91.05509	$C_{10}H_9N$	6-Methylquinoline	+	+
2	4.109	169.07701 $[M + H]^+$	−0.93	169.07664, 168.06664, 125.06008	$C_8H_9FN_2O$	3-Fluoro-N′-hydroxy-4-methylbenzenecarboximidamide	+	+
3	4.130	328.11978 $[M + H]^+$	1.7	135.08096, 131.04948, 121.06533, 105.07039, 103.05487	$C_{19}H_{13}N_5O$	3-Methyl-6-oxo-1-phenyl-4-(3-pyridinyl)-6,7-dihydro-1H-pyrazolo[3,4-b]pyridine-5-carbonitrile	-	+
4	4.695	212.11839 $[M + H]^+$	3.33	195.09239, 167.07089, 119.06101, 94.06588, 77.03954	$C_{13}H_{13}N_3$	N,N′-Diphenylguanidine	+	+
5	5.639	198.12854 $[M + H]^+$	4.13	181.10182, 166.07823, 106.06578, 91.05499, 79.05508	$C_{14}H_{15}N$	Dibenzylamine	+	+
6	5.643	433.11496 $[M + H]^+$	4.29	271.06094, 243.06612, 215.07085, 153.01874, 91.05487	$C_{21}H_{20}O_{10}$	unknow	-	+
7	7.662	319.082 $[M + H]^+$	2.43	273.07654, 245.08148, 167.03447, 163.03957, 123.04462	$C_{16}H_{14}O_7$	Padmatin	+	-
8	9.081	284.12912 $[M + H]^+$	3.45	164.07121, 147.04465, 121.06541, 119.04980, 103.05490	$C_{17}H_{17}NO_3$	Paprazine	+	+
9	10.059	255.06584 $[M + H]^+$	2.53	227.07100, 199.07602, 153.07043, 137.02394, 119.04977	$C_{15}H_{10}O_4$	Daidzein	-	+
10	10.986	303.08786 $[M + H]^+$	4.41	257.08173, 229.08670, 167.03453, 163.03960, 135.04468,	$C_{16}H_{14}O_6$	unknow	+	-
11	11.570	181.05023 $[M + H]^+$	3.87	163.11285, 135.04454, 107.08614, 91.05480, 67.05514	$C_9H_8O_4$	4-oxo-4,5,6,7-tetrahydrobenzo[b]furan-3-carboxylic acid	+	+
12	11.571	193.05038 $[M + H]^+$	3.95	165.05515, 137.06024, 109.06550, 91.05497, 68.99794	$C_{10}H_8O_4$	5,7-Dihydroxy-4-methylcoumarin	+	-
13	11.792	271.0614 $[M + H]^+$	4.82	243.06604, 215.07108, 169.06546, 153.01881, 121.02882	$C_{15}H_{10}O_5$	Genistein	-	+

Table 1. *Cont.*

No.	t/min	Precursor Ion (m/z)	Error /ppm	Fragment Ions (m/z)	Molecular Formula	Identification	Raw PCH	Processed PCH
14	12.214	295.22791 $[M + H]^+$	3.44	227.21707, 151.11235, 135.11743, 95.08626, 67.05510	$C_{18}H_{30}O_3$	13(S)-HOTrE	+	-
15	12.276	202.21744 $[M + H]^+$	4.76	184.20668, 85.10194, 71.08640, 62.06094, 57.07082	$C_{12}H_{27}NO$	N,N-Dimethyldecylamine N-oxide	+	-
16	12.468	287.09293 $[M + H]^+$	4.05	167.03462, 147.04471, 119. 04974, 91.05491, 68.99786	$C_{16}H_{14}O_5$	unknow	+	-
17	12.631	211.08733 $[M + H]^+$	3.51	193.07681, 192.06796, 165.07085, 115.05492, 105.07058	$C_{13}H_{10}N_2O$	5,7-Dihydro-6H-dibenzo[d,f][1,3]diazepin-6-one	+	+
18	13.128	315.08737 $[M + H]^+$	3.37	313.07111, 286.08279, 241.08656, 213.09215, 198.06728	$C_{17}H_{14}O_6$	Aflatoxin B2	+	+
19	13.941	207.06589 $[M + H]^+$	3.41	179.07083, 164.04680, 148.05188, 133.06537, 108.04921	$C_{11}H_{10}O_4$	Scoparone	+	+
20	14.844	301.10815 $[M + H]^+$	4.29	179.03517, 137.06004, 121.06537, 122.06859, 91.05498	$C_{17}H_{16}O_5$	unknow	+	+
21	16.153	437.19495 $[M + Na]^+$	3.48	119.08613, 117.07032, 91.05495, 79.05503	$C_{24}H_{30}O_6$	Bis(4-ethylbenzylidene)sorbitol	+	-
22	18.692	219.17557 $[M + H]^+$	4.22	201.16444, 163.11180, 135.08084, 123.11755, 81.07075	$C_{15}H_{22}O$	Nootkatone	+	-
23	19.537	320.25732 $[M + H]^+$	−3.2	95.08607	$C_{20}H_{33}NO_2$	unknow	+	+
24	20.040	235.16977 $[M + H]^+$	2.26	179.10745, 180.11069, 123.04454, 57.07079	$C_{15}H_{22}O_2$	3,5-di-tert-Butyl-4-hydroxybenzaldehyde	-	+
25	20.122	359.14987 $[M + H]^+$	2.67	341.13843, 235.09723, 219.06676, 175.07613, 137.06026	$C_{20}H_{22}O_6$	Matairesinol	+	+
26	20.372	279.23309 $[M + H]^+$	4.77	137.13301, 123.11738, 109.10186, 95.08624, 93.07060	$C_{18}H_{30}O_2$	α-Eleostearic acid	+	+
27	21.366	355.28586 $[M + H]^+$	4.09	337.27499, 263.23770, 245.22723, 161.13303, 133.10173	$C_{21}H_{38}O_4$	1-Linoleoyl glycerol	-	+

Table 1. *Cont.*

No.	t/min	Precursor Ion (m/z)	Error /ppm	Fragment Ions (m/z)	Molecular Formula	Identification	Raw PCH	Processed PCH
28	21.390	478.32339 $[M + H]^+$	3.48	434.26038, 390.19824, 329.20343, 285.13789, 258.12930	$C_{33}H_{39}N_3$	unknow	+	+
29	21.411	227.21759 $[M + H]^+$	4.25	241.19543, 221.15463, 171.11710, 161.13290, 151.11226	$C_{18}H_{30}O_3$	9-Oxo-10(E),12(E)-octadecadienoic acid	+	+
30	21.554	279.16013 $[M + H]^+$	3.73	205.08685, 167.03441, 150.02721, 149.92388, 121.02893	$C_{16}H_{22}O_4$	Dibutyl phthalate	+	+
31	21.829	284.33249 $[M + H]^+$	4.63	60.08168, 57.07076	$C_{19}H_{41}N$	Cetrimonium	+	+
32	22.269	338.34314 $[M + H]^+$	4.05	321.31607, 303.30536, 212.20132, 149.13286, 135.11731	$C_{22}H_{43}NO$	Erucamide	+	+
33	22.533	324.29108 $[M + H]^+$	4.24	306.28064, 263.23862, 245.22665, 179.18025, 147.11760	$C_{20}H_{37}NO_2$	Linoleoyl Ethanolamide	+	+
34	23.066	293.24863 $[M + H]^+$	3.83	261.22220, 151.11247, 123.11769, 109.10181, 81.07066	$C_{19}H_{32}O_2$	9(Z),11(E),13(E)-Octadecatrienoic Acid methyl ester	+	+
35	23.434	300.29102 $[M + H]^+$	4.38	283.26355, 123.11771, 109.10162, 95.08624, 85.10188	$C_{18}H_{37}NO_2$	Palmitoyl ethanolamide	+	+
36	23.873	326.30673 $[M + H]^+$	4.23	309.27997, 135.11768, 121.10165, 83.08633	$C_{20}H_{39}NO_2$	Oleoyl ethanolamide	+	+
37	23.891	311.16556 $[M + H]^+$	4.47	255.10117, 203.10631, 177.05539, 161.09665, 135.04459	$C_{20}H_{22}O_3$	Avobenzone	+	+
38	24.121	161.06029 $[M + H]^+$	3.26	133.06534, 118.04193, 105.07047, 79.05498, 66.04727	$C_{10}H_{10}O_3$	4-Methoxycinnamic acid	+	-
39	24.357	256.26447 $[M + H]^+$	3.85	130.12366, 116.10772, 102.09198, 95.08604, 88.07630	$C_{16}H_{33}NO$	Hexadecanamide	+	+

Compounds were identified according to their characteristic fragment ions compared to reported references and standards. For example, compound **9** (flavonoid) showed an $[M + H]^+$ ion at m/z 255.06584 ($C_{15}H_{10}O_4$) and exhibited fragment ions at m/z 227.07100 by the losses of CO. It also produced fragment ions at 137.0230 ($1.3A^+$) and 119.04977 by fragmentation pathways of Retro-Diels-Alder (RDA). It can be definitively confirmed to be daidzein through comparison with the literature and reference standard. The hypothesized fragmentation process of compound **9** is displayed in Figure 2A. Compound **19** (coumarin) yielded an $[M + H]^+$ ion at m/z 207.06589 with t formula $C_{11}H_{10}O_4$, and then it lost a molecule of CO, CH_3 and H_2O to form a $[M + H - CO]^+$ fragment ion of m/z 17907032, $[M + H - CO - CH_3]^+$ fragment ion of m/z 164.04680 and $[M + H - CO - CH_3 - H_2O]^+$ fragment ion of m/z 148.05188. Compound **19** was unambiguously confirmed as scoparone through comparison with the literature and standard substance. The hypothesized cleavage mode of compound **19** is shown in Figure 2B. Compound **39** displayed an $[M + H]^+$ ion at m/z 256.26447 with the formula $C_{16}H_{33}NO$, and it exhibited characteristic fragment ions at m/z 130.12366, 102.09198, and 88.07630, indicating successive losses of C_9H_{18}, C_2H_4, and CH_2. Compound **39** was tentatively inferred to be hexadecanamide, and the hypothesized fragmentation pattern is shown in Figure 2C.

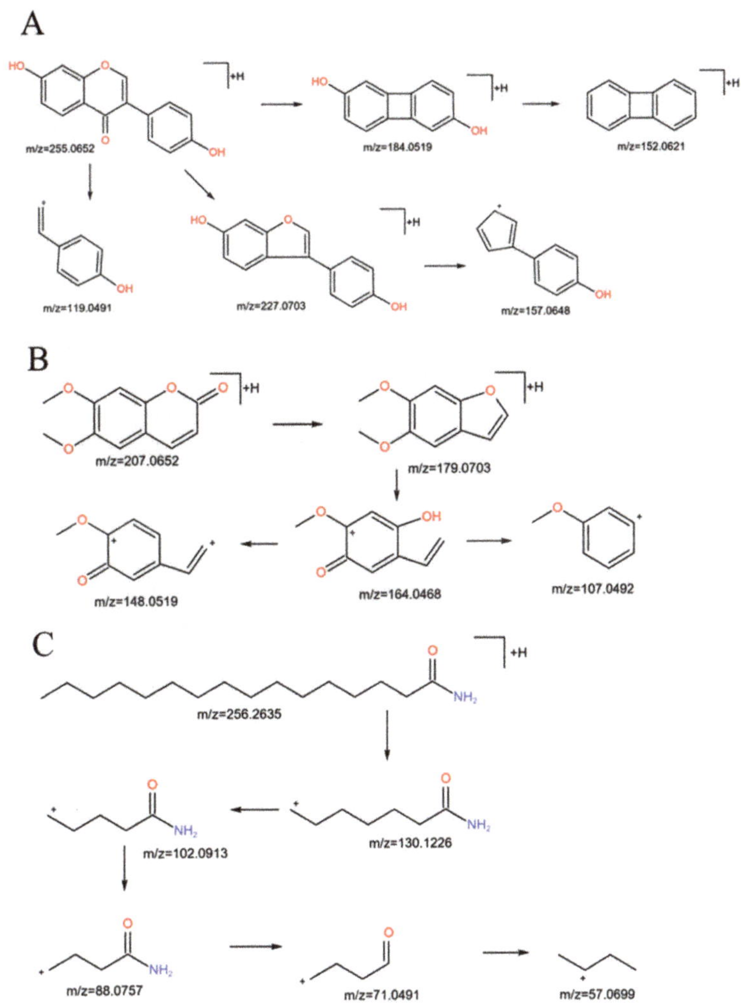

Figure 2. The hypothesized fragmentation pathway of compound **9** (**A**), compound **19** (**B**) and compound **39** (**C**).

2.3. Comparison of Chemical Profiles of Raw and Processed PCH

The newly established UPLC-Q-Exactive-MS/MS method was used to qualitatively and semi-quantitatively compare the chemical fingerprints between raw and processed

PCH samples. As shown in Figure 1, it was notable that the quantity and intensity of lower polarity compounds increased significantly in processed samples, while the medium polarity compounds clearly decreased and even disappeared after processing. There were six newly generated compounds (peaks 3, 6, 9, 13, 24 and 27) in processed PCH samples. The intensities of peaks 25 and 29 were markedly increased in processed PCH samples, while peaks 17, 18, 19, 28 and 37 were sharply decreased. Peaks 7, 10, 12, 14,15, 16, 21, 22, 28 and 38 of processed samples disappeared completely. The results indicated that processing could have caused some chemical changes in PCH during processing with black beans, which consequently changed the chemical fingerprints of the PCH samples.

2.4. Multivariate Statistical Analysis and Discovering Potential Chemical Markers

Due to chemical complexity, the differences between raw and processed PCH samples were unclear according to UHPLC-Q-Orbitrap-MS chromatograms. To distinguish the raw and processed PCH samples, a cluster heatmap was first carried out using TBtools software (TIBCO Software Inc., Palo Alto, CA, USA). In the cluster heatmap, different colors represent different content, with bright green representing higher content and dark red representing lower content. As shown in Figure 3A, the results of the cluster heatmap showed that all the samples were distributed into two major clusters. One cluster was regarded as the raw samples (G1–G10), and another cluster was considered to be the processed samples (P1–P10). It should be noted that the content of these compounds (peaks 1, 9, 11, 13, 15, 18, 19, 20, 24, 27, 28 and 39) are more than other compounds, and they may be the key compounds of differences between raw and processed PCH samples, suggesting that the chemical component and/or contents of compounds may change remarkably in the procedure of processing with black beans.

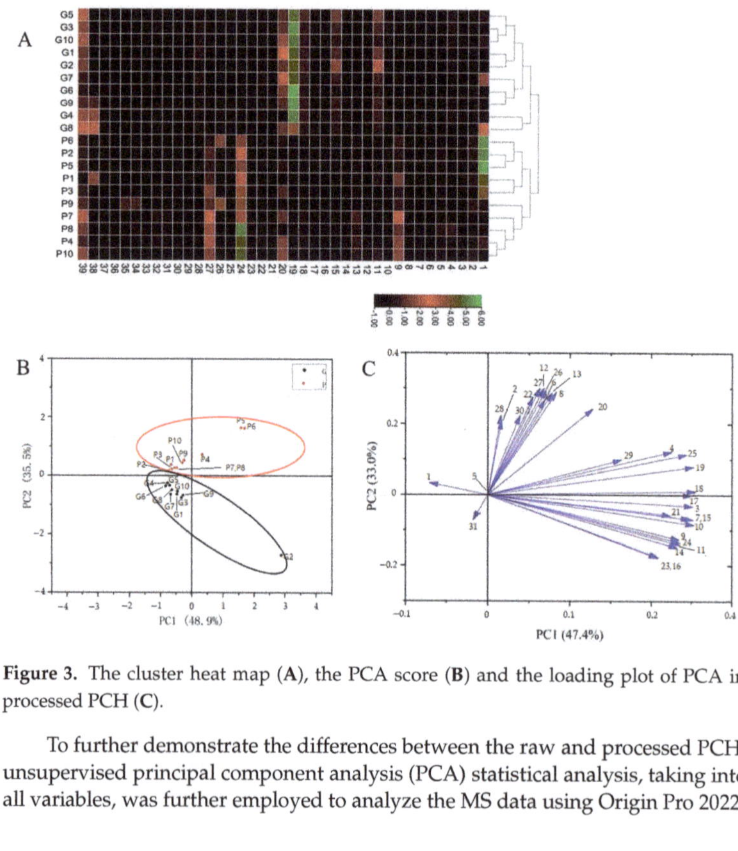

Figure 3. The cluster heat map (**A**), the PCA score (**B**) and the loading plot of PCA in raw and processed PCH (**C**).

To further demonstrate the differences between the raw and processed PCH samples, unsupervised principal component analysis (PCA) statistical analysis, taking into account all variables, was further employed to analyze the MS data using Origin Pro 2022 software

(Microcal Software, Northampton, MA, USA). The first and second principal components explained 48.9% and 35.5% of the total variance, respectively, meaning that the analytical platform was stable. The scatter plots of PCA indicated that all the samples were divided quite obviously into raw and processed PCH groups (Figure 3B). The raw samples (G1–G10) were distributed below the plot, while the processed samples (P1–P10) were at the upper of the plot, suggesting that the procedures of processing with black beans induced remarkable changes in the chemical component and/or contents of compounds in PCH. The loading plot of PCA is shown in Figure 3C. The spot will be far from the original point, meaning that it contributes the most to the differences between raw and processed PCH samples. Thus, these peaks had a significant contribution to showing the differences between raw and processed PCH samples except peaks 1, 5 and 31 because they were far away from the center in the model, and these compounds could be considered as the different markers for the raw and processed PCH samples. Therefore, these results demonstrated great differences in chemical constituents between the raw and processed PCH.

2.5. Polysaccharides Content of PCH

In this study, glucose was used as a standard. The equation of the regression line was $y = 28.37x + 0.0011$ with r = 0.9997, and the polysaccharide content was obtained from a regression line of glucose in the 2–16 μg/mL range, showing good linearity. The RSD values of the precision and repeatability tests were within 3.0% and 2.3%, respectively. The RSD values of the stability were no more than 2.8% within 12 h post reaction. The recovery rate of glucose was 103.0%, and the RSD was within 3.4%, indicating an accurate and reliable method. As listed in Table 2, the polysaccharide content of raw samples ranged from 8.46% to 14.88%, and the average value of polysaccharide content was 12.19%. However, the content of polysaccharides in processed samples ranged from 29.40% to 35.12%, and the average value of polysaccharide content was 32.28%. The results indicated that the processed samples suggested an obvious increase in polysaccharide content. The variation showed that processing with black beans has an obvious effect on polysaccharide content in PCH, which may induce some compounds with glycoside structures to degrade to oligosaccharides, consistent with earlier studies [30,31].

Table 2. Polysaccharides and Saponin content and information of Sample.

No.	Collecting Location	Collection Year	Classification	Total Saccharide Content (%)	Total Saponin Content (%)
G1	Meishan, Sichuan	June 2020	Raw	8.46 ± 0.65	5.02 ± 0.42
G2	Meishan, Sichuan	June 2020	Raw	14.04 ± 1.07	5.97 ± 0.33
G3	Meishan, Sichuan	June 2020	Raw	8.54 ± 0.24	3.64 ± 0.65
G4	Mianyang, Sichuan	October 2020	Raw	14.88 ± 0.86	4.73 ± 0.49
G5	Mianyang, Sichuan	October 2020	Raw	13.45 ± 0.75	5.59 ± 0.55
G6	Baise, Guangxi	May 2021	Raw	14.59 ± 1.32	4.72 ± 0.63
G7	Baise, Guangxi	May 2021	Raw	10.84 ± 0.87	4.31 ± 0.61
G8	Baise, Guangxi	May 2021	Raw	10.35 ± 0.53	5.28 ± 0.73
G9	Heshan, Guangdong	June 2021	Raw	14.48 ± 1.16	5.22 ± 0.48
G10	Heshan, Guangdong	June 2021	Raw	12.25 ± 0.88	5.63 ± 0.61
P1	Meishan, Sichuan	June 2020	processed	32.12 ± 2.21	8.50 ± 0.46
P2	Meishan, Sichuan	June 2020	processed	32.27 ± 1.86	8.20 ± 0.50
P3	Meishan, Sichuan	June 2020	processed	29.40 ± 0.90	7.86 ± 0.62
P4	Mianyang, Sichuan	October 2020	processed	33.78 ± 1.69	9.41 ± 0.85
P5	Mianyang, Sichuan	October 2020	processed	35.12 ± 2.11	7.82 ± 0.46
P6	Baise, Guangxi	May 2021	processed	32.37 ± 2.30	6.09 ± 0.49
P7	Baise, Guangxi	May 2021	processed	32.02 ± 0.95	7.58 ± 0.82
P8	Baise, Guangxi	May 2021	processed	32.90 ± 0.70	6.71 ± 0.35
P9	Heshan, Guangdong	June 2021	processed	30.57 ± 1.06	6.63 ± 0.44
P10	Heshan, Guangdong	June 2021	processed	32.21 ± 0.82	8.09 ± 0.37

2.6. Saponin Content of PCH

Saponins are also the main bioactive compounds in PCH [24]. In this study, diosgenin was used as a standard. The equation of the regression line was $y = 38.693x - 0.1676$ with $r = 0.9998$, and the saponin content was calculated by a standard curve of diosgenin ranging between 6 and 16 µg/mL. The RSD values of the precision and repeatability tests were less than 2.1% and 2.6%, respectively. The RSD values of the stability were within 2.7% in 12 h. The recovery rate was 103.2%, and the RSD value was within 2.7%. These results showed that the approach could be suitable for total saponin content measure. As shown in Table 2, the saponin content of raw samples ranged between 3.64% and 8.50%, and the average value of saponin content was 5.01%. However, the saponin content of processed samples ranged from 6.09% to 9.41%, and the average value of saponin content was 7.69%. The results revealed that there were a few differences among the raw and processed samples, and processing caused a change in some chemical structures and showed a rising trend in saponin content.

2.7. Quantitative Analysis of Monosaccharide by GC-MS

2.7.1. Method Validation

Method validation for the determination of monosaccharide results is listed in Table 3. The results of the linear equation, LOQs and LODs for the five monosaccharides showed a good linearity ($r > 0.9990$) for each reference standard within a certain concentration range. For the precision test, the intraday and inter-day RSDs were no more than 1.81% and 2.52%, respectively, which indicated that the instrument was in order. The RSD values of the repeatability were within 2.40%, and the established method was reproducible. The stability showed that the sample solution had a steady repetition at 24 h at room temperature, and the RSD values were less than 1.84%. Moreover, the average recoveries ranged from 94.47% to 103.23%, and the RSD values of recovery were no more than 1.39%. The results indicated that the newly established approach can be suitable for quantitative analysis.

Table 3. The results of linear regression, LOQs and LODs.

Compound	Regression Equation	Linearity Range (µg/mL)	r	LOQ (µg/mL)	LOD (µg/mL)
rhamnose	$y = 3615416637x - 5554691$	2.528~25.275	0.9996	8.173×10^{-1}	3.342×10^{-1}
arabinose	$y = 1777422855x - 2702648$	5.065~50.650	0.9996	6.573×10^{-1}	2.471×10^{-1}
mannose	$y = 1501991081x - 3075994$	5.085~50.850	0.9995	5.438×10^{-1}	4.184×10^{-1}
glucose	$y = 1051468840x - 2269974$	25.225~504.500	0.9995	9.935×10^{-1}	7.488×10^{-1}
fructose	$y = 357575280x - 1347648$	25.450~254.500	0.9988	4.126	8.836×10^{-1}

2.7.2. Monosaccharide Composition and Levels Differ between Raw and Processed PCH

GC–MS was utilized to determine monosaccharide composition and levels in different PCH samples. The monosaccharide composition analysis revealed that the extracts of both raw and processed PCH samples primarily contained five kinds of monosaccharides, including rhamnose, arabinose, mannose, glucose and fructose, by comparison with the reference standards. The monosaccharide content hydrolyzed from the extracts is listed in Figure 4. After processing, the contents of arabinose, glucose and fructose hydrolyzed from extracts were 0.7114%, 7.3248% and 53.3072%, respectively, and were reduced by 0.25%, 3.46% and 12.70%, respectively, while the contents of rhamnose and mannose were 0.2161% and 2.2363%, respectively, and increased by 0.03% and 1.03%, respectively. Specifically, fructose was the most abundant monosaccharide hydrolyzed product in both raw and processed PCH. The monosaccharide composition and level results showed that processing had no obvious impact on the type of monosaccharides present in the extract but changed their content and ratio, which could further induce chemical structure variations, consistent with previous studies [2,24].

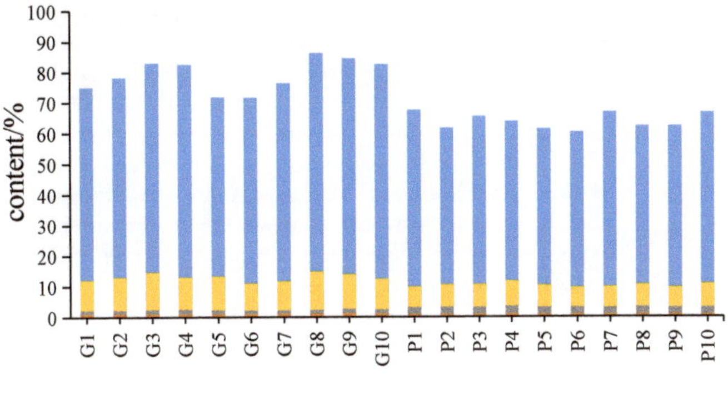

Figure 4. The content of monosaccharide in extracts of the raw (G) and processed (P) samples of PCH.

2.8. Antioxidant Activity Test Results

To compare the antioxidant capacity of raw and processed samples, antioxidant assays were performed (Figure 5A–C). DPPH and ABTS are stable free radical commonly used to evaluate the total antioxidant capacity of antioxidant materials. DPPH and ABTS were therefore used to evaluate the radical scavenging activity of the raw and processed PCH. As shown in Figure 5A,B, the raw and processed samples both displayed DPPH and ABTS radical-scavenging activities in dose-dependent at concentrations from 0 to 2.0 mg/mL. In addition, the antioxidant capacity of the processed sample was significantly stronger than that of the raw sample ($p < 0.05$). The IC_{50} values of the raw and processed samples for DPPH were 1.0 ± 0.12 and 0.65 ± 0.10 mg/mL, respectively, while the IC_{50} values for ABTS were 0.65 ± 0.07 and 0.25 ± 0.04 mg/mL, respectively. We also determined the reducing activity of the raw and processed samples by using the FRAP assay (Figure 5C). Both the raw and processed samples also showed dose-dependent scavenging activity, and the processed samples showed a higher FRAP reducing capacities than the raw samples. These results indicated that the processing with black beans had an obvious influence on the antioxidant activity of PCH, and antioxidant activity of processing with black beans sample showed significantly stronger than that of the raw sample. These studies are consistent with recent studies [3,9,30] that showed that the antioxidant activity of PCH was enhanced during steam processing.

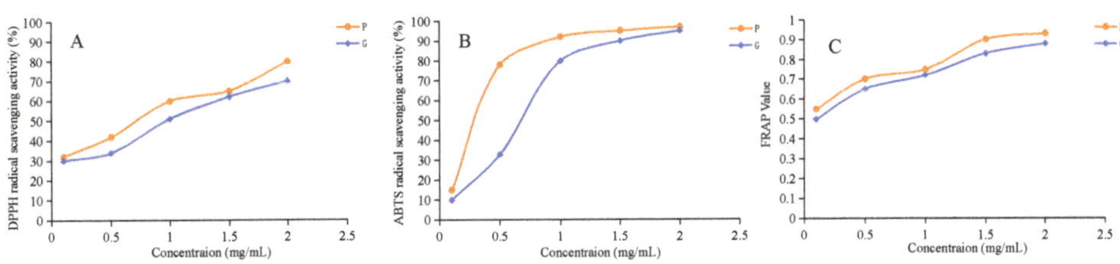

Figure 5. DPPH radical scavenging activity (**A**), ABTS radical scavenging activity (**B**) and Ferric reducing antioxidant power (**C**). G, the extract of the raw sample; P, the extract of processed sample.

2.9. Antidiabetic Activity In Vitro

In addition to antioxidant effects, PCH has also exhibited an antihyperglycemic effect in previous studies [3]. The inhibition activity of α-glucosidase and α-amylase activities of the raw and processed sample extracts was studied in this work. As shown in Figure 6A,B, the inhibitory effects of raw, processed and acarbose (positive control) on α-glucosidase and α-amylase activity showed an increase with a dose-dependent manner. As expected, acarbose showed the highest inhibition of α-glucosidase and α-amylase.

The extracts of raw and processed PCH also showed significant inhibition against α-glucosidase and α-amylase. Lower IC_{50} values indicate stronger inhibition. The IC_{50} values of the extracts from raw and processed samples against α-glucosidase were 5.58 ± 0.22 and 1.29 ± 0.12 mg/mL (Figure 6A), respectively, and the extracts of raw and processed samples also had strong α-amylase inhibition with IC_{50} values of 0.80 ± 0.09 and 0.48 ± 0.05 mg/mL (Figure 6B), respectively, indicating that processed extracts had higher inhibition against both α-glucosidase and α-amylase than the raw sample ($p < 0.05$). The results indicated that the extracts of raw and processed PCH showed significant antidiabetic activity in vitro. After processing, the extracts exhibited a higher antidiabetic activity. However, further in vivo studies are needed.

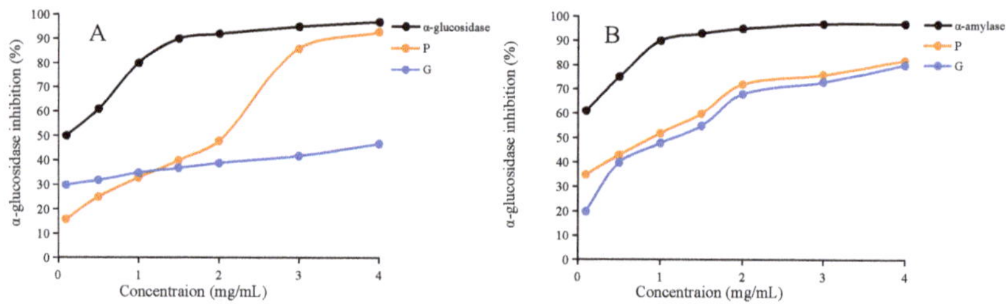

Figure 6. The α-glucosidase inhibitory activity (**A**) and the α-amylase inhibitory activity (**B**). G, the extract of the raw sample; P, the extract of processed sample.

3. Materials and Methods

3.1. Materials and Chemicals

LC/MS grade of methanol, acetonitrile and formic acid from Merck (Darmstadt, Germany) were purchased. Ultra-pure water was obtained from Watsons Food and Beverage Company (Guangzhou, China). Reference substances, including D-rhamnose (Rha), arabinose (Ara), D-mannose (Man), D-glucose (Glu), D-galactose (Gal) and D-fructose (Fru) from Chengdu Kangbang Biotechnology Co., Ltd. (Chengdu, China) were purchased. Diosgenin was obtained from the National Institute for the Control of Pharmaceutical and Biological Products (Beijing, China). Daidzein and scoparone were purchased from Chengdu Maide Biotechnology Co., Ltd. (Chengdu, China). Each compound was 98% purity as indicated by the supplier.

2,2-diphenyl-1-picrylhydrazyl (DPPH) and acarbose were supplied by Chengdu Kangbang Biotechnology Co. Ltd. (Chengdu, China). Trifluoroacetic acid, hydroxylamine hydrochloride, pyridine, acetic anhydride and trichloromethane from Chroma-Biotechnology Co., Ltd. (Chengdu, China) were purchased. 2,20-azinobis (3-ethylbenzothia-zo-line-6-sulfonicacid) diammonium salt (ABTS) was supplied by Beijing Soleibao Technology Co., Ltd. (Beijing, China). Phosphate buffer solution (PBS), 3,5-dinitrosalicylic acid (DNS) reagent, α-glucosidase and α-amylase were supplied by Shanghai Maclean Biochemical Technology Co., Ltd. (Shanghai, China). Ferric chloride ($FeCl_3$), potassium ferricyanide ($K_3[Fe(CN)_6]$), trichloroacetic acid (Cl_3CCOOH), potassium persulfate ($K_2S_2O_8$), perchloric acid, anthrone, sulfuric acid, glacial acetic acid and soluble starch were supplied by Tianjin Zhiyuan Chemical Reagent Co., Ltd. (Tianjin, China). All other chemicals were analytically pure.

Black beans were obtained from the Chengdu Hehuachi medicinal herbs market. Ten batches of *Polygonatum cyrtonema* Hua samples were collected from Sichuan, Guangxi and Guangdong provinces in China and identified by Professor Yuan Liu (School of Pharmacy, Southwest Minzu University, China). All the samples were stored in the herbarium of the School of Pharmacy of Southwest Minzu University (Chengdu, China). Sample information is presented in Table 3.

3.2. Processing Methods of PCH

The raw PCH: The fresh samples were dried in a drying oven at 45 °C until complete dryness and then cut into thin slices. Processed PCH: Black beans (100 g) were boiled in a pot, and 1 L of juice was obtained. Raw PCH (10 kg) was soaked in 1 L of black bean juice for 12 h and then steamed for the outside, until the insides were moist black. Finally, samples were cut into thick slices and dried in a drying oven at 45 °C.

3.3. UHPLC-Q-Exactive-MS/MS Analysis

3.3.1. Preparation of Sample Solution

All the batches of raw and processed PCH samples were ground to powder and sieved through a 10 mesh sieve. Then, the dried powder (1.0 g) was suspended in 25 mL of methanol, sonicated for 60 min at room temperature, filtered and dried. The residue was dissolved in methanol and fixed in a 5 mL volumetric flask. The extracted solution was filtered through a 0.22 μm microporous filters prior to UHPLC-Q-Exactive-MS/MS analysis.

3.3.2. Ultra Performance Liquid Chromatography

The extracts were analysed by a Thermo Scientific™ Vanquish™ Flex UHPLC (Thermo Fisher Scientific Inc., Waltham, MA, USA) coupled with a binary gradient pump, automatic sampler, and diode array detector (DAD). Chromatographic separation was achieved using a Hypersil GOLD C18 column (2.1 × 100 mm, 1.8 μm). The mobile phases were 0.1% aqueous formic acid (A) and acetonitrile (B) with a flow rate of 0.3 mL/min. The optimized gradient elution conditions were set as follows: 0–3 min, 10–20% B; 3–8 min, 20% B; 8–10 min, 20–46% B; 10–15 min, 46% B; 15–20 min, 46–70% B; 20–25 min, 70–100% B. The column temperature was maintained at 30 °C and the injection volume was 5 μL.

3.3.3. MS Conditions

After chromatographic separation, a Thermo UHPLC-Q-Exactive Orbitrap mass spectrometer was connected to the Thermo Scientific™ Vanquish™ Flex UHPLC system via a heating of the electrospray ionization (HESI) source for mass spectrometry. Positive ionization mode was employed to obtain data in range of m/z 100 to 1500 Da with 0.2 s scan time in a 30 min analysis period. The mass data was obtained under the following parameters: a source temperature of 100 °C and a desolvation temperature of 350 °C. The capillary voltage was 3.8 kV (positive mode); the sheath gas pressure was 3.5 MPa; the auxiliary gas pressure was 1.0 MPa, and the collision energy was 40 eV. Finally, Xcalibur 13.0 software (Thermo Fisher Scientific Inc., USA) was used to acquired and analysed the mass data.

3.4. Preparation and Determination of Polysaccharides Content

3.4.1. Sample Preparation

The dried powder (100 g) was mixed with 1500 mL 80% ethanol and then sonicated for 1 h at 60 °C in water bath. The extraction step was repeated 3 times. The ultrasonic power was set at 250 W. The combined supernatant was filtered using a Buchner funnel and then concentrated with a rotary evaporator (Büchi R-100, Essen, Germany) under reduced pressure. To completely remove the solvent, the concentrated extracts further underwent lyophilization for 48 h using an FD-1B-80 lyophilizer (Boyikang, Beijing, China). Then, the PCH extracts were kept at 4 °C.

3.4.2. Determination of Polysaccharides Content

The polysaccharide content of PCH was measured using an anthrone-sulfuric acid method described in the Chinese Pharmacopoeia 2020 Edition (Volume I) [19]. The dried PCH extracts were dissolved with purified water to the concentration of 1 mg/mL. An amount of 0.2 mL of extract solutions and 1.8 mL purified water was mixed in the test tube with a stopper. After that 8.0 mL of 2% anthrone-sulfuric acid was drawn slowly in an ice bath and shaken ten times. The test tube was then placed into a boiling water bath for heating 10 min and rapidly cooled down to the indoor temperature. The absorbance of 200 μL postreaction solutions was determined at a wavelength of 582 nm in 96-well plates using a Varioskan LUX2 automatic microplate reader (Thermo Fisher Scientific Inc., USA). For this, 0.2 mL deionized water replaced the extract solution as a blank. The results were calculated by using g of polysaccharides per 100 g of sample.

3.4.3. Determination of Saponin Content

The saponin content of PCH was measured in the light of reports by Wang et al. [32], with a few modifications. In brief, the dried PCH extracts were dissolved with methanol to the concentration of 1 mg/mL. Aliquots of 0.4 mL were added to the test tube with a stopper and vaporized solvent at 80 °C in a water bath. After that, 0.2 mL 5% vanillin-glacial acetic acid was added to the test tube, followed by 0.5 mL perchloric acid in an ice bath. The solutions were vortexed and heated at 60 °C for 15 min, cooled in an ice bath for 2 min and finally, 4 mL of glacial acetic acid was added and incubated for 5 min. The absorbance of 200 μL postreaction solutions was determined at a wavelength of 452 nm in 96-well plates. For this, 0.4 mL methanol replaced the extract solution as a blank. The results were calculated by using g of saponin per 100 g of sample.

3.5. GC-QQQ-MSMS Analysis

3.5.1. Hydrolysis of Extracts

Ten milligrams of PCH extracts from Section 3.4.1 and 2 mL of 2 mol/L trichloroacetic acid were mixed in a 10 mL ampoule bottle. After that, the extracts were hydrolysed for 3 h at normal temperature. The hydrolysates were cleaned at 40 °C by using a Termovap Sample Concentrator (Hangzhou Aosheng Instrument Co., LTD, Hangzhou, China) and kept at 4 °C.

3.5.2. Derivatization of Saccharide

The hydrolysis products were mixed with 10 mg hydroxylamine hydrochloride and 0.5 mL pyridine and reacted for 30 min at 90 °C in a thermostat water bath. After that, the hydrolysis products were removed, cooled down to normal temperature and mixed with 1 mL acetic anhydride. After reacting for 30 min, the hydrolysis products were cooled down to normal temperature. The solvent was cleaned up at 50 °C by using a Termovap Sample Concentrator. The residue was dissolved in 4 mL trichloromethane. The solution was filtered through a 0.22 μm microporous filter before GC-MS analysis.

3.5.3. Preparation of Standard Solutions

Derivatization of monosaccharide (rhamnose, arabinose, mannose, glucose, galactose, fructose) was prepared as described in Section 3.5.2, and the concentration of each monosaccharide standard stock solution was 2.5 mg/mL. Then, a series of appropriate concentration mixed standard solutions were achieved from the stock solution by adding an appropriate volume of chloroform. The mixed standard solutions were kept at 4 °C and filtered using a 0.22 μm microporous filter for further GC-MS analysis.

3.5.4. Validation Method

The quantitative GC-MS a method was validated in terms of linearity, precision, repeatability, stability and recovery. The linear relationship was achieved by precisely injecting appropriate concentrations of monosaccharide standard stock solution into the GC-MS system to draw the linear equation, correlation coefficient and linearity range. The detection limit (LOD) and quantification limit (LOQ) were assessed on the basis of signal-to-noise ratios (S/N) of 3:1 and 10:1. The precision was achieved by determining six injections of the same standard solutions. The intraday and inter-day tests were used to evaluate repeatability. The intraday test was analysed by injecting the same solution for six times on the same day. The inter-day test was determined three times a day for three consecutive days. Stability test was evaluated by injecting the same solution at 0, 2, 4, 8, 12, 24 h, respectively. The test was achieved by adding the corresponding components of 80%, 100% and 120% into the sample to ensure recovery. Each level was evaluated for three times. The relative standard deviation (RSD) of relative retention time (RRT) of compound was calculated to analyze the precision, repeatability, stability and recovery of the method.

3.5.5. GC-MS Conditions

GC-MS analysis was carried out using a Agilent7000D Triple Quadrupole GC/MS (Agilent Technologies, Santa Clara, CA, USA). The separations were performed using an HP-5 m capillary column (30 m × 0.25 mm, 0.25 μm). The temperature of GC injector was 250 °C, the carrier gas was helium (99.99% purity), the split ratio was set at 20:1, and the injection volume was set to 1 μL each time. The oven temperature programs were as follows: the initial temperature of 150 °C, then the temperature was increased to 200 °C at a rate of 8 °C/min (held for 1 min). The temperature was subsequently increased to 260 °C at a rate of 10 °C/min (held 1 min).

Mass spectrometry was carried out on an electrospray ionization (ESI) source. The ion source temperature and quadrupole temperature were set at 230 °C and 150 °C, respectively. The solvent delay was 3 min, and the scan range was m/z 30–600.

3.6. Determination of Antioxidant Capacity

A weighed sample of 10 g powder was sonicated with 25 mL of 80% ethanol for 60 min, repeated for three times. After that, the solution was filtered, concentrated and freeze-dried for standby application. For antioxidant activity assay, raw and processed solution was further diluted to a series concentration in the range of 0.1 to 1.5 mg/mL. The antioxidant activity of the raw and processed PCH samples was evaluated via DPPH assays, ABTS assay and FRAP assay.

3.6.1. DPPH Assay

The DPPH activity was assessed based on the method reported by Blois [33] with the same modifications. For the DPPH assay, 1 mL sample solution was mixed with 1 mL of the DPPH solution (0.1 mmol/mL) in a test tube. After that, the mixed solutions were left in darkness and then incubated for 30 min. Finally, 200 μL mixed solutions were placed in 96-well plates, and the absorbance at 517 nm was determined. The DPPH radical scavenging activity was calculated as Equation (1):

$$\text{Scavenging activity} = (1 - \frac{A_s - A_c}{A_b}) \times 100\% \quad (1)$$

where A_s represented the absorbance of 1 mL sample extract with 1 mL DPPH solution; A_c is 1 mL sample extract with 1 mL 70% ethanol; and A_b is 1 mL 70% ethanol with 1 mL DPPH solution. The concentration of the radical content reduced by 50% was defined as the IC_{50} value. The IC_{50} values were measured by equation.

3.6.2. Assay of ABTS

The ABTS activity was evaluated according to the method mentioned by Moon et al. [34] with some modifications. For the ABTS assay, ABTS (384.076 mg) and $K_2S_2O_8$ (50.064 mg) were dissolved in water and fixed in a 100 mL volumetric flask. The configured solution was kept in the darkness at normal temperature for 12 h. Phosphate buffer solution (pH = 7.4) was used to diluted the configured $ABTS^+$ solution, and then an absorbance of 0.70 ± 0.02 was determined at a wavelength of 734 nm for $ABTS^+$ analysis. Then, 0.5 mL sample solution was added to 1.5 mL of $ABTS^+$ solution. After reaction for 10 min in the darkness, 200 μL mixed solutions was put into 96-well plates and determined at a wavelength of 734 nm. The DPPH radical scavenging activity was converted by Equation (2):

$$\text{Scavenging activity } (\%) = (1 - \frac{A_s}{A_0}) \times 100\% \quad (2)$$

where A_s represents the absorbance of 0.5 mL sample solution plus1.5 mL $ABTS^+$ solution; A_0 represents 0.5 mL 70% ethanol plus 1.5 mL $ABTS^+$ solution. The IC_{50} values were calculated using regression analysis.

3.6.3. Assay of FRAP

The FRAP assay was determined as the procedure described by Moon et al. [34] with slight modifications. For analysis, 0.5 mL sample solution, 2.5 mL of phosphate buffer solution (pH = 6.6) and 2.5 mL of 1% $K_3[Fe(CN)_6]$ were mixed in 10 mL test tube and reacted at 50 °C for 20 min. After that, 2.5 mL of 10% Cl_3CCOOH was added and reacted at room temperature for 10 min. Finally, 200 μL of mixed solution was placed in a 96-well plate, and the absorbance was determined at 700 nm. The solution without sample was used as blank control.

All antioxidant assays (DPPH, ABTS, FRAP) were achieved by using Thermo Fisher Scientific Varioskan LUX2 automatic microplate reader (Thermo Fisher Scientific Inc., USA) equipment. All the tests were repeated three times, and the average mean was calculated.

3.7. *Inhibition of Hypoglycemic Activity*

The hypoglycemic activity was measured in vitro via the inhibition of α-glucosidase and α-amylase. The PCH extracts from Section 3.4.1 were diluted with purified water to concentrations of 0.1, 0.5, 1, 1.5, 2, 3, and 4 mg/mL.

3.7.1. α-Glucosidase Inhibitory Assay

The inhibition of α-glucosidase assay was evaluated according to a previous method reported by Zaharudin et al. [35], with a few modifications. In brief, P-nitrophenol (PNPG) was set as a substrate in the test. A total of 10 μL of sample solution and 50 μL of phosphate buffer solutions (pH = 6.8) were mixed, and then 10 μL of 4 U/mL α-glucosidase enzyme was added as the sample group. Then, 60 μL of phosphate buffer solution (pH = 6.8) was mixed with 10 μL of sample solution as a sample control. Subsequently, 50 μL of phosphate buffer solution, 10 μL of α-glucosidase and 10 μL of distilled water were mixed as the enzyme group, and 60 μL of phosphate buffer solution was mixed with 10 μL of purified water as an enzyme control. All of the test reactions occurred at room temperature for 10 min, and then 50 μL of 5.0 mmol/L PNPG was added. The mixture was vortexed and reacted for 30 min at 37 °C. After that, the reaction was terminated by adding 100 μL 1 mol/L sodium carbonate solution. The absorbance value was recorded at a wavelength of 405 nm. The positive control was acarbose. The percent inhibitory was calculated as Equation (3):

$$\text{Inhibitory } (\%) = (1 - \frac{OD\text{sample} - OD\text{sample control}}{OD\text{enzyme} - OD\text{enzyme control}}) \times 100\% \quad (3)$$

3.7.2. α-Amylase Inhibitory Assay

The inhibition of α-amylase was measured based on the method mentioned by Yonemoto et al. [36] with slight modifications. In brief, 30 μL of sample solution and 30 μL of 10 U/mL α-amylase were added to the test tube and incubated at 37 °C for 15 min. Next, 30 μL of 1% soluble starch solution was added and reacted for 15 min. Then, the reaction was terminated by adding 50 μL of DNS reagent, and boiling at 100 °C for 10 min, then cooling down to room temperature. After that, the reaction mixture was diluted by adding 840 μL of phosphate buffer solution (pH = 6.8). Finally, 200 μL of the mixture was placed in a 96-well plate, and the absorbance was recorded at a wavelength of 540 nm. The percent inhibition was calculated using Equation (4):

$$\text{Inhibitory } (\%) = (1 - \frac{OD_C - OD_D}{OD_A - OD_B}) \times 100\% \quad (4)$$

The positive control was acarbose. Where OD_C is the absorbance of sample solution or acarbose; OD_D is the absorbance of phosphate buffer solution instead of α-amylase; OD_A is the absorbance of purified water replaced sample solution; and OD_B is the absorbance of distilled water and phosphate buffer solution instead of sample solution and α-amylase, respectively. The IC_{50} values were calculated by using regression analysis.

All hypoglycemic activity tests (α-glucosidase and α-amylase) were determined on Thermo Fisher Scientific Varioskan LUX2 automatic microplate reader (Thermo Fisher Scientific Inc., USA) equipment. All tests were repeated in triplicate, and the results were averaged.

3.8. Statistical Analysis

UHPLC-Q-Orbitrap-MS data were converted into the excel format using Xcalibur 13.0 software to align retention time and peak integration. The peak area information of all tentatively assigned compounds was used to further multivariate statistical analysis containing cluster heatmap and principal components analysis (PCA) by using TBtools and Origin Pro 2022, respectively. All values were presented as the means ± standard errors (SE) of triplicates. One-way analysis of variance (ANOVA) from SPSS statistical software (version 20.0, SPSS Inc., Chicago, IL, USA) was employed to assess differences in mean values among groups. $p < 0.05$ was considered statistically significant.

4. Conclusions

In this study, we investigated and compared the chemical compositions and biological activities of raw and processed PCH with black beans. The results revealed significant differences in chemical composition, polysaccharide content, saponin content and biological activities between the two samples. The polysaccharide and saponin content in PCH processed with black beans were notably higher compared to the raw samples. The average values of polysaccharide and saponin content in the raw samples were 12.19% and 5.01%, respectively. However, the average values of polysaccharide and saponin content in the samples processed with black beans were 32.18% and 7.69%, respectively. Furthermore, bioactivity assays demonstrated that the processed PCH with black beans exhibited stronger antioxidant and antidiabetic activities compared to the raw material. The IC_{50} values for DPPH assay were 1.0 ± 0.12 mg/mL for the raw PCH and 0.65 ± 0.10 mg/mL for the processed PCH with black beans. Similarly, the IC_{50} values for ABTS were 0.65 ± 0.07 mg/mL and 0.25 ± 0.04 mg/mL for the raw and processed samples, respectively. In terms of α-glucosidase inhibitory activity, the IC_{50} values were 5.58 ± 0.22 mg/mL and 1.29 ± 0.12 mg/mL for the raw and processed PCH, respectively. For α-amylase inhibitory activity, the IC_{50} values were 0.80 ± 0.09 mg/mL and 0.48 ± 0.04 mg/mL for the raw and processed samples, respectively. These results suggest that processing PCH with black bean juice can alter its chemical composition, content, and subsequently impact its antioxidant and antidiabetic activities. Additionally, this study highlights the significance of black bean-processing for PCH and provides a foundation for its further development as a functional food. However, it is essential to conduct further in vivo studies to demonstrate the role of black bean-processing in PCH's activities.

Author Contributions: This work was conceived by Z.Z.; J.Z. performed the UHPLC-Q-Orbitrap-MS analysis and drafted the manuscript. J.W. conducted the GC-MS analysis. L.Y. conducted the bioactive experiment. Y.W. identified the chemical compounds. W.J. conducted the statistical analysis. J.L. revised the manuscript. All authors have read and agreed to the published version of the manuscript.

Funding: This work was supported by the National Natural Science Foundation of China (31870314), the Sichuan Provincial Regional Innovation Cooperation Project (2023YFQ0084), the Science and Technology Major Project of Tibetan Autonomous Region of China (XZ202201ZD0001G06), Southwest Minzu University Basic Scientific Research Business Special Fund project for Central Universities (NO. ZYN2023033) and the Technology Research and Development Project of Qinzhou Science and Technology Bureau (20198509).

Institutional Review Board Statement: Not applicable.

Informed Consent Statement: Not applicable.

Data Availability Statement: Not applicable.

Conflicts of Interest: The authors have declared no conflict of interest.

Sample Availability: Not applicable.

References

1. Zhao, P.; Zhao, C.C.; Li, X.; Gao, Q.Z.; Huang, L.Q.; Xiao, P.G.; Gao, W.Y. The genus *Polygonatum*: A review of ethnopharmacology, phytochemistry and pharmacology. *J. Ethnopharmacol.* **2018**, *214*, 274–291. [CrossRef] [PubMed]
2. Jin, J.; Lao, J.; Zhou, R.R.; He, W.; Qin, Y.; Zhong, C.; Xie, J.; Liu, H.; Wan, D.; Zhang, S.H.; et al. Simultaneous Identification and Dynamic Analysis of Saccharides during Steam Processing of Rhizomes of *Polygonatum cyrtonema* by HPLC–QTOF–MS/MS. *Molecules* **2018**, *23*, 2855. [CrossRef]
3. Chen, S.C.; Yang, C.S.; Chen, J.J. Main Bioactive Components and Their Biological Activities from Natural and Processed Rhizomes of *Polygonum sibiricum*. *Antioxidants* **2022**, *11*, 1383. [CrossRef] [PubMed]
4. Xie, Y.; Jiang, Z.W.; Yang, R.; Ye, Y.Y.; Pei, L.X.; Xiong, S.; Wang, S.C.; Wang, L.S.; Liu, S. Polysaccharide-rich extract from *Polygonatum sibiricum* protects hematopoiesis in bone marrow suppressed by triple negative breast cancer. *Biomed. Pharmacother.* **2021**, *137*, 111338. [CrossRef] [PubMed]
5. Tai, Y.; Sun, Y.M.; Zou, X.; Pan, Q.; Lan, Y.D.; Huo, Q.; Zhu, J.W.; Guo, F.; Zheng, C.Q.; Wu, C.Z.; et al. Effect of *Polygonatum odoratum* extract on human breast cancer MDA-MB-231 cell proliferation and apoptosis. *Exp. Ther. Med.* **2016**, *12*, 2681–2687. [CrossRef] [PubMed]
6. Zhao, H.; Wang, Q.L.; Hou, S.B.; Chen, G. Chemical constituents from the rhizomes of *Polygonatum sibiricum* Red. and anti-inflammatory activity in RAW264.7 macrophage cells. *Nat. Prod. Res.* **2019**, *33*, 2359–2362. [CrossRef]
7. Singla, R.K.; De, R.; Efferth, T.; Mezzetti, B.; Sahab Uddin, M.; Sanusi; Ntie-Kang, F.; Wang, D.; Schultz, F.; Kharat, K.R.; et al. The International Natural Product Sciences Taskforce (INPST) and the power of Twitter networking exemplified through #INPSThashtag analysis. *Phytomedicine* **2023**, *108*, 154520. [CrossRef]
8. Wang, F.; Wang, B.; Wang, L.; Xiong, Z.Y.; Gao, W.; Li, P.; Li, H.J. Discovery of discriminatory quality control markers for Chinese herbal medicines and related processed products by combination of chromatographic analysis and chemometrics methods: *Radix Scutellariae* as a case study. *J. Pharm. Biomed. Anal.* **2017**, *138*, 70–79. [CrossRef]
9. Teng, H.H.; Zhang, Y.; Jin, C.S.; Wang, T.S.; Huang, S.Z.; Li, L.; Xie, S.Z.; Wu, D.L.; Xu, F.Q. Polysaccharides from steam-processed *Polygonatum cyrtonema* Hua protect against d-galactose-induced oxidative damage in mice by activation of Nrf2/HO-1 signaling. *J. Sci. Food Agric.* **2023**, *103*, 779–791. [CrossRef]
10. Xu, Y.D.; Ye, Y.L.; Liu, C.; Chen, B.Y.; Ji, J.; Sun, J.D.; Zhang, Y.Z.; Sun, X.L. Positive effects of steamed *Polygonatum sibiricum* polysaccharides including a glucofructan on fatty acids and intestinal microflora. *Food Chem.* **2022**, *402*, 134068. [CrossRef]
11. Wang, Y.Q.; Shi, L.; Lv, L.L.; Huang, N.N.; Guo, X.; Cai, T.T.; Li, X.Y.; Sun, R. Study on Acute Toxicity of Different Processed Water Extract of *Polygoni multiflori* in Mice. *Chin. J. Pharmacovigil.* **2018**, *14*, 599–602.
12. Zhao, X.M.; Li, X.Y.; Sun, R.; Huang, N.N.; Wang, Y.Q.; Guo, X.; Cai, T.T.; Lv, L.L. Study on Acute Toxicity of Different Processed Ethanol Extract of *Polygoni multiflori* in Mice. *Chin. J. Pharmacovigil.* **2017**, *14*, 603–606.
13. Sichuan Food and Drug Administration. *Processing Standard of Chinese Herbal Pieces in Sichuan Province*; Sichuan Science and Technology Press: Sichuan, China, 2015.
14. Zheng, S.Y. Protective effect of *Polygonatum sibiricum* Polysaccharide on D-galactose-induced aging rats model. *Sci. Rep.* **2020**, *10*, 2246. [CrossRef]
15. He, L.L.; Yan, B.X.; Yao, C.Y.; Chen, X.Y.; Li, L.W.; Wu, Y.J.; Song, Z.J.; Song, S.S.; Zhang, Z.F.; Luo, P. Oligosaccharides from *Polygonatum cyrtonema* Hua: Structural characterization and treatment of LPS-induced peritonitis in mice. *Carbohydr. Polym.* **2021**, *255*, 117392. [CrossRef]
16. Tang, C.; Yu, Y.M.; Qi, Q.L.; Wu, X.D.; Wang, J.; Tang, S.A. Steroidal saponins from the rhizome of *Polygonatum sibiricum*. *J. Asian Nat. Prod. Res.* **2019**, *21*, 197–206. [CrossRef]
17. Hu, C.Y.; Xu, D.P.; Wu, Y.M.; Ou, S.Y. Triterpenoid saponins from the rhizome of *Polygonatum sibiricum*. *J. Asian Nat. Prod. Res.* **2010**, *12*, 801–808. [CrossRef] [PubMed]
18. Chen, H.; Li, Y.J.; Li, X.F.; Sun, Y.J.; Li, H.W.; Su, F.Y.; Cao, Y.G.; Zhang, Y.L.; Zheng, X.K.; Feng, W.S. Homoisoflavanones with estrogenic activity from the rhizomes of *Polygonatum sibiricum*. *J. Asian Nat. Prod. Res.* **2018**, *20*, 92–100. [CrossRef] [PubMed]

19. Chinese Pharmacopoeia Commission. *Pharmacopoeia of the People's Republic of China*; China Medical Science Press: Beijing, China, 2020; Volume 1.
20. Xue, S.J.; Wang, L.L.; Chen, S.Q.; Cheng, Y.X. Simultaneous Analysis of Saccharides between Fresh and Processed *Radix rehmanniae* by HPLC and UHPLC-LTQ-Orbitrap-MS with Multivariate Statistical Analysis. *Molecules* **2018**, *23*, 541. [CrossRef] [PubMed]
21. Zhang, X.F.; Chen, J.; Yang, J.L.; Shi, Y.P. UPLC-MS/MS analysis for antioxidant components of *Lycii fructus* based on spectrum-effect relationship. *Talanta* **2018**, *180*, 389–395. [CrossRef]
22. Sharma, S.; Joshi, R.; Kumar, D. Quantitative analysis of flavonols, flavonol glycoside and homoisoflavonoids in *Polygonatum verticillatum* using UHPLC-DAD-QTOF-IMS and evaluation of their antioxidant potential. *Phytochem. Anal.* **2020**, *31*, 333–339. [CrossRef] [PubMed]
23. Yu, X.A.; Ge, A.H.; Zhang, L.; Li, J.; An, M.R.; Gao, J.; He, J.; Gao, X.M.; Chang, Y.X. Influence of different processing times on the quality of *Polygoni Multiflora Radix* by metabolomics based on ultra high performance liquid chromatography with quadrupole time-of-flight mass spectrometry. *J. Sep. Sci.* **2017**, *40*, 1928–1941. [CrossRef]
24. Jiang, T.; Wu, T.; Gao, P.Y.; Wang, L.X.; Yang, X.Y.; Chen, X.X.; Chen, Y.Y.; Yue, C.Y.; Liang, K.Q.; Tang, L.Y.; et al. Research on Processing-Induced Chemical Variations in *Polygonatum cyrtonema* Rhizome by Integrating Metabolomics and Glycomics. *Molecules* **2022**, *27*, 5869. [CrossRef] [PubMed]
25. Wang, Y.H.; Wang, X.R.; Ji, T.; Lu, C.L.; Wu, D.J.; Shi, F.F.; Zhou, J.H. The study of oligosaccharides in loquat honey during maturation by gas chromatography-mass spectrometry. *Food Ferment. Ind.* **2022**, *18*, 235–243.
26. Ruiz-Matute, A.I.; Brokl, M.; Soria, A.C.; Sanz, M.L.; Martínez-Castro, I. Gas chromatographic–mass spectrometric characterisation of tri- and tetrasaccharides in honey. *Food Chem.* **2009**, *120*, 637–642. [CrossRef]
27. Ma, X.B.; Wu, Y.J.; Li, Y.; Huang, Y.F.; Liu, Y.; Luo, P.; Zhang, Z.F. Rapid discrimination of *Notopterygium incisum* and *Notopterygium franchetii* based on characteristic compound profiles detected by UHPLC-QTOF-MS/MS coupled with multivariate analysis. *Phytochem. Anal.* **2020**, *31*, 355–365. [CrossRef] [PubMed]
28. Jiang, Y.L.; Xu, Z.J.; Cao, Y.F.; Wang, F.; Chu, C.; Zhang, C.; Tao, Y.; Wang, P. HPLC fingerprinting-based multivariate analysis of chemical components in *Tetrastigma hemsleyanum* Diels et Gilg: Correlation to their antioxidant and neuraminidase inhibition activities. *J. Pharm. Biomed. Anal.* **2021**, *205*, 114314. [CrossRef]
29. Han, J.; Xu, K.; Yan, Q.X.; Sui, W.W.; Zhang, H.T.; Wang, S.J.; Zhang, Z.; Wei, Z.Y.; Han, F. Qualitative and quantitative evaluation of *Flos puerariae* by using chemical fingerprint in combination with chemometrics method. *J. Pharm. Anal.* **2022**, *12*, 489–499. [CrossRef] [PubMed]
30. Li, Q.Y.; Zeng, J.; Gong, P.X.; Wu, Y.C.; Li, H.J. Effect of steaming process on the structural characteristics and antioxidant activities of polysaccharides from *Polygonatum sibiricum* rhizomes. *Glycoconj. J.* **2021**, *38*, 561–572. [CrossRef] [PubMed]
31. Zhu, S.Q.; Liu, P.; Wu, W.X.; Li, D.; Shang, E.X.; Sheng, G.; Qian, D.W.; Yan, H.; Wang, W.; Duan, J.A. Multi-constituents variation in medicinal crops processing: Investigation of nine cycles of steam-sun drying as the processing method for the rhizome of *Polygonatum cyrtonema*. *J. Pharm. Biomed. Anal.* **2022**, *209*, 114497. [CrossRef]
32. Wang, D.; Zhang, H.; Liu, J.L.; Man, Q.; Dong, X.R.; Liu, D.M.; Huang, L.X. Comparative Analysis of Bioactive Components of *Polygonatum cyrtonema* Hua from Different Areas. *Agric. Sci. Technol.* **2020**, *21*, 50–57.
33. Blois, M.S. Antioxidant determinations by the use of a stable free radical. *Nature* **1958**, *181*, 1199–1200. [CrossRef]
34. Moon, J.K.; Shibamoto, T. Antioxidant assays for plant and food components. *J. Agric. Food Chem.* **2009**, *57*, 1655–1666. [CrossRef] [PubMed]
35. Zaharudin, N.; Staerk, D.; Dragsted, L.O. Inhibition of α-glucosidase activity by selected edible seaweeds and fucoxanthin. *Food Chem.* **2019**, *270*, 481–486. [CrossRef] [PubMed]
36. Yonemoto, R.; Shimada, M.; Gunawan-Puteri, M.; Kato, E.; Kawabata, J. α-Amylase inhibitory triterpene from *Abrus precatorius* leaves. *J. Agric. Food Chem.* **2014**, *62*, 8411–8414. [CrossRef] [PubMed]

Disclaimer/Publisher's Note: The statements, opinions and data contained in all publications are solely those of the individual author(s) and contributor(s) and not of MDPI and/or the editor(s). MDPI and/or the editor(s) disclaim responsibility for any injury to people or property resulting from any ideas, methods, instructions or products referred to in the content.

molecules

MDPI

Article

Evaluation of the Biological Properties of an Optimized Extract of *Polygonum cuspidatum* Using Ultrasonic-Assisted Extraction

Gabriela Fletes-Vargas [1,2], Rogelio Rodríguez-Rodríguez [2,3,*], Neith Pacheco [4], Alejandro Pérez-Larios [1] and Hugo Espinosa-Andrews [2,*]

[1] Laboratorio de Nanomateriales, Agua y Energía, Departamento de Ingenierías, Centro Universitario de Los Altos, Universidad de Guadalajara, Tepatitlán de Morelos 47600, Mexico; ana.fletes3623@alumnos.udg.mx (G.F.-V.); alarios@cualtos.udg.mx (A.P.-L.)

[2] Unidad de Tecnología Alimentaria, Centro de Investigación y Asistencia en Tecnología y Diseño del Estado de Jalisco, Zapopan 45019, Mexico

[3] Departamento de Ciencias Naturales y Exactas, Centro Universitario de los Valles (CUVALLES), Universidad de Guadalajara, Ameca 46600, Mexico

[4] Centro de Investigación y Asistencia en Tecnología y Diseño del Estado de Jalisco CIATEJ, A.C. Subsede Sureste, Parque Científico Tecnológico de Yucatán, Mérida 97302, Mexico; npacheco@ciatej.mx

* Correspondence: rogelio.rodriguez4085@academicos.udg.mx (R.R.-R.); hespinosa@ciatej.mx (H.E.-A.)

Abstract: Phytochemicals are natural compounds found in plants that have potential health benefits such as antioxidants, anti-inflammatory and anti-cancer properties, and immune reinforcement. *Polygonum cuspidatum* Sieb. et Zucc. is a source rich in resveratrol, traditionally consumed as an infusion. In this study, *P. cuspidatum* root extraction conditions were optimized to increase antioxidant capacity (DPPH, ABTS+), extraction yield, resveratrol concentration, and total polyphenolic compounds (TPC) via ultrasonic-assisted extraction using a Box–Behnken design (BBD). The biological activities of the optimized extract and the infusion were compared. The optimized extract was obtained using a solvent/root powder ratio of 4, 60% ethanol concentration, and 60% ultrasonic power. The optimized extract showed higher biological activities than the infusion. The optimized extract contained 16.6 mg mL^{-1} resveratrol, high antioxidant activities (135.1 µg TE mL^{-1} for DPPH, and 230.4 µg TE mL^{-1} for ABTS+), TPC (33.2 mg GAE mL^{-1}), and extraction yield of 12.4%. The EC_{50} value (effective concentration 50) of the optimized extract was 0.194 µg mL^{-1}, which revealed high cytotoxic activity against the Caco-2 cell line. The optimized extract could be used to develop functional beverages with high antioxidant capacity, antioxidants for edible oils, functional foods, and cosmetics.

Keywords: antioxidant capacity; resveratrol; *Polygonum cuspidatum*; ultrasonic-assisted extraction; response surface methodology; infusion

Citation: Fletes-Vargas, G.; Rodríguez-Rodríguez, R.; Pacheco, N.; Pérez-Larios, A.; Espinosa-Andrews, H. Evaluation of the Biological Properties of an Optimized Extract of *Polygonum cuspidatum* Using Ultrasonic-Assisted Extraction. *Molecules* **2023**, *28*, 4079. https://doi.org/10.3390/molecules28104079

Academic Editors: Eliza Oprea, Arunaksharan Narayanankutty and Ademola C. Famurewa

Received: 1 April 2023
Revised: 29 April 2023
Accepted: 10 May 2023
Published: 13 May 2023

Copyright: © 2023 by the authors. Licensee MDPI, Basel, Switzerland. This article is an open access article distributed under the terms and conditions of the Creative Commons Attribution (CC BY) license (https://creativecommons.org/licenses/by/4.0/).

1. Introduction

Nutraceutical is a term derived from "nutrition" and "pharmaceutics", used for compounds isolated from herbal products with biological activity. These compounds provide health benefits, especially for preventing and treating diseases such as cancer, diabetes, cardiovascular and neurological disorders [1,2]. These diseases are associated with the generation and accumulation of reactive oxygen species (ROS) produced by cellular oxidative stress [3,4]. Antioxidant compounds can inhibit or decrease oxidation processes that affect biomolecules such as proteins, lipids, and DNA [5]. Antioxidants can protect cells against oxidation by blocking the initiation phase of radical production or neutralizing radicals. Commonly, herbal plants contain antioxidant properties due to the presence of bioactive compounds. *Polygonum cuspidatum Siebold & Zucc.* belongs to the Polygonaceae family and grows widely in Asia and North America. It has been used for centuries in China and Japan as an herbal medicine to treat inflammatory diseases, hepatitis, tumors, diarrhea, arthralgia, chronic bronchitis, amenorrhea, hypertension, neuroprotector, and hypercholesterolemia. In addition, its ethanolic extracts have estrogenic and antiviral activities against hepatitis B viruses and SARS-CoV-2 omicron [6,7]. The root of *P. cuspidatum* contains many secondary metabolites with biological efficacy. These compounds have been identified as stilbenes, including resveratrol, piceid, and emodin.

Resveratrol is one of the most highly investigated antioxidant molecules [8]. Resveratrol (3,5,4′-trihydroxy-stilbene) is a polyphenolic molecule found in many foods such as grapes, mulberries, peanuts, cereals, vegetables, flowers, and roots [9,10]. Resveratrol is a secondary metabolite that confers protection against pathogenic attack, UV radiation and environmental stress, heavy metals, and in some cases, climate change [11]. Resveratrol reduces the formation of intracellular ROS and oxidative damage, thereby providing several biological activities: anti-inflammatory, antioxidant, anti-aging, anti-tumor, and anti-mutagenic [12]. However, the therapeutic potential and bioavailability of resveratrol are limited due to its low water solubility [13]. Piceid is a stilbenoid with a neurological protection effect that has been reported at concentrations 10 or 15 times higher than resveratrol [14]. In addition, its bioavailability is lower compared to resveratrol due to intestinal cells absorbing piceid slowly, and this process requires glycosidases [15]. Emodin is an anthraquinone located in the rhizome, and quercetin is a flavonoid in leaves and stems. Polyphenols are commonly found in flowers [16,17].

Traditionally, all vegetable parts of *P. cuspidatum* are consumed as tea beverages or infusions with medicinal aims. However, the extraction of phenolic compounds from *P. cuspidatum* can be performed using Soxhlet extraction, which has the disadvantage that the extraction is performed at elevated temperatures for a long time, which can degrade the polyphenols [18]. In addition, polyphenol extraction can use organic solvents such as acetone, ethanol, methanol, and ethyl acetate; however, organic solvents may not be efficient [19]. Recently, ultrasound-assisted extraction technology (UAE) has been employed to increase extraction yields and, in some cases, to perform more selective extractions. UAE is an innovative extraction technique in which the sample can be mixed with organic solvents at a controlled temperature, reducing the extraction time [20]. The release of phytochemicals is due to the rupture of the cell walls by ultrasound waves, a phenomenon called cavitation. Extraction of polyphenols with UAE is higher in yield (6–35%) and more time-saving than other traditional techniques [21,22].

Response surface methodology (RSM) is a mathematical and statistical technique widely used to investigate multiple parameters and their possible interactions to optimize processes [23]. RSM reduces the number of experimental runs and the time required to investigate the optimal conditions for extraction [24]. Kuo et al. [18] optimized the extraction conditions of phenolic compounds from *P. cuspidatum* using multiple RSM. The authors reported that temperature and ethanol concentration impacted the extraction yields of the bioactive compounds. In addition, using supercritical carbon dioxide technology, Ruan et al. [25] reported that *P. cuspidatum* extracts showed high scavenging capacity. However, there are few reports about the biological activities of ethanolic extracts of *P. cuspidatum*.

This study aimed to optimize *P. cuspidatum* root extraction conditions to increase antioxidant capacity (DPPH, ABTS+), extraction yield, resveratrol concentration, and total polyphenolic compound content via UAE using a Box–Behnken design (BBD). The independent variables were solvent/root powder ratio, ethanol concentration, and ultrasonic power. We hypothesized that the interaction of the independent variables would increase the antioxidant capacity of the extract, extraction yield, polyphenolic compound content, and resveratrol concentration compared to the infusion. Thus, the biological activities of the optimized extract and the traditional infusion were compared.

2. Results

2.1. Model Fitting from Extracts of *P. cuspidatum* Root

Table 1 shows the experimental values of antioxidant capacity (DPPH and ABTS+), TPC, resveratrol concentration, and extraction yield obtained by UAE from the interaction variables: solvent/root-powder ratio (X_1), ethanol concentration (X_2), and ultrasonic power (X_3). According to the experimental values, the scavenging capacities for DPPH and ABTS+ ranged from 51.1 to 135.1 μg TE mL^{-1} and 119.5 to 230.4 μg TE mL^{-1}, respectively. The data obtained for total polyphenolic compounds ranged from 5.8 to 33.2 mg GAE mL^{-1}. The resveratrol content in crude extract and the extraction yield ranged from 12.0 to 16.7 mg mL^{-1} and 2.74% to 12.43%, respectively. A reduced quadratic polynomial was used to predict the experimental data, shown in Equation (1).

Table 1. Experimental values of antioxidant activity, total polyphenolic compounds, resveratrol concentration, and extraction yield obtained by UAE with a Box–Behnken design.

Experimental Run	Independent Variables			Responses				
				Antioxidant Capacity (μg TE mL^{-1})		TPC	Resveratrol	Extraction Yield
	X_1 ($mL\ g^{-1}$)	X_2 (%)	X_3 (W)	DPPH	ABTS+	(mg GAE mL^{-1})	Concentration (mg mL^{-1})	(%)
1	7	45	129	103.5	190.3	18.9	16.3	6.6
2	7	45	129	116.2	198.4	18.9	16.5	8.5
3	7	45	129	126.7	214.7	27.9	16.3	9.3
4	7	45	129	109.0	194.3	19.4	16.4	7.5
5	7	45	129	109.7	195.3	19.1	16.5	7.3
6	4	45	107	132.9	222.2	25.0	16.5	9.9
7	7	30	150	52.5	126.6	8.0	12.7	3.6
8	10	45	150	103.5	188.2	16.8	16.3	5.6
9	10	60	107	103.3	190.4	19.3	16.2	6.6
10	7	30	129	51.1	121.2	6.8	12.0	2.7
11	7	60	150	126.7	215.8	29.3	16.6	9.4
12	4	45	107	133.8	229.0	27.6	16.7	10.1
13	10	30	150	51.3	119.5	5.8	12.4	2.9
14	4	60	107	135.1	230.4	33.2	16.6	12.4
15	4	30	129	60.3	129.4	8.8	13.6	3.3
16	7	60	129	129.4	215.8	29.1	16.5	9.3
17	10	45	129	99.9	197.1	17.7	16.2	5.24

Table 2 shows the ANOVA results for the responses and interactions of the independent variables. Regarding antioxidant activities ($ABTS^+$ and DPPH), variables X_1 and $X_2{}^2$ showed negative effects, while variable X_2 had highly significant positive effects ($p < 0.001$). Meanwhile, variable X_3 did not show a significant effect; however, it contributed positively to the antioxidant activities. The analysis indicated that the determinant coefficients (R^2) were 0.9612 for $ABTS^+$ and 0.9593 for DPPH. The regression model explained 96.12% and 95.93% of the responses for $ABTS^+$ and DPPH, respectively ($p < 0.05$). The TPC showed a linear impact for variables X_1 and X_2, with an $R^2 = 0.8581$, able to explain 85.81% of the fitted regression model ($p < 0.05$). Ethanol concentration was the most critical parameter for increasing the resveratrol content in the extract. The lineal variable X_2 showed a highly significant positive effect ($p < 0.001$), while variable $X_2{}^2$ showed a highly significant negative effect ($p < 0.001$). According to the F-value, the impact of the linear X_2 is higher than the quadratic X_2, resulting in a positive effect on resveratrol extraction. Variables X_1, X_3, X_1X_2, X_2X_3, significantly impacted the resveratrol content, showing an excellent coefficient of correlation of the predicted model ($R^2 = 0.9850$). The extraction yield indicated a linear effect for X_1 and X_3 and a quadratic impact for X_2, an $R^2 = 0.8993$. Data analysis of the extraction yield showed that the mathematical model could predict the effect of the interactions for variables X_1, X_2, and X_3 on UAE from *P. cuspidatum*.

Table 2. ANOVA results of the correlated polynomial model for antioxidant activity, TPC, resveratrol content, and extraction yield.

Response	Source	Sum of Squares	df	Mean Square	*F*-Value	*p*-Value
ABTS+	X_1	1676.21	1	1676.21	18.75	0.0123 *
	X_2	15,815.3	1	15,815.3	176.94	0.0002 **
	X_3	55.6513	1	55.6513	0.62	0.4742
	${X_2}^2$	5082.14	1	5082.14	56.86	0.0017 *
	Lack of fit	555.787	8	69.4734	0.78	0.6481
	Pure error	357.52	4	89.38		
	Cor Total	23,542.6	16			
	$R_2 = 0.9612$					
	$R_{2adju} = 0.9483$					
DPPH	X_1	1352.0	1	1352.0	17.16	0.0143 *
	X_2	9751.06	1	9751.06	123.80	0.0004 **
	X_3	1.90125	1	1.90125	0.02	0.8841
	${X_2}^2$	2934.15	1	2934.15	37.25	0.0036 *
	Lack of fit	341.858	8	42.7323	0.54	0.7860
	Pure error	315.068	4	78.767		
	Cor Total	14,696.0	16			
	$R^2 = 0.9552$					
	${R^2}_{adju} = 0.9403$					
TPC	X_1	153.125	1	153.125	9.80	0.0351
	X_2	830.281	1	830.281	53.16	0.0019 *
	X_3	3.00125	1	3.00125	0.19	0.6838
	Lack of fit	100.61	9	11.1789	0.72	0.6901
	Pure error	62.472	4	15.618		
	Cor Total	1149.49	16			
	$R^2 = 0.8581$					
	${R^2}_{adju} = 0.8253$					
Resveratrol content	X_1	0.66125	1	0.66125	66.12	0.0012 *
	X_2	28.88	1	28.88	2888.00	0.0000 **
	X_3	0.10125	1	0.10125	10.12	0.0335 *
	X_1X_2	0.16	1	0.16	16.00	0.0161 *
	X_2X_3	0.09	1	0.09	9.0	0.0399 *
	${X_1}^2$	0.0796053	1	0.0796053	7.96	0.0478
	${X_2}^2$	14.2164	1	14.2164	1421.64	0.0000 **
	${X_3}^2$	0.0532895	1	0.0532895	5.33	0.0822
	Lack of fit	0.505	4	0.12625	12.62	0.0154 *
	Pure error	0.04	4	0.01		
	Cor Total	44.8424	16			
	$R^2 = 0.9878$					
	${R^2}_{adju} = 0.9756$					
Extraction yield	X_1	29.6835	1	29.6835	26.46	0.0068 *
	X_2	79.317	1	79.317	70.72	0.0011 *
	X_3	0.09245	1	0.09245	0.08	0.7883
	${X_2}^2$	9.45017	1	9.45017	8.43	0.0440 *
	Lack of fit	8.78113	8	1.09764	0.98	0.5486
	Pure error	4.48648	4	1.12162		
	Cor Total	131.811	16			
	$R^2 = 0.8993$	29.6835				
	${R^2}_{adju} = 0.8658$					

* Statistical significative ($p < 0.05$), ** high statistical significative ($p < 0.001$).

2.2. *Response Surface Plot Analysis*

The 3D response surface plots show the impacts of independent variables on the antioxidant capacity, TPC, resveratrol content, and extraction yield of *P. cuspidatum* (Figures 1 and 2). Results revealed high antioxidant activities with increasing ethanol concentration (from 45% to 60%) and decreasing solvent/root-powder ratio (from 4 mL g^{-1} to 7 mL g^{-1}), reaching maximum values of 135.1 μg TE mL^{-1} (IC_{50} = 78 μg TE mL^{-1}) and 230.4 μg TE mL^{-1} (IC_{50} = 158 μg TE mL^{-1}) for DPPH and ABTS+, respectively.

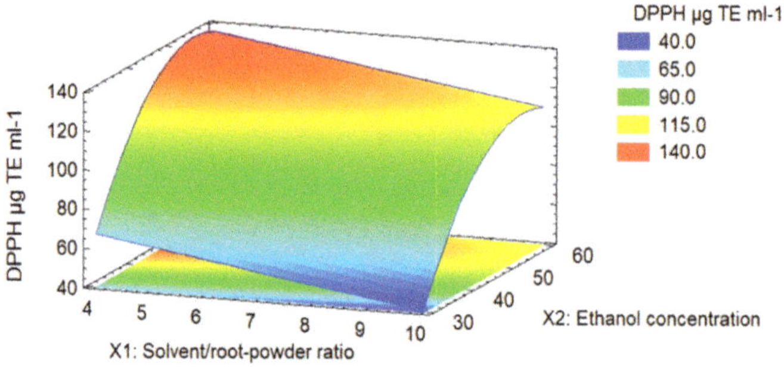

(**A**)

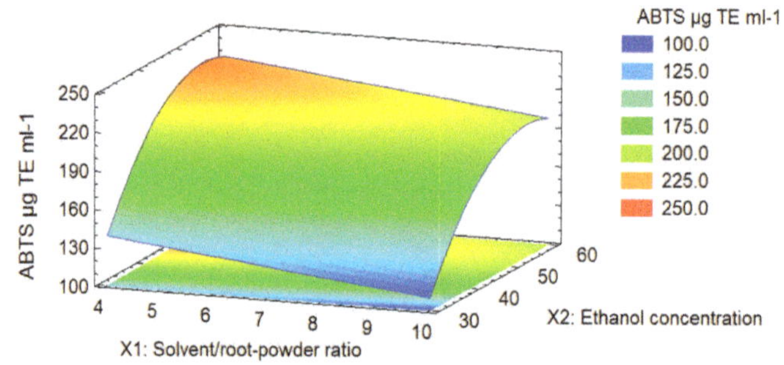

(**B**)

Figure 1. Surface plot of the combined effect of the independent variables on antioxidant activities: (**A**) DPPH (X_1X_2) and (**B**) $ABTS^+$ radical (X_1X_2). X_1 = solvent/root-powder ratio and X_2 = ethanol concentration.

TPC showed a maximum concentration of 33.2 mg GAE mL^{-1} at higher ethanol concentrations (45% to 60%) and low solvent/root-powder ratios (4 mL g^{-1} to 7 mL g^{-1}) (Figure 2A).

Resveratrol content (Figure 2B,C) showed a highly significant impact ($p < 0.001$), with resveratrol content increasing to 16.7 mg mL^{-1} when the ethanol concentration was higher than 50%. These results suggest that solvent/root-powder ratio, ultrasonic power, and interactions impacted the resveratrol concentration. The maximum extraction yield (Figure 2D), corresponding to 12.43%, reached a lower solvent/root-powder ratio (4 mL g^{-1} to 7 mL g^{-1}) and high ethanol concentration (45 to 60%).

The regression coefficients to calculate the predicted response for the antioxidant activities of ABTS+ and DPPH, TPC, resveratrol content, and extraction yield were performed using a reduced second-order polynomial equation (Table 3).

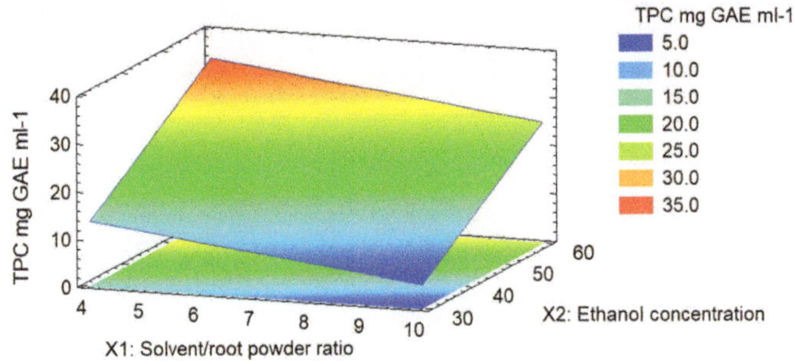

(A)

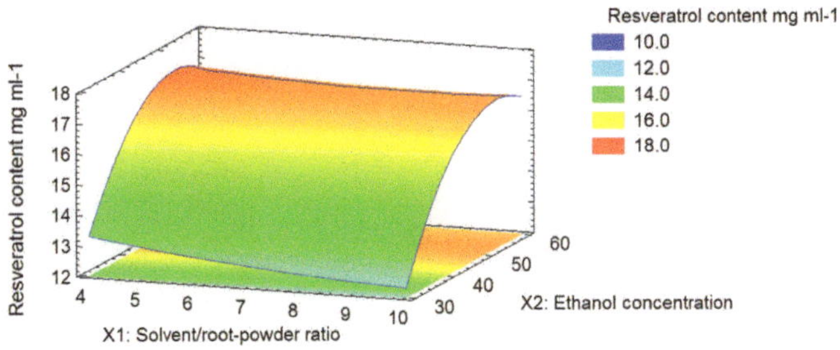

(B)

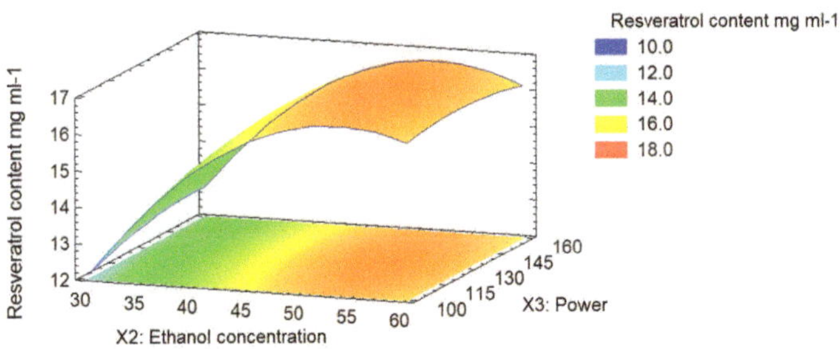

(C)

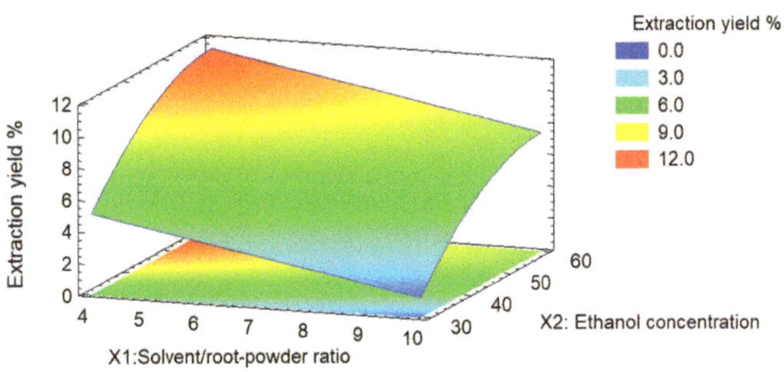

(D)

Figure 2. 3D plots for interaction variables for (**A**) TPC (X_1X_2), (**B**) resveratrol content (X_1X_2) y (**C**) resveratrol content (X_2X_3), and (**D**) extraction yield (X_2X_3). X_1 = solvent/root-powder ratio, X_2 = ethanol concentration, and X_3 = ultrasonic power.

Table 3. Reduced polynomial equations for antioxidant capacity, TPC, resveratrol content, and extraction yield.

Response	Reduced Equations
$ABTS^+$	$ABTS+ = -223.86 - 4.82X_1 + 16.82X_2 + 0.12X_3 - 0.15X_2^2$
DPPH	$DPPH = -193.35 - 4.33X_1 + 12.86X_2 - 0.02X_3 - 0.12X_2^2$
TPC	$TPC = -4.51 - 1.46X_1 + 0.68X_2 + 0.03X_3$
Resveratrol content	$\text{Resveratrol content} = -10.40 - 0.51X_1 + 0.89X_2 + 0.088X_3 + 0.015X_1^2 + 4.4 \times 10^{-3}X_1 X_2 - 8 \times 10^{-3}X_2^2 - 4.6 \times 10^{-4} X_2 X_3 - 2.4 \times 10^{-4} X_3^2$
Extraction yield	$\text{Extraction yield} = -11.26 - 0.64X_1 + 0.81X_2 + 0.017X_3 - 6.64 \times 10^{-3}X_2^2$

X_1 = solvent/root-powder ratio, X_2 = ethanol concentration, X_3 = ultrasonic power.

The optimal predicted value was obtained in experimental run 14, corresponding to a solvent/root-powder ratio of 4, 60% ethanol concentration, and 128.5 W of ultrasonic power, corresponding to 58 KJ/g. The predicted values for the antioxidant activity of DPPH and ABTS+ corresponded to 136.625 μg TE mL^{-1} and 227.575 μg TE mL^{-1}, respectively. For TPC, the expected value was 34.068 mg GAE mL^{-1}, and resveratrol content and extraction yield were 15.56 mg mL^{-1} and 11.36%, respectively. The predicted desirability value for optimal extraction of *P. cuspidatum* was 0.9661 (Figure 3). The experimental conditions were validated using an independent experiment, finding 136.2 μg TE mL^{-1}, 195.4 μg TE mL^{-1}, 29.95 mg GAE mL^{-1}, 16.72 mg mL^{-1}, and 12.3% for DPPH, ABTS+, TPC, resveratrol, and extraction yield, respectively.

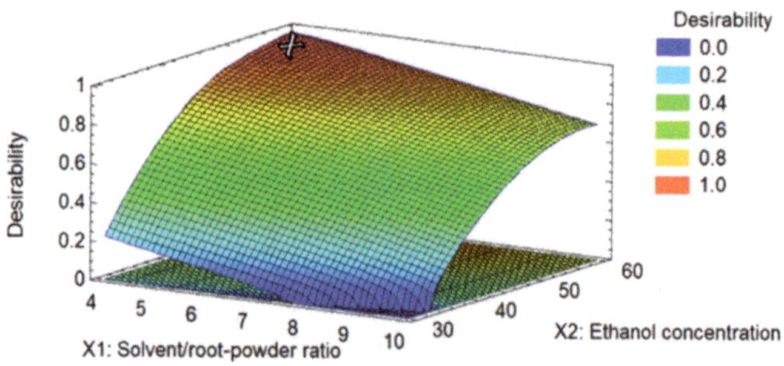

Figure 3. Desirability 3D plot; the cross mark represents the desirability for optimal extraction conditions.

Analysis of the infusion of *P. cuspidatum* root powder indicates that the values for the antioxidant activities of DPPH and ABTS+ were 103.75 μg TE mL^{-1} and 147.78 μg TE mL^{-1}, respectively. In addition, the TPC value was 0.4024 mg GAE mL^{-1}. The resveratrol content and extraction yield were 0.044 mg mL^{-1} and 0.340%, respectively.

2.3. Cytotoxic Assay and EC_{50} Value

The optimized extract was diluted (1:110) in MEM (Minimal Eagle's Medium), and the infusion was used to directly compare the cytotoxicity on colorectal cancer cells for 24 h. The results obtained for the optimized extract demonstrate that the viability of Caco-2 cells decreases at low concentrations in a dose-depending manner (1.24 μg mL^{-1} to 0.03 μg mL^{-1}), and the EC_{50} value corresponded to 0.125 ± 0.008 (R^2 = 0.9913) (Figure 4A). In contrast, the infusion extraction of *P. cuspidatum* showed low viability with an estimated EC_{50} of 0.03 ± 0.002 μg mL^{-1} (R^2 = 0.9892) (Figure 4B). Figure 5A shows that Caco-2 cells treated with optimized extraction were smaller and had low cell confluency compared to untreated cells (Figure 5C). Cells treated with the infusion of *P. cuspidatum* show higher confluency and apoptotic bodies, characteristic of the induction of apoptotic cell death (Figure 5B), compared to those treated with the optimized extract (Figure 5C).

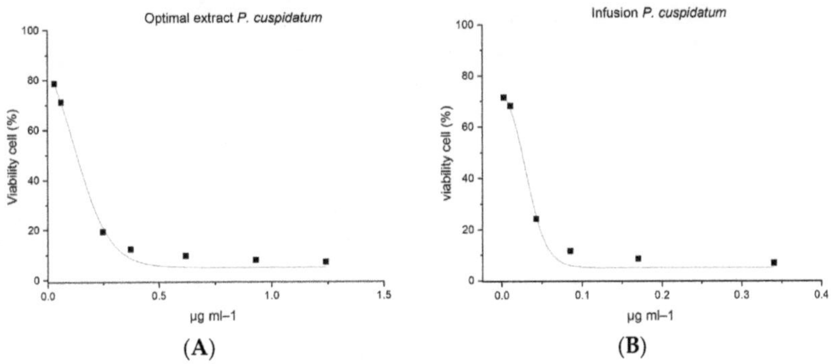

Figure 4. Effect of *P. cuspidatum* extracts on viability of cells at different concentrations for 24 h. (**A**) Caco-2 cells treated with diluted optimized extract, (**B**) Caco-2 cells treated with direct infusion. Data show mean ± SEM from three independent experiments.

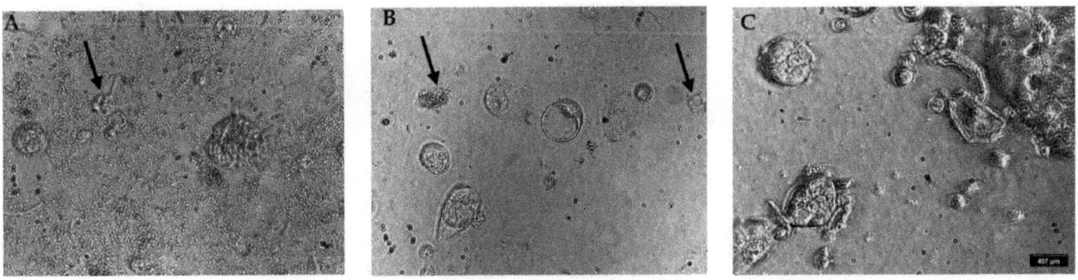

Figure 5. Microphotography of Caco-2 cells treated with *P. cuspidatum* extracts after 24 h: (**A**) small cells after the treatment with optimized extract; (**B**) apoptotic cells induced by aqueous infusion and (**C**) non-treated cells (negative control). Arrows indicate the cells at different treatments.

2.4. Compounds Identified by UPLC-Mass Spectrometry (MS)

Figure 6 shows the chromatogram of the optimized extract from *P. cuspidatum* obtained through MS analysis. The spectra revealed two main peaks attributed to stilbene compounds: piceid (RT = 5.71) and resveratrol (RT = 7.55). Thus, these results indicate that piceid and resveratrol are the major bioactive components in the optimized extract.

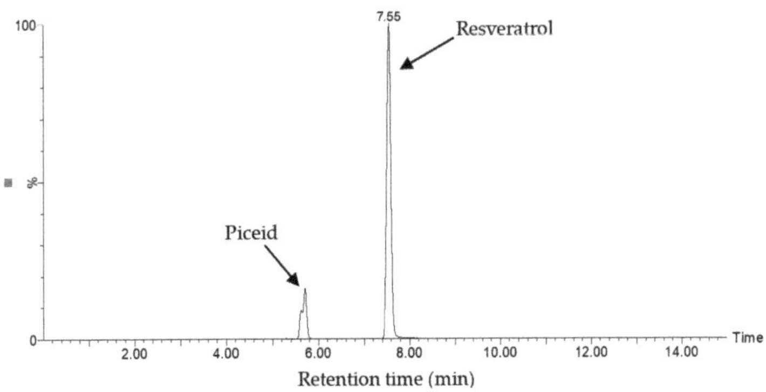

Figure 6. Chromatogram of the optimized extract from *P. cuspidatum* obtained by mass spectrometry (MS).

3. Discussion

Bioactive compounds have medicinal benefits and are extracted from fresh or dried plants using different extraction methods and solvents [26]. Solvents impact the recovery and quality of bioactive compounds, but can degrade them at high temperatures [27]. Water is used to prepare beverages with an affinity for hydrophilic molecules, such as phenolic compounds, proteins, and carbohydrates. Studies have reported that the low dielectric constant and low polarity of hot water increase the diffusion of compounds, improving the extraction of lipophilic bioactive compounds. On the other hand, ethanol is used to extract bioactive compounds, such as polyphenols, flavonoids, alkaloids, and terpenes [28,29]. Several factors, such as plant type, ethanol concentration, and extraction time, impact the extraction yield. The ethanol–water concentration increases the polarity of the solvent, improving the efficiency of the extraction of nonpolar phytochemicals [30].

Usually, resveratrol is extracted with 95% ethanol from the root of *P. cuspidatum* using a refluxing method followed by liquid–liquid extraction with organic solvents [25]. Kuo et al. [18] found that when applying a temperature of 70 °C, 60% ethanol concentration, and 120 W ultrasonic power, the amount of resveratrol was 3.9 mg/g. Jia et al. [31] reported high flavonoid concentration (94.5%) using UAE- and CCRD-extracted flavonoids from *P. cuspidatum* using an ethanol concentration of 60%, solid–liquid ratio of 1:20 g mL^{-1}, extraction temperature of 45 °C, extraction time of 34 min, and ultrasonic power of 80 W. Ruan et al. [32] reported an extraction yield value for resveratrol and emodin of 2.564 ± 0.121 mg mL^{-1} and 2.804 ± 0.108 mg mL^{-1}, respectively, and a high scavenging capacity. Our results showed that resveratrol content at the optimized extract conditions was higher than those reported in the literature.

The standard methods to evaluate the antioxidant capacity of foods employ the stable radicals DPPH and the cation ABTS+. Results obtained from antioxidant assays reported the necessary antioxidant concentrations to reduce radicals [33]. Becze et al. [34] obtained extracts with a high antioxidant capacity of 34.623 µg AAE mL^{-1} and 182.4 µL of resveratrol content from *F. regala* leaves. UAE of *Arachis repens*, known as peanut ass, showed high concentrations of trans-resveratrol and total polyphenolics, demonstrating a high level of DPPH free radical inhibition (70%). Polyphenolic compounds can reduce the Folin–Ciocalteu reagent under alkaline conditions, which yields a colored product [35]. El Moussaoui et al. [36] evaluated the antioxidant activity and total polyphenols of extract from *Whitania frutescens* L., and found that roots are 67 times richer in polyphenols (53.3 ± 1.2 mg GAE/g).

The optimized ethanolic extract showed higher cytotoxic activity than the infusion extract cont due to the higher resveratrol concentration found in the ethanolic extract. Similar results were found by Youmbi and coworkers [37]. They found that crude extract obtained from the leaves of *Brucea antidysenterica* induces cell death mediated by caspases on drug-resistant cancer cells, such as CCRF-CEM, a human leukemia cell line, at a low concentration (from 12.42 µg mL^{-1} to 38.70 µg mL^{-1}). Other authors reported that ethanol extracts of *P. cuspidatum* demonstrated efficacy on the loss of viability of cells in oral, breast, and ovarian cancer at lower concentrations, EC_{50}: <50 µg mL^{-1} [17,38,39].

On the other hand, tea or infusion is widely used for its beneficial properties and low cost. Traditionally, *P. cuspidatum* is commonly used as part of traditional Chinese medicine, and is prepared as a herbal infusion in boiling water for a few minutes. Kosovic et al. [40] evaluated the stability of resveratrol from *Vitis vinifera* L. at high temperatures, finding a better release of stilbenes such as trans-resveratrol. However, if a prolonged time was applied, the polyphenolic compounds could be degraded. The effect of green tea on rat hepatocyte cells was evaluated by Schmidt et al. [41], and the results show that cell viability decreased at 500 µg mL^{-1} of the extract. Aqueous extract of *Reynoutris japonica* Houtt root, a Polygonaceae family, was not cytotoxic in SK-Hep1 and Huh7 cell lines, but *in vitro* results indicated that at a concentration of 20 µg mL^{-1} inhibited wound recovery and invasion in Huh7 cells [42].

The main compounds in *P. cuspidatum* root extracts have been investigated using chromatographical analysis. Yi et al. [43] reported that the *P. cuspidatum* rhizome is rich in stilbenes (piceid and resveratrol) and anthraquinones (emodin-8-*O*-β-D-glucoside, and emodin). These compounds are commonly used as an indicator of quality assessment for herbal extracts. Vastano et al. [44] evaluated two varieties of *P. cuspidatum* root extracts (Hu Zhang and Mexican Bamboo). They identified three stilbene glucosides: piceatannol glucoside, resveratroloside, and piceid. By comparison, our results showed that the optimized extract of *P. cuspidatum* root contains piceid and resveratrol as the major bioactive compounds.

4. Materials and Methods

4.1. Plant Material and Reagents

Dried root powder of *P. cuspidatum* with particle size 80 mesh (~0.177 mm) was obtained from Herbal Mexico (Tlalnepantla de Baz, Mexico). Pure ethyl alcohol, 6-hydroxy-2,5,7,8-tetramethyl chroman-2-carboxylic acid (Trolox), 2,2-diphenyl-1-picrylhydrazyl (DPPH), 2,2′-Azino-bis(3-ethylbenzothiazoline-6-sulfonic acid) diammonium salt (ABTS+), sodium persulfate, Folin–Ciocualteu phenol reagent, and resveratrol standard (purity > 99%) were purchased from Sigma-Aldrich (St. Louis, MO, USA). Gallic acid and sodium carbonate (Na_2CO_3) were obtained from Jalmek (San Nicolás de Los Garza, NL, Mexico). MEM and Fetal Bovine Serum (FBS) were purchased from Biowest, and Tripsine-EDTA solution and Resazurin sodium salt from Sigma-Aldrich.

4.2. Optimization of Ultrasonic-Assisted Extraction (UAE)

P. cuspidatum powder was dispersed in ethanol, according to Table 1. Then, dispersions were sonicated using an Elmasonic p70H ultrasonic bath (37 kHz, Elma Schmidbauer GmbH, Singen, Germany) following the conditions of the Box–Behnken design. After applying the ultrasound, the samples were centrifuged at 4000 rpm for 15 min. The supernatant was collected and stored at −20 °C until analysis. As a control, an infusion of 10 g of root powder was dispersed into 200 mL of boiling water; afterward, the solution was filtrated to remove sediment and stored at −20 °C until analysis.

4.3. Experiment Design Strategy and Statistical Analysis

RSM optimized the UAE extraction parameters for DPPH, ABTS+, TPC, resveratrol concentration, and extraction yield from *P. cuspidatum*. In a three-level, three-factorial Box–Behnken design, including five replicates of the central point, 17 runs were analyzed in random order (Table 4). The independent variables were solvent/root-powder ratio (X_1, mL g^{-1}), ethanol concentration (X_2, %), and ultrasonic power (X_3, W). Temperature and extraction time were fixed at 45 °C and 30 min, respectively.

Table 4. Levels of variables for experimental design.

Independent Variable	Coded Levels		
	−1	0	1
X_1	4	7	10
X_2	30	45	60
X_3	107	128.5	150

RSM was applied to obtain the optimal UAE conditions for each raw material. A second-order polynomial equation derived from the RSM was used to calculate the predicted response (Equation (1)):

$$Y = \beta_0 + \sum_{i=1}^{k} \beta_i X_i + \sum_{i=1}^{k} \beta_{ii} X_{ii} + \sum_{i>j}^{k} \beta_{ij} X_j + E \quad (1)$$

Y is the response variable, X_i is the coded or uncoded value for the factors evaluated, β_0 is a constant, β_i is the main effect of the coefficient for each variable, β_{ij} represents the interaction effect coefficients, and E is the error of the model. ANOVA evaluated significant interactions of the model ($p < 0.05$). The coefficient of determination was quantified (R^2 and adjusted R^2). The predicted values for antioxidant activity (DPPH and ABTS+), TPC, resveratrol content, and extraction yield were maximized to establish the optimal UAE via RSM. An additional extraction was conducted using the optimal predicted conditions to verify the model's suitability. Applied energy was estimated according to Strieder et al. [45]: energy = [ultrasonic power × time extraction]/mass of the sample. The regression coefficients were used to generate 3D surface plots from the fitted polynomial equation. Also, the regression coefficients were used to visualize the relationship between the response variable and experimental levels and predict the optimum conditions.

4.4. *Determination of Extraction Yield*

The extraction yield (%) was quantified by vaporization utilizing 2 mL of crude extract in an oven at 40 °C until the dry mass was obtained. The results were shown as a mass of total extractable solids per 100 g of dry plant material (%).

4.5. *Resveratrol Quantification Using Spectrophotometry UV–Vis*

UV–Vis spectroscopy was used to assess the concentration of resveratrol in the crude extracts. Samples were compared with a standard curve of resveratrol in ethyl alcohol. Ethyl alcohol was utilized as a blank for background correction. Absorbance was read at 312 nm using a multiwell plate reader (TECAN infinite pro-200, Trading AG, Steinhausen, Switzerland).

4.6. *In Vitro Antioxidant Capacity*

4.6.1. DPPH Radical Scavenging Assay

For the DPPH stable radical scavenging assay, samples (ethanolic samples 1:400, infusion 1:2), blank (ethyl alcohol), or Trolox standard were added to a 96-well microplate. Subsequently, DPPH solution was added to each well and allowed to stand for 30 min in the dark before reading absorbance at 517 nm. All experiments were realized in triplicate. The capacity to scavenge the DPPH radical was presented as µg of Trolox equivalent (TE) mL^{-1} [46].

4.6.2. ABTS+ Assay

For the ABTS+ test, in a 96-well microplate the sample (ethanolic samples 1:400, infusion 1:2), blank (ethyl alcohol pure), or Trolox standard were added. Subsequently, the ABTS+ working solution was added to each well and allowed to stand for 5 min in the dark before reading absorbance at 734 nm [22]. All experiments were realized in triplicate. The results for antioxidant activity in the ABTS+ radical test were presented as µg of Trolox equivalent (TE) mL^{-1} [46].

4.6.3. Total Polyphenolic Compounds

Polyphenols were measured using Folin–Ciocalteu's method. The samples (ethanolic samples 1:400, infusion 1:2) and blank or gallic standard curves were placed in a 96-well plate. Next, 0.5 N Folin–Ciocalteu phenol reagent and 75 µL Na_2CO_3 solution (0.1 mg mL^{-1}) were added. Absorbance was measured at 765 nm and the results expressed as mg GAE per mL [47].

4.7. *Caco-2 Cell Culture and Cytotoxic Assay*

The cytotoxic assay for the optimal extraction of *P. cuspidatum* was evaluated using a resazurin reduction assay [48]. The optimal extract was diluted with MEM to obtain different concentrations. The Caco-2 cell line (HTB-37 ATCC), a human colorectal adenocarcinoma, was cultured in MEM supplemented with 10% FBS and 1% penicillin/streptomycin. Briefly, cells were incubated until reaching ~80% confluence and seeded in a 96-well plate at 5×10^3 cells per well at 37 °C and 5% CO_2. After 24 h, MEM was replaced with 100 μL of the extracts at different concentrations and incubated for 24 h (37 °C and 5% CO_2). After that, 20 μL of resazurin solution (0.2 mg mL^{-1}) was added and incubated for 3 h at 37 °C. The fluorescence emitted by resorufin, metabolized from resazurin by viable cells, was measured (excitation 560 nm, emission 590 nm). A negative control (MEM medium) and positive control (DMSO 10%) were used. Percentage viability and effective concentration 50 (EC_{50}) of crude extracts and the infusion of *P. cuspidatum* were calculated using OriginPro 2018 Software (Origin Lab Corporation, Northampton, MA, USA). The cell viability was calculated as shown:

$$\% \text{ cell viability} = I_s / I_c \times 100\%$$

where: I_s is the absorbance of the cells exposed to the extracts and I_c is the absorbance of the cells without exposure to the extracts

4.8. Chromatography Analysis by Mass Spectrometry (MS/MS)

Mass spectrometry (MS) analysis was performed using a Waters Xevo TQ-S instrument (Waters, Milford, MA, USA). The conditions used for identifying the bioactive compounds from the optimized extract by comparing fingerprint and MS data were those previously reported [49,50]. Resveratrol and piceid were identified using multiple reaction monitoring (MRM), a sensitive method of targeted MS. For acquisition and data processing, MassLynx V4.1 software was employed (Waters, Milford, MA, USA).

4.9. Statistical Analysis

The results obtained were analyzed using ANOVA and Tukey's means comparison analysis. Data from the Box–Behnken experimental design were examined using the least square multiple regression methodology to fit the polynomial models for UAE optimization. Data analysis and response surfaces were conducted using Statgraphics Centurion XVI. I software (Statistical Graphics Corp., Manugistics, Inc., Cambridge, MA, USA). Significance was established at $p \leq 0.05$. All experiments were conducted in triplicate.

5. Conclusions

Ultrasonic-assisted extraction from *P. cuspidatum* roots was effectively developed. The extraction efficiency, antioxidant activity, resveratrol content, and total polyphenolic compounds could be further improved through BBD experimental design and response surface methodology. A second-order model was obtained to describe the relationship between the responses and their interactions with independent variables X_1 (solvent/root powder ratio), X_2 (ethanol concentration), and X_3 (ultrasonic power at 45 °C for 30 min). The evaluated parameters affected antioxidant activity (DPPH, ABTS+), total polyphenolic compounds, resveratrol content, and extraction yields from the crude extract from *P. cuspidatum* root. The optimized extract showed higher cytotoxic activity than the traditional infusion of *P. cuspidatum* root against Caco-2 colorectal cancer cells. Due to the complex mixture in the optimized extract of *P. cuspidatum*, future research is required to identify and quantify all the bioactive compounds and to determine the mechanism by which this extract induces cell death.

Author Contributions: Conceptualization, G.F.-V. and R.R.-R.; methodology, H.E.-A., G.F.-V., R.R.-R., H.E.-A. and N.P.; validation, G.F.-V. and H.E.-A.; formal analysis, G.F.-V. and H.E.-A.; investigation, G.F.-V., R.R.-R. and H.E.-A.; resources, R.R.-R. and H.E.-A.; writing—original draft preparation, G.F.-V., R.R.-R. and H.E.-A.; supervision, A.P.-L., R.R.-R. and H.E.-A. All authors have read and agreed to the published version of the manuscript.

Funding: This research received no external funding.

Data Availability Statement: Data will be made available on request.

Acknowledgments: The authors thank CIATEJ A.C. and G.F.-V. thanks CONACYT for scholarship grant number 778177.

Conflicts of Interest: The authors declare no conflict of interest.

Sample Availability: Not applicable.

References

1. Kalra, E.K. Nutraceutical-Definition and Introduction. *AAPS PharmSci* **2003**, *5*, 27–28. [CrossRef]
2. Nasri, H.; Baradaran, A.; Shirzad, H.; Rafieian-Kopaei, M. New Concepts in Nutraceuticals as Alternative for Pharmaceuticals. *Int. J. Prev. Med.* **2014**, *5*, 1487. [PubMed]
3. Sachdeva, V.; Roy, A.; Bharadvaja, N. Current Prospects of Nutraceuticals: A Review. *Curr. Pharm. Biotechnol.* **2020**, *21*, 884–896. [CrossRef] [PubMed]
4. Prasad, S.; Gupta, S.C.; Tyagi, A.K. Reactive Oxygen Species (ROS) and Cancer: Role of Antioxidative Nutraceuticals. *Cancer Lett.* **2017**, *387*, 95–105. [CrossRef] [PubMed]
5. Barba-Ostria, C.; Carrera-Pacheco, S.E.; Gonzalez-Pastor, R.; Heredia-Moya, J.; Mayorga-Ramos, A.; Rodríguez-Pólit, C.; Zúñiga-Miranda, J.; Arias-Almeida, B.; Guamán, L.P. Evaluation of Biological Activity of Natural Compounds: Current Trends and Methods. *Molecules* **2022**, *27*, 4490. [CrossRef]
6. HSU, C.-Y.; CHAN, Y.-P.; CHANG, J. Antioxidant Activity of Extract from *Polygonum cuspidatum*. *Biol. Res.* **2007**, *40*. [CrossRef] [PubMed]
7. Lin, S.; Wang, X.; Tang, R.W.-L.; Lee, H.C.; Chan, H.H.; Choi, S.S.A.; Dong, T.T.-X.; Leung, K.W.; Webb, S.E.; Miller, A.L. The Extracts of *Polygonum cuspidatum* Root and Rhizome Block the Entry of SARS-CoV-2 Wild-Type and Omicron Pseudotyped Viruses via Inhibition of the S-Protein and 3CL Protease. *Molecules* **2022**, *27*, 3806. [CrossRef] [PubMed]
8. Bononi, I.; Tedeschi, P.; Mantovani, V.; Maietti, A.; Mazzoni, E.; Pancaldi, C.; Brandolini, V.; Tognon, M. Antioxidant Activity of Resveratrol Diastereomeric Forms Assayed in Fluorescent-Engineered Human Keratinocytes. *Antioxidants* **2022**, *11*, 196. [CrossRef]
9. Kotta, S.; Mubarak Aldawsari, H.; Badr-Eldin, S.M.; Alhakamy, N.A.; Md, S. Coconut Oil-Based Resveratrol Nanoemulsion: Optimization Using Response Surface Methodology, Stability Assessment and Pharmacokinetic Evaluation. *Food Chem.* **2021**, *357*, 129721. [CrossRef]
10. Syahdi, R.R.; Nadyana, R.; Putri, R.H.; Santi, R.; Mun'im, A. Application of green extraction methods to resveratrol extraction from peanut (*Arachis hypogaea* L.) skin. *Int. J. Appl. Pharm.* **2020**, 38–42. [CrossRef]
11. Guamán-Balcázar, M.C.; Setyaningsih, W.; Palma, M.; Barroso, C.G. Ultrasound-Assisted Extraction of Resveratrol from Functional Foods: Cookies and Jams. *Appl. Acoust.* **2016**, *103*, 207–213. [CrossRef]
12. Jin, S.; Gao, M.; Kong, W.; Yang, B.; Kuang, H.; Yang, B.; Fu, Y.; Cheng, Y.; Li, H. Enhanced and Sustainable Pretreatment for Bioconversion and Extraction of Resveratrol from Peanut Skin Using Ultrasound-Assisted Surfactant Aqueous System with Microbial Consortia Immobilized on Cellulose. *3 Biotech* **2020**, *10*, 293. [CrossRef]
13. Chimento, A.; De Amicis, F.; Sirianni, R.; Sinicropi, M.; Puoci, F.; Casaburi, I.; Saturnino, C.; Pezzi, V. Progress to Improve Oral Bioavailability and Beneficial Effects of Resveratrol. *Int. J. Mol. Sci.* **2019**, *20*, 1381. [CrossRef] [PubMed]
14. Zhang, J.; Zhou, L.; Zhang, P.; Liu, T.; Yang, G.; Lin, R.; Zhou, J. Extraction of Polydatin and Resveratrol from *Polygonum cuspidatum* Root: Kinetics and Modeling. *Food Bioprod. Process.* **2015**, *94*, 518–524. [CrossRef]
15. Signorelli, P.; Ghidoni, R. Resveratrol as an Anticancer Nutrient: Molecular Basis, Open Questions and Promises. *J. Nutr. Biochem.* **2005**, *16*, 449–466. [CrossRef]
16. Dong, J.; Wang, H.; Wan, L.; Hashi, Y.; Chen, S. Identification and Determination of Major Constituents in Polygonum Cuspidatum Sieb. et Zucc. by High Performance Liquid Chromatography/Electrospray Ionization-Ion Trap-Time-of-Flight Mass Spectrometry. *Se Pu* **2009**, *27*, 425–430. [PubMed]
17. Ke, J.; Li, M.-T.; Xu, S.; Ma, J.; Liu, M.-Y.; Han, Y. Advances for Pharmacological Activities of *Polygonum cuspidatum*—A Review. *Pharm. Biol.* **2023**, *61*, 177–188. [CrossRef]
18. Kuo, C.-H.; Chen, B.-Y.; Liu, Y.-C.; Chang, C.-M.; Deng, T.-S.; Chen, J.-H.; Shieh, C.-J. Optimized Ultrasound-Assisted Extraction of Phenolic Compounds from *Polygonum cuspidatum*. *Molecules* **2013**, *19*, 67–77. [CrossRef] [PubMed]
19. Stalikas, C.D. Extraction, Separation, and Detection Methods for Phenolic Acids and Flavonoids. *J. Sep. Sci.* **2007**, *30*, 3268–3295. [CrossRef] [PubMed]
20. Martins Strieder, M.; Keven Silva, E.; Angela, A.; Meireles, M. Specific Energy: A New Approach to Ultrasound-Assisted Extraction of Natural Colorants. *Food Public Health* **2019**, *9*, 45–52. [CrossRef]
21. Irfan, S.; Ranjha, M.M.A.N.; Nadeem, M.; Safdar, M.N.; Jabbar, S.; Mahmood, S.; Murtaza, M.A.; Ameer, K.; Ibrahim, S.A. Antioxidant Activity and Phenolic Content of Sonication- and Maceration-Assisted Ethanol and Acetone Extracts of Cymbopogon Citratus Leaves. *Separations* **2022**, *9*, 244. [CrossRef]
22. Ranjha, M.M.A.N.; Irfan, S.; Lorenzo, J.M.; Shafique, B.; Kanwal, R.; Pateiro, M.; Arshad, R.N.; Wang, L.; Nayik, G.A.; Roobab, U.; et al. Sonication, a Potential Technique for Extraction of Phytoconstituents: A Systematic Review. *Processes* **2021**, *9*, 1406. [CrossRef]

23. Kadam, S.U.; Tiwari, B.K.; Smyth, T.J.; O'Donnell, C.P. Optimization of Ultrasound Assisted Extraction of Bioactive Components from Brown Seaweed Ascophyllum Nodosum Using Response Surface Methodology. *Ultrason Sonochem.* **2015**, *23*, 308–316. [CrossRef]
24. Aguilar-Hernández, G.; de los Vivar-Vera, M.Á.; de García-Magaña, M.L.; González-Silva, N.; Pérez-Larios, A.; Montalvo-González, E. Ultrasound-Assisted Extraction of Total Acetogenins from the Soursop Fruit by Response Surface Methodology. *Molecules* **2020**, *25*, 1139. [CrossRef]
25. Sun, B.; Zheng, Y.-L.; Yang, S.-K.; Zhang, J.-R.; Cheng, X.-Y.; Ghiladi, R.; Ma, Z.; Wang, J.; Deng, W.-W. One-Pot Method Based on Deep Eutectic Solvent for Extraction and Conversion of Polydatin to Resveratrol from *Polygonum cuspidatum*. *Food Chem.* **2021**, *343*, 128498. [CrossRef]
26. Vinatoru, M. An Overview of the Ultrasonically Assisted Extraction of Bioactive Principles from Herbs. *Ultrason Sonochem.* **2001**, *8*, 303–313. [CrossRef] [PubMed]
27. Kumar, A.; Kumar, M.; Jose, A.; Tomer, V.; Oz, E.; Proestos, C.; Zeng, M.; Elobeid, T.; Oz, F. Major Phytochemicals: Recent Advances in Health Benefits and Extraction Method. *Molecules* **2023**, *28*, 887. [CrossRef]
28. Thavamoney, N.; Sivanadian, L.; Tee, L.H.; Khoo, H.E.; Prasad, K.N.; Kong, K.W. Extraction and Recovery of Phytochemical Components and Antioxidative Properties in Fruit Parts of Dacryodes Rostrata Influenced by Different Solvents. *J. Food Sci. Technol.* **2018**, *55*, 2523–2532. [CrossRef] [PubMed]
29. Babbar, N.; Oberoi, H.S.; Sandhu, S.K.; Bhargav, V.K. Influence of Different Solvents in Extraction of Phenolic Compounds from Vegetable Residues and Their Evaluation as Natural Sources of Antioxidants. *J. Food Sci. Technol.* **2014**, *51*, 2568–2575. [CrossRef] [PubMed]
30. Aboulghazi, A.; Bakour, M.; Fadil, M.; Lyoussi, B. Simultaneous Optimization of Extraction Yield, Phenolic Compounds and Antioxidant Activity of Moroccan Propolis Extracts: Improvement of Ultrasound-Assisted Technique Using Response Surface Methodology. *Processes* **2022**, *10*, 297. [CrossRef]
31. Jia, W.; Chen, Z.; Zhao, Y.; Li, K.; Tichnell, B.; Tang, Z.; Ruso, J.M.; Liu, Z. The Study of Ultrasound-Assisted Extraction of Flavonoids from *Polygonum cuspidatum* Sieb. et Zucc. *J. Mater. Res.* **2019**, *34*, 3254–3262. [CrossRef]
32. Ruan, N.; Jiao, Z.; Tang, L. Response Surface Methodology to Optimize Supercritical Carbon Dioxide Extraction of *Polygonum cuspidatum*. *J. AOAC Int.* **2022**, *105*, 272–281. [CrossRef]
33. Sadowska-Bartosz, I.; Bartosz, G. Evaluation of The Antioxidant Capacity of Food Products: Methods, Applications and Limitations. *Processes* **2022**, *10*, 2031. [CrossRef]
34. Becze, A.; Babalau-Fuss, V.L.; Varaticeanu, C.; Roman, C. Optimization of High-Pressure Extraction Process of Antioxidant Compounds from Feteasca Regala Leaves Using Response Surface Methodology. *Molecules* **2020**, *25*, 4209. [CrossRef]
35. Piyaratne, P.S.; LeBlanc, R.; Myracle, A.D.; Cole, B.J.W.; Fort, R.C. Extraction and Purification of (E)-Resveratrol from the Bark of Conifer Species. *Processes* **2022**, *10*, 647. [CrossRef]
36. El Moussaoui, A.; Jawhari, F.Z.; Almehdi, A.M.; Elmsellem, H.; Fikri Benbrahim, K.; Bousta, D.; Bari, A. Antibacterial, Antifungal and Antioxidant Activity of Total Polyphenols of *Withania frutescens* L. *Bioorg. Chem.* **2019**, *93*, 103337. [CrossRef] [PubMed]
37. Youmbi, L.M.; Makong, Y.S.D.; Mbaveng, A.T.; Tankeo, S.B.; Fotso, G.W.; Ndjakou, B.L.; Wansi, J.D.; Beng, V.P.; Sewald, N.; Ngadjui, B.T.; et al. Cytotoxicity of the Methanol Extracts and Compounds of Brucea Antidysenterica (Simaroubaceae) towards Multifactorial Drug-Resistant Human Cancer Cell Lines. *BMC Complement. Med. Ther.* **2023**, *23*, 48. [CrossRef]
38. Wang, Y.; Horng, C.; Hsieh, M.; Chen, H.; Huang, Y.; Yang, J.; Wang, G.; Chiang, J.; Chen, H.; Lu, C.; et al. Autophagy and Apoptotic Machinery Caused by *Polygonum cuspidatum* Extract in Cisplatin-resistant Human Oral Cancer CAR Cells. *Oncol. Rep.* **2019**, *41*, 2549–2557. [CrossRef] [PubMed]
39. Pan, B.; Shi, X.; Ding, T.; Liu, L. Unraveling the Action Mechanism of *Polygonum cuspidatum* by a Network Pharmacology Approach. *Am. J. Transl. Res.* **2019**, *11*, 6790–6811.
40. Kosović, E.; Topiař, M.; Cuřínová, P.; Sajfrtová, M. Stability Testing of Resveratrol and Viniferin Obtained from *Vitis vinifera* L. by Various Extraction Methods Considering the Industrial Viewpoint. *Sci. Rep.* **2020**, *10*, 5564. [CrossRef]
41. Schmidt, M.; Schmitz, H.-J.; Baumgart, A.; Guédon, D.; Netsch, M.I.; Kreuter, M.-H.; Schmidlin, C.B.; Schrenk, D. Toxicity of Green Tea Extracts and Their Constituents in Rat Hepatocytes in Primary Culture. *Food. Chem. Toxicol.* **2005**, *43*, 307–314. [CrossRef]
42. Kim, B.R.; Ha, J.; Lee, S.; Park, J.; Cho, S. Anti-Cancer Effects of Ethanol Extract of Reynoutria Japonica Houtt. Radix in Human Hepatocellular Carcinoma Cells via Inhibition of MAPK and PI3K/Akt Signaling Pathways. *J. Ethnopharmacol.* **2019**, *245*, 112179. [CrossRef] [PubMed]
43. Yi, T.; Zhang, H.; Cai, Z. Analysis of Rhizoma Polygoni Cuspidati by HPLC and HPLC-ESI/MS. *Phytochem. Anal.* **2007**, *18*, 387–392. [CrossRef] [PubMed]
44. Vastano, B.C.; Chen, Y.; Zhu, N.; Ho, C.-T.; Zhou, Z.; Rosen, R.T. Isolation and Identification of Stilbenes in Two Varieties of *Polygonum c uspidatum*. *J. Agric. Food Chem.* **2000**, *48*, 253–256. [CrossRef]
45. Strieder, M.M.; Silva, E.K.; Meireles, M.A.A. Specific Energy: A New Approach to Ultrasound-Assisted Extraction of Natural Colorants. *Probe (Adelaide)* **2019**, *23*, 30.
46. Medina-Torres, N.; Espinosa-Andrews, H.; Trombotto, S.; Ayora-Talavera, T.; Patrón-Vázquez, J.; González-Flores, T.; Sánchez-Contreras, Á.; Cuevas-Bernardino, J.C.; Pacheco, N. Ultrasound-Assisted Extraction Optimization of Phenolic Compounds from Citrus Latifolia Waste for Chitosan Bioactive Nanoparticles Development. *Molecules* **2019**, *24*, 3541. [CrossRef]

47. Fonseca-Hernández, D.; Lugo-Cervantes, E.D.C.; Escobedo-Reyes, A.; Mojica, L. Black Bean (*Phaseolus vulgaris* L.) Polyphenolic Extract Exerts Antioxidant and Antiaging Potential. *Molecules* **2021**, *26*, 6716. [CrossRef]
48. O'Brien, J.; Wilson, I.; Orton, T.; Pognan, F. Investigation of the Alamar Blue (Resazurin) Fluorescent Dye for the Assessment of Mammalian Cell Cytotoxicity. *Eur. J. Biochem.* **2000**, *267*, 5421–5426. [CrossRef]
49. Guerrero, R.F.; Valls-Fonayet, J.; Richard, T.; Cantos-Villar, E. A Rapid Quantification of Stilbene Content in Wine by Ultra-High Pressure Liquid Chromatography—Mass Spectrometry. *Food Control* **2020**, *108*, 106821. [CrossRef]
50. Jiménez-Morales, K.; Castañeda-Pérez, E.; Herrera-Pool, E.; Ayora-Talavera, T.; Cuevas-Bernardino, J.C.; García-Cruz, U.; Pech-Cohuo, S.C.; Pacheco, N. Ultrasound-Assisted Extraction of Phenolic Compounds from Different Maturity Stages and Fruit Parts of Cordia Dodecandra A. DC.: Quantification and Identification by UPLC-DAD-ESI-MS/MS. *Agriculture* **2022**, *12*, 2127. [CrossRef]

Disclaimer/Publisher's Note: The statements, opinions and data contained in all publications are solely those of the individual author(s) and contributor(s) and not of MDPI and/or the editor(s). MDPI and/or the editor(s) disclaim responsibility for any injury to people or property resulting from any ideas, methods, instructions or products referred to in the content.

molecules

MDPI

Article

In Vivo Neuropharmacological Effects of Neophytadiene

Maria L. Gonzalez-Rivera [1,†], Juan Carlos Barragan-Galvez [1,†], Deisy Gasca-Martínez [2], Sergio Hidalgo-Figueroa [3], Mario Isiordia-Espinoza [4] and Angel Josabad Alonso-Castro [1,*]

1 Departamento de Farmacia, División de Ciencias Naturales y Exactas, Universidad de Guanajuato, Guanajuato 36200, Mexico
2 Unidad de Análisis Conductual, Instituto de Neurobiología, Campus UNAM-Juriquilla, Juriquilla 76230, Mexico
3 CONACyT-División de Biología Molecular, Instituto Potosino de Investigación Científica y Tecnológica A.C., San Luis Potosí 78216, Mexico
4 Instituto de Investigación en Ciencias Médicas, Departamento de Clínicas, División de Ciencias Biomédicas, Centro Universitario de los Altos, Universidad de Guadalajara, Tepatitlán de Morelos 47620, Mexico
* Correspondence: angeljosabad@ugto.mx; Tel.: +52-(442)-2381036 (ext. 340365)
† These authors contributed equally to this work.

Abstract: Neophytadiene (NPT) is a diterpene found in the methanolic extracts of *Crataeva nurvala* and *Blumea lacera*, plants reported with anxiolytic-like activity, sedative properties, and antidepressant-like actions; however, the contribution of neophytadiene to these effects is unknown. This study determined the neuropharmacological (anxiolytic-like, antidepressant-like, anticonvulsant, and sedative) effects of neophytadiene (0.1–10 mg/kg p.o.) and determined the mechanisms of action involved in the neuropharmacological actions using inhibitors such as flumazenil and analyzing the possible interaction of neophytadiene with GABA receptors using a molecular docking study. The behavioral tests were evaluated using the light–dark box, elevated plus-maze, open field, hole-board, convulsion, tail suspension, pentobarbital-induced sleeping, and rotarod. The results showed that neophytadiene exhibited anxiolytic-like activity only to the high dose (10 mg/kg) in the elevated plus-maze and hole-board tests, and anticonvulsant actions in the 4-aminopyridine and pentylenetetrazole-induced seizures test. The anxiolytic-like and anticonvulsant effects of neophytadiene were abolished with the pre-treatment with 2 mg/kg flumazenil. In addition, neophytadiene showed low antidepressant effects (about 3-fold lower) compared to fluoxetine. On other hand, neophytadiene had no sedative or locomotor effects. In conclusion, neophytadiene exerts anxiolytic-like and anticonvulsant activities with the probable participation of the GABAergic system.

Keywords: neophytadiene; anxiety; convulsion; diterpene

Citation: Gonzalez-Rivera, M.L.; Barragan-Galvez, J.C.; Gasca-Martínez, D.; Hidalgo-Figueroa, S.; Isiordia-Espinoza, M.; Alonso-Castro, A.J. In Vivo Neuropharmacological Effects of Neophytadiene. *Molecules* **2023**, *28*, 3457. https://doi.org/10.3390/molecules28083457

Academic Editors: Eliza Oprea, Arunaksharan Narayanankutty and Ademola C. Famurewa

Received: 3 March 2023
Revised: 3 April 2023
Accepted: 11 April 2023
Published: 14 April 2023

Copyright: © 2023 by the authors. Licensee MDPI, Basel, Switzerland. This article is an open access article distributed under the terms and conditions of the Creative Commons Attribution (CC BY) license (https://creativecommons.org/licenses/by/4.0/).

1. Introduction

In the world, neurological disorders are the leading cause of disability and the second leading cause of death (8.8 million people died in 2019). Only in the USA and Europe, the annual cost of treating mental disorders is USD 1.7 trillion [1]. More than 50 million and 322 million individuals suffer from epilepsy and depression, respectively [2]. Multiple investigations have indicated that anxiety and depression might simultaneously occur [3]. Scientific research has focused on studying medicinal plants and the discovery of potentially bioactive compounds for treating neurological disorders [4]. Several potential compounds isolated from plant extracts have been assayed as possible treatments for anxiety, depression, and epilepsy [4]. For instance, methanolic extracts from the leaves of *Blumea lacera* and *Crataeva nurvala* showed anxiolytic-like, sedative, and antidepressant activities with inhibition of locomotor activity in mice. Several bioactive metabolites were identified in these two studies, and one of the compounds was neophytadiene (NPT), a diterpene (Figure 1) [3,5]. Diterpenes with anticonvulsant effects remained to be analyzed, such as phytol, which increased the latency of onset clonic and tonic seizures and the survival rate with minimal motor impairment in a model of convulsion induced by pentylenetetrazole (PTZ) [4,6].

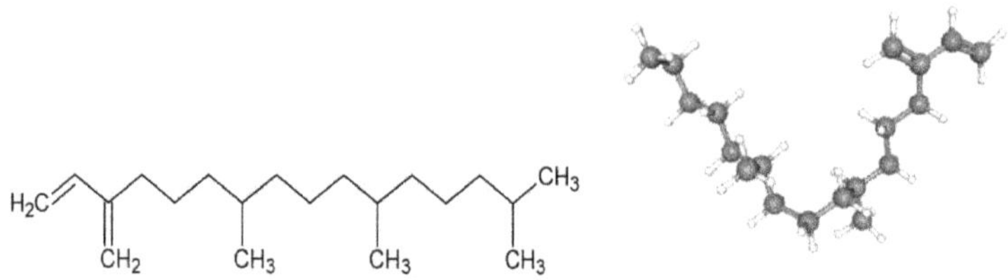

Figure 1. Chemical structure of neophytadiene shown in 2D structure and 3D conformer (NPT).

Medicinal plants containing NPT are used in the treatment of headaches, rheumatism, and some skin diseases, whereas NPT has shown analgesic, antipyretic, anti-inflammatory, and antioxidant properties [7–12]. The anti-inflammatory effect of NPT is attributed to reducing nitric oxide (NO) production and decreasing mRNA levels of proinflammatory cytokines (IL-6 and IL-10) and inhibiting the production of tumor necrosis factor α (TNFα), nuclear factor kB (NF-kB), and downregulated production of inducible nitric synthase (iNOS) in an in vitro assay [13]. NPT also showed antimicrobial activity against gram-positive and gram-negative bacteria, and fungal species [7,14,15].

Few studies have evaluated the pharmacological effects of NPT. To our knowledge, no scientific research has focused the analysis of NPT on mental disorders and seizure protection. Therefore, this research focused on evaluating the anxiolytic-like, antidepressant-like, anticonvulsant, and sedative effects of NPT using in vivo models.

2. Results

2.1. Anxiolytic-like Activity of NPT and Its Possible Mechanism of Action

In the light–dark box test (LDBT), the three tested doses of NPT did not show statistical significance ($p > 0.05$) in the time and the number of entries in the light compartment compared to the vehicle group, whereas the reference drug clonazepam (CNZ) raised ($p < 0.05$) the time and the number of entries in the light compartment (Figure 2a,b).

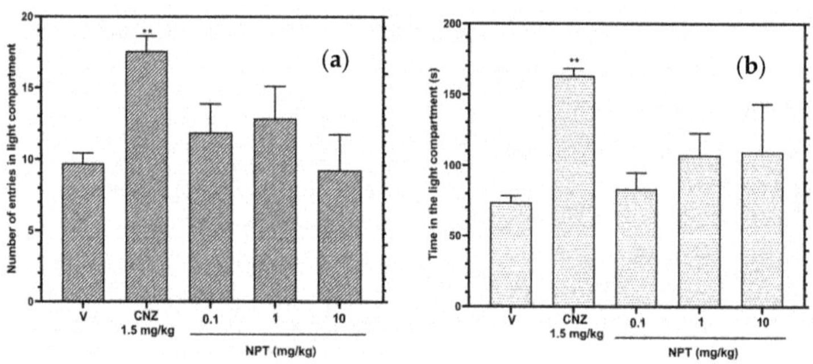

Figure 2. Anxiolytic-like actions of NPT (0.1–10 mg/kg p.o.) in the LDBT were determined by the time in the light compartment (**a**) and the number of crossings into the light compartment (**b**). The reference drug was 1.5 mg/kg p.o. clonazepam (CNZ). Bars represent mean values (±SEM) for the experimental group. $n = 8$, ** $p < 0.05$ compared to the vehicle group (indicated as V).

In the elevated plus-maze test (EPMT), only the high dose of NPT (10 mg/kg), compared to the vehicle group, increased ($p < 0.05$) the time in open arms and decreased ($p < 0.05$) the number of entries and time in closed arms (Figure 3b,d); however, this same dose did not affect the number of entries in open arms (Figure 3a). The anxiolytic-like effect presented by NPT was lower than that presented by CNZ.

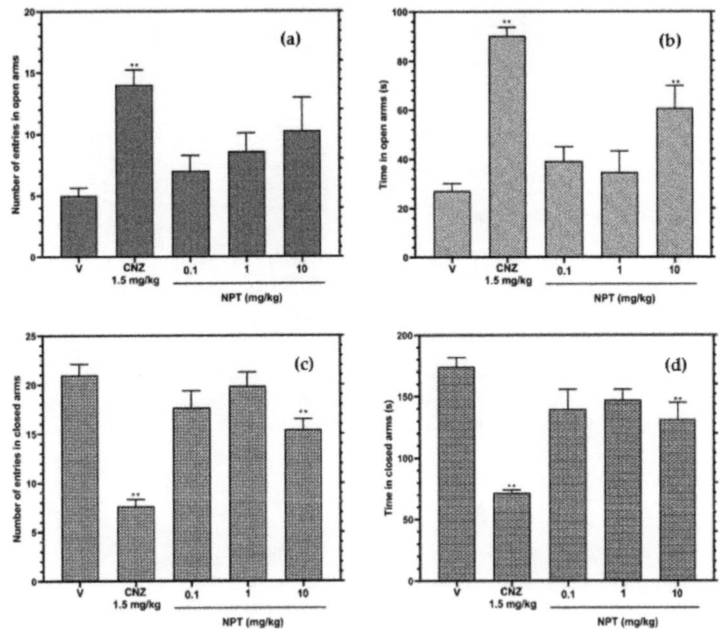

Figure 3. Anxiolytic-like effects of NPT (0.1–10 mg/kg p.o.) were assessed by the number of admissions in open (**a**) and closed arms (**c**) and the time elapsed in open (**b**) and closed (**d**) arms in the EPMT for 5 min. The reference drug was 1.5 mg/kg p.o. clonazepam (CNZ). The results of all experimental groups are reported as the mean values (±SEM). $n = 8$, ** $p < 0.05$ regarding to the vehicle group (indicated as V).

In the open field test (OFT) (Table 1), no dose of NPT elicited anxiolytic-like activity compared to 1.5 mg/kg of the reference drug CNZ, which decreased the total distance and increased the resting time, time in the center squares, and distance in center squares ($p < 0.05$ vs. vehicle group).

Table 1. Effects of NPT in the OFT.

Treatment	Total Distance (cm)	Resting Time (s)	Time in Center Squares (s)	Distance in Center Squares (cm)
Vehicle	2070 ± 58.69	69.59 ± 2.45	16.08 ± 0.53	345.6 ± 13.85
CNZ 1.5 mg/kg	640.5 ± 16.57 ***	170.0 ± 4.15 ***	56.28 ± 1.54 ***	748.3 ± 12.59 ***
NPT 0.1 mg/kg	2081 ± 130.7	70.95 ± 11.56	19.94 ± 3.30	352.9 ± 55.97
NPT 1 mg/kg	2160 ± 295.6	73.99 ± 14.95	19.63 ± 3.40	348.5 ± 57.42
NPT 10 mg/kg	2252 ± 312.8	93.86 ± 15.71	14.53 ± 4.14	367.6 ± 81.51

CNZ-Clonazepam, NPT-neophytadiene. *** $p < 0.0001$ vs. vehicle group.

In the hole-board test (HBT), only 10 mg/kg NPT increased ($p < 0.05$) the occurrence of head dipping by 2.6-fold compared to the vehicle group (Figure 4a). The anxiolytic-like activity of NPT was close to the effect shown by the reference drug CNZ (3.3-fold). On the other hand, Figure 4b shows that the pre-treatments with the antagonists of GABA receptor bicuculline (1 mg/kg) and flumazenil (2 mg/kg) abolished the anxiolytic-like effects shown by NPT, showing similar results to the vehicle group ($p > 0.05$). These findings suggest that 10 mg/kg of NPT shows an anxiolytic-like effect.

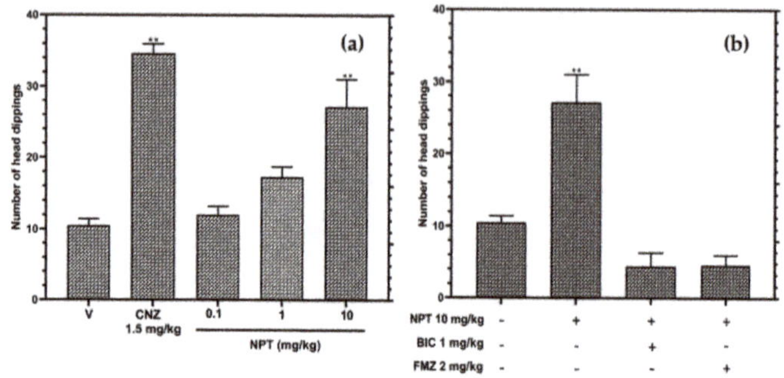

Figure 4. Anxiolytic-like actions of NPT (0.1–10 mg/kg p.o.) in the HBT (**a**), and the possible anxiolytic mechanism of NPT (**b**). Bicuculline (BIC) and flumazenil (FMZ) were used as antagonists. The reference drug was 1.5 mg/kg p.o. clonazepam (CNZ). The results of all experimental groups are reported as the mean values (±SEM). $n = 8$, ** $p < 0.05$ regarding to the vehicle group (indicated as V).

A molecular docking study evaluated the possible interaction of NPT in $GABA_A$ receptors (Figure 5). NPT exhibits the same preference in binding site as flumazenil (Figure 5A–C). Interestingly, flumazenil and NPT can form hydrophobic interactions in aspartic acid residue at position 56 (Asp56), tyrosine residue at position 58 (Tyr58), phenylanalanine residue at position 77 (Phe77), alanine residue at position 79 (Ala79), histidine residue at position 102 (His102), threonine residue at position 142 (Thr142), tyrosine residue at position 210 (Tyr210), and tyrosine residue at position 260 (Tyr260) of $GABA_A$ receptors (Figure 5B,C). Flumazenil showed an additional H-bond with Thr142. The binding energy for flumazenil (redocked) was −9.4 Kcal/mol, whereas the binding energy for NPT was −7.1 Kcal/mol.

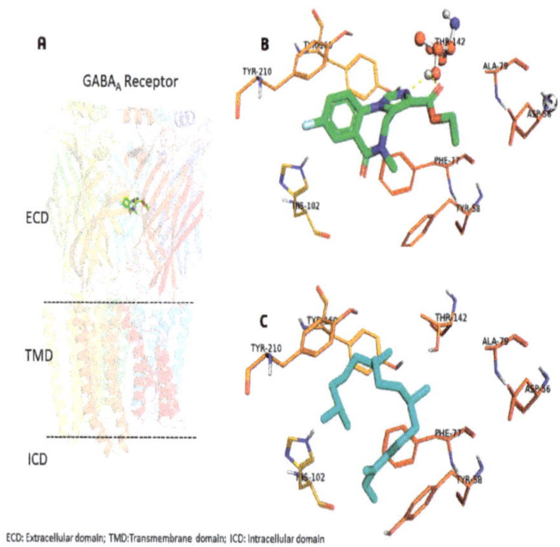

Figure 5. (**A**) Heteropentameric $GABA_A$ receptor (PDB id: 6D6T) displayed in cartoon representation with competitive antagonist flumazenil (green color). (**B**) Flumazenil interacting with the binding pocket with hydrophobic residues, H-bond is shown in dashed line, and Thr142 is represented in ball and stick. (**C**) Neophytadiene (cyan color) is surrounded by hydrophobic residues in the pocket.

2.2. *Anticonvulsant Activity and Possible Mechanism of Action*

NPT (0.1–10 mg/kg) decreased the duration of the convulsion and decreased the mortality rate in a dose-dependent compared to the vehicle group in the pentylenetetrazole (PTZ)-induced convulsion test (Table 2). It is important to mention that two mice treated with 10 mg/kg NPT did not show convulsions. The pretreatment with 2 mg/kg flumazenil abolished the anticonvulsant effects of 10 mg/kg NPT, increasing the mortality rate (Table 2, upper section). In the 4-aminopyridine (4-AP)-induced convulsion, the three doses of NPT decreased the mortality rate compared to the vehicle group; the dose of 10 mg/kg had a significant ($p < 0.05$) effect on the onset of convulsion and the 0.1 mg/kg and 1.0 mg/kg doses significantly decreased the duration of convulsion (Table 2, bottom section). This data shows that NPT exhibits anticonvulsant effects and reduces the mortality for seizures.

Table 2. Anticonvulsant activity of NPT in mice treated with 90 mg/kg pentylenetetrazole (PTZ) or 12 mg/kg 4-aminipyridine (4-AP) to induce convulsions.

Treatment	Onset of Convulsion (s)	Duration of Convulsion (s)	Mortality (%)
	90 mg/kg PTZ		
Vehicle	69.34 ± 2.11	165.6 ± 3.77	100.00
CNZ 1.5 mg/kg	0.00 ± 0.00 ***	0.00 ± 0.00 ***	0.00
NPT 0.1 mg/kg	109.10 ± 13.12	47.71 ± 6.58 ***	14.00
NPT 1 mg/kg	97.86 ± 7.01	36.43 ± 6.61 ***	57.14
NPT 10 mg/kg ¥	73.43 ± 23.73	21.29 ± 6.50 ***	42.85
NPT 10 mg/kg + Flumazenil 2 mg/kg	100.9 ± 5.50	13.43 ± 2.88 ***	85.71
	12 mg/kg 4-AP		
Vehicle	129.1 ± 6.95	38.63 ± 2.54	100.00
CNZ 1.5 mg/kg	258.1 ± 11.93 ***	15.50 ± 1.91 ***	50.00
NPT 0.1 mg/kg	176.9 ± 32.51	14.00 ± 2.78 ***	85.71
NPT 1 mg/kg	162.5 ± 9.74	18.88 ± 2.41 ***	87.50
NPT 10 mg/kg	191.3 ± 14.96 *	32.38 ± 4.79	75.00

CNZ-Clonazepam, NPT-neophytadiene. *** $p< 0.0001$ and * $p < 0.05$ vs. vehicle group. ¥ Two mice did not convulse.

2.3. *Antidepressant Effects*

In the tail suspension test (TST), the immobility time of the groups treated with NPT decreased significantly ($p < 0.05$) by 22.3% (0.1 mg/kg), 17.7% (1 mg/kg), and 19.1% (10 mg/kg) compared to the vehicle group, but this antidepressant-like activity was lower compared to the group treated with 20 mg/kg of the reference drug FLX (66.2%) (Figure 6). Therefore, no further antidepressant-like assays were performed.

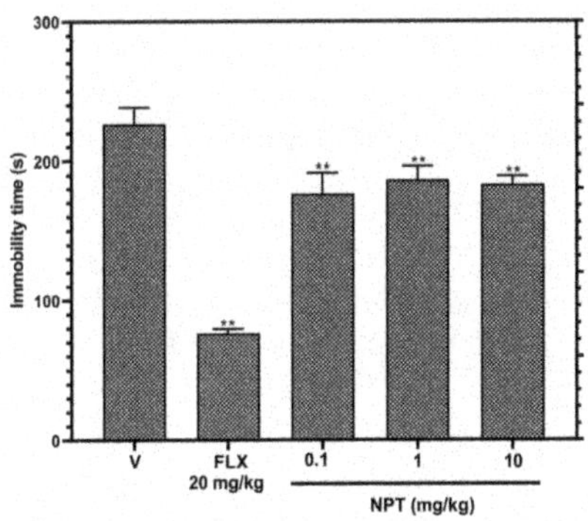

Figure 6. Antidepressant-like effects of NPT. Effects of NPT (0.1, 1 and 10 mg/kg p.o.) on immobility time were determined in the tail suspension proof. Fluoxetine (FLX) was used as the positive control. Bars represent mean values (±SEM) for the experimental group. $n = 8$, ** $p < 0.05$ compared to the vehicle group (indicated as V).

2.4. Sedative Effects and Locomotor Actions

The mice treated with NPT showed no significant ($p > 0.05$) difference in the beginning of sleep and length of sleep in the model of sedation produced with pentobarbital, showing the same results with the vehicle group and opposite effects to the group treated with the reference drug CNZ (Figure 7).

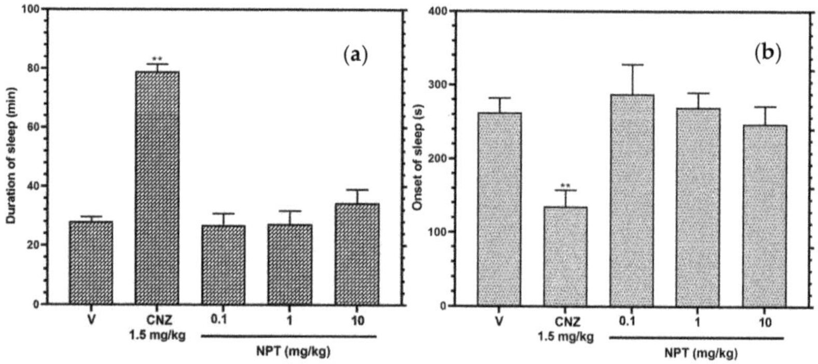

Figure 7. Sedative effects of NPT. The action of NPT (0.1, 1 and 10 mg/kg p.o.) was determined by the beginning of sleep (**a**) and length of sleep (**b**) in the pentobarbital-generated sleeping test. Clonazepam (CNZ) was the anxiolytic drug at 1.5 mg/kg p.o. The results of all experimental groups are reported as the mean values (±SEM). $n = 8$, ** $p < 0.05$ regarding to the vehicle group (indicated as V).

For the rotarod test (RT), the three doses of NPT did not affect the motor coordination of the mice compared to 5 mg/kg CNZ (reference drug), which decreased the time in the rotarod by 37.4% (60 min) and 57.5% (120 min) (Figure 8).

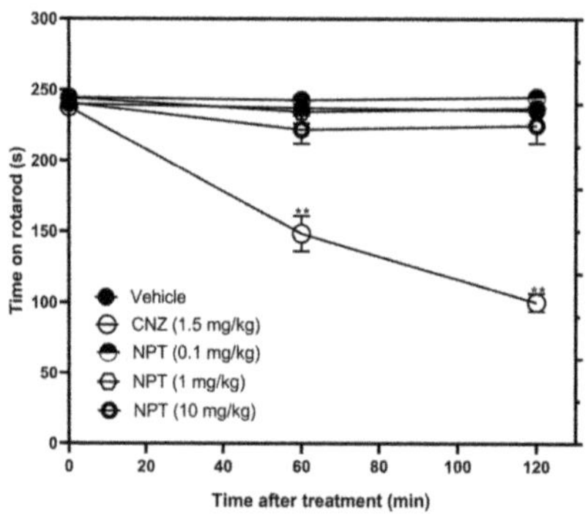

Figure 8. Locomotor actions of NPT. Effects of NPT (0.1–10 mg/kg p.o.) on time on rotarod were measured. Clonazepam (CNZ) was the anxiolytic drug at 1.5 mg/kg p.o. The results of all experimental groups are reported as the mean values (±SEM). $n = 8$, ** $p < 0.05$ regarding to the vehicle group (indicated as V).

3. Discussion

There are many plants and plant-derived compounds used for treating different ailments, but their pharmacological effects are still unknown. New treatments for mental disorders are necessary due to these affections decrease the quality of life, increase the cost of stays in the hospital, and diminish productivity at work [16]. For this reason, this study evaluated the neuropharmacological activity of NPT and provided evidence of anxiolytic-like and anticonvulsant actions of this diterpene.

The behavioral tests used for anxiety evaluation were LDBT, EPMT, OFT, and HBT. These assays are based on the conflict that each animal experiences between exploring a new environment and rodents' natural aversion to bright, elevated, and open spaces [17]. CNZ was used as a positive control due to the anxiolytic activity demonstrated in clinical increase trials [18]. CNZ showed in all experimental tests related to anxiety that the mice overcame the aversion to bright and new, open spaces. CNZ belongs to the benzodiazepine family and is a positive allosteric modulator on $GABA_A$ receptors that hyperpolarizes neurons by increasing the opening of Cl^- channel, reducing the excitability of neurons [19].

Only 10 mg/kg NPT presented anxiolytic activity in 2 (EPMT and HBT) of the 4 models of anxiety in mice. In the EPMT, NPT increased the rodent's time elapsed in an unprotected area (open arms) and decreased the number of entries and the time in a protected place (closed arms); however, these effects were not comparable to those shown by CNZ. In the HBT, NPT increased the number of head dips in the hole board with an anxiolytic activity such as that elicited by CNZ. Other diterpenes such as 3,4-secoisopimar-4(18),7,15-triene-3-oic acid (CMP1) and rosmanol, intraperitoneally administered and tested at similar doses as NPT, both showed anxiolytic-like activities by increasing the time and the number of entries in open arms in the EPMT [20,21]. In addition, other diterpenes, such as tilifodiolide (TFD), tested at 50 and 100 mg/kg p.o., and neo-clerodane 7-keto-neoclerodan-3,13-dien-18,19:15,16-diolide, tested at 10 mg/kg p.o., both showed anxiolytic-like effects in the EPMT [22,23].

Furthermore, the anxiolytic-like mechanism of NPT (10 mg/kg) was evaluated using the GABA inhibitors flumazenil and bicuculline in the HBT. The results showed that NPT effects are reversed in the individual co-administrations of these drugs. These actions were due to the antagonism produced by flumazenil and bicuculline on the $GABA_A$ receptors [24–26]. Flumazenil is a potent benzodiazepine receptor antagonist that competitively inhibits the activity of the benzodiazepine recognition site of the subunits α_2 and α_3 of the $GABA_A$ receptor implicated in anxiolytic-like and anticonvulsant effects [27]. Based on the molecular docking study, the hypothesis is that neophytadiene has the potential to occupy the benzodiazepine site and exerts anxiolytic-like and anticonvulsant effects through the GABAergic system without inducing impairments in motor coordination or sedation. According to random molecular docking, this study found that NPT has an affinity for the same flumazenil binding site and not for other available $GABA_A$ receptor binding sites.

Furthermore, these findings would explain the anticonvulsant effects of NPT presented in the convulsion models induced with PTZ, where 10 mg/kg NPT decreased the duration of convulsion and the mortality rate by at least 50%. The convulsion mechanism for 4-AP is induced by selectively blocking voltage-activated K^+ channel [28] where the anticonvulsant action of NPT was different, only decreasing seizure-related mortality 15%, but interestingly, NPT showed a partial anticonvulsant effect in onset and duration of convulsion. A possible theory is that NPT could also interact on K^+ channels but with a lower affinity than seen for $GABA_A$ receptors.

The mechanism of action found by NPT has not been the same as other diterpenes. For instance, the anxiolytic-like effects of TFD and rosmanol failed or were slightly reversed with the same inhibitor used in this study (flumazenil), demonstrating that the anxiolytic-like activity of TFD and rosmanol is probably not related to the participation of the GABAergic system [21,22].

Few diterpenes with anticonvulsant effects have been reported. For instance, diterpene lactone reported in leaf extracts from *Leonotis leonorus* showed protection against PTZ-induced seizures by up to 50% at the high dose of 400 mg/kg (i.p.); however, the mechanism of the anticonvulsant activity was not elucidated [29]. The diterpene phytol, at the high dose of 250 mg/kg (i.p.), suppressed the PTZ-induced convulsions with minimal locomotor impairments. The anticonvulsant activity was abolished with the coadministration of flumazenil, demonstrating the possible participation of the GABAergic system [6]. In this study, 10 mg/kg p.o. NPT protected mice and delayed the onset of convulsions in the

PTZ-induced convulsion test. As shown in the docking study, GABAergic participation is the possible mechanism of the anticonvulsant activity shown by NPT.

The tail suspension test is a model consisting in recreating a despair behavior reflected by a failure of mice in the persistent attempt to escape from an inescapable situation [30]. NPT (0.1, 1, and 10 mg/kg p.o.) moderately reduced the time the mice remained immobile, indicating a low antidepressant-like effect (17.7–22.3%), which was the same regardless of the increase in its dose, whereas 20 mg/kg fluoxetine showed an antidepressant effect of 66.2%, being nearly 3-fold higher than that of NPT. These data indicate that NPT would be a compound with low effectiveness for treating depression, but it could be synergized in combination with other diterpenes, such as the diterpene TFD, which is reported to have a better antidepressant effect at doses of 10, 50, and 100 mg/kg p.o. [22].

In neurological studies, diterpenes such as CMP1 and phyllocladane (16*R*)-16,17-dihydroxyphyllocladan-3-one) have been reported to have no effect on motor coordination [20,31]. In a similar outcome, NPT lacked locomotor coordination impairment and sedative effects in the rotarod and pentobarbital-induced sleeping tests, respectively. This is important because the anxiolytic-like activity of NPT is not attributed to its sedative properties like benzodiazepines [17].

4. Materials and Methods

4.1. Drugs

Neophytadiene was obtained from Santa Cruz Biotechnology (Santa Cruz, CA, USA) and had a purity of at least 95% according to the manufacturer). Clonazepam (CNZ), pentylenetetrazole (PTZ), flumazenil, 4-aminopyridine, and bicuculline were acquired from Sigma-Aldrich (St. Louis, MO, USA). Pentobarbital was purchased from Pisa Farmaceutica (Mexico City, Mexico).

4.2. Animals

Male CD-1 mice, weighing 38 ± 2 g within 8–10 weeks old, were provided by the Institute of Neurobiology at the Autonomous University of Mexico in Juriquilla, Queretaro, Mexico. The experiments were conducted in the Bioterium at the University of Guanajuato and in the Behavioral Unit of the Neurobiology at UNAM. Animals were housed in groups of 5 per polycarbonate cage, maintained on a 12 h: 12 h light-dark schedule with free access to water and a standard diet (LabDiet 5001). After 1 week of adaptation, the mice were randomly distributed into groups of 8 animals and used only once. All tests were conducted between 08:00 a.m. and 02:00 p.m. Experimental protocols were approved by the Ethics Committee of the University of Guanajuato (CIBIUG-P03-2020) before the beginning of all experiments. Behavioral assessments were conducted in quiet areas by observers unaware of the treatment administered.

4.3. Pharmacological Treatment

Neophytadiene (NPT) and all drugs were prepared in 0.5% (w/v) carboxymethyl cellulose (CMC)-physiological saline solution and administered orally (p.o.). Each experimental group consisted of a total of eight mice randomly distributed. The treatments were administered 1 h before each evaluation. Groups of mice were treated with NPT (0.1, 1, and 10 mg/kg) or positive controls 1.5 mg/kg clonazepam (CNZ) and 20 mg/kg fluoxetine (FLX) [32]. The doses of NPT were selected considering the percentage present in *Blumea lacera* leaves [3] and the results of preliminary studies carried out in our laboratory.

4.4. Light–Dark Box Test (LDBT)

The dimensions of the light–dark box were one-third of the dark compartment (illumination of 40 lux) and two-thirds of the light compartment (illumination of 400 lux) with an exterior size of 46 × 27 × 30 cm (l × b × h). Each mouse was placed into the dark section, and the crossing between the two sections and the time spent in the light section were recorded for 5 min [33]. The parameters evaluated in the light section were time in the compartment and the number of entries.

4.5. Elevated Plus-Maze Test (EPMT)

The methodology and apparatus used were those described by Lister [34]. The test began with the placement of a mouse in the center of the elevated plus-maze, positioned in front of the closed arm. The animal was allowed to roam freely for 5 min, and the entire trial was recorded. The parameters determined were the number of entries in the open and closed arms and the time spent in the open and closed arms. The increase in time and number of entries in the open arms were indicative of anxiolytic-like activity.

4.6. Open Field Test (OFT)

A mouse was settled in the center of a 42 × 42 × 42 polyvinyl chloride box. The movement around the central and peripherical areas of the box was recorded with a camera [35]. The total distance, the resting time, the time in central squares, and the distance in center squares were the parameters evaluated for 5 min.

4.7. Hole–Board Test (HBT)

The test consisted of a wooden box (42 × 42 × 30 cm) with 4 equidistant holes 3 cm in diameter in the floor. The test began with the placement of a mouse in the center of the hole-board. The mouse was allowed to explore the hole-board and the total number of head-dipping was counted for 5 min [36]. The anxiolytic-like mechanism of action of NPT was determined using 1 mg/kg bicuculline (BIC) ($GABA_A$ antagonist) and 2 mg/kg flumazenil (FMZ) ($GABA_A$ antagonist). These antagonists were administered intraperitoneally 15 min before the administration of the dose of 10 mg/kg NPT. The hole-board test was performed after 45 min of NPT administration. The data collected were total distance, resting time, time in central squares, and the distance in central squares.

4.8. Molecular Docking Study of Neophytadiene

The $GABA_A$ receptor in complex with flumazenil was used in the docking study and it was obtained from the Protein Data Bank (www.wwpdb.org, accessed on 10 February 2023) [37] with the following accession code: 6D6T with the resolution of 3.86 Å. The molecular structure of neophytadiene (PubChem id: 10446) was constructed in MOE 2022.02 [38]. The MOE 2022.02 software was used to protonate the $GABA_A$ receptor and neophytadiene structures. The molecular docking of neophytadiene was performed using AutoDock Vina 1.2.0 [39]. The grid box was fixed at the following coordinates: X = 119.613, Y = 169.091, and Z = 154.264 (centered covering the flumazenil coordinates), and the proportions of the grid were 25 × 20 × 20 points separated by 1 Å. We explored the possibility of binding with random molecular docking, centering at the following coordinates and grid dimensions: X = 127.392, Y = 167.341, and Z = 127.574, and grid size of 80 × 70 × 80 points. The pose with the lowest energy of binding was extracted for further analysis. We used the PyMOL software (The PyMOL Molecular Graphics System, Version 2.0 Schrödinger, LLC.) to visualize the complex structures.

4.9. Convulsion Test

Mice were injected with PTZ (90 mg/kg i.p.) or 4-Aminopyridine (4-AP) (12 mg/kg i.p.) (prepared in 0.9% saline solution). After 1 h, each animal received NPT or CNZ orally and individually placed in acrylic cylinders (23 × 27 × 6.5 cm). The onset and the duration of tonic–clonic convulsions and mortality of the mice were recorded [40]. All experiments were carried out between 10:00 a.m. and 02:00 p.m. For an additional test, mice were pretreated with 2 mg/kg i.p. flumazenil (FMZ) (antagonist of the $GABA_A$ receptor) [41] 15 min before the administration of NPT 10 mg/kg and 45 min after the administration of 90 mg/kg PTZ.

4.10. Tail Suspension Test (TST)

The procedure consisted of wrapping adhesive tape around the mice's tails, placing it 1 cm from the tip of the tail, and the suspension of every mouse 50 cm above the floor on the table's edge. The immobility time(s) was evaluated for 6 min [42].

4.11. Pentobarbital-Induced Sleeping Test

The mice were administered a single dose of NPT (0.1, 1, 10 mg/kg) orally and 60 min after, the sodic pentobarbital (50 mg/kg) was injected intraperitoneally. Each mouse was observed for the onset and duration of sleep [43].

4.12. Rotarod Test

Before carrying out the test, the mice were individually placed in the rotarod apparatus (Harvard Apparatus, Barcelona, Spain) for 3 consecutive days so that they could adapt. The animals included in the test were those that managed to walk for 4 minutes at 4 rpm on the device [44]. The motor coordination test was performed at 60 and 120 min after the oral administration of the different doses of NPT. The mice were placed on the rotarod at the same speed (4 rpm) and the time spent on the apparatus was recorded [18]. The duration of the test was 4 min. The decrease in the time on the rotarod is indicative of impaired motor coordination.

4.13. Statistical Analysis of Data

All the data were reported as mean ± standard error of the mean (SEM) and analyzed by one-way or two-way analysis of variance followed by Dunnett's multiple comparison test and Tukey's multiple comparisons test, respectively. The statistical program used was GraphPad Prism (version 8.0.1). A probability value of $p < 0.05$ was considered a statistical difference.

5. Conclusions

The present investigation revealed that NPT has anxiolytic-like activity and anticonvulsant effects without sedative-locomotor effects in short-term studies. The experimental outcomes and the molecular docking analysis propose the possible participation of the GABAergic system in the anxiolytic-like and anticonvulsant activity provided by neophytadiene. The neuropharmacological actions of NPT will be assessed in subacute and chronic administration in murine models of Parkinson´s disease and other neurodegenerative diseases. The combination of NPT with other anxiolytic drugs will be assessed using isobologram studies.

Author Contributions: Conceptualization, A.J.A.-C., M.L.G.-R. and J.C.B.-G.; methodology, M.L.G.-R., J.C.B.-G., S.H.-F. and D.G.-M.; formal analysis, M.L.G.-R., J.C.B.-G., A.J.A.-C., S.H.-F. and M.I.-E.; investigation, M.L.G.-R., J.C.B.-G. and A.J.A.-C.; writing—original draft preparation, M.L.G.-R., J.C.B.-G., S.H.-F. and A.J.A.-C.; writing—review and editing, M.L.G.-R., J.C.B.-G. and A.J.A.-C.; supervision, A.J.A.-C., M.I.-E. and D.G.-M. All authors have read and agreed to the published version of the manuscript.

Funding: This research received no external funding.

Institutional Review Board Statement: This study was conducted in accordance with the NIH Guide for Treatment and Care for Laboratory Animals [45] and the Official Mexican Norm NOM 062-ZOO-1999 (technical instructions for the production, care, and use of laboratory animals). The protocol for the use of animals was approved by the Research Bioethics Committee of the University of Guanajuato ((CIBIUG-P03-2020)). The animals were sacrificed in a CO_2 chamber.

Informed Consent Statement: Not applicable.

Data Availability Statement: Data is contained within the article.

Acknowledgments: M.L.G.-R (CVU 705887) and J.C.B.-G (CVU 489981) received a postdoctoral fellow from CONACYT. Thanks to Edgar Bolaños-Aquino for his technical assistance.

Conflicts of Interest: The authors declare that they have no conflict of interest.

Sample Availability: Not applicable.

References

1. Feigin, V.L. The Evolution of Neuroepidemiology: Marking the 40-Year Anniversary of Publishing Studies on Epidemiology of Neurological Disorders. *Neuroepidemiology* **2022**, *56*, 2–3. [CrossRef]
2. Rauf, A.; Rahman, M.M. Potential therapeutics against neurological disorders: Natural products-based drugs. *Front. Pharmacol.* **2022**, *13*, 3178. [CrossRef] [PubMed]
3. Hossen, M.A.; Ali Reza, A.S.M.; Amin, M.B.; Nasrin, M.S.; Khan, T.A.; Rajib, M.H.R.; Tareq, A.M.; Haque, M.A.; Rahman, M.A.; Haque, M.A. Bioactive metabolites of *Blumea lacera* attenuate anxiety and depression in rodents and computer-aided model. *Food Sci. Nutr.* **2021**, *9*, 3836–3851. [CrossRef]
4. Kaur, J.; Famta, P.; Famta, M.; Mehta, M.; Satija, S.; Sharma, N.; Vyas, M.; Khatik, G.L.; Chellappan, D.K.; Dua, K. Potential anti-epileptic phytoconstituents: An updated review. *J. Ethnopharmacol.* **2021**, *268*, 113565. [CrossRef] [PubMed]
5. Moniruzzaman, M.; Mannan, M.A.; Hossen Khan, M.F.; Abir, A.B.; Afroze, M. The leaves of *Crataeva nurvala* Buch-Ham. modulate locomotor and anxiety behaviors possibly through GABAergic system. *BMC Complement. Altern. Med.* **2018**, *18*, 283. [CrossRef]
6. Tchekalarova, J.; Freitas, R.M. Effect of Diterpene Phytol on Pentlylentetrazol and Maximal Electroshock Seizure Models: Possible Role of GABAergic Mechanism. *Pharmacologia* **2014**, *5*, 351–356. Available online: https://scialert.net/abstract/?doi=pharmacologia.2014.351.356 (accessed on 20 February 2023).
7. Ngobeni, B.; Mashele, S.S.; Malebo, N.J.; Watt, E.V.D.; Manduna, I.T. Disruption of microbial cell morphology by *Buxus macowanii*. *BMC Complement. Med. Ther.* **2020**, *20*, 266. [CrossRef]
8. Swamy, M.K.; Arumugan, G.; Kaur, R.; Ghasemzadeh, A.; Yusoff, M.M.; Sinniah, U.R. GS-MS Based Metabolite Profiling, Antioxidant and Antimicrobial Propierties of Different Solvent Extracts of Malaysian *Plectranthus amboinicus* Leaves. *Evid.-Based Complement. Altern. Med.* **2017**, *2017*, 1517683. [CrossRef]
9. Kazemi, M. Phenolic profile, antioxidant capacity and anti-inflammatory activity of *Anethum graveolens* L. essential oil. *Natural. Prod. Res.* **2015**, *29*, 551–553. [CrossRef]
10. Chandan, G.; Kumar, C.; Verma, M.K.; Satti, N.K.; Saini, A.K.; Saini, R.V. *Datura stramonium* essential oil composition and it's immunostimulatory potential against colon cancer cells. *3 Biotech* **2020**, *10*, 451. [CrossRef] [PubMed]
11. Sajid, M.; Khan, M.R.; Ljaz, M.U.; Ismail, H.; Bhatti, M.Z.; Shah, S.A.; Ali, S.; Tareen, M.U.; Alotaibi, S.S.; Albogami, S.M.; et al. Evaluation of Phytochemistry and Pharmacological Properties of *Alnus nitida*. *Molecules* **2022**, *27*, 4582. [CrossRef]
12. Shah, M.D.; Maran, B.A.V.; Shaleh, S.R.M.; Zuldin, W.H.; Gnanaraj, C.; Yong, Y.S. Therapeutic Potential and Nutraceutical Profiling of North Bornean Seaweeds: A Review. *Mar. Drugs* **2022**, *20*, 101. [CrossRef] [PubMed]
13. Bhardwaj, M.; Sali, V.K.; Mani, S.; Vasanthi, H.R. Neophytadiene from *Turbinaria ornate* Suppresses LPS-Induced Inflammatory Response in RAW 264.7 Macrophages and Sprague Dawley Rats. *Inflammation* **2020**, *43*, 937–950. [CrossRef] [PubMed]
14. Al-Rajhi, A.M.H.; Qanash, H.; Almuhayawi, M.S.; Al Jaouni, S.K.; Bakri, M.M.; Ganash, M.; Salama, H.M.; Selim, S.; Abdelghany, T.M. Molecular Interaction Studies and Phytochemical Characterization of *Mentha pulegium* L. Constituents with Multiple Biological Utilities as Antioxidant, Antimicrobial, Anticancer and Anti-Hemolytic Agents. *Molecules* **2022**, *27*, 4824. [CrossRef] [PubMed]
15. Venkata, R.B.; Samuel, L.A.; Pardha, S.M.; Narashimha, R.B.; Naga, V.K.A.; Sudhakar, M.; Radhakrishnan, T.M. Antibacterial, antioxidant activity and GS-MS Analysis of *Eupatorium odoratum*. *Asian J. Pharm. Clin. Res.* **2012**, *5* (Suppl. S2), 99–106.
16. Calvo, M.I.; Cavero, R.Y. Medicinal plants used for neurological and mental disorders in Navarra and their validation from official sources. *J. Ethnopharmacol.* **2015**, *169*, 263–268. [CrossRef] [PubMed]
17. Rosso, M.; Wirz, R.; Loretan, A.V.; Sutter, N.A.; Pereira da Cunha, C.T.; Jaric, I.; Würbel, H.; Voelkl, B. Reliability of common mouse behavioral test of anxiety: A systematic review and meta-analysis on the effects of anxiolytics. *Neurosci. Biobehav. Rev.* **2022**, *143*, 104928. [CrossRef]
18. Bustos-Gómez, C.I.; Gasca-Martínez, D.; Yáñez-Barrientos, E.; Hidalgo-Figueroa, S.; Gonzalez-Rivera, M.L.; Barragan-Galvez, J.C.; Zapata-Morales, J.R.; Isiordia-Espinoza, M.; Corrales-Escobosa, A.R.; Alonso-Castro, A.J. Neuropharmacological Activities of *Ceiba aesculifolia* (Kunth) Britten & Baker f (Malvaceae). *Pharmaceuticals* **2022**, *15*, 1580. [CrossRef]
19. Basit, H.; Kahwaji, C.I. *Clonazepam. [Updated 2022 Sep 1]*; StatPearls Publishing: Treasure Island, FL, USA, 2022. Available online: https://www.ncbi.nlm.nih.gov/books/NBK556010/ (accessed on 20 February 2023).
20. Maione, F.; Bonito, M.C.; Colucci, M.; Cozzolino, V.; Bisio, A.; Romussi, G.; Cicala, C.; Pieretti, S.; Mascolo, N. First evidence for an anxiolytic effect of a diterpenoid from *Salvia cinnabarina*. *Nat. Prod. Commun.* **2009**, *4*, 469–472. [CrossRef]
21. Abdelhalim, A.; Karim, N.; Chebib, M.; Aburjai, T.; Khan, I.; Johnston, G.A.; Hanrahan, J. Antidepressant, Anxiolytic and Antinociceptive Activities of Constituents from *Rosmarinus officinalis*. *J. Pharm. Pharm. Sci.* **2015**, *18*, 448–459. [CrossRef]
22. Alba-Betancourt, C.; Sánchez-Recillas, A.; Alonso-Castro, A.J.; Esquivel-Juárez, D.; Zapata-Morales, J.R.; Yáñez-Pérez, V.; Álvarez-Camacho, D.; Medina-Rivera, Y.E.; González-Chávez, M.M.; Gasca-Martínez, D.; et al. Antidiarrheal, vasorelaxant, and neuropharmacological actions of the diterpene tilifodiolide. *Drug Dev. Res.* **2019**, *80*, 981–991. [CrossRef] [PubMed]

23. Ortiz-Mendoza, N.; Zavala-Ocampo, L.M.; Martínez-Gordillo, M.J.; González-Trujano, M.E.; Peña, F.A.B.; Rodríguez, I.J.B.; Chávez, J.A.R.; Dorazco-González, A.; Aguirre-Hernández, E. Antinociceptive and anxiolytic-like effects of a neo-clerodane diterpene from *Salvia semiatrata* aerial parts. *Pharm. Biol.* **2020**, *58*, 620–629. [CrossRef] [PubMed]
24. Gallo, A.T.; Hulse, G.K. A theory of the anxiolytic action of flumazenil in anxiety disorders. *J. Psychopharmacol.* **2022**, *36*, 439–448. [CrossRef] [PubMed]
25. Tang, X.H.; Diao, Y.G.; Ren, Z.Y.; Zang, Y.Y.; Zhang, G.F.; Wang, X.M.; Duan, G.F.; Shen, J.C.; Hashimoto, K.; Zhou, Z.Q.; et al. A role of GABAA receptor $\alpha1$ subunit in the hippocampus for rapid-acting antidepressant-like effects of ketamine. *Neuropharmacology* **2023**, *225*, 109383. [CrossRef] [PubMed]
26. Fathalizade, F.; Baghani, M.; Khakpai, F.; Fazli-Tabei, S.; Zarrindast, M.R. GABA-ergic agents modulated the effects of histamine on the behavior of male mice in the elevated plus maze test. *Exp. Physiol.* **2022**, *107*, 233–242. [CrossRef]
27. Sivilotti, M.L. Flumazenil, naloxone and the 'coma coktail'. *Br. J. Clin. Pharmacol.* **2016**, *81*, 428–436. [CrossRef]
28. Zhao, M.; McGarry, L.M.; Ma, H.; Harris, S.; Berwick, J.; Yuste, R.; Schwartz, T.H. Optical triggered seizures using a caged 4-Aminopyridine. *Front. Neurosci.* **2015**, *9*, 25. [CrossRef]
29. Muhizi, T.; Green, I.R.; Amabeoku, G.J.; Bienvenu, E. An Anticonvulsant Diterperne Lactone Isolated From the Leaves of *Leonotis leonorus* (L) R.BR. *East Cent. Afr. J. Pharm. Sci.* **2005**, *8*, 54–61. [CrossRef]
30. Alonso-Castro, A.J.; Gasca-Martínez, D.; Cortez- Mendoza, L.V.; Alba-Betancourt, C.; Ruiz-Padilla, A.J.; Zapata-Morales, J.R. Evaluation of the neuropharmacological effects of Gardenin A in mice. *Drug Dev. Res.* **2020**, *81*, 600–608. [CrossRef]
31. Wasowski, C.; Marder, M. Central nervous system activities of two diterpenes isolated from *Aloysia virgata*. *Phytomedicine* **2011**, *18*, 393–401. [CrossRef]
32. Alonso-Castro, A.J.; Alba-Betancourt, C.; Rocha-González, E.; Ruiz-Arredondo, A.; Zapata-Morales, J.R.; Gasca-Martínez, D.; Pérez-Gutiérrez, S. Neuropharmacological effects of D-pinitol and its possible mechanisms of action. *J. Food Biochem.* **2019**, *43*, e13070. [CrossRef] [PubMed]
33. Bourin, M.; Hascoët, M. The mouse light/dark box test. *Eur. J. Pharmacol.* **2003**, *463*, 55–65. [CrossRef] [PubMed]
34. Lister, R.G. The use of a plus-maze to measure anxiety in the mouse. *Phychopharmacology* **1987**, *92*, 180–185. [CrossRef] [PubMed]
35. Kraeuter, A.K.; Guest, P.C.; Sarnyai, Z. The Open Field Test for Measuring Locomotor Activity and Anxiety—Like Behavior. In *Preclinical Models*; Guest, P., Ed.; Methods in Molecular Biology; Humana Press: New York, NY, USA, 2019; Volume 1916. [CrossRef]
36. File, S.E.; Wardill, A.G. Validity of head-dipping as a measure of exploration in a modified hole-board. *Psychopharmacologia* **1975**, *44*, 53–59. [CrossRef]
37. Berman, H.M.; Henrick, K.; Nakamura, H. Announcing the worldwide Protein Data Bank. *Nat. Struct. Biol.* **2003**, *10*, 980. [CrossRef] [PubMed]
38. Molecular Operating Environment (MOE). 2022.02 Chemical Computing Group ULC,1010 Sherbooke St. West, Suite #910, Montreal, QC, Canada, H3A 2R7, 2022. Available online: https://www.chemcomp.com/Products.htm (accessed on 10 February 2023).
39. Eberhardt, J.; Santos-Martins, D.; Tillack, A.F.; Forli, S. AutoDock Vina 1.2.0: New Docking Methods, Expanded Force Field, and Python Bindings. *J. Chem. Inf. Model.* **2021**, *61*, 3891–3898. [CrossRef]
40. Velluci, S.V.; Webster, R.A. Antagonism of caffeine—Induced seizures in mice by Ro15-1788. *Eur. J. Pharmacol.* **1984**, *97*, 289–293. [CrossRef] [PubMed]
41. Kandeda, A.K.; Taiwe, G.S.; Ayissi, R.E.M.; Moutchida, C. An aqueous extract of *Canarium schweinfurthii* attenuates seizures and potentiates sleep in mice: Evidence for involvement of GABA Pathway. *Biomed. Pharmacother.* **2021**, *142*, 111973. [CrossRef]
42. Castagné, V.; Moser, P.; Roux, S.; Porsolt, R.D. Rodent Models of Depression: Forced Swim and Tail Suspension Behavioral Despair Tests in Rats and Mice. *Curr. Protoc. Neurosci.* **2011**, *55*, 8.10A.1–8.10A.14. [CrossRef]
43. Wolfman, C.; Viola, H.; Marder, M.; Wasowski, C.; Ardenghi, P.; Izquierdo, I.; Paladini, A.C.; Medina, J.H. Anxioselective properties of 6,3'-dinitroflavone, a high-affinity benzodiazepine receptor ligand. *Eur. J. Pharmacol.* **1996**, *318*, 23–30. [CrossRef]
44. Deacon, R.M.J. Measuring Motor Coordination in Mice. *J. Vis. Exp.* **2013**, *75*, e2609. [CrossRef]
45. National Research Council (US) Committee for the Update of the Guide for the Care and Use of Laboratory Animals. *Guide for the Care and Use of Laboratory Animals*, 8th ed.; National Academies Press: Washington, DC, USA, 2011.

Disclaimer/Publisher's Note: The statements, opinions and data contained in all publications are solely those of the individual author(s) and contributor(s) and not of MDPI and/or the editor(s). MDPI and/or the editor(s) disclaim responsibility for any injury to people or property resulting from any ideas, methods, instructions or products referred to in the content.

molecules

MDPI

Article

Evaluation of the Antihypertensive Activity of Eggplant Acetylcholine and γ-Aminobutyric Acid in Spontaneously Hypertensive Rats

Wenhao Wang [1], Shohei Yamaguchi [2,3], Masahiro Koyama [2] and Kozo Nakamura [1,3,4,*]

[1] Department of Science and Technology, Graduate School of Medicine, Science and Technology, Shinshu University, 8304 Minamiminowa, Nagano 399-4598, Japan
[2] Wellnas Co., Ltd., 1-28-5 Koenjiminami, Suginami-Ku, Tokyo 166-0003, Japan
[3] Department of Bioscience and Biotechnology, Faculty of Agriculture, Shinshu University, 8304 Minamiminowa, Nagano 399-4598, Japan
[4] Institute of Agriculture, Academic Assembly, Shinshu University, 8304 Minamiminowa, Nagano 399-4598, Japan
* Correspondence: knakamu@shinshu-u.ac.jp

Abstract: Daily consumption of eggplant powder containing 2.3 mg acetylcholine (ACh) is known to alleviate hypertension and improve mental status. However, eggplant powder used in clinical trials also contains the antihypertensive compound γ-aminobutyric acid (GABA). Although our previous study indicated that the main antihypertensive compound in eggplant is ACh, given that GABA amounts in eggplant do not reach the effective dosage, the effects of GABA on the antihypertensive effect of eggplant remain unclear. It is necessary to establish whether there is a synergistic effect between GABA and ACh and whether GABA in eggplant exerts antihypertensive effects. Consequently, here we sought to evaluate the effects of GABA on the antihypertensive effects of eggplant. We used a probability sum (q) test to investigate the combined effects of ACh and GABA and prepared eggplant powder with very low ACh content for oral administration in animals. ACh and GABA exhibited additive effects but the GABA content in eggplants was not sufficient to promote a hypotensive effect. In conclusion, ACh is the main component associated with the antihypertensive effects of eggplant but GABA within eggplants has a minimal effect in this regard. Thus, compared with GABA, ACh could be a more effective functional food constituent for lowering blood pressure.

Keywords: eggplant; acetylcholine; γ-aminobutyric acid; spontaneously hypertensive rat; hypertension; functional food

Citation: Wang, W.; Yamaguchi, S.; Koyama, M.; Nakamura, K. Evaluation of the Antihypertensive Activity of Eggplant Acetylcholine and γ-Aminobutyric Acid in Spontaneously Hypertensive Rats. *Molecules* **2023**, *28*, 2835. https://doi.org/10.3390/molecules28062835

Academic Editors: Arunaksharan Narayanankutty, Ademola C. Famurewa and Eliza Oprea

Received: 9 February 2023
Revised: 17 March 2023
Accepted: 19 March 2023
Published: 21 March 2023

Copyright: © 2023 by the authors. Licensee MDPI, Basel, Switzerland. This article is an open access article distributed under the terms and conditions of the Creative Commons Attribution (CC BY) license (https://creativecommons.org/licenses/by/4.0/).

1. Introduction

Hypertension is a prominent risk factor that increases the likelihood of developing brain, heart, kidney, and other diseases [1]. Worldwide, an estimated 1.13 billion individuals are living with hypertension [2]. The main causes of hypertension are a combination of genetic and environmental factors, among the latter of which are diet, lack of exercise, smoking, and alcohol consumption. Research on improving blood pressure via dietary modification has received increasing attention in recent years, particularly in the context of antihypertensive nutraceuticals and functional foods [3]. Consequently, the discovery of a broader range of functional foods with antihypertensive properties could contribute to diversifying the dietary options and thus contribute to enhancing the health status of hypertensive individuals.

Eggplant (*Solanum melongena* L.), a common vegetable consumed daily worldwide, is an abundant source of minerals and dietary fiber; contains vitamins B_6, C, and B_9; and is low in protein and calories [4]. Furthermore, eggplants contain functional compounds such as chlorogenic acid, which has anti-inflammatory and antioxidant properties [5,6];

γ-aminobutyric acid (GABA), with antihypertensive effects [7,8]; and nasunin with anti-angiogenic activity [9,10].

Acetylcholine (ACh) is an ester compound comprising quaternary ammonium cation choline and acetic acid that functions as a neurotransmitter, not only in the nervous system of mammals but also in that of many other organisms [11,12]. However, given that orally administered ACh is rapidly broken down by cholinesterase in the body and thus not readily absorbed, it has long been used as an intravenous drug. In this regard, we have previously conducted single and repeat oral studies of dried eggplant powder using spontaneously hypertensive rats (SHRs) and confirmed a significant antihypertensive effect. We speculate that the substance promoting the hypotensive effect is ACh [13], the content of which in eggplant was found to 2900-fold higher than that in other cultivated crops [14]. This finding confirmed, for the first time, that orally administered ACh has potential utility as a novel functional food compound. On the basis of the results of animal studies, we conducted a randomized, placebo-controlled trial on hypertensive patients and accordingly found that the daily intake of ACh-rich eggplant powder (ACh content: 2.3 mg) contributed to a significant reduction in blood pressure and enhanced mental status (stress and mood) [15]. The eggplant powder used in the clinical trial contained GABA in addition to ACh. The daily GABA intake in this clinical trial was 7.65 mg [15], which is lower than the previously reported effective dose of 10 mg [7], although GABA may have been regarded to be a common antihypertensive compound in lower effective dosages [7,16,17]. On the basis of these findings, we assume ACh to be the major antihypertensive factor in eggplant powder. However, we have yet to sufficiently establish the role of GABA in the antihypertensive effect of eggplant consumption, i.e., whether GABA acts synergistically with ACh or whether the GABA content in eggplant is unassociated with the hypotensive effect. Accordingly, establishing the contribution of GABA in this regard is important with respect to determining whether the minimum effective intake of eggplant should take only the ACh content into account, which would reduce the cost of testing the contents of different functional compounds in products during the commercialization of eggplant products.

Consequently, in this study we sought to evaluate the effects of eggplant GABA on the antihypertensive effect of eggplants. We prepared an eggplant powder with low ACh content and the same level of GABA as in the eggplant powder used in the previous clinical trial and investigated the effects of oral administration to evaluate the blood pressure-lowering effects. In addition, we evaluated the interaction between ACh and GABA on the blood pressure-lowering effect in SHRs using the probability sum (q) test, a common analytical approach used to determine the combined effect of two medications. On the basis of the findings of this study, we identified the main factor in eggplants that exerts hypotensive effects and established the association between the GABA content of eggplants and it antihypertensive effects. Moreover, we highlight the merits of ACh compared with GABA as a hypotensive constituent in eggplants.

2. Results

2.1. Combined Effects of ACh and GABA

To determine whether GABA influences the hypotensive effect of ACh, we assessed the combined effects of ACh and GABA using the probability sum (q) test, the results of which obtained for rats treated with ACh, GABA, or a combination of ACh and GABA are shown in Table 1. At 6 h post-administration, rats characterized by a reduction in systolic blood pressure (SBP) of more than 20 mmHg compared with that prior to administration (0 h) were defined as responders. Of the six rats in the ACh group, two were found to have reductions in SBP of more than 20 mmHg, (i.e., a probability of $P_{ACh} = 2/6 = 1/3$). Similar responses were recorded among the GABA group rats ($P_{GABA} = 2/6 = 1/3$), whereas three of the six rats in the ACh + GABA group showed a reduction in SBP greater than 20 mmHg ($P_{ACh + GABA} = 3/6 = 1/2$). Substituting the above probabilities into the q test equation [$q = P_{ACh + GABA}/(P_{ACh} + P_{GABA} - P_{ACh} \times P_{GABA})$] yielded a q-value of 0.90. We accordingly established that the combination of ACh and GABA had an additive effect on the hypotensive effect in experimental rats, which tends to indicate an absence of any synergistic effect between GABA and ACh with respect to lowering blood pressure.

Table 1. The *q* test results for SHRs following acute ACh and GABA treatment (*n* = 6 in each group).

Group	Medication	Dose (μmol/kg BW)	Identification Number	SBP (mmHg)			Probability	*q*-Value
				Before	After	Change		
ACh	ACh	10	A1	178	157	−21	1/3	0.90
			A2	184	176	−9		
			A3	186	163	−23		
			A4	187	206	19		
			A5	195	191	−5		
			A6	168	161	−7		
GABA	GABA	5	G1	179	170	−8	1/3	
			G2	170	186	16		
			G3	190	169	−22		
			G4	157	147	−11		
			G5	188	166	−22		
			G6	157	145	−11		
ACh + GABA	ACh + GABA	10 + 5	AG1	188	194	6	1/2	
			AG2	191	176	−15		
			AG3	176	175	−1		
			AG4	176	144	−32		
			AG5	191	171	−21		
			AG6	206	165	−41		

ACh: acetylcholine; GABA: γ-aminobutyric acid; BW: body weight; SBP: systolic blood pressure. $q = P_{ACh+GABA}/(P_{ACh} + P_{GABA} - P_{ACh} \times P_{GABA})$; P_{ACh}: probability of the ACh group, P_{GABA}: probability of the GABA group, $P_{ACh+GABA}$: probability of the ACh + GABA group.

2.2. Quantification of the ACh and GABA Contents in Samples

To determine whether the GABA contained within eggplants has a hypotensive effect, we prepared eggplant powder with a very low ACh content (whereas the GABA concentration remained unmodified). ACh and GABA concentrations in the ACh-reduced eggplant powder (ACh-reduced eggplant sample) and the normally prepared eggplant powder (eggplant control sample) are shown in Table 2. The GABA content of the ACh-reduced eggplant sample was comparable to that of the control sample, whereas the ACh content was only 0.22% of that of the control sample. The doses of the ACh-reduced eggplant samples used in the oral administration studies could, therefore, be adjusted to give the same GABA dosage in both samples, whereas the ACh dosage delivered in the ACh-reduced eggplant sample was lower than in the eggplant control sample. Furthermore, to confirm the antihypertensive effects of GABA in eggplants, we subjected rats to three treatments, namely, the ACh-reduced eggplant preparation, positive control (eggplant control sample), and negative control (pure water).

Table 2. Contents of ACh and GABA in eggplant samples (mg/g DW; *n* = 3).

Sample	ACh	GABA
ACh-reduced eggplant sample (ACh-reduced eggplant powder)	0.0055 ± 0.00070	6.5 ± 0.28
Eggplant control sample (unmodified eggplant powder)	2.5 ± 0.12	8.1 ± 0.10

ACh: acetylcholine; GABA: γ-aminobutyric acid; DW: dry weight.

2.3. Acute Antihypertensive Effect

The changes in the SBP of rats before and after the oral administration of eggplant powder suspensions are shown in Figure 1. Compared with rats in the negative control group (pure water), those in the positive control group (eggplant control sample) showed a declining trend of SBP at 3 and 6 h post-administration and significantly lower values at 9 h ($p < 0.05$), thereby confirming the short-term hypotensive effect of the eggplant control sample. Comparatively, compared with rats in the negative control group, those in the subject group (ACh-reduced eggplant sample) showed no significant difference following administration the eggplant preparation. Furthermore, fluctuations in the SBP of subject group rats were similar to those observed in the negative control group rats and no hypotensive effect was detected.

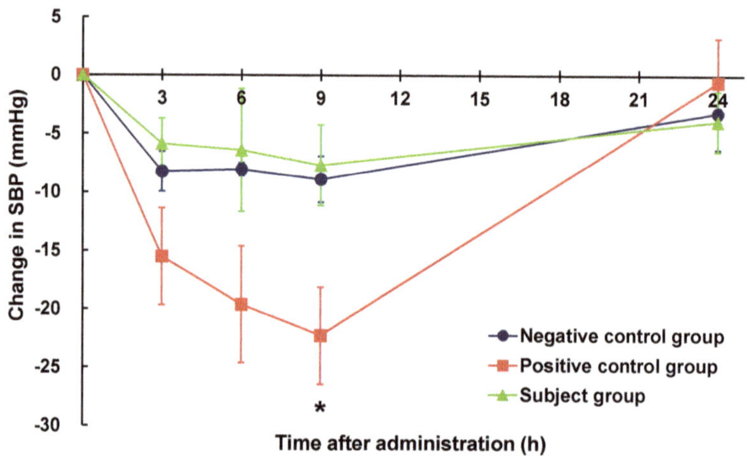

Figure 1. Changes in the SBP of SHRs after a single oral administration of eggplant suspension. SBP: systolic blood pressure. •: negative control group (n = 6), ■: positive control group (n = 6), ▲: subject group (n = 6). The negative control group was administered pure water, the positive control group was administered an eggplant control sample, and the subject group was administered an ACh-reduced eggplant sample. * $p < 0.05$ versus the negative control group, evaluated using Student's t-test (homoscedasticity) and Welch's t-test (heteroscedasticity).

2.4. Chronic Antihypertensive Effect

Among rats in the positive control and subject groups, we detected no significant differences with respect to food intake, water intake, or body weight compared with those in the negative control group (Figure 2a–c). Figure 2d depicts the sequential changes in blood pressure, which indicates that the blood pressure of subject group rats (ACh-reduced eggplant sample) did not differ significantly from that of rats in the negative control group (pure water). Contrastingly, compared with the negative control groups rats, those in the positive control group (eggplant control sample) showed significantly lower values on days 14 ($p < 0.05$) and 21 ($p < 0.01$) following the commencement of administration, thereby confirming the inhibitory effect of eggplant on blood pressure elevation. These results thus provide evidence to indicate that the GABA contained in eggplant samples does not have any pronounced antihypertensive effect. Overall, the condition of rats in the treatment was good during the repeated oral administration study.

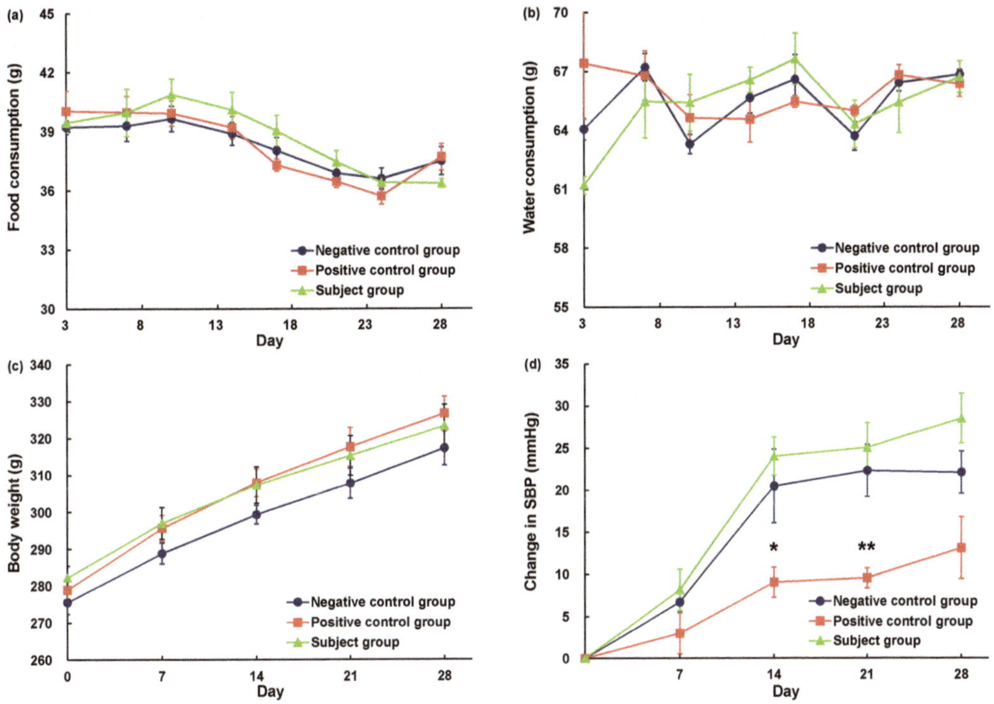

Figure 2. Daily food consumption (**a**), water consumption (**b**), body weight (**c**), and change in the SBP (**d**) of SHRs during chronic administration of eggplant powder suspension. SBP: systolic blood pressure. •: negative control group (n = 6), ■: positive control group (n = 6), ▲: subject group (n = 6). The negative control group was administered pure water, the positive control group was administered an eggplant control sample, and the subject group was administered an ACh-reduced eggplant sample. * $p < 0.05$, ** $p < 0.01$ versus the negative control group, evaluated using Student's t-test (homoscedasticity) and Welch's t-test (heteroscedasticity).

3. Discussion

In this study, we investigated the combined effects of ACh and GABA to determine the effect of GABA on the antihypertensive properties of eggplant. On the basis of our findings in a previous study, we had speculated that ACh is probably the main antihypertensive constituent in eggplants and that the content of GABA within eggplant is in all likelihood insufficiently high to have any appreciable effects in this regard [13]. To confirm this inference, we prepared ACh-reduced eggplant powder and indeed verified that the main antihypertensive compound in eggplant is ACh.

The results of our combined effect study revealed that the combination of ACh and GABA had an additive effect on the reduction in blood pressure in SHRs. The mechanisms underlying the observed reductions in blood pressure in response to the oral administration of ACh and GABA are assumed to be as follows. Following administration, ACh acts on the M_3 muscarinic ACh receptor in the gastrointestinal tract without being absorbed, thereby promoting afferent vagal (parasympathetic) nerve activity, resulting in a reduction in sympathetic nervous activity associated with an autonomic reflex, suppression of noradrenaline (NAD) release, and lowering of acute and chronic blood pressure [18]. Autonomic reflexes are believed to involve afferent vagal (parasympathetic) stimulation transmitted to the rostral lateral medulla (RVLM), thereby inhibiting RVLM activity and thus a reduction in RVLM stimulation of sympathetic nerve activity [19,20]. In contrast, orally administered GABA is absorbed by the body [21], wherein it binds to $GABA_b$ receptors in ganglionic or presynaptic (peripheral nerves) sites. In response to the activation of $GABA_b$, K^+ channels open, thereby bringing neurons closer to a K^+ equilibrium potential, and as a consequence, there are reductions in both action potential frequency and neurotransmitter release [22]. This accordingly has the effect of attenuating sympathetic conduction, inhibiting NAD secretion, and reducing blood pressure [23,24]. On the basis of these putative blood pressure-lowering mechanisms of orally administered ACh and GABA, it is assumed that ACh does not influence the binding of GABA to the $GABA_b$ receptor at ganglionic sites or presynaptic (peripheral nerves), and thus the oral administration of ACh is considered unlikely to enhance the sympathoinhibitory effects of GABA. As the blood-brain barrier is impermeable to GABA [25], orally delivered GABA is unable to enter the central nervous system and has no appreciable effect on the hypotensive effect of oral ACh. In summary, on the basis of the established mechanistic activities, it is assumed that orally administered ACh and GABA do not mutually enhance their respective sympathetic depressant effects, which explains our findings that the combined effects of these two compounds are additive. In this regard, it has previously been found that the effects of phenobarbital and spironolactone combination therapy on UDP-glucuronosyltransferase are additive, as these two compounds act via independent mechanisms [26]. Similarly, combination therapy using pravastatin and olmesartan has been shown to have additive effects in reducing myocardial infarction via different mechanisms [27]. These examples thus indicate that the additive effects of combined medications are probably attributable to the independent mechanisms of the individual preparations. Accordingly, we reason that our findings revealing that the combined effects of ACh and GABA are additive rather than synergistic provide evidence to indicate that the GABA contained within eggplants does not influence the antihypertensive effect of ACh.

In this study, we prepared a lyophilized powder of ACh-reduced eggplant to determine the antihypertensive effect of GABA in eggplant. In the preparation of the ACh-reduced eggplant samples, in which eggplant samples were stored frozen at −20 °C for 15 months after harvest, ACh was reduced to a level that is 0.22% of that in a corresponding eggplant control sample, which was prepared by freeze-drying immediately after harvest. The reduction in the ACh content of eggplant during long-term frozen storage at −20 °C is believed to be associated with the gradual degradation of ACh by cholinesterase, which is naturally present in eggplants [28].

For the single oral dose study, we established three experimental groups, namely, a subject group (the ACh-reduced eggplant sample), positive control group (the eggplant control sample), and negative control group (pure water), and the hypotensive effects of ACh-reduced eggplant or eggplant control sample were evaluated based on a comparison with the negative control group (pure water). The subject group showed values similar to those of the negative control group, whereas the positive control group confirmed the short-term hypotensive effect of eggplant, as demonstrated in a previous study [13].

In the repeated oral administration study, rats were administered either ACh-reduced or unmodified eggplant preparations over the mid- to long-terms and evaluated in vivo by comparing the changes in general conditions and blood pressure with those of rats in the negative control group. Rats in the subject group were found to show blood pressure changes comparable to those recorded in the negative control group rats, whereas measurements obtained for the positive control group rats confirmed the mid- to long-term inhibition of blood pressure increases that occur during aging, as previously demonstrated [13].

In contrast to the significantly lower levels of ACh in ACh-reduced eggplant samples, the GABA content in these samples was similar to that in eggplant control samples. Our studies on SHRs orally administered the ACh-reduced eggplant and eggplant control samples revealed that blood pressure was not lowered in the subject group rats and only those in the positive control group showed a hypotensive effect, thereby providing evidence that ACh is the main hypotensive substance in eggplants and that GABA in eggplant has no apparent hypotensive effect at the assessed dose. The blood pressure-lowering low dose of

GABA for SHRs is reported to be 19 μmol/kg body weight (BW) [29], which would explain our findings indicating that the ACh-reduced eggplant sample (GABA dose: 4.6×10^{-3} μmol/kg BW) assessed in the present study lacked antihypertensive effects, as the GABA content in this preparation was below the assumed effective dose. In addition to ACh and GABA, chlorogenic acid has been identified as a further functional compound in eggplants that has blood pressure regulatory effects [13]. It has previously been demonstrated that the chlorogenic acid content of eggplants does not change markedly after 15 days of refrigerated storage at 4 °C [30], and thus we assume that the chlorogenic acid content in the ACh-reduced eggplant sample remains relative stable during prolonged storage. This also indirectly confirms that the chlorogenic acid content of eggplants would not reach an effective dose.

In clinical trials, the effective doses of GABA for lowering blood pressure have been found to range from 10 to 80 mg/day [7,16,17], whereas tranquilizer-like effects have been observed at concentrations between 26.4 and 70 mg/day [31–33]. Furthermore, the recommended daily intake of GABA in several dietary supplements worldwide is 100 mg, consumed in divided doses [34]. In contrast, it has been established that ACh can contribute to reducing blood pressure and improve psychological status at concentrations as low as 2.3 mg/day [15]. Moreover, given that orally administered ACh is not absorbed but instead lowers blood pressure via the regulatory effects of sympathetic activity, there are no potential side effects from excessive intake [18], thereby indicating ACh to be a superior novel functional compound. In addition, our findings in this study indicate that only the ACh content and not the GABA content needs to be evaluated to determine the hypotensive effects of eggplant, thereby simplifying investigations of the hypotensive effects of different eggplant varieties. This conclusion can be generalized in that it is based on significant results from placebo-controlled parallel-group comparison studies using SHRs, which have several human applications and are highly extrapolatable. However, further research is needed to confirm the validity of this conclusion.

4. Materials and Methods

4.1. Chemicals

An Arium 611 ultrapure water system (Sartorius, Göttingen, Germany) was used to provide ultrapure water with a specific resistance of 18.2 MΩ × cm.

Reagents used in the ACh and GABA combination study: ACh chloride was obtained from Kanto Chemical Co., Inc. (Tokyo, Japan) and GABA was obtained from Tokyo Chemical Industry Co., Ltd. (Tokyo, Japan). The median lethal dose of ACh and GABA is 2500 mg/kg and 12,680 mg/kg, respectively.

Reagents used for ACh and GABA quantification: methanol (high-performance liquid chromatography grade), 1 mol/L hydrochloric acid (HCl), and formic acid were obtained from Nacalai Tesque, Inc. (Kyoto, Japan).

4.2. Animals and Ethics Statement

SHR/Izm rats (male 10 weeks old; Japan SLC, Inc., Shizuoka, Japan) were used in the combined treatment effect experiment (ACh and GABA combination effect study). SHR/NCrlCrlj rats (male 14 weeks old; Charles River Laboratories Japan, Inc., Kanagawa, Japan) were used in tests of a single oral administration and SHR/NCrlCrlj rats (male 10 weeks old; Charles River Laboratories Japan, Inc., Kanagawa, Japan) were used in tests of chronic oral administration (validation experiment of the blood pressure-lowering effect of GABA in eggplant powder containing a hypotensive effective amount of ACh). SHRs were maintained in cages under a 12-h light/dark cycle at 23 ± 4 °C with a humidity of 50% ± 20%. These rats had unrestricted access to faucet water and laboratory feed (MF; Oriental Yeast Co., Ltd., Tokyo, Japan). In total, 54 rats were used in experiments. In the ACh and GABA combination effect study, we used 18 rats, which were allocated to one of the following three groups (n = 6 per group): ACh (numbered A1 to A6), GABA (numbered G1 to G6), and ACh + GABA (numbered AG1 to AG6) groups. In the validation experiment of the blood pressure-lowering effect of GABA in eggplant powder containing a hypotensive effective amount of ACh, we used 36 rats. For the measurement of the acute blood pressure-lowering effects, we allocated 18 rats to the following three groups ($n = 6$ per group): negative control, positive control, and subject groups. For the measurement of the chronic blood pressure-lowering effect, we allocated 18 rats to the following three groups ($n = 6$ per group): negative control, positive control, and subject groups. All experimental procedures conducted in this study were approved by The Animal Care Committee of the Faculty of Shinshu University (approval number: 290061).

In addition, we used two strains of SHR in this study. In the ACh and GABA combination efficacy study, we selected SHR/Izm based on the fact that we needed to determine the minimum effective dosages of ACh and GABA, and a literature search revealed that the minimum effective dosages of ACh and GABA have only been assessed in this strain.

In the validation experiment of the blood pressure-lowering effect of GABA in eggplant powder containing a hypotensive effective amount of ACh, we set the ACh and GABA concentrations in the samples based on the ACh content of eggplant in a previous study using SHR/NCrlCrlj rats [13], and thus we selected this strain for consistency.

Regarding the age of the selected experimental animals, that of SHR/Izm was determined based on the variation in blood pressure with age, described in the literature for this strain, which indicated that blood pressure undergoes a significant upward trend from 10 weeks of age [35].

In the validation experiment of the blood pressure-lowering effect of GABA in eggplant powder containing a hypotensive effective amount of ACh, we used rats of 14 weeks of age in the acute antihypertensive effect experiment, which is consistent with that of the animals used in a previous study [13]. We selected rats of 10 weeks of age in the chronic antihypertensive effect trial in order to match the age of the SHR/NCrlCrlj rats at the end of the trial with that of the SHR/NCrlCrlj rats in the acute antihypertensive effect trial.

4.3. Sample Preparation

ACh and GABA combination effect study: The dosages administered to rats in the three study groups, namely, ACh, GABA, and ACh + GABA, were 10 μmol/kg BW (ACh group), 5 μmol/kg BW (GABA group), and 10 + 5 μmol/kg BW (ACh + GABA group). The concentrations of GABA administered were determined based on reference to the hypotensive effect dose-dependent range for SHR/Izm rats (0.5–50 μmol/kg BW) and by the findings of our preliminary experiments [36]. ACh concentrations were set at the lowest hypotensive effect dose (10 μmol/kg BW) for SHR/Izm rats [37].

Validation experiment of the blood pressure-lowering effect of GABA in eggplant powder containing a hypotensive effective amount of ACh: Eggplant (Tosataka) fruits were harvested from Kochi Agricultural Research Center (Kochi Prefecture, Japan) in July 2019 for the preparation of an ACh-reduced eggplant sample and in December 2020 for the preparation of an eggplant control sample. The eggplant fruits (edible parts), with the calyx removed, were cut into 2 to 3 cm cubes. In our preliminary experiments, we established that the ACh content in eggplant declined by up to 71% when stored at −20 °C for 12 weeks (Figure 3; Experimental method: 21 samples of Tosataka eggplants were cut into 2 to 3 cm pieces and stored in a freezer at −20 °C. Three samples were taken out at 0, 2, 4, 6, 8, 10, and 12 weeks to quantify the ACh content). Thus, to ensure that the ACh content in eggplant was reduced to a very low level, we extended the storage period by freezing 2 to 3 cm cubes of eggplant at −20 °C for 15 months. The ACh-reduced eggplant material was freeze-dried in a freeze-dryer (FDU-2110; EYELA Tokyo Rikakikai Co., Ltd., Tokyo, Japan) and subsequently powdered using a mill mixer (1 min, 28,000 rpm: Wonder Crusher WC-3; Osaka Chemical Co., Ltd., Osaka, Japan) to produce an ACh-reduced eggplant powder. Eggplant control samples were prepared by cutting eggplant fruits into 2 to 3 cm squares immediately after harvesting and then freeze-drying and powdering as previously described. The lyophilized powders were suspended in pure water to prepare the respective samples. The dosages are shown in Table 3.

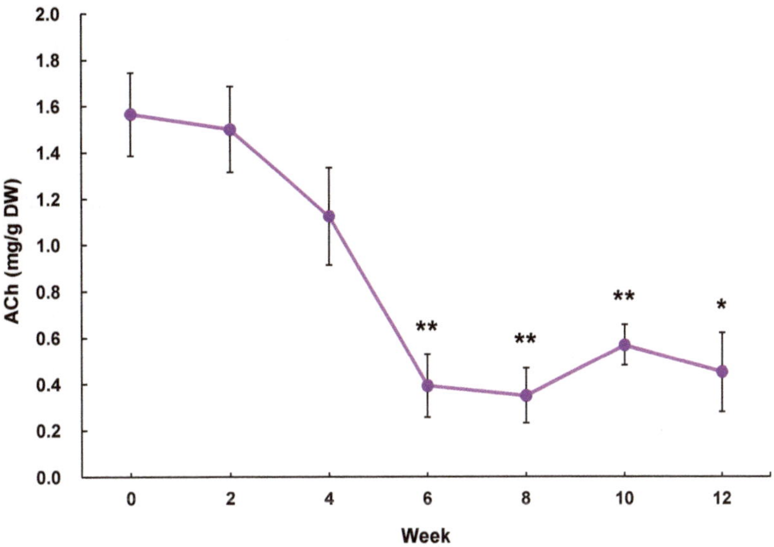

Figure 3. Changes in the ACh content in eggplant (Tosataka) fruit stored at −20 °C. * $p < 0.05$, ** $p < 0.01$ versus week 0, evaluated using Student's *t*-test. ACh: acetylcholine.

Table 3. Dose setting for validation experiment of the blood pressure-lowering effect of GABA in eggplant powder containing a hypotensive effective amount of ACh.

Experiment	Sample	Dose (μg/kg BW)	ACh (mol/kg BW)	GABA (mol/kg BW)
Single oral administration	Eggplant control sample	59	1.0×10^{-9}	4.6×10^{-9}
	ACh-reduced eggplant sample	73	2.7×10^{-12}	
Chronic oral administration	Eggplant control sample	590	1.0×10^{-8}	4.6×10^{-8}
	ACh-reduced eggplant sample	730	2.7×10^{-11}	

ACh: acetylcholine; GABA: γ-aminobutyric acid; BW: body weight.

4.4. Quantification of ACh and GABA in Eggplants

Lyophilized eggplant powder (approximately 10 mg) was added to a 1.5 mL tube containing 50 mmol/L HCl (200 μL), vortex shaken for 3 min, and centrifuged (3 min, 1000× *g*, 25 ± 5 °C) to collect the supernatant. To the remaining residue, we added 200 μL of 50 mmol/L HCl, followed by vortex shaking (3 min), centrifugation (3 min, 1000× *g*, 25 ± 5 °C), and collection of the supernatant. These steps were repeated twice. All collected supernatants were homogeneously mixed (approximately 600 μL) and made up to a final volume of 1 mL with 50% (*v*/*v*) methanol containing 0.01% formic acid, followed by thorough mixing. After filling up to 1 mL, the samples to be measured were diluted 50-fold with the 50% (*v*/*v*) methanol containing 0.01% formic acid and quantification was performed using the standard addition method and liquid chromatography-tandem mass spectrometry (LC-MS/MS).

The quantification of ACh and GABA was performed based on previously described methodology [13]. The analytical systems used were Nexera-I LC-2040C 3D (UPLC) and LCMS-8045 (MS) (Shimadzu Co., Kyoto, Japan) and the column used was a YMC-Triart PFP column (4.6 mm × 250 mm, 5 μm). The chromatographic conditions were as follows: mobile phase of 50% (*v*/*v*) methanol containing 0.01% formic acid at a flow rate of 0.50 mL/min, a separation temperature of 40 °C, an injection volume of 1 μL, and an analysis time of 18 min. Multiple reaction monitoring data were optimized in positive electrospray ionization (ESI) mode (ESI [+] MRM). LC-MS/MS analysis employed multiple reaction monitoring, a desolvation line temperature of 250 °C, an interface temperature of 300 °C, a heat block temperature of 400 °C, drying and heating gases flowed at 10 L/h, and nebulizer gas flowed at 3.0 L/min. The voltage settings were as follows: Q1 Pre-Bias (V) was set to −10.0 (ACh) and −11.0 (GABA); collision energy (V) was set to −14.0 (ACh) and −13.0 (GABA); and Q3 Pre-Bias (V) was set to −17.0 (ACh) and −17.0 (GABA). Monitoring was performed at the following mass-to-charge ratio (m/z) transitions: 146.15 → 87.10 (ACh) and 104.15 → 87.20 (GABA). The methodology adopted has been described in our previous study, in which we used the standard addition method to measure the concentrations of ACh and GABA [38].

4.5. Probability Sum Test

To determine the combined effects of ACh and GABA, we used a modified probability sum test [39] (abbreviated as the "q test"), which is based on classic probability analysis and has been suggested to be appropriate for evaluating the synergism of a combination of two medications [39,40]. SHR/Izm rats were acclimated for 1 week prior to taking blood pressure measurements and subsequent grouping, thereby ensuring that there were no significant differences among the groups with respect to baseline blood pressures. The sample size of each group was set to six based on our previous experimental experience [13]. The 10-week-old male SHR/Izm rats were allocated to one of the following three study groups: ACh group (10 µmol/kg BW), GABA group (5 µmol/kg BW), and ACh + GABA group (10 + 5 µmol/kg BW). The rats were initially fasted overnight and at 09:00 the following day were orally administered the indicated samples using a feeding needle (gavage). Prior to and 6 h after oral administration, we measured SBP using the tail-cuff method (BP-98A; Softron Co., Tokyo, Japan). On the basis of clinical experience, the rats were classified as responders when their SBP dropped by more than 20 mmHg [41].

The formula used to determine the q-value is as follows:

$$q = P_{A+B}/(P_A + P_B - P_A \times P_B)$$

where P (probability) represents the percentage of responders in each group (P = number of responders/all rats in the group), and A and B represent samples A and B, respectively. P_{A+B} is the real percentage of the responders in combinations with A and B. The expected response rate is ($P_A + P_B - P_A \times P_B$), where ($P_A + P_B$) represents the sum of probabilities when samples A and B are used alone and ($P_A \times P_B$) is the probability of rats responding to both samples when used alone (i.e., assuming the two medications act independently). In this study, samples A and B corresponded to ACh and GABA, respectively. The combination was considered antagonistic when $q < 0.85$, synergistic when $q > 1.15$, and additive when $q = 0.85$–1.15 [39,40].

4.6. Measurement of the Acute Antihypertensive Effect

SHR/NCrlCrlj rats (male 14 weeks old) were acclimated for 1 week prior to taking blood pressure measurements and subsequent grouping, as described in the previous section. The three assessed groups were as follows: a negative control group (pure water, $n = 6$), a positive control group (eggplant control sample, $n = 6$), and a subject group (ACh-reduced eggplant sample, $n = 6$). Doses were set at 0.073 mg/kg BW (ACh: 2.7×10^{-12} mol/kg BW; GABA: 4.6×10^{-9} mol/kg BW) of the ACh-reduced eggplant sample and 0.059 mg/kg BW (ACh: 10^{-9} mol/kg BW; GABA: 4.6×10^{-9} mol/kg BW) of the eggplant control sample. The eggplant control sample contained an ACh dose of 10^{-9} mol/kg, which is consistent with the dosage used in our previous study [13]. The ACh-reduced eggplant sample was designed to contain the same content of GABA as the eggplant control sample. The rats in each group were fasted overnight and at 09:00 the following day were orally administered the indicated samples with a feeding needle. To determine the changes in blood pressure, SBP was measured prior to and 3, 6, 9, and 24 h after oral administration using the tail-cuff method.

4.7. Measurement of the Chronic Antihypertensive Effect

SHR/NCrlCrlj rats (male 10 weeks old) were acclimated for 1 week prior to taking blood pressure measurements and subsequent grouping, as described in Section 4.5. The three assessed groups were as follows: a negative control group (pure water, $n = 6$), a positive control group (eggplant control sample, $n = 6$), and a subject group (ACh-reduced eggplant sample, $n = 6$). Doses were set at 0.73 mg/kg BW (ACh: 2.7×10^{-11} mol/kg BW; GABA: 4.6×10^{-8} mol/kg BW) of the ACh-reduced eggplant sample and 0.59 mg/kg BW (ACh: 10^{-8} mol/kg BW; GABA: 4.6×10^{-8} mol/kg BW) of the eggplant control sample. Given that assessments of the chronic antihypertensive effect were conducted without fasting, the doses of ACh and GABA in chronic oral administration were set at 10-fold higher than those used in single oral administration in order to prevent feed intake adversely influencing absorption of the sample. The oral administration of samples to rats was repeated daily for 28 days. SBP was measured prior to administration and on days 7, 14, 21, and 28 during the administration period using the tail-cuff method. Blood pressure was measured between 09:00 and 12:00, and the indicated samples were orally administered using a feeding needle (gavage) at 18:00 each day. The changes in SBP were defined as the blood pressures measured after oral administration (days 7, 14, 21, and 28) minus the blood pressure measured prior to administration (day 0). Throughout the study period, water and food consumption were assessed twice weekly, and body weights was measured once each week.

4.8. Statistical Analysis

All experimental results are presented as the means $\pm$ standard error. Differences were considered significant at $p < 0.05$ and highly significant at $p < 0.01$. The F-test was used to confirm the homogeneity of variance of the data. Student's t-test was used to compare the means of groups with homoscedastic data, whereas Welch's t-test was used to compare the means of groups with heteroscedastic data. Analyses were performed using Microsoft Excel 365 MSO 2212 Build 16.0.15928.20196.

5. Conclusions

In this study, we evaluated the effects of GABA on the hypotensive effects of eggplant using SHRs and confirmed the main constituents associated with the antihypertensive effects of eggplants. The results of the combined effect study of ACh and GABA revealed that the effects of ACh and GABA were additive rather than synergistic. To assess the antihypertensive effects of GABA in eggplant, we orally administered SHRs with ACh-reduced eggplant samples containing almost no ACh but unaltered levels of GABA, along with eggplant control samples containing ACh and GABA, to examine blood pressure changes. The results revealed that only the eggplant control samples had an antihypertensive effect. On the basis of these findings, we speculate that in clinical studies, GABA makes a negligible contribution to the detected antihypertensive effects of eggplant, as the contents of GABA in eggplant powder are less than the effective dose and this compound shows no synergistic effect with ACh. Accordingly, we concluded that only the ACh content needs to be considered when establishing an appropriate beneficial intake of eggplant.

Overall, the blood pressure-lowering dose of ACh is lower than that of GABA, a functional compound commonly used for its blood pressure-lowering effects. Moreover, ACh is effective in small amounts without being absorbed by the body. Consequently, ACh is considered a promising novel and safe functional food constituent. On the basis of our findings in this study, we intend to further assess efficacious use of foods containing the functional constituent ACh and to search for other promising ACh-containing agricultural, forestry, and fishery food resources, thereby contributing to a healthier society through the diversification of ACh intake options. In this study, we focused on evaluating the effects of the GABA content of eggplant on its antihypertensive effects; however, eggplants also contain other antihypertensive compounds, such as chlorogenic acid, and although the levels of chlorogenic acid in eggplants do not appear to reach those considered necessary for an appreciable therapeutic effect, its plausible influence on the antihypertensive effects of eggplant needs to be further investigated (for example, to establish whether it has any synergistic effects with ACh).

Author Contributions: Writing—original draft, W.W.; investigation, W.W.; funding acquisition, K.N.; conceptualization, K.N.; supervision, K.N.; project administration, K.N.; writing—review and editing, W.W., S.Y., M.K. and K.N. All authors have read and agreed to the published version of the manuscript.

Funding: This work was supported by the Special Scheme Project on Vitalizing Management Entities of Agriculture, Forestry, and Fisheries from the National Agriculture and Food Research Organization (NARO) Bio-oriented Technology Research Advancement Institution, Project Number: 16932813, Principal Investigator: Nakamura, K. The funding source had no role in the study design; in the collection, analysis, and interpretation of data; in the writing of the report; or in the decision to submit the article for publication.

Institutional Review Board Statement: The Animal Care Committee of the Faculty of Shinshu University approved all experimental procedures (approval number: 290061).

Informed Consent Statement: Not applicable.

Data Availability Statement: The submitted manuscript contains all of the data, models, and codes generated or utilized during this study.

Acknowledgments: We thank the Kochi Agricultural Research Center for providing us with eggplant fruits.

Conflicts of Interest: S. Yamaguchi is a shareholder-employee in Wellnas Co., Ltd. M. Koyama is a shareholder-CEO in Wellnas Co., Ltd. K. Nakamura is an inventor of pending (WO2015147251 and WO2018070545) and awarded (Japanese Patent No. 6192167 and Indonesia Patent No. P000063624) patents associated with this work and a shareholder in Wellnas Co., Ltd. The funder played no role in the design of this study; in the collection, analysis, or interpretation of data; in the writing of the manuscript; or in the decision to publish the results. There are no other potential conflicts of interest to declare.

Sample Availability: Samples other than the ACh-reduced eggplant sample are available from the authors. The ACh-reduced eggplant sample has been used up.

References

1. World Health Organization. A Global Brief on Hypertension: Silent killer, Global Public Health Crisis: World Health Day 2013 (No. WHO/DCO/WHD/2013.2). World Health Organization. 2013. Available online: https://www.who.int/publications/i/item/a-global-brief-on-hypertension-silent-killer-global-public-health-crisis-world-health-day-2013 (accessed on 8 February 2023).
2. NCD Risk Factor Collaboration (NCD-RisC). Worldwide trends in blood pressure from 1975 to 2015: A pooled analysis of 1479 population-based measurement studies with 19·1 million participants. *Lancet* **2017**, *389*, 37–55. [CrossRef]
3. Chen, Z.Y.; Peng, C.; Jiao, R.; Wong, Y.M.; Yang, N.; Huang, Y. Anti-hypertensive nutraceuticals and functional foods. *J. Agric. Food Chem.* **2009**, *57*, 4485–4499. [CrossRef]
4. Gürbüz, N.; Uluişik, S.; Frary, A.; Frary, A.; Doğanlar, S. Health benefits and bioactive compounds of eggplant. *Food Chem.* **2018**, *268*, 602–610. [CrossRef]
5. Singh, A.P.; Luthria, D.; Wilson, T.; Vorsa, N.; Singh, V.; Banuelos, G.S.; Pasakdee, S. Polyphenols content and antioxidant capacity of eggplant pulp. *Food Chem.* **2009**, *114*, 955–961. [CrossRef]
6. Yun, N.; Kang, J.W.; Lee, S.M. Protective effects of chlorogenic acid against ischemia/reperfusion injury in rat liver: Molecular evidence of its antioxidant and anti-inflammatory properties. *J. Nutr. Biochem.* **2012**, *23*, 1249–1255. [CrossRef]
7. Inoue, K.; Shirai, T.; Ochiai, H.; Kasao, M.; Hayakawa, K.; Kimura, M.; Sansawa, H. Blood-pressure-lowering effect of a novel fermented milk containing γ-aminobutyric acid (GABA) in mild hypertensives. *Eur. J. Clin. Nutr.* **2003**, *57*, 490–495. [CrossRef]
8. Horie, H.; Ando, A.; Saito, T. The contents of γ-amino butyric acid in eggplant and its accumulation with heat treatment. *Nippon Shokuhin Kagaku Kogaku Kaishi* **2013**, *60*, 661–664. (In Japanese, with English Abstract) [CrossRef]
9. Matsubara, K.; Kaneyuki, T.; Miyake, T.; Mori, M. Antiangiogenic activity of nasunin, an antioxidant anthocyanin, in eggplant peels. *J. Agric. Food Chem.* **2005**, *53*, 6272–6275. [CrossRef]
10. Casati, L.; Pagani, F.; Braga, P.C.; Lo Scalzo, R.L.; Sibilia, V. Nasunin, a new player in the field of osteoblast protection against oxidative stress. *J. Funct. Foods* **2016**, *23*, 474–484. [CrossRef]
11. Horiuchi, Y.; Kimura, R.; Kato, N.; Fujii, T.; Seki, M.; Endo, T.; Kato, T.; Kawashima, K. Evolutional study on acetylcholine expression. *Life Sci.* **2003**, *72*, 1745–1756. [CrossRef]

12. Yamada, T.; Fujii, T.; Kanai, T.; Amo, T.; Imanaka, T.; Nishimasu, H.; Wakagi, T.; Shoun, H.; Kamekura, M.; Kamagata, Y.; et al. Expression of acetylcholine (ACh) and ACh-synthesizing activity in Archaea. *Life Sci.* **2005**, *77*, 1935–1944. [CrossRef]
13. Yamaguchi, S.; Matsumoto, K.; Koyama, M.; Tian, S.; Watanabe, M.; Takahashi, A.; Miyatake, K.; Nakamura, K.; Nakamura, K. Antihypertensive effects of orally administered eggplant (*Solanum melongena*) rich in acetylcholine on spontaneously hypertensive rats. *Food Chem.* **2019**, *276*, 376–382. [CrossRef]
14. Wang, W.; Yamaguchi, S.; Koyama, M.; Tian, S.; Ino, A.; Miyatake, K.; Nakamura, K. LC-MS/MS analysis of choline compounds in Japanese-cultivated vegetables and fruits. *Foods* **2020**, *9*, 1029. [CrossRef]
15. Nishimura, M.; Suzuki, M.; Takahashi, R.; Yamaguchi, S.; Tsubaki, K.; Fujita, T.; Nishihira, J.; Nakamura, K.; Nakamura, K. Daily ingestion of eggplant powder improves blood pressure and psychological state in stressed individuals: A randomized placebo-controlled study. *Nutrients* **2019**, *11*, 2797. [CrossRef]
16. Matsubara, F.; Ueno, H.; Tadano, K.; Suyama, T.; Imaizumi, K.; Suzuki, T.; Magata, K.; Kikuchi, N.; Muneyuki, K.; Nakamichi, N.; et al. Effects of GABA supplementation on blood pressure and safety in adults with mild hypertension. *Jpn. Pharmacol. Ther.* **2002**, *30*, 963–972.
17. Shimada, M.; Hasegawa, T.; Nishimura, C.; Kan, H.; Kanno, T.; Nakamura, T.; Matsubayashi, T. Anti-hypertensive effect of γ-aminobutyric acid (GABA)-rich Chlorella on high-normal blood pressure and borderline hypertension in placebo-controlled double blind study. *Clin. Exp. Hypertens.* **2009**, *31*, 342–354. [CrossRef]
18. Yamaguchi, S.; Hayasaka, Y.; Suzuki, M.; Wang, W.; Koyama, M.; Nagasaka, Y.; Nakamura, K. Antihypertensive mechanism of orally administered acetylcholine in spontaneously hypertensive rats. *Nutrients* **2022**, *14*, 905. [CrossRef]
19. Buckey, J.C.; Homick, J.L. *The Neurolab Spacelab Mission: Neuroscience Research in Space: Results from the STS-90, Neurolab Spacelab Mission*; Government Printing Office: Washington, DC, USA, 2003; Volume 535.
20. Katsurada, K.; Ogoyama, Y.; Imai, Y.; Patel, K.P.; Kario, K. Renal denervation based on experimental rationale. *Hypertens. Res.* **2021**, *44*, 1385–1394. [CrossRef]
21. Yamatsu, A.; Yamashita, Y.; Pandharipande, T.; Maru, I.; Kim, M. Effect of oral γ-aminobutyric acid (GABA) administration on sleep and its absorption in humans. *Food Sci. Biotechnol.* **2016**, *25*, 547–551. [CrossRef]
22. Misgeld, U.; Bijak, M.; Jarolimek, W. A physiological role for $GABA_B$ receptors and the effects of Baclofen in the mammalian central nervous system. *Prog. Neurobiol.* **1995**, *46*, 423–462. [CrossRef]
23. Kimura, M.; Hayakawa, K.; Sansawa, H. Involvement of γ-aminobutyric acid (GABA) B receptors in the hypotensive effect of systemically administered GABA in spontaneously hypertensive rats. *Jpn. J. Pharmacol.* **2002**, *89*, 388–394. [CrossRef]
24. Hayakawa, K.; Kimura, M.; Kamata, K. Mechanism underlying γ-aminobutyric acid-induced antihypertensive effect in spontaneously hypertensive rats. *Eur. J. Pharmacol.* **2002**, *438*, 107–113. [CrossRef]
25. Kuriyama, K.; Sze, P.Y. Blood-brain barrier to H3-γ-aminobutyric acid in normal and amino oxyacetic acid-treated animals. *Neuropharmacology* **1971**, *10*, 103–108. [CrossRef]
26. Mottino, A.D.; Guibert, E.E.; Rodríguez Garay, E.A.R. Additive effect of combined spironolactone and phenobarbital treatment on hepatic bilirubin UDP-glucuronyltransferase. *Biochem. Pharmacol.* **1989**, *38*, 851–853. [CrossRef]
27. Lee, T.M.; Lin, M.S.; Chou, T.F.; Chang, N.C. Additive effects of combined blockade of AT1 receptor and HMG-CoA reductase on left ventricular remodeling in infarcted rats. *Am. J. Physiol. Heart Circ. Physiol.* **2006**, *291*, H1281–H1289. [CrossRef]
28. Fluck, R.A.; Jaffe, M.J. Cholinesterases from plant tissues VI. Preliminary characterization of enzymes from *Solanum melongena* L. and *Zea mays* L. *Biochim. Biophys. Acta* **1975**, *410*, 130–134. [CrossRef]
29. Yang, N.C.; Jhou, K.Y.; Tseng, C.Y. Antihypertensive effect of mulberry leaf aqueous extract containing γ-aminobutyric acid in spontaneously hypertensive rats. *Food Chem.* **2012**, *132*, 1796–1801. [CrossRef]
30. Galani, J.H.; Patel, J.S.; Patel, N.J.; Talati, J.G. Storage of fruits and vegetables in refrigerator increases their phenolic acids but decreases the total phenolics, anthocyanins and vitamin C with subsequent loss of their antioxidant capacity. *Antioxidants* **2017**, *6*, 59. [CrossRef]
31. Sonoda, H. Functionality of the GABA derived from lactic acid bacterium fermentation-Effect on climacteric disturbance and presenile mental disorder. *Food Style 21* **2001**, *5*, 92–96.
32. Okada, T.; Sugishita, T.; Murakami, T.; Murai, H.; Saikusa, T.; Horino, T.; Onoda, A.; Kajimoto, O.; Takahashi, R.; Takahashi, T.; et al. Effect of the defatted rice germ enriched with GABA for sleeplessness, depression, autonomic disorder by oral administration. *Nippon Shokuhin Kagaku Kogaku Kaishi J. Jpn. Soc. Food Sci. Technol.* **2000**, *47*, 596–603. [CrossRef]
33. Horie, K.; Higashiguchi, S.; Yokogoshi, H. Influence of GABA upon immunity and mental health. *Food Style 21* **2003**, *7*, 64–68. (In Japanese)
34. Oketch-Rabah, H.A.; Madden, E.F.; Roe, A.L.; Betz, J.M. United States Pharmacopeia (USP) safety review of gamma-aminobutyric acid (GABA). *Nutrients* **2021**, *13*, 2742. [CrossRef]
35. Fukuda, S.; Tsuchikura, S.; Iida, H. Age-related changes in blood pressure, hematological values, concentrations of serum biochemical constituents and weights of organs in the SHR/Izm, SHRSP/Izm and WKY/Izm. *Exp. Anim.* **2004**, *53*, 67–72. [CrossRef]
36. Hayakawa, K.; Kimura, M.; Kasaha, K.; Matsumoto, K.; Sansawa, H.; Yamori, Y. Effect of a γ-aminobutyric acid-enriched dairy product on the blood pressure of spontaneously hypertensive and normotensive Wistar-Kyoto rats. *Br. J. Nutr.* **2004**, *92*, 411–417. [CrossRef]

37. Yamaguchi, S.; Matsumoto, K.; Wang, W.; Nakamura, K. Differential antihypertensive effects of oral doses of acetylcholine between spontaneously hypertensive rats and normotensive rats. *Foods* **2021**, *10*, 2107. [CrossRef]
38. Wang, W.; Yamaguchi, S.; Suzuki, A.; Wagu, N.; Koyama, M.; Takahashi, A.; Takada, R.; Miyatake, K.; Nakamura, K.; Miyatake, K.; et al. Investigation of the distribution and content of acetylcholine, a novel functional compound in eggplant. *Foods* **2021**, *10*, 81. [CrossRef]
39. Jin, Z.J. About the evaluation of drug combination. *Acta Pharmacol. Sin.* **2004**, *25*, 146–147.
40. Su, D.F.; Xu, L.P.; Miao, C.Y.; Xie, H.H.; Shen, F.M.; Jiang, Y.Y. Two useful methods for evaluating antihypertensive drugs in conscious freely moving rats. *Acta Pharmacol. Sin.* **2004**, *25*, 148–151.
41. Committee for Medicinal Products for Human Use. *Guideline on Clinical Investigation of Medicinal Products in the Treatment of Hypertension*; European Medicines Agency: Amsterdam, The Netherlands, 2016.

Disclaimer/Publisher's Note: The statements, opinions and data contained in all publications are solely those of the individual author(s) and contributor(s) and not of MDPI and/or the editor(s). MDPI and/or the editor(s) disclaim responsibility for any injury to people or property resulting from any ideas, methods, instructions or products referred to in the content.

molecules

MDPI

Article

Evaluation of the Chemical Composition of Selected Varieties of *L. caerulea* var. *kamtschatica* and *L. caerulea* var. *emphyllocalyx*

Józef Gorzelany [1], Oskar Basara [1,2,*], Ireneusz Kapusta [3], Korfanty Paweł [4] and Justyna Belcar [1,2]

1 Department of Food and Agriculture Production Engineering, University of Rzeszow, 4 Zelwerowicza Street, 35-601 Rzeszów, Poland; justyna.belcar@op.pl (J.B.)
2 Doctoral School of the University of Rzeszów, st Rejtana 16C, 35-959 Rzeszów, Poland
3 Department of Food Technology and Human Nutrition, University of Rzeszow, 4 Zelwerowicza Street, 35-601 Rzeszów, Poland; ikapusta@ur.edu.pl
4 Fruit Plant Nursery 'Korfanty', 36-207 Grabownica Starzeńska, Poland; korfanty.farm@wp.pl
* Correspondence: oskarb@dokt.ur.edu.pl

Abstract: *Lonicera caerulea* fruits are a rich source of vitamins, organic acids, and phenolic compounds, which are characterised by their health-promoting properties. The content of bioactive compounds in this fruit may vary depending on the cultivar and the harvest date. The fruits of the *L. caerulea* var. *kamtschatica* cultivars 'Duet' and 'Aurora' and the *L. caerulea* var. *emphyllocalyx* cultivars 'Lori', 'Colin' and 'Willa' were used in this study. *L. emphyllocalyx* fruit, especially the cultivar 'Willa', was characterised as having a higher acidity by an average of 29.96% compared to *L. kamtschatica*. The average ascorbic acid content of the *L. kamtschatica* fruit was 53.5 mg·100 g^{-1} f.w., while *L. emphyllocalyx* fruit had an average content that was 14.14% lower. The antioxidant activity (determined by DPPH, FRAP, and ABTS) varied according to the cultivar and the species of fruit analysed. The total polyphenol content differed significantly depending on the cultivar analysed; fruits of the *L. emphyllocalyx* cultivar 'Willa' were characterised by the lowest content of total polyphenols—416.94 mg GAE·100 g^{-1} f.w.—while the highest content of total polyphenols—747.85 GAE·100 g^{-1} f.w.—was found in the fruits of the *L. emphyllocalyx* cultivar 'Lori'. *Lonicera caerulea* fruits contained 26 different phenolic compounds in their compositions, of which the highest content was characterised by cyanidin 3-*O*-glucoside (average: 347.37 mg·100 g^{-1}). On the basis of this study, it appears that both *L. kamtschatica* fruits and *L. emphyllocalyx* fruits, especially of the cultivars 'Lori' and 'Willa', can be used in food processing.

Keywords: *Lonicera*; chemical composition; quality fruits; organic acids; sugars; polyphenolic profile; antioxidant activity

Citation: Gorzelany, J.; Basara, O.; Kapusta, I.; Paweł, K.; Belcar, J. Evaluation of the Chemical Composition of Selected Varieties of *L. caerulea* var. *kamtschatica* and *L. caerulea* var. *emphyllocalyx*. *Molecules* **2023**, *28*, 2525. https://doi.org/10.3390/molecules28062525

Academic Editors: Arunaksharan Narayanankutty, Ademola C. Famurewa and Eliza Oprea

Received: 14 February 2023
Revised: 7 March 2023
Accepted: 8 March 2023
Published: 9 March 2023

Copyright: © 2023 by the authors. Licensee MDPI, Basel, Switzerland. This article is an open access article distributed under the terms and conditions of the Creative Commons Attribution (CC BY) license (https://creativecommons.org/licenses/by/4.0/).

1. Introduction

The blue honeysuckle (*Lonicera caerulea* L.) belongs to the genus *Lonicera* (*Caprifoliaceae*), which contains more than 200 species, native to the cold lands of the Far East and central Asia. Most of them are ornamental plants; only about 17 are edible fruit-producing species [1]. *Lonicera caerulea* has numerous varieties, several of which are widely cultivated, originating from Russia (*L. caerulea var. edulis*, *L. caerulea var. kamtschatica*, *L. caerulea var. altaica*, *L. caerulea var. boczkarnikovae*) and the island of Hokkaido in Japan (*L. caerulea* var. *emphyllocalyx*). They are long-living (25–30 years) shrubs that can reach 0.8–3.0 m height. They need an outside pollinator to bear fruits one year after planting. After three years, approx. 500 g of fruit can be obtained from one plant. Berries are dark blue or dark purple in colour, with a size of 1.5–3.0 cm, and a mostly cylindrical shape with a wax coating. They are plants that are very resistant to frost, able to withstand temperatures up to −46 °C [2]. *Lonicera caerulea* var. *kamtschatica* is a variety of honeysuckle, commonly known as Kamchatka berry, which is one of the most popular fruits in Poland, the Czech Republic, Canada, and Russia. Several varieties native to the species have been selected, which differ, among other things, in terms of flowering time, prolificacy, and content of bioactive substances [3]. A lesser-known variety is *Lonicera caerulea var. emphyllocalyx* (Maxim.) Nakai, commonly referred to as Japanese haskap. This is native to the island of Hokkaido in northern Japan and is also cultivated in China, Korea, or Russia [4]. The word 'haskap', from the Japanese hasukappu, hascup, haskappu, or hasakapu, in the language of the natives literally means a lot of small objects on the tops of the branches. In the regions where the plant originates, the fruit is recognised as a medicine and immune booster, used to treat stomach ailments and protect against many diseases. The native Ainu people of the island of Hokkaido called the fruit the 'elixir of life', making it an important part of their diet. They were eaten fresh, preserved with sugar and salt and were also used to prepare spirits and stockpile valuable raw material for winter [5,6].

The *L. caerulea* fruit is rich in sugars, organic acids, and polyphenols; this has a significant impact not only on sensory perception, but also on the health-promoting properties of the fruit. The content of polyphenols, the main group showing biological activity, can vary depending on the cultivar and the harvest date. The cultivation conditions (soil type and fertilisation method) do not significantly affect the content of active compounds; fruit from different sites will be characterised by similar contents of bioactive compounds [6]. *Lonicera caerulea* berries are characterised by a high content of phenolic compounds, ranging from 428.14 to 622.52 mg GAE·100 g^{-1} f.w. [7,8]. *L. caerulea* is an important source of phenolic compounds such as anthocyanins, flavonoids, proanthocyanidins and phenolic acids [6], the predominant group of phenolic compunds being anthocyanins, mainly cyanidin 3-*O*-glucoside. This is a widely used compound in the plant world, accounting for about 79–92% of the the anthocyanin content of *L. caerulea* fruit [9]. Fruits are a valuable source of vitamin C (ascorbic acid), at around 30.5–186.6 mg·100 g^{-1} [10]. The fruit is a source of minerals, including magnesium (79–163 mg·kg), potassium (3000–5000 mg·kg), phosphorus (486–2252 mg·kg), and calcium (1077 mg·kg) [11]. The berries are characterised by anti-inflammatory, antimicrobial, and antioxidant properties [6]. The fruits of *L. caerulea* has many uses in food processing. Pulp made from *L. caerulea* berries has a high juice yield and tiny seeds, making it excellent source of fruit juice [12]. Due to the very dense colour of the fruit juice, it is suitable for the production of food products that require the addition of juice. It also has high antioxidant properties [13,14]. Canned food and spreads are a relatively good way to improve the utility value of *L. caerulea*; however, due to the high content of sugars, the amount of added sugar should be controlled during the production stage [15]. The processing of fruits might change the levels of bioactive compounds. Fresh-processed foods have more health-promoting properties than the thermally treated foods. The concentration of some compounds might increase after water loss caused by fruit processing methods, e.g., heating. After freeze drying or pressing the contents of some compounds might change [15].

The purpose of this study was to compare the chemical compositions of the fruits of *L. caerulea* var. *kamtschatica* ('Duet' and 'Aurora' cultivars) and *L. caerulea* var. *emphyllocalyx* ('Lori', 'Colin', and 'Willa' cultivars) and the potential use of *L. emphyllocalyx* fruit in food processing.

2. Results and Discussion

2.1. pH and Acidity of Fruits

The organic acid content of the fruit decreases with successive stages of ripening. The organic acids contained in the fruit are perishable and, under the influence of various factors, e.g., temperature, can change in terms of concentration in the plant material [16,17]. *L. caerulea* fruit are rich in organic acids (e.g., malic acid, citric acid, quinic acid and fumaric acid; [17]). The contents of individual organic acids significantly influence the taste qualities of ripe fruit and the acceptability of the consumer. The average pH values of the *L. kamtschatica* and the *L. emphyllocalyx* fruits were similar: 3.07–3.32 and 3.13–3.52, respectively (Figure 1).

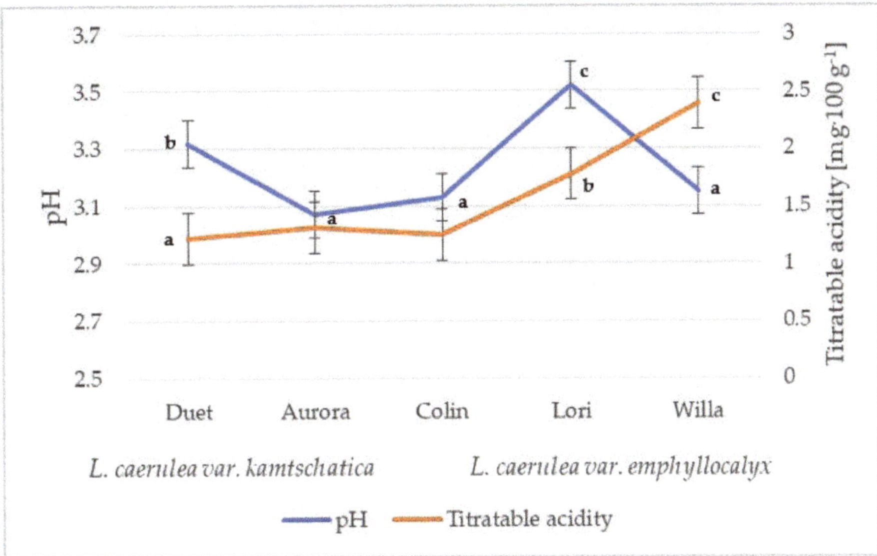

Figure 1. pH and total acidity of *L. caerulea* var. *kamtschatica* and *L. caerulea* var. *emphyllocalyx* fruits (data are expressed as mean values ($n = 3$) ± SD; SD—standard deviation). Mean values with different letters are significantly different ($p < 0.05$).

These results are comparable to those obtained by other authors. In a study by Gerbrand et al. [18], the fruit pH of *Lonicera caerulea* ranged from 2.42 to 3.10; in an experiment of Auzanneau et al. [19], the fruit pH of the same species ranged from 2.70 to 3.30, depending on the growing year and harvest date. Compared to Miyashita et al. [20] and Oszmiański et al. [21], we found a lower average pH value in *L. kamtschatica* and *L. emphyllocalyx* berries: 2.60 and 2.65, respectively. The pH value is diversified compared to other species grown in Poland, e.g., saskatoon berry (4.12–5.03) [22], red currant (3.20 and 3.27) [23], sea buckthorn (3.02–3.19) [24], highbush blueberry fruit (2.76–3.33) [25], raspberry (3.72) and mulberry (5.17) [26]. In general, both *L. kamtschatica* and *L. emphyllocalyx* are rich in organic acids that give a taste resembling blueberries, with a distinct hint of sourness.

The difference in acidity between varieties may have been caused by different climatic conditions. The acidity of the *L. kamtschatica* fruit was at the level of 1.22–1.31 $g \cdot 100\ g^{-1}$, while the *L. emphyllocalyx* fruit, especially the 'Willa' cultivar, had a higher acidity by 29.96% on average compared to the *L. kamtschatica* fruit (Figure 1). The results in this study are similar to a study by MacKenzie et al. [27], where the total acidity in *L. caerulea* fruit ranged from 1.6 to 3.0 $g \cdot 100\ g^{-1}$, depending on the year of cultivation. The obtained results are comparable with berries grown in Poland, e.g., cranberry (1.56–1.60 $g \cdot 100\ g^{-1}$) [28], saskatoon berry (0.3–1.5 $g \cdot 100\ g^{-1}$) [22], red currant (0.7–1.6 $g \cdot 100\ g^{-1}$) [29], mulberry fruit (0.26 $g \cdot 100\ g^{-1}$) and raspberry (0.63 $g \cdot 100\ g^{-1}$) [26].

2.2. Contents of Ascorbic Acid and Antioxidant Activity in L. kamtschatica and L. emphyllocalyx Fruit

Ascorbic acid is a powerful antioxidant that is required for the activation of many enzymes. This compound is essential for the functioning of the human body, influencing the immune and circulatory systems, accelerating wound healing, slowing skin ageing, and regulating collagen production [30]. The average ascorbic acid content of the *L. kamtschatica* fruit was 53.5 $mg \cdot 100\ g^{-1}$ f.w., while the *L. emphyllocalyx* fruit had an average ascorbic acid content of 45.93 $mg \cdot 100\ g^{-1}$ f.w. (Table 1). The content ascorbic acid of *L. caerulea* fruit differs from that reported by Celli et al. [10]; we obtained results ranging from 30.5 to 186.6 mg·100 g^{-1} f.w. According to Jurnikova et al. [31], the content of ascorbic acid in *Lonicera caerulea*, higher than 70 mg·100 g^{-1} f.w., is associated with a lower accumulation of phenolics. The ascorbic acid content of *L. kamtschatica* and *L. emphyllocalyx* fruit was at a similar level to popular berries grown in Poland, e.g., strawberries (average: 50.1 mg·100 g^{-1}), raspberries (average: 30.6 mg·100 g^{-1} f.w.), blueberries (average: 60.1 mg·100 g^{-1} f.w.) [32,33], plum blackthorn (21.94 mg·100 g^{-1} f.w.), blackberry (33.85 mg·100 g^{-1} f.w.) [33]; red currant (31.2 mg·100 g^{-1}–44.1 mg·100 g^{-1} f.w.) [23], sea buckthorn (4.0–9.1 mg·100 g^{-1} f.w.) [24]. According to Senica et al. [14], spreads and smoothies made out of *L. caerulea* berries increase the concentration of ascorbic acid by 100%. Heating plant material before crushing may influence the stability of ascorbic acid, preserving its contents. This can be used in the food industry in the production of *L. caerulea* products [14].

Table 1. Ascorbic acid content and antioxidant activity of *L. kamtschatica* and *L. emphyllocalyx* fruit.

		Lonicera kamtschatica		*Lonicera emphyllocalyx*		
		'Duet'	'Aurora'	'Colin'	'Lori'	'Willa'
Ascorbic acid content [mg·100 g^{-1} f.w.]		44.40 ^{b} ± 0.4	62.60 ^{e} ± 0.6	49.70 ^{c} ± 0.7	51.30 ^{d} ± 0.3	36.80 ^{a} ± 0.5
Antioxidant activity	DPPH [% inhibition]	89.62 ^{d} ± 0.01	68.68 ^{a} ± 0.08	72.11 ^{b} ± 0.01	79.02 ^{b} ± 0.02	68.97 ^{a} ± 0.04
	FRAP [μM Fe^{2+}·g^{-1} f.w.]	37.67 ^{e} ± 0.07	30.52 ^{a} ± 0.02	36.77 ^{d} ± 0.07	33.17 ^{b} ± 0.07	35.22 ^{c} ± 0.02
	ABTS [mM TE·100 g^{-1} f.w.]	2.26 ^{b} ± 0.06	1.97 ^{a} ± 0.07	1.91 ^{a} ± 0.01	2.12 ^{b} ± 0.02	2.21 ^{c} ± 0.04

Data are expressed as mean values (n = 3) ± SD; SD—standard deviation. Mean values within rows with different letters are significantly different ($p < 0.05$).

Lonicera caerulea fruit extract containing anthocyanins is a highly effective antioxidant agent, effectively reducing reactive oxygen species (ROS), which are produced by immune cells as a result of inflammation. Persistent inflammation can lead to cell damage and chronic disease. The fruit extract further reduces lipid peroxidation, which affects cellular damage under oxidative stress conditions, which can contribute to the reduction in diseases related to oxidative stress [34]. The antioxidant activity of *L. kamtschatica* and *L. emphyllocalyx* fruit, determined by the DPPH method, was at the level of 68.68–89.62% inhibition, of which the *L. kamtschatica* cultivar 'Duet' had the highest free radical scavenging activity (on average 13.4–30.4% more than the other cultivars analysed). These results are comparable to those obtained by Kula et al. [2] and Khattab et al. [35], in which the antiradical values of *Lonicera caerluea* were, respectively, 85% and 78.70%. The antioxidant activity of *L. caerulea* fruits is significantly higher compared to other species, e.g., sea buckthorn (74%), bilberry (37–91%), and garden rhubarb (48–98%) [36]. The iron reduction capacity (FRAP method) of *L. kamtschatica* range from 30.52 to 37.67 μM Fe^{2+}·g^{-1} f.w. and *L. emphyllocalyx* range from 30.52 to 37.67 μM Fe^{2+}·g^{-1} f.w. Results comparable to this study were obtained by Rupasinghe et al. [7], with an antioxidant value (FRAP) ranging from 27.96 to 46.90 μM Fe^{2+}·g^{-1}. According to studies, *L. caerulea* had more antioxidative properties (FRAP) compared to other species grown in Poland, e.g., strawberries (8.00 μM TE·g^{-1} f.w), blackberries (15.03 μM TE·g^{-1} f.w.), highbush blueberry (16.24 μM TE·g^{-1} f.w.), elderberry (29.56 mM Fe^{2+}·g^{-1} f.w.), blackthorn (14.74 mM Fe^{2+}·g^{-1} f.w.) and wild strawberry (10.99 Fe^{2+}·g^{-1} f.w.) [7,33]. *Lonicera caerulea* fruits are characterised by a higher antioxidant capacity. The antioxidant activity of the fruits of *L. kamtschatica* and *L. emphyllocalyx*, determined by the ABTS method, ranged from 1.91 to 2.21 mM TE·100 g^{-1} f.w. As reported by Rop et al. [37], the ABTS antiradical activity of ABTS of *Lonicera caerulea* was 0.30 μM·g^{-1}. The values obtained in this study are different to other species cultivated in Poland, e.g., highbush blueberry (16.87 mM TE·g^{-1} f.w.), elderberry (15.88 mM TE·g^{-1} f.w.), blackberry (9.55 mM TE·g^{-1} f.w.), strawberries (5.61 TE·g^{-1} f.w.) [7,33], red currant (11.83–12.59 μM TE·g^{-1} d.m.) [13], mulberry

(0.14 mM TE·100 g^{-1} f.w.) and raspberry (0.08 mM TE·100 g^{-1} f.w.) [26]. The high level of antioxidant capacity in *L. kamtschatica* and *L. emphyllocalyx* makes them very valuable in terms of bioactivity. Research on the antioxidant activity of extracts with *L. caerulea* revealed that the fruits of this plant are characterised by stronger antioxidant properties from other berries commonly regarded as effective antioxidants. Considering the relationship between modern-day diseases and long-term oxygen stress, strong antioxidant properties may indicate the potential importance of this fruit, not only in prophylaxis, but also in the treatment of many diseases [2], thus making *L. caerulea* products more valuable for food industry.

2.3. Polyphenolic Compound Content in L. kamtschatica and L. emphyllocalyx Fruit

The phenolic compounds, mainly anthocyanins, contained in *L. caerulea* fruit extract exhibit anti-inflammatory effects. They reduce cellular damage under conditions of oxidative stress in in vitro cultures of rat microsomes and reduce ROS production in cultures of pro-inflammatory gingival fibroblasts [6,38]. The contents of total polyphenols differed significantly depending on the cultivar analysed; fruits of the *L. emphyllocalyx* cultivar 'Willa' were characterised by the lowest total polyphenol content—416.94 mg GAE·100 g^{-1} f.w.—while the highest total polyphenol content of total polyphenols—747.85 GAE·100 g^{-1} f.w.—was also found in fruits of *L. emphyllocalyx* but of the cultivar 'Lori' (Table 2). The results obtained in this study are comparable to those reported by Rop et al. [37]; the content of total polyphenols in the *L. kamtschatica* fruit ranged from 575 to 903 mg GAE·100 g^{-1} f.w, while in the study by Oszmiański et al. [21], the polyphenol content of the *L. kamtschatica* fruit was 12.29 g·100 g^{-1} f.w. The analysed fruits of *L. emphyllocalyx* and *L. kamtschatica* were characterised by significantly higher total polyphenol contents compared to fruits grown in Poland: raspberries—445.5 mg GAE·100 g^{-1} f.w.; strawberries—238.0 mg GAE·100 g^{-1} f.w.; sea buckthorn—302.72 mg GAE·100 g^{-1} f.w. [6,39,40]; blackberry—247.25 mg GAE·100 g^{-1} f.w.; blackthorn—402.67 mg GAE·100 g^{-1} f.w.; highbush blueberry—424.72 mg GAE·100 g^{-1} f.w.; and elderberry—535.98 mg GAE·100 g^{-1} f.w. [33].

The analysis of phenolic compounds using the UPLC-PDA-MS/MS method allowed for the determination of the differences between the contents of individual groups of polyphenolic compounds contained in the fruit of the analysed cultivars of *L. kamtschatica* and *L. emphyllocalyx* (Table 2). Berries of the *L. kamtschatica* cultivars 'Duet' and 'Aurora', and *L. emphyllocalyx* cultivars 'Lori', 'Willa' and 'Colin' were characterised by different contents of individual polyphenolic compounds. The total content of phenolic compounds depends, among other things, on the cultivar, the degree of fruit maturity, and also on the harvest date. The identification of compounds was carried out based on retention time (Rt), MS and MS/MS with available publications [41–45]. The extract prepared from the fruit contained 26 different compounds in its composition, including 14 anthocyanins, 8 flavonoids, 2 phenolic acids and 1 flavan-3-ol (Table 2).

The highest anthocyanin content in the analysed fruits was found in the berries *of L. kamtschatica* 'Duet', at 456.3 mg·100 g^{-1} (Table 2). Among the phenolic compounds found in the fruits of *L. kamtschatica* (509.29–597.29 mg GAE·100 g^{-1}) and *L. emphyllocalyx* (416.94–747.85 mg GAE·100 g^{-1}), anthocyanins represented, on average, 94% of all polyphenols, and the main representative was cyanidin 3-*O*-glucoside—C3G (the compound represented, on average, 82.2% of the total anthocyanin content detected in the fruits). Among the varieties analysed, the *L. kamtschatica* fruits of the 'Duet' variety contained the highest concentration of C3G in their composition—382.18 mg·100 g^{-1} (Table 2). The results obtained in this study are comparable to those obtained by Rupasinghe et al. [9]; the C3G content of *Lonicera caerulea* fruit ranged from 68 to 649 mg·100 g^{-1}. C3G content was comparable to that reported by Khattab et al. [35], and the C3G content in the fruit of the cultivars 'Tundra', 'Berry Blue' and 'Indigo gem' reached 79–88% of the total anthocyanin content. The C3G content of the *Lonicera caerulea* fruit was significantly higher compared to the strawberry fruit (3.7 mg·100 g^{-1}), blueberry (3.0 mg·100 g^{-1}), and the cranberry (0.7 mg·100 g^{-1}) [9]. C3G is the most prevalent anthocyanin in edible fruits and has been shown to have anti-inflammatory, antioxidant, chemotherapeutic, and epigenetic effects [9]. Fruits with a larger diameter and harvested at the optimal harvest time will have a higher anthocyanin content due to the larger skin area [15]. In the study of Senica et al. [15], spreads and smoothies made out of *L. caerulea* had higher concentrations of C3G (6.48, 5.00 mg·100 g^{-1}) in comparison to fresh fruit. The fruits of this species play important roles in a wide range of physiological processes, e.g., protective effect against UV radiation for skin, protection against pathogenic strains, etc..

Table 2. Individual phenolic compounds identified by UPLC-PDA-MS/MS in *L. kamtschatica* i *L. emphyllocalyx*.

No.	Compund [mg·100 g^{-1}]	Rt	λ_{max}	$[M\text{-}H]^{+/-}$ *m/z*		*Lonicera emphyllocalyx*			*Lonicera kamtschatica*	
		min	nm	MS	MS/MS	'Lori'	'Colin'	'Willa'	'Aurora'	'Duet'
				Anthocyanins						
1.	(+)Catechin-Cyanidin-3-*O*-glucoside	1.99	280, 522	737^{+}	449, 287	1.30 ^{a} ± 0.10	1.21 ^{a} ± 0.04	1.35 ^{a} ± 0.31	1.56 ^{a} ± 0.23	1.41 ^{a} ± 0.21
2.	Pelargonidin 3-O-rutinoside	2.01	279, 520	579^{+}	271	1.12 ^{bc} ± 0.10	0.87 ^{ab} ± 0.05	0.67 ^{a} ± 0.03	0.63 ^{a} ± 0.44	1.30 ^{c} ± 0.31
3.	(−)Epicatechin-Cyanidin-3-*O*-glucoside	2.08	279, 520	737^{+}	449, 287	1.67 ^{b} ± 0.22	1.05 ^{a} ± 0.01	1.14 ^{a} ± 0.19	1.18 ^{a} ± 0.14	2.09 ^{b} ± 0.48
4.	Cyanidin 3,5-*O*-diglucoside	2.20	276, 512	611^{+}	287	26.31 ^{c} ± 1.48	17.50 ^{ab} ± 0.56	20.40 ^{bc} ± 2.42	14.00 ^{a} ± 1.54	40.49 ^{d} ± 6.62
5.	Malvidin 3-*O*-glucoside	2.33	279, 522	493^{+}	331	1.70 ^{d} ± 0.27	1.36 ^{bc} ± 0.10	1.05 ^{b} ± 0.20	0.72 ^{a} ± 0.10	1.47 ^{cd} ± 0.14
6.	Cyanidin 3-*O*-glucoside	2.61	279, 515	449^{+}	287	355.73 ^{b} ± 8.40	259.41 ^{a} ± 2.03	364.89 ^{b} ± 33.45	374.62 ^{b} ± 18.00	382.18 ^{b} ± 28.34
7.	Cyanidin 3-*O*-rutinoside	2.93	279, 517	595^{+}	287	33.73 ^{d} ± 0.20	32.06 ^{cd} ± 1.82	30.67 ^{c} ± 3.07	16.06 ^{b} ± 1.00	12.48 ^{a} ± 0.45
8.	Cyanidin 3-*O*-galactoside	3.05	278, 521	449^{+}	287	0.12 ^{a} ± 0.03	0.82 ^{d} ± 0.09	0.63 ^{c} ± 0.06	0.37 ^{b} ± 0.01	0.38 ^{b} ± 0.01
9.	Pelagonidin 3-*O*-glucoside	3.16	277, 502	433^{+}	271	1.40 ^{a} ± 0.04	2.05 ^{b} ± 0.13	2.23 ^{b} ± 0.37	3.24 ^{c} ± 0.27	1.28 ^{a} ± 0.14
10.	Pelargonidin 3-*O*-galactoside	3.28	277, 505	433^{+}	271	0.32 ^{b} ± 0.04	0.83 ^{c} ± 0.04	0.38 ^{b} ± 0.03	0.76 ^{c} ± 0.01	0.19 ^{a} ± 0.06
11.	Peonidin 3-*O*-glucoside	3.46	279, 517	463^{+}	301	21.42 ^{c} ± 0.55	14.27 ^{ab} ± 0.06	16.15 ^{b} ± 2.00	16.29 ^{b} ± 1.56	12.57 ^{a} ± 0.40
12.	Cyanidin 3-(6″-acetylo)-glucoside	3.59	279, 519	591^{+}	449, 287	2.52 ^{c} ± 0.02	2.75 ^{d} ± 0.17	1.69 ^{b} ± 0.10	0.79 ^{a} ± 0.04	0.63 ^{a} ± 0.06
13.	Cyanidin 3-*O*-rutinoside-5-*O*-dlucoside	3.79	279, 522	757^{+}	595, 287	0.13 ^{b} ± 0.03	0.15 ^{b} ± 0.03	0.43 ^{c} ± 0.05	0.39 ^{c} ± 0.02	0.07 ^{a} ± 0.01
14.	Delphinidin 3-*O*-glucoside-pentoside	3.77	279, 522	597^{+}	303	0.26 ^{c} ± 0.07	0.16 ^{b} ± 0.00	0.26 ^{c} ± 0.01	0.37 ^{d} ± 0.01	0.09 ^{a} ± 0.02
				Phenolic acids						
15.	Neochlorogenic acid	2.25	288sh, 324	353^{-}	191	0.68 ^{b} ± 0.02	0.42 ^{a} ± 0.01	0.76 ^{b} ± 0.10	1.18 ^{c} ± 0.01	1.28 ^{d} ± 0.02
16.	Chlorogenic acid	2.87	288sh, 324	353^{-}	191	6.92 ^{a} ± 0.26	8.82 ^{b} ± 0.48	9.94 ^{b} ± 1.41	13.44 ^{c} ± 0.39	9.71 ^{b} ± 0.81
				Flavon-3-ols						
17.	Procyanidin dimer B-type	3.11	279	577^{-}	289	2.10 ^{b} ± 0.25	1.46 ^{a} ± 0.05	2.35 ^{b} ± 0.22	1.97 ^{b} ± 0.10	2.32 ^{b} ± 0.36
				Flavonoids						
18.	Quercetin 3-*O*-rutinoside-7-*O*-rhamnoside	4.03	255, 350	755^{-}	609, 301	0.30 ^{b} ± 0.00	0.45 ^{c} ± 0.02	0.30 ^{b} ± 0.02	0.56 ^{d} ± 0.01	0.21 ^{a} ± 0.04
19.	Quercetin 3-*O*-arabinoside-glucoside	4.22	255, 355	595^{-}	301	4.44 ^{e} ± 0.20	3.06 ^{d} ± 0.06	0.88 ^{a} ± 0.16	2.02 ^{c} ± 0.07	1.12 ^{b} ± 0.01
20.	Quercetin 3-*O*-rutinoside	4.54	255, 355	609^{-}	301	3.04 ^{a} ± 0.09	2.65 ^{a} ± 0.15	4.75 ^{b} ± 0.66	14.08 ^{c} ± 0.29	5.32 ^{b} ± 0.11
21.	Quercetin 3-*O*-glucoside	4.72	255, 355	463^{-}	301	1.88 ^{c} ± 0.02	1.32 ^{b} ± 0.02	0.98 ^{a} ± 0.15	2.46 ^{d} ± 0.25	1.92 ^{c} ± 0.27
22.	Quercetin 3-*O*-pentoside	4.88	255, 355	433^{-}	301	0.15 ^{a} ± 0.01	0.20 ^{ab} ± 0.03	0.17 ^{a} ± 0.07	0.29 ^{b} ± 0.05	0.87 ^{c} ± 0.11
23.	Quercetin 3-*O*-rhamnoside	3.95	255, 355	447^{-}	301	0.11 ^{a} ± 0.01	0.43 ^{c} ± 0.06	0.23 ^{b} ± 0.04	0.53 ^{d} ± 0.04	0.44 ^{c} ± 0.02
24.	Kaempferol 3-*O*-rutinoside	5.09	265, 347	593^{-}	285	0.06 ^{a} ± 0.00	0.07 ^{a} ± 0.01	0.08 ^{a} ± 0.02	0.41 ^{b} ± 0.05	0.04 ^{a} ± 0.02
25.	3,4–di-*O*-caffeoyl-quinic acid	5.25	288sh, 324	515^{-}	353	1.90 ^{b} ± 0.15	1.87 ^{b} ± 0.08	3.23 ^{c} ± 0.38	2.92 ^{c} ± 0.02	1.22 ^{a} ± 0.02
26.	Quercetin 3-O-(6″-acetylo)-glucoside	5.50	255, 333	505^{-}	463, 301	0.92 ^{b} ± 0.01	1.06 ^{c} ± 0.05	0.66 ^{a} ± 0.06	1.11 ^{c} ± 0.04	0.89 ^{b} ± 0.13
	Total polyphenols content [mg GAE·100 g f.w.]					747.85 ^{e} ± 0.05	522.06 ^{c} ± 0.06	416.94 ^{a} ± 0.04	597.29 ^{d} ± 0.04	506.29 ^{b} ± 0.06

Data are expressed as mean values (n = 3) ± SD; SD—standard deviation. Mean values within rows with different letters are significantly different ($p < 0.05$).

The total flavonoid contents ranged from 11.11 mg·100 g^{-1} in the *L. emphyllocalyx* cv. cultivar 'Colin' to 24.4 mg·100 g^{-1} in the case of the *L. kamtschatica* cv. cultivar 'Duet' (Table 2). In the *L. kamtschatica* and *L. emphyllocalyx* fruits analysed, the average content of flavonoids accounted for 3% of all phenolic compounds and the main representative was quercetin 3-*O*-rutinoside, which represented an average of 41.6% of all flavonoids (Table 2). Quercetin 3-*O*-rutinoside content in our study was significantly higher compared to the results of Oszmiański et al. [21]; the *L. kamtschatica* fruits of the cultivar 'Wojtek' were characterised by a quercetin 3-*O*-rutinoside content of 0.21 mg·100 g^{-1} f.w. Quercetin 3-*O*-rutinoside shows a protective effect on the liver or blood vessels and has anti-inflammatory and antidiabetic properties [46].

Two phenolic acids, chlorogenic acid and neochlorogenic acid, were determined in the composition of the *L. kamtschatica* and *L. emphyllocalyx* fruits. The fruits of *L. kamtschatica* cultivars were characterised by a higher content of phenolic acids than those of *L. emphyllocalyx* cultivars. Of the *L. kamtschatica* and *L. emphyllocalyx* cultivars analysed, the *L. kamtschatica* cultivar 'Duet' was characterised by the highest contents of phenolic acids, on average 14.62 mg·100 g^{-1}, which was 51% higher than in the fruit of the *L. emphyllocalyx* cultivar 'Colin' (Table 2). Chlorogenic acid, with an average of 9.77 mg·100 g^{-1}, was present in the analysed fruits at a significantly higher concentration than neochlorogenic acid (Table 2). The chlorogenic acid contained in the fruits analysed constituted 91.8% of all phenolic acids, which is comparable to the results of Kithama et al. [47], who determined that, in the cultivars 'Aurora', 'Evie', 'Larissa' and 'Rebecca', the content of chlorogenic acid made up 95% of all phenolic acids contained in *L. kamtschatica* fruits.

A chemical compound from the flavan-3-ol group—B-type procyanidins—was present among the fruits tested of *L. kamtschatica* and *L. emphyllocalyx* cultivars. The highest concentration of B-type procyanidins was determined in the berries of the *L. emphyllocalyx* cultivar 'Willa'—an average of 2.35 mg·100 g^{-1}, which was on average 62.1% higher than in the fruit of the cultivar 'Colin' (Table 2). In a study by Raudonė [48], *L. kamtschatica* fruits of the cultivars 'Wojtek', 'Indigo Gem', 'Iga', 'Leningradskij Velikan', 'Nimfa', 'Amphora', 'Tola' and 'Tundra' contained from 9.76 mg·100 g^{-1} f.w. to 27.1 mg·100 g^{-1} f.w. of procyanidin type B procyanidin in their compositions. *L. kamtschatica* fruits of the cultivar 'Wojtek' in the study by Oszmiański et al. [21] were characterised by a procyanidin dimer content of 7.05 mg·100 g^{-1} f.w.

2.4. Sugar Content of *L. kamtschatica* and *L. emphyllocalyx* Fruit

Of the *L. kamtschatica* berry cultivars analysed, the 'Duet' cultivar had the highest total sugar content, with an average of 4968 mg·100 g^{-1}, and glucose was the dominant sugar in the fruit of all the cultivars (Table 3). A significant difference in the contents of total sugars was observed between the fruits of the *L. kamtschatica* and the *L. emphyllocalyx* cultivars. The *L. kamtschatica* fruits of the cultivar 'Aurora' were characterised by the highest glucose content, on average 2963 mg·100 g^{-1}, while the significantly lower glucose content between the analysed cultivars was characteristic of *L. emphyllocalyx* fruits of the cultivar 'Willa' (Table 3). However, this cultivar was characterised by a high fructose content (comparable to that of the *L. kamtschatica* cultivar 'Duet') and significantly higher compared to the other cultivars (Table 3). The sucrose content of the fruits analysed ranged from 1325 mg·100 g^{-1} (*L. emphyllocalyx* cultivar 'Willa') to 2750 mg·100 g^{-1} (*L. kamtschatica* cultivar 'Aurora'; Table 3). The sugar content results are comparable to those obtained by Senica et al. [49], who obtained a total sugar content ranging from 1557.37 to 2585.45 mg·100 g^{-1} f.w. The results obtained by Gołba et al. [11] were significantly lower compared to those of this study, the total sugar content of the *Lonicera caerulea* fruit ranged from 1500 mg·100 g^{-1} to 2585 mg·100 g^{-1}. The average fructose content of *Lonicera caerulea* reported by Sharma et al. [17] was comparable with this study and ranged from 1047.53 to 1363.67 mg·100 g^{-1} f.w., but the content of glucose was much smaller and ranged from 750.89 to 1129.35 mg·100 g^{-1} f.w. A range of 3.72–126.12 mg·100 g^{-1} f.w. of sucrose has also been reported by Cheng et al. [50]; these values are comparable to those obtained in the experiment. As reported by Wojdyło et al. [51], Sorbitol was previously found in Polish *L. kamtschatica* cultivars, with a concentration ranging from 0.1 to 0.4 mg·100 g^{-1} f.w. The sugar content of *L. kamtschatica* and *L. emphyllocalyx* fruit depends on environmental conditions, light intensity, fruit maturity and species, among other factors [16].

Table 3. Total sugar content of the fruits of *L. kamtschatica* and *L. emphyllocalyx*.

	Lonicera kamtschatica		*Lonicera emphyllocalyx*		
	'Aurora'	'Duet'	'Lori'	'Colin'	'Willa'
Fructose content [mg·100 g^{-1}]	1503.00 ^{b} ± 5	2109.00 ^{c} ± 2	1179.00 ^{a} ± 9	1163.00 ^{a} ± 4	2087.00 ^{c} ± 4
Glucose content [mg·100 g^{-1}]	2963.00 ^{c} ± 3	2666.00 ^{b} ± 1	2943.00 ^{c} ± 5	2795.00 ^{bc} ± 5	2125.00 ^{a} ± 3
Sucrose content [mg·100 g^{-1}]	275.00 ^{c} ± 4	192.00 ^{b} ± 6	192.00 ^{b} ± 1	171.00 ^{b} ± 5	132.00 ^{a} ± 5
Total sugar content [mg·100 g^{-1}]	4741.00 ^{bc} ± 0.4	4968.50 ^{c} ± 0.4	4315.50 ^{ab} ± 0.8	4129.50 ^{a} ± 0.1	4345.50 ^{ab} ± 0.9

Data are expressed as mean values (n = 3) ± SD; SD—standard deviation. Mean values within rows with different letters are significantly different ($p < 0.05$).

3. Materials and Methods

3.1. Material

Fruits of the *L. caerulea* var. *kamtschatica* cultivars 'Duet' and 'Aurora' were obtained from a nursery crop located in Tyczyn (49°57′52″ N 22°2′47″ E, Subcarpathian Voivodship, Poland) in the year 2022. Fruits of the *L. caerulea* var. *emphyllocalyx* cultivars 'Lori', 'Willa' and 'Colin' were obtained from 'Korfanty' (49°41′41″ N 22°5′3″ E, Grabownica Starzeńska, Subcarpathian Voivodeship, Poland) in 2022. Both species were grown in containers filled with a peat substrate containing sand and perlite in a ratio of 20:1:1, with the addition of fertilizer Osmocote Exact 3–4 m (ICL, Sydney, Autralia) at concentration of 2.0 kg for 1 m^3 of substrate.

The average monthly temperatures in the period from March to June in Tyczyn were, respectively, 3.3, 7.0, 14.8, and 19.8 °C, and in Grabownica Starzeńska were, respectively, 2.2, 6.0, 13.7, and 18.5 °C. The average monthly rainfall values in the period from March to June in Tyczyn were, respectively, 20 mm, 60 mm, 50 mm, and 20 mm, and in Grabownica Starzeńska the average monthly rainfall in the period from March to June was 50 mm.

The fruits of the analysed cultivars were harvested by hand at the stage of their harvest maturity (first decade of June), 1000 g each. Immediately after harvest, the fruits were subjected to chemical analysis.

3.2. Determination of pH and Acidity

The total acidity (as citric acid) and the pH of the *L. kamtschatica* and *L. emphyllocalyx* fruit were determined through the potentiometric titration of the sample for analysis with a standard 0.1 M NaOH solution at pH = 8.1 using a TitroLine 5000 (SI Analytcs, Weilheim, Germany) according to the method given in PN-EN 12147:2000 [52]. The results are expressed as g of citric acid per 100 g of fruit. The analyses were performed in triplicate.

3.3. Determination of the Contents of Bioactive Compounds in Fruit and Determination of Their Antioxidant Activity

Vitamin C (ascorbic acid) was determined according to PN-A-04019:1998 [53]. Total polyphenol content (mg GAE·100 g^{-1} f.w.) was determined using the Folin–Ciocalteu method according to the methodology described by Bakowska-Barczak et al. [8]. The identification of the polyphenolic profile in *L. kamtschatica* and *L. emphyllocalyx* fruit was determined according to the methodology reported by Gorzelany et al. [24].

The ability of the fruit to reduce iron ions (FRAP method) was determined according to the methodology given by Rupasinghe et al. [26], and the results are given in μM Fe^{2+}·g^{-1} f.w. The antioxidant activity of the fruit was determined using DPPH methods according to the methodology given by Jurčaga et al. [54], and the result is expressed as % inhibition of DPPH radicals, and through the ABTS method according to the methodology given by Gawroński et al. [1], and the results are expressed in mM TE·100 g^{-1} f.w. All analyses were performed in triplicate.

3.4. Determination of Sugars in *L. kamtschatica* and *L. emphyllocalyx* Fruit

The sugar content was measured using the HPLC method with refractive index detection. The chromatographic equipment SYKAM (Sykam GmbH, Eresing, Germany), consisting of sample injector S5250, pump system S1125, column oven S4120 and RI detector S3590, was used. Separation was carried out using a Cosmosil Sugar-D column 250 × 4.6 mm (Nacalai, Kyoto, Japan). The separation was achieved with a mobile phase of 70% of acetonitrile in water in isocratic mode. The flow rate was 0.5 mL/min at column temperature set at 30 °C. The volume of injected sample was 20 μL and 15 min was needed to complete the analysis. Samples before injection were centrifuged at 5000 rpm for 10 min using Centrifuge 5430 (Eppendorf, Hamburg, Germany) and diluted with mobile phase 1:4 (*v*/*v*). All determinations were performed in triplicate.

3.5. Statistical Analysis

Using Statistica 13.3. software (TIBCO Software Inc., Tulsa, OK, USA), a statistical analysis of the results obtained was performed that included the analysis of variance (ANOVA) and NIR significance test at a significance level of $\alpha = 0.05$.

4. Conclusions

Based on this study, differences in fruit composition were found in both individual species and the *L. caerulea* cultivars. The fruits of the *L. kamtschatica* cultivar 'Aurora' contained the highest amount of ascorbic acid in their composition, approximately 22% more compared to 'Lori', which had the highest content of this compound among *L. emphyllocalyx* cultivars. The highest total polyphenol content was found in the *L. emphyllocalyx* cultivar 'Lori', while the predominant polyphenolic compound was cyanidin 3-*O*-glucoside. On the basis of the results obtained, it can be concluded that the *L. emphyllocalyx* cultivars 'Lori' and 'Willa' can find applications in various food industries. Despite the diversity of chemical composition, the fruits have high antioxidant properties compared to the well-known varieties 'Duet' and 'Aurora'. The variety 'Lori' has the highest content of phenolic compounds among the tested varieties. 'Willa' has the highest concentration of C3G comparable to the widely known varieties of *L. kamtschatica* and the highest sugar content among *L. emphyllocalyx*. The use of *Lonicera* as an ingredient of functional foods, natural colourant or source of natural antioxidant seems to be promising. Compared to other types of berries grown in Poland, e.g., raspberries and red currant, *L. kamtschatica* and *L. emphyllocalyx* can be a good source of bioactive substances and sugar, making them good in food industry and processing, including juices, wines, spreads and dried fruits.

Author Contributions: Conceptualisation, O.B. and J.G.; methodology, J.B., I.K. and O.B.; validation, J.B., O.B. and J.G.; formal analysis, J.B. and K.P.; investigation, O.B. and K.P.; writing—original draft preparation, O.B. and J.B.; writing—review and editing, J.G., O.B. and I.K.; visualisation, J.B.; supervision, O.B. and J.B.; project administration, O.B. and J.B.; funding acquisition, J.G. and J.B. All authors have read and agreed to the published version of the manuscript.

Funding: This research received no external funding.

Institutional Review Board Statement: Not applicable.

Informed Consent Statement: Not applicable.

Data Availability Statement: Not applicable.

Conflicts of Interest: The authors declare no conflict of interest.

Sample Availability: Samples of the compounds used in research are available from the authors.

References

1. Gawronski, J.; Hortynski, J.; Kaczmarska, E.; Dyduch-Sieminska, M.; Marecki, W.; Witorozec, A. Evaluation of phenotypic and genotypic diversity of some Polish and Russian blue honeysuckle (*Lonicera caerulea* L.) cultivars and clones. *Acta Sci. Polonorum. Hortorum Cultus* **2014**, *13*, 157–169.
2. Kula, M.; Krauze-Baranowska, M. Blue Honeysuckle (*Lonicera caerulea* L.)—The current state of phytochemical research and biological activity. *Post Fitoter.* **2016**, *17*, 111–118.
3. Kucharska, A.Z.; Sokół-Łętowska, A.; Oszmiański, J.; Piórecki, N.; Fecka, I. Iridoids, Phenolic Compounds and Antioxidant Activity of Edible Honeysuckle Berries (*Lonicera caerulea* var. *kamtschatica* Sevast.). *Molecules* **2017**, *22*, 405. [CrossRef]
4. Maxine, T.M. Introducing haskap, Japanese Blue honeysuckle. *J. Am. Pomol. Soc.* **2006**, *60*, 164–168.
5. Minami, M.; Takase, H.; Nakamura, M.; Makino, T. Methanol extract of *Lonicera caerulea* var. *emphyllocalyx* fruit has anti-motility and anti-biofilm activity against enteropathogenic *Escherichia coli*. *Drug Discov. Ther.* **2019**, *19*, 335–342. [CrossRef]
6. Ochmian, D.; Skupień, K.; Grajkowski, J.; Smolik, M.; Ostrowska, K. Chemical Composition and Physical Characteristics of Fruits of Two Cultivars of Blue Honeysuckle (*Lonicera caerulea* L.) in Relation to their Degree of Maturity and Harvest Date. *Not. Bot. Horti. Agrobot.* **2012**, *40*, 155–162. [CrossRef]
7. Rupasinghe, H.P.V.; Yu, L.J.; Bhullar, K.S.; Bors, B. Short Communication: Haskap (*Lonicera caeruela*): A new berry crop with high antioxidant capacity. *Can. J. Plant Sci.* **2012**, *21*, 4334. [CrossRef]
8. Bakowska-Barczak, A.M.; Marianchuk, M.; Kolodziejczak, P. Survey of bioactive components in Western Canadian berries. *Can. J. Physiol. Pharmacol.* **2007**, *85*, 1139–1152. [CrossRef]
9. Rupasinghe, H.P.V.; Arumuggam, N.; Amararathna, M.; De Silva, A.B.K.H. The potential health benefits of haskap (*Lonicera caerulea* L.): Role of cyanidin-3-O-glucoside. *J. Funct. Foods* **2018**, *44*, 24–39. [CrossRef]
10. Celli, G.B.; Ghanem, A.; Brooks, M.S.L. Haskap Berries (*Lonicera caerulea* L.)—A Critical Review of Antioxidant Capacity and Health-Related Studies for Potential Value-Added Products. *Food Bioprocess Technol.* **2014**, *7*, 1541–1554. [CrossRef]
11. Khattab, R.; Ghanem, A.; Brooks, M.S.L. Quality of dried haskap berries (*Lonicera caerulea* L.) as affected by prior juice extraction, osmotic treatment, and drying conditions. *Dry. Technol.* **2017**, *35*, 375–391. [CrossRef]
12. Gołba, M.; Sokół-Łętowska, A.; Kucharska, A. Health Properties and Composition of Honeysuckle Berry *Lonicera caerulea* L. An U date on Recent Studies. *Molecules* **2020**, *25*, 749. [CrossRef]
13. Belyaeva, O.V.; Sergeeva, I.Y.; Belyaeva, E.E.; Chernobrovkina, E.V. Study of antioxidant activity of juices and beverages from blue honeysuckle and black chokeberry. *IOP Conf. Ser. Earth Environ. Sci.* **2021**, *640*, 052008. [CrossRef]
14. Grobelna, A.; Kalisz, S.; Kieliszek, M. The effect of the addition of blue honeysuckle berry juice to apple juice on the selected quality characteristics, anthocyanin stability, and antioxidant properties. *Biomolecules* **2019**, *9*, 744. [CrossRef]
15. Senica, M.; Stampar, F.; Mikulic-Petkovsek, M. Different extraction processes affect the metabolites in blue honeysuckle (*Lonicera caerulea* L. subsp. *edulis*) food products. *Turk. J. Agric. For.* **2019**, *43*, 576–585. [CrossRef]
16. Jurikova, T.; Matuškovič, J.; Gazdik, Z. Effect of irrigation on intensity of respiration and study of sugar and organic acids content in different development stages of *Lonicera kamtschatica* and *Lonicera edulis* berries. *HortScience* **2009**, *36*, 14–20.
17. Sharma, A.; Hae-Jeung, L. *Lonicera caerulea*: An updated account of its phytoconstituents and health-promoting activities. *Trends Food Sci. Technol.* **2021**, *107*, 130–149. [CrossRef]
18. Gerbrand, E.M.; Bors, R.H.; Meyer, D.; Wilen, R.; Chibbar, R. Fruit quality of Japanese, Kuril and Russian blue honeysuckle (*Lonicera caerulea* L.) germplasm compared to blueberry, raspberry and strawberry. *Euphytica* **2020**, *216*, 59. [CrossRef]
19. Auzanneau, N.; Weber, P.; Kosińska-Cagnazzo, A.; Audlauer, W. Bioactive compounds and antioxidant capacity of *Lonicera caerulea* berries: Comparison of seven cultivars over three harvesting years. *J. Food Compos. Anal.* **2018**, *66*, 81–89. [CrossRef]
20. Miyashita, T.; Hoshino, Y. Intersecific hybridization in *Lonicera caerulea* and *Lonicera gracilipes*: The occurrence of green/albino plants by reciprocal crossing. *Sci. Hortic.* **2010**, *125*, 692–699. [CrossRef]
21. Oszmiański, J.; Wojdyło, A.; Lachowicz, S. Effect of dried powder preparation process on polyphenolic content and antioxidant activity of blue honeysuckle berries (*Lonicera caerulea L. var. kamtschatica*). *LWT Food Sci. Technol.* **2015**, *64*, 214–222. [CrossRef]
22. Gorzelany, J.; Kapusta, I.; Zardzewiały, M.; Belcar, J. Effects of ozone application on microbiological stability and content of sugars and bioactive compounds in the fruit of the saskatoon berry (*Amelanchier alnifolia* Nutt.). *Molecules* **2022**, *27*, 6446. [CrossRef]
23. Kuźniar, P.; Belcar, J.; Zardzewiały, M.; Basara, O.; Gorzelany, J. Effect of Ozonation on the Mechanical, Chemical, and Microbiological Properties of Organically Grown Red Currant (*Ribes rubrum* L.) Fruit. *Molecules* **2022**, *27*, 8231. [CrossRef]
24. Gorzelany, J.; Basara, O.; Kuźniar, P.; Pawłowska, M.; Belcar, J. Effect of ozone treatment on mechanical and chemical properties of sea-buckthorn (*Hippophae rhamnoides* L.) fruit. *Acta Univ. Cibinensis Ser. E Food Technol.* **2022**, *2*, 183–194. [CrossRef]
25. Cheng, K.; Peng, B.; Yuan, F. Volatile composition of eight blueberry cultivars and their relationship with sensory attributes. *Flavour Fragr. J.* **2020**, *35*, 443–453. [CrossRef]
26. Lee, Y.; Lee, J.-H.; Kim, S.-D.; Chang, M.-S.; Jo, I.-S.; Kim, S.-J.; Hwang, K.T.; Jo, H.-B.; Kim, J.-H. Chemical composition, functional constituents, and antioxidant activities of berry fruits produced in Korea. *J. Korean Soc. Food Sci. Nutr.* **2015**, *44*, 1295–1303. [CrossRef]

27. MacKenzie, J.O.; Eiford, E.M.A.; Subramanian, J.; Brandt, R.W.; Stone, K.E.; Sulivan, J.A. Performance of five haskap (*Lonicera caeurlea* L.) cultivars and effect of hexanal on postharvest quality. *Can. J. Plant Sci.* **2018**, *98*, 432–443. [CrossRef]
28. Gorzelany, J.; Belcar, J.; Kuźniar, P.; Niedbała, G.; Pentoś, K. Modelling of Mechanical Properties of Fresh and Stored Fruit of Large Cranberry Using Multiple Linear Regression and Machine Learning. *Agriculture* **2022**, *12*, 200. [CrossRef]
29. Djordjević, B.S.; Djurovic, D.B.; Zec, G.D.; Meland, M.O.; Fotiric Aksic, M.M. Effect of shoot age on biological and chemical properties of red currant (*Ribes rubrum* L.) cultivars. *Folia Hort.* **2020**, *32*, 291–305. [CrossRef]
30. Du, J.; Cullen, J.J.; Buettner, G.R. Ascorbic acid: Chemistry, biology and the treatment of cancer. *Biochim. Biophys. Acta* **2012**, *1826*, 443–457. [CrossRef]
31. Jurnikova, T.; Rop, O.; Mlcek, J.; Sochor, J.; Balla, S.; Szekeres, L.; Hegedusova, A.; Hubalek, J.; Adam, V.; Kizek, R. Phenolic profiles of edible honeysuckle berries (Genus *Lonicera*) and Their Biological Effects. *Molecules* **2012**, *17*, 61–79. [CrossRef]
32. Martinsen, B.K.; Aaby, K.; Skrede, G. Effect of temperature on stability of anthocyanins, ascorbic acid and color in strawberry and raspberry jams. *Food Chem.* **2020**, *316*, 126297. [CrossRef]
33. Jabłońska-Ryś, E.; Zalewska-Korona, M.; Kalbarczyk, J. Antioxidant capacity, ascorbic acid and phenolics content in wild edible fruits. *J. Fruit Ornam. Plant Res.* **2009**, *17*, 115–120.
34. Paliková, I.; Valentová, K.; Oborná, I.; Ulrichová, J. Protectivity of blue honeysuckle extract against oxidative human endothelilal cells and rat hepatocyte damage. *J. Agric. Food Chem.* **2009**, *57*, 6584–6589. [CrossRef]
35. Khattab, R.; Brooks, M.S.-L.; Ghanem, A. Phenolic analyses of haskap berries (*Lonicera caerulea* L.): Spectrophotometry Versus High Performance Liquid Chromatography. *Int. J. Food Prop.* **2016**, *19*, 1708–1725. [CrossRef]
36. Raudsepp, P.; Anton, D.; Roasto, M.; Meremäe, K.; Pedastsarr, P.; Mäesaar, M.; Raal, A.; Laikoja, K.; Püssa, T. The antioxidative and antimicrobial properties of the blue honeysuckle (*Lonicera caerulea* L.), Siberian rhubarb (*Rheum rhaponticum* L.) and some other plants, compared to ascorbic acid and sodium nitrate. *Food Control* **2013**, *31*, 129–135. [CrossRef]
37. Rop, O.; Řezníček, V.; Mlček, J.; Juríková, T.; Balík, J.; Sochor, J.; Kramářová, D. Antioxidant and radical oxygen species scavenging activities of 12 cultivars of blue honeysuckle fruit. *Hort. Sci.* **2011**, *38*, 63–70. [CrossRef]
38. Zdarilová, A.; Svobodvaá, A.R.; Chytilová, K.; Simánek, V.; Ulrichová, J. Polyphenolic fraction of Lonicera caerulea L. fruits reduces oxidative stress and inflammatory markers induced by lipopolysaccharide in gingival fibroblasts. *Food Chem. Toxicol.* **2010**, *48*, 1555–1561. [CrossRef]
39. Kant, V.; Mehta, M.; Varshneya, C. Antioxidant potential and total phenolic contents of seabuckthorn (*Hippophae rhamnoides)* pomace. *Free. Radic. Antioxid.* **2012**, *2*, 79–86. [CrossRef]
40. Vasco, C.; Ruales, J.; Kamal-Eldin, A. Total phenolic compound and antioxidant capacities of major fruits from Ecuador. *Food Chem.* **2008**, *4*, 816–823. [CrossRef]
41. Oszmiański, J.; Lachowicz, S.; Gorzelany, J.; Matłok, N. The effect of different maturity stages on phytochemical composition and antioxidant capacity of cranberry cultivars. *Eur. Food Res. Technol.* **2017**, *244*, 705–719. [CrossRef]
42. Truong, V.D.; McFeeters, R.F.; Thompson, R.T.; Dean, L.L.; Shofran, B. Phenolic Acid Content and Composition in Leaves and Roots of Common Commercial Sweet Potato (*Ipomea batatas* L.) Cultivars in the United States. *J. Food Sci.* **2007**, *72*, 343–349. [CrossRef]
43. Lee, S.; Kim, S.H.; Koo, B.; Kim, H.-B.; Jo, Y.-Y.; Kweon, H.; Ju, W.-T. Flavonoids analysis in leaves and fruits of Korean mulberry cultivar, Baekokwang having white fruits. *Int. J. Ind. Entomol.* **2020**, *41*, 45–50. [CrossRef]
44. Ju, W.-T.; Kwon, O.-C.; Kim, H.-B.; Sung, G.-B.; Kim, H.-W.; Kim, Y.-S. Qualitative and quantitative analysis of flavonoids from 12 species of Korean mulberry leaves. *J. Food Sci. Technol.* **2018**, *55*, 1789–1796. [CrossRef]
45. Wu, X.; Prior, R. Systematic Identification and Characterization of Anthocyanins by HPLC-ESI-MS/MS in Common Foods in the United States: Fruits and Berries. *J. Agric. Food Chem.* **2005**, *53*, 2589–2599. [CrossRef]
46. Semwal, R.; Joshi, S.K.; Semwal, R.B.; Semwal, D.K. Health benefits and limiatation of rutin—A natural flavonoid with high nutraceutical value. *Phytochem. Lett.* **2021**, *46*, 119–128. [CrossRef]
47. Kithama, A.; De Silva, H.; Rupasinghe, H.P.V. Polyphenols composition and anti-diabetic properties in vitro oh haskap (*Lonicera caerulea* L.) beries in relation to cultivar harvesting date. *J. Food Compos. Anal.* **2019**, *88*, 103402. [CrossRef]
48. Raudonė, L.; Liaudanskas, M.; Vilkickytė, G.; Kviklys, D.; Žwikas, V.; Viškelis, P. Phenolic profiles, antioxidant activity and phenotypic characterization of *Lonicera caerulea* L. berries, cultivated in Lithuania. *Antioxidants* **2021**, *10*, 115. [CrossRef]
49. Senica, M.; Stampar, F.; Mikulic-Petkovsek, M. Blue honeysuckle (*Lonicera cearulea* L. subs. Edulis) berry; A rich source of some nutrients and their differences among four different cultivars. *Sci. Hortic.* **2018**, *19*, 215–221. [CrossRef]
50. Cheng, Z.; Bao, Y.; Li, Z.; Wang, J.; Wang, M.; Wang, S.; Wang, Y.; Wang, Y.; Li, B. *Lonicera caerulea* (Haskap berries): A review of development traceability, functional value, product development status, future opportunities and challenges. *Crit. Rev. Food Sci. Nutr.* **2022**, 1–25. [CrossRef]
51. Wojdyło, A.; Jáuregui, P.N.N.; Carbonell-Barrachina, A.A.; Oszmiański, J.; Golis, T. Varability of Phytochemical Properties and content of bioactive compounds in *Lonicera caerulea* L. var. *kamtschatica* Berries. *J. Agric. Food Chem.* **2013**, *61*, 12072–12084. [CrossRef]

52. *PN-EN 12147:2000*; Fruit and Vegetable Juices—Determination of Titrable Acidity. Polish Committee for Standardization: Warsaw, Poland, 2000.
53. *PN-A-04019:1998*; Food Products—Determination of Vitamin C Content. Polish Committee for Standardization: Warsaw, Poland, 1998.
54. Jurčaga, L.; Bobko, M.; Kolesárová, A.; Bobková, A.; Demianová, A.; Haščík, P.; Belej, L.; Mendelová, A.; Bučko, O.; Kročko, M.; et al. Blackcurrant (*Ribes nigrum* L.) and Kamchatka Honeysuckle (*Lonicera caerulea* var. *Kamtschatica*) Extract Effects on Technological Properties, Sensory Quality, and Lipid Oxidation of Raw-Cooked Meat Product (Frankfurters). *Foods* **2021**, *10*, 2957. [CrossRef]

Disclaimer/Publisher's Note: The statements, opinions and data contained in all publications are solely those of the individual author(s) and contributor(s) and not of MDPI and/or the editor(s). MDPI and/or the editor(s) disclaim responsibility for any injury to people or property resulting from any ideas, methods, instructions or products referred to in the content.

molecules

Article

Ganoderic Acid A and Its Amide Derivatives as Potential Anti-Cancer Agents by Regulating the p53-MDM2 Pathway: Synthesis and Biological Evaluation

Yi Jia, Yan Li, Hai Shang, Yun Luo and Yu Tian *

Institute of Medicinal Plant Development, Chinese Academy of Medical Sciences & Peking Union Medical College, Beijing 100193, China
* Correspondence: ytian@implad.ac.cn

Abstract: The mechanisms of action of natural products and the identification of their targets have long been a research hotspot. Ganoderic acid A (GAA) is the earliest and most abundant triterpenoids discovered in *Ganoderma lucidum*. The multi-therapeutic potential of GAA, in particular its anti-tumor activity, has been extensively studied. However, the unknown targets and associated pathways of GAA, together with its low activity, limit in-depth research compared to other small molecule anti-cancer drugs. In this study, GAA was modified at the carboxyl group to synthesize a series of amide compounds, and the in vitro anti-tumor activities of the derivatives were investigated. Finally, compound **A2** was selected to study its mechanism of action because of its high activity in three different types of tumor cell lines and low toxicity to normal cells. The results showed that **A2** could induce apoptosis by regulating the p53 signaling pathway and may be involved in inhibiting the interaction of MDM2 and p53 by binding to MDM2 (K_D = 1.68 μM). This study provides some inspiration for the research into the anti-tumor targets and mechanisms of GAA and its derivatives, as well as for the discovery of active candidates based on this series.

Keywords: Ganoderic acid A; derivatives; p53; MDM2; tumor; mechanism of action

Citation: Jia, Y.; Li, Y.; Shang, H.; Luo, Y.; Tian, Y. Ganoderic Acid A and Its Amide Derivatives as Potential Anti-Cancer Agents by Regulating the p53-MDM2 Pathway: Synthesis and Biological Evaluation. *Molecules* **2023**, *28*, 2374. https://doi.org/10.3390/molecules28052374

Academic Editor: Claudio Ferrante

Received: 8 February 2023
Revised: 1 March 2023
Accepted: 2 March 2023
Published: 4 March 2023

Copyright: © 2023 by the authors. Licensee MDPI, Basel, Switzerland. This article is an open access article distributed under the terms and conditions of the Creative Commons Attribution (CC BY) license (https://creativecommons.org/licenses/by/4.0/).

1. Introduction

Ganoderma lucidum is the dry fruiting body of *Ganoderma lucidum* Karst and *Ganoderma sinensis*, which belong to the genus Ganoderma of the family Polyporaceae. The chemical composition of *Ganoderma lucidum* is complex. There are currently about 400 known compounds, the majority of which are triterpenoids, polysaccharides, nucleosides, sterols, and other compounds, of which more than 300 are triterpenoids. Ganoderic acid A (GAA, Figure 1), one of the most prominent and highly concentrated triterpenes from *Ganoderma lucidum*, exhibits a variety of biological properties, including anti-tumor [1], anti-inflammatory [2–4], anti-depressant [5,6], neuroprotection [7,8], anti-fibrosis [9], liver protection [10,11], improvement of glucose and lipid metabolism and myocardial protection [12–14], etc., which can be used as a potential resource for drug development. The anti-tumor activity is one of the earliest discovered activities of GAA, which has received the most attention since then [1]. Many researchers have investigated the anti-tumor activities of *Ganoderma lucidum* triterpenoids and predicted their anti-tumor pathway. Studies have shown that GAA can inhibit tumor growth through a variety of signaling pathways. For example, GAA has good cytotoxicity on human glioblastoma by inducing apoptosis, autophagy and inhibiting PI3K/AKT signaling pathways [15]; it can inhibit the expression of *KDR* mRNA and protein, induce apoptosis of human glioma cell U251 cell, and inhibit its proliferation and invasion [16]. However, there is no relevant literature that clearly indicates the possible anti-tumor target of GAA.

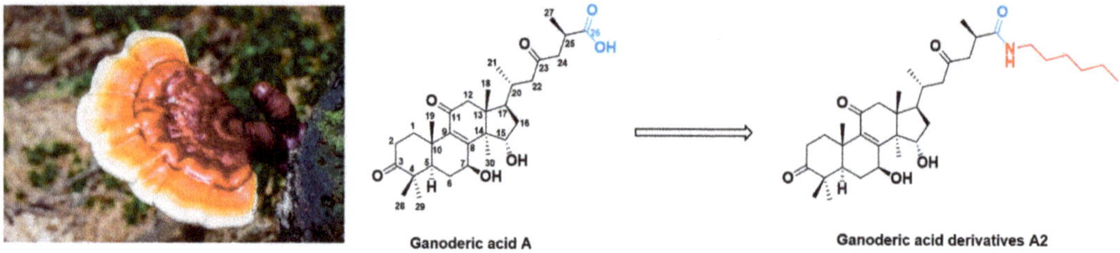

Figure 1. The structure of GAA and compound **A2**.

Cancer and the MDM2-p53 signaling pathway are closely related. p53 is a tumor suppressor gene. When cells are damaged by a variety of causes such as DNA damage, ribosomal stress, the expression of the p53 protein is activated to repair damaged cells or to directly induce apoptosis if the DNA damage is already too severe. p53 is essential for a number of processes that occur throughout life, including DNA damage repair, cell cycle arrest, metabolism, senescence, and apoptosis [17]. If too much p53 protein is produced during certain physiological processes, cell function is impaired or the tendency to form tumors is increased. Therefore, the expression of murine double minute 2 (MDM2) protein in the downstream signaling pathway will increase when p53 protein accumulates in normal cells. To achieve the balance and stability of p53 protein levels in cells, MDM2 can interact with the transcriptional activation domain of the p53 to form the p53-MDM2 complex, which suppresses the transcriptional activity of p53. When a cell is stressed, MDM2 expression decreases, p53 expression increases, and the increase in p53 induces MDM2 expression at the transcriptional level, creating a negative feedback regulatory loop (Figure 2) [18]. MDM2 can also act as an E3 ubiquitin ligase, targeting p53 protein and inducing its ubiquitination and degradation to maintain low levels of p53 protein [19]. p53 has long been an intriguing cancer target [20]. Individuals carrying certain inherited loss-of-function mutations in p53 have a 50% chance of developing cancer by the age of 30 and a 90% chance of developing cancer by the age of 70 [21]. Mice knocked out of p53 quickly develop tumors. Up to 50% of cancers have mutations in both copies of p53 [22]. Drugs that can reactivate the tumor suppressing ability of p53 may therefore have a powerful anti-cancer effect. However, it is more difficult to activate proteins than to inhibit them, so the interaction of MDM2 with p53 provides an opportunity to activate p53 by inhibiting the interaction of MDM2 to exert anti-tumor effects.

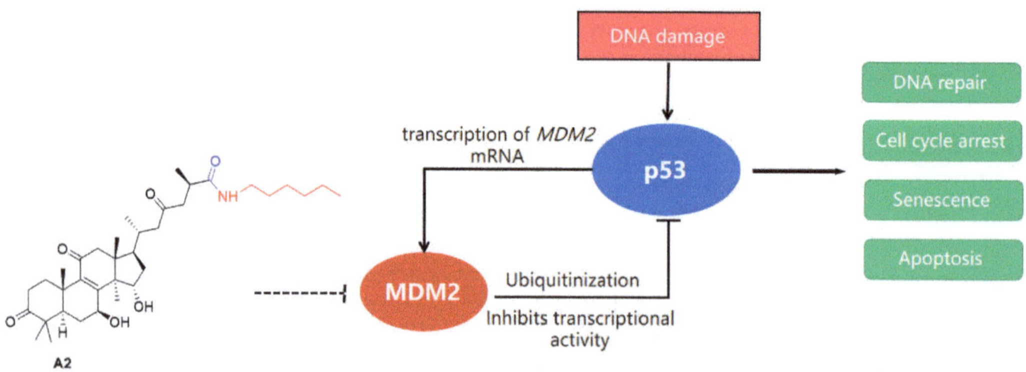

Figure 2. MDM2-p53 negative feedback regulatory loop.

After summarizing the relevant literature, we discovered that GAA may interact with the p53-MDM2 pathway. For example, Xu Bin et al. found that GAA inhibited LNCaP in a concentration-dependent manner. Real-time experiments showed that GAA promoted the apoptosis-promoting genes bad and p53 [23]. Tang Wen et al. found that 95-D cells expressing wild-type p53 protein were 3.3 times more sensitive to ganoderic acid T than H1299 cells that did not express p53 protein [24]. Other studies suggest that GAA and sterols with similar structures may have some affinity for the MDM2. Froufe et al. found that some *Ganoderma lucidum* triterpenoids have potential affinities with MDM2 protein through virtual screening prediction, including ganoderic acid A (K_i = 147 nM) and ganoderic acid F (K_i = 212 nM) [25]. Staszczak et al. summarized the role of secondary metabolites in fungi on the ubiquitin–protesome system, in which sterols have certain interactions with MDM2, indicating that such structure has advantages in interactions with MDM2 [26]. All these results suggest that GAA is likely to be related to the p53-MDM2 pathway. However, considering the low anti-tumor effect of GAA and the size of the pocket of MDM2, and there is no relevant literature highlighting the anti-cancer activities of synthetic GAA derivatives on potential MDM2-p53 interaction inhibitions, we decided to simply modify the structure of GAA at the carboxyl group to improve its anti-tumor activity and reduce possible pharmacokinetic problems caused by the free carboxyl group, and investigated the effects of the different GAA amide derivatives on the MDM2-p53 pathway.

In this study, GAA was modified to determine the in vitro anti-tumor activity of these derivatives on different tumor cell lines, and compound **A2** (Figure 1), which has good activity in different cell lines and low toxicity to normal cells, was selected to investigate the relevant mechanism. First, we investigated the effect of **A2** on cell apoptosis and the expression of proteins related to the MDM2-p53 pathway by flow cytometry and Western blot experiment. Next, in silico target fishing and molecular docking was performed to investigate the binding potential of **A2** and MDM2. We then used a surface plasmon resonance (SPR) experiment to show that **A2** has a certain binding affinity with MDM2 in vitro. It was speculated that **A2** might play a role in increasing p53 protein levels by binding to MDM2 to inhibit the interaction of MDM2 and p53. This work is valuable in further demonstrating the potential of GAA and its amide derivatives as MDM2-p53 binding inhibitors and in developing candidates with anti-tumor activity.

2. Results and Discussion

2.1. Chemistry

Based on the structure of GAA, we retained its core structure of tetracyclic triterpenoids, and introduced a series of amino groups to modify GAA at the carboxyl site. As shown in Scheme 1, GAA was treated with amino compounds, 2-(1*H*-benzotriazole-1-yl)-1,1,3,3-tetramethyluronium tetrafluoroborate (TBTU), and *N*,*N*-diisopropylethylamine (DIPEA) to obtain GAA derivatives [27,28]. **A1–A12** refers to the amide derivatives formed with fatty amine, aniline, benzylamine, phenylethylamine and other different types of primary amine compound. **A13–A15** refers to the derivatives formed with piperazine compounds.

Except for the low yield of substituted aniline, the yields of the other compounds are 70~98%, which is easy to obtain. See Methods and Materials for the detailed synthesis and purification methods of all compounds. After being substituted by different amine fragments, the hydrogen signal of amide bond appears at 7–5 ppm. The methylene peak of amine fragments is mostly distributed at 4.5–3 ppm. The chemical shift of hydrogen signal in GAA itself does not change very much. All new compounds were identified by ^{1}H-NMR, ^{13}C-APT and HRMS spectroscopy. The corresponding spectra are presented in the Supplementary Materials.

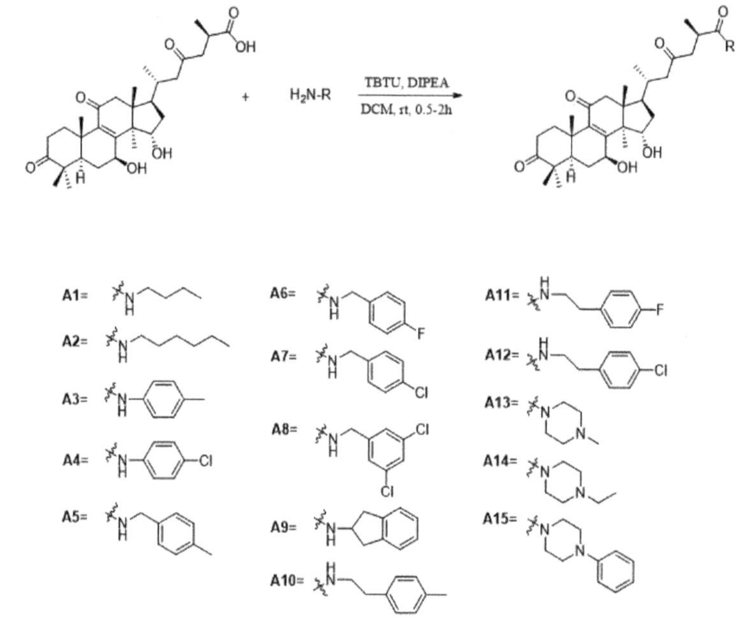

Scheme 1. The synthesis of GAA derivatives. Reagents and conditions: TBTU, DIPEA, DCM, rt, 0.5–2 h.

2.2. *In Vitro Anti-Proliferation Activity*

2.2.1. The Anti-Proliferation Activity on MCF-7

MCF-7 is a commonly used tumor cell line. Previous studies have shown that GAA has some anti-tumor activity against MCF-7, and there is a high expression of MDM2-p53 in MCF-7. The anti-tumor activities of GAA derivatives on MCF-7 were tested for 48 h, and the results were shown in Table 1 and Figure 3. The results showed that compounds **A2, A6, A7, A8, A9, A15** had significant anti-proliferation activities on MCF-7 cell line compared with GAA. Among all derivatives, **A6** has the strongest anti-proliferation effect, and its inhibition rate of MCF-7 at 50 μM can reach 63.64%.

Overall, among aliphatic amines, anilines, benzylamines, phenylethylamines, and (hetero) cyclic amines, benzylamine derivatives (**A6, A7, A8**) were significantly more potent than other substituted compounds. Among the aliphatic amines (**A1, A2**), the chain length of six carbon atoms is better than that of four carbon atoms. Among benzylamine compounds, the activities of electron withdrawing groups on benzene ring (**A6, A7, A8**) (cell viability at 50 μM less than 50%) are better than that of electron donating group (**A5**), and 3,5-diCl double substitution is better than 4-Cl single substitution, indicating that the position and amount of electron withdrawing groups can affect the activities of GAA derivatives. Compared with anilines (**A3, A4**), benzylamines (**A6, A7, A8**) and phenylethylamines (**A10, A11, A12**) substituted compounds, the activities of benzylamines are better, which also indicated that the chain length of substituents may affect their activities. At the same time, the introduction of the common anti-tumor fragment indene can improve the anti-proliferative activity of GAA (**A9**). The introduction of *N*-methyl or *N*-ethyl piperazine with strong hydrophilicity can't improve the anti-tumor activity of GAA, but *N*-phenyl piperazine can improve activity (**A13, A14, A15**), indicating that the anti-proliferation activities of GAA derivatives may have certain requirements for hydrophobicity.

Table 1. The viability of MCF-7 breast cancer cells. Mean values ± SD from three independent experiments (n = 3) performed in duplicate are presented.

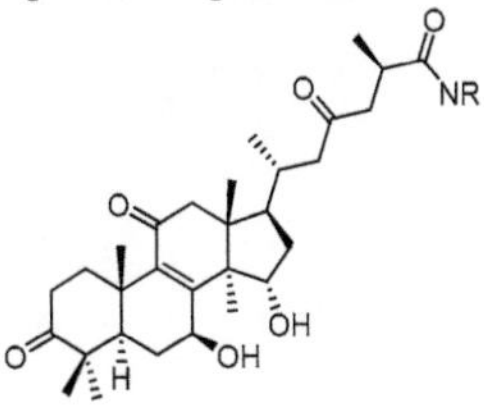

Compound	R	Viability of Cells (% of Control)		
		25 μM	50 μM	100 μM
A1	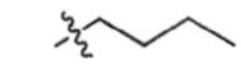	96.6 ± 6.2	114.3 ± 1.2	78.7 ± 3.4
A2	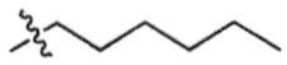	66.76 ± 6.2	59.4 ± 4.7	50.37 ± 1.7
A3	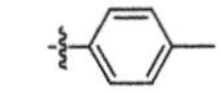	110.9 ± 2.2	114.3 ± 1.2	78.7 ± 3.9
A4	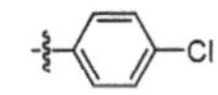	116.4 ± 4.8	120.7 ± 2.0	108.9 ± 1.1
A5	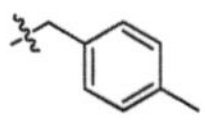	115.0 ± 2.9	107.4 ± 6.7	99.5 ± 1.0
A6	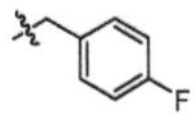	77.1 ± 4.2	36.4 ± 2.6	26.4 ± 1.5
A7	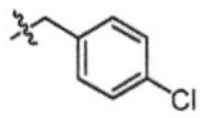	73.5 ± 4.0	46.6 ± 2.5	75.1 ± 13.2
A8	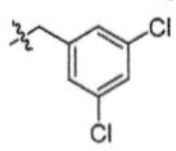	63.3 ± 8.3	45.4 ± 3.9	39.4 ± 5.8
A9	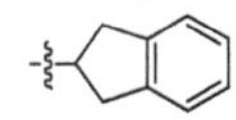	87.5 ± 7.6	53.3 ± 6.9	40.3 ± 7.8
A10	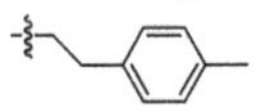	120.8 ± 4.8	118.3 ± 6.0	104.1 ± 2.2
A11	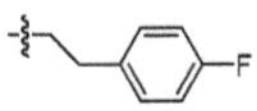	100.5 ± 2.5	101.8 ± 0.2	58.4 ± 5.5
A12	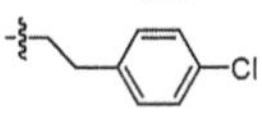	119.0 ± 3.4	110.5 ± 6.0	102.9 ± 3.8
A13	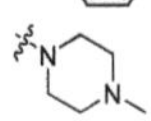	113.6 ± 2.1	107.3 ± 9.1	102.7 ± 3.5
A14	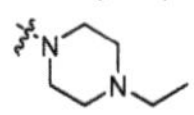	106.2 ± 4.7	102.2 ± 5.9	96.2 ± 5.0
A15	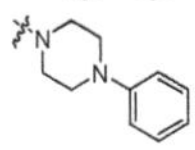	74.3 ± 5.2	70.4 ± 1.5	66.5 ± 1.4
GAA		109.9 ± 2.1	86.2 ± 5.2	83.6 ± 9.4

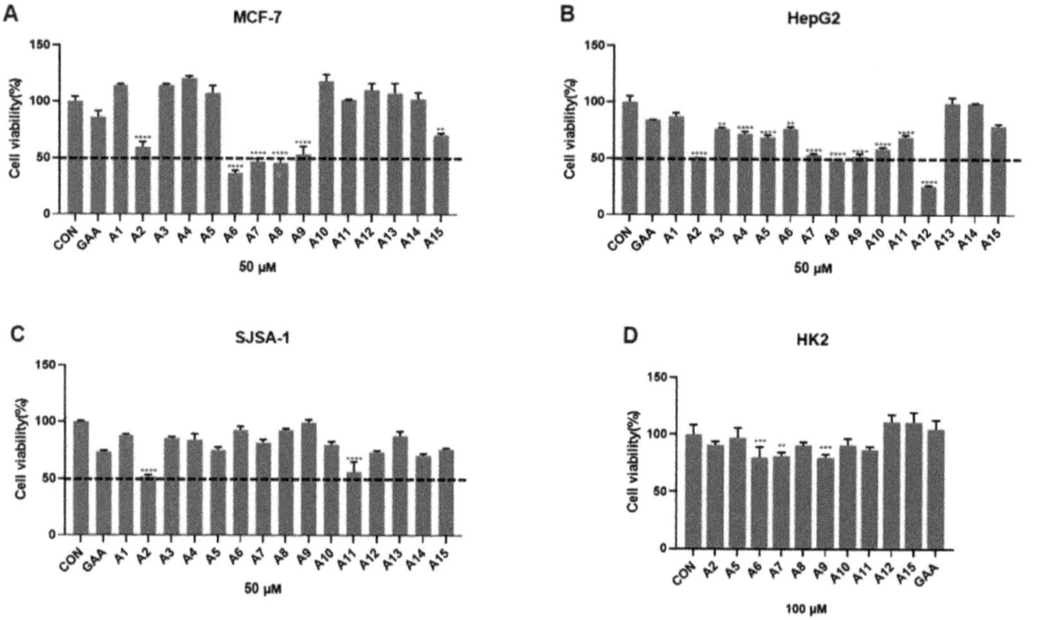

Figure 3. The anti-proliferation effect of GAA derivatives on different cell lines. (**A**) The viability of MCF-7 cells treated for 48 h with 50 μM concentration of synthesized compounds; (**B**) The viability of HepG2 cells treated for 48 h with 50 μM concentration of synthesized compounds; (**C**) The viability of SJSA-1 cells treated for 72 h with 50 μM concentration of synthesized compounds; (**D**) The viability of HK2 cells treated for 48 h with 100 μM concentration of tested compounds. All data are presented as mean ± SD of three independent experiments. CON refers to cells treated with 0.1% DMSO; ** $p < 0.01$, *** $p < 0.01$, **** $p < 0.0001$ vs GAA by ANOVA.

2.2.2. The Anti-Proliferation Activity on SJSA-1, HepG2 and HK2

To investigate the selectivity of these derivatives towards different tumor cell lines, we also selected HepG2 and osteosarcoma cell line SJSA-1 cells to evaluate the anti-proliferation activity of the derivatives. The results were shown in Table 2 and Figure 3. The results showed that the inhibitory effect of this series of derivatives on HepG2 was overall better than that on MCF-7 on the whole. Except for compounds **A2** and **A11**, the effects of other compounds on SJSA-1 were not strong. In HepG2 cell line, compounds **A2**, **A7**, **A8** and **A9** still have potent anti-proliferation activity, whereas **A6** and **A15**, which were better in MCF-7, have weaker anti-proliferation effect on HepG2. However, **A12** had strong selectivity on HepG2, and the inhibition rate of this cell below 50 μM can reach 74.37%. In SJSA-1 cell line, compound **A2** still showed potent inhibition, whereas **A11** showed some selectivity for SJSA-1, and it was found that GAA had better anti-tumor activity for SJSA-1 than for HepG2 and MCF-7.

Table 2. The viability of HepG2, SJSA-1, HK-2 cells at different concentration. Mean values ± SD from three independent experiments (n = 3) performed in duplicate are presented.

Compound	Viability of Cells (% of Control)					
	HepG2		SJSA-1		HK-2	
	25 μM	50 μM	50 μM	100 μM	50 μM	100 μM
A1	90.5 ± 1.1	87.2 ± 3.1	87.9 ± 0.9	74.0 ± 3.8	-	-
A2	78.8 ± 1.7	48.4 ± 2.1	51.4 ± 1.7	28.5 ± 3.4	97.1 ± 4.1	90.9 ± 3.0
A3	90.1 ± 1.1	76.3 ± 1.1	85.3 ± 1.1	83.3 ± 3.0	-	-
A4	87.7 ± 1.9	72.0 ± 1.9	83.9 ± 5.4	57.4 ± 4.2	-	-
A5	78.5 ± 3.0	69.1 ± 1.9	75.0 ± 2.6	64.5 ± 1.2	110.4 ± 13.1	97.0 ± 8.8
A6	84.0 ± 0.8	76.6 ± 1.6	92.6 ± 3.1	65.8 ± 3.2	89.6 ± 6.9	80.2 ± 9.0
A7	73.8 ± 1.2	53.4 ± 1.0	81.6 ± 2.8	77.1 ± 4.6	86.2 ± 3.7	81.3 ± 3.3
A8	54.2 ± 2.0	48.1 ± 1.5	92.4 ± 1.4	80.4 ± 4.7	107.8 ± 14.3	91.0 ± 2.4
A9	64.2 ± 1.4	52.4 ± 2.1	99.2 ± 2.6	89.6 ± 3.8	79.0 ± 2.9	80.1 ± 2.8
A10	57.9 ± 0.5	58.8 ± 1.4	80.0 ± 2.7	75.6 ± 4.8	110.9 ± 9.6	91.1 ± 5.5
A11	78.5 ± 3.0	69.1 ± 2.0	56.1 ± 8.7	41.9 ± 3.1	89.8 ± 9.6	86.9 ± 2.7
A12	29.1 ± 0.4	25.6 ± 0.8	73.5 ± 1.2	65.5 ± 1.5	124.1 ± 2.6	111.2 ± 6.5
A13	104.4 ± 2.0	99.3 ± 4.9	87.6 ± 3.8	77.6 ± 3.5	-	-
A14	95.6 ± 1.9	99.3 ± 0.2	70.4 ± 1.6	47.3 ± 2.1	-	-
A15	77.1 ± 2.1	99.3 ± 0.2	75.9 ± 0.9	83.5 ± 0.8	123.3 ± 5.5	111.0 ± 8.7
GAA	83.0 ± 1.5	84.0 ± 0.1	73.5 ± 1.1	74.0 ± 4.0	117.8 ± 6.3	104.9 ± 7.6

2.3. **A2** *Induces Apoptosis in SJSA-1 Cells*

We next examined the effect of **A2** (24 h incubation, at concentrations of 12.5, 25, 50 μM) on the SJSA-1 which **A2** showed the highest anti-proliferation potency among all the cell lines. Cells were stained with Annexin V-FITC and propidium iodide. The results are shown in the Figure 4. The results showed that different concentrations of **A2** could induce different degrees of apoptosis in SJSA-1 cells. At low concentrations, the proportion of cells undergoing early apoptosis increased slightly from 11.6% (12.5 μM) to 12.3% (25 μM). However, the proportion of apoptosis cells increased significantly at 50 μM (18.7%), while the proportion of late apoptosis remained essentially unchanged with increasing of concentration. The results indicated that **A2** can induce cell apoptosis in a dose-dependent manner.

2.4. **A2** *Effects MDM2-P53 Signaling Pathway*

2.4.1. **A2** Effects the Expression of p53 Protein, MDM2 and Bcl-2/Bax

In the introduction section, we introduced that the concept that the MDM2-p53 pathway can induce cell apoptosis by up-regulating the expression of p53 protein to inhibit the proliferation of tumor cells, and by blocking the interaction of MDM2 and p53, the activation of p53 results in transcription of *MDM2* mRNA, leading to robust MDM2 protein accumulation [29–31]. In order to verify the effect of **A2** on this pathway, we examined the effect of **A2** on the protein level of MDM2, p53 protein and Bcl-2/Bax related to apoptosis, as shown in Figure 5. The results showed that after treatment of MCF-7 cells with **A2** for 24 h, both MDM2 and p53 protein showed an increasing trend at 50 μM. The level of Bcl-2/Bax decreased which was consistent with the apoptosis of MCF-7 cells induced by **A2**. We also investigated the effect of **A2** on the SJSA-1 cell line which overexpresses MDM2. Compared with MCF-7, the expression of MDM2 and p53 protein in this cell line increased in a dose-dependent manner which may be the reason for the best anti-proliferation effect on SJSA-1 among all three cell lines. This experiment demonstrated that **A2** can affect the MDM2-p53 pathway to induces apoptosis.

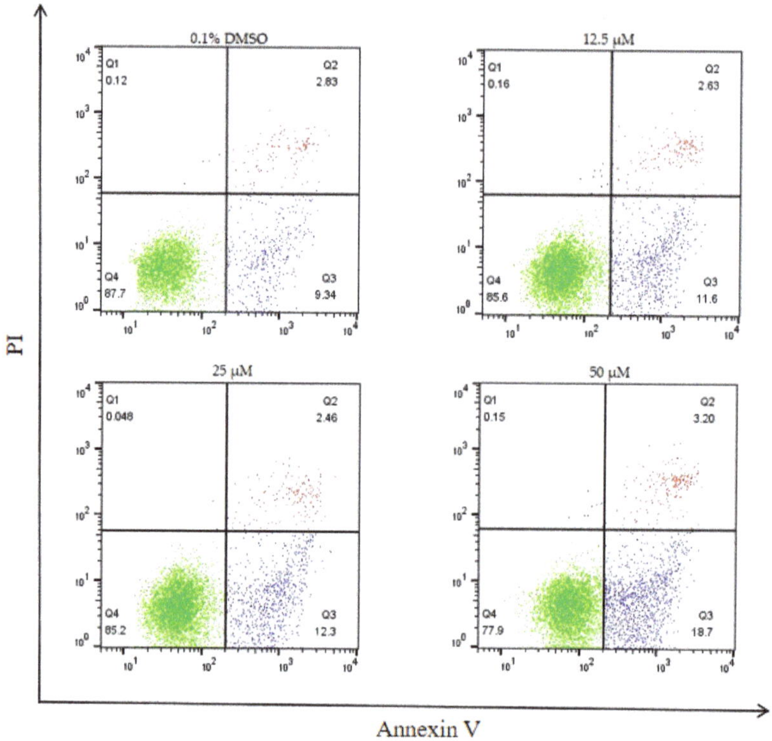

Figure 4. Flow cytometry analysis of SJSA-1 cells after 24 h incubation with **A2** at concentration of 12.5 μM, 25 μM, 50 μM and subsequent staining with Annexin V and propidium iodide. Q3 and the percentage means the proportion of the early phase apoptosis cells and Q2 and the percentage means the proportion of the late phase apoptosis cells.

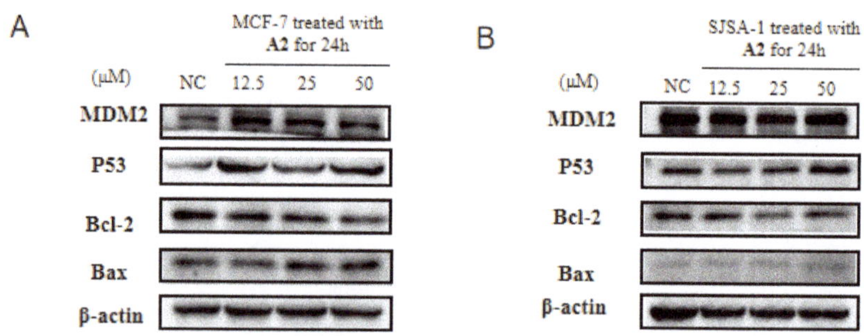

Figure 5. The effect of **A2** on the relative protein expression. The protein expression in MCF-7 (**A**) and SJSA-1 (**B**) treated with A2 for 24 h with a concentration of 12.5, 25 and 50 μM. NC refers to 0.1% DMSO.

2.4.2. GAA and **A2** Have In Vitro Binding Affinity with MDM2

In order to speculate whether **A2** effects the p53-MDM2 pathway by binding with MDM2 to inhibit the interaction between MDM2 and p53, we carried out in silico and in vitro binding experiments. First, we performed computer simulation to conduct target fishing of GAA and found that MDM2 interacts with GAA in silico (FitValue 0.79). We used the S-value to evaluate the binding degree of the compound and MDM2 in the molecular docking experiments. The higher the absolute value of this number, the stronger the binding force. Molecular docking (see Figure 6) revealed that the hydroxyl-H of GAA interacts with Met58 in MDM2 (S-value: −6.49). When **A2** was docked to MDM2, it was found that, the core of **A2** was in the opposite direction compared to GAA. In addition to the interaction with Met58 similar to GAA, the *n*-hexyl is well anchored in the hydrophobic pocket and the methylene has some hydrophobic interaction with His92 (S-value: −7.22). To verify whether GAA and **A2** have a certain binding ability with MDM2 in vitro, we used surface plasmon resonance (SPR) experiment to investigate the interaction between GAA, **A2** and MDM2 (see Figure 7). The K_D of GAA and MDM2 is 12.73 μM, indicating that they do have some affinity. At the same time, **A2** which has a stronger anti-proliferation activity has a stronger binding affinity with MDM2 than GAA, with a K_D of 1.68 μM. These results demonstrated that **A2** can affect the MDM2-p53 pathway to induces apoptosis probably by inhibiting the interaction of MDM2 and p53.

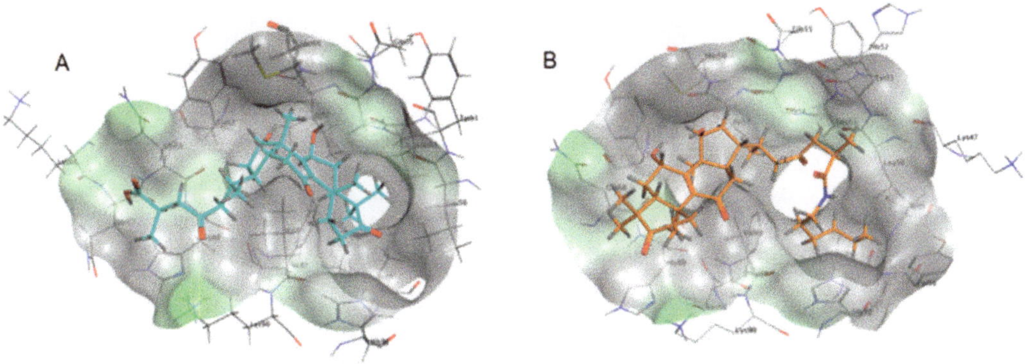

Figure 6. Molecular docking results of MDM2 (PBD ID: 4j3e) with GAA (**A**) and **A2** (**B**). The pose of GAA in MDM2 is that the carboxyl term exposed to the solvent, the ring core is located in the pocket, and the hydroxyl-H has hydrogen bond interaction with Met58. The pose of **A2** in MDM2 is opposite to that of GAA. The longer hydrophobic *n*-hexyl anchors into the pocket and has hydrophobic interaction with His92. The ring part faces out of the solvent and retains the interaction with Met58.

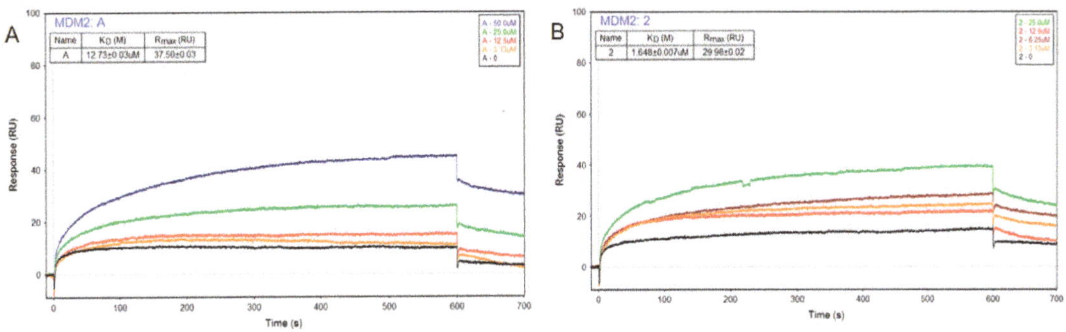

Figure 7. The binding curve of MDM2 with GAA (**A**) and **A2** (**B**).

We then investigated the effects of compounds with significantly higher activity than GAA in three cell lines on HK2, which is a normal cell line used to assess cytotoxicity. The results are shown in Table 2 and Figure 3. The results showed that at high concentration, benzylamine compounds **A6**, **A7** and **A9** with anti-tumor fragments had some toxicity to HK2 cells, whereas the other compounds with stronger activity had lower cytotoxicity to HK2 cells. To sum up, this series of GAA derivatives showed some selectivity in different cell lines, and have the potential to be developed as various tumor inhibitors. Given the strong anti-proliferation effect of derivative **A2** in various cell lines and its low effect on normal cells, **A2** was selected to investigate its anti-proliferation mechanism.

3. Materials and Methods

3.1. Chemistry

Unless otherwise stated, all reagents and solvents were obtained from commercial sources were used without further purification. GAA were purchased from Biopurify (Chengdu, China). Flash column chromatography was performed on Biotage Isolera Four (Sweden). ^{1}H NMR and ^{13}C-APT spectra were recorded on a Bruker AvanceIII 600MHz spectrometer (Germany). HRMS was performed on a Thermo Fisher LTQ Orbitrap XL (United States).

3.1.1. Synthesis of (*n*-butyl)-(7β,15α,25R)-7,15-Dihydroxy-3,11,23-Trioxolanost-8-en-26-oic Amide (**A1**)

To a solution of GAA (1eq., 50 mg, 0.01 mmol) in DCM (5 mL), *n*-butylamine (2 eq.), TBTU (1.5 eq.) and DIPEA (1.5 eq.) were added. The resulting reaction mixture was stirred at room temperature for 1 h and monitored by TLC. Upon completion, the reaction was then quenched with water and extracted with DCM. The organic layer was washed twice with water, dried over anhydrous sodium sulfate, filtered and concentrated under reduced pressure. The crude material was purified by column chromatography using dichloromethane and methanol (10:1, *v*/*v*) as mobile phase to obtain target molecule as white powder (yield 80.6%). mp: 120.7–121.5 °C. ^{1}H-NMR (600 MHz, $CDCl_3$) δ: 5.81 (t, *J* = 5.6 Hz, 1H, CON*H*), 4.80–4.77 (m, 1H, H-15), 4.63–4.60 (m, 1H, H-7), 4.16–4.15 (m, 1H, OH-15), 3.55–3.52 (m, 1H, OH-7), 3.22–3.19 (m, 2H, CONHCH_2), 2.90–2.79 (m, 2H, H-24a, H-1b), 2.78–2.68 (m, 2H, H-12a, H-25), 2.54–2.45 (m, 3H, H-22a, H-12b, H-24b), 2.44–2.37 (m, 2H, H-2), 2.26–2.19 (m, 1H, H-22b), 2.07–2.01 (m, 1H, H-6a), 2.01–1.95 (m,1H, H-20), 1.83–1.78 (m, 3H, H-17, H-16), 1.74–1.66 (m, 2H, H-5, H-6b), 1.52–1.42 (m, 3H, H-1a, CH_2), 1.37–1.30 (m, 2H, CH_2), 1.27 (s, 3H, CH_3), 1.25 (s, 3H, CH_3), 1.14 (d, *J* = 7.0 Hz, 3H, CH_3), 1.12 (s, 3H, CH_3), 1.10 (s, 3H, CH_3), 0.99 (s, 3H, CH_3), 0.91 (t, *J* = 7.3 Hz, 3H, CH_2CH_3), 0.87 (d, *J* = 6.4 Hz, 3H, CH_3). ^{13}C-APT (150 MHz, $CDCl_3$) δ: 217.3, 209.7, 199.6, 175.5, 159.3, 140.1, 72.3, 68.8, 53.9, 51.7, 49.8, 48.7, 48.1, 47.2, 46.7, 46.6, 39.3, 37.9, 36.3, 36.0, 35.5, 34.3, 32.7, 31.6, 28.9, 27.3, 20.7, 20.0, 19.6, 19.5, 19.4, 18.0, 17.3, 13.8. HRMS calculated for $C_{34}H_{53}NO_6Na$ $[M + Na]^+$ *m/z* 594.3765, found 594.3746.

3.1.2. Synthesis of (*n*-hexyl)-(7β,15α,25R)-7,15-Dihydroxy-3,11,23-Trioxolanost-8-en-26-oic Amide (**A2**)

The title compound was obtained from 1-hexanamine following similar synthesis procedure of **A1** (white powder, yield 65.6%). mp: 129.9–131.3 °C. ^{1}H-NMR (600 MHz, $CDCl_3$) δ: 5.78 (t, *J* = 5.6 Hz, 1H, CON*H*), 4.79–4.77 (m, 1H, H-15), 4.64–4.60 (m, 1H, H-7), 4.06–3.96 (m, 1H, OH), 3.50–3.35 (m, 1H, OH), 3.21–3.17 (m, 2H, CONHCH_2), 2.91–2.80 (m, 2H, H-24a, H-1b), 2.78–2.68 (m, 2H, H-12a, H-25), 2.54–2.45 (m, 3H, H-22a, H-12b, H-24b), 2.44–2.37 (m, 2H, H-2), 2.26–2.19 (m, 1H, H-22b), 2.09–2.02 (m, 1H, H-6a), 2.01–1.95 (m,1H, H-20), 1.84–1.76 (m, 3H, H-17, H-16), 1.74–1.64 (m, 2H, H-5, H-6b), 1.51–1.43 (m, 3H, H-1a, CH_2), 1.35–1.23 (m, 12H, $CH_2 \times 3$, $CH_3 \times 2$), 1.25 (s, 3H, CH_3), 1.14 (d, *J* = 7.1 Hz, 3H, CH_3), 1.12 (s, 3H, CH_3), 1.10 (s, 3H, CH_3), 0.99 (s, 3H, CH_3), 0.89−0.86 (m, 6H, $2\times CH_3$). ^{13}C-APT (150 MHz, $CDCl_3$) δ: 217.2, 209.7, 199.6, 175.5, 159.3, 140.2, 72.3, 68.8, 54.0, 51.7, 49.9, 48.7, 48.1, 47.2, 46.8, 46.6, 39.6, 38.0, 36.4, 36.0, 35.5, 34.3, 32.8, 31.5, 29.5, 29.0, 27.4, 26.5, 22.6, 20.7, 19.6, 19.5, 19.4, 18.0, 17.3, 14.1. HRMS calculated for $C_{36}H_{57}NO_6Na$ $[M + Na]^+$ *m/z* 622.4078, found 622.4070.

3.1.3. Synthesis of (4-Methylphenyl)-(7β,15α,25R)-7,15-Dihydroxy-3,11,23-Trioxolanost-8-en-26-oic Amide (**A3**)

The title compound was obtained from *p*-toluidine following similar synthesis procedure of **A1** (white powder, yield 30.3%). mp: 174.3–175.5 °C. ^{1}H-NMR (600 MHz, $CDCl_3$) δ: 7.76 (s, 1H, CON*H*), 7.29 (d, *J* = 8.2 Hz, 2H, Ph-2, 6-H), 7.02 (d, *J* = 8.2 Hz, 2H, Ph-2,

6-H), 4.66–4.64 (m, 1H, H-15), 4.51–4.50 (m, 1H, H-7), 4.07–4.03 (m, 1H, OH), 3.46-3.32 (m, 1H, OH), 2.92–2.80 (m, 2H, H-24a, H-1b), 2.79–2.72 (m, 1H, H-25), 2.67–2.62 (d, *J* = 16.1 Hz, H-12a), 2.44–2.37 (m, 4H, H-22a, H-12b, H-24b, H-2a), 2.36–2.30 (m, 1H, H-2b), 2.23 (s, 3H, Ph-CH_3), 2.21–2.14 (m, 1H, H-22b), 1.96–1.92 (m, 1H, H-6a), 1.92–1.87 (m,1H, H-20), 1.76-1.67 (m, 3H, H-17, H-16), 1.62–1.58 (m, 2H, H-5, H-6b), 1.42–1.34 (m, 1H, H-1a), 1.18 (s, 3H, CH_3), 1.16 (d, *J* = 6.7 Hz, 3H, CH_3), 1.15 (s, 3H, CH_3), 1.03 (s, 3H, CH_3), 1.01 (s, 3H, CH_3), 0.89 (s, 3H, CH_3), 0.78 (d, *J* = 6.7 Hz, 3H, CH_3). ^{13}C-APT (150 MHz, $CDCl_3$) δ: 216.4, 209.2, 198.7, 172.9, 158.3, 139.0, 134.2, 133.0, 128.4, 119.0, 71.2, 67.7, 52.9, 50.6, 48.8, 47.6 47.0, 46.3, 45.7, 45.6, 36.9, 35.7, 35.1, 34.4, 33.2, 31.8, 27.8, 26.3, 19.9, 19.6, 18.6, 18.4, 18.4, 16.9, 16.2. HRMS calculated for $C_{37}H_{51}NO_6Na$ $[M + Na]^+$ *m/z* 628.3609, found 628.3594.

3.1.4. Synthesis of (4-Chlorophenyl)-(7β,15α,25R)-7,15-Dhydroxy-3,11,23-Trioxolanost-8-en-26-oic Amide (**A4**)

The title compound was obtained from *p*-chloroaniline following similar synthesis procedure of **A1** (white powder, yield 26.3%). mp: 184.9–185.3 °C. ^{1}H-NMR (600 MHz, $CDCl_3$) δ: 7.78 (s, 1H, CON*H*), 7.37 (d, *J* = 8.6 Hz, 2H, Ph-2, 6-H), 7.17 (d, *J* = 8.7 Hz, 2H, Ph-2, 6-H), 4.69–4.66 (m, 1H, H-15), 4.55–4.52 (m, 1H, H-7), 2.93–2.80 (m, 2H, H-24a, H-1b), 2.79–2.71 (m, 1H, H-25), 2.69–2.63 (d, *J* = 16.1 Hz, H-12a), 2.47–2.38 (m, 4H, H-22a, H-12b, H-24b, H-2a), 2.36–2.30 (m, 1H, H-2b), 2.21–2.14 (m, 1H, H-22b), 1.99–1.94 (m, 1H, H-6a), 1.94–1.88 (m,1H, H-20), 1.74–1.69 (m, 3H, H-17, H-16), 1.64–1.57(m, 2H, H-5, H-6b), 1.42–1.35(m, 1H, H-1a), 1.19–1.55 (m, 9H, 3 × CH_3), 1.04 (s, 3H, CH_3), 1.01 (s, 3H, CH_3), 0.89 (s, 3H, CH_3), 0.78 (d, *J* = 6.2 Hz, 3H, CH_3). ^{13}C-APT (150 MHz, $CDCl_3$) δ: 216.3, 209.3, 198.5, 173.0, 158.1, 139.1, 135.5, 128.1, 127.9, 120.0, 71.3, 67.8, 52.8, 50.6, 48.7, 47.6, 47.0, 46.4, 45.7, 45.6, 36.9, 35.6, 35.2, 34.4, 33.2, 31.8, 27.9, 26.3, 19.6, 18.6, 18.4, 18.3, 16.8, 16.2. HRMS calculated for $C_{36}H_{48}ClNO_6Na$ $[M + Na]^+$ *m/z* 648.3062, found 648.3049.

3.1.5. Synthesis of (4-methylbenzyl)-(7β,15α,25R)-7,15-Dihydroxy-3,11,23-Trioxolanost-8-en-26-oic Amide (**A5**)

The title compound was obtained from 4-methylphenyl following similar synthesis procedure of **A1** (white powder, yield 91.9%). mp: 208.7–209.0 °C. ^{1}H-NMR (600 MHz, $CDCl_3$) δ: 7.06 (m, 4H, Ph-H), 6.09 (t, *J* = 7.2 Hz, 1H, CON*H*), 4.70–4.67 (m, 1H, H-15), 4.53–4.51 (m, 1H, H-7), 4.29–4.28 (m, 2H, CONHCH_2), 4.21–4.24 (m, 1H, OH), 3.64–3.45 (m, 1H, OH), 2.87–2.79 (m, 1H, H-1b), 2.78–2.73 (m, 1H, H-24a), 2.72-2.64 (m, 2H, H-25, H-12a), 2.45–2.30 (m, 5H, H-22a, H-12b, H-24b, H-2), 2.26 (s, 3H, Ph-CH_3), 2.20–2.12 (m, 1H, H-22b), 1.97–1.93 (m, 1H, H-6a), 1.92–1.87 (m,1H, H-20), 1.76–1.69 (m, 3H, H-17, H-16), 1.64–1.56 (m, 2H, H-5, H-6b), 1.43–1.34 (m, 1H, H-1a), 1.19 (s, 3H, CH_3), 1.18 (s, 3H, CH_3), 1.10 (d, *J* = 7.11 Hz, 3H, CH_3), 1.03 (s, 3H, CH_3), 1.01 (s, 3H, CH_3), 0.90 (s, 3H, CH_3), 0.78 (d, *J* = 6.4 Hz, 3H, CH_3). ^{13}C-APT (150 MHz, $CDCl_3$) δ: 216.4, 208.6, 198.7, 174.6, 158.5, 139.0, 136.1, 134.0, 128.3, 126.5, 71.2, 67.7, 52.9, 50.6, 48.8, 47.6, 47.0, 46.1, 45.7, 45.6, 42.3, 36.9, 35.2, 34.9, 34.5, 33.3, 31.7, 27.8, 26.3, 20.1, 19.6, 18.6, 18.5, 18.4, 17.0, 16.2. HRMS calculated for $C_{38}H_{53}NO_6Na$ $[M + Na]^+$ *m/z* 642.3765, found 642.3752.

3.1.6. Synthesis of (4-Fluorobenzyl)-(7β,15α,25R)-7,15-Dihydroxy-3,11,23-Trioxolanost-8-en-26-oic Amide (**A6**)

The title compound was obtained from *p*-fluorobenzylamine following similar synthesis procedure of **A1** (white powder, yield 90.2%). mp: 178.4–179.3 °C. ^{1}H-NMR (600 MHz, $CDCl_3$) δ: 7.16 (m, 2H, Ph-2, 6-H), 6.93 (m, 2H, Ph-3, 5-H), 6.18 (m, 1H, CON*H*), 4.70–4.67 (m, 1H, H-15), 4.53–4.52 (m, 1H, H-7), 4.34–4.26 (m, 2H, CONHCH_2), 4.12–3.98 (m, 1H, OH), 3.56–3.37 (m, 1H, OH), 2.88–2.80 (m, 1H, H-1b), 2.78–2.64 (m, 3H, H-24a, H-12a, H-25), 2.47–2.30 (m, 5H, H-22a, H-12b, H-24b, H-2), 2.18–2.11 (m, 1H, H-22b), 1.99–1.93 (m, 1H, H-6a), 1.92–1.87 (m,1H, H-20), 1.75–1.69 (m, 3H, H-17, H-16), 1.64–1.58 (m, 2H, H-5, H-6b), 1.43–1.35 (m, 1H, H-1a), 1.19 (s, 3H, CH_3), 1.17 (s, 3H, CH_3), 1.10 (d, *J* = 7.4 Hz, 3H, CH_3), 1.04 (s, 3H, CH_3), 1.01 (s, 3H, CH_3), 0.91 (s, 3H, CH_3), 0.78 (d, *J* = 6.4 Hz, 3H, CH_3). ^{13}C-APT (150 MHz, $CDCl_3$) δ: 216.3, 208.7, 198.6, 174.7, 161.9, 158.3, 139.1, 132.97, 132.95, 128.18, 128.13, 114.55, 114.41, 71.3, 67.8, 52.9, 50.6, 48.7, 47.6, 47.0, 46.1, 45.7, 45.6, 41.8, 36.9, 35.2,

34.9, 34.4, 33.2, 31.7, 28.7, 27.9, 26.3, 19.6, 18.6, 18.5, 18.4, 17.0, 16.2. HRMS calculated for $C_{37}H_{51}FNO_6$ $[M + H]^+$ *m/z* 624.3695, found 624.3686.

3.1.7. Synthesis of (4-Chlorobenzyl)-(7β,15α,25R)-7,15-Dihydroxy-3,11,23-Trioxolanost-8-en-26-oic Amide (**A7**)

The title compound was obtained from *p*-chlorobenzylamine following similar synthesis procedure of **A1** (white powder, yield 94.1%). mp: 188.6–189.4 °C. ^{1}H-NMR (600 MHz, $CDCl_3$) δ: 7.21 (d, *J* = 8.5 Hz, 2H, Ph-2, 6-H), 7.11 (d, *J* = 8.5 Hz, 2H, Ph-3, 5-H), 6.28 (m, 1H, CON*H*), 4.67 (m, 1H, H-15), 4.52 (m, 1H, H-7), 4.34–4.25 (m, 2H, CONHCH_2), 4.22–4.04 (m, 1H, OH), 3.74–3.38 (m, 1H, OH), 2.87–2.80 (m, 1H, H-1b), 2.78–2.64 (m, 3H, H-24a, H-12a, H-25), 2.46–2.28 (m, 5H, H-22a, H-12b, H-24b, H-2), 2.18-2.11 (m, 1H, H-22b), 2.00–1.93 (m, 1H, H-6a), 1.93–1.86 (m, 1H, H-20), 1.75–1.67 (m, 3H, H-17, H-16), 1.64–1.58 (m, 2H, H-5, H-6b), 1.44–1.33 (m, 1H, H-1a), 1.19 (s, 3H, CH_3), 1.16 (s, 3H, CH_3), 1.10 (d, *J* = 7.0 Hz, 3H, CH_3), 1.04 (s, 3H, CH_3), 1.01 (s, 3H, CH_3), 0.90 (s, 3H, CH_3), 0.78 (d, *J* = 7.0 Hz, 3H, CH_3). ^{13}C-APT (150 MHz, $CDCl_3$) δ: 216.4, 208.7, 198.7, 174.8, 158.3, 139.0, 135.8, 132.1, 127.8, 127.7, 71.2, 67.7, 52.9, 50.6, 48.7, 47.6, 47.0, 46.1, 45.7, 45.6, 41.8, 36.9, 35.2, 34.7, 34.4, 33.2, 31.7, 27.8, 26.3, 19.6, 18.6, 18.5, 18.4, 17.0, 16.2. HRMS calculated for $C_{37}H_{50}ClNO_6Na$ $[M + Na]^+$ *m/z* 662.3219, found 662.3206.

3.1.8. Synthesis of (3,5-Dichlorobenzyl)-(7β,15α,25R)-7,15-Dihydroxy-3,11,23-Trioxolanost-8-en-26-oic Amide (**A8**)

The title compound was obtained from 3,5-dichlorobenzylamine following similar synthesis procedure of **A1** (white powder, yield 96.5%). mp: 182.6–183.8 °C. ^{1}H-NMR (600 MHz, $CDCl_3$) δ: 7.17 (m, 1H, Ph-4-H), 7.08 (m, 2H, Ph-2, 6-H), 6.49 (m, 1H, CON*H*), 4.67 (m, 1H, H-15), 4.53 (m, 1H, H-7), 4.40–4.19 (m, 2H, CONHCH_2), 2.89–2.80 (m, 1H, H-1b), 2.79–2.64 (m, 3H, H-24a, H-12a, H-25), 2.46–2.36 (m, 4H, H-22a, H-12b, H-24b, H-2a), 2.34–2.27 (m, 1H, H-2b), 2.19–2.12 (m, 1H, H-22b), 1.97–1.89 (m, 2H, H-6a, H-20), 1.76–1.68 (m, 3H, H-17, H-16), 1.64–1.56 (m, 2H, H-5, H-6b), 1.42–1.33 (m, 1H, H-1a), 1.19 (s, 3H, CH_3), 1.16 (s, 3H, CH_3), 1.11 (d, *J* = 6.9 Hz, 3H, CH_3), 1.04 (s, 3H, CH_3), 1.01 (s, 3H, CH_3), 0.90 (s, 3H, CH_3), 0.77 (d, *J* = 6.2 Hz, 3H, CH_3). ^{13}C-APT (150 MHz, $CDCl_3$) δ: 208.8, 198.8, 175.1, 158.5, 140.9, 139.0, 134.0, 126.4, 124.7, 71.2, 67.7, 52.9, 50.6, 48.5, 47.6, 47.0, 45.6, 41.3, 37.6, 36.9, 35.1, 34.9, 34.5, 33.2, 31.7, 27.8, 26.3, 19.6, 18.6, 18.5, 18.4, 17.0, 16.2. HRMS calculated for $C_{37}H_{49}Cl_2NO_6Na$ $[M + Na]^+$ *m/z* 696.2829, found 696.2818.

3.1.9. Synthesis of (2,3-Dihydro-1*H*-inden-2-yl)-(7β,15α,25R)-7,15-Dihydroxy-3,11,23-Trioxolanost-8-en-26-oic Amide (**A9**)

The title compound was obtained from 2-aminoindane HCl following similar synthesis procedure of **A1** (white powder, yield 94.4%). mp: 199.7–200.3 °C. ^{1}H-NMR (600 MHz, $CDCl_3$) δ: 7.23–7.22 (m, 2H, Ph-H), 7.19–7.17 (m, 2H, Ph-H), 6.10 (d, *J* = 7.8 Hz, 1H, CON*H*), 4.78 (m, 1H, H-15), 4.68–4.64 (m, 1H, CONHC*H*), 4.63–4.60 (m, 1H, H-7), 4.20–4.12 (m, 1H, OH), 3.62–3.52 (m, 1H, OH), 3.31–3.27 (m, 2H, CONHCHCH_2), 2.88–2.82 (m, 2H, CONHCHCH_2), 2.89–2.72 (m, 5H, H-1b, H-24a, H-12a, H-25, CONHCHCH_2), 2.69–2.61 (m, 1H, CONHCHCH_2), 2.53–2.35 (m, 5H, H-22a, H-12b, H-24b, H-2), 2.23–2.19 (m, 1H, H-22b), 2.07–2.01 (m, 1H, H-6a), 2.00–1.95 (m, 1H, H-20), 1.84–1.76 (m, 3H, H-17, H-16), 1.75–1.65 (m, 2H, H-5, H-6b), 1.50–1.42 (m, 1H, H-1a), 1.27 (s, 3H, CH_3), 1.26 (s, 3H, CH_3), 1.12–1.11 (d, *J* = 7.3 Hz, 6H, 2 × CH_3), 1.09 (s, 3H, CH_3), 0.99 (s, 3H, CH_3), 0.89 (d, *J* = 6.3 Hz, 3H, CH_3). ^{13}C-APT (150 MHz, $CDCl_3$) δ: 217.1, 209.5, 199.5, 175.3, 159.0, 140.8, 140.7, 140.2, 126.8, 126.7, 124.8, 124.7, 72.4, 68.8, 53.9, 51.7, 50.5, 49.8, 48.7, 48.1, 47.2, 46.8, 46.6, 43.4, 40.1, 40.0, 37.9, 36.4, 35.9, 35.5, 34.3, 32.7, 29.0, 27.3, 20.7, 19.6, 19.5, 19.4, 17.9, 17.3. HRMS calculated for $C_{39}H_{53}NO_6Na$ $[M + Na]^+$ *m/z* 654.3765, found 654.3760.

3.1.10. Synthesis of (4-Methylphenethyl)-(7β,15α,25R)-7,15-Dihydroxy-3,11,23-Trioxolanost-8-en-26-oic Amide (**A10**)

The title compound was obtained from 2-(4-methylphenyl) ethanamine following similar synthesis procedure of **A1** (white powder, yield 85.6%). mp: 221.9–223.0 °C. ^{1}H-NMR

(600 MHz, $CDCl_3$) δ: 7.12 (d, *J* = 8.0 Hz, 2H, Ph-H), 7.09 (d, *J* = 8.0 Hz, 2H, Ph-H), 5.82–5.77 (m, 1H, CONH), 4.80–4.76 (m, 1H, H-15), 4.63–4.59 (m, 1H, H-7), 4.18–4.12 (m, 1H, OH-7), 3.55–3.39 (m, 3H, $CONHCH_2$, OH-15), 2.88–2.80 (m, 2H, H-1b, H-24a), 2.78–2.64 (m, 4H, H-12a, H-25, CH_2Ph), 2.52–2.46 (m, 3H, H-22a, H-12b, H-24b), 2.42–2.36 (m, 2H, H-2), 2.33 (s, 1H, Ph-CH_3), 2.25–2.18 (m, 1H, H-22b), 2.07–2.00 (m, 1H, H-6a), 2.00–1.95 (m, 1H, H-20), 1.84–1.77 (m, 3H, H-17, H-16), 1.72–1.66 (m, 2H, H-5, H-6b), 1.50–1.42 (m, 1H, H-1a), 1.27 (s, 3H, CH_3), 1.25 (s, 3H, CH_3), 1.11–1.09 (m, 9H, 3 ×CH_3), 0.99 (s, 3H, CH_3), 0.87 (d, *J* = 6.5 Hz, 2H). ^{13}C-APT (150 MHz, $CDCl_3$) δ: 217.2, 209.5, 199.6, 175.6, 159.3, 140.2, 136.1, 135.6, 129.3, 128.7, 72.4, 68.8, 53.9, 51.7, 49.9, 48.7, 48.1, 47.0, 46.8, 46.6, 40.8, 38.0, 36.4, 36.0, 35.5, 35.2, 34.3, 32.8, 29.0, 27.4, 21.1, 20.7, 19.6, 19.5, 19.4, 18.0, 17.3. HRMS calculated for $C_{39}H_{55}NO_6Na$ $[M + Na]^+$ *m/z* 656.3922, found 656.3910.

3.1.11. Synthesis of (4-Fluorophenethyl)-(7β,15α,25R)-7,15-Dihydroxy-3,11,23 -Trioxolanost-8-en-26-oic Amide (**A11**)

The title compound was obtained from 4-fluorophenethylamine hydrochloride following similar synthesis procedure of **A1** (white powder, yield 97.2%). mp: 188.1–189.0 °C. ^{1}H-NMR (600 MHz, $CDCl_3$) δ: 7.09 (m, 2H, Ph-H), 6.92 (m, 2H, Ph-H), 5.83 (m, 1H, CONH), 4.50 (m, 1H, H-15), 4.54 (m, 1H, H-7), 3.42–3.34 (m, 2H, $CONHCH_2$), 2.81–2.73 (m, 2H, H-1b, H-24a), 2.72–2.56 (m, 4H, H-12a, H-25, CH_2Ph), 2.45–2.37 (m, 3H, H-22a, H-12b, H-24b), 2.36–2.28 (m, 2H, H-2), 2.19–2.10(m, 1H, H-22b), 1.99–1.94 (m, 1H, H-6a), 1.93–1.87 (m, 1H, H-20), 1.77–1.70 (m, 3H, H-17, H-16), 1.67–1.59 (m, 2H, H-5, H-6b), 1.44–1.35(m, 1H, H-1a), 1.20 (s, 3H, CH_3), 1.18 (s, 3H, CH_3), 1.04–1.02 (m, 9H, 3 × CH_3), 0.91 (s, 3H, CH_3), 0.79 (d, *J* = 6.4 Hz, 2H). ^{13}C-APT (150 MHz, $CDCl_3$) δ: 216.4, 208.6, 198.6, 174.7, 161.4, 159.8, 158.4, 139.0, 133.39, 133.37, 129.2, 129.2, 114.4, 114.3, 71.2, 67.7, 52.9, 50.6, 48.8, 47.6, 47.0, 46.0, 45.7, 39.7, 36.9, 35.2, 34.9, 34.4, 33.8, 33.2, 31.7, 27.9, 26.3, 19.6, 18.6, 18.5, 18.4, 17.0, 16.2. HRMS calculated for $C_{38}H_{52}FNO_6Na$ $[M + Na]^+$ *m/z* 660.3671, found 660.3650.

3.1.12. Synthesis of (4-Chlorophenethyl)-(7β,15α,25R)-7,15-Dihydroxy-3,11,23-Trioxolanost-8-en-26-oic Amide (**A12**)

The title compound was obtained from 4-chlorobenzeneethanamine following similar synthesis procedure of **A1** (white powder, yield 95.5%). mp: 195.4–196.1 °C. ^{1}H-NMR (600 MHz, $CDCl_3$) δ: 7.20 (m, 2H, Ph-H), 7.07 (m, 2H, Ph-H), 5.78 (m, 1H, CONH), 4.70 (m, 1H, H-15), 4.55 (m, 1H, H-7), 3.99 (s, 1H, OH), 3.67 (m, 3H, OH, $CONHCH_2$), 2.81–2.73 (m, 2H, H-1b, H-24a), 2.72–2.55 (m, 4H, H-12a, H-25, CH_2Ph), 2.45–2.37 (m, 3H, H-22a, H-12b, H-24b), 2.35–2.28 (m, 2H, H-2), 2.18–2.09(m, 1H, H-22b), 2.00–1.94 (m, 1H, H-6a), 1.93–1.87 (m, 1H, H-20), 1.78–1.70 (m, 3H, H-17, H-16), 1.66–1.59 (m, 2H, H-5, H-6b), 1.45–1.30 (m, 1H, H-1a), 1.19 (s, 3H, CH_3), 1.18 (s, 3H, CH_3), 1.04–1.02 (m, 9H, 3 × CH_3), 0.91 (s, 3H, CH_3), 0.79 (d, *J* = 6.5 Hz, 2H). ^{13}C-APT (150 MHz, $CDCl_3$) δ: 216.4, 208.6, 198.6, 174.7, 158.3, 139.1, 136.2, 131.3, 129.1, 127.7, 71.3, 67.7, 52.9, 50.6, 48.8, 47.6, 47.0, 46.0, 45.7, 45.6, 39.5, 36.9, 35.2, 34.9, 34.5, 34.0, 33.3, 31.7, 27.9, 26.3, 19.6, 18.6, 18.5, 18.4, 17.0, 16.2. HRMS calculated for $C_{38}H_{53}ClNO_6$ $[M + H]^+$ *m/z* 654.3556, found 654.3547.

3.1.13. Synthesis of (4-Methylpiperazin-1-yl)-(7β,15α,25R)-7,15-Dihydroxy-3,11,23 -Trioxolanost-8-en-26-oic Amide (**A13**)

The title compound was obtained from methylpiperazine following similar synthesis procedure of **A1** (white powder, yield 92.7%). mp: 173.2–174.9 °C. ^{1}H-NMR (600 MHz, $CDCl_3$) δ: 4.75–4.72 (m, 1H, H-15), 4.58–4.56 (m, 1H, H-7), 3.74–3.68 (m, 2H, $CONCH_2$), 3.61−3.54 (m, 2H, $CONCH_2$), 3.22–3.12 (1H, H-25), 3.01–2.95 (m, 1H, H-1b), 2.83–2.77 (m, 1H, H-24a), 2.72 (d, *J* = 16.5 Hz, 1H, H-12a), 2.64–2.51 (m, 2H, CH_2N), 2.51–2.33 (m, 10H, H-24b, H-12b, H-22a, H-2, NCH_3, CH_2N), 2.25–2.16 (m, 1H, H-22b), 2.02–1.97 (m, 1H, H-6a), 1.97–1.91 (m, 1H, H-20), 1.82–1.73 (m, 3H, H-17, H-16), 1.71–1.62 (m, 2H, H-5, H-6b), 1.48–1.39 (m, 1H, H-1a), 1.24 (s, 3H, CH_3), 1.22 (s, 3H, CH_3), 1.09–1.06 (m, 9H, 3 × CH_3), 0.98 (s, 3H, CH_3), 0.84 (d, *J* = 6.5 Hz, 3H, CH_3). ^{13}C-APT (150 MHz, $CDCl_3$) δ: 216.4, 208.8, 198.7, 173.1, 158.7, 138.9, 71.1, 67.6, 53.9, 53.5, 52.9, 50.7, 48.7, 47.6, 47.1, 46.2, 45.7, 45.6, 44.6,

44.1, 40.3, 36.9, 35.1, 34.5, 33.3, 31.8, 29.8, 27.8, 26.3, 19.7, 18.6, 18.5, 18.4, 16.5, 16.3. HRMS calculated for $C_{35}H_{55}N_2O_6$ $[M + H]^+$ *m/z* 599.4055, found 599.4036.

3.1.14. Synthesis of (4-Ethylpiperazin-1-yl)-(7β,15α,25R)-7,15-Dihydroxy-3,11,23 -Trioxolanost-8-en-26-oic Amide (**A14**)

The title compound was obtained from 1-ethylpiperazine following similar synthesis procedure of A1 (white powder, yield 98.1%). mp: 198.5–199.4 °C. ^{1}H-NMR (600 MHz, $CDCl_3$) δ: 4.72−4.70 (m, 1H, H-15), 4.56−4.54 (m, 1H, H-7), 3.82–3.72 (m, 2H, $CONCH_2$), 3.66−3.50 (m, 2H, $CONCH_2$), 3.18–3.09 (1H, H-25), 2.98–2.90 (m, 1H, H-1b), 2.81–2.72 (m, 1H, H-24a), 2.72–2.66 (m, 3H, H-12a, CH_2), 2.66–2.31 (m, 9H, H-24b, H-12b, H-22a, H-2, CH_2N), 2.22–2.12 (m, 1H, H-22b), 1.99–1.94 (m, 1H, H-6a), 1.94–1.86 (m, 1H, H-20), 1.77–1.70 (m, 3H, H-17, H-16), 1.49–1.60 (m, 2H, H-5, H-6b), 1.45–1.35 (m, 1H, H-1a), 1.20 (s, 3H, CH_3), 1.18 (s, 3H, CH_3), 1.09 (t, *J* = 7.0 Hz, 3H, NCH_2CH_3), 1.05–1.02 (m, 9H, 3 × CH_3), 0.92 (s, 3H, CH_3), 0.80 (d, *J* = 6.2 Hz, 3H, CH_3). ^{13}C-APT (150 MHz, $CDCl_3$) δ: 216.4, 208.9, 198.6, 173.2, 158.4, 139.0, 71.2, 67.7, 52.9, 51.5, 51.2, 51.0, 50.7, 48.8, 47.6, 47.1, 46.2, 45.7, 45.6, 43.7, 40.0, 36.9, 35.2, 34.5, 33.3, 31.8, 29.7, 27.9, 26.4, 19.7, 18.6, 18.5, 18.4, 16.5, 16.3. HRMS calculated for $C_{36}H_{57}N_2O_6$ $[M + H]^+$ *m/z* 613.4211, found 613.4205.

3.1.15. Synthesis of (4-Phenylpiperazin-1-yl)-(7β,15α,25R)-7,15-Dihydroxy-3,11,23 -Trioxolanost-8-en-26-oic Amide (**A15**)

The title compound was obtained from 1-phenylpiperazine following similar synthesis procedure of **A1** (white powder, yield 89.9%). mp: 229.4–229.9 °C. ^{1}H-NMR (600 MHz, $CDCl_3$) δ: 7.23−7.21 (m, 2H, Ph-H), 6.88−6.83 (m, 3H, Ph-H), 4.72−4.69 (m, 1H, H-15), 4.55−4.53 (m, 1H, H-7), 3.75−3.71 (m, 2H, $CONCH_2$), 3.65−3.62 (m, 2H, $CONCH_2$), 3.26–3.16 (m, 2H, CH_2), 3.17–3.10(1H, H-25), 3.10–3.05 (m, 2H, CH_2), 3.00–2.92 (m, 1H, H-1b), 2.80–2.74 (m, 1H, H-24a), 2.78 (d, *J* = 16.5 Hz, 1H, H-12a), 2.45–2.39 (m, 3H, H-22a, H-12b, H-24b), 2.38–2.31 (m, 2H, H-2), 2.22–2.14 (m, 1H, H-22b), 2.00–1.94 (m, 1H, H-6a), 1.94–1.88 (m, 1H, H-20), 1.78–1.71 (m, 3H, H-17, H-16), 1.66–1.58 (m, 2H, H-5, H-6b), 1.43–1.35 (m, 1H, H-1a), 1.20 (s, 3H, CH_3), 1.18 (s, 3H, CH_3), 1.06 (d, *J* = 7.1 Hz, 3H, CH_3), 1.04 (s, 3H, CH_3), 1.02 (s, 3H, CH_3), 0.92 (s, 3H, CH_3), 0.80 (d, *J* = 6.4 Hz, 3H, CH_3). ^{13}C-APT (150 MHz, $CDCl_3$) δ: 216.3, 208.6, 198.6, 173.1, 158.2, 149.8, 139.1, 128.2, 119.5, 115.5, 71.3, 67.7, 52.9, 50.6, 48.8, 48.7, 48.4, 47.6, 47.1, 46.2, 45.7, 45.6, 44.6, 40.9, 37.6, 36.9, 35.3, 34.4, 33.2, 31.7, 29.9, 27.9, 26.3, 19.6, 18.6, 18.5, 18.4, 16.6, 16.2. HRMS calculated for $C_{40}H_{57}N_2O$ $[M + H]^+$ *m/z* 661.4211, found 661.4203.

3.2. Cell Culture

The cell lines present in this study were obtained from Procell Life Science & Technology Co. Ltd. MCF-7, HepG2 and SJSA-1 cells were cultured in DMEM medium (DMEM, Gibco) supplemented with 10% fetal bovine serum (FBS, Gibco), 1% penicillin-streptomycin (Hyclone) at 37°C in a humid environment with 5% CO_2. HK-2 cells were cultured in DMEM/F12 (1:1) medium and placed in incubators in the same environment.

3.3. Cell Viability Assay

Cell viability was determined by MTT assay. MCF-7, HepG2, SJSA-1 and HK-2 cells (6×10^3 cells /well) were seeded in 96-well plates with serum-free medium for 24 h. Then MCF-7, HepG2, SJSA-1 cells were treated with 0.1% DMSO, 25, 50, 100 μM of GAA derivatives for 48 h (MCF-7, HepG2, HK2) or 72 h (SJSA-1). After 48 or 72 h, 10 μL MTT (5 mg/mL, Beyotime) was added and incubated at 37 °C for 4 h. Then 100 μL of lysate was added. After complete dissolution of the crystal, the absorbance was measured at 540 nm and expressed as the average percentage of absorbance between treated and control cells. The value for control cells was set at 100%. Cell survival rate was calculated as the ratio of the absorbance of the cells and negative control after minus the blank absorbance respectively.

3.4. Target Fishing and Molecular Docking by In Silico Approaches

The binding targets of GAA were predicted using Discovery Studio 2016 v16.1 (BIOVIA Software Inc., San Diego, CA, USA), a software suite for the computational analysis of data relevant to Life Sciences research. To predict the probable targets of GAA, we used Ligand Profiler protocol which maps a set of pharmacophores, including Pharma DB by default. The ligand GAA was prepared by the Specifying Ligands parameter protocol. After setting parameters, the job was run, and the results were gained for three days. To explore the potential binding mode of GAA and **A2** with MDM2 protein (PDB code: 4j3e), a molecular modeling research was performed with docking program named Induced-Fit, a refinement method in another software MOE. To eliminate any bond length and bond angle biases, the ligands (GAA and **A2**) were subjected to the "energy minimize" prior to docking. The binding affinities (*S*-values) in MOE were used to evaluate the interactions between MDM2 and ligands. The scores (binding affinities) were obtained based on the virtual calculation of various interactions of ligands with the targeted receptor.

3.5. Surface Plasmon Resonance (SPR) Assay

GAA derivatives bound to MDM2 protein were assayed with a molecular interaction analyzer (PALL FORTEBIO, USA). MDM2 protein (5 mg/mL, Protintech) was immobilized on a PCH sensor chip (Octet) and preactivated with EDC/NHS mixture for 420 s at a flow rate of 10 μL/min. **A2** was diluted to 100, 50, 25, 12.5, 6.25, 3.13, 0 μM with PBST buffer containing 1% DMSO. The binding time was 600 s, and the flow rate was 20 μL/min. The dissociation time was 180 s, and the affinity constant K_D value was obtained by computer fitting and steady-state analysis.

3.6. Flow Cytometric Analysis of the Apoptosis Rate with Annexin V-FITC/PI Staining

To determine the apoptosis rate, an Annexin V-FITC/PI double staining apoptosis assay kit (Beyotime) was used to detect apoptotic cells by flow cytometry (BD FACSALOBUR), according to the manufacturer's instructions. Briefly, SJSA-1 cells were treated with 0.1% DMSO, 12.5, 25, and 50 μM of **A2** for 24 h. After harvesting, the cells were incubated with 5 μL Annexin V-FITC for 15 min and 10 μL PI for 5 min at 4 °C under dark conditions. Flow cytometry was then performed to analyze the apoptosis rate. Data were analyzed by using BD FACSDiva 8.0.1.

3.7. Western Blot Analysis of Protein Expression

For Western blot analysis, MCF-7 and SJSA-1 cells were treated with different concentrations of **A2** for 24 h. The total cell protein was extracted, and proteins were isolated using 10% SDS-PAGE gel system. The proteins on the gel were transferred to PVDF membrane, blocked in 5% BSA at room temperature for 2 h, incubated in primary antibody dilution at 4 °C overnight, and washed with TBST for 3 times, 10 min each. Then, they were transferred to dilute release solution of secondary antibody and incubated at room temperature for 2 h. ECL chemiluminescence development solution (Beyotime, BeyoECL star) was added uniformly and detected on gel imaging system (Clinx ChemiScope, China). Antibodies for blotting were MDM2 (abcam, ab16895), P53 (Proteintech, 10442-1-AP), Bcl-2 (CST, 15071S), Bax (CST, 2772T) and β-actin (abcam, ab8226).

3.8. Statistical Analysis

The results are expressed as the means ± standard deviation. A one-way AVONA and t-test were used for comparison of differences between groups, and GraphPad Prism 8.0 software was used for graph and statistical analysis. Statistical significance was set at $p < 0.05$.

4. Conclusions

Natural products are rich in beneficial scaffolds that have been used in anti-tumor, anti-inflammatory, neuroprotective and other aspects. However, these natural products have the problem of unclear targets and weak activity. Therefore, if we can determine the relevant mechanism of the action of natural products and identify specific pathways and targets, we can improve their activity based on adding appropriate interaction with binding amino acid residues in the active pocket of the target. In this study, we modified *Ganoderma lucidum* triterpenoid compound GAA and evaluated anti-proliferative effects of these derivatives in different tumor cell lines. Finally, compound **A2** was selected for further investigation of its mechanism. The results showed that **A2** could induce apoptosis by interfering with the MDM2-p53 pathway. Target fishing and SPR experiments suggested that **A2** might play a role by binding to MDM2 and blocking its inhibition of p53. Although these compounds may have weaker anti-tumor activity than other small molecule anti-tumor drugs, this study may provide insights into finding the target of GAA and developing new natural product anti-cancer compounds. If we can confirm the specific targets of GAA in different diseases, we can carry out target-based rational design of GAA to greatly improve its efficacy and provide an excellent scaffold for the development of new drugs.

5. Patents

In order to protect the structure and activity of compounds in a timely manner, the patent Preparation method of Ganoderic A amide derivatives useful as anti-tumor drugs, China CN112574272 A 2021-03-30, refers to the synthesis of the derivatives and simple in vitro cell anti-proliferation screening. In subsequent studies, the activity of the derivatives in other cell lines was found and the mechanism was investigated. The relevant experimental results are presented in this article.

Supplementary Materials: The following supporting information can be downloaded at: https://www.mdpi.com/article/10.3390/molecules28052374/s1, ^{1}H-NMR, ^{13}C-APT and HRMS spectroscopy of all new compounds.

Author Contributions: Conceptualization, Y.T.; methodology, Y.J., Y.L. (Yan Li) and Y.L. (Yun Luo); software, Y.J. and Y.T.; validation, Y.J., Y.T. and H.S.; formal analysis, Y.J.; investigation, Y.J.; resources, Y.L. (Yan Li), H.S., and Y.L. (Yun Luo); data curation, Y.J.; writing—original draft preparation, Y.J.; writing—review and editing, Y.T.; visualization, Y.J.; supervision, Y.T.; project administration, Y.T.; funding acquisition, Y.T. All authors have read and agreed to the published version of the manuscript.

Funding: We are grateful for the financial support of the National Natural Sciences Foundation of China (Grant No. 82173710), the CAMS Innovation Fund for Medical Science (CIFMS) (Grant No. 2021-I2M-1-028, 2022-I2M-2-002), the National Key Research and Development Plan of China (Grant No. 2019YFC1710504), and the Beijing Natural Science Foundation (Grant No. 7192129).

Institutional Review Board Statement: Not applicable.

Informed Consent Statement: Not applicable.

Data Availability Statement: Not applicable.

Acknowledgments: We are grateful for the National Natural Sciences Foundation of China, National Key Research and Development Plan of China, the CAMS Innovation Fund for Medical Science (CIFMS), and the Beijing Natural Science Foundation.

Conflicts of Interest: The authors declare no conflict of interest.

References

1. Zhao, R.; He, Y. Network pharmacology analysis of the anti-cancer pharmacological mechanisms of *Ganoderma lucidum* extract with experimental support using Hepa1-6-bearing C57 BL/6 mice. *J. Ethnopharmacol.* **2018**, *210*, 287–295. [CrossRef] [PubMed]
2. Zhang, Y.; Wang, X.; Yang, X.; Yang, X.; Xue, J.; Yang, Y. Ganoderic Acid A to Alleviate Neuroinflammation of Alzheimer's Disease in Mice by Regulating the Imbalance of the Th17/Tregs Axis. *J. Agric. Food Chem.* **2021**, *69*, 14204–14214. [CrossRef] [PubMed]
3. Zheng, S.; Ma, J.; Zhao, X.; Yu, X.; Ma, Y. Ganoderic Acid A Attenuates IL-1β-Induced Inflammation in Human Nucleus Pulposus Cells Through Inhibiting the NF-κB Pathway. *Inflammation* **2021**, *45*, 851–862. [CrossRef]
4. Jia, Y.; Zhang, D.; Li, H.; Luo, S.; Xiao, Y.; Zhou, F.; Wang, C.; Feng, L.; Wang, G.; Wu, P.; et al. Activation of FXR by ganoderic acid A promotes remyelination in multiple sclerosis via anti-inflammation and regeneration mechanism. *Biochem. Pharmacol.* **2021**, *185*, 114422. [CrossRef]

5. Zhang, L.; Zhang, L.; Sui, R. Ganoderic Acid A-Mediated Modulation of Microglial Polarization is Involved in Depressive-Like Behaviors and Neuroinflammation in a Rat Model of Post-Stroke Depression. *Neuropsych. Dis. Treat.* **2021**, *17*, 2671–2681. [CrossRef]
6. Bao, H.; Li, H.; Jia, Y.; Xiao, Y.; Luo, S.; Zhang, D.; Han, L.; Dai, L.; Xiao, C.; Feng, L.; et al. Ganoderic acid A exerted antidepressant-like action through FXR modulated NLRP3 inflammasome and synaptic activity. *Biochem. Pharmacol.* **2021**, *188*, 114561. [CrossRef]
7. Qi, L.; Liu, S.; Liu, Y.; Li, P.; Xu, X. Ganoderic Acid A Promotes Amyloid-β Clearance (In Vitro) and Ameliorates Cognitive Deficiency in Alzheimer's Disease (Mouse Model) through Autophagy Induced by Activating Axl. *Int. J. Mol. Sci.* **2021**, *22*, 5559. [CrossRef]
8. Yu, Z.; Jia, W.; Liu, C.; Wang, H.; Yang, H.; He, G.; Chen, R.; Du, G. Ganoderic acid A protects neural cells against NO stress injury in vitro via stimulating β adrenergic receptors. *Acta Pharmacol. Sin.* **2020**, *41*, 516–522. [CrossRef] [PubMed]
9. Ma, J.; Zhang, Y.; Tian, Z. Anti-oxidant, anti-inflammatory and anti-fibrosis effects of ganoderic acid A on carbon tetrachloride induced nephrotoxicity by regulating the Trx/TrxR and JAK/ROCK pathway. *Chem. Biol. Interact.* **2021**, *344*, 109529. [CrossRef]
10. Liu, F.; Shi, K.; Dong, J.; Jin, Z.; Wu, Y.; Cai, Y.; Lin, T.; Cai, Q.; Liu, L.; Zhang, Y. Ganoderic acid A attenuates high-fat-diet-induced liver injury in rats by regulating the lipid oxidation and liver inflammation. *Arch. Pharm. Res.* **2020**, *43*, 744–754. [CrossRef] [PubMed]
11. Li, X.; Li, Y.; Song, H. Ganoderic acid A against cyclophosphamide-induced hepatic toxicity in mice. *J. Biochem. Mol. Toxic.* **2019**, *33*, e22271. [CrossRef]
12. Guo, W.; Guo, J.; Liu, B.; Lu, J.; Chen, M.; Liu, B.; Bai, W.; Rao, P.; Ni, L.; Lv, X. Ganoderic acid A from *Ganoderma lucidum* ameliorates lipid metabolism and alters gut microbiota composition in hyperlipidemic mice fed a high-fat diet. *Food Funct.* **2020**, *11*, 6818–6833. [CrossRef] [PubMed]
13. Zhu, J.; Jin, J.; Ding, J.; Li, S.; Cen, P.; Wang, K.; Wang, H.; Xia, J. Ganoderic Acid A improves high fat diet-induced obesity, lipid accumulation and insulin sensitivity through regulating SREBP pathway. *Chem. Biol. Interact.* **2018**, *290*, 77–87. [CrossRef] [PubMed]
14. Zhang, Y.; Shi, K.; Lin, T.; Xia, F.; Cai, Y.; Ye, Y.; Liu, L.; Liu, F. Ganoderic acid A alleviates myocardial ischemia-reperfusion injury in rats by regulating JAK2/STAT3/NF-κB pathway. *Int. Immunopharmacol.* **2020**, *84*, 106543. [CrossRef] [PubMed]
15. Cheng, Y.; Xie, P. Ganoderic acid A holds promising cytotoxicity on human glioblastoma mediated by incurring apoptosis and autophagy and inactivating PI3K/AKT signaling pathway. *J. Biochem. Mol. Toxic.* **2019**, *33*, e22392. [CrossRef] [PubMed]
16. Liu, H.; Zheng, K.; Shan, X. Effects of Ganoderic Acid A on Proliferation, Apoptosis and Invasion of Human Glioma U251 Cells. *Chin. J. Comp. Med.* **2016**, *26*, 64–69.
17. Kanapathipillai, M. Treating p53 Mutant Aggregation-Associated Cancer. *Cancers* **2018**, *10*, 154. [CrossRef]
18. Zhao, Y.; Yu, H.; Hu, W. The regulation of MDM2 oncogene and its impact on human cancers. *Acta. Biochim. Biophys. Sin.* **2014**, *46*, 180–189. [CrossRef] [PubMed]
19. Li, B.; Cheng, Q.; Li, Z.; Chen, J. p53 inactivation by MDM2 and MDMX negative feedback loops in testicular germ cell tumors. *Cell Cycle* **2010**, *9*, 1411–1420. [CrossRef]
20. Nakamura, M.; Obata, T.; Daikoku, T.; Fujiwara, H. The Association and Significance of p53 in Gynecologic Cancers: The Potential of Targeted Therapy. *Int. J. Mol. Sci.* **2019**, *20*, 5482. [CrossRef] [PubMed]
21. Munisamy, M.; Mukherjee, N.; Thomas, L.; Pham, A.; Shakeri, A.; Zhao, Y.; Kolesar, J.; Rao, P.; Rangnekar, V.M.; Rao, M. Therapeutic opportunities in cancer therapy: Targeting the p53-MDM2/MDMX interactions. *Am. J. Cancer Res.* **2021**, *11*, 5762–5781. [PubMed]
22. Hassin, O.; Oren, M. Drugging p53 in cancer: One protein, many targets. *Nat. Rev. Drug Discov.* **2023**, *22*, 127–144. [CrossRef]
23. Xu, B.; Jia, W.; Wang, Z.; Zhang, H.; Wu, D.; Tang, C.; Yang, Y.; Liu, D.; Zhang, J.; Wang, W. Inhibitory effect and mechanism of ganoderic acid A on the growth of prostate cancer LNCaP cells. *Mycosystema* **2019**, *38*, 717–727. [CrossRef]
24. Tang, W.; Wu, Y.; Wang, Y.; Sun, P.; Ouyang, J. P53 protein participates in the process of ganoderic acid inhibiting the proliferation of cancer cells. *Sci. Technol. Food Ind.* **2015**, *36*, 193–196. [CrossRef]
25. Froufe, H.; Abreu, R.; Ferreira, I. Virtual screening of low molecular weight mushrooms compounds as potential Mdm2 inhibitors. *J. Enzym. Inhib. Med. Chem.* **2013**, *28*, 569–575. [CrossRef]
26. Staszczak, M. Fungal Secondary Metabolites as Inhibitors of the Ubiquitin-Proteasome System. *Int. J. Mol. Sci.* **2021**, *22*, 13309. [CrossRef]
27. Walczak, J.; Maksymilian, D.; Ziomkowska, M.; Śliwka-Kaszyńska, M.; Daśko, M.; Trzonkowski, P.; Cholewiński, G. Novel amides of mycophenolic acid and some heterocyclic derivatives as immunosuppressive agents. *J. Enzym. Inhib. Med. Chem.* **2022**, *37*, 2725–2741. [CrossRef] [PubMed]
28. Nassim, E.; Wang, M. Synthesis and biological evaluation of naloxone and naltrexone-derived hybrid opioids. *Med. Chem.* **2012**, *8*, 683–689. [CrossRef]
29. Li, Y.; Yang, J.; Aguilar, A.; McEachern, D.; Przybranowski, S.; Liu, L.; Yang, C.; Wang, M.; Han, X.; Wang, S. Discovery of MD-224 as a First-in-Class, Highly Potent, and Efficacious Proteolysis Targeting Chimera Murine Double Minute 2 Degrader Capable of Achieving Complete and Durable Tumor Regression. *J. Med. Chem.* **2019**, *62*, 448–466. [CrossRef] [PubMed]

30. Wurz, R.; Cee, V. Targeted Degradation of MDM2 as a New Approach to Improve the Efficacy of MDM2-p53 Inhibitors. *J. Med. Chem.* **2019**, *62*, 445–447. [CrossRef] [PubMed]
31. Lodi, G.; Gentili, V.; Casciano, F.; Romani, A.; Zauli, G.; Secchiero, P.; Zauli, E.; Simioni, C.; Beltrami, S.; Fernandez, M.; et al. Cell cycle block by p53 activation reduces SARS-CoV-2 release in infected alveolar basal epithelial A549-hACE2 cells. *Front. Pharmacol.* **2022**, *13*, 1018761. [CrossRef] [PubMed]

Disclaimer/Publisher's Note: The statements, opinions and data contained in all publications are solely those of the individual author(s) and contributor(s) and not of MDPI and/or the editor(s). MDPI and/or the editor(s) disclaim responsibility for any injury to people or property resulting from any ideas, methods, instructions or products referred to in the content.

molecules

Article

Extraction of Antioxidant Compounds from Brazilian Green Propolis Using Ultrasound-Assisted Associated with Low- and High-Pressure Extraction Methods

Thiago Dantas Teixeira [1], Bruna Aparecida Souza Machado [2,*], Gabriele de Abreu Barreto [2], Jeancarlo Pereira dos Anjos [2], Ingrid Lessa Leal [3], Renata Quartieri Nascimento [4], Katharine Valéria Saraiva Hodel [2] and Marcelo Andrés Umsza-Guez [1,*]

1 Department of Biotechnology, Health Science Institute, Federal University of Bahia, Salvador 40170-115, Brazil
2 SENAI Institute of Innovation (ISI) in Health Advanced Systems (CIMATEC ISI SAS), Senai Cimatec University Center, Salvador 41650-010, Brazil
3 Department of Food and Beverages, Applied Research Laboratory of Biotechnology and Food, Senai Cimatec University Center, Salvador 41650-010, Brazil
4 Postgraduate Program in Biotechnology-Northeast Biotechnology Network (RENORBIO), Institute of Health Sciences, Federal University of Bahia, Salvador 40170-115, Brazil
* Correspondence: brunam@fieb.org.br (B.A.S.M.); marcelo.umsza@ufba.br (M.A.U.-G.)

Abstract: The demand for bee products has been growing, especially regarding their application in complementary medicine. *Apis mellifera* bees using *Baccharis dracunculifolia* D.C. (*Asteraceae*) as substrate produce green propolis. Among the examples of bioactivity of this matrix are antioxidant, antimicrobial, and antiviral actions. This work aimed to verify the impact of the experimental conditions applied in low- and high-pressure extractions of green propolis, using sonication (60 kHz) as pretreatment to determine the antioxidant profile in the extracts. Total flavonoid content (18.82 ± 1.15–50.47 ± 0.77 mgQE·g^{-1}), total phenolic compounds (194.12 ± 3.40–439.05 ± 0.90 mgGAE·g^{-1}) and antioxidant capacity by DPPH (33.86 ± 1.99–201.29 ± 0.31 μg·mL^{-1}) of the twelve green propolis extracts were determined. By means of HPLC-DAD, it was possible to quantify nine of the fifteen compounds analyzed. The results highlighted formononetin (4.76 ± 0.16–14.80 ± 0.02 mg·g^{-1}) and *p*-coumaric acid (<LQ—14.33 ± 0.01 mg·g^{-1}) as majority compounds in the extracts. Based on the principal component analysis, it was possible to conclude that higher temperatures favored the release of antioxidant compounds; in contrast, they decreased the flavonoid content. Thus, the obtained results showed that samples pretreated with 50 °C associated with ultrasound displayed a better performance, which may support the elucidation of the use of these conditions.

Keywords: formononetin; *p*-coumaric acid; rutin; antioxidant capacity; alcoholic extraction; supercritical extraction; PCA

Citation: Teixeira, T.D.; Machado, B.A.S.; Barreto, G.d.A.; dos Anjos, J.P.; Leal, I.L.; Nascimento, R.Q.; Hodel, K.V.S.; Umsza-Guez, M.A. Extraction of Antioxidant Compounds from Brazilian Green Propolis Using Ultrasound-Assisted Associated with Low- and High-Pressure Extraction Methods. *Molecules* **2023**, *28*, 2338. https://doi.org/10.3390/molecules28052338

Academic Editors: Arunaksharan Narayanankutty, Ademola C. Famurewa and Eliza Oprea

Received: 6 February 2023
Revised: 24 February 2023
Accepted: 27 February 2023
Published: 3 March 2023

Copyright: © 2023 by the authors. Licensee MDPI, Basel, Switzerland. This article is an open access article distributed under the terms and conditions of the Creative Commons Attribution (CC BY) license (https://creativecommons.org/licenses/by/4.0/).

1. Introduction

The demand for apiculture products has been growing, especially regarding their application in complementary medicine [1], as is the case of propolis. The composition of propolis is strongly associated with its botanical and geographic origin [2,3], but generally its centesimal composition is treated and described in a generic way [4]. *Apis mellifera* bees using *Baccharis dracunculifolia* D.C. (*Asteraceae*) as a substrate produce green propolis, classified by Park et al. [5] along with other types of Brazilian propolis. *B. dracunculifolia*, also widely known as "Alecrim do campo", is native to the Southeast and South regions of Brazil and has been the subject of different investigations for ethnomedicinal, phytochemical and pharmacological purposes [6].

Over the years, records in the literature have been demonstrating the bioactive potential of this resin [7]. Among the examples of bioactivity of this matrix are the antioxidant, antimicrobial, anti-inflammatory, antiparasitic, antiviral, and antitumor actions [8–11]. In their studies, Silveira et al. [12] showed that the antiviral action of green propolis was promising in the treatment of patients with COVID-19, reducing hospitalization time and the development of kidney damage (a common sequela in patients with the disease). Silva-Beltrán et al. [13] verified the efficacy of the extract of green propolis as well as of some individual compounds present therein against human coronavirus 229E. Sokolonski et al. [14] showed the antifungal action of green propolis against *Candida albicans* isolates. The bioactivity of this matrix is mainly due to the presence of phenolic compounds (flavonoids, phenolic acids and their esters) in its chemical composition [15].

These bioactive compounds present in different types of propolis can be extracted using different methods and solvents [10], which can result in different chemical profiles of extracts [16]. Conventional methods for extraction of biocomposites, such as Soxhlet extraction, have some disadvantages, such as the high consumption of organic solvents, degradation of bioactive compounds by exposure to high temperatures, and the amount of time required to perform these techniques. On the other hand, the non-conventional methods are characterized by shorter operational time, low environmental impact, besides allowing the obtainment of extracts with greater purity [17,18].

Among the non-conventional methods, the extraction with supercritical fluid presents desirable characteristics in what concerns the extraction of thermosensitive compounds. Once it allows the use of low temperatures, providing a minor degradation of sample constituents, besides eliminating eventual problems with residual solvents [19–21], the suppression of the solvent/extract separation step becomes possible. These properties are of fundamental importance for the extraction of natural products, where the quality of the final product depends directly on the integrity of the biocomposites present in it [22,23].

Efforts have been made to improve conventional methods, applying alternatives that improve the yield and reduce the time and costs of the extraction step [24,25]. One alternative is the increasingly common use of sonication as a pretreatment in extraction processes of bioactive compounds in plant matrices [26,27]. The propagation of mechanical ultrasound waves provokes the phenomenon of acoustic cavitation in the sample, which induces a series of compressions and rarefactions in the solvent molecules, leading to bubble formation on the solute surface [28]. These bubbles implode, generating an increased interaction between solute and solvent due to the increased penetrability through the open channels on the sample surface [29].

In this context, the present work aimed to verify the impact of different experimental conditions applied to low-pressure (ethanolic) and high-pressure (supercritical) extractions of green propolis, using sonication (60 kHz) as pretreatment, as well as to determine the profile of phenolic compounds in the extracts obtained.

2. Results and Discussion

2.1. Antioxidant Profile of Green Propolis Extracts (Total Phenolic Compounds, Flavonoid Content and Antioxidant Capacity)

Propolis is the third most important component of bee products. This product is highly rich in bioactive compounds such as phenolic compounds, esters, flavonoids, terpenes, among other important organic compounds [1]. Figure 1 presents the results for the content of total phenolic compounds (TPC), flavonoids (FT) and antioxidant capacity (AC) of the extracts of different green propolis samples obtained by the two extraction methods as per Figure 1c (conventional ethanolic/low pressure (LPE) and supercritical/high pressure (SFE)). In general, a significant variation ($p < 0.05$) was observed for TPC, FT and CA among the green propolis extracts obtained by different methods. TPC showed a variation of 57.5% among the extracts (186.81 ± 0.32 to 439.05 ± 0.90 mgGAE·g^{-1}, ESC samples and B20, respectively) (Figure 1a), while FT varied by 63% (18.82 ± 1.15 to 50.47 ± 0.77 mgQE·g^{-1}, samples C30 and UESC, respectively) (Figure 1b), whereas AC (IC_{50}) varied by 83% (33.86 ± 1.99 to 201.29 ± 0.31 μg.mL^{-1}, samples B10 and ESC, respectively. The mean and standard deviation of the values obtained in each of the analyses are shown in Table S1 (Supplementary Materials).

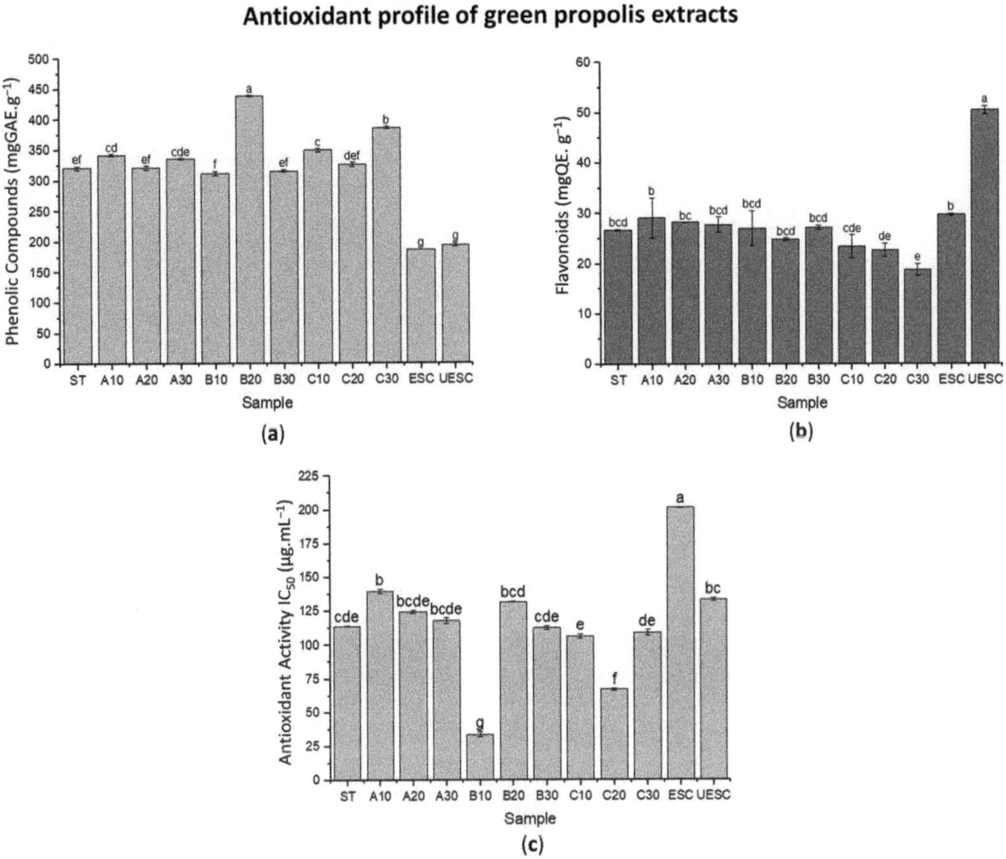

Figure 1. Antioxidant profile of green propolis extracts: (**a**) Determination of total phenolic compounds (mgGAE·g^{-1}); (**b**) flavonoids (mgQE·g^{-1}) and (**c**) antioxidant capacity to DPPH (IC_{50}) of different green propolis extracts obtained by ethanolic (LPE) and supercritical extraction (SFE). Error bars represent the standard deviation ($n = 3$). Values presenting the same letter do not show significant differences ($p > 0.05$) in Tukey's test at 95% confidence.

Importantly, the extracts obtained by LPE showed a 29% variation in TPC (311.74 ± 3.43 to 439.05 ± 0.90 mgGAE·g^{-1}, samples B10 and B20, respectively) while those obtained by SFE varied by approximately 4% (186.81 ± 0.32 to 194.12 ± 3.40 mgGAE·g^{-1}, samples ESC and UESC, respectively). Therefore, on average, LPE extracts showed 45% higher TPC than SFE. Since the low polarity of supercritical CO_2 provides it with a high power to solubilize compounds with similar polarity, such as waxes, this may explain the lower TPC yield in the extracts obtained by SFE. Waxes are poorly soluble in ethanol, which contributes to the increased interaction of this solvent with the phenolic compounds present in the propolis sample, increasing the yield in this type of extraction. In contrast, CO_2 displays weaker interaction with the phenolic compounds, leading to a decrease in the extraction yield. Although solvent polarity is an important parameter, this is not the only preponderant characteristic in the extraction process, since aspects such as solute/solvent interaction are of equal importance [30]. In their studies using *B. dracunculifolia*, Casagrande et al. [31] evidenced that the extracted phenolic compounds are strongly influenced by the concentration of ethanol (40%, 60% and 80%) used in the extraction solution. Devequi-Nunes et al. [8] evidenced that the brown, green and red propolis extracts obtained by LPE showed higher TPC content than their respective extracts obtained by SFE, and the values of both green propolis extracts corroborate those obtained in the present study (374.10 and 174.31 mgGAE·g^{-1}, respectively).

The UESC sample (60 KHz, 50 °C and 20 min) showed a TPC value (194.12 ± 3.40 mgGAE·g^{-1}) of approximately 4% higher than the ESC (186.81 ± 0.32 mgGAE·g^{-1}). Of the LPE extracts, the sample with the highest TPC was B20 (439.05 ± 0.90 mgGAE·g^{-1}), pretreated with ultrasound under the same conditions as the UESC extract, and this was 27% higher than the ST sample (320.97 ± 3.07 mgGAE·g^{-1}). De Souza et al. [25] detected a similar effect in obtaining CFT in grape seed oil, where samples pretreated with ultrasound obtained higher TPC concentrations. This effect can be explained by the rupture of the material due to acoustic cavitation caused by sonication, forming pores in its structure [32]. Due to the pores opened by acoustic cavitation, phenolic compounds are easily released from the matrix [26]. In their studies, Taddeo et al. [33] obtained a 28% increase in the yield of biocompounds in Italian propolis extracts using conventional solvent extraction combined with ultrasound exposure which corresponds with the yield reported in the present study for samples obtained by LPE.

Regarding the FT, a different behavior was observed, where the LPE samples obtained 36% lower yield (between 18.82 ± 1.15 and 29.14 ± 3.98 mgQE·g^{-1}) when compared to the SFE extracts (between 29.68 ± 0.26 and 50.47 ± 0.77 mgQE·g^{-1}). Among the samples obtained by LPE, the one that obtained the highest FT concentration was A10 (60 KHz, 25 °C and 10 min) (29.14 ± 3.98 mgQE·g^{-1}). In contrast, sample C30 (60 KHz, 75 °C and 30 min) (18.82 ± 1.15 mgQE·g^{-1}) obtained 35% lower yield and was pretreated at higher temperature and treatment time conditions, which can be explained by the thermosensitivity of flavonoids [34]. Since sample C30 had three times the temperature and exposure time conditions of sample A10, it is possible that degradation of flavonoid molecules occurred in the sample. In a study by Liu, Wang and Cai [35], it was shown that the extraction of flavonoids in *Scutellaria baicalensis* (Chinese medicinal plant) had decreased yield in extractions conducted with temperature higher than 60 °C due to loss of activity and degradation of flavonoids.

The UESC sample presented flavonoid concentration 1.7 times higher than its control (ESC). The authors De Andrade et al. [24] indicate that the pretreatment of grape skin samples with ultrasound improves the yield of FT. This can be justified because the ultrasound waves stimulate the formation of small bubbles subjected to rapid compression and expansion, causing rapid local increase in temperature and pressure, which facilitates the solubilization of compounds present in the matrix [26]. In contrast, sample B20, pretreated under the same conditions as UESC (60 KHz, 50 °C and 20 min), was 50.8% lower than this sample obtained by SFE considering the concentration of FT (Figure 1b). Saito et al. [36] observed the same pattern in the amount of flavonoids, where supercritical extracts of red propolis showed these constituents in larger amounts when compared to ethanolic extracts. Similar behavior was also reported by Martinez-Correa et al. [37], who observed higher amount of flavonoids in supercritical extract of *Eugenia uniflora* when compared to ethanolic and aqueous extracts of this matrix. This fact indicates that supercritical extraction has greater selectivity in obtaining flavonoids.

The research for natural matrices with high content of compounds with antioxidant capacity has increased considerably in recent years, especially due to the potential benefits that these components present considering biological environments [38,39]. Within this perspective, it was observed that the IC_{50} value was lower in ethanolic sample B10 (60 KHz, 50 °C and 10 min) (33.86 ± 1.99 μg.mL^{-1}). This value was approximately 2.8 times lower than the extract with the highest antioxidant capacity reported by Zhang et al. [40] for Brazilian green propolis (93.51 μg.mL^{-1}). Considering that, the IC_{50} expresses the sample concentration required to neutralize the DPPH radical by 50%; the lower this value, the higher the antioxidant potential of the sample. Within this perspective, the sample with the lowest antioxidant capacity, represented by the highest IC_{50} value, was the supercritical

ESC (201.29 ± 0.31 μg.mL^{-1}). These data can be explained since the TPC of the B10 extract is approximately 1.6 times higher than that of the ESC sample. This indicates that the phenolic compounds comprising this pretreated sample, or even the presence of other compounds released in the pretreatment, contribute to the antioxidant potential [41].

Additionally, in the samples obtained by SFE, it was possible to notice that the UESC sample had higher AC (IC_{50}: 133.17 ± 1.09 μg.mL^{-1}) when compared to its ESC control (IC50: 201.29 ± 0.31 μg.mL^{-1}). This effect is consistent with the FT concentration of these samples, since in the pretreated sample the concentration of these compounds was approximately two times higher than that in the control sample. Flavonoids are biocompounds that are oxidized by free radicals, resulting in a more stable and less reactive radical, providing these compounds with the antioxidant potential [42].

Considering the potential application of samples with higher antioxidant capacity, such as sample B10, there is a great need for the use of matrices with these properties in scientific and technological studies. For example, it has been shown that the presence of antioxidant components in propolis samples is related to the increase in its anti-aging [43], anti-inflammatory [44], and anti-tumor activities [45], among others. From the technological development point of view, the presence of propolis extracts with antioxidant properties in food packaging has contributed to the increase in the shelf life of products [46–48]. This perspective reinforces the need for studies such as this one, which have, among their objectives, the aim of elucidation of the antioxidant property of natural matrices.

Figure 2 presents the correlation of extraction methods with respect to the content of phenolic compounds (mgGAE·g^{-1}), flavonoids (mgQE·g^{-1}) and DPPH radical scavenging capacity (IC_{50}, μg.mL^{-1}) through Principal Component Analysis (PCA) in order to detect the principal component that best describes the highlighted influences of this study. The principal components (PC1 x-axis: 64.34% of score value and PC2 y-axis: 25.95% of score value, corresponding to 90.29% of total cumulative variance) differentiate the extraction methods according to TPC, FT and AC properties. Figures S1 and S2 show the graph of loadings PC1 and PC2, respectively.

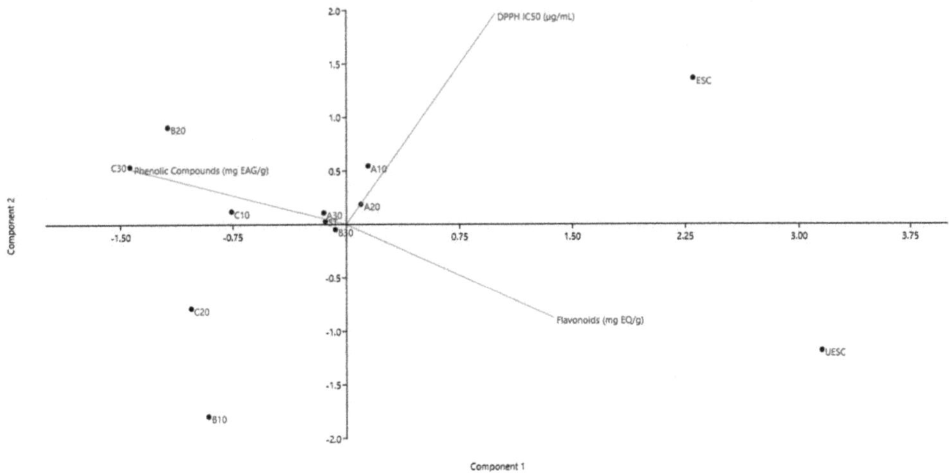

Figure 2. Principal component analysis of the samples obtained by the different extraction methods (PC scores) according to the number of loadings (content of phenolic compounds, flavonoids, and DPPH radical scavenging ability).

Treatments such as A10 and A20 were grouped in the right quadrant (positive) along with the DPPH variable (Figure 2), indicating that samples with lower temperature and time of ultrasound exposure had lower antioxidant capacity due to the high IC_{50} value. The ESC treatment, despite using higher temperature and time, was also grouped in the same quadrant due to its high IC_{50} value; however, the lower phenolic content of this sample explains its low antioxidant capacity. Inversely located on the left side (negative), samples B10 and C20 were grouped together (Figure 2), indicating that the increase in temperature may be directly related to favoring the extraction of compounds with antioxidant potential, since these samples had the lowest IC_{50} values. On the right side (negative) the UESC treatment was grouped along with the flavonoids variable (Figure 2). This may indicate that the supercritical extraction associated with ultrasound exposure directly contributed to flavonoid extraction due to the high concentration obtained in this extract (Figure 1b).

Conversely, on the opposite left side (positive and negative) are samples C10, C20 and C30, that, despite sonication, showed the lowest flavonoid contents (Figure 1b). This shows a trend of higher temperatures associated with a longer exposure time being able to affect flavonoid extraction. Observing the left side (upper), it was noted that the B20 treatment was the one that best represented the variable TPC along with C10 and C30 (Figure 2), since they obtained higher concentrations of these compounds (Figure 1a). On the right side (positive and negative) the ESC and UESC treatments are positioned. From this, it is possible to note a tendency that ethanol extraction associated with sonication favors the extraction of phenolic compounds.

2.2. Quantification of Compounds by HPLC

Among the fifteen compounds analyzed in the different extracts, it was possible to identify and quantify nine compounds in most samples. Among them are formononetin (from 4.76 ± 0.16 to 14.80 ± 0.02 $mg \cdot g^{-1}$); *p*-coumaric acid (<LQ—14.33 ± 0.01 $mg \cdot g^{-1}$); quercetin (from 0.67 ± 0.02 to 2.45 ± 0.00 $mg \cdot g^{-1}$); gallic acid (<LQ—2.78 ± 0.01 $mg \cdot g^{-1}$); kaempferol (from 0.34 ± 0.01 to 2.51 ± 0.03 $mg \cdot g^{-1}$); caffeic acid (0.19—3.02 ± 0.30 $mg \cdot g^{-1}$); catechin (from 0.52 ± 0.00 to 1.37 ± 0.03 $mg \cdot g^{-1}$); epicatechin (from 0.22 ± 0.03 to 0.98 ± 0.01 $mg \cdot g^{-1}$) and rutin (from 0.15 ± 0.01 to 10.00 ± 0.03 $mg \cdot g^{-1}$) (Table 1). Determining and quantifying the bioactive compounds of propolis is of great importance, since each type of propolis has unique characteristics, and when its main components are determined, the type of propolis can be targeted for specific therapeutic indications [49].

Table 1. Quantification of the nine major phenolic compounds by HPLC in different green propolis extracts obtained by ultrasound-assisted LPE and SFE.

Sample	Compounds ($mg \cdot g^{-1}$)								
	Quercetin	Gallic Acid	Formononetin	Kaempferol	*p*-Coumaric Acid	Caffeic Acid	Catechin	Epicatechin	Rutin
ST	2.30	0	13.37	1.17	14.03	1.21	1.05	0.51	1.93
A10	2.45	1.37	9.20	1.43	7.29	1.39	1.13	0.55	1.65
A20	2.37	0.68	6.04	1.01	7.53	1.69	1.37	0.98	0.21
A30	1.25	1.12	7.60	1.32	7.17	2.48	1.01	0.33	0.03
B10	1.01	1.48	14.80	0.88	14.33	1.48	0.71	0.22	2.71
B20	1.37	2.78	7.77	0.91	8.10	2.19	0.91	0.48	0.56
B30	2.24	0.44	6.66	0.78	0.00	3.02	1.16	11.49	0.48
C10	0.69	2.66	14.10	0.98	13.41	1.86	0.90	0.27	0.15
C20	0.67	2.62	14.31	0.93	12.56	1.76	0.83	0.39	1.42
C30	1.67	<LQ	11.90	0.89	12.20	2.06	0.93	0.66	1.75
ESC	2.06	2.15	8.39	0.34	7.87	0.23	0.66	0.51	4.17
UESC	0.68	1.79	4.76	2.51	0.00	0.19	0.52	0.31	14.99

In most samples obtained by LPE, the majority compounds were formononetin (from 4.76 ± 0.16 to 14.80 ± 0.02 $mg \cdot g^{-1}$) and *p*-coumaric acid (<LQ—14.33 ± 0.01 $mg \cdot g^{-1}$) (Table 1). Formononetin is an isoflavone commonly found in red propolis samples that has been reported in the literature for its fungicidal, antioxidant, gastroprotective, and dyslipidemic regulating properties [50–52]. Despite being considered as a biomarker of red propolis, formonononetin was reported as a majority compound in most extracts obtained from green propolis, indicating the possibility that bees of the *Apis mellifera* species also occasionally collect plants containing this biocompound [53]. The presence of formononetin

has also been reported in samples of green propolis from the state of Minas Gerais, located in southeastern Brazil [54]. These findings reinforce the idea of the chemical complexity of propolis, which has encouraged the publication of studies aimed at its quality control and standardization [55]. *p*-Coumaric acid and its prenylated derivatives are phenolic acids widely known as biomarkers of Brazilian green propolis [56,57]. Ferreira et al. [58] showed that this biocompound can act directly and indirectly in mitigating inflammatory processes. Celińska-Janowicz et al. [59] showed that *p*-coumaric acid has the ability to cause apoptosis in tongue squamous cell carcinoma cells (CAL-27). Many of these effects are directly associated with its AC [60].

Among the extracts obtained by LPE, those with the highest AC were B10 (IC_{50}: 33.86 ± 1.99 μg.mL^{-1}) and C20 (IC_{50}: 67.1 ± 0.86 μg.mL^{-1}). From the results of the chromatographic analyses, it was possible to note the high concentration of formononetin and *p*-coumaric acid simultaneously in these samples, which may have directly influenced the ability to neutralize the DPPH free radical. In their in vitro study to determine the antioxidant potential of formononetin, Vishnuvathan et al. [61] concluded that the ability of this substance to neutralize the DPPH radical increased in a concentration-dependent relationship. Shen et al. [62] detected a similar relationship in an in vitro study of the antioxidant potential of *p*-coumaric acid against the DPPH radical.

In the samples obtained by SFE, it was possible to observe that the ESC sample presented formononetin (8.39 ± 0.03 mg·g^{-1}) and *p*-coumaric acid (7.87 ± 0.03 mg·g^{-1}) as the majority compounds. In contrast, the UESC sample showed lower levels of these compounds (4.76 ± 0.16 mg·g^{-1} and <LQ, respectively), but the high amount of rutin (10.00 ± 0.03 mg·g^{-1}) may have directly interfered with the antioxidant potential of this sample (Table 2). Selvaraj et al. [63] concluded that rutin has antioxidant capacity against both DPPH and ABTS [2,2′-azino-bis (3-ethylbenzothiazolin) 6-sulfonic acid] radicals, the antioxidant potential of which grows in a concentration-dependent relationship. In a study by Silva-Beltrán et al. [13], it was shown that some of the phenolic compounds present in green propolis have antiviral properties against human coronavirus (HCoV 229-E), where quercetin reduced the cytopathogenicity of the virus by up to 90%, followed by caffeic acid (80–90%) and rutin (75%).

Table 2. Pretreatment conditions for each green propolis extract obtained by ethanol (EtOH) (LPE) and supercritical (SCO_2) extraction (SFE).

Sample	Treatment
ST	LPE Control
A10	25 °C, 10 min.
A20	25 °C, 20 min.
A30	25 °C, 30 min.
B10	50 °C, 10 min.
B20	50 °C, 20 min.
B30	50 °C, 30 min.
C10	75 °C, 10 min.
C20	75 °C, 20 min.
C30	75 °C, 30 min.
ESC	Control SFE
UESC	SFE, 50 °C, 20 min.

The PCA presented in Figure 3 was performed in order to evaluate the influence of the extraction methods with respect to the content of the nine phenolic compounds quantified by HPLC (Table 1). The first two components (PC1 and PC2) explained 94.30% of the data, demonstrating that the influence of the extraction methods on the concentration of these compounds is strong. PC1 *x*-axis had the highest score value (89.59%), while PC2 *y*-axis had the lowest score value (4.71%). Observing the right side (positive), it is possible to notice the grouping of treatments B10, C10, C20 and C30 together with the variable formononetin and *p*-coumaric acid. On the other hand, on the left side (negative) are opposed treatments such as A10, A20 and A30 that have lower content of these biocompounds, which may be an indication that higher temperatures favored the extraction of these phenolic compounds.

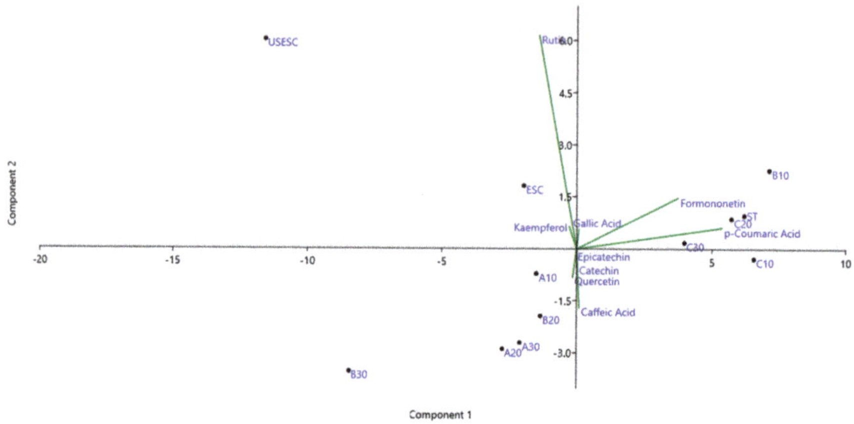

Figure 3. Principal component analysis of the samples obtained by different extraction methods according to the amount of phenolic compounds by HPLC.

Both the ESC and UESC samples were positioned to the left-positive side, showing that it is possible that supercritical extraction and/or pretreatment will have disfavored the extraction of the caffeic acid compound. However, the samples obtained by SFE occupied this position along with the variable rutin due to the higher concentration obtained for this compound. It is possible, analyzing the control sample (ESC), that the pretreatment with ultrasound may have been responsible for the release of kaempferol and rutin in the UESC sample. Since most of the samples obtained by LPE are opposed to the rutin variable, it may be an indication that SFE may have been more selective for obtaining this compound.

3. Materials and Methods

3.1. Reagents

Ethanol (HPLC grade) and acetic acid (HPLC grade) were purchased from Merck Co. (Darmstadt, Germany), methanol (HPLC grade) and DMSO (dimethyl sulfoxide) were purchased from Sigma-Aldrich Chemical Co. (St. Louis, MO, USA). Cellulose ester membrane filters of 0.45 μm (SLCR025NS, Millipore 1Co. Bedford, MA, USA) were used. The carbon dioxide (CO_2) used in the extraction presented 99.9% purity (White Martins Gases Industrials—São Paulo, Brazil). The gallic acid (CAS number 149-91-7), catechin (CAS number 7295-85-4), epicatechin (CAS number 490-46-0), trans-cinnamic acid (CAS number 140-10-3), narigenin (CAS number 67604-48-2), caffeic acid (CAS number 331-39-5), *p*-coumaric acid (CAS number 501-98-4), resveratrol (CAS number 501-36-0), formononetin (CAS number 485-72-3), rutin hydrate (CAS number 207671-50-9), quercetin (CAS number 117-39-5), kaempferol (CAS number 520-18-3), and myricetin (CAS number 529-44- 2) were purchased from Sigma-Aldrich Chemical Co. (St. Louis, MO, USA), and trans-ferulic acid (CAS number 537-98-4) was purchased from Fluka (Buchs, Switzerland).

3.2. Obtaining Extracts from Green Propolis

The sample of green propolis used in this study was acquired from an apiary in the city of Carmo da Mata (Minas Gerais, Brazil). The sample was processed in a mill (Cadence, Santa Catarina, Brazil) to obtain a diameter between 52 and 92 μm, thus facilitating the extraction process and the uniformity of the material. The sample was kept at −10 °C in inert atmosphere conditions (N2) in a fractional form to avoid the oxidation of the material.

3.2.1. Low Pressure Extraction (LPE)

Propolis extracts were prepared from the homogenization of 2 g of raw green propolis in 15 mL of ethanol (80%). The system was pretreated in an ultrasonic bath at a frequency of 60 KHz, varying the temperature conditions (A-25 °C, B-50 °C and C-75 °C) and exposure time (A, B, C—10, 20 and 30 min, respectively) (Table 2). Then, the system was stored in the dark for 7 days, with manual shaking for 5 min every 24 h. The extract was recovered by centrifugation (SIGMA Centrifuge 2–16 KL, USA) at 5000 rpm (10 °C) for 11 min, and the supernatant was transferred to glass test tubes (15 × 160 mm) and kept at 45 °C until complete evaporation of the solvent. A control (ST sample) was obtained, without ultrasonic bath pretreatment. All extracts were kept at 5 °C until use [64].

3.2.2. Supercritical Fluid Extraction (SFE)

To obtain propolis extracts by supercritical extraction, an SFT-110 Supercritical Fluid Extractor (Supercritical Fluid Technologies, Inc., Newark, NJ, USA) was used. In each experiment, the extraction cell was composed of 5 g of ground green propolis with 5.5 mL of ethanol (80%) as co-solvent, as well as wool and glass beads. The system was pretreated in an ultrasonic bath at the frequency of 60 KHz, 50 °C for 20 min, since this treatment condition obtained higher TPC yield in the extracts obtained by LPE (Figure 1). The extraction conditions were the following: pressure—350 bar; temperature—50 °C; CO_2 flow—6 g.min^{-1}. The extraction time was approximately 2.5 h. The extracts, collected in glass vials, were stored under the shelter of light, in inert atmospheric conditions (N_2) to avoid degradation of the constituents. A control (ESC sample) was obtained without pretreatment in an ultrasonic bath. The extracts were kept at 5 °C until the time of use [64].

3.3. Determination of Total Phenolic Compounds

The analyses for determination of total phenolic compounds in green propolis extracts were performed by the Folin–Ciocalteu's spectrophotometric method using gallic acid as standard [65]. Ethanol 80% was used to solubilize the extracts to obtain a concentration of 20 mg.mL^{-1}. Then, 0.5 mL of the extract solution was withdrawn and mixed with 2.5 mL of aqueous Folin–Ciocalteu solution (10%) and 2.0 mL of 7.5% sodium carbonate. The solution was placed in a thermoregulated bath at 50 °C for 5 min and then the absorbance was measured in a spectrophotometer (Lambda 25 UV/vis Systems—PerkinElmer, Washington, DC, USA) at 765 nm. The results of the concentrations of total phenolic compounds were compared with an analytical curve of gallic acid (mgGAE·g^{-1}) (y = 0.0104x + 0.0688; R2 = 0.9976) under the same conditions. All analyses were performed in triplicate and expressed as gallic acid equivalents (mgGAE·g^{-1}).

3.4. Determination of Flavonoid Content

The determination of the total flavonoid content of the green propolis extracts was performed in a spectrophotometer (Lambda 25 UV/vis Systems—PerkinElmer, Washington, DC, USA) at 415 nm. A solution with the propolis extracts was prepared using 2.0% aluminum chloride in methanol in 1:1 (*v*/*v*) solution [66]. The same procedure was performed using known solutions of quercetin standard to prepare an analytical curve (y = 0.0311x + 0.0259; R2 = 0.9987). In addition, a blank sample was prepared under the same conditions and the amount of flavonoids was expressed as quercetin equivalents (QE) (mgQE·g^{-1}). All analyses were performed in triplicate.

3.5. Determination of the Antioxidant Capacity: DPPH Method

The evaluation of the antioxidant capacity of the extracts was performed with 1,1-diphenyl-2-picrylhydrazyl (DPPH), according to the methodology described by Brand-Williams et al. [67]. The green propolis extracts were diluted to five concentrations in triplicate. Then, 1.0 mL of each dilution was transferred to a test tube containing 3.0 mL of ethanolic DPPH solution (0.004%). After 30 min of incubation in the dark and at room temperature, the reduction in the DPPH free radical was measured by reading the absorbance in a spectrophotometer (Lambda 25 UV/vis Systems—PerkinElmer, Washington, DC, USA) at 517 nm. A blank sample was prepared using ethanol instead of the sample. The IC_{50} value (concentration in $\mu g.mL^{-1}$ required of the extract to sequester 50% of the DPPH radical) was calculated using the straight line equation obtained by constructing the curve based on the concentrations of the extracts (Table S2).

3.6. Chromatographic Analysis of Green Propolis Extracts

The phenolic compounds (gallic acid, catechin, epicatechin, trans-cinnamic acid, naringenin, caffeic acid, *p*-coumaric acid, resveratrol, formononetin, rutin hydrate, quercetin, kaempferol, myricetin, O-dianiside, and trans-ferulic acid) were analyzed from the green propolis extracts. First, 10 mg of green propolis extracts obtained from the different extraction methods were prepared and dissolved in ethanol (HPLC grade). A 0.45 µm cellulose ester membrane filter (Micropore) was used to filter the samples, before injection into the chromatographic system. The chromatographic analyses were performed using a High-Performance Liquid Chromatography (HPLC) equipment (Shimadzu, Tokyo, Japan), composed of a pump (LC-20AT), supplied with an automatic injector (SIL-20AHT), degasser (DGU-205), diode array detector (DAD) (SPD-M20A) and a column oven (CTO-20A). The method used to promote chromatographic separation was adapted from Daugsch (2007) and Machado et al. (2015). A NUCLEODUR® 100-5 C18 column (150 × 4 mm ID) (5 µm) was used in conjunction with a ZORBAX Eclipse Plus C18 pre-column (4.6 × 12.5 mm) from Agilent.

Chromatographic analysis was performed with an elution gradient using a mobile phase of 5% acetic acid (solvent A) and acetonitrile (solvent B), for 0–35 min (0–92% B); 35 to 40 min (92–0% B); 40 to 42 min (0% B), for a total time of 42 min. The furnace was operated at 40°C. The injection volume was 20 µL, analyzed in triplicate, and the chromatographic acquisition was set at wavelengths in the 280 nm, 300 nm, and 320 nm regions (Table S3). Fifteen phenolic compounds were analyzed, of which analytical curves were constructed from dilutions of a 40 $mg.L^{-1}$ stock solution containing all analytes dissolved in methanol. From the stock solution, dilutions were prepared for the construction of the analytical curves in a range of 0.5–15.0 $mg.L^{-1}$. The limit of quantification ranged from 0.076 to 0.380 $mg.L^{-1}$. The detection limit was in a range of 0.023 to 0.056 $mg.L^{-1}$.

3.7. Statistical Analysis

The results were evaluated using ANOVA (one-way) analysis of variance and Tukey's test to identify whether the changes in the parameters evaluated were significant at 95% confidence level. The principal component analysis (PCA) was performed to evaluate the influence of extraction methods with respect to the content of phenolic compounds ($mgGAE \cdot g^{-1}$), flavonoids ($mgQE \cdot g^{-1}$) and DPPH radical scavenging capacity (IC_{50}, $\mu g.mL^{-1}$), using the software PAST (Paleontological Statistics, Oslo, Norway) version 3.26. Since the averages for the aforementioned characterization tests use different units of measurement, the data were normalized in the range of 0 to 1.

4. Conclusions

The results showed that samples pretreated with medium temperature (50 °C) associated with ultrasound had higher antioxidant capacity (IC_{50} up to 33.86 ± 1.99 $\mu g.mL^{-1}$) and total phenolic content (concentrations up to 439.05 ± 0.90 $mgGAE \cdot g^{-1}$). Based on the principal component analysis, it was observed that the samples obtained by SFE tended to have higher flavonoid contents (concentrations up to 50.47 ± 0.77 $mgQE \cdot g^{-1}$), an effect that was enhanced when this extraction technique was associated with sonication. In contrast, treatments with higher temperature (75 °C) showed lower total phenolic values (concentrations up to 23.38 ± 2.3 $mgQE \cdot g^{-1}$). In addition, by means of PCA, it was possible to observe that extracts obtained by SFE favored the obtainment of phenolic compounds such as rutin and kaempferol, with the SFE method being more selective in obtaining rutin (concentrations up to 10.00 ± 0.03 $mg \cdot g^{-1}$), an effect that was also strengthened with the use of sonication. Thus, it is hoped that these findings can support the elucidation of the use of pretreatment and type of extractive methods in the process of obtaining natural extracts composed of bioactive molecules.

Supplementary Materials: The following supporting information can be downloaded at: https://www.mdpi.com/article/10.3390/molecules28052338/s1, Table S1: Antioxidant profile of green propolis extracts based on determination of total phenolic compounds, flavonoids, and antioxidant activity to DPPH (IC_{50}); Figure S1: Graph of loadings for principal component 1; Figure S2: Graph of loadings for principal component 2; Table S2: Linear equations and their respective correlation coefficients used to calculate the IC_{50} of green propolis extracts; Table S3: HPLC-DAD data for the nine main quantified phenolic compounds: wavelength (λ), retention time (rt), concentration range (CR), limit of quantification (LQ) and limit of detection (LOD).

Author Contributions: Conceptualization, B.A.S.M., J.P.d.A. and M.A.U.-G.; Data curation, T.D.T., G.d.A.B., J.P.d.A., I.L.L., R.Q.N. and K.V.S.H.; Formal analysis, T.D.T., G.d.A.B., J.P.d.A. and I.L.L.; Funding acquisition, B.A.S.M.; Investigation, T.D.T., G.d.A.B., J.P.d.A., I.L.L. and K.V.S.H.; Methodology, B.A.S.M., J.P.d.A. and M.A.U.-G.; Project administration, B.A.S.M., J.P.d.A. and M.A.U.-G.; Resources, B.A.S.M., J.P.d.A. and M.A.U.-G.; Supervision, B.A.S.M., J.P.d.A. and M.A.U.-G.; Validation, B.A.S.M., J.P.d.A. and M.A.U.-G.; Visualization, K.V.S.H. and M.A.U.-G.; Writing—original draft, T.D.T., B.A.S.M., J.P.d.A., I.L.L., K.V.S.H. and M.A.U.-G.; Writing—review and editing, T.D.T., B.A.S.M., J.P.d.A., K.V.S.H. and M.A.U.-G. All authors have read and agreed to the published version of the manuscript.

Funding: This research received no external funding.

Institutional Review Board Statement: Not applicable.

Informed Consent Statement: Not applicable.

Data Availability Statement: Data are contained within the article.

Acknowledgments: The authors would like to thank Serviço Nacional de Aprendizagem Industrial—SENAI/CIMATEC (Bahia). M.A.U.G and B.A.S.M are Technological Development fellow from CNPq (Proc. 304747/2020-3 and Proc. 306041/2021-9, respectively). T.D.T. was supported by the Bahia State Research Support Foundation (FAPESB-Bol1771/2018), in the Institutional Scientific Initiation Scholarship Program (PIBIC).

Conflicts of Interest: The authors declare no conflict of interest.

Sample Availability: Samples of the compounds are available from the authors.

References

1. Pasupuleti, V.R.; Sammugam, L.; Ramesh, N.; Gan, S.H. Honey, Propolis, and Royal Jelly: A Comprehensive Review of Their Biological Actions and Health Benefits. *Oxid. Med. Cell. Longev.* **2017**, *2017*, 1259510. [CrossRef]
2. De Groot, A.C. Propolis: A review of properties, applications, chemical composition, contact allergy, and other adverse effects. *Dermatitis* **2013**, *24*, 263–282. [CrossRef]
3. El-Guendouz, S.; Lyoussi, B.; Miguel, M.G. Insight on Propolis from Mediterranean Countries: Chemical Composition, Biological Activities and Application Fields. *Chem. Biodivers.* **2019**, *16*, e1900094. [CrossRef]
4. Salatino, A.; Salatino, M.L.F. Scientific note: Often quoted, but not factual data about propolis composition. *Apidologie* **2021**, *52*, 312–314. [CrossRef]
5. Park, Y.K.; Alencar, S.M.; Aguiar, C.L. Botanical Origin and Chemical Composition of Brazilian Propolis. *J. Agric. Food Chem.* **2002**, *50*, 2502–2506. [CrossRef]
6. Gazim, Z.C.; Valle, J.S.; Carvalho dos Santos, I.; Rahal, I.L.; Silva, G.C.C.; Lopes, A.D.; Ruiz, S.P.; Faria, M.G.I.; Piau, R., Jr.; Gonçalves, D.D. Ethnomedicinal, phytochemical and pharmacological investigations of Baccharis dracunculifolia DC. (ASTERACEAE). *Front. Pharmacol.* **2022**, *13*, 1048688. [CrossRef]
7. Bobiş, O. Plants: Sources of Diversity in Propolis Properties. *Plants* **2022**, *11*, 2298. [CrossRef]
8. Devequi-Nunes, D.; Machado, B.A.S.; de Barreto, G.A.; Rebouças Silva, J.; da Silva, D.F.; da Rocha, J.L.C.; Brandão, H.N.; Borges, V.M.; Umsza-Guez, M.A. Chemical characterization and biological activity of six different extracts of propolis through conventional methods and supercritical extraction. *PLoS ONE* **2018**, *13*, e0207676. [CrossRef]
9. Khayrani, A.C.; Irdiani, R.; Aditama, R.; Pratami, D.K.; Lischer, K.; Ansari, M.J.; Chinnathambi, A.; Alharbi, S.A.; Almoallim, H.S.; Sahlan, M. Evaluating the potency of Sulawesi propolis compounds as ACE-2 inhibitors through molecular docking for COVID-19 drug discovery preliminary study. *J. King Saud Univ.—Sci.* **2021**, *33*, 101297. [CrossRef]

10. Silva-Beltrán, N.P.; Umsza-Guez, M.A.; Ramos Rodrigues, D.M.; Gálvez-Ruiz, J.C.; de Paula Castro, T.L.; Balderrama-Carmona, A.P. Comparison of the Biological Potential and Chemical Composition of Brazilian and Mexican Propolis. *Appl. Sci.* **2021**, *11*, 11417. [CrossRef]
11. Hodel, K.V.S.; Machado, B.A.S.; Sacramento, G.d.C.; Maciel, C.A.d.O.; Oliveira-Junior, G.S.; Matos, B.N.; Gelfuso, G.M.; Nunes, S.B.; Barbosa, J.D.V.; Godoy, A.L.P.C. Active Potential of Bacterial Cellulose-Based Wound Dressing: Analysis of Its Potential for Dermal Lesion Treatment. *Pharmaceutics* **2022**, *14*, 1222. [CrossRef]
12. Silveira, M.A.D.; De Jong, D.; Berretta, A.A.; Galvão, E.B.d.S.; Ribeiro, J.C.; Cerqueira-Silva, T.; Amorim, T.C.; Conceição, L.F.M.R.d. Gomes, M.M.D.; Teixeira, M.B.; et al. Efficacy of Brazilian green propolis (EPP-AF®) as an adjunct treatment for hospitalized COVID-19 patients: A randomized, controlled clinical trial. *Biomed. Pharmacother.* **2021**, *138*, 111526. [CrossRef]
13. Silva-Beltrán, N.P.; Galvéz-Ruíz, J.C.; Ikner, L.A.; Umsza-Guez, M.A.; de Paula Castro, T.L.; Gerba, C.P. In vitro antiviral effect of Mexican and Brazilian propolis and phenolic compounds against human coronavirus 229E. *Int. J. Environ. Health Res.* **2022**, 1–13. [CrossRef]
14. Santos, L.M.; Fonseca, M.S.; Sokolonski, A.R.; Deegan, K.R.; Araújo, R.P.C.; Umsza-Guez, M.A.; Barbosa, J.D.V.; Portela, R.D.; Machado, B.A.S. Propolis: Types, composition, biological activities, and veterinary product patent prospecting. *J. Sci. Food Agric.* **2020**, *100*, 1369–1382. [CrossRef]
15. Woźniak, M.; Mrówczyńska, L.; Waśkiewicz, A.; Rogoziński, T.; Ratajczak, I. Phenolic profile and antioxidant activity of propolis extracts from Poland. *Nat. Prod. Commun.* **2019**, *14*, 1–7. [CrossRef]
16. Sforcin, J.M. Biological Properties and Therapeutic Applications of Propolis. *Phyther. Res.* **2016**, *30*, 894–905. [CrossRef]
17. Liu, X.; Ou, H.; Gregersen, H. Ultrasound-assisted supercritical CO_2 extraction of cucurbitacin E from Iberis amara seeds. *Ind. Crops Prod.* **2020**, *145*, 112093. [CrossRef]
18. Manjare, S.D.; Dhingra, K. Supercritical fluids in separation and purification: A review. *Mater. Sci. Energy Technol.* **2019**, *2*, 463–484. [CrossRef]
19. Machado, B.A.S.; Pereira, C.G.; Nunes, S.B.; Padilha, F.F.; Umsza-Guez, M.A. Supercritical Fluid Extraction Using CO_2: Main Applications and Future Perspectives. *Sep. Sci. Technol.* **2013**, *48*, 2741–2760. [CrossRef]
20. Valadez-Carmona, L.; Ortiz-Moreno, A.; Ceballos-Reyes, G.; Mendiola, J.A.; Ibáñez, E. Valorization of cacao pod husk through supercritical fluid extraction of phenolic compounds. *J. Supercrit. Fluids* **2018**, *131*, 99–105. [CrossRef]
21. Espinosa-Pardo, F.A.; Martinez, J.; Martinez-Correa, H.A. Extraction of bioactive compounds from peach palm pulp (*Bactris gasipaes*) using supercritical CO_2. *J. Supercrit. Fluids* **2014**, *93*, 2–6. [CrossRef]
22. Machado, B.A.S.; de Abreu Barreto, G.; Costa, A.S.; Costa, S.S.; Silva, R.P.D.; da Silva, D.F.; Brandão, H.N.; da Rocha, J.L.C.; Nunes, S.B.; Umsza-Guez, M.A.; et al. Determination of Parameters for the Supercritical Extraction of Antioxidant Compounds from Green Propolis Using Carbon Dioxide and Ethanol as Co-Solvent. *PLoS ONE* **2015**, *10*, e0134489. [CrossRef]
23. dos Santos, L.C.; Bitencourt, R.G.; dos Santos, P.; de Tarso Vieira e Rosa, P.; Martínez, J. *Solubility of passion fruit (Passiflora edulis Sims*) seed oil in supercritical CO_2. *Fluid Phase Equilib.* **2019**, *493*, 174–180. [CrossRef]
24. de Andrade, R.B.; Machado, B.A.S.; Barreto, G.D.A.; Nascimento, R.Q.; Corrêa, L.C.; Leal, I.L.; Tavares, P.P.L.G.; Ferreira, E.D.S.; Umsza-Guez, M.A. Syrah grape skin residues has potential as source of antioxidant and anti-microbial bioactive compounds. *Biology* **2021**, *10*, 1262. [CrossRef]
25. Souza, R.D.C.D.; Machado, B.A.S.; Barreto, G.D.A.; Leal, I.L.; dos Anjos, J.P.; Umsza-Guez, M.A. Effect of Experimental Parameters on the Extraction of Grape Seed Oil Obtained by Low Pressure and Supercritical Fluid Extraction. *Molecules* **2020**, *25*, 1634. [CrossRef] [PubMed]
26. Oroian, M.; Ursachi, F.; Dranca, F. Influence of ultrasonic amplitude, temperature, time and solvent concentration on bioactive compounds extraction from propolis. *Ultrason. Sonochem.* **2020**, *64*, 105021. [CrossRef]
27. de Oliveira Reis, J.H.; Barreto, G.d.A.; Cerqueira, J.C.; dos Anjos, J.P.; Andrade, L.N.; Padilha, F.F.; Druzian, J.I.; Machado, B.A.S. Evaluation of the antioxidant profile and cytotoxic activity of red propolis extracts from different regions of northeastern Brazil obtained by conventional and ultrasound-assisted extraction. *PLoS ONE* **2019**, *14*, e0219063. [CrossRef]
28. Dent, M.; Verica, D.-U.; Garofulić, I.; Bosiljkov, T.; Ježek, D.; Brncic, M. Comparison of Conventional and Ultrasound Assisted Extraction Techniques on Mass Fraction of Phenolic Compounds from sage (*Salvia officinalis* L.). *Chem. Biochem. Eng. Q.* **2015**, *29*, 475–484. [CrossRef]
29. Medina-Torres, N.; Ayora, T.; Andrews, H.; Sanchez, A.; Pacheco López, N. Ultrasound Assisted Extraction for the Recovery of Phenolic Compounds from Vegetable Sources. *Agronomy* **2017**, *7*, 47. [CrossRef]
30. Biscaia, D.; Ferreira, S.R.S. Propolis extracts obtained by low pressure methods and supercritical fluid extraction. *J. Supercrit. Fluids* **2009**, *51*, 17–23. [CrossRef]
31. Casagrande, M.; Zanela, J.; Wagner, A.; Busso, C.; Wouk, J.; Iurckevicz, G.; Montanher, P.F.; Yamashita, F.; Malfatti, C.R.M. Influence of time, temperature and solvent on the extraction of bioactive compounds of Baccharis dracunculifolia: In vitro antioxidant activity, antimicrobial potential, and phenolic compound quantification. *Ind. Crops Prod.* **2018**, *125*, 207–219. [CrossRef]
32. Vidal, A.R.; Cansian, R.L.; de Oliveira Mello, R.; Kubota, E.H.; Demiate, I.M.; Zielinski, A.A.F.; Dornelles, R.C.P. Effect of ultrasound on the functional and structural properties of hydrolysates of different bovine collagens. *Food Sci. Technol.* **2020**, *40*, 346–353. [CrossRef]

33. Taddeo, V.A.; Epifano, F.; Fiorito, S.; Genovese, S. Comparison of different extraction methods and HPLC quantification of prenylated and unprenylated phenylpropanoids in raw Italian propolis. *J. Pharm. Biomed. Anal.* **2016**, *129*, 219–223. [CrossRef] [PubMed]
34. De Castro, D.S.; De Oliveira, T.K.B.; Lemos, D.M.; Rocha, A.P.T.; Almeida, R.D. Efeito da temperatura sobre a composição físico-química e compostos bioativos de farinha de taro obtida em leito de jorro. *Brazilian J. Food Technol.* **2017**, *20*, e2016060. [CrossRef]
35. Liu, Y.; Wang, H.; Cai, X. Optimization of the extraction of total flavonoids from Scutellaria baicalensis Georgi using the response surface methodology. *J. Food Sci. Technol.* **2015**, *52*, 2336–2343. [CrossRef]
36. Saito, É.; Sacoda, P.; Paviani, L.C.; Paula, J.T.; Cabral, F.A. Conventional and supercritical extraction of phenolic compounds from Brazilian red and green propolis. *Sep. Sci. Technol.* **2021**, *56*, 3119–3126. [CrossRef]
37. Martinez-Correa, H.A.; Magalhães, P.M.; Queiroga, C.L.; Peixoto, C.A.; Oliveira, A.L.; Cabral, F.A. Extracts from pitanga (*Eugenia uniflora* L.) leaves: Influence of extraction process on antioxidant properties and yield of phenolic compounds. *J. Supercrit. Fluids* **2011**, *55*, 998–1006. [CrossRef]
38. Jaiswal, V.; Lee, H.-J. Antioxidant Activity of Urtica dioica: An Important Property Contributing to Multiple Biological Activities. *Antioxidants* **2022**, *11*, 2494. [CrossRef]
39. Dumanović, J.; Nepovimova, E.; Natić, M.; Kuča, K.; Jaćević, V. The Significance of Reactive Oxygen Species and Antioxidant Defense System in Plants: A Concise Overview. *Front. Plant Sci.* **2021**, *11*, 2106. [CrossRef]
40. Zhang, C. Identification of Free Radical Scavengers from Brazilian Green Propolis Using Off-Line HPLC-DPPH Assay and LC-MS. *J. Food Sci.* **2017**, *82*, 1602–1607. [CrossRef]
41. Cottica, S.M.; Sabik, H.; Antoine, C.; Fortin, J.; Graveline, N.; Visentainer, J.V.; Britten, M. Characterization of Canadian propolis fractions obtained from two-step sequential extraction. *LWT—Food Sci. Technol.* **2015**, *60*, 609–614. [CrossRef]
42. Wang, X.; Ding, G.; Liu, B.; Wang, Q. Flavonoids and antioxidant activity of rare and endangered fern: Isoetes sinensis. *PLoS ONE* **2020**, *15*, e0232185. [CrossRef]
43. An, J.Y.; Kim, C.; Park, N.R.; Jung, H.S.; Koo, T.; Yuk, S.H.; Lee, E.H.; Cho, S.H. Clinical Anti-aging Efficacy of Propolis Polymeric Nanoparticles Prepared by a Temperature-induced Phase Transition Method. *J. Cosmet. Dermatol.* **2022**, *21*, 4060–4071. [CrossRef] [PubMed]
44. Pahlavani, N.; Malekahmadi, M.; Firouzi, S.; Rostami, D.; Sedaghat, A.; Moghaddam, A.B.; Ferns, G.A.; Navashenaq, J.G.; Reazvani, R.; Safarian, M.; et al. Molecular and cellular mechanisms of the effects of Propolis in inflammation, oxidative stress and glycemic control in chronic diseases. *Nutr. Metab.* **2020**, *17*, 65. [CrossRef]
45. Barreto, G.d.A.; Cerqueira, J.C.; Reis, J.H.d.O.; Hodel, K.V.S.; Gama, L.A.; Anjos, J.P.; Minafra-Rezende, C.S.; Andrade, L.N.; Amaral, R.G.; Pessoa, C.d.Ó.; et al. Evaluation of the Potential of Brazilian Red Propolis Extracts: An Analysis of the Chemical Composition and Biological Properties. *Appl. Sci.* **2022**, *12*, 11741. [CrossRef]
46. Costa, S.S.; Druzian, J.I.; Machado, B.A.S.; De Souza, C.O.; Guimaraes, A.G. Bi-functional biobased packing of the cassava starch, glycerol, licuri nanocellulose and red propolis. *PLoS ONE* **2014**, *9*, e112554. [CrossRef]
47. Skowron, K.; Kwiecińska-Piróg, J.; Grudlewska, K.; Gryń, G.; Wiktorczyk, N.; Balcerek, M.; Załuski, D.; Wałecka-Zacharska, E.; Kruszewski, S.; Gospodarek-Komkowska, E. Antilisterial Activity of Polypropylene Film Coated with Chitosan with Propolis and/or Bee Pollen in Food Models. *Biomed Res. Int.* **2019**, *2019*, 7817063. [CrossRef] [PubMed]
48. Suriyatem, R.; Auras, R.; Rachtanapun, C.; Rachtanapun, P. Biodegradable Rice Starch/Carboxymethyl Chitosan Films with Added Propolis Extract for Potential Use as Active Food Packaging. *Polymers* **2018**, *10*, 954. [CrossRef]
49. Hochheim, S.; Guedes, A.; Faccin-Galhardi, L.; Rechenchoski, D.Z.; Nozawa, C.; Linhares, R.E.; Filho, H.H.D.S.; Rau, M.; Siebert, D.A.; Micke, G.; et al. Determination of phenolic profile by HPLC–ESI-MS/MS, antioxidant activity, in vitro cytotoxicity and anti-herpetic activity of propolis from the Brazilian native bee Melipona quadrifasciata. *Rev. Bras. Farmacogn.* **2019**, *29*, 339–350. [CrossRef]
50. Marques das Nevesa, M.V.; Sarmento da Silva, T.M.; Lima, E.D.O.; Leitao da Cunha, E.V.; Oliveira, E.D.J. Isoflavone formononetin from red propolis acts as a fungicide against Candida sp. *Braz. J. Microbiol.* **2016**, *47*, 159–166. [CrossRef] [PubMed]
51. Mendonça, M.A.A.d.; Ribeiro, A.R.S.; Lima, A.K.d.; Bezerra, G.B.; Pinheiro, M.S.; Albuquerque-Júnior, R.L.C.d.; Gomes, M.Z.; Padilha, F.F.; Thomazzi, S.M.; Novellino, E.; et al. Red Propolis and Its Dyslipidemic Regulator Formononetin: Evaluation of Antioxidant Activity and Gastroprotective Effects in Rat Model of Gastric Ulcer. *Nutrients* **2020**, *12*, 2951. [CrossRef] [PubMed]
52. Li, Z.; Dong, X.; Zhang, J.; Zeng, G.; Zhao, H.; Liu, Y.; Qiu, R.; Mo, L.; Ye, Y. Formononetin protects TBI rats against neurological lesions and the underlying mechanism. *J. Neurol. Sci.* **2014**, *338*, 112–117. [CrossRef] [PubMed]
53. Lopez, B.G.-C.; Schmidt, E.M.; Eberlin, M.N.; Sawaya, A.C.H.F. Phytochemical markers of different types of red propolis. *Food Chem.* **2014**, *146*, 174–180. [CrossRef] [PubMed]
54. Figueiredo, Y.G.; Mendonça, H.O.P.; Oliveira-Junior, A.H.; Ramos, A.L.C.C.; Paula, A.C.C.F.F.; Marcucci, M.C.; Parreira, A.G.; Araujo, R.L.B.; Augustini, R.; Melo, J.O.F. Green Propolis Chemical Profile Determination by Mass Spectrometry with Paper Spray Ambient Ionization (PSMS) in Positive Mode. Available online: https://proceedings.science/slaca/slaca-2021/papers/green-propolis-chemical-profile-determination-by-mass-spectrometry-with-paper-sp?lang=en# (accessed on 3 February 2023).
55. Contieri, L.S.; de Souza Mesquita, L.M.; Sanches, V.L.; Viganó, J.; Martinez, J.; da Cunha, D.T.; Rostagno, M.A. Standardization proposal to quality control of propolis extracts commercialized in Brazil: A fingerprinting methodology using a UHPLC-PDA-MS/MS approach. *Food Res. Int.* **2022**, *161*, 111846. [CrossRef] [PubMed]

56. Rodrigues, D.M.; De Souza, M.C.; Arruda, C.; Pereira, R.A.S.; Bastos, J.K. The Role of Baccharis dracunculifolia and its Chemical Profile on Green Propolis Production by Apis mellifera. *J. Chem. Ecol.* **2020**, *46*, 150–162. [CrossRef]
57. Salatino, A.; Fernandes-Silva, C.C.; Righi, A.A.; Salatino, M.L.F. Propolis research and the chemistry of plant products. *Nat. Prod. Rep.* **2011**, *28*, 925. [CrossRef]
58. Ferreira, J.C.; Reis, M.B.; Coelho, G.D.P.; Gastaldello, G.H.; Peti, A.P.F.; Rodrigues, D.M.; Bastos, J.K.; Campo, V.L.; Sorgi, C.A.; Faccioli, L.H.; et al. Baccharin and p-coumaric acid from green propolis mitigate inflammation by modulating the production of cytokines and eicosanoids. *J. Ethnopharmacol.* **2021**, *278*, 114255. [CrossRef]
59. Celińska-Janowicz, K.; Zaręba, I.; Lazarek, U.; Teul, J.; Tomczyk, M.; Pałka, J.; Miltyk, W. Constituents of Propolis: Chrysin, Caffeic Acid, p-Coumaric Acid, and Ferulic Acid Induce PRODH/POX-Dependent Apoptosis in Human Tongue Squamous Cell Carcinoma Cell (CAL-27). *Front. Pharmacol.* **2018**, *9*, 336. [CrossRef]
60. Boo, Y.C. p-Coumaric Acid as An Active Ingredient in Cosmetics: A Review Focusing on its Antimelanogenic Effects. *Antioxidants* **2019**, *8*, 275. [CrossRef]
61. Vishnuvathan, V.J.; Lakshmi, K.S.; Srividya, A.R. STUDY OF ANTIOXIDANT ACTIVITY OF FORMONONETIN BY IN VITRO METHOD. *Int. J. Pharm. Pharm. Sci.* **2017**, *9*, 273. [CrossRef]
62. Shen, Y.; Song, X.; Li, L.; Sun, J.; Jaiswal, Y.; Huang, J.; Liu, C.; Yang, W.; Williams, L.; Zhang, H.; et al. Protective effects of p-coumaric acid against oxidant and hyperlipidemia-an in vitro and in vivo evaluation. *Biomed. Pharmacother.* **2019**, *111*, 579–587. [CrossRef] [PubMed]
63. Selvaraj, K.; Chowdhury, R.; Bhattacharjee, C. Isolation and structural elucidation of flavonoids from aquatic fern Azolla microphylla and evaluation of free radical scavenging activity. *Int. J. Pharm. Pharm. Sci.* **2013**, *5*, 743–749.
64. Machado, B.A.S.; Silva, R.P.D.; de Abreu Barreto, G.; Costa, S.S.; Silva, D.F.D.; Brandão, H.N.; Rocha, J.L.C.d.; Dellagostin, O.A.; Henriques, J.A.P.; Umsza-Guez, M.A.; et al. Chemical Composition and Biological Activity of Extracts Obtained by Supercritical Extraction and Ethanolic Extraction of Brown, Green and Red Propolis Derived from Different Geographic Regions in Brazil. *PLoS ONE* **2016**, *11*, e0145954. [CrossRef] [PubMed]
65. Singleton, V.; Rossi, J. Colorimetry of total phenolics with phosphomolybdic-phosphotungstic acid reagents. *Am. J. Enol. Vitic.* **1965**, *16*, 144–158.
66. Marcucci, M.C.; Ferreres, F.; García-Viguera, C.; Bankova, V.S.; De Castro, S.L.; Dantas, A.P.; Valente, P.H.M.; Paulino, N. Phenolic compounds from Brazilian propolis with pharmacological activities. *J. Ethnopharmacol.* **2001**, *74*, 105–112. [CrossRef]
67. Brand-Williams, W.; Cuvelier, M.E.; Berset, C. Use of a free radical method to evaluate antioxidant activity. *LWT—Food Sci. Technol.* **1995**, *28*, 25–30. [CrossRef]

Disclaimer/Publisher's Note: The statements, opinions and data contained in all publications are solely those of the individual author(s) and contributor(s) and not of MDPI and/or the editor(s). MDPI and/or the editor(s) disclaim responsibility for any injury to people or property resulting from any ideas, methods, instructions or products referred to in the content.

molecules

Article

Study on the Comprehensive Phytochemicals and the Anti-Ulcerative Colitis Effect of *Saussurea pulchella*

Yunhe Liu [1], Caixia Wang [1], Junzhe Wu [1], Luying Tan [1], Peng Gao [1], Sinuo Wu [1], Daohao Tang [1], Qianyun Wang [1], Cuizhu Wang [1,2], Pingya Li [1,2] and Jinping Liu [1,2,*]

1 School of Pharmaceutical Sciences, Jilin University, Changchun 130021, China
2 Research Center of Natural Drugs, Jilin University, Changchun 130021, China
* Correspondence: liujp@jlu.edu.cn; Tel./Fax: +86-431-85619803

Abstract: Background: *Saussurea pulchella (SP)* is a traditional medicinal plant that is widely used in folk medicine because of its diverse biological activities, particularly its anti-inflammatory effects. However, the alleviation effect of SP on ulcerative colitis (UC) has not yet been realized. Purpose: To investigate the chemical composition and therapeutic effect of SP extract against UC. Methods: First, qualitative and quantitative analysis of SP 75% ethanol extract was performed by UPLC-Q/TOF-MS. Second, a dextran sodium sulfate (DSS) model of UC mice was developed to study the effects of SP on the symptoms, inflammatory factors, oxidative stress indexes and colon histopathology. Third, an integration of network pharmacology with metabolomics was performed to investigate the key metabolites, biological targets and metabolisms closely related to the effect of SP. Results: From the SP ethanol extract, 149 compounds were identified qualitatively and 20 were determined quantitatively. The SP could dose-dependently decrease the DAI score, spleen coefficient and the levels of TNF-α, IL-6, iNOS, MPO and MDA; increase the colon length, GSH level and SOD activity; and protect the intestinal barrier in the UC mice. Moreover, 10 metabolite biomarkers,18 targets and 5 metabolisms were found to play crucial roles in the treatment of UC with SP. Conclusions: SP 75% ethanol extract could effectively alleviate the progression of UC and, therefore, could be classified as a novel natural treatment for UC.

Keywords: *Saussurea pulchella*; chemical composition; ulcerative colitis; metabolomics; network pharmacology

Citation: Liu, Y.; Wang, C.; Wu, J.; Tan, L.; Gao, P.; Wu, S.; Tang, D.; Wang, Q.; Wang, C.; Li, P.; et al. Study on the Comprehensive Phytochemicals and the Anti-Ulcerative Colitis Effect of *Saussurea pulchella*. *Molecules* **2023**, *28*, 1526. https://doi.org/10.3390/molecules28041526

Academic Editors: Arunaksharan Narayanankutty, Ademola C. Famurewa and Eliza Oprea

Received: 18 January 2023
Revised: 2 February 2023
Accepted: 3 February 2023
Published: 4 February 2023

Copyright: © 2023 by the authors. Licensee MDPI, Basel, Switzerland. This article is an open access article distributed under the terms and conditions of the Creative Commons Attribution (CC BY) license (https://creativecommons.org/licenses/by/4.0/).

1. Introduction

Ulcerative colitis (UC), an inflammatory disease, has increasing prevalence worldwide [1]. The typical symptoms of UC include bloody diarrhea, tenesmus and abdominal pain [2]. At present, drug intervention is the main method to treat UC. The main first-line drugs are aminosalicylic acids, glucocorticoids, immunomodulators and biological drugs. However, there are also some disadvantages, such as dose-dependent toxicity, drug dependence, irreversible complications or high cost [3]. The research and development of safer, more effective and more economical drugs has become a hotspot in research. With the tide of global natural medicines, products that have the effect of clearing heat and removing dampness, such as *Coptis chinensis*, *Glycyrrhiza uralensis*, *Croton crassifolius* and Shaoyao Decoction, have been used to treat UC in clinic. To our satisfaction, the curative effects were unique and remarkable, and the toxicity was also low [4–8].

Saussurea pulchella (SP), with the functions of dispelling wind, clearing heat, removing dampness and relieving pain [9], is widely distributed in northeastern Asia, particularly in China, Japan, Korea and Russia [10–12]. In Korea, SP had been used as folk medicine for the treatment of inflammation, hypertension, hepatitis and arthritis [12]. In Russia, SP has been used to treat diarrhea [13]. While, in our country, it has been widely used in folk medicine for treating rheumatoid arthritis, hepatitis, diarrhea and other diseases [14,15].

Modern pharmacological research has also demonstrated that the ethanol extract of SP and the sesquiterpenes from SP both exerted obvious anti-inflammatory effects [12,16,17]. It is noteworthly that, *Saussurea lappa*, which is in the same family and has the same genus as SP has also been reported to have the anti-ulcecrative effect. For example, the sesquiterpenes isolated from *Saussurea lappa* methanol extract could alleviate HCl/ethanol-induced gastric mucosal lesions in gastric ulcer rats [18]. In gastric ulceration rats and duodenal ulceration rats, the ethyl acetate extract of *Saussurea lappa* showed good anti-ulcer activity [19] However, there has been no report about the effect of SP on ulcerative colitis. It has been reported that SP contains a variety of secondary metabolites, such as phenylpropanoids, flavonoids and terpenoids [20]. However, the chemical composition of SP is still not clear, namely, there is no literature report on the comprehensive phytochemicals analysis or the quantitative assay of the main chemical components of SP.

This work is intended to investigate the chemical components and anti-ulcerative colitis activity of SP. Firstly, by using UPLC-Q/TOF-MS, the collective understanding of the chemical components of SP was refined based on the UNIFI platform, and the content assay of the main phenylpropanoids and flavonoids in SP was also performed. Secondly, a dextran sulfate sodium (DSS)-induced UC mouse model was established to evaluate the anti-ulcerative colitis activity of SP by examining the biochemical indicators, disease activity index, histopathological changes, etc. Thirdly, to identify the key metabolite biomarkers, targets and metabolisms linked with the effect of SP, an integrated analysis of metabolomics and network pharmacology was carried out. This research is conducive to illustrating the chemical components of SP and to providing a theoretical basis for expanding the application of SP. In addition, the study could also provide a potential natural medicine with good anti-ulcerative colitis activity.

2. Results

2.1. Comprehensive Phytochemical Analysis

2.1.1. Qualitative Analysis

The base peak intensity (BPI) chromatograms of the SP 75% ethanol extract are shown in Figure 1. A total of 149 components were identified or tentatively identified (Table 1). Among them, 35 components were identified through a comparison with the reference substances, while other components were preliminarily identified through accurate molecular weight and typical mass fragment analysis. It is also worth mentioning that 139 of the 149 components were identified from SP for the first time. The identification of these phytochemicals highlights the structural diversity of secondary metabolites in SP.

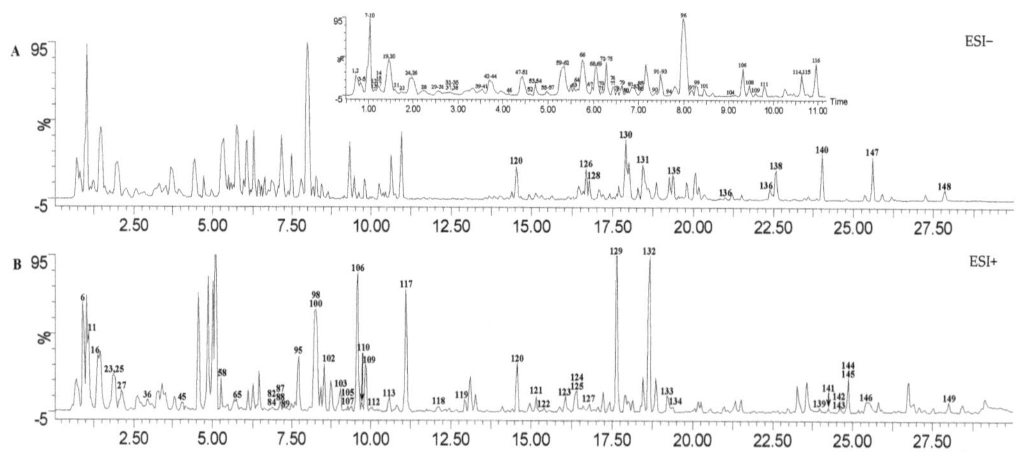

Figure 1. The base peak intensity (BPI) chromatograms of SP in ESI− and ESI+ modes.

Table 1. Compounds identified from SP by UPLC-Q/TOF-MS.

NO.	t_R (min)	Formula	Theoretical Mass (Da)	Calculated Mass (Da)	Mass Error (ppm)	MS^E Fragmentation	Identification	Ref.
1 *	0.71	$C_7H_{12}O_6$	192.0634	192.0643	4.69	191.0570[M–H]$^-$, 173.0443[M–H–H_2O]$^-$	Quinic acid	S
2 *	0.79	$C_{16}H_{18}O_9$	354.0951	354.0961	2.82	353.0988[M–H]$^-$, 191.0569[M–H–$C_9H_6O_3$]$^-$, 179.0338[M–H–$C_7H_{10}O_5$]$^-$, 135.0451[M–H–$C_8H_{10}O_7$]$^-$	Chlorogenic acid	S
3 *	0.82	$C_6H_{12}O_6$	180.0634	180.0637	1.67	179.0564[M–H]$^-$, 131.0358[M–H–H_2O–CH_2O]$^-$, 103.0404[M–H–H_2O–$C_2H_5O_2$]$^-$	D-Galactose	[21]
4 *	0.83	$C_{30}H_{38}O_{15}$	638.2211	638.2232	3.29	683.2214[M–H]$^-$, 489.1377[M–H–Fuc]$^-$, 458.1828[M–H–$C_9H_7O_4$]$^-$, 161.0201[M–H–Glu–Fuc–$C_8H_{11}O_2$]$^-$	Sucrose	[22]
5 *	0.83	$C_{12}H_{22}O_{11}$	342.1162	342.1179	4.97	341.1106[M–H]$^-$, 179.0571[M–H–Glu]$^-$, 161.0465[M–H–OFru]$^-$	Cistanoside C	[23]
6 *	0.94	$C_{10}H_{17}NO_3$	199.1208	199.1213	2.51	200.1286[M+H]$^+$, 168.1017[M+H–CH_3O]$^+$, 126.0930[M+H–C_2H_4–H_2O–CH_3O]$^+$, 122.0978[M+H–H_2O–$C_2H_3O_2$]$^+$, 94.0687[M+H–C_2H_4–H_2O–$C_2H_3O_2$]$^+$	Tussilagine	[24]
7 *	0.96	$C_{14}H_{18}O_9$	330.0951	330.0961	3.03	329.0889[M–H]$^-$, 167.0356[M–H–Glu]$^-$	Mudanoside A	[25]
8 *	0.98	$C_{13}H_{16}O_{10}$	332.0743	332.0754	3.31	331.0681[M–H]$^-$, 168.0068[M–H–Glu]$^-$, 124.0173[M–H–Glu–CO_2]$^-$	Glucogallin	[26]
9 *	1.01	$C_7H_6O_4$	154.0266	154.0264	−1.30	153.0192[M–H]$^-$, 109.0281[M–H–HCOOH]$^-$	Protocatechuic acid	S
10 *	1.03	$C_{11}H_{12}O_6$	240.0634	240.0643	3.75	285.0625[M+HCOO]$^-$, 239.0564[M–H]$^-$, 149.0597[M–H–2HCOOH]$^-$, 108.0518[M–H–$C_4H_2O_5$]$^-$	Eucomic acid	[27]
11 *	1.08	$C_{15}H_{21}NO_7$	327.1318	327.1323	1.53	328.1395[M+H]$^+$, 310.1288[M+H–H_2O]$^+$, 292.1183[M+H–2H_2O]$^+$, 264.1229[M+H–H_2O–HCOO]$^+$, 166.0867[M+H–$C_6H_{10}O_5$]$^+$	Fructose-phenylalanine	[28]
12 *	1.13	$C_{14}H_{20}O_9$	332.1107	332.1121	4.22	331.1049[M–H]$^-$, 168.0431[M–H–Glu]$^-$, 154.0237[M–H–Glu–CH_3]$^-$, 139.0028[M–H–Glu–2CH_3]$^-$, 137.0246[M–H–Glu–CH_3O]$^-$	Leonuriside A	[29]
13 *	1.16	$C_{16}H_{18}O_9$	354.0951	354.0964	3.75	353.0887[M–H]$^-$, 191.0568[M–H–$C_9H_6O_3$]$^-$, 135.0456[M–H–$C_8H_{10}O_7$]$^-$	Neochlorogenic acid	S
14 *	1.20	$C_{15}H_{18}O_9$	342.0951	342.0965	4.09	341.0892[M–H]$^-$, 179.0342[M–H–Glu]$^-$, 135.0446[M–H–Glu–CO_2]$^-$	Phoeniceoside	[30]
15 *	1.24	$C_{15}H_{18}O_8$	326.1002	326.1011	2.76	325.0938[M–H]$^-$, 163.0413[M–H–Glu]$^-$, 119.0513[M–H–Glu–HCOOH]$^-$	Melilotoside	[31]
16 *	1.30	$C_{18}H_{18}O_5$	314.1154	314.1162	3.18	315.1235[M+H]$^+$, 193.0875[M+H–$C_7H_6O_2$]$^+$, 147.0451[M+H–CH_3O–$C_8H_9O_2$]$^+$, 137.0622[M+H–$C_{10}H_{10}O_3$]$^+$	*p*-Hydroxyphenethyl ferulate	CFM-ID
17 *	1.31	$C_7H_6O_4$	154.0266	154.0270	2.60	153.0197[M–H]$^-$, 109.0293[M–H–HCOOH]$^-$	3,4-Dihydroxybenzoic acid	[32]
18 *	1.45	C_9H_7NO	145.0528	145.0523	−3.45	146.0595[M+H]$^+$, 118.0655[M+H–CHO]$^+$	Indole-3-aldehyde	[33]
19 *	1.47	$C_{27}H_{28}N_2O_4$	444.2049	444.2035	−3.16	443.1962[M–H]$^-$, 252.1025[M–H–$C_{11}H_{13}O_2N$]$^-$	Cryptochlorogenic acid	S
20 *	1.47	$C_{16}H_{18}O_9$	354.0951	354.0962	3.11	353.0889[M–H]$^-$, 307.0824[M–H–HCOOH]$^-$, 191.0566[M–H–$C_9H_6O_3$]$^-$, 146.0587[M–H–$C_9H_6O_3$–HCOOH]$^-$	Aurantiamide acetate	[34]

Table 1. *Cont.*

NO.	t_R (min)	Formula	Theoretical Mass (Da)	Calculated Mass (Da)	Mass Error (ppm)	MS^E Fragmentation	Identification	Ref.
21 *	1.67	$C_{10}H_{12}O_4$	196.0736	196.0742	3.06	241.0724[M+HCOO]$^-$, 195.0661[M–H]$^-$, 179.0721[M–H–H$_2$O]$^-$, 165.0563[M–H–CH$_3$O]$^-$	Acetosyringone	[35]
22 *	1.76	$C_{18}H_{26}O_9$	386.1577	386.1590	3.37	431.1572[M+HCOO]$^-$, 385.1503[M–H]$^-$, 223.0995[M–H–Glu]$^-$, 135.0467[M–H–OGlu–C$_4$H$_7$O]$^-$	Methylsyringin	[36]
23	1.78	$C_{19}H_{27}NO_6$	365.1838	365.1844	1.62	366.1917[M+H]$^+$, 330.1704[M+H–2H$_2$O]$^+$, 262.1437[M+H–H$_2$O–C$_4$H$_7$O$_2$]$^+$	Pulchellamine B	[37]
24 *	1.82	$C_{14}H_{18}O_7$	298.1053	298.1067	4.70	343.1039[M+HCOO]$^-$, 164.0695[M–H–C$_8$H$_7$O$_2$]$^-$, 133.0303[M–H–Glu]$^-$, 121.0300[M–H–Glu–CH$_3$]$^-$	Ameliaroside	[38]
25 *	1.88	$C_{20}H_{27}NO_6$	377.1838	377.1846	2.12	378.1919[M+H]$^+$, 360.1802[M+H–H$_2$O]$^+$, 332.1862[M+H–HCOOH]$^+$, 314.1749[M+H–HCOOH–H$_2$O]$^+$, 227.1060[M+H–2H$_2$O–C$_5$H$_9$NO$_2$]$^+$	Calophyllamine A	[39]
26	1.97	$C_{17}H_{24}O_9$	372.1420	372.1433	3.49	417.1445[M+HCOO]$^-$, 371.1351[M–H]$^-$, 209.0821[M–H–Glu]$^-$, 194.0586[M–H–CH$_3$–Glu]$^-$, 151.0409[M–H–Glu–C$_3$H$_5$O]$^-$	Syringin	[26]
27 *	2.08	$C_{20}H_{20}O_8$	388.1158	388.1156	−0.51	411.1048 [M+Na]$^+$, 389.1232[M+H]$^+$, 371.1133[M+H–H$_2$O]$^+$, 167.0720[M+H–C$_{11}$H$_{10}$O$_5$]$^+$	6α-Catechyl-2α-guaicyl-3,7-dioxabicyclo [3.3.0]octan-4-one	[40]
28 *	2.26	$C_7H_6O_3$	138.0317	138.0323	4.35	137.0251[M–H]$^-$, 109.0302[M–H–CHO]$^-$	Protocatechuic aldehyde	S
29 *	2.36	$C_{15}H_{20}O_8$	328.1158	328.1168	3.05	327.1095[M–H]$^-$, 165.0562[M–H–Glu]$^-$, 147.0453[M–H–H$_2$O–Glu]$^-$	Paeonoside	[41]
30 *	2.48	$C_{17}H_{24}O_{10}$	388.1369	388.1387	4.61	387.1305[M–H]$^-$, 371.0989[M–H–CH$_3$]$^-$, 207.0664[M–H–OGlu]$^-$, 192.0432[M–H–Glu–CH$_3$O]$^-$	Geniposide	[42]
31 *	2.57	$C_{16}H_{18}O_8$	338.1002	338.1010	2.37	337.0931[M–H]$^-$, 191.0562[M–H–C$_9$H$_7$O$_2$]$^-$, 163.0402[M–H–C$_7$H$_{11}$O$_5$]$^-$	3-*p*-Coumaroylquinic acid	[43]
32 *	2.61	$C_8H_8O_4$	168.0423	168.0426	1.79	167.0373[M–H]$^-$, 123.0355[M–H–HCOOH]$^-$, 108.0216[M–H–HCOOH–CH$_3$]$^-$, 93.0343[M–H–HCOOH–CH$_3$O]$^-$	Vanillic acid	[44]
33 *	2.77	$C_9H_{10}O_5$	198.0528	198.0551	1.51	197.0449[M–H]$^-$, 179.0345[M–H–H$_2$O]$^-$, 135.0444[M–H–H$_2$O–HCOOH]$^-$	Syringic acid	[45]
34 *	2.80	$C_{17}H_{26}O_7$	342.1679	342.1692	3.80	387.1664[M+HCOO]$^-$, 341.1608[M–H]$^-$, 163.1127[M–H–OGlu]$^-$	Jasmolone glucoside	CFM-ID
35 *	2.81	$C_9H_8O_4$	180.0423	180.0422	−0.56	179.0340[M–H]$^-$, 135.0438[M–H–HCOOH]$^-$	Caffeic acid	S
36 *	2.82	$C_{20}H_{27}NO_6$	377.1838	377.1831	−1.92	378.1904[M+H]$^+$, 332.1854[M+H–HCOOH]$^+$, 257.1408[M+H–2H$_2$O–CH$_2$–C$_3$H$_3$O$_2$]$^+$, 235.0971[M+H–C$_3$H$_3$O$_2$–C$_4$H$_8$N]$^+$, 206.0939[M+H–C$_3$H$_3$O$_2$–C$_5$H$_8$NO]$^+$	Lanicepomine A	[37]
37 *	2.98	$C_{13}H_{18}O_6$	270.1103	270.1111	2.96	315.1113[M+HCOO]$^-$, 269.1029[M–H]$^-$, 161.0455[M–H–C$_7$H$_8$O]$^-$	Benzyl β-D-glucoside	[46]
38 *	3.06	$C_{15}H_{16}O_6$	292.0947	292.0958	3.77	337.0930[M+HCOO]$^-$, 291.0873[M–H]$^-$, 163.0414[M–H–H$_2$O–C$_2$HO–C$_4$H$_5$O]$^-$	Cnidimol D	[47]

Table 1. *Cont.*

NO.	t_R (min)	Formula	Theoretical Mass (Da)	Calculated Mass (Da)	Mass Error (ppm)	MS^E Fragmentation	Identification	Ref.
39 *	3.50	$C_{11}H_{14}O_5$	226.0841	226.0844	1.34	225.0770[M–H]$^-$, 195.0663[M–H–CH$_3$O]$^-$, 180.0427[M–H–C$_2$H$_5$O]$^-$, 149.0240[M–H–CH$_3$O–C$_2$H$_5$O]$^-$	3-Hydroxy-1-(4-hydroxy-3,5-dimethoxyphenyl)propan-1-one	[48]
40 *	3.52	$C_{26}H_{34}O_{12}$	538.2050	538.2058	1.49	583.2031[M+HCOO]$^-$, 537.1982[M–H]$^-$, 375.1454[M–H–Glu]$^-$, 357.1342[M–H–Glu–H$_2$O]$^-$, 151.0407[M–H–Glu–C$_{12}$H$_{16}$O$_4$]$^-$	Medusaside A	[49]
41 *	3.58	$C_{17}H_{20}O_9$	368.1107	368.1110	0.81	367.1037[M–H]$^-$, 191.0564[M–H–CH$_3$–C$_9$H$_5$O$_3$]$^-$, 161.0241[M–H–C$_8$H$_{14}$O$_6$]$^-$, 135.0450[M–H–C$_9$H$_{12}$O$_7$]$^-$	Methyl 3-caffeoylquinate	[50]
42 *	3.70	$C_{19}H_{32}O_8$	388.2097	388.2116	4.89	433.2099[M+HCOO]$^-$, 387.2030[M–H]$^-$, 225.1501[M–H–Glu]$^-$, 153.0920[M–H–C$_4$H$_7$O–Glu]$^-$	Icariside B8	CFM-ID
43 *	3.72	$C_{26}H_{34}O_{12}$	538.2050	538.2068	3.34	583.2039[M+HCOO]$^-$, 537.1980[M–H]$^-$, 375.1451[M–H–Glu]$^-$, 153.0927[M–H–C$_{17}$H$_{20}$O$_{10}$]$^-$	Medusaside B	[49]
44 *	3.73	$C_{19}H_{30}O_8$	386.1941	386.1949	2.07	431.1961 [M+HCOO]$^-$, 385.1970[M–H]$^-$, 223.1344[M–H–Glu]$^-$, 205.1231[M–H–Glu–H$_2$O]$^-$	Saussureoside B	[51]
45	4.21	$C_{20}H_{29}NO_6$	379.1995	379.2002	1.85	380.2075[M+H]$^+$, 334.2013[M+H–HCOOH]$^+$, 316.1910[M+H–HCOOH–H$_2$O]$^+$, 215.1075[M+H–2H$_2$O–C$_6$H$_{11}$NO$_2$]$^+$	Pulchellamine E	[37]
46 *	4.29	$C_9H_{10}O_3$	166.0630	166.0635	3.01	165.0562[M–H]$^-$, 147.0452[M–H–H$_2$O]$^-$	Phloretic acid	[52]
47 *	4.37	$C_{26}H_{34}O_{12}$	538.2050	538.2069	3.53	537.1986[M–H]$^-$, 375.1451[M–H–Glu]$^-$, 327.1240[M–H–Glu–H$_2$O–CH$_3$O]$^-$, 297.1136[M–H–Glu–H$_2$O-2CH$_3$O]$^-$, 225.1250[M–H–Glu–C$_8$H$_7$O$_3$]–	Lanicepside A	[53]
48 *	4.44	$C_8H_8O_2$	136.0524	136.0527	2.21	135.0455[M–H]$^-$, 120.0213[M–H–CH$_3$]$^-$, 92.0267[M–H–C$_2$H$_3$O]$^-$	Curculigoside C	[54]
49 *	4.44	$C_{22}H_{26}O_{12}$	482.1424	482.1439	3.11	481.1346[M–H]$^-$, 197.0455[M–H–Glu– C$_7$H$_5$O$_2$]$^-$, 121.0295[M–H–Glu–C$_9$H$_9$O$_5$]$^-$	*p*-Hydroxyacetophenone	[55]
50 *	4.49	$C_{21}H_{34}O_9$	430.2203	430.2220	3.95	429.2127[M–H]$^-$,401.1817[M–H–C$_2$H$_4$]$^-$, 267.1603[M–H–Glu]$^-$	4α(15),11β(13)-Tetrahydroridentin B-1 -glucoside	[56]
51 *	4.56	$C_{26}H_{34}O_{12}$	538.2050	538.2067	3.16	537.2064[M–H]$^-$, 327.1240[M–H–Glu–H$_2$O–CH$_3$O]$^-$, 195.0664[M–H–Glu–C$_{10}$H$_{11}$O$_3$]$^-$, 161.0464[M–H–C$_{20}$H$_{24}$O$_7$]$^-$	Citrusin A	[57]
52 *	4.64	$C_{26}H_{34}O_{12}$	538.2050	538.2070	3.72	583.2048[M+HCOO]$^-$, 537.1982[M–H]$^-$, 375.1442[M–H–Glu]$^-$, 327.1245[M–H–Glu–H$_2$O–CH$_3$O]$^-$, 179.0561[M–H–C$_{20}$H$_{22}$O$_6$]$^-$	Lanicepside B	[53]
53 *	4.66	$C_{32}H_{42}O_{16}$	682.2473	682.2499	3.81	727.2481[M+HCOO]$^-$, 681.2411[M–H]$^-$, 519.1877[M–H–Glu]$^-$, 339.1242[M–H–2Glu–H$_2$O]$^-$	Pinoresinol diglucoside	S
54 *	4.73	$C_{27}H_{36}O_{13}$	568.2156	568.2166	1.76	613.2188[M+HCOO]$^-$, 567.2092[M–H]$^-$, 521.2040[M–H–H$_2$O–CH$_3$O]$^-$, 405.1565[M–H–Glu]$^-$, 195.0662[M–H–Glu–C$_{11}$H$_{13}$O$_4$]$^-$	Citrusin B	CFM-ID

Table 1. *Cont.*

NO.	t_R (min)	Formula	Theoretical Mass (Da)	Calculated Mass (Da)	Mass Error (ppm)	MS^E Fragmentation	Identification	Ref.
55 *	5.01	$C_{26}H_{36}O_{11}$	524.2258	524.2240	−3.43	523.2167[M–H]$^-$, 507.1880[M–H–CH$_3$]$^-$, 361.1690[M–H–Glu]$^-$, 346.1771[M–H–Glu–CH$_3$]$^-$, 315.1331[M–H–Glu–CH$_3$–CH$_3$O]$^-$	(-)-Secoisolariciresinol-4-O-β-D-glucoside	[58]
56 *	5.01	$C_{17}H_{20}O_9$	368.1107	368.1116	2.44	367.1033[M–H]$^-$, 179.0346[M–H–C$_8$H$_{12}$O$_5$]$^-$, 161.0247[M–H–C$_8$H$_{14}$O$_6$]$^-$, 135.0472[M–H–C$_9$H$_{12}$O$_7$]$^-$	Methyl 4-caffeoylquinate	[59]
57 *	5.08	$C_{19}H_{24}O_8$	380.1471	380.1484	3.42	425.1466[M+HCOO]$^-$, 379.1404[M–H]$^-$, 343.1188[M–H–2H$_2$O]$^-$	15-Hydroxyjanerin	CFM-ID
58	5.11	$C_{21}H_{31}NO_6$	393.2151	393.2163	3.05	416.2095[M+Na]$^+$, 394.2241[M+H]$^+$, 378.1929[M+H–CH$_3$]$^+$, 342.1711[M+H–CH$_3$–2H$_2$O]$^+$, 262.1448[M+H–C$_6$H$_{12}$O$_2$–H$_2$O]$^+$, 228.1161[M+H–C$_6$H$_{12}$NO$_2$–2H$_2$O]$^+$	Pulchellamine G	[37]
59 *	5.21	$C_{28}H_{38}O_{13}$	582.2312	582.2303	−1.55	581.2230[M–H]$^-$, 419.1720[M–H–Glu]$^-$, 389.1603[M–H–Glu–CH$_3$O]$^-$, 373.1298[M–H–Glu–CH$_3$–CH$_3$O]$^-$	Lyoniresinol-3α-glucoside	[60]
60 *	5.28	$C_{21}H_{18}O_{12}$	462.0798	462.0806	1.73	461.0734[M–H]$^-$, 285.0404[M–H–Gluac]$^-$, 151.0049[M–H–Gluac–C$_8$H$_6$O$_2$]$^-$, 132.0210[M–H–Gluac–C$_7$H$_4$O$_4$]$^-$	Luteolin 7-glucuronide	S
61	5.32	$C_{27}H_{30}O_{16}$	610.1534	610.1530	−0.66	609.1457[M–H]$^-$, 461.0731[M–H–Rha]$^-$, 300.0281[M–H–Glu–Rha]$^-$	Rutin	S
62 *	5.57	$C_{21}H_{20}O_{12}$	464.0955	464.0978	4.96	463.0906[M–H]$^-$, 300.0280[M–H–Glu]$^-$, 151.0041[M–H–Glu–C$_8$H$_5$O$_3$]$^-$, 150.0328[M–H–Glu–C$_7$H$_4$O$_4$]$^-$	Isoquercitroside	S
63 *	5.62	$C_{19}H_{22}O_5$	330.1467	330.1475	2.42	375.1447[M+HCOO]$^-$, 329.1395[M–H]$^-$, 297.1131[M–H–CH$_3$–H$_2$O]$^-$, 282.0899[M–H–CH$_3$–CH$_2$–H$_2$O]$^-$, 226.0641[M–H–H$_2$O–CH$_2$–C$_4$H$_7$O]$^-$	Aguerin B	[61]
64 *	5.66	$C_{27}H_{34}O_{12}$	550.2050	550.2069	3.45	595.2042[M+HCOO]$^-$, 549.1984[M–H]$^-$, 519.1876[M–H–CH$_3$O]$^-$, 387.1454[M–H–Glu]$^-$	Saussurenoside	[62]
65 *	5.72	$C_{15}H_{10}O_7$	302.0427	302.0435	2.65	303.0508[M+H]$^+$, 178.0272[M+H– C$_6$H$_5$O$_3$]$^+$, 153.0195[M+H–C$_8$H$_5$O$_3$]$^+$, 108.0216[M+H–H2O–C$_9$H$_5$O$_4$]$^+$	Isoetin	[63]
66 *	5.76	$C_{25}H_{24}O_{12}$	516.1268	516.1287	3.68	515.1204[M–H]$^-$, 353.0885[M–H–C$_9$H$_6$O$_3$]$^-$, 335.0776[M–H–C$_9$H$_9$O$_4$]$^-$, 191.0570[M–H–2C$_9$H$_6$O$_3$]$^-$, 179.0353[M–H–C$_{16}$H$_{16}$O$_8$]$^-$	1,4-Dicaffeoylquinic acid	S
67 *	5.96	$C_{27}H_{30}O_{15}$	594.1585	594.1598	2.19	593.1515[M–H]$^-$, 285.0407[M–H–Rut]$^-$	Luteolin-7-rutinoside	[64]
68 *	6.03	$C_{25}H_{24}O_{12}$	516.1268	516.1271	0.58	515.1198[M–H]$^-$, 353.0878[M–H–C$_9$H$_6$O$_3$]$^-$, 191.0561[M–H–2C$_9$H$_6$O$_3$]$^-$, 179.0352[M–H–C$_{16}$H$_{16}$O$_8$]$^-$	1,5-Dicaffeoylquinic acid	S
69 *	6.06	$C_{28}H_{32}O_{16}$	624.1690	624.1720	4.81	623.1647[M–H]$^-$, 351.0735[M–H–ORha–C$_6$H$_4$O$_2$]$^-$, 315.0530[M–H–Rut]$^-$	Narcisin	S
70 *	6.15	$C_{27}H_{30}O_{14}$	578.1636	578.1653	2.94	577.1580[M–H]$^-$, 269.0474[M–H–Neo]$^-$	Rhoifolin	[65]
71 *	6.18	$C_{21}H_{18}O_{11}$	446.0849	446.0868	4.26	445.0791[M–H]$^-$, 284.0322[M–H–Glu]$^-$, 269.0464[M–H– OGlu]$^-$	Rhein-8-glucoside	[66]
72	6.29	$C_{22}H_{26}O_8$	418.1628	418.1639	2.63	417.1567[M–H]$^-$, 402.1271[M–H–CH$_3$]$^-$, 387.1080[M–H–2CH$_3$]$^-$, 181.0521[M–H–C$_{13}$H$_{11}$O$_4$]$^-$	Syringaresinol	S

Table 1. *Cont.*

NO.	t_R (min)	Formula	Theoretical Mass (Da)	Calculated Mass (Da)	Mass Error (ppm)	MS^E Fragmentation	Identification	Ref.
73	6.29	$C_{21}H_{20}O_{11}$	448.1006	448.1021	3.35	447.0938$[M-H]^-$, 301.0375$[M-H-Rha]^-$, 283.0255$[M-H-Rha-H_2O]^-$, 151.0043$[M-H-Rha-C_8H_5O_3]^-$	Quercitrin	S
74	6.31	$C_{26}H_{32}O_{12}$	536.1894	536.1901	1.31	535.1823$[M-H]^-$, 501.1768$[M-H-2H_2O]^-$, 355.1188$[M-H-Glu-H_2O]^-$, 151.0405$[M-H-Glu-C_{12}H_{13}O_4]^-$	1-Hydroxypinoresinol-1 -glucoside	[67]
75 *	6.36	$C_{21}H_{20}O_{10}$	432.1056	432.1075	4.40	477.1057$[M+HCOO]^-$, 431.0993$[M-H]^-$, 285.0405$[M-H-Rha]^-$, 161.0464$[M-H-C_{15}H_{10}O_5]^-$	Afzelin	[68]
76 *	6.45	$C_{25}H_{24}O_{12}$	516.1268	516.1292	4.65	515.1219$[M-H]^-$, 353.0891$[M-H-C_9H_6O_3]^-$, 191.0579$[M-H-2C_9H_6O_3]^-$, 179.0359$[M-H-C_{16}H_{16}O_8]^-$	4,5-Dicaffeoylquinic acid	S
77 *	6.47	$C_{21}H_{20}O_{10}$	432.1056	432.1071	3.47	431.0988$[M-H]^-$, 269.0461$[M-H-Glu]^-$	Cosmosiin	[68]
78 *	6.57	$C_{26}H_{32}O_{11}$	520.1945	520.1931	−2.71	565.1913$[M+HCOO]^-$, 519.1851$[M-H]^-$, 357.1323$[M-H-Glu]^-$, 151.0387$[M-H-Glu-C_{12}H_{13}O_3]^-$	Pinoresinol 4- glucoside	S
79 *	6.63	$C_9H_{16}O_4$	188.1049	188.1050	0.53	187.0977$[M-H]^-$, 143.1081$[M-H-HCOOH]^-$, 125.0968$[M-H-H_2O-HCOOH]^-$	Azelaic acid	[69]
80 *	6.67	$C_{22}H_{22}O_{11}$	462.1162	462.1151	−2.38	461.1078$[M-H]^-$, 446.0853$[M-H-CH_3]^-$, 298.0472$[M-H-Glu]^-$, 283.0244$[M-H-Glu-CH_3]^-$	Thermopsoside	[70]
81 *	6.82	$C_{34}H_{30}O_{15}$	678.1585	678.1606	3.10	677.1513$[M-H]^-$, 515.1194$[M-H-C_9H_6O_3]^-$, 497.1098$[M-H-C_9H_8O_4]^-$, 353.0881$[M-H-2C_9H_6O_3]^-$, 179.0346$[M-H-C_{25}H_{22}O_{11}]^-$	1,3,5-Tricaffeoylquinic acid	[71]
82 *	6.89	$C_{20}H_{26}O_8$	394.1628	394.1641	3.30	417.1533$[M+Na]^+$, 395.1713$[M+H]^+$, 359.1508$[M+H-2H_2O]^+$, 350.1378$[M+H-C_2H_5O]^+$, 327.1243$[M+H-2H_2O-CH_3O]^+$, 229.0776$[M+H-C_2H_5O-H_2O-C_4H_7O_3]^+$	Methoxyjanerin	[72]
83 *	6.93	$C_{20}H_{26}O_6$	362.1729	362.1737	2.21	361.1664$[M-H]^-$, 346.1428$[M-H-CH_3]^-$, 327.1231$[M-H-H_2O-CH_3]^-$, 315.1247$[M-H-CH_3-CH_3O]^-$, 165.0563$[M-H-C_{10}H_{13}O_3-CH_3]^-$	Secoisolariciresinol	S
84 *	7.00	$C_{21}H_{22}O_7$	386.1366	386.1361	−1.29	387.1434$[M+H]^+$, 163.0400$[M+H-C_{10}H_{10}O_4-2CH_3]^+$, 135.0453$[M+H-C_{13}H_{16}O_5]^+$	Conicaol B	[73]
85 *	7.06	$C_{26}H_{32}O_{11}$	520.1945	520.1959	2.69	519.1876$[M-H]^-$, 357.1345$[M-H-Glu]^-$, 342.1116$[M-H-Glu-CH_3]^-$, 121.0305$[M-H-Glu-C_{13}H_{15}O_4]^-$	Matairesinoside	S
86 *	7.09	$C_{20}H_{28}O_7$	380.1835	380.1852	4.47	425.1827$[M+HCOO]^-$, 379.1772$[M-H]^-$, 221.0840$[M-H-C_8H_{14}O_3]^-$, 209.0834$[M-H-C_9H_{14}O_3]^-$	Elemacarmanin	CFM-ID
87 *	7.10	$C_{18}H_{22}O_6$	334.1416	334.1421	1.50	357.1353$[M+Na]^+$, 335.1502$[M+H]^+$, 317.1404$[M+H-H_2O]^+$, 137.0614$[M+H-H_2O-C_8H_9O-C_2H_3O_2]^+$	7α-Hydroxygerin	[74]
88 *	7.14	$C_{20}H_{22}O_4$	326.1518	326.1524	1.84	327.1597$[M+H]^+$, 203.1089$[M+H-C_7H_8O_2]^+$, 189.0924$[M+H-C_7H_7O_2-CH_3]^+$, 137.0614$[M+H-C_{12}H_{14}O_2]^+$	Dehydrodiisoeugenol	[75]

Table 1. *Cont.*

NO.	t_R (min)	Formula	Theoretical Mass (Da)	Calculated Mass (Da)	Mass Error (ppm)	MS^E Fragmentation	Identification	Ref.
89 *	7.28	$C_{22}H_{24}O_8$	416.1471	416.1469	−0.48	417.1542[M+H]$^+$, 399.1435[M+H–H$_2$O]$^+$, 358.1362[M+H–C$_2$H$_2$O$_2$]$^+$, 137.0613[M+H–H$_2$O–C$_{14}$H$_{14}$O$_5$]$^+$	Acetoxypinoresinol	CFM-ID
90 *	7.35	$C_{30}H_{34}O_{10}$	554.2152	554.2171	3.43	553.2089[M–H]$^-$, 535.1990[M–H–H$_2$O]$^-$, 357.1352[M–H–H$_2$O–C$_{10}$H$_{10}$O$_3$]$^-$, 181.0877[M–H–C$_{20}$H$_{20}$O$_7$]$^-$	Lappaol E	[76]
91 *	7.49	$C_{18}H_{18}O_3$	282.1256	282.1267	3.90	327.1249[M+HCOO]$^-$, 239.0726[M–H–C$_3$H$_6$]$^-$, 197.0626[M–H–C$_3$H$_5$–C$_2$H$_3$–H$_2$O]$^-$, 163.0405[M–H–C$_9$H$_{10}$]$^-$	Obovatol	[77]
92 *	7.50	$C_{20}H_{20}O_5$	340.1311	340.1321	2.94	339.1248[M–H]$^-$, 324.1008[M–H–CH$_3$]$^-$, 293.0825[M–H–CH$_3$–CH$_3$O]$^-$, 265.0519[M–H–H$_2$O–C$_4$H$_8$]$^-$	Licocoumarone	[78]
93 *	7.51	$C_{20}H_{22}O_6$	358.1416	358.1425	2.51	357.1342[M–H]$^-$, 342.1117[M–H–CH$_3$]$^-$, 151.0405[M–H–C$_{12}$H$_{16}$O$_3$]$^-$, 136.0538[M–H–C$_{12}$H$_{11}$O$_3$–H$_2$O]$^-$	Pinoresinol	S
94 *	7.68	$C_{30}H_{34}O_{10}$	554.2152	554.2168	2.89	553.2095[M–H]$^-$,535.1954[M–H–H$_2$O]$^-$, 517.1888[M–H–2H$_2$O]$^-$	Lappaol C	[79]
95 *	7.72	$C_{17}H_{20}O_4$	288.1362	288.1376	4.86	311.1268[M+Na]$^+$, 289.1457[M+H]+, 230.1312[M+H–C$_2$H$_3$O$_2$]$^+$, 202.1370[M+H–C$_2$H$_3$O$_2$–CO]$^+$	8α-Acetoxydehydrocostuslactone	[80]
96 *	8.00	$C_{27}H_{34}O_{11}$	534.2101	534.2118	3.18	579.2090[M+HCOO]$^-$, 533.2035[M–H]$^-$, 371.1512[M–H–Glu]$^-$, 356.1280[M–H–Glu–CH$_3$]$^-$, 136.0535[M–H–Glu–C$_{13}$H$_{14}$O$_4$]$^-$, 121.0306[M–H–Glu–CH$_3$–C$_{13}$H$_{14}$O$_4$]$^-$	Arctiin	S
97 *	8.18	$C_{15}H_{10}O_6$	286.0477	286.0488	3.85	285.0415[M–H]$^-$, 151.0044[M–H–C$_8$H$_6$O$_2$]$^-$, 133.0308[M–H–C$_7$H$_4$O$_4$]$^-$, 107.0144[M–H–C$_9$H$_6$O$_4$]$^-$	Luteolin	S
98 *	8.23	$C_8H_8O_2$	136.0524	136.0521	−2.21	137.0613[M+H]$^+$, 122.0364[M+H–CH$_3$]$^+$, 94.0407[M+H–C$_2$H$_3$O]$^+$	Phenyl acetate	[81]
99 *	8.24	$C_{34}H_{30}O_{15}$	678.1585	678.1614	4.28	677.1521[M–H]$^-$, 515.1210[M–H–C$_9$H$_6$O$_3$]$^-$, 353.0895[M–H–2C$_9$H$_6$O$_3$]$^-$, 335.0788[M–H–C$_9$H$_7$O$_3$–C$_9$H$_7$O$_4$]$^-$, 179.0352[M–H–C$_{25}$H$_{22}$O$_{11}$]$^-$	3,4,5–Tricaffeoylquinic acid	[82]
100 *	8.26	$C_{21}H_{24}O_6$	372.1573	372.1583	2.69	373.1656[M+H]$^+$, 355.1549[M+H–H$_2$O]$^+$, 137.0617[M+H–C$_{13}$H$_{16}$O$_4$]$^+$, 122.0386[M+H–C$_{14}$H$_{19}$O$_4$]$^+$	Phillygenin	S
101 *	8.46	$C_{30}H_{36}O_9$	540.2359	540.2381	4.07	585.2352[M+HCOO]$^-$, 539.2308[M–H]$^-$, 521.2194[M–H–H$_2$O]$^-$, 509.2192[M–H–CH$_3$O]$^-$, 371.1505[M–H–CH$_3$O–C$_8$H$_9$O$_2$]$^-$, 297.1145[M–H–H$_2$O–C$_{12}$H$_{16}$O$_4$]$^-$	Sesquipinsapol B	[83]
102 *	8.54	$C_{16}H_{12}O_7$	316.0583	316.0596	4.11	317.0668[M+H]$^+$, 302.0429 [M+H–CH$_3$]$^+$, 168.0062[M+H–CH$_3$–C$_8$H$_6$O$_2$]$^+$, 140.0506[M+H–C$_9$H$_5$O$_4$]$^+$	Eupafolin	S
103 *	9.04	$C_{15}H_{24}O_2$	236.1776	236.1786	4.23	237.1858[M+H]$^+$, 219.1771[M+H–H$_2$O]$^+$, 108.0945[M+H–C$_7$H$_{13}$O$_2$]$^+$, 92.0631[M+H–C$_3$H$_7$O–H$_2$O–C$_5$H$_8$]$^+$	Eudesma-4(14),11(13)-diene-3β,12-diol	[84]
104 *	9.08	$C_{31}H_{36}O_{10}$	568.2308	568.2328	3.52	567.2256[M–H]$^-$, 535.1982[M–H–CH$_3$O]$^-$, 517.1888[M–H–H$_2$O–CH$_3$O]$^-$, 191.0714[M–H–C$_{20}$H$_{24}$O$_7$]$^-$	Lappaol D	[79]

Table 1. *Cont.*

NO.	t_R (min)	Formula	Theoretical Mass (Da)	Calculated Mass (Da)	Mass Error (ppm)	MS^E Fragmentation	Identification	Ref.
105 *	9.28	$C_{34}H_{37}N_3O_6$	583.2682	583.2684	0.34	584.2757[M+H]$^+$, 438.2385[M+H–$C_9H_6O_2$]$^+$, 292.2026[M+H–2$C_9H_6O_2$]$^+$, 275.1765[M+H–C_9H_6O2–$C_9H_9NO_2$]$^+$, 147.0453[M+H–$C_{25}H_{31}N_3O_4$]$^+$	*N*1,*N*5,*N*10-Tri-*p*-coumaroylspermidine	[85]
106 *	9.33	$C_{15}H_{10}O_5$	270.0528	270.0539	4.07	269.0456[M–H]$^-$, 151.0039[M–H–C_8H_6O]$^-$, 117.0356[M–H–$C_7H_8O_4$]$^-$, 107.0145[M–H–$C_9H_6O_3$]$^-$	Apigenin	S
107	9.38	$C_{26}H_{30}N_2O_6$	466.2104	466.2087	−3.65	489.1989[M+Na]$^+$, 467.2160[M+H]$^+$, 321.1205[M+H–CH_3–C_9H_9N]$^+$, 303.1119[M+H–H_2O–CH_3–C_9H_9N]$^+$, 265.1430[M+H–$C_{11}H_{10}N_2O_2$]$^+$, 202.0747[M+H–$C_{15}H_{21}O_4$]$^+$	Pulchellamine F	[37]
108 *	9.47	$C_{20}H_{22}O_6$	358.1416	358.1431	4.19	357.1348[M–H]$^-$, 342.1113[M–H–CH_3]$^-$, 179.0718[M–H–$C_{10}H_{10}O_3$]$^-$, 165.0563[M–H–$C_{10}H_9O_3$–CH_3]$^-$, 122.0370[M–H–$C_{13}H_{15}O_4$]$^-$	Matairesinol	S
109 *	9.61	$C_{16}H_{12}O_6$	300.0634	300.0640	2.00	299.0568[M–H]$^-$, 284.0330[M–H–CH_3]$^-$, 256.0384[M–H–C_2H_3O]$^-$, 161.0246[M–H–$C_7H_6O_3$]$^-$	Hispidulin	S
110 *	9.71	$C_{18}H_{22}O_5$	318.1467	318.1476	2.83	341.1378[M+Na]$^+$, 319.1556[M+H]$^+$, 287.1297[M+H–CH_3O]$^+$, 189.0917[M+H–C_2H_4O–$C_4H_6O_2$]$^+$	Gerin	[74]
111 *	9.81	$C_{18}H_{32}O_5$	328.2250	328.2260	3.05	327.2228[M–H]$^-$, 291.1969[M–H–2H_2O]$^-$, 229.1455[M–H–$C_6H_{10}O$]$^-$, 183.1392[M–H–H_2O–$C_7H_{10}O_2$]$^-$, 171.1040[M–H–$C_9H_{16}O_2$]$^-$	Malyngic acid	CFM-ID
112	10.05	$C_{16}H_{28}O_2$	252.2089	252.2099	3.96	275.2001[M+Na]$^+$, 253.2178[M+H]$^+$, 219.1756[M+H–H_2O–CH_3]$^+$, 149.0969[M+H–CH_3–$C_5H_{11}O$]$^+$	7δ-Methoxy-4(14)-oppositen-1β-ol	[86]
113 *	10.61	$C_{15}H_{22}O_2$	234.1620	234.1623	1.28	235.1705[M+H]$^+$, 177.1273[M+H–H_2O–C_3H_4]$^+$, 163.1480[M+H–$C_3H_2O_2$]$^+$, 121.0663[M+H–H_2O–C_7H_{12}]$^+$	Germacra-1(10),4,11(13)-trien-12-oic acid	[87]
114 *	10.65	$C_{18}H_{34}O_5$	330.2406	330.2417	3.33	329.2335[M–H]$^-$, 229.1447[M–H–$C_6H_{12}O$]$^-$, 211.1343[M–H–$C_6H_{12}O$–H_2O]$^-$, 99.0814[M–H–$C_{12}H_{22}O_4$]$^-$	9,12,13-TriHOME	CFM-ID
115 *	10.69	$C_{30}H_{32}O_9$	536.2046	536.2063	3.17	535.2021[M–H]$^-$, 505.1877[M–H–CH_3O]$^-$, 490.1633[M–H–CH_3–CH_3O]$^-$	Lappaol A	[88]
116 *	10.96	$C_{21}H_{24}O_6$	372.1573	372.1587	3.76	371.1501[M–H]$^-$, 356.1264[M–H–CH_3]$^-$, 136.0528[M–H–$C_{13}H_{15}O_4$]$^-$, 121.0094[M–H–$C_{13}H_{15}O_4$–CH_3]$^-$, 83.0144[M–H–$C_9H_{11}O_2$–$C_8H_9O_2$]$^-$	Arctigenin	S
117 *	11.14	$C_{21}H_{22}O_6$	370.1416	370.1421	1.35	371.1493[M+H]$^+$, 219.0652[M+H–$C_9H_{12}O_2$]$^+$, 151.0766[M+H–$C_{12}H_{12}O_4$]$^+$, 137.0606[M+H–$C_{13}H_{14}O_4$]$^+$, 107.0500[M+H–$C_{13}H_{14}O_4$–CH_3O]$^+$	(+)-7,8-Didehydroarctigenin	[89]
118 *	12.10	$C_{15}H_{20}O_2$	232.1463	232.1472	3.88	233.1545[M+H]$^+$, 203.1084[M+H–2CH_3]$^+$, 189.1630[M+H–CO_2]$^+$, 149.1335[M+H–$C_4H_4O_2$]$^+$	Costunolide	[87]

Table 1. *Cont.*

NO.	t_R (min)	Formula	Theoretical Mass (Da)	Calculated Mass (Da)	Mass Error (ppm)	MS^E Fragmentation	Identification	Ref.
119 *	12.94	$C_{15}H_{22}O_2$	234.1620	234.1625	2.14	235.1699[M+H]$^+$, 161.1320[M+H–$C_3H_6O_2$]$^+$, 133.1022[M+H–$C_5H_{10}O_2$]$^+$, 121.1026[M+H–$C_6H_{10}O_2$]$^+$, 81.0712[M+H–$C_9H_{14}O_2$]$^+$	Costic acid	[90]
120 *	14.58	$C_{42}H_{46}O_{12}$	742.2989	742.2978	−1.48	765.2856[M+Na]$^+$, 743.3051[M+H]$^+$, 725.2928[M+H–H_2O]$^+$, 707.2841[M+H–2H_2O]$^+$, 151.0763[M+H–$C_{33}H_{36}O_{10}$]$^+$, 137.0601[M+H–$C_{13}H_{15}O_4$–$C_{21}H_{23}O_6$]$^+$	Diarctigenin	[91]
121 *	15.17	$C_{42}H_{46}O_{12}$	742.2989	742.2991	0.27	765.2867[M+Na]$^+$, 743.3063[M+H]$^+$, 725.2951[M+H–H_2O]$^+$, 707.2834[M+H–2H_2O]$^+$, 151.0465[M+H–$C_{33}H_{36}O_{10}$]$^+$, 137.0612[M+H–$C_{34}H_{38}O_{10}$]$^+$	Conicaol A	[91]
122 *	15.41	$C_{28}H_{50}O_2$	418.3811	418.3830	4.54	441.3725[M+Na]$^+$, 419.3898[M+H]+, 259.2380[M+H–H_2O–$C_9H_{18}O$]$^+$, 151.1500[M+H–H_2O–$C_{17}H_{30}O$]$^+$, 95.0880[M+H–H_2O–$C_{21}H_{38}O$]$^+$	Ergostane-3,24-diol	CFM-ID
123 *	16.06	$C_{15}H_{18}O_2$	230.1307	230.1315	3.48	231.1388[M+H]$^+$, 203.1441[M+H–CO]$^+$, 121.1028[M+H–$C_6H_6O_2$]$^+$, 105.0718[M+H–$C_7H_{10}O_2$]$^+$	Dehydrocostus lactone	S
124 *	16.35	$C_{26}H_{48}NO_7P$	517.3168	517.3181	2.51	518.3254[M+H]$^+$, 184.0744[M+H– $C_{21}H_{34}O_3$]$^+$, 104.1100[M+H–$C_{21}H_{35}O_6P$]$^+$, 86.0986[M+H– $C_{21}H_{37}O_7P$]$^+$	LPC (18:3)	CFM-ID
125 *	16.40	$C_{15}H_{22}O$	218.1671	218.1680	4.13	219.1757[M+H]$^+$, 203.1444[M+H–CH_3]$^+$, 162.1419[M+H–C_3H_5O]$^+$	Germacra-1(10),4,11(13)-trien-12-al	[92]
126 *	16.68	$C_{18}H_{36}O_4$	316.2614	316.2628	4.43	315.2545[M–H]$^-$, 297.2453[M–H–H_2O]$^-$, 171.1031[M–H–C_9H_{18}–H_2O]$^-$, 141.1291[M–H–$C_9H_{16}O_2$–H_2O]$^-$	9,10-Dihydroxystearic acid	CFM-ID
127 *	16.72	$C_{16}H_{30}O_3$	270.2195	270.222	2.59	293.2116[M+Na]$^+$, 269.2124[M+H]$^+$, 165.1651[M+H–$C_4H_8O_3$]$^+$, 121.1025[M+H–$C_7H_{16}O_3$]$^+$, 95.0869[M+H–$C_9H_{18}O_3$]$^+$	(6Z)-2-Hydroxy-6-hexadecenoic acid	CFM-ID
128 *	16.78	$C_{18}H_{30}O_3$	294.2195	294.2207	4.08	293.2134[M−H]$^-$, 275.2035[M−H−H_2O]$^-$, 249.2230[M−H−HCOOH]$^-$, 195.1401[M− H–$C_6H_{10}O$]$^-$	13-oxo-9,11-Octadecadienoic acid	CFM-ID
129 *	17.64	$C_{26}H_{50}NO_7P$	519.3325	519.3336	2.12	520.3408[M+H]$^+$, 184.0744[M+H– $C_{21}H_{36}O_3$]$^+$, 104.1101[M+H–$C_{21}H_{37}O_6P$]$^+$, 86.1006[M+H– $C_{21}H_{39}O_7P$]$^+$	LPC (18:2)	CFM-ID
130 *	17.92	$C_{18}H_{32}O_3$	296.2351	296.2361	3.38	295.2288[M–H]$^-$, 277.2180[M–H_2O]$^-$, 250.2309[M–HCOOH]$^-$	Coronaric acid	CFM-ID
131 *	18.46	$C_{18}H_{30}O_3$	294.2195	294.2203	2.72	293.2131[M−H]$^-$, 275.2042[M−H−H_2O]$^-$, 249.2230[M−H −HCOOH]$^-$, 113.0973[M−H–$C_{11}H_{16}O_2$]$^-$	9-Oxo-10,12-Octadecadienoic acid	S
132 *	18.65	$C_{24}H_{50}NO_7P$	495.3325	495.3337	2.42	496.3409[M+H]$^+$, 184.0742[M+H– $C_{19}H_{36}O_3$]$^+$, 104.1100[M+H–$C_{19}H_{37}O_6P$]$^+$, 86.1006[M+H– $C_{19}H_{39}O_7P$]$^+$	LPC (16:0)	S
133 *	19.26	$C_{26}H_{52}NO_7P$	521.3481	521.3486	0.96	522.3559[M+H]$^+$, 184.0745[M+H– $C_{21}H_{38}O_3$]$^+$, 104.1101[M+H–$C_{21}H_{39}O_6P$]$^+$, 86.1005[M+H– $C_{21}H_{41}O_7P$]$^+$	LPC (18:1)	S

Table 1. *Cont.*

NO.	t_R (min)	Formula	Theoretical Mass (Da)	Calculated Mass (Da)	Mass Error (ppm)	MS^E Fragmentation	Identification	Ref.
134 *	19.30	$C_{16}H_{22}O_4$	278.1518	278.1526	2.88	301.1419[M+Na]$^+$, 279.1571[M+H]$^+$, 149.0245[M+H–C_4H_9–C_4H_9O]$^+$, 121.0305[M+H–C_4H_9–$C_5H_9O_2$]$^+$	Dibutyl phthalate	[93]
135 *	19.38	$C_{18}H_{34}O_3$	298.2508	298.2519	3.69	297.2446[M–H]$^-$, 279.2335[M–H–H_2O]$^-$, 253.2542[M–H–HCOOH]$^-$	Ricinoleic acid	CFM-ID
136 *	21.00	$C_{30}H_{48}O_4$	472.3553	472.3575	4.66	471.3492[M–H]$^-$, 427.3588[M–H–HCOOH]$^-$, 411.3273[M–H–HCOOH–CH_3]$^-$	Macrocarpoic acid A	[94]
137 *	22.39	$C_{18}H_{30}O_2$	278.2246	278.2259	4.67	277.2176[M–H]$^-$, 259.2076[M–H–H_2O]$^-$, 109.0661[M–H–$C_{11}H_{18}$–H_2O]$^-$	Linolenic acid	S
138 *	22.58	$C_{16}H_{32}O_3$	272.2351	272.2386	2.21	271.2314[M–H]$^-$, 225.2255[M–H–HCOOH]$^-$, 223.2086[M–H–2H_2O–CH_3]$^-$, 197.1904[M–H–2H_2O–C_3H_7]$^-$	3-Hydroxyhexadecanoic acid	CFM-ID
139 *	23.98	$C_{30}H_{48}O$	424.3705	424.3700	−1.18	425.3773[M+H]$^+$, 205.1942[M+H–$C_{15}H_{24}O$]$^+$, 189.1644[M+H–$C_{16}H_{28}O$]$^+$, 161.1335[M+H–$C_{18}H_{32}O$]$^+$	Lupenone	[95]
140 *	24.04	$C_{18}H_{32}O_2$	280.2402	280.2412	3.57	279.2329[M–H]$^-$, 261.2229 [M –H–H_2O]$^-$	Linoleic acid	S
141 *	24.25	$C_{30}H_{48}O$	424.3705	424.3695	−2.36	425.3767[M+H]$^+$, 409.3454[M+H–CH_3]$^+$, 217.1953[M+H–$C_{14}H_{24}O$]$^+$, 137.1337[M+H–$C_{20}H_{32}O$]$^+$	Amyrone	[96]
142 *	24.62	$C_{30}H_{48}O_2$	440.3654	440.3651	−0.68	441.3724[M+H]$^+$, 231.2112[M+H– $C_{13}H_{21}O$]$^+$, 187.1493[M+H–$C_{16}H_{28}O$–H_2O]$^+$	Ptiloepoxide	[97]
143 *	24.63	$C_{30}H_{48}O$	424.3705	424.3693	−2.83	425.3765[M+H]$^+$, 205.1954[M+H– $C_{15}H_{24}O$]$^+$, 189.1640[M+H–$C_{16}H_{28}O$]$^+$	Taraxasterone	[98]
144 *	24.82	$C_{30}H_{48}O_2$	440.3654	440.3658	0.91	441.3731[M+H]$^+$, 423.3611[M+H–H_2O]$^+$, 191.1803[M+H–$C_{16}H_{26}O_2$]$^+$, 123.1191[M+H–$C_{21}H_{34}O_2$]$^+$	11α-Hydroxyurs-12-en-3-one	[99]
145 *	24.87	$C_{35}H_{36}N_4O_5$	592.2686	592.2696	1.69	593.2769[M+H]$^+$, 533.2556[M+H–$C_2H_4O_2$]$^+$	Pheophorbide A	[100]
146 *	25.44	$C_{30}H_{48}O_2$	440.3654	440.3644	−2.27	441.3717[M+H]$^+$, 189.1638[M+H–$C_{16}H_{28}O_2$]$^+$, 135.1184 [M+H–$C_{20}H_{34}O_2$]$^+$	11-Oxo-amyrin	[101]
147 *	25.60	$C_{16}H_{32}O_2$	256.2402	256.2409	2.73	255.2326[M–H]$^-$, 237.2208[M–H–H_2O]$^-$	Hexadecanoic acid	CFM-ID
148 *	27.87	$C_{28}H_{48}O_2$	416.3654	416.3669	3.60	461.3661[M+HCOO]$^-$, 415.3591[M–H]$^-$, 281.2849[M–H–$C_9H_{10}O$]$^-$	β-Tocopherol	CFM-ID
149 *	28.02	$C_{24}H_{38}O_4$	390.2770	390.2770	0.00	413.2662[M+Na]$^+$, 391.2847[M+H]$^+$, 149.0247[M+H–C_8H_{17}–$C_8H_{17}O$]$^+$	Dioctyl phthalate	[102]

S: compared with the reference compounds. CFM-ID: compared with the CFM-ID 4.0 [103]. * identified from SP for the first time.

According to the types of chemical structure, these identified compounds could be divided into phenylpropanoids, flavonoids, terpenoids, organic acids and other types. The structures are listed in Figure 2.

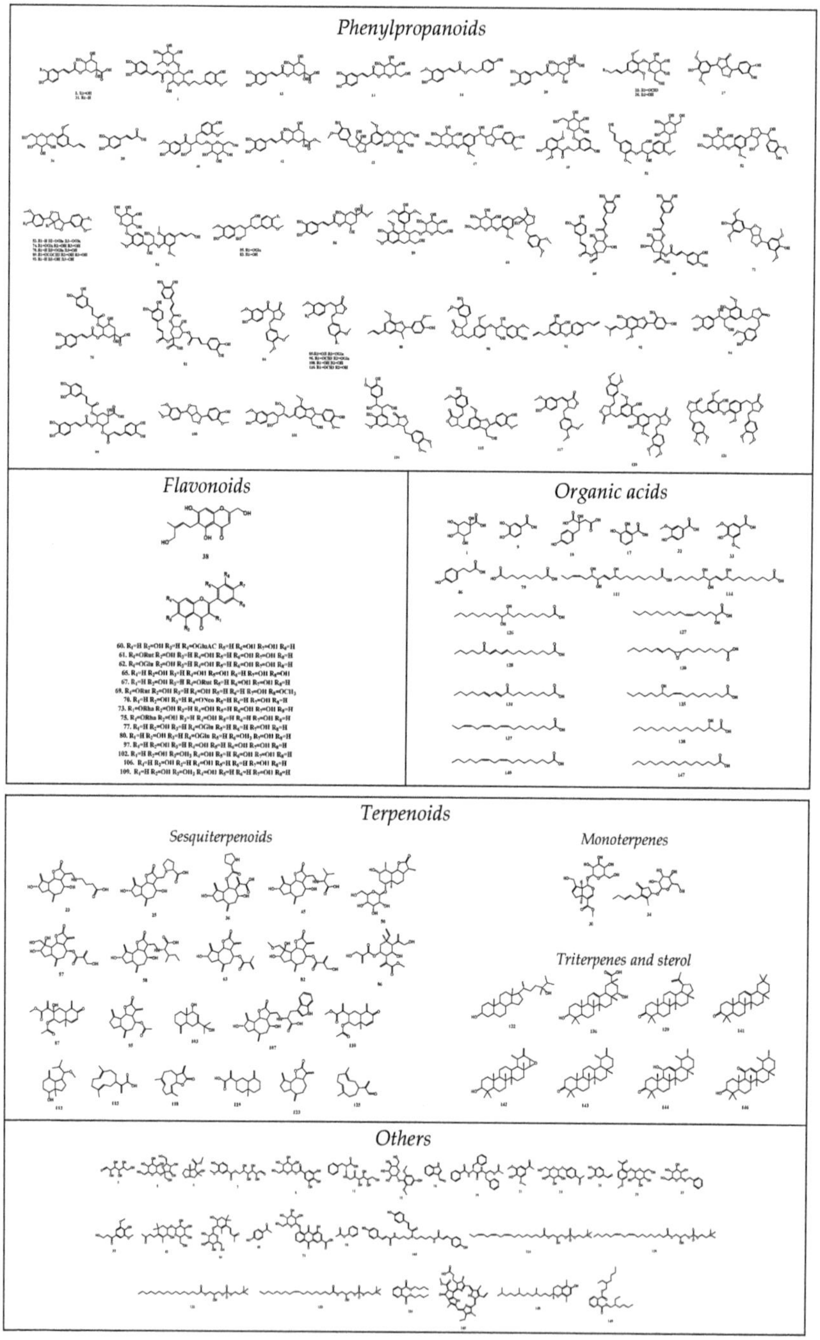

Figure 2. Chemical structures of components identified in SP.

2.1.2. Quantitative Analysis

Methodological verification The RSDs of accuracy and precision, displayed in Table S1, were all less than 3.0%. The average recoveries of 20 compounds were all more than 95%. The LOD, LOQ and linear relationships are presented in Table 1. The detection and quantitation limits of the 20 components were within the appropriate ranges, and the standard curves exhibited good linearity over the corresponding ranges. The results showed that the method could be used for the quantitative assay of the main polyphenols of SP ethanol extract.

Quantitative Analysis results The contents of all of the compounds are summarized in Table 2. The results showed that 20 polyphenols accounted for 33.2% of the ethanol extract of SP. Among them, the chemical components with high contents were narcisin (6.94%), rutin (6.86%), arctiin (5.42%), chlorogenic acid (4.60%), apigenin (4.10%), 1,4-dicaffeoylquinic acid (2.04%) and pinoresinol (1.12%).

Table 2. Contents of 20 polyphenols in ethanol extract of SP.

No.	Compound	Regression Equations	R^2	linearity Range ($\mu g \cdot mL^{-1}$)	LOD ($\mu g \cdot mL^{-1}$)	LOQ ($\mu g \cdot mL^{-1}$)	Content (%)
2	Chlorogenic acid	$y = 35.49x + 468.5$	0.9991	1~100	0.20	1.0	4.60
13	Neochlorogenic acid	$y = 47.005x + 5.7739$	0.9993	0.1~10	0.04	0.1	0.13
60	Luteolin 7-glucuronide	$y = 155.56x - 19.818$	0.9992	0.1~10	0.05	0.1	0.07
61	Rutin	$y = 27.813x + 1119$	0.9993	1~100	0.33	1.0	6.86
62	Isoquercitroside	$y = 895.09x - 133.08$	0.9995	0.1~10	0.02	0.1	0.20
66	1,4-Dicaffeoylquinic acid	$y = 174.39x + 1766.7$	0.9997	1~100	0.33	1.0	2.04
68	1,5-Dicaffeoylquinic acid	$y = 471.27x + 438.81$	0.9991	0.1~10	0.03	0.1	0.14
69	Narcisin	$y = 56.783x + 1673.6$	0.9991	1~100	0.20	1.0	6.94
72	Syringaresinol	$y = 879.44x - 136.7$	0.9992	0.05~5	0.02	0.05	0.006
73	Quercitrin	$y = 3142.4x + 151.2$	0.9993	0.05~5	0.03	0.05	0.026
76	4.5-Dicaffeoylquinic acid	$y = 522.11x + 15.755$	0.9994	0.1~10	0.02	0.1	0.52
78	Pinoresinol 4-glucoside	$y = 61.543x + 55.902$	0.9998	0.1~10	0.05	0.1	0.16
85	Matairesinoside	$y = 702.5x - 75.75$	0.9993	0.1~10	0.02	0.1	0.08
93	Pinoresinol	$y = 24.76x + 1.1$	0.9998	1~100	0.20	1.0	1.12
96	Arctiin	$y = 1631x + 4406.1$	0.9994	0.5~50	0.10	0.5	5.42
97	Luteolin	$y = 3162x + 331$	0.9996	0.05~5	0.02	0.05	0.01
102	Eupafolin	$y = 17011x + 24.231$	0.9990	0.05~5	0.02	0.05	0.03
106	Apigenin	$y = 226.54x + 5077.6$	0.9996	0.5~50	0.20	0.5	4.10
108	Matairesinol	$y = 1519.7x + 132.26$	0.9997	0.05~5	0.01	0.05	0.01
116	Arctigenin	$y = 39.1x - 32.2$	0.9995	0.1~10	0.03	0.1	0.76

2.2. *Alleviated Ulcerative Colitis Activity*

2.2.1. Body Weights, Clinical Signs Observations and DAI

Throughout the experiment, the mice in the control group had normal weight growth, clinical signs and DAI. In contrast, the mice in the model group developed obvious anorexia and weight loss. As for the mice intervened with CNY or SP, the weight loss and clinical signs were alleviated to various degrees. From the fifth day of administration, SP dose-dependently reduced the DSS-mediated increase in the DAI scores during the disease progression compared with the model mice. By the seventh day, the UC model mice became more symptomatic with the increasing DSS induction time, as evidenced by loose stools, blood in the stool and the DAI scores. On the tenth day, the weights in all of the administration groups, except the SPL group, were higher than those in the model group ($p < 0.05$, Figure 3A), and the DAI scores in all of the administration groups, except the SPL group, were lower than those in the model group ($p < 0.01$, Figure 3B).

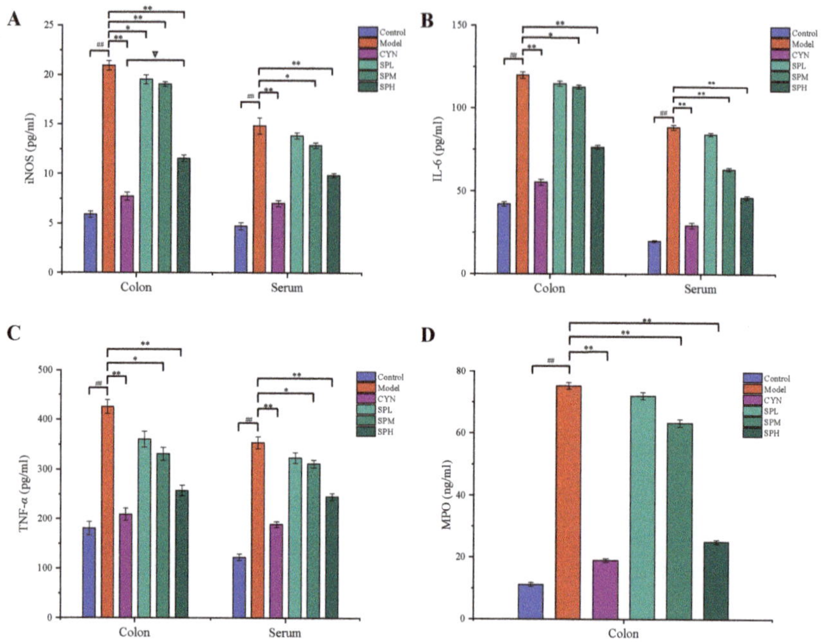

Figure 3. Effect of SP on (**A**) body weight, (**B**) disease activity index, (**C**) colon length and (**D**) spleen coefficient. (Data are expressed as the means ± S.E.M., $n = 10$). Compared with control group, ## $p < 0.01$; compared with model group, * $p < 0.05$, ** $p < 0.01$; compared with CYN group, ∇ $p > 0.05$).

2.2.2. Colon Length and Spleen Coefficient

As shown in Figure 3C, the colon of the model group mice was significantly shorter than that of the control group, indicating that the colon tissue had been damaged ($p < 0.01$). Colon damage could be reduced with the oral administration of SP or CYN. Compared to the model group, a significant increase ($p < 0.01$) in colon length was observed in the CYN, SPM and SPH groups, and high doses of SP provided a similar effect to CYN ($p > 0.05$). Compared to the control group, the spleen coefficient increased significantly in the model group, indicating that the UC mice exhibited inflammatory responses. In contrast, after CYN or three dosages of SP, the spleen coefficients were significantly decreased ($p < 0.01$). The above results are shown in Figure 3D. The above results showed that both SP and the positive drug could decrease the inflammatory responses in UC mice.

2.2.3. Measurement of Cytokines and MPO Contents

As shown in Figure 4, the levels of TNF-α, IL-6 and iNOS in the serum or colon, and the level of MPO in the colon, were all considerably higher in the model group compared to the control group. While being treated with CYN and SP, the levels of the above cytokines all significantly decreased compared with the model group. In addition, in terms of modulating TNF-α and MPO, the SP at a high dose showed similar effects to CYN, the positive control drug.

2.2.4. Measurement of Oxidative Stress Indexes Levels

Lipid peroxidation is associated with ulcerative colitis due to oxidative damage. The activated free radicals will deplete the antioxidant level in the colon and aggravate ulcerative colitis. As demonstrated in Figure 5, compared with the control group, the GSH and SOD levels of the model group mice significantly decreased, while the MDA level significantly increased. However, compared with the model group, after the intervention of CYN or SPH, the levels of GSH and SOD in the mice were significantly increased, while the level of MDA was significantly reduced.

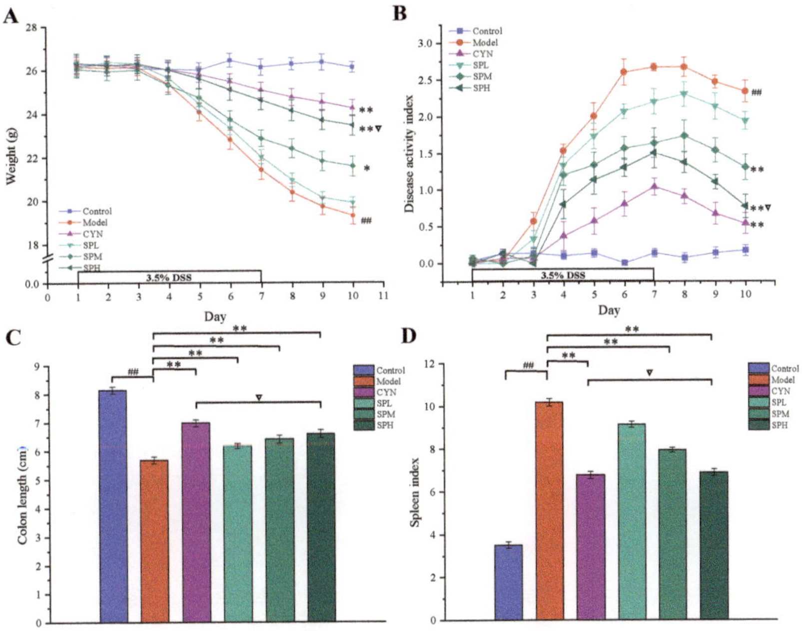

Figure 4. Effect of SP on (**A**) iNOS, (**B**) IL-6, (**C**) TNF-α and (**D**) MPO. (Data are expressed as the means ± S.E.M., $n = 10$). Compared with control group, $^{\#\#}$ $p < 0.01$; compared with model group, ** $p < 0.01$; compared with CYN group, ∇ $p > 0.05$).

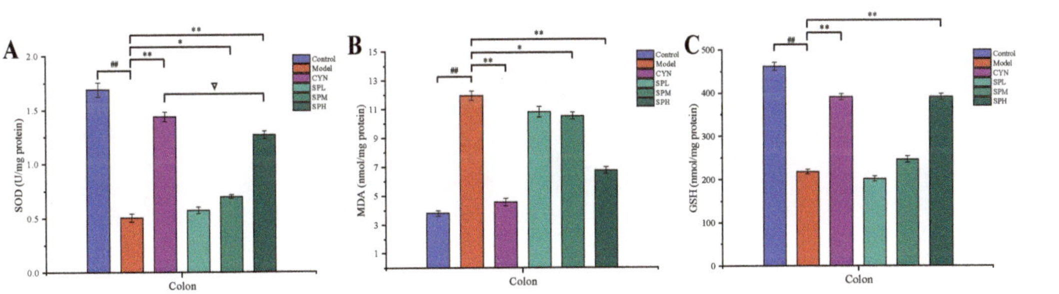

Figure 5. Effect of SP on (**A**) SOD, (**B**) MDA and (**C**) GSH. (Data are expressed as the means ± S.E.M., $n = 10$). Compared with control group, $^{\#\#}$ $p < 0.01$; compared with model group, * $p < 0.05$, ** $p < 0.01$; compared with CYN group, ∇ $p > 0.05$).

2.2.5. Histopathology

The typical H&E staining photos are list in Figure 6. In the control group, the normal whole colonic structure and mucosal epithelium was visible. Severe mucosal damage and edema in the submucosal region and goblet cell were found in the model group. Compared with the model group, the inflammatory cell infiltration in the SP and CYN groups was decreased, the epithelial damage was recovered, and the colonic tissues were relatively complete, indicating that the inflammatory symptoms of the colonic tissue in each group were alleviated to various degrees after CYN or SP intervention.

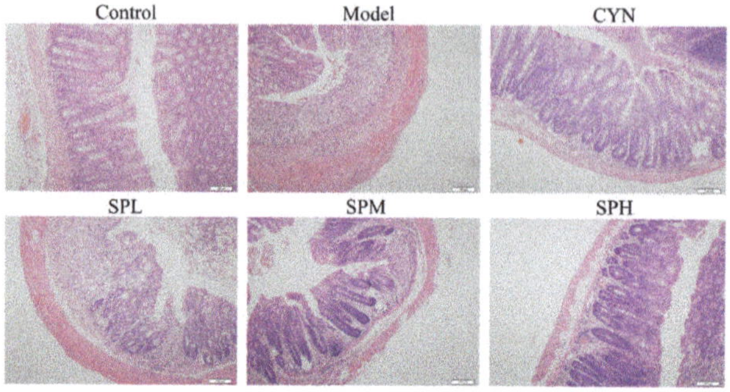

Figure 6. The typical H&E staining photos of the colon section.

2.2.6. Transmission Electron Microscopy Analysis

To confirm the effect of SP on the intestinal microvilli, transmission electron microscopy analysis was performed. The results are shown in Figure 7. In the control group, the villi of the colonic epithelial cells were neatly arranged and fully formed. However, various degrees of villous shedding and disorder were observed in the model group. Meanwhile, vacuolar degeneration was seen in mitochondria. For the SPH group, the villi arranged neatly without obvious shedding and organelle morphology is intact and normal.

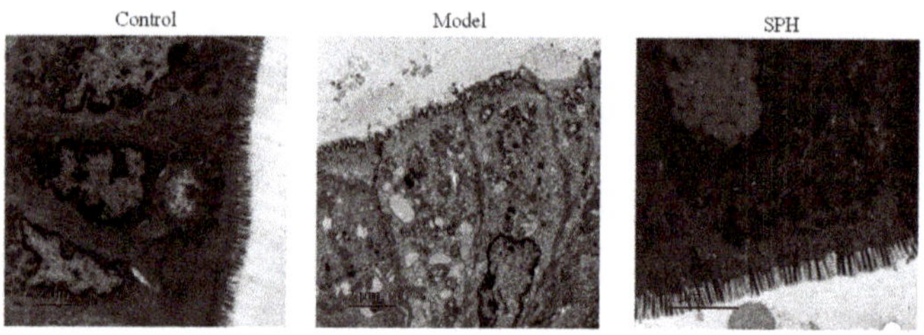

Figure 7. The transmission electron microscopy photos in colon section.

2.3. *Metabolomics*

2.3.1. Validation and Determination

The m/z-RT pairs in the ESI+ mode and ESI− mode included 132.0865-0.67, 274.2741-12.64, 362.3267-12.98, 104.1092-17.08, 496.3401-17.58; and 286.8602-0.59, 191.0191-0.78, 329.2319-10.40, 233.1547-16.68, 452.2766-17.94, respectively. The RSDs of the peak intensity and RT for the system stability, precision, reproducibility and sample stability were calculated and are listed in Table S2; they were all less than 3.0%. It was indicated that the established method with good precision, reproducibility and stability could be applied to assay the serum and colon samples. The detected representative base peak intensity (BPI) chromatograms of the serum and colon samples are shown in Figure S1.

2.3.2. Multivariate Statistical Analyses of Serum and Colon Metabolomics

The metabolomic study was performed in both the ESI+ and in ESI− modes. A satisfactory level of system stability was also shown by the clustered QC samples in the PCA results (Figure 8A). The tested serum or colon samples from the control, model or SPH groups were clustered, respectively. The samples from the three groups were located in different regions, indicating that the metabolic disturbances in the three groups were differential. In order to achieve maximal separation between two groups, the OPLS-DA models were then established (Figure 8B). The separation between the control group and model group, or between the SPH group and model group, were achieved with satisfactory R2Y values and Q2 values. Moreover, the permutation test (Figure 8C) also showed that all of the Q2-values to the left were lower than the original points to the right, indicating that the OPLS-DA models were valid. Volcano maps (Figure 8D) were further performed to screen the differentiated metabolites. As a result, a total of 21 metabolites were identified and given the red color. Moreover, the generated ROC curves (Figure 9A,B) analyzed the above 21 metabolites, and the AUC values (all greater than 0.8) and p values (all less than 0.01) are listed in Table S3. All of them have the potential to be used as UC diagnostic biomarkers, according to the ROC analysis between the model group and the control group. The analysis of the ROC curves between the model and SPH groups showed that these metabolites contributed to the effects of SPH in UC treatment.

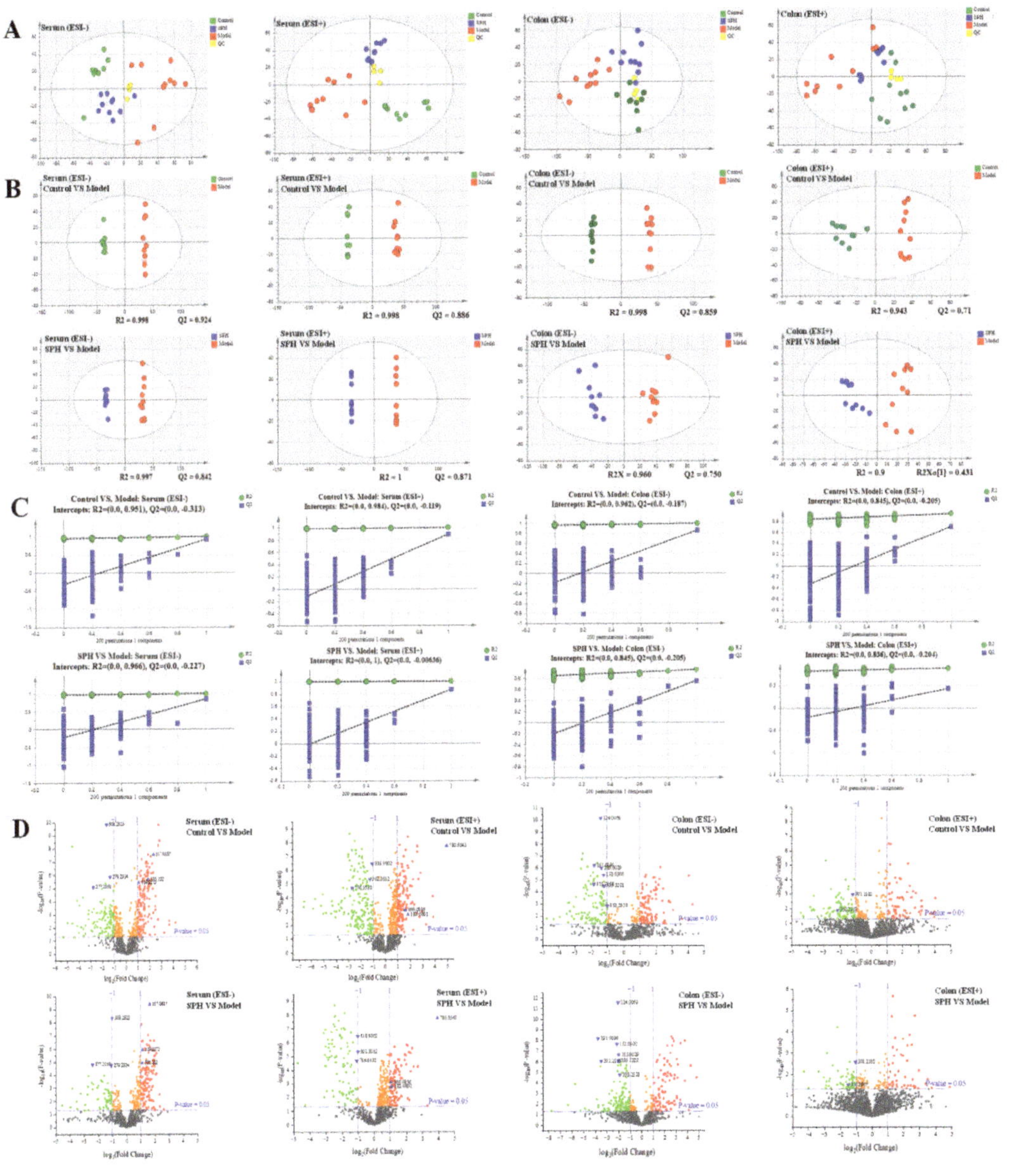

Figure 8. The PCA score (**A**), OPLS-DA score (**B**), permutations test (**C**) and volcano plots (**D**) of serum and colon metabolic profiling.

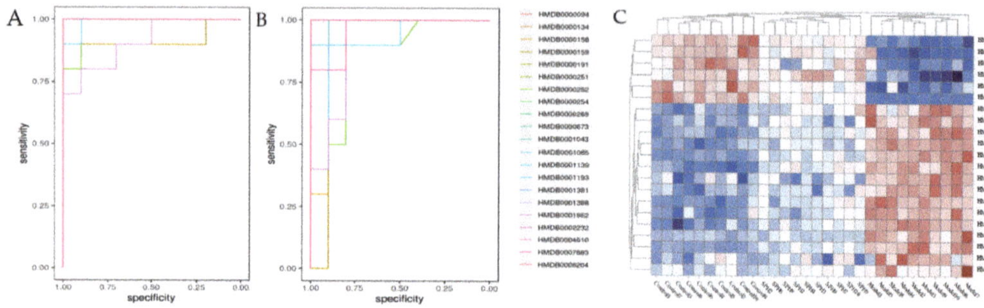

Figure 9. The predictive receiver operating characteristic (ROC) curves (**A**: between control group and model group, **B**: between model group and SPH group) and the heatmap (**C**) of the identified potential biomarkers in each group.

2.3.3. Biomarkers Screening and Pathway Enrichment

As potential biomarkers, 21 endogenous metabolites were identified (Table 3). After that, these potential biomarkers from different groups were visualized and mapped on the heat map (Figure 9C). From blue to red, the colors indicated increasing abundance of the metabolites.

Table 3. The information of identified metabolites in serum and colon.

No.	RT/min	Measured Mass (Da)	Mass Error (ppm)	Adducts	Biomarkers	Sources	Pathway	HMDB ID	Change Trend	
									M/C	D/M
1 *	0.60	132.0300	0.23	M-H	L-Aspartic acid	Colon	ASGM, HD	HMDB0000191	↑	↓
2 #	0.64	124.0069	0.08	M-H	Taurine	Colon	THM	HMDB0000251	↑	↓
3 *	0.68	115.0029	−0.17	M-H	Fumarate	Colon	ASGM, TM, CC	HMDB0000134	↑	↓
4 #	0.79	191.0190	−0.10	M-H	Citrate	Colon	ASGM, CC	HMDB0000094	↑	↓
5 #	0.8	117.0187	−0.09	M-H	Succinate	Serum	CC	HMDB0000254	↓	↑
6 *	0.8	182.0802	−0.82	M+H	L-Tyrosine	Serum	PTTB, TM, PM	HMDB0000158	↓	↑
7 *	0.98	166.0856	−0.72	M+H	L-Phenylalanine	Serum	PTTB, PM	HMDB0000159	↓	↑
8 #	10.51	353.2328	0.00	M-H	Prostaglandin F2a	Colon	AM	HMDB0001139	↑	↓
9 *	10.53	301.2182	0.46	M+H	all-trans-Retinoic acid	Colon	RM	HMDB0001852	↑	↓
10 #	12.81	318.3002	−0.19	M+H	Phytosphingosine	Serum	SM	HMDB0004610	↑	↓
11 #	12.97	300.2907	0.13	M+H	Sphingosine	Colon	SM	HMDB0000252	↑	↓
12 *	13.68	351.2158	−0.37	M-H	Prostaglandin H2	Colon	AM	HMDB0001381	↑	↓
13 *	14.36	335.2222	0.00	M-H	Leukotriene B4	Colon	AM	HMDB0001085	↑	↓
14 #	15.06	302.3052	−0.23	M+H	Sphinganine	Serum	SM	HMDB0000269	↑	↓
15 #	16.33	335.2220	0.00	M-H	5(S)-HpETE	Serum	AM	HMDB0001193	↓	↑
16 *	18.16	319.2272	−0.03	M-H	8,9-EET	Serum	AM	HMDB0002232	↓	↑
17 *	21.69	277.2158	−0.36	M-H	α-Linolenic acid	Serum	ALM	HMDB0001388	↑	↓
18 *	22.81	303.2323	−0.03	M-H	Arachidonate	Serum	AM	HMDB0001043	↑	↓
19 *	23.11	279.2324	0.00	M-H	Linoleic acid	Serum	LM	HMDB0000673	↑	↓
20 #	26.60	780.5543	0.00	M+H	PC(18:3/18:2)	Serum	LM, ALM	HMDB0008204	↓	↑
21 #	27.54	756.5532	0.17	M+H	PC(14:0/20:4)	Serum	AM	HMDB0007883	↑	↓

* Metabolites validated with standards. # Metabolites confirmed by MS data. "↑" represents the content was up-regulated. "↓" represents the content was down-regulated.

The MetaboAnalyst analysis revealed that the 21 potential biomarkers were mainly associated with 11 potential metabolisms with impact values above 0.10 (Table 4).

Table 4. The results from metabolic pathways of differential metabolites.

Pathway Name	Match Status	*p*	−log (*p*)	Holm *p*	FDR	Impact
Phenylalanine, tyrosine and tryptophan biosynthesis (PTTB)	2/4	1.09×10^{-3}	2.9613	0.0885	0.0230	1.0000
Linoleic acid metabolism (LM)	2/5	1.81×10^{-3}	2.7431	0.1445	0.0304	1.0000
Arachidonic acid metabolism (AM)	7/36	2.22×10^{-7}	6.6532	1.87×10^{-5}	1.87×10^{-5}	0.5861
Taurine and hypotaurine metabolism (THM)	1/8	0.1065	0.9727	1.0000	0.7141	0.4286
Phenylalanine metabolism (PM)	3/12	4.75×10^{-4}	3.3234	0.0389	0.0133	0.3571
alpha-Linolenic acid metabolism (ALM)	2/13	0.0132	1.8803	1.0000	0.1230	0.3333
Retinol metabolism (RM)	1/16	0.2021	0.6944	1.0000	0.9433	0.2275
Alanine, aspartate and glutamate metabolism (ASGM)	4/28	4.61×10^{-4}	3.3361	0.0383	0.0133	0.2260
Sphingolipid metabolism (SM)	3/21	0.0026	2.5772	0.2065	0.0318	0.2028
Tyrosine metabolism (TM)	2/42	0.1142	0.9423	1.0000	0.7141	0.1644
Citrate cycle (TCA cycle) (CC)	3/20	0.0023	2.6403	0.1809	0.0318	0.1529

2.4. Network Pharmacology

The intersection of 1532 SP-related targets and 4920 UC-related targets provided a total of 373 core targets. Inflammatory factors, such as IL-6, TNF, NOS2 and MPO, determined in the study of the anti-UC activity, are also included in these targets. Among the various targets, enzymes (137 species) accounted for the greatest fraction (36.73%), followed by kinases (16.89%).

Next, the interactions of 149 compounds on 373 core targets were examined, and the SP-core targets network was built, as shown in Figure 10, which illustrated a network with 535 nodes and 11,387 edges. On one hand, 116 of the components' degrees were greater than the average degree, which is 65. Among these 116 components, there were 17 components that had been quantified determined. On the other hand, the degrees of the phenylpropanoids and flavonoids, being 75 and 74, respectively, were greater than the other structure type's component degrees.

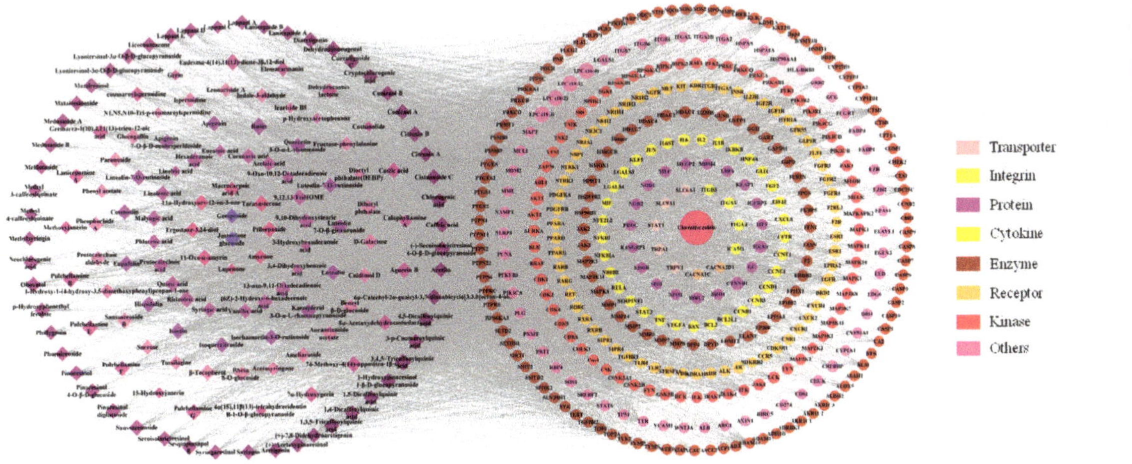

Figure 10. The network of SP-core targets and PPI network.

Based on the aforementioned topology analysis, the components with high degree values (indicating that more targets were related) might be regarded as potential active components. In addition, the PPI network was also developed to identify potential targets for SP against UC.

2.5. Integrated Analysis Involving Metabolomics and Network Pharmacology

The integrated analysis was performed based on the 373 potential targets obtained from the network pharmacology and the 21 potential biomarkers identified from the metabolomics in order to further confirm the key targets, biomarkers and pathways. The "biomarkers-targets-pathways" network was then constructed (Figure 11). Through matching analysis, there were ten biomarkers (succinate, L-phenylalanine, L-tyrosine, fumarate, PC(18:3/18:2), citrate, arachidonate, linoleic acid, 5(S)-HpETE, 8,9-EET), 18 targets (ADH1B, AKR1C1, ALDH2, ALOX5, CBR1, COMT, CYP1A1, CYP1A2, CYP1B1, DBH, EPHX1, GSTP1, HSD11B2, MIF, MPO, PNMT, PTGS1, PTGS2) and five pathways (arachidonic acid metabolism, citrate cycle, linoleic acid metabolism, sphingolipid metabolism and tyrosine metabolism) that were closely connected in the constructed network. In addition, the metabolic network, including these ten key biomarkers and their metabolic sites, is shown in Figure 12.

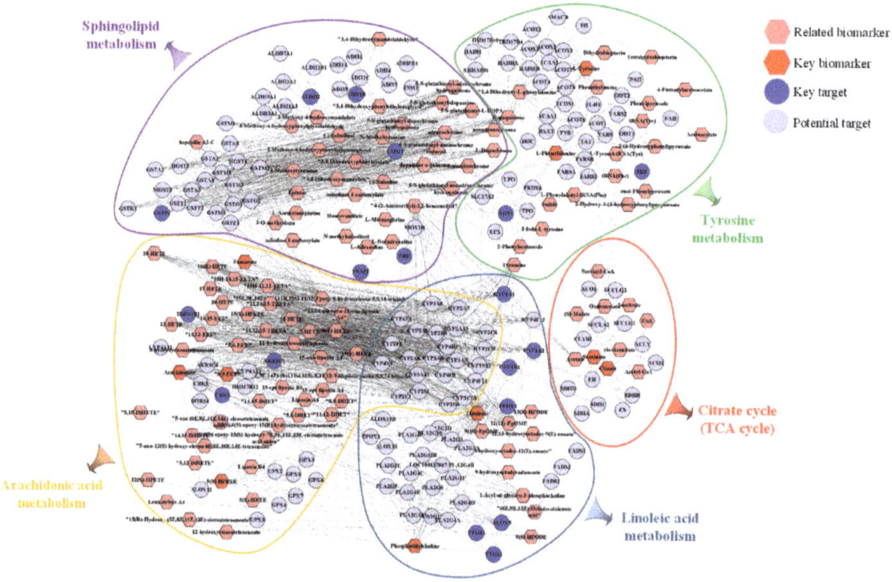

Figure 11. The network of "biomarkers-targets-pathways.

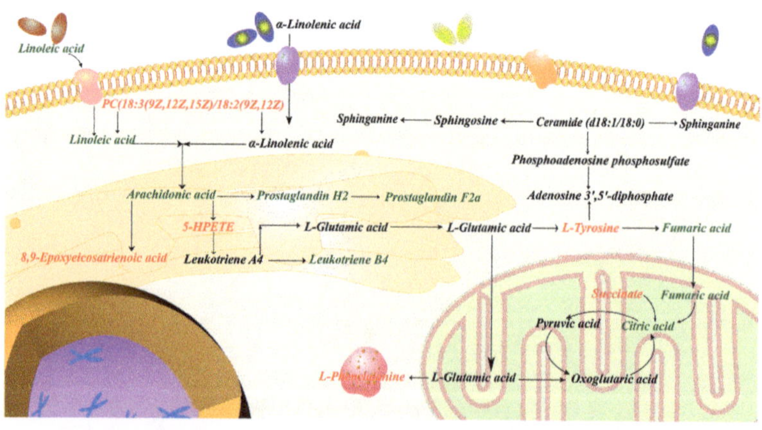

Figure 12. Metabolic network including key biomarkers and their metabolic sites.

3. Discussion

In this study, the chemical composition and pharmacological effect of relieving ulcerative colitis with SP 75% ethanol extract were investigated for the first time. It sheds fresh light on the medical significance of SP as a viable candidate for alleviating UC symptoms.

Both the qualitative and quantitative characteristics of SP extract were determined by UPLC-Q/TOF-MS. A total of 149 components were identified. It was reported that both the phenylpropanoids and flavonoids have anti-UC effects [104–107]. Therefore, we quantitatively assayed the twelve phenylpropanoids and eight flavonoids in the SP extract. In addition, a total of 116 components (17 of them were quantified) with degrees greater than the average degree were screened as the potential active components in network pharmacology. Interestingly, the degrees of the phenylpropanoids and flavonoids were higher than the other structure types, which suggested that these two kinds of substances contributed the most to the pharmacological activity of SP. The above chemical composition research results provided the material basis for the pharmacological activity of SP.

As DSS consumption could damage the intestinal epithelium chemically, expose the lamina propria to lumen antigens and intestinal bacteria, and trigger an inflammatory and immunological response in the gut [108], an experimental model of UC in mice was established, and induced by using DSS in the present pharmacological activity study. This model exhibits very similar clinical symptoms to human UC [109]. Firstly, bodyweight loss, DAI score, shortened colon length and spleen coefficient are frequently regarded as inflammatory signs to evaluate UC progression. It is also believed that colonic MPO activity is directly connected to the degree of neutrophil infiltration, which could cause the tissue damage at the site of UC inflammation. Our current investigation demonstrated that the intervention by SP may significantly reduce the above indexes in a dose-dependent way. Secondly, TNF-α triggers a wide range of inflammatory genes and encourages the production of pro-inflammatory cytokines [110]; IL-6 promotes neutrophil infiltration and results in tissue necrosis [111]; and iNOS produces excessive inflammatory mediators [112]. Namely, these mediators play a crucial role in the development of intestinal damage. Our findings also reinforced the significance of these inflammatory factors in the incidence and progression of ulcerative colitis, and also demonstrated that SP could drastically lower the iNOS, TNF-α and IL-6 levels in UC mice. Thirdly, oxidative stress is also involved in the pathogenesis of ulcerative colitis, with compelling evidence that the increased formation of reactive oxygen species damages cellular macromolecules and jeopardizes epithelial cell integrity. GSH, SOD and MDA are the most significant typical indicators for evaluating oxidative stress. To our satisfaction, SP treatment could dramatically reduce MDA concentrations, raise GSH levels and enhance SOD activity. Fourthly, the histopathology and transmission electron microscopy examination of colonic tissue are also important indexes to investigate the protective effect of SP on the intestinal barrier. As we expected, H&E staining and TEM revealed that SP could reduce the damage to the colonic intestinal barrier.

In order to further assess the effectiveness of SP and to investigate the relevant mechanisms, metabolomics analysis was carried out in this work. A total of 21 potential metabolite biomarkers and 11 metabolisms were identified to be closely related to the effect of SP. Network pharmacology analysis was then performed to screen out the active components (such as phenylpropanoids and flavonoids) and 373 potential biological targets. Aiming to establish the connection network between the biological targets and metabolites, integrated analysis, by merging metabolomics with network pharmacology, was finally employed. As a result, 10 metabolites out of 21 potential biomarkers were discovered to have a direct link with 18 biological targets among the 373 potential targets. Specifically, these ten metabolites were involved in five metabolisms. Three of these five pathways were lipid metabolism (arachidonic acid metabolism, linoleic acid metabolism and sphingolipid metabolism). Lipids influence the immune response by acting as intracellular and intercellular signaling molecules. It has been reported that lipid metabolism was expected to have a significant role in the pathophysiology of UC [113]. When colitis develops, the citrate cycle is disturbed, which reduces the amount of energy that the gut receives through aerobic breakdown. Tyrosine plays a critical role in the metabolism and development of both humans and animals and is linked to immunological activation and inflammation. To summarize, these 10 biomarkers, 18 targets and 5 metabolisms were thought to be critical in the therapeutic effect of SP on UC. It is believed that the substantial pharmacological effects of SP are due to its multi-target mechanism.

4. Materials and Methods

4.1. Materials and Reagents

The SP was collected in Shipeng Village, Panshi City, Jilin Province, China, in mid-September 2021. It was authenticated by Prof. Pingya Li as the whole herb of SP and was then air-dried. The specimen was preserved in the Natural Drugs Research Center of Jilin University.

The methanol and acetonitrile, of LC-MS grade, were bought from Fisher Chemical Company. The formic acid for UPLC was purchased from Sigma-Aldrich Company. All of the other chemicals were of analytical purity.

The phillygenin, pinoresinol diglucoside, luteolin 7-glucuronide, pinoresinol 4-glucoside, pinoresinol, isoquercitroside, 1,5-dicaffeoylquinic acid, matairesinoside, matairesinol and sec-oisolarieiresinol were purchased from ChemFaces. Chlorogenic acid, syringic acid, hispidulin, neochlorogenic acid, protocatechuic aldehyde, protocatechuic acid, 4,5-dicaffeoylquinic acid, 1,4-dicaffeoylquinic acid, eupafolin, quinic acid, cryptochlorogenic acid, rutin, caffeic acid, narcisin, quercitrin, arctiin, syringaresinol, apigenin, arctigenin, luteolin, dehydrocostus lactone, LPC (16:0), LPC (18:1), linolenic acid, linoleic acid and 9-oxo-10,12-octadecadienoic acid were purchased from Chengdu HerbSubstance Co., Ltd.

The DSS (MW: 40,000 Da) was purchased from Macklin Inc. Mouse MPO, the TNF-α and IL-6 ELISA kits were obtained from MultiSciences (Lianke) Biotech, Co., Ltd. The Mouse iNOS ELISA kit was purchased from Shanghai zcibio technology Co.,Ltd. The SOD, MDA, GSH assay kits were purchased from Nanjing Jiancheng Bioengineering Institute. Changyanning Tablet (Batch No. 2003044) was produced by Jiangxi Kang'enbei Traditional Chinese Medicine Co., Ltd.

4.2. Animals

Adult male BALB/c mice (22 $\pm$ 2 g) were bought from YISI Experimental Animal Technology Co., Ltd. (Changchun, China, License serial number: 202100040595). All of the mice were fed in the Observation Facility of Animal Experiment in Barrier Environment (SPF level, School of Basic Medicine, Jilin University) maintained under relative humidity (60 $\pm$ 5%) and standard temperature (25 $\pm$ 2 °C) with a 12 h light/dark cycle. After one week of acclimation, the mice were stochastically assigned to different experimental groups. In accordance with the Guide for Institutional Animal Care and Use of Laboratory Animals, the mice were kept in facilities approved by the Association for Institutional Animal Care and Use Committee of Jilin University.

4.3. Sample Preparation

Ethanol extract of SP: The dried whole herb of SP (1.0 kg) was extracted with 75% ethanol (10 L) for three times (3 h per time). The extracts were combined, and the ethanol was recovered by vacuum distillation, the obtained dried residue (ethanol extract of SP, 73.2 g) was stored at room temperature for further study.

Test solution for qualitative analysis: Ethanol extract was dissolved in methanol to obtain the solution at a concentration of 3.0 mg·mL^{-1}.

Test solutions for quantitative analysis: (1) Ethanol extract was dissolved in methanol to obtain the solution at a concentration of 3.0 mg·mL^{-1}; (2) Ethanol extract (70 mg) was suspended in water (30 mL), then extracted for three times with n-hexane (50 mL) and ethyl acetate (50 mL), respectively. The ethyl acetate layer was combined and recovered to dryness. The dried residue was then dissolved in methanol (1 mL) for test.

Test solution for pharmacological activity test: Ethanol extract was suspended in 0.5% sodium carboxymethylcellulose (CMC-Na) to prepare the solutions with the concentrations of 12.0, 6.0, 3.0 mg·mL^{-1}.

4.4. UPLC-Q/TOF-MS

A Waters Acquity UPLC system connected to a Waters Xevo G2-XS QTOF mass spectrometer (Waters Co., Milford, MA, USA) was used to perform chromatographic separations and mass spectrometry detections via electrospray ionization interface. UPLC-MS/MS method was conducted as previously reported [114]. The details are shown in the Supporting Information.

4.5. Comprehensive Phytochemical Analysis

4.5.1. Qualitative Analysis

Firstly, an independent database was created in addition to the Traditional Medicine Library within the UNIFI platform [30]. Namely, the chemical compositions reported for the *Saussurea* species were searched in online databases, including China National Knowledge Infrastructure (CNKI), Web of Science, ChemSpider, Medline and PubMed, and were gathered to form the database, including the names, chemical structures and molecular formulas of the components being acquired. Secondly, the MS raw data compressed by Waters Compression and Archival Tool v1.10, were imported into the UNIFI software (Waters, Manchester, UK) and were automatically analyzed by the workflow. The main parameters for the workflow were as follows: the minimum peak area was 200; the peak intensities of low and high energy were 200 and 1000 counts, respectively; the acceptable difference of retention time of reference substance was in the range of ±0.1 min. Both positive adducts (+H and +Na) and negative adducts (−H and +COOH) were selected in the analysis. The components that matched the evaluation criteria were screened quickly and were listed. Thirdly, the results were refined with a filter (mass error of the molecular weight or the typical fragments in the range of ±5 ppm, response value >5000). Finally, following the above conditions, the compound was identified by comparing the retention time and accurate molecular weight with the reference substance or by comparing the representative MS fragmentation patterns with the literature.

4.5.2. Quantitative Analysis

The quantitative analysis of the SP ethanol extract was performed on polyphenols with representative skeletons, including 12 phenylpropanoids and 8 flavonoids, using UPLC-Q/TOF-MS. Three standard stock solutions (I~III) of mixtures were prepared in methanol: solution I contained chlorogenic acid, pinoresinol, luteolin, syringaresinol, 1,5-dicaffeoylquinic acid and pinoresinol 4-glucoside; solution II contained matairesinol, neochlorogenic acid, luteolin 7-glucuronide, isoquercitroside, 4,5-dicaffeoylquinic acid, matairesinoside and eupafolin; solution III contained rutin, 1,4-dicaffeoylquinic acid, narcisin, quercitrin, arctiin, apigenin and arctigenin.

Before the assay, a series of standard working solutions were created by properly diluting the stock solution. The external calibration method was used for the quantitative analysis. The validation of the method was as follows:

Calibration curves Each concentration of the mixed three standard solutions was injected and analyzed. The calibration curves were constructed by plotting the peak areas versus the concentrations.

Limits of detection and quantification The standard stocks were diluted with methanol to appropriate concentrations. The LOD and LOQ for each analyte were determined at S/N of about 3 and 10, respectively.

Precision and accuracy The method's precision was assessed by intra- and inter-day variations. The standard solution was analyzed five times in a single day to calculate the intra-day precision, and the sample was analyzed multiple times over the course of six days to determine the inter-day precision. The recovery test was conducted to assess the method's accuracy.

4.6. *Alleviated Ulcerative Colitis Activity*

4.6.1. Experimental Design

In this study, Changyanning Tablet was used as a positive control drug [115]. After being fed adaptively for one week, the mice were randomly assigned to six groups (n = 10) consisting of control group, model group, Changyanning tablet group (CYN, 1.2 g·kg^{-1}), low, middle and high dosages of SP ethanol extract groups (SPL, 30 mg·kg^{-1}; SPM, 60 mg·kg^{-1}; SPH, 120 mg·kg^{-1}). From day 1 to day 7, the mice in control group were given normal water, while other five groups drank DSS aqueous solution (3.5%, *w/v*) *ad libitum* to induce UC model. From day 4 to day 10, the mice in the control and model groups were intragastrically administered with 0.5% sodium carboxymethylcellulose (CMC-Na) solution once a day, while the mice in the other groups were separately intragastrically administered with CYN or SP CMC-Na solution once a day. The volume of administration was all 10 mL/kg. All the mice were sacrificed on day 11 after fasting for 12 h, the blood and tissues were collected and explored for biochemical and histological changes.

4.6.2. Body Weights, Clinical Signs Observations and Disease Activity Index (DAI)

On a daily basis, all mice were weighted and their general clinical signs including fecal characteristics and blood stool were recorded throughout the study period. DAI, obtained on the basis of the scores of weight loss, fecal characteristics and blood stool [116], was used to obtain a quantitative assessment.

4.6.3. Sample Collection and Preparation

The blood obtained through eyeball enucleation was coagulated for half an hour and centrifuged (4000 rpm) for 15 min to obtain the serum samples for biochemical index determination. In addition, serum samples from control group, model group and SPH group were also used for metabolomic study.

After blood collection, the spleen and colon were flushed with PBS solution. The colon length (in terms of centimeters) and spleen coefficient (spleen weight (mg)/body weight (g)) were then calculated or measured for assessing the degree of inflammatory reaction. Then, the colons from each group were used to perform biochemical parameter determination, histological evaluation (fixed in 10% neutral-buffered formalin) and electron microscopy examinations (fixed in 2.5% glutaraldehyde). Moreover, the colons from the control, model and SPH groups were also used for the metabolomic study.

4.6.4. Measurement of Cytokines and Myeloperoxidase (MPO) Contents

The homogenized colon samples were centrifuged for 10 min at 13,000 rpm at 4 °C after homogenization in PBS. TNF-α, iNOS and IL-6 levels in serum samples and in colon homogenate samples were measured using ELISA kits. In order to assess the activity of the neutrophils infiltrated into the colonic lamina, the MPO level in the colon homogenate sample was also evaluated using an ELISA kit.

4.6.5. Measurement of Oxidative Stress Indexes Levels

According to the kit's instructions, the activities of MDA, SOD and GSH in the colon homogenate samples were assessed.

4.6.6. Histological Analysis

The colon tissue was sectioned, deparaffinized, hydrated and H&E stained after being fixed in 10% neutral-buffered formalin and paraffin-embedded. Photographs were taken of the colonic slides under a microscope.

4.6.7. Transmission Electron Microscopy Examination

The fixed colon tissue was post-fixed in 1% OsO_4, and then dehydrated through a graded ethanol series and embedded in epoxy resin. Uranyl acetate and lead citrate were used to counterstain ultrathin sections. Transmission electron microscopy (FEI Tecnai Spirit, USA) was used for observation and photography.

4.6.8. Statistical Analysis

The SPSS 20.0 software was used for statistical analysis. The results were presented as Mean S.E.M. A one-way analysis of variance (ANOVA), followed by a Tukey test, was used to determine statistically significant difference ($p < 0.05$).

4.7. Metabolomics

Serum and colon samples of three groups of mice (control, model and SPH) were collected for metabolomic analysis (n = 10 mice in each group). The method for the metabolomic and data processing was conducted as previously reported [116]. The details are shown in the Supporting Information.

4.8. Network Pharmacology

The network pharmacology study was continued in order to explain the interactions between the phytochemicals and the pharmacological activity, and to predict the potential targets closely associated with the effect of SP from a comprehensive perspective. The method for network pharmacology was conducted as previously reported [117]. The details are shown in the Supporting Information.

4.9. Integrated Analysis Involving Metabolomics and Network Pharmacology

The potential biomarkers obtained from the metabolomics study and the potential targets obtained from the network pharmacology were used to perform the integrated analysis. Then, the "biomarkers-targets" correlation network was then constructed by using MetScape plugin (Cytoscape) based on the Metascape database (http://metascape.org/ (accessed on 11 October 2022)), DAVID database (https://david.ncifcrf.gov/ (accessed on 11 October 2022)) and Reactome database (https://reactome.org/ (accessed on 11 October 2022)). Finally, the intersection of the metabolisms from the integrated analysis and the metabolisms from the metabolomic study were screened out.

5. Conclusions

In the present study, the chemical composition and the pharmacological effect of SP 75% ethanol extract were investigated. A total of 149 components were qualitatively identified or tentatively identified from SP 75% ethanol extract. Among these, 139 components were identified from SP for the first time. Wherein, 12 phenylpropanoids and 8 flavonoids were quantitatively assayed and accounted for 33.2% of the ethanol extract of SP. The components with high contents were narcisin (6.94%), rutin (6.86%), arctiin (5.42%), chlorogenic acid (4.60%), apigenin (4.10%), 1,4-dicaffeoylquinic acid (2.04%) and pinoresinol (1.12%). Network pharmacology analysis showed that the phenylpropanoids and flavonoids contributed the most to the pharmacological activity of SP. By using the DSS-induced UC model mice, it was proven that SP 75% ethanol extract could dose-dependently alleviate bodyweight loss; decrease DAI score, spleen coefficient, levels of TNF-α, IL-6, iNOS, MPO and MDA; increase the colon length, GSH levels and SOD activity; and protect the intestinal barrier. A total of 10 biomarkers, 18 targets and 5 metabolisms were screened out to play vital roles in the therapeutic effect of SP on UC. To summarize, the SP 75% ethanol extract containing phenylpropanoids and flavonoids has a good anti-UC pharmacological effect, and it might be a viable candidate for alleviating UC symptoms.

Supplementary Materials: The following supporting information can be downloaded at: https://www.mdpi.com/article/10.3390/molecules28041526/s1, Figure S1: The representative BPI chromatograms of serum and colon samples of control, model and SPH groups in negative modes (A-F) and in positive modes (G–L).; Table S1: Precision and accuracy of 20 investigated analytes by UPLC-Q/TOF-MS; Table S2: The RSDs(%)of peak area and RT in validation tests; Table S3: The AUCs and *p* values of the biomarkers in different ROC curves; Table S4. The AUCs and p values of the biomarkers in different ROC curves.

Author Contributions: Y.L.: Methodology, Software, Writing and Original draft. C.W. (Caixia Wang): Software and visualization. J.W.: Software and Data curation. L.T.: Methodology, Investigation and Formal analysis. P.G.: Methodology and Formal analysis. S.W.: Methodology. D.T.: Methodology. Q.W.: Investigation and Methodology. C.W. (Cuizhu Wang): Investigation, Methodology and Supervision. P.L.: Conceptualization and Funding acquisition. J.L.: Conceptualization, Data curation, Funding acquisition, Investigation, Supervision, Validation, Resources, Writing—review & editing. All authors have read and agreed to the published version of the manuscript.

Funding: This work was supported by the Science and Technology Innovation Center Project of Jilin Province (No. 20210502005ZP), Industry Independent Innovation Capacity Special project of Jilin Province (No. 2020C038-3) and Young and Middle-aged Science and Technology Innovation Leaders and Team Projects of Jilin Province (No. 20200301002RQ).

Institutional Review Board Statement: In accordance with the Guide for Institutional Animal Care and Use of Laboratory Animals, the mice were kept in facilities approved by the Association for Institutional Animal Care and Use Committee of Jilin University (No.20210060).

Informed Consent Statement: Not applicable.

Data Availability Statement: The data presented in this study are available on request from the corresponding author.

Conflicts of Interest: All the authors declare that there are no conflicts of interest associated with this publication and there is no significant financial support for this work that could have influenced its outcome.

Sample Availability: Samples of the compounds are available on request from the corresponding author.

Abbreviations

BPI	base peak intensity
CMC-Na	sodium carboxymethylcellulose
CYN	changyanning tablet
DAI	disease active index
DSS	dextran sodium sulfate
ESI	electron spray ionization
GSH	glutathione
H&E	hematoxylin-eosin staining
IL-6	lnterleukin-6
iNOS	inducible nitric oxide synthase
LOD	limits of detection
LOQ	limits of quantification
LPC	lysophosphatidylcholine
MDA	malondialdehyde
MPO	myeloperoxidase
OPLS-DA	orthogonal projections to latent structures discriminant analysis
PBS	phosphate buffered saline
PCA	principal component analysis
QC	quality control
QTOF-MS	quadrupole time of flight-mass spectrometry
ROC	receiver operating characteristic

RSD	relative standard deviation
RT	retention time
S.E.M.	standard error of the mean
SOD	superoxide dismutase
SP	*Saussurea pulchella*
TEM	transmission electron microscopy
TNF-α	tumor necrosis factor-α
UC	ulcerative colitis
VIP	variable importance for the projection

References

1. Ungaro, R.; Mehandru, S.; Allen, P.B.; Peyrin-Biroulet, L.; Colombel, J.F. Ulcerative colitis. *Lancet* **2017**, *389*, 1756–1770. [CrossRef] [PubMed]
2. Gajendran, M.; Loganathan, P.; Jimenez, G.; Catinella, A.P.; Ng, N.; Umapathy, C.; Ziade, N.; Hashash, J.G. A comprehensive review and update on ulcerative colitis. *Dis. Mon.* **2019**, *65*, 100851. [CrossRef] [PubMed]
3. Niu, W.; Chen, X.; Xu, R.; Dong, H.; Yang, F.; Wang, Y.; Zhang, Z.; Ju, J. Polysaccharides from natural resources exhibit great potential in the treatment of ulcerative colitis: A review. *Carbohydr. Polym.* **2021**, *254*, 117189. [CrossRef] [PubMed]
4. Xiao, H.; Du, M. Clinical Research Progress of Traditional Chinese Medicine in Ulcerative Colitis. *Pop. Sci. Technol.* **2020**, *22*, 84–87. [CrossRef]
5. Guo, L.; Jiang, X.; Li, J.; Zhang, C.; Li, J.; Chen, J.; Huang, B. Research progress of traditional Chinese medicine for ulcerative colitis. *China Mod. Med.* **2020**, *34*, 26–30. [CrossRef]
6. Jiang, S.; Shen, X.; Xuan, S.; Yang, B.; Ruan, Q.; Cui, H.; Zhao, Z.; Jin, J. Serum and colon metabolomics study reveals the anti-ulcerative colitis effect of Croton crassifolius Geisel. *Phytomedicine* **2021**, *87*, 153570. [CrossRef]
7. Wei, Y.-Y.; Fan, Y.-M.; Ga, Y.; Zhang, Y.-N.; Han, J.-C.; Hao, Z.-H. Shaoyao decoction attenuates DSS-induced ulcerative colitis, macrophage and NLRP3 inflammasome activation through the MKP1/NF-κB pathway. *Phytomedicine* **2021**, *92*, 153743. [CrossRef]
8. Bai, Y.; Hong, S.; Yue, L.; Wang, Y.-Q. Clinical observations on 100 cases of ulcerative colitis treated with the method of clearing away heat, expelling dampness, promoting blood circulation and healing ulcer. *J. Tradit. Chin. Med.* **2010**, *30*, 98–102.
9. Zhong Hua Ben Cao Commission. *Chinese Materia Medica (Zhong Hua Ben Cao)*; Shanghai Science and Technology Press: Shanghai, China, 1999.
10. Nishiuchi, T.; Yamasaki, N. Saussurea Pulchella as a New Cut Flower. In Proceedings of the IV International Symposium on New Floricultural Crops, Chania, Greece, 22–27 May 1999; Volume 541, pp. 247–252.
11. Воробьева, А.Н.; Басаргин, Д.Д. Особенности строения эпидермы листа Saussurea pulchella (Fisch.) Fisch. и S. *Neopulchella Lipsch.* **2013**, *3*, 38–45.
12. Lee, D.-S.; Choi, H.-G.; Woo, K.W.; Kang, D.-G.; Lee, H.-S.; Oh, H.; Lee, K.R.; Kim, Y.-C. Pulchellamin G, an amino acid-sesquiterpene lactone, from Saussurea pulchella suppresses lipopolysaccharide-induced inflammatory responses via heme oxygenase-1 expression in murine peritoneal macrophages. *Eur. J. Pharmacol.* **2013**, *715*, 123–132. [CrossRef]
13. Basargin, D.D.; Tsiklauri, G.C. The phenolic compounds of Saussurea pulchella (Fisch.) Fisch. *Rastit. Resur.* **1990**, *26*, 68–71.
14. Wu, Z. *Compendium of New China Herbal*; Xinhua: Beijing, China, 1990.
15. Ye, Z.; LI, L.; Fu, S.; Liu, J.; Li, P.; Liu, Y. Study on the Anti-hepatocellular Activity of Saussurea pulchella. *Spec. Wild Econ. Anim. Plant Res.* **2021**, *43*, 10–14. [CrossRef]
16. Diao, E.; Wang, G.; Gao, M. Clinical Observation on Rheumatoid Arthritis Treated by Saussurea pulchella. *Chin. J. Tradit. Med. Sci. Technol.* **2000**, *1*, 42–43.
17. Wang, G.; Nie, J.; Diao, E. Study on the anti-inflammatory effect of chrysanthemum. *Chin. J. Tradit. Med. Sci. Technol.* **2000**, *1*, 39–40.
18. Matsuda, H.; Kageura, T.; Inoue, Y.; Morikawa, T.; Yoshikawa, M. Absolute stereostructures and syntheses of saussureamines A, B, C, D and E, amino acid–sesquiterpene conjugates with gastroprotective effect, from the roots of Saussurea lappa. *Tetrahedron* **2000**, *56*, 7763–7777. [CrossRef]
19. Sutar, N.; Garai, R.; Sharma, U.S.; Singh, N.; Roy, S.D. Antiulcerogenic activity of Saussurea lappa root. *Int. J. Pharm. Life Sci.* **2011**, *2*, 516–520.
20. Zhao, T.; Li, S.-J.; Zhang, Z.-X.; Zhang, M.-L.; Shi, Q.-W.; Gu, Y.-C.; Dong, M.; Kiyota, H. Chemical constituents from the genus Saussurea and their biological activities. *Heterocycl. Commun.* **2017**, *23*, 331–358. [CrossRef]
21. Korul'Kina, L.; Shul'ts, E.; Zhusupova, G.; Abilov, Z.A.; Erzhanov, K.; Chaudri, M. Biologically active compounds from Limonium gmelinii and L. popovii I. *Chem. Nat. Compd.* **2004**, *40*, 465–471. [CrossRef]
22. Lin, H.; Zhu, H.; Tan, J.; Wang, H.; Wang, Z.; Li, P.; Zhao, C.; Liu, J. Comparative analysis of chemical constituents of Moringa oleifera leaves from China and India by ultra-performance liquid chromatography coupled with quadrupole-time-of-flight mass spectrometry. *Molecules* **2019**, *24*, 942. [CrossRef]
23. Yang, Z.-Y.; Lu, D.-Y.; Yao, S.; Zhang, R.-R.; Jiang, Z.-J.; Ma, Z.-G. Chemical fingerprint and quantitative analysis of Cistanchedeserticola by HPLC-DAD-ESI-MS. *J. Food Drug Anal.* **2013**, *21*, 50–57.

24. Smyrska-Wieleba, N.; Wojtanowski, K.K.; Mroczek, T. Comparative HILIC/ESI-QTOF-MS and HPTLC studies of pyrrolizidine alkaloids in flowers of Tussilago farfara and roots of Arnebia euchroma. *Phytochem. Lett.* **2017**, *20*, 339–349. [CrossRef]
25. He, J.; Dong, Y.; Liu, X.; Wan, Y.; Gu, T.; Zhou, X.; Liu, M. Comparison of chemical compositions, antioxidant, and anti-photoaging activities of Paeonia suffruticosa flowers at different flowering stages. *Antioxidants* **2019**, *8*, 345. [CrossRef]
26. Kumar, S.; Chandra, P.; Bajpai, V.; Singh, A.; Srivastava, M.; Mishra, D.; Kumar, B. Rapid qualitative and quantitative analysis of bioactive compounds from Phyllanthus amarus using LC/MS/MS techniques. *Ind. Crops Prod.* **2015**, *69*, 143–152. [CrossRef]
27. Llorach, R.; Favari, C.; Alonso, D.; Garcia-Aloy, M.; Andres-Lacueva, C.; Urpi-Sarda, M. Comparative metabolite fingerprinting of legumes using LC-MS-based untargeted metabolomics. *Food Res. Int.* **2019**, *126*, 108666. [CrossRef]
28. Hau, J.; Devaud, S.; Blank, I. Detection of Amadori compounds by capillary electrophoresis coupled to tandem mass spectrometry. *Electrophoresis* **2004**, *25*, 2077–2083. [CrossRef]
29. Otsuka, H.; Takeuchi, M.; Inoshiri, S.; Sato, T.; Yamasaki, K. Phenolic compounds from Coix lachryma-jobi var. ma-yuen. *Phytochemistry* **1989**, *28*, 883–886. [CrossRef]
30. Lin, L.-Z.; Sun, J.; Chen, P.; Harnly, J. UHPLC-PDA-ESI/HRMS/MS n analysis of anthocyanins, flavonol glycosides, and hydroxycinnamic acid derivatives in red mustard greens (Brassica juncea Coss variety). *J. Agric. Food Chem.* **2011**, *59*, 12059–12072. [CrossRef]
31. Yang, Y.-L.; Al-Mahdy, D.A.; Wu, M.-L.; Zheng, X.-T.; Piao, X.-H.; Chen, A.-L.; Wang, S.-M.; Yang, Q.; Ge, Y.-W. LC-MS-based identification and antioxidant evaluation of small molecules from the cinnamon oil extraction waste. *Food Chem.* **2022**, *366*, 130576. [CrossRef]
32. Bartsch, M.; Bednarek, P.; Vivancos, P.D.; Schneider, B.; von Roepenack-Lahaye, E.; Foyer, C.H.; Kombrink, E.; Scheel, D.; Parker, J.E. Accumulation of isochorismate-derived 2, 3-dihydroxybenzoic 3-O-β-D-xyloside in Arabidopsis resistance to pathogens and ageing of leaves. *J. Biol. Chem.* **2010**, *285*, 25654–25665. [CrossRef]
33. Mujahid, M.; Sasikala, C.; Ramana, C.V. Aniline-induced tryptophan production and identification of indole derivatives from three purple bacteria. *Curr. Microbiol.* **2010**, *61*, 285–290. [CrossRef]
34. Shakya, R.; Navarre, D.A. Rapid screening of ascorbic acid, glycoalkaloids, and phenolics in potato using high-performance liquid chromatography. *J. Agric. Food Chem.* **2006**, *54*, 5253–5260. [CrossRef]
35. Liu, L.; Cui, Z.-X.; Zhang, Y.-B.; Xu, W.; Yang, X.-W.; Zhong, L.-J.; Zhang, P.; Gong, Y. Identification and quantification analysis of the chemical constituents from Mahonia fortune using Q-Exactive HF Mass Spectrometer and UPLC–ESI-MS/MS. *J. Pharm. Biomed. Anal.* **2021**, *196*, 113903. [CrossRef]
36. Antunes, A.C.; Acunha, T.d.S.; Perin, E.C.; Rombaldi, C.V.; Galli, V.; Chaves, F.C. Untargeted metabolomics of strawberry (Fragaria x ananassa 'Camarosa') fruit from plants grown under osmotic stress conditions. *J. Sci. Food Agric.* **2019**, *99*, 6973–6980. [CrossRef]
37. Yang, M.C.; Choi, S.U.; Choi, W.S.; Kim, S.Y.; Lee, K.R. Guaiane sesquiterpene lactones and amino acid-sesquiterpene lactone conjugates from the aerial parts of Saussurea pulchella. *J. Nat. Prod.* **2008**, *71*, 678–683. [CrossRef]
38. Kammerer, B.; Kahlich, R.; Biegert, C.; Gleiter, C.H.; Heide, L. HPLC-MS/MS analysis of willow bark extracts contained in pharmaceutical preparations. *Phytochem. Anal.* **2005**, *16*, 470–478. [CrossRef]
39. Milutinović, V.; Niketić, M.; Krunić, A.; Nikolić, D.; Petković, M.; Ušjak, L.; Petrović, S. Sesquiterpene lactones from the methanol extracts of twenty-eight Hieracium species from the Balkan Peninsula and their chemosystematic significance. *Phytochemistry* **2018**, *154*, 19–30. [CrossRef]
40. Duan, H.; Takaishi, Y.; Momota, H.; Ohmoto, Y.; Taki, T. Immunosuppressive constituents from Saussurea medusa. *Phytochemistry* **2002**, *59*, 85–90. [CrossRef]
41. Xu, S.j.; Yang, L.; Zeng, X.; Zhang, M.; Wang, Z.t. Characterization of compounds in the Chinese herbal drug Mu-Dan-Pi by liquid chromatography coupled to electrospray ionization mass spectrometry. *Rapid Commun. Mass Spectrom.* **2006**, *20*, 3275–3288. [CrossRef]
42. Saravanakumar, K.; Park, S.; Sathiyaseelan, A.; Kim, K.-N.; Cho, S.-H.; Mariadoss, A.V.A.; Wang, M.-H. Metabolite profiling of methanolic extract of Gardenia jaminoides by LC-MS/MS and GC-MS and its anti-diabetic, and anti-oxidant activities. *Pharmaceuticals* **2021**, *14*, 102. [CrossRef]
43. Anttonen, M.J.; Karjalainen, R.O. High-performance liquid chromatography analysis of black currant (*Ribes nigrum* L.) fruit phenolics grown either conventionally or organically. *J. Agric. Food Chem.* **2006**, *54*, 7530–7538. [CrossRef]
44. Fang, N.; Yu, S.; Prior, R.L. LC/MS/MS characterization of phenolic constituents in dried plums. *J. Agric. Food Chem.* **2002**, *50*, 3579–3585. [CrossRef]
45. Owen, R.; Haubner, R.; Hull, W.; Erben, G.; Spiegelhalder, B.; Bartsch, H.; Haber, B. Isolation and structure elucidation of the major individual polyphenols in carob fibre. *Food Chem. Toxicol.* **2003**, *41*, 1727–1738. [CrossRef]
46. Tolonen, A.; Hohtola, A.; Jalonen, J. Comparison of electrospray ionization and atmospheric pressure chemical ionization techniques in the analysis of the main constituents from Rhodiola rosea extracts by liquid chromatography/mass spectrometry. *J. Mass Spectrom.* **2003**, *38*, 845–853. [CrossRef]
47. Hu, H.; Yau, L.-F.; Peng, J.; Hu, B.; Li, J.; Li, Y.; Huang, H. Comparative Research of Chemical Profiling in Different Parts of Fissistigma oldhamii by Ultra-High-Performance Liquid Chromatography Coupled with Hybrid Quadrupole-Orbitrap Mass Spectrometry. *Molecules* **2021**, *26*, 960. [CrossRef]

48. Duan, L.; Xiong, H.; Du, Y.; Wang, Z.; Li, Y.; Zhao, S.; Chen, J.; Si, D.; Pan, H. High-throughput LC–MS method for the rapid characterisation and comparative analysis of multiple ingredients of four hawthorn leaf extracts. *Phytochem. Anal.* **2022**, *33*, 635–643. [CrossRef]
49. Fan, C.-Q.; Yue, J.-M. Biologically active phenols from Saussurea medusa. *Bioorganic Med. Chem.* **2003**, *11*, 703–708. [CrossRef]
50. Jaiswal, R.; Kuhnert, N. How to identify and discriminate between the methyl quinates of chlorogenic acids by liquid chromatography–tandem mass spectrometry. *J. Mass Spectrom.* **2011**, *46*, 269–281. [CrossRef]
51. Xie, H.; Wang, T.; Matsuda, H.; Morikawa, T.; Yoshikawa, M.; Tani, T. Bioactive constituents from Chinese natural medicines. XV. Inhibitory effect on aldose reductase and structures of saussureosides A and B from Saussurea medusa. *Chem. Pharm. Bull.* **2005**, *53*, 1416–1422. [CrossRef]
52. Knust, U.; Erben, G.; Spiegelhalder, B.; Bartsch, H.; Owen, R.W. Identification and quantitation of phenolic compounds in faecal matrix by capillary gas chromatography and nano-electrospray mass spectrometry. *Rapid Commun. Mass Spectrom.* **2006**, *20*, 3119–3129. [CrossRef]
53. Zhou, Z.W.; Yin, S.; Wang, X.N.; Fan, C.Q.; Li, H.; Yue, J.M. Two new lignan glycosides from Saussurea laniceps. *Helv. Chim. Acta* **2007**, *90*, 951–956. [CrossRef]
54. Mellegård, H.; Stalheim, T.; Hormazabal, V.; Granum, P.; Hardy, S. Antibacterial activity of sphagnum acid and other phenolic compounds found in Sphagnum papillosum against food-borne bacteria. *Lett. Appl. Microbiol.* **2009**, *49*, 85–90. [CrossRef]
55. Wu, D.; Wang, H.; Tan, J.; Wang, C.; Lin, H.; Zhu, H.; Liu, J.; Li, P.; Yin, J. Pharmacokinetic and metabolism studies of curculigoside C by UPLC-MS/MS and UPLC-QTOF-MS. *Molecules* **2018**, *24*, 21. [CrossRef]
56. Yang, D.S.; Whang, W.K.; Kim, I.H. The constituents of Taraxacum hallaisanensis roots. *Arch. Pharmacal Res.* **1996**, *19*, 507–513. [CrossRef]
57. Rodríguez-Pérez, C.; Quirantes-Piné, R.; Fernández-Gutiérrez, A.; Segura-Carretero, A. Comparative characterization of phenolic and other polar compounds in Spanish melon cultivars by using high-performance liquid chromatography coupled to electrospray ionization quadrupole-time of flight mass spectrometry. *Food Res. Int.* **2013**, *54*, 1519–1527. [CrossRef]
58. Liang, Y.-H. Lignans and flavonoids from rhizome of Drynaria fortunei. *Chin. Tradit. Herb. Drugs* **2011**, *24*, 25–30.
59. Michalska, A.; Wojdyło, A.; Bogucka, B. The influence of nitrogen and potassium fertilisation on the content of polyphenolic compounds and antioxidant capacity of coloured potato. *J. Food Compost. Anal.* **2016**, *47*, 69–75. [CrossRef]
60. Cao, Y.; Gu, C.; Zhao, F.; Tang, Y.; Cui, X.; Shi, L.; Xu, L.; Yin, L. Therapeutic effects of Cyathula officinalis Kuan and its active fraction on acute blood stasis rat model and identification constituents by HPLC-QTOF/MS/MS. *Pharmacogn. Mag.* **2017**, *13*, 693. [PubMed]
61. Ha, T.J.; Jang, D.S.; Lee, J.R.; Lee, K.D.; Lee, J.; Hwang, S.W.; Jung, H.J.; Nam, S.H.; Park, K.H.; Yang, M.S. Cytotoxic effects of sesquiterpene lactones from the flowers ofHemisteptia lyrata B. *Arch. Pharmacal Res.* **2003**, *26*, 925–928. [CrossRef]
62. Kuo, Y.-H.; Way, S.-T.; Wu, C.-H. A new triterpene and a new lignan from Saussurea japonica. *J. Nat. Prod.* **1996**, *59*, 622–624. [CrossRef]
63. Yang, N.; Wang, H.; Lin, H.; Liu, J.; Zhou, B.; Chen, X.; Wang, C.; Liu, J.; Li, P. Comprehensive metabolomics analysis based on UPLC-Q/TOF-MS E and the anti-COPD effect of different parts of Celastrus orbiculatus Thunb. *RSC Adv.* **2020**, *10*, 8396–8420. [CrossRef]
64. Han, Y.; Zhou, M.; Wang, L.; Ying, X.; Peng, J.; Jiang, M.; Bai, G.; Luo, G. Comparative evaluation of different cultivars of Flos Chrysanthemi by an anti-inflammatory-based NF-κB reporter gene assay coupled to UPLC-Q/TOF MS with PCA and ANN. *J. Ethnopharmacol.* **2015**, *174*, 387–395. [CrossRef] [PubMed]
65. Gattuso, G.; Caristi, C.; Gargiulli, C.; Bellocco, E.; Toscano, G.; Leuzzi, U. Flavonoid glycosides in bergamot juice (Citrus bergamia Risso). *J. Agric. Food Chem.* **2006**, *54*, 3929–3935. [CrossRef]
66. Ye, M.; Han, J.; Chen, H.; Zheng, J.; Guo, D. Analysis of phenolic compounds in rhubarbs using liquid chromatography coupled with electrospray ionization mass spectrometry. *J. Am. Soc. Mass Spectrom.* **2007**, *18*, 82–91. [CrossRef] [PubMed]
67. Wei, L.; Mei, Y.; Zou, L.; Chen, J.; Tan, M.; Wang, C.; Cai, Z.; Lin, L.; Chai, C.; Yin, S. Distribution patterns for bioactive constituents in pericarp, stalk and seed of Forsythiae fructus. *Molecules* **2020**, *25*, 340. [CrossRef] [PubMed]
68. Dai, X.; Zhuang, J.; Wu, Y.; Wang, P.; Zhao, G.; Liu, Y.; Jiang, X.; Gao, L.; Xia, T. Identification of a flavonoid glucosyltransferase involved in 7-OH site glycosylation in tea plants (Camellia sinensis). *Sci. Rep.* **2017**, *7*, 5926. [CrossRef] [PubMed]
69. Flores, R.M.; Doskey, P.V. Evaluation of multistep derivatization methods for identification and quantification of oxygenated species in organic aerosol. *J. Chromatogr.* **2015**, *1418*, 1–11. [CrossRef]
70. Fu, S.; Arráez-Roman, D.; Segura-Carretero, A.; Menéndez, J.A.; Menéndez-Gutiérrez, M.P.; Micol, V.; Fernández-Gutiérrez, A. Qualitative screening of phenolic compounds in olive leaf extracts by hyphenated liquid chromatography and preliminary evaluation of cytotoxic activity against human breast cancer cells. *Anal. Bioanal. Chem.* **2010**, *397*, 643–654. [CrossRef]
71. Lin, L.-Z.; Harnly, J.M. Identification of hydroxycinnamoylquinic acids of arnica flowers and burdock roots using a standardized LC-DAD-ESI/MS profiling method. *J. Agric. Food Chem.* **2008**, *56*, 10105–10114. [CrossRef]
72. Singh, P.; Bhala, M. Guaianolides from Saussurea candicans. *Phytochemistry* **1988**, *27*, 1203–1205. [CrossRef]
73. Fan, C.-Q.; Zhu, X.-Z.; Zhan, Z.-J.; Ji, X.-Q.; Li, H.; Yue, J.-M. Lignans from Saussurea conica and their NO production suppressing activity. *Planta Med.* **2006**, *72*, 590–595. [CrossRef]
74. Wang, X.R.; Wu, Q.X.; Shi, Y.P. Terpenoids and sterols from Saussurea cauloptera. *Chem. Biodivers.* **2008**, *5*, 279–289. [CrossRef] [PubMed]

75. Zhang, Y.-B.; Yang, X.-B.; Yang, X.-W.; Xu, W.; Li, F.; Gonzezal, F.J. Liquid chromatography with tandem mass spectrometry: A sensitive method for the determination of dehydrodiisoeugenol in rat cerebral nuclei. *Molecules* **2016**, *21*, 321. [CrossRef]
76. Ichihara, A.; Numata, Y.; Kanai, S.; Sakamura, S. New sesquilignans from Arctium lappa L. The structure of lappaol C, D and E. *Agric. Biol. Chem.* **1977**, *41*, 1813–1814. [CrossRef]
77. Joo, J.; Lee, D.; Wu, Z.; Shin, J.H.; Lee, H.S.; Kwon, B.M.; Huh, T.L.; Kim, Y.W.; Lee, S.J.; Kim, T.W. In vitro metabolism of obovatol and its effect on cytochrome P450 enzyme activities in human liver microsomes. *Biopharm. Drug Dispos.* **2013**, *34*, 195–202. [CrossRef] [PubMed]
78. Freund, D.M.; Martin, A.C.; Cohen, J.D.; Hegeman, A.D. Direct detection of surface localized specialized metabolites from Glycyrrhiza lepidota (American licorice) by leaf spray mass spectrometry. *Planta* **2018**, *247*, 267–275. [CrossRef]
79. Aabideen, Z.U.; Mumtaz, M.W.; Akhtar, M.T.; Mukhtar, H.; Raza, S.A.; Touqeer, T.; Saari, N. Anti-obesity attributes; UHPLC-QTOF-MS/MS-based metabolite profiling and molecular docking insights of Taraxacum officinale. *Molecules* **2020**, *25*, 4935. [CrossRef]
80. Bohlmann, F.; Singh, P.; Jakupovic, J.; Huneck, S. Further guaianolides from Saussurea species. *Planta Med.* **1985**, *51*, 74–75. [CrossRef]
81. Ratnam, K.J.; Reddy, R.S.; Sekhar, N.; Kantam, M.L.; Figueras, F. Sulphated zirconia catalyzed acylation of phenols, alcohols and amines under solvent free conditions. *J. Mol. Catal. A Chem.* **2007**, *276*, 230–234. [CrossRef]
82. Zheleva-Dimitrova, D.; Gevrenova, R.; Zaharieva, M.M.; Najdenski, H.; Ruseva, S.; Lozanov, V.; Balabanova, V.; Yagi, S.; Momekov, G.; Mitev, V. HPLC-UV and LC–MS analyses of acylquinic acids in Geigeria alata (DC) Oliv. & Hiern. and their contribution to antioxidant and antimicrobial capacity. *Phytochem. Anal.* **2017**, *28*, 176–184.
83. Barrero, A.F.; Haidour, A.; Dorado, M.M. Sesquipinsapols A and B: Two sesquilignans from Abies pinsapo. *Nat. Prod. Lett.* **1993**, *2*, 255–262. [CrossRef]
84. Fan, C.Q.; Zhan, Z.J.; Li, H.; Yue, J.M. Eudesmane-Type Sesquiterpene Derivatives from Saussurea conica. *Helv. Chim. Acta* **2004**, *87*, 1446–1451. [CrossRef]
85. Lee, J.H.; Lee, S.J.; Park, S.; Kim, H.K.; Jeong, W.Y.; Choi, J.Y.; Sung, N.-J.; Lee, W.S.; Lim, C.-S.; Kim, G.-S. Characterisation of flavonoids in Orostachys japonicus A. Berger using HPLC–MS/MS: Contribution to the overall antioxidant effect. *Food Chem.* **2011**, *124*, 1627–1633. [CrossRef]
86. Choi, S.U.; Yang, M.C.; Lee, K.H.; Kim, K.H.; Lee, K.R. Lignan and terpene constituents from the aerial parts of Saussurea pulchella. *Arch. Pharmacal Res.* **2007**, *30*, 1067–1074. [CrossRef]
87. Liu, Q.; Majdi, M.; Cankar, K.; Goedbloed, M.; Charnikhova, T.; Verstappen, F.W.; De Vos, R.C.; Beekwilder, J.; Van der Krol, S.; Bouwmeester, H.J. Reconstitution of the costunolide biosynthetic pathway in yeast and Nicotiana benthamiana. *PLoS ONE* **2011**, *6*, e23255. [CrossRef] [PubMed]
88. Aboutabl, E.A.; El Mahdy, M.E.; Sokkar, N.M.; Sleem, A.A.; Shams, M.M. Bioactive lignans and other phenolics from the roots, leaves and seeds of Arctium lappa L. grown in Egypt. *Egypt. Pharm. J.* **2012**, *11*, 59.
89. Matsumoto, T.; Hosono-Nishiyama, K.; Yamada, H. Antiproliferative and apoptotic effects of butyrolactone lignans from Arctium lappa on leukemic cells. *Planta Med.* **2006**, *72*, 276–278. [CrossRef] [PubMed]
90. Nguyen, D.T.; Göpfert, J.C.; Ikezawa, N.; MacNevin, G.; Kathiresan, M.; Conrad, J.; Spring, O.; Ro, D.-K. Biochemical conservation and evolution of germacrene A oxidase in Asteraceae. *J. Biol. Chem.* **2010**, *285*, 16588–16598. [CrossRef]
91. Huh, J.; Lee, C.-M.; Lee, S.; Kim, S.; Cho, N.; Cho, Y.-C. Comprehensive Characterization of Lignans from Forsythia viridissima by UHPLC-ESI-QTOF-MS, and Their NO Inhibitory Effects on RAW 264.7 Cells. *Molecules* **2019**, *24*, 2649. [CrossRef]
92. de Kraker, J.-W.; Franssen, M.C.; de Groot, A.; Shibata, T.; Bouwmeester, H.J. Germacrenes from fresh costus roots. *Phytochemistry* **2001**, *58*, 481–487. [CrossRef]
93. Roy, R.N.; Laskar, S.; Sen, S. Dibutyl phthalate, the bioactive compound produced by Streptomyces albidoflavus 321.2. *Microbiol. Res.* **2006**, *161*, 121–126. [CrossRef]
94. Piacente, S.; Santos, L.C.D.; Mahmood, N.; Pizza, C. Triterpenes from Maytenus macrocarpa and evaluation of their anti-HIV activity. *Nat. Prod. Commun.* **2006**, *1*, 1934578X0600101201. [CrossRef]
95. Na, M.; Kim, B.Y.; Osada, H.; Ahn, J.S. Inhibition of protein tyrosine phosphatase 1B by lupeol and lupenone isolated from Sorbus commixta. *J. Enzym. Inhib. Med. Chem.* **2009**, *24*, 1056–1059. [CrossRef] [PubMed]
96. Frankenberger, L.; Mora, T.D.; de Siqueira, C.D.; Filippin-Monteiro, F.B.; de Moraes, M.H.; Biavatti, M.W.; Steindel, M.; Sandjo, L.P. UPLC-ESI-QTOF-MS2 characterisation of Cola nitida resin fractions with inhibitory effects on NO and TNF-α released by LPS-activated J774 macrophage and on Trypanosoma cruzi and Leishmania amazonensis. *Phytochem. Anal.* **2018**, *29*, 577–589. [CrossRef] [PubMed]
97. Menichini, F.; Di Benedetto, R.; Delle Monache, F. A triterpene epoxide and a guaianolide from Ptilostemmon gnaphaloides. *Phytochemistry* **1996**, *41*, 1377–1379. [CrossRef]
98. Pütter, K.M.; van Deenen, N.; Müller, B.; Fuchs, L.; Vorwerk, K.; Unland, K.; Bröker, J.N.; Scherer, E.; Huber, C.; Eisenreich, W. The enzymes OSC1 and CYP716A263 produce a high variety of triterpenoids in the latex of Taraxacum koksaghyz. *Sci. Rep.* **2019**, *9*, 5942. [CrossRef]
99. Luis, J.G.; Andrés, L.S. New ursane type triterpenes from Salvia mellifera greene. *Nat. Prod. Lett.* **1999**, *13*, 187–194. [CrossRef]
100. Pop, R.M.; Weesepoel, Y.; Socaciu, C.; Pintea, A.; Vincken, J.-P.; Gruppen, H. Carotenoid composition of berries and leaves from six Romanian sea buckthorn (*Hippophae rhamnoides* L.) varieties. *Food Chem.* **2014**, *147*, 1–9. [CrossRef]

101. Rontani, J.-F.; Charrière, B.; Menniti, C.; Aubert, D.; Aubert, C. EIMS Fragmentation and MRM quantification of autoxidation products of α-and β-amyrins in natural samples. *Rapid Commun. Mass Spectrom.* **2018**, *18*, 1599–1607. [CrossRef]
102. Takatori, S.; Kitagawa, Y.; Kitagawa, M.; Nakazawa, H.; Hori, S. Determination of di (2-ethylhexyl) phthalate and mono (2-ethylhexyl) phthalate in human serum using liquid chromatography-tandem mass spectrometry. *J. Chromatogr. B* **2004**, *804*, 397–401. [CrossRef]
103. Wang, F.; Dana, A.; Tian, S.; Eponine, O.; Vasuk, G.; Russell, G.; Thomaso, M.; Davids, W. CFM-ID 4.0—A web server for accurate MS-based metabolite identification. *Nucleic Acids Res.* **2022**, *50*, W165–W174. [CrossRef]
104. Wu, X.; Yang, Y.; Dou, Y.; Ye, J.; Bian, D.; Wei, Z.; Tong, B.; Kong, L.; Xia, Y.; Dai, Y. Arctigenin but not arctiin acts as the major effective constituent of Arctium lappa L. fruit for attenuating colonic inflammatory response induced by dextran sulfate sodium in mice. *Int. Immunopharmacol.* **2014**, *23*, 505–515. [CrossRef] [PubMed]
105. Zeng, J.; Zhang, D.; Wan, X.; Bai, Y.; Yuan, C.; Wang, T.; Yuan, D.; Zhang, C.; Liu, C. Chlorogenic Acid Suppresses miR-155 and Ameliorates Ulcerative Colitis through the NF-κB/NLRP3 Inflammasome Pathway. *Mol. Nutr. Food Res.* **2020**, *64*, 2000452. [CrossRef] [PubMed]
106. Liu, Y.; Huang, W.; Ji, S.; Wang, J.; Luo, J.; Lu, B. Sophora japonica flowers and their main phytochemical, rutin, regulate chemically induced murine colitis in association with targeting the NF-κB signaling pathway and gut microbiota. *Food Chem.* **2022**, *393*, 133395. [CrossRef] [PubMed]
107. Marquez-Flores, Y.K.; Villegas, I.; Cárdeno, A.; Rosillo, M.Á.; Alarcon-de-la-Lastra, C. Apigenin supplementation protects the development of dextran sulfate sodium-induced murine experimental colitis by inhibiting canonical and non-canonical inflammasome signaling pathways. *J. Nutr. Biochem.* **2016**, *30*, 143–152. [CrossRef] [PubMed]
108. Bilsborough, J.; Fiorino, M.F.; Henkle, B.W. Select animal models of colitis and their value in predicting clinical efficacy of biological therapies in ulcerative colitis. *Expert Opin. Drug Discov.* **2021**, *16*, 567–577. [CrossRef]
109. Randhawa, P.K.; Singh, K.; Singh, N.; Jaggi, A.S. A review on chemical-induced inflammatory bowel disease models in rodents. *Korean J. Physiol. Pharmacol.* **2014**, *18*, 279–288. [CrossRef]
110. Vilcek, J.; Lee, T.H. Tumor necrosis factor.: New insights into the molecular mechanisms of its multiple actions. *J. Biol. Chem.* **1991**, *266*, 7313–7316. [CrossRef]
111. Luo, J.; Cao, J.; Jiang, X.; Cui, H. Effect of low molecular weight heparin rectal suppository on experimental ulcerative colitis in mice. *Biomed. Pharmacother.* **2010**, *64*, 441–445. [CrossRef]
112. Itzkowitz, S.H. Molecular biology of dysplasia and cancer in inflammatory bowel disease. *Gastroenterol. Clin.* **2006**, *35*, 553–571. [CrossRef]
113. Diab, J.; Hansen, T.; Goll, R.; Stenlund, H.; Ahnlund, M.; Jensen, E.; Moritz, T.; Florholmen, J.; Forsdahl, G. Lipidomics in ulcerative colitis reveal alteration in mucosal lipid composition associated with the disease state. *Inflamm. Bowel Dis.* **2019**, *25*, 1780–1787. [CrossRef]
114. Liu, Y.; Wang, Z.; Wang, C.; Si, H.; Yu, H.; Li, L.; Fu, S.; Tan, L.; Li, P.; Liu, J. Comprehensive phytochemical analysis and sedative-hypnotic activity of two Acanthopanax species leaves. *Food Funct.* **2021**, *12*, 2292–2311. [CrossRef]
115. Liu, Z.; Peng, Y.; Ma, P.; Fan, L.; Zhao, L.; Wang, M.; Li, X. An integrated strategy for anti-inflammatory quality markers screening of traditional Chinese herbal medicine Mume Fructus based on phytochemical analysis and anti-colitis activity. *Phytomedicine* **2022**, *99*, 154002. [CrossRef] [PubMed]
116. Zhou, B.; Liu, J.; Wang, Y.; Wu, F.; Wang, C.; Wang, C.; Liu, J.; Li, P. Protective Effect of Ethyl Rosmarinate against Ulcerative Colitis in Mice Based on Untargeted Metabolomics. *Int. J. Mol. Sci.* **2022**, *23*, 1256. [CrossRef] [PubMed]
117. Lin, H.; Wang, C.; Yu, H.; Liu, Y.; Tan, L.; He, S.; Li, Z.; Wang, C.; Wang, F.; Li, P. Protective effect of total Saponins from American ginseng against cigarette smoke-induced COPD in mice based on integrated metabolomics and network pharmacology. *Biomed. Pharmacother.* **2022**, *149*, 112823. [CrossRef] [PubMed]

Disclaimer/Publisher's Note: The statements, opinions and data contained in all publications are solely those of the individual author(s) and contributor(s) and not of MDPI and/or the editor(s). MDPI and/or the editor(s) disclaim responsibility for any injury to people or property resulting from any ideas, methods, instructions or products referred to in the content.

molecules

MDPI

Article

Proximate Composition and Nutritional Values of Selected Wild Plants of the United Arab Emirates

Mohammad Shahid, Rakesh Kumar Singh and Sumitha Thushar *

Division of Crop Diversification, International Center for Biosaline Agriculture, Dubai P.O. Box 14660, United Arab Emirates
* Correspondence: s.thushar@biosaline.org.ae

Abstract: Wild plants supply food and shelter to several organisms; they also act as important sources of many nutrients and pharmaceutical agents for mankind. These plants are widely used in traditional medicinal systems and folk medicines. The present study analyzed the nutritional and proximate composition of various compounds in selected wild plants available in the UAE, viz., *Chenopodium murale* L., *Dipterygium glaucum* Decne., *Heliotropium digynum* Asch. ex C.Chr., *Heliotropium kotschyi* Gürke., *Salsola imbricata* Forssk., *Tribulus pentandrus* Forssk., *Zygophyllum qatarense* Hadidi. The predominant amino acids detected in the plants were glycine, threonine, histidine, cysteine, proline, serine, and tyrosine; the highest quantities were observed in *H. digynum* and *T. pentandrus*. The major fatty acids present were long-chain saturated fatty acids; however, lauric acid was only present in *S. imbricata*. The presence of essential fatty acids such as oleic acid, α-Linoleic acid, and linolenic acid was observed in *H. digynum*, *S. imbricata*, and *H. kotschyi*. These plants also exhibited higher content of nutrients such as carbohydrates, proteins, fats, ash, and fiber. The predominant vitamins in the plants were vitamin B complex and vitamin C. *C. murale* had higher vitamin A, whereas vitamin B complex was seen in *T. pentandrus* and *D. glaucum*. The phosphorus and zinc content were high in *T. pentandrus*; the nitrogen, calcium, and potassium contents were high in *H. digynum*, and *D. glaucum*. Overall, these plants, especially *H. digynum* and *T. pentandrus* contain high amounts of nutritionally active compounds and important antioxidants including trace elements and vitamins. The results from the experiment provide an understanding of the nutritional composition of these desert plant species and can be better utilized as important agents for pharmacological drug discovery, food, and sustainable livestock production in the desert ecosystem.

Keywords: wild plants; amino acid content; fatty acid profile; mineral content; vitamin composition

Citation: Shahid, M.; Singh, R.K.; Thushar, S. Proximate Composition and Nutritional Values of Selected Wild Plants of the United Arab Emirates. *Molecules* **2023**, *28*, 1504. https://doi.org/10.3390/molecules28031504

Academic Editors: Arunaksharan Narayanankutty, Ademola C. Famurewa and Eliza Oprea

Received: 30 November 2022
Revised: 19 January 2023
Accepted: 20 January 2023
Published: 3 February 2023

Copyright: © 2023 by the authors. Licensee MDPI, Basel, Switzerland. This article is an open access article distributed under the terms and conditions of the Creative Commons Attribution (CC BY) license (https://creativecommons.org/licenses/by/4.0/).

1. Introduction

The body needs six nutrients for proper functioning and overall health. These include carbohydrates, proteins, fats, water, vitamins, and minerals. In many developing countries, starvation and undernourishment are on the rise because of population explosion, scarcity of productive land, and soaring food costs. The desert ecosystem has diverse environmental constraints, including low rainfall, drought, high soil salinity, extreme temperatures, intense sunshine, low moisture retention, and infertile soil, all of which cumulatively impact plant growth and development. Despite these ecological restrictions, the natural flora is well-adapted to these adverse environmental conditions [1]. In the wild, there are many plants of high nutritional value that can be used to feed the ever-growing human population and help to secure nutritional security. Finding a diet from the wild has been closely related to humans for hundreds of thousands of years [2]. Around 12,000 years ago, before the invention of agriculture, people depended on Wild Edible Plants (WEPs) and hunting for survival. WEPs are found in a variety of botanical types, including herbs, vines, bushes, grasses, and trees, both annual and perennial species [3]. Approximately 30,000 edible plant species found around the world have the potential to be used as food or feed. This wild edible flora can potentially convert the food systems into more nourishing, sustainable, and resilient to abiotic stresses [4]. The WEPs provide both micro- and macronutrients that enrich the dietary quality, which are inexpensive sources of nutrition for different people worldwide [5,6]. Wild food use is still thriving, particularly in isolated, economically deprived places of the world [7,8] and during crop failures [9].

In addition to food and fodder, wild plants are also important sources of pharmacologically important agents such as antioxidant and anti-inflammatory compounds. Among these, the major ones include phytochemicals such as phenols, flavonoids, tannins, alkaloids, terpenoids, and saponins [10]. These compounds can break the reactive chain reactions such as free radicals and thereby inhibit oxidative insults to the body [11]. In addition, these compounds are known to inhibit enzymes and signaling cascades involved in inflammatory reactions [12,13]. Apart from these, phytochemicals are associated with the prevention and curing of various diseases such as cardiovascular diseases, neurodegenerative disorders, and various forms of cancers [14,15]. In addition to this, plants contain major vitamins such as vitamin B complex and vitamin C; these vitamins are central molecules involved in various physiological activities of the body and play a central role in redox balance [16,17]. Further, trace elements such as zinc, selenium, etc. are also known for their biologically important roles; considering their importance in the normal physiological functioning of the body and disease prevention, these plants are also considered to be important in medicine and food [18–20]. Hence, wild plants are important agents that provide various bioactive and pharmacologically important compounds to humans.

The United Arab Emirates (UAE), located in the southeast of the Arabian Peninsula, has an 83,600 square km area. The country has four major landforms: sand, salt flat, gravel, and mountains, which harbor a great diversity of flora that has adapted to diverse rough and mild environments found here. Most of the country is comprised of the desert, which is part of a vast sea of sand called Rub' al Khali or Empty Quarter. Most plants here are either xerophytic (adapted to dry arid habitats) or halophytic (salt-tolerant). The UAE has about 820 plant species growing in different regions of the country [21]. Karim et al. [22] have screened 170 different plants in the UAE (trees, shrubs, and grasses) which are salt resistant along with their description, distribution, and uses. Native plants of the UAE were used in traditional medicine long ago in the UAE but in ethnobotanical literature most of the species found in the UAE are rare. Zayed Complex for Herbal Research and Traditional Medicine, a publication about medicinal plants that enlisted only 29 species, including some non-indigenous [23] species. Sakkir et al. [24] analyzed the medicinal status of UAE flora, and they listed a total of 132 plants (nearly 20% of total species) that are traditionally used in the UAE for their medicinal properties.

Wild plant species thus play a significant role in worldwide nutrition and food security [25,26] and the use of healthy edible wild flora can be investigated to support the nutritional/medicinal needs around the globe. A nutritional values database for WEPs needs to be systematically created to enhance dietary diversity and control hunger [10,20]. Information on wild plants' edibility and medicinal properties is generally scarce, and data on their nutritional composition is insignificant [2,4,5,27]. In the present study, we aimed to evaluate the selected wild plants of the United Arab Emirates, viz., *Chenopodium murale, Dipterygium glaucum, Heliotropium digynum, Heliotropium kotschyi, Salsola imbricata, Tribulus pentandrus* and *Zygophyllum qatarense* which were found to grow abundantly in the UAE under the harsh environments for their nutritional and proximate composition.

Tribulus is characterized by buttercup-like yellow flowers and flora references of the region report five species in the United Arab Emirates (UAE), namely *T. macropterous* Boiss. (perennial), *T. omanense* (perennial), *T. parvispinos* Presl (annual), *T. pentandrus* Forssk. (annual/perennial) and *T. terrestris* L (annual, occasionally biennial). Among these, *T. pentandrus*, a well-distributed and less investigated species, were selected for the study. The native range of *T. pentandrus* (Arabic name: Zahar) belongs to the family Zygophyllaceae in the Arabian Peninsula. It is a perennial and grows primarily in the desert or dry shrubland biome(s) [28]. The native range of *C. murale* (Arabic name: Abu' efein) belongs to the family

Amaranthaceae in Macaronesia, Europe, Medit. to NE tropical Africa and Sri Lanka. It is an annual herb reaching 70 cm in height and grows primarily in the temperate biome(s) with an erect stem which is usually red or red-streaked green and leafy with green foliage. It has environmental uses, such as fodder, medicine, and food [29,30]. *D. glaucum* (Arabic name: Safrawi) belongs to the family Capparaceae and is very common in the UAE, along the Arabian gulf coast, often very close to beach lines, also on saline sand inland, except for the southern part of Abu Dhabi emirate. The plant prefers the deep sand of dunes and similar habitats where it can form fairly dense stands. Globally, the plant has been known from northern Sudan and Egypt east of the Nile through the Arabian Peninsula to the Desert areas of NW India in the provinces of Rajasthan, Gujarat, and Pakistan. This plant is much grazed by livestock and often very stunted, also a preferred feeding plant of the rare and endangered Houbara Bustard. The plant is used in some countries to treat respiratory diseases [31,32].

Heliotropium, with its different species, is considered a valuable medicinal plant worldwide. Genus *Heliotropium* belongs to Boraginaceae, a large family of dicotyledonous angiosperms which includes 16 genera, and 170 species present in the Mediterranean basin and the Middle East and extending through Europe and Tropical Africa [33]. The southwestern region of the Arabian Peninsula is considered a part of the floristic hotspot where there are many genera that have not received proper attention, e.g., *Heliotropium* L. [34]. The native range of *H. digynum* (Arabic name: Kary, Jery) is N. Africa to Iraq and the Arabian Peninsula. It is a perennial subshrub and grows primarily in the desert or dry shrubland biome(s). It is common in the UAE and widespread in sandy deserts, between dunes and sandy plains, and in the shade of Eucalyptus plantations. The plant is used potentially as a feed source for camels [35]. *H. kotschyi* (Arabic name: Ramram) is common and widespread along coasts; it is tolerant to salinity. By far it is the most common heliotrope in the UAE, abundant in all zones except steeper and higher mountains; it thrives in semi-saline flats and alluvial gravels and is not found in deep mobile sands. A poultice of leaves is used to treat blisters and snake bites [36–38]. *S. imbricata* (Arabic name: Ghadrib) belonging to the family Amaranthaceae is a perennial halophytic shrub that grows in deserts and arid regions of the Arabian Peninsula, southwestern Asia, and North Africa. *S. imbricata* can also be used as a model plant to study the cross-tolerance for salt and drought stress and improve the stress resistance in many other plant species [39]. The species has traditionally been used as a vermifuge and for treating certain skin disorders [40]. Five triterpene glycosides have been isolated from the roots of *S. imbricata,* with two of them being new glycoside derivatives not previously known [41]. It is used for producing alkali, eaten by camels only, with crushed leaves with a strong fishy smell and taste.

Z. qatarens (Arabic name: Hadidi) belonging to the family Zygophyllaceae is a salt-tolerant dwarf shrub with multiple stems that grows in the Arabian Peninsula and is both drought and salt tolerant. It has tiny, fleshy leaves with paired leaflets that are deciduous and store water, dropping off in stressful conditions, and can survive a leafless state for years [42,43]. It typically grows in coarse, stony, or sandy soils at the edge of salt flats, around salt marshes, and in the sand that accumulates on the base of depressions. It also grows in non-salty locations, on calcareous soils, and around the fringes of dune areas, and often dominates plant communities in these locations [44]. On well-drained sandy soil on coastal plains, it may cover 75% of the ground surface, and this plant community is probably the most encountered around the western side of the Persian Gulf [45]. The plants are found growing in association with several species of soil microfungi, regularly with *Cladosporium sphaerospermum*, but also sometimes with *Penicillium citrinum* and *Aspergillus fumigatus* [46]. Aqueous extract of the plant is documented to produce a lowering of blood pressure and acts as a diuretic and antipyretic, local anesthetic, with antihistamine activity, stimulation, and depression of isolated amphibian heart, relaxation of the isolated intestine, contraction of the uterus, and vasodilation. The extract antagonized acetylcholine action on skeletal muscle and acted additively to the muscle relaxant effect of d-tubocurarine. The juice from fresh leaves and stems is known to be used as an abrasive cleanser and as a remedy for the treatment of certain skin diseases. It is a rich source of water, and a treasure of salts made as food for camels [47].

Although little/no reports have been published on the nutritional value of these selected native plants from the UAE instead of *C. murale* [48], to contribute to the growing body of knowledge on this subject, we analyzed these seven plant specimens from the UAE for their lipid, fatty acid, protein, amino acid, and trace mineral content. It is expected that the results of the present study may yield important information regarding the utility of various plants present in the wilderness of the UAE.

2. Results and Discussion

2.1. Amino Acid Composition of Plants

The results revealed the presence of 17 various amino acids including eight essentials, in the aerial parts of the selected wild plant species (Table 1) from the United Arab Emirates (UAE). Different plant species exhibited varying concentrations of different essential and non-essential amino acids. *C. morale*, *D. glaucum*, *H. digynum*, and *T. pentandrus* contain all the 17 analyzed amino acids, while some amino acids are not detected in the selected plants due to their distribution profile being less optimal, viz., *H. kotschyi* (glutamic acid), *Z. qatarense* (histidine, cystine), and *S. imbricata* (glutamic acid, histidine, and cystine). The analysis shows that *H. digynum* has the maximum amount of amino acids (7.598 g/100 g) compared to the other six wild flora. The species contain higher amounts of glutamic acid, followed by aspartic acid, leucine, lysine, and alanine. For the amino acids' concentration, the second in line is *T. pentandrus* (5.069 g/100 g), which is about 33% less than the first. On the other hand, *S. imbricata* has the lowest quantity of amino acids (1.21 g/100 g), which is six times less than *H. digynum*.

Table 1. Free amino acids contents (g/100 g) in the leaves of seven wild species from the United Arab Emirates.

Amino Acid	*Chenopodium murale* L.	*Dipterygium glaucum* Decne.	*Heliotropium digynum* Asch. ex C.Chr.	*Heliotropium kotschyi* Gürke	*Salsola imbricata* Forssk.	*Tribulus pentandrus* Forssk.	*Zygophyllum qatarense* Hadidi	Human Adult Requirements, mg/kg per Day
Alanine	0.103	0.286	0.424	0.12	0.059	0.27	0.07	-
Arginine	0.122	0.433	0.359	0.151	0.144	0.275	0.093	-
Aspartic acid	0.213	0.672	1.35	0.43	0.163	0.588	0.133	-
Valine *	0.099	0.27	0.397	0.14	0.097	0.274	0.062	10
Glutamic acid	0.234	0.702	1.499	0.268		0.762	0.206	-
Glycine	0.099	0.28	0.368	0.118	0.08	0.336	0.065	-
Threonine *	0.092	0.252	0.365	0.117	0.096	0.329	0.066	7
Isoleucine *	0.082	0.208	0.296	0.103	0.049	0.142	0.056	10
Leucine *	0.149	0.385	0.571	0.185	0.083	0.481	0.107	14
Histidine *	0.073	0.134	0.168	0.08	ND	0.100	ND	8–12
Cystine +	0.023	0.041	0.029	ND	ND	0.03	ND	13
Methionine +	0.047	0.067	0.021	0.031	0.02	0.031	0.021	
Proline	0.087	0.245	0.343	0.116	0.076	0.321	0.078	-
Lysine *	0.106	0.287	0.455	0.124	0.064	0.267	0.056	12
Serine	0.09	0.243	0.355	0.115	0.072	0.394	0.051	-
Tyrosine #	0.107	0.178	0.224	0.113	0.135	0.219	0.081	14
Phenylalanine *,#	0.134	0.295	0.374	0.132	0.072	0.244	0.088	

Human daily requirements are recommended by WHO, 1985; *—essential amino acids; #—tyrosine + phenylalanine; +—cystine + methionine; ND—not detectable.

H. digynum also has the highest concentration (2.647 g/100 g) of essential amino acids (histidine, isoleucine, leucine, lysine, methionine, phenylalanine, threonine, tryptophan, and valine), followed by *D. glaucum* (1.898 g/100 g), with the difference of more than 28% between the two species. The lowest amount of essential amino acids was recorded in *Z.*

qatarense (0.456 g/100 g). In all the plant species analyzed, tryptophan was unable to reach the limit of detection.

These amino acids are important in the normal functioning of the body and its regular activities. The amino acids that are normally not synthesized in the body are termed Essential Amino Acids (EAA); these plants act as the sources of various essential amino acids [49,50]. Further, the lower levels of these amino acids in the body can lead to complications such as metabolic diseases and other disorders [2,51,52]. EAA deficiency will lead to growth slowdown and development in children, the development of diseases, and the destruction of cells in adults. Among these amino acids, some are bioactive (lysine, isoleucine, leucine, valine, threonine, phenylalanine, and tyrosine), and others are antioxidants (histidine, methionine, and cysteine) [53]. *H. digynum* exhibited the highest glutamic acid content of 19.73%, and *S. imbricata* did not show a detectable amount. Glutamic acid acts as a neurotransmitter for the central nervous system, the brain, and the spinal cord. It supports the brain to correct the body's physiological imbalances [54]. Serine is necessary for the muscles' development and the immune system's maintenance. It is important in the synthesis of RNA and DNA within the cells. Serine detected ranges from 4.1% to 7.7% in *Z. qatarense* and *T. pentandrus,* respectively. The values for alanine were between 4.8% (*S. imbricata*) and 5.6% (*Z. qatarense*). Alanine plays an important role in the transfer of nitrogen in the body and glucose that the body uses as energy and strengthens the immune system by producing antibodies. It also regulates the discharge of toxic substances. Proline ranges from 4.5% (*H. digynum*) to 6.3% (*T. pentandrus* & *Z. qatarense*). Proline will help to slow down the production of collagen thus facilitating improved skin texture and leading to the slowdown of the aging process. Proline plays an important role in curative treatment to avoid problems in the cartilage, tendons, and muscles of the heart [55]. Arginine values fluctuate between 4.7% (*H. digynum*) and 11.9% (*S. imbricata*). Arginine boosts the immune system thereby delaying the growth of cancerous tumors. Arginine plays a major role in the detoxification of the liver by neutralizing ammonia and reducing the toxicity of alcohol. In addition, it is frequently used in the treatment of infertility in men. Glycine hinders muscle degeneration; it improves glycogen storage and releases glucose to meet energy needs. Glycine values are ranges from 4.8% (*H. digynum*) to 6.6% (*T. pentandrus* and *S. imbricata*). Histidine is used in the treatment of cardiovascular disease with a physiological antioxidant role it plays on the free radicals (hydroxyl radical and singlet oxygen), and the values fluctuate from 1.9% to 3.9% in *T. pentandrus* and *C. murale.* Histidine cannot be detected in a traceable amount in *S. imbricata* & *Z. qatarense.* According to WHO, the daily need for this amino acid is 12 mg/kg or 840 mg to 70 kg of body weight. It is important to mention that methionine is an antioxidant with high sulfur content [25] that helps prevent deficiencies in hair cells, skin, and nails. Furthermore, this amino acid protects against greasy clusters around the liver and the arteries that cause obstructions. It promotes the detoxification of harmful agents such as lead and other heavy metals. Additionally, cysteine works as a powerful antioxidant to eliminate harmful toxins [56]. Therefore, these two-sulfur amino acid deficiencies would cause physiological disturbances, even the risks of contracting degenerative diseases [53]. Cysteine cannot be detected in *H. kotschyi, Z. qatarense,* and *S. imbricata.* All the wild plants except *S. imbricata* (8.0%) exhibited a range of 5% valine. Valine is used in the treatment of liver and gallbladder diseases and promotes brain vigor [56]. Threonine ranges from 4.8% (*H. digynum*) to 7.9% (*S. imbricata*). Threonine helps in maintaining the balanced intake of proteins in the body and is also an important part of the formation of dental enamel, collagen, and elastin. In the deficiency, the isoleucine causes physical and mental disorders. The highest range of leucine, isoleucine, valine, and phenylalanine was detected in *T. pentandrus* (9.5%), *Z. qatarense* (4.5%), *S. imbricata* (8.01%) and *C. murale* (7.2%). Leucine acts with isoleucine and valine to promote muscle, skin, and bone function [55]. In addition, it is recognized that the brain uses phenylalanine to produce norepinephrine, a chemical that transmits signals between nerve cells. Additionally, it promotes alertness and vitality, regulates human mood, and reduces pain. This amino acid is also used in the treatment of arthritis, depression, painful menstruation, migraine, obesity, Parkinson's disease, and schizophrenia [55,56].

2.2. *Fatty Acid Composition of Plants*

The biochemical analysis of the leaves of seven wild species indicates the presence of 11 fatty acids in the present study (Table 2). All the plants contain C14:0 (myristic acid), C16:0 (palmitic acid), C18:0 (stearic acid), C18:1 (oleic acid), and C18:2 (linoleic acid) in their leaves. At the same time, C16:1 (palmitoleic acid) has been detected in *D. glaucum, H. digynum, S. imbricata, T. pentandrus*, and *Z. qatarense*. C17:0 (heptadecanoic acid) was reported in *C. murale, D. glaucum, H. kotschyi, T. pentandrus, Z. qatarense*. C18:3 ω3 (α-Linolenic acid) was in all plants except *C. murale*. On the other hand, C12:0 (lauric acid) was found in *S. imbricata*, and C20:1 (eicosenoic acid) in *H. kotschyi* only.

Table 2. Fatty acid compositions (% of total fat) of the United Arab Emirates wild plants.

Fatty Acid	TP	HD	DG	SI	HI	ZQ	CM
C12:0	ND	ND	ND	ND	0.93	ND	ND
C14:0	2.58	2.15	2.33	0.19	1.18	2.21	3.26
C16:0	32.4	30.7	24.3	15.7	19.8	28.2	32.8
C16:1	ND	1.18	2.03	ND	0.34	1.35	0.43
C18:0	48.5	17.2	16.1	7.61	12.6	16.9	50.4
C17:0	0.6	0.38	ND	0.06	ND	0.12	0.88
C18:1	10.5	23	27.9	51.8	35.5	25.7	8.58
C18:2 ω6	3.99	15	16.3	24	28.2	15.6	2.8
C18:3 ω3	ND	8.74	9.88	0.21	0.58	8.99	0.56
C20:0	1.09	1.25	1.16	0.34	0.84	1.18	ND
C20:1	ND	ND	ND	0.13	ND	ND	ND

TP—*Tribulus pentandrus*; HD—*Heliotropium digynum*; DG—*Digiterygium glaucum*; SI—*Salsola imbricata*; HI—*Heliotropium kotschyi*; ZQ—*Zygophyllum qatarense*; CM—*Chenopodium murale*; ND—not detectable.

T. pentandrus had higher levels of palmitic acid and oleic acid; likewise, *H. digynum* contained higher amounts of palmitic acid, oleic acid, α-Linoleic acid, and linolenic acid. The predominant fatty acids in *D. glaucum* include palmitic acid, stearic acid, and oleic acid. In *S. imbricate*, palmitic acid, oleic acid, and α-Linoleic acid were the major fatty acids. *H. kotschyi* includes about 50% oleic acid as the primary fatty acid, whereas *Z. qatarense* and *C. murale* contain stearic acid, the primary fatty acid (50.4%).

The leaves of wild plants used for food have lower oil contents but are rich in essential fatty acids, e.g., linoleic acid (C18:2 ω6) and α-linolenic acid (C18:3 ω3) [57,58]. The human body cannot produce either α-linolenic acid or linoleic acid. These fatty acids are essential in regulating various metabolisms and have beneficial impacts against cardiovascular disease [27], non-alcoholic fatty liver disease, hyperlipidemia, and even cancers [59–61]. α-linolenic acid (ALA), an omega-3 amino, helps reduce inflammation in the human body. The results of the current study suggest that the leaf lipids of wild plant species, including the experiment (Table 2), are rich in essential fatty acids (18:2 ω6 and 18:3 ω3). Hence, the consumption of these plants may yield essential fatty acids.

The medium chain saturated fatty acids such as lauric acid are important in health and are known to have several pharmacological effects including anticancer properties [3,9,27] and hypolipidemic effects [4,5]. Oleic acid and α-Linoleic acid are important essential fatty acids, and plants act as important sources of these fatty acids. The essential fatty acids are important in the management of diseases including non-alcoholic fatty liver disease, hyperlipidemia and even cancers [6,62].

2.3. Nutrient Contents of Plants

The proximate composition of selected seven wild plants is described in Table 3. The predominant nutrients in different plants include carbohydrates, proteins, fats, ash, and fiber content. The highest level of carbohydrates, saturated fat, poly-unsaturated fats, fiber, and energy were present in *D. glaucum*. The highest protein content was observed in *H. digynum* (7.21 g/100 g) and the lowest in *C. murale* (3.09 g/100 g). The presence of high protein content suggests its nutritive superiority over other traditionally consumed crops. For total carbohydrates, *S. imbricata* (3.7 g/100 g) showed the least amount, while with 10.7 g/100 g, *D. glaucum* was at the top. The protein composition in the fresh leaves of the seven plants ranged from 1.75 to 7.21 g/100 g, and these values are close to the estimates of certain orphan leafy vegetables described in another report [25,63,64]. Diets that provide about 12% of their calorific amount from proteins are good protein sources [50]. Hence, these wild plants can have a significant role in supplying inexpensive and readily available proteins for people living in rural areas. The protein content in *C. murale* from Spain is reported to contain 4.35 ± 0.41 g [48], which is close to the values that we obtained in samples from the UAE. Plant proteins are cellular functional macromolecules that are required to perform a wide range of functions as enzymatic activities and managing transport across cellular membranes. The variation in protein content present in different species which belong to the same genus was previously reported [65].

Table 3. Nutritional composition of the different plant species.

Parameter	TP	HD	DG	SI	HI	ZQ	CM WHO	Recommended Rates
Saturated Fats (g/100 g)	0.34	0.31	0.45	0.13	0.11	0.2	0.34	10% of total kcal/day
Mono-Unsaturated Fats (g/100 g)	0.19	0.21	0.21	0.13	0.23	ND	0.19	-
Poly-Unsaturated Fats (g/100 g)	0.17	0.18	0.21	0.10	0.11	ND	0.17	-
Energy (kcal/100 g)	95	98	124	25	57	32	22	18–25 (Kcal/kg bwt.)/day
Fat (g/100 g)	0.7	0.7	0.87	0.36	0.45	0.23	0.76	-
Carbohydrates (total) (g/100 g)	16.5	15.8	22.7	3.7	10	5.8	0.7	130 g/day
Proteins (g/100 g)	5.72	7.21	6.22	1.87	3.2	1.57	3.09	0.75 g/kg/day
Salt (g/100 g)	0.042	0.087	0.071	3.61	0.594	1.08	1	5 g/day
Ash (total) (g/100 g)	6.57	3.9	3.5	7.6	4.1	7.78	6.98	-
Moisture (g/100 g)	70.5	72.4	66.7	86.5	82.2	84.6	88.5	-
Crude Fiber (%)	0.071	0.032	0.152	0.02	0.06	0.042	0.018	-
Ash (Insoluble in acids) (g/100 g)	1.9	1.2	1	2.2	1.2	2.3	2.1	-
Neutral Detergent Fiber (NDF) (%)	13.9	17	20	5.24	10	5.56	3.94	-
Acid Detergent Fiber (ADF) (%)	7.78	10.7	12.2	0.717	5.67	1.89	1.69	-

TP—*Tribulus pentandrus*; HD—*Heliotropium digynum*; DG—*Digiterygium glaucum*; SI—*Salsola imbricata*; HI—*Heliotropium kotschyi*; ZQ—*Zygophyllum qatarense*; CM—*Chenopodium murale*; bwt—body weight; ND—not detectable.

The total fat content varied between 0.23 and 0.87 g/100 g in these plants, similar to the results of several studies that concluded that leafy vegetables are inferior resources of fats [4]. Though a sizable part of the fat present in the plant aerial parts is saturated, most of the lipids are unsaturated, either mono or poly. Unsaturated fats that are fluid at room temperature are useful as they can reduce blood cholesterols, relieve inflammation, soothe heartbeats, and play several other beneficial roles. Food supplying 1–2% of its caloric energy as fat is appropriate for humans, as excess fat intake results in cardiac diseases [50]. Fats are also a source of essential fatty acids such as linoleic and linolenic acid, which the body cannot synthesize and can only be acquired from diets. They are vital for managing inflammation, blood clotting, and brain function and development. The absorption of fat-soluble vitamins such as carotene and vitamin A in the body is also augmented by the presence of lipids [8].

The total carbohydrate content in the studied wild plant range was 3.5–22.7 g/100 g. There is a big difference (more than six times) for this important dietary compound among the seven species. A study in India [66] showed a lesser carbohydrate content in various green leafy vegetables consumed by some tribesmen there. Other research [67] also registered lower values for carbohydrate contents in wild plants eaten in eastern parts of India. On the other hand, the carbohydrates of the wild plant were comparable to those described by studies in Bangladesh [51], northeast India [68,69], and Pakistan [70]. The optimal daily carbohydrate needed for humans is 130 g. It indicated that 7.0–14.5% of the daily requirement could be achieved through the consumption of 100 g of these dried plants.

Ash contains all the important dietary ingredients, especially minerals, micro and macronutrients that are very significant for the normal physiological functions of the body. Ash comprises inorganic matter of the plant that contains oxides and salts, including anions, e.g., phosphates, sulfates, chlorides, and other cations and halides such as calcium, sodium, potassium, magnesium, manganese, and iron [8]. The ash content shows the aggregate of minerals in the food. This study found the highest ash in *Z. qatarense* (7.78 g/100 g) and the lowest in *D. glaucum* (3.5 g/100 g). The ash present in these wild plants is similar to some commonly used wild vegetables in Bangladesh, and India [51,68,69]. The ash content in *C. murale* collected from the UAE showed higher value as compared to the report from those collected from Spain (4.23 $\pm$ 0.28 g/100 mg) [48]. The quantity and composition of ash left over after burning of plant material varies significantly according to the plant's age, time, organ to organ [71]. This must be evaluated further.

Moisture content is the amount of water present in a substance. Water is an important part of food. Around 20% of the total water intake is by diet moisture [32]. When foods are consumed, their moisture content is soaked up by the body. All the plant species under investigation had moisture content varying from 66.7% to 88.5% in fresh weight. The reasonably high moisture contents show that there might also be a higher activity of water-soluble enzymes [72]. The moisture content is determined mainly by the humidity, temperature, and harvest time of the plant species. The higher moisture contents in plants of less humid and dry regions might be due to their higher water retention capabilities because they have xeric nature and xerophytes store water and have sunken stomata to avoid transpiration of water. The maximum moisture content among the seven studied plants was in *C. murale* (88.5 g/100 g), and the lowest was in *D. glaucum* (66.7 g/100 g). *C. murale* having the highest moisture value makes it more prone to a decline in nutrients, since foods with high moisture content are more vulnerable to perishability. The present study also supports the above observation. The moisture content values of *C. murale* collected from the UAE in the present study is in correlation with samples collected from Spain (82.02 $\pm$ 3.01) [48].

Out of seven wild plants studied, the highest nutritive value regarding energy was noted in *D. glaucum* (124 Kcal/100 g) and the lowest in *C. murale* (22 Kcal/100 g). The energy was reported in the samples of *C. murale* (33 $\pm$ 19 Kcal) is comparable to the samples from the UAE used for the present study [48]. It has been reported that *D. glaucum* is used as herbal medicinal and camel's favorite food source in the UAE. The camel prefers the plant especially when it is in flowering and seeding stages as a good source of revitalizing camel body. The tiny flowers of the plants work as a strong laxative agent. The stalk with flowers is a source of flashing the digestive system of the livestock as well as a human being. It is also given to camels suffering weakness and a lower desire for food [73]. Dietary diversification is important to improve the intake of critical nutrients. The low-fat content (0.54 to 2.37%) in fresh weight of aerial parts of the plants suggests that the plants can be used as a valuable source of good dietary practices and may be advised to individuals with overweight or obesity problems. Therefore, high protein, carbohydrate, and nutritional composition can form an ideal diet for children, breastfeeding mothers, and adults. Further toxicological studies need to perform the effective utilization of these plants as dietary supplements.

2.4. Vitamin and Mineral Composition

Vitamins are considered important nutrients in foods and carry out specific functions essential for health though their daily requirements are minute. The daily recommended intake of vitamin A, C, and riboflavin (B2) for pregnant women and children are (800 and 400 μg, 55 and 30 mg, 1.4 and 0.5 mg), respectively [74]. The predominant vitamins in the plants were the vitamin B complex. At the same time, a higher level of vitamin A was found in *C. murale*. The amounts of vitamin B2 and B3 were maximum in *T. pentandrus*, whereas vitamin B6 concentration was higher in *D. glaucum* compared to other species analyzed for the study (Table 4). B vitamins have an essential role in maintaining good well-being and directly influence energy levels, brainwork, and cell metabolism. Vitamin B complex may assist prevent microbial infections and help support cell health.

Table 4. Composition of vitamin and minerals in the plants.

Parameter	TP	HD	DG	SI	HI	ZQ	CM
Vitamin A (free Retinol) (mg/100 g)	0.02	0.02	ND	ND	ND	ND	0.08
Vitamin B2 (Riboflavin) (mg/Kg)	6.29	ND	1.41	ND	ND	ND	ND
Vitamin B3 (Niacin) (mg/100 g)	0.37	0.095	0.21	0.08	ND	0.09	0.03
Vitamin B6 (Pyridoxin) (mg/100 g)	0.06	0.03	0.10	0.03	ND	0.03	0.03
L-ascorbic acid (Vitamin C) (mg/100 g)	20	20	20	20	20	20	20
Phosphorus (P) (mg/Kg)	1154	502	922	137	280	133	762
Nitrogen (N) (mg/Kg)	9157	11,532	9958	2996	5117	2514	4939
Zinc (Zn) (mg/Kg)	642	87.6	1.92	27.3	70	0.233	36.7
Calcium (Ca) (mg/Kg)	4595	1710	9662	6100	8333	8825	3753
Potassium (K) (mg/Kg)	6665	2050	7476	4395	4509	438	9406

TP—*Tribulus pentandrus*; HD—*Heliotropium digynum*; DG—*Digiterygium glaucum*; SI—*Salsola imbricata*; HI—*Heliotropium kotschyi*; ZQ—*Zygophyllum qatarense*; CM—*Chenopodium murale*; ND—not detectable.

The essential minerals potassium (K), Calcium (Ca), and trace mineral zinc (Zn) of the seven wild plants are shown in Table 4 ranged between 438 and 9410 mg/kg in fresh plant leaves. D. glaucum shows the highest concentration, while Z. qatarense has the lowest amount; it plays a major role in our body to help in keeping normal levels of water within the cells. The Ca concentration varied between 1710 and 9662 mg/kg, while in some of the important leafy vegetables (cabbage, lettuce, and spinach), it varies between 390 and 730 mg/kg [8]. The outcome from this present study is roughly similar to the wild edible plants consumed in parts of rural Bangladesh [51] but better than the wild flora eaten as food in Pakistan [75]. Based on the finding, these wild plants may be a good Ca source for our food. This essential mineral is important for blood clotting and the normal working of the heart muscles [76].

Zn plays an important role in stabilizing macromolecular synthesis and structure. The metal ion's role in the synthesis of the nucleotides (DNA and RNA) is well known, and both RNA and DNA polymerases are zinc-dependent enzymes. The highest Zn concentration was recorded for the studied seven plants in H. digynum (87.6 mg/kg) and the lowest in Z. qatarense (0.233 mg/kg). The results are close to the levels reported in some wild plants in India [77] and all the minerals quantified in C. murale from the UAE showed an increased amount as compared to samples from Spain [48].

Nitrogen (N) and phosphorus (P) are two main plant minerals, and their deficiencies frequently restrict floral development. Plants utilize N for leaf growth and its green color, while they use P to help form new roots and produce flowers, fruits, and seeds. The mineral also helps plants in fighting against diseases. The biochemical analysis of the plants reveals that H. digynum contains the highest amount of N (11,532 mg/kg), while Z. qatarense has the lowest (2514 mg/kg). For P, T. pentandrus has the maximum amount (1154 mg/kg), and with just 133 mg/kg, Z. qatarense is at the bottom. N is a vital part of the food that helps synthesize amino acids in the body. The amino acids are the building block of protein, and nitrogen is a part of their structure. P is an essential element of bone health and cellular activities in the body.

The micronutrients, e.g., trace elements and vitamins, are important in the normal functioning of the body [7–9]; they act as essential components of various enzymes including those regulate body antioxidant status [57,78]. In addition, these trace elements and vitamins act as inhibitors of oxidative stress and inflammation [25,26]; subsequently, these compounds prevent the development of various diseases including cancer and cardiovascular diseases [18]. Different habitats could directly affect the amounts of bioactive products accumulated in plants. Abd El Gawad et al. [79] compared phenolics content as well as the antioxidant activity of different Heliotropium species collected from coastal and inland habitats.

3. Materials and Methods

3.1. Collection and Preparation of Samples

Aerial parts of *C. murale* L., *D. glaucum* Decne., *H. digynum* Asch. ex C.Chr., *H. kotschyi* Gürke, *S. imbricata* Forssk., *T. pentandrus* Forssk., and *Z. qatarens* Hadidi were collected from the plants growing naturally at the ICBA campus (25.0947° N, 55.3899° E), Dubai, United Arab Emirates. All seven species are native to the United Arab Emirates (UAE) and are also found in various regions of Africa, Asia, and Europe (Table 5 and Figure 1). Other related information on the wild flora has been provided in the table. The samples were washed thoroughly 2–3 times with running tap water and then once with sterile water and used for further analysis. The detail of analysis is provided in Supplementary Materials.

Table 5. Information on wild plant species analyzed for their nutritional composition.

Species	Family	Local Name	Life Cycle	Habitat	Native Range	Uses
Chenopodium murale L.	Amaranthaceae	Abu' efei	Annual	Plantation, fallow fields	Europe, N Africa, Arabian Peninsula, SW Asia	Vegetable, fodder
Dipterygium glaucum Decne.	Cappraceae	Safrawi	Perennial	Sandplains	NE Africa, Arabian Peninsula, Iran, and S Asia	Fodder, medicine
Heliotropium digynum Asch. ex C.Chr.	Boraginaceae	Kary, Jery	Perennial	Sandplains	N Africa, W Asia including Arabian Peninsula	Fodder
Heliotropium kotschyi Gürke	Boraginaceae	Ramram	Perennial	Sandplains, gravels	NE Africa, Arabian Peninsula and parts of SW Asia	Medicine
Salsola imbricata Forssk	Amaranthaceae	Ghadraf	Perennial	Saline sand, disturbed land	N Africa, Arabian Peninsula and SW Asia	Fodder, medicine
Tribulus pentandrus Forssk.	Zygophyllaceae	Shersir	Perennial	Sandplains, valleys	N Africa, SW Asia	Fodder
Zygophyllum qatarense Hadidi	Zygophyllaceae	Haram	Perennial	Sandplains, coastal areas	Arabian Peninsula	Fodder

Figure 1. (**a**) *Chenopodium murale* L. (**b**) *Heliotropium digynum* Asch. ex C.Chr. (**c**) *Salsola imbricata* Forssk. (**d**) *Dipterygium glaucum* Decne. (**e**) *Heliotropium kotschyi* Gürke. (**f**) *Tribulus pentandrus* Forssk. (**g**) *Zygophyllum qatarense* Hadidi.

The soils at ICBA are sandy in texture, that is, fine sand (sand 98%, silt 1%, and clay 1%), calcareous (50–60% $CaCO_3$ equivalents), porous (45% porosity). The saturation percentage of the soil is 26, with very high drainage capacity, while its saturated extract's electrical conductivity (ECe) is 1.2 dS m^{-1}. In line with American Soil Taxonomy [36], the soil is categorized as typic torripsamments, carbonatic, and hyperthermic [80].

3.2. Nutritional Analysis

Nutritional analyses followed the methodology of the Association of Official Analytical Chemicals [81]. Moisture content was determined by the difference between fresh matter and dry matter; fat was determined by the Soxhlet extraction method; protein (PS) was measured by the Kjeldahl method; crude fiber (CF) content was determined using the neutral detergent reagent method [82]; and total carbohydrate (CHO) content was estimated by the difference between 100 and the sum of the percentages of moisture, protein, total lipid and ash contents [83]. Total ash content (Ash) was analyzed after burning the plants in a muffle furnace. The micronutrients potassium (K^+), calcium (Ca^{2+}), phosphorus (P^{3-}), iron (Fe^{2+}) and zinc (Zn^{2+}), were analyzed using the Inductive Coupled Plasma (ICP) spectrometer and atomic absorption [83].

3.3. Amino Acid and Vitamin Analysis

The amino acids were determined using Sykam Amino Acid Analyzer (Sykam GmbH, Eresing, Germany).

3.4. Fatty Acid Composition

The fat extracted from the samples were further analyzed for the fatty acids composition using gas–liquid chromatography (GLC) and gas chromatography mass spectrometry. More information about the methodologies applied and equipment used for the chemical analyses is provided in the Supplementary Materials.

3.5. Trace Elemental Analysis and Proximate Composition

The trace element composition of the plant was estimated according to the standard methods. Briefly, the plant tissue was reflexed with concentrated nitric acid and perchloric acid; the samples were then analyzed by inductively coupled argon plasma atomic emission spectroscopy (ICP-AES). More information about the methodologies applied and equipment used for the chemical analyses is provided in the Supplementary Materials.

4. Conclusions

A biochemical analysis of seven wild plants aims to unravel the rich dietary composition of indigenous plants and their capacity to be utilized as alternative sources of nutrients and nutritional supplements. The nutritional composition information of wild plant species will also help promote the use of more biodiverse foods and feed for healthy diets and the pharmaceutical industry in the UAE and elsewhere. The analyzed plant species could be an excellent substitute for other commonly eaten vegetables because of their superior nutrient content, and they should be tested for their toxicological properties. Unfortunately, many precious desert plants including the selected samples for the present study are depleting for numerous reasons. This flora really needs to be well studied, documented, and the status and risk factors ought to be identified. Most of these plants selected in the present study are used by the local population as fodder. However, no detailed study carried out an analysis of the nutritional quality of these plants for better utilization. Understanding the relative importance and preference of different species is also crucial for the sustainable management of the local forage resources and can help animal husbandry technicians to optimize the selection of useful fodder species and to improve the livestock system efficiency in the desert ecosystem.

Supplementary Materials: The following supporting information can be downloaded at: https://www.mdpi.com/article/10.3390/molecules28031504/s1. Detailed methodologies for the analysis of the chemical composition of selected wild plants of the United Arab Emirates [81–83].

Author Contributions: Conceptualization, M.S. and R.K.S.; methodology and original draft preparation, S.T. and M.S.; review and editing, M.S., R.K.S. and S.T.; fund acquisition, R.K.S. All authors have read and agreed to the published version of the manuscript.

Funding: The current study funded by International Center for Biosaline Agriculture core fund.

Institutional Review Board Statement: Not applicable.

Informed Consent Statement: Not applicable.

Data Availability Statement: The data may be shared upon a valid request.

Acknowledgments: Authors gratefully acknowledge ICBA management for funding the project.

Conflicts of Interest: The authors declare no conflict of interest.

References

1. Madouh, T.A.; Al-Sabbagh, T.A. Nutritional quality and adaptation of several native plant species of Kuwait to the farming system as potential livestock feed. *Int. J. Env. Earth Sci.* **2021**, *2*, 1–8.
2. FAO. *The Second Report on the State of the World's Plant Genetic Resources for Food and Agriculture*; Food and Agriculture Organization of the United Nations: Rome, Italy, 2010.
3. Carvalho, A.M.; Barata, A.M. The Consumption of Wild Edible Plants. In *Wild Plants Mushrooms Nuts Functional Food Properties and Applications*; John Wiley & Sons, Inc.: Hoboken, NJ, USA, 2016; pp. 159–198.
4. Powell, B.; Thilsted, S.; Ickowitz, A.; Termote, C.; Sunderland, T.; Herforth, A. Improving diets with wild and cultivated biodiversity from across the landscape. *Food Secur.* **2015**, *7*, 535–554. [CrossRef]
5. Ogle, B.M. Wild Vegetables and Micronutrient Nutrition: Studies on the Significance of Wild Vegetables in Women's Diets in Vietnam. Ph.D. Thesis, Acta Universitatis Upsaliensis, Uppsala, Sweden, 2001.
6. Jones, A.D. Critical review of the emerging research evidence on agricultural biodiversity, diet diversity, and nutritional status in low- and middle-income countries. *Nutr. Rev.* **2017**, *75*, 769–782. [CrossRef] [PubMed]
7. Wunder, S.; Börner, J.; Shively, G.; Wyman, M. Safety nets, gap filling, and forests: A global-comparative perspective. *World Dev.* **2014**, *64* (Suppl. 1), S29–S42. [CrossRef]
8. FAO; WFP; IFAD. *The State of Food Insecurity in the World 2012 Economic Growth Is Necessary but Not Sufficient to Accelerate Reduction of Hunger and Malnutrition*; FAO: Rome, Italy, 2012.
9. Pinela, J.; Carvalho, A.; Ferreira, I.C.F.R. Wild edible plants: Nutritional and toxicological characteristics, retrieval strategies and importance for today's society. *Food Chem. Toxicol.* **2017**, *110*, 165–188. [CrossRef] [PubMed]
10. Yu, M.; Gouvinhas, I.; Rocha, J.; Barros, A.I.R.N.A. Phytochemical and antioxidant analysis of medicinal and food plants towards bioactive food and pharmaceutical resources. *Sci Rep.* **2021**, *11*, 10041. [CrossRef]
11. Zhang, Y.J.; Gan, R.Y.; Li, S.; Zhou, Y.; Li, A.N.; Xu, D.P.; Li, H.B. Antioxidant Phytochemicals for the prevention and treatment of chronic diseases. *Molecules* **2015**, *20*, 21138–21156. [CrossRef]
12. Narayanankutty, A. Natural Products as PI3K/Akt Inhibitors: Implications in Preventing hepatocellular carcinoma. *Curr. Mol. Pharm.* **2021**, *14*, 760–769. [CrossRef]
13. Job, J.T.; Rajagopal, R.; Alfarhan, A.; Narayanankutty, A. *Borassus flabellifer* Linn haustorium methanol extract mitigates fluoride-induced apoptosis by enhancing Nrf2/Haeme oxygenase 1—Dependent glutathione metabolism in intestinal epithelial cells. *Drug Chem. Toxicol.* **2022**, *45*, 2269–2275. [CrossRef]
14. Narayanankutty, A. PI3K/Akt/mTOR pathway as a therapeutic target for colorectal cancer: A review of preclinical and clinical evidence. *Curr. Drug Targets* **2019**, *20*, 1217–12-26. [CrossRef]
15. Vinayak, N.; Arunaksharan, N.; Anusree, N. Heat Shock Proteins (HSPs): A Novel Target for Cancer Metastasis Prevention. *Curr. Drug Targets* **2019**, *20*, 727–737. [CrossRef]
16. Kennedy, D.O. B Vitamins and the Brain: Mechanisms, Dose and efficacy—A review. *Nutrients* **2016**, *8*, 68. [CrossRef] [PubMed]
17. Padayatty, S.J.; Levine, M. Vitamin C: The known and the unknown and Goldilocks. *Oral Dis.* **2016**, *22*, 463–493. [CrossRef] [PubMed]
18. AlFadhly, N.K.Z.; Alhelfi, N.; Altemimi, A.B.; Verma, D.K.; Cacciola, F.; Narayanankutty, A. Trends and technological advancements in the possible food applications of spirulina and their health benefits: A review. *Molecules* **2022**, *27*, 5584. [CrossRef]
19. Scheiermann, E.; Puppa, M.A.; Rink, L.; Wessels, I. Zinc status impacts the epidermal growth factor receptor and downstream protein expression in A549 cells. *Int. J. Mol. Sci.* **2022**, *23*, 2270. [CrossRef]
20. Wooten, D.J.; Sinha, I.; Sinha, R. Selenium induces pancreatic cancer cell death alone and in combination with gemcitabine. *Biomedicines* **2022**, *10*, 149. [CrossRef]
21. Shahid, M.; Rao, N.K. New flowering plant species records for the United Arab Emirates. *Tribulus* **2016**, *24*, 131–136.
22. Karim, F.; Dakheel, A.J. *Salt Tolerant Plants of the United Arab Emirates*; International Center for Biosaline Agriculture (ICBA): Dubai, United Arab Emirates, 2006.
23. El-Ghonemy, A. *Encyclopedia of Medicinal Plants of the United Arab Emirates*; University of United Arab Emirates Press: Abu Dhabi, United Arab Emirates, 1993.
24. Sakkir, S.; Kabshawi, M.; Mehairbi, M. Medicinal plants diversity and their conservation status in the United Arab Emirates (UAE). *J. Med. Plants Res.* **2012**, *6*, 1304–1322.
25. Toledo, A.; Burlingame, B. Biodiversity and nutrition: A common path toward global food security and sustainable development. *J. Food Compos. Anal.* **2006**, *19*, 477–483. [CrossRef]
26. Anderson, E.N.; Pearsall, D.M.; Hunn, E.S.; Turner, N.J. (Eds.) *Ethnobiology*; John Wiley & Sons, Inc.: Hoboken, NJ, USA, 2011. [CrossRef]
27. Ali, A.; Deokule, S.S. Studies on Nutritional Values of Some Wild Edible Plants from Iran and India. *Pak. J. Nutr.* **2009**, *8*, 26–31.
28. Available online: https://powo.science.kew.org/taxon/urn:lsid:ipni.org:names:873398-1 (accessed on 2 November 2022).
29. Available online: https://powo.science.kew.org/taxon/urn:lsid:ipni.org:names:77121007-1 (accessed on 2 November 2022).
30. Available online: https://en.wikipedia.org/wiki/Chenopodiastrum_murale (accessed on 7 September 2022).
31. Available online: https://www.ddcr.org/florafauna/Detail.aspx?Class=Plants&Referrer=Capparaceae%20(Shrubs)&Subclass=Shrubs&Id=221 (accessed on 11 September 2022).
32. Available online: https://www.gbif.org/species/3873162 (accessed on 11 September 2022).

33. Selvi, F.; Bigazzi, M. Leaf surface and anatomy in Boraginaceae tribe Boraginaceae with respect to ecology and taxonomy. *Flora* **2001**, *196*, 269–285. [CrossRef]
34. Myers, N.; Mittermeier, R.A.; Mittermeier, C.G.; Da Fonseca, G.A.; Kent, J. Biodiversity hotspots for conservation priorities. *Nature* **2000**, *403*, 853. [CrossRef]
35. Available online: https://www.ddcr.org/florafauna/Detail.aspx?Class=Plants&Referrer=Boraginaceae%20(Shrubs)&Subclass=Shrubs&Id=264 (accessed on 17 October 2022).
36. Available online: https://www.ddcr.org/florafauna/Detail.aspx?Class=Plants&Referrer=Boraginaceae%20(Shrubs)&Subclass=Shrubs&Id=220 (accessed on 17 October 2022).
37. Available online: http://data.rbge.org.uk/herb/E00558486 (accessed on 24 October 2022).
38. Available online: https://www.gbif.org/occurrence/574689181 (accessed on 11 September 2022).
39. Alam, H.; Zamin, M.; Adnan, M.; Shah, A.N.; Alharby, H.F.; Bamagoos, A.A.; Alabdallah, N.M.; Alzahrani, S.S.; Alharbi, B.M.; Saud, S.; et al. Exploring suitability of *Salsola imbricata* (Fetid Saltwort) for salinity and drought conditions: A step toward sustainable landscaping under changing climate. *Front. Plant Sci.* **2022**, *13*, 900210. [CrossRef] [PubMed]
40. Khan, M.A.; Benno, B.; Hans-Jörg, B.; German, K.S. *Sabkha Ecosystems: Volume II: West and Central Asia*; Springer Science & Business Media: Berlin/Heidelberg, Germany, 2006; pp. 143–147. ISBN 978-1-4020-5071-8.
41. Hamed, A.I.; Milena, M.; Mohamed, S.G.; Usama, A.M.; Moatz, T.M.; Angela, P.; Sonia, P. Triterpene saponins from Salsola imbricata. *Phytochem. Lett.* **2011**, *4*, 353–356. [CrossRef]
42. Retief, E. *Zygophyllum*; South African National Biodiversity Institute: Pretoria, South Africa, 2009; Available online: https://www.plantzafrica.com (accessed on 11 September 2022).
43. Available online: https://web.archive.org/web/20160122050413/http://www.arkive.org/zygophyllum/zygophyllum-qatarense/ (accessed on 11 September 2022).
44. Ghazanfar, S.A.; Fisher, M. *Vegetation of the Arabian Peninsula*; Springer Science & Business Media: Berlin/Heidelberg, Germany, 2013; ISBN 978-94-017-3637-4.
45. Abdulaziz, H.; Abuzinada Hans-Jörg, B.; Krupp, F.; Böer, B.; Al Abdessalaam, Z.T. *Protecting the Gulf's Marine Ecosystems from Pollution*; Springer Science & Business Media: Berlin/Heidelberg, Germany, 2008; pp. 32–33. ISBN 978-3-7643-7947-6.
46. Mandeel, Q.A. Microfungal community associated with rhizosphere soil of *Zygophyllum qatarense* in arid habitats of Bahrain. *J. Arid Environ.* **2002**, *50*, 665–681. [CrossRef]
47. Available online: https://arkbiodiv.com/2017/01/08/the-strong-and-resilient-plant-of-desert-zygophyllum-zygophyllum-qatarense/ (accessed on 28 December 2022).
48. Guerrero, J.L.G.; Isasa, M.E.T. Nutritional composition of leaves of *Chenopodium* species (*C. album* L., *C. murale* L. and *C. opulifolium* Shraeder). *Int. J. food Sci. Nutr.* **2009**, *5*, 321–327.
49. Pelletier, D.L.; Frongillo EAJr Schroeder, D.G.; Habicht, J.P. The effects of malnutrition on child mortality in developing countries. *Bull World Health Organ.* **1995**, *73*, 443–448. [PubMed]
50. Kris-Etherton, P.M.; Krummel, D.; Russell, M.E.; Dreon, D.; Mackey, S.; Borchers, J.; Wood, P.D. The effect of diet on plasma lipids, lipoproteins, and coronary heart disease. *J. Am. Diet Assoc.* **1988**, *88*, 1373–1400. [CrossRef]
51. Sheela, K.; Kamal, G.; Nath, D.; Vijayalakshmi Geeta, M.Y.; Roopa, B.P. Proximate composition of underutilized green leafy vegetables in southern Karnataka. *J. Human Ecol.* **2004**, *15*, 227–229. [CrossRef]
52. Aletor, V.A.; Aladetimi, O.O. Compositional evaluation of some cowpea varieties and some under-utilized edible legumes in Nigeria. *Nahrung* **1989**, *33*, 999–1007. [CrossRef]
53. Okada, Y.; Okada, M. Scavenging effect of water-soluble proteins in broad beans on free radicals and active oxygen species. *J. Agric. Food Chem.* **1998**, *42*, 401–406. [CrossRef] [PubMed]
54. Shabert, J.K.; Winslow, C.; Lacey, J.M.; Wilmore, D.W. Glutamine-antioxidant supplementation increases body cell mass in aids patients with weight loss: A randomized, double-blind controlled trial. *Nutrition* **1999**, *15*, 860–864. [CrossRef] [PubMed]
55. Onuegbu, N.C.; Adedokun, I.I.; Kabuo, N.O.; Nwosu, J.N. Amino acid profile and micronutrient composition of the African Pear (*Dacryodes edulis*) pulp. *Pak. J. Nutr.* **2011**, *10*, 555–557. [CrossRef]
56. Williamson, P.; Bevers, E.M.; Sineets, E.F.; Comfurius, P.; Schlegel, R.A.; Zwaal, R.F.A. Continuous analysis of the mechanism of activated transbilayer lipid movement in platelets. *Biochemistry* **1995**, *34*, 10448–10455. [CrossRef] [PubMed]
57. Kuhnlein, H.V. Nutrient values in indigenous wild plant greens and roots used by the Nuxalk people of Bella Coola, British Columbia. *J. Food Composit. Anal.* **1990**, *3*, 38–46. [CrossRef]
58. Sridhar, R.; Lakshminarayana, G. Lipid classes, fatty acids, and tocopherols of leaves of six edible plant species. *J. Agric. Food Chem.* **1993**, *41*, 61–63. [CrossRef]
59. Kaur, N.; Chugh, V.; Gupta, A.K. Essential fatty acids as functional components of foods—A review. *J. Food Sci. Technol.* **2014**, *51*, 2289–2303. [CrossRef]
60. Simopoulos, A.P. Essential fatty acids in health and chronic disease. *Am. J. Clin. Nutr.* **1999**, *70*, 560s–569s. [CrossRef] [PubMed]
61. Aloskar, L.V.; Kakkar, K.K.; Chakra, O.J. *Second Supplement to Glossary of Indian Medicinal Plants with Active Principles, Part-I (A–K), 1965–1981*; NISC, CSIR: New Delhi, India, 1992; pp. 265–266.
62. Trugo, L.C.; Von Baer, D.; Von Bayer, E. Lupin. In *Encyclopedia of Food Sciences and Nutrition*, 2nd ed.; Cabellero, B., Ed.; Academic Press: Oxford, UK, 2003.

63. Saha, J.; Biswal, A.K.; Deka, S.C. Chemical composition of some underutilized green leafy vegetables of Sonitpur district of Assam, India. *Int. Food Res. J.* **2015**, *22*, 1466–1473.
64. Vishwakarma, K.L.; Dubey, V. Nutritional analysis of indigenous wild edible herbs used in Eastern Chattisgarh, India. *Emir. J. Food Agric.* **2011**, *23*, 554–560.
65. Seal, T.; Chaudhuri, K. Effect of solvent extraction system on the antioxidant activity of some selected wild leafy vegetables of Meghalaya state in India. *Int. J. Pharm. Sci. Res.* **2015**, *4*, 1046–1051.
66. Seal, T.; Chaudhuri, K. Nutritional analysis of some selected wild edible plants consumed by the tribal people of Meghalaya state in India. *Int. J. Food Sci. Nutr.* **2016**, *1*, 39–43.
67. Khan, N.; Sultana, A.; Tahir, N.; Jamila, N. Nutritional composition, vitamins, minerals and toxic heavy metals analysis of *Trianthema portulacastrum* L., a wild edible plant from Peshawar, Khyber Pakhtunkhwa, Pakistan. *Afr. J. Biotechnol.* **2013**, *12*, 6079–6085.
68. Vermani, A.; Navneet, P.; Chauhan, A. Physicochemical analysis of ash of some medicinal plants growing in Uttarakhand. India. *Nat. Sci.* **2006**, *8*, 88–91.
69. Iheanacho, K.; Udebuani, A.C. Nutritional composition of some leafy vegetables consumed in Imo State, Nigeria. *J. Appl. Sci. Environ. Manag.* **2009**, *13*, 35. [CrossRef]
70. Available online: https://arkbiodiv.com/2019/05/23/plants-that-are-liked-such-as-icecream-by-the-camels-part-4/ (accessed on 28 December 2022).
71. FAO/WHO Expert Consultation on Human Vitamin and Mineral Requirements. Available online: https://www.fao.org/3/y2809e/y2809e.pdf (accessed on 30 December 2022).
72. Stadlmayr, B.; Nilsso, E.; Mouille, B.; Medhammar, E.; Burlingame, B.; Charrondiere, R. Nutrition indicator for biodiversity on food composition. A report on the progress of data availability. *J. Food Compos. Anal.* **2011**, *24*, 692–698. [CrossRef]
73. Sundriyal, M.; Sundriyal, R.C. Wild edible plants of the Sikkim Himalaya: Nutritive values of selected species. *Econ. Bot.* **2004**, *58*, 286–299. [CrossRef]
74. Saikia, P.; Deka, D.C. Mineral content of some wild green leafy vegetables of North-East India. *J. Chem. Pharm. Res.* **2013**, *5*, 117–121.
75. Abdus Satter, M.M.; Khan, M.M.R.R.L.; Jabin, S.A.; Abedin, N.; Islam, M.F.; Shaha, B. Nutritional quality and safety aspects of wild vegetables consume in Bangladesh. *Asian Pac. J. Trop. Biomed.* **2016**, *6*, 125–131. [CrossRef]
76. Abd-Elgawad, A.M.; Elshamy, A.I.; Alrowaily, S.L.; Rowaily, L.; El-Amier, Y.A. Habitat affects the chemical profile, allelopathy, and antioxidant properties of essential oils and phenolphenolic-enriched of the invasive plant *Heliotropium curassavicum*. *Plants* **2019**, *8*, 482. [CrossRef]
77. Shahid, S.A.; Dakheel, A.J.; Mufti, K.A.; Shabbir, G. Automated in-situ soil salinity logging in irrigated agriculture. *Eur. J. Sci. Res.* **2009**, *26*, 288–297.
78. Pastor-Cavada, E.; Juan, R.; Pastor, J.E.; Alaiz, M.; Vioque, J. Protein and amino acid composition of select wild legume species of tribe Fabaceae. *Food Chem.* **2014**, *163*, 97–102. [CrossRef] [PubMed]
79. Niyukuri, J.; Raiti, J.; Ntakarutimana, V.; Hafidi, A. Lipid composition and antioxidant activities of some underused wild plant seeds from Burundi. *Food Sci. Nutr.* **2021**, *9*, 111–122. [CrossRef] [PubMed]
80. Szymansky, C.M.; Muscolo, A.; Yeo, M.; Colville, L.; Clatworthy, I.; Salge, T.; Seal, C.E. Elemental localisation and a reduced glutathione redox state protect seeds of the halophyte Suaeda maritima from salinity during over-wintering and germination. *Environ. Exp. Bot.* **2021**, *190*, 104569. [CrossRef]
81. AOAC. Official Methods of Analysis. In *Agricultural Chemical, Contaminants, Drugs*, 15th ed.; Helrich, K., Ed.; AOAC: Arlington, VA, USA, 1990; Available online: https://law.resource.org/pub/us/cfr/ibr/002/aoac.methods.1.1990.pdf (accessed on 28 December 2022).
82. Siew, W.L.; Tang, T.S.; Tan, Y.A. *PORIM Test Methods Vol. 1*; Palm Oil Research Institute of Malaysia: Kuala Lumpur, Malaysia, 1995.
83. Shukla, A.; Vats, S.; Shukla, R.K. Phytochemical screening, proximate analysis and antioxidant activity of *Dracaena reflexa* Lam. Leaves. *Ind. J. Pharm.Sci.* **2015**, *77*, 640–644. [CrossRef]

Disclaimer/Publisher's Note: The statements, opinions and data contained in all publications are solely those of the individual author(s) and contributor(s) and not of MDPI and/or the editor(s). MDPI and/or the editor(s) disclaim responsibility for any injury to people or property resulting from any ideas, methods, instructions or products referred to in the content.

molecules

MDPI

Article

Screening of Active Ingredients from Wendan Decoction in Alleviating Palmitic Acid-Induced Endothelial Cell Injury

Nan Xu [1,2,†], Muhammad Ijaz [1,†], Haiyan Shi [3,4], Muhammad Shahbaz [2,5], Meichao Cai [6], Ping Wang [2,7], Xiuli Guo [1,*] and Lei Ma [2,6,*]

1 Department of Pharmacology, School of Pharmaceutical Science, Shandong University, Jinan 250012, China
2 Laboratory of Chinese Medicine Preparation, Shandong Academy of Chinese Medicine, Jinan 250014, China
3 Department of Clinical Pharmacy, The First Affiliated Hospital of Shandong First Medical University, Jinan 250014, China
4 Shandong Provincial Qianfoshan Hospital, Jinan 250014, China
5 Department of Radiology, Qilu Hospital Affiliated to Shandong University, Jinan 250012, China
6 School of Pharmacy, Shandong University of Traditional Chinese Medicine, Jinan 250355, China
7 State Key Laboratory of Precision Measurement Technology and Instruments, Tianjin University, Tianjin 300072, China
* Correspondence: guoxl@sdu.edu.cn (X.G.); 18615208870@163.com (L.M.)
† These authors contributed equally to this work.

Abstract: (1) Objective: Traditional Chinese medicine (TCM) plays an important role in the treatment of numerous illnesses. As a classic Chinese medicine, Wendan Decoction (WDD) encompasses a marvelous impact on the remedy of hyperlipidemia. It is known that hyperlipidemia leads to cardiovascular injury, therefore anti-vascular endothelial cell injury (AVECI) may be an underlying molecular mechanism of WDD in the cure of hyperlipidemia. However, there is no relevant research on the effect of WDD on vascular endothelial cells and its pharmacodynamic substances. Therefore, the purpose of this study was to investigate the protective effect of WDD on vascular endothelial cells. (2) Methods: The chemical constituents of WDD were determined by LC-MS/MS technology. The protective effects of 16 batches of WDD on samples from human umbilical vein endothelial cells (HUVECs) were evaluated. Finally, gray relation analysis (GRA) and partial least squares regression (PLSR) were used to analyze the potential correlation between chemical ingredients and AVECI. (3) Results: The results indicated that WDD had apparent protective effect on endothelial cells, and pharmacological properties in 16 batches of WDD tests were apparently discrepant. The GRA and PLSR showed that trigonelline, liquiritin, hesperidin, hesperetin, scopoletin, morin, quercetin, isoliquiritigenin, liquiritigenin and formononetin may be the active ingredients of AVECI in WDD. (4) Conclusions: WDD has a protective effect on endothelial cell injury induced by palmitic acid, which may be related to its component content. This method was suitable for the search of active components in classical TCM.

Keywords: Wendan decoction; active ingredients; screening; anti-vascular endothelial cell injury

Citation: Xu, N.; Ijaz, M.; Shi, H.; Shahbaz, M.; Cai, M.; Wang, P.; Guo, X.; Ma, L. Screening of Active Ingredients from Wendan Decoction in Alleviating Palmitic Acid-Induced Endothelial Cell Injury. *Molecules* **2023**, *28*, 1328. https://doi.org/10.3390/molecules28031328

Academic Editors: Arunaksharan Narayanankutty, Ademola C. Famurewa and Eliza Oprea

Received: 26 December 2022
Revised: 20 January 2023
Accepted: 24 January 2023
Published: 30 January 2023

Copyright: © 2023 by the authors. Licensee MDPI, Basel, Switzerland. This article is an open access article distributed under the terms and conditions of the Creative Commons Attribution (CC BY) license (https://creativecommons.org/licenses/by/4.0/).

1. Introduction

In recent years, significant changes have taken place in diet and lifestyle. The incidence of metabolic diseases, including hyperlipidemia, has increased, and patients have high levels of triglyceride, phospholipid and total cholesterol [1]. Hyperlipidemia can significantly increase the risk of other metabolic diseases and reduce the quality of life. Studies have shown that the level of lipid peroxidation in patients with hyperlipidemia has increased, and the mechanism of oxidative stress was involved in abnormal lipid metabolism [2]. A large number of clinical and basic research data at home and abroad have proved that hyperlipidemia, especially hypercholesterolemia is one of the important risk factors of atherosclerosis and cardio-cerebrovascular diseases. It is a process of chronic pathological changes, mediated by a variety of factors and links, in which there is always an interaction between blood and blood vessel wall. Therefore, the study of the effect of dyslipidemia on blood cells and vascular endothelial cells and their interaction has important theoretical and application value for understanding the mechanism of the occurrence and development of atherosclerosis and putting forward effective counter measures.

HUVECs are monolayer flat cells, lining the inner surface of the vascular system, and are important components of vascular intima. They can synthesize, secrete and express a variety of antithrombotic and thrombogenic substances. Endothelium injury is regarded as a reason of pathogenesis and clinical sign of hyperlipidemic vascular injury, and a large number of essential and auxiliary metabolic disarrangements have been connected to endothelial injury [3–6]. Experimental and clinical research has indisputably associated the endothelium with the pathogenesis of hyperlipidemia. Therefore, the clinical evaluation of endothelial cells may be considered to evaluate vascular wellbeing (or illness) [7].

TCM has played a basic role in treatment of various ailments since its long-standing history of clinical utilization [8]. In China, hyperlipidemia has become an important factor endangering public health, and shows a younger trend. However, long-term use of hypolipidemic chemical drugs has greater side effects [9]. Therefore, it is urgent to find lipid-lowering drugs with good safety and low toxicity. The TCM meets this demand. It is an important task to research and develop new lipid-lowering TCM preparations, which are of great significance and broad prospects. Through the study, it was found that the extracts of millet leaves can inhibit the apoptosis of vascular endothelial cells induced by hyperlipidemia in model rats by regulating the expression of apoptotic proteins Bax and Bcl-2. Bax and Bcl-2 regulate two key proteins of apoptosis, and they regulate the susceptibility of cells to apoptosis. The increased expression of Bax leads to cell death, while the effect of Bcl-2 is the reverse. The extract of millet leaves can significantly up-regulate the expression of Bcl-2 protein and down-regulate the expression of Bax protein, so as to protect against vascular endothelial oxidative damage caused by hyperlipidemia [10]. Paeonol regulated HUVEC development via downregulating miR-338-3p expression. Interestingly, miR-338-3p targeted TET2 and inhibited TET2 expression. MiR-338-3p modulated ox-LDL-treated VEC growth through suppressing TET2 expression. We demonstrated that paeonol attenuated the effect of ox-LDL on the development of mice HUVEC via modulating miR-338-3p/TET2 axis [11]. Vast evidence is now available that non-esterified FFA, usually typified by PA, abruptly triggers endothelial cell apoptosis in vitro1 [12,13]. HUVEC could not be maintained longer than 2 days even at low PA (or other FFA) concentrations that are compatible with their normal levels in human blood.

WDD dates from "Bei ji qian jin yao fang (Essential Recipes for Emergent Use Worth A Thousand Gold)" (written by Sun Simiao of the Tang dynasty), and was recorded in the Catalogue of Classical Prescriptions (first batch) issued by the state administration of traditional medicine of China in 2018 (NATCM, 2018). This was one of the primary 100 classic Chinese pharmaceutical medicines discharged and consisted of Pineilia ternata (Thunb.) Breit. (Ban-Xia, BX), Glycyrrhiza uralensis Fisch. (Gan-Cao, GC), Citrus reticulata Blanco (Chen-Pi, CP), Phyllostachys nigra (Lodd.) Munro var. henonis (Mitf.) Stapf ex Rendle (Zhu-Ru, ZR), Citrus aurantium L. (Zhi-Shi, ZS), Zingiber officinale Rose. (Sheng-Jiang, SJ). This prescription is the basic prescription of "expectorant". Clinically, it is generally utilized within the treatment of diabetes, chronic kidney disease and hyperlipidemi [14–16]. The role of vascular complications in the development and progression of hyperlipidemia was well documented in the pulmonary vascular dysfunction, endothelial dysfunction and atherosclerosis [17,18]. Therefore, we infer that the efficacy of WDD is linked to the protection of endothelial cells. However, there is no research on the effect of WDD on endothelial cells, and the pharmacodynamic substances are not clear.

The identification method of chemical constituents in WDD refer to a previous study [19]. In all, a total of 20 ingredients related to hyperlipidemia were elected as quantitative chemical markers [20–24]. In this study, 16 batches of 20 chemical components of WDD were analyzed by ultra-performance liquid chromatography-triple quadrupole tandem mass spectrometry (UPLC-QQQ-MS/MS)), and the optimal concentration of WDD in the treatment of AVECI was selected by MTT method. Then the 10 active components of WDD for the treatment of AVECI were determined by two mathematical analysis modes (GRA and PLSR).

2. Result

2.1. Content Determination of Components in WDD

The single mass spectra of 20 compounds were studied by pouring the relevant standard solutions into the mass spectrometer in both positive and negative ion modes. The optimization results are shown in Table 1. The chemical structures of 20 chemical ingredients and IS were shown in Figure 1.

Table 1. List of quantified 20 compounds of WDD and IS parameters.

NO.	Compound	Formula	Mass	t_r (min)	Polarity	Precursor Ion (*m*/*z*)	Product Ion (*m*/*z*)	Collision Energy (V)	Radio Frequency Lens (V)
1	Trigonelline	$C_7H_7NO_2$	137.14	0.84	+	138.01	92.13	19.91	97
2	Synephrine	$C_9H_{13}NO_2$	167.21	0.97	+	168.03	150.13	8.16	49
3	Isoquercitrin	$C_{21}H_{20}O_{12}$	464.38	3.69	+	465.17	229.05	44.22	106
4	Liquiritin	$C_{21}H_{22}O_9$	418.39	3.69	-	417.13	135.04	31.41	147
5	Naringin	$C_{27}H_{32}O_{14}$	580.53	3.71	-	579.2	459.16	22.99	234
6	Hesperidin	$C_{28}H_{34}O_{15}$	610.56	3.73	-	609.21	286.08	41.73	170
7	Hesperetin	$C_{16}H_{14}O_6$	302.28	3.74	+	303.11	177.13	16.84	133
8	Scopoletin	$C_{10}H_8O_4$	192.17	3.89	-	190.97	103.99	25.05	83
9	Daidzein	$C_{15}H_{10}O_4$	254.24	4.08	-	253.04	223.07	33.81	151
10	Naringenin	$C_{15}H_{12}O_5$	272.25	4.51	-	271.03	119.05	26.52	124
11	Glycyrrhizic acid	$C_{42}H_{62}O_{16}$	822.93	4.55	+	823.52	453.39	19.87	171
12	Morin	$C_{15}H_{10}O_7$	302.24	4.59	-	301.04	125.04	20.25	137
13	Quercetin	$C_{15}H_{10}O_7$	302.24	4.61	-	301.06	151.04	21.09	149
14	Isoliquiritigenin	$C_{15}H_{12}O_4$	256.25	4.78	-	254.97	134.97	15.7	104
15	Liquiritigenin	$C_{15}H_{12}O_4$	256.25	4.79	-	254.98	135.04	15.24	118
16	Formononetin	$C_{16}H_{12}O_4$	268.26	4.96	-	267.03	223.04	32.21	127
17	Nobiletin	$C_{21}H_{22}O_8$	402.4	5.43	+	403.17	327.13	32.17	182
18	Tangeretin	$C_{20}H_{20}O_7$	372.37	6.03	+	373.17	358.13	19.07	167
19	Obacunone	$C_{26}H_{30}O_7$	454.51	6.05	+	455.22	161.07	23.28	137
20	Glycyrrhetinic acid	$C_{30}H_{46}O_4$	470.68	8.29	-	469.36	355.33	46.45	261
21	Furosemide	$C_{12}H_{11}ClN_2O_5S$	330.74	4.49	-	328.95	204.99	21.51	113

WDD = Wendan decoction; IS = Internal standard.

The contents of 20 compounds in 16 batches of WDD were determined by UPLC-QQQ-MS/MS within 10 min. The degree speed had major upgrades over the HPLC method, so ensuring that we may checkout still more cases in one day. Figure 2 shows the chromatographic behavior of 20 compounds and IS in WDD. The specific outcomes are shown in Table 2.

Method Validation

The 20 primitive courses of action were precisely debilitated with 50% (*v*/*v*) methanol-water. The slopes, intercepts and correlation coefficients (r^2) were obtained by weighted (1/x) linear regression. The estimation ranges (S/N $\geq$ 10), lower limit of quantitation (LLOQ) and straight twists of diverse chemical fixings are tabulated in Table 3. The relative standard deviations (RSD) of precision, repeatability and stability were all less than 15%, and the single solution was stable at room temperature for 24 h and the recovery rate was within acceptable limits.

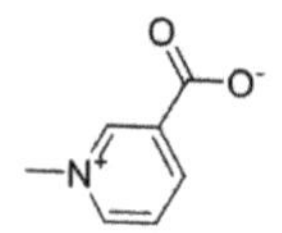

NO.1 Trigonelline

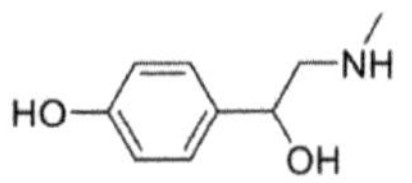

NO.2 Synephrine

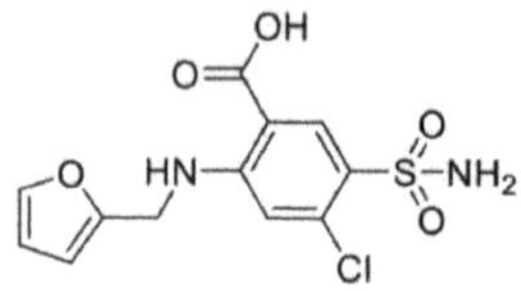

NO.21 Furosemide

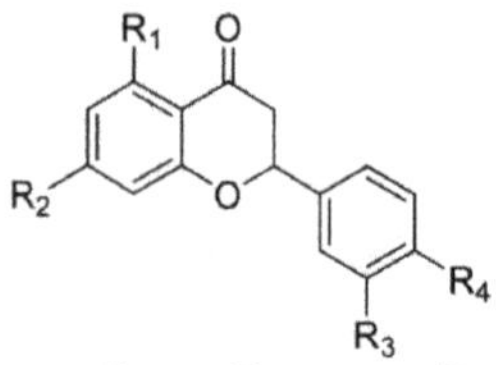

NO.	R1	R2	R3	R4
4 Liquiritin	H	OH	H	OGlc
5 Naringin	OH	OGlc(2-1)Rha	H	OH
6 Hesperidin	OH	OGlc(6-1)Rha	OH	OCH_3
7 Hesperetin	OH	OH	OH	OCH_3
10 Naringenin	OH	OH	H	OH
15 Liquiritigenin	H	OH	H	OH

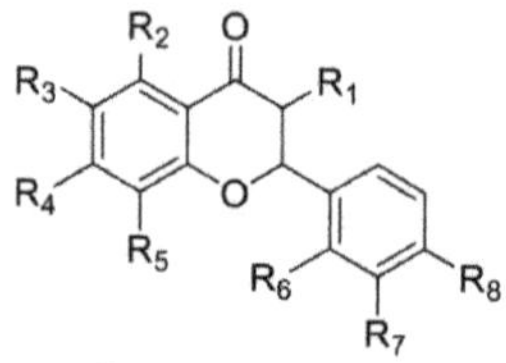

NO.	R1	R2	R3	R4	R5	R6	R7	R8
3 Isoquercitrin	OGlc	OH	H	OH	H	H	OH	OH
12 Morin	OH	OH	H	OH	H	OH	H	OH
13 Quercetin	OH	OH	H	OH	H	H	OH	OH
17 Nobiletin	H	OCH_3	OCH_3	OCH_3	OCH_3	H	OCH_3	OCH_3
18 Tangeretin	H	OCH_3	OCH_3	OCH_3	OCH_3	H	H	OCH_3

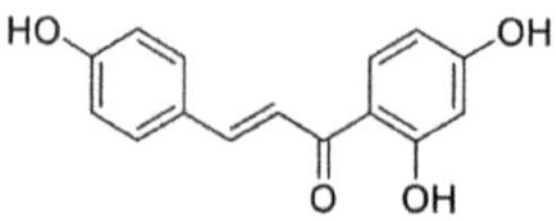

NO.14 Isoliquiritigenin

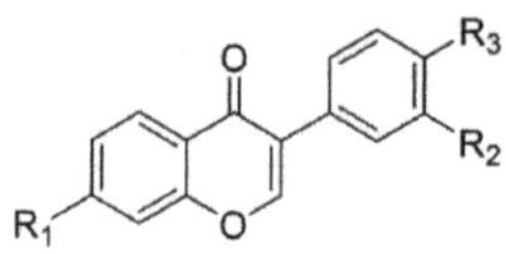

NO.	R1	R2	R3
9 Daidzein	OH	H	OH
16 Formononetin	OH	H	OCH_3

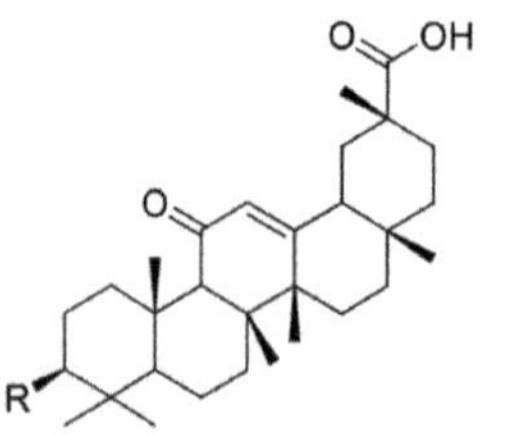

NO.	R
11 Glycyrrhizic acid	O-2GluA
20 Glycyrrhetinic acid	OH

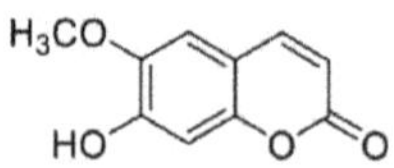

NO.8 Scopoletin

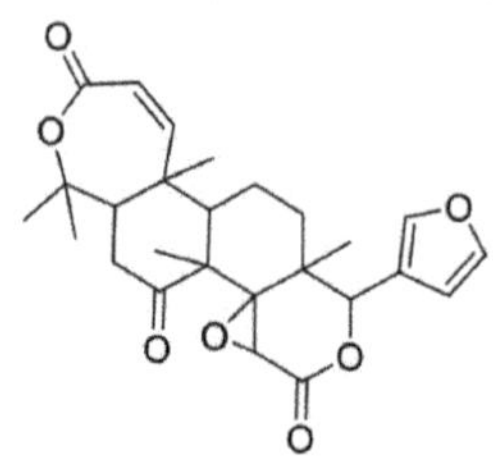

NO.19 Obacunone

Figure 1. Structures of 20 chemicals analyzed in WDD and IS.

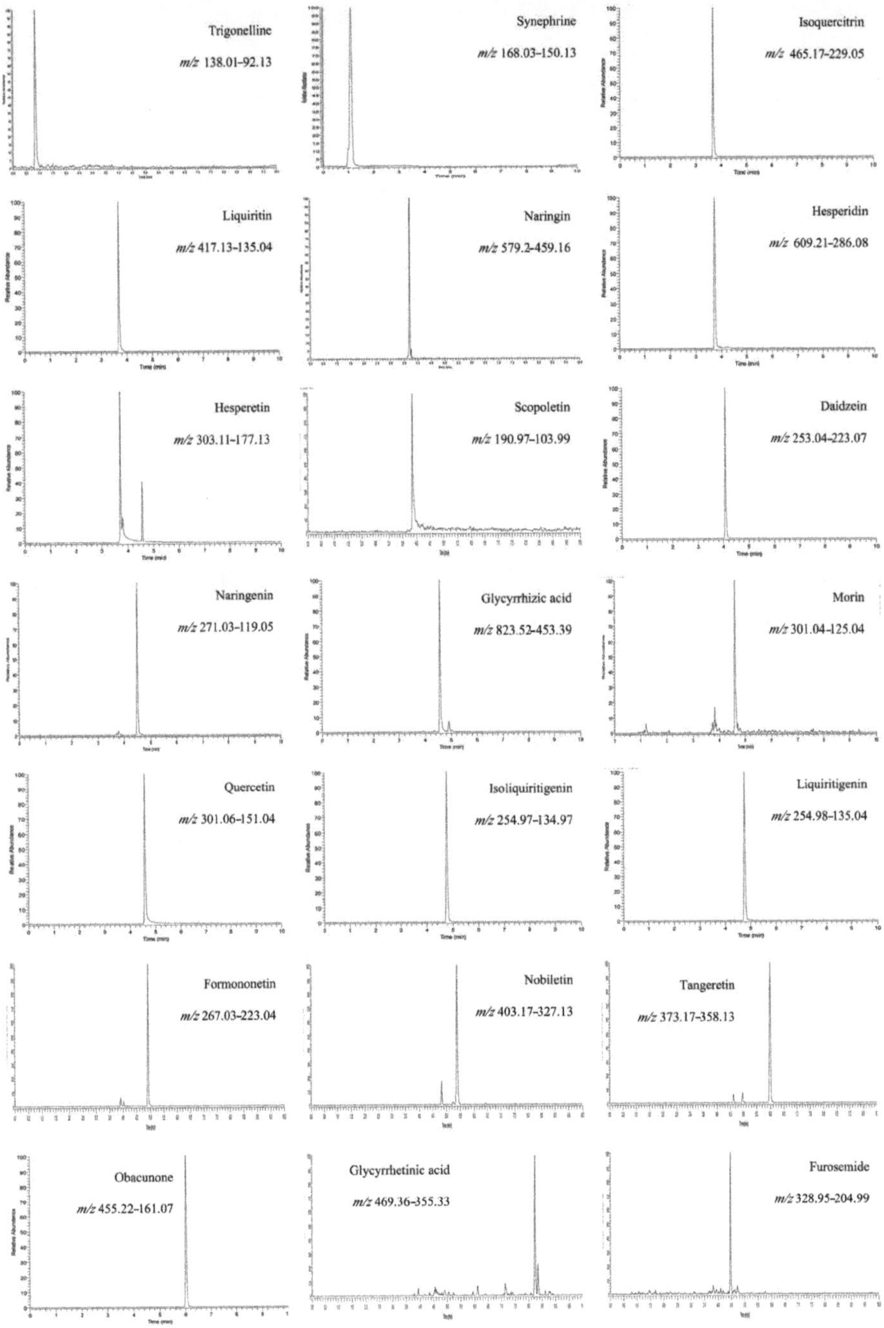

Figure 2. Chromatographic behavior of 20 compounds and IS.

Table 2. The mean of the 20 chemical contents of WDD ($n = 3$, μg/g).

NO.	Compound	S1	S2	S3	S4	S5	S6	S7	S8	S9	S10	S11	S12	S13	S14	S15	S16
1	Trigonelline	1081.13 ± 115.55	1214.56 ± 98.98	968.28 ± 78.62	986.33 ± 54.57	2132.75 ± 248.64	1667.38 ± 253.09	2644.48 ± 250.17	1542.58 ± 182.98	2016.93 ± 213.55	2241.31 ± 238.07	1969.98 ± 67.36	798.42 ± 34.41	567.43 ± 71.72	1778.26 ± 45.44	2003.50 ± 273.90	1674.62 ± 261.21
2	Synephrine	13147.38 ± 1846.14	17359.46 ± 1855.39	18033.69 ± 165.49	12952.94 ± 1978.05	38277.40 ± 4178.07	36066.95 ± 4630.68	68086.80 ± 9517.01	18511.95 ± 2834.50	25981.23 ± 3620.37	58408.02 ± 3153.10	36739.46 ± 2112.75	27157.50 ± 965.88	6773.77 ± 960.51	37962.47 ± 1573.09	28646.49 ± 2927.31	38754.01 ± 4227.24
3	Isoquercitrin	123.85 ± 6.77	107.03 ± 2.40	123.63 ± 4.37	33.89 ± 3.42	80.13 ± 8.19	58.89 ± 6.21	95.03 ± 11.35	94.24 ± 2.64	100.42 ± 13.38	78.47 ± 13.19	96.81 ± 8.38	78.29 ± 6.24	91.47 ± 12.04	84.97 ± 4.08	60.87 ± 5.10	94.16 ± 10.12
4	Liquiritin	489.62 ± 58.75	1729.21 ± 240.56	601.38 ± 15.65	440.53 ± 33.29	995.53 ± 51.83	624.55 ± 59.79	1754.29 ± 222.48	764.55 ± 94.22	1717.53 ± 206.92	639.79 ± 57.67	621.43 ± 82.80	657.53 ± 43.81	774.28 ± 56.84	1159.37 ± 20.32	663.45 ± 53.62	1135.81 ± 163.75
5	Naringin	4293.99 ± 694.84	544.19 ± 57.61	2330.00 ± 62.10	43.33 ± 5.88	183.49 ± 21.41	493.23 ± 77.71	133.77 ± 22.64	1972.70 ± 308.36	1907.97 ± 298.14	155.73 ± 15.23	4980.87 ± 716.43	856.40 ± 27.96	1003.64 ± 121.72	2221.39 ± 43.38	966.65 ± 106.91	251.75 ± 39.35
6	Hesperidin	3874.78 ± 652.37	4160.72 ± 528.16	3091.37 ± 148.23	1195.08 ± 80.82	3060.41 ± 269.90	2626.72 ± 259.04	3575.26 ± 467.00	3227.13 ± 494.79	3102.85 ± 409.92	2480.56 ± 231.93	4758.26 ± 689.58	3177.08 ± 196.28	3868.27 ± 317.84	2890.70 ± 92.02	3181.91 ± 368.53	3125.96 ± 179.77
7	Hesperetin	5837.87 ± 264.68	3048.85 ± 303.57	2239.34 ± 96.43	814.26 ± 68.23	1993.62 ± 77.28	1807.70 ± 127.78	2145.84 ± 144.19	2330.46 ± 243.40	2295.72 ± 329.59	1834.12 ± 181.45	7156.59 ± 316.95	2489.32 ± 129.60	2786.83 ± 294.40	2006.73 ± 48.25	3308.82 ± 266.13	1839.25 ± 192.44
8	Scopoletin	150.36 ± 3.34	76.50 ± 12.71	69.38 ± 4.00	34.08 ± 5.08	64.44 ± 7.10	60.95 ± 1.04	44.94 ± 5.80	56.11 ± 6.83	86.36 ± 4.00	34.28 ± 4.00	237.55 ± 13.04	57.43 ± 2.46	64.86 ± 10.42	69.80 ± 4.72	69.86 ± 11.88	45.06 ± 3.03
9	Daidzein	14.06 ± 0.94	17.63 ± 1.57	9.12 ± 0.78	1.95 ± 0.28	9.00 ± 0.75	6.81 ± 0.91	6.31 ± 0.81	13.53 ± 1.14	12.57 ± 0.84	10.28 ± 1.19	6.89 ± 0.71	8.58 ± 1.20	9.99 ± 1.24	6.57 ± 0.83	9.20 ± 1.35	5.17 ± 0.10
10	Naringenin	2600.18 ± 217.74	379.36 ± 18.67	784.54 ± 27.63	35.69 ± 0.54	113.35 ± 12.30	206.63 ± 18.81	71.18 ± 4.94	425.68 ± 35.30	525.10 ± 70.04	227.03 ± 19.17	3259.16 ± 241.86	206.70 ± 8.61	246.75 ± 10.36	397.22 ± 10.40	153.62 ± 22.52	179.03 ± 12.65
11	Glycyrrhizic acid	3070.18 ± 175.29	4269.03 ± 199.04	3065.04 ± 77.75	1164.14 ± 183.71	3355.01 ± 426.61	2576.06 ± 255.23	4157.77 ± 211.33	3826.32 ± 335.32	3899.42 ± 542.40	3107.87 ± 378.10	2768.07 ± 250.76	2558.80 ± 180.03	2942.77 ± 268.86	3080.26 ± 131.21	3006.57 ± 333.22	3113.40 ± 136.04
12	Morin	3813.18 ± 383.83	708.37 ± 28.98	844.89 ± 111.80	105.58 ± 13.89	110.88 ± 16.48	563.48 ± 63.58	170.98 ± 23.36	539.10 ± 49.68	550.87 ± 49.07	311.26 ± 26.17	4019.50 ± 136.56	374.92 ± 20.03	449.90 ± 49.18	727.34 ± 59.61	379.26 ± 31.03	201.83 ± 22.11
13	Quercetin	2902.92 ± 420.22	464.34 ± 27.99	576.66 ± 67.05	81.35 ± 11.27	129.22 ± 16.14	455.71 ± 31.50	145.68 ± 12.20	378.52 ± 26.58	395.13 ± 41.98	252.14 ± 31.94	3107.47 ± 277.25	301.80 ± 24.18	344.56 ± 42.39	599.26 ± 51.07	304.66 ± 20.80	231.75 ± 22.44
14	Isoliquiritigenin	123.33 ± 1.54	107.88 ± 0.74	80.49 ± 2.98	17.05 ± 1.62	22.02 ± 1.40	42.85 ± 3.26	34.17 ± 5.03	142.57 ± 16.27	129.15 ± 11.10	65.46 ± 4.92	199.95 ± 8.58	16.57 ± 0.55	19.10 ± 0.64	61.64 ± 1.80	26.24 ± 1.75	17.41 ± 0.71
15	Liquiritigenin	392.14 ± 7.16	356.60 ± 0.22	263.01 ± 5.46	54.58 ± 5.33	67.60 ± 3.23	133.48 ± 17.50	113.34 ± 10.75	466.33 ± 60.63	408.01 ± 56.91	213.66 ± 28.54	649.33 ± 28.39	51.91 ± 2.90	58.18 ± 4.37	201.42 ± 3.37	85.16 ± 9.88	57.06 ± 6.40
16	Formononetin	146.95 ± 8.45	139.01 ± 3.45	107.96 ± 3.51	17.32 ± 2.04	110.97 ± 8.81	66.66 ± 9.18	55.63 ± 5.85	150.10 ± 11.65	92.85 ± 9.92	92.21 ± 7.64	115.78 ± 2.23	73.09 ± 3.82	81.79 ± 9.80	89.48 ± 4.13	74.02 ± 6.13	65.94 ± 5.70
17	Nobiletin	4129.53 ± 49.24	1783.50 ± 20.77	2515.31 ± 77.18	1850.06 ± 156.62	1954.99 ± 191.13	3735.75 ± 474.17	4389.54 ± 507.72	2561.99 ± 276.92	5054.44 ± 659.23	2330.66 ± 252.21	2747.98 ± 84.84	3001.93 ± 283.54	3325.35 ± 186.39	5093.37 ± 125.05	3912.02 ± 401.66	2206.13 ± 167.75
18	Tangeretin	2239.72 ± 63.67	692.66 ± 23.77	781.88 ± 35.83	764.14 ± 108.30	775.59 ± 80.63	1638.23 ± 199.89	1865.95 ± 205.54	750.53 ± 69.41	2278.55 ± 273.85	674.31 ± 71.45	1527.77 ± 102.02	1410.63 ± 114.69	1628.44 ± 71.47	2597.79 ± 53.12	1789.81 ± 206.46	684.47 ± 51.29
19	Obacunone	1826.45 ± 219.79	1392.87 ± 138.75	346.31 ± 12.17	178.50 ± 2.65	1142.22 ± 153.81	540.02 ± 80.23	725.34 ± 55.98	428.33 ± 51.34	303.89 ± 51.81	329.90 ± 10.00	1818.21 ± 207.43	481.09 ± 58.54	580.77 ± 17.44	224.79 ± 29.90	441.34 ± 26.66	298.26 ± 20.22
20	Glycyrrhetinic acid	40.62 ± 4.81	32.24 ± 1.36	19.98 ± 2.46	2.86 ± 0.20	14.83 ± 2.20	7.85 ± 0.78	11.66 ± 1.95	34.82 ± 4.88	31.69 ± 2.53	13.99 ± 0.90	80.30 ± 11.80	3.27 ± 0.21	4.79 ± 0.42	13.26 ± 0.93	3.68 ± 0.27	6.10 ± 0.46

WDD = Wendan Decoction.

Table 3. Regression data, LLOQs, precision, stabilities and repeatabilities for the 20 components of WDD.

NO.	Compound	Regression Equation	r^2	LLOQ ng/mL	Linear Range ng/mL	Precision RSD, $n = 6$	Stability RSD, 24 h, $n = 6$	Repeatability RSD, $n = 6$	Recovery % mean ± SD
1	Trigonelline	$Y = 0.00299197X + 0.0378053$	0.9904	120	120–60,000	13.98	7.44	8.54	94.48 ± 7.50
2	Synephrine	$Y = 0.0900787X + 0.463625$	0.9911	8	8–4000	13.79	6.91	11.87	104.74 ± 8.29
3	Isoquercitrin	$Y = 0.00046952X - 0.00753226$	0.9907	140	140–70,000	14.75	13.05	10.49	94.13 ± 6.43
4	Liquiritin	$Y = 0.00497658X - 0.00447113$	0.9968	6	6–3000	8.83	3.63	11.53	93.91 ± 5.71
5	Naringin	$Y = 0.00275605X - 0.0152202$	0.9913	12	12–6000	11.31	6.63	12.38	102.78 ± 7.37
6	Hesperidin	$Y = 0.00185764X - 0.0162377$	0.9926	24	24–12,000	10.96	5.71	13.37	100.97 ± 6.38
7	Hesperetin	$Y = 0.00747729X + 0.0165338$	0.9924	60	60–30,000	11.36	5.54	5.87	96.00 ± 5.06
8	Scopoletin	$Y = 0.00145419X - 0.0126843$	0.9917	16	16–8000	12.2	7.17	7.38	101.27 ± 8.60
9	Daidzein	$Y = 0.00364942X + 0.0178624$	0.9948	4	4–2000	9.82	13.04	12.79	111.84 ± 10.65
10	Naringenin	$Y = 0.0144335X - 0.0633108$	0.993	14	14–7000	6.82	7.71	6.62	93.33 ± 5.30
11	Glycyrrhizic acid	$Y = 0.0419993X - 0.197051$	0.9905	10	10–5000	7.60	6.45	6.21	96.27 ± 9.18
12	Morin	$Y = 0.0005182X + 0.130608$	0.9932	20	20–10,000	10.6	13.25	11.03	97.35 ± 7.92
13	Quercetin	$Y = 0.00131174X - 0.00216878$	0.9921	26	26–13,000	11.65	7.66	8.00	98.59 ± 10.04
14	Isoliquiritigenin	$Y = 0.0370696X - 0.106264$	0.9910	6	6–3000	8.00	7.50	4.77	111.43 ± 1.22
15	Liquiritigenin	$Y = 0.0114821X + 0.0776073$	0.9992	4	4–2000	10.33	6.54	8.89	89.62 ± 3.19
16	Formononetin	$Y = 0.0197593X - 0.0374138$	0.9903	4	4–2000	7.41	7.49	5.89	112.36 ± 3.13
17	Nobiletin	$Y = 0.047895X - 0.0133269$	0.9942	1	1–500	8.52	6.06	9.76	92.59 ± 3.64
18	Tangeretin	$Y = 0.188519X + 0.0417842$	0.9948	0.4	0.4–200	8.74	6.05	9.44	99.47 ± 8.74
19	Obacunone	$Y = 0.000297133X - 0.00452012$	0.9902	50	50–25,000	11.95	13.62	13.09	90.47 ± 7.57
20	Glycyrrhetinic acid	$Y = 0.000755862X - 0.00133086$	0.9946	60	60–30,000	11.89	6.39	12.95	102.81 ± 10.84

LLOQ = lower limit of quantification; WDD = Wendan Decoction.

2.2. *Cell Experiments*

2.2.1. Screening of PA Concentration by MTT

To determine the optimal inhibitory concentration of PA on HUVEC cells, the results were expressed as inhibition rate.

$$\text{Inhibition rate (\%)} = (OD_{sample} - OD_{blank})/OD_{blank} \times 100\% \quad (1)$$

OD means Optical Density. The tests were carried out in triplicate. The data are shown as the mean ± SD. The p value of < 0.05 was considered significant. The inhibition rate of PA extracted from HUVEC are shown in Figure 3. The inhibition rates were as follows: 0 μmol/L PA(Blank): 0%, 50 μmol/L PA: 5.50%, 100 μmol/L PA: 14.05%, 150 μmol/L PA: 29.50%, 200 μmol/L PA: 45.12%, 250 μmol/L PA: 54.10%, 300 μmol/L PA: 62.34%. The experimental results showed that cell proliferation was dose-dependent with PA concentration compared with the control group. Therefore, 200 μmol/L PA concentration was used for subsequent tests.

2.2.2. Cell Protective Effect of WDD

Inhibition rates were utilized as the benchmark, the inhibitory impact of the ideal concentration of WDD on HUVEC damage was decided.

$$\text{Inhibition rate (\%)} = (OD_{sample} - OD_{model})/OD_{model} \times 100\% \quad (2)$$

The results showed that there were no major changes within the cell state of the control bunch. When the concentration of each bunch of WDD was 0.25 mg/mL, 0.5 mg/mL and 1 mg/mL, the cell condition had clear hyperplasia, and when the consistence was 2 mg/mL or over, it had inhibitory impact on the cell. Subsequently, 1 mg/mL concentration was chosen for ensuing tests. The results of filtrating HUVEC cells shielded by different consistencies of WDD by MTT strategy are shown in Table 4.

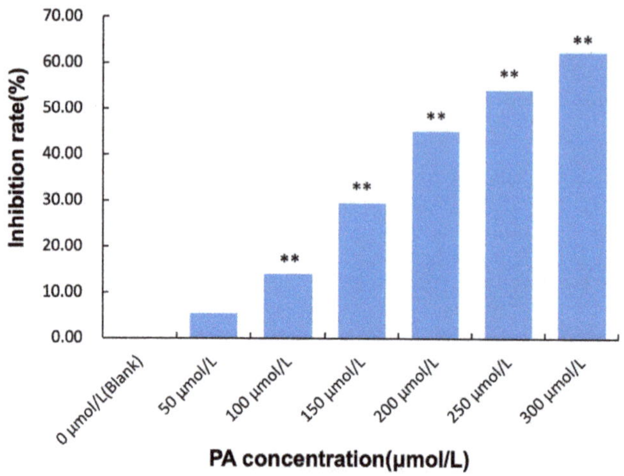

Figure 3. The inhibition rate of PA extracted from HUVEC. **: Compared with blank group, $p < 0.01$.

Table 4. The inhibition rates of 16 batches of WDD (Mean ± SD, $n = 3$,).

Samples	0.25 mg/mL	0.5 mg/mL	1 mg/mL	2 mg/mL	4 mg/mL
S1	0.27 ± 0.16	0.38 ± 0.12	0.57 ± 0.05	0.24 ± 0.09	−0.03 ± 0.17
S2	0.19 ± 0.08	0.25 ± 0.10	0.53 ± 0.24	0.07 ± 0.02	−0.14 ± 0.12
S3	0.28 ± 0.20	0.33 ± 0.08	0.50 ± 0.13	0.10 ± 0.23	−0.10 ± 0.03
S4	0.19 ± 0.14	0.36 ± 0.07	0.59 ± 0.21	0.08 ± 0.23	−0.01 ± 0.16
S5	0.25 ± 0.25	0.32 ± 0.12	0.52 ± 0.16	0.08 ± 0.25	−0.10 ± 0.15
S6	0.18 ± 0.04	0.39 ± 0.26	0.58 ± 0.08	0.08 ± 0.06	−0.08 ± 0.21
S7	0.18 ± 0.04	0.39 ± 0.26	0.58 ± 0.08	0.08 ± 0.06	−0.08 ± 0.21
S8	0.18 ± 0.10	0.31 ± 0.11	0.61 ± 0.23	0.10 ± 0.18	−0.08 ± 0.25
S9	0.20 ± 0.15	0.30 ± 0.18	0.52 ± 0.13	0.17 ± 0.20	−0.22 ± 0.08
S10	0.12 ± 0.15	0.25 ± 0.11	0.48 ± 0.14	0.08 ± 0.11	−0.10 ± 0.11
S11	0.17 ± 0.19	0.26 ± 0.04	0.53 ± 0.12	0.22 ± 0.10	−0.05 ± 0.16
S12	0.12 ± 0.08	0.29 ± 0.14	0.47 ± 0.16	0.12 ± 0.10	−0.12 ± 0.18
S13	0.12 ± 0.10	0.26 ± 0.12	0.47 ± 0.06	0.06 ± 0.07	−0.11 ± 0.11
S14	0.09 ± 0.13	0.29 ± 0.21	0.47 ± 0.18	0.23 ± 0.05	0.04 ± 0.12
S15	0.20 ± 0.09	0.33 ± 0.15	0.52 ± 0.21	0.13 ± 0.15	−0.11 ± 0.13
S16	0.17 ± 0.13	0.33 ± 0.19	0.52 ± 0.26	0.16 ± 0.08	0.00 ± 0.13

WDD = Wendan Decoction.

2.3. *Statistical Analysis*

The results of GRA and PLSR are shown in Figure 4 and Table 5. It can be seen from Figure 4 that synephrine, isoquercitrin, naringin, daidzein, naringenin, glycyrrhizic acid, nobiletin, tangeretin, obacunone, and glycyrrhetinic acid had a negative correlation with the in vitro efficacy results. The ingredients of trigonelline, liquiritin, hesperidin, hesperetin, scopoletin, morin, quercetin, isoliquiritigenin, liquiritigenin and formononetin had a positive relationship with the in vitro adequacy results. Observing the GRA results, it can be seen that these were the primary ten compounds with great advancing impact on pharmacodynamic parameters in vitro.

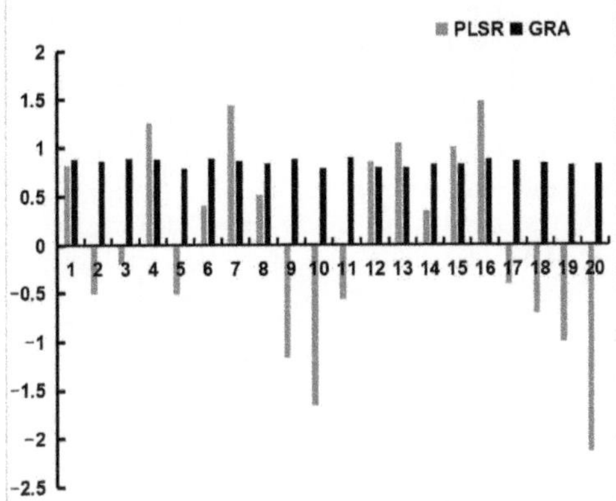

Figure 4. GRA and PLSR results for WDD compounds. GRA = Gray relation analysis; PLSR = Partial least squares regression; WDD = Wendan Decoction.

Table 5. GRA and PLSR results for WDD compounds.

NO.	Compounds	GRA	PLSR	NO.	Compounds	GRA	PLSR
1	Trigonelline	0.88	0.822	11	Glycyrrhizic acid	0.901	−0.565
2	Synephrine	0.869	−0.507	12	Morin	0.795	0.856
3	Isoquercitrin	0.888	−0.193	13	Quercetin	0.798	1.044
4	Liquiritin	0.886	1.259	14	Isoliquiritigenin	0.828	0.349
5	Naringin	0.789	−0.513	15	Liquiritigenin	0.828	1.001
6	Hesperidin	0.892	0.404	16	Formononetin	0.883	1.477
7	Hesperetin	0.865	1.438	17	Nobiletin	0.864	−0.403
8	Scopoletin	0.843	0.521	18	Tangeretin	0.838	−0.716
9	Daidzein	0.885	−1.167	19	Obacunone	0.819	−1.003
10	Naringenin	0.785	−1.659	20	Glycyrrhetinic acid	0.829	−2.138

GRA = Gray relation analysis; PLSR = partial least squares regression; WDD = Wendan Decoction.

3. Discussion

Since the prehistoric age, human beings have used natural products, such as animals and marine organisms, in medicines to alleviate and treat diseases. On the basis of fossil records, the past records of human use of flora as medicines may be traced back at least 60,000 years. In the wake of the progress of theoretical background, therapeutic theory and related technologies, as well as comprehension of the life sciences, a limpid comprehension of the active ingredients of TCM has proved to be possible [25]. TCM includes a wide application prospect within the treatment of complicated infections such as cancer and diabetes due to its points of interest in combined direction component of "multi-components, multi-targets and multi-approaches" [26]. An assortment of tissues, organs or different targets take part within the entire preparation of complicated affliction and shaping a complicated network regulation mechanism [27]. It has complex components and multiple targets, which is in line with the characteristics of TCM [28].

The inquiry about the viable components of TCM as a rule begins with the efficient division of chemical components, and its organic movement or pharmacodynamic impact was screened to clarify its compelling components [29], this strategy is time consuming. In this research, we used UPLC-QQQ-MS/MS technology to study the content of WDD. Combined with MTT method and mathematical statistical analysis, we screened 10 ingredients of WDD to protect endothelial cells. The results showed that WDD was concentration dependent in protecting endothelial cells. However, when the concentration was 2 mg/mL, the protective ability decreased. This may be because when the consistencies change, the condition of molecules in the liquor will also change. This may bring about the formation of complex structures or changes within the pharmacokinetics of the drugs and thereby reducing binding of drug to targets [30,31].

The agreeable impacts of "multicomponent, multichannel and multi-target" makes it difficult to inquire about the exercises of its pharmacodynamic materials [32]. At the same time, this is also one of the advantages of TCM. The 17 miRNAs in HUVECs that were raised by formononetin; the impact was biggest within the circumstances of miR-375, which was situated in an elevation backward-looking intergenic region between the cryba2 and Ccdc108 [33]. The move to communicate with miR-375 in HUVECs can increment hyperplasia and diminish apoptosis. The truth that formononetin incites the miR-375/RASD1/ERα input circuit in HUVECs, which at that pathway point is activated, may clarify why the isoflavone energized hyperplasia and limited apoptosis in HUVECs [34]. In addition, formononetin regulates ERK1/2 and P38 MAPK pathways, participates in the overexpression of EGR-1 transcription factor, and promotes endothelial cell repair and wound healing. The hesperidin hindered the expression of ERK, P38 MAPK and PI3K/AKT in VEGF-induced HUVECs. Hesperidin also inhibited the growth of aortic ring microvessels in mice in vitro. In conclusion, hesperidin inhibits endothelial angiogenesis by inhibiting PI3K/AKT, ERK and P38 MAPK signaling pathways [35]. In vivo studies further showed that isoliquiritigenin can inhibit the growth and angiogenesis of breast cancer, and inhibit VEGF/VEGFR-2 signal transduction, and increase the rate of apoptosis with little toxicity [36]. Isoliquiritigenin inhibited the expression of VCAM-1, e-selectin and PECAM-1 proteins induced by pro-inflammatory cytokines at the transcriptional level and weakened the interaction between THP-1 monocytes and endothelial cells. In this study, the decrease of VCAM-1 suggested the effect of isoliquiritigenin on endothelial function [37].

It is becoming evident that low-dose liquiritigenin can clear endocellular ROS, control or repress ROS interceded flag pathway, and ensure an arrangement of cells such as neurons, HUVECs, macrophages, hippocampal neurons, osteoblasts and hepatocytes [38]. Low-dose liquiritigenin antagonized HUVECs by inhibiting ROS mediated signal pathway [39]. HMEC-1 cells were cultivated in Matrigel for 24 h to observe the tube-formation. They were suddenly vascularized beneath Matrigel condition. The quercetin overwhelmingly quelled cells, practicality, relocation, VEGF expression and encouraged apoptosis in HMEC-1 cells. Moreover, concealment of miR-216a was found in HMEC-1 cells after quercetin incitement, in the interim miR-216a over expression abrogated the capacities of quercetin in HMEC-1 cells. The quercetin also deactivated PI3K/AKT and JAK/STAT pathways through altering miR-216a [40].

In conclusion, some chemical ingredients can act on the same pathway and target to treat diseases, but they can also act on different pathway targets. The ensemble curative effect of TCM is the outcome of the comprehensive action of the compounds. The active constituents of WDD are special compounds with multi-component, multiple targets and comprehensive regulation. Compared to a single-target strategy, the focal points of multi-objective strategy are becoming self-evident. In order to find out the dynamic substances of WDD, it was vital to understand the pharmacodynamics of its potential dynamic compositions. Therefore, we deduced that these active constituents are part of the effective constituents of WDD. The technique is highly efficient, high-speed, and economical. The quality marker (Q-marker) is the standard to evaluate the quality of traditional Chinese medicine compound, which should not be based on the ingredients of a few traditional Chinese medicines but should reflect the situation of the whole compound [41]. These ingredients can be used as reference standards for quality evaluation of WDD.

Limitations

The research still has some shortcomings, such as small selection of the components. It still cannot fully represent the efficacy of WDD.

4. Materials and Methods

4.1. Materials

The 16 batches, comprising of WDD counting of BX, GC, ZS, CP, SJ, ZR were included in this research. These 16 batches of Chinese homegrown drugs come from different pharmaceutical markets of China. All homegrown solutions were distinguished by Teacher Jin Guangqian, TCM recognizable proof master of Shandong Institute of TCM. All the herbs met the necessities of Ch. P (2020 Version) as shown in Table 6.

Table 6. Details of 16 batches of WDD.

Samples	BX	ZR	ZS	CP	GC
S1	Shandong	Guangdong	Jiangxi	Sichuan	Baotou, Inner Mongolia
S2	Shandong	Zhejiang	Chongqing	Fujian	Chifeng, Inner Mongolia
S3	Shandong	Hubei	Sichuan	Guangdong	Ordos, Inner Mongolia
S4	Shandong	Sichuan	Hunan	Hunan	Gansu
S5	Sichuan	Guangdong	Chongqing	Guangdong	Gansu
S6	Sichuan	Zhejiang	Jiangxi	Hunan	Ordos, Inner Mongolia
S7	Sichuan	Hubei	Hunan	Sichuan	Chifeng, Inner Mongolia
S8	Sichuan	Sichuan	Sichuan	Fujian	Baotou, Inner Mongolia
S9	Gansu	Guangdong	Sichuan	Hunan	Chifeng, Inner Mongolia
S10	Gansu	Zhejiang	Hunan	Guangdong	Baotou, Inner Mongolia
S11	Gansu	Hubei	Jiangxi	Fujian	Gansu
S12	Gansu	Sichuan	Chongqing	Sichuan	Ordos, Inner Mongolia
S13	Guangxi	Guangdong	Hunan	Fujian	Ordos, Inner Mongolia
S14	Guangxi	Zhejiang	Sichuan	Sichuan	Gansu
S15	Guangxi	Hubei	Chongqing	Hunan	Baotou, Inner Mongolia
S16	Guangxi	Sichuan	Jiangxi	Guangdong	Chifeng, Inner Mongolia

WDD = Wendan decoction; BX = Breit; ZR = Stapf ex Rendle; ZS = Citrus aurantium L.; CP = Citrus reticulata Blanco; GC = Glycyrrhiza uralensis Fisch.

Hesperidin (wkq20030407), tangeretin (wkq20041611), naringin (wkq21020606), nobiletin (wkq18020111), isoliquiritigenin (wkq18033006), formononetin (wkq18022712), naringenin (wkq21022409), glycyrrhetinic acid (wkq16070701) and glycyrrhizic acid (wkq16032502) were purchased from Sichuan Weikeqi Biotechnology Co., LTD., China. Daidzein (PS000251), isoquercitrin (PS001042), hesperetin (PS000219), trigonelline (PS000427), morin (PS020344), scopoletin (PS010525), liquiritigenin (PS010083) and liquiritin (PS012028) were provided by Chengdu Pusi Biotechnology Co., LTD., Sichuan Province, China. Obacunone (111923-201102) and furosemide (internal standard, IS) (100544-201503) were obtained from National Institutes for Food and Drug Control. Quercetin (100081-200406) and synephrine (0727-200105) were obtained by China National Institute for the Control of Pharmaceutical and Biological Products. All the reference substances were HPLC $\geq$ 98%. Acetonitrile for UPLC-QQQ-MS/MS was acquired from Merck KGaA (Darmstadt, Germany).

HUVEC cell line was purchased from Wuxi Bohe Biomedical Technology Co., LTD. (Jiangsu, China). RPMI-1640 basal medium was obtained from Hyclone Laboratory Media (Logan, UT, USA). Pancreatin was obtained from Shanghai Beyotime Biotechnology Co., LTD, China. The detection kit of MTT was bought from Sigma Company, Saint Louis, MO, USA. Penicillin-streptomycin was purchased from biological sharp, from Shanghai, China. Dimethyl sulfoxide (DMSO) was obtained from Invitrogen (Carlsbad, CA, USA).

4.2. Content Determination of Components in WDD

The preparation of medicinal materials was based on previous studies [42]. It was prepared according to the compatibility of medicinal materials and previous method [43]. Briefly, 10.0 g of BX, 10.0 g of ZS, 10.0 g of ZR, 15.0 g of CP, 20.0 g of SJ, and 5.0 g of GC were included to 840 mL of immaculate water and drenched for 30 min. The blend was decocted on high flame until bubbling, and after that decocted on low flame for 2 h. The above operations were repeated to obtain WDD solution and then concentrated with a rotary evaporator (Temperature: 60 °C; Rotational speed: 40–80 rpm). After merging and filtering, the supernatant was collected for concentration by Vacuum Freezing and then dehydrated in a drying oven at −50 °C for 72 h. This produced a dry powder. It was kept in a desiccator before use, and other batches of WDD powder were prepared using the same method.

The freeze-dried powder of WDD extricate (0.5000 g) was precisely weighed and ultrasonically broken up in methanol (30 min, at 25 °C). The arrangements were blended and balanced to a volume of 5 mL and diluted to the proper consistence. The test arrangements were sifted through a 0.22 μm channel film and the filtrate was stored at 4 °C for later UPLC-QQQ-MS/MS examination.

4.2.1. Preparation of Standard Solution

All standard solutions were weighed accurately and diluted in methanol to form a reserve solution, which was stored at 4 °C. All reserve fluids were diluted to the desired concentration and then mixed immediately prior to analysis.

4.2.2. UPLC-QQQ-MS/MS Conditions

LC–MS/MS investigations were carried out employing a Thermo Scientific Vanquish LC system connected to a TSQ Quantum Ultra triple quadrupole mass spectrometer (Thermo Scientific, Orlando, FL, USA) equipped with a heated electrospray ionisation source (H-ESI). Xcalibur 4.4 Software (Thermo Scientific) was used for method setup, data processing and reporting. The mass spectrometer was operated with a H-ESI interface handled in both positive and negative ionization modes and was utilized for the multiple reaction monitoring. Instrument settings were: positive ion: 3600 v, negative ion: 3400 v, sheath gas 0.3 mL/min, aux gas 0.3 mL/min, ion transfer tube temp 345 °C, vaporizer temp 350 °C.

Chromatographic fractionation was accomplished on a Thermo Hypersil GOLD C18 column (1.9 μm, 100 × 2.1 mm) at 30 °C, and versatile stage comprised 0.1% formic corrosive and water for dissolvable A and acetonitrile for dissolvable B at a stream rate of 0.3 mL/min. The gradient elution conditions of mobile phase B are as follows: 0–0.5 min: 5%, 0.5–2 min: 5–8%, 2–2.1 min: 40%, 2.1–4 min: 40–50%, 4–6 min: 50–60%, 6–6.1 min: 60–70%, 6.1–8 min: 70–80%, 8–8.1 min: 80–5%, 8.1–10 min: 5%. The sample size for analysis was 3 μL.

4.2.3. Methodology Validation

The standard curve was established in the method validation part with internal standard method. The precision, stability, repeatability and recovery were investigated. The precision was calculated by 6 consecutive injections of the same sample solution, and repeatability was evaluated by six samples from the same source. The sample solution was put into the sample bottle at room temperature and injected for analysis at different times (0, 4, 8, 12, 16, 24 h) after preparation to observe the sample stability at room temperature. The measured WDD content was precisely weighed and added equal to 100% of the measured reference substance in WDD to determine the recovery of the sample.

4.3. Cell Experiments

4.3.1. Screening of Palmitic Acid (PA) Concentration by MTT

The HUVECs were stored in a RPMI-1640 basal medium containing 10% FBS and other essential development variables in a humidified incubator with the climate containing 5% CO_2 at 37 °C. The PA was broken down in DMSO (200 mM) as a stock solution and after that diluted with a medium containing RPMI-1640 basal medium with 10% FBS for later use (A: 0, B: 50 μmol/L, C: 100 μmol/L, D: 150 μmol/L, E: 200 μmol/L, F: 250 μmol/L, G: 300 μmol/L). The control treatment: cells were cultured with RPMI-1640 basal medium with 10% FBS. The experimental group added different concentrations of PA with the climate containing 5% CO_2 at 37 °C. All the treatments were performed in triplicate.

According to the above grouping, 10 μL of MTT was included in each well with the air brake containing 5% CO_2 at 37 °C for 3–4 h in a dark incubator. After the liquid absorption, 150 μL DMSO was added and shaken at the shaking table at room temperature for 10 min. Finally, the optical density (OD) value at the same time point was calibrated at 492 nm wavelength by enzyme labeling instrument, and the measured OD value was used for cell viability analysis.

4.3.2. Cell Protective Effect of WDD

The 16 batches of WDD powder were absolutely weighed. The volume was settled to 1 mL by PBS, and the microscopic organisms were sifted by 0.22 μm channel layer. Later the cells were included, and each batch of concentrated arrangement was weakened to 5 concentrations of 0.25 mg/mL, 0.5 mg/mL, 1 mg/mL, 2 mg/mL, 4 mg/mL by RPMI-1640 basal medium with 10% FBS. They were isolated into control group and the WDD treatment group. The cells treatment strategy was rehashed. All groups were in triplicate.

4.4. Statistical Analysis

4.4.1. GRA

GRA's significant concept is to select whether the filiation is close by characterising the geometric resemblance between the reference data course of action and a couple of comparison data columns. It mirrors the relationship degree between bends [44]. The data reflecting the pharmacodynamic behavior of WDD system were taken as the reference series, and the quantitative data of each component that might affect the pharmacodynamic behavior of WDD system were taken as the comparison series. GRA program (Grey Modeling_V3.0) was utilized to analyze, and the related coefficients between different components and in vitro pharmacodynamics evaluation records were obtained, which were sorted according to GRA the calculation condition.

4.4.2. PLSR

PLSR could be a measurement strategy which brings the calculated variable amount and watched variable into an unused room to look out for a direct relapse demonstrate [45]. To encourage the unequivocal reliance between constituents and healing impact of WDD, the substance of 20 fundamental compositions in this study were contention and ampleness data were subordinate components. PLSR was obtained to analyze the information of 1 mg/mL and the regression coefficient.

5. Conclusions

A fast, sensitive and dependable LC-/MSMS strategy, combined with MTT and scientific factual examination strategy, was set up for the screening of numerous dynamic fixings of WDD with AVECI action. The results show that the extracts of WDD significantly increased the proliferation of endothelial cells and protected endothelial cells. However, the interaction between them can be additive, synergistic and antagonistic, and the mechanism remains unclear. More appropriate models should be included in the follow-up to conduct a comprehensive pharmacodynamics evaluation of WDD, laying a foundation for accurately and comprehensively revealing the active component group of WDD, so as to provide the basis for the rational use of WDD.

Author Contributions: N.X., M.I., H.S., M.C., M.S., P.W., X.G., L.M. were responsible for Conceptualization, Study design, Project development, N.X., M.I., M.C., and L.M. performed Data acquisition, analysis and interpretation of data, N.X., M.I., H.S., M.S., P.W. and L.M. wrote the main manuscript, edited and revised the manuscript. All authors have read and agreed to the published version of the manuscript.

Funding: This study was supported by the following grants: Natural Science Foundation of Shandong Province, No. ZR2020KH017; Shandong Traditional Chinese Medicine Science and Technology Project, No. 2021Z048,Q-2022084; National Natural Science Foundation of China, No. 82074052; Collaborative

Innovation Center project of TCM quality control and whole industry chain construction in colleges and universities of Shandong Province No. CYLXTCX2021-02.

Institutional Review Board Statement: Not applicable.

Informed Consent Statement: Not applicable. This study was not involving humans or animals.

Data Availability Statement: The datasets during and/or analysed during the current study available from the corresponding author on reasonable request.

Acknowledgments: The authors wish to thank Zeeshan Farhaj for his assistance in proofreading the manuscript.

Conflicts of Interest: The authors declare that they have no competing interests.

References

1. Kumar, S.; Mittal, A.; Babu, D.; Mittal, A. Herbal Medicines for Diabetes Management and its Secondary Complications. *Curr. Diabetes Rev.* **2021**, *17*, 437–456. [CrossRef] [PubMed]
2. Boudina, S.; Sena, S.; Sloan, C.; Tebbi, A.; Han, Y.H.; O'Neill, B.T.; Cooksey, R.C.; Jones, D.; Holland, W.L.; McClain, D.A.; et al. Early mitochondrial adaptations in skeletal muscle to diet-induced obesity are strain dependent and determine oxidative stress and energy expenditure but not insulin sensitivity. *Endocrinology* **2012**, *153*, 2677–2688. [CrossRef] [PubMed]
3. Poredos, P.; Jezovnik, M.K. Testing endothelial function and its clinical relevance. *J. Atheroscler. Thromb.* **2013**, *20*, 1–8. [CrossRef] [PubMed]
4. Villalta, S.A.; Lang, J.; Kubeck, S.; Kabre, B.; Szot, G.L.; Calderon, B.; Wasserfall, C.; Atkinson, M.A.; Brekken, R.A.; Pullen, N.; et al. Inhibition of VEGFR-2 reverses type 1 diabetes in NOD mice by abrogating insulitis and restoring islet function. *Diabetes* **2013**, *62*, 2870–2878. [CrossRef] [PubMed]
5. Engin, A. Endothelial Dysfunction in Obesity. *Adv. Exp. Med. Biol.* **2017**, *960*, 345–379. [CrossRef] [PubMed]
6. Taylor, F.C.; Dunstan, D.W.; Fletcher, E.; Townsend, M.K.; Larsen, R.N.; Rickards, K.; Maniar, N.; Buman, M.; Dempsey, P.C.; Joham, A.E.; et al. Interrupting Prolonged Sitting and Endothelial Function in Polycystic Ovary Syndrome. *Med. Sci. Sport. Exerc.* **2021**, *53*, 479–486. [CrossRef]
7. Incalza, M.A.; D'Oria, R.; Natalicchio, A.; Perrini, S.; Laviola, L.; Giorgino, F. Oxidative stress and reactive oxygen species in endothelial dysfunction associated with cardiovascular and metabolic diseases. *Vasc. Pharmacol.* **2018**, *100*, 1–19. [CrossRef]
8. Zhou, W.; Wang, J.; Wu, Z.; Huang, C.; Lu, A.; Wang, Y. Systems pharmacology exploration of botanic drug pairs reveals the mechanism for treating different diseases. *Sci. Rep.* **2016**, *6*, 36985. [CrossRef]
9. Zhang, Y.; Zhang, Y.; Liang, J.; Kuang, H.X.; Xia, Y.G. Exploring the effects of different processing techniques on the composition and biological activity of Platycodon grandiflorus (Jacq.) A.DC. by metabonomics and pharmacologic design. *J. Ethnopharmacol.* **2022**, *289*, 114991. [CrossRef]
10. Yang, N.; Liu, F.L. Effect of Extracts from *Broussonetia papyrifera* (L.) Leaves on Blood Lipid and the Expression of Bax and Bcl-2 in Vascular Endothelial Cells in Rats with Hyperlipidemia. *Guangming J. Chin. Med.* **2019**, *34*, 1659–1661.
11. Yu, Y.; Yan, R.; Chen, X.; Sun, T.; Yan, J. Paeonol suppresses the effect of ox-LDL on mice vascular endothelial cells by regulating miR-338-3p/TET2 axis in atherosclerosis. *Mol. Cell. Biochem.* **2020**, *475*, 127–135. [CrossRef] [PubMed]
12. Su, J.; Zhou, H.; Tao, Y.; Guo, J.; Guo, Z.; Zhang, S.; Zhang, Y.; Huang, Y.; Tang, Y.; Dong, Q.; et al. G-CSF protects human brain vascular endothelial cells injury induced by high glucose, free fatty acids and hypoxia through MAPK and Akt signaling. *PLoS ONE* **2015**, *10*, e0120707. [CrossRef] [PubMed]
13. Chen, P.; Liu, H.; Xiang, H.; Zhou, J.; Zeng, Z.; Chen, R.; Zhao, S.; Xiao, J.; Shu, Z.; Chen, S.; et al. Palmitic acid-induced autophagy increases reactive oxygen species via the Ca^{2+}/PKCα/NOX_4 pathway and impairs endothelial function in human umbilical vein endothelial cells. *Exp. Ther. Med.* **2019**, *17*, 2425–2432. [CrossRef] [PubMed]
14. Li, Y.B.; Zhang, W.H.; Liu, H.D.; Liu, Z.; Ma, S.P. Protective effects of Huanglian Wendan Decoction aganist cognitive deficits and neuronal damages in rats with diabetic encephalopathy by inhibiting the release of inflammatory cytokines and repairing insulin signaling pathway in hippocampus. *Chin. J. Nat. Med.* **2016**, *14*, 813–822. [CrossRef]
15. Huang, Y.M.; Xu, J.H.; Ling, W.; Li, Y.; Zhang, X.X.; Dai, Z.K.; Sui, Y.; Zhao, H.L. Efficacy of the wen dan decoction, a Chinese herbal formula, for metabolic syndrome. *Altern. Ther. Health Med.* **2015**, *21*, 54–67.
16. Zhang, X.; Wang, Y.; Liu, L.; Jiang, H.; Wang, J.; Xiao, Y.; Wang, J. Efficacy of Wen-Dan Decoction in the treatment of patients with coronary heart disease: A protocol for systematic review and meta-analysis. *Medicine* **2022**, *101*, e28041. [CrossRef]
17. Willson, C.; Watanabe, M.; Tsuji-Hosokawa, A.; Makino, A. Pulmonary vascular dysfunction in metabolic syndrome. *J. Physiol.* **2019**, *597*, 1121–1141. [CrossRef]
18. Murni, I.K.; Sulistyoningrum, D.C.; Susilowati, R.; Julia, M. Risk of metabolic syndrome and early vascular markers for atherosclerosis in obese Indonesian adolescents. *Paediatr. Int. Child Health* **2020**, *40*, 117–123. [CrossRef]
19. Hong, F.; Yan, Y.; Zhao, L.; Yu, B.Y.; Chai, C.Z. Analysis on ingredients and metabolites of Gegen Decoction absorbed into blood based on UHPLC-Q-TOF-MS. *China J. Chin. Mater. Med.* **2021**, *46*, 5944–5952. [CrossRef]

20. Kumar, R.; Akhtar, F.; Rizvi, S.I. Protective effect of hesperidin in Poloxamer-407 induced hyperlipidemic experimental rats. *Biol. Futur.* **2021**, *72*, 201–210. [CrossRef]
21. Wang, F.; Zhao, C.; Tian, G.; Wei, X.; Ma, Z.; Cui, J.; Wei, R.; Bao, Y.; Kong, W.; Zheng, J. Naringin Alleviates Atherosclerosis in $ApoE^{-/-}$ Mice by Regulating Cholesterol Metabolism Involved in Gut Microbiota Remodeling. *J. Agric. Food Chem.* **2020**, *68*, 12651–12660. [CrossRef] [PubMed]
22. George, M.Y.; Menze, E.T.; Esmat, A.; Tadros, M.G.; El-Demerdash, E. Naringin treatment improved main clozapine-induced adverse effects in rats; emphasis on weight gain, metabolic abnormalities, and agranulocytosis. *Drug Dev. Res.* **2021**, *82*, 980–989. [CrossRef] [PubMed]
23. Wang, Y.; Zhao, H.; Li, X.; Wang, Q.; Yan, M.; Zhang, H.; Zhao, T.; Zhang, N.; Zhang, P.; Peng, L.; et al. Formononetin alleviates hepatic steatosis by facilitating TFEB-mediated lysosome biogenesis and lipophagy. *J. Nutr. Biochem.* **2019**, *73*, 108214. [CrossRef] [PubMed]
24. Qiu, M.; Huang, K.; Liu, Y.; Yang, Y.; Tang, H.; Liu, X.; Wang, C.; Chen, H.; Xiong, Y.; Zhang, J.; et al. Modulation of intestinal microbiota by glycyrrhizic acid prevents high-fat diet-enhanced pre-metastatic niche formation and metastasis. *Mucosal Immunol.* **2019**, *12*, 945–957. [CrossRef] [PubMed]
25. Liu, Y.; Yang, S.; Wang, K.; Lu, J.; Bao, X.; Wang, R.; Qiu, Y.; Wang, T.; Yu, H. Cellular senescence and cancer: Focusing on traditional Chinese medicine and natural products. *Cell Prolif.* **2020**, *53*, e12894. [CrossRef]
26. Liu, Z.; Guo, F.; Wang, Y.; Li, C.; Zhang, X.; Li, H.; Diao, L.; Gu, J.; Wang, W.; Li, D.; et al. BATMAN-TCM: A Bioinformatics Analysis Tool for Molecular mechANism of Traditional Chinese Medicine. *Sci. Rep.* **2016**, *6*, 21146. [CrossRef]
27. Wang, Z.Y.; Yang, S.Y.; Lyu, W.; Ding, J.B.; Ma, J.L.; Yang, J. Analysis of spectrum-activity relationship among antioxidant parts of Lycii Fructus using three different chemometrics methods. *China J. Chin. Mater. Med.* **2021**, *46*, 3377–3387. [CrossRef]
28. Li, S.; Xu, T.; Liu, S.; Liu, Z.; Pi, Z.; Song, F.; Jin, Y. Exploring the potential pharmacodynamic material basis and pharmacologic mechanism of the Fufang-Xialian-Capsule in chronic atrophic gastritis by network pharmacology approach based on the components absorbed into the blood. *R. Soc. Open Sci.* **2018**, *5*, 171806. [CrossRef]
29. Wang, Y.L.; Liu, Y.T.; Tao, Y.Y.; Liu, C.H.; Qu, H.Y.; Zhou, H.; Yang, T. Research ideas and method on screening active components of traditional Chinese medicine against hepatotoxicity with mitochondria as target. *China J. Chin. Mater. Med.* **2021**, *46*, 306–311. [CrossRef]
30. Tian, X.; Wang, P.; Li, T.; Huang, X.; Guo, W.; Yang, Y.; Yan, M.; Zhang, H.; Cai, D.; Jia, X.; et al. Self-assembled natural phytochemicals for synergistically antibacterial application from the enlightenment of traditional Chinese medicine combination. *Acta Pharm. Sin. B* **2020**, *10*, 1784–1795. [CrossRef]
31. Amtaghri, S.; Akdad, M.; Slaoui, M.; Eddouks, M. Traditional Uses, Pharmacological, and Phytochemical Studies of Euphorbia: A Review. *Curr. Top. Med. Chem.* **2022**, *22*, 1553–1570. [CrossRef] [PubMed]
32. Wang, K.X.; Du, G.H.; Qin, X.M.; Gao, L. 1H-NMR-based metabolomics reveals the biomarker panel and molecular mechanism of hepatocellular carcinoma progression. *Anal. Bioanal. Chem.* **2022**, *414*, 1525–1537. [CrossRef] [PubMed]
33. Wei, J.; Lu, Y.; Wang, R.; Xu, X.; Liu, Q.; He, S.; Pan, H.; Liu, X.; Yuan, B.; Ding, Y.; et al. MicroRNA-375: Potential cancer suppressor and therapeutic drug. *Biosci. Rep.* **2021**, *41*, BSR20211494. [CrossRef]
34. Chen, J.; Zhang, X.; Wang, Y.; Ye, Y.; Huang, Z. Differential ability of formononetin to stimulate proliferation of endothelial cells and breast cancer cells via a feedback loop involving MicroRNA-375, RASD1, and ERα. *Mol. Carcinog.* **2018**, *57*, 817–830. [CrossRef]
35. Kim, G.D. Hesperetin Inhibits Vascular Formation by Suppressing of the PI3K/AKT, ERK, and p38 MAPK Signaling Pathways. *Prev. Nutr. Food Sci.* **2014**, *19*, 299–306. [CrossRef]
36. Wang, C.; Chen, Y.; Wang, Y.; Liu, X.; Liu, Y.; Li, Y.; Chen, H.; Fan, C.; Wu, D.; Yang, J. Inhibition of COX-2, mPGES-1 and CYP4A by isoliquiritigenin blocks the angiogenic Akt signaling in glioma through ceRNA effect of miR-194-5p and lncRNA NEAT1. *J. Exp. Clin. Cancer Res. CR* **2019**, *38*, 371. [CrossRef] [PubMed]
37. Park, S.M.; Ki, S.H.; Han, N.R.; Cho, I.J.; Ku, S.K.; Kim, S.C.; Zhao, R.J.; Kim, Y.W. Tacrine, an oral acetylcholinesterase inhibitor, induced hepatic oxidative damage, which was blocked by liquiritigenin through GSK3-beta inhibition. *Biol. Pharm. Bull.* **2015**, *38*, 184–192. [CrossRef]
38. Zeng, J.; Chen, Y.; Ding, R.; Feng, L.; Fu, Z.; Yang, S.; Deng, X.; Xie, Z.; Zheng, S. Isoliquiritigenin alleviates early brain injury after experimental intracerebral hemorrhage via suppressing ROS- and/or NF-κB-mediated NLRP3 inflammasome activation by promoting Nrf2 antioxidant pathway. *J. Neuroinflammation* **2017**, *14*, 119. [CrossRef]
39. Shi, C.; Wu, H.; Xu, K.; Cai, T.; Qin, K.; Wu, L.; Cai, B. Liquiritigenin-Loaded Submicron Emulsion Protects Against Doxorubicin-Induced Cardiotoxicity via Antioxidant, Anti-Inflammatory, and Anti-Apoptotic Activity. *Int. J. Nanomed.* **2020**, *15*, 1101–1115. [CrossRef]
40. Wang, X.; Xue, X.; Wang, H.; Xu, F.; Xin, Z.; Wang, K.; Cui, M.; Qin, W. Quercetin inhibits human microvascular endothelial cells viability, migration and tube-formation in vitro through restraining microRNA-216a. *J. Drug Target.* **2020**, *28*, 609–616. [CrossRef]
41. Xiong, Y.; Hu, Y.; Li, F.; Chen, L.; Dong, Q.; Wang, J.; Gullen, E.A.; Cheng, Y.C.; Xiao, X. Promotion of quality standard of Chinese herbal medicine by the integrated and efficacy-oriented quality marker of Effect-constituent Index. *Phytomed. Int. J. Phytother. Phytopharm.* **2018**, *45*, 26–35. [CrossRef] [PubMed]
42. Chen, R.S.; Li, Y.; Lu, Y.; Li, L.J. A textual research on the key information of the famous classical formula Wendan Decoction based on ancient bibliometric analysis. *J. Nanjing Univ. TCM* **2021**, *37*, 439–445.

43. Zhou, L.; Shi, H.Y.; Shi, Y.P.; Wang, P.; Xu, N. Optimization of extraction technology of Classical Mingfang Wendan decoction by fingerprint combined with chemometrics. *Chin. J. Hosp. Pharm.* **2020**, *40*, 2214–2219.
44. Rehman, E.; Ikram, M.; Feng, M.T.; Rehman, S. Sectoral-based CO2 emissions of Pakistan: A novel Grey Relation Analysis (GRA) approach. *Environ. Sci. Pollut. Res. Int.* **2020**, *27*, 29118–29129. [CrossRef] [PubMed]
45. Foodeh, R.; Ebadollahi, S.; Daliri, M.R. Regularized Partial Least Square Regression for Continuous Decoding in Brain-Computer Interfaces. *Neuroinformatics* **2020**, *18*, 465–477. [CrossRef] [PubMed]

Disclaimer/Publisher's Note: The statements, opinions and data contained in all publications are solely those of the individual author(s) and contributor(s) and not of MDPI and/or the editor(s). MDPI and/or the editor(s) disclaim responsibility for any injury to people or property resulting from any ideas, methods, instructions or products referred to in the content.

molecules

MDPI

Review

Long Non-Coding RNAs as Novel Targets for Phytochemicals to Cease Cancer Metastasis

Sadegh Rajabi [1,*], Huda Fatima Rajani [2], Niloufar Mohammadkhani [3], Andrés Alexis Ramírez-Coronel [4,5,6,7], Mahsa Maleki [8], Marc Maresca [9,*] and Homa Hajimehdipoor [10,*]

1 Traditional Medicine and Materia Medica Research Center, Shahid Beheshti University of Medical Sciences, Tehran 1434875451, Iran
2 Department of Immunology, Max Rady College of Medicine, University of Manitoba, Winnipeg, MB R3E0T5, Canada
3 Department of Clinical Biochemistry, School of Medicine, Shahid Beheshti University of Medical Sciences, Tehran 1985717443, Iran
4 Research Group in Educational Statistics, National University of Education (UNAE), Azogues 030101, Ecuador
5 Azogues Campus Nursing Career, Health and Behavior Research Group (HBR), Psychometry and Ethology Laboratory, Catholic University of Cuenca, Cuenca 010107, Ecuador
6 Psychology Department, University of Palermo, Buenos Aires 1175ABT, Argentina
7 Epidemiology and Biostatistics Research Group, CES University, Medellín 050022, Colombia
8 Department of Molecular Genetics, Faculty of Biological Sciences, Tarbiat Modares University, Tehran 331-14115, Iran
9 Aix Marseille University, CNRS, Centrale Marseille, iSm2, 13397 Marseille, France
10 Traditional Medicine and Materia Medica Research Center and Department of Traditional Pharmacy, School of Traditional Medicine, Shahid Beheshti University of Medical Sciences, Tehran 1516745811, Iran
* Correspondence: sadegh.rajabi2017@gmail.com (S.R.); m.maresca@univ-amu.fr (M.M.); hajimehd@sbmu.ac.ir (H.H.)

Abstract: Metastasis is a multi-step phenomenon during cancer development leading to the propagation of cancer cells to distant organ(s). According to estimations, metastasis results in over 90% of cancer-associated death around the globe. Long non-coding RNAs (LncRNAs) are a group of regulatory RNA molecules more than 200 base pairs in length. The main regulatory activity of these molecules is the modulation of gene expression. They have been reported to affect different stages of cancer development including proliferation, apoptosis, migration, invasion, and metastasis. An increasing number of medical data reports indicate the probable function of LncRNAs in the metastatic spread of different cancers. Phytochemical compounds, as the bioactive agents of plants, show several health benefits with a variety of biological activities. Several phytochemicals have been demonstrated to target LncRNAs to defeat cancer. This review article briefly describes the metastasis steps, summarizes data on some well-established LncRNAs with a role in metastasis, and identifies the phytochemicals with an ability to suppress cancer metastasis by targeting LncRNAs.

Keywords: long non-coding RNA; LncRNAs; phytochemical; cancer; metastasis; therapy

Citation: Rajabi, S.; Rajani, H.F.; Mohammadkhani, N.; Ramírez-Coronel, A.A.; Maleki, M.; Maresca, M.; Hajimehdipoor, H. Long Non-Coding RNAs as Novel Targets for Phytochemicals to Cease Cancer Metastasis. *Molecules* **2023**, *28*, 987. https://doi.org/10.3390/molecules28030987

Academic Editors: Arunaksharan Narayanankutty and Zhaojun Wei

Received: 23 November 2022
Revised: 31 December 2022
Accepted: 11 January 2023
Published: 18 January 2023

Copyright: © 2023 by the authors. Licensee MDPI, Basel, Switzerland. This article is an open access article distributed under the terms and conditions of the Creative Commons Attribution (CC BY) license (https://creativecommons.org/licenses/by/4.0/).

1. Introduction

Cancer metastasis is a multi-step process that results in the spread of cancer cells to distant tissues and organs beyond the primary site [1]. This phenomenon is responsible for more than 90% of cancer-related mortality in the world [2]. According to estimations, about 50% of cancer patients already have clinically detectable metastases at the time of initial diagnosis. However, metastasis is most often due to the recurrence of the disease mainly after definitive treatment [3].

Long non-coding RNAs (LncRNAs) are members of the non-coding RNAs family that are more than 200 base pairs in length. These molecules are transcribed from various regions of the genome such as introns, exons, and intergenic connections. To date, nearly

30,000 LncRNAs have been recognized in humans and mice, but the function of only some of them has been recognized [4]. The main role of these RNA molecules is the regulation of gene expression by acting on a variety of intracellular processes from transcription to translation and interfering with signaling pathways [5]. As LncRNAs regulate gene expression, they have critical effects on the proliferation, apoptosis, migration, invasion, and metastasis of cancer cells [6]. A growing body of evidence suggests the potential role of LncRNAs in different steps tumor metastasis, which is the critical stage in cancer progression leading to decreased patient survival [7]. Therefore, targeting these novel and important molecules in cancers may enable the development of treatment methods for this disease.

Phytochemicals are natural bioactive ingredients of a variety of plants with beneficial health effects beyond basic nutrition [8]. They exhibit a number of desirable biological activities including anti-cancer, anti-inflammatory, anti-oxidant, and antimicrobial effects in vitro and in vivo [9]. Phytochemicals exert anti-cancer effects through different mechanisms. They induce cell death in cancer cells, target specific molecules in cellular pathways, modulate oxidative stress, and prevent tumors angiogenesis, which hinders metastasis [10]. A variety of phytochemicals have been shown to inhibit the metastatic propagation of cancer cells via several mechanisms. For example, curcumin is a polyphenol derived from *Curcuma longa* with ability to hamper metastasis of cancer cells by inhibiting transcription factors, cell adhesion molecules, cell surface markers, and epithelial-mesenchymal transition (EMT) [11]. As before mentioned, LncRNAs have a key role in the metastatic spread of cancer cells [7]. Therefore, targeting these regulatory RNA molecules by phytochemicals is of great importance in the treatment of cancer. This review paper summarizes the metastasis process and the well-established LncRNAs involved in this process. It particularly provides a list of phytochemicals that have been used to target LncRNAs for cancer therapy.

2. Metastasis and Its Compartments

Metastasis formation involves different steps that are summarized in Figure 1.

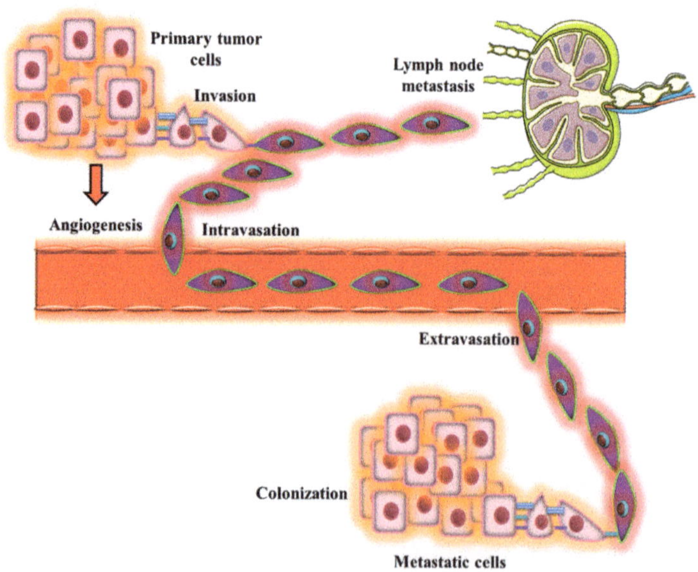

Figure 1. Different steps of cancer cell metastasis. Primary tumor cells invade the adjacent tissue to migrate away from the primary site. Afterward, they intravasate into blood vessels or enter lymph nodes and get transported to the other organs. Circulating tumor cells then extravasate into the secondary organ and start to proliferate to form a colony in the metastatic site.

2.1. Invasion

The invasion of tumor cells occurs due to the loss of attachment and polarity, modification of extracellular matrix (ECM), and alteration of migratory characteristics in these cells. Before metastasis, tumor cells are located in the luminal space, which is different from the stromal compartment [12]. The metastatic tumor cells undergo the EMT process to be ready for invasion and detachment from the primary tumor. This process is achieved by activating a number of cell signaling pathways that are triggered by the transforming growth factor (TGF-β), Wnt ligand, and tyrosine kinase receptors. Genetic alterations also support the activation of EMT-related transcription factors involved in the modulation of cell adhesion and polarity. The lncRNAs affect the expression of these transcription factors [13]. Post-transcriptional modifications and splicing also facilitate this process [14,15]. For example, loss of expression or downregulation of E-cadherin, a cell-cell adhesion protein, is associated with the recruitment and accumulation of circulating tumor cells (CTC) and their intravasation into the blood vessels [16]. Expression of E-cadherin is suppressed by the action of EMT-related factors such as Snail, Slug, Zeb1, and Zeb2, which bind to the promoter region of E-cadherin and lead to the formation of repressive chromatin structure [17]. Mesenchymal features of tumor cells are characterized by a bipolar structure with higher mobility, stemness, and invasiveness. Vimentin, epithelial cytokeratin 8, LC3B, alanine aminopeptidase, occludin, fibronectin, mastermind-like protein 1, myocardin-related transcription factors, some LncRNAs, and various signaling molecules are known to contribute to the EMT [18–24].

2.2. Angiogenesis

Tumor angiogenesis differs from normal angiogenesis in terms of endothelial cell mitogens and chemo-attractants. Tumor neovascularization is characterized by the invasion of the basement membrane toward the primary blood vessel resulting in vessel growth. With the growth of the tumor, primary tumor cells go farther from the blood vessels, the process that leads to hypoxia in the tumor tissue [25]. Hypoxia instigates the production of factors such as vascular endothelial growth factor (VEGF) and fibroblast growth factor (FGF), which promote angiogenesis. Production of some enzymes causes the degradation of the basement membrane of the capillary followed by migration and subsequent proliferation of epithelial cells. Overproduction of VEGF makes tumor vessels leaky and highly permeable leading to increased fluidity in the tumor microenvironment and interstitial pressure [26].

2.3. Intravasation

Intravasation is a crucial step in tumor propagation in which tumor cells penetrate vessel walls and enter circulation. This process leads to the dissemination of CTCs into circulation and their movement toward the metastatic site. The entry of tumor cells into lymph vessels is relatively easier than blood vessels because lymph vessels are devoid of the endothelial junction [27]. Through intravasation, tumor cells break through a dense ECM to enter the vessel [28]. Amoeboid intravasation is facilitated by the Rho/ROCK signaling pathway that leads to the formation of blebs. The production of VEGF increases the permeability of endothelial cells for tumor cells [29]. Furthermore, the tumor microenvironment of metastasis plays an important role in the recruitment of tumor cells through chemotactic signals. It has been reported that the presence of $CD68^+$ macrophages and $CD31^+$ endothelial cells in the vicinity of breast cancer cells instigates hematogenous metastasis [30,31].

2.4. Tumor Cell in Circulation

Due to the loss of ECM adhesion, tumor cells encounter a great amount of stress in circulation. Tumor cells need to survive in circulation in order to extravasate and disseminate in a distant organ. These tumor cells form aggregates with blood cells in the circulation; however, in smaller vessels like capillaries, aggregates are modified into chains to allow the passage of the cluster. This helps tumor cells to overcome mechanical shear stress in the bloodstream [32,33]. The expression of mutated pannexin-1, a membrane channel, supports mechanical stress and inhibits the apoptosis of CTCs [34].

Loss of adhesion to the ECM, in an integrin-dependent way, can trigger anoikis, which is a programmed cell death induced by the detachment of metastatic cancer cell from the ECM. The tumor cells have to develop anoikis resistance to survive in circulation [35]. To escape anoikis, the tumor cells tend to activate several cell signaling pathways by a zinc-finger transcription factor, FAK, phosphatase and tensin homolog, tyrosine kinases, insulin-like growth factor, and PI3K/Akt [36,37]. Furthermore, these cells escape from immune cells in circulation by the secretion of immunoregulatory molecules that protects them from natural killer cells [38]. Vascular cell adhesion molecule 1 (VCAM1) and vascular adhesion protein 1 (VAP1) also recruit macrophages by expressing tissue factor, which results in blood clotting and facilitates tumor cell survival [39].

2.5. *Extravasation*

Following successful survival in circulation, tumor cells extravasate to a secondary organ. The process involves the adhesion of these cells to the endothelium of blood vessels, alteration in the endothelial barrier to cross it, and migration into underlying tissue. This can or cannot be preceded by the proliferation and differentiation of cancer cells in the blood vessels. Disruption of the endothelial cell-to-cell barrier is an important step in this process [40]. Tumor cells adhere to endothelium by producing several cell adhesion molecules like cadherins, selectins, and integrins [41–43]. Ligand-receptor interaction may also contribute to transendothelial migration. Homophilic interactions have been also reported to facilitate extravasation. Jouve et al. showed that expression of CD146 in melanoma and endothelial cells supports metastasis into the lungs, through increased production of VEGF-2 [44]. Hemodynamic shear stress increases the production of reactive oxidation species and extracellular signal-regulated kinases that promote the migration of tumor cells [45].

2.6. *Colonization*

After detachment from the primary tumor, metastatic cancer cells infiltrate and colonize different organs. The gap between infiltration and colonization is latency. The prolonged period of latency implies greater malignant evolution of disseminated tumor cells and/or their microenvironment before colonization [1]. When metastasis is fast, like in lung cancer and pancreatic adenocarcinoma, there is a little-to-no capacity for metastatic cells to evolve. Common organs for cancer cell colonization are the liver, brain, bone marrow, and lungs due to circulation patterns [1]. This organotropic feature of cancer cells is favored by the upregulation of cell adhesion molecules such as metadherin that can specifically bind to the pulmonary vasculature to help the CTC to enter the lung tissue [46]. Furthermore, sinusoids in the capillaries of bone marrow have fenestrated endothelia for the passage of blood cells. These structures allow the CTC to enter bone marrow [47]. It has been shown that the transcription factor SNAI2 in glioma gives the tumor cells an ability to metastasize into multi-organs [48]. Once metastatic tumor cells reach the target organ, they undergo a process called mesenchymal-to-endothelial transition (MET) for the localization and proliferation in the metastatic organ. Loss of mesenchymal phenotype gives macrometastatic colonies a capacity to overcome growth arrest during the EMT process. Several genes are known to be involved in the formation of metastatic colonies. For example, an inhibitor of DNA-binding (Id) renders tumor cells self-renewable properties and induces MET and pulmonary colonization in breast cancer cells [49]. Inhibition of the Paired Related Homeobox 1 (Prrx1) gene is also important for tumor cells to obtain stem cell characteristics and metastatic colonization [50].

3. The LncRNAs Involved in Metastasis

Figure 2 shows different LncRNA molecules that are implicated in the invasion, migration, and metastasis processes of a variety of cancers.

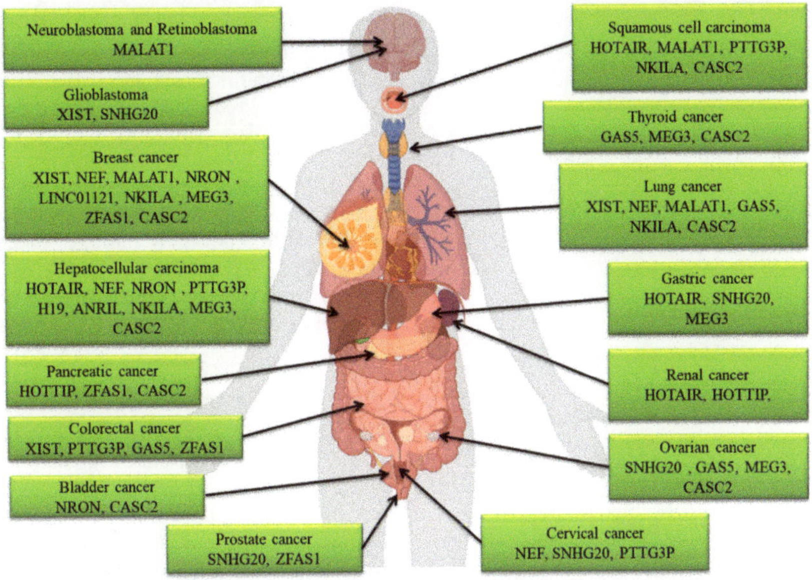

Figure 2. LncRNAs with a role in the invasion, migration, and metastasis of different types of cancers.

3.1. *ANRIL*

The *ANRIL* LncRNA was first reported in melanoma with a 403 kb deletion at the *CDKN2A/B* locus (9p21.3) [51]. Due to its important location, near *CDKN2A/B*, it has been numerously studied for the inheritance of several diseases. Tumor suppressor proteins, including p15, p16, and cyclin-dependent kinase, are encoded by *CDKN2A* and *CDKN2B*. These loci are silenced in nearly 40% of human cancers where *ANRIL* mediates oncogenic effects such as cell proliferation, adhesion, and metastasis. Increased expression of *ANRIL* is also associated with chemoresistance [52].

Hua et al. reported that high expression of *ANRIL* in human hepatocellular carcinoma (HCC) tissue is positively associated with histologic grade, cell proliferation, and poor survival rate [53]. Down-regulated *ANRIL* has an opposite effect and increases radiosensitivity and expression of miR-125a in nasopharyngeal carcinoma cells [54]. A recent study has shown that an increased expression of *ANRIL* in multiple myeloma inhibits bortezomib-induced apoptosis via *PTEN* promoter [55]. The lack of data on the role of *ANRIL* in metastatic characteristics of tumors urges more investigations to unravel its effects on this cancer hallmark.

3.2. *CASC2*

The LncRNA cancer susceptibility candidate 2 (*CASC2*) is a novel tumor suppressor with the ability to hamper invasion, migration, and metastasis in HCC cells by suppressing EMT [56]. In pancreatic cancer, *CASC2* is involved in the suppression of invasion and metastasis by upregulating *PTEN* and downregulating *miR-21* [57]. In breast cancer, *CASC2* suppresses cell proliferation and metastasis by targeting two different mechanisms, which involve the TGF-β signaling and *miR-96-5p/synoviolin* pathways [58,59]. Under-expression of *CASC2* is correlated with the serous histological subtype, lymph node metastasis, poor histological grade, and large tumor size in ovarian cancer samples [60]. The LncRNA *CASC2* acts as a tumor suppressor in esophageal squamous cell carcinoma by inhibiting proliferation, migration, and invasion in these cancer cells [61]. The role of *CASC2* as a tumor suppressor has also been established in several cancer types, including thyroid cancer, lung cancer, bladder cancer, osteosarcoma, and oral squamous cell carcinoma by suppressing the proliferation and metastasis [62–65].

3.3. *GAS5*

Growth arrest-specific 5 (*GAS5*) is a LncRNA with the ability to induce cell death by binding glucocorticoid receptors. In colorectal cancer cells, it binds *YAP* and *YTH* N6-Methyladenosine RNA Binding Protein 3 (*YTHDF3*) to inhibit cancer progression [66]. The *GAS5* promotes apoptosis in triple-negative breast cancer, which is highly metastatic breast cancer, by binding *miR-378a* [67]. This LncRNA also has anti-invasive effects on ovarian cancer by suppressing *miR-96-5p* and promoting the PTEN/mTOR signaling pathway [68]. In melanoma cancer cells, *GAS5* inhibits metastasis by reducing the expression of MMP-7 and 9, which are two important markers of cancer metastasis [69]. It also prohibits EMT in osteosarcoma cancer [70]. In a recent study, Xu et al. reported that reduced expression levels of GAS5 in papillary thyroid carcinoma decreased tumor cell growth, migration, and lymph node metastasis of cancer cells via the IFNγ/STAT1 signaling pathway [71].

3.4. *HOTAIR*

The HOX antisense intergenic RNA (*HOTAIR*) is a LncRNA that is transcribed from the antisense strand of *HOX* gene cluster with the ability to bind the chromatin modification complex. It regulates gene expression in a trans-regulatory fashion. Through enhancer of zeste homolog 2 (*EZH2*), lysine-specific histone demethylase 1A (*LSD1*), and polycomb repressive complex 2 (*PRC2*), *HOTAIR* silences gene expression and histone methylation. Its positive role in the promotion of metastasis, invasion, and tumor cell proliferation by epigenetic regulation of several metastatic genes and protein products has been extensively studied [72,73]. The *HOTAIR* is also upregulated in cancer-associated fibroblast (CAF) due to increased secretion of TGF-β1 [74]. The CAFs play an important role in metastasis, invasion, and drug resistance of different cancers [74].

The *HOTAIR* also suppresses miR-122 expression which instigates activation of cyclin G1 and subsequent cancerous response in HCC [75]. A decrease in the expression of *miR-122* is associated with the progression of HCC by targeting several genes involved in EMT and angiogenesis. These genes include cyclin G1, insulin-like growth factor-1, WNT1, pyruvate kinase M2, and A disintegrin and metalloprotease 10 (ADAM10) [76,77]. Drug resistance in HCC, which is characterized by the overexpression of TGF-β1, p glycoprotein, and breast cancer resistance protein, is associated with upregulation of *HOTAIR*. This is also associated with the promotion of metastasis in HCC cells [78].

A study by Yang et al. showed that *HOTAIR* mediates SNAP23 phosphorylation, activation of mammalian target of rapamycin (mTOR) signaling cascade, and secretion of exosomes in HCC [79]. Exosomes contain various mRNA, miRNA, LncRNA, and some other non-coding RNAs. Tumor cells use exosomes to help the spread and progression of the tumor [79]. The Collagen alpha-1(V) chain gene is upregulated during gastric cancer progression and immune infiltration. These are mediated by *HOATIR* overexpression and subsequent downregulation of miR-1277-5p [80]. Metastasis of squamous cell carcinoma is also supported by the upregulation of *HOTAIR*, which induces tumor invasion and stimulates EMT [81]. A recent study suggested that the knockdown of *HOTAIR* decreases angiogenesis, proliferation, and migration of renal carcinoma. The HOTAIR competitively binds *miR-126* and regulates the expression of epidermal growth factor-like domain multiple 7 (EGFLD7) and metastasis in these cells [82].

3.5. *HOTTIP*

The HOXA Distal Transcript Antisense RNA (*HOTTIP*) is a LncRNA located at the 5′ end of *HOXA* gene cluster with the ability to facilitate the transcription of these genes upon recruitment of WD repeat domain 5/ mixed lineage leukemia (WDR5/MLL). Activation of the *HOXA13* gene promotes tumorigenesis in the tissue by downregulating *miR-30b* [83,84]. Renal cancer is marked with increased expression of *HOTTIP*, which also is an indicator of poor prognoses such as metastasis, increased tumor size, vascular invasion, and reduced overall survival rate [85]. Furthermore, *HOTTIP* upregulates insulin-like growth factor-2 (IGF-2), which has a role in tumor progression [85]. In pancreatic cancer, *HOTTIP* is also upregulated and imposes tumorigenic effects by the promotion of cancer growth, proliferation, migration, and metastasis [86]. The *HOXA9* binds WDR5 and activates the Wnt/β-catenin pathway, which promotes cancer cell progression and the EMT process in pancreatic cancer cells. Stemness of pancreatic cancer cells is regulated by increased expression of *HOTTIP*, as a result of the production of stem cell factors such as NANOG, OCT4, and SOX2 [87]. The *HOTTIP* increases the resistance of pancreatic cancer cells to cisplatin by inhibiting *miR-137*, which increases the resistance of pancreatic cancer cells to cisplatin. Silencing of *HOTTIP* in these cells induces apoptosis and suppresses the growth and metastasis of pancreas tumor [88].

3.6. *H19*

The *H19* LncRNA, located on chromosome 11p15.5, is expressed in fetal and adult periods and is associated with the differentiation of skeletal muscle cells. Its expression is upregulated in hypoxic stress through the p53/HIF1-α signaling pathway. Furthermore, several oncogenes like *ZEB1, HER2, CALN1, MYC,* and *STAT3/EZH2/Catenin* are upregulated with the expression of *H19* LncRNA [89]. It also increases cell viability, motility, growth, migration, invasion, metastasis, EMT, autophagy, cell cycle progression, colony formation, and glucose metabolism [90,91]. It promotes the development of cancer-mediated chronic infection in HCC [92], contributes to EMT in papillary thyroid carcinoma [93], and increases estrogen-mediated cell survival and proliferation in breast cancer [94].

3.7. *LINC01121*

Long intergenic noncoding RNA 01121 (*LINC01121*) is expressed LncRNA with the ability to act as upstream regulator of SIX Homeobox 2 (*SIX2*) gene [95]. The *LINC01121* is substantially overexpressed in breast cancer cell lines compared with healthy breast epithelial cells [96]. Downregulation of *LINC01121* in breast tumors is associated with the inhibition of cell proliferation, cell cycle progression, migration, and invasion in breast cancer cells [96]. High-mobility group protein 2 (HMGA2) is a target gene of miR-150-5p and is significantly overexpressed in breast tumors [97]. This gene encodes a protein with the ability to enhance the proliferation and metastasis of breast cancer cells. The *miR-150-5p* contributes to the suppression of triple-negative breast cancer metastasis through impeding HMGA2 expression [98]. Further studies revealed that *LINC01121* could indirectly upregulate HMGA2 protein expression through the interaction with miR-150-5p [96].

3.8. *MALAT1*

Metastasis associated with lung adenocarcinoma transcript 1 (*MALAT1*), as the name indicates, was primarily known for its role in the survival rate of patients with non-small-cell lung cancer [99]. It is an excellent predictor of tumor invasion and progression [100]. This LncRNA is highly expressed in various cancer types and exerts its tumorigenic effects by blocking the PI3K/Akt pathway and increasing matrix metalloproteinase-9 (MMP-9) [101,102]. Moreover, in neuroblastoma and retinoblastoma cells, upregulation of MALAT1 activates mitogen-activated protein kinase (MAPK) along with the peroxisome proliferator-activated receptor (PPAR), P53-dependent signaling, and the Wnt/β-catenin pathway [103]. This is mediated by increased expression of miR-124 and subsequent activation of the mentioned pathways via slug knockdown [104]. Its expression in tumor tissues is a biomarker for the development and progression of cancer. For instance, it can help to determine the stage and invasiveness of the tumor [105]. It is also positively associated with lung cancer metastasis and resistance to gefitinib and doxorubicin [106]. Xiang et al. showed that the induction of EMT is accompanied by upregulation of TGF-β1 because of *MALAT1* expression in endothelial progenitor cells. The LncRNA MALAT1 regulates TGF-β receptor 2 and the SMAD3 signaling pathway in these cells [107]. On the contrary, Kim and colleagues reported that *MALAT1*, in breast cancer, has a metastasis-suppressing role, which is facilitated by various pro-metastatic transcription factors of transcriptional enhancer associated domain (TEAD) family [108]. Overexpression of TEAD proteins is associated with the activation of several genes responsible for tumor growth and metastasis such as Yes-associated protein 1 (YAP) and Transcriptional co-activator with PDZ-binding motif (TAZ), which are involved in the hippo pathway [109,110], the pathway that is essential for angiogenesis and tissue regeneration.

3.9. MEG3

Maternally expressed gene 3 (*MEG3*) is a LncRNA located on chromosome 14q32.3 and is downregulated in human cancers [111]. Wang et al. showed that *MEG3* was remarkably reduced in patients with metastatic papillary thyroid carcinoma. In addition, they revealed that downregulated *MEG3* had a direct correlation with lymph-node metastasis [112]. Jiao et al. revealed that MEG-3 functions as a suppressor of gastric carcinoma cell growth, invasion, and migration. They suggested that *MEG3* suppresses migratory features of gastric cancer cells by modulating EMT in these cancer cells [113]. Overexpression of *MEG3* in human osteosarcoma cell line, MG63, leads to a significant decrease in proliferation and invasion as well as a remarkable increase in apoptosis [114]. The *MEG3* acts as a metastasis suppressor in melanoma by a mechanism that involves miR-21/E-cadherin axis [115]. It is downregulated in melanoma tissues and cell lines and its level is markedly associated with poor prognosis in patients with this disease [116]. In ovarian cancer, *MEG3* prohibits the tumor progression by acting on *miR-30e-3p* and laminin alpha4 [117]. Additionally, *MEG3* suppresses the metastatic progression of several cancers such as breast cancer, lung cancer, and HCC [91,118,119].

3.10. NEF

Neighboring enhancer of FOXA2 (*NEF*) is a LncRNA known for its tumor-suppressive role. Several preclinical studies have shown that the upregulation of *NEF* inhibits cancer progression [120,121]. In triple-negative breast cancer, *NEF* is downregulated because of the upregulation of *miR-155* [122]. It is also reported to suppress metastasis in HCC and inhibits cell invasion and migration in osteosarcoma by downregulating *miR-21* [123]. In cervical cancer, the reduced expression of *NEF* is a characteristic of patients with reduced survival rates. Downregulation of *NEF* is associated with increased production of TGF-β1 that induces metastasis of cancer cells [124,125]. This is likely to be achieved by inhibition of the Wnt/β-catenin pathway [126]. Chang et al. reported that serum concentration of *NEF* is negatively correlated with the stage of non-small-cell lung cancer [127].

3.11. NKILA

Nuclear Factor-κB Interacting LncRNA (*NKILA*) is an inflammation-induced LncRNA molecule that has been recognized in triple-negative breast cancer cells after exposing them to tumor necrosis factor (TNF)-α and interleukin-1β (IL-1β) [128]. Downregulated expression of *NKILA* is associated with metastasis and invasiveness in breast cancer patients [128]. In HCC, *NKILA* suppresses the metastatic spread of the tumor by inhibiting NF-κB/Slug-mediated EMT in these tumor cells [129]. This LncRNA also hampers the invasion and migration of tongue squamous cell carcinoma cells by blocking the EMT process in these cancer cells [130]. A study on non-small cell lung cancer showed downregulated *NKILA* in tumor samples. The results indicated that *NKILA* suppresses these cells by acting on the NF-κB/Snail signal pathway [131]. The effect of *NKILA* on the inhibition of migration and invasion of malignant melanoma cells was also shown to be achieved by the regulation of the NF-kB signaling pathway [132]. Several studies have affirmed the key role of *NKILA* in the invasion, migration, and metastasis of different cancers such as esophageal squamous cell carcinoma, laryngeal cancer, and head and neck cancer [133–135].

3.12. NRON

The nuclear factor of activated T-cells (NFAT) is a transcription factor present in the extracts of T-cells with a role in cancer invasion, angiogenesis, and differentiation [136,137]. Non-coding repressor of NFAT (*NRON*) is a LncRNA that represses NFAT by inhibiting the transfer of this factor between the nucleus and cytoplasm (nucleocytoplasmic shuttling). By reducing the transfer of NFAT to the nucleus, *NRON* reduces the proliferation and invasion of vascular endothelial cells [138]. The *NRON* levels are markedly increased in some cancers such as bladder cancer, where it facilitates proliferation, migration, differentiation, and metastasis of cancer cells [136]. The LncRNA *NRON* is downregulated in breast tumors compared to healthy tissues [139]. Reduction in *NRON* levels in breast cancer is associated with increased cell invasion and differentiation and reduced apoptosis [140]. Nonetheless, its increased levels had the opposite effect marked by reduced levels of CCND1, CDK4, and Bcl-2 and an increase in Bax and *miR-302b* leading to the inhibition of cancer progression and metastasis [140].

3.13. *PTTG3P*

Pituitary tumor-transforming 3 (*PTTG3P*) is a LncRNA with a confined protein-coding capacity and implications in tumorigenesis of various cancer types. It is noticeably homologous to its parental gene, *PTTG1* [141]. In the resected cervical cancer (CC) tissue, *PTTG3P* and *PTTG1* had considerably higher expression levels in comparison with their paired adjacent healthy counterparts. Furthermore, the invasiveness of CC cells was enhanced by *PTTG3P* through SNAIL upregulation and E-cadherin downregulation [141]. High expression levels of *PTTG3P*, *PTTG1*, and *PTTG2* have been observed in esophageal squamous cell carcinoma (ESCC) patients and cell lines. In addition, there has been a correlation between TNM stage, tumor depth, and lymph node invasion with the elevated expression of *PTTG3P* in ESCC [142]. Interestingly, the results of an in vitro *PTTG3P* gain-of-function study demonstrated that the invasion and migration of ESCC cells were stimulated because of the increased expression of *PTTG3P*. Conclusively, it is suggested that *PTTG3P* functions as an oncogene in ESCC [143]. The considerable up-regulation of *PTTG3P* in colorectal cancer (CRC) tissues is correlated with distant and lymph node metastasis. It has been indicated that the motility of CRC cells might be promoted by *PTTG3P* through downregulation of miR-155-5P [144]. Huang et al. investigated the oncogenic function of PTTG3P in HCC and revealed that *PTTG3P* expression was considerably increased in HCC patients. They showed that *PTTG3P* upregulation was positively associated with TNM stage, tumor size, and poor survival of patients [145]. Further experiments demonstrated that recombinant overexpression of *PTTG3P* leads to enhanced cell proliferation, migration, and invasion in vitro as well as augmented metastasis and tumorigenesis in vivo. Conversely, opposite influences were recorded following the knockdown of *PTTG3P*. Mechanistically, overexpressed *PTTG3P* leads to the activation of PI3K/AKT and its downstream signals, including cell apoptosis, cell cycle progression, EMT markers, and up-regulation of *PTTG1* [145].

3.14. *SNHG20*

Small nucleolar RNA host gene 20 (*SNHG20*) is primarily known for its role in HCC. However, research has shown its role in the pathogenesis of a variety of cancers including bladder, lung, bone, colorectal cancer, and ovarian cancer. For example, the upregulation of SNHG20 is positively correlated with the activation of Wnt/β-catenin signaling, leading to the proliferation and invasion of ovarian cancer cells [146]. This is supported by increased production of cyclin-dependent kinase inhibitor 1 (p21), cyclin D1, N-cadherin, and vimentin. Guo et al. revealed that *SNHG20* downregulates *miR-140-5p* and ADAM10 to activate the MEK/ERK signaling pathway, which leads to subsequent cell proliferation, invasion, and differentiation [147]. In gastric cancer patients, *SNHG20* is correlated with the size of the tumor, lymphatic metastasis, and a lower overall survival rate [148]. Furthermore, overexpression of *SNHG20* activates the PI3K/Akt/mTOR signaling pathway, contributing to tumor progression and stemness in glioblastoma [149]. Several studies have also shown its upregulation in prostate cancer, osteosarcoma, and laryngeal squamous cell carcinoma [150–152].

3.15. *XIST*

LncRNA x-inactive specific transcript (*XIST*) inactivates the X chromosome by accumulation near the transcriptional loci of different proteins and contributes to gene silencing [153]. Knockdown of this LncRNA suppresses the growth, proliferation, migration, and invasion of some tumor cells. In glioblastoma stem cells, this effect is mediated by the upregulation of miR-152, indicating a negative correlation between *XIST* and miR-152 [154]. There are several pathways by which *XIST* exerts its effect on different types of cancer such as non-small cell lung cancer, breast cancer, and colorectal cancer [155,156]. In glioma cells, silencing of *XIST* suppresses metastasis and angiogenesis because of increased expression of *miR-429* [157].

A recent study by Xu et al. reported that silencing of *XIST* inhibits lung cancer cell growth by allowing the transcription of p53 and NLR family pyrin domain containing 3 (NLRP3), which suppresses the function of SMAD2 to inhibit its translocation to the nucleus. This LncRNA upregulates the transcription of Bcl2 and reduces that of E-cadherin, which results in the detachment of tumor cells from the primary tissue and its metastasis to distant organs [158]. The LncRNA *XIST* also competes with *miR-744*, which activates the Wnt/β-catenin signaling pathway, leading to tumor progression, invasion, and migration [159].

The LncRNA *XIST* promotes metastasis of breast cancer through different pathways and mechanisms. It inhibits the action of miR-125b leading to an increase in the production of NOD-like receptor family CARD domain containing 5 (NLRC5), which is a known inducer of metastasis in breast cancer [160]. However, Xing et al. showed that loss of *XIST* is associated with metastasis of breast cancer to the brain, via activation of the EMT process and c-Met [161]. Similar findings have been suggested by Zheng et al. [162].

The role of *XIST* in tumor progression and metastasis has also been reported in gastric cancer. It suppresses *miR-101* and modulates the function of EZH2. Additionally, it targets TGF-β1 by repressing *miR-185*, metastasis-associated in colon cancer 1 gene (MACC) via suppression of *miR-497* and JAK expression through competing with *miR-337* [163–166].

3.16. *ZFAS1*

The ZNFX1 antisense RNA 1 (*ZFAS1*) is a novel LncRNA transcribed in the antisense orientation of zinc finger NFX1-type containing 1(ZNFX1). The LncRNA *ZFAS1* is upregulated in several cancers and may contribute to the development and progression of these cancers [167]. In prostate cancer, knocking down *ZFAS1* suppresses the migration and invasion of these cancer cells by inhibiting EMT [168]. Upregulation of *ZFAS1* induces colorectal cancer cell migration, invasion, and metastasis and is positively correlated with TNM stage these tumors [169]. The *ZFAS1* also stimulates proliferation and metastasis in pancreatic cancer cells by acting on *miR-497-5p* [170]. In colorectal cancer, the upregulated level of *ZFAS1* is directly associated with poor prognosis and promotes invasion and metastasis [169]. The *ZFAS1* is involved in colorectal cancer progression by inducing vascular endothelial growth factor A (VEGFA), which is one of the important inducers of angiogenesis in tumors [171]. Moreover, *ZFAS1* acts as a tumor suppressor in breast cancer and its downregulated level is associated with augmented proliferation and metastatic breast tumors [172]. Some data affirm that *ZFAS1* is a major modulator of the EMT process in colon adenocarcinoma [173].

4. Phytochemicals That Target LncRNAs to Cease Metastasis

During the last decade, numerous bioactive compounds have been studied for their potential activities against metastasis through modulating LncRNAs (Table 1, Figure 3). Some of the more frequent cancer models in which these phytochemicals have been examined to regulate metastasis by acting on LncRNAs include breast cancer, hepatocellular carcinoma, and prostate cancer [174].

Table 1. Phytochemicals targeting LncRNAs to inhibit cancer metastasis.

Phytochemicals	Plant Source	Concentration	LncRNAs	Alteration	Cancer Type	Cancer Model	Ref.
Betulinic acids	Outer bark of a variety of tree species like white-barked birch	PLC/PRF/5 cell line: IC_{50} = 63.04 µM for 48 h MHCC97L cell line: IC_{50} = 40.02 for 48 h Mice: 10 mg/kg/day	*MALAT1*	Down-regulated	Hepatocellular Carcinoma	BALB/c nude mice, PLC/PRF/5 and MHCC97L cell lines	[175]
Bharangin	*Pygmacopremna herbacea*	0, 1, 2.5, and 5 µM for 24 h	*GAS-5* *MEG3* *H19*	Up-regulated Up-regulated Down-regulated	Breast Cancer	MCF-7 cell line	[176]
Curcumin	*Curcuma longa* (turmeric)	0, 5, 15, and 20 µM for 48 h	*H19*	Down-regulated	Breast Cancer	MCF-7/TAMR * cell line	[177]
Curcumin	*Curcuma longa* (turmeric)	5 to 10 µM for 24 h	*HOTAIR*	Down-regulated	Renal Cell Carcinoma	769-P-HOTAIR and 786-0 cell lines	[178]
DNC *	*Curcuma longa* (turmeric)	0–25 µM for 48 h	*MEG3* *HOTAIR*	Up-regulated Down-regulated	Hepatocellular Cancer	HuH-7 and HepG2 cell lines HuH-7 cell line	[179]
Genistein	Soybean	25 µM for 48 h	*HOTAIR*	Down-regulated	Prostate Cancer	PC3, DU145 cell lines	[180]
IDET *	*Elephantopus scaber* Linn	1, 2.5 and 5 µM for 24 h	*NKILA* *GAS-5* *H19* *HOTAIR* *ANRIL*	Up-regulated Up-regulated Down-regulated Up-regulated Up-regulated	Breast Cancer	MDA-MB-231 cell line	[129,181]
Pterostilbene	Grapes, blueberries, and peanuts	0, 1, 5, 25, and 50 µM for 24 h	*MEG3* *HOTAIR* *LINC01121* *PTTG3P*	Up-regulated Down-regulated Down-regulated Down-regulated	Breast Cancer	MCF7 cell line	[182]
Resveratrol	Berries, grapes, peanuts, pistachio, plums, and white hellebore	IC_{50} = 55 µM for 24 h	*MALAT1*	Down-regulated	Colorectal Cancer Cells	LoVo cell line	[183]
Sanguinarine	*Sanguinaria canadensis* (Bloodroot)	0–5 µM for 24 h	*CASC2*	Up-regulated	Epithelial Ovarian Cancer	SKOV3 cell line	[184]
Silibinin	*Silybum marianum* (Seeds of milk thistle)	10 µM for 24 h	*HOTAIR* *ZFAS1*	Down-regulated	Bladder Cancer	T24, UM-UC-3 cell lines	[185]

* IDET: Isodeoxyelephantopin, MCF-7/TAMR: MCF 7/tamoxifen-resistant cell, DNC: Dendrosomal curcumin (Nanocurcumin).

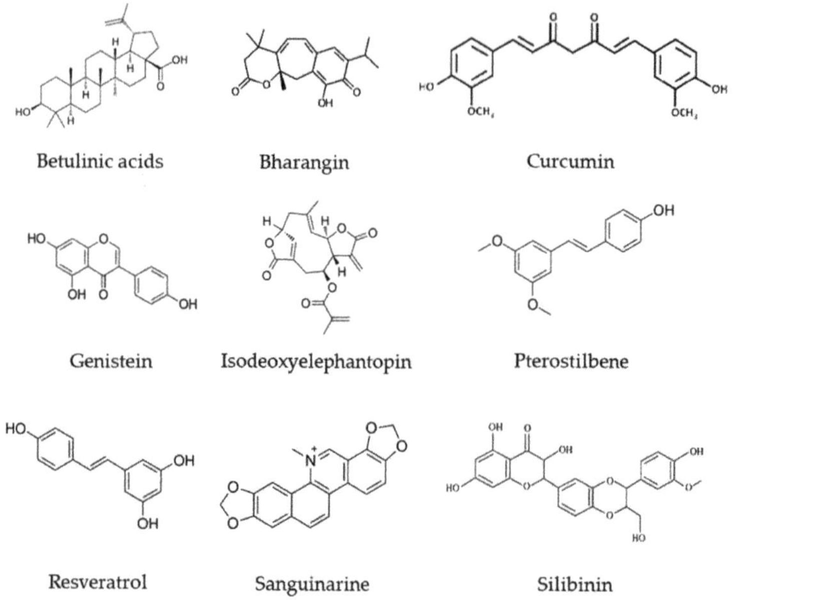

Figure 3. Chemical structure of phytochemicals involved in the invasion, migration, and metastasis of various cancers.

Betulinic acid (BA) is a natural component derived from the outer bark of a variety of tree species like white-barked birch [186]. A recent in vivo and in vitro study conducted to evaluate the effects of BA on HCC cells revealed that this triterpenoid can suppress the progression and invasion of these cells through inhibition of *MALAT1* expression. These effects were shown to be exerted in a dose-dependent manner. Moreover, Hematoxylin and eosin staining and immunohistochemical (IHC) observations were applied to visualize tumor tissues in a BALB/c nude mice model of HCC. In the mice treated with BA, a well-defined layer of tumor tissues without notable invasion was recognized, while the untreated group showed observable invasion in their tumor tissues. This was suggestive of the inhibitory role of BA on HCC cell invasion. Likewise, Ki67 expression, as a cell proliferation marker, was suppressed in BA-treated mice compared with untreated animals [187]. The biological activities of BA warrant more studies to prove an eventual translation into clinical settings, which can lead to the identification of a novel therapeutic approach for cancer patients [188]. Its potential anti-cancer effects led to the conduction of a phase I/II clinical trial to investigate the effect of 20% BA ointment on the treatment of moderate to severe forms of dysplastic nevi with no reported results. However, its challenging extraction process and poor water solubility limit its potential application as an anti-cancer drug [175].

Bharangin is a diterpenoid quinone methide, which is derived from the roots of the *Pygmacopremna herbacea* plant [189]. The results of an investigation revealed that MDA-MB-231 cells treated with bharangin could reduce the migration capacity of these cells compared with non-treated cells. The expression of tumor suppressor LncRNAs including *GAS-5* and *MEG3* was significantly augmented in bharangin-treated cells, while this diterpenoid down-regulated oncogenic *H19* LncRNA. They also reported that bharangin had a remarkable potential to inhibit the activation of okadaic acid-induced NF-κB in breast cancer cells [190].

Curcumin (diferuloylmethane) is an ingredient in yellow spice turmeric (*Curcuma longa*) [191]. This polyphenol has different biological activities against numerous human diseases, including cancer. In a recent study, curcumin effectively reduced *H19*-induced EMT in MCF-7/TAMR cells by downregulating N-cadherin and upregulation of E-cadherin, which are two well-known EMT biomarkers. Moreover, wound healing and transwell assays demonstrated that curcumin considerably reduces the migration and invasion of these breast cancer cells [176]. Accordingly, available data from an in vitro study support the key role of curcumin in the suppression of *HOTAIR*-induced migration of renal cell carcinoma (RCC) cell lines [192]. Two RCC cell lines were utilized including 769-P-*HOTAIR* and 769-P-vector cells with high and stable *HOTAIR* expression. The migration capacity of 769-P-HOTAIR cells was substantially higher than that of 769-P-vector cells. Interestingly, curcumin prohibited the migration of 769-P-HOTAIR cells in a concentration-dependent manner [192]. Zamani and colleagues studied the effect of dendrosomal curcumin (DNC) on *MEG3* expression in HCC cells. They observed that DNC effectively augments the expression levels of *MEG3* via upregulation of mir-29a and mir-185 [177]. However, more investigations are needed to confirm the effect of DNC-induced *MEG3* upregulation in the suppression of metastasis in this cancer.

Several clinical trials have been performed to assess the effectiveness of curcumin in treating different cancer types. According to the literature, the nitric oxide (NO) level is associated with different stages of malignancies and increased levels of NO have been reported in leukemic patients [178]. Therefore, Ghalaut et al. conducted a clinical study to assess the effectiveness of imatinib alone or in combination with turmeric powder on the levels of NO in 50 patients with chronic myeloid leukemia (CML). Twenty-five patients were treated with imatinib (400 mg twice a day) alone, and 25 subjects received imatinib in combination with turmeric powder (5 g three times/day) for 42 days. A more significant decrease in the serum levels of NO in the group with a combined treatment suggested that turmeric powder can be used as an adjuvant in reducing NO levels and may be effective in the treatment of CML [179].

Another randomized, double-blind placebo-controlled clinical trial evaluated the effect of curcumin (4 g daily) on free light-chain ratio response and bone turnover in patients with monoclonal gammopathy of undetermined significance (MGUS) and smoldering multiple myeloma (SMM). The data showed that curcumin could reduce disease procession in these patients [193]. Mahammedi et al. conducted a Phase II trial for examining the efficacy of docetaxel/prednisone for six cycles in combination with curcumin (6 mg per day) in 30 patients with metastatic prostate cancer. Their observations revealed a prostate-specific antigen (PSA) response in 59% of patients and significant efficacy of curcumin in treating cancer with a high response rate, well tolerability, and patient acceptability [194]. The safety profile and tolerability of curcumin was explored in a Phase I/II trial in metastatic colorectal cancer patients. The results showed that oral curcumin (2 g daily) with 12 cycles of 5-fluorouracil, folinic acid, and oxaliplatin chemotherapy regimen is safe and tolerable [195].

In another Phase II open-label clinical trial, the immunomodulatory efficacy of 100 mg of curcuminoids (extracted from *Curcuma longa* root) was assessed for tumor-induced inflammation in seven patients with endometrial carcinoma. The levels of inflammatory biomarkers in the patients who received this regimen were significantly suppressed and this may indicate curcumin-based compounds as supplementary regimens in endometrial carcinoma [196]. A Phase 1 clinical study on the chemopreventive potential of curcumin (4 g daily for 4 weeks) in colorectal cancer was conducted in 2010. Forty patients were enrolled to be participated in this study. No results have yet been reported for this study [197]. A Phase I study aimed to evaluate the short-term effects of supplementation with a turmeric extract, Curcumin C3 Complex®, on the biomarkers of head and neck squamous cell carcinoma (HNSCC). The tumor samples' adjacent tissues were used to measure the concentrations of curcumin and its metabolites in patients. The results revealed that this curcumin derivative could be used as a cancer preventing agent in smokers and tobacco users who are at risk of oral cancer [198]. A Phase II trial on effect of a curcumin derivative (Meriva®, 500 mg twice daily) was conducted for chemotherapy-treated breast cancer patients undergoing radiotherapy. The activity of nuclear factor-κB and its downstream modulators were quantified after treatment of patients with curcumin. No final results have yet been released [199].

Genistein, as a flavonoid compound derived from soybeans, targets an oncogenic LncRNA *HOTAIR* to suppress the migration and invasion of prostate tumor cells [200]. The tumor suppressor miR-34a, which binds to the *HOTAIR* mRNA sequence, participates in the anti-metastatic mechanism of genistein. Notably, genistein increases the expression of *miR-34a*, which in turn downregulates *HOTAIR* expression and thus suppresses the cell movement capacity of prostate cancer cells [201]. As reported in the literature, genistein changes the levels of phosphorylated tyrosine residues in cellular proteins. Accordingly, a Phase I clinical trial was conducted to determine the pharmacokinetic of two isoflavone preparations, PTI G-2535 and PTI G-4660 (which contained 43% and 90% genistein, respectively), in 13 patients with metastatic prostate cancer. The study also evaluated the toxicity and levels of protein-tyrosine phosphorylation in peripheral blood samples of the patients. Moreover, cohorts of four patients were administered genistein at three doses of 2, 4, or 8 mg/kg daily. The toxicity test results showed that one case with a treatment-related rash. Besides, a significant increase in tyrosine was identified in blood samples of the patients. This may suggest a potential anti-metastatic activity for genistein [202]. In another clinical study, Miltyk et al. investigated the probable genotoxic effect of a purified soy unconjugated isoflavone mixture containing in genistein, daidzein, and glycitein on 20 men with prostate cancer. The patients received 300 mg genistein for 28 days and then with 600 mg/d for another 56 days. Fluorescence in situ hybridization technique was used to measure genotoxicity markers in peripheral lymphocytes. Based on their data, no remarkable toxic changes were observed in genistein-treated patients. Therefore, the authors reported no toxic effects for the mentioned isoflavone mixture despite the in vitro genotoxicity that has been reported in the literature [203].

In a recent study, the impact of IDET, a sesquiterpene lactone extracted from *Elephantopus scaber* [180], on breast cancer cell migration was examined [204]. Data reported from scratch (wound healing) assay showed that IDET could significantly prevent the invasiveness of MDA-MB-231 cells. The healthy breast epithelia abundantly express the LncRNA *NKILA* and long GAS5, but their low expression correlates with metastasis of breast cancer [128,205]. Interestingly, IDET administration remarkably enhanced the expression levels of these LncRNAs [204]. Furthermore, the expression of oncogenic LncRNA *H19*, which is constitutively expressed in various tumor types like breast cancer, was significantly down-regulated due to the treatment of MDA-MB-231 cells with IDET [204]. This suggests that the up-regulation of tumor suppressor LncRNAs and down-regulation of oncogenic LncRNAs by IDET may contribute to motility suppression of MDA-MB-231cells. Likewise, elevated expression levels of oncogenic LncRNAs such as *ANRIL* and *HOTAIR* were observed in the serum samples and clinical tumor tissues of breast cancer patients compared to their paired healthy controls [206]. Nevertheless, the results of a study by Verma and colleagues showed the increased expression of these oncogenic LncRNAs by IDET. This may indicate a compensatory mechanism in response to the suppressed expression of other oncogenic LncRNAs and upregulation of tumor suppressor LncRNAs [204].

As described before, EMT is considered a vital stage in the metastatic propagation of all cancer types. Huang et al. showed that treatment of MCF7 cells with a phytochemical compound (extracted from different plants including grapes, blueberries, and peanuts) known as Pterostilbene [181] impedes EMT through downregulation of *HOTAIR*, *LINC01121*, and *PTTG3P*, as well as upregulation of *MEG3* [207].

A polyphenolic phytoalexin, known as resveratrol, is extracted from a variety of herbs, including berries, grapes, peanuts, pistachio, plums, and white hellebore [208]. In an investigation conducted by Ji et al., in situ hybridization confirmed that there are significantly higher *MALAT1* expression levels in tumor tissues compared to adjacent normal tissues. Besides, they found a statistically significant correlation between the extent of tumor metastasis and invasion with *MALAT1* expression. They also demonstrated that resveratrol can remarkably suppress migration and invasion of human colon cancer cell line LoVo through *MALAT1*-mediated Wnt/β-catenin signaling and its downstream targets in a dose-dependent manner. Overexpression of *MALAT1* using recombinant lentiviralbased experiment confirmed that this oncogenic LncRNA impedes the inhibitory impact of resveratrol on migration and invasion of LoVo cells [182]. Several clinical trials have been done to explore the effects of resveratrol or resveratrol-reached plant extracts on different types of cancer. A phase I trial evaluated the safety, tolerability, and dose determination of muscadine grape skin extract, which contains resveratrol, in men with recurrent prostate cancer (BRPC). Of 14 patients, seven remained in the study and received 4000 mg of the extract. According to the results, the extract led to a delayed disease recurrence by lengthening the PSA doubling time by 5.3 months. The safety assessments showed four patients with gastrointestinal symptoms, including grade 1 flatulence, grade 1 soft stool, and grade 1 eructation [209]. Nguyen et al. designed a clinical trial to study the effects of freeze-dried grape powder (GP) (containing resveratrol and resveratrol derived from plants) on the expression of factors involved in the Wnt pathway in 8 colorectal cancer patients. Treatment of the patients with GP (80 g/day containing 0.07 mg of resveratrol) for 14 days downregulated the expression of the Wnt target genes within regular mucosa of the patients' samples. According to the results, the authors suggested GP or resveratrol as colon cancer preventing compounds [183]. In another study, Patel et al. treated 20 colorectal cancer patients with resveratrol at 0.5 or 1 g/day for eight days. Then they quantified the expression of proliferation marker Ki-67 in tumor tissues. The results showed a 5% decline in the expression levels of Ki-67 in tumor tissues, indicating tumor suppressing activity of resveratrol in colorectal cancer patients [210]. Howells et al., in a Phase I randomized, double-blind pilot clinical trial, studied the effect of a resveratrol derivative (SRT501) on colorectal cancer patients with hepatic metastases. They clarified that SRT501 at a dose of 5 g/day for two weeks upregulated caspase-3 within liver tissue. This may suggest a pro-apoptotic activity for this resveratrol derivative in this cancer type [211]. Popat et al. conducted a Phase II trial to assess the possible activity of another resveratrol (SRT501) in combination with bortezomib in patients with relapsed and or refractory multiple myeloma. This resveratrol formulation was administered to 24 participants at a dose of 5 g/day for 20 days in a 21-day cycle up to 12 cycles. The results of the study indicated an unacceptable safety profile and minimal efficacy in these patients [212]. In a randomized placebo controlled clinical trial, Kjaer et al. treated 66 patients with prostate hyperplasia with two doses of resveratrol (150 or 1000 mg/day) for 4 months. Their data revealed that resveratrol treatment significantly decreased serum levels of androstenedione, dehydroepiandrosterone, and dehydroepiandrosterone-sulphate, but no remarkable effect was observed in prostate sizes [213].

Sanguinarine is an alkaloid derived from Bloodroot (*Sanguinaria canadensis*) with a significant inhibitory activity against the migratory ability of ovarian epithelial cancer cells [214]. In an in vitro experiment conducted by Zhang et al. to evaluate the potential effects of sanguinarine on human ovarian SKOV3 cells, this alkaloid inhibited the viability, migration, and invasion of these cells and increased apoptosis as well. Interestingly, *CASC2* is induced by this alkaloid and silencing *CASC2* rescues the antitumor effects of sanguinarine. This process was suggested to be mediated through *CASC2*–EIF4A3 signaling and/or PI3K/AKT/mTOR and NF-κB signal transductions [215].

Silibinin, a bioactive component isolated from the seeds of milk thistle (*Silybum marianum*), also has inhibitory potential against bladder cancer [216]. The *HOTAIR* expression is augmented by KRAS [184] and the PI3K pathways [217], and silibinin imposes its inhibitory effects on *HOTAIR* and *ZFAS1* by decreasing the activity of actin cytoskeleton and PI3K/Akt signal transductions in bladder cancer cells [218]. In a clinical study, Barrera et al. treated two patients with brain metastases from non-small cell lung cancer (NSCLC) with silibinin-based nutraceutical (Legasil). They found that Legasil treatment could significantly improve the clinical and radiological data of these patients. They also observed that silibinin treatment of the patients not only suppressed progressive brain metastases and reduced peritumoral brain edema but also did not affect the size of NSCLC tumors. The authors suggested that the combination of brain radiotherapy and Legasil may be a promising regimen to reduce brain edema and can provide local control and time for seeking other potential therapies for these patients [219]. Siegel et al. conducted a phase I study of silibinin phosphatidylcholine to determine the maximum tolerated dose per day of the compound in patients with advanced HCC. The serum levels of silibinin and silibinin glucuronide were increased within 1 to 3 weeks but all three patients died within 23–69 days of enrolling into the trial and no remarkable data were found in this study [185]. Flaig et al. enrolled 12 patients with prostate cancer to a trial study to estimate the tissue and blood effects of high-dose silibinin-phytosome in prostate cancer. Six patients were treated with silibinin at a single dose of 13 g/day for 20 days and six additional subjects were served as a control. The results revealed that high-dose silibinin led to high blood levels transiently, but low concentrations of the compound were observed in prostate tissue, indicating a weak penetration of silibinin into the prostate tissue [220].

5. Conclusions

Recent data shows that epigenetic role-players such as LncRNAs play key roles in regulating the malignant transformation and progression of cancers. Given that LncRNAs have pivotal roles in the modulation of a variety of cellular processes, more explorations are needed to unravel their possible mechanism of action in these processes. As has been numerously reported in the literature, phytochemical compounds from natural plants show potential effects on LncRNAs. Phytochemical compounds have been thus demonstrated to modulate the balance of expression of both oncogenic and antitumor LncRNAs, resulting in an anti-metastatic and anticancer effect (Figure 4). However, it appears that direct cellular target molecules of the phytochemicals and their exact mechanism of action are not known. Although well documented, the anti-metastatic effect of phytochemical compounds demands more preclinical and clinical studies to confirm their potential and further identify their molecular mechanism(s) of action. The evidence reviewed herein implies that targeted therapies using cancer-related LncRNAs could lead to the development of novel and effective treatment strategies for different types of cancer. Due to the important roles of LncRNAs in different cellular processes, phytochemicals that target these molecules may also boost the sensitivity of tumors to therapeutic methods.

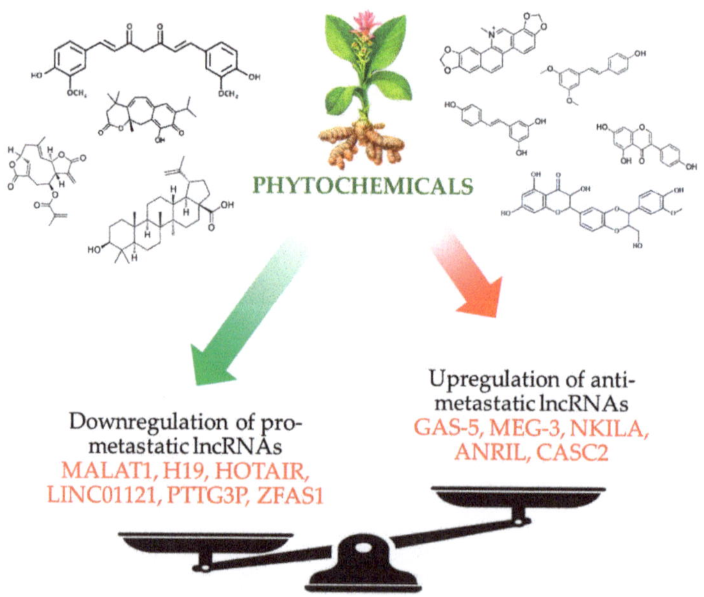

Figure 4. Schematic representation of the effects of phytochemicals on the balance between the expressions of pro- and anti-metastatic LncRNAs.

Funding: This research received no external funding.

Institutional Review Board Statement: Not applicable.

Informed Consent Statement: Not applicable.

Data Availability Statement: Not applicable.

Conflicts of Interest: The authors declare no conflict of interest.

References

1. Fares, J.; Fares, M.Y.; Khachfe, H.H.; Salhab, H.A.; Fares, Y. Molecular principles of metastasis: a hallmark of cancer revisited. *Signal Transduct. Target. Ther.* **2020**, *5*, 28. [CrossRef] [PubMed]
2. Dillekås, H.; Rogers, M.S.; Straume, O. Are 90% of deaths from cancer caused by metastases? *Cancer Med.* **2019**, *8*, 5574–5576. [CrossRef] [PubMed]
3. Riggio, A.I.; Varley, K.E.; Welm, A.L. The lingering mysteries of metastatic recurrence in breast cancer. *Br. J. Cancer* **2021**, *124*, 13–26. [CrossRef] [PubMed]
4. Chen, J.; Wang, Y.; Wang, C.; Hu, J.F.; Li, W. LncRNA Functions as a New Emerging Epigenetic Factor in Determining the Fate of Stem Cells. *Front. Genet.* **2020**, *11*, 277. [CrossRef] [PubMed]
5. Statello, L.; Guo, C.J.; Chen, L.L.; Huarte, M. Gene regulation by long non-coding RNAs and its biological functions. *Nat. Rev. Mol. Cell Biol.* **2021**, *22*, 96–118. [CrossRef]
6. Jiang, M.C.; Ni, J.J.; Cui, W.Y.; Wang, B.Y.; Zhuo, W. Emerging roles of lncRNA in cancer and therapeutic opportunities. *Am. J. Cancer Res.* **2019**, *9*, 1354–1366.
7. Chen, S.; Shen, X. Long noncoding RNAs: functions and mechanisms in colon cancer. *Mol. Cancer* **2020**, *19*, 167. [CrossRef]
8. Leitzmann, C. Characteristics and Health Benefits of Phytochemicals. *Forsch. Komplementmed.* **2016**, *23*, 69–74. [CrossRef]
9. Kochman, J.; Jakubczyk, K.; Antoniewicz, J.; Mruk, H.; Janda, K. Health Benefits and Chemical Composition of Matcha Green Tea: A Review. *Molecules* **2020**, *26*, 85. [CrossRef]
10. Choudhari, A.S.; Mandave, P.C.; Deshpande, M.; Ranjekar, P.; Prakash, O. Phytochemicals in Cancer Treatment: From Preclinical Studies to Clinical Practice. *Front. Pharmacol.* **2019**, *10*, 1614. [CrossRef]
11. Deng, Y.I.; Verron, E.; Rohanizadeh, R. Molecular Mechanisms of Anti-metastatic Activity of Curcumin. *Anticancer Res.* **2016**, *36*, 5639–5647. [CrossRef] [PubMed]
12. Ye, X.; Weinberg, R.A. Epithelial-Mesenchymal Plasticity: A Central Regulator of Cancer Progression. *Trends Cell Biol.* **2015**, *25*, 675–686. [CrossRef] [PubMed]
13. Jiang, Y.; Cao, W.; Wu, K.; Qin, X.; Wang, X.; Li, Y.; Yu, B.; Zhang, Z.; Wang, X.; Yan, M.; et al. LncRNA LINC00460 promotes EMT in head and neck squamous cell carcinoma by facilitating peroxiredoxin-1 into the nucleus. *J. Exp. Clin. Cancer Res.* **2019**, *38*, 365. [CrossRef]
14. Lamouille, S.; Subramanyam, D.; Blelloch, R.; Derynck, R. Regulation of epithelial-mesenchymal and mesenchymal-epithelial transitions by microRNAs. *Curr. Opin. Cell Biol.* **2013**, *25*, 200–207. [CrossRef]
15. Harvey, S.E.; Xu, Y.; Lin, X.; Gao, X.D.; Qiu, Y.; Ahn, J.; Xiao, X.; Cheng, C.J.R. Coregulation of alternative splicing by hnRNPM and ESRP1 during EMT. *RNA* **2018**, *24*, 1326–1338. [CrossRef] [PubMed]
16. Na, T.-Y.; Schecterson, L.; Mendonsa, A.M.; Gumbiner, B.M. The functional activity of E-cadherin controls tumor cell metastasis at multiple steps. *Proc. Natl. Acad. Sci. USA* **2020**, *117*, 5931–5937. [CrossRef]
17. Nagaishi, M.; Nakata, S.; Ono, Y.; Hirata, K.; Tanaka, Y.; Suzuki, K.; Yokoo, H.; Hyodo, A.J.J.o.C.N. Tumoral and stromal expression of Slug, ZEB1, and ZEB2 in brain metastasis. *J. Clin. Neurosci* **2017**, *46*, 124–128. [CrossRef]
18. Gasparics, Á.; Sebe, A. MRTFs- master regulators of EMT. *Dev. Dyn.* **2018**, *247*, 396–404. [CrossRef]
19. Shariat Razavi, S.M.; Forghanifard, M.M.; Kordi-Tamandani, D.M.; Abbaszadegan, M.R. MAML1 regulates EMT markers expression through NOTCH-independent pathway in breast cancer cell line MCF7. *Biochem. Biophys. Res. Commun.* **2019**, *510*, 376–382. [CrossRef]
20. Wang, J.-Y.; Wu, T.; Ma, W.; Li, S.; Jing, W.-J.; Ma, J.; Chen, D.-M.; Guo, X.-J.; Nan, K.-J. Expression and clinical significance of autophagic protein LC3B and EMT markers in gastric cancer. *Cancer Manag. Res.* **2018**, *10*, 1479–1486. [CrossRef]
21. Yamanaka, C.; Wada, H.; Eguchi, H.; Hatano, H.; Gotoh, K.; Noda, T.; Yamada, D.; Asaoka, T.; Kawamoto, K.; Nagano, H.; et al. Clinical significance of CD13 and epithelial mesenchymal transition (EMT) markers in hepatocellular carcinoma. *Jpn. J. Clin. Oncol.* **2017**, *48*, 52–60. [CrossRef] [PubMed]
22. Milano, A.; Mazzetta, F.; Valente, S.; Ranieri, D.; Leone, L.; Botticelli, A.; Onesti, C.E.; Lauro, S.; Raffa, S.; Torrisi, M.R.; et al. Molecular Detection of EMT Markers in Circulating Tumor Cells from Metastatic Non-Small Cell Lung Cancer Patients: Potential Role in Clinical Practice. *Anal. Cell. Pathol. (Amst.)* **2018**, *2018*, 3506874. [CrossRef] [PubMed]
23. Pearson, G.W. Control of Invasion by Epithelial-to-Mesenchymal Transition Programs during Metastasis. *J. Clin. Med.* **2019**, *8*, 646. [CrossRef] [PubMed]
24. Gao, G.; Liu, X.; Lin, Y.; Liu, H.; Zhang, G.J.E.R.M.P.S. LncRNA CASC9 promotes tumorigenesis by affecting EMT and predicts poor prognosis in esophageal squamous cell cancer. *Eur. Rev. Med. Pharmacol. Sci.* **2018**, *22*, 422–429. [PubMed]

25. Lugano, R.; Ramachandran, M.; Dimberg, A. Tumor angiogenesis: causes, consequences, challenges and opportunities. *Cell. Mol. Life Sci.* **2020**, *77*, 1745–1770. [CrossRef] [PubMed]
26. Geindreau, M.; Ghiringhelli, F.; Bruchard, M. Vascular Endothelial Growth Factor, a Key Modulator of the Anti-Tumor Immune Response. *Int. J. Mol. Sci.* **2021**, *22*, 4871. [CrossRef]
27. Zavyalova, M.V.; Denisov, E.V.; Tashireva, L.A.; Savelieva, O.E.; Kaigorodova, E.V.; Krakhmal, N.V.; Perelmuter, V.M. Intravasation as a Key Step in Cancer Metastasis. *Biochemistry (Mosc.)* **2019**, *84*, 762–772. [CrossRef]
28. Majidpoor, J.; Mortezaee, K. Steps in metastasis: an updated review. *Med. Oncol.* **2021**, *38*, 3. [CrossRef]
29. Zakaria, M.A.; Rajab, N.F.; Chua, E.W.; Selvarajah, G.T.; Masre, S.F. Roles of Rho-associated kinase in lung cancer (Review). *Int. J. Oncol.* **2021**, *58*, 185–198. [CrossRef]
30. Neophytou, C.M.; Panagi, M.; Stylianopoulos, T.; Papageorgis, P. The Role of Tumor Microenvironment in Cancer Metastasis: Molecular Mechanisms and Therapeutic Opportunities. *Cancers (Basel)* **2021**, *13*, 2053. [CrossRef]
31. Ma, Y.; Li, Y.; Guo, P.; Zhao, J.; Qin, Q.; Wang, J.; Liang, Z.; Wei, D.; Wang, Z.; Shen, J.; et al. Endothelial Cells Potentially Participate in the Metastasis of Triple-Negative Breast Cancer. *J. Immunol. Res.* **2022**, *2022*, 5412007. [CrossRef] [PubMed]
32. Strilic, B.; Offermanns, S. Intravascular Survival and Extravasation of Tumor Cells. *Cancer Cell* **2017**, *32*, 282–293. [CrossRef] [PubMed]
33. Headley, M.B.; Bins, A.; Nip, A.; Roberts, E.W.; Looney, M.R.; Gerard, A.; Krummel, M.F. Visualization of immediate immune responses to pioneer metastatic cells in the lung. *Nature* **2016**, *531*, 513–517. [CrossRef] [PubMed]
34. Furlow, P.W.; Zhang, S.; Soong, T.D.; Halberg, N.; Goodarzi, H.; Mangrum, C.; Wu, Y.G.; Elemento, O.; Tavazoie, S.F. Mechanosensitive pannexin-1 channels mediate microvascular metastatic cell survival. *Nat. Cell Biol.* **2015**, *17*, 943–952. [CrossRef]
35. Buchheit, C.L.; Weigel, K.J.; Schafer, Z.T. Cancer cell survival during detachment from the ECM: multiple barriers to tumour progression. *Nat. Rev. Cancer* **2014**, *14*, 632–641. [CrossRef]
36. Guadamillas, M.C.; Cerezo, A.; Del Pozo, M.A.J.J.o.c.s. Overcoming anoikis–pathways to anchorage-independent growth in cancer. *J. Cell Sci.* **2011**, *124*, 3189–3197. [CrossRef]
37. Paoli, P.; Giannoni, E.; Chiarugi, P.J.B.e.B.A.-M.C.R. Anoikis molecular pathways and its role in cancer progression. *Biochim. Biophys. Acta* **2013**, *1833*, 3481–3498. [CrossRef]
38. Morvan, M.G.; Lanier, L.L. NK cells and cancer: you can teach innate cells new tricks. *Nat. Rev. Cancer* **2016**, *16*, 7–19. [CrossRef]
39. Ferjančič, Š.; Gil-Bernabé, A.M.; Hill, S.A.; Allen, P.D.; Richardson, P.; Sparey, T.; Savory, E.; McGuffog, J.; Muschel, R.J. VCAM-1 and VAP-1 recruit myeloid cells that promote pulmonary metastasis in mice. *Blood* **2013**, *121*, 3289–3297. [CrossRef]
40. Hänggi, K.; Vasilikos, L.; Valls, A.F.; Yerbes, R.; Knop, J.; Spilgies, L.M.; Rieck, K.; Misra, T.; Bertin, J.; Gough, P.J.; et al. RIPK1/RIPK3 promotes vascular permeability to allow tumor cell extravasation independent of its necroptotic function. *Cell Death Dis.* **2017**, *8*, e2588. [CrossRef]
41. Aragon-Sanabria, V.; Pohler, S.E.; Eswar, V.J.; Bierowski, M.; Gomez, E.W.; Dong, C.J.S.r. VE-cadherin disassembly and cell contractility in the endothelium are necessary for barrier disruption induced by tumor cells. *Sci. Rep.* **2017**, *7*, 1–15.
42. Häuselmann, I.; Roblek, M.; Protsyuk, D.; Huck, V.; Knopfova, L.; Grässle, S.; Bauer, A.T.; Schneider, S.W.; Borsig, L.J.C.r. Monocyte induction of E-selectin–mediated endothelial activation releases VE-cadherin junctions to promote tumor cell extravasation in the metastasis cascade. *Cancer Res.* **2016**, *76*, 5302–5312. [CrossRef] [PubMed]
43. Gakhar, G.; Navarro, V.N.; Jurish, M.; Lee, G.Y.; Tagawa, S.T.; Akhtar, N.H.; Seandel, M.; Geng, Y.; Liu, H.; Bander, N.H.; et al. Circulating tumor cells from prostate cancer patients interact with E-selectin under physiologic blood flow. *PLoS ONE* **2013**, *8*, e85143. [CrossRef] [PubMed]
44. Jouve, N.; Bachelier, R.; Despoix, N.; Blin, M.G.; Matinzadeh, M.K.; Poitevin, S.; Aurrand-Lions, M.; Fallague, K.; Bardin, N.; Blot-Chabaud, M.; et al. CD146 mediates VEGF-induced melanoma cell extravasation through FAK activation. *Int. J. Cancer* **2015**, *137*, 50–60. [CrossRef] [PubMed]
45. Ma, S.; Fu, A.; Chiew, G.G.Y.; Luo, K.Q. Hemodynamic shear stress stimulates migration and extravasation of tumor cells by elevating cellular oxidative level. *Cancer Lett.* **2017**, *388*, 239–248. [CrossRef]
46. Urabe, F.; Patil, K.; Ramm, G.A.; Ochiya, T.; Soekmadji, C. Extracellular vesicles in the development of organ-specific metastasis. *J. Extracell. Vesicles* **2021**, *10*, e12125. [CrossRef] [PubMed]
47. Kusumbe, A.P. Vascular niches for disseminated tumour cells in bone. *J. Bone Oncol.* **2016**, *5*, 112–116. [CrossRef]
48. Peng, L.; Fu, J.; Chen, Y.; Ming, Y.; He, H.; Zeng, S.; Zhong, C.; Chen, L. Transcription factor SNAI2 exerts pro-tumorigenic effects on glioma stem cells via PHLPP2-mediated Akt pathway. *Cell Death Dis.* **2022**, *13*, 516. [CrossRef]
49. Stankic, M.; Pavlovic, S.; Chin, Y.; Brogi, E.; Padua, D.; Norton, L.; Massagué, J.; Benezra, R. TGF-β-Id1 signaling opposes Twist1 and promotes metastatic colonization via a mesenchymal-to-epithelial transition. *Cell Rep.* **2013**, *5*, 1228–1242. [CrossRef]
50. Ocaña, O.H.; Córcoles, R.; Fabra, Á.; Moreno-Bueno, G.; Acloque, H.; Vega, S.; Barrallo-Gimeno, A.; Cano, A.; Nieto, M.A. Metastatic Colonization Requires the Repression of the Epithelial-Mesenchymal Transition Inducer Prrx1. *Cancer Cell* **2012**, *22*, 709–724. [CrossRef]
51. Kong, Y.; Hsieh, C.H.; Alonso, L.C. ANRIL: A lncRNA at the CDKN2A/B Locus With Roles in Cancer and Metabolic Disease. *Front. Endocrinol. (Lausanne)* **2018**, *9*, 405. [CrossRef] [PubMed]
52. Iacobucci, I.; Ferrari, A.; Lonetti, A.; Papayannidis, C.; Paoloni, F.; Trino, S.; Storlazzi, C.T.; Ottaviani, E.; Cattina, F.; Impera, L.; et al. CDKN2A/B alterations impair prognosis in adult BCR-ABL1-positive acute lymphoblastic leukemia patients. *Clin. Cancer Res.* **2011**, *17*, 7413–7423. [CrossRef] [PubMed]

53. Hua, L.; Wang, C.Y.; Yao, K.H.; Chen, J.T.; Zhang, J.J.; Ma, W.L. High expression of long non-coding RNA ANRIL is associated with poor prognosis in hepatocellular carcinoma. *Int. J. Clin. Exp. Pathol.* **2015**, *8*, 3076–3082.
54. Hu, X.; Jiang, H.; Jiang, X. Downregulation of lncRNA ANRIL inhibits proliferation, induces apoptosis, and enhances radiosensitivity in nasopharyngeal carcinoma cells through regulating miR-125a. *Cancer Biol. Ther.* **2017**, *18*, 331–338. [CrossRef]
55. Yang, L.H.; Du, P.; Liu, W.; An, L.K.; Li, J.; Zhu, W.Y.; Yuan, S.; Wang, L.; Zang, L. LncRNA ANRIL promotes multiple myeloma progression and bortezomib resistance by EZH2-mediated epigenetically silencing of PTEN. *Neoplasma* **2021**, *68*, 788–797. [CrossRef] [PubMed]
56. Wang, Y.; Liu, Z.; Yao, B.; Li, Q.; Wang, L.; Wang, C.; Dou, C.; Xu, M.; Liu, Q.; Tu, K. Long non-coding RNA CASC2 suppresses epithelial-mesenchymal transition of hepatocellular carcinoma cells through CASC2/miR-367/FBXW7 axis. *Mol. Cancer* **2017**, *16*, 123. [CrossRef]
57. Zhang, H.; Feng, X.; Zhang, M.; Liu, A.; Tian, L.; Bo, W.; Wang, H.; Hu, Y. Long non-coding RNA CASC2 upregulates PTEN to suppress pancreatic carcinoma cell metastasis by downregulating miR-21. *Cancer Cell Int.* **2019**, *19*, 18. [CrossRef] [PubMed]
58. Zhang, Y.; Zhu, M.; Sun, Y.; Li, W.; Wang, Y.; Yu, W. Upregulation of lncRNA CASC2 Suppresses Cell Proliferation and Metastasis of Breast Cancer via Inactivation of the TGF-β Signaling Pathway. *Oncol. Res.* **2019**, *27*, 379–387. [CrossRef]
59. Gao, Z.; Wang, H.; Li, H.; Li, M.; Wang, J.; Zhang, W.; Liang, X.; Su, D.; Tang, J. Long non-coding RNA CASC2 inhibits breast cancer cell growth and metastasis through the regulation of the miR-96-5p/SYVN1 pathway. *Int. J. Oncol.* **2018**, *53*, 2081–2090. [CrossRef]
60. Xue, Z.; Zhu, X.; Teng, Y. Long non-coding RNA CASC2 inhibits progression and predicts favorable prognosis in epithelial ovarian cancer. *Mol. Med. Rep.* **2018**, *18*, 5173–5181. [CrossRef]
61. Sun, K.; Zhang, G. Long noncoding RNA CASC2 suppresses esophageal squamous cell carcinoma progression by increasing SOCS1 expression. *Cell Biosci.* **2019**, *9*, 90. [CrossRef] [PubMed]
62. Liu, Q.Y.; Gao, L.Y.; Xu, L.; Zhang, X.L.; Zhang, L.J.; Gong, X.L.; Luo, S.B.; Zhao, R.; Cheng, R.C. CASC2 inhibits the growth, migration, and invasion of thyroid cancer cells through sponging miR-18a-5p/FIH1 axis. *Kaohsiung J. Med. Sci.* **2021**, *37*, 268–275. [CrossRef] [PubMed]
63. Wang, D.; Gao, Z.M.; Han, L.G.; Xu, F.; Liu, K.; Shen, Y. Long noncoding RNA CASC2 inhibits metastasis and epithelial to mesenchymal transition of lung adenocarcinoma via suppressing SOX4. *Eur. Rev. Med. Pharmacol. Sci.* **2020**, *24*, 7210. [PubMed]
64. Xing, H.B.; Qiu, H.M.; Li, Y.; Dong, P.F.; Zhu, X.M. Long noncoding RNA CASC2 alleviates the growth, migration and invasion of oral squamous cell carcinoma via downregulating CDK1. *Eur. Rev. Med. Pharmacol. Sci.* **2020**, *24*, 10916.
65. Ba, Z.; Gu, L.; Hao, S.; Wang, X.; Cheng, Z.; Nie, G. Downregulation of lncRNA CASC2 facilitates osteosarcoma growth and invasion through miR-181a. *Cell Prolif.* **2018**, *51*, e12409. [CrossRef] [PubMed]
66. Ni, W.; Yao, S.; Zhou, Y.; Liu, Y.; Huang, P.; Zhou, A.; Liu, J.; Che, L.; Li, J. Long noncoding RNA GAS5 inhibits progression of colorectal cancer by interacting with and triggering YAP phosphorylation and degradation and is negatively regulated by the m(6)A reader YTHDF3. *Mol. Cancer* **2019**, *18*, 143. [CrossRef] [PubMed]
67. Zheng, S.; Li, M.; Miao, K.; Xu, H. lncRNA GAS5-promoted apoptosis in triple-negative breast cancer by targeting miR-378a-5p/SUFU signaling. *J. Cell. Biochem.* **2020**, *121*, 2225–2235. [CrossRef]
68. Dong, Q.; Long, X.; Cheng, J.; Wang, W.; Tian, Q.; Di, W. LncRNA GAS5 suppresses ovarian cancer progression by targeting the miR-96-5p/PTEN axis. *Ann. Transl. Med.* **2021**, *9*, 1770. [CrossRef]
69. Chen, L.; Yang, H.; Xiao, Y.; Tang, X.; Li, Y.; Han, Q.; Fu, J.; Yang, Y.; Zhu, Y. LncRNA GAS5 is a critical regulator of metastasis phenotype of melanoma cells and inhibits tumor growth in vivo. *Onco Targets Ther.* **2016**, *9*, 4075–4087. [CrossRef]
70. Wang, Y.; Ren, X.; Yuan, Y.; Yuan, B.S. Downregulated lncRNA GAS5 and Upregulated miR-21 Lead to Epithelial-Mesenchymal Transition and Lung Metastasis of Osteosarcomas. *Front. Cell Dev. Biol.* **2021**, *9*, 707693. [CrossRef]
71. Xu, Y.; Lu, J.; Lou, N.; Lu, W.; Xu, J.; Jiang, H.; Ye, G. Long noncoding RNA GAS5 inhibits proliferation and metastasis in papillary thyroid carcinoma through the IFN/STAT1 signaling pathway. *Pathol. Res. Pract.* **2022**, *233*, 153856. [CrossRef] [PubMed]
72. Xiao, Q.; Zheng, F.; Tang, Q.; Wu, J.-J.; Xie, J.; Huang, H.-D.; Yang, X.-B.; Hann, S.S.J.C.P. Biochemistry. Repression of PDK1-and LncRNA HOTAIR-mediated EZH2 gene expression contributes to the enhancement of atractylenolide 1 and erlotinib in the inhibition of human lung cancer cells. *Cell Physiol. Biochem.* **2018**, *49*, 1615–1632. [CrossRef] [PubMed]
73. Hajjari, M.; Salavaty, A. HOTAIR: an oncogenic long non-coding RNA in different cancers. *Cancer Biol. Med.* **2015**, *12*, 1–9.
74. Ren, Y.; Jia, H.H.; Xu, Y.Q.; Zhou, X.; Zhao, X.H.; Wang, Y.F.; Song, X.; Zhu, Z.Y.; Sun, T.; Dou, Y.; et al. Paracrine and epigenetic control of CAF-induced metastasis: the role of HOTAIR stimulated by TGF-ß1 secretion. *Mol. Cancer* **2018**, *17*, 5. [CrossRef] [PubMed]
75. Nakao, K.; Miyaaki, H.; Ichikawa, T. Antitumor function of microRNA-122 against hepatocellular carcinoma. *J. Gastroenterol.* **2014**, *49*, 589–593. [CrossRef]
76. Fornari, F.; Gramantieri, L.; Giovannini, C.; Veronese, A.; Ferracin, M.; Sabbioni, S.; Calin, G.A.; Grazi, G.L.; Croce, C.M.; Tavolari, S.; et al. MiR-122/cyclin G1 interaction modulates p53 activity and affects doxorubicin sensitivity of human hepatocarcinoma cells. *Cancer Res.* **2009**, *69*, 5761–5767. [CrossRef]
77. Bai, S.; Nasser, M.W.; Wang, B.; Hsu, S.H.; Datta, J.; Kutay, H.; Yadav, A.; Nuovo, G.; Kumar, P.; Ghoshal, K. MicroRNA-122 inhibits tumorigenic properties of hepatocellular carcinoma cells and sensitizes these cells to sorafenib. *J. Biol. Chem.* **2009**, *284*, 32015–32027. [CrossRef]

78. Kong, J.; Qiu, Y.; Li, Y.; Zhang, H.; Wang, W. TGF-β1 elevates P-gp and BCRP in hepatocellular carcinoma through HOTAIR/miR-145 axis. *Biopharm. Drug Dispos.* **2019**, *40*, 70–80. [CrossRef]
79. Yang, L.; Peng, X.; Li, Y.; Zhang, X.; Ma, Y.; Wu, C.; Fan, Q.; Wei, S.; Li, H.; Liu, J. Long non-coding RNA HOTAIR promotes exosome secretion by regulating RAB35 and SNAP23 in hepatocellular carcinoma. *Mol. Cancer* **2019**, *18*, 78. [CrossRef]
80. Wei, Z.; Chen, L.; Meng, L.; Han, W.; Huang, L.; Xu, A. LncRNA HOTAIR promotes the growth and metastasis of gastric cancer by sponging miR-1277-5p and upregulating COL5A1. *Gastric Cancer* **2020**, *23*, 1018–1032. [CrossRef]
81. Tao, D.; Zhang, Z.; Liu, X.; Zhang, Z.; Fu, Y.; Zhang, P.; Yuan, H.; Liu, L.; Cheng, J.; Jiang, H. LncRNA HOTAIR promotes the invasion and metastasis of oral squamous cell carcinoma through metastasis-associated gene 2. *Mol. Carcinog.* **2020**, *59*, 353–364. [CrossRef] [PubMed]
82. Guo, Y.P.; Wang, Z.F.; Li, N.; Lei, Q.Q.; Cheng, Q.; Shi, L.G.; Zhou, S.L.; Wang, X.H.; Sun, Y.; Kong, L.F. Suppression of lncRNA HOTAIR alleviates RCC angiogenesis through regulating miR-126/EGFL7 axis. *Am. J. Physiol Cell Physiol* **2021**, *320*, C880–c891. [CrossRef] [PubMed]
83. Wong, C.H.; Li, C.H.; He, Q.; Chan, S.L.; Tong, J.H.-M.; To, K.-F.; Lin, L.-z.; Chen, Y.J.C.l. Ectopic HOTTIP expression induces noncanonical transactivation pathways to promote growth and invasiveness in pancreatic ductal adenocarcinoma. *Cancer Lett.* **2020**, *477*, 1–9. [CrossRef] [PubMed]
84. Lin, C.; Wang, Y.; Zhang, S.; Yu, L.; Guo, C.; Xu, H.J.O. Transcriptional and posttranscriptional regulation of HOXA13 by lncRNA HOTTIP facilitates tumorigenesis and metastasis in esophageal squamous carcinoma cells. *Oncogene* **2017**, *36*, 5392–5406. [CrossRef]
85. Wang, Q.; Wu, G.; Zhang, Z.; Tang, Q.; Zheng, W.; Chen, X.; Chen, F.; Li, Q.; Che, X. Long non-coding RNA HOTTIP promotes renal cell carcinoma progression through the regulation of the miR-615/IGF-2 pathway. *Int. J. Oncol.* **2018**, *53*, 2278–2288. [CrossRef]
86. Cheng, Y.; Jutooru, I.; Chadalapaka, G.; Corton, J.C.; Safe, S. The long non-coding RNA HOTTIP enhances pancreatic cancer cell proliferation, survival and migration. *Oncotarget* **2015**, *6*, 10840–10852. [CrossRef]
87. Fu, Z.; Chen, C.; Zhou, Q.; Wang, Y.; Zhao, Y.; Zhao, X.; Li, W.; Zheng, S.; Ye, H.; Wang, L.; et al. LncRNA HOTTIP modulates cancer stem cell properties in human pancreatic cancer by regulating HOXA9. *Cancer Lett.* **2017**, *410*, 68–81. [CrossRef]
88. Yin, F.; Zhang, Q.; Dong, Z.; Hu, J.; Ma, Z. Lncrna hottip participates in cisplatin resistance of tumor cells by regulating mir-137 expression in pancreatic cancer. *Onco Targets Ther.* **2020**, *13*, 2689. [CrossRef]
89. Yang, J.; Qi, M.; Fei, X.; Wang, X.; Wang, K. LncRNA H19: A novel oncogene in multiple cancers. *Int. J. Biol. Sci.* **2021**, *17*, 3188–3208. [CrossRef]
90. Pan, Y.; Zhang, Y.; Liu, W.; Huang, Y.; Shen, X.; Jing, R.; Pu, J.; Wang, X.; Ju, S.; Cong, H.; et al. LncRNA H19 overexpression induces bortezomib resistance in multiple myeloma by targeting MCL-1 via miR-29b-3p. *Cell Death Dis.* **2019**, *10*, 106. [CrossRef]
91. Zheng, Q.; Lin, Z.; Xu, J.; Lu, Y.; Meng, Q.; Wang, C.; Yang, Y.; Xin, X.; Li, X.; Pu, H.; et al. Long noncoding RNA MEG3 suppresses liver cancer cells growth through inhibiting β-catenin by activating PKM2 and inactivating PTEN. *Cell Death Dis.* **2018**, *9*, 253. [CrossRef] [PubMed]
92. Gamaev, L.; Mizrahi, L.; Friehmann, T.; Rosenberg, N.; Pappo, O.; Olam, D.; Zeira, E.; Bahar Halpern, K.; Caruso, S.; Zucman-Rossi, J.; et al. The pro-oncogenic effect of the lncRNA H19 in the development of chronic inflammation-mediated hepatocellular carcinoma. *Oncogene* **2021**, *40*, 127–139. [CrossRef] [PubMed]
93. Liang, W.Q.; Zeng, D.; Chen, C.F.; Sun, S.M.; Lu, X.F.; Peng, C.Y.; Lin, H.Y. Long noncoding RNA H19 is a critical oncogenic driver and contributes to epithelial-mesenchymal transition in papillary thyroid carcinoma. *Cancer Manag. Res.* **2019**, *11*, 2059–2072. [CrossRef]
94. Sun, H.; Wang, G.; Peng, Y.; Zeng, Y.; Zhu, Q.-N.; Li, T.-L.; Cai, J.-Q.; Zhou, H.-H.; Zhu, Y.-S. H19 lncRNA mediates 17β-estradiol-induced cell proliferation in MCF-7 breast cancer cells. *Oncol. Rep.* **2015**, *33*, 3045–3052. [CrossRef]
95. Hadziselimovic, F.; Verkauskas, G.; Vincel, B.; Stadler, M.B. Testicular expression of long non-coding RNAs is affected by curative GnRHa treatment of cryptorchidism. *Basic Clin. Androl* **2019**, *29*, 18. [CrossRef] [PubMed]
96. Wang, Z.; Wang, P.; Cao, L.; Li, F.; Duan, S.; Yuan, G.; Xiao, L.; Guo, L.; Yin, H.; Xie, D.; et al. Long Intergenic Non-Coding RNA 01121 Promotes Breast Cancer Cell Proliferation, Migration, and Invasion via the miR-150-5p/HMGA2 Axis. *Cancer Manag. Res.* **2019**, *11*, 10859–10870. [CrossRef] [PubMed]
97. Yang, E.; Cisowski, J.; Nguyen, N.; O'Callaghan, K.; Xu, J.; Agarwal, A.; Kuliopulos, A.; Covic, L. Dysregulated protease activated receptor 1 (PAR1) promotes metastatic phenotype in breast cancer through HMGA2. *Oncogene* **2016**, *35*, 1529–1540. [CrossRef]
98. Sun, M.; Song, C.X.; Huang, H.; Frankenberger, C.A.; Sankarasharma, D.; Gomes, S.; Chen, P.; Chen, J.; Chada, K.K.; He, C.; et al. HMGA2/TET1/HOXA9 signaling pathway regulates breast cancer growth and metastasis. *Proc. Natl. Acad. Sci. USA* **2013**, *110*, 9920–9925. [CrossRef]
99. Cheng, Y.; Imanirad, P.; Jutooru, I.; Hedrick, E.; Jin, U.H.; Rodrigues Hoffman, A.; Leal de Araujo, J.; Morpurgo, B.; Golovko, A.; Safe, S. Role of metastasis-associated lung adenocarcinoma transcript-1 (MALAT-1) in pancreatic cancer. *PLoS ONE* **2018**, *13*, e0192264. [CrossRef]
100. Eißmann, M.; Gutschner, T.; Hämmerle, M.; Günther, S.; Caudron-Herger, M.; Groß, M.; Schirmacher, P.; Rippe, K.; Braun, T.; Zörnig, M.; et al. Loss of the abundant nuclear non-coding RNA MALAT1 is compatible with life and development. *RNA Biol.* **2012**, *9*, 1076–1087. [CrossRef]

101. Yuan, J.; Xu, X.J.; Lin, Y.; Chen, Q.Y.; Sun, W.J.; Tang, L.; Liang, Q.X. LncRNA MALAT1 expression inhibition suppresses tongue squamous cell carcinoma proliferation, migration and invasion by inactivating PI3K/Akt pathway and downregulating MMP-9 expression. *Eur. Rev. Med. Pharmacol. Sci.* **2019**, *23*, 198–206. [PubMed]
102. Dong, Y.; Liang, G.; Yuan, B.; Yang, C.; Gao, R.; Zhou, X. MALAT1 promotes the proliferation and metastasis of osteosarcoma cells by activating the PI3K/Akt pathway. *Tumour Biol.* **2015**, *36*, 1477–1486. [CrossRef] [PubMed]
103. Chen, L.; Feng, P.; Zhu, X.; He, S.; Duan, J.; Zhou, D. Long non-coding RNA Malat1 promotes neurite outgrowth through activation of ERK/MAPK signalling pathway in N2a cells. *J. Cell Mol. Med.* **2016**, *20*, 2102–2110. [CrossRef] [PubMed]
104. Liu, S.; Yan, G.; Zhang, J.; Yu, L. Knockdown of Long Noncoding RNA (lncRNA) Metastasis-Associated Lung Adenocarcinoma Transcript 1 (MALAT1) Inhibits Proliferation, Migration, and Invasion and Promotes Apoptosis by Targeting miR-124 in Retinoblastoma. *Oncol. Res.* **2018**, *26*, 581–591. [CrossRef]
105. Li, Z.-X.; Zhu, Q.-N.; Zhang, H.-B.; Hu, Y.; Wang, G.; Zhu, Y.-S. MALAT1: a potential biomarker in cancer. *Cancer Manag. Res.* **2018**, *10*, 6757–6768. [CrossRef]
106. Yang, T.; Li, H.; Chen, T.; Ren, H.; Shi, P.; Chen, M. LncRNA MALAT1 Depressed Chemo-Sensitivity of NSCLC Cells through Directly Functioning on miR-197-3p/p120 Catenin Axis. *Mol. Cells* **2019**, *42*, 270–283.
107. Xiang, Y.; Zhang, Y.; Tang, Y.; Li, Q. MALAT1 Modulates TGF-β1-Induced Endothelial-to-Mesenchymal Transition through Downregulation of miR-145. *Cell Physiol. Biochem.* **2017**, *42*, 357–372. [CrossRef]
108. Kim, J.; Piao, H.L.; Kim, B.J.; Yao, F.; Han, Z.; Wang, Y.; Xiao, Z.; Siverly, A.N.; Lawhon, S.E.; Ton, B.N.; et al. Long noncoding RNA MALAT1 suppresses breast cancer metastasis. *Nat. Genet.* **2018**, *50*, 1705–1715. [CrossRef]
109. Boopathy, G.T.K.; Hong, W. Role of Hippo Pathway-YAP/TAZ Signaling in Angiogenesis. *Front. Cell Dev. Biol.* **2019**, *7*, 49. [CrossRef]
110. Huh, H.D.; Kim, D.H.; Jeong, H.-S.; Park, H.W. Regulation of TEAD Transcription Factors in Cancer Biology. *Cells* **2019**, *8*, 600. [CrossRef]
111. Ghafouri-Fard, S.; Taheri, M. Maternally expressed gene 3 (MEG3): A tumor suppressor long non coding RNA. *Biomed. Pharmacother.* **2019**, *118*, 109129. [CrossRef] [PubMed]
112. Wang, C.; Yan, G.; Zhang, Y.; Jia, X.; Bu, P. Long non-coding RNA MEG3 suppresses migration and invasion of thyroid carcinoma by targeting of Rac1. *Neoplasma* **2015**, *62*, 541–549. [CrossRef] [PubMed]
113. Jiao, J.; Zhang, S. Long non-coding RNA MEG-3 suppresses gastric carcinoma cell growth, invasion and migration via EMT regulation. *Mol. Med. Rep.* **2019**, *20*, 2685–2693. [CrossRef]
114. Shi, Y.; Lv, C.; Shi, L.; Tu, G. MEG3 inhibits proliferation and invasion and promotes apoptosis of human osteosarcoma cells. *Oncol. Lett.* **2018**, *15*, 1917–1923. [CrossRef] [PubMed]
115. Wu, L.; Zhu, L.; Li, Y.; Zheng, Z.; Lin, X.; Yang, C. LncRNA MEG3 promotes melanoma growth, metastasis and formation through modulating miR-21/E-cadherin axis. *Cancer Cell Int.* **2020**, *20*, 12. [CrossRef] [PubMed]
116. Long, J.; Pi, X. lncRNA-MEG3 Suppresses the Proliferation and Invasion of Melanoma by Regulating CYLD Expression Mediated by Sponging miR-499-5p. *Biomed. Res. Int.* **2018**, *2018*, 2086564. [CrossRef] [PubMed]
117. Liu, Y.; Xu, Y.; Ding, L.; Yu, L.; Zhang, B.; Wei, D. LncRNA MEG3 suppressed the progression of ovarian cancer via sponging miR-30e-3p and regulating LAMA4 expression. *Cancer Cell Int.* **2020**, *20*, 181. [CrossRef]
118. Zhang, C.Y.; Yu, M.S.; Li, X.; Zhang, Z.; Han, C.R.; Yan, B. Overexpression of long non-coding RNA MEG3 suppresses breast cancer cell proliferation, invasion, and angiogenesis through AKT pathway. *Tumour Biol.* **2017**, *39*, 1010428317701311. [CrossRef]
119. Zhao, Y.; Zhu, Z.; Shi, S.; Wang, J.; Li, N. Long non-coding RNA MEG3 regulates migration and invasion of lung cancer stem cells via miR-650/SLC34A2 axis. *Biomed. Pharmacother.* **2019**, *120*, 109457. [CrossRef]
120. Wan, X.; Xiang, J.; Zhang, Q.; Bian, C. Downregulation of lnRNA-NEF is involved in the postoperative cancer distant recurrence in prostate carcinoma patients. *J. Cell. Biochem.* **2019**, *120*, 9601–9607. [CrossRef]
121. Liang, W.-C.; Ren, J.-L.; Wong, C.-W.; Chan, S.-O.; Waye, M.M.-Y.; Fu, W.-M.; Zhang, J.-F. LncRNA-NEF antagonized epithelial to mesenchymal transition and cancer metastasis via cis-regulating FOXA2 and inactivating Wnt/β-catenin signaling. *Oncogene* **2018**, *37*, 1445–1456. [CrossRef]
122. Song, X.; Liu, Z.; Yu, Z. LncRNA NEF is downregulated in triple negative breast cancer and correlated with poor prognosis. *Acta Biochim. Biophys. Sin. (Shanghai)* **2019**, *51*, 386–392. [CrossRef] [PubMed]
123. Yang, Q.; Yu, H.; Yin, Q.; Hu, X.; Zhang, C. lncRNA-NEF is downregulated in osteosarcoma and inhibits cancer cell migration and invasion by downregulating miRNA-21. *Oncol. Lett.* **2019**, *17*, 5403–5408. [CrossRef] [PubMed]
124. Ju, W.; Luo, X.; Zhang, N. LncRNA NEF inhibits migration and invasion of HPV-negative cervical squamous cell carcinoma by inhibiting TGF-β pathway. *Biosci. Rep.* **2019**, *39*, 20180878. [CrossRef] [PubMed]
125. Xie, J.; Liu, Y.; Du, X.; Wu, Y. TGF-β1 promotes the invasion and migration of papillary thyroid carcinoma cells by inhibiting the expression of lncRNA-NEF. *Oncol. Lett.* **2019**, *17*, 3125–3132. [CrossRef]
126. Zhang, J.; Hu, S.L.; Qiao, C.H.; Ye, J.F.; Li, M.; Ma, H.M.; Wang, J.H.; Xin, S.Y.; Yuan, Z.L. LncRNA-NEF inhibits proliferation, migration and invasion of esophageal squamous-cell carcinoma cells by inactivating wnt/β-catenin pathway. *Eur. Rev. Med. Pharmacol. Sci.* **2018**, *22*, 6824–6831.
127. Chang, L.; Xu, W.; Zhang, Y.; Gong, F. Long non-coding RNA-NEF targets glucose transportation to inhibit the proliferation of non-small-cell lung cancer cells. *Oncol. Lett.* **2019**, *17*, 2795–2801. [CrossRef]

128. Liu, B.; Sun, L.; Liu, Q.; Gong, C.; Yao, Y.; Lv, X.; Lin, L.; Yao, H.; Su, F.; Li, D.; et al. A cytoplasmic NF-κB interacting long noncoding RNA blocks IκB phosphorylation and suppresses breast cancer metastasis. *Cancer Cell* **2015**, *27*, 370–381. [CrossRef]
129. Chen, R.; Cheng, Q.; Owusu-Ansah, K.G.; Song, G.; Jiang, D.; Zhou, L.; Xu, X.; Wu, J.; Zheng, S. NKILA, a prognostic indicator, inhibits tumor metastasis by suppressing NF-κB/Slug mediated epithelial-mesenchymal transition in hepatocellular carcinoma. *Int. J. Biol. Sci.* **2020**, *16*, 495–503. [CrossRef]
130. Huang, W.; Cui, X.; Chen, J.; Feng, Y.; Song, E.; Li, J.; Liu, Y. Long non-coding RNA NKILA inhibits migration and invasion of tongue squamous cell carcinoma cells via suppressing epithelial-mesenchymal transition. *Oncotarget* **2016**, *7*, 62520–62532. [CrossRef]
131. Lu, Z.; Li, Y.; Wang, J.; Che, Y.; Sun, S.; Huang, J.; Chen, Z.; He, J. Long non-coding RNA NKILA inhibits migration and invasion of non-small cell lung cancer via NF-κB/Snail pathway. *J. Exp. Clin. Cancer Res.* **2017**, *36*, 54. [CrossRef] [PubMed]
132. Bian, D.; Gao, C.; Bao, K.; Song, G. The long non-coding RNA NKILA inhibits the invasion-metastasis cascade of malignant melanoma via the regulation of NF-kB. *Am. J. Cancer Res.* **2017**, *7*, 28–40. [PubMed]
133. Lu, Z.; Chen, Z.; Li, Y.; Wang, J.; Zhang, Z.; Che, Y.; Huang, J.; Sun, S.; Mao, S.; Lei, Y.; et al. TGF-β-induced NKILA inhibits ESCC cell migration and invasion through NF-κB/MMP14 signaling. *J. Mol. Med. (Berl)* **2018**, *96*, 301–313. [CrossRef] [PubMed]
134. Yang, T.; Li, S.; Liu, J.; Yin, D.; Yang, X.; Tang, Q. lncRNA-NKILA/NF-κB feedback loop modulates laryngeal cancer cell proliferation, invasion, and radioresistance. *Cancer Med.* **2018**, *7*, 2048–2063. [CrossRef]
135. Luo, X.; Qiu, Y.; Jiang, Y.; Chen, F.; Jiang, L.; Zhou, Y.; Dan, H.; Zeng, X.; Lei, Y.L.; Chen, Q. Long non-coding RNA implicated in the invasion and metastasis of head and neck cancer: possible function and mechanisms. *Mol. Cancer* **2018**, *17*, 14. [CrossRef]
136. Dai, Z.-T.; Xiang, Y.; Wang, Y.; Bao, L.-Y.; Wang, J.; Li, J.-P.; Zhang, H.-M.; Lu, Z.; Ponnambalam, S.; Liao, X.-H. Prognostic value of members of NFAT family for pan-cancer and a prediction model based on NFAT2 in bladder cancer. *Aging* **2021**, *13*, 13876–13897. [CrossRef]
137. Yao, Z.; Xiong, Z.; Li, R.; Liang, H.; Jia, C.; Deng, M. Long non-coding RNA NRON is downregulated in HCC and suppresses tumour cell proliferation and metastasis. *Biomed. Pharmacother.* **2018**, *104*, 102–109. [CrossRef]
138. Pan, M.-G.; Xiong, Y.; Chen, F.J.C.m.m. NFAT gene family in inflammation and cancer. *Curr. Mol. Med.* **2013**, *13*, 543–554. [CrossRef]
139. Mao, Q.; Li, L.; Zhang, C.; Sun, Y.; Liu, S.; Li, Y.; Shen, Y.; Liu, Z. Long non coding RNA NRON inhibited breast cancer development through regulating miR-302b/SRSF2 axis. *Am. J. Transl. Res.* **2020**, *12*, 4683–4692.
140. Niu, L.; Fan, Q.; Yan, M.; Wang, L. LncRNA NRON down-regulates lncRNA snaR and inhibits cancer cell proliferation in TNBC. *Biosci. Rep.* **2019**, *39*, 20190468. [CrossRef]
141. Guo, X.C.; Li, L.; Gao, Z.H.; Zhou, H.W.; Li, J.; Wang, Q.Q. The long non-coding RNA PTTG3P promotes growth and metastasis of cervical cancer through PTTG1. *Aging (Albany NY)* **2019**, *11*, 1333–1341. [CrossRef] [PubMed]
142. Grzechowiak, I.; Graś, J.; Szymańska, D.; Biernacka, M.; Guglas, K.; Poter, P.; Mackiewicz, A.; Kolenda, T. The Oncogenic Roles of PTTG1 and PTTG2 Genes and Pseudogene PTTG3P in Head and Neck Squamous Cell Carcinomas. *Diagnostics (Basel, Switzerland)* **2020**, *10*, 606. [CrossRef] [PubMed]
143. Zhang, Z.; Shi, Z. The Pseudogene PTTG3P Promotes Cell Migration and Invasion in Esophageal Squamous Cell Carcinoma. *Open Med. (Wars)* **2019**, *14*, 516–522. [CrossRef] [PubMed]
144. Liu, N.; Dou, L.; Zhang, X. LncRNA PTTG3P Sponge Absorbs microRNA-155-5P to Promote Metastasis of Colorectal Cancer. *Onco Targets Ther.* **2020**, *13*, 5283–5291. [CrossRef] [PubMed]
145. Huang, J.L.; Cao, S.W.; Ou, Q.S.; Yang, B.; Zheng, S.H.; Tang, J.; Chen, J.; Hu, Y.W.; Zheng, L.; Wang, Q. The long non-coding RNA PTTG3P promotes cell growth and metastasis via up-regulating PTTG1 and activating PI3K/AKT signaling in hepatocellular carcinoma. *Mol. Cancer* **2018**, *17*, 93. [CrossRef] [PubMed]
146. He, S.; Zhao, Y.; Wang, X.; Deng, Y.; Wan, Z.; Yao, S.; Shen, H. Up-regulation of long non-coding RNA SNHG20 promotes ovarian cancer progression via Wnt/β-catenin signaling. *Biosci. Rep.* **2018**, *38*, 20170681. [CrossRef] [PubMed]
147. Guo, H.; Yang, S.; Li, S.; Yan, M.; Li, L.; Zhang, H. LncRNA SNHG20 promotes cell proliferation and invasion via miR-140-5p-ADAM10 axis in cervical cancer. *Biomed. Pharmacother.* **2018**, *102*, 749–757. [CrossRef]
148. Cui, N.; Liu, J.; Xia, H.; Xu, D. LncRNA SNHG20 contributes to cell proliferation and invasion by upregulating ZFX expression sponging miR-495-3p in gastric cancer. *J. Cell. Biochem.* **2019**, *120*, 3114–3123. [CrossRef]
149. Gao, X.F.; He, H.Q.; Zhu, X.B.; Xie, S.L.; Cao, Y. LncRNA SNHG20 promotes tumorigenesis and cancer stemness in glioblastoma via activating PI3K/Akt/mTOR signaling pathway. *Neoplasma* **2019**, *66*, 532–542. [CrossRef]
150. Zhang, J.; Ju, C.; Zhang, W.; Xie, L. LncRNA SNHG20 is associated with clinical progression and enhances cell migration and invasion in osteosarcoma. *IUBMB Life* **2018**, *70*, 1115–1121. [CrossRef]
151. Wu, X.; Xiao, Y.; Zhou, Y.; Zhou, Z.; Yan, W. lncRNA SNHG20 promotes prostate cancer migration and invasion via targeting the miR-6516-5p/SCGB2A1 axis. *Am. J. Transl. Res.* **2019**, *11*, 5162–5169. [PubMed]
152. Li, Y.; Xu, J.; Guo, Y.; Yang, B. LncRNA SNHG20 promotes the development of laryngeal squamous cell carcinoma by regulating miR-140. *Eur. Rev. Med. Pharmacol. Sci.* **2019**, *23*, 3401–3409. [PubMed]
153. Pandya-Jones, A.; Markaki, Y.; Serizay, J.; Chitiashvili, T.; Mancia Leon, W.R.; Damianov, A.; Chronis, C.; Papp, B.; Chen, C.K.; McKee, R.; et al. A protein assembly mediates Xist localization and gene silencing. *Nature* **2020**, *587*, 145–151. [CrossRef]

154. Yao, Y.; Ma, J.; Xue, Y.; Wang, P.; Li, Z.; Liu, J.; Chen, L.; Xi, Z.; Teng, H.; Wang, Z.; et al. Knockdown of long non-coding RNA XIST exerts tumor-suppressive functions in human glioblastoma stem cells by up-regulating miR-152. *Cancer Lett.* **2015**, *359*, 75–86. [CrossRef]
155. Liu, J.; Yao, L.; Zhang, M.; Jiang, J.; Yang, M.; Wang, Y. Downregulation of LncRNA-XIST inhibited development of non-small cell lung cancer by activating miR-335/SOD2/ROS signal pathway mediated pyroptotic cell death. *Aging (Albany NY)* **2019**, *11*, 7830–7846. [CrossRef] [PubMed]
156. Yang, X.; Zhang, S.; He, C.; Xue, P.; Zhang, L.; He, Z.; Zang, L.; Feng, B.; Sun, J.; Zheng, M. METTL14 suppresses proliferation and metastasis of colorectal cancer by down-regulating oncogenic long non-coding RNA XIST. *Mol. Cancer* **2020**, *19*, 46. [CrossRef] [PubMed]
157. Cheng, Z.; Li, Z.; Ma, K.; Li, X.; Tian, N.; Duan, J.; Xiao, X.; Wang, Y. Long Non-coding RNA XIST Promotes Glioma Tumorigenicity and Angiogenesis by Acting as a Molecular Sponge of miR-429. *J. Cancer* **2017**, *8*, 4106–4116. [CrossRef]
158. Xu, X.; Zhou, X.; Chen, Z.; Gao, C.; Zhao, L.; Cui, Y. Silencing of lncRNA XIST inhibits non-small cell lung cancer growth and promotes chemosensitivity to cisplatin. *Aging* **2020**, *12*, 4711–4726. [CrossRef]
159. Wang, J.; Cai, H.; Dai, Z.; Wang, G. Down-regulation of lncRNA XIST inhibits cell proliferation via regulating miR-744/RING1 axis in non-small cell lung cancer. *Clin. Sci. (Lond.)* **2019**, *133*, 1567–1579. [CrossRef]
160. Zong, Y.; Zhang, Y.; Hou, D.; Xu, J.; Cui, F.; Qin, Y.; Sun, X. The lncRNA XIST promotes the progression of breast cancer by sponging miR-125b-5p to modulate NLRC5. *Am. J. Transl. Res.* **2020**, *12*, 3501–3511.
161. Xing, F.; Liu, Y.; Wu, S.-Y.; Wu, K.; Sharma, S.; Mo, Y.-Y.; Feng, J.; Sanders, S.; Jin, G.; Singh, R.; et al. Loss of XIST in Breast Cancer Activates MSN-c-Met and Reprograms Microglia via Exosomal miRNA to Promote Brain Metastasis. *Cancer Res.* **2018**, *78*, 4316–4330. [CrossRef] [PubMed]
162. Zheng, R.; Lin, S.; Guan, L.; Yuan, H.; Liu, K.; Liu, C.; Ye, W.; Liao, Y.; Jia, J.; Zhang, R. Long non-coding RNA XIST inhibited breast cancer cell growth, migration, and invasion via miR-155/CDX1 axis. *Biochem. Biophys. Res. Commun.* **2018**, *498*, 1002–1008. [CrossRef] [PubMed]
163. Ma, L.; Zhou, Y.; Luo, X.; Gao, H.; Deng, X.; Jiang, Y. Long non-coding RNA XIST promotes cell growth and invasion through regulating miR-497/MACC1 axis in gastric cancer. *Oncotarget* **2017**, *8*, 4125–4135. [CrossRef]
164. Zheng, W.; Li, J.; Zhou, X.; Cui, L.; Wang, Y. The lncRNA XIST promotes proliferation, migration and invasion of gastric cancer cells by targeting miR-337. *Arab J. Gastroenterol.* **2020**, *21*, 199–206. [CrossRef] [PubMed]
165. Zhang, Q.; Chen, B.; Liu, P.; Yang, J. XIST promotes gastric cancer (GC) progression through TGF-β1 via targeting miR-185. *J. Cell. Biochem.* **2018**, *119*, 2787–2796. [CrossRef]
166. Chen, D.-l.; Ju, H.-q.; Lu, Y.-x.; Chen, L.-z.; Zeng, Z.-l.; Zhang, D.-s.; Luo, H.-y.; Wang, F.; Qiu, M.-z.; Wang, D.-s.; et al. Long non-coding RNA XIST regulates gastric cancer progression by acting as a molecular sponge of miR-101 to modulate EZH2 expression. *J. Exp. Clin. Cancer Res.* **2016**, *35*, 142. [CrossRef]
167. He, A.; He, S.; Li, X.; Zhou, L. ZFAS1: A novel vital oncogenic lncRNA in multiple human cancers. *Cell Prolif* **2019**, *52*, e12513. [CrossRef]
168. Pan, J.; Xu, X.; Wang, G. lncRNA ZFAS1 Is Involved in the Proliferation, Invasion and Metastasis of Prostate Cancer Cells Through Competitively Binding to miR-135a-5p. *Cancer Manag. Res.* **2020**, *12*, 1135–1149. [CrossRef]
169. Wang, W.; Xing, C. Upregulation of long noncoding RNA ZFAS1 predicts poor prognosis and prompts invasion and metastasis in colorectal cancer. *Pathol. Res. Pract.* **2016**, *212*, 690–695. [CrossRef]
170. Rao, M.; Xu, S.; Zhang, Y.; Liu, Y.; Luan, W.; Zhou, J. Long non-coding RNA ZFAS1 promotes pancreatic cancer proliferation and metastasis by sponging miR-497-5p to regulate HMGA2 expression. *Cell Death Dis.* **2021**, *12*, 859. [CrossRef]
171. Chen, X.; Zeng, K.; Xu, M.; Hu, X.; Liu, X.; Xu, T.; He, B.; Pan, Y.; Sun, H.; Wang, S. SP1-induced lncRNA-ZFAS1 contributes to colorectal cancer progression via the miR-150-5p/VEGFA axis. *Cell Death Dis.* **2018**, *9*, 982. [CrossRef] [PubMed]
172. Fan, S.; Fan, C.; Liu, N.; Huang, K.; Fang, X.; Wang, K. Downregulation of the long non-coding RNA ZFAS1 is associated with cell proliferation, migration and invasion in breast cancer. *Mol. Med. Rep.* **2018**, *17*, 6405–6412. [CrossRef]
173. O'Brien, S.J.; Fiechter, C.; Burton, J.; Hallion, J.; Paas, M.; Patel, A.; Patel, A.; Rochet, A.; Scheurlen, K.; Gardner, S.; et al. Long non-coding RNA ZFAS1 is a major regulator of epithelial-mesenchymal transition through miR-200/ZEB1/E-cadherin, vimentin signaling in colon adenocarcinoma. *Cell Death Discov* **2021**, *7*, 61. [CrossRef] [PubMed]
174. Mishra, S.; Verma, S.S.; Rai, V.; Awasthee, N.; Chava, S.; Hui, K.M.; Kumar, A.P.; Challagundla, K.B.; Sethi, G.; Gupta, S.C. Long non-coding RNAs are emerging targets of phytochemicals for cancer and other chronic diseases. *Cell. Mol. Life Sci.* **2019**, *76*, 1947–1966. [CrossRef] [PubMed]
175. Coricovac, D.; Dehelean, C.A.; Pinzaru, I.; Mioc, A.; Aburel, O.M.; Macasoi, I.; Draghici, G.A.; Petean, C.; Soica, C.; Boruga, M.; et al. Assessment of Betulinic Acid Cytotoxicity and Mitochondrial Metabolism Impairment in a Human Melanoma Cell Line. *Int. J. Mol. Sci.* **2021**, *22*, 4870. [CrossRef]
176. Cai, J.; Sun, H.; Zheng, B.; Xie, M.; Xu, C.; Zhang, G.; Huang, X.; Zhuang, J. Curcumin attenuates lncRNA H19-induced epithelial-mesenchymal transition in tamoxifen-resistant breast cancer cells. *Mol. Med. Rep.* **2021**, *23*, 13.
177. Zamani, M.; Sadeghizadeh, M.; Behmanesh, M.; Najafi, F. Dendrosomal curcumin increases expression of the long non-coding RNA gene MEG3 via up-regulation of epi-miRs in hepatocellular cancer. *Phytomedicine* **2015**, *22*, 961–967. [CrossRef] [PubMed]
178. Khan, F.H.; Dervan, E.; Bhattacharyya, D.D.; McAuliffe, J.D.; Miranda, K.M.; Glynn, S.A. The Role of Nitric Oxide in Cancer: Master Regulator or NOt? *Int. J. Mol. Sci.* **2020**, *21*, 9393. [CrossRef]

179. Ghalaut, V.S.; Sangwan, L.; Dahiya, K.; Ghalaut, P.S.; Dhankhar, R.; Saharan, R. Effect of imatinib therapy with and without turmeric powder on nitric oxide levels in chronic myeloid leukemia. *J. Oncol. Pharm. Pract.* **2012**, *18*, 186–190. [CrossRef]
180. Hiradeve, S.M.; Rangari, V.D. Elephantopus scaber Linn.: A review on its ethnomedical, phytochemical and pharmacological profile. *J. Appl. Biomed.* **2014**, *12*, 49–61. [CrossRef]
181. Ren, G.; Rimando, A.M.; Mathews, S.T. AMPK activation by pterostilbene contributes to suppression of hepatic gluconeogenic gene expression and glucose production in H4IIE cells. *Biochem. Biophys. Res. Commun.* **2018**, *498*, 640–645. [CrossRef] [PubMed]
182. Ji, Q.; Liu, X.; Fu, X.; Zhang, L.; Sui, H.; Zhou, L.; Sun, J.; Cai, J.; Qin, J.; Ren, J.; et al. Resveratrol inhibits invasion and metastasis of colorectal cancer cells via MALAT1 mediated Wnt/β-catenin signal pathway. *PLoS ONE* **2013**, *8*, e78700. [CrossRef] [PubMed]
183. Nguyen, A.V.; Martinez, M.; Stamos, M.J.; Moyer, M.P.; Planutis, K.; Hope, C.; Holcombe, R.F. Results of a phase I pilot clinical trial examining the effect of plant-derived resveratrol and grape powder on Wnt pathway target gene expression in colonic mucosa and colon cancer. *Cancer Manag. Res.* **2009**, *1*, 25–37. [PubMed]
184. Chen, D.; Sun, Q.; Cheng, X.; Zhang, L.; Song, W.; Zhou, D.; Lin, J.; Wang, W. Genome-wide analysis of long noncoding RNA (lncRNA) expression in colorectal cancer tissues from patients with liver metastasis. *Cancer Med.* **2016**, *5*, 1629–1639. [CrossRef] [PubMed]
185. Siegel, A.B.; Narayan, R.; Rodriguez, R.; Goyal, A.; Jacobson, J.S.; Kelly, K.; Ladas, E.; Lunghofer, P.J.; Hansen, R.J.; Gustafson, D.L.; et al. A phase I dose-finding study of silybin phosphatidylcholine (milk thistle) in patients with advanced hepatocellular carcinoma. *Integr. Cancer Ther.* **2014**, *13*, 46–53. [CrossRef]
186. Lou, H.; Li, H.; Zhang, S.; Lu, H.; Chen, Q. A Review on Preparation of Betulinic Acid and Its Biological Activities. *Molecules* **2021**, *26*, 5583. [CrossRef]
187. Chen, F.; Zhong, Z.; Tan, H.Y.; Guo, W.; Zhang, C.; Cheng, C.S.; Wang, N.; Ren, J.; Feng, Y. Suppression of lncRNA MALAT1 by betulinic acid inhibits hepatocellular carcinoma progression by targeting IAPs via miR-22-3p. *Clin. Transl. Med.* **2020**, *10*, e190. [CrossRef] [PubMed]
188. Jiang, W.; Li, X.; Dong, S.; Zhou, W. Betulinic acid in the treatment of tumour diseases: Application and research progress. *Biomed. Pharmacother.* **2021**, *142*, 111990. [CrossRef]
189. Gupta, S.C.; Kannappan, R.; Kim, J.; Rahman, G.M.; Francis, S.K.; Raveendran, R.; Nair, M.S.; Das, J.; Aggarwal, B.B. Bharangin, a diterpenoid quinonemethide, abolishes constitutive and inducible nuclear factor-κB (NF-κB) activation by modifying p65 on cysteine 38 residue and reducing inhibitor of nuclear factor-κB α kinase activation, leading to suppression of NF-κB-regulated gene expression and sensitization of tumor cells to chemotherapeutic agents. *Mol. Pharmacol.* **2011**, *80*, 769–781. [PubMed]
190. Awasthee, N.; Rai, V.; Verma, S.S.; Sajin Francis, K.; Nair, M.S.; Gupta, S.C. Anti-cancer activities of Bharangin against breast cancer: Evidence for the role of NF-κB and lncRNAs. *Biochim. Biophys. Acta Gen. Subj.* **2018**, *1862*, 2738–2749. [CrossRef] [PubMed]
191. Gupta, S.C.; Patchva, S.; Koh, W.; Aggarwal, B.B. Discovery of curcumin, a component of golden spice, and its miraculous biological activities. *Clin. Exp. Pharmacol. Physiol.* **2012**, *39*, 283–299. [CrossRef] [PubMed]
192. Pei, C.S.; Wu, H.Y.; Fan, F.T.; Wu, Y.; Shen, C.S.; Pan, L.Q. Influence of curcumin on HOTAIR-mediated migration of human renal cell carcinoma cells. *Asian Pac. J. Cancer Prev.* **2014**, *15*, 4239–4243. [CrossRef] [PubMed]
193. Golombick, T.; Diamond, T.H.; Manoharan, A.; Ramakrishna, R. Monoclonal gammopathy of undetermined significance, smoldering multiple myeloma, and curcumin: a randomized, double-blind placebo-controlled cross-over 4g study and an open-label 8g extension study. *Am. J. Hematol.* **2012**, *87*, 455–460. [CrossRef] [PubMed]
194. Mahammedi, H.; Planchat, E.; Pouget, M.; Durando, X.; Curé, H.; Guy, L.; Van-Praagh, I.; Savareux, L.; Atger, M.; Bayet-Robert, M.; et al. The New Combination Docetaxel, Prednisone and Curcumin in Patients with Castration-Resistant Prostate Cancer: A Pilot Phase II Study. *Oncology* **2016**, *90*, 69–78. [CrossRef]
195. Howells, L.M.; Iwuji, C.O.O.; Irving, G.R.B.; Barber, S.; Walter, H.; Sidat, Z.; Griffin-Teall, N.; Singh, R.; Foreman, N.; Patel, S.R.; et al. Curcumin Combined with FOLFOX Chemotherapy Is Safe and Tolerable in Patients with Metastatic Colorectal Cancer in a Randomized Phase IIa Trial. *J. Nutr.* **2019**, *149*, 1133–1139. [CrossRef]
196. Tuyaerts, S.; Rombauts, K.; Everaert, T.; Van Nuffel, A.M.T.; Amant, F. A Phase 2 Study to Assess the Immunomodulatory Capacity of a Lecithin-based Delivery System of Curcumin in Endometrial Cancer. *Front. Nutr.* **2018**, *5*, 138. [CrossRef]
197. Asher, G.N. Curcumin Biomarkers. ClinicalTrials.gov Identifier: NCT01333917; 2013. Available online: https://clinicaltrials.gov/ct2/show/record/NCT01333917?recrs=e&id=NCT01333917&rank=1 (accessed on 24 April 2018).
198. Nathan, A.C. Neck Cancer. ClinicalTrials. gov Identifier: NCT01160302; 2010. Available online: https://clinicaltrials.gov/ct2/show/NCT01160302?recrs=e&id=NCT01160302&rank=1 (accessed on 2 March 2016).
199. Miller, A.H. Phase II Study of Curcumin vs Placebo for Chemotherapy-Treated Breast Cancer Patients Undergoing Radiotherapy. ClinicalTrials.gov Identifier: NCT01740323; 2012. Available online: https://clinicaltrials.gov/ct2/show/NCT01740323?recrs=e&cond=curcumin%2C+cancer&draw=2&rank=14 (accessed on 29 Jun 2020).
200. Spagnuolo, C.; Russo, G.L.; Orhan, I.E.; Habtemariam, S.; Daglia, M.; Sureda, A.; Nabavi, S.F.; Devi, K.P.; Loizzo, M.R.; Tundis, R.; et al. Genistein and cancer: current status, challenges, and future directions. *Adv. Nutr.* **2015**, *6*, 408–419. [CrossRef]
201. Chiyomaru, T.; Yamamura, S.; Fukuhara, S.; Yoshino, H.; Kinoshita, T.; Majid, S.; Saini, S.; Chang, I.; Tanaka, Y.; Enokida, H.; et al. Genistein inhibits prostate cancer cell growth by targeting miR-34a and oncogenic HOTAIR. *PLoS ONE* **2013**, *8*, e70372. [CrossRef]

202. Takimoto, C.H.; Glover, K.; Huang, X.; Hayes, S.A.; Gallot, L.; Quinn, M.; Jovanovic, B.D.; Shapiro, A.; Hernandez, L.; Goetz, A.; et al. Phase I pharmacokinetic and pharmacodynamic analysis of unconjugated soy isoflavones administered to individuals with cancer. *Cancer Epidemiol. Biomarkers Prev.* **2003**, *12*, 1213–1221.
203. Miltyk, W.; Craciunescu, C.N.; Fischer, L.; Jeffcoat, R.A.; Koch, M.A.; Lopaczynski, W.; Mahoney, C.; Jeffcoat, R.A.; Crowell, J.; Paglieri, J.; et al. Lack of significant genotoxicity of purified soy isoflavones (genistein, daidzein, and glycitein) in 20 patients with prostate cancer. *Am. J. Clin. Nutr.* **2003**, *77*, 875–882. [CrossRef]
204. Verma, S.S.; Rai, V.; Awasthee, N.; Dhasmana, A.; Rajalaksmi, D.S.; Nair, M.S.; Gupta, S.C. Isodeoxyelephantopin, a Sesquiterpene Lactone Induces ROS Generation, Suppresses NF-κB Activation, Modulates LncRNA Expression and Exhibit Activities Against Breast Cancer. *Sci. Rep.* **2019**, *9*, 17980. [CrossRef] [PubMed]
205. Ding, Y.X.; Duan, K.C.; Chen, S.L. Low expression of lncRNA-GAS5 promotes epithelial-mesenchymal transition of breast cancer cells in vitro. *Nan Fang Yi Ke Da Xue Xue Bao* **2017**, *37*, 1427–1435. [PubMed]
206. Ding, W.; Ren, J.; Ren, H.; Wang, D. Long Noncoding RNA HOTAIR Modulates MiR-206-mediated Bcl-w Signaling to Facilitate Cell Proliferation in Breast Cancer. *Sci. Rep.* **2017**, *7*, 17261. [CrossRef] [PubMed]
207. Huang, Y.; Du, J.; Mi, Y.; Li, T.; Gong, Y.; Ouyang, H.; Hou, Y. Long Non-coding RNAs Contribute to the Inhibition of Proliferation and EMT by Pterostilbene in Human Breast Cancer. *Front. Oncol.* **2018**, *8*, 629. [CrossRef] [PubMed]
208. Gupta, S.C.; Kannappan, R.; Reuter, S.; Kim, J.H.; Aggarwal, B.B. Chemosensitization of tumors by resveratrol. *Ann. N. Y. Acad. Sci.* **2011**, *1215*, 150–160. [CrossRef] [PubMed]
209. Paller, C.J.; Rudek, M.A.; Zhou, X.C.; Wagner, W.D.; Hudson, T.S.; Anders, N.; Hammers, H.J.; Dowling, D.; King, S.; Antonarakis, E.S.; et al. A phase I study of muscadine grape skin extract in men with biochemically recurrent prostate cancer: Safety, tolerability, and dose determination. *Prostate* **2015**, *75*, 1518–1525. [CrossRef]
210. Patel, K.R.; Brown, V.A.; Jones, D.J.; Britton, R.G.; Hemingway, D.; Miller, A.S.; West, K.P.; Booth, T.D.; Perloff, M.; Crowell, J.A.; et al. Clinical pharmacology of resveratrol and its metabolites in colorectal cancer patients. *Cancer Res.* **2010**, *70*, 7392–7399. [CrossRef]
211. Howells, L.M.; Berry, D.P.; Elliott, P.J.; Jacobson, E.W.; Hoffmann, E.; Hegarty, B.; Brown, K.; Steward, W.P.; Gescher, A.J. Phase I randomized, double-blind pilot study of micronized resveratrol (SRT501) in patients with hepatic metastases–safety, pharmacokinetics, and pharmacodynamics. *Cancer Prev. Res. (Phila)* **2011**, *4*, 1419–1425. [CrossRef]
212. Popat, R.; Plesner, T.; Davies, F.; Cook, G.; Cook, M.; Elliott, P.; Jacobson, E.; Gumbleton, T.; Oakervee, H.; Cavenagh, J. A phase 2 study of SRT501 (resveratrol) with bortezomib for patients with relapsed and or refractory multiple myeloma. *Br. J. Haematol.* **2013**, *160*, 714–717. [CrossRef]
213. Kjaer, T.N.; Ornstrup, M.J.; Poulsen, M.M.; Jørgensen, J.O.; Hougaard, D.M.; Cohen, A.S.; Neghabat, S.; Richelsen, B.; Pedersen, S.B. Resveratrol reduces the levels of circulating androgen precursors but has no effect on, testosterone, dihydrotestosterone, PSA levels or prostate volume. A 4-month randomised trial in middle-aged men. *Prostate* **2015**, *75*, 1255–1263. [CrossRef]
214. Croaker, A.; King, G.J.; Pyne, J.H.; Anoopkumar-Dukie, S.; Liu, L. Sanguinaria canadensis: Traditional Medicine, Phytochemical Composition, Biological Activities and Current Uses. *Int. J. Mol. Sci.* **2016**, *17*, 1414. [CrossRef] [PubMed]
215. Zhang, S.; Leng, T.; Zhang, Q.; Zhao, Q.; Nie, X.; Yang, L. Sanguinarine inhibits epithelial ovarian cancer development via regulating long non-coding RNA CASC2-EIF4A3 axis and/or inhibiting NF-κB signaling or PI3K/AKT/mTOR pathway. *Biomed. Pharmacother.* **2018**, *102*, 302–308. [CrossRef] [PubMed]
216. Gazák, R.; Walterová, D.; Kren, V. Silybin and silymarin–new and emerging applications in medicine. *Curr. Med. Chem.* **2007**, *14*, 315–338. [CrossRef] [PubMed]
217. Chen, L.L.; Carmichael, G.G. Decoding the function of nuclear long non-coding RNAs. *Curr. Opin. Cell Biol.* **2010**, *22*, 357–364. [CrossRef]
218. Imai-Sumida, M.; Chiyomaru, T.; Majid, S.; Saini, S.; Nip, H.; Dahiya, R.; Tanaka, Y.; Yamamura, S. Silibinin suppresses bladder cancer through down-regulation of actin cytoskeleton and PI3K/Akt signaling pathways. *Oncotarget* **2017**, *8*, 92032–92042. [CrossRef] [PubMed]
219. Bosch-Barrera, J.; Sais, E.; Cañete, N.; Marruecos, J.; Cuyàs, E.; Izquierdo, A.; Porta, R.; Haro, M.; Brunet, J.; Pedraza, S.; et al. Response of brain metastasis from lung cancer patients to an oral nutraceutical product containing silibinin. *Oncotarget* **2016**, *7*, 32006–32014. [CrossRef]
220. Flaig, T.W.; Glodé, M.; Gustafson, D.; van Bokhoven, A.; Tao, Y.; Wilson, S.; Su, L.J.; Li, Y.; Harrison, G.; Agarwal, R.; et al. A study of high-dose oral silybin-phytosome followed by prostatectomy in patients with localized prostate cancer. *Prostate* **2010**, *70*, 848–855. [CrossRef]

Disclaimer/Publisher's Note: The statements, opinions and data contained in all publications are solely those of the individual author(s) and contributor(s) and not of MDPI and/or the editor(s). MDPI and/or the editor(s) disclaim responsibility for any injury to people or property resulting from any ideas, methods, instructions or products referred to in the content.

molecules

MDPI

Review

Maslinic Acid: A New Compound for the Treatment of Multiple Organ Diseases

Yan He [1,†], Yi Wang [1,†], Kun Yang [1], Jia Jiao [2], Hong Zhan [2], Youjun Yang [1], De Lv [3], Weihong Li [1,*] and Weijun Ding [1,*]

1 Department of Fundamental Medicine, Chengdu University of Traditional Chinese Medicine, 1166 Liutai Avenue, Chengdu 611137, China
2 Department of Clinical Medicine, Chengdu University of Traditional Chinese Medicine, 1166 Liutai Avenue, Chengdu 611137, China
3 Affiliated Hospital of Chengdu University of Traditional Chinese Medicine, Chengdu 611137, China
* Correspondence: lwh@cdutcm.edu.cn (W.L.); dingweijun@cdutcm.edu.cn (W.D.); Tel.: +86-28-6180-0219 (W.L. & W.D.); Fax: +86-28-6180-0225 (W.L. & W.D.)
† These authors contributed equally to this work.

Abstract: Maslinic acid (MA) is a pentacyclic triterpene acid, which exists in many plants, including olive, and is highly safe for human beings. In recent years, it has been reported that MA has anti-inflammatory, antioxidant, anti-tumor, hypoglycemic, neuroprotective and other biological activities. More and more experimental data has shown that MA has a good therapeutic effect on multiple organ diseases, indicating that it has great clinical application potential. In this paper, the extraction, purification, identification and analysis, biological activity, pharmacokinetics in vivo and molecular mechanism of MA in treating various organ diseases are reviewed. It is hoped to provide a new idea for MA to treat various organ diseases.

Keywords: Maslinic acid; extraction; purification; identification; biological activity; pharmacokinetics; treatment of multiple organ diseases

Citation: He, Y.; Wang, Y.; Yang, K.; Jiao, J.; Zhan, H.; Yang, Y.; Lv, D.; Li, W.; Ding, W. Maslinic Acid: A New Compound for the Treatment of Multiple Organ Diseases. *Molecules* **2022**, 27, 8732. https://doi.org/10.3390/molecules27248732

Academic Editors: Arunaksharan Narayanankutty, Ademola C. Famurewa, Eliza Oprea and Arjun H. Banskota

Received: 31 October 2022
Accepted: 7 December 2022
Published: 9 December 2022

Publisher's Note: MDPI stays neutral with regard to jurisdictional claims in published maps and institutional affiliations.

Copyright: © 2022 by the authors. Licensee MDPI, Basel, Switzerland. This article is an open access article distributed under the terms and conditions of the Creative Commons Attribution (CC BY) license (https://creativecommons.org/licenses/by/4.0/).

1. Introduction

Maslinic acid (MA) is widely found in a variety of plants including olive [1,2], loquat leaves [3,4], red dates [5,6], eucalyptus [7], crape myrtle [8], sage [9], plantain [10], prunella vulgaris [11], spiny leaf dong [12], etc. Olive oil is rich in MA. Olive oil is the traditional edible oil of people in Mediterranean coastal countries, with good nutritional and health care effects. In addition, it is also widely used in beauty care, which also shows that MA is very safe for the human body. The research shows that the MA content of extra virgin olive oil in different regions is 20–98 mg/kg, the MA content of high acid value olive oil in different regions is 212–356 mg/kg, the MA content in olive residue is the highest, and the MA content in olive residue in different regions is 212–1485 mg/kg [13]. Therefore, olive pomace is often used as raw material for MA extraction.

Pure MA is a white powder, molecular formula: $C_{30}H_{48}O_4$, molecular weight: 472.70, melting point: 266–268 °C. It is insoluble in water and petroleum ether, soluble in ethanol and methanol, and soluble in ethyl acetate, benzene and chloroform. The molecular formula of MA is shown in Figure 1. Research has shown that high concentration of MA does not cause any discomfort or death in the treatment of animal disease models [14]. Therefore, MA is considered as a natural compound with low toxicity and side effects.

In the past few decades, people have become more and more interested in natural compounds with health benefits. MA has been extracted from natural plants, such as jujube, olive and hawthorn [15]. Subsequently, researchers studied its pharmacological activity and potential molecular mechanism for disease treatment and toxicological effect [15], and found that MA has beneficial effects in medicine [15,16], health food [15–17], animal feed [18,19] and other fields. The biological activity of MA has been systematically summarized by scholars, but the extraction of MA and the therapeutic mechanism of MA for various organ diseases have not been discussed [15]. In this paper, we will comprehensively and systematically review the research progress on the extraction, isolation, identification, bioavailability and bioactivity of MA, and discuss the preventive and therapeutic effects of MA on diseases related to brain, lung, heart, stomach, intestine and kidney organs. The purpose of writing this review is to provide useful experimental data and research ideas for researchers interested in MA research.

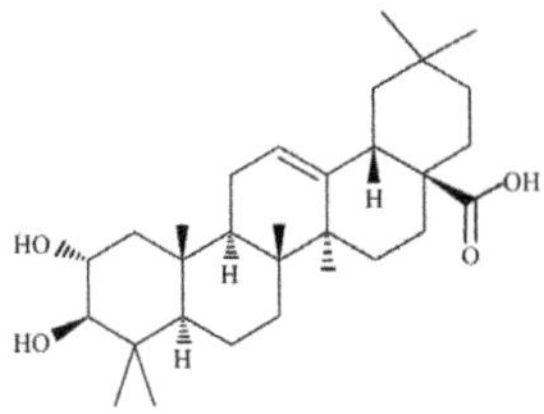

Figure 1. Structural formula of MA.

2. Extraction of MA

2.1. Extraction

Solvent extraction technology is the most classical and convenient extraction method, and also the most commonly used method for extracting natural ingredients. The results have shown that MA [20–24] can be successfully extracted from olive, apple, jujube and other plants with methanol, ethanol and ethyl acetate. Fernandez-Pastor et al. [24] determined that the optimal conditions for extracting MA from olive were as follows: "at T = 65–70 °C, t = 30 min, ethyl acetate (EtOAc) was used as the solvent, and the ratio of material to liquid was 1:40, 1:20 and 1:10 respectively". The solvent extraction of MA has the disadvantages of time-consuming, high-energy consumption, low extraction efficiency and destruction of biological activity. Therefore, it is necessary to develop an economical and effective MA extraction method.

In order to improve the extraction efficiency of MA, microwave assisted extraction (MAE) is widely used in MA extraction [20,24]. Compared with traditional solvent extraction, MAE has the advantages of low organic solvent consumption, high recovery and low cost [24]. Xie et al. [20] determined the best extraction conditions of microwave assisted extraction: "ethanol as the extraction solvent, extraction time of 5 min, liquid–solid ratio of 30 mL/g, extraction temperature of 50 °C, and microwave power of 600 w". Compared with traditional solvent extraction, MAE not only greatly shortened the extraction time of MA, but the extraction solution was heated evenly and rapidly, so the extraction efficiency of MA was greatly improved.

In addition, in order to improve the yield of MA and optimize the experimental conditions for MA extraction, ultrasonic-assisted extraction was developed and used. Wu et al. [25] successfully extracted MA from a variety of Chinese medicinal materials by using ultrasonic-assisted dispersion liquid–liquid microextraction (UA-DLLME) with chloroform as the extraction solvent and acetone as the dispersion solvent. Gómez-Cruz et al. [26] used ultrasound-assisted extraction for 16 min to extract MA from olive pomace. Song et al. [27] determined the optimal conditions for ultrasonic assisted extraction of MA from jujube to be: "temperature 55.14 °C, ethanol concentration 86.57%, time 34.41 min, liquid–solid ratio 39.33 mL/g, and the average content of MA extracted was 265.568 (μg/g DW)". Xie et al. [20] determined the optimal conditions for ultrasonic assisted extraction of MA from olive residue to be: "extraction temperature 50 °C, liquid–solid ratio 30 (mL/g), ultrasonic frequency 60 kHz, extraction 5 min, and ultrasonic intensity 135.6 W/cm^2". Goulasal et al. [28] determined that ethanol or methanol–ethanol mixture is the most effective solvent for ultrasonic extraction of MA from olive dregs. Ultrasound-assisted extraction is not only convenient, simple and time-consuming, but also allows extraction at a lower temperature, which can avoid the risk of damaging MA activity at high temperatures. In recent years, supercritical fluid extraction (SFE) with both liquid and gas properties has also emerged. It has the characteristics of high diffusivity, low viscosity and good solubility. The experimental results show the optimum conditions for supercritical extraction of MA to be: "CO_2 as the medium, extraction pressure 22 MPa, extraction temperature 60 °C, material liquid ratio 1:40, extraction time 3 h". Under these conditions, the extraction yield of MA was the highest [29–31].

Solvent extraction, microwave-assisted extraction, ultrasonic-assisted extraction and supercritical fluid extraction are compared. Solvent extraction of MA has the disadvantages of time-consuming, high-energy consumption, low extraction efficiency and destruction of biological activity. MAE greatly shortens the extraction time of MA, and the extraction solution is heated evenly and rapidly. Ultrasound-assisted extraction is not only convenient, simple and time-saving, but also can be performed at a lower temperature, which can avoid the risk of destroying MA activity at high temperatures. SFE is the only method that does not consume organic solvents, but is time-consuming and inefficient.

2.2. *Purification and Separation of MA*

The separation and purification procedures are summarized as follows: In short, the MA liquid extract is first concentrated, and then dried until the weight of the solid sediment remains unchanged, so that the MA crude extract can be obtained. The crude extract of MA was extracted with ethanol (100% v/v ethanol, 24 h, 10% w/v solid load) to obtain ethanol extract [32]. Finally, the ethanol extract was purified by HPLC. Water containing 5% acetonitrile (A) and 100% acetonitrile (B) was used as the mobile phase on the chromatographic column for gradient elution. The flow rate of the mobile phase was 1 mL min^{-1}, and the column temperature was set at 30 °C [25]. It should be noted that in order to prevent bubbles from escaping from the system and affecting the mobile phase, the mobile phase used for HPLC needs to be ultrasonically treated in advance to be completely degassed [33]. The elution procedure is as follows: 0–5 min, 65–90% B; 5–15 min, 90–92% B; 15–35 min, 92–92.5% B; 35–38 min, 92.5–100% B. The final steps include recrystallization with methanol, filtration, freeze drying and finally obtaining purified MA.

2.3. *Identification of MA*

2.3.1. Colorimetric Method

MA is a pentacyclic triterpene acid compound with 2,3 dihydroxy groups, which has the characteristic color reaction of terpenoids. MA is a 12-alkene structure with slow color development. The main color reaction is as follows: (1) MA reacts with acetic anhydride concentrated sulfuric acid: the sample is dissolved in acetic anhydride, then a few drops of concentrated sulfuric anhydride (1:20) are added, which can produce yellow → red → purple → blue and finally fade, and this color change could identify the sample as MA. (2) MA reacts with trichloroacetic acid: the sample chloroform solution is dropped on filter paper, then 1 drop of 25% TCA solution is added and heated to 100 °C for color development; if red to purple color was generated, this color change could identify the sample as MA. (3) MA reacts with chloroform concentrated sulfuric acid: the sample is dissolved in chloroform. After adding concentrated sulfuric acid, chloroform is red or blue, and the sulfuric acid layer is green fluorescent, so the sample can be identified as MA. (4) MA reacts with antimony pentachloride: the chloroform or ethanol solution of the sample is dropped on the filter paper, and then 20% antimony pentachloride chloroform solution is sprayed on the filter paper. If it is heated, it will show blue, gray blue and gray purple color spots, and this color change could identify the sample as MA. (5) Glacial acetic acid–acetyl chloride reaction: The sample is dissolved in glacial acetic acid, and then a few drops of acetyl chloride and a few grains of zinc oxide are added. After the sample solution is slightly heated, if the color changed to light red or purplish red, this color change could identify the sample as MA [34,35].

2.3.2. Chromatography

In addition to colorimetric methods, we can also use chromatographic methods to identify MA. Guo et al. [6] used high-performance liquid chromatography (HPLC) coupled with evaporative light-scattering detection (ELSD) to identify MA in date palm leaves. Yang et al. [36] also analyzed the content of hawthorneic acid in plum blossoms using high-performance liquid chromatography coupled with evaporative light-scattering detection (HPLC-ELSD). Perez-Camino et al. [13] used gas chromatography (GC) to alkalize MA samples, which were then injected into GC to obtain gas chromatograms of MA alkylated derivatives, and this method was used to identify MA in olive oil. In GC, MA cannot be injected directly into the chromatograph due to its low volatility and high molecular weight, and must be derivatized prior to gas chromatographic analysis. The use of HPLC hinders the detection of MA because it has a saturated backbone, does not show fluorescence and has a very low UV absorption. Therefore, this method may not be sensitive enough to detect low doses of hawthorneic acid. Sánchez-González et al. [37] developed a new detection technique, liquid chromatography–atmospheric pressure chemical ionization mass spectrometry (LC-APCI-MS), for the analysis of MA in plasma. Compared to GC, LC-APCI-MS does not depend on specific groups in the molecule, thus reducing the derivatization step and avoiding sample consumption. Also, it can be used for highly specific detection without long chromatographic separations compared to HPLC. Therefore, LC-APCI-MS may be a relatively effective and sensitive method to determine the concentration of MA. In conclusion, the extraction, purification and identification of MA are shown in Figure 2.

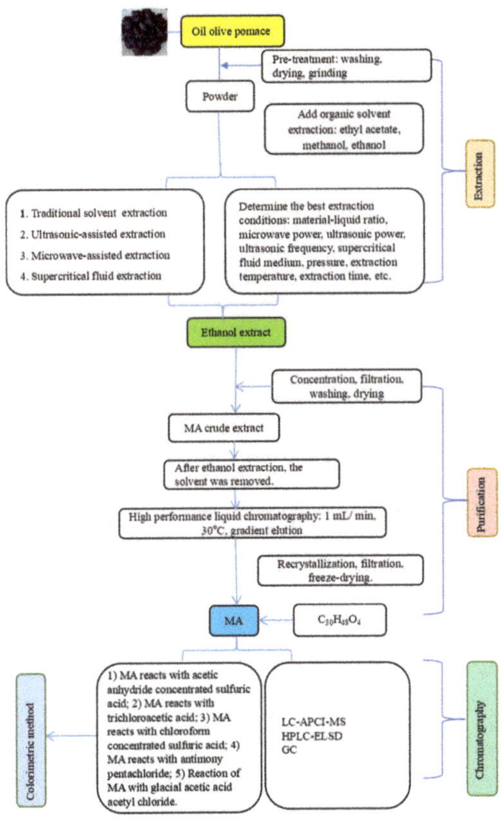

Figure 2. Extraction, purification and identification of MA.

3. Biological Activity of MA

3.1. Hypoglycemic Effect

MA has been widely studied in reducing blood glucose. As a hypoglycemic drug, MA has the following advantages: (1) non-toxic, fewer side effects; (2) treatment of various complications of diabetes, such as nonalcoholic fatty liver disease caused by diabetes [38], diabetes nephropathy [39] and other complications, which seriously endanger human health. In addition, MA can also act as an inhibitor of glycogen phosphorylase, inhibit the activity of glycogen phosphorylase, reduce glycogen decomposition and reduce blood sugar [40]. It has also been shown that MA can reduce the activity of carbohydrate hydrolase and glucose related transporters in the small intestine during the digestion and absorption of carbohydrates in the small intestine, and reduce the absorption of glucose [41]. After a large amount of glucose intake, MA can also increase insulin sensitivity by activating the AMPK/Sirtl pathway, and promote the transformation and storage of blood glucose [38,40,42].

3.2. Antioxidant Effect

There are two kinds of antioxidant systems in the body. One is the non-enzymatic antioxidant system, including ergot thiophene, vitamin C, vitamin E, glutathione and trace elements. The other is the enzyme antioxidant system, including CAT, SOD and GPX. Oxidative stress is a negative effect of free radicals in the body and is considered to be one of the important factors leading to aging and disease [43]. Free radicals and lipid peroxides produced in the process of oxidative stress can cause cross-linking and polymerization of life macromolecules such as proteins and nucleic acids, leading to protein damage, inhibition of RNA synthesis and damage to normal cell structure, which may interfere with cell cycle, chromosome separation, hormone synthesis, cell proliferation and apoptosis, and even lead to dysfunction of oocytes [44]. It is reported that MA has an antioxidant effect and can improve oxidative damage caused by oxidative stress [31]. It was found that MA mainly prevented the generation of oxidative stress by preventing oxidative stress caused by excessive H_2O_2, thereby reducing the production level of ROS [31], inhibiting the ability of low-density lipoprotein (LDL) peroxidation [45] and inhibiting the production of nitric oxide (NO) induced by lipopolysaccharide (LPS) [46]. However, it is worth noting that although MA has antioxidant properties, MA can lead to an increase in reactive oxygen species in cells at higher doses, further inducing cell damage and apoptosis [47]. The results showed that in MA-treated B16F10 melanoma cells, ROS levels were decreased at MA concentrations less than ($IC_{50/8}$) and increased at MA concentrations greater than ($IC_{50/8}$) [48].

3.3. Neuroprotective Effect

The nervous system plays a leading role in maintaining the homeostasis of the internal environment, maintaining the integrity and unity of the body, and its coordination and balance with the external environment. When the nervous system is injured, a series of diseases will appear, such as cerebrovascular disease, peripheral neuropathy, spinal cord disease, extrapyramidal disease, neuromuscular junction disease, paroxysmal disease, etc., which not only damage human health, but also bring a heavy psychological and economic burden to patients' families and society. Some studies have shown that MA can significantly reduce ischemia-induced neuronal apoptosis, reduce the level of MDA, and increase SOD activity in the cortex and hippocampus; MA can inhibit NF in the astrocyte κB signal pathway [49]. MA can also reduce the glutamate toxicity of primary neurons in the cerebral cortex in a dose-dependent manner, and improve the survival rate of neurons [50]. At the same time, MA can also regulate extracellular glutamate concentration by increasing the expression of glutamate transporter in astrocytes, which may provide neuroprotection [50]. After the occurrence of nerve injury, MA can also promote Akt activity, increase the growth of nerve axons and synapses, and restore the loss of nerve synapses [48].

3.4. Anti-Inflammatory Effect

Inflammation refers to the pathological process in which the body mainly defends against the stimulation of inflammatory factors. Acute inflammatory reaction is one of the most active and important molecular processes in the human body. It clears damaged tissues, restores homeostasis, and protects the host from the threat of pathogens or damage. It is now believed that uncontrolled inflammation is pathological and related to many diseases, such as cardiovascular disease, asthma, neurodegenerative disease, diabetes and obesity [51]. Although the current anti-inflammatory treatment can treat the uncomfortable signs and symptoms caused by inflammation, it may lead to severe immunosuppression and opportunistic infection. Natural compounds have become the focus of anti-inflammation because of their extensive biological activity, low toxicity and weak side effects. Interleukin-1β (IL-1β) is one of the key mediators in stimulating inflammatory response. It was found that in IL-1β-induced inflammatory reaction, MA can effectively pass the PI3K/AKT/NF-κ channel B [52], HMGB1/TLR4/NF-κB pathway [53] and p-STAT-1 pathway [54], inhibit the expression of various inflammatory factors and thus inhibit the inflammatory response.

3.5. Anti-Tumor Effect

At present, radiotherapy, chemotherapy and surgery are the main treatments for tumors. However, some advanced and well differentiated patients have poor sensitivity to radiotherapy, while conventional chemotherapy drugs have large side effects and are prone to drug resistance, which significantly reduces the quality of life of patients. Traditional Chinese medicine (TCM) has many advantages in treating tumors, such as rich drug sources, flexible prescriptions and small side effects. It has become one of the research hotspots of anti-tumor drugs to search for anti-tumor active components from traditional Chinese medicine and then study their mechanism of action. Studies have found that MA has a wide range of anti-tumor activities. MA has good anti-tumor activity in a variety of tumors, including colon cancer [55–58], lung cancer [59,60], colorectal cancer [61,62], pancreatic cancer [59,63,64], kidney cancer [65], prostate cancer [66], bladder cancer [67], gallbladder cancer [68], papilloma [69], gastric cancer [70], human neuroblastoma [71], cardiac cancer [72], etc. The mechanisms of MA in treating tumors include inhibiting cell proliferation [63], blocking cell cycle [48,73], inducing cancer cell apoptosis [56] and inhibiting angiogenesis [74]. The biological activity of MA is summarized in Figure 3.

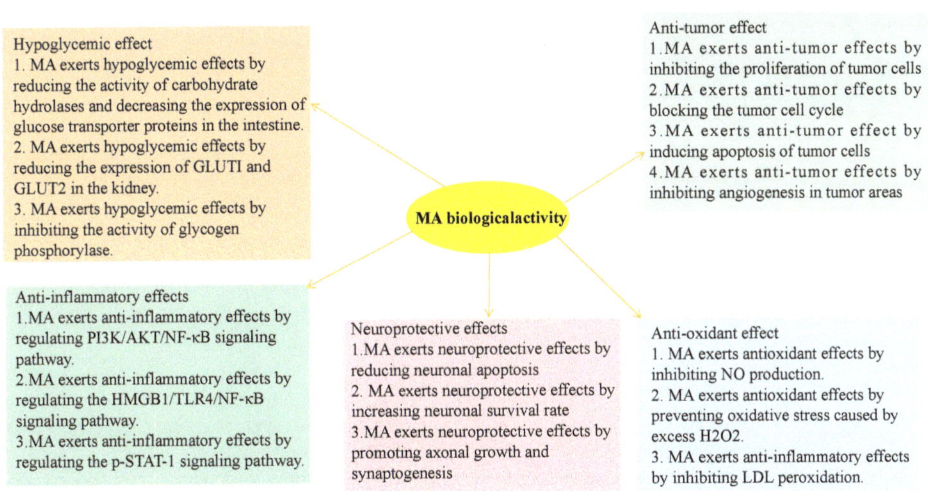

Figure 3. Biological activity of MA.

4. Bioavailability of MA

Although MA has so many biological activities, its in vivo efficacy in the human body also depends on its absorption and metabolism, so it is important to understand the bioavailability of MA. Bioavailability is the fraction of an orally administered dose of intact drug that reaches the systemic circulation, taking into account absorption and local metabolic degradation. The NUTRAOLEUM study [75] was used to evaluate the pharmacokinetics of MA in the human body in a single dose of high triterpene-acid content olive oil. After repeated administration of MA, its plasma concentration gradually accumulated in the dry period of one week. After a single dose of 30 mL MA for 24 h, it increased by 6.7 ng/mL from 1.8 ng/mL at baseline, and reached 21.5 ng/mL three weeks later. The bioavailability based on Cmax and AUC0-10 is 7 times higher than that of OA. The oral bioavailability of MA is about 6.25%.

Studies in rodents have shown that [75] MA is rapidly absorbed orally, reaching a peak at 0.51 h with a bioavailability of 5.13%, which is consistent with the reported values for other pentacyclic triterpenoids; after entering the blood MA was then widely distributed in tissues with central and peripheral distribution volumes of 8.41 L/70 kg and 63.6 L/70 kg, respectively. The reason for the low oral availability of MA may be that MA is affected by the mechanical action of the digestive tract, digestive enzymes, pH and other foods consumed together during oral administration. After oral administration of triterpenes for 4 weeks, the concentration of triterpenes in organs was measured, and the results showed that some triterpenes including MA had the highest concentrations in the liver, followed by the kidneys [76]. It is suggested that the liver may be the main organ for MA storage. Furthermore, MA is able to penetrate the blood–brain barrier after being absorbed and metabolized [76], a finding that supports the possibility of using MA for brain protection.

5. Therapeutic Effects of MA

5.1. MA Treatment of Brain Diseases

The therapeutic effect of MA on brain diseases is mainly through anti-inflammatory [49], antioxidant [77] and neuroprotective bioactivity [50]. MA is mainly used to treat Alzheimer's disease (AD), epilepsy, stroke, astrocytoma, etc.

5.1.1. MA Treatment of Alzheimer's Disease

AD is a neurodegenerative disease whose pathogenesis is still unclear. It is currently believed that impaired brain neurotrophic factor (BDNF) signaling may be one of the pathogenic mechanisms leading to neurodegenerative diseases [78]. BDNF is a neuroprotective agent that is very important for neuronal survival and neurotransmission, and can prevent and improve neurodegeneration in AD [78]. A study has shown that BDNF can prevent and improve memory cognitive impairment in AD [79]. In addition, BDNF has been reported to be critical for memory consolidation and enhancement 9 to 12 h after memory acquisition [80,81]. Bae et al. [82] used a scopolamine-induced mouse AD model and then treated the mice with MA (concentration of 1 mg/kg and 3 mg/kg). The results showed that the mRNA expression level of BDNF was increased, neuronal apoptosis was reduced, and the cognitive function and memory cognitive impairment of the mice were improved 9 and 12 h after treatment. Thus, MA may be a potential drug for the treatment of AD.

5.1.2. MA Treatment of Epilepsy

Epilepsy is a chronic brain dysfunction caused by excessive discharge of neurons in the brain, and both experimental and clinical evidence suggests that inflammatory stress in the brain is associated with epilepsy, especially in the hippocampus [83,84]. Key inflammatory mediators contribute to the progression of epileptic seizures, such as IL-1β, IL-6, TNF-α and prostaglandin E2 (PGE2) [85,86]. Excessive production of reactive oxygen species (ROS) can also lead to seizures associated with neuronal depolarization and exacerbate neurological dysfunction [87]. In addition, glutamate excitotoxicity is another important factor that induces epilepsy [88]. Therefore, any drug that reduces inflammation, oxidation or glutamate toxicity in the hippocampus can reduce the severity of epilepsy.

Wang et al. [89] used kainic acid (KA) to induce epileptiform behavior in mice, which were at the same time given 20 or 40 mg/kg/d MA (MA) treatment; the results showed that MA preadministration reduces the production of IL-1β, IL-6, TNF-α and PGE2, which are inflammatory factors. Also, MA preadministration maintains the activity of glutathione peroxide (GPX) and glutathione reductase (GR), thereby reducing oxidized glutathione (GSSG) production and preserving glutathione (GSH), reducing ROS production, enhancing hippocampal glutamine synthetase activity, decreasing glutamate levels, and increasing glutamine levels, thereby alleviating epileptiform behavior. These findings suggest that MA may be an effective therapeutic agent for seizure relief.

5.1.3. MA Treatment of Ischemic Stroke

Ischemic stroke is a cerebrovascular disease that causes ischemia and hypoxia in brain tissue due to impaired blood supply to the brain, resulting in necrosis, softening and eventually infarct foci. Neural regeneration and brain repair in the injured area after ischemic stroke mainly rely on axonal remodeling and dendritic plasticity. MA has a better preventive effect on ischemic stroke. Qian et al. [90] induced an ischemic stroke model by oxygen–glucose deprivation, and the experimental results showed that MA treatment normalizes the expression/activation of caspase3 and caspase9, and increases the ratio of Bcl-2/Bax. In addition, MA inhibited NO production and iNOS expression induced by oxygen–glucose deprivation. These results indicate that MA has beneficial effects on hypoxic neurons by inhibiting the activation of iNOS. It indicates that MA has a better preventive effect on ischemic stroke and can reduce the necrosis of neurons. Guan et al. [91] found that MA also has a good therapeutic effect on the stage of stroke. MA treatment can enhance the expression of glial glutamate transporter GLT-1 at the protein and mRNA levels, leaving extracellular glutamate at a low concentration, thus playing a protective role in nerve cells during the ischemic stroke. In addition, Qian et al. [92] found that the presence of MA prolonged the therapeutic time window of MK-801, and the combination of the two may be more effective in the treatment of acute ischemic stroke. For the recovery period of stroke, MA can also have a therapeutic effect. After MA was given to mice treated with reperfusion after ischemic stroke for a period of time, it was found that synaptophysin levels increased significantly, nerve axon damage recovered significantly, and synapse growth and synaptogenesis increased [93]. The above experimental results indicate that MA has therapeutic effects on ischemic stroke through multi-stage and multi-pathway effects.

5.1.4. MA Treatment of Malignant Astrocytoma

Malignant astrocytoma is a common primary brain tumor. Traditional Chinese medicine has many advantages in the treatment of tumors, such as rich sources of drugs, flexible prescriptions and small side effects. It has become one of the research hotspots of anti-tumor drugs to find anti-tumor active components from traditional Chinese medicine and study their mechanism of action. It has been found that MA has a good therapeutic effect on malignant astrocytoma.

Martín et al. [77] incubated 1321N1 cells with different concentrations of MA (1, 5, 25 and 50 μmol/L) for 24 h. It was found that MA led to an increase in intracellular reactive oxygen species, loss of mitochondrial membrane integrity and subsequent apoptosis of tumor cells. Therefore, MA may be a promising new drug for the treatment of astrocytoma.

5.2. MA Treatment of Lung Diseases

MA can treat lung diseases through various anti-inflammatory pathways [54,94] and, at the same time, it can also treat lung cancer by inhibiting cancer cell proliferation and inducing cancer cell apoptosis [59,60].

5.2.1. MA Treatment of Lung Cancer

Lung cancer is one of the most common cancers in the world. After many treatments, the five-year survival rate of lung cancer is still very low [95]. Therefore, it is very important to find drugs that can induce apoptosis of lung cancer cells.

Hsia et al. [59] incubated human lung cancer A549 cells with different concentrations of MA (4, 8, 16, 32 and 64 μmol/L) and found that MA can mediate the apoptosis pathway of tumor cells and HIF-1 α, and cause apoptosis of cancer cells. Bai et al. [60] incubated human lung cancer A549 cells with different concentrations of MA (0, 9, 12, 15, 18, and 21 μg/mL) for 24 h, and found that MA can regulate Smac expression and reduce c-IAP1, c-IAP2, XIAP and survivin expression to regulate the expression of caspase-3, caspase-8 and caspase-9, thereby inducing cancer cell apoptosis. These findings prove that MA can induce apoptosis of lung adenocarcinoma cells, and thus have a therapeutic effect on lung cancer.

5.2.2. MA Treatment of Lung System Damage

Lung injury is lung parenchyma injury caused by different factors, and its pathological mechanism is mainly related to oxidative stress and inflammatory injury. Therefore, anti-inflammatory and antioxidant drugs may have beneficial effects on lung injury. Jeong et al. [94] induced lung injury in mice using PM2.5 (10 mg/kg) generated by intratracheal instillation of diesel and, 30 min later, MA (0.2, 0.4, 0.6 and 0.8 mg/kg) was injected via the tail vein. It was found that MA has the ability to modulate the TLR4-MyD88 and mTOR autophagy pathways to counteract PM2.5-induced lung injury. Lee et al. [54] injected LPS (15 mg/kg i.p.) into the peritoneal cavity to induce lung injury, followed by intravenous injection of MA (0.07–0.7 mg/kg) six hours after the injection, and found that MA can down-regulate NF-κB and p-STAT-1 to regulate iNOS, exerting an anti-inflammatory effect. These results suggest that MA may be a potential natural compound for the treatment of lung injury.

5.3. MA Treatment of Heart Disease

MA plays a therapeutic role in heart-related diseases through its anti-hyperlipidemia, inhibition of lipid peroxidation (LPO), and anti-blood-glucose and antioxidant effects.

5.3.1. MA Treatment of Pathological Cardiac Hypertrophy

Physiological cardiac hypertrophy is an adaptive response to physiological and pathological stimuli and is designed to cope with increased workload [96]. However, pathological cardiac hypertrophy is a major risk factor for cardiomyopathy, heart failure and sudden cardiac death [97]. The mechanism of pathological myocardial hypertrophy is that cardiomyocyte hypertrophy enters a decompensated phase, resulting in cell death, myocardial fibrosis, sarcomere structure changes, metabolic reprogramming, protein damage, etc., which ultimately lead to poor cardiac remodeling and heart failure [98]. N6-methyladenosine (m6A) is the most prevalent internal modification of mammalian messenger RNA (mRNA) [99], and m6A plays an important role in cardiac biological processes and the pathogenesis of cardiovascular disease [100,101]. At the same time, there is increasing evidence that the extracellular regulated protein kinase (ERK) and protein kinase B (PKB/AKT) signaling pathways are involved in the process of cardiac hypertrophy [102,103]. Therefore, pharmacological intervention with respect to these molecules could facilitate the development of treatments for cardiac hypertrophy.

Fang et al. [104] found that MA treatment significantly inhibited the hypertrophy of NMCMs induced by Ang II, and the dose did not affect the cell viability of NCMCs. At the same time, the researchers established a mouse model of transverse aortic contraction (TAC) to simulate cardiac hypertrophy caused by pressure overload, and then injected MA (30 mg/kg/d) intraperitoneally for 14 days. MA administration reduced the level of total RNA m6A methylation and METTL3 in TAC stressed hearts. In vivo and in vitro, MA can significantly improve myocardial hypertrophy, myocardial fibrosis and cardiac function. Liu et al. [105] ligated the aorta of mice to induce cardiac hypertrophy caused by pressure overload. After operation, the mice took MA (20 mg/kg) orally for 4 weeks, and the hypertrophic cardiomyocytes induced by PE were treated with MA in vitro. It was found that MA alleviated the pressure-overload-induced cardiac hypertrophy in vivo and PE-induced cardiac hypertrophy in vitro by reducing the phosphorylation of AKT and ERK signaling pathways. These results suggest that MA may be a potential drug for the treatment of cardiac hypertrophy.

5.3.2. MA Treatment of Acute Myocardial Infarction

Ischemic heart disease has become one of the most important causes of death worldwide [96]. Abnormal lipid metabolism is the most important risk factor for atherosclerosis, and high triglycerides are an independent marker of increased risk of ischemic events [106]. Oxyphosphatase (PON) prevents oxidation of low-density lipoprotein cholesterol (LDL-C) and protects cell membranes from free radical damage to prevent atherosclerosis [107], and is an important target for the treatment of ischemic cardiomyopathy. In addition, xanthine oxidase (XO) is an enzyme that produces free radicals, which can produce superoxide, hydrogen peroxide, NADH and NO, and eventually lead to some cardiovascular diseases (CVD). Therefore, XO is a potential drug target for the treatment of CVD [108].

Hussain Shaik et al. [109] used MA (15 mg/kg) to intervene in a myocardial infarction model induced by isoproterenol (85 mg/kg). It is found that MA can increase PON activity, reduce LDL-C level and inhibit LPO activity, thus playing a protective role in the heart. Another experiment by Shaik et al. [110] found that MA can relieve myocardial infarction (MI) by inhibiting XO. Therefore, MA can be considered as a potential natural drug for the treatment of MI without obvious adverse reactions.

5.3.3. MA Treatment of Diabetic Heart Disease

Diabetes is a global epidemic with a strong association with cardiovascular disease [111]. Diabetes heart disease is due to the imbalance of sugar and lipid metabolism related to diabetes, which leads to increased oxidative stress and activation of multiple inflammatory pathways, ultimately leading to tissue damage, heart remodeling, heart dysfunction, etc. [112].

Khathi et al. [41] found that the IC_{50} values for sucrase, α-amylase and α-glucosidase were lower for different concentrations of MA than for acarbose, suggesting that MA was effective in controlling blood glucose. Hung et al. [113] injected streptozotocin (50 mg/kg) into the tail vein of mice to induce diabetes in mice (fasting blood glucose level $\geq$ 200 mg/dL), and then fed mice with 0.1% MA and 0.2% MA diet for 12 consecutive weeks to study the cardioprotective effect of MA in diabetic mice. This study found that MA can protect the hearts of diabetic mice by alleviating glycosyl damage and coagulation disorders, inhibiting NF-κB and MAPK pathways, and reducing oxidative stress. While inhibiting oxidative stress, MA can also regulate blood sugar and improve heart function without obvious adverse reactions. MA can be used to treat diabetes heart disease [19].

5.4. MA Treatment of Liver Diseases

MA effectively inhibits fat synthesis and restores hepatic glycogen to normal levels [114,115]. In addition, MA increases the activity of antioxidant enzymes through anti-inflammatory and antioxidant effects during liver injury, improves liver function [38], and can also reduce the production of inflammatory factors and protect the liver. Finally, MA can also inhibit the metastasis and invasion of liver cancer [116].

5.4.1. MA Treatment of Acute Liver Injury

Acute liver injury is a life-threatening syndrome with high morbidity [117], which is often caused by viruses, bacteria, drugs and toxins [118,119], and the pathological mechanisms involved are complex and yet to be elucidated. However, there is increasing evidence that inflammation and oxidative stress are involved in the pathogenesis of liver injury [120]. For example, the detrimental effects of alcohol in alcoholic acute liver injury are mainly attributed to the large production of reactive oxygen species (ROS) and acetaldehyde in ethanol metabolism, which deplete glutathione (GSH) and lead to free-radical-mediated apoptosis [121,122]. In addition, ethanol and its metabolites enhance the formation of inflammatory cytokines such as IL-6 and TNF-α [123,124], partly due to increased stimulation of oxidative stress, which can lead to cellular factor imbalances and immune disorders, and further impair liver function. Therefore, any drug that can inhibit inflammation and oxidative stress has the potential to reduce liver damage [125].

Yan et al. [126] found that MA can significantly inhibit CYP2E1 and NF-κB and MAPK pathways, thereby reducing downstream oxidation and inflammatory factors (such as NO, TNF- α And PGE2, etc.), and finally reduce alcohol induced liver injury. This indicates that MA can effectively prevent alcohol-induced acute liver injury. Wang et al. [127] gave MA before LPS/D-gal-induced liver injury, and found that MA inhibited NF-κB and activate the Nrf2 signaling pathway to play an anti-inflammatory and antioxidant role, thereby reducing liver damage. Relevant experiments show that MA can reduce liver injury induced by various factors through anti-inflammatory and antioxidant effects, so these findings may support MA as an effective preventive agent for acute liver injury.

5.4.2. MA Treatment of Liver Cancer

Liver cancer, also known as hepatocellular carcinoma (HCC), is a hypervascular tumor characterized by massive angiogenesis [128]. It is reported that hypoxia is the main pathophysiological condition to promote angiogenesis, and hypoxia-inducible factor (HIF-1R) regulates the basic adaptive response of cancer cells to hypoxia. Vascular endothelial growth factor (VEGF) [129], urokinase plasminogen activator (uPA) [130], reactive oxygen species (ROS) and nitric oxide (NO) [131,132] play an important role in cancer angiogenesis. Therefore, any drug that can inhibit angiogenesis may be beneficial to the treatment of HCC.

Lin et al. [116] cultured hepatoma Hep3B, Huh7 and HA22T cells in vitro, and treated these hepatocellular carcinoma cells with 2 or 4 μmol/L MA for 72 h. It was found that 4 μmol/L MA significantly inhibited the mRNA expression of angiogenic factors HIF-1R, VEGF, IL-8 and uPA, and reduced ROS by maintaining GSH levels and reducing NO production, while MA also reduced the invasion and migration of three types of hepatocellular carcinoma cells. Thus, MA is an effective anti-angiogenic agent that delays the invasion and migration of hepatocellular carcinoma cells.

5.4.3. MA Treatment of Nonalcoholic Fatty Liver Disease

Nonalcoholic fatty liver disease develops from abnormal lipid metabolism. Excessive intake of free fatty acids interferes with lipid storage and metabolism, and also forms excessive accumulation of triglycerides and lipids in the liver, eventually leading to chronic hepatitis with abnormal liver function and liver metabolic disorders [133]. If patients fail to maintain an appropriate body weight and a balanced diet, the disease may progressively worsen and progress to nonalcoholic steatohepatitis, irreversible fibrosis or cirrhosis, and even hepatocellular carcinoma [134].

Liou et al. [114] induced nonalcoholic fatty liver disease in mice by feeding mice with HFD (60% fat, *w*/*w*) for four weeks. After successful modeling, mice were injected intraperitoneally with MA (10 mg/kg and 20 mg/kg) twice a week for 12 weeks. Experiments found that MA can reduce liver fat infiltration by inhibiting the expression of genes involved in liver adipogenesis, and restore liver glycogen levels and reduce triglycerides and total cholesterol by enhancing ATGL and Sirt1 expression, and AMPK phosphorylation. It is suggested that MA can prevent obesity-induced nonalcoholic fatty liver disease by regulating the Sirt1/AMPK signaling pathway.

5.5. MA Treatment of Gastric Diseases

MA mainly protects the stomach by enhancing gastric mucosal protective factor [135] and inhibiting the activity of enzymes such as H[+], K[+]-ATP [136].

5.5.1. MA Treatment of Gastric Ulcer

Gastric ulcer is one of the most common gastric diseases in the world, which is mainly related to the destruction of gastric mucosa by aggressive factors (hydrochloric acid, pepsin, bile reflux and reactive oxygen species) [137,138]. Da Rosa et al. [136] pre-administered MA (1–10 mg/kg) to mice and, 1 h after oral administration or 30 min after intraperitoneal treatment, mice were given ethanol/HCl (60%/0.3 M, 10 mL) /kg, po) or indomethacin (80 mg/kg, orally) to induce gastric injury. The results showed that MA pretreatment can effectively reduce the area of gastric injury by more than 90%, and the mechanism may be the inhibition of H[+], K[+]-ATPase activities. This experiment shows that MA can protect gastric mucosa and thus play a therapeutic role in gastric ulcer.

5.5.2. MA Treatment of Gastric Cancer

One study found that MA was able to inhibit IL-6 expression, induce JAK and STAT3 phosphorylation, and down-regulate STAT3-mediated protein Bad, Bcl-2 and Bax expression to treat gastric cancer [70].

5.6. MA Treatment of Intestinal Diseases

In the intestine, MA can inhibit carbohydrate hydrolysis and reduce glucose absorption, and can also induce apoptosis in colorectal cancer cells.

MA Treatment of Colorectal Cancer

Colorectal cancer (CRC) is one of the five major cancers worldwide, and its conventional treatment is mainly surgery, radiotherapy and chemotherapy [139], but there are problems such as recurrence and significant side effects [140]. Therefore, there is an urgent need to develop a potent drug with low side effects to treat CRC.

In 2006, Reyes et al. [141] found that MA exerts anti-proliferative and pro-apoptotic effects in human colon cancer cell lines HT29 and Caco-2. In recent years, it has been found that, in human colon cancer cells HT29, MA caused G0/G1 phase cell cycle arrest and induced tumor apoptosis through the JNK-Bid-mediated mitochondrial apoptotic pathway and p53 activation [53]. In p53-deficient Caco-2 colon cancer cells, MA was also able to rapidly activate caspase-8 and caspase-3, resulting in late activation of caspase-9 while Bax protein expression levels remained unchanged [56,57]. This suggests that MA can exert anti-tumor effects by activating the endogenous mitochondrial apoptotic pathway or exogenous apoptotic pathway. In addition, MA can also exert anti-tumor effects by inducing cytoskeletal changes in HT29 human colon cancer cells [58]. In vivo experiments have also demonstrated that MA can reduce intestinal tumorigenesis in ApcMin/+ mice by inhibiting the formation of intestinal polyps [142]. It has also been found that MA dose-dependently regulates the AMPK-mTOR pathway, thereby inhibiting SW480 and HCT116 colon cancer cell viability to exert anti-tumor effects [61]. At the same time, MA at high concentrations in the intestine can be useful for better prevention of colon cancer [143]. However, the oral bioavailability of MA is low, about 5% [144]; the relative abundance of MA metabolites in intestinal contents is higher than that in plasma or urine. Therefore, oral MA may be a more effective drug delivery method to prevent colon cancer.

5.7. MA Treatment of Kidney Disease

MA protects the kidneys through anti-inflammatory, hypoglycemic and antioxidant effects, and can also inhibit the development of renal cell carcinoma by inhibiting the proliferation and generation of blood vessels.

5.7.1. MA Treatment of Diabetic Nephropathy

Diabetic nephropathy occurs in approximately 40% of people with diabetes [145]. Diabetes nephropathy has the strongest correlation with the death of diabetes patients [146], and has increased the cardiovascular incidence rate and mortality of diabetes patients [147].

Studies have shown that MA inhibits the expression of oxidative stress markers and inflammatory factors in the kidney of diabetic mice, and activates the AMPK/SIRT1 signaling pathway to affect renal metabolism, thereby protecting renal function [42]. Pre-diabetes is the intermediate stage between normal blood sugar and diabetes. The risk of developing into diabetes is very high, and the risk of developing into cardiovascular disease, kidney disease and death will also increase [148]; for example, renal insufficiency can occur at an early stage of impaired glucose metabolism [149]. Based on the significant cost of diabetic complications, it is essential to reverse hyperglycemia and complications prior to pre-diabetes. Preliminary studies have found that MA attenuates renal oxidative stress, reduces urinary podocin mRNA expression and attenuates renal insufficiency [150]. However, molecular mechanisms need to be identified to elucidate the mechanisms by which MA improves renal function. Prevention of diabetic nephropathy should require prevention of renal reabsorption of sodium ions in addition to strict glycemic control. Studies have shown that intravenous infusion of MA derivative phenylhydrazine (PH-MA) significantly increased renal Na+ and lithium excretion fractions, significantly increased glomerular filtration rate (GFR), reduced plasma aldosterone levels and improved diabetic renal function [151]. MA can also reduce the expression of glucose transporter protein 1 (GLUT1) and glucose transporter protein 2 (GLUT2) in the kidneys of diabetic rats when increasing the renal excretion of Na+, thereby reducing the blood sugar level [39]. MA not only reduces the function of renal excretion of Na+, but also can lower blood sugar and improve renal function. MA may be a potential drug for the treatment of diabetes nephropathy.

5.7.2. MA Treatment of Acute Kidney Injury

Acute kidney injury (AKI) is the sudden loss of excretory renal function, mainly due to renal ischemia/reperfusion injury (IRI) following predisposing factors such as renal transplantation and renal surgery [152,153], and leads to renal irreversible damage [154]. After IRI, a large amount of reactive oxygen species (ROS) and calcium overload are produced to activate cell apoptosis, necrosis and necrotic apoptosis, activate inflammatory reaction, and finally lead to renal structural damage and long-term tissue damage [155].

One study showed that MA inhibits IRI-induced AKI injury through NF-κB and MAPK signaling pathways [156]. Inflammation is important for the appearance, progression, exacerbation and prognosis of IRI [157], and MA has a therapeutic effect on IRI-related inflammation, indicating that MA is a promising drug for treating AKI.

Renal cell carcinoma (RCC) is a highly metastatic, heterogeneous disease that is resistant to conventional treatment modalities [155,158]. At present, it is still a refractory cancer, because RCC is a solid tumor with the highest degree of vascularization [159]. MA has antiproliferative and antiangiogenic effects. In metastatic renal cell carcinoma cell lines, MA inhibited the proliferation of cancer cells by reducing nuclear antigen expression, anti-proliferation and anti-colony production in proliferating cells, and down-regulating VEGF in vascular endothelial cells and PCNA in RCC to inhibit angiogenesis and proliferation [65]. It is worth noting that, unlike the mechanism of MA against other tumors, its anti-RCC mechanism of action is mediated by inhibiting proliferation rather than inducing apoptosis. The treatment mechanism of MA in treating diseases of various organs are summarized, as shown in Table 1.

Table 1. The treatment mechanism of MA in treating diseases of various organs are summarized.

Organs	In Vivo/In Vitro	Diseases	Treatment Mechanism	References
Brain				
	In vivo	AD	MA promotes the expression of BDNF, reduces the apoptosis of neurons, improves the memory and cognitive impairment of mice caused by cholinergic system damage, and enhances the cognitive function of mice	[82]
	In vivo	Epilepsy	MA can reduce the production of inflammatory factors, reduce the level of glutamate in the hippocampus, improve the antioxidant capacity of the hippocampus and thus improve the production of epileptic behavior	[89]
	In vitro	Ischemic stroke	MA can block the cell necrosis induced by hypoxia, reduce the necrosis of neurons, effectively prevent the damage of cell bodies and neurites, and increase the survival rate of neurons	[90]
	In vivo	Ischemic stroke	MA prolonged the therapeutic time window of MK-801 from 1 h to 3 h. MA and MK-801 jointly increased the level of glutamate transporter GLT-1 in astrocytes and promoted astrocytes to regulate glutamate excitotoxicity, thus playing a therapeutic role in ischemia	[92]
	In vivo	Ischemic stroke	MA can significantly prevent axon injury, promote axon regeneration and increase the expression of synaptophysin after 7 days of ischemia	[93]
	In vivo	Ischemic stroke	MA treatment can enhance the expression of glial glutamate transporter GLT-1 at the protein and mRNA levels, leaving extracellular glutamate at a low concentration, thus playing a protective role in nerve cells during stroke ischemia	[91]
	In vitro	Astrocytoma (1321N1 cells)	MA can induce apoptosis of 1321N1 cell line	[77]
Lung				
	In vitro	Lung cancer (A549 cells)	MA treatment mediates mitochondrial apoptosis pathway and HIF-1 α pathway induced apoptosis of A549 cells	[59]
	In vitro	Lung cancer (A549 cells)	MA can promote the expression of caspase-3, caspase-8 and caspase-9 by regulating the expression of Smac and reducing the expression of c-IAP1, c-IAP2, XIAP and survivin, thereby inducing apoptosis of A549 cells	[60]
	In vivo	Lung damage	MA antagonizes lung injury caused by diesel PM2.5 by regulating TLR4-MyD88 and mTOR autophagy pathway	[94]
	In vivo	Lung injury	MA exerts anti-inflammatory effects by down-regulating NF-κB and p-STAT-1 to regulate iNOS	[54]

Table 1. *Cont.*

Organs	In Vivo/In Vitro	Diseases	Treatment Mechanism	References
Heart				
	In vitro	Myocardial hypertrophy (NMCMs, H9C2 cells)	MA treatment significantly inhibited Ang-II-induced hypertrophy of NMCMs, and the dose did not affect the cell viability of H9C2 and NCMCs	[104]
	In vivo	Myocardial hypertrophy	MA can significantly improve myocardial hypertrophy, myocardial fibrosis and cardiac function, probably through the METTL3-mediated m 6A methylation pathway	[104]
	In vivo	Myocardial hypertrophy	MA reduces stress-overload-induced cardiac hypertrophy in vivo by reducing phosphorylation of AKT and ERK signaling pathways	[105]
	In vivo	Myocardial infarction	MA provides cardioprotection by increasing PON activity, reducing LDL-C levels and inhibiting lipid peroxidation (LPO)	[109]
	In vivo	Myocardial infarction	MA can inhibit the enzyme xanthine oxidase XO to relieve myocardial infarction	[110]
Liver				
	In vivo	Acute liver injury	MA inhibits CYP2E1, NF-κB and MAPK pathways, reducing the production of downstream oxidative and inflammatory factors (such as NO, TNF-α and PGE2), ultimately reducing alcohol-induced hepatotoxicity	[126]
	In vivo	Acute liver injury	MA exerts anti-inflammatory and antioxidant effects by inhibiting NF-κB and activating the Nrf2 signaling pathway, thereby providing protection against LPS/D-gal-induced liver injury	[127]
	In vitro	Liver cancer (hepatocellular carcinoma Hep3B, Huh7 and HA22T cells)	MA significantly inhibits angiogenesis and delays the metastasis and invasion of liver cancer cells	[116]
	In vitro	Fatty liver disease	MA can reduce hepatic fat infiltration, restore liver glycogen levels and reduce triglyceride and total cholesterol levels by inhibiting the expression of genes involved in hepatic fat formation	[114]
Stomach				
	In vivo	Gastric ulcer	MA pretreatment effectively reduces the area of gastric damage, inhibits H[+] and K[+]-ATPase activity, and provides gastroprotection	[136]
	In vivo	Gastric cancer	MA was able to inhibit IL-6 expression, induce JAK and STAT3 phosphorylation, and down-regulate STAT3-mediated protein Bad, Bcl-2 and Bax expression to treat gastric cancer	[70]

Table 1. *Cont.*

Organs	In Vivo/In Vitro	Diseases	Treatment Mechanism	References
Intestine				
	In vitro	Colorectal cancer (HCT116, SW480 cells)	MA mainly induces apoptosis of colorectal cancer cells and inhibits proliferation and migration of colorectal tumors, and induces apoptosis to play an anti-tumor role	[61]
Kidney				
	In vivo	Diabetic nephropathy	MA activation of renal AMPK/SIRT1 signaling pathway improves diabetic nephropathy	[42]
	In vivo	Diabetic nephropathy	MA increases renal excretion of Na+ and can also lower blood glucose values	[151]
	In vivo	Renal cell carcinoma	MA inhibited the proliferation of cancer cells by reducing nuclear antigen expression, anti-proliferation and anti-colony production in proliferating cells, and down-regulating VEGF in vascular endothelial cells and PCNA in RCC to inhibit angiogenesis and proliferation	[65]
	In vivo	Acute kidney injury	MA inhibits IRI-induced AKI injury via NF-κB and MAPK signaling pathways	[156]

6. Conclusions and Future Perspectives

This paper systematically reviewed the biological activity and clinical research progress of MA in recent years. The use of natural compounds to treat diseases has become increasingly popular, because these natural compounds have almost no side effects. As a natural compound, MA has a wide range of biological activities, low toxicity and side effects. As a natural medicine, it has been paid more and more attention in clinical treatment of diseases.

MA has a variety of biological activities such as antioxidant, anti-inflammatory, anti-tumor, hypoglycemic and neuroprotective activities, so it has a good preventive and therapeutic effect on various organ-related diseases, such as some cancers, ischemic diseases, diabetes, stroke, and other acute and chronic diseases caused by oxidative stress or inflammation.

MA is a pentacyclic triterpene acid, which is widely found in medicinal plants such as olive, jujube, hawthorn and loquat leaves. As MA is particularly rich in olive peel, most of the current studies use olive residue to extract MA. MA extraction is usually by the SE extraction method, but it is always time-consuming and inefficient. Although UAE, MAE and SFE can better overcome these problems as new extraction methods, they cannot be widely promoted due to their immature processes and costs.

Although the research on MA has been so comprehensive, there are still many questions to be solved: (1) MA can promote the expression of GLT-1 in astrocytes and reduce the level of extracellular glutamate, but the pathway of MA activating GLT-1 is still unclear. (2) At present, the extraction process of MA is either high energy consumption, time consuming, low efficiency, immature process or high cost, so it is urgent to develop new extraction methods. (3) At present, the in vivo research of MA mostly adopts the method of gavage and intraperitoneal injection. However, clinical medication is generally oral and intravenous infusion, so whether MA can achieve the expected effect through intravenous administration requires further research. (4) MA has antioxidant biological activity, but at higher doses, which in turn exhibits pro-oxidative effects leading to an increase in intracellular reactive oxygen species. (5) MA, as a natural compound, can be used for the prevention and treatment of various diseases, but research into its toxicological effects and side effects under different circumstances is still lacking.

Overall, MA is of great interest in modern scientific research due to its beneficial effects in numerous diseases. MA has a wide range of biological activities, weak side effects and other properties, which makes it have great development potential in the fields of modern farming, functional health care and medical treatment. We believe that the research on MA will become more and more comprehensive, and the practical application of MA will become more and more extensive. It will usher in a broad market and broad prospects, and go out of the laboratory to enter the food and medicine fields.

Author Contributions: Y.H., Y.W., W.D. and W.L. designed this review; K.Y., J.J., H.Z., Y.Y. and D.L. participated in the literature collection for this review; Y.H. and Y.W. wrote the first draft; W.D. and W.L. revised and reviewed the first draft. All authors have read and agreed to the published version of the manuscript.

Funding: This review was supported by grants from the National Natural Science Foundation of China (No. 82074151) and the Sichuan Youth Science and Technology Innovation Experiment Group (2020JDTD0022).

Institutional Review Board Statement: Not applicable.

Informed Consent Statement: Not applicable.

Data Availability Statement: Not applicable.

Conflicts of Interest: The authors declare no conflict of interest.

References

1. Stiti, N.; Triki, S.; Hartmann, M.A. Formation of triterpenoids throughout *Olea europaea* fruit ontogeny. *Lipids* **2007**, *42*, 55–67. [CrossRef] [PubMed]
2. García, A.; Brenes, M.; Dobarganes, M.; Romero, C.; Ruiz-Méndez, M. Enrichment of pomace olive oil in triterpenic acids during storage of "Alpeorujo" olive paste. *Eur. J. Lipid Sci. Technol.* **2008**, *110*, 1136–1141. [CrossRef]
3. Sun, Z.; Li, Z.; Zuo, L.; Wang, Z.; Zhou, L.; Shi, Y.; Kang, J.; Zhu, Z.; Zhang, X. Qualitative and quantitative determination of YiXinShu Tablet using ultra high performance liquid chromatography with Q Exactive hybrid quadrupole orbitrap high-resolution accurate mass spectrometry. *J. Sep. Sci.* **2017**, *40*, 4453–4466. [CrossRef] [PubMed]
4. Ho, H.; Lin, W.; Kitanaka, S.; Chang, C.; Wu, J. Analysis of bioactive triterpenes in *Eriobotrya japonica* Lindl. by high-performance liquid chromatography. *Yao Wu Shi Pin Fen Xi = J. Food Drug Anal.* **2020**, *16*, 5. [CrossRef]
5. Guo, S.; Duan, J.A.; Tang, Y.P.; Yang, N.Y.; Qian, D.W.; Su, S.L.; Shang, E.X. Characterization of triterpenic acids in fruits of ziziphus species by HPLC-ELSD-MS. *J. Agric. Food Chem.* **2010**, *58*, 6285–6289. [CrossRef] [PubMed]
6. Guo, S.; Duan, J.A.; Tang, Y.; Qian, Y.; Zhao, J.; Qian, D.; Su, S.; Shang, E. Simultaneous qualitative and quantitative analysis of triterpenic acids, saponins and flavonoids in the leaves of two Ziziphus species by HPLC-PDA-MS/ELSD. *J. Pharm. Biomed. Anal.* **2011**, *56*, 264–270. [CrossRef]
7. Savina, A.; Sokol'skaya, T.; Fesenko, D. Maslinic acid from the leaves of *Eucalyptus viminalis*. *Chem. Nat. Compd.* **1983**, *19*, 114–115. [CrossRef]
8. Zong, W.; Xia, W.; Cui, B. Determination of corosolic and maslinic acids in *Lagerstroemia speciosa* leaves by TLC/HPLC method. *Pharm. Chem. J.* **2007**, *41*, 222–224. [CrossRef]
9. Pan, Z.H.; Wang, Y.Y.; Li, M.M.; Xu, G.; Peng, L.Y.; He, J.; Zhao, Y.; Li, Y.; Zhao, Q.S. Terpenoids from *Salvia trijuga*. *J. Nat. Prod.* **2010**, *73*, 1146–1150. [CrossRef]
10. Tarvainen, M.; Suomela, J.; Kallio, H.; Yang, B. Triterpene Acids in Plantago major: Identification, Quantification and Comparison of Different Extraction Methods. *Chromatographia* **2010**, *71*, 279–284. [CrossRef]
11. Lee, I.K.; Kim, D.H.; Lee, S.Y.; Kim, K.R.; Choi, S.U.; Hong, J.K.; Lee, J.H.; Park, Y.H.; Lee, K.R. Triterpenoic acids of *Prunella vulgaris* var. lilacina and their cytotoxic activities in vitro. *Arch. Pharmacal Res.* **2008**, *31*, 1578–1583. [CrossRef] [PubMed]
12. Guinda, A.; Rada, M.; Delgado, T.; Castellano, J. Pentacyclic triterpenic acids from *Argania spinosa*. *Eur. J. Lipid Sci. Technol.* **2011**, *113*, 231–237. [CrossRef]
13. Pérez-Camino, M.C.; Cert, A. Quantitative determination of hydroxy pentacyclic triterpene acids in vegetable oils. *J. Agric. Food Chem.* **1999**, *47*, 1558–1562. [CrossRef] [PubMed]
14. Sánchez-González, M.; Lozano-Mena, G.; Juan, M.E.; García-Granados, A.; Planas, J.M. Assessment of the safety of maslinic acid, a bioactive compound from *Olea europaea* L. *Mol. Nutr. Food Res.* **2013**, *57*, 339–346. [CrossRef]
15. Lozano-Mena, G.; Sánchez-González, M.; Juan, M.; Planas, J. Maslinic acid, a natural phytoalexin-type triterpene from olives—A promising nutraceutical? *Molecules* **2014**, *19*, 11538–11559. [CrossRef]
16. Jing, Z.; Rui, W.; Ruihua, L.; Hao, Y.; Hengtong, F. Review of the Biological Activity of Maslinic Acid. *Curr. Drug Targets* **2021**, *22*, 1496–1506. [CrossRef]

17. Yap, W.; Lim, Y. Mechanistic Perspectives of Maslinic Acid in Targeting Inflammation. *Biochem. Res. Int.* **2015**, *2015*, 279356. [CrossRef]
18. Rufino-Palomares, E.; Reyes-Zurita, F.; Fuentes-Almagro, C.; de la Higuera, M.; Lupiáñez, J.; Peragón, J. Proteomics in the liver of gilthead sea bream (*Sparus aurata*) to elucidate the cellular response induced by the intake of maslinic acid. *Proteomics* **2011**, *11*, 3312–3325. [CrossRef]
19. Fernández-Navarro, M.; Peragón, J.; Esteban, F.; de la Higuera, M.; Lupiáñez, J. Maslinic acid as a feed additive to stimulate growth and hepatic protein-turnover rates in rainbow trout (*Onchorhynchus mykiss*). *Comp. Biochem. Physiol. Toxicol. Pharmacol. CBP* **2006**, *144*, 130–140. [CrossRef]
20. Xie, P.; Huang, L.; Zhang, C.; Deng, Y.; Wang, X.; Cheng, J. Enhanced extraction of hydroxytyrosol, maslinic acid and oleanolic acid from olive pomace: Process parameters, kinetics and thermodynamics, and greenness assessment. *Food Chem.* **2019**, *276*, 662–674. [CrossRef]
21. Guinda, A.; Rada, M.; Delgado, T.; Gutiérrez-Adánez, P.; Castellano, J.M. Pentacyclic triterpenoids from olive fruit and leaf. *J. Agric. Food Chem.* **2010**, *58*, 9685–9691. [CrossRef]
22. Giménez, E.; Juan, M.E.; Calvo-Melià, S.; Planas, J.M. A sensitive liquid chromatography-mass spectrometry method for the simultaneous determination in plasma of pentacyclic triterpenes of *Olea europaea* L. *Food Chem.* **2017**, *229*, 534–541. [CrossRef] [PubMed]
23. Fernández-Hernández, A.; Martinez, A.; Rivas, F.; García-Mesa, J.A.; Parra, A. Effect of the solvent and the sample preparation on the determination of triterpene compounds in two-phase olive-mill-waste samples. *J. Agric. Food Chem.* **2015**, *63*, 4269–4275. [CrossRef] [PubMed]
24. Fernandez-Pastor, I.; Fernandez-Hernandez, A.; Perez-Criado, S.; Rivas, F.; Martinez, A.; Garcia-Granados, A.; Parra, A. Microwave-assisted extraction versus Soxhlet extraction to determine triterpene acids in olive skins. *J. Sep. Sci.* **2017**, *40*, 1209–1217. [CrossRef] [PubMed]
25. Wu, H.; Li, G.; Liu, S.; Liu, D.; Chen, G.; Hu, N.; Suo, Y.; You, J. Simultaneous determination of six triterpenic acids in some Chinese medicinal herbs using ultrasound-assisted dispersive liquid-liquid microextraction and high-performance liquid chromatography with fluorescence detection. *J. Pharm. Biomed. Anal.* **2015**, *107*, 98–107. [CrossRef] [PubMed]
26. Gómez-Cruz, I.; Contreras, M.D.M.; Romero, I.; Castro, E. Sequential Extraction of Hydroxytyrosol, Mannitol and Triterpenic Acids Using a Green Optimized Procedure Based on Ultrasound. *Antioxidants* **2021**, *10*, 1781. [CrossRef]
27. Song, L.; Zhang, L.; Xu, L.; Ma, Y.; Lian, W.; Liu, Y.; Wang, Y. Optimized Extraction of Total Triterpenoids from Jujube (*Ziziphus jujuba* Mill.) and Comprehensive Analysis of Triterpenic Acids in Different Cultivars. *Plants* **2020**, *9*, 412. [CrossRef] [PubMed]
28. Goulas, V.; Manganaris, G.A. Towards an efficient protocol for the determination of triterpenic acids in olive fruit: A comparative study of drying and extraction methods. *Phytochem. Anal.* **2012**, *23*, 444–449. [CrossRef] [PubMed]
29. De Melo, M.; Silvestre, A.; Silva, C. Supercritical fluid extraction of vegetable matrices: Applications, trends and future perspectives of a convincing green technology. *J. Supercrit. Fluids* **2014**, *92*, 115–176. [CrossRef]
30. Benavides, A.; Martín-Álvarez, P.; Vázquez, L.; Reglero, G.; Señoráns, F.; Ibáñez, E. Optimization of Countercurrent Supercritical Fluid Extraction of Minor Components from Olive Oil. *Curr. Anal. Chem.* **2013**, *10*, 78–85. [CrossRef]
31. Xynos, N.; Papaefstathiou, G.; Psychis, M.; Argyropoulou, A.; Aligiannis, N.; Skaltsounis, A. Development of a green extraction procedure with super/subcritical fluids to produce extracts enriched in oleuropein from olive leaves. *J. Supercrit. Fluids* **2012**, *67*, 89–93. [CrossRef]
32. Gómez-Cruz, I.; Contreras, M.D.M.; Romero, I.; Castro, E. Optimization of Microwave-Assisted Water Extraction to Obtain High Value-Added Compounds from Exhausted Olive Pomace in a Biorefinery Context. *Foods* **2022**, *11*, 2002. [CrossRef] [PubMed]
33. Nile, S.H.; Nile, A.; Liu, J.; Kim, D.H.; Kai, G. Exploitation of apple pomace towards extraction of triterpenic acids, antioxidant potential, cytotoxic effects, and inhibition of clinically important enzymes. *Food Chem. Toxicol.* **2019**, *131*, 110563. [CrossRef] [PubMed]
34. Yu, Q.S. *Triterpenoid Chemistry*; Chemical Industry Press: Beijing, China, 2008; p. 75.
35. Wu, L.J. *Natural Pharmaceutical Chemistry*, 4th ed.; People's Medical Publishing House: Beijing, China, 2004; pp. 292–293.
36. Yang, J.; Hu, Y.-J.; Yu, B.-Y.; Qi, J. Integrating qualitative and quantitative characterization of Prunellae Spica by HPLC-QTOF/MS and HPLC-ELSD. *Chin. J. Nat. Med.* **2016**, *14*, 391–400. [CrossRef] [PubMed]
37. Sánchez-González, M.; Lozano-Mena, G.; Juan, M.E.; García-Granados, A.; Planas, J.M. Liquid chromatography-mass spectrometry determination in plasma of maslinic acid, a bioactive compound from *Olea europaea* L. *Food Chem.* **2013**, *141*, 4375–4381. [CrossRef]
38. Akinnuga, A.M.; Siboto, A.; Khumalo, B.; Sibiya, N.H.; Ngubane, P.; Khathi, A. Bredemolic Acid Ameliorates Selected Liver Function Biomarkers in a Diet-Induced Prediabetic Rat Model. *Can. J. Gastroenterol. Hepatol.* **2020**, *2020*, 2475301. [CrossRef]
39. Mkhwanazi, B.N.; Serumula, M.R.; Myburg, R.B.; Van Heerden, F.R.; Musabayane, C.T. Antioxidant effects of maslinic acid in livers, hearts and kidneys of streptozotocin-induced diabetic rats: Effects on kidney function. *Ren. Fail.* **2014**, *36*, 419–431. [CrossRef]
40. Liu, J.; Sun, H.; Duan, W.; Mu, D.; Zhang, L. Maslinic acid reduces blood glucose in KK-Ay mice. *Biol. Pharm. Bull.* **2007**, *30*, 2075–2078. [CrossRef]

41. Khathi, A.; Serumula, M.R.; Myburg, R.B.; Van Heerden, F.R.; Musabayane, C.T. Effects of *Syzygium aromaticum*-derived triterpenes on postprandial blood glucose in streptozotocin-induced diabetic rats following carbohydrate challenge. *PLoS ONE* **2013**, *8*, e81632. [CrossRef]
42. Gao, H.; Wu, H. Maslinic acid activates renal AMPK/SIRT1 signaling pathway and protects against diabetic nephropathy in mice. *BMC Endocr. Disord.* **2022**, *22*, 25. [CrossRef]
43. Liu, W.; Wang, J.; Zhang, Z.; Xu, J.; Xie, Z.; Slavin, M.; Gao, X. In vitro and in vivo antioxidant activity of a fructan from the roots of *Arctium lappa* L. *Int. J. Biol. Macromol.* **2014**, *65*, 446–453. [CrossRef] [PubMed]
44. Rodriguez-Rodriguez, R. Oleanolic acid and related triterpenoids from olives on vascular function: Molecular mechanisms and therapeutic perspectives. *Curr. Med. Chem.* **2015**, *22*, 1414–1425. [CrossRef] [PubMed]
45. Xu, H.X.; Zeng, F.Q.; Wan, M.; Sim, K.Y. Anti-HIV triterpene acids from Geum japonicum. *J. Nat. Prod.* **1996**, *59*, 643–645. [CrossRef] [PubMed]
46. Márquez Martín, A.; de la Puerta Vázquez, R.; Fernández-Arche, A.; Ruiz-Gutiérrez, V. Supressive effect of maslinic acid from pomace olive oil on oxidative stress and cytokine production in stimulated murine macrophages. *Free. Radic. Res.* **2006**, *40*, 295–302. [CrossRef] [PubMed]
47. Bast, A.; Haenen, G.R. Ten misconceptions about antioxidants. *Trends Pharmacol. Sci.* **2013**, *34*, 430–436. [CrossRef]
48. Mokhtari, K.; Rufino-Palomares, E.E.; Pérez-Jiménez, A.; Reyes-Zurita, F.J.; Figuera, C.; García-Salguero, L.; Medina, P.P.; Peragón, J.; Lupiáñez, J.A. Maslinic Acid, a Triterpene from Olive, Affects the Antioxidant and Mitochondrial Status of B16F10 Melanoma Cells Grown under Stressful Conditions. *Evid. Based Complementary Altern. Med.* **2015**, *2015*, 272457. [CrossRef]
49. Huang, L.; Guan, T.; Qian, Y.; Huang, M.; Tang, X.; Li, Y.; Sun, H. Anti-inflammatory effects of maslinic acid, a natural triterpene, in cultured cortical astrocytes via suppression of nuclear factor-kappa B. *Eur. J. Pharmacol.* **2011**, *672*, 169–174. [CrossRef]
50. Qian, Y.; Guan, T.; Tang, X.; Huang, L.; Huang, M.; Li, Y.; Sun, H.; Yu, R.; Zhang, F. Astrocytic glutamate transporter-dependent neuroprotection against glutamate toxicity: An in vitro study of maslinic acid. *Eur. J. Pharmacol.* **2011**, *651*, 59–65. [CrossRef]
51. Serhan, C.N. Treating inflammation and infection in the 21st century: New hints from decoding resolution mediators and mechanisms. *FASEB J.* **2017**, *31*, 1273–1288. [CrossRef]
52. Chen, Y.L.; Yan, D.Y.; Wu, C.Y.; Xuan, J.W.; Jin, C.Q.; Hu, X.L.; Bao, G.D.; Bian, Y.J.; Hu, Z.C.; Shen, Z.H.; et al. Maslinic acid prevents IL-1β-induced inflammatory response in osteoarthritis via PI3K/AKT/NF-κB pathways. *J. Cell. Physiol.* **2021**, *236*, 1939–1949. [CrossRef]
53. Li, Q.; Xu, M.; Li, Z.; Li, T.; Wang, Y.; Chen, Q.; Wang, Y.; Feng, J.; Yin, X.; Lu, C. Maslinic Acid Attenuates Ischemia/Reperfusion Injury-Induced Myocardial Inflammation and Apoptosis by Regulating HMGB1-TLR4 Axis. *Front. Cardiovasc. Med.* **2021**, *8*, 768947. [CrossRef] [PubMed]
54. Lee, W.; Kim, J.; Park, E.K.; Bae, J.S. Maslinic Acid Ameliorates Inflammation via the Downregulation of NF-κB and STAT-1. *Antioxidants* **2020**, *9*, 106. [CrossRef] [PubMed]
55. Reyes-Zurita, F.J.; Pachón-Peña, G.; Lizárraga, D.; Rufino-Palomares, E.E.; Cascante, M.; Lupiáñez, J.A. The natural triterpene maslinic acid induces apoptosis in HT29 colon cancer cells by a JNK-p53-dependent mechanism. *BMC cancer* **2011**, *11*, 154. [CrossRef] [PubMed]
56. Reyes-Zurita, F.J.; Rufino-Palomares, E.E.; Medina, P.P.; Leticia García-Salguero, E.; Peragón, J.; Cascante, M.; Lupiáñez, J.A. Antitumour activity on extrinsic apoptotic targets of the triterpenoid maslinic acid in p53-deficient Caco-2 adenocarcinoma cells. *Biochimie* **2013**, *95*, 2157–2167. [CrossRef]
57. Reyes-Zurita, F.J.; Rufino-Palomares, E.E.; García-Salguero, L.; Peragón, J.; Medina, P.P.; Parra, A.; Cascante, M.; Lupiáñez, J.A. Maslinic Acid, a Natural Triterpene, Induces a Death Receptor-Mediated Apoptotic Mechanism in Caco-2 p53-Deficient Colon Adenocarcinoma Cells. *PLoS ONE* **2016**, *11*, e0146178. [CrossRef]
58. Rufino-Palomares, E.E.; Reyes-Zurita, F.J.; García-Salguero, L.; Mokhtari, K.; Medina, P.P.; Lupiáñez, J.A.; Peragón, J. Maslinic acid, a triterpenic anti-tumoural agent, interferes with cytoskeleton protein expression in HT29 human colon-cancer cells. *J. Proteom.* **2013**, *83*, 15–25. [CrossRef]
59. Hsia, T.C.; Liu, W.H.; Qiu, W.W.; Luo, J.; Yin, M.C. Maslinic acid induces mitochondrial apoptosis and suppresses HIF-1α expression in A549 lung cancer cells under normoxic and hypoxic conditions. *Molecules* **2014**, *19*, 19892–19906. [CrossRef]
60. Bai, X.; Zhang, Y.; Jiang, H.; Yang, P.; Li, H.; Zhang, Y.; He, P. Effects of maslinic acid on the proliferation and apoptosis of A549 lung cancer cells. *Mol. Med. Rep.* **2016**, *13*, 117–122. [CrossRef]
61. Wei, Q.; Zhang, B.; Li, P.; Wen, X.; Yang, J. Maslinic Acid Inhibits Colon Tumorigenesis by the AMPK-mTOR Signaling Pathway. *J. Agric. Food Chem.* **2019**, *67*, 4259–4272. [CrossRef]
62. Juan, M.E.; Lozano-Mena, G.; Sánchez-González, M.; Planas, J.M. Reduction of Preneoplastic Lesions Induced by 1,2-Dimethylhydrazine in Rat Colon by Maslinic Acid, a Pentacyclic Triterpene from *Olea europaea* L. *Molecules* **2019**, *24*, 1266. [CrossRef]
63. Li, C.; Yang, Z.; Zhai, C.; Qiu, W.; Li, D.; Yi, Z.; Wang, L.; Tang, J.; Qian, M.; Luo, J.; et al. Maslinic acid potentiates the anti-tumor activity of tumor necrosis factor alpha by inhibiting NF-kappaB signaling pathway. *Mol. Cancer* **2010**, *9*, 73. [CrossRef] [PubMed]
64. Tian, Y.; Xu, H.; Farooq, A.A.; Nie, B.; Chen, X.; Su, S.; Yuan, R.; Qiao, G.; Li, C.; Li, X.; et al. Maslinic acid induces autophagy by down-regulating HSPA8 in pancreatic cancer cells. *Phytother. Res.* **2018**, *32*, 1320–1331. [CrossRef] [PubMed]

65. Thakor, P.; Song, W.; Subramanian, R.B.; Thakkar, V.R.; Vesey, D.A.; Gobe, G.C. Maslinic Acid Inhibits Proliferation of Renal Cell Carcinoma Cell Lines and Suppresses Angiogenesis of Endothelial Cells. *J. Kidney Cancer VHL* **2017**, *4*, 16–24. [CrossRef] [PubMed]
66. Park, S.Y.; Nho, C.W.; Kwon, D.Y.; Kang, Y.H.; Lee, K.W.; Park, J.H. Maslinic acid inhibits the metastatic capacity of DU145 human prostate cancer cells: Possible mediation via hypoxia-inducible factor-1α signalling. *Br. J. Nutr.* **2013**, *109*, 210–222. [CrossRef]
67. Zhang, S.; Ding, D.; Zhang, X.; Shan, L.; Liu, Z. Maslinic acid induced apoptosis in bladder cancer cells through activating p38 MAPK signaling pathway. *Mol. Cell. Biochem.* **2014**, *392*, 281–287. [CrossRef]
68. Yu, Y.; Wang, J.; Xia, N.; Li, B.; Jiang, X. Maslinic acid potentiates the antitumor activities of gemcitabine in vitro and in vivo by inhibiting NF-κB-mediated survival signaling pathways in human gallbladder cancer cells. *Oncol. Rep.* **2015**, *33*, 1683–1690. [CrossRef]
69. Cho, J.; Rho, O.; Junco, J.; Carbajal, S.; Siegel, D.; Slaga, T.J.; DiGiovanni, J. Effect of Combined Treatment with Ursolic Acid and Resveratrol on Skin Tumor Promotion by 12-O-Tetradecanoylphorbol-13-Acetate. *Cancer Prev. Res.* **2015**, *8*, 817–825. [CrossRef]
70. Wang, D.; Tang, S.; Zhang, Q. Maslinic acid suppresses the growth of human gastric cells by inducing apoptosis via inhibition of the interleukin-6 mediated Janus kinase/signal transducer and activator of transcription 3 signaling pathway. *Oncol. Lett.* **2017**, *13*, 4875–4881. [CrossRef]
71. Liu, Y.; Lu, H.; Dong, Q.; Hao, X.; Qiao, L. Maslinic acid induces anticancer effects in human neuroblastoma cells mediated via apoptosis induction and caspase activation, inhibition of cell migration and invasion and targeting MAPK/ERK signaling pathway. *AMB Express.* **2020**, *10*, 104. [CrossRef]
72. Chang, T.; Li, X.; Chen, X.; Zhang, L.; Yang, F.; Li, Z.; Li, J. Maslinic acid activates mitochondria-dependent apoptotic pathway in cardiac carcinoma. *Clin. Investig. Med.* **2014**, *37*, E217–E224. [CrossRef]
73. Yap, W.H.; Khoo, K.S.; Lim, S.H.; Yeo, C.C.; Lim, Y.M. Proteomic analysis of the molecular response of Raji cells to maslinic acid treatment. *Phytomedicine* **2012**, *19*, 183–191. [CrossRef] [PubMed]
74. Lin, C.; Yan, S.; Yin, M. Corrigendum to "Inhibitory effects of maslinic acid upon human esophagus, stomach and pancreatic cancer cells" [J. Funct. Foods 11 (2014) 581–588]. *J. Funct. Foods* **2017**, *33*, 446. [CrossRef]
75. De la Torre, R.; Carbó, M.; Pujadas, M.; Biel, S.; Mesa, M.D.; Covas, M.I.; Expósito, M.; Espejo, J.A.; Sanchez-Rodriguez, E.; Díaz-Pellicer, P.; et al. Pharmacokinetics of maslinic and oleanolic acids from olive oil—Effects on endothelial function in healthy adults. A randomized, controlled, dose-response study. *Food Chem.* **2020**, *322*, 126676. [CrossRef] [PubMed]
76. Yin, M.C.; Lin, M.C.; Mong, M.C.; Lin, C.Y. Bioavailability, distribution, and antioxidative effects of selected triterpenes in mice. *J. Agric. Food Chem.* **2012**, *60*, 7697–7701. [CrossRef]
77. Martín, R.; Carvalho-Tavares, J.; Carvalho, J.; Ibeas, E.; Hernández, M.; Ruiz-Gutierrez, V.; Nieto, M.L. Acidic triterpenes compromise growth and survival of astrocytoma cell lines by regulating reactive oxygen species accumulation. *Cancer Res.* **2007**, *67*, 3741–3751. [CrossRef]
78. Jerónimo-Santos, A.; Fonseca-Gomes, J.; Guimarães, D.A.; Tanqueiro, S.R.; Ramalho, R.M.; Ribeiro, J.A.; Sebastião, A.M.; Diógenes, M.J. Brain-derived neurotrophic factor mediates neuroprotection against Aβ-induced toxicity through a mechanism independent on adenosine 2A receptor activation. *Growth Factors* **2015**, *33*, 298–308. [CrossRef]
79. Ismail, N.A.; Leong Abdullah, M.F.I.; Hami, R.; Ahmad Yusof, H. A narrative review of brain-derived neurotrophic factor (BDNF) on cognitive performance in Alzheimer's disease. *Growth Factors* **2020**, *38*, 210–225. [CrossRef]
80. Kim, D.H.; Lee, Y.; Lee, H.E.; Park, S.J.; Jeon, S.J.; Jeon, S.J.; Cheong, J.H.; Shin, C.Y.; Son, K.H.; Ryu, J.H. Oroxylin A enhances memory consolidation through the brain-derived neurotrophic factor in mice. *Brain Res. Bull.* **2014**, *108*, 67–73. [CrossRef]
81. Kim, D.H.; Kim, J.M.; Park, S.J.; Lee, S.; Shin, C.Y.; Cheong, J.H.; Ryu, J.H. Hippocampal extracellular signal-regulated kinase signaling has a role in passive avoidance memory retrieval induced by GABAA Receptor modulation in mice. *Neuropsychopharmacology* **2012**, *37*, 1234–1244. [CrossRef]
82. Bae, H.J.; Kim, J.; Kim, J.; Goo, N.; Cai, M.; Cho, K.; Jung, S.Y.; Kwon, H.; Kim, D.H.; Jang, D.S.; et al. The effect of maslinic acid on cognitive dysfunction induced by cholinergic blockade in mice. *Br. J. Pharmacol.* **2020**, *177*, 3197–3209. [CrossRef]
83. Friedman, A.; Dingledine, R. Molecular cascades that mediate the influence of inflamation on epilepsy. *Epilepsia* **2011**, *52* (Suppl. S3), 33–39. [CrossRef] [PubMed]
84. Vezzani, A.; French, J.; Bartfai, T.; Baram, T.Z. The role of inflammation in epilepsy. *Nat. Rev. Neurol.* **2011**, *7*, 31–40. [CrossRef] [PubMed]
85. Huang, H.L.; Lin, C.C.; Jeng, K.C.; Yao, P.W.; Chuang, L.T.; Kuo, S.L.; Hou, C.W. Fresh green tea and gallic acid ameliorate oxidative stress in kainic acid-induced status epilepticus. *J. Agric. Food Chem.* **2012**, *60*, 2328–2336. [CrossRef] [PubMed]
86. Teocchi, M.A.; Ferreira, A.; da Luz de Oliveira, E.P.; Tedeschi, H.; D'Souza-Li, L. Hippocampal gene expression dysregulation of Klotho, nuclear factor kappa B and tumor necrosis factor in temporal lobe epilepsy patients. *J. Neuroinflammation* **2013**, *10*, 53. [CrossRef] [PubMed]
87. Kovac, S.; Domijan, A.M.; Walker, M.C.; Abramov, A.Y. Seizure activity results in calcium- and mitochondria-independent ROS production via NADPH and xanthine oxidase activation. *Cell Death Dis.* **2014**, *5*, e1442. [CrossRef]
88. Ravizza, T.; Balosso, S.; Vezzani, A. Inflammation and prevention of epileptogenesis. *Neurosci. Lett.* **2011**, *497*, 223–230. [CrossRef]
89. Wang, Z.H.; Mong, M.C.; Yang, Y.C.; Yin, M.C. Asiatic acid and maslinic acid attenuated kainic acid-induced seizure through decreasing hippocampal inflammatory and oxidative stress. *Epilepsy Res.* **2018**, *139*, 28–34. [CrossRef] [PubMed]

90. Qian, Y.; Guan, T.; Tang, X.; Huang, L.; Huang, M.; Li, Y.; Sun, H. Maslinic acid, a natural triterpenoid compound from *Olea europaea*, protects cortical neurons against oxygen-glucose deprivation-induced injury. *Eur. J. Pharmacol.* **2011**, *670*, 148–153. [CrossRef]
91. Guan, T.; Qian, Y.; Tang, X.; Huang, M.; Huang, L.; Li, Y.; Sun, H. Maslinic acid, a natural inhibitor of glycogen phosphorylase, reduces cerebral ischemic injury in hyperglycemic rats by GLT-1 up-regulation. *J. Neurosci. Res.* **2011**, *89*, 1829–1839. [CrossRef]
92. Qian, Y.; Tang, X.; Guan, T.; Li, Y.; Sun, H. Neuroprotection by Combined Administration with Maslinic Acid, a Natural Product from *Olea europaea*, and MK-801 in the Cerebral Ischemia Model. *Molecules* **2016**, *21*, 1093. [CrossRef]
93. Qian, Y.; Huang, M.; Guan, T.; Chen, L.; Cao, L.; Han, X.J.; Huang, L.; Tang, X.; Li, Y.; Sun, H. Maslinic acid promotes synaptogenesis and axon growth via Akt/GSK-3β activation in cerebral ischemia model. *Eur. J. Pharmacol.* **2015**, *764*, 298–305. [CrossRef] [PubMed]
94. Jeong, S.Y.; Kim, J.; Park, E.K.; Baek, M.C.; Bae, J.S. Inhibitory functions of maslinic acid on particulate matter-induced lung injury through TLR4-mTOR-autophagy pathways. *Environ. Res.* **2020**, *183*, 109230. [CrossRef] [PubMed]
95. Siegel, R.L.; Miller, K.D.; Fuchs, H.E.; Jemal, A. Cancer statistics, 2022. *CA Cancer J. Clin.* **2022**, *72*, 7–33. [CrossRef] [PubMed]
96. Shan, D.; Guo, S.; Wu, H.K.; Lv, F.; Jin, L.; Zhang, M.; Xie, P.; Wang, Y.; Song, Y.; Wu, F.; et al. Cardiac Ischemic Preconditioning Promotes MG53 Secretion through H_2O_2-Activated Protein Kinase C-δ Signaling. *Circulation* **2020**, *142*, 1077–1091. [CrossRef] [PubMed]
97. Zhu, L.; Li, C.; Liu, Q.; Xu, W.; Zhou, X. Molecular biomarkers in cardiac hypertrophy. *J. Cell. Mol. Med.* **2019**, *23*, 1671–1677. [CrossRef] [PubMed]
98. Nakamura, M.; Sadoshima, J. Mechanisms of physiological and pathological cardiac hypertrophy. *Nat. Rev. Cardiol.* **2018**, *15*, 387–407. [CrossRef]
99. Cao, G.; Li, H.B.; Yin, Z.; Flavell, R.A. Recent advances in dynamic m6A RNA modification. *Open Biol.* **2016**, *6*, 160003. [CrossRef]
100. Dorn, L.E.; Lasman, L.; Chen, J.; Xu, X.; Hund, T.J.; Medvedovic, M.; Hanna, J.H.; van Berlo, J.H.; Accornero, F. The N6-Methyladenosine mRNA Methylase METTL3 Controls Cardiac Homeostasis and Hypertrophy. *Circulation* **2019**, *139*, 533–545. [CrossRef]
101. Qin, Y.; Li, L.; Luo, E.; Hou, J.; Yan, G.; Wang, D.; Qiao, Y.; Tang, C. Role of m6A RNA methylation in cardiovascular disease (Review). *Int. J. Mol. Med.* **2020**, *46*, 1958–1972. [CrossRef]
102. Ma, Z.G.; Yuan, Y.P.; Zhang, X.; Xu, S.C.; Wang, S.S.; Tang, Q.Z. Piperine Attenuates Pathological Cardiac Fibrosis Via PPAR-γ/AKT Pathways. *EBioMedicine* **2017**, *18*, 179–187. [CrossRef]
103. Ma, Z.G.; Dai, J.; Zhang, W.B.; Yuan, Y.; Liao, H.H.; Zhang, N.; Bian, Z.Y.; Tang, Q.Z. Protection against cardiac hypertrophy by geniposide involves the GLP-1 receptor/AMPKα signalling pathway. *Br. J. Pharmacol.* **2016**, *173*, 1502–1516. [CrossRef] [PubMed]
104. Fang, M.; Deng, J.; Zhou, Q.; Hu, Z.; Yang, L. Maslinic acid protects against pressure-overload-induced cardiac hypertrophy by blocking METTL3-mediated m6A methylation. *Aging* **2022**, *14*, 2548–2557. [CrossRef] [PubMed]
105. Liu, Y.L.; Kong, C.Y.; Song, P.; Zhou, H.; Zhao, X.S.; Tang, Q.Z. Maslinic acid protects against pressure overload-induced cardiac hypertrophy in mice. *J. Pharmacol. Sci.* **2018**, *138*, 116–122. [CrossRef] [PubMed]
106. Bhatt, D.L.; Steg, P.G.; Miller, M.; Brinton, E.A.; Jacobson, T.A.; Ketchum, S.B.; Doyle, R.T.; Juliano, R.A.; Jiao, L.; Granowitz, C.; et al. Cardiovascular Risk Reduction with Icosapent Ethyl for Hypertriglyceridemia. *N. Engl. J. Med.* **2019**, *380*, 11–22. [CrossRef]
107. Cervellati, C.; Vigna, G.B.; Trentini, A.; Sanz, J.M.; Zimetti, F.; Dalla Nora, E.; Morieri, M.L.; Zuliani, G.; Passaro, A. Paraoxonase-1 activities in individuals with different HDL circulating levels: Implication in reverse cholesterol transport and early vascular damage. *Atherosclerosis* **2019**, *285*, 64–70. [CrossRef]
108. Sagor, M.A.; Tabassum, N.; Potol, M.A.; Alam, M.A. Xanthine Oxidase Inhibitor, Allopurinol, Prevented Oxidative Stress, Fibrosis, and Myocardial Damage in Isoproterenol Induced Aged Rats. *Oxidative Med. Cell. Longev.* **2015**, *2015*, 478039. [CrossRef]
109. Hussain Shaik, A.; Rasool, S.N.; Abdul Kareem, M.; Krushna, G.S.; Akhtar, P.M.; Devi, K.L. Maslinic acid protects against isoproterenol-induced cardiotoxicity in albino Wistar rats. *J. Med. Food* **2012**, *15*, 741–746. [CrossRef]
110. Shaik, A.H.; Shaik, S.R.; Shaik, A.S.; Daoud, A.; Salim, M.; Kodidhela, L.D. Analysis of maslinic acid and gallic acid compounds as xanthine oxidase inhibitors in isoprenaline administered myocardial necrotic rats. *Saudi J. Biol. Sci.* **2021**, *28*, 2575–2580. [CrossRef]
111. Huang, D.; Refaat, M.; Mohammedi, K.; Jayyousi, A.; Al Suwaidi, J.; Abi Khalil, C. Macrovascular Complications in Patients with Diabetes and Prediabetes. *BioMed Res. Int.* **2017**, *2017*, 7839101. [CrossRef]
112. Tan, Y.; Zhang, Z.; Zheng, C.; Wintergerst, K.A.; Keller, B.B.; Cai, L. Mechanisms of diabetic cardiomyopathy and potential therapeutic strategies: Preclinical and clinical evidence. *Nat. Rev. Cardiol.* **2020**, *17*, 585–607. [CrossRef]
113. Hung, Y.C.; Yang, H.T.; Yin, M.C. Asiatic acid and maslinic acid protected heart via anti-glycative and anti-coagulatory activities in diabetic mice. *Food Funct.* **2015**, *6*, 2967–2974. [CrossRef] [PubMed]
114. Liou, C.J.; Dai, Y.W.; Wang, C.L.; Fang, L.W.; Huang, W.C. Maslinic acid protects against obesity-induced nonalcoholic fatty liver disease in mice through regulation of the Sirt1/AMPK signaling pathway. *FASEB J.* **2019**, *33*, 11791–11803. [CrossRef] [PubMed]
115. Guillen, N.; Acín, S.; Surra, J.C.; Arnal, C.; Godino, J.; García-Granados, A.; Muniesa, P.; Ruiz-Gutiérrez, V.; Osada, J. Apolipoprotein E determines the hepatic transcriptional profile of dietary maslinic acid in mice. *J. Nutr. Biochem.* **2009**, *20*, 882–893. [CrossRef] [PubMed]
116. Lin, C.C.; Huang, C.Y.; Mong, M.C.; Chan, C.Y.; Yin, M.C. Antiangiogenic potential of three triterpenic acids in human liver cancer cells. *J. Agric. Food Chem.* **2011**, *59*, 755–762. [CrossRef]

117. Mohr, A.M.; Lavery, R.F.; Barone, A.; Bahramipour, P.; Magnotti, L.J.; Osband, A.J.; Sifri, Z.; Livingston, D.H. Angiographic embolization for liver injuries: Low mortality, high morbidity. *J. Trauma-Inj. Infect. Crit. Care* **2003**, *55*, 1077–1081; discussion 1081–1802. [CrossRef]
118. Andrade, R.; Lucena, M.; Fernández, M.; Pelaez, G.; Pachkoria, K.; García-Ruiz, E.; García-Muñoz, B.; González-Grande, R.; Pizarro, A.; Durán, J.; et al. Drug-Induced Liver Injury: An Analysis of 461 Incidences Submitted to the Spanish Registry over a 10-Year Period. *Gastroenterology* **2005**, *129*, 512–521. [CrossRef]
119. Mahler, H.; Pasi, A.; Kramer, J.M.; Schulte, P.; Scoging, A.C.; Bär, W.; Krähenbühl, S. Fulminant liver failure in association with the emetic toxin of Bacillus cereus. *N. Engl. J. Med.* **1997**, *336*, 1142–1148. [CrossRef]
120. Su, G.L. Lipopolysaccharides in liver injury: Molecular mechanisms of Kupffer cell activation. *Am. J. Physiol. Gastrointest. Liver Physiol.* **2002**, *283*, G256–G265. [CrossRef]
121. Koch, O.R.; Pani, G.; Borrello, S.; Colavitti, R.; Cravero, A.; Farrè, S.; Galeotti, T. Oxidative stress and antioxidant defenses in ethanol-induced cell injury. *Mol. Asp. Med.* **2004**, *25*, 191–198. [CrossRef]
122. Lluis, J.M.; Colell, A.; García-Ruiz, C.; Kaplowitz, N.; Fernández-Checa, J.C. Acetaldehyde impairs mitochondrial glutathione transport in HepG2 cells through endoplasmic reticulum stress. *Gastroenterology* **2003**, *124*, 708–724. [CrossRef]
123. McClain, C.J.; Song, Z.; Barve, S.S.; Hill, D.B.; Deaciuc, I. Recent advances in alcoholic liver disease. IV. Dysregulated cytokine metabolism in alcoholic liver disease. *Am. J. Physiol. Gastrointest. Liver Physiol.* **2004**, *287*, G497–G502. [CrossRef] [PubMed]
124. Naveau, S.; Balian, A.; Capron, F.; Raynard, B.; Fallik, D.; Agostini, H.; Grangeot-Keros, L.; Portier, A.; Galanaud, P.; Chaput, J.C.; et al. Balance between pro and anti-inflammatory cytokines in patients with acute alcoholic hepatitis. *Gastroenterol. Clin. Biol.* **2005**, *29*, 269–274. [CrossRef] [PubMed]
125. Tipoe, G.L.; Leung, T.M.; Liong, E.C.; Lau, T.Y.; Fung, M.L.; Nanji, A.A. Epigallocatechin-3-gallate (EGCG) reduces liver inflammation, oxidative stress and fibrosis in carbon tetrachloride (CCl4)-induced liver injury in mice. *Toxicology* **2010**, *273*, 45–52. [CrossRef] [PubMed]
126. Yan, S.L.; Yang, H.T.; Lee, H.L.; Yin, M.C. Corrigendum to "Protective effects of maslinic acid against alcohol-induced acute liver injury in mice" [Food and Chem. Toxicol. 74 (2014 Dec) 149–55]. *Food Chem. Toxicol.* **2017**, *106 Pt A*, 570. [CrossRef]
127. Wang, Y.Y.; Diao, B.Z.; Zhong, L.H.; Lu, B.L.; Cheng, Y.; Yu, L.; Zhu, L.Y. Maslinic acid protects against lipopolysaccharide/d-galactosamine-induced acute liver injury in mice. *Microb. Pathog.* **2018**, *119*, 49–53. [CrossRef]
128. Pang, R.; Poon, R.T. Angiogenesis and antiangiogenic therapy in hepatocellular carcinoma. *Cancer Lett.* **2006**, *242*, 151–167. [CrossRef]
129. Finn, R.S.; Zhu, A.X. Targeting angiogenesis in hepatocellular carcinoma: Focus on VEGF and bevacizumab. *Expert Rev. Anticancer. Ther.* **2009**, *9*, 503–509. [CrossRef]
130. Andreasen, P.A.; Kjøller, L.; Christensen, L.; Duffy, M.J. The urokinase-type plasminogen activator system in cancer metastasis: A review. *Int. J. Cancer* **1997**, *72*, 1–22. [CrossRef]
131. Morbidelli, L.; Donnini, S.; Ziche, M. Role of nitric oxide in the modulation of angiogenesis. *Curr. Pharm. Des.* **2003**, *9*, 521–530. [CrossRef]
132. Wu, W.S. The signaling mechanism of ROS in tumor progression. *Cancer Metastasis Rev.* **2006**, *25*, 695–705. [CrossRef]
133. Kitade, H.; Chen, G.; Ni, Y.; Ota, T. Nonalcoholic Fatty Liver Disease and Insulin Resistance: New Insights and Potential New Treatments. *Nutrients* **2017**, *9*, 387. [CrossRef]
134. Hazlehurst, J.M.; Woods, C.; Marjot, T.; Cobbold, J.F.; Tomlinson, J.W. Non-alcoholic fatty liver disease and diabetes. *Metab. Clin. Exp.* **2016**, *65*, 1096–1108. [CrossRef] [PubMed]
135. Ishikawa, T.; Donatini, R.S.; Diaz, I.E.; Yoshida, M.; Bacchi, E.M.; Kato, E.T. Evaluation of gastroprotective activity of *Plinia edulis* (Vell.) Sobral (Myrtaceae) leaves in rats. *J. Ethnopharmacol.* **2008**, *118*, 527–529. [CrossRef] [PubMed]
136. Da Rosa, R.L.; Nesello, L.; Mariano, L.N.B.; Somensi, L.B.; Campos, A.; Pinheiro, A.M.; Costa, S.; Rial, M.; Tozzo, M.; Cechinel-Filho, V.; et al. Gastroprotective activity of the methanol extract from peels of *Plinia edulis* (Vell.) Sobral fruits and its isolated triterpenes: Maslinic and ursolic acids. *Naunyn-Schmiedebergs Arch. Pharmacol.* **2018**, *391*, 95–101. [CrossRef] [PubMed]
137. Laine, L.; Takeuchi, K.; Tarnawski, A. Gastric mucosal defense and cytoprotection: Bench to bedside. *Gastroenterology* **2008**, *135*, 41–60. [CrossRef]
138. Rozza, A.L.; Cesar, D.A.; Pieroni, L.G.; Saldanha, L.L.; Dokkedal, A.L.; De-Faria, F.M.; Souza-Brito, A.R.; Vilegas, W.; Takahira, R.K.; Pellizzon, C.H. Antiulcerogenic Activity and Toxicity of Bauhinia holophylla Hydroalcoholic Extract. *Evid. Based Complementary Altern. Med.* **2015**, *2015*, 439506. [CrossRef]
139. Njor, S.H.; Friis-Hansen, L.; Andersen, B.; Søndergaard, B.; Linnemann, D.; Jørgensen, J.C.R.; Roikjær, O.; Rasmussen, M. Three years of colorectal cancer screening in Denmark. *Cancer Epidemiol.* **2018**, *57*, 39–44. [CrossRef]
140. Eng, C. Toxic effects and their management: Daily clinical challenges in the treatment of colorectal cancer. *Nat. Rev. Clin. Oncol.* **2009**, *6*, 207–218. [CrossRef]
141. Reyes, F.J.; Centelles, J.J.; Lupiáñez, J.A.; Cascante, M. (2Alpha,3beta)-2,3-dihydroxyolean-12-en-28-oic acid, a new natural triterpene from Olea europea, induces caspase dependent apoptosis selectively in colon adenocarcinoma cells. *FEBS Lett.* **2006**, *580*, 6302–6310. [CrossRef]
142. Sánchez-Tena, S.; Reyes-Zurita, F.J.; Díaz-Moralli, S.; Vinardell, M.P.; Reed, M.; García-García, F.; Dopazo, J.; Lupiáñez, J.A.; Günther, U.; Cascante, M. Maslinic acid-enriched diet decreases intestinal tumorigenesis in Apc(Min/+) mice through transcriptomic and metabolomic reprogramming. *PLoS ONE* **2013**, *8*, e59392. [CrossRef]

143. Lozano-Mena, G.; Sánchez-González, M.; Parra, A.; Juan, M.E.; Planas, J.M. Identification of gut-derived metabolites of maslinic acid, a bioactive compound from *Olea europaea* L. *Mol. Nutr. Food Res.* **2016**, *60*, 2053–2064. [CrossRef]
144. Sánchez-González, M.; Colom, H.; Lozano-Mena, G.; Juan, M.E.; Planas, J.M. Population pharmacokinetics of maslinic acid, a triterpene from olives, after intravenous and oral administration in rats. *Mol. Nutr. Food Res.* **2014**, *58*, 1970–1979. [CrossRef] [PubMed]
145. Furuichi, K.; Shimizu, M.; Okada, H.; Narita, I.; Wada, T. Clinico-pathological features of kidney disease in diabetic cases. *Clin. Exp. Nephrol.* **2018**, *22*, 1046–1051. [CrossRef] [PubMed]
146. Flemming, N.B.; Gallo, L.A.; Forbes, J.M. Mitochondrial Dysfunction and Signaling in Diabetic Kidney Disease: Oxidative Stress and Beyond. *Semin. Nephrol.* **2018**, *38*, 101–110. [CrossRef] [PubMed]
147. Morimoto, K.; Matsui, M.; Samejima, K.; Kanki, T.; Nishimoto, M.; Tanabe, K.; Murashima, M.; Eriguchi, M.; Akai, Y.; Iwano, M.; et al. Renal arteriolar hyalinosis, not intimal thickening in large arteries, is associated with cardiovascular events in people with biopsy-proven diabetic nephropathy. *Diabet. Med.* **2020**, *37*, 2143–2152. [CrossRef]
148. Echouffo-Tcheugui, J.B.; Selvin, E. Prediabetes and What It Means: The Epidemiological Evidence. *Annu. Rev. Public Health* **2021**, *42*, 59–77. [CrossRef]
149. De Nicola, L.; Conte, G.; Minutolo, R. Prediabetes as a Precursor to Diabetic Kidney Disease. *Am. J. Kidney Dis.* **2016**, *67*, 817–819. [CrossRef]
150. Akinnuga, A.M.; Siboto, A.; Khumalo, B.; Sibiya, N.H.; Ngubane, P.; Khathi, A. Ameliorative Effects of Bredemolic Acid on Markers Associated with Renal Dysfunction in a Diet-Induced Prediabetic Rat Model. *Oxidative Med. Cell. Longev.* **2020**, *2020*, 2978340. [CrossRef]
151. Mkhwanazi, B.N.; van Heerden, F.R.; Mavondo, G.A.; Mabandla, M.V.; Musabayane, C.T. Triterpene derivative improves the renal function of streptozotocin-induced diabetic rats: A follow-up study on maslinic acid. *Ren. Fail.* **2019**, *41*, 547–554. [CrossRef]
152. Andreucci, M.; Faga, T.; Pisani, A.; Sabbatini, M.; Russo, D.; Michael, A. Prevention of contrast-induced nephropathy through a knowledge of its pathogenesis and risk factors. *Sci. World J.* **2014**, *2014*, 823169. [CrossRef]
153. Cooper, J.E.; Wiseman, A.C. Acute kidney injury in kidney transplantation. *Curr. Opin. Nephrol. Hypertens.* **2013**, *22*, 698–703. [CrossRef] [PubMed]
154. Dong, Y.; Zhang, Q.; Wen, J.; Chen, T.; He, L.; Wang, Y.; Yin, J.; Wu, R.; Xue, R.; Li, S.; et al. Ischemic Duration and Frequency Determines AKI-to-CKD Progression Monitored by Dynamic Changes of Tubular Biomarkers in IRI Mice. *Front. Physiol.* **2019**, *10*, 153. [CrossRef]
155. Han, S.J.; Lee, H.T. Mechanisms and therapeutic targets of ischemic acute kidney injury. *Kidney Res. Clin. Pract.* **2019**, *38*, 427–440. [CrossRef]
156. Sun, W.; Choi, H.S.; Kim, C.S.; Bae, E.H.; Ma, S.K.; Kim, S.W. Maslinic Acid Attenuates Ischemia/Reperfusion-Induced Acute Kidney Injury by Suppressing Inflammation and Apoptosis through Inhibiting NF-κB and MAPK Signaling Pathway. *Front. Pharmacol.* **2022**, *13*, 807452. [CrossRef] [PubMed]
157. Dellepiane, S.; Marengo, M.; Cantaluppi, V. Detrimental cross-talk between sepsis and acute kidney injury: New pathogenic mechanisms, early biomarkers and targeted therapies. *Crit. Care* **2016**, *20*, 61. [CrossRef] [PubMed]
158. Lim, S.H.; Hwang, I.G.; Ji, J.H.; Oh, S.Y.; Yi, J.H.; Lim, D.H.; Lim, H.Y.; Lee, S.J.; Park, S.H. Intrinsic resistance to sunitinib in patients with metastatic renal cell carcinoma. *Asia-Pac. J. Clin. Oncol.* **2017**, *13*, 61–67. [CrossRef] [PubMed]
159. Rini, B.I.; Small, E.J. Biology and clinical development of vascular endothelial growth factor-targeted therapy in renal cell carcinoma. *J. Clin. Oncol.* **2005**, *23*, 1028–1043. [CrossRef]

molecules

MDPI

Article

Root Bark Extract of *Oroxylum indicum* Vent. Inhibits Solid and Ascites Tumors and Prevents the Development of DMBA-Induced Skin Papilloma Formation

Seema Menon [1,2,*], Jawaher J. Albaqami [3], Hamida Hamdi [3,4], Lincy Lawrence [1], Menon Kunnathully Divya [1], Liya Antony [1], Jose Padikkala [1], Shaji E. Mathew [1,2] and Arunaksharan Narayanankutty [1,5,*]

1. Department of Biochemistry and Plant Biotechnology, Amala Cancer Research Centre, Amala Nagar, Thrissur 680555, India
2. Department of Zoology, Kodungallur Kunjikuttan Thampuran Memorial Government College, Pullut, Kodungallur, Thrissur 680663, India
3. Department of Biology, College of Science, Taif University, P.O. Box 11099, Taif 21944, Saudi Arabia
4. Zoology Department, Faculty of Science, Cairo University, Giza 12613, Egypt
5. Division of Cell and Molecular Biology, PG & Research Department of Zoology, St. Joseph's College (Autonomous), Devagiri, Calicut 673008, India

* Correspondence: seematty2@gmail.com (S.M.); arunaksharann@devagiricollege.org (A.N.)

Abstract: *Oroxylum indicum* is a traditionally used plant in Ayurvedic and folk medicines. The plant is useful for the management of gastrointestinal diseases as well as skin diseases. In the present study, we analyzed the antitumor potential of *O. indicum* in Dalton's lymphoma ascites tumor cells (DLA) and Ehrlich ascites carcinoma (EAC)-induced solid and ascites tumors. Further, the potential of *O. indicum* extract (OIM) on skin papilloma induction by dimethyl benz(a) anthracene (DMBA) and croton oil was evaluated. The chemical composition of the extract was analyzed using UPLC-Q-TOF-MS. The predominant compounds present in the extract were demethoxycentaureidin 7-*O*-rutinoside, isorhamnetin-3-*O*-rutinoside, baicalein-7-*O*-glucuronide, 5,6,7-trihydroxyflavone, 3-Hydroxy-3′,4′,5′-trimethoxyflavone, 5,7-dihydroxy-3-(4-methoxyphenyl) chromen-4-one, and 4′-Hydroxy-5,7-dimethoxyflavanone. Treatment with high-dose OIM enhanced the percentage of survival in ascites tumor-bearing mice by 34.97%. Likewise, high and low doses of OIM reduced the tumor volume in mice by 61.84% and 54.21%, respectively. Further, the skin papilloma formation was brought down by the administration of low- and high-dose groups of OIM (by 67.51% and 75.63%). Overall, the study concludes that the *Oroxylum indicum* root bark extract is a potentially active antitumor and anticancer agent.

Keywords: *Oroxylum indicum*; antioxidant activity; dimethyl benz(a) anthracene; skin papilloma; antitumor activity

Citation: Menon, S.; Albaqami, J.J.; Hamdi, H.; Lawrence, L.; Divya, M.K.; Antony, L.; Padikkala, J.; Mathew, S.E.; Narayanankutty, A. Root Bark Extract of *Oroxylum indicum* Vent. Inhibits Solid and Ascites Tumors and Prevents the Development of DMBA-Induced Skin Papilloma Formation. *Molecules* **2022**, *27*, 8459. https://doi.org/10.3390/molecules27238459

Academic Editor: Kyoko Nakagawa-Goto

Received: 5 November 2022
Accepted: 28 November 2022
Published: 2 December 2022

Publisher's Note: MDPI stays neutral with regard to jurisdictional claims in published maps and institutional affiliations.

Copyright: © 2022 by the authors. Licensee MDPI, Basel, Switzerland. This article is an open access article distributed under the terms and conditions of the Creative Commons Attribution (CC BY) license (https://creativecommons.org/licenses/by/4.0/).

1. Introduction

Cancer is characterized by the abnormal and uncontrolled proliferation of cells, which possess the ability to metastasize or invade other parts of the body [1,2]. The invading cells form a subset of neoplasms, which either form a tumor lump or a diffusely distributed cell population [3,4]. Chemotherapy is one of the common modalities of cancer treatment. Chemotherapeutic agents eliminate the rapidly proliferating cells, including hematopoietic bone marrow cells, hair follicles, digestive tract epithelial cells, and reproductive tract cells, apart from the cancer cells, which essentially are the prime targets [5]. This poses serious side effects that affect vital organs, such as the heart, lungs, kidneys, and digestive organs [6]. Clinically used chemotherapeutics are also facing issues of resistance by various molecular pathways [7–9]. Hence, there is a need for novel anticancer drug candidates to overcome the issues associated with present-day chemotherapeutics [10].

Oroxylum indicum Vent. is one of the ten plants whose roots are used as an ingredient of the *Dashamoola* combination in Ayurveda [11,12]. Different parts of the tree also find use in folklore medicine to cure ailments such as urinary infections, bronchitis, leukoderma, diarrhea, nasopharyngeal cancers, oral cancers, etc., as reviewed by Deka, et al. [13], Mao [14], and Preety and Sharma [15]. The root bark possesses anti-ulcer [16], immunomodulatory [17], antioxidant [18], and hepatoprotective [19] properties. The antiproliferative potential of different fractions of the root bark has also been reported [20] in human breast carcinoma cells. Still, thorough research on its anticancer properties in in vivo models has not been conducted.

Animals, including mouse models, reiterate the human equivalent of different malignancies [21]. Anticancer agents of plant origin bring about the inhibition/suppression of carcinogenesis, thereby acting as chemopreventive agents, or they pose toxicity to the cells in already developed tumors, thereby reducing tumor burden [22,23]. Ehrlich ascites carcinoma (EAC) is a rapidly proliferating, experimental transplantable tumor maintained in outbred mice by a series of intraperitoneal passages. It was originally identified as a murine mammary adenocarcinoma, which later was adapted to ascites form [24,25], and has been exploited for many chemotherapeutic studies [26–28]. The antitumor activity of any agent against Ehrlich ascites can be assessed by cytological examination of the ascites cells post-treatment, calculating the increase in survival time and/or by measuring the number of ascites formed after treatment [24]. Our study relied on the evaluation of the antitumor activity of *Oroxylum indicum* Vent. root bark, in EAC cells, in terms of its effect on the increase in the average life span.

Earlier models of evaluation of antitumor activity included murine models induced with ascitic leukemia using different types of cancer cell lines [29,30], but were not adequate enough for the identification of therapeutic agents against solid tumors [31,32]. In our study, the antitumor efficacy of the extract was evaluated using the solid tumor model in mice, inoculated intramuscularly with Dalton's lymphoma ascites (DLA) cell line, which has been known to be in use for many similar investigations before [33,34]. The 7, 12-dimethyl-benz[a] anthracene (DMBA) acts as an inducer of the two-step mutagenesis and the TPA present in croton oil promotes the malignant conversion of mouse skin tumors [35]. Overall, the present study evaluates the antitumor and anticarcinogenic potential of *O. indicum* Vent. root bark in multiple models in mice.

2. Results

2.1. Phytochemical Analysis

2.1.1. Qualitative Phytochemical Analysis

Figure 1 shows the total ion chromatogram of the OIM extract subjected to UPLC-Q-TOF-MS analysis. The compounds provisionally identified (based on previous fragmentation data from the literature) are enlisted in Table 1.

Table 1. The compounds tentatively identified from *O. indicum* Vent. wild root bark (OIM) extract by UPLC-Q-TOF-MS analysis.

Peak No.	RT (Min)	*m/z*	Molecular Weight (kDa)	Molecular Formula	Name of the Compound
1	0.79	562.2056	–	–	Unidentified
2	2.92	487.1551	–	–	Unidentified
3	3.70	639.2048	–	–	Unidentified
4	3.90	637.1891	638.184685	$C_{29}H_{34}O_{16}$	Demethoxycentaureidin 7-*O*-rutinoside
5	4.02	653.2208	–	–	Unidentified
6	4.13	623.2098	624.16903	$C_{28}H_{32}O_{16}$	Isorhamnetin-3-*O*-rutinoside (Narcissin)
7	4.36	607.2140	–	–	Unidentified
8	4.52	547.1558	–	–	Unidentified
9	4.75	651.2419	–	–	Unidentified

Table 1. *Cont.*

Peak No.	RT (Min)	*m/z*	Molecular Weight (kDa)	Molecular Formula	Name of the Compound
10	4.99	445.1238	446.08491	$C_{21}H_{18}O_{11}$	Baicalein-7-*O*-glucuronide (Baicalin)
11	5.65	269.0502	270.05282	$C_{15}H_{10}O_5$	5,6,7-Trihydroxyflavone (Baicalein)
12	5.82	327.2242	328.094688	$C_{18}H_{16}O_6$	3-Hydroxy-3′,4′,5′-trimethoxyflavone
13	6.02	283.0659	284.068473	$C_{16}H_{12}O_5$	5,7-Dihydroxy-3-(4-methoxyphenyl)chromen-4-one (Biochanin A)
14	7.05	299.2069	300.099774	$C_{17}H_{16}O_5$	4′-Hydroxy-5,7-dimethoxyflavanone
15	7.23	295.2332	296.104859	$C_{18}H_{16}O_4$	6-Ethoxy-3(4′-hydroxyphenyl)-4-methylcoumarin
16	7.80	311.1749	–	–	Unidentified
17	8.16	325.1904	–	–	Unidentified

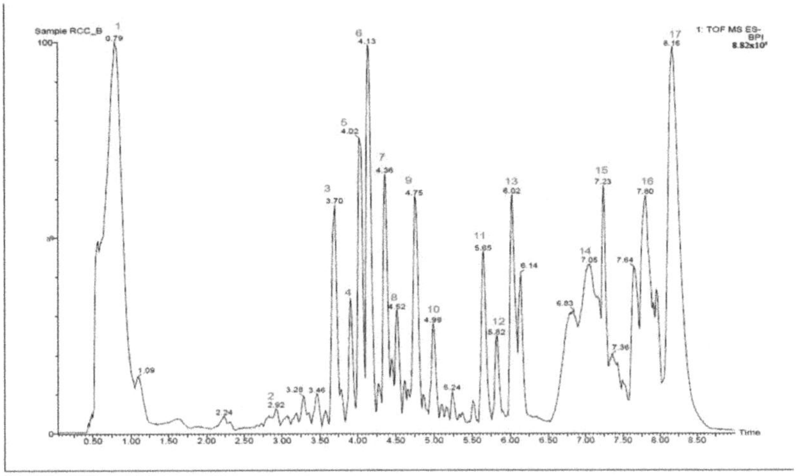

Figure 1. Total ion chromatogram of *O. indicum* Vent. wild root bark (OIM) extract subjected to UPLC-Q-TOF-MS analysis. The compounds corresponding to the peak numbers are listed in Table 1.

2.1.2. HPLC-Based Quantification of Chrysin and Baicalein

The quantitative determination of chrysin and baicalein by HPLC is shown in the chromatogram (Supplementary Figure S1). The quantity of baicalein was 1.794 ± 0.23 mg/g of *O. indicum* powder. The quantity of chrysin was 0.725 ± 0.08 mg/g of the powder of *O. indicum* (Table 2).

Table 2. The concentration of baicalein (Bcl) and chrysin (Chr) expressed as mg/g dry wt. of powder in *O. indicum* Vent. medium.

Compound	Concentration
Baicalein	1.794 ± 0.23
Chrysin	0.725 ± 0.08

2.2. Cytotoxicity of O. indicum Extract

The methanolic extract of *Oroxylum indicum* Vent. showed a profound decline in the viability of DLA and EAC cell lines in a dose-dependent pattern (Figure 2). No cell death was observed in the control. With OIM extract concentrations at 259.07 and 252.53 µg/mL respectively, 50% death of DLA and EAC cells occurred. The extract was found to be non-toxic to spleen cells up to the highest concentration used (500 µg/mL).

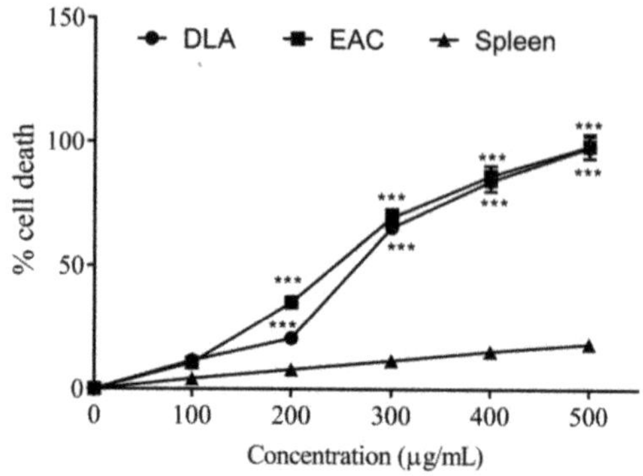

Figure 2. In vitro cytotoxicity of different concentrations of methanolic extract of *O. indicum* Vent. root bark (OIM) on DLA and EAC cell lines and normal spleen cells. *** Indicates a significant variation with $p < 0.001$ with the cytotoxicity in spleen cells.

2.3. Effect of OIM Extract on Ascites Tumor

All animals developed ascites tumors upon receiving intraperitoneal inoculation of EAC cells. When the untreated control group sustained only an average survival period of 16.67 ± 1.86 days, treatment with OIM extract at 400 and 200 mg/kg increased the average life span to 22.5 ± 2.43 ($p < 0.01$) and 20.67 ± 2.94 ($p < 0.05$) days, respectively. OIMH treatment elevated the percentage increase of survival by 34.97%. The animals under standard drug cyclophosphamide survived for 23.67 ± 2.16 days (percentage increase in survival—41.99%) (Figure 3).

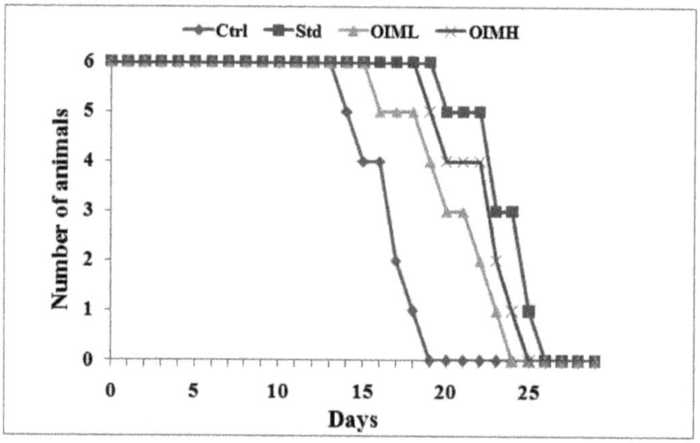

Figure 3. Effect of methanolic extract of *O. indicum* Vent. (OIM) on the average survival period of ascites tumor-bearing animals. Ctrl—untreated negative control; Std—positive control, treated with cyclophosphamide (i.p.) at 10 mg/kg b.wt.; OIML—low-dose group treated with OIM at dosage 200 mg/kg, and OIMH—high-dose group treated with OIM at dosage 400 mg/kg b.wt.

2.4. Effect of OIM Extract on Solid Tumor

All animals who received intramuscular DLA injection developed tumors by the sixth day of induction, which grew in volume in successive days. Yet, compared to the final volume recorded on the 30th day in the control group (2.07 ± 0.23 cm^3), *O. indicum Vent.* extract-treated groups showed a marked decrease ($p < 0.01$) in tumor volume up to 0.79 ± 0.03 cm^3 in OIMH and 0.95 ± 0.09 cm^3 in OIML. OIMH and OIML treatment brought about 61.84% and 54.21% inhibition of tumor growth, respectively, closer to that affected by the cyclophosphamide treatment in the standard animal group (71.05%) (Figure 4).

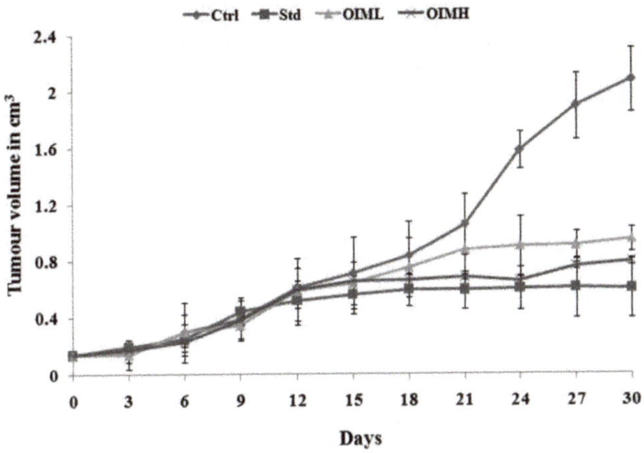

Figure 4. Effect of methanolic extract of *O. indicum* Vent. (OIM) on tumor volume in solid tumor-bearing animals. Values are expressed as mean ± SD for 6 animals. Ctrl—untreated negative control; Std—positive control, treated with cyclophosphamide (i.p.) at 10 mg/kg b.wt; OIML—low-dose group treated with OIM at dosage 200 mg/kg b.wt., and OIMH—high-dose group treated with OIM at dosage 400 mg/kg b.wt.

2.5. *Effect of OIM Extract on Hematological Parameters*

Though it was evident that animals treated with cyclophosphamide exhibited substantial anti-tumor activity in solid tumor models, there was a gradual reduction in the hemoglobin concentration (Figure 5a) and total leucocyte count of the animals, which was not obviously seen in the control, OIML, or OIMH groups (Figure 5b).

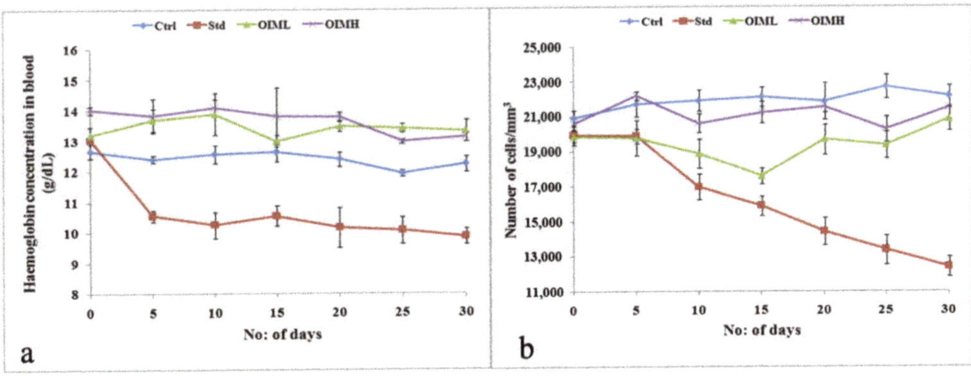

Figure 5. Comparison of hematological parameters between cyclophosphamide-treated animals and the other study groups during solid tumor induction. (**a**) Percentage hemoglobin (g/dL) and (**b**) total leucocyte count (no: of cells/mm^3) of Ctrl—tumor-induced group, having received no treatment; Std—cyclophosphamide (i.p.) treated group at dosage 10 mg/kg b.wt; OIML—low-dose group treated with OIM at dosage 200 mg/kg b.wt., and OIMH—high-dose group treated with OIM at dosage 400 mg/kg b.wt.

2.6. *Effect of OIM Extract on Skin Papilloma*

The application of either DMBA or croton oil alone did not induce papilloma in any of the animals. Meanwhile, DMBA plus croton oil application in untreated control animals led to full-blown papilloma development (100% incidence). However, OIM application at low and high regimes reduced the incidence of a tumor to 83.3 and 66.6%, respectively

(Table 1). Papilloma onset was registered in the control group in the 6th week of the topical application itself, whereas it was delayed in treated groups—9th week in OIML and 12th week in OIMH (Figure 6). Table 1 shows that the average number of papillomas per animal was significantly ($p < 0.01$) lower in treated mice compared to the control. OIM treatment was found to inhibit papilloma development in low- and high-dose groups by 67.51% and 75.63%, respectively.

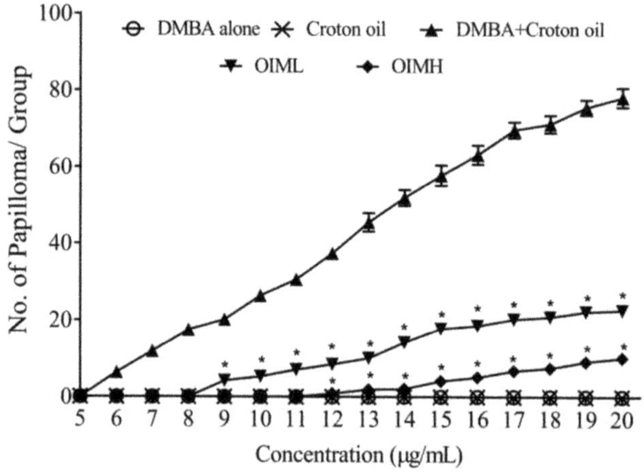

Figure 6. Effect of methanolic extract of *O. indicum* (OIM) on skin tumorigenesis in the DMBA–croton oil-induced animals. Group I: animals induced with 470 nmol of DMBA alone, in 200 acetone; Group II: animals induced 1% croton oil alone, in 200 acetone; Group III: animals induced with DMBA plus croton oil, which served as negative control; Group IV: [(DMBA + croton oil + OIM 5% in 200 μL distilled water)]; and Group V: [(DMBA + croton oil + OIM 10% in 200 μL distilled water)]. * indicate significant variation at $p < 0.05$ with DMBA + croton oil group.

3. Discussion

Cell-based screening assays are widely employed in drug discovery systems for determining the anticancer properties of both synthetic and plant-derived compounds, and the viability measurements of cancer cell lines in the presence of the tested compound are preliminary steps in drug screening. In the study, we found significant in vitro anticancer properties of OIM extract using short-term cytotoxicity assays in DLA and EAC cell lines, where non-toxic effects were noticed in normal splenocytes. Methanolic and aqueous extracts of *O. indicum* bark were previously demonstrated to induce cytotoxicity in MDA-MB-435S cell lines, as determined by the XTT assay [36]. The anti-proliferative ability of the methanolic extracts of fruits from *O. indicum* was studied on HL-60 cell lines, with baicalein identified as an active principle in the extract [37]. The flavonoid (already reported to be present in the root bark [38]) may be responsible for its capability to induce mortality in DLA and EAC cell lines, as uncovered by the present study.

Followingly, the antitumor activity of *O. indicum* Vent. root bark extract was evaluated using EAC-induced ascites tumor models. Prolongation of the life span of animals is a reliable criterion for adjudging an anticancer agent. Though death is inevitable, OIM in high doses showed a significant increase in the survival rate of mice, compared to untreated control. However, OIM extract at a low dose (200 mg/kg b.wt) increased the life span of tumor-bearing animals by merely 23.99%, but significantly different from the survival rate of untreated animals. Similar results of anti-tumor activity in EAC-induced tumors were previously reported by Samudrala, et al. [39] with *Alternanthera brasiliana,* and Senthil Kumar et al. [40], with *Prosopis glandulosa*. In EAC-induced models, the survival times of mice and the final volumes of the tumors were positively correlated, revealing the progressive growth pattern. Tumor growth is also enhanced by the amount of ascitic fluid in the mouse peritoneal cavity, which is an inflammatory exudate produced in response to EAC cell growth [41,42]. In all EAC-induced animals of our study, there was a gradual increase noticed in ascites fluid volume, pertinent to the above concept.

Solid tumors (e.g., sarcomas, carcinomas, and lymphomas) are abnormal, localized masses of tissues devoid of cysts or liquid areas, of either benign or malignant nature [43,44]. Dalton's lymphoma spontaneously originates in the thymus of murine animals in the transplantable form [45–47] and, hence, widely used in cancer research, with tumor volume measurement as a reproducible parameter. The current study also revealed that oral administration of OIM extract brought about a gradual, but significant reduction in tumor volume, which was evident from the 12th day of induction onwards up to day 30, signifying the anticancer property of the extract, consistent with previous anticancer studies of plant extracts [48,49] using DLA induced tumors.

The study reveals the significant anti-carcinogenic efficacy of the root bark extract of *O. indicum* Vent. in suppressing papilloma genesis, which is induced by the application of DMBA and croton oil. In the typical two-stage chemical carcinogenesis system in the mouse skin, a mutation in *Hras1* induced by a low dose of 7,12-dimethyl benz(a)anthracene is promoted by repeated application of TPA, which acts as a tumor promoter. During the tumor promotion stage, due to the altered expression of genes whose encoded protein products take part in hyperproliferation, tissue remodeling, and inflammation, the initiated cells are subjected to selective clonal expansion, forming visible tumors [50]. This facilitates the initiation and promotion stages to be distinctly viewed both operationally and mechanistically [51].

The inhibition of tumorigenesis, as revealed from the study, has occurred in the promotion stage. Topical application of OIM extract showed a dose-dependent inhibition of skin tumorigenesis, with a delay of three and six weeks respectively, in high and low-dose groups, compared to the full-blown skin tumors that developed in control mice, having received no external treatment. Chemoprevention, as stated by Wattenberg [52], can occur either during initiation or during a promotion; the mediators preventing the former are categorized as blocking agents, and those inhibiting the latter as suppressing agents. Chemoprevention can be best accomplished during the promotion stage of carcinogenesis because of its reversible nature which requires more time and a higher incidence of exposure, unlike initiation or progression [53,54].

The relationship between inflammation and papilloma-genesis has been previously established. During human papillomavirus infection, the host inflammatory response is postulated to promote lesion progression through the direct participation of inflammatory cells. Under the influence of the microbial genome or epigenetic factors, somatic cells undergo changes mediated by autocrine and paracrine signals indicative of the association between chronic inflammation and cancer [55,56]. In addition to being a component of *Dashamoola* formulation, there is also mention of the usage of *O. indicum* Vent. as a single ingredient in treating rheumatoid arthritis and inflammation [57,58]. The methanolic extract root bark of *O. indicum* Vent. has been previously reported by the authors indicating its potential anti-inflammatory properties in acute and chronic paw edema models [59] in mice. In light of the above facts, it is conclusive that the inhibitive action of OIM extract may be due to its suppressive effect on inflammation.

In the context of serious side effects posed by chemotherapeutic agents used in cancer therapy, a surge has been observed in developing complementary medication from plant sources, which are repositories of bioactive compounds. In the antitumor study conducted by us, cyclophosphamide was used to treat the positive control group, the compound being a common chemotherapeutic agent [60], which is also known to cause myelosuppression and anemia [61]. In our study, it was observed that, though cyclophosphamide treatment increased the average life span in ascites tumor models and reduced solid tumors, the animals developed anemia and suffered a decrease in total leucocyte count, as revealed from the hematological examination. Contrarily, the study reveals that OIM extract exerts simultaneous antitumor and myeloprotective activities, similar to other plant extracts rich in bioactive compounds [62,63], possibly due to the synergistic effect of these substances present in them.

The root bark of *O. indicum* has been found to contain chrysin, baicalein, biochanin-A (flavonoids), and ellagic acid (phenolic compound) [38]. With previously reported anti-cancer activities of baicalein [64,65], chrysin [66,67], biochanin-A [68], and ellagic acid [69], it may be concluded that the antitumor activity of the extract is due to the combined activity of these bioactive components, exerting anti-carcinogenic and cytotoxic effects.

4. Materials and Methods

4.1. Reagents and Chemicals

7,12-dimethyl-benz[a] anthracene of an analytical grade was used for the study (Sigma Aldrich, MO, USA); the hematological test kit manufactured by Agappe Diagnostics (Ernakulam, Kerala, India) was used for hemoglobin estimation. The reagents and chemicals that were used in the study were of analytic grade.

4.2. Collection of Oroxylum Indicum and Methanol Extraction

The roots of *Oroxylum indicum* Vent. were collected from Thrissur and authenticated (KFRI/SILVA/GEN/06/11). The peeled bark was dried, powdered, and extracted using methanol as solvent. The methanol extract was selected based on preliminary studies and also considering the previous studies that methanol extracts most of the phytochemicals [70,71]. The concentrated extract was dissolved in the deionized water for the animal model studies and cytotoxicity analysis.

4.3. UPLC-Q-TOF-MS Analysis

Ultra-high-pressure liquid chromatographic (UPLC) analysis of the extract was performed using an Acquity UPLC H class (Waters) system equipped with an autosampler and a diode-array detector (DAD). The mobile phase (methanol and 0.1% aqueous formic acid) was allowed to percolate through a BEH C18 column (with specifications, 50 mm × 2.1 mm × 1.7 μm, purchased from Waters, USA) using gradient elution (0–5 min, 5% acetonitrile; 5–7 min, 95% methanol; 8–9 min 5% methanol) at a flow rate of 0.3 mL/min. Detection was achieved at a wavelength of 210–400 nm. The MS and MS/MS data were retrieved from Xevo G2 (Waters, Massachusetts, USA) Quadrupole-Time-of-Flight (Q-TOF) system. Mass spectrometric operations were carried out by the methods of Felipe, et al. [72].

The quantification of flavonoids chrysin and baicalein were also carried out using the same solvent and running conditions using HPLC SCL-10A VP (Shimadzu, Kyoto, Japan). The standard compounds were initially dissolved in 0.1 mg/mL concentration and loaded in the HPLC system. Similarly, the extract was analyzed and compared with the standards, and the final concentration was estimated using SCL-VP software.

4.4. Animals

BALB/c mice weighing 20–25 g were grouped and the in vivo acute toxicity analysis of *O. indicum* Vent. root bark extract was conducted for 14 days according to OECD guideline-423 [73]. Based on the acute toxicity studies conducted by the authors as per OECD—423 guidelines, OIM extract was found to be having no observed adverse effect limit (NOAEL) till 5 g/ kg [59]. For further studies, 400 mg/kg b.wt. of OIM was selected as the high dose (OIMH) and 200 mg/kg of OIM as the low dose (OIML).

4.5. Cell Lines

Dalton's lymphoma ascites tumor cells (DLA) and Ehrlich ascites carcinoma (EAC) cells were aspirated aseptically in PBS (0.1 M, pH 7.4). The wash was repeated thrice and the cell count was set as 10^7 cells/mL and used for cytotoxicity and antitumor studies [74].

4.6. Cytotoxic Analysis of O. indicum on DLA and EAC Cells

About 0.1 mL of the DLA/EAC cell suspension was dispensed into test tubes containing OIM extract, at concentrations ranging from 100 to 500 μg/mL. After the incubation for 3 h, cell viability was estimated by trypan blue exclusion [75]. The half-maximal cell death (IC_{50}) was also estimated graphically. Spleen cells isolated from BALB/c mice were used as normal [76].

4.7. Antitumor Study

4.7.1. Ascites Tumor Model

The ascites tumor was induced in BALB/c mice by intra-peritoneal injection of 1×10^6 EAC cells/animal. Animals were divided into four groups, each with 6 members. Group I served as the untreated negative control. Group II served as the standard or positive control group treated with cyclophosphamide (i.p.) at a dosage of 10 mg/kg b.wt. Groups III and IV were the extract-treated groups receiving oral administration of methanol extract of *O. indicum* Vent. root bark (OIM) extract at low (OIML—200 mg/kg b.w.) and high doses (OIMH—400 mg/kg b.w.), respectively. The drug treatment was conducted for 10 days after 24 h of induction and the percentage increase in lifespan was calculated [77].

4.7.2. Solid Tumor Model

To induce a solid tumor, the intramuscular injection was given to the hind limb with 1×10^6 cells/animal. The grouping of animals and drug treatment protocol followed was the same as that of the ascites tumor model. The tumor development was measured using vernier calipers every 3rd day after induction, up to day 30) and the volume was calculated [78].

4.8. Hematological Parameters

The total WBC count and hemoglobin level of the animals in the solid tumor study were tracked every third day to compare the effect of cyclophosphamide and OIM extract on hematological parameters. The caudal vein blood was drawn into heparinized tubes and the total WBC count was determined by a standard procedure using the hemocytometer [79] and the hemoglobin level was estimated using a standard kit procedure (cyan methemoglobin method) (Agappe Diagnostics Kit).

4.9. DMBA–Croton Oil-Induced Papilloma

The study was conducted using male BALB/c mice. Three animals were accommodated per cage to avoid fighting and skin aberrations that may result; aggressive members were maintained separately. A circular area of diameter 2 cm on the dorsal side of each mouse was shaved. These mice were grouped as followed ($n = 6$);

Group I: 470 nmol of DMBA /mouse in 200 μL acetone as a single dose on day 1.

Group II: 1% croton oil/mouse in 200 μL acetone, (2 weeks after DMBA application, in a frequency of two times per week for 6 continuous weeks).

Group III: DMBA (same regime as for group I) plus croton oil (same regime as for group II). This group served as a control.

Group IV and V: [(DMBA + croton oil—same regime as for group III)] + *Oroxylum indicum* Vent. methanolic extract low dose (OIML)—5% in 200 μL distilled water and high dose (OIMH)—10% in 200 μL, respectively, 30 min before each croton oil application

Throughout the study period of 20 weeks, all animals were monitored for their food intake and toxicity symptoms. Weekly formation of papillomas, the percentage incidences of papillomas, the cumulative count of papillomas developed, the number of papillomas developed per animal, and the delay in tumor onset in each group were noted [80].

4.10. Statistical Analysis

The data representation followed a mean ± SD format with six animals per group. The comparison was done using one-way ANOVA followed by the Dunnett post hoc test using Graph Pad Prism 7.0.

5. Conclusions

The study concludes that the root bark extract of *O. indicum* is a rich source of various flavonoid compounds that are capable of inducing strong anticancer properties. In addition, the extract improved the life expectancy and tumor burden in the mice without inducing toxic insults, such as what occurs in cytotoxic anticancer chemotherapeutic agents.

Supplementary Materials: The following supporting information can be downloaded at: https://www.mdpi.com/article/10.3390/molecules27238459/s1, Figure S1: HPLC chromatograms of wild extract and the flavonoids—baicalein (Bcl) and chrysin (Chr).

Author Contributions: Conceptualization, S.M., J.J.A., H.H. and J.P.; methodology, S.M., J.J.A., H.H. and J.P., software, S.M., L.L. and M.K.D.; validation, S.M., L.L., M.K.D. and A.N.; formal analysis, S.M., J.J.A., H.H., J.P., L.L., M.K.D., L.A., S.E.M. and A.N.; investigation, S.M., J.J.A., H.H. and J.P.; resources, S.M., L.L., M.K.D., L.A., S.E.M. and A.N.; data curation, S.M., L.L., M.K.D., L.A., S.E.M. and A.N.; writing—original draft preparation, S.M., J.J.A., H.H. and J.P.; writing—review and editing, S.M., L.L., M.K.D., L.A., S.E.M. and A.N.; visualization, S.M., S.E.M. and A.N.; supervision, S.M. and A.N.; project administration, S.M., J.J.A., H.H. and J.P.; funding acquisition J.J.A. and H.H. All authors have read and agreed to the published version of the manuscript.

Funding: S.M. acknowledge the Council of Scientific and Industrial Research, Government of India, for the financial support in the form of the Senior Research Fellowship (09/869 (0008)/2011-EMR-I).

Institutional Review Board Statement: All animal experiments conducted during the present study received prior permission from the Institutional Animal Ethics Committee, Amala Cancer Research Centre (approval no. ACRC/IAEC/15/02-(01)) and strictly followed the guidelines of the Committee for the Purpose of Control and Supervision of Experiments on Animals (CPCSEA) constituted by the Animal Welfare Division, Government of India.

Informed Consent Statement: Not applicable.

Data Availability Statement: Data can be made available upon valid request.

Acknowledgments: The authors are thankful to the Amala Cancer Research Centre, Thrissur, for providing the necessary facilities.

Conflicts of Interest: The authors declare no conflict of interest.

References

1. Alyami, H.S.; Naser, A.Y.; Dahmash, E.Z.; Alyami, M.H.; Belali, O.M.; Assiri, A.M.; Rehman, A.; Alsaleh, A.M.; Alsaleh, H.A.; Hussein, S.H.; et al. Clinical and Therapeutic Characteristics of Cancer Patients in the Southern Region of Saudi Arabia: A Cross-Sectional Study. *Int. J. Environ. Res. Public Health* **2021**, *18*, 6654. [CrossRef] [PubMed]
2. Hanahan, D. Hallmarks of Cancer: New Dimensions. *Cancer Discov.* **2022**, *12*, 31–46. [CrossRef]
3. Sutton, T.L.; Patel, R.K.; Anderson, A.N.; Bowden, S.G.; Whalen, R.; Giske, N.R.; Wong, M.H. Circulating Cells with Macrophage-like Characteristics in Cancer: The Importance of Circulating Neoplastic-Immune Hybrid Cells in Cancer. *Cancers* **2022**, *14*, 3871. [CrossRef] [PubMed]
4. Senga, S.S.; Grose, R.P. Hallmarks of cancer-the new testament. *Open Biol.* **2021**, *11*, 20. [CrossRef] [PubMed]
5. Wu, D.; Pusuluri, A.; Vogus, D.; Krishnan, V.; Shields, C.W.; Kim, J.; Razmi, A.; Mitragotri, S. Design principles of drug combinations for chemotherapy. *J. Control. Release* **2020**, *323*, 36–46. [CrossRef] [PubMed]
6. van den Boogaard, W.M.C.; Komninos, D.S.J.; Vermeij, W.P. Chemotherapy Side-Effects: Not All DNA Damage Is Equal. *Cancers* **2022**, *14*, 627. [CrossRef] [PubMed]
7. Zhou, J.; Kang, Y.; Chen, L.; Wang, H.; Liu, J.; Zeng, S.; Yu, L. The Drug-Resistance Mechanisms of Five Platinum-Based Antitumor Agents. *Front. Pharmacol.* **2020**, *11*, 343. [CrossRef]
8. Rezayatmand, H.; Razmkhah, M.; Razeghian-Jahromi, I. Drug resistance in cancer therapy: The Pandora's Box of cancer stem cells. *Stem Cell Res. Ther.* **2022**, *13*, 181. [CrossRef]
9. Bahar, E.; Han, S.Y.; Kim, J.Y.; Yoon, H. Chemotherapy Resistance: Role of Mitochondrial and Autophagic Components. *Cancers* **2022**, *14*, 1462. [CrossRef]

10. Dehelean, C.A.; Marcovici, I.; Soica, C.; Mioc, M.; Coricovac, D.; Iurciuc, S.; Cretu, O.M.; Pinzaru, I. Plant-Derived Anticancer Compounds as New Perspectives in Drug Discovery and Alternative Therapy. *Molecules* **2021**, *26*, 1109. [CrossRef]
11. Singh, R.S.; Ahmad, M.; Wafai, Z.A.; Seth, V.; Moghe, V.V.; Upadhyaya, P. Anti-inflammatory effects of Dashmula, an Ayurvedic preparation, versus Diclofenac in animal models. *J. Chem. Pharm. Res.* **2011**, *3*, 882–888.
12. Bhalerao, P.P.; Pawade, R.B.; Joshi, S. Evaluation of analgesic activity of Dashamoola formulation by using experimental models of pain. *Indian J. Basic Appl. Med. Res.* **2015**, *4*, 245–255.
13. Deka, D.C.; Kumar, V.; Prasad, C.; Kumar, K.; Gogoi, B.J.; Singh, L.; Srivastava, R.B. *Oroxylum indicum*—A medicinal plant of North East India: An overview of its nutritional, remedial, and prophylactic properties. *J. Appl. Pharm. Sci.* **2013**, *3* (Suppl. 1), S104–S112.
14. Mao, A.A. *Oroxylum indicum* Vent.—A potential anticancer medicinal plant. *Indian J. Tradit. Knowl.* **2002**, *1*, 17–21.
15. Preety, A.; Sharma, S. A review on *Oroxylum indicum* (L.) Vent: An important medicinal tree. *Int. J. Res. Biol. Sci.* **2016**, *6*, 7–12.
16. Khandhar, M.; Shah, M.; Santani, D.; Jain, S. Antiulcer Activity of the Root Bark of *Oroxylum indicum* against Experimental Gastric Ulcers. *Pharm. Biol.* **2006**, *44*, 363–370. [CrossRef]
17. Zaveri, M.; Gohil, P.; Jain, S. Immunostimulant activity of n-butanol fraction of root bark of *Oroxylum indicum* Vent. *J. Immunotoxicol.* **2006**, *3*, 83–99. [CrossRef]
18. Mishra, S.L.; Sinhamahapatra, P.K.; Nayak, A.; Das, R.; Sannigrahi, S. In vitro Antioxidant Potential of Different Parts of *Oroxylum indicum*: A Comparative Study. *Indian J. Pharm. Sci.* **2010**, *72*, 267–269.
19. Sastry, A.V.S.; Sastry, V.G.; Mallikarjun, P.; Srinivas, K. Chemical and Pharmacological Evaluation of Aqueous Extract of Root Bark of "*Oroxylum Indicum*" Vent. *Int. J. Pharm. Technol.* **2011**, *3*, 1796–1806.
20. Dhru, B.; Bhatt, D.; Jethva, K.; Zaveri, M. In vitro Cytotoxicity Studies of the Anti-Cancer Potential of Fractions of Root Bark of *Oroxylum Indicum* in Human Breast Carcinoma Cells. *Int. J. Pharm. Sci. Rev. Res.* **2016**, *38*, 18–21.
21. Yee, N.S.; Ignatenko, N.; Finnberg, N.; Lee, N.; Stairs, D. Animal models of cancer biology. *Cancer Growth Metastasis* **2015**, *8* (Suppl. 1), 115–118. [CrossRef] [PubMed]
22. Ruby, A.J.; Kuttan, G.; Dinesh Babu, K.; Rajasekharan, K.N.; Kuttan, R. Anti-tumour and antioxidant activity of natural curcuminoids. *Cancer Lett.* **1995**, *94*, 79–83. [CrossRef] [PubMed]
23. Zhang, H.-M.; Zhao, L.; Li, H.; Xu, H.; Chen, W.-W.; Tao, L. Research progress on the anticarcinogenic actions and mechanisms of ellagic acid. *Cancer Biol. Med.* **2014**, *11*, 92–100. [PubMed]
24. Jaganathan, S.K.; Mondhe, D.; Wani, Z.A.; Pal, H.C.; Mandal, M. Effect of Honey and Eugenol on Ehrlich Ascites and Solid Carcinoma. *J. Biomed. Biotechnol.* **2010**, *2010*, 5. [CrossRef]
25. Ozaslan, M.; Karagoz, I.D.; Kilic, I.H.; Guldur, M.E. Ehrlich ascites carcinoma. *Afr. J. Biotechnol.* **2011**, *10*, 2375–2378.
26. Osman, A.-M.M.; Alqahtani, A.A.; Damanhouri, Z.A.; Al-Harthy, S.E.; ElShal, M.F.; Ramadan, W.S.; Kamel, F.; Osman, M.A.M.; Khan, L.M. Dimethylsulfoxide excerbates cisplatin-induced cytotoxicity in Ehrlich ascites carcinoma cells. *Cancer Cell Int.* **2015**, *15*, 104. [CrossRef]
27. Islam, F.; Khatun, H.; Ghosh, S.; Ali, M.M.; Khanam, J.A. Bioassay of Eucalyptus extracts for anticancer activity against Ehrlich ascites carcinoma (eac) cells in Swiss albino mice. *Asian Pac. J. Trop. Biomed.* **2012**, *2*, 394–398. [CrossRef]
28. Gayatri, S.; Maheswara Reddy, C.U.; Chitra, K.; Parthasarathy, V. Assessment of in vitro cytotoxicity and in vivo antitumor activity of Sphaeranthus amaranthoides burm.f. *Pharmacogn. Res.* **2015**, *7*, 198–202. [CrossRef]
29. Skipper, H.E.; Schabel, F.M., Jr.; Wilcox, W.S. Experimental Evaluation of Potential Anticancer Agents. Xiii. On the Criteria and Kinetics Associated with "Curability" of Experimental Leukemia. *Cancer Chemother. Rep.* **1964**, *35*, 3–111.
30. Teicher, B.A. Tumor models for efficacy determination. *Mol. Cancer Ther.* **2006**, *5*, 2435–2443. [CrossRef]
31. Schein, P.S.; Scheffler, B. Barriers to efficient development of cancer therapeutics. *Clin. Cancer Res.* **2006**, *12 Pt 1*, 3243–3248. [CrossRef]
32. Talmadge, J.E.; Singh, R.K.; Fidler, I.J.; Raz, A. Murine Models to Evaluate Novel and Conventional Therapeutic Strategies for Cancer. *Am. J. Pathol.* **2007**, *170*, 793–804. [CrossRef] [PubMed]
33. Babu, T.D.; Kuttan, G.; Padikkala, J. Cytotoxic and anti-tumour properties of certain taxa of Umbelliferae with special reference to *Centella asiatica* (L.) Urban. *J. Ethnopharmacol.* **1995**, *48*, 53–57. [CrossRef] [PubMed]
34. Natesan, S.; Badami, S.; Dongre, S.H.; Godavarthi, A. Antitumor Activity and Antioxidant Status of the Methanol Extract of Careya arborea Bark Against Dalton's Lymphoma Ascites-Induced Ascitic and Solid Tumor in Mice. *J. Pharmacol. Sci.* **2007**, *103*, 12–23. [CrossRef] [PubMed]
35. Indra, A.K.; Castaneda, E.; Antal, M.C.; Jiang, M.; Messaddeq, N.; Meng, X.; Loehr, C.V.; Gariglio, P.; Kato, S.; Wahli, W.; et al. Malignant transformation of DMBA/TPA-induced papillomas and nevi in the skin of mice selectively lacking retinoid-X-receptor alpha in epidermal keratinocytes. *J. Investig. Dermatol.* **2007**, *127*, 1250–1260. [CrossRef]
36. Kumar, D.R.N.; George, V.C.; Suresh, P.K.; Kumar, R.A. Cytotoxicity, Apoptosis Induction and Anti-Metastatic Potential of *Oroxylum indicum* in Human Breast Cancer Cells. *Asian Pac. J. Cancer Prev.* **2012**, *13*, 2729–2734. [CrossRef]
37. Roy, M.K.; Nakahara, K.; Na, T.V.; Trakoontivakorn, G.; Takenaka, M.; Isobe, S.; Tsushida, T. Baicalein, a flavonoid extracted from a methanolic extract of *Oroxylum indicum* inhibits proliferation of a cancer cell line in vitro via induction of apoptosis. *Pharmazie* **2007**, *62*, 149–153.
38. Zaveri, M.; Khandhar, A.; Jain, S. Quantification of Baicalein, Chrysin, Biochanin-A and Ellagic Acid in Root Bark of *Oroxylum indicum* by RP- HPLC with UV Detection. *Eurasian J. Anal. Chem.* **2008**, *3*, 245–257.

39. Samudrala, P.K.; Augustine, B.B.; Kasala, E.R.; Bodduluru, L.N.; Barua, C.; Lahkar, M. Evaluation of antitumor activity and antioxidant status of Alternanthera brasiliana against Ehrlich ascites carcinoma in Swiss albino mice. *Pharmacogn. Res.* **2015**, *7*, 66–73. [CrossRef]
40. Senthil Kumar, R.; Rajkapoor, B.; Perumal, P.; Dhanasekaran, T.; Alvin Jose, M.; Jothimanivannan, C. Antitumor Activity of Prosopis glandulosa Torr. on Ehrlich Ascites Carcinoma (EAC) Tumor Bearing Mice. *Iran. J. Pharm. Res.* **2011**, *10*, 505–510.
41. Hartveit, F. The Survival Time of Mice with Ehrlich's Ascites Carcinoma related to the Sex and Weight of the Mouse, and the Blood Content of the Tumour. *Br. J. Cancer* **1961**, *15*, 336–341. [CrossRef] [PubMed]
42. Hartveit, F. The growth of Ehrlich's ascites carcinoma in C3H mice and in mice of an unrelated closed colony. Variation in survival time. *Br. J. Cancer* **1966**, *20*, 813–817. [CrossRef] [PubMed]
43. Mohan, H. *Textbook of Pathology*, 4th ed.; Jaypee Publications: New Delhi, India, 2002.
44. Patil, T.B.; Shrikhande, S.V.; Kanhere, H.A.; Saoji, R.R.; Ramadwar, M.R.; Shukla, P.J. Solid pseudopapillary neoplasm of the pancreas: A single institution experience of 14 cases. *HPB Off. J. Int. Hepato Pancreato Biliary Assoc.* **2006**, *8*, 148–150. [CrossRef]
45. Klein, G. Comparative studies of mouse tumors with respect to their capacity for growth as "ascites tumors" and their average nucleic acid content per cell. *Exp. Cell Res.* **1951**, *2*, 518–573. [CrossRef]
46. Goldie, H.; Dingman Felix, M. Growth Characteristics of Free Tumor Cells Transferred Serially in the Peritoneal Fluid of the Mouse. *Cancer Res.* **1951**, *11*, 73–80. [PubMed]
47. Koiri, R.K.; Mehrotra, A.; Trigun, S.K. Dalton's Lymphoma as a Murine Model for Understanding the Progression and Development of T-Cell Lymphoma and Its Role in Drug Discovery. *Int. J. Immunother. Cancer Res.* **2017**, *3*, 001–006.
48. Naik, A.V.; Dessai, S.N.; Sellappan, K. Antitumour activity of *Annona muricata* L. leaf methanol extracts against Ehrlich Ascites Carcinoma and Dalton's Lymphoma Ascites mediated tumours in Swiss albino mice. *Libyan J. Med.* **2021**, *16*, 1846862. [CrossRef]
49. Jeena, K.; Liju, V.B.; Kuttan, R. Antitumor and cytotoxic activity of ginger essential oil (*Zingiber officinale* Roscoe). *Int. J. Pharm. Pharm. Sci.* **2015**, *7*, 341–344.
50. Abel, E.L.; DiGiovanni, J. Multistage Carcinogenesis. In *Chemical Carcinogenesis*; Penning, T.M., Ed.; Humana Press: Totowa, NJ, USA, 2011; pp. 27–51.
51. Abel, E.L.; Angel, J.M.; Kiguchi, K.; DiGiovanni, J. Multi-stage chemical carcinogenesis in mouse skin: Fundamentals and applications. *Nat. Protoc.* **2009**, *4*, 1350–1362. [CrossRef]
52. Wattenberg, L.W. Chemoprevention of cancer. *Cancer Res.* **1985**, *45*, 1–8. [CrossRef]
53. DiGiovanni, J. Multistage carcinogenesis in mouse skin. *Pharmacol. Ther.* **1992**, *54*, 63–128. [CrossRef] [PubMed]
54. Sporn, M.B. Approaches to prevention of epithelial cancer during the preneoplastic period. *Cancer Res.* **1976**, *36 Pt 2*, 2699–2702.
55. Boccardo, E.; Lepique, A.P.; Villa, L.L. The role of inflammation in HPV carcinogenesis. *Carcinogenesis* **2010**, *31*, 1905–1912. [CrossRef] [PubMed]
56. Fernandes, J.V.; Fernandes, T.A.; de Azevedo, J.C.; Cobucci, R.N.; de Carvalho, M.G.; Andrade, V.S.; de Araújo, J.M. Link between chronic inflammation and human papillomavirus-induced carcinogenesis (Review). *Oncol. Lett.* **2015**, *9*, 1015–1026. [CrossRef] [PubMed]
57. Chunekar, K.C.; Pandey, G.S. *Bhavaprakasha Nighantu*, 10th ed.; Chaukhamba Bharati Academy: Varanasi, India, 1999; pp. 283–285.
58. Kirtikar, K.R.; Basu, B.D. *Indian Medicinal Plants*, 2nd ed.; Periodical Experts: Delhi, India, 1975; Volume 4, pp. 1839–1841.
59. Menon, S.; Lawrence, L.; Vipin, P.S.; Padikkala, J. Phytochemistryand evaluation of in vivo antioxidant and anti-inflammatory activities of *Oroxylum indicum* Vent. Root bark. *J. Chem. Pharm. Res.* **2015**, *7*, 767–775.
60. Groopman, J.E.; Itri, L.M. Chemotherapy-Induced Anemia in Adults: Incidence and Treatment. *JNCI J. Natl. Cancer Inst.* **1999**, *91*, 1616–1634. [CrossRef] [PubMed]
61. Thews, O.; Kelleher, D.K.; Vaupel, P. Erythropoietin restores the anemia-induced reduction in cyclophosphamide cytotoxicity in rat tumors. *Cancer Res.* **2001**, *61*, 1358–1361.
62. Alam, B.; Majumder, R.; Akter, S.; Lee, S.-H. Piper betle extracts exhibit antitumor activity by augmenting antioxidant potential. *Oncol. Lett.* **2015**, *9*, 863–868. [CrossRef]
63. Mondal, A.; Singha, T.; Maity, T.K.; Pal, D. Evaluation of Antitumor and Antioxidant Activity of Melothria heterophylla (Lour.) Cogn. *Indian J. Pharm. Sci.* **2013**, *75*, 515–522.
64. Liu, H.; Dong, Y.; Gao, Y.; Du, Z.; Wang, Y.; Cheng, P.; Chen, A.; Huang, H. The Fascinating Effects of Baicalein on Cancer: A Review. *Int. J. Mol. Sci.* **2016**, *17*, 1681–1698. [CrossRef]
65. Peng, Y.; Li, Q.; Li, K.; Zhao, H.; Han, Z.; Li, F.; Sun, M.; Zhang, Y. Antitumor activity of baicalein on the mice bearing U14 cervical cancer. *Afr. J. Biotechnol.* **2011**, *10*, 14169–14176.
66. Kasala, E.R.; Bodduluru, L.N.; Madana, R.M.; V, A.K.; Gogoi, R.; Barua, C.C. Chemopreventive and therapeutic potential of chrysin in cancer: Mechanistic perspectives. *Toxicol. Lett.* **2015**, *233*, 214–225. [CrossRef] [PubMed]
67. Khoo, B.Y.; Chua, S.L.; Balaram, P. Apoptotic Effects of Chrysin in Human Cancer Cell Lines. *Int. J. Mol. Sci.* **2010**, *11*, 2188. [CrossRef] [PubMed]
68. Peterson, G.; Barnes, S. Genistein and biochanin A inhibit the growth of human prostate cancer cells but not epidermal growth factor receptor tyrosine autophosphorylation. *Prostate* **1993**, *22*, 335–345. [CrossRef]
69. Losso, J.N.; Bansode, R.R.; Trappey, A.; Bawadi, H.A.; Truax, R. In vitro anti-proliferative activities of ellagic acid. *J. Nutr. Biochem.* **2004**, *15*, 672–678. [CrossRef] [PubMed]

70. Harminder; Singh, V.; Chaudhary, A.K. A Review on the Taxonomy, Ethnobotany, Chemistry and Pharmacology of *Oroxylum indicum* Vent. *Indian J. Pharm. Sci.* **2011**, *73*, 483–490. [PubMed]
71. Mohamat, S.A.; Shueb, R.H.; Che Mat, N.F. Anti-viral Activities of *Oroxylum indicum* Extracts on Chikungunya Virus Infection. *Indian J. Microbiol.* **2018**, *58*, 68–75. [CrossRef]
72. Felipe, D.F.; Brambilla, L.Z.S.; Porto, C.; Pilau, E.J.; Cortez, D.A.G. Phytochemical Analysis of Pfaffia glomerata Inflorescences by LC-ESI-MS/MS. *Molecules* **2014**, *19*, 15720–15734. [CrossRef]
73. Organization of Economic Co-operation and Development. *The OECD Guideline for Testing of Chemicals: 423 Acute Oral Toxicity—Acute Toxic Class Method*; Organization of Economic Co-Operation and Development: Paris, France, 2001.
74. Gothoskar, S.V.; Ranadive, K.J. Anticancer screening of SAN-AB: An extract of marking nut, Semecarpus anacardium. *Indian J. Exp. Biol.* **1971**, *9*, 372–375.
75. Kerschbaum, H.H.; Tasa, B.A.; Schürz, M.; Oberascher, K.; Bresgen, N. Trypan Blue—Adapting a Dye Used for Labelling Dead Cells to Visualize Pinocytosis in Viable Cells. *Cell. Physiol. Biochem. Int. J. Exp. Cell.Physiol.Biochem. Pharmacol.* **2021**, *55*, 171–184.
76. Strober, W. Trypan Blue Exclusion Test of Cell Viability. *Curr. Protoc. Immunol.* **2015**, *111*, A3.B.1–A3.B.3. [CrossRef] [PubMed]
77. Saleh, N.; Allam, T.; Korany, R.M.S.; Abdelfattah, A.M.; Omran, A.M.; Abd Eldaim, M.A.; Hassan, A.M.; El-Borai, N.B. Protective and Therapeutic Efficacy of Hesperidin versus Cisplatin against Ehrlich Ascites Carcinoma-Induced Renal Damage in Mice. *Pharmaceuticals* **2022**, *15*, 294. [CrossRef] [PubMed]
78. Smina, T.P.; Mathew, J.; Janardhanan, K.K. Ganoderma lucidum total triterpenes attenuate DLA induced ascites and EAC induced solid tumours in Swiss albino mice. *Cell Mol. Biol.* **2016**, *62*, 55–59. [PubMed]
79. Chung, J.; Ou, X.; Kulkarni, R.P.; Yang, C. Counting White Blood Cells from a Blood Smear Using Fourier Ptychographic Microscopy. *PLoS ONE* **2015**, *10*, e0133489. [CrossRef] [PubMed]
80. Kong, Y.H.; Xu, S.P. Salidroside prevents skin carcinogenesis induced by DMBA/TPA in a mouse model through suppression of inflammation and promotion of apoptosis. *Oncol. Rep.* **2018**, *39*, 2513–2526. [CrossRef] [PubMed]

molecules

Article

Chemical Composition, Antioxidant, Anti-Bacterial, and Anti-Cancer Activities of Essential Oils Extracted from *Citrus limetta* Risso Peel Waste Remains after Commercial Use

Arunaksharan Narayanankutty [1,*], Naduvilthara U. Visakh [2], Anju Sasidharan [1], Berin Pathrose [2,*], Opeyemi Joshua Olatunji [3,4,*], Abdullah Al-Ansari [5], Ahmed Alfarhan [5] and Varsha Ramesh [6]

1 Division of Cell and Molecular Biology, PG & Research Department of Zoology, St. Joseph's College (Autonomous), Calicut 673008, India
2 Department of Agricultural Entomology, College of Agriculture, Kerala Agricultural University, Thrissur 680656, India
3 African Genome Center, Mohammed VI Polytechnic University, Ben Guerir 43150, Morocco
4 Traditional Thai Medical Research and Innovation Center, Faculty of Traditional Thai Medicine, Prince of Songkla University, Hat Yai 90110, Thailand
5 Department of Botany and Microbiology, College of Science, King Saud University, P.O. Box 2455, Riyadh 11451, Saudi Arabia
6 Department of Biotechnology, Deakin University, Geelong, VIC 3217, Australia
* Correspondence: arunaksharann@devagiricollege.org (A.N.); berin.pathrose@kau.in (B.P.); Joshua.OLATUNJI@um6p.ma (O.J.O.)

Abstract: Citrus plants are widely utilized for edible purposes and medicinal utility throughout the world. However, because of the higher abundance of the antimicrobial compound D-Limonene, the peel waste cannot be disposed of by biogas production. Therefore, after the extraction of D-Limonene from the peel wastes, it can be easily disposed of. The D-Limonene rich essential oil from the *Citrus limetta* risso (CLEO) was extracted and evaluated its radical quenching, bactericidal, and cytotoxic properties. The radical quenching properties were DPPH radical scavenging (11.35 ± 0.51 μg/mL) and ABTS scavenging (10.36 ± 0.55 μg/mL). There, we observed a dose-dependent antibacterial potential for the essential oil against pathogenic bacteria. Apart from that, the essential oil also inhibited the biofilm-forming properties of *E. coli*, *P. aeruginosa*, *S. enterica*, and *S. aureus*. Further, cytotoxicity was also exhibited against estrogen receptor-positive (MCF7) cells (IC_{50}: 47.31 ± 3.11 μg/mL) and a triple-negative (MDA-MB-237) cell (IC_{50}: 55.11 ± 4.62 μg/mL). Upon evaluation of the mechanism of action, the toxicity was mediated through an increased level of reactive radicals of oxygen and the subsequent release of cytochrome C, indicative of mitotoxicity. Hence, the D-Limonene rich essential oil of *C. limetta* is useful as a strong antibacterial and cytotoxic agent; the antioxidant properties exhibited also increase its utility value.

Keywords: *Citrus limetta*; antioxidant activity; essential oil; antibacterial activity; anticancer activity

Citation: Narayanankutty, A.; Visakh, N.U.; Sasidharan, A.; Pathrose, B.; Olatunji, O.J.; Al-Ansari, A.; Alfarhan, A.; Ramesh, V. Chemical Composition, Antioxidant, Anti-Bacterial, and Anti-Cancer Activities of Essential Oils Extracted from *Citrus limetta* Risso Peel Waste Remains after Commercial Use. *Molecules* **2022**, *27*, 8329. https://doi.org/10.3390/molecules27238329

Academic Editor: Alessandra Guerrini

Received: 20 October 2022
Accepted: 25 November 2022
Published: 29 November 2022

Copyright: © 2022 by the authors. Licensee MDPI, Basel, Switzerland. This article is an open access article distributed under the terms and conditions of the Creative Commons Attribution (CC BY) license (https://creativecommons.org/licenses/by/4.0/).

1. Introduction

Citrus fruits are widely consumed and, as a result, the waste products in the form of peels are accumulating [1]. Sustainable waste management is highly dependent on the conversion of waste materials into biogas or other forms of energy [2]. However, because of the presence of the antibacterial compound D-Limonene, the peels of citrus fruits are less utilizable in biogas production. Studies have indicated that the D-Limonene present in citrus peels inhibits the anaerobic digestion of the waste by preventing the growth of bacterial colonies [3,4]. In addition, studies have also indicated that the extraction of the D-Limonene and other bactericidal compounds from the citrus peel enhances the anaerobic digestion and subsequent conversion of these waste products into biogas [5,6]. Hence, the extraction of D-limonene from the citrus peel is of high significance and, also, the compound has commercial and pharmacological uses [7,8].

Among the major citrus species, *Citrus maxima, Citrus limetta, Citrus limon, Citrus aurantifolia,* and *Citrus reticulata* are widely used and consumed in world markets [9]. The major method of extraction of bioactive D-Limonene from peel waste is in the form of essential oils [10–12]. The essential oil extracted from the *Citrus reticulata,* the commonly found variety among citrus fruits, was found to inhibit the growth of microbes [13,14], as well fungal strains such as *Penicillium italicum* and *P. digitatum* [15]. The radical quenching and bactericidal potentials of the essential oil is also reported that subsequently resulted in its wound-healing properties [16]. Similarly, antibacterial activities are also reported for essential oil isolated from *C. maxima* by different extraction methods [17]; likewise, this essential oil was also effective against pests of stored products [18]. The growth inhibitory potential against *Aspergillus flavus,* the producer of the common food toxicant aflatoxin, was also observed [19]. The *Citrus limetta* essential oil exhibited strong larvicidal activity against the *Anopheles* and *Aedes* mosquitoes [20]; the essential oil was also effective against skin diseases with underlining inflammation or oxidative stress in cell culture models and animals [21]. The essential oil of *Citrus medica L. var. sarcodactylis* was found to be a good antibacterial agent by altering the membrane integrity of the different microbial strains [22]. Anticancer activities are also attributed to citrus-derived essential oils. A study by Yang, et al. [23] indicated that the gannan navel orange (*C. sinensis*) exerts antiproliferative potential against human prostate and lung cancer cells. *Citrus medica* is another plant that is shown to have anti-neoplastic activities against colorectal cancer cells [24]. The *C. limon* essential oil-nanoemulsion has been found to exert apoptosis in A549 cells under in vitro conditions [25]. A study by Elansary, et al. [26] compared the anticancer potential of *C. aurantifolia, C. limon,* and *C. paradisi* against various cancer cell lines. Among these, *C. paradisi,* was most effective and capable of inducing apoptosis.

The utility of various essential oils extracted from peels, fruits, and leaves of different citrus plants is available. However, the novelty of the work is that the source of essential oil used for the analysis was the peel waste of commercially used *Citrus limetta,* and it was collected from juice shops. Further, we analyzed the constituent compounds and potentials of the essential oil as an antioxidant, bactericidal, and cytotoxic agent.

2. Results

2.1. The Average Yield and Volatile Content in the Peel Essential Oil of Citrus limetta

The average yield of essential oils from the waste peels of *C. limetta* was 0.63%. The GC–MS chromatogram of essential oil isolated from the peels of *Citrus limetta* (Figure 1) shown the occurrence of D-limonene and α-myrcene as chief components (Table 1).

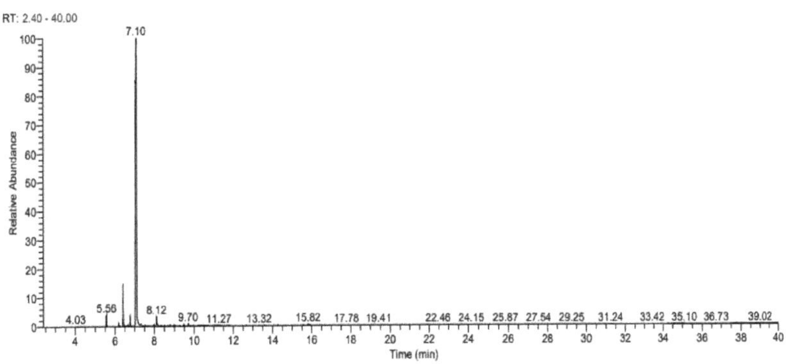

Figure 1. The total ion chromatogram of the peel of the *C. limetta* essential oil.

Table 1. Chemical constituents of the peel of the *C. limetta* essential oil.

Peak No.	Retention Time	Component	Retention Index		%Relative Area [a]
			Calculated	Library	
1	5.56	α-pinene	939	938	1.17 ± 0.96
2	6.41	α-myrcene	985	983	4.85 ± 0.67
3	6.76	3-carene	1011	1012	1.20 ± 0.08
4	7.09	D-limonene	1029	1029	85.71 ± 0.41
5	8.12	Linalyl acetate	1237	1133	1.64 ± 0.37
6	8.97	Citronellal	1153	1141	0.36 ± 0.06
7	9.48	Terpinen-4-ol	1175	1176	0.36 ± 0.28
8	9.70	α-Terpineol	1178	1180	0.44 ± 0.03
9	10.31	cis-p-mentha-1(7),8-dien-2-ol	1190	1195	0.43 ± 0.43
10	15.82	Guaia-1(10),11-diene	1490	1488	0.32 ± 0.91

[a] Relative area (peak area relative to the total peak area).

2.2. Anti-Radical Activities

The anti-radical abilities of the peel of the CLEO indicated a higher activity compared to the major compound S-limonene. On contrary, the ascorbic acid (standard compound) was more active in scavenging DPPH free radicals and ferric-reducing properties. However, the lipid peroxidation inhibition and hydrogen peroxide scavenging were high in the CLEO ($p < 0.05$). The ABTS radical scavenging potential was found to be similar in both CLEO and ascorbic acid (Table 2).

Table 2. Radical quenching abilities of essential oil from *Citrus limetta* essential oil (CLEO) are expressed in terms of IC_{50} (μg/mL).

	DPPH Radical Scavenging	ABTS Radical Scavenging	H_2O_2 Radical Scavenging	Ferric Reducing Antioxidant Power	Lipid Peroxidation Inhibition
CLEO	11.35 ± 0.51 *	10.36 ± 0.55 *	8.28 ± 0.35 *	8.67 ± 0.21	30.19 ± 0.27 *
D-limonene	48.49 ± 0.22	41.22 ± 0.13	20.67 ± 0.34	19.08 ± 0.33	58.16 ± 0.43
Ascorbic acid	9.57 ± 0.75 *	11.08 ± 2.11 *	19.62 ± 1.60	3.41 ± 0.29 *	65.98 ± 1.95

(* $p < 0.05$).

2.3. Cytotoxicity of the *C. limetta* Essential Oil, D-Limonene, and Cyclophosphamide

The essential oil treatment-induced dose-dependent cytotoxicity against MCF7 and MDAMB-231 cell lines (Figure 2).

The cytotoxicity of the CLEO, D-Limonene and cyclophosphamide were indicated as the IC_{50} values in Table 3. The anticancer activity of the CLEO was high against MCF7 cells; whereas, in the MDAMB231 cells, the IC_{50} value was high. However, the standard drug, cyclophosphamide, had significantly higher activity compared to the CLEO (Table 3). However, the cytotoxicity of D-Limonene was low compared to the essential oil. The morphological changes are indicated in the Figures S1 and S2 (Supplementary Materials).

Table 3. Anticancer activity expressed in terms of IC_{50} (μg/mL) value of the *Citrus limetta* peel essential oil.

	MCF-7	MDAMB231
CLEO	47.31 ± 3.11	55.11 ± 4.62
D-Limonene	392.57 ± 5.29	428.33 ± 4.61
Cyclophosphamide	10.02 ± 0.38	9.37 ± 0.25

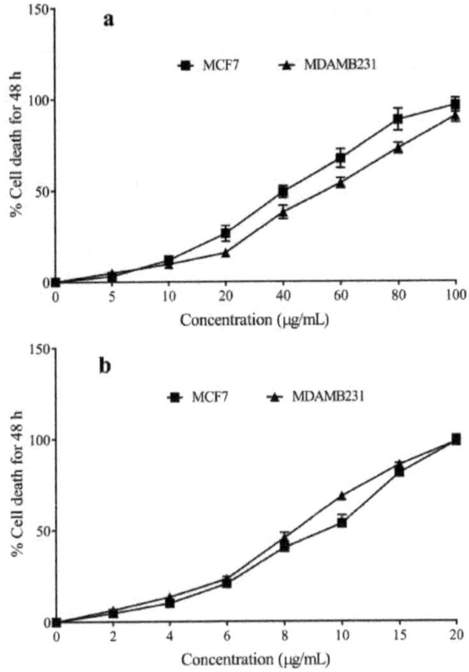

Figure 2. Anticancer activity of the *Citrus limetta* essential oil (**a**) and cyclophosphamide (**b**) was analyzed against MCF7 and MDAMB-231 cells.

The mechanism of action was estimated in terms of the reactive oxygen species generated in the cells and also based on the release of mitochondrial cytochrome C release. There observed a noteworthy elevation in the ROS levels of cells treated with the IC_{50} value equivalent dose of different compounds (Figure 3) and a subsequent increase in the release of cytochrome C levels.

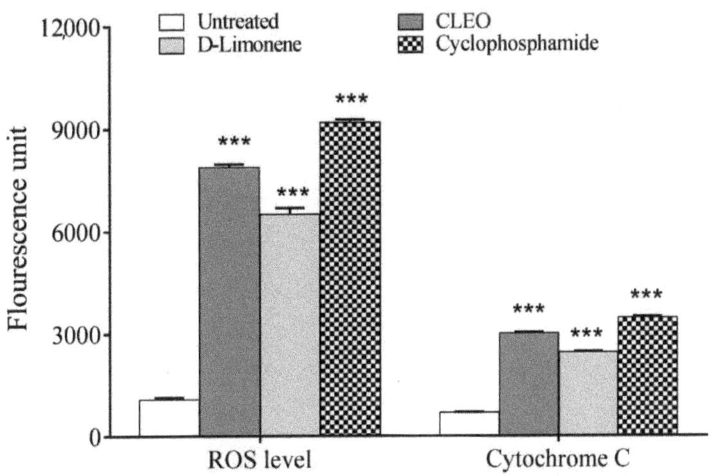

Figure 3. The mechanistic bases of action of the Citrus limetta peel essential oil along with the standard anticancer drug, cyclophosphamide, at their respective IC_{50} values. (*** indicate $p < 0.001$).

2.4. Bactericidal Properties of the CLEO

The bactericidal activity of the CLEO was indicated in Table 4. The disc diffusion assay observed significant inhibition of bacterial growth in the *Citrus limetta* essential oil treatment. The highest activity was observed against *Staphylococcus aureus* (Table 4). The CLEO showed moderate activities against other bacterial strains.

Table 4. Bactericidal efficacy of the *Citrus limetta* (CLEO) essential oil, D-Limonene, and gentamicin (GM) by disc diffusion assay in MHA plates.

Strain	Zone of Inhibition (mm)		
	CLEO	D-Limonene	GM
Escherichia coli	13.5 ± 0.4	16.7 ± 0.2	22.5 ± 0.1
Pseudomonas aeruginosa	16.8 ± 0.2	18.7 ± 0.2	19.5 ± 0.2
Staphylococcus aureus	17.1 ± 0.5	20.9 ± 0.3	23.0 ± 0.1
Salmonella enterica	15.9 ± 0.3	17.1 ± 0.4	19.5 ± 0.3

Further, the MIC value of the CLEO was estimated against the same microbial species (Table 5). The CLEO had the lowest level of MIC value against *E. coli* (0.50 ± 0.03 mg/mL) and *S. aureus* (0.50 ± 0.02 mg/mL). The standard antibiotic gentamicin had been more effective, as shown in Table 5.

Table 5. Minimum inhibitory concentrations (mg/mL) of the *Citrus limetta* (CLEO) essential oil and standard gentamicin (GM) against selected microbial strains.

Bacteria	MIC Concentration (mg/mL)		
	CLEO	D-Limonene	GM
Escherichia coli	0.50 ± 0.03 *	0.0625 ± 0.02	0.0312 ± 0.01
Pseudomonas aeruginosa	0.75 ± 0.03	0.0312 ± 0.01	0.0312 ± 0.01
Staphylococcus aureus	0.50 ± 0.02	1.25 ± 0.1	1.5 ± 0.3
Salmonella enterica	0.625 ± 0.03 *	0.0312 ± 0.00	0.0312 ± 0.01

(* $p < 0.05$).

The anti-biofilm formation activity of CLEO and gentamicin was also determined. As indicated in Table 6, we observed significant anti-biofilm formation (0.5 mg/mL) for CLEO compared to that of the standard antibiotic gentamicin.

Table 6. The percentage inhibition of antibiofilm formation activity of the *Citrus limetta* (CLEO) essential oil and standard gentamicin (GM) against selected microbial strains (at 0.5 mg/mL).

	Percentage Inhibition		
	CLEO	D-Limonene	GM
Escherichia coli	90.6 ± 1.6	100	100
Pseudomonas aeruginosa	92.19 ± 1.2	100	100
Staphylococcus aureus	95.6 ± 2.1	100	100
Salmonella enterica	93.8 ± 1.5	100	100

3. Discussion

Various products from Citrus plants are widely consumed fruits and source for various nutrients and pharmacologically active agents. However, fruits also contribute to large quantities of waste products, as in other agriculture sectors [27,28]. The predominant waste products from various citrus plants include their peel wastes; these waste products later decay and lead to pollution at various levels [2]. However, these waste peels also emerge as important sources of biological and pharmacologically active essential oils [5,6]. Hence, the present study analyzed the potentials of citrus peel-derived essential oils as anticancer and antibacterial agents using in vitro experimental models.

The gas chromatography analysis indicated the presence of *D-limonene* as major component in CLEO. The predominant compound in the essential oil was *D-limonene*. The highest level was observed in *Citrus limetta, Citrus reticulata,* and *Citrus limon*. Limonene is an important bioactive compound that is shown to have strong antibacterial and antifungal properties, and thereby acts as a potent agent against microbial diseases [29–31]. The D-Limonene is also reported to have significant anticancer potentials; the mechanistic basis of action is proven to be mediated by autophagy and apoptosis in various cancer cells [32–34]. The compounds α-pinene and α-myrcene are are the other minor constituents present in the essential oil; they are also shown to have potent antimicrobial, anti-inflammatory, and antitumor properties [35,36]. The presence of these compounds at a lower level is noted in various citrus essential oils prepared from leaves or fruits [37–41].

Furthermore, the bactericidal properties of the *C. limetta* essential oil and its bioactive compound D-Limonene are also observed. The bacterial strains tested are pathogenic to animals and humans [42,43]. Previous studies using different citrus essential oils also indicated the antibacterial potentials [41,44–46]. A previous study has indicated that the inhibition zone of *Citrus* spp. was in the range of 5.8–21.0 mm for *E. coli* and the same was 5.0–10.0 mm for *Lactbacillus plantarum* [47]. A study has also observed that the inhibition zones were in the range of 14.0–26.0 mm for various bacteria, and the minimum inhibitory concentration value was varied between 0.039 and 2.5 mg/mL [48]. The bactericidal properties are ascribed to *D*-limonene, present in essential oils [29–31]. D-Limonene in the peels is also known to cause issues in the anaerobic degradation during biogas [3,4].

The CLEO induced anticancer effects in MCF7 and MDAMB231 cells. Previous reports have indicated that the citrus essential oil IC_{50} values against human lung cancer cells are estimated to be 17.53–45.74 μg/mL [23,49]. The MCF7 cells are considered to be estrogen receptor-positive cells and MDAMB231 is a triple-negative breast cancer cell [50,51]. Hence, it can be possible that essential oils exert anticancer properties in both types of breast cancer cells. This will open up a new source of anticancer agents against different types of breast cancers. The bioactive compounds present in the *Citrus limetta* essential oil, such as limonene [32,52], citral [53,54], and terpineol [55,56], are strong anti-proliferative and apoptotic agents in cancer cells. Hence, the bioactive compounds present in the CLEO might be accountable for these activities.

4. Materials and Methods

4.1. Essential Oil Extraction from the Peel Waste of Citrus limetta

Peels of *Citrus limetta* were obtained from juice shops in Kerala, India (10.5276° N, 76.2144° E). After washing, the peels were extracted by hydro-distillation in a Clevenger-type apparatus for 4–5 h (100 °C). The final yield from the peels (CLEO) was represented as mg of CLEO obtained per gram of fresh peels (%, *V/w*). The dehydration of CLEO was performed using sodium sulfate (AR) and kept in amber-colored bottles in refrigerated conditions.

4.2. Analysis of the Component Chemicals in CLEO

The chemical constituents are analyzed using a TSQ 8000 Evo GC-MS system (Thermoscientific, Waltham, MA, USA) furnished with an autosampling system and TG-5MS column as described by our previously published method [18]. The individual constituents were derived by matching the MS spectra of the NIST library. Further, a blank run was performed following each sample analysis to avoid contamination. The retention index (Kovats index) of individual compounds was calculated by the co-injection of the n-alkene mixture (C_7-C_{30}) passed through the column with maintaining the same conditions followed by essential oil chemical characterization. The calculated RI of each constituent was compared with their library RI.

4.3. *Quenching Abilities of Citrus limetta Peel Essential Oil against Various Free Radicals*

The radical quenching abilities were estimated using different models. Initially, different concentrations of the *Citrus limetta* essential oil were prepared in Tween 80 (0–100 µg/mL), likewise, the D-Limonene and ascorbic acid was also dissolved in dimethyl sulfoxide. The DPPH radical scavenging was estimated according to the procedures prescribed by House, et al. [57]. ABTS- quenching was performed by the methods mentioned by Baliyan, et al. [58]. The quenching of peroxide radicals was carried out using H_2O_2 as mentioned by Al-Amiery, et al. [59].

4.4. *Anti-Proliferative Effect of the Citrus limetta Peel Essential Oil*

The estrogen receptor-positive human breast cancer cell (MCF7) and human triple-negative breast cancer cell (MDAMB231) was received from the NCCS cell repository (Pune, Maharashtra, India) and cultured in complete DMEM media. The cytotoxicity analysis was carried out using an MTT assay as described earlier [60]. The cell death was measured spectrophotometrically at 570 nm and expressed as percentage using the following formula (Equation (1)):

$$\%\ \text{Cell death} = \frac{\text{OD of Control} - \text{OD of Sample}}{\text{OD of Control}} \times 100 \qquad (1)$$

4.5. *Effect of the CLEO on ROS Level and Cytochrome C Release*

The mechanism of action was determined in terms of the cytochrome-C release and reactive oxygen species levels; these changes in essential oil-treated cells were determined by commercially available kits, as described in our previous article [61]. The cells were treated with the respective IC_{50} value doses of different citrus peel essential oils for the mechanistic basis of action.

4.6. *Analysis of Antibacterial Activity*

4.6.1. Bacterial Maintenance

The bacterial colonies of *Pseudomonas aeruginosa, Escherichia coli, Staphylococcus aureus,* and *Salmonella enterica* were procured from MTCC, Chandigarh, India. The bacteria were initially grown under standard atmospheric conditions; the procedures strictly adhered to the methods described in a previous study [62].

4.6.2. Inhibition Zone Formation by *C. limetta* Essential Oil Treatment

The aforementioned bacteria were cultured using LB broth. For the disc diffusion assay, the MHA agar plate was inoculated with individual bacterial strains. Further, the *C. limetta* essential oil (10 µL), D-Limonene, and gentamicin were applied to Whatman No.1 filter paper (8 mm diameter) and placed in the MHA agar plates. The inhibition zones for each were determined after 24 h [63].

4.6.3. *C. limetta* Essential Oil Minimum Inhibitory Concentrations (MIC)

The MIC value was estimated by the methods described by the standard methods described previously [64–66]. The different bacteria were set to 5×10^5 CFU/mL density using a spectrophotometer. From this, about 50 µL was transferred to the individual wells of a 96-well plate together with the *Citrus limetta* essential oil, gentamicin, and D-Limonene. The media was then supplemented with 2,3,5-triphenyl tetrazolium chloride (TTC) (10 µL). The MIC concentration was determined to be the lowest concentration without pink color.

4.6.4. Analysis of Biofilm Formation Inhibition by the *Citrus limetta* Essential Oil

The inhibition of biofilm formation by the *Citrus limetta* essential oil was carried out by the methods of Selim, et al. [67]. Briefly, the assay was carried out using a 96-well plate containing growing cells that were incubated with 5% concentrations of the *Citrus limetta* essential oil; the cells after 24h were stained with crystal violet.

4.7. Statistical Analysis

The final values of radical quenching assay, cytotoxicity analysis, and bactericidal studies were shown as mean ± standard deviation. These assays were repeated three times and each was performed in triplicate.

5. Conclusions

The agro-waste products of *Citrus limetta* plants are their peels; the results indicated that the peels are important sources of aromatic essential oils and, also, the predominant compound *D*-limonene, α-pinene, and α-myrcene. Results also indicated the radical quenching potential of the *C. limetta* essential oil against different types of radicals. The bactericidal properties of the essential oil were also significant; however, they were less than that of the D-Limonene and gentamicin. Likewise, the essential oil is found to inhibit the biofilm formation properties of different bacteria. The cytotoxic effect of the *C. limetta* essential oil was noticed against breast cancer cells of different receptors, specificity. Furthermore, the mechanism of action is found to be mediated through reactive oxygen species-mediated mitochondrial toxicity. Hence, the essential oil from the peel wastes of *Citrus limetta* is found to be pharmacologically active and emerges as a potential antibacterial and cytotoxic agent.

Supplementary Materials: The following supporting information can be downloaded at: https://www.mdpi.com/article/10.3390/molecules27238329/s1, Figure S1. The untreated MCF7 cells (a) and the cytotoxicity of the *Citrus limetta* essential oil (b), D-Limonene (c), and cyclophosphamide (d); Figure S2. The untreated MDA-MB-231 cells (a) and the cytotoxicity of the *Citrus limetta* essential oil (b), D-Limonene (c), and cyclophosphamide (d).

Author Contributions: Conceptualization, O.J.O., A.A.-A. and A.N.; methodology, A.N., A.S. and B.P.; software, N.U.V.; validation, N.U.V., O.J.O. and A.N.; formal analysis, A.N. and A.S.; investigation, N.U.V.; resources, V.R., A.A. and A.N.; data curation, B.P., A.A.-A. and A.N.; writing—original draft preparation, A.S., N.U.V. and A.N.; writing—review and editing, A.N., V.R., A.A. and B.P.; visualization, V.R., A.A.-A. and B.P.; supervision, A.N. and A.A.; project administration, A.N., B.P., V.R. and A.A.-A.; funding acquisition O.J.O. and A.A.-A. All authors have read and agreed to the published version of the manuscript.

Funding: The authors are grateful for the seed grant from Mohammed VI Polytechnic University, Morocco and the study was funded by King Saud University, Riyadh, Saudi Arabia through Researchers Supporting Project No. RSP 2022/11.

Institutional Review Board Statement: Not applicable.

Informed Consent Statement: Not applicable.

Data Availability Statement: Data may be made available upon valid request.

Acknowledgments: The authors are grateful for the seed grant from Mohammed VI Polytechnic University, Morocco. The authors also acknowledge the funding support from Researchers Supporting Project Number (RSP 2022/11), King Saud University, Riyadh, Saudi Arabia. DBT-STAR (Project number: BT/HRD/11/09/2020) scheme supported Infrastructural development in St. Joseph's College (Autonomous), Devagiri, Calicut.

Conflicts of Interest: The authors declare no conflict of interest.

Samples of the compounds are available from the authors.

References

1. Wikandari, R.; Nguyen, H.; Millati, R.; Niklasson, C.; Taherzadeh, M.J. Improvement of Biogas Production from Orange Peel Waste by Leaching of Limonene. *Biomed. Res. Int.* **2015**, *2015*, 494182. [CrossRef] [PubMed]
2. Duque-Acevedo, M.; Belmonte-Ureña, L.J.; Cortés-García, F.J.; Camacho-Ferre, F. Agricultural waste: Review of the evolution, approaches and perspectives on alternative uses. *Glob. Ecol. Conserv.* **2020**, *22*, e00902. [CrossRef]
3. Mizuki, E.; Akao, T.; Saruwatari, T. Inhibitory effect of Citrus unshu peel on anaerobic digestion. *Biol. Wastes* **1990**, *33*, 161–168. [CrossRef]

4. Millati, R.; Permanasari, E.D.; Sari, K.W.; Cahyanto, M.N.; Niklasson, C.; Taherzadeh, M.J. Anaerobic digestion of citrus waste using two-stage membrane bioreactor. *IOP Conf. Ser. Mater. Sci. Eng.* **2018**, *316*, 012063. [CrossRef]
5. Kumar Sarangi, P.; Subudhi, S.; Bhatia, L.; Saha, K.; Mudgil, D.; Prasad Shadangi, K.; Srivastava, R.K.; Pattnaik, B.; Arya, R.K. Utilization of agricultural waste biomass and recycling toward circular bioeconomy. *Environ. Sci. Pollut. Res. Int.* **2022**, *13*, 022–20669. [CrossRef]
6. Ravindran, R.; Hassan, S.S.; Williams, G.A.; Jaiswal, A.K. A Review on Bioconversion of Agro-Industrial Wastes to Industrially Important Enzymes. *Bioengineering* **2018**, *5*, 93. [CrossRef]
7. Calabrò, P.S.; Pontoni, L.; Porqueddu, I.; Greco, R.; Pirozzi, F.; Malpei, F. Effect of the concentration of essential oil on orange peel waste biomethanization: Preliminary batch results. *Waste Manag.* **2016**, *48*, 440–447. [CrossRef]
8. Ruiz, B.; Flotats, X. Citrus essential oils and their influence on the anaerobic digestion process: An overview. *Waste Manag.* **2014**, *34*, 2063–2079. [CrossRef]
9. USDA Foreign Agricultural Service. *Citrus: World Markets and Trade*; USDA Foreign Agricultural Service: Washington, DC, USA, 2020.
10. Deng, M.; Jia, X.; Dong, L.; Liu, L.; Huang, F.; Chi, J.; Ma, Q.; Zhao, D.; Zhang, M.; Zhang, R. Structural elucidation of flavonoids from Shatianyu (*Citrus grandis* L. Osbeck) pulp and screening of key antioxidant components. *Food Chem.* **2022**, *366*, 130605. [CrossRef]
11. Somanathan Karthiga, R.; Sukhdeo, S.V.; Madhugiri Lakshminarayan, S.; Mysuru Nanjarajurs, S. Efficacy of Citrus maxima fruit segment supplemented paranthas in STZ induced diabetic rats. *J. Food Sci.* **2021**, *86*, 2091–2102. [CrossRef]
12. Tsai, M.L.; Lin, C.D.; Khoo, K.A.; Wang, M.Y.; Kuan, T.K.; Lin, W.C.; Zhang, Y.N.; Wang, Y.Y. Composition and Bioactivity of Essential Oil from Citrus grandis (L.) Osbeck 'Mato Peiyu' Leaf. *Molecules* **2017**, *22*, 2154. [CrossRef]
13. Yabalak, E.; Erdogan Eliuz, E.A.; Nazli, M.D. Evaluation of Citrus reticulata essential oil: Chemical composition and antibacterial effectiveness incorporated gelatin on E. coli and S. aureus. *Int. J. Environ. Health Res.* **2022**, *32*, 1261–1270. [CrossRef]
14. Song, X.; Liu, T.; Wang, L.; Liu, L.; Li, X.; Wu, X. Antibacterial Effects and Mechanism of Mandarin (Citrus reticulata L.) Essential Oil against Staphylococcus aureus. *Molecules* **2020**, *25*, 4956. [CrossRef]
15. Tao, N.; Jia, L.; Zhou, H. Anti-fungal activity of Citrus reticulata Blanco essential oil against Penicillium italicum and Penicillium digitatum. *Food Chem.* **2014**, *153*, 265–271. [CrossRef]
16. Ishfaq, M.; Akhtar, B.; Muhammad, F.; Sharif, A.; Akhtar, M.F.; Hamid, I.; Sohail, K.; Muhammad, H. Antioxidant and Wound Healing Potential of Essential Oil from Citrus reticulata Peel and Its Chemical Characterization. *Curr. Pharm. Biotechnol.* **2021**, *22*, 1114–1121. [CrossRef]
17. Thavanapong, N.; Wetwitayaklung, P.; Charoenteeraboon, J. Comparison of Essential Oils Compositions of Citrus maxima Merr. Peel Obtained by Cold Press and Vacuum Stream Distillation Methods and of Its Peel and Flower Extract Obtained by Supercritical Carbon Dioxide Extraction Method and Their Antimicrobial Activity. *J. Essent. Oil Res.* **2010**, *22*, 71–77. [CrossRef]
18. Visakh, N.U.; Pathrose, B.; Narayanankutty, A.; Alfarhan, A.; Ramesh, V. Utilization of Pomelo (Citrus maxima) Peel Waste into Bioactive Essential Oils: Chemical Composition and Insecticidal Properties. *Insects* **2022**, *13*, 480. [CrossRef]
19. Singh, P.; Shukla, R.; Prakash, B.; Kumar, A.; Singh, S.; Mishra, P.K.; Dubey, N.K. Chemical profile, antifungal, antiaflatoxigenic and antioxidant activity of Citrus maxima Burm. and Citrus sinensis (L.) Osbeck essential oils and their cyclic monoterpene, DL-limonene. *Food Chem. Toxicol.* **2010**, *48*, 1734–1740. [CrossRef]
20. Kumar, S.; Warikoo, R.; Mishra, M.; Seth, A.; Wahab, N. Larvicidal efficacy of the Citrus limetta peel extracts against Indian strains of Anopheles stephensi Liston and Aedes aegypti L. *Parasitol. Res.* **2012**, *111*, 173–178. [CrossRef]
21. Maurya, A.K.; Mohanty, S.; Pal, A.; Chanotiya, C.S.; Bawankule, D.U. The essential oil from Citrus limetta Risso peels alleviates skin inflammation: In-vitro and in-vivo study. *J. Ethnopharmacol.* **2018**, *212*, 86–94. [CrossRef]
22. Li, Z.H.; Cai, M.; Liu, Y.S.; Sun, P.L.; Luo, S.L. Antibacterial Activity and Mechanisms of Essential Oil from Citrus medica L. var. sarcodactylis. *Molecules* **2019**, *24*, 1577. [CrossRef] [PubMed]
23. Yang, C.; Chen, H.; Chen, H.; Zhong, B.; Luo, X.; Chun, J. Antioxidant and Anticancer Activities of Essential Oil from Gannan Navel Orange Peel. *Molecules* **2017**, *22*, 1391. [CrossRef] [PubMed]
24. Fitsiou, E.; Pappa, A. Anticancer Activity of Essential Oils and Other Extracts from Aromatic Plants Grown in Greece. *Antioxidants* **2019**, *8*, 290. [CrossRef] [PubMed]
25. Yousefian Rad, E.; Homayouni Tabrizi, M.; Ardalan, P.; Seyedi, S.M.R.; Yadamani, S.; Zamani-Esmati, P.; Haghani Sereshkeh, N. Citrus lemon essential oil nanoemulsion (CLEO-NE), a safe cell-depended apoptosis inducer in human A549 lung cancer cells with anti-angiogenic activity. *J. Microencapsul.* **2020**, *37*, 394–402. [CrossRef] [PubMed]
26. Elansary, H.O.; Abdelgaleil, S.A.M.; Mahmoud, E.A.; Yessoufou, K.; Elhindi, K.; El-Hendawy, S. Effective antioxidant, antimicrobial and anticancer activities of essential oils of horticultural aromatic crops in northern Egypt. *BMC Complement. Altern. Med.* **2018**, *18*, 214. [CrossRef]
27. Sharma, K.; Garg, V.K. Vermicomposting of Waste: A Zero-Waste Approach for Waste Management. In *Sustainable Resource Recovery and Zero Waste Approaches*; Taherzadeh, M.J., Bolton, K., Wong, J., Pandey, A., Eds.; Elsevier: Amsterdam, The Netherlands, 2019; pp. 133–164. [CrossRef]
28. Gomes-Araújo, R.; Martínez-Vázquez, D.G.; Charles-Rodríguez, A.V.; Rangel-Ortega, S.; Robledo-Olivo, A. Bioactive Compounds from Agricultural Residues, Their Obtaining Techniques, and the Antimicrobial Effect as Postharvest Additives. *Int. J. Food Sci.* **2021**, *2021*, 9936722. [CrossRef]

29. Vuuren, S.F.v.; Viljoen, A.M. Antimicrobial activity of limonene enantiomers and 1,8-cineole alone and in combination. *Flavour Fragr. J.* **2007**, *22*, 540–544. [CrossRef]
30. Thakre, A.; Zore, G.; Kodgire, S.; Kazi, R.; Mulange, S.; Patil, R.; Shelar, A.; Santhakumari, B.; Kulkarni, M.; Kharat, K.; et al. Limonene inhibits Candida albicans growth by inducing apoptosis. *Med. Mycol.* **2017**, *56*, 565–578. [CrossRef]
31. Han, Y.; Sun, Z.; Chen, W. Antimicrobial Susceptibility and Antibacterial Mechanism of Limonene against Listeria monocytogenes. *Molecules* **2019**, *25*, 33. [CrossRef]
32. Araújo-Filho, H.G.; Dos Santos, J.F.; Carvalho, M.T.B.; Picot, L.; Fruitier-Arnaudin, I.; Groult, H.; Quintans-Júnior, L.J.; Quintans, J.S.S. Anticancer activity of limonene: A systematic review of target signaling pathways. *Phytother. Res.* **2021**, *35*, 4957–4970. [CrossRef]
33. Yu, X.; Lin, H.; Wang, Y.; Lv, W.; Zhang, S.; Qian, Y.; Deng, X.; Feng, N.; Yu, H.; Qian, B. D-limonene exhibits antitumor activity by inducing autophagy and apoptosis in lung cancer. *Onco Targets Ther.* **2018**, *11*, 1833–1847. [CrossRef]
34. Zhou, J.; Azrad, M.; Kong, L. Effect of Limonene on Cancer Development in Rodent Models: A Systematic Review. *Front. Sustain. Food Syst.* **2021**, *5*, 407. [CrossRef]
35. Zhou, H.; Tao, N.; Jia, L. Antifungal activity of citral, octanal and α-terpineol against Geotrichum citri-aurantii. *Food Control* **2014**, *37*, 277–283. [CrossRef]
36. Zielińska, A.; Martins-Gomes, C.; Ferreira, N.R.; Silva, A.M.; Nowak, I.; Souto, E.B. Anti-inflammatory and anti-cancer activity of citral: Optimization of citral-loaded solid lipid nanoparticles (SLN) using experimental factorial design and LUMiSizer®. *Int. J. Pharm.* **2018**, *553*, 428–440. [CrossRef]
37. Oliveira, A.; Fernandes, C.C.; Santos, L.S.; Candido, A.; Magalhaes, L.G.; Miranda, M.L.D. Chemical composition, in vitro larvicidal and antileishmanial activities of the essential oil from Citrus reticulata Blanco fruit peel. *Braz. J. Biol. = Rev. Brasleira Biol.* **2021**, *83*, e247539. [CrossRef]
38. Ambrosio, C.M.S.; Diaz-Arenas, G.L.; Agudelo, L.P.A.; Stashenko, E.; Contreras-Castillo, C.J.; da Gloria, E.M. Chemical Composition and Antibacterial and Antioxidant Activity of a Citrus Essential Oil and Its Fractions. *Molecules* **2021**, *26*, 2888. [CrossRef]
39. Farahmandfar, R.; Tirgarian, B.; Dehghan, B.; Nemati, A. Changes in chemical composition and biological activity of essential oil from Thomson navel orange (Citrus sinensis L. Osbeck) peel under freezing, convective, vacuum, and microwave drying methods. *Food Sci. Nutr.* **2020**, *8*, 124–138. [CrossRef]
40. Abdel-Kawy, M.A.; Michel, C.G.; Kirollos, F.N.; Hussien, R.A.A.; Al-Mahallawi, A.M.; Sedeek, M.S. Chemical composition and potentiation of insecticidal and fungicidal activities of Citrus trifoliata L. fruits essential oil against Spodoptera littoralis, Fusarium oxysporum and Fusarium solani via nano-cubosomes. *Nat. Prod. Res.* **2021**, *35*, 2438–2443. [CrossRef]
41. Teneva, D.; Denkova-Kostova, R.; Goranov, B.; Hristova-Ivanova, Y.; Slavchev, A.; Denkova, Z.; Kostov, G. Chemical composition, antioxidant activity and antimicrobial activity of essential oil from Citrus aurantium L zest against some pathogenic microorganisms. *Z. Naturforschung. C J. Biosci.* **2019**, *74*, 105–111. [CrossRef]
42. Goncalves Mendes Neto, A.; Lo, K.B.; Wattoo, A.; Salacup, G.; Pelayo, J.; DeJoy, R., 3rd; Bhargav, R.; Gul, F.; Peterson, E.; Albano, J.; et al. Bacterial infections and patterns of antibiotic use in patients with COVID-19. *J. Med. Virol.* **2021**, *93*, 1489–1495. [CrossRef]
43. Murray, C.J.L.; Ikuta, K.S.; Sharara, F.; Swetschinski, L.; Robles Aguilar, G.; Gray, A.; Han, C.; Bisignano, C.; Rao, P.; Wool, E.; et al. Global burden of bacterial antimicrobial resistance in 2019: A systematic analysis. *Lancet* **2022**, *399*, 629–655. [CrossRef] [PubMed]
44. Fancello, F.; Petretto, G.L.; Marceddu, S.; Venditti, T.; Pintore, G.; Zara, G.; Mannazzu, I.; Budroni, M.; Zara, S. Antimicrobial activity of gaseous Citrus limon var pompia leaf essential oil against Listeria monocytogenes on ricotta salata cheese. *Food Microbiol.* **2020**, *87*, 103386. [CrossRef] [PubMed]
45. Al-Aamri, M.S.; Al-Abousi, N.M.; Al-Jabri, S.S.; Alam, T.; Khan, S.A. Chemical composition and in-vitro antioxidant and antimicrobial activity of the essential oil of Citrus aurantifolia L. leaves grown in Eastern Oman. *J. Taibah Univ. Med. Sci.* **2018**, *13*, 108–112. [CrossRef] [PubMed]
46. Mandalari, G.; Bennett, R.N.; Bisignano, G.; Trombetta, D.; Saija, A.; Faulds, C.B.; Gasson, M.J.; Narbad, A. Antimicrobial activity of flavonoids extracted from bergamot (Citrus bergamia Risso) peel, a byproduct of the essential oil industry. *J. Appl. Microbiol.* **2007**, *103*, 2056–2064. [CrossRef] [PubMed]
47. Raspo, M.A.; Vignola, M.B.; Andreatta, A.E.; Juliani, H.R. Antioxidant and antimicrobial activities of citrus essential oils from Argentina and the United States. *Food Biosci.* **2020**, *36*, 100651. [CrossRef]
48. Ben Hsouna, A.; Ben Halima, N.; Smaoui, S.; Hamdi, N. Citrus lemon essential oil: Chemical composition, antioxidant and antimicrobial activities with its preservative effect against Listeria monocytogenes inoculated in minced beef meat. *Lipids Health Dis.* **2017**, *16*, 146. [CrossRef]
49. Narang, N.; Jiraungkoorskul, W. Anticancer Activity of Key Lime, Citrus aurantifolia. *Pharm. Rev.* **2016**, *10*, 118–122.
50. Hero, T.; Buhler, H.; Kouam, P.N.; Priesch-Grzeszowiak, B.; Lateit, T.; Adamietz, I.A. The Triple-negative Breast Cancer Cell Line MDA-MB 231 Is Specifically Inhibited by the Ionophore Salinomycin. *Anticancer Res.* **2019**, *39*, 2821–2827. [CrossRef]
51. Shi, Y.; Ye, P.; Long, X. Differential Expression Profiles of the Transcriptome in Breast Cancer Cell Lines Revealed by Next Generation Sequencing. *Cell. Physiol. Biochem.* **2017**, *44*, 804–816. [CrossRef]
52. Mukhtar, Y.M.; Adu-Frimpong, M.; Xu, X.; Yu, J. Biochemical significance of limonene and its metabolites: Future prospects for designing and developing highly potent anticancer drugs. *Biosci. Rep.* **2018**, *38*, BSR20181253. [CrossRef]

53. Nordin, N.; Yeap, S.K.; Rahman, H.S.; Zamberi, N.R.; Abu, N.; Mohamad, N.E.; How, C.W.; Masarudin, M.J.; Abdullah, R.; Alitheen, N.B. In vitro cytotoxicity and anticancer effects of citral nanostructured lipid carrier on MDA MBA-231 human breast cancer cells. *Sci. Rep.* **2019**, *9*, 1614. [CrossRef]
54. White, B.; Evison, A.; Dombi, E.; Townley, H.E. Improved delivery of the anticancer agent citral using BSA nanoparticles and polymeric wafers. *Nanotechnol. Sci. Appl.* **2017**, *10*, 163–175. [CrossRef]
55. Liu, S.; Zhao, Y.; Cui, H.F.; Cao, C.Y.; Zhang, Y.B. 4-Terpineol exhibits potent in vitro and in vivo anticancer effects in Hep-G2 hepatocellular carcinoma cells by suppressing cell migration and inducing apoptosis and sub-G1 cell cycle arrest'. *J. B.U.ON. Off. J. Balkan Union Oncol.* **2021**, *26*, 294.
56. Hassan, S.B.; Gali-Muhtasib, H.; Goransson, H.; Larsson, R. Alpha terpineol: A potential anticancer agent which acts through suppressing NF-kappaB signalling. *Anticancer Res.* **2010**, *30*, 1911–1919.
57. House, N.C.; Puthenparampil, D.; Malayil, D.; Narayanankutty, A. Variation in the polyphenol composition, antioxidant, and anticancer activity among different Amaranthus species. *S. Afr. J. Bot.* **2020**, *135*, 408–412. [CrossRef]
58. Baliyan, S.; Mukherjee, R.; Priyadarshini, A.; Vibhuti, A.; Gupta, A.; Pandey, R.P.; Chang, C.M. Determination of Antioxidants by DPPH Radical Scavenging Activity and Quantitative Phytochemical Analysis of Ficus religiosa. *Molecules* **2022**, *27*, 1326. [CrossRef]
59. Al-Amiery, A.A.; Al-Majedy, Y.K.; Kadhum, A.A.; Mohamad, A.B. Hydrogen Peroxide Scavenging Activity of Novel Coumarins Synthesized Using Different Approaches. *PLoS ONE* **2015**, *10*, e0132175. [CrossRef]
60. Narayanankutty, A.; Gopinath, M.K.; Vakayil, M.; Ramavarma, S.K.; Babu, T.D.; Raghavamenon, A.C. Non-enzymatic conversion of primary oxidation products of Docosahexaenoic acid into less toxic acid molecules. *Spectrochim. Acta Part A Mol. Biomol. Spectrosc.* **2018**, *203*, 222–228. [CrossRef]
61. Kim, Y.O.; Narayanankutty, A.; Kuttithodi, A.M.; Kim, H.-J.; Na, S.W.; Kunnath, K.; Rajagopal, R.; Alfarhan, A. Azima tetracantha Leaf Methanol Extract Inhibits Gastric Cancer Cell Proliferation through Induction of Redox Imbalance and Cytochrome C Release. *Appl. Sci.* **2022**, *12*, 120. [CrossRef]
62. Balouiri, M.; Sadiki, M.; Ibnsouda, S.K. Methods for in vitro evaluating antimicrobial activity: A review. *J. Pharm. Anal.* **2016**, *6*, 71–79. [CrossRef]
63. Walia, S.; Mukhia, S.; Bhatt, V.; Kumar, R.; Kumar, R. Variability in chemical composition and antimicrobial activity of Tagetes minuta L. essential oil collected from different locations of Himalaya. *Ind. Crops Prod.* **2020**, *150*, 112449. [CrossRef]
64. European Committee for Antimicrobial Susceptibility Testing (EUCAST) of the European Society of Clinical Microbiology and Infectious Diseases (ESCMID). E.C.f.A.S.T.E.o.t.E.S.o.C.M.a.I.D. Determination of minimum inhibitory concentrations (MICs) of antibacterial agents by agar dilution. *Clin. Microbiol. Infect.* **2000**, *6*, 509–515. [CrossRef] [PubMed]
65. Campana, R.; Tiboni, M.; Maggi, F.; Cappellacci, L.; Cianfaglione, K.; Morshedloo, M.R.; Frangipani, E.; Casettari, L. Comparative Analysis of the Antimicrobial Activity of Essential Oils and Their Formulated Microemulsions against Foodborne Pathogens and Spoilage Bacteria. *Antibiotics* **2022**, *11*, 447. [CrossRef] [PubMed]
66. Aljeldah, M.M. Antioxidant and Antimicrobial Potencies of Chemically-Profiled Essential Oil from Asteriscus graveolens against Clinically-Important Pathogenic Microbial Strains. *Molecules* **2022**, *27*, 3539. [CrossRef]
67. Selim, S.; Almuhayawi, M.S.; Alqhtani, H.; Al Jaouni, S.K.; Saleh, F.M.; Warrad, M.; Hagagy, N. Anti-Salmonella and Antibiofilm Potency of Salvia officinalis L. Essential Oil against Antibiotic-Resistant Salmonella enterica. *Antibiotics* **2022**, *11*, 489. [CrossRef]

molecules

MDPI

Article

Antioxidant, Antimicrobial, Cytotoxicity, and Larvicidal Activities of Selected Synthetic Bis-Chalcones

Aswathi Moothakoottil Kuttithodi [1], Divakaran Nikhitha [1], Jisha Jacob [1], Arunaksharan Narayanankutty [2,*], Manoj Mathews [3], Opeyemi Joshua Olatunji [4,*], Rajakrishnan Rajagopal [5], Ahmed Alfarhan [5] and Damia Barcelo [6]

1 Molecular Microbial Ecology Lab, PG and Research Department of Zoology, St. Joseph's College (Autonomous), Devagiri, Calicut 680 555, Kerala, India
2 Division of Cell and Molecular Biology, PG and Research Department of Zoology, St. Joseph's College (Autonomous), Devagiri, Calicut 673 008, Kerala, India
3 PG and Research Department of Chemistry, St. Joseph's College (Autonomous), Devagiri, Calicut 673 008, Kerala, India
4 African Genome Center, Mohammed VI Polytechnic University, Ben Guerir 43150, Morocco
5 Department of Botany and Microbiology, College of Science, King Saud University, P.O. Box 2455, Riyadh 11451, Saudi Arabia
6 Water and Soil Research Group, Department of Environmental Chemistry, IDAEA-CSIC, Jordi Girona 18–26, 08034 Barcelona, Spain
* Correspondence: arunaksharann@devagiricollege.org (A.N.); joshua.olatunji@um6p.ma (O.J.O.)

Abstract: Plants are known to have numerous phytochemicals and other secondary metabolites with numerous pharmacological and biological properties. Among the various compounds, polyphenols, flavonoids, anthocyanins, alkaloids, and terpenoids are the predominant ones that have been explored for their biological potential. Among these, chalcones and bis-chalcones are less explored for their biological potential under in vitro experiments, cell culture models, and animal studies. In the present study, we evaluated six synthetic bis-chalcones that were different in terms of their aromatic cores, functional group substitution, and position of substitutions. The results indicated a strong antioxidant property in terms of DPPH and ABTS radical-scavenging potentials and ferric-reducing properties. In addition, compounds **1**, **2**, and **4** exhibited strong antibacterial activities against *Escherichia coli*, *Pseudomonas aeruginosa*, *Staphylococcus aureus*, and *Salmonella enteritidis*. The disc diffusion assay values were indicative of the antibacterial properties of these compounds. Overall, the study indicated the antioxidant and antimicrobial properties of the compounds. Our preliminary studies point to the potential of this class of compounds for further in vivo investigation.

Keywords: bis-chalcones; antioxidant activity; antimicrobial activity; larvicidal activity; chemical structure

Citation: Kuttithodi, A.M.; Nikhitha, D.; Jacob, J.; Narayanankutty, A.; Mathews, M.; Olatunji, O.J.; Rajagopal, R.; Alfarhan, A.; Barcelo, D. Antioxidant, Antimicrobial, Cytotoxicity, and Larvicidal Activities of Selected Synthetic Bis-Chalcones. *Molecules* **2022**, *27*, 8209. https://doi.org/10.3390/molecules27238209

Academic Editor: Ionel Mangalagiu

Received: 15 October 2022
Accepted: 18 November 2022
Published: 25 November 2022

Publisher's Note: MDPI stays neutral with regard to jurisdictional claims in published maps and institutional affiliations.

Copyright: © 2022 by the authors. Licensee MDPI, Basel, Switzerland. This article is an open access article distributed under the terms and conditions of the Creative Commons Attribution (CC BY) license (https://creativecommons.org/licenses/by/4.0/).

1. Introduction

Plants are an important component of the biosphere that is essential for the sustainability of entire ecosystems [1]. They act as the primary source of food for other organisms, which are primarily composed of carbohydrates [2]. Apart from that, these plants are also home to a wide variety of compounds. Phytochemicals are important components obtained from plants by various methods and they are used as food and medicinal compounds by various populations in India and around the world [3]. Most compounds of plant origin are useful for their biological and pharmacological properties. [4]. A majority of these are allelochemicals that are known to repel various pests that attach to these plants [5,6]. Various biochemicals such as flavonoids, alkaloids, saponins and phenolic compounds are a useful part of medicine [7]. Compounds are extracted from almost all parts of a plant including the root, bark, flower, leaf, etc. Fatty oils, both essential and non-essential, and many active compounds can be seen in different parts of a single plant. Some produce medicinal effects whereas some produce toxic effects.

Plant products including primary and secondary metabolites are widely applied in the pharmacological field and nutritional aspects. These biochemical products are "side-tracks" or secondary metabolites that are essential in plant growth and development, protection, attraction or signaling. The main chemical groups of bioactive compounds in plants include polyphenols, flavonoids, anthocyanins, tannins, and chalcones. Flavonoids are an important class of bioactive compounds of plant origin. Flavonoid compounds possess antioxidant, antimicrobial, antiviral, and antitumor activities. The intake of large amounts of flavonoids can help prevent cancer and heart disease. Structurally, they are further classified into chalcones, flavones, isoflavones, flavanols, and anthocyanins. A chalcone is a compound that consists of two aromatic rings linked by an unsaturated α, β-ketone. Chalcones are a class of bioactive plant metabolites that are equipped with numerous biological and pharmaceutical benefits to humans. Plants belonging to the Leguminosae, Asteraceae and Moraceae families are rich in natural chalcones [8,9]. The radical-quenching abilities of the chalcones isolated from different plants have been widely reported [10,11]. The *Angelica keiskei*-derived chalcones have been shown to block the activities of cysteine proteases of the COVID virus [12]. A pharmacological analysis has indicated the anti-radical and anti-edematous properties of these *A. keiskei* chalones [13]. Later, studies by Shin et al. [14] found that *A. keiskei* chalcones inhibit cytokine production in macrophages. Further, Enoki et al. [15] indicated that 4-hydroxyderricin (4-HD) and xanthoangelol, the two major chalcones in the plant, inhibit the development of diabetes via peroxisome-proliferator-activated receptor-gamma activation in mice. Likewise, the extracted chalcones from different species of *Artocarpus* inhibited platelet aggregation in vitro [16]. Ngameni et al. [17] indicated the potential of *Dorstenia turbinata* chalcones to inhibit brain tumor cell proliferation and invasion by blocking MMP2. The bioavailability of chalcones, i.e., the proportion of chalcone that enters the circulation when introduced into the body and its ability to have an effect, is low. The issue itself has important implications for the pharmaceutical applications of chalcones or their derivative molecules. [18]. Structurally, dibenzylidines or bis-chalcones also belong to the chalcone family [19]. Synthetic chalcones and bis-chalcones have therefore become more important than the naturally present chalcones. However, the medical properties of bis-chalcones are less known. The chalcones are also known for the inhibition of microbial populations, especially bacteria and fungi [20,21]. Further, these compounds are known to inhibit diseases associated with oxidative damage and inflammation [22].

Plant chalcones are flavonoid derivatives formed in the biosynthetic pathway of flavonols. They do not accumulate in plants in larger quantities; hence, the availability of these compounds is too limited from plant sources. Another challenge that limits the potential of chalcones is their low half-life. In this work, we synthesized six bis-chalcone compounds whose structures are shown in Figure 1. Compounds **1** to **3** are cyclohexanone derivatives while compounds 4 to 6 are cyclopentanone derivatives. We report the properties of these synthetic bis-chalcones as antioxidant molecules and antibacterial and larvicidal agents.

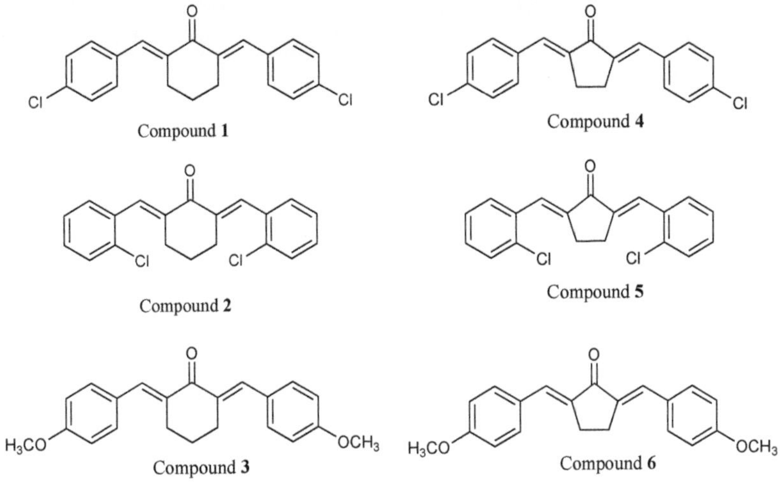

Figure 1. Molecular structures of synthetic chalcones **1** to **6**.

2. Results and Discussion

2.1. Characterization of Compounds

Phytochemicals are important compounds that are present in different plant parts at different concentrations [23]. These compounds are produced as part of the metabolic pathway or as secondary metabolites. Chalcones are one of the few bioactive compounds in plants [18]; they are known to have strong pharmacological activities against infectious and chronic diseases [24]. However, these molecules have a lower biological availability in humans. To overcome this issue, various synthetic chalcones/bis-chalcones have been chemically synthesized and studied [25,26]. In the present study, we evaluated six synthetic bis-chalcones for their biological efficacies. The synthetic chalcones were characterized by FT-IR and NMR spectroscopy techniques and high-quality figures are included in Supplementary Figures S1–S13.

2.2. Antiradical Potentials of Various Synthetic Bis-Chalcones

The synthetic bis-chalcones exhibited strong antioxidant activity; the scavenging of DPPH, nitric oxide, and ABTS radicals was high in compound **1** and compound **2**, whereas compounds **5** and **6** were the least active (Table 1). The FRAP assay identified the EC50 values of compounds A, B, and C; the highest activity was shown by compound A (EC50 of 1.35 ± 0.10 μg/mL). This was followed by compound B (5.24 ± 0.21 μg/mL) and compound C (12.4 ± 0.20 μg/mL). The results indicated the antioxidant properties of the selected bis-chalcones, especially for compounds **1**, **2**, and compound **4**. As can be seen from the structure, all of these molecules have the –Cl group as a substitution in their aromatic core unit. Strong radical-scavenging, reducing and enzyme inhibitory properties of these molecules can be attributed to the presence of the chlorine substitution in these bis-chalcones. Our results showed the ability of bis-chalcones to strongly inhibit the DPPH radical and reduce ferric ions. Antioxidant abilities eventually help to reduce oxidative radicals from various biological systems and thereby prevent the development of oxidative stress [27,28]. Oxidative damage induced by free radicals results in the progression of degenerative disorders including diabetes, obesity, cardiovascular problems, and neoplasia [29,30]. Hence, synthetic bis-chalcones may be helpful to prevent the development and progression of various oxidative-stress-associated diseases.

Table 1. Antiradical properties of different synthetic bis-chalcones.

	DPPH (IC_{50} µg/mL)	ABTS (IC_{50} µg/mL)	Nitric Oxide (IC_{50} µg/mL)	FRAP (EC_{50} µg/mL)
(2E,6E)-2,6-bis(4-methoxybenzylidene) cyclohexanone (compound **1**)	18.41 ± 1.45	18.63 ± 1.41	28.87 ± 1.49	1.35 ± 0.10
(2E,6E)-2,6-bis(4-chlorobenzylidene) cyclohexanone (compound **2**)	19.92 ± 1.52	21.57 ± 1.55	26.04 ± 1.61	5.24 ± 0.21
(2E,6E)-2,6-bis(2-chlorobenzylidene) cyclohexanone (compound **3**)	27.75 ± 2.50	26.47 ± 1.42	34.30 ± 2.55	12.40 ± 0.20
(2E,5E)-2,5-bis(4-(tetrahydro-2H-pyran-2-yloxy) benzylidene) cyclopentanone (compound **4**)	25.42 ± 1.39	22.18 ± 1.29	29.15 ± 1.72	4.34 ± 0.11
2,5-bis(4-hydroxybenzylidene) cyclopentanone (compound **5**)	36.49 ± 1.55	42.10 ± 2.27	45.67 ± 3.04	15.61 ± 0.30
4-(tetrahydro-2H-pyran-2-yloxy) benzaldehyde (compound **6**)	35.47 ± 1.64	46.17 ± 3.23	49.09 ± 3.11	16.20 ± 0.24

2.3. *Synthetic Bis-Chalcones as Antimicrobial Agents*

The selected synthetic bis-chalcones exhibited varying toxicity against the bacterial strains (Table 2); compound **2** was the most active against *Escherichia coli* (22.5 ± 0.2 mm). Similarly, compound **2** and compound **4** had LC50 values of 57.6 ± 3.2 and 69.7 ± 2.4 µg/mL. Previous studies have also indicated their bactericidal properties against various pathogenic microbial organisms [31–33]. Emerging studies have also indicated that chalcones and their derivatives prevent biofilm formation in bacterial colonies [34–36] and enhance antibiotic sensitivity [37,38]. The bacterial strains used are pathogenic to humans and animals as well as known to be lethal in many conditions [39,40]. The infection of *P. aeruginosa* has been widely associated with patients of cancers, organ transplantation and HIV [41]. Additionally, there have also been raising concerns about antibiotic resistance over the years. Likewise, *S. aureus* and *S. enteritidis* are also associated with infections of the digestive tract in foodborne diseases [42,43]. Hence, the inhibitory potential of bis-chalcones on various microbes may indicate their potential as antibiotic agents for future use.

Table 2. Efficacy of synthetic bis-chalcones against Gram-positive and Gram-negative bacterial strains as indicated by the zone of inhibition by the disc diffusion method.

Bacteria	Compounds with Zone of Inhibition (mm)					
	1	2	3	4	5	6
Escherichia coli	18.4 ± 0.1	22.5 ± 0.2 *	16.7 ± 0.3	20.6 ± 0.3 *	17.7 ± 0.3	16.4 ± 0.1
Pseudomonas aeruginosa	19.8 ± 0.2 *	19.1 ± 0.2 *	14.7 ± 0.2	20.1 ± 0.3	16.9 ± 0.2	15.0 ± 0.3
Staphylococcus aureus	18.5 ± 0.1	20.1 ± 0.3 *	13.7 ± 0.1	19.5 ± 0.2 *	15.1 ± 0.2	14.6 ± 0.3
Salmonella enteritidis	18.2 ± 0.1	19.0 ± 0.3 *	15.4 ± 0.2	18.6 ± 0.1 *	16.0 ± 0.2	17.4 ± 0.1

2.4. *Larvicidal Activity of Synthetic Bis-Chalcones*

Apart from their antioxidant properties, it was also noted that the bis-chalcones exhibited larvicidal properties. The larvicidal property of the synthetic bis-chalcones revealed stronger activity in compound **1** (45.27 ± 2.34 µg/mL), compound **2** (59.81 ± 2.09 µg/mL), and compound **4** (56.46 ± 3.07 µg/mL) (Table 3). Limited studies are available on the potential of synthetic chalcones or their derivatives against mosquito larvae. Previous studies by Targanski et al. [44], Begum et al. [45] and Pasquale et al. [46] are among the few that indicated the potential of chalcones against *Aedes aegypti*. Mosquitoes are important vectors of various diseases including arboviral diseases [47], Chikungunya [48], and dengue [49]. Further, the recent literature has indicated that mosquitoes are also involved in the spreading of Zika viral infections [50]. Therefore, the beneficial larvicidal potential of synthetic bis-chalcones may be helpful in the management of infectious diseases.

Table 3. Larvicidal properties of different synthetic bis-chalcones against *Aedes albopictus*.

Compound	LC_{50} (μg/mL)
(2E,6E)-2,6-bis(4-methoxybenzylidene) cyclohexanone (compound **1**)	45.27 ± 2.34
(2E,6E)-2,6-bis(4-chlorobenzylidene) cyclohexanone (compound **2**)	59.81 ± 2.09
(2E,6E)-2,6-bis(2-chlorobenzylidene) cyclohexanone (compound **3**)	99.04 ± 2.18
(2E,5E)-2,5-bis(4-(tetrahydro-2H-pyran-2-yloxy) benzylidene) cyclopentanone (compound **4**)	56.46 ± 3.07
2,5-bis(4-hydroxybenzylidene) cyclopentanone (compound **5**)	89.22 ± 3.12
4-(tetrahydro-2H-pyran-2-yloxy) benzaldehyde (compound **6**)	79.18 ± 2.69

The cytotoxicity evaluation was performed against two human breast cancer cells, MCF-7 and MDA-MB-231 (Table 4). The MCF-7 cell expresses receptors for estrogen, epidermal growth factor, and progesterone [51–53]. On the contrary, MDA-MB-231 is a cell with basal expression for these three receptors and is often considered to be "triple negative" [54–56]. The chalcones were more toxic towards the MCF-7 cells compared to MDA-MB-231 in terms of the IC_{50} values. Previous studies have also reported the anticancer potentials of chalcones and their derivatives in various cell and animal models [57,58]. In addition, the fact that the synthetic chalcones are in a purified form means their applicability in medicine is higher than that of crude extracts or isolated plant products.

Table 4. Cytotoxicity evaluation of the synthetic bis-chalcones against human cancer cells and results are expressed as IC_{50} values.

Compound	LC_{50} (μg/mL)	
	MCF-7	**MDA-MB-231**
(2E,6E)-2,6-bis(4-methoxybenzylidene) cyclohexanone (compound **1**)	86.13 ± 3.45	128.66 ± 3.62
(2E,6E)-2,6-bis(4-chlorobenzylidene) cyclohexanone (compound **2**)	79.51 ± 2.85	97.64 ± 3.15
(2E,6E)-2,6-bis(2-chlorobenzylidene) cyclohexanone (compound **3**)	132.49 ± 3.71	160.54 ± 5.22
(2E,5E)-2,5-bis(4-(tetrahydro-2H-pyran-2-yloxy) benzylidene) cyclopentanone (compound **4**)	71.09 ± 2.34	89.62 ± 2.18
2,5-bis(4-hydroxybenzylidene) cyclopentanone (compound **5**)	103.56 ± 2.48	141.05 ± 4.84
4-(tetrahydro-2H-pyran-2-yloxy) benzaldehyde (compound **6**)	109.82 ± 4.10	155.32 ± 5.03

Further studies are also necessary to ascertain the safety aspects of these synthetic bis-chalcones against various non-target organisms, freshwater fishes and germinating seeds. This will ensure a bis-chalcone-based synthetic pesticide, which is an efficient alternative to the existing antioxidant supplement, antimicrobial compounds, and larvicidal systems. Additionally, further studies on the functional food aspect of the compounds will ensure the possible potential of these selected compounds as a nutraceutical against various diseases.

3. Materials and Methods

3.1. *Chemicals and Reagents*

The chemicals, reagents, and analytical-grade solvents were obtained from Sigma-Aldrich and local chemical companies. Bulk solvents used for purification or extraction and other general purposes were purified and dried before use by following standard procedures. Unless otherwise specified, chemicals or reagents were used as received from the suppliers without further purification. Chromatography was performed using

Silica gel (60–120 and 100–200 mesh size). The progress and completion of the reaction were monitored by thin-layer chromatography (TLC). For this purpose, aluminum sheets pre-coated with silica gel (Merck, Kieselgel60, F254) were used.

3.2. Instruments Used for the Study

IR spectra were recorded on a Nicolet iS5 Thermo Fischer Scientific FT-IR spectrometer (Waltham, MA, USA). The spectral positions are given in the wavenumber (cm^{-1}) unit. ^1H and ^{13}C NMR spectra of the compounds in $CDCl_3$ were recorded using Bruker AMX-400 (400 MHz) spectrometer (Billerica, MA, USA). For ^1H NMR spectra, the chemical shifts (δ) are reported in parts per million (ppm) relative to tetramethylsilane (TMS) as an internal standard. Coupling constants (J) are given in Hz. The spectrophotometric measurements were taken using UV 1280 Shimadzu UV/Visible spectrophotometer (Kyoto, Japan).

3.3. Synthesis and Characterization of Bis-Chalcones

The target bis-chalcone compounds with cyclohexanone and cyclopentanone cores were synthesized as outlined in Figure 2. The required chemicals such as ortho-chloro benzaldehyde, para-chloro benzaldehyde, para-methoxy benzaldehyde, cyclohexanone and cyclopentanone were obtained from Aldrich and used without further purification. Solvents and other reagents were obtained from local sources. The base-catalyzed Claisen–Schmidt condensation reaction of substituted benzaldehydes with cyclohexanone and cyclopentanone yielded the target compounds. In a typical reaction, the double mixed-aldol condensation reaction between a ketone and substituted benzaldehyde compound was carried out [16]. The ketone has α-hydrogens (on both sides) and thus can be deprotonated to give a nucleophilic enolate anion. The alkoxide produced is protonated by the solvent, giving a β-hydroxy ketone, which undergoes base-catalyzed dehydration. The elimination process is particularly fast in this case because the alkene is stabilized by conjugation not only to the carbonyl but also to the benzene. In this synthesis, two equivalents of the substituted benzaldehyde compound were used such that the aldol condensation could occur on both sides of the ketone. The aldehyde carbonyl is more reactive than that of the ketone and therefore reacts rapidly with the anion of the ketone to give a β-hydroxy ketone, which easily undergoes base-catalyzed dehydration (Figure 2). The molecular structures of all the target compounds were confirmed by standard spectroscopic methods of analysis. Detailed synthetic procedures for the compounds along with their characterization data are given as supplementary information.

3.4. Radical Generation Inhibition and Reducing Potential of the Synthetic Bis-Chalcones

The antiradical activity of the synthetic bis-chalcones was estimated against various radical generators including DPPH, ABTS, and nitric oxide according to the standard protocols mentioned by Lalhminghlui and Jagetia [59]. The reducing potential was assessed as the ferric reduction ability of the compounds as described by Youn et al. [60].

3.5. Antibacterial Activity of the Synthetic Bis-Chalcones by Disc Diffusion Method

The antibacterial activity was estimated against *Escherichia coli, Pseudomonas aeruginosa, Staphylococcus aureus*, and *Salmonella enteritidis* strains. Briefly, the bacteria were grown in LB broth, and for the disc diffusion assay they were plated in an MHA agar plate (5 mm thickness). The plate was immersed with an 8 mm filter-paper disc corresponding to 20 μg/mL. The plates were incubated for 24 h in a bacteriological incubator and the inhibition zone was determined [61].

3.6. Analysis of the Larvicidal Activity of Synthetic Bis-Chalcones

The larvae of *Aedes albopictus* (third instar) were collected from the maintained culture; different concentrations of the synthetic bis-chalcones were added to glass jars (500 mL) and fifty larvae were put in each. The larvae were observed for 24 h for mortality. The percentage of death in each treatment was determined and the LC_{50} value was determined by probit analysis.

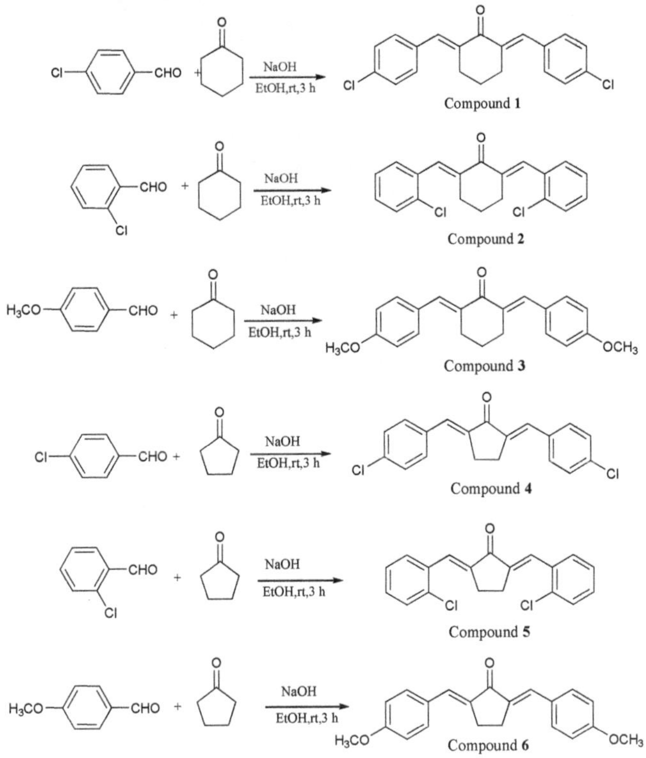

Figure 2. Synthetic route for the target compounds.

3.7. Cytotoxicity Analysis of Synthetic Bis-Chalcones

The MCF-7 and MDA-MB-231 cells were procured from the National Centre for Cell Science (Pune, Maharashtra). The cells were maintained according to the methods suggested by the supplier. The cytotoxicity was estimated using the MTT assay as described by the studies of Khanapure et al. [62]. The IC_{50} value was estimated using probit analysis using GraphPad Prism.

3.8. Statistical Analysis

The experimental results were initially sorted using spreadsheet software and statistical analysis was performed using GraphPad Prism version 7.0 (La Jolla, CA, USA). All the data were represented as mean ± SD for every experiment.

4. Conclusions

In this study, we synthesized six bis-chalcone compounds based on cyclohexanone and cyclopentanone cores. Their molecular structures were characterized by spectroscopic methods. The bis-chalones were different in their aromatic cores, functional group substitution, and position of substitutions. The compounds with a cyclohexanone core and –Cl substitution exhibited better antioxidant properties. In addition, compounds **1**, **2**, and **4** exhibited strong antibacterial activities against *Escherichia coli, Pseudomonas aeruginosa, Staphylococcus aureus*, and *Salmonella enteritidis*. Apart from their antioxidant and antibacterial properties, the compounds also exhibited larvicidal properties. The effective control of mosquito populations by killing the larvae of the *Aedes* mosquito indicates their potential application in preventing infectious diseases and their vectors. The cytotoxic effect of these compounds is also indicative of their antineoplastic potentials. Our preliminary studies highlight the potential of bis-chalcones as pharmacologically active compounds and therefore further research with structural modifications with polar and non-polar substitutions is currently underway.

Supplementary Materials: The following supporting information can be downloaded at: https://www.mdpi.com/article/10.3390/molecules27238209/s1, Supplementary Figure S1: FT-IR spectrum of compound **1**; Supplementary Figure S2: ^{1}H NMR Spectrum of compound **1** in CDCl3; Supplementary Figure S3: FT-IR spectrum of compound **2**; Supplementary Figure S4: ^{1}H NMR Spectrum of compound **2** in CDCl3; Supplementary Figure S5: FT-IR spectrum of compound **3**; Supplementary Figure S6: ^{1}H NMR Spectrum of compound **3** in CDCl3; Supplementary Figure S7: FT-IR spectrum of compound **4**; Supplementary Figure S8: ^{1}H NMR Spectrum of compound **4** in CDCl3; Supplementary Figure S9: FT-IR spectrum of compound **5**; Supplementary Figure S10: ^{1}H NMR Spectrum of compound **5** in CDCl3; Supplementary Figure S11: 13C NMR Spectrum of compound **5** in CDCl3; Supplementary Figure S12: FT-IR spectrum of compound **6**; Supplementary Figure S13: ^{1}H NMR Spectrum of compound **6** in CDCl3.

Author Contributions: A.M.K., analysis, manuscript preparation, experimentation; D.N. analysis, manuscript preparation, experimentation R.R. analysis, manuscript preparation, experimentation, A.N.: Study design, methodology, experimentation, analysis, funding acquisition, manuscript editing. M.M., Study design, methodology, experimentation, analysis, funding acquisition, manuscript editing J.J., experimentation, supervision, manuscript editing O.J.O. Study design, methodology, experimentation, analysis, funding acquisition, manuscript editing and A.A.: Study design, methodology, experimentation, analysis, funding acquisition, manuscript editing. D.B.: Study design, analysis, manuscript editing. All authors have read and agreed to the published version of the manuscript.

Funding: The authors are grateful for the seed grant from Mohammed VI Polytechnic University, Morocco and funding support from Researchers Supporting Project Number (RSP2022R465), King Saud University, Riyadh, Saudi Arabia.

Institutional Review Board Statement: Not applicable.

Informed Consent Statement: Not applicable.

Data Availability Statement: The data may be shared upon valid request.

Acknowledgments: The authors are grateful for the seed grant from Mohammed VI Polytechnic University, Morocco. The authors also acknowledge the funding support from Researchers Supporting Project Number (RSP2022R465), King Saud University, Riyadh, Saudi Arabia. DBT-STAR (Project number: BT/HRD/11/09/2020) scheme supported Infrastructural development in St. Joseph's College (Autonomous), Devagiri, Calicut.

Conflicts of Interest: The authors declare no conflict of interest.

Sample Availability: Samples of the compounds are available from the authors.

References

1. Nieto, G. How Are Medicinal Plants Useful When Added to Foods? *Medicines* **2020**, *7*, 58. [CrossRef]
2. Vatansever, R.; Ozyigit, I.; Filiz, E. Essential and Beneficial Trace Elements in Plants, and Their Transport in Roots: A Review. *Appl. Biochem. Biotechnol.* **2017**, *181*, 464–482. [CrossRef]
3. Choudhari, A.S.; Mandave, P.C.; Deshpande, M.; Ranjekar, P.; Prakash, O. Phytochemicals in Cancer Treatment: From Preclinical Studies to Clinical Practice. *Front. Pharmacol.* **2020**, *10*, 1614. [CrossRef]
4. Ashokkumar, K.; Murugan, M.; Dhanya, M.K.; Pandian, A.; Warkentin, T.D. Phytochemistry and therapeutic potential of black pepper [Piper nigrum (L.)] essential oil and piperine: A review. *Clin. Phytosci.* **2021**, *7*, 52. [CrossRef]
5. Seca, A.M.L.; Pinto, D.C.G.A. Biological Potential and Medical Use of Secondary Metabolites. *Medicines* **2019**, *6*, 66. [CrossRef] [PubMed]
6. Zhang, S.; Zhang, L.; Zou, H.; Qiu, L.; Zheng, Y.; Yang, D.; Wang, Y. Effects of Light on Secondary Metabolite Biosynthesis in Medicinal Plants. *Front. Plant Sci.* **2021**, *11*, 497. [CrossRef]
7. Cory, H.; Passarelli, S.; Szeto, J.; Tamez, M.; Mattei, J. The Role of Polyphenols in Human Health and Food Systems: A Mini-Review. *Front. Nutr.* **2018**, *5*, 87. [CrossRef] [PubMed]

8. Orlikova, B.; Tasdemir, D.; Golais, F.; Dicato, M.; Diederich, M. Dietary chalcones with chemopreventive and chemotherapeutic potential. *Genes. Nutr.* **2011**, *6*, 125–147. [CrossRef] [PubMed]
9. Wu, X.; Zhang, S.; Liu, X.; Shang, J.; Zhang, A.; Zhu, Z.; Zha, D. Chalcone synthase (CHS) family members analysis from eggplant (*Solanum melongena* L.) in the flavonoid biosynthetic pathway and expression patterns in response to heat stress. *PLoS ONE* **2020**, *15*, e0226537. [CrossRef] [PubMed]
10. Oldoni, T.L.C.; Cabral, I.S.R.; d'Arce, M.A.B.R.; Rosalen, P.L.; Ikegaki, M.; Nascimento, A.M.; Alencar, S.M. Isolation and analysis of bioactive isoflavonoids and chalcone from a new type of Brazilian propolis. *Sep. Purif. Technol.* **2011**, *77*, 208–213. [CrossRef]
11. Escobar-Ramos, A.; Lobato-Garcia, C.E.; Zamilpa, A.; Gomez-Rivera, A.; Tortoriello, J.; Gonzalez-Cortazar, M. Homoisoflavonoids and Chalcones Isolated from Haematoxylum campechianum L. with Spasmolytic Activity. *Molecules* **2017**, *22*, 1405. [CrossRef]
12. Park, J.Y.; Ko, J.A.; Kim, D.W.; Kim, Y.M.; Kwon, H.J.; Jeong, H.J.; Kim, C.Y.; Park, K.H.; Lee, W.S.; Ryu, Y.B. Chalcones isolated from Angelica keiskei inhibit cysteine proteases of SARS-CoV. *J. Enzym. Inhib. Med. Chem.* **2016**, *31*, 23–30. [CrossRef]
13. Xue, Y.; Liu, Y.; Luo, Q.; Wang, H.; Chen, R.; Liu, Y.; Li, Y. Antiradical Activity and Mechanism of Coumarin–Chalcone Hybrids: Theoretical Insights. *J. Phys. Chem. A* **2018**, *122*, 8520–8529. [CrossRef]
14. Shin, J.E.; Choi, E.J.; Jin, Q.; Jin, H.G.; Woo, E.R. Chalcones isolated from Angelica keiskei and their inhibition of IL-6 production in TNF-alpha-stimulated MG-63 cell. *Arch. Pharmacal Res.* **2011**, *34*, 437–442. [CrossRef]
15. Enoki, T.; Ohnogi, H.; Nagamine, K.; Kudo, Y.; Sugiyama, K.; Tanabe, M.; Kobayashi, E.; Sagawa, H.; Kato, I. Antidiabetic activities of chalcones isolated from a Japanese Herb, Angelica keiskei. *J. Agric. Food Chem.* **2007**, *55*, 6013–6017. [CrossRef]
16. Jantan, I.; Mohd Yasin, Y.H.; Jamil, S.; Sirat, H.; Basar, N. Effect of prenylated flavonoids and chalcones isolated from Artocarpus species on platelet aggregation in human whole blood. *J. Nat. Med.* **2010**, *64*, 365–369. [CrossRef]
17. Ngameni, B.; Touaibia, M.; Patnam, R.; Belkaid, A.; Sonna, P.; Ngadjui, B.T.; Annabi, B.; Roy, R. Inhibition of MMP-2 secretion from brain tumor cells suggests chemopreventive properties of a furanocoumarin glycoside and of chalcones isolated from the twigs of Dorstenia turbinata. *Phytochemistry* **2006**, *67*, 2573–2579. [CrossRef]
18. Salehi, B.; Quispe, C.; Chamkhi, I.; El Omari, N.; Balahbib, A.; Sharifi-Rad, J.; Bouyahya, A.; Akram, M.; Iqbal, M.; Docea, A.O.; et al. Pharmacological Properties of Chalcones: A Review of Preclinical Including Molecular Mechanisms and Clinical Evidence. *Front. Pharmacol.* **2021**, *11*, 592654. [CrossRef]
19. Zhuang, C.; Zhang, W.; Sheng, C.; Zhang, W.; Xing, C.; Miao, Z. Chalcone: A Privileged Structure in Medicinal Chemistry. *Chem Rev.* **2017**, *117*, 7762–7810. [CrossRef]
20. Okolo, E.N.; Ugwu, D.I.; Ezema, B.E.; Ndefo, J.C.; Eze, F.U.; Ezema, C.G.; Ezugwu, J.A.; Ujam, O.T. New chalcone derivatives as potential antimicrobial and antioxidant agent. *Sci. Rep.* **2021**, *11*, 21781. [CrossRef]
21. Gupta, D.; Jain, D.K. Chalcone derivatives as potential antifungal agents: Synthesis, and antifungal activity. *J. Adv. Pharm. Technol. Res.* **2015**, *6*, 114–117. [CrossRef] [PubMed]
22. Thapa, P.; Upadhyay, S.P.; Suo, W.Z.; Singh, V.; Gurung, P.; Lee, E.S.; Sharma, R.; Sharma, M. Chalcone and its analogs: Therapeutic and diagnostic applications in Alzheimer's disease. *Bioorg. Chem.* **2021**, *108*, 29. [CrossRef]
23. Kucharíková, A.; Kusari, S.; Sezgin, S.; Spiteller, M.; Čellárová, E. Occurrence and Distribution of Phytochemicals in the Leaves of 17 In vitro Cultured Hypericum spp. Adapted to Outdoor Conditions. *Front. Plant Sci.* **2016**, *7*, 1616. [CrossRef] [PubMed]
24. Forni, C.; Facchiano, F.; Bartoli, M.; Pieretti, S.; Facchiano, A.; D'Arcangelo, D.; Norelli, S.; Valle, G.; Nisini, R.; Beninati, S.; et al. Beneficial Role of Phytochemicals on Oxidative Stress and Age-Related Diseases. *Biomed. Res. Int.* **2019**, *2019*, 8748253. [CrossRef] [PubMed]
25. Liargkova, T.; Hadjipavlou-Litina, D.J.; Koukoulitsa, C.; Voulgari, E.; Avgoustakis, C. Simple chalcones and bis-chalcones ethers as possible pleiotropic agents. *J. Enzym. Inhib. Med. Chem.* **2016**, *31*, 302–313. [CrossRef]
26. Tajudeen Bale, A.; Mohammed Khan, K.; Salar, U.; Chigurupati, S.; Fasina, T.; Ali, F.; Wadood, A.; Taha, M.; Sekhar Nanda, S.; Ghufran, M.; et al. Chalcones and bis-chalcones: As potential α-amylase inhibitors; synthesis, in vitro screening, and molecular modelling studies. *Bioorg. Chem.* **2018**, *79*, 179–189. [CrossRef] [PubMed]
27. Pisoschi, A.M.; Pop, A. The role of antioxidants in the chemistry of oxidative stress: A review. *Eur. J. Med. Chem.* **2015**, *97*, 55–74. [CrossRef] [PubMed]
28. Sharifi-Rad, M.; Anil Kumar, N.V.; Zucca, P.; Varoni, E.M.; Dini, L.; Panzarini, E.; Rajkovic, J.; Tsouh Fokou, P.V.; Azzini, E.; Peluso, I.; et al. Lifestyle, Oxidative Stress, and Antioxidants: Back and Forth in the Pathophysiology of Chronic Diseases. *Front. Physiol.* **2020**, *11*, 694. [CrossRef]
29. Moylan, J.S.; Reid, M.B. Oxidative stress, chronic disease, and muscle wasting. *Muscle Nerve* **2007**, *35*, 411–429. [CrossRef]
30. Liguori, I.; Russo, G.; Curcio, F.; Bulli, G.; Aran, L.; Della-Morte, D.; Gargiulo, G.; Testa, G.; Cacciatore, F.; Bonaduce, D.; et al. Oxidative stress, aging, and diseases. *Clin. Interv. Aging* **2018**, *13*, 757–772. [CrossRef]
31. Xu, M.; Wu, P.; Shen, F.; Ji, J.; Rakesh, K.P. Chalcone derivatives and their antibacterial activities: Current development. *Bioorg. Chem.* **2019**, *91*, 103133. [CrossRef] [PubMed]
32. Božić, D.D.; Milenković, M.; Ivković, B.; Ćirković, I. Antibacterial activity of three newly-synthesized chalcones & synergism with antibiotics against clinical isolates of methicillin-resistant Staphylococcus aureus. *Indian J. Med. Res.* **2014**, *140*, 130–137. [PubMed]
33. Koudokpon, H.; Armstrong, N.; Dougnon, T.V.; Fah, L.; Hounsa, E.; Bankolé, H.S.; Loko, F.; Chabrière, E.; Rolain, J.M. Antibacterial Activity of Chalcone and Dihydrochalcone Compounds from *Uvaria chamae* Roots against Multidrug-Resistant Bacteria. *BioMed Res. Int.* **2018**, *2018*, 1453173. [CrossRef] [PubMed]

34. Satokata, A.A.C.; Souza, J.H.; Silva, L.L.O.; Santiago, M.B.; Ramos, S.B.; Assis, L.R.; Theodoro, R.S.; Oliveira, L.R.; Regasini, L.O.; Martins, C.H.G. Chalcones with potential antibacterial and antibiofilm activities against periodontopathogenic bacteria. *Anaerobe* **2022**, *76*, 102588. [CrossRef]
35. Kunthalert, D.; Baothong, S.; Khetkam, P.; Chokchaisiri, S.; Suksamrarn, A. A chalcone with potent inhibiting activity against biofilm formation by nontypeable Haemophilus influenzae. *Microbiol. Immunol.* **2014**, *58*, 581–589. [CrossRef]
36. Bozic, D.D.; Milenkovic, M.; Ivkovic, B.; Cirkovic, I. Newly-synthesized chalcones-inhibition of adherence and biofilm formation of methicillin-resistant Staphylococcus aureus. *Braz. J. Microbiol.* **2014**, *45*, 263–270. [CrossRef]
37. Uchil, A.; Murali, T.S.; Nayak, R. Escaping ESKAPE: A chalcone perspective. *Results Chem.* **2021**, *3*, 100229. [CrossRef]
38. Hellewell, L.; Bhakta, S. Chalcones, stilbenes and ketones have anti-infective properties via inhibition of bacterial drug-efflux and consequential synergism with antimicrobial agents. *Access Microbiol.* **2020**, *2*, acmi000105. [CrossRef]
39. Goncalves Mendes Neto, A.; Lo, K.B.; Wattoo, A.; Salacup, G.; Pelayo, J.; DeJoy, R., III; Bhargav, R.; Gul, F.; Peterson, E.; Albano, J.; et al. Bacterial infections and patterns of antibiotic use in patients with COVID-19. *J. Med. Virol.* **2021**, *93*, 1489–1495. [CrossRef]
40. Murray, C.J.L.; Ikuta, K.S.; Sharara, F.; Swetschinski, L.; Robles Aguilar, G.; Gray, A.; Han, C.; Bisignano, C.; Rao, P.; Wool, E.; et al. Global burden of bacterial antimicrobial resistance in 2019: A systematic analysis. *Lancet* **2022**, *399*, 629–655. [CrossRef]
41. Qin, S.; Xiao, W.; Zhou, C.; Pu, Q.; Deng, X.; Lan, L.; Liang, H.; Song, X.; Wu, M. Pseudomonas aeruginosa: Pathogenesis, virulence factors, antibiotic resistance, interaction with host, technology advances and emerging therapeutics. *Signal Transduct. Target. Ther.* **2022**, *7*, 022–01056. [CrossRef] [PubMed]
42. Nakao, J.H.; Talkington, D.; Bopp, C.A.; Besser, J.; Sanchez, M.L.; Guarisco, J.; Davidson, S.L.; Warner, C.; McIntyre, M.G.; Group, J.P.; et al. Unusually high illness severity and short incubation periods in two foodborne outbreaks of Salmonella Heidelberg infections with potential coincident Staphylococcus aureus intoxication. *Epidemiol. Infect.* **2018**, *146*, 19–27. [CrossRef] [PubMed]
43. Akbar, A.; Anal, A.K. Prevalence and antibiogram study of Salmonella and Staphylococcus aureus in poultry meat. *Asian Pac. J. Trop. Biomed.* **2013**, *3*, 163–168. [CrossRef] [PubMed]
44. Targanski, S.K.; Sousa, J.R.; de Pádua, G.M.; de Sousa, J.M.; Vieira, L.C.; Soares, M.A. Larvicidal activity of substituted chalcones against Aedes aegypti (Diptera: Culicidae) and non-target organisms. *Pest Manag. Sci.* **2021**, *77*, 325–334. [CrossRef]
45. Begum, N.A.; Roy, N.; Laskar, R.A.; Roy, K. Mosquito larvicidal studies of some chalcone analogues and their derived products: Structure–activity relationship analysis. *Med. Chem. Res.* **2011**, *20*, 184–191. [CrossRef]
46. Pasquale, G.; Romanelli, G.P.; Autino, J.C.; García, J.; Ortiz, E.V.; Duchowicz, P.R. Quantitative Structure–Activity Relationships of Mosquito Larvicidal Chalcone Derivatives. *J. Agric. Food Chem.* **2012**, *60*, 692–697. [CrossRef]
47. Weetman, D.; Kamgang, B.; Badolo, A.; Moyes, C.L.; Shearer, F.M.; Coulibaly, M.; Pinto, J.; Lambrechts, L.; McCall, P.J. Aedes Mosquitoes and Aedes-Borne Arboviruses in Africa: Current and Future Threats. *Int. J. Environ. Res. Public Health* **2018**, *15*, 220. [CrossRef]
48. Aragão, C.F.; Pinheiro, V.C.S.; Nunes Neto, J.P.; Silva, E.; Pereira, G.J.G.; Nascimento, B.; Castro, K.D.S.; Maia, A.M.; Catete, C.P.; Martins, L.C.; et al. Natural Infection of Aedes aegypti by Chikungunya and Dengue type 2 Virus in a Transition Area of North-Northeast Brazil. *Viruses* **2019**, *11*, 1126. [CrossRef]
49. Gloria-Soria, A.; Armstrong, P.M.; Powell, J.R.; Turner, P.E. Infection rate of Aedes aegypti mosquitoes with dengue virus depends on the interaction between temperature and mosquito genotype. *Proc. Biol. Sci.* **2017**, *284*, 1864.
50. Mourya, D.T.; Gokhale, M.D.; Majumdar, T.D.; Yadav, P.D.; Kumar, V.; Mavale, M.S. Experimental Zika virus infection in Aedes aegypti: Susceptibility, transmission & co-infection with dengue & chikungunya viruses. *Indian J. Med. Res.* **2018**, *147*, 88–96.
51. Hegde, S.M.; Kumar, M.N.; Kavya, K.; Kumar, K.M.; Nagesh, R.; Patil, R.H.; Babu, R.L.; Ramesh, G.T.; Sharma, S.C. Interplay of nuclear receptors (ER, PR, and GR) and their steroid hormones in MCF-7 cells. *Mol. Cell. Biochem.* **2016**, *422*, 109–120. [CrossRef] [PubMed]
52. Fog, C.K.; Christensen, I.J.; Lykkesfeldt, A.E. Characterization of a human breast cancer cell line, MCF-7/RU58R-1, resistant to the pure antiestrogen RU 58,668. *Breast Cancer Res. Treat.* **2005**, *91*, 133–144. [CrossRef] [PubMed]
53. Leung, E.; Kannan, N.; Krissansen, G.W.; Findlay, M.P.; Baguley, B.C. MCF-7 breast cancer cells selected for tamoxifen resistance acquire new phenotypes differing in DNA content, phospho-HER2 and PAX2 expression, and rapamycin sensitivity. *Cancer Biol. Ther.* **2010**, *9*, 717–724. [CrossRef] [PubMed]
54. Mielczarek, L.; Krug, P.; Mazur, M.; Milczarek, M.; Chilmonczyk, Z.; Wiktorska, K. In the triple-negative breast cancer MDA-MB-231 cell line, sulforaphane enhances the intracellular accumulation and anticancer action of doxorubicin encapsulated in liposomes. *Int. J. Pharm.* **2019**, *558*, 311–318. [CrossRef]
55. Al-Bader, M.; Ford, C.; Al-Ayadhy, B.; Francis, I. Analysis of estrogen receptor isoforms and variants in breast cancer cell lines. *Exp. Ther. Med.* **2011**, *2*, 537–544. [CrossRef]
56. Kalinina, T.S.; Kononchuk, V.V.; Gulyaeva, L.F. Expression of estrogen-, progesterone-, and androgen-responsive genes in MCF-7 and MDA-MB-231 cells treated with o,p'-DDT, p,p'-DDT, or endosulfan. *J. Biochem. Mol. Toxicol.* **2021**, *35*, 1–8. [CrossRef]
57. Constantinescu, T.; Lungu, C.N. Anticancer Activity of Natural and Synthetic Chalcones. *Int. J. Mol. Sci.* **2021**, *22*, 11306. [CrossRef]
58. Ouyang, Y.; Li, J.; Chen, X.; Fu, X.; Sun, S.; Wu, Q. Chalcone Derivatives: Role in Anticancer Therapy. *Biomolecules* **2021**, *11*, 894. [CrossRef]
59. Lalhminghlui, K.; Jagetia, G.C. Evaluation of the free-radical scavenging and antioxidant activities of Chilauni, Schima wallichii Korth in vitro. *Future Sci. OA* **2018**, *4*, FSO272. [CrossRef]

60. Youn, J.S.; Kim, Y.J.; Na, H.J.; Jung, H.R.; Song, C.K.; Kang, S.Y.; Kim, J.Y. Antioxidant activity and contents of leaf extracts obtained from Dendropanax morbifera LEV are dependent on the collecting season and extraction conditions. *Food Sci. Biotechnol.* **2018**, *28*, 201–207. [CrossRef]
61. Walia, S.; Mukhia, S.; Bhatt, V.; Kumar, R.; Kumar, R. Variability in chemical composition and antimicrobial activity of Tagetes minuta L. essential oil collected from different locations of Himalaya. *Ind. Crops Prod.* **2020**, *150*, 112449. [CrossRef]
62. Khanapure, S.; Jagadale, M.; Bansode, P.; Choudhari, P.; Rashinkar, G. Anticancer activity of ruthenocenyl chalcones and their molecular docking studies. *J. Mol. Struct.* **2018**, *1173*, 142–147. [CrossRef]

molecules

Review

Collection of Hairy Roots as a Basis for Fundamental and Applied Research

Anna Yurievna Stepanova *, Maria Viktorovna Malunova, Evgeny Aleksandrovich Gladkov, Sergey Viktorovich Evsyukov, Dmitry Viktorovich Tereshonok and Aleksandra Ivanovna Solov'eva

K.A. Timiryazev Institute of Plant Physiology, Russian Academy of Sciences, 127276 Moscow, Russia
* Correspondence: step_ann@mail.ru; Tel.: +7-(499)-678-53-97

Abstract: Due to population growth, instability of climatic conditions, and reduction of the areas of natural ecosystems, it becomes necessary to involve modern biotechnological approaches to obtain highly productive plant material. This statement applies both to the creation of plant varieties and the production of new pharmaceutical raw materials. Genetic transformation of valuable medicinal plants using *Agrobacterium rhizogenes* ensures the production of stable and rapidly growing *hairy roots* cultures that have a number of advantages compared with cell culture and, above all, can synthesize root-specific substances at the level of the roots of the intact plant. In this regard, special attention should be paid to the collection of hairy roots of the Institute of Plant Physiology RAS, Russian Academy of Sciences, the founder of which was Dr. Kuzovkina I.N. Currently, the collection contains 38 hairy roots lines of valuable medicinal and forage plants. The review discusses the prospects of creating a hairy roots collection as a basis for fundamental research and commercial purposes.

Keywords: hairy roots; secondary metabolites; collection

Citation: Stepanova, A.Y.; Malunova, M.V.; Gladkov, E.A.; Evsyukov, S.V.; Tereshonok, D.V.; Solov'eva, A.I. Collection of Hairy Roots as a Basis for Fundamental and Applied Research. *Molecules* **2022**, *27*, 8040. https://doi.org/10.3390/molecules27228040

Academic Editors: Arunaksharan Narayanankutty, Ademola C. Famurewa and Eliza Oprea

Received: 29 October 2022
Accepted: 16 November 2022
Published: 19 November 2022

Publisher's Note: MDPI stays neutral with regard to jurisdictional claims in published maps and institutional affiliations.

Copyright: © 2022 by the authors. Licensee MDPI, Basel, Switzerland. This article is an open access article distributed under the terms and conditions of the Creative Commons Attribution (CC BY) license (https://creativecommons.org/licenses/by/4.0/).

1. Introduction

Plants produce a wide range of substances useful to mankind which are used in the food industry as biologically active additives, feed additives, medicines, as well as flavorings and food colorings [1–4]. Approximately 25% of the world pharmaceutical market is products obtained from plants [4]. However, often high-value secondary metabolites are synthesized by plants in small amounts under natural conditions. For example, about 10,000 kg of dry bark of *Taxus* sp. is required to obtain 1 kg of taxol [5]. Taxol (paclitaxel) is the first drug from the taxan group, which entered clinical practice and firmly took the leading position in the treatment of the most frequent malignant tumors—breast cancer, ovarian cancer, and non-small-cell lung cancer [6–9]. In addition, Taxol has shown promising results in the treatment of Kaposi sarcoma [10]. Taxol is also being studied for non-cancerous diseases that require microtubule stabilization to avoid cell proliferation and angiogenesis, such as psoriasis, and for the treatment of Alzheimer's or Parkinson's disease [11,12]. Taxol's annual turnover by 2000 was $1.5 billion [13]. To obtain 1 kg of vinblastine and 1 g of vincristine, which are also widely used anticancer drugs, 530 kg of fresh *Catharanthus roseus* (L.) G. Don. leaves is needed [14]. Vinblastine and vincristine are recommended for the treatment of rapidly proliferating neoplasms (hematosarcoma, myeloma, acute leukemia, etc.), breast cancer, neuroblastoma, chorionepithelioma, and lymphogranulomatosis [15].

Due to the global problem of instability of climatic conditions and the shortage of plant raw materials, it is necessary to use biotechnological approaches to obtain it in sufficient quantities, namely, the cultivation of plant cells and plant organs in vitro. In this regard, special attention is drawn to biotechnological approaches associated with the cultivation of cultures in vitro: undifferentiated—callus and cell suspension cultures and differentiated—adventitious and hairy roots.

The biotechnology of higher plants developed in the middle of the last century. Biotechnology in most cases is understood as an application of callus and suspension cultures, which, as a rule, are combined under the general name "tissue culture". The first patent for the production of substances using plant tissue culture was obtained in 1956 [16]. Even today, the greatest number of publications is devoted to the direction associated with the use of in vitro cultured cells, namely, undifferentiated cultures. Numerous reviews discuss the production of metabolites important for humans from suspension and callus cultures [1,17–19]. Some authors have noted that the content of secondary metabolites in the resulting suspension and callus cultures was higher than in intact plants [4,19,20]. However, the opposite result was obtained in most cases [21–26]. Difficulties arise during large-scale cultivation due to the instability of the synthesis of substances by undifferentiated cultures in a liquid medium [19,27]. In addition, the synthesis of some pharmacologically important substances may not be possible in undifferentiated cultures [17,27]. In this regard, attention is drawn to differentiated cultures: shoots, adventitious, and hairy roots (transformed roots). Adventitious roots are induced on media with high auxin and low cytokinin content. However, despite the studies that showed the promise of using adventitious [28–30] (untransformed) roots, most studies noted the disadvantages of their use, primarily slow growth and termination of the synthesis of target substances [27,31]. On the contrary, hairy roots have rapid growth on a hormone-free medium and a high and stable synthesis of essential substances, which has been repeatedly shown [31–38]. Therefore, the technology of hairy roots is very promising.

In our group of specialized root metabolism at the Institute of Plant Physiology, the first studies were carried out with non-transformed adventitious roots, but with the appearance of the possibility of obtaining hairy roots, these studies were discontinued [39]. Thus, the technology of hairy roots in Russia, as well as in other countries, arose later than cellular biotechnology and the culture of isolated roots. Nevertheless, the number of works related to hairy roots is steadily increasing.

This review discusses obtaining hairy roots as producers of valuable metabolites, as well as the prospects of creating a collection of hairy roots consisting of different types of crops for fundamental and commercial purposes.

2. The History of the Development of the Hairy Roots Trend in the World

The history of the emergence of hairy roots as an object of scientific research began in the late 19th to early 20th century and is associated primarily with the American phytopathologist Smith, who studied the formation of crown galls and hairy roots in a number of fruit plants [40]. In November 1908, cultures of bacteria capable of infecting new plants were isolated from the hairy roots of an apple tree, which is described in detail in a large-scale 215-page work by American authors, accompanied by many high-quality photos to document the experimental material obtained [40]. The same paper describes numerous experiments in which bacteria isolated from one plant species successfully infected another and formed similar hairy roots in that one. However, the original causes of the modification of plant organisms under the influence of crown gall bacteria in some and hairy roots in others remained unclear.

The first work that showed that the formation of hairy roots is caused by the transfer of genetic material from *Agrobacterium rhizogenes* dates back to 1982 [41]. *Agrobacterium rhizogenes* is a Gram-negative bacillus, a symbiotic bacterium that currently has been renamed (also named *Rhizobium rhizogenes*). The article drew a parallel between two types of pathogenic bacteria—*Agrobacterium tumefaciens* and *Agrobacterium rhizogenes* and concluded that the Ri-plasmid of *A. rhizogenes*, as well as the Ti-plasmid of A. tumefaciens, can be a vector for the transfer of genetic material. Starting from this date, researchers began to consider hairy roots not just as a neoplasm resulting from the attack of a pathogenic bacterium, but also as a promising model for studying the features of secondary metabolism and, ultimately, as producers of natural products.

In 1997, the monograph "Hairy Roots: Culture and Applications" (Doran P.M., Harwood Academic) was published on the status of research activities in the field of hairy root biotechnology at that time [42]. The monograph outlined laboratory protocols for the initiation and cultivation of hairy roots; described the use of hairy roots as producers of secondary metabolites; as an expression system for the production of antibodies; considerations are given for their large-scale cultivation in bioreactors. Studies on phytoremediation were also presented. The monograph combines the work of scientists from many fields, from genetics and molecular biology to horticulture, medicine, environmental research, and biotechnology.

According to 2006 data since the time of the first publications, the roots of more than 140 plant species belonging to 40 families have been introduced into in vitro culture using methods of genetic transformation of roots [43]. The number of introduced species is only presently increasing [34]. However, hairy roots are mainly induced in dicotyledonous plants. Obtaining hairy roots in monocotyledonous plants is difficult since the infection of such plants with *A. rhizogenes* is a very rare event in nature. One of the reasons for this may be the lack of production of wounder phenolic compounds, such as acetosyringone [44]. It acts as a chemotactic agent at very low concentrations and activates *vir*-genes on Ri-plasmids, which initiate the infection process for the transfer of T-DNA [45,46]. Nevertheless, there are works on obtaining hairy roots in representatives of the monocotyledonous class, such as *Alstroemeria* [47], *Chlorophytum borivilianum* [48], *Zea mays* [49], *Crocus sativus* [44], etc. These works are of great interest since a significant number of medicinal plants are monocotyledonous. One of the ways to overcome the difficulty of agrobacterial transformation for monocotyledonous plants is to use the microparticle bombardment method, since there is no limitation to the range of hosts with this method. It was developed in 1990 by Sanford and coworkers [50]. Indeed, in some cases, the microparticle bombardment method has been successful enough to produce hairy roots in both monocotyledonous and dicotyledonous plants [51–53]. Despite this, most authors tend to use Agrobacterium even in the case of monocotyledonous plants, since this method is simpler and does not require expensive equipment. The advantage of agrobacterial transformation over the microparticle bombardment method is the integration of one copy of the transgene in most cases, the low incident of transgene silencing, and the ability for long DNA segment transfer [47,52,54].

For 40 years since the publication of the first works, hairy roots have been used as producers of secondary metabolites, such as alkaloids, anthocyanins [37,55], flavonoids, ginsenosides, stilbenes, lignans, terpenoids, and shikonin [56]; as well as recombinant proteins such as vaccine [57,58], monoclonal antibodies [59], and therapeutic proteins [60]. Studies on the possibility of using hairy roots in phytoremediation are being conducted [61,62]. However, to the greatest extent, the hairy roots are used as a source of pharmacologically valuable secondary metabolites. It should be noted that, in most cases, the content of medicinal substances in hairy roots is at the level or higher than their content in the roots of intact plants (Table 1). One of the reasons for the high synthesis of various secondary metabolites is the presence of *rol*-genes, primarily *rol*B and *rol*C [63,64]. Bulgakov et al. showed that *rol*B affects the expression patterns of MYB factors controlling the early steps of flavonoid biosynthesis [64].

Hairy roots can synthesize a number of secondary metabolites that are not typical for the intact roots of plants [65]. It has been shown that naphthochinon lawson, which accumulated in the aerial parts of *Lawsonia inermis* L., was not found in plant roots and adventitious roots cultivated in vitro, but was presented in hairy roots [66]. It was considered that only artemisin accumulates in the aboveground part of the *Artemisia annua* plant [67], but it was shown later that hairy roots can also produce artemisin [68].

In the case of an insufficient level of synthesis of secondary metabolites in root cultures, either elicitors or genetic engineering methods are used to increase their content [69]. The use of elicitors is well highlighted in a recently published review covering the period from 2010–2022 [21]. According to this review, methyl jasmonate acts as the main elicitor and the content of secondary metabolites in the hairy roots of various families was significantly increased with it, but the greatest effect was shown in representatives of the genus *Lamiaceae* [21].

The introduction of genes controlling the synthesis of valuable secondary metabolites makes it possible to obtain hairy roots with a stable synthesis of substances regardless of the action of external factors. For example, the introduction of transcription factors *WRKY1,2* [70,71], *MYB98* [72], key genes involved in the tanshinone biosynthetic pathway *HMGH* and *DXR* [73], *GGPPS* and *DXSII* [20] increased the synthesis of tanshinones in the hairy roots of *Salvia miltiorrhiza* by 1.4–21 times [74]; overexpressing CrORCA4 in *Catharanthus roseus* increased tabersonine synthesis by 40 times [75]. The introduction of the maize transcription factor *ZmLc* and *Arabidopsis* transcription factor *AtPAP1* made it possible to increase the content of three main flavones (baicalin, baicalein, and wogonin) in hairy roots of *Scutellaria baicalensis* by 322% and 532%, respectively, by a comprehensively upregulating flavonoid biosynthesis of pathway genes [76]. Through the introduction of curcumin biosynthetic pathway genes, it was possible to increase the level of curcumin and its glycosides in *Atropa belladonna* hairy roots [77]. The number of such works is steadily increasing.

All of the aforementioned makes hairy roots technology a powerful tool for both fundamental and applied research.

Table 1. The benefits of using hairy roots. The examples of studies in which the content of secondary metabolites in hairy roots is at the level or higher than in untransformed tissues.

Species	Analyzed Metabolites	Metabolites Content in Hairy Roots	Metabolites in Plants	Metabolites in Callus Suspension	Reference
Atropa belladonna	Total alkaloids contents	1.1–8 mg/gDW	Intact roots—0.3 mg/g DW		[78]
Atropa belladonna	Total alkaloids contents	1.32%	In untransformed roots—0.8%		[79]
Artemisia dubui	Artemisin	0.603–0.753%.	In untransformed roots—0.001%		[80]
Artemisia sp.	Artemisinin and its co-products	1.02 mg/g DW	In untransformed roots—up to 0.687 mg/g DW		[81]
Panax ginseng	Ginsenoside	Total content 5.44 mg/g DW	Total content in untransformed roots—4.55 mg/g DW		[82]
Panax ginseng	Saponins	2–2.4 times higher compared with native root			[83]
Panax ginseng	Ginsenoside	2.88% of dry weight when cultured in 1/8 MS medium for 8 weeks	2.56% of dry weight (cultivated roots were 5 years old)		[84]
Rubia yunnanensis	Rubiaceae-type cyclopeptides (RAs)	Amount of (RAs) in 1/2 MS liquid medium—4.611 μg/g DW	In plants in vitro—0.331 μg/g and 4.096 μg/g DW for shoots and roots, respectively. Amount of RAs in seed-borne plants—80.296 μg/g, quinones—7409 μg/g DW	In calli—1.082 μg/g DW	[85]
Rubia yunnanensis	Quinones	5067 μg/g DW	24–132 μg/g DW in plants in vitro; in seed-borne plants 7409 μg/g DW	In calli—338 μg/g DW	[85]
Rubia cordifolia Linn	Total phenolic contents	139.7 mg/g DW	41.02 mg/g DW of field grown roots	-	[86]
Rubia cordifolia Linn	Alizarin	In 5.16-fold than normal roots of field grown plants	5.16 times lower in the roots of the field-grown plants compared with the hairy roots		[86]
Withania sominifera	Withanolide A	157.4 μg/g DW	57.9 μg/g DW	-	[87]

3. The Collection of Hairy Roots of the Institute of Plant Physiology

In Russia, the technology of obtaining hairy roots is associated with the name of Kuzovkina I.N. The first object was the hairy roots of *Peganum harmala* L. obtained in 1987 [88,89] (Table 2).

Peganum harmala L. (Zygophyllaceae) is a perennial herbaceous plant common in the Mediterranean region of Europe, Central Asia, and southern South America [90,91]. It refers to medicinal plants widely used in folk and traditional Chinese medicine for the treatment of various human diseases [90–94]. The therapeutic effect of extracts from seeds and vegetative parts of the plant is explained by the presence of two classes of alkaloids—quinazoline and indole β-carboline type. Alkaloids of the first class (peganine and its derivatives) are found only in the aerial part, and β-carboline (the main ones are harmine and harmaline) are found only in the roots (Figure 1).

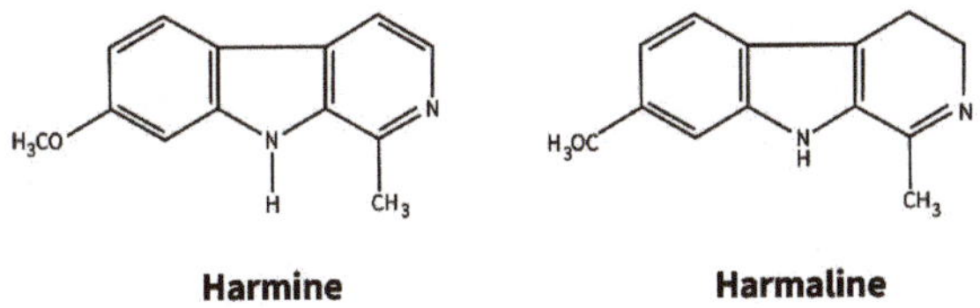

Figure 1. The main β-carboline alkaloids in roots of *Peganum harmala*.

According to the literature, *Peganum harmala* has antibacterial [95], anti-inflammatory [96], anti-fungal [97], and antitumor effects [92], and is used to treat hypertension [98], cough [99], diabetes [100], jaundice [94], malaria [101], tremor paralysis, Parkinson's disease, and Alzheimer's disease [92,102]. Despite the wide spectrum of action, the medicine uses peganin hydrochloride (ampoules and tablets) for the treatment of myopathy and myasthenia gravis and harmine hydrochloride for the treatment of encephalitis, tremor paralysis, and Parkinson's disease [103].

It was shown by our group that the content of harmine and harmalol (β-carboline alkaloids) in the transformed roots was 30 and 4.3 times higher than in callus, respectively, which indicates the promise of the obtained roots [89]. Calli and hairy roots of *P. harmala* are still maintained in the collection of the IPP RAS (Figure 2). Because the content of harmine in the hairy roots was predominant and only slightly inferior to its content in the roots of intact plants, the resulting hairy roots of *Peganum harmala* can be a source for obtaining an important group of pharmacologically valuable alkaloids which include harmine.

Figure 2. Calli (**left**) and hairy roots (**right**) of *Peganum harmala* obtained in 1987. The photos were taken in 2021.

After, hairy roots of *Ruta graveolens* L. (Rutaceae) were obtained (Figure 3). *Ruta graveolens* (common rue, rue) is a plant with a very rich composition of secondary metabolites (about 200) belonging to various groups of low molecular weight compounds [104]. *Ruta graveolens* contains coumarins, alkaloids, volatile oils, flavonoids, and phenolic acids [104]. Rue has been known as a medicinal plant since ancient times. It has already been used to treat various diseases since the time of Hippocrates [105]. There are numerous reports on the use of rue herbs in the folk medicine of various countries, namely, in Indian and Chinese medicine. Rue has been used for a long time as an analgesic, to eliminate eye problems, and to improve the condition of patients with rheumatism and dermatitis [104,106]. In Russian folk medicine, it was used for heart diseases, disorders of the nervous system, and as an effective abortifacient [107]. Currently, the antiviral, antibacterial, and fungicidal effects of rue have been proven [106]. Along with pronounced pharmacological properties, rue is also used as an essential oil plant. Rue essential oil finds application in the perfumery and food industry [108].

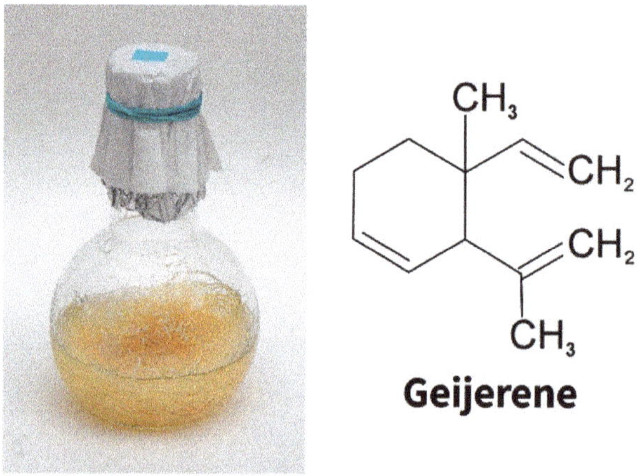

Figure 3. The hairy roots of *Ruta graveolens* (**left**). The main component of the essential oil of common rue is geijarene (**right**).

The peculiarity of this plant is that the synthesis of these substances is organ-specific, i.e., the root and aerial parts of the plant form various secondary metabolites, and the main part of them is characterized by a kind of fluorescence. With in vitro cultures at its disposal, it is possible to study the spatial distribution of metabolites. From 1969 to 1976, calli of obtained *Ruta graveolens* were obtained by Kuzovkina (Figure 4) to study the biogenesis of furocoumarins and acridone alkaloids. However, it became possible to study the spatial organization of low molecular weight metabolites only with the production of hairy roots of *Ruta graveolens* in 1991 [109,110].

Figure 4. The examples of *Ruta graveolens* calli obtained during the period 1969–1976. The photos were taken in 2021.

Intensively branching genetically transformed roots forming a large number of root apexes turned out to be a convenient object for studying the function of the so-called border cells, which are root cap cells that separate from the root tip during its growth. They are the first to come into direct contact with various representatives of the rhizosphere. Hairy roots of common rue retain the ability to synthesize the essential oil typical of intact plant roots, the main component of which is sesquiterpene geijerene [110]. Interestingly, geijerene (Figure 3) belongs to the number of root-specific volatile metabolites that attract entomopathogenic nematodes eating the larvae of a dangerous root pest of young citrus plants (*Rutaceae*), the weevil *Diaprepes abbreviate* [111].

Thus, hairy roots can be a model not only for the spatial study of secondary metabolites but also for the study of allelopathic relationships between roots and soil micro- and macrobiota.

When creating the collection, the main attention was paid to those plants in which the roots were found to contain low molecular weight metabolites that are of practical interest and are used in the medical and food industries. For this purpose, hairy roots of *Glycyrrhiza uralensis, Rauwolfia serpentina, Rubia tinctorum, Rhodiola rosea,* various species of the genus *Scutellaria*, and other most important medicinal plants were obtained [112].

Rauwolfia serpentina Benth. (Apocynaceae) has pronounced pharmacological properties. The main use of *R. serpentina* was for snake and insect bites, fever, cholera, diarrhea, as a mild sedative for children, and in Java, it was used as an anthelmintic [113–115]. To date, it has been shown that *Rauwolfia serpentina* has antibacterial [116], antifungal [117], anti-inflammatory [118], antidiabetic, mosquito larvicidal, antihistamine, antidiarrheal [119], hypoglycemic and hypolipidemic, anticancer, as well as sedative [120] and hepatoprotective activities [113,121]. Currently, more than 80 alkaloids have been isolated from this plant [122]. However, reserpine is a pharmacologically more potent alkaloid [122]. On the basis of *Rauwolfia serpentina* alkaloids, the drug "Reserpine" against arterial hypertension and "Ajmalin"—an antiarrhythmic agent was released. The hairy roots of *Rauwolfia serpentina* were obtained in 1990 and are cultivated to the present (Figure 5).

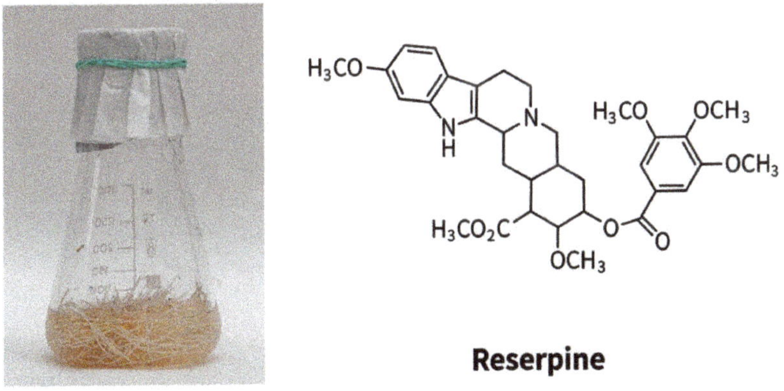

Figure 5. The hairy roots of *Rauwolfia serpentina* (**left**) and the main alkaloid (**right**).

Rubia tinctorum L. (Rubiaceae) is also a medicinal plant, according to modern studies, containing about 250 compounds with different chemical classes [123]. The main components are anthraquinones and their derivatives such as alizarin and purpurin, which have a diuretic, antispasmodic, and laxative effect [124]. At present, the antitumor, hepatoprotective and antidiabetic properties of the roots of common madder have been shown [123,125,126]. Besides that, the roots of common madder have been used as a dye for dyeing fabrics and applying patterns since ancient times [127,128].

The hairy roots of *Rubia tinctorum* were obtained in 1991 and are cultivated to the present in the collection of hairy roots of the Institute of Plant Physiology (Figure 6).

Figure 6. Hairy roots of *Glycyrrhiza uralensis* (**left**) and *Rubia tinctorum* (**right**). The photo was taken in 2021.

Glycyrrhiza uralensis (licorice) (Fabaceae) is a plant that has found its application both in the food and pharmaceutical industries. Licorice contains many different substances, among which glycyrrhizin and glycyrrhetic acid can be considered the main ones. The plant has been known since ancient times. Licorice was considered a panacea for many diseases in oriental medicine. The Egyptians, Greeks, and Romans recommended licorice as a remedy to help fight physical stress and fatigue [129]. In our country (Russia), licorice has always been the largest harvesting object, as well as the subject of raw material exports. Biologically active constituents of licorice have antiviral and anti-inflammatory effects; it has recently been shown that it can also be used to treat alcoholic liver damage [130–132]. Licorice root is also used in the food industry as a sweetener, flavor, and aroma enhancer [129,133].

The hairy roots of *Glycyrrhiza uralensis* were obtained in 1990 and are cultivated to the present in the collection of hairy roots of the Institute of Plant Physiology (Figure 6).

The genetic stability of hairy roots and their ability to maintain the synthesis of low molecular weight metabolites under in vitro conditions at the level of the roots of the whole plant formed the basis for studies conducted with the roots of valuable medicinal plants of the genus *Scutellaria*. The roots of Baikal skullcap, which have been steadily growing for more than 25 years, synthesize flavones typical of the roots of this plant—glucuronides: baicalin and wogonoside, and aglycones: baicalein and wogonin [134–138] (Figure 7).

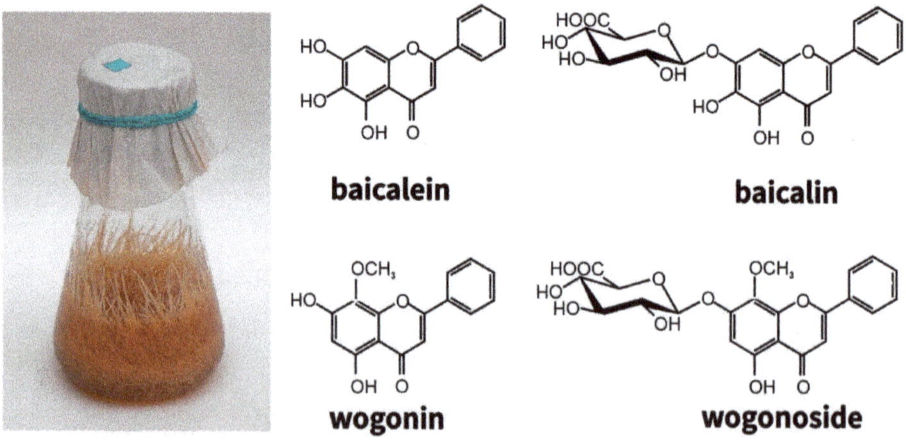

Figure 7. Hairy roots of *Scutellaria baicalensis* obtained in 1996 (**left**). The main root-specific flavones (**right**). The photo was taken in 2021.

Although the content of major metabolites in hairy roots is three times lower than in the roots of intact plants, this disadvantage is compensated by the rapid growth and year-round cultivation of hairy roots. However, if the dominant metabolite of the *Scutellaria baicalensis* roots of intact plants is the glucuronide baicalin and, accordingly, its aglycone baicalein, then the main flavone in hairy roots is the mono-O-methylated glucuronide wogonoside and its aglycone wogonin [134,135,138]. Later it was suggested that the increase in the content of monomethylated flavones is associated with the climatic features of the region (Dauria, Russia), where the plants were taken for transformation [139]. Nevertheless, when obtaining other members of the genus *Scutellaria (Scutellaria andrachnoides* (together with Kyrgyz colleagues), *Scutellaria przewalskii, Scutellaria pycnoclada,* and *Scutellaria lateriflora* taken from different climatic zones, increased content of mono-O-methylated flavones was also shown [136,140]. This suggests more subtle regulatory mechanisms that are currently being investigated. The practical significance of such a ratio of flavones in skullcap hairy roots was assessed after the publication of Japanese researchers in 2009, who showed that the aglycone wogonin selectively induces apoptosis only in cancer cells, while not affecting normal cells [141]. Its antitumor activity was confirmed by in vivo studies, which opened up the possibility of the clinical application of wogonin [142–144]. Wogonin has been confirmed to be effective against neurodegenerative diseases, including Alzheimer's disease [145]. The anti-coronavirus properties of wogonin have also been recently discovered [146–148]. Because the aglycone wogonin has a pronounced therapeutic effect, but the main methylated flavone in the hairy roots of *S. baicalensis* obtained by us is wogonoside glucuronide, the latter must be hydrolyzed. Some publications propose different ways of obtaining aglycones from glucuronides, one is hydrolysis using β-glucuronidase from various microorganisms. For instance, 90% of glucuronides were hydrolyzed to aglycones within 3 h with the help of β-glucuronidase from *Lactobacillus delbruecki* [149]. At the same time, it is known that the Baikal skullcap contains its own β-glucuronidase (sGUS) [150,151]. We have conducted a number of studies of the relationship between the content of basic flavones and the activity of the sGUS both in the maintained *S. baicalensis* hairy roots strain and in the undifferentiated callus and suspension cultures obtained from it [26,77]. As a result, it was shown that the activity of sGUS during the cultivation cycle of *S. baicalensis* hairy roots correlated with the content of wogonin more than with the content of baicalein. We have also demonstrated that the content of wogonin increases in response to mechanical stress and is presumably associated with the protection of plants from biotic stress, in particular, from insect pests. It should be noted that undifferentiated cultures (calli and suspensions) of *S. baicalensis* had a different ratio of flavones with a predominance of the baicalin/baicalein pair, in contrast to hairy roots. Interestingly, sGUS activity in hairy roots was 10 times higher than in undifferentiated cultures [137]. In the future, we plan to study the possibilities of sGUS activation and an increase in the level of biologically active flavones-aglycones.

Thus, in our studies conducted on the hairy roots of members of the genus *Scutellaria* from the collection of hairy roots of the Institute of Plant Physiology, it was shown that they contain the same set of flavones as the intact roots. However, the roots of intact plants and hairy roots differ in the ratio of flavones.

This makes our objects unique both for research and commercial use.

Another rare medicinal plant from which both hairy roots and undifferentiated cells were obtained was *Rhodiola quadrifida* Pall. (Crassulaceae), known for its medicinal properties [152–154]. It belongs to the alpine species and its range is rapidly shrinking [155]. It should be pointed out that the content of the main metabolites (salidroside and rosavin) was higher in undifferentiated callus cultures than in hairy roots [156]. The results are of considerable interest for further research.

At present, the collection of hairy roots includes 38 hairy roots strains belonging to 25 plant species and 16 lines of callus cultures (Table 2).

The collection mainly maintains hairy roots of medicinal plant species (84%), and the two families that are predominantly represented are *Fabaceae* and *Lamiaceae* (Figure 8).

Table 2. Some strains in the hairy roots collection of Institute of Plant Physiology.

Family	Species Name	Number of Lines	Type of the Culture	The Year Obtained	Origin of the Culture	Secondary Metabolites	Note	Possible Use	Literature References in Which This Strain Is Mentioned
Apocynaceae	*Rauwolfia serpentina* L. (Benth.)	1	Hairy roots	1990	Leaves of juvenile plants	Contain indole alkaloids (vomilenine, vinorine, perakine) higher than in suspension culture.	The total alkaloid content was 3 times lower compared with the suspension culture. There is no raucaffricine alkaloid, which predominates in suspension culture.	Pharmacology, medicine	[112]
Caryophyllaceae	*Silene vulgaris* L.	1	Hairy roots	2002	-	-	-	Phytoremediation (accumulation of heavy metals)	-
Crassulaceae	*Rhodiola quadrifida Pall*	2	Hairy roots	2017	Cotyledons and hypocotyls	Contain salidroside, rosavin	Tyrosol and rosarin are missing. The content is significantly lower than in callus tissue.	Pharmacology, preserving of rare and endangered species	[156]
	Rhodiola quadrifida Pall	4	Calli	2019	Hairy roots	Contain salidroside, rosavin, rosin	Thyrosol and rosarin are absent		
Fabaceae	*Hedysarum* sp.	5	Hairy roots	2001–2019	Juvenile seedlings	Isoflavones (ononine)	-	Pharmacology, medicine	[157]

Table 2. *Cont.*

Family	Species Name	Number of Lines	Type of the Culture	The Year Obtained	Origin of the Culture	Secondary Metabolites	Note	Possible Use	Literature References in Which This Strain Is Mentioned
	Glycyrrhiza uralensis L.	1	Hairy roots	1990	Hypocotyle of juvenile plant	Phenolic derivatives	-	Pharmacology, medicine	-
	Lupinus polyphyllus L.	1	Hairy roots	1990	Hypocotyle of juvenile plant	Isoflavone glycoside		Biotechnology, feed additives	
	Ononis sp. L.	4	Hairy roots	1993–1994	Hypocotyle of juvenile plant	isoflavonoids			
	Sophora korolkovii Koehne	1	Hairy roots	2004	Hypocotyle of juvenile plant	Phenolic compounds			
	Trifolium repens L.	1	Hairy roots	1991	Hypocotyle of juvenile plant	A model object for studying arbuscular mycorrhizae			
	Medicago sativa L.	2	Hairy roots	2013	Hypocotyle of juvenile plant	Phytoremediation			
Lamiaceae	*Scutellaria baicalensis Georgi*	3	Hairy roots	1993, 2018, 2021	Leaves of juvenile plant	Contains flavones (baicalin, baicalein, wogonin, wogonoside)	The content of methylated flavones is higher than in the roots of intact plants	Pharmacology, medicine, food industry	[135–138,158]
	Scutellaria baicalensis Georgi	2	Calli		Hairy roots	Contains flavones (baicalin, baicalein, wogonin, wogonoside)	The main flavones are baicalin and baicalein, as in the roots of intact plants.		

Table 2. *Cont.*

Family	Species Name	Number of Lines	Type of the Culture	The Year Obtained	Origin of the Culture	Secondary Metabolites	Note	Possible Use	Literature References in Which This Strain Is Mentioned
	Scutellaria androchnoides	1	Hairy roots	2006	Cotyledon and hypocotyl	The dominant compounds are acteoside (phenylethanoids) and the four main flavones of representatives of *Scutellaria* sp.: baicalin, wogonoside, wogonin and baicalein	The content of acteoside is 10 times higher, and the content of methylated flavone wogonoside is 3 times higher compared with the roots of intact plants	Pharmacology, medicine	[140]
	Scutellaria androchnoides	1	Calli	2006	Hairy roots	The dominant metabolites are acteoside and the methylated flavones wogonoside and wogonin	The content of acteoside is 2.5 times, wogonoside 1.5 times higher than in the roots of intact plants	Pharmacology, medicine	
	Scutellaria przewalskii	1	Hairy roots	2014, 2020	Cotyledons, hypocotyls of sterile-grown plants	Based on HPLC data, 17 flavones were found, among which the main metabolites are baicalin and wogonoside glucuronides	The content of the main metabolites is higher than in all of the above representatives of *Scutellaria*	Pharmacology, medicine	[136]

Table 2. *Cont.*

Family	Species Name	Number of Lines	Type of the Culture	The Year Obtained	Origin of the Culture	Secondary Metabolites	Note	Possible Use	Literature References in Which This Strain Is Mentioned
	Scutellaria lateriflora	2	Hairy roots	2020	Hypocotyls of sterile-grown plants	Contains flavones (baicalin, baicalein, wogonin, wogonoside)	The content of flavones is 4.57 mg/g DW, that is lower than in all the species of hairy roots studied by us	Pharmacology, medicine	[136]
	Scutellaria pycnoclada	8	Hairy roots	2020	Hypocotyls of sterile-grown plants	In contrast to other lines of skullcaps, the ratio of main flavones is close to that of the roots of intact plants		Pharmacology, medicine	[136]
Linaceae	*Linum usitatissimum* L.	1	Hairy roots	1995	Cotyledons of sterile-grown seedlings	Cyanogenic glycosides, lignan		Pharmacology, medicine	
Rubiaceae	*Rubia tictorum* L.	1	Hairy roots	1991		Anthraquinones	The content of anthraquinones is 2.5% by dry weight. In the roots and rhizomes of an intact plant—5.2%	Pharmacology, medicine	[159]
Rutaceae	*Ruta graveolens* L.	1	Hairy roots	1991		Hypocotyle of a juvenile plant	Acridon alkaloids	Study of the spatial distribution of acridone alkaloids	[109]

Table 2. *Cont.*

Family	Species Name	Number of Lines	Type of the Culture	The Year Obtained	Origin of the Culture	Secondary Metabolites	Note	Possible Use	Literature References in Which This Strain Is Mentioned
	Ruta graveolens L.	8	Calli	1969, 1970, 1978, 1980, 1999	Stem of whole plants, hypocotyle of juvenile seedlings, roots of juvenile seedling	Acridonalkaloids		Study of the biosynthesis of acridone alkaloids	[160,161]
Zygophyllaceae	*Peganum harmala* L.	1	Hairy roots	1988	Stem of a juvenile plant	β-carbolinealkaloids (harmine, harmalol, harmaline), serotonin		Study of the distribution of secondary metabolites in plant roots. Pharmacology	[52]
	Peganum harmala L.	1	Calli	1988	Spontaneous callus formation on a juvenile plant stem	β-carboline alkaloids, serotonin			

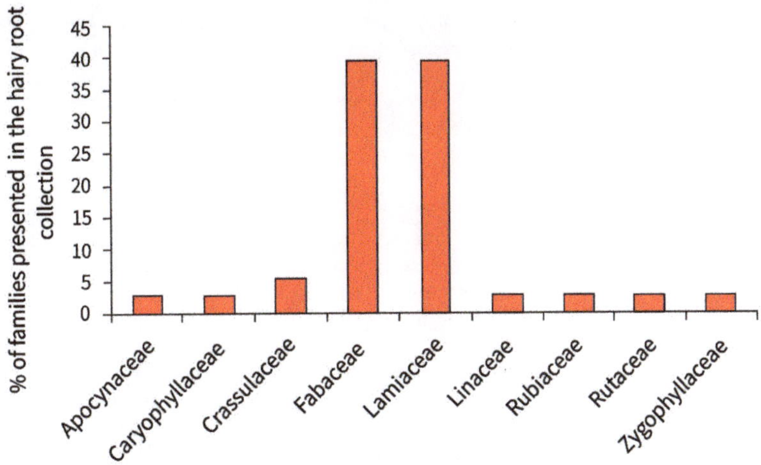

Figure 8. The percentage of plant families represented in the collection of hairy roots.

The collection contains 14 genera of plants of the *Fabaceae* family including both forage plants (*Lupinus polyphyllus* L., *Medicago sativa* L., and *Trifolium repens*) and medicinal plants (*Sophora korolkovii* Koehne, *Glycyrrhiza uralensis* L., *Ononis* sp., and *Hedysarum* sp.). The *Lamiaceae* family is only represented in the collection by the genus *Scutellaria*, which is widely known for its medicinal properties. The number of lines obtained from each species of the genus *Scutellaria* is from 1 to 8, so the total number of lines of hairy roots belonging to the *Lamiaceae* family is equal to the number of lines of the *Fabaceae* family. The remaining families are represented by single species, each of which is maintained by one line of hairy roots.

For comparative studies, undifferentiated callus cultures are maintained in the collection (Table 2), however, they grow much slower than hairy roots (Figure 9).

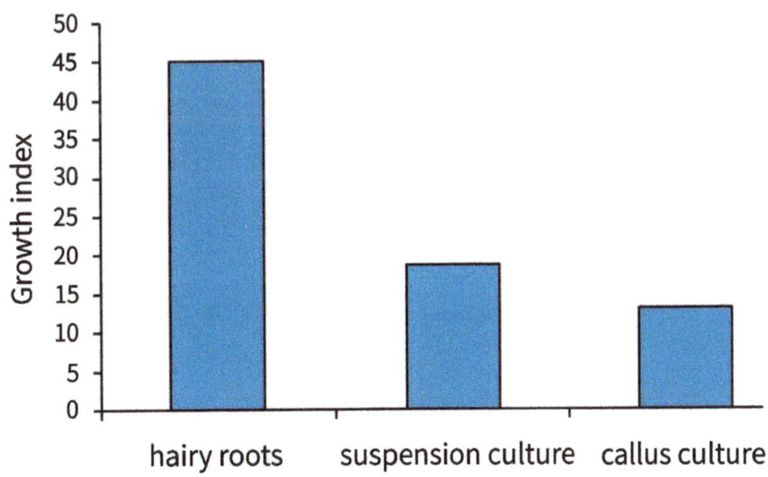

Figure 9. Growth index of the *Scutellaria baicalensis* Georgi in in vitro cultures.

The hairy roots are continuously cultivated in our collection. Surely, the permanent cultivation of hairy roots is laborious, but cryopreservation of hairy roots is not an easy task, requiring the development of an individual protocol for each line of hairy root cultures [26]. Nevertheless, due to the genetic stability of hairy roots, permanent transplants do not affect the level of synthesis of metabolites.

It should be noted that the presence of the collection makes it possible to conduct a wide range of studies that cannot be done with one or more lines at one's disposal (Table 2). The results obtained can be used in practical activities for the production of biologically active substances in the pharmaceutical industry. The use of hairy roots is becoming increasingly important due to the reduction of the areas of medicinal plants and the production of new strains of hairy roots.

4. Conclusions

Hairy roots are a unique in vitro system capable of rapid growth on hormone-free media and significant synthesis of secondary metabolites characteristic of both underground and aboveground parts of plants. Since they are differentiated structures, they can be used as a model to study the spatial distribution of secondary metabolites in plant roots. The collection of hairy roots can be not only a tool for conducting various fundamental and applied research but also a way to preserve rare and endangered species. It is also extremely interesting to reveal the features of the synthesis of secondary metabolites in differentiated and undifferentiated cultures obtained from them, as well as on media of different compositions. The latter can be of great practical importance. The prospects for the development of the collection of hairy roots include the expansion of the range of plant families, from which they will be obtained, including at the expense of "stable" monocotyledonous species.

Author Contributions: Conceptualization, A.Y.S. and M.V.M.; writing and editing, A.Y.S., A.I.S., E.A.G., D.V.T. and S.V.E. All authors have read and agreed to the published version of the manuscript.

Funding: This research was funded by the Ministry of Education and Science of the Russian Federation (Topic No122042600086-7).

Institutional Review Board Statement: Not applicable.

Informed Consent Statement: Not applicable.

Data Availability Statement: The data presented in this study are available on request from the corresponding author.

Conflicts of Interest: The authors declare no conflict of interest.

References

1. Chandran, H.; Meena, M.; Barupal, T.; Sharma, K. Plant tissue culture as a perpetual source for production of industrially important bioactive compounds. *Biotechnol. Rep.* **2020**, *26*, e00450. [CrossRef] [PubMed]
2. Nasri, H.; Baradaran, A.; Shirzad, H.; Kopaei, M.R. New concepts in nutraceuticals as alternative for pharmaceuticals. *Int. J. Prev. Med.* **2014**, *5*, 1487–1499. [PubMed]
3. Quaglio, D.; Talib, W.H.; Daoud, S.; Mahmod, A.I.; Hamed, R.A.; Awajan, D.; Feras Abuarab, S.; Odeh, L.H.; Khater, S.; Al Kury, L.T. Plants as a source of anticancer agents: From bench to bedside. *Molecules* **2022**, *27*, 4818. [CrossRef]
4. Ramachandra Rao, S.; Ravishankar, G.A. Plant cell cultures: Chemical factories of secondary metabolites. *Biotechnol. Adv.* **2002**, *20*, 101–153. [CrossRef]
5. Barrales-Cureño, H.J.; Valdivia, A.C.R.; Hernández, M.S. Increased production of taxoids in suspension cultures of Taxus globosa after elicitation. *Futur. Pharmacol.* **2022**, *2*, 45–54. [CrossRef]
6. Shen, F.; Long, D.; Yu, T.; Chen, X.; Liao, Y.; Wu, Y.; Lin, X. Vinblastine differs from Taxol as it inhibits the malignant phenotypes of NSCLC cells by increasingthe phosphorylation of Op18/stathmin. *Oncol. Rep.* **2017**, *37*, 2481–2489. [CrossRef]
7. Gallego-Jara, J.; Lozano-Terol, G.; Sola-Martínez, R.A.; Cánovas-Díaz, M.; de Diego Puente, T. A Compressive Review about Taxol®: History and Future Challenges. *Molecules* **2020**, *25*, 5986.
8. Hennessy, B.T.; Coleman, R.L.; Markman, M. Ovarian cancer. *Lancet* **2009**, *374*, 1371–1382. [CrossRef]
9. Wani, M.C.; Taylor, H.L.; Wall, M.E.; Coggon, P.; Mcphail, A.T. Plant Antitumor Agents.VI.The Isolation and Structure of Taxol, a Novel Antileukemic and Antitumor Agent from Taxus brevifolia2. *J. Am. Chem. Soc.* **1971**, *93*, 2325–2327. [CrossRef]
10. Weaver, B.A. How Taxol/paclitaxel kills cancer cells. *Mol. Biol. Cell* **2014**, *25*, 2677–2681. [CrossRef]
11. Gozes, I.; Harrington, C.; Lowery, L.A.; Coffey, E.T.; Varidaki, A.; Hong, Y. Repositioning Microtubule Stabilizing Drugs for Brain Disorders. *Front. Cell. Neurosci.* **2018**, *1*, 226. [CrossRef]

12. Zhang, B.; Maiti, A.; Shively, S.; Lakhani, F.; McDonald-Jones, G.; Bruce, J.; Lee, E.B.; Xie, S.X.; Joyce, S.; Li, C.; et al. Microtubule-binding drugs offset tau sequestration by stabilizing microtubules and reversing fast axonal transport deficits in a tauopathy model. *Proc. Natl. Acad. Sci. USA* **2004**, *102*, 227–231. [CrossRef]
13. Success Story: Taxol. Available online: https://dtp.cancer.gov/timeline/flash/success_stories/S2_taxol.htm (accessed on 7 November 2022).
14. Barrales-Cureño, H.J.; Reyes, C.R.; García, I.V.; Valdez, L.G.L.; De Jesús, A.G.; Ruíz, J.A.C.; Herrera, L.M.S.; Caballero, M.C.C.; Magallón, J.A.S.; Perez, J.E.; et al. Alkaloids of pharmacological importance in Catharanthus roseus. *Alkaloids-Their. Importance Nat. Hum. Life* **2019**, *1*, 18. [CrossRef]
15. Coderch, C.; Morreale, A.; Gago, F. Tubulin-based Structure-affinity Relationships for Antimitotic Vinca Alkaloids. *Anticancer. Agents Med. Chem.* **2012**, *12*, 219–225. [CrossRef]
16. Heilig, M.L. United States Patent Office. *ACM SIGGRAPH Comput. Graph.* **1994**, *28*, 131–134. [CrossRef]
17. Mehrotra, S.; Rahman, L.U.; Kukreja, A.K. An extensive case study of hairy-root cultures for enhanced secondary-metabolite production through metabolic-pathway engineering. *Biotechnol. Appl. Biochem.* **2010**, *56*, 161–172. [CrossRef]
18. Yue, W.; Ming, Q.L.; Lin, B.; Rahman, K.; Zheng, C.J.; Han, T.; Qin, L.P. Medicinal plant cell suspension cultures: Pharmaceutical applications and high-yielding strategies for the desired secondary metabolites. *Crit. Rev. Biotechnol.* **2016**, *36*, 215–232. [CrossRef]
19. Fazili, M.A.; Bashir, I.; Ahmad, M.; Yaqoob, U.; Geelani, S.N. In vitro strategies for the enhancement of secondary metabolite production in plants: A review. *Bull. Natl. Res. Cent.* **2022**, *46*, 35. [CrossRef]
20. Yazaki, K. Lithospermum erythrorhizon cell cultures: Present and future aspects. *Plant Biotechnol.* **2017**, *34*, 131. [CrossRef] [PubMed]
21. Alcalde, M.A.; Perez-Matas, E.; Escrich, A.; Cusido, R.M.; Palazon, J.; Bonfill, M. Biotic Elicitors in Adventitious and Hairy Root Cultures: A Review from 2010 to 2022. *Molecules* **2022**, *27*, 5253. [CrossRef]
22. Królicka, A.; Staniszewska, I.; Bielawski, K.; Maliński, E.; Szafranek, J.; Lojkowska, E. Establishment of hairy root cultures of Ammi majus. *Plant Sci.* **2001**, *160*, 259–264. [CrossRef]
23. Purwianingsih, W.; Hidayat, R.Y.; Rahmat, A. Increasing anthraquinone compounds on callus leaf *Morinda citrifolia* (L.) by elicitation method using chitosan shell of shrimps (*Penaeus monodon*). *J. Phys. Conf. Ser.* **2019**, *1280*, 022001. [CrossRef]
24. Yamada, Y.; Endo, T. Tropane alkaloid production in cultured cells of *Duboisia leichhardtii*. *Plant Cell Rep.* **1984**, *3*, 186–188. [CrossRef] [PubMed]
25. Hashimoto, T.; Yamada, Y. Scopolamine production in suspension cultures and redifferentiated roots of Hyoscyamus niger. *Planta Med.* **1983**, *47*, 195–199. [CrossRef] [PubMed]
26. Häkkinen, S.T.; Moyano, E.; Cusido, R.M.; Oksman-Caldentey, K.M. Exploring the Metabolic Stability of Engineered Hairy Roots after 16 Years Maintenance. *Front Plant Sci.* **2016**, *7*, 1486. [CrossRef] [PubMed]
27. Thi, L.; Tien, T. Root Cultures for Secondary Products. *Plant Roots* **2020**, 425–574. [CrossRef]
28. Khanam, M.N.; Anis, M.; Bin Javed, S.; Mottaghipisheh, J.; Csupor, D. Adventitious root culture-an alternative strategy for secondary metabolite production: A review. *Agronomy* **2022**, *12*, 1178. [CrossRef]
29. Praveen, N.; Manohar, S.H.; Naik, P.M.; Nayeem, A.; Jeong, J.H.; Murthy, H.N. Production of andrographolide from adventitious root cultures of Andrographis paniculata. *Curr. Sci.* **2009**, *2009*, 694–697.
30. Ho, T.T.; Lee, J.D.; Jeong, C.S.; Paek, K.Y.; Park, S.Y. Improvement of biosynthesis and accumulation of bioactive compounds by elicitation in adventitious root cultures of Polygonum multiflorum. *Appl. Microbiol. Biotechnol.* **2018**, *102*, 199–209. [CrossRef]
31. Shi, M.; Liao, P.; Nile, S.H.; Georgiev, M.I.; Kai, G. Biotechnological Exploration of Transformed Root Culture for Value-Added Products. *Trends Biotechnol.* **2021**, *39*, 137–149. [CrossRef]
32. Sharma, P.; Padh, H.; Shrivastava, N. Hairy root cultures: A suitable biological system for studying secondary metabolic pathways in plants. *Eng. Life Sci.* **2013**, *13*, 62–75. [CrossRef]
33. Srivastava, S.; Srivastava, A.K. Hairy root culture for mass-production of high-value secondary metabolites. *Crit. Rev. Biotechnol.* **2007**, *27*, 29–43. [CrossRef]
34. Gutierrez-Valdes, N.; Häkkinen, S.T.; Lemasson, C.; Guillet, M.; Oksman-Caldentey, K.M.; Ritala, A.; Cardon, F. Hairy Root Cultures—A Versatile Tool With Multiple Applications. *Front. Plant Sci.* **2020**, *11*, 33. [CrossRef]
35. Rekha, K.; Thiruvengadam, M. Secondary Metabolite Production in Transgenic Hairy Root Cultures of Cucurbits. *Transgenesis Second. Metab.* **2017**, *267*, 267–293. [CrossRef]
36. Giri, A.; Narasu, M.L. Transgenic hairy roots: Recent trends and applications. *Biotech. Adv.* **2000**, *18*, 1–22. [CrossRef]
37. Morey, K.J.; Peebles, C.A.M. Hairy roots: An untapped potential for production of plant products. *Front. Plant Sci.* **2022**, *13*, 2808. [CrossRef]
38. Sun, J.; Ma, L.; San, K.Y.; Peebles, C.A.M. Still stable after 11 years: A Catharanthus roseus Hairy root line maintains inducible expression of anthranilate synthase. *Biotechnol. Prog.* **2017**, *33*, 66–69. [CrossRef]
39. Smirnov, A.M. *Rost i Metabolizm Izolirovannyh Kornej v Steril'noj Kul'ture*; Ratner, E.I., Ed.; Izdatel'stvo "Nauka": Moscow, Russia, 1970.
40. Smith, E.F.; Brown, N.A.; Townsend, C.O. *Crown-Gall of Plants: Its Cause and Remedy*; US Government Printing Office: Washington, DC, USA, 1911; pp. 213, 1–215.
41. Chilton, M.D.; Tepfer, D.A.; Petit, A.; David, C.; Casse Delbart, F.; Tempé, J. *Agrobacterium rhizogenes* inserts T-DNA into the genomes of the host plant root cells. *Nature* **1982**, *295*, 432–434. [CrossRef]
42. Doran, P.M. *Hairy Roots: Culture and Applications*; Doran, P.M., Ed.; CRC Press: Sydney, Australia, 2020. ISBN 9781000725339.

43. Kuzovkina, I.N.; Schneider, B. Genetically transformed root cultures—Generation, properties and application in plant sciences. *Prog. Bot.* **2006**, *67*, 275–314. [CrossRef]
44. Sharma, S.; Singh, Y.; Verma, P.K.; Vakhlu, J. Establishment of *Agrobacterium rhizogenes*-mediated hairy root transformation of *Crocus sativus* L. *3 Biotech* **2021**, *11*, 1–8. [CrossRef]
45. Rahimi, K.; Haghbeen, K.; Marefatjo, J.; Jazii, F.R.; Sheikhani, R. Successful production of hairy root of *Valeriana sisymbriifolium* by *Agrobacterium rhizogenes*. *Biotechnology* **2008**, *7*, 200–204. [CrossRef]
46. Chandra, S.; Chandra, R. Engineering secondary metabolite production in hairy roots. *Phytochem. Rev.* **2011**, *10*, 371–395. [CrossRef]
47. Akutsu, M.; Ishizaki, T.; Sato, H. Transformation of the monocotyledonous Alstroemeria by *Agrobacterium tumefaciens*. *Plant Cell Rep.* **2004**, *22*, 561–568. [CrossRef] [PubMed]
48. Bathoju, G.; Rao, K.; Giri, A. Production of sapogenins (stigmasterol and hecogenin) from genetically transformed hairy root cultures of Chlorophytum borivilianum (Safed musli). *Plant Cell Tissue Organ Cult.* **2017**, *131*, 369–376. [CrossRef]
49. Xu, H.; Zhou, X.; Lu, J.; Wang, J.; Wang, X. Hairy roots induced by *Agrobacterium rhizogenes* and production of regenerative plants in hairy root cultures in maize. *Sci. China Ser. C Life Sci.* **2006**, *49*, 305–310. [CrossRef] [PubMed]
50. Sanford, J.C. Biolistic plant transformation. *Physiol. Plant.* **1990**, *79*, 206–209. [CrossRef]
51. Yasybaeva, G.; Vershinina, Z.; Kuluev, B.; Mikhaylova, E.; Baymiev, A.; Chemeris, A. Biolistic-mediated plasmid-free transformation for induction of hairy roots in tobacco plants. *Plant Root* **2017**, *11*, 33–39. [CrossRef]
52. Vogt, T.; Liu, Q.; John Loake, G.; Hou, W.; Shakya, P.; Franklin, G. A Perspective on Hypericum perforatum Genetic Transformation. *Front. Plant Sci.* **2016**, *1*, 879. [CrossRef]
53. Christou, P. Strategies for variety-independent genetic transformation of important cereals, legumes and woody species utilizing particle bombardment. *Euphytica* **1995**, *85*, 13–27. [CrossRef]
54. Barampuram, S.; Zhang, Z.J. Recent advances in plant transformation. *Methods Mol. Biol.* **2011**, *701*, 1–35. [CrossRef]
55. Barba-Espín, G.; Chen, S.T.; Agnolet, S.; Hegelund, J.N.; Stanstrup, J.; Christensen, J.H.; Müller, R.; Lütken, H. Ethephon-induced changes in antioxidants and phenolic compounds in anthocyanin-producing black carrot hairy root cultures. *J. Exp. Bot.* **2020**, *71*, 7030–7045. [CrossRef]
56. Wawrosch, C.; Zotchev, S.B. Production of bioactive plant secondary metabolites through in vitro technologies-status and outlook. *Appl. Microbiol. Biotechnol.* **2021**, *105*, 6649–6668. [CrossRef]
57. Massa, S.; Paolini, F.; Marino, C.; Franconi, R.; Venuti, A. Bioproduction of a therapeutic vaccine against human papillomavirus in tomato hairy root cultures. *Front. Plant Sci.* **2019**, *10*, 452. [CrossRef]
58. Skarjinskaia, M.; Ruby, K.; Araujo, A.; Taylor, K.; Gopalasamy-Raju, V.; Musiychuk, K.; Chichester, J.A.; Palmer, G.A.; de la Rosa, P.; Mett, V.; et al. Hairy Roots as a Vaccine Production and Delivery System. *Adv. Biochem. Eng. Biotechnol.* **2013**, *134*, 115–134. [CrossRef]
59. Donini, M.; Marusic, C. Hairy roots as bioreactors for the production of biopharmaceuticals. In *Hairy Roots: An Effective Tool of Plant Biotechnology*; Springer: Singapore, 2018; pp. 213–225. ISBN 9789811325625.
60. Zhang, N.; Wright, T.; Wang, X.; Karki, U.; Savary, B.J.; Xu, J. Engineering 'designer' glycomodules for boosting recombinant protein secretion in tobacco hairy root culture and studying hydroxyproline-O-glycosylation process in plants. *Plant Biotechnol. J.* **2019**, *17*, 1130–1141. [CrossRef]
61. Majumder, A.; Ray, S.; Jha, S. Hairy Roots and Phytoremediation. *Bioprocess. Plant Vitr. Syst.* **2016**, 1–24. [CrossRef]
62. Moola, A.K.; Balasubramanian, P.; Satish, L.; Shamili, S.; Ramesh, M.; Kumar, T.S.; Kumari, B.D.R.; Moola, A.K.; Kumar, T.S.; Kumari, B.D.R.; et al. Hairy Roots as a Source for Phytoremediation. In *Strategies and Tools for Pollutant Mitigation*; Springer: Cham, Switzerland, 2021; pp. 29–47. [CrossRef]
63. Bulgakov, V.P. Functions of rol genes in plant secondary metabolism. *Biotechnol. Adv.* **2008**, *26*, 318–324. [CrossRef]
64. Bulgakov, V.P.; Veremeichik, G.N.; Grigorchuk, V.P.; Rybin, V.G.; Shkryl, Y.N. The rolB gene activates secondary metabolism in Arabidopsis calli via selective activation of genes encoding MYB and bHLH transcription factors. *Plant Physiol. Biochem. PPB* **2016**, *102*, 70–79. [CrossRef]
65. Kim, Y.; Wyslouzil, B.E.; Weathers, P.J. Secondary metabolism of hairy root cultures in bioreactors. *Vitr. Cell. Dev. Biol.-Plant* **2002**, *38*, 1–10. [CrossRef]
66. Bakkali, A.T.; Jaziri, M.; Foriers, A.; Vander Heyden, Y.; Vanhaelen, M.; Homès, J. Lawsone accumulation in normal and transformed cultures of henna, Lawsonia inermis. *Plant Cell. Tissue Organ Cult.* **1997**, *51*, 83–87. [CrossRef]
67. Wallaart, T.E.; Pras, N.; Quax, W.J. Isolation and identification of dihydroartemisinic acid hydroperoxide from Artemisia annua: A novel biosynthetic precursor of artemisinin. *J. Nat. Prod.* **1999**, *62*, 1160–1162. [CrossRef] [PubMed]
68. Patra, N.; Srivastava, A.K. Artemisinin production by plant hairy root cultures in gas- and liquid-phase bioreactors. *Plant Cell Rep.* **2016**, *35*, 143–153. [CrossRef] [PubMed]
69. Bhaskar, R.; Xavier, L.S.E.; Udayakumaran, G. Biotic elicitors: A boon for the in-vitro production of plant secondary metabolites. *Plant Cell Tiss Organ Cult.* **2022**, *149*, 7–24. [CrossRef]
70. Cao, W.; Wang, Y.; Shi, M.; Hao, X.; Zhao, W.; Wang, Y.; Ren, J.; Kai, G. Transcription factor SmWRKY1 positively promotes the biosynthesis of tanshinones in *Salvia miltiorrhiza*. *Front. Plant Sci.* **2018**, *9*, 554. [CrossRef] [PubMed]
71. Deng, C.; Hao, X.; Shi, M.; Fu, R.; Wang, Y.; Zhang, Y.; Zhou, W.; Feng, Y.; Makunga, N.P.; Kai, G. Tanshinone production could be increased by the expression of SmWRKY2 in *Salvia miltiorrhiza* hairy roots. *Plant Sci.* **2019**, *284*, 1–8. [CrossRef]

72. Hao, X.; Pu, Z.; Cao, G.; You, D.; Zhou, Y.; Deng, C.; Shi, M.; Nile, S.H.; Wang, Y.; Zhou, W.; et al. Tanshinone and salvianolic acid biosynthesis are regulated by SmMYB98 in *Salvia miltiorrhiza* hairy roots. *J. Adv. Res.* **2020**, *23*, 1–12. [CrossRef]
73. Shi, M.; Luo, X.; Ju, G.; Yu, X.; Hao, X.; Huang, Q.; Xiao, J.; Cui, L.; Kai, G. Increased accumulation of the cardio-cerebrovascular disease treatment drug tanshinone in *Salvia miltiorrhiza* hairy roots by the enzymes 3-hydroxy-3-methylglutaryl CoA reductase and 1-deoxy-d-xylulose 5-phosphate reductoisomerase. *Funct. Integr. Genom.* **2014**, *14*, 603–615. [CrossRef]
74. Shi, M.; Luo, X.; Ju, G.; Li, L.; Huang, S.; Zhang, T.; Wang, H.; Kai, G. Enhanced Diterpene Tanshinone Accumulation and Bioactivity of Transgenic *Salvia miltiorrhiza* Hairy Roots by Pathway Engineering. *J. Agric. Food Chem.* **2016**, *64*, 2523–2530. [CrossRef]
75. Paul, P.; Singh, S.K.; Patra, B.; Sui, X.; Pattanaik, S.; Yuan, L. A differentially regulated AP2/ERF transcription factor gene cluster acts downstream of a MAP kinase cascade to modulate terpenoid indole alkaloid biosynthesis in Catharanthus roseus. *New Phytol.* **2017**, *213*, 1107–1123. [CrossRef]
76. Park, C.H.; Xu, H.; Yeo, H.J.; Park, Y.E.; Hwang, G.S.; Park, N., II; Park, S.U. Enhancement of the flavone contents of *Scutellaria baicalensis* hairy roots via metabolic engineering using maize Lc and Arabidopsis PAP1 transcription factors. *Metab. Eng.* **2021**, *64*, 64–73. [CrossRef]
77. Singh, S.; Pandey, P.; Akhtar, M.Q.; Negi, A.S.; Banerjee, S. A new synthetic biology approach for the production of curcumin and its glucoside in Atropa belladonna hairy roots. *J. Biotechnol.* **2021**, *328*, 23–33. [CrossRef]
78. Bonhomme, V.; Laurain-Mattar, D.; Lacoux, J.; Fliniaux, M.A.; Jacquin-Dubreuil, A. Tropane alkaloid production by hairy roots of *Atropa belladonna* obtained after transformation with *Agrobacterium rhizogenes* 15834 and *Agrobacterium tumefaciens* containing rol A, B, C genes only. *J. Biotechnol.* **2000**, *81*, 151–158. [CrossRef]
79. Jung, G.; Tepfer, D. Use of genetic transformation by the Ri T-DNA of *Agrobacterium rhizogenes* to stimulate biomass and tropane alkaloid production in *Atropa belladonna* and *Calystegia sepium* roots grown in vitro. *Plant Sci.* **1987**, *50*, 145–151. [CrossRef]
80. Mannan, A.; Shaheen, N.; Arshad, W.; Qureshi, R.A.; Zia, M.; Mirza, B. Hairy roots induction and artemisinin analysis in *Artemisia dubia* and *Artemisia indica*. *Afr. J. Biotechnol.* **2010**, *7*, 3288–3292. [CrossRef]
81. Drobot, K.O.; Matvieieva, N.A.; Ostapchuk, A.M.; Kharkhota, M.A.; Duplij, V.P. Study of artemisinin and sugar accumulation in Artemisia vulgaris and Artemisia dracunculus "hairy" root cultures. *Prep. Biochem. Biotechnol.* **2017**, *47*, 776–781. [CrossRef]
82. Mallol, A.; Cusidó, R.M.; Palazón, J.; Bonfill, M.; Morales, C.; Piñol, M.T. Ginsenoside production in different phenotypes of Panax ginseng transformed roots. *Phytochemistry* **2001**, *57*, 365–371. [CrossRef]
83. Yoshikawa, T.; Furuya, T. Saponin production by cultures of Panax ginseng transformed with *Agrobacterium rhizogenes*. *Plant Cell Rep.* **1987**, *6*, 449–453. [CrossRef]
84. Washida, D.; Shimomura, K.; Nakajima, Y.; Takido, M.; Kitanaka, S. Ginsenosides in hairy roots of a panax hybrid. *Phytochemistry* **1998**, *49*, 2331–2335. [CrossRef]
85. Miao, Y.; Hu, Y.; Yi, S.; Zhang, X.; Tan, N. Establishment of hairy root culture of Rubia yunnanensis Diels: Production of Rubiaceae-type cyclopeptides and quinones. *J. Biotechnol.* **2021**, *341*, 21–29. [CrossRef]
86. Kudale, S.; Ghatge, S.; Desai, N. Quantification of Phytochemicals in hairy root cultures of *Rubia cordifolia Linn*. *Int. J. Adv. Res.* **2015**, *3*, 903–913.
87. Murthy, H.N.; Dijkstra, C.; Anthony, P.; White, D.A.; Davey, M.R.; Power, J.B.; Hahn, E.J.; Paek, K.Y. Establishment of Withania somnifera Hairy Root Cultures for the Production of Withanolide A. *J. Integr. Plant Biol.* **2008**, *50*, 975–981. [CrossRef] [PubMed]
88. Berlin, J.; Rügenhagen, C.; Greidziak, N.; Kuzovkina, I.N.; Witte, L.; Wray, V. Biosynthesis of serotonin and β-carboline alkaloids in hairy root cultures of *Peganum harmala*. *Phytochemistry* **1993**, *33*, 593–597. [CrossRef]
89. Kuzovkina, I.N.; Gohar, A.; Alterman, I.E. Production of ß-Carboline Alkaloids in Transformed Root Cultures of *Peganum harmala* L. *Zeitschrift Fur Naturforsch.-Sect. C J. Biosci.* **1990**, *45*, 727–728. [CrossRef]
90. Sharifi-Rad, J.; Quispe, C.; Herrera-Bravo, J.; Semwal, P.; Painuli, S.; Özçelik, B.; Hacihasanoğlu, F.E.; Shaheen, S.; Sen, S.; Acharya, K.; et al. Peganum spp.: A Comprehensive Review on Bioactivities and Health-Enhancing Effects and Their Potential for the Formulation of Functional Foods and Pharmaceutical Drugs. *Oxid. Med. Cell. Longev.* **2021**, *2021*, 1–20. [CrossRef] [PubMed]
91. Cheng, X.-M.; Zhao, T.; Yang, T.; Wang, C.-H.; Bligh, S.W.A.; Wang, Z.-T.; Key, M. HPLC Fingerprints Combined with Principal Component Analysis, Hierarchical Cluster Analysis and Linear Discriminant Analysis for the Classifi cation and Diff erentiation of *Peganum* sp. Indigenous to China. *Phytochem. Anal* **2010**, *21*, 279–289. [CrossRef]
92. Zhang, Y.; Shi, X.; Xie, X.; Laster, K.V.; Pang, M.; Liu, K.; Hwang, J.; Kim, D.J. Harmaline isolated from *Peganum harmala* suppresses growth of esophageal squamous cell carcinoma through targeting mTOR. *Phyther. Res.* **2021**, *35*, 6377–6388. [CrossRef]
93. Zhu, Z.; Zhao, S.; Wang, C. Antibacterial, Antifungal, Antiviral, and Antiparasitic Activities of *Peganum harmala* and Its Ingredients: A Review. *Molecules* **2022**, *27*, 4161. [CrossRef]
94. Li, S.; Cheng, X.; Wang, C. A review on traditional uses, phytochemistry, pharmacology, pharmacokinetics and toxicology of the genus Peganum. *J. Ethnopharmacol.* **2017**, *203*, 127–162. [CrossRef]
95. Khadraoui, N.; Essid, R.; Jallouli, S.; Damergi, B.; Ben Takfa, I.; Abid, G.; Jedidi, I.; Bachali, A.; Ayed, A.; Limam, F.; et al. Antibacterial and antibiofilm activity of *Peganum harmala* seed extract against multidrug-resistant Pseudomonas aeruginosa pathogenic isolates and molecular mechanism of action. *Arch. Microbiol.* **2022**, *204*, 133. [CrossRef]
96. Abbas, M.W.; Hussain, M.; Qamar, M.; Ali, S.; Shafiq, Z.; Wilairatana, P.; Mubarak, M.S. Antioxidant and Anti-Inflammatory Effects of *Peganum harmala* Extracts: An In Vitro and In Vivo Study. *Molecules* **2021**, *26*, 6084. [CrossRef]

97. Sarpeleh, A.; Sharifi, K.; Sonbolkar, A. Evidence of antifungal activity of wild rue (*Peganum harmala* L.) on phytopathogenic fungi. *J. Plant Dis. Prot.* **2009**, *116*, 208–213. [CrossRef]
98. Samaha, A.A.; Fawaz, M.; Salami, A.; Baydoun, S.; Eid, A.H. Antihypertensive Indigenous Lebanese Plants: Ethnopharmacology and a Clinical Trial. *Biomolecules* **2019**, *9*, 292. [CrossRef]
99. Liu, W.; Cheng, X.; Wang, Y.; Li, S.; Zheng, T.; Gao, Y.; Wang, G.; Qi, S.; Wang, J.; Ni, J.; et al. In vivo evaluation of the antitussive, expectorant and bronchodilating effects of extract and fractions from aerial parts of *Peganum harmala* linn. *J. Ethnopharmacol.* **2015**, *162*, 79–86. [CrossRef]
100. Abd El Baky, H.H.; Ahemd, A.A.; Mekawi, E.M.; Ibrahem, E.A.; Shalapy, N.M. The anti-diabetic and anti-lipidemic effects of *Peganum harmala* seeds in diabetic rats. *Der Pharm. Lett.* **2016**, *8*, 1–10.
101. Astulla, A.; Zaima, K.; Matsuno, Y.; Hirasawa, Y.; Ekasari, W.; Widyawaruyanti, A.; Zaini, N.C.; Morita, H. Alkaloids from the seeds of *Peganum harmala* showing antiplasmodial and vasorelaxant activities. *J. Nat. Med.* **2008**, *62*, 470–472. [CrossRef]
102. Nasibova, T.; Garaev, E. Potential anti-Alzheimer alkaloids of *Peganum harmala*. *Alzheimer's Dement.* **2021**, *17*, e056722. [CrossRef]
103. Kempster, P.; Ma, A. Parkinson's disease, dopaminergic drugs and the plant world. *Front. Pharmacol.* **2022**, *13*, 3216. [CrossRef]
104. Malik, S.; Moraes, D.F.C.; do Amaral, F.M.M.; Ribeiro, M.N.S. *Ruta graveolens*: Phytochemistry, Pharmacology, and Biotechnology. *Ref. Ser. Phytochem.* **2017**, *4*, 177–204. [CrossRef]
105. Pollio, A.; De Natale, A.; Appetiti, E.; Aliotta, G.; Touwaide, A. Continuity and change in the Mediterranean medical tradition: *Ruta* spp. (rutaceae) in Hippocratic medicine and present practices. *J. Ethnopharmacol.* **2008**, *116*, 469–482. [CrossRef]
106. Jinous Asgarpanah Phytochemistry and pharmacological properties of *Ruta graveolens* L. *J. Med. Plants Res.* **2012**, *6*, 3942–3949. [CrossRef]
107. Shikov, A.N.; Pozharitskaya, O.N.; Makarov, V.G.; Wagner, H.; Verpoorte, R.; Heinrich, M. Medicinal Plants of the Russian Pharmacopoeia; their history and applications. *J. Ethnopharmacol.* **2014**, *154*, 481–536. [CrossRef] [PubMed]
108. Semerdjieva, I.B.; Burducea, M.; Astatkie, T.; Zheljazkov, V.D.; Dincheva, I. Essential oil composition of ruta graveolens l. fruits and hyssopus officinalis subsp. aristatus (godr.) nyman biomass as a function of hydrodistillation time. *Molecules* **2019**, *24*, 4047. [CrossRef] [PubMed]
109. Kuzovkina, I.; Al'terman, I.; Schneider, B. Specific accumulation and revised structures of acridone alkaloid glucosides in the tips of transformed roots of *Ruta graveolens*. *Phytochemistry* **2004**, *65*, 1095–1100. [CrossRef] [PubMed]
110. Kuzovkina, I.N.; Szarka, S.; Héthelyi, É.; Lemberkovics, E.; Szöke, É. Composition of essential oil in genetically transformed roots of *Ruta graveolens*. *Russ. J. Plant Physiol.* **2009**, *56*, 846–851. [CrossRef]
111. Ali, J.G.; Alborn, H.T.; Stelinski, L.L. Constitutive and induced subterranean plant volatiles attract both entomopathogenic and plant parasitic nematodes. *J. Ecol.* **2011**, *99*, 26–35. [CrossRef]
112. Falkenhagen, H.; Stockigt, J.; Kuzovkina, I.N.; Alterman, I.E.; Kolshorn, H. Indole alkaloids from "hairy roots" of *Rauwolfia serpentina*. *Can. J. Chem.* **1993**, *71*, 2201–2203. [CrossRef]
113. Lobay, D. Rauwolfia in the Treatment of Hypertension. *Integr. Med. A Clin. J.* **2015**, *14*, 40.
114. Sourabh, P. Ethnomedicinal Uses and Cultivation of *Rauvolfia serpentina*. *Recent Adv. Med. Plants Their Cultiv.* **2012**, *40*, 153–159.
115. Arjariya, A.; Chaurasia, K. Some Medicinal Plants among the Tribes of Chhatarpur District (M.P.) India. *Ecoprint An Int. J. Ecol.* **1970**, *16*, 43–50. [CrossRef]
116. Alshahrani, M.Y.; Rafi, Z.; Alabdallah, N.M.; Shoaib, A.; Ahmad, I.; Asiri, M.; Zaman, G.S.; Wahab, S.; Saeed, M.; Khan, S. A Comparative Antibacterial, Antioxidant, and Antineoplastic Potential of *Rauwolfia serpentina* (L.) Leaf Extract with Its Biologically Synthesized Gold Nanoparticles (R-AuNPs). *Plants* **2021**, *10*, 2278. [CrossRef]
117. Singh, H.K.; Charan, A.A.; Charan, A.I.; Prasad, S.M. Antifungal and antibacterial activity of methanolic, ethanolic and acetonic leaf extracts of sarpagandha (*Rauwolfia serpentina*). *J. Pharmacogn. Phytochem.* **2017**, *6*, 152–156.
118. Rao, B.G.; Rao, P.U.; Rao, E.S.; Rao, T.M.; Praneeth, D.V.S. Evaluation of in-vitro antibacterial activity and anti-inflammatory activity for different extracts of *Rauvolfia tetraphylla* L. root bark. *Asian Pac. J. Trop. Biomed.* **2012**, *2*, 818–821. [CrossRef]
119. Ezeigbo, I.I.; Ezeja, M.I.; Madubuike, K.G.; Ifenkwe, D.C.; Ukweni, I.A.; Udeh, N.E.; Akomas, S.C. Antidiarrhoeal activity of leaf methanolic extract of *Rauwolfia serpentina*. *Asian Pac. J. Trop. Biomed.* **2012**, *2*, 430. [CrossRef]
120. Weerakoon, S.W.; Arambewela, L.S.R.; Premakumara, G.A.S.; Ratnasooriya, W.D. Sedative activity of the crude extract of *Rauvolfia densiflora*. *Pharm. Biol.* **1998**, *36*, 360–361. [CrossRef]
121. Gupta, A.K.; Chitme, H.; Dass, S.K.; Misra, N. Hepatoprotective Activity of *Rauwolfia serpentina* Rhizome in Paracetamol Intoxicated Rats. *J. of Pharmacol. Toxicol.* **2006**, *1*, 82–88. [CrossRef]
122. Kaur, J.; Gulati, S. Therapeutic potential of *Rauwolfia serpentina*. *Indian J. Adv.* **2017**, *2*, 99–104.
123. Eltamany, E.E.; Nafie, M.S.; Khodeer, D.M.; El-Tanahy, A.H.H.; Abdel-Kader, M.S.; Badr, J.M.; Abdelhameed, R.F.A. *Rubia tinctorum* root extracts: Chemical profile and management of type II diabetes mellitus. *RSC Adv.* **2020**, *10*, 24159–24168. [CrossRef]
124. Taha, K.; Abu, M. A Natural Anthraquinone Plants with Multi-Pharmacological Activities. *Texas J. Med. Sci.* 2022; 10, 23–32.
125. Kalyoncu, F.; Cetin, B.; Saglam, H. Antimicrobial activity of common madder (*Rubia tinctorum* L.). *Phyther. Res.* **2006**, *20*, 490–492. [CrossRef]
126. Wang, W.; Zhang, J.; Qi, W.; Su, R.; He, Z.; Peng, X. Alizarin and Purpurin from *Rubia tinctorum* L. Suppress Insulin Fibrillation and Reduce the Amyloid-Induced Cytotoxicity. *ACS Chem. Neurosci.* **2021**, *12*, 2182–2193. [CrossRef]

127. Ahmed, H.E.; Tahoun, I.F.; Elkholy, I.; Shehata, A.B.; Ziddan, Y. Identification of natural dyes in rare Coptic textile using HPLC- DAD and mass spectroscopy in museum of Faculty of Arts, Alexandria University, Egypt. *Dye. Pigment.* **2017**, *145*, 486–492. [CrossRef]
128. Karapanagiotis, I.; Abdel-kareem, O.; Kamaterou, P.; Mantzouris, D. Identification of dyes in coptic textiles from the museum of faculty of archaeology, cairo university. *Heritage* **2021**, *4*, 3147–3156. [CrossRef]
129. Palagina, M.V.; Abramova, G.A. Solodka ural'skaya i ee ispol'zovanie v pishchevoj i farmacevticheskoj promyshlennosti. *Nov. v Pishchevyh Tekhnologiyah* **2005**, *1*, 77–87.
130. Zhu, L.; Xie, S.; Geng, Z.; Yang, X.; Zhang, Q. Evaluating the Potential of *Glycyrrhiza uralensis* (Licorice) in Treating Alcoholic Liver Injury: A Network Pharmacology and Molecular Docking Analysis Approach. *Processes* **2022**, *10*, 1808. [CrossRef]
131. Chen, H.; Zhang, X.; Feng, Y.; Rui, W.; Shi, Z.; Wu, L. Bioactive components of *Glycyrrhiza uralensis* mediate drug functions and properties through regulation of CYP450 enzymes. *Mol. Med. Rep.* **2014**, *10*, 1355–1362. [CrossRef] [PubMed]
132. Zhang, Q.; Ye, M. Chemical analysis of the Chinese herbal medicine Gan-Cao (licorice). *J. Chromatogr. A* **2009**, *1216*, 1954–1969. [CrossRef]
133. Deutch, M.R.; Grimm, D.; Wehland, M.; Infanger, M.; Krüger, M. Bioactive Candy: Effects of Licorice on the Cardiovascular System. *Foods* **2019**, *8*, 495. [CrossRef]
134. Kuzovkina, I.N.; Guseva, A.V.; Kovács, D.; Szöke, É.; Vdovitchenko, M.Y. Flavones in genetically transformed *Scutellaria baicalensis* roots and induction of their synthesis by elicitation with methyl jasmonate. *Russ. J. Plant Physiol.* **2005**, *52*, 77–82. [CrossRef]
135. Olina, A.V.; Solovyova, A.I.; Solovchenko, A.E.; Orlova, A.V.; Stepanova, A.Y. Physiologically active flavones contentin *Scutellaria baicalensis* georgiinvitro cultures. *Biotekhnologiya* **2017**, *33*, 29–37. [CrossRef]
136. Stepanova, A.Y.; Solov'eva, A.I.; Malunova, M.V.; Salamaikina, S.A.; Panov, Y.M.; Lelishentsev, A.A. Hairy roots scutellaria spp. (lamiaceae) as promising producers of antiviral flavones. *Molecules* **2021**, *26*, 3927. [CrossRef]
137. Solov'eva, A.I.; Evsyukov, S.V.; Sidorov, R.A.; Stepanova, A.Y. Correlation of endogenous β-glucuronidase activity with differentiation of in vitro cultures of *Scutellaria baicalensis*. *Acta Physiol. Plant.* **2020**, *42*, 106. [CrossRef]
138. Dikaya, V.S.; Solovyeva, A.I.; Sidorov, R.A.; Solovyev, P.A.; Stepanova, A.Y. The Relationship Between Endogenous β-Glucuronidase Activity and Biologically Active Flavones-Aglycone Contents in Hairy Roots of Baikal Skullcap. *Chem. Biodivers.* **2018**, *15*, e1700409. [CrossRef]
139. Elkin, Y.N.; Kulesh, N.I.; Stepanova, A.Y.; Solovieva, A.I.; Kargin, V.M.; Manyakhin, A.Y. Methylated flavones of the hairy root culture *Scutellaria baicalensis*. *J. Plant Physiol.* **2018**, *231*, 277–280. [CrossRef]
140. Kuzovkina, I.N.; Prokof'eva, M.Y.; Umralina, A.R.; Chernysheva, T.P. Morphological and biochemical characteristics of genetically transformed roots of *Scutellaria andrachnoides*. *Russ. J. Plant Physiol.* **2014**, *61*, 697–706. [CrossRef]
141. Li-Weber, M. New therapeutic aspects of flavones: The anticancer properties of Scutellaria and its main active constituents Wogonin, Baicalein and Baicalin. *Cancer Treat. Rev.* **2009**, *35*, 57–68. [CrossRef]
142. Sharifi-Rad, J.; Herrera-Bravo, J.; Salazar, L.A.; Shaheen, S.; Abdulmajid Ayatollahi, S.; Kobarfard, F.; Imran, M.; Imran, A.; Custódio, L.; Dolores López, M.; et al. The Therapeutic Potential of Wogonin Observed in Preclinical Studies. *Evid.-Based Complement. Altern. Med.* **2021**, *2021*, 9935451. [CrossRef]
143. Lin, C.C.; Lin, J.J.; Wu, P.P.; Lu, C.C.; Chiang, J.H.; Kuo, C.L.; Ji, B.C.; Lee, M.H.; Huang, A.C.; Chung, J.G. Wogonin, a natural and biologically-active flavonoid, influences a murine WEHI-3 leukemia model in vivo through enhancing populations of T-And B-cells. *Vivo* **2013**, *27*, 733–738.
144. Qi, Q.; Peng, J.; Liu, W.; You, Q.; Yang, Y.; Lu, N.; Wang, G.; Guo, Q. Toxicological studies of wogonin in experimental animals. *Phyther. Res.* **2009**, *23*, 417–422. [CrossRef]
145. Huang, D.S.; Yu, Y.C.; Wu, C.H.; Lin, J.Y. Protective Effects of Wogonin against Alzheimer's Disease by Inhibition of Amyloidogenic Pathway. *Evid. Based. Complement. Alternat. Med.* **2017**, *2017*, 3545169. [CrossRef]
146. Tronina, T.; Mrozowska, M.; Bartmańska, A.; Popłoński, J.; Sordon, S.; Huszcza, E. Simple and rapid method for wogonin preparation and its biotransformation. *Int. J. Mol. Sci.* **2021**, *22*, 8973. [CrossRef]
147. Tong, T.; Wu, Y.Q.; Ni, W.J.; Shen, A.Z.; Liu, S. The potential insights of Traditional Chinese Medicine on treatment of COVID-19. *Chin. Med.* **2020**, *15*, 51. [CrossRef]
148. Huang, Y.F.; Bai, C.; He, F.; Xie, Y.; Zhou, H. Review on the potential action mechanisms of Chinese medicines in treating Coronavirus Disease 2019 (COVID-19). *Pharmacol. Res.* **2020**, *158*, 104939. [CrossRef] [PubMed]
149. Ku, S.; Zheng, H.; Soo Park, M.; Eog Ji, G. Optimization of β-Glucuronidase Activity from *Lactobacillus delbrueckii* Rh2 and Its Use for Biotransformation of Baicalin and Wogonoside. *J. Korean Soc. Appl. Biol. Chem.* **2011**, *54*, 275–280. [CrossRef]
150. Levy, G.A. Baicalinase, a plant beta-glucuronidase. *Biochem. J.* **1954**, *58*, 462–469. [CrossRef] [PubMed]
151. Ikegami, F.; Matsunae, K.; Hisamitsu, M.; Murakoshi, I.; Kurihara, T.; Yamamoto, T. Purification and properties of a plant beta-D-glucuronidase form Scutellaria root. *Biol. Pharm. Bull.* **1995**, *18*, 1531–1534. [CrossRef] [PubMed]
152. Yoshikawa, M.; Shimada, H.; Shimoda, H.; Matsuda, H.; Yamahara, J.; Murakami, N. Rhodiocyanosides A and B, New Antiallergic Cyanoglycosides from Chinese Natural Medicine "Si Lie Hong Jing Tian", The Underground Part of Rhodiola Quadrifida (Pall.) Fisch. Et Mey. *Chem. Pharm. Bull.* **1995**, *43*, 1245–1247. [CrossRef]
153. Yoshikawa, M.; Shimada, H.; Shimoda, H.; Murakami, N.; Yamahara, J.; Matsuda, H. Bioactive constituents of Chinese natural medicines. II. Rhodiolae radix. (1). Chemical structures and antiallergic activity of rhodiocyanosides A and B from the underground part of *Rhodiola quadrifida* (Pall.) Fisch. et Mey. (Crassulaceae). *Chem. Pharm. Bull.* **1996**, *44*, 2086–2091. [CrossRef]

154. Wiedenfeld, H.; Dumaa, M.; Malinowski, M.; Furmanowa, M.; Narantuya, S. Erratum: Phytochemical and analytical studies of extracts from *Rhodiola rosea* and *Rhodiola quadrifida*. *Pharmazie* **2007**, *62*, 308–311.
155. You, J.; Qin, X.; Ranjitkar, S.; Lougheed, S.C.; Wang, M.; Zhou, W.; Ouyang, D.; Zhou, Y.; Xu, J.; Zhang, W.; et al. Response to climate change of montane herbaceous plants in the genus Rhodiola predicted by ecological niche modelling. *Sci. Rep.* **2018**, *8*, 5879. [CrossRef]
156. Stepanova, A.; Malunova, M.; Salamaikina, S.; Selimov, R.; Solov'eva, A. Establishment of *Rhodiola quadrifida* Hairy Roots and Callus Culture to Produce Bioactive Compounds. *Phyton (B. Aires).* **2021**, *90*, 543–552. [CrossRef]
157. Vdovitchenko, M.Y.; Kuzovkina, I.N.; Paetz, C.; Schneider, B. Formation of phenolic compounds in the roots of *Hedysarum theinum* cultured in vitro. *Russ. J. Plant Physiol.* **2007**, *54*, 536–544. [CrossRef]
158. Kuzovkina, I.N.; Guseva, A.V.; Alterman, I.E.; Karnachuk, R.A. Flavonoid production in transformed scutellaria baicalensis roots and ways of its regulation. *Russ. J. Plant Physiol.* **2001**, *48*, 448–452. [CrossRef]
159. Bányai, P.; Kuzovkina, I.N.; Kursinszki, L.; Szoke, É. HPLC analysis of alizarin and purpurin produced by *Rubia tinctorum* L. hairy root cultures. *Chromatographia* **2006**, *63*, S111–S114. [CrossRef]
160. Baumert, A.; Gröger, D.; Kuzovkina, I.N.; Reisch, J. Secondary metabolites produced by callus cultures of various Ruta species. *Plant Cell Tissue Organ Cult.* **1992**, *28*, 159–162. [CrossRef]
161. Baumert, A.; Kuzovkina, I.N.; Krauss, G.; Hieke, M.; Gröger, D. Biosynthesis of rutacridone in tissue cultures of *Ruta graveolens* L. *Plant Cell Rep.* **1982**, *1*, 168–171. [CrossRef]

molecules

Article

The Chemical Composition and Antimitotic, Antioxidant, Antibacterial and Cytotoxic Properties of the Defensive Gland Extract of the Beetle, *Luprops tristis* Fabricius

Ovungal Sabira [1], Attuvalappil Ramdas Vignesh [1], Anthyalam Parambil Ajaykumar [1,*], Sudhir Rama Varma [2], Kodangattil Narayanan Jayaraj [3,*], Merin Sebastin [1], Kalleringal Nikhila [1], Annet Babu [1], Vazhanthodi Abdul Rasheed [1], Valiyaparambil Sivadasan Binitha [4], Zeena koldath Vasu [1] and Madathilpadi Subrahmanian Sujith [1]

1 Division of Biomaterial Sciences, Department of Zoology, Sree Neelakanta Government Sanskrit College, Pattambi 679306, India
2 Clinical Sciences Department, Centre for Medical and Bio-Allied Health Sciences Research, Ajman University, Ajman P.O. Box 346, United Arab Emirates
3 Basic Sciences Department, Centre for Medical and Bio-Allied Health Sciences Research, Ajman University, Ajman P.O. Box 346, United Arab Emirates
4 Department of Zoology, Sree Narayana College, Nattika, Thrissur 680566, India
* Correspondence: ajaykumar@sngscollege.org (A.P.A.); j.narayanan@ajman.ac.ae (K.N.J.)

Citation: Sabira, O.; Vignesh, A.R.; Ajaykumar, A.P.; Varma, S.R.; Jayaraj, K.N.; Sebastin, M.; Nikhila, K.; Babu, A.; Rasheed, V.A.; Binitha, V.S.; et al. The Chemical Composition and Antimitotic, Antioxidant, Antibacterial and Cytotoxic Properties of the Defensive Gland Extract of the Beetle, *Luprops tristis* Fabricius. *Molecules* **2022**, *27*, 7476. https://doi.org/10.3390/molecules27217476

Academic Editors: Arunaksharan Narayanankutty, Ademola C. Famurewa and Eliza Oprea

Received: 2 September 2022
Accepted: 28 October 2022
Published: 2 November 2022

Publisher's Note: MDPI stays neutral with regard to jurisdictional claims in published maps and institutional affiliations.

Copyright: © 2022 by the authors. Licensee MDPI, Basel, Switzerland. This article is an open access article distributed under the terms and conditions of the Creative Commons Attribution (CC BY) license (https://creativecommons.org/licenses/by/4.0/).

Abstract: The unpredictable invasion of the Mupli beetle, *Luprops tristis* Fabricius (Coleoptera: Tenebrionidae), makes areas uninhabitable to humans. These beetles produce a strong-smelling, irritating secretion as a defence mechanism, which causes blisters on contact with human skin. In the current study, gas chromatography high-resolution mass spectrometry (GC-HRMS) analysis of the defensive gland extract of the Mupli beetle revealed the presence of compounds such as 2,3,dimethyl-1,4-benzoquinone, 1,3-dihydroxy-2-methylbenzene, 2,5-dimethyl hydroquinone, tetracosane, oleic acid, hexacosane, pentacosane, 7-hexadecenal and tert-hexadecanethiol. The defensive gland extracts showed considerable antibacterial activity on Gram-negative and Gram-positive bacteria in an agar diffusion assay. The chromosomal aberration analysis using root tips of *Allium cepa* L. exposed to the defensive secretion showed chromosomal aberrations such as disturbed metaphase, sticky chromosomes and chromosomal breakage. The antioxidant activity of the extract was determined using a radical scavenging (DPPH) assay. A cytotoxic assay of the defensive gland extract against Dalton's lymphoma ascites (DLA) cell line showed anticancer properties. In the present study, the defensive gland extract of the Mupli beetle, *L. tristis*, which is generally perceived as a nuisance insect to humans, was found to have beneficial biological activities.

Keywords: *Luprops tristis*; Mupli beetle; GC-HRMS analysis; defensive gland; Coleoptera

1. Introduction

Coleoptera is the largest order of insects, and Tenebrionidae is a prominent family within this order. Most species of Tenebrionidae are found in rotten wood, under logs and the bark of old trees. Both the adults and larvae feed on plant materials, decaying vegetation, etc. [1]. Based on the structure of mouth parts, male genitalia and the occurrence of a defensive gland, two broad evolutionary lineages of tenebrionid beetles are recognized; in these, tentyrioids lack defensive glands, and the tenebrionoid lineage possesses abdominal defensive glands. Many Coleopteran families possess abdominal defensive glands, but a comparative investigation was performed only in Carabidae and Tenebrionidae [2]. Comparative investigations of the compounds in defensive gland extract reveal that it has systematic value without focusing on its biological applications [3]. Gas–liquid chromatography of the chemical constituents of the defensive gland secretion of different species of

Tenebrionidae indicates the presence of toluquinone, ethylquinone and benzoquinone [4]. In addition to quinones, previously unidentified 4-methyl ketones and unsaturated ketones are also identified in the defensive secretion [5].

Despite the fact that many studies on the biochemical components of the defensive gland secretions of beetles have been carried out in various countries [6], studies on Indian beetles are comparatively rare [7]. There are no available data on the biomolecules present in the defensive secretion of the members of the Coleopteran genus, *Luprops*. The experimental organism *L. tristis* is a darkling beetle, which produces an odoriferous secretion when it gets disturbed that causes skin blisters in humans. The defensive glands are invaginations of the intersegmental membrane between the seventh and eighth sternites of the beetle, which open backward and everted on pressing the abdomen. When the beetle is disturbed, this gland is ruptured by rubbing with the hind tarsus for the release of the secretion as a part of a defence mechanism against predators [8]. Biomolecules of both plant and animal origin with antimitotic, antibacterial, cytotoxic and antioxidant properties are demonstrated to have important therapeutic uses. The biologically significant metabolites in animals have attracted much attention from the scientific community. The main objective of the current study is to identify the chemical components of the defensive secretion of the Mupli beetle, *L. tristis*. In addition, we have analysed bioassays to examine the antimitotic, antibacterial, free radical scavenging and cytotoxic properties of the defensive glandular extract of the beetle, *L. tristis*.

2. Results

2.1. GC-HRMS Analysis

The adult beetle possesses two tiny defensive glands that measure about 0.8–0.9 mm in size (Figure 1). The chemical composition of the defensive gland secretion of *L. tristis* was analysed using the gas chromatography high-resolution mass spectrometry (GC-HRMS) technique. The GC-HRMS data are presented in Figure 2. The results show that the defensive secretion of *L. tristis* consists of 2,3-dimethyl-1,4-benzoquinone, 1,3-dihydroxy-2-methylbenzene, 2,5-dimethylhydroquinone, tetracosane, oleic acid, hexacosane, pentacosane, 7-hexadecenal and tert-hexadecanethiol (Figure 3 and Table 1).

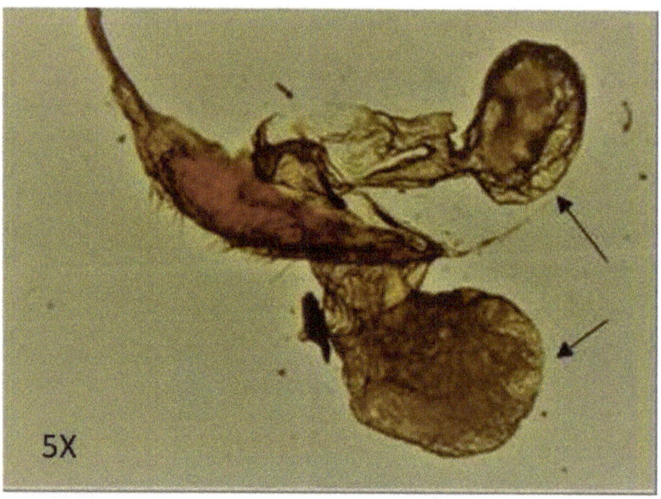

Figure 1. The defensive gland of the Mupli beetle, *Luprops tristis*. Arrows indicate the reservoir region of the gland.

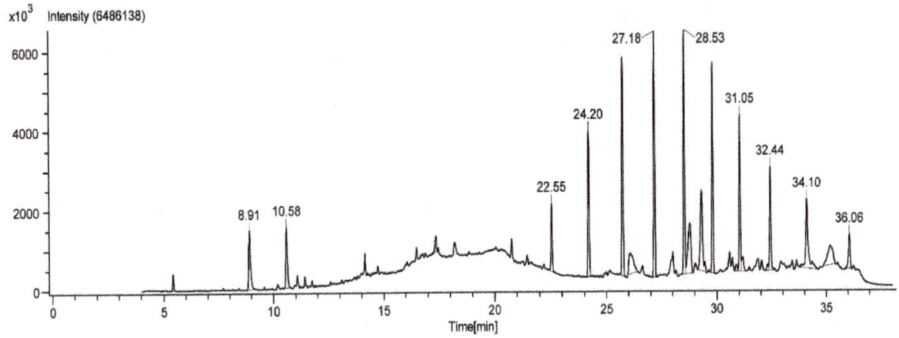

Figure 2. Gas chromatography high-resolution mass spectrometry (GC-HRMS) chromatogram of the defensive gland extracts of the beetle, *L. tristis*.

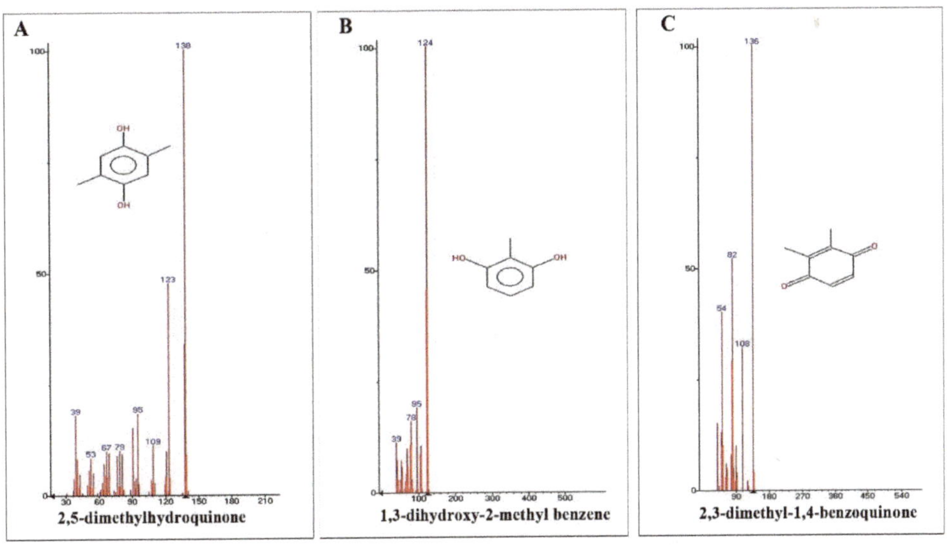

Figure 3. GC-HRMS analysis of identified quinone and phenolic compounds in the defensive gland extract of *L. tristis*. Data show the presence of 2,5-dimethylhydroquinone (**A**), 1,3-dihydroxy-2-methylbenzene (**B**) and 2,3-dimethyl-1,4-benzoquinone (**C**).

Table 1. Compounds in the defensive gland extract of *L. tristis*.

Sl. No	Compound Name	Molecular Formula	Retention Time (min)
1	2,3-dimethyl-1,4-benzoquinone	$C_8H_8O_2$	5.43
2	1,3-dihydroxy-2-methylbenzene	$C_7H_8O_2$	8.91
3	2,5-dimethyl hydroquinone	$C_8H_{10}O_2$	10.58
4	tetracosane	$C_{24}H_{50}$	25.74
5	oleic acid	$C_{18}H_{34}O_2$	26.05
6	hexacosane	$C_{26}H_{54}$	29.81
7	pentacosane	$C_{25}H_{52}$	31.05
8	7-hexadecenal	$C_{16}H_{30}O$	32.31
9	tert-hexadecanethiol	$C_{16}H_{34}S$	35.18

2.2. *Antimitotic Activity*

Various chromosomal aberrations were observed in the mitotic chromosomes when different concentrations of defensive secretion of *L. tristis* were applied to growing root meristem cells of *A. cepa* (Figure 4, Table 2). The frequency of chromosomal aberrations gradually increased in the experimental group along with an increasing concentration of the defensive gland extract (Table 2). The defensive extract hindered the normal mitotic cell division process. The aberrations are in the form of disturbed metaphase, sticky chromosomes and chromosomal breakage. However, the control group showed normal mitotic cell division without any aberrations.

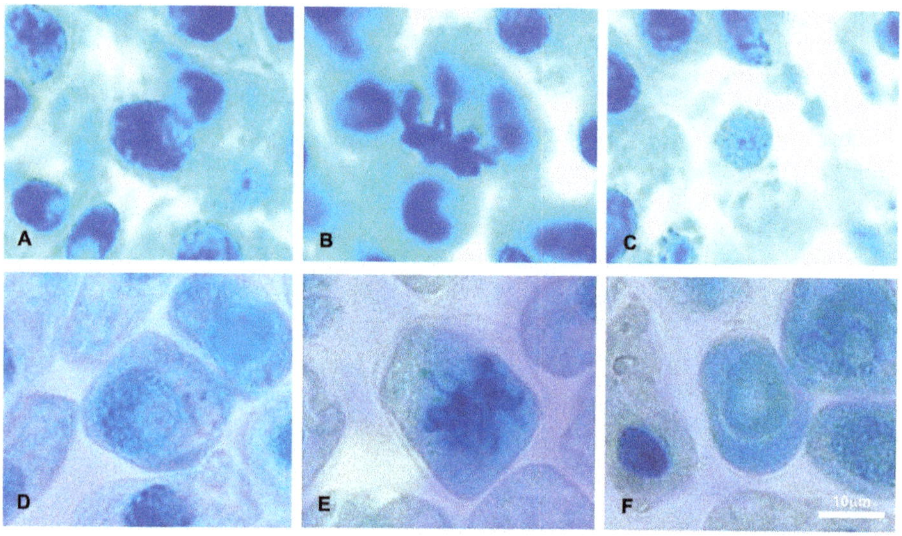

Figure 4. The effect of the defensive gland extract on dividing *A. cepa* cells. (**A**) Normal prophase, (**B**) normal metaphase, (**C**) normal nuclei, (**D**) disintegrated prophase, (**E**) abnormal metaphase and (**F**) disintegrated nuclei.

Table 2. Chromosome aberrations observed in the root meristem cells of *A. cepa* treated with different concentrations of the defensive gland extract of *L. tristis*.

Treatment	Concentration (μL)	Mitotic Index (% ± SD)	Chromosomal Aberration (% ± SD)
Control (distilled water)	100	11.29 ± 0.32	NIL
	200	12.65 ± 0.25	NIL
	300	14.44 ± 0.37	NIL
	400	14.55 ± 0.32	NIL
	500	15.58 ± 0.30	NIL
Experiment (defensive gland extract)	100	9.50 ± 0.50	11.17 ± 0.29
	200	6.57 ± 0.27	13.61 ± 0.27
	300	5.64 ± 0.24	14.46 ± 0.42
	400	3.57 ± 0.36	15.35 ± 0.36
	500	3.33 ± 0.33	16.33 ± 0.33

2.3. *Antioxidant Activity*

The DPPH, ABTS radical scavenging, peroxide scavenging and ferric reducing antioxidant power assays were used to investigate the antiradical scavenging efficiency of the defensive secretion of *L. tristis*. The half-maximal inhibitory concentration (IC50) for the defensive gland extract was 108.3 ± 2.1 μg/mL, 95.6 ± 1.5 μg/mL, 69.4 ± 3.5 μg/mL and 34.1 ± 1.0 μg/mL for the DPPH, ABTS radical scavenging, peroxide scavenging and iron reducing antioxidant capacity assays, respectively (Table 3).

Table 3. The IC_{50} values of antioxidant activities exhibited by the defensive secretion of *L. tristis*.

Sl. No	Assay	IC_{50} Value (μg/mL)
1	DPPH radical scavenging	108.3 ± 2.1
2	ABTS radical scavenging	95.6 ± 1.5
3	H_2O_2 scavenging	69.4 ± 3.5
4	Ferric reducing antioxidant power	34.1 ± 1.0

2.4. Antibacterial Activity

In vitro antibacterial activity of the defensive gland secretion was carried out following the agar disc diffusion assay method. Two different bacterial strains were used for the analysis. The results of the disc diffusion assay are presented in Figure 5. The data show that the defensive secretion of *L. tristis* is a very effective antimicrobial agent against the bacteria tested. The zone of inhibition against the bacteria *Staphylococcus aureus* and *Escherichia coli* is found to be 11 mm and 9 mm, respectively (Figure 5).

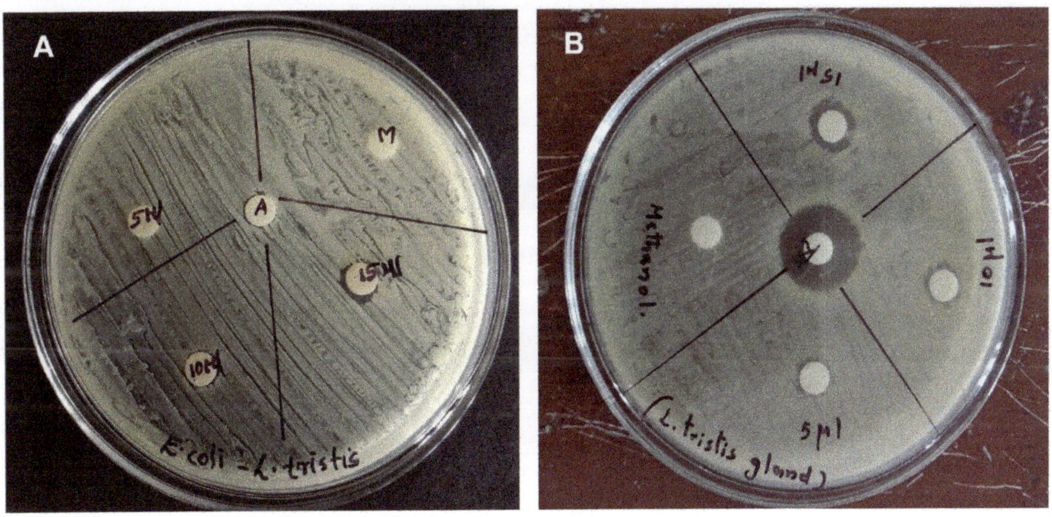

Figure 5. In vitro antibacterial analysis of *L. tristis* defensive secretion on *E. coli* (**A**) and *S. aureus* (**B**).

2.5. Cytotoxicity Assay

The anticancer property of the defensive gland extract of the *L. tristis* beetle was estimated, and the result of the analysis is presented in Figure 6. A cytotoxicity assay using Dalton's lymphoma ascites (DLA) cells revealed the anticancer property of the defensive secretion of *L. tristis*. A gradual increase in cytotoxicity was observed with an increase in the concentration (10 μL–200 μL) of the extract.

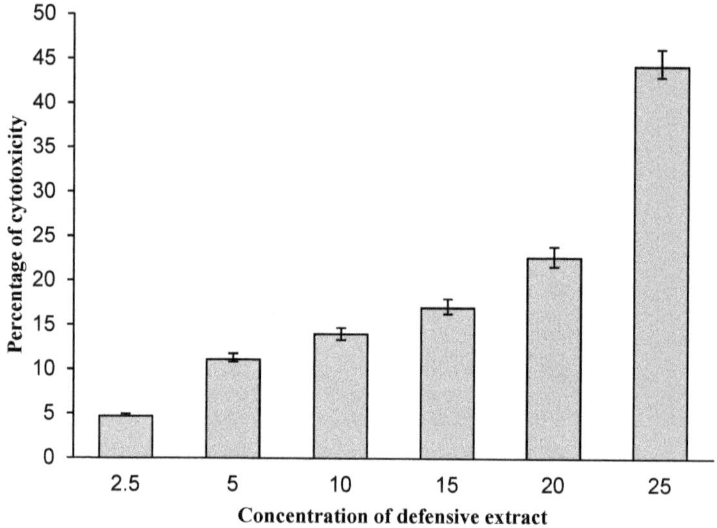

Figure 6. Cytotoxic activity of defensive gland extract of *L. tristis*.

3. Discussion

The current investigation into the chemical composition of the defensive gland secretion of *L. tristis* showed the presence of 2,3-dimethyl-1,4-benzoquinone, 1,3-dihydroxy-2-methylbenzene, 2,5-dimethylhydroquinone, tetracosane, oleic acid, hexacosane, pentacosane, 7-hexadecenal and tert-hexadecanethiol. Quinone compounds are one of the major biomolecules present in the defensive secretion of beetles [9–11]. Earlier studies by Klaus and Thomas (1987) demonstrated two quinone chemicals—methyl-p-benzoquinone and ethyl-p-benzoquinone—which are important components of the defensive secretion of the tenebrionid beetle, *Blaps mucronata* L. They also identified quinone molecules in the defensive gland which induced restlessness in other insects [12]. The significant characteristic of the hydroquinone molecule is its ability to irritate the mucous membranes, skin and eyes of human beings. It is moderately harmful when ingested or absorbed through the skin. Therefore, it is inferred that the quinones and polyphenolic molecules (2,5-dimethyl hydroquinone, 2,3,dimethyl-1,4-benzoquinone and 1,3-dihydroxy-2-methylbenzene) present in the defensive secretion of *L. tristis* are responsible for the skin blistering and irritation in people who are exposed to it. Sonja et al. (2013) analysed the biochemical components in the defensive secretions of three ground beetles—*Abax parallelepipedus* (Piller and Mitter pacher; 1783), *Calosoma sycophanta* L. and *Carabus ullrichii* Gerner—and showed all the three samples contained methacrylic, tiglic and isobutyric acids [13]. Analysis of the chemical components of the defensive gland of both land and water bombardier beetles shows they possess hot quinones, while adephagans contain weak and strong alkaloids, steroids, phenols, carboxylic acids and terpenes [14]. The GC-MS analysis of the defensive secretion of the tenebrionid beetle indicates the presence of toluquinone, ethylquinone and a relatively lower quantity of benzoquinone [4]. In addition to quinones, previously unidentified 4-methyl ketones and unsaturated ketones also are identified in the defensive secretion [5]. The *L. tristis* beetle—a member of the Tenebrionidae family—also contains similar types of quinone chemicals. However, we were unable to identify ketone molecules (4-methyl ketones) in the defensive secretion. Pentacosane and heptacosane are the other chemicals found in the defensive gland extract of *L. tristis*, and these hydrocarbons have also been found in the sting glands of the Braconid wasp, *Bracon hebetor* Say [15]. The hydrocarbon tetracosane, identified in *L. tristis*, was identified in the cuticle of the parasite butterfly *Phengaris nausithous* (Bergstrasser, 1779) [16]. Another significant molecule discovered in *L. tristis* is oleic acid, which is also present in the defensive secretion gland system-1 of the rove beetle, *Deleaster dichrous* (Gravenhorst, 1802) [17]. The sex pheromone 7-hexadecenal, which was isolated in the ovipositor of the female *Heliothis virescens* Fabricius, was discovered in the defensive gland of *L. tristis*. However, it needs to be confirmed experimentally that the chemical acts as a sex pheromone in *L. tristis*. Comparable to the above, the defensive extract of *D. dichrous* has been shown to contain the molecules, sec-butyl decanoate, sec-butyl dodecanoate, sec-butyl (Z)-7-tetradecenoate and isopropyl (Z)-7-tetradecenoate, which are sex pheromone components of the western grape leaf skeletonizer, *Harrisina brillians* (Barnes and McDunnough, 1910). This suggests that the defensive gland secretions act as pheromones or precursors of pheromones in insects [18].

The defensive secretion of the Mupli beetle induced chromosomal aberrations in the dividing cells of *A. cepa*. It is assumed that the defensive secretion enters the cells and stimulates various types of chromosomal damage. The mutagenic research demonstrated that certain heteropteran bug secretions tested on the root tip cells of *Allium* induced significant damage during mitotic division. The scent components of the insect also induced chromosomal abnormalities such as disorientation of chromosomes at metaphase due to non-formation of spindles, unequal separation of chromosomes, formation of anaphase bridges, tripolar grouping and unoriented chromosomes [19]. In the current study, various radical scavenging assays (DPPH, ABTS radical scavenging, peroxide scavenging, and ferric reducing antioxidant power) using the defensive secretion of *L. tristis* revealed promising antioxidant activity. Earlier studies by Liu et al. (2012) using the ethanolic whole-body extract of *Holotrichia parallela* Motschulsky showed that the main component responsible for the antioxidant properties of the defensive gland extracts may be the presence of quinone and phenolic compounds such as methyl-p-benzoquinone and ethyl-p-benzoquinone [20]. We also isolated quinone and phenolic compounds, 2,3-dimethyl-1,4-benzoquinone, 1,3-dihydroxy-2-methylbenzene, 2,5-dimethylhydroquinone from the *L. tristis* defensive gland extract, which may be the reasonable cause of improved antiradical scavenging activity. Earlier research works also studied the antioxidant properties of different extracts of the insects, *Tenebrio molitor* L and *Gryllus bimaculatus* (de Geer) [21–23].

Compounds with a quinone group are known to induce a variety of physiological actions, including antibacterial, antifungal, antiviral, antimicrobial and anticancer effects [24]. The quinone and polyphenolic molecules present in the defensive extract of *L. tristis* are the probable reason for the antimicrobial activity against the bacteria *E. coli* and *S. aureus*. Similar outcomes were obtained with the millipede *Pachyiulus hungaricus* Karsch (1881) defensive gland extract, which contains high amounts of benzoquinones and hydroquinones and demonstrated substantial antibacterial activity against seven bacterial strains, including *E. coli* and *S. aureus* [25]. Previous studies have revealed that the defensive gland extract from several ground beetle species and the secretions of carrion beetles have antibacterial activities [26]. The pygidial gland secretion of the woodland caterpillar hunter, *Calosoma sycophanta*, was used to identify the antibacterial activities against human bacterial pathogens [10]. Antimicrobial activity of exocrine glandular secretion of *Chrysomela* larvae showed that insect antimicrobial peptide complexes prevent resistance development in bacteria [11]. A cytotoxicity assay using DLA cells revealed the anticancer property of the defensive secretion of *L. tristis*. Similar observations were noticed with the pygidial gland secretions of four ground beetle species (Coleoptera: Carabidae); the results have shown an inhibition of tumour and non-tumour cell proliferation by the antiproliferative effect on the tested cell line [14]. Defensive secretion of *Ulomoides dermestoides* (Fairmaire, 1893) on A549 cells also showed intense cytotoxic activity [27]. Similar to the aforementioned finding, the defensive secretion of *L. tristis* also showed anticancer activity.

4. Materials and Methods

4.1. Collection of Experimental Organism

The experimental insect *L. tristis* was collected from various localities of Kerala, India, using the hand-picking method from crevices of buildings and rubber plantations. The collected beetles were kept in perforated plastic containers with lids so as to ensure the availability of proper air, temperature and humidity. They were then brought to the laboratory for the extraction of the defensive gland secretion.

4.2. Collection of Defensive Secretion from the Gland

Insects of both sexes were used for milking the defensive secretion. The beetles were held between the thumb and index finger after displacing and exposing the elytra and terga and then the defensive gland was located. The terminal part of the abdomen was cleaned with cotton soaked in deionized water. A thin needle was used to irritate the beetle, and the defensive glands were extruded out by gently pressing the abdomen. Much care was taken to extrude the gland out and prevent cross-contamination from other substances such as faecal matter. The tip of the extruded gland was immersed in a solvent contained in an Eppendorf tube (500 μL capacity), and then the gland was broken with a sharp needle. The extract was collected in the Eppendorf tubes containing 300 μL of methanol (HPLC grade) for GC-HRMS analysis. Deionized water was used to collect the extract for the DPPH assay, antimicrobial assay and cytotoxicity experiments. The collected extract was centrifuged at 5000 rpm for 2 min to get rid of any remaining tissue debris. The filtrate was then used for further analysis.

4.3. GC-HRMS Analysis

The chemical composition of the defensive gland secretion of the beetle was analysed using a gas chromatograph coupled with mass spectrometer (Jeol, AccuTOF GCV, Agilent) [28]. The methanolic extract of the defensive secretion was used for GC-HRMS analysis. (GC-MS data is available in Supplementary Materials).

4.4. Antimitotic Activity

The antimitotic activity of the defensive extract was examined using healthy, young, uniformly sized *Allium* bulbs. The bulb tissues grown in 300 μL of the defensive secretion were used as experimental groups, and tissues grown in distilled water were used as control groups. The root tips of both the control group and the experimental group were examined using the squash preparation method. The chromosomal aberrations were examined with a microscope (Leica DMi1 inverted microscope) equipped with a camera. The images of the chromosomes were captured at a magnification of 40×.

4.5. In Vitro Antioxidant Activity Analysis

4.5.1. DPPH Radical Scavenging Assay

The antioxidant activity of the extracts was studied by a DPPH assay using ascorbic acid as the standard. Various concentrations of the defensive gland extract were prepared (20 μL, 40 μL, 60 μL, 80 μL and 100 μL) by adding 370 μL DPPH and made up to 2000 μL by adding distilled water. Five different concentrations of ascorbic acid were also prepared. The solutions were kept in the dark for 30 min and analysed using a UV absorption spectrophotometer.

4.5.2. ABTS Radical Scavenging Assay

The defensive extract of the beetle was mixed with 1 mL of a working solution of the ABTS + radical and incubated for 20 min. The absorbance of each concentration and the control was measured at a wavelength of 734 nm using spectrophotometric analysis [29].

4.5.3. Hydrogen Peroxide Scavenging Assay

Different concentrations of the defensive extract of the beetle were mixed with phosphate buffer (0.1 M, pH 7.4) that contained hydrogen peroxide (25 mM). The change in absorbance from the start to the end of 5 min was measured at 230 nm [30].

4.5.4. Ferric Reducing Antioxidant Power

A previous method was used for performing the FRAP experiment. As a FRAP reagent, 300 mM sodium acetate buffer (pH 3.6, 10 mL) was added to 10 mM TPTZ solution in 40 mM hydrochloric acid (1 mL) and 20 mM iron (III) chloride (1 mL). The FRAP reagent was used in a water bath at 37 °C. The sample (20 L) and FRAP reagent (150 L) were mixed, and the absorbance at 593 nm was measured immediately [31].

4.6. Antibacterial Activity

In vitro antibacterial activity of the gland extract of the Mupli beetle was studied by the agar disc diffusion assay method. To test the antibacterial activity, *S. aureus* and *E. coli*—two prominent Gram-positive and Gram-negative bacteria—were selected. These pathogens cause various diseases in human beings. The Mueller–Hinton agar (MHA) well diffusion method was employed to investigate the antibacterial activity. Spore suspension of bacteria was added to a sterile Mueller–Hinton medium before solidification. It was then poured into sterile Petri dishes (9 cm in diameter) and spread using a cotton swab. Various concentrations (5 μL, 10 μL and 15 μL) of the defensive gland extract were pipetted into sterile discs of 6 mm and placed at the centre of the Petri dish. The antibiotic kanamycin was used as a positive control in the experiment. The Petri dish was incubated for 16 h at 37 °C. The zone of inhibition was analysed to estimate the antibacterial effect [32].

4.7. Anticancer Activity

The gland extract of the beetle was studied for short-term in vitro cytotoxicity using DLA cells. The tumour cells aspirated from the peritoneal cavity of tumour-bearing mice were washed thrice with phosphate-buffered saline (PBS) or normal saline. Cell viability was determined using the trypan blue exclusion method. Viable cell suspension (1×10^6 cells in 0.1 mL) was added to tubes containing various concentrations of the gland extract, and the volume was made up to 1 mL using PBS. The control tube contained only the cell suspension. The assay mixture was incubated for 3 h at 37 °C. After the incubation, the cell suspension was mixed with 0.1 mL of 1% trypan blue, kept for 2–3 min and then loaded on a haemocytometer. Dead cells take up the blue colour of trypan blue, while live cells do not take up the dye. The number of stained and unstained cells was counted individually, and the following formula was used to estimate the percentage of cytotoxicity.

Percentage of cytotoxicity = number of dead cells/number of live cells + number of dead cells × 100 [33].

4.8. Statistical Analysis and Data Representation

The results obtained from the cytotoxicity assay were represented as percentage of cytotoxicity, and a Microsoft Excel programme was used to plot the results graphically. The concentration of the sample required to produce 50% scavenging activity (IC_{50}) was analysed from the graph through linear regression analysis. The results obtained from the different scavenging assay are represented as percentage ± standard deviation.

5. Conclusions

In the current study, the components of the defensive gland extract of the Mupli beetle, *L. tristis*, and its biomedical applications were analysed. The defensive gland extract consists of compounds such as 2,3-dimethyl-1,4-benzoquinone, 1,3-dihydroxy-2-methylbenzene, 2,5-dimethylhydroquinone, tetracosane, oleic acid, hexacosane, pentacosane, 7-hexadecenal and tert-hexadecanethiol. Much effort has been made to find substances from natural sources that can act as effective antimicrobial agents, which has earned great attention as an essential medical need, particularly for pathogenic bacteria. The antibacterial effects of the defensive gland extract of *L. tristis* on the harmful bacteria *S. aureus* and *E. coli* were analysed for the first time in the present study. The defensive gland extract of *L. tristis* also exhibited antioxidant properties. This ability to scavenge free radicals makes this extract advantageous, particularly in illness conditions where there is an abundance of free radical species. The active compounds in the defensive gland extract of *L. tristis* can eliminate these free radicals. The potential anticancer and antimitotic properties of the defensive gland extract imply that the molecules in the beetle are a potentially beneficial source of natural products, and the compounds offer chances for the development of novel chemotherapeutics. In this study, the defensive gland extract of the Mupli beetle, *L. tristis*, which is usually considered a nuisance insect by the human society, was found to have beneficial biochemical properties.

Supplementary Materials: The following supporting information can be downloaded at: https://www.mdpi.com/article/10.3390/molecules27217476/s1.

Author Contributions: Conceptualization, writing—original draft preparation, A.P.A., K.N.J. and O.S.; methodology, writing—original draft preparation, A.P.A., K.N.J., A.R.V., V.S.B. and M.S.; software, writing—review and editing, A.B., V.A.R. and S.R.V.; validation, project administration, critical revision, A.P.A., K.N.J., Z.k.V., S.R.V. and K.N.; writing—review and editing, resources, A.P.A., K.N.J. and M.S.S. All authors have read and agreed to the published version of the manuscript.

Funding: This research received no external funding.

Institutional Review Board Statement: Not applicable.

Informed Consent Statement: Not applicable.

Data Availability Statement: The data generated and analysed during the current study are available from the corresponding author upon reasonable request.

Acknowledgments: We thank the Sophisticated Analytical Instrument Facility (SAIF), IIT Bombay, for the GC-HRMS analysis.

Conflicts of Interest: The authors declare no conflict of interest.

References

1. Tschinkel, W.R. A Comparative Study of the Chemical Defensive System of Tenebrionid Beetles III. Morphology of the Glands. *J. Morphol.* **1975**, *145*, 355–370. [CrossRef] [PubMed]
2. Tschinkel, W.R. A Comparative Study of the Chemical Defensive System of Tenebrionid Beetles: Chemistry of the Secretions. *J. Insect Physiol.* **1975**, *21*, 753–783. [CrossRef]
3. Tschinkel, W.R. A Comparative Study of the Chemical Defensive System of Tenebrionid Beetles. Defensive Behavior and Ancillary Features1,2. *Ann. Entomol. Soc. Am.* **1975**, *68*, 439–453. [CrossRef]
4. Kanehisa, K. Comparative Study of the Abdominal Defensive Systems in Tenebrionid Beetles. *Ber. Ohara Inst. Für Landwirtsch. Biol. Okayama Univ.* **1978**, *17*, 47–55.
5. Zvereva, E.L.; Kozlov, M.V. The Costs and Effectiveness of Chemical Defenses in Herbivorous Insects: A Meta-Analysis. *Ecol. Monogr.* **2016**, *86*, 107–124. [CrossRef]
6. Brown, W.V.; Doyen, J.T.; Moore, B.P.; Lawrence, J.F. Chemical Composition and Taxonomic Significance of Defensive Secretions of Some Australian Tenebrionidae (Coleoptera). *Aust. J. Entomol.* **1992**, *31*, 79–89. [CrossRef]
7. Bao, T.; Zhang, X.; Walczyńska, K.S.; Wang, B.; Rust, J. Earliest Mordellid-like Beetles from the Jurassic of Kazakhstan and China (Coleoptera: Tenebrionoidea). *Proc. Geol. Assoc.* **2019**, *130*, 247–256. [CrossRef]
8. Abhitha, P.; Vinod, K.V.; Sabu, T.K. Defensive Glands in the Adult and Larval Stages of the Darkling Beetle, *Luprops Tristis*. *J. Insect Sci.* **2010**, *10*, 7. [CrossRef]
9. Blum, M.S.; Crewe, R.M.; Pasteels, J.M. Defensive Secretion of *Lomechusa strumosa*, a Myrmecophilous Beetle1,2. *Ann. Entomol. Soc. Am.* **1971**, *64*, 975–976. [CrossRef]
10. Markarian, H.; Florentine, G.J.; Pratt, J.J. Quinone Production of Some Species of Tribolium. *J. Insect Physiol.* **1978**, *24*, 785–790. [CrossRef]
11. Gross, J.; Podsiadlowski, L.; Hilker, M. Antimicrobial activity of exocrine glandular secretion of chrysomela larvae. *J. Chem. Ecol.* **2002**, *28*, 317–331. [CrossRef] [PubMed]
12. Peschke, K.; Eisner, T. Defensive secretion of the tenebrionid beetle, *Blaps mucronata:* Physical and chemical determinants of effectiveness. *J. Comp. Physiol. A* **1987**, *161*, 377–388. [CrossRef] [PubMed]
13. Lečić, S.; Ćurčić, S.; Vujisić, L.; Ćurčić, B.; Curcic, N.; Nikolić, Z.; Anđelković, B.; Milosavljević, S.; Tešević, V.; Makarov, S. Defensive secretions in three ground-beetle species (Insecta: Coleoptera: Carabidae). *Ann. Zool. Fenn.* **2014**, *51*, 285–300. [CrossRef]

14. Nenadić, M.; Soković, M.; Calhelha, R.C.; Ferreira, I.C.F.R.; Ćirić, A.; Vesović, N.; Ćurčić, S. Inhibition of tumour and non-tumour cell proliferation by pygidial gland secretions of four ground beetle species (coleoptera: Carabidae). *Biologia* **2018**, *73*, 787–792. [CrossRef]
15. Fukushima, J.; Kuwahara, Y.; Yamada, A.; Suzuki, T. New Non-Cyclic Homo-Diterpene from the Sting Glands of *Bracon hebetor* Say (Hymenoptera: Braconidae). *Agric. Biol. Chem.* **1990**, *54*, 809–810. [CrossRef]
16. Solazzo, G.; Seidelmann, K.; Moritz, R.F.A.; Settele, J. Tetracosane on the Cuticle of the Parasitic Butterfly *Phengaris (Maculinea) nausithous* Triggers the First Contact in the Adoption Process by *Myrmica rubra* Foragers. *Physiol. Entomol.* **2015**, *40*, 10–17. [CrossRef]
17. Dettner, K.; Schwinger, G.; Wunderle, P. Sticky Secretion from Two Pairs of Defensive Glands of Rove Beetle *Deleaster dichrous* (Grav.) (Coleoptera: Staphylinidae): Gland Morphology, Chemical Constituents, Defensive Functions, and Chemotaxonomy. *J. Chem. Ecol.* **1985**, *11*, 859–883. [CrossRef]
18. Pfeiffer, L.; Ruther, J.; Hofferberth, J.; Stökl, J. Interference of Chemical Defence and Sexual Communication can Shape the Evolution of Chemical Signals. *Sci. Rep.* **2018**, *8*, 321. [CrossRef]
19. Surender, P.; Mogili, T.; Thirupathi, D.; Janaiah, C.; Vidyavati. Effect of scent components on somatic cells of *Allium sativum* L. *Curr. Sci.* **1987**, *56*, 964–967.
20. Liu, S.; Sun, J.; Yu, L.; Zhang, C.; Bi, J.; Zhu, F.; Qu, M.; Yang, Q. Antioxidant Activity and Phenolic Compounds of *Holotrichia parallela* Motschulsky Extracts. *Food Chem.* **2012**, *134*, 1885–1891. [CrossRef]
21. Tang, Y.; Debnath, T.; Choi, E.-J.; Kim, Y.W.; Ryu, J.P.; Jang, S.; Chung, S.U.; Choi, Y.-J.; Kim, E.-K. Changes in the Amino Acid Profiles and Free Radical Scavenging Activities of *Tenebrio molitor* Larvae Following Enzymatic Hydrolysis. *PLoS ONE* **2018**, *13*, e0196218. [CrossRef] [PubMed]
22. Hong, F.; Yang, F.; Liu, C.; Gao, Q.; Wan, Z.; Gu, F.; Wu, C.; Ma, Z.; Zhou, J.; Yang, P. Influences of Nano-TiO2 on the Chloroplast Aging of Spinach under Light. *Biol. Trace Elem. Res.* **2005**, *104*, 249–260. [CrossRef]
23. Hwang, B.B.; Chang, M.H.; Lee, J.H.; Heo, W.; Kim, J.K.; Pan, J.H.; Kim, Y.J.; Kim, J.H. The Edible Insect *Gryllus bimaculatus* Protects against Gut-Derived Inflammatory Responses and Liver Damage in Mice after Acute Alcohol Exposure. *Nutrients* **2019**, *11*, 857. [CrossRef] [PubMed]
24. Beheshti, A.; Norouzi, P.; Ganjali, M.R. A simple and robust model for predicting the reduction potential of quinones family; electrophilicity index effect. *Int. J. Electrochem. Sci.* **2012**, *7*, 11.
25. Stanković, S.; Dimkić, I.; Vujisić, L.; Pavković-Lučić, S.; Jovanović, Z.; Stević, T.; Sofrenić, I.; Mitić, B.; Tomić, V. Chemical Defence in a Millipede: Evaluation and Characterization of Antimicrobial Activity of the Defensive Secretion from *Pachyiulus hungaricus* (Karsch, 1881) (Diplopoda, Julida, Julidae). *PLoS ONE* **2016**, *11*, e0167249. [CrossRef]
26. Hoback, W.W.; Bishop, A.A.; Kroemer, J.; Scalzitti, J.; Shaffer, J.J. Differences among antimicrobial properties of carrion beetle secretions reflect phylogeny and ecology. *J. Chem. Ecol.* **2004**, *30*, 719–729. [CrossRef] [PubMed]
27. Crespo, R.; Villaverde, M.L.; Girotti, J.R.; Güerci, A.; Juárez, M.P.; de Bravo, M.G. Cytotoxic and Genotoxic Effects of Defence Secretion of *Ulomoides dermestoides* on A549 Cells. *J. Ethnopharmacol.* **2011**, *136*, 204–209. [CrossRef]
28. Honda, K. GC-MS and 13C-NMR Studies on the Biosynthesis of Terpenoid Defensive Secretions by the Larvae of Papilionid Butterflies (Luehdorfia and Papilio). *Insect Biochem.* **1990**, *20*, 245–250. [CrossRef]
29. Anusmitha, K.M.; Aruna, M.; Job, J.T.; Narayanankutty, A.; Pb, B.; Rajagopal, R.; Alfarhan, A.; Barcelo, D. Phytochemical Analysis, Antioxidant, Anti-Inflammatory, Anti-Genotoxic, and Anticancer Activities of Different *Ocimum* Plant Extracts Prepared by Ultrasound-Assisted Method. *Physiol. Mol. Plant Pathol.* **2022**, *117*, 101746. [CrossRef]
30. Malayil, D.; Jose, B.; Narayanankutty, A.; Ramesh, V.; Rajagopal, R.; Alfarhan, A. Phytochemical Profiling of *Azima tetracantha* Lam. Leaf Methanol Extract and Elucidation of Its Potential as a Chain-Breaking Antioxidant, Anti-Inflammatory and Anti-Proliferative Agent. *Saudi J. Biol. Sci.* **2021**, *28*, 6040–6044. [CrossRef]
31. Malayil, D.; House, N.C.; Puthenparambil, D.; Job, J.T.; Narayanankutty, A. *Borassus flabellifer* Haustorium Extract Prevents Pro-Oxidant Mediated Cell Death and LPS-Induced Inflammation. *Drug Chem. Toxicol.* **2022**, *45*, 1716–1722. [CrossRef] [PubMed]
32. Es, B. Antibacterial Potential of *Luprops tristis*–The Nuisance Rubber Plantation Pest from Western Ghats of India. *Int. J. Agric. Innov. Res.* **2014**, *3*, 1–6.
33. Ahn, M.Y.; Ryu, K.S.; Lee, Y.W.; Kim, Y.S. Cytotoxicity and L-Amino Acid Oxidase Activity of Crude Insect Drugs. *Arch. Pharm. Res.* **2000**, *23*, 477–481. [CrossRef] [PubMed]

molecules

MDPI

Review

Application of Metabolomics in Fungal Research

Guangyao Li [1], Tongtong Jian [2], Xiaojin Liu [1], Qingtao Lv [1], Guoying Zhang [1,*] and Jianya Ling [1,3,*]

1 School of Pharmacy, Shandong University of Traditional Chinese Medicine, Jinan 250355, China
2 Hospital of Shandong University of Traditional Chinese Medicine, Jinan 250355, China
3 State Key Laboratory of Microbial Technology, Shandong University, Qingdao 266237, China
* Correspondence: zhangguoying2000@126.com (G.Z.); lingjian-ya@sdu.edu.cn (J.L.)

Abstract: Metabolomics is an essential method to study the dynamic changes of metabolic networks and products using modern analytical techniques, as well as reveal the life phenomena and their inherent laws. Currently, more and more attention has been paid to the development of metabolic histochemistry in the fungus field. This paper reviews the application of metabolomics in fungal research from five aspects: identification, response to stress, metabolite discovery, metabolism engineering, and fungal interactions with plants.

Keywords: metabolomics; fungi; application

Citation: Li, G.; Jian, T.; Liu, X.; Lv, Q.; Zhang, G.; Ling, J. Application of Metabolomics in Fungal Research. *Molecules* **2022**, *27*, 7365. https://doi.org/10.3390/molecules27217365

Academic Editors: Arunaksharan Narayanankutty, Ademola C. Famurewa and Eliza Oprea

Received: 22 September 2022
Accepted: 25 October 2022
Published: 29 October 2022

Publisher's Note: MDPI stays neutral with regard to jurisdictional claims in published maps and institutional affiliations.

Copyright: © 2022 by the authors. Licensee MDPI, Basel, Switzerland. This article is an open access article distributed under the terms and conditions of the Creative Commons Attribution (CC BY) license (https://creativecommons.org/licenses/by/4.0/).

1. Introduction

Metabolomics is an emerging omics technology following genomics, proteomics, and transcriptomics. The concept of metabolomics first originated from metabolomic profiling proposed in 1971 by Devaux et al. [1]. Metabolomics was put forward as a group of metabolites in organisms by Oliver et al. in 1998 [2]. Nicholson raised the concept of metabolomics on the basis of a statistical analysis of NMR spectroscopic data from mouse urine. These data are defined as "a quantitative measurement of the dynamic multiparametric metabolic response of living systems to pathophysiological stimuli or genetic modifications" [3]. Traditional metabolomics is divided into targeted metabolomics and untargeted metabolomics. Targeted metabolomics is the measurement of a defined set of chemically characterized and biochemically annotated metabolites, usually focusing on one or more relevant metabolic pathways [4]. Recently, it has been subdivided and further developed into widely targeted metabolomics, pseudotargeted metabolomics, quasi-targeted metabolomics, LM precision targeted metabolomics, etc. Although the above methods all use the MRM mode for mass spectrometric data acquisition, widely targeted metabolomics and pseudotargeted metabolomics are performed by qualitatively and relatively quantifying the target through substances in the local library (established on the basis of partial standards, untargeted data, literature data, etc.), LM precision targeted metabolomics can absolutely characterize the substances corresponding to all standards. Even in combination with external standard methods, absolute quantification of metabolites in a sample can be achieved. However, untargeted metabolomics analyzes all measurable metabolites in a sample. The aim was to measure metabolites in the samples whenever possible [5]. According to different research objects and purposes, metabonomics can be divided into four levels: metabolic fingerprinting analysis, metabolic target analysis, metabolic profiling, and metabolomics [6].

Metabolomics, through modern instrumental analytical methods with high throughput, sensitivity, and resolution, combined with chemometric methods, analyzes the change law of metabolites after stimulation or interference in biological systems. Its focus is more on small-molecule metabolites with relative molecular weights of less than 1000 in biological tissues or cells and is often used to study plant and microbial systems [7–9]. The existence time of fungi on the Earth is unknown, and no definite conclusions can be drawn about their origin. Fungal cells do not contain chloroplasts and plastids; they are typical heterotrophs with parasitic or saprophytic patterns. The latest research speculates that there are as many as six million species of fungi worldwide, of which more than 600 are closely related to humans. They can participate in the formation of the human micro-ecosystem as resident fungi or cause diseases as pathogens. Fungi profoundly affect human health, agriculture, biodiversity, natural ecology, industry, biomedicine, etc. [10–12]. The application of metabolomics to different research fields of fungi, such as the classification and identification of fungi, metabolic pathways of fungi, the discovery of fungal natural products, and plant–fungal interactions [13–18], can help researchers entirely mine the potential of fungi. This article mainly reviews the research methods of metabolomics and the latest progress of metabolomics in various research fields of fungi (Table 1), aiming to promote further research on fungal metabolomics.

Table 1. Application of metabolomics in fungal research in recent years.

Species	Techniques	Nos. of Metabolites	Main Metabolites	Involved Pathway	Ref.
		Fungal response to stress			
Agaricus subrufescens	UHPLC–MS/MS	38	Ergosterol, agaritine, pyroglutamic acid, vitamin B3, sphingolipids		[19]
Phanerochaete chrysosporium	GC–MS	53	Veratryl alcohol, threonate, and erythronate		[20]
Alternaria sp. MG1	GC–TOF-MS	239	Amino acid, carbohydrate, xenobiotics, and lipid	PPPN biosynthesis pathway	[21]
Cryptococcus neoformans	GC–TOF-MS		Amino acids, carbohydrates	Amino acid and carbohydrate metabolism	[22]
Pleurotus ostreatus	LC–Q/TOF-MS	59	Sucrose, dextrin, adenine, uracil, L-glutamine, and L-lysine	glutathione metabolism, sphingolipid metabolism, and some amino-acid metabolism	[23]
Volvariella volvacea	LC–Q/TOF-MS	547	Organic acids, fatty acids, amino acids, carbohydrate metabolites, nucleotides	Amino-acid metabolism, carbohydrate metabolism, the TCA cycle, energy metabolism	[24]
Aspergillus aculeatus	GC–MS	42	Amino acids, organic acids, sugars, fatty acids, and sugar alcohol		[25]
Aspergillus flavus	LC–MS/MS	135		Tricarboxylic acid cycle, amino acid biosynthesis, protein degradation, absorption, mineral absorption	[26]
Aspergillus niger	GC–MS		Mannitol and gluconic acid	Mannitol cycle	[27]
Aspergillus niger	LC–MS/MS	68	Triacylglycerol, monoacylglycerol, hydroxy-triacylglycerol	Glycerolipid metabolism	[28]
Ganoderma lucidum	GC–MS and LC–MS/MS	154/70	L-Malic acid, alpha-hydroxycholesterol, zymosterol, ergosterol	Protein digestion, absorption, purine metabolism, unsaturated fatty acids, fatty-acid biosynthesis, purine metabolism	[29]
Cunninghamella echinulata	LC–MS/MS		Protein and amino acid	Purine, amino-acid, TCA, and sugar metabolism	[30]
Schizochytrium limacinum SR21	GC–MS	30	Fatty acids, amino acids, organic acids, carbohydrates, alcohols, squalene, cholesterol	Mevalonate, lipid synthesis, and pentose phosphate pathway	[31]
Industrial yeast	GC–MS	59	Trehalose, glycerin acid, fatty acids	TCA cycle, fatty-acid synthesis, glycolysis pathway, arginine metabolism, etc.	[32]
		Discovery of fungal natural products			

Table 1. *Cont.*

Species	Techniques	Nos. of Metabolites	Main Metabolites	Involved Pathway	Ref.
Ganoderma lucidum and *Cordyceps sinensis*	HPTLC–MS	6	Thymine, uracil, adenine, cytosine, guanine and guanosine		[33]
Ophiocordyceps sinensis	UHPLC–Q-TOF-IMS	345	Tyrosyl-phenylalanine, 2-phenylethyl beta-D-glucopyranoside and 3′,5′-odimethylmyricetin 3-O-beta-D-2″,3″-diacetylglucopyranoside		[34]
Cordyceps militaris	GC–MS	39	Amino acid, nucleosides, organic acids, and sugars	Nucleotide, carbohydrate, and amino-acid metabolism	[35]
Ophiocordyceps sinensis and *Cordyceps militaris*	LC–TOF-MS	100	Amino acids, unsaturated fatty acid, peptides, mannitol, adenosine, and succinoadenosine		[36]
Cordyceps sinensis and *Cordyceps militaris*	LC–MS	39	L-Tyrosine, 9,10-dihydroxy-12Z-octadecenoic acid and (−)-riboflavin	Histidine metabolism	[37]
Cordyceps militaris	LC–ESI-IT-MS/MS and GC–EI-IT-MS		Soyasaponin, pyroglutamic acid, isoflavone methyl-glycosides		[38]
Trametes versicolor and *Ganoderma applanatum*		57	N-(4-Methoxyphenyl)formamide 2-O-β-D-xyloside and N-(4-methoxyphenyl)formamide 2-O-β-D-xylobioside		[39]
Aspergillus oryzae and *Zygosaccharomyces rouxii*	UHPLC–Q-TOF-MS	32	N-Formyl-l-aspartate, imidazoleacetic acid, taurine, glycocholate, phenylpyruvate	Histidine metabolism, phenylalanine, adenosine kinase, phosphatidylserine synthase homo sapiens, phosphatidylethanolamine scramblase	[40]
Agaricus bisporus	UPLC–Q-TOF-MS	40	Organic acids, trehalose	Fatty-acid biosynthesis, tyrosine metabolism, and citrate cycle	[41]
Flammulina filiformis	HILIC–ESI(±)-QTOF-MS, LC–MS/MS	107	Melanin, l-dopa (3,4-dihydroxy-l-phenylalanine)	Phenylpropanoid biosynthesis and tyrosine metabolism	[42]
Aspergillus terreus	LC–HRMS	18	Quinones, isocoumarins, polyketides		[43]
Morchella sp.	UPLC–Q-TOF-MS	50	Fatty acids, peptides		[44]
Penicillium restrictum MMS417	UPLC–IT/TOF-MS/MS		Pyran-2-ones		[45]
			Fungal metabolic engineering		
Saccharomyces cerevisiae	GC–EI-MS		Geranyl diphosphate, farnesyl diphosphate, geranylgeranyl diphosphate, squalene, lanosterol, and ergosterol	Isoprenoid pathway	[46]
Aspergillus nidulans	LC–MS		Fellutamide B, antibiotic 1656-G, and antibiotic 3127		[47]
Aspergillus nidulans	UHPLC–ESI-HRMS	6	Orcinol, phenoxyacetic acid, orsellinic acid, monodictyphenone, gentisic acid, and caffeic acid	Glycine, serine, and threonine metabolic pathway, glycolysis, and TCA cycle	[48]
Fusarium verticillioides and *Streptomyces* sp.	LC–ESI-QqQ	36	Amino acids, saccharides, nucleotides, organic acids, phenol, lipid, and amine	Protein synthesis, Krebs cycle	[49]

Table 1. *Cont.*

Species	Techniques	Nos. of Metabolites	Main Metabolites	Involved Pathway	Ref.
Fusarium verticillioides	GC–MS	46	Arabitol, mannitol, and trehalose	Fumonisin biosynthesis and trehalose biosynthesis	[50]
Fusarium graminearum	LC–MS	22	N-Ethyl anthranilic acid, N-phenethylacetamide, tricinolone and tricinolonoic acid, fusarins, zearalenones, and fusaristatin A		[51]
Fusarium graminearum	NMR–GC-FID–MS	45	Sugars, amino acids, organic acids, choline metabolites	Inhibiting glycolysis, tricarboxylic acid cycle	[52]
Aspergillus nidulans	GC–EI-MS	86	Carbohydrates, amino acids, and carboxylic and lipid acids, purines and pyrimidines	Amino-acid and carbohydrate metabolism	[53]
		Plant–fungal interaction			
Diaporthe phaseolorum, Trichoderma spirale	NMR	20	Threonine, malic acid, and N-acetyl-mannosamine		[54]
Pisolithus tinctorius	NMR, FT-ICR	61	Carbohydrates, organic acids, tannins, long-chain fatty acids, monoacylglycerols, gamma-aminobutyric acid (GABA), and terpenoids		[55]
Fusarium verticillioides	UPLC–Q-TOF/MS		Isoflavones, jasmonic acid	Phenylpropanoid, flavone metabolic,	[56]
Armillaria luteobubalina	GC–MS	117	Sugars, sugar alcohols, amines, or amino acids	D-Threitol synthesis	[57]
Tilletia controversa	LC–MS	62	9-HODE, prostaglandin D3, caffeic acid, pyroglutamic acid, tetracosanoic acid		[58]
Penicillium digitata	UHPLC–Q-TOF/MS	85	amino acids, lipids, fatty acids, TCA metabolites, galactose metabolites, carbohydrate metabolites, nucleic acids, amino sugars, and nucleotide sugars	Amino-acid, lipid, fatty-acid, and purine metabolism, and TCA cycle	[59]
Trichoderma fungi	HRMAS NMR		γ-Aminobutyric acid, acetylcholine, and amino acids		[60]

2. Fungal Metabolomic Approaches

The technical route for metabolomic research in fungi mainly involves three main processes: sample preparation, data collection and processing, and analysis (Figure 1). Sample preparation can affect not only the observed metabolite content but also the biological interpretation of the data. Therefore, appropriate sample collection and preparation steps are required to avoid interfering with the efficient metabolomics analysis. Currently, studies on sample preparation in biological fluids, tissues, mammalian cells, and plants have been relatively comprehensive, but there are few reviews related to fungal sample preparation strategies [61–64]. An ideal sample preparation method for metabolomics should be as simple, rapid, highly selective, and reproducible as possible and capable of quenching to determine the actual metabolic composition at the sampling time. General sample preparation steps include rapid sampling, quenching, and sample extraction [65].

2.1. Rapid Sampling

Rapid sampling from fermentation tanks or shake flasks is the first step in performing fungal sample preparation, since 1969 when Harrison et al. first attempted rapid sampling from small-scale laboratory bioreactors [66], to Iversen and Theobald et al., who developed iterations of sampling systems with minimal dead volume for manual feeding [67,68]. Fast sampling systems are currently characterized by motorized sampling, high frequency, negligible dead areas, and efficient inactivation [69–71]. The technology developed by van Gulik used adenine nucleotide as an indicator to analyze its dynamic response to changing glucose concentration and quickly sample yeast metabolites. It has a very high sampling frequency, which can also ensure long-term sterility [72]. Although the sampling method developed by Hannes Link et al., which directly injects fungi such as yeast into high-resolution mass spectrometers for real-time metabolomic analysis, achieves automated detection of target compounds [73], this approach is expensive and impractical for the majority of researchers.

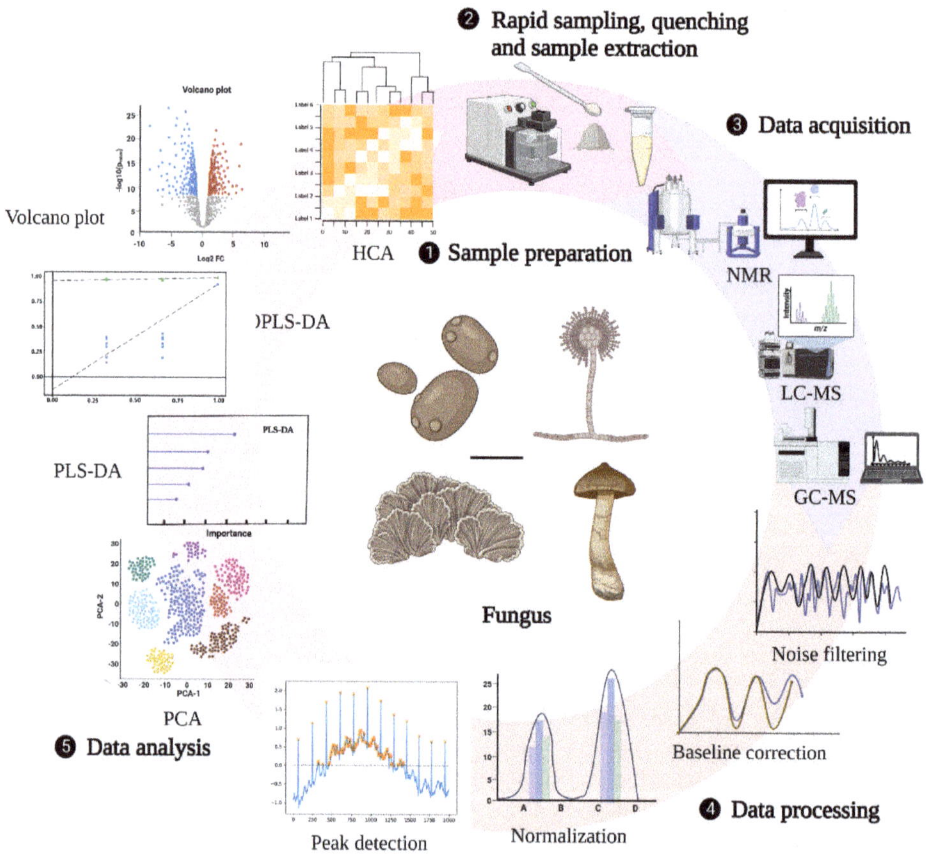

Figure 1. General technical route of metabolomics. Created with BioRender.com (accessed on 28 October 2022).

2.2. Quenching

The purpose of quenching is to rapidly stop various metabolic activities within the cells, inactivate enzymes, keep the metabolite content stable, and reduce the degradation of metabolites. The cold methanol quenching method is a standard method used in the past, and the research on cell quenching mainly focused on different methanol content and quenching temperature. Specifically, 60% (v/w) methanol quenching at −40 °C is one of the standard methods for microbial metabolomics research. However, this method causes the leakage of cellular metabolites [74,75]. So far, cold methanol use for quenching has remained a controversial topic. Due to the advantage of a high extraction efficiency of representative intracellular metabolites by cold methanol quenching, some researchers have continuously improved the original method of cold methanol quenching proposed by de Koning et al. According to the degree of metabolite leakage depending on the exposure time, temperature, and nature of the methanol solution, under fast sampling equipment, setting the sample/quencher liquid (v/v) ratio to 1:5 or less for quenching in pure methanol at ≤ -40 °C effectively prevents metabolite leakage [76]. Of course, other quenching methods are also under continuous development. Rapid filtration of liquid nitrogen is more suitable for cell quenching than cold methanol quenching, with minimal damage to cell integrity and improved recovery of intracellular metabolites [77]. However, ice crystals produced by the freezing of liquid nitrogen can potentially damage the cell membrane, which also leads to the leakage of metabolites inside the cell. Quenching using strong acids such as perchloric acid can lead to the degradation of some compounds in a strong acid environment with severe metabolite reduction [78]. It seems that the perfect quenching method is challenging to achieve. Although Meinert et al. believe that the quenching method of methanol quenching solution (60%, −40 °C) has no metabolite leakage, we can still find that the metabolite investigated using this method is insufficient and far from reaching the level of no leakage for all detection indicators [79]. Moreover, most of the quenching studies focused on Gram-positive bacteria [80], Gram-negative bacteria [81], yeast cells [82,83], etc., whereas research on filamentous fungi is rare.

2.3. Sample Extraction

How to extract the quenching sample is a critical step in the sample preparation stage. Furthermore, the ideal extraction should not change the metabolites' physical properties and chemical structure and should maximize the sequestration of the metabolite content. Current extraction methods can presumably be grouped into physical and chemical categories. Physical methods often use homogenizers, ultrasonic, microwave, and other tools. Chemical methods mostly use acid, base, water, methanol, ethanol, and different ratios of water and organic solvents, mixed solvents, etc. Researchers can choose different solvents depending on different samples. As mentioned above, strong acids and bases can cause severe metabolite leakage resulting in a low recovery rate of final metabolites. Although this is a classical extraction method, using strong acids and bases such as perchloric acid is still not recommended. Solvents with the highest extraction efficiency can be selected experimentally. Three sample preparation methods and five solvent mixtures of *Mortierella alpina* were evaluated using gas chromatography/mass spectrometry (GC–MS) [84]. The results showed better reproducibility and recovery of lyophilized. Methanol/water (1:1) was more effective in extracting metabolites of *Mortierella alpina*. Compared with biphasic extraction at different pH with methanol extraction, which is easy and fast and suitable for the extraction of metabolites from *Phanerochaete chrysosporium*, biphasic extraction at different pH is more suitable for target analysis [85]. Of note, despite the high efficiency and recovery of metabolites extracted by supercritical fluids, partially unstable metabolites may undergo decomposition due to the pressure ranging between 200 and 500 bar [86].

2.4. Instrumental Analysis Methods

Commonly used metabolomics analysis methods require collecting raw data after sample quenching and extraction. Currently, several analytical methods exist for qualitative and quantitative analysis of metabonomic extracts in metabonomic research. The commonly used analytical methods for fungal metabolomics include gas chromatography/mass spectrometry (GC–MS), liquid chromatography/mass spectrometry (LC–MS), nuclear magnetic resonance spectroscopy/mass spectrometry (NMR–MS), capillary electrophoresis/mass spectrometry (CE–MS), and matrix-assisted laser desorption ionization mass spectrometry (MALDI-MS). GC–MS is often used to analyze substances with excellent thermal stability and volatility, allowing simultaneous analysis of sugars, amino acids, phosphorylated metabolites, organic acids, lipids, amines, and other compounds. It exhibits extraordinary robustness, excellent separation ability, high selectivity, effective sensitivity, and reproducibility [87]. However, there are also some disadvantages, as nonvolatile compounds require derivatization. The main derivatization methods are silanization, acylation, alkylation, and esterification. Koek et al. established a GC–MS spectrometry metabonomic analysis technology suitable for microorganisms and verified the method with various microorganisms [88]. The results showed that the method has good repeatability, effective reproducibility, and fast linear regression characteristics. It can be used for the metabonomic analysis of various components of microorganisms, such as alcohols, aldehydes, amino acids, fatty acids, organic acids, sugars, purines, pyrimidines, and aromatic compounds. HPLC reduces the complexity of samples and offers several advantages, such as simple preparation, high sensitivity, signal reproducibility, minimal noise, and high qualitative and quantitative ability. It is helpful for thermally labile compounds, nonvolatile compounds, polar compounds, and compounds that are macromolecules. With the development of high-performance liquid chromatography (HPLC) and ultra-performance liquid chromatography (UHPLC), the resolution of peaks was improved, and the speed of analysis was accelerated [89]. NMR techniques have the advantages of high reproducibility, accurate quantification, simple sample preparation, measurable analytes in various solvents, clear identification of unknown metabolites, and complete metabolite detection. The disadvantage is low sensitivity, which severely limits the use of NMR in metabolomics [90]. Capillary electrophoresis techniques are relatively new and less applied analytical methods, mainly for studying molecules, but they are preferred when dealing with highly polar, charged metabolites [91]. They allow rapid and high-resolution analysis of charged metabolites such as nucleic acids, amino acids, carboxylic acids, and sugar phosphates. Each analytical tool has advantages and disadvantages. A single analytical platform tool cannot directly and precisely characterize or quantify thousands of small-molecule metabolites involved in fungal metabolic processes. The right combination of tools is often needed depending on the experimental situation to better analyze the target fungi.

2.5. Data Processing and Analysis Methods

The original data obtained by the analytical instrument cannot provide a clean and comparable metabolite spectrum. Therefore, the original data must be preprocessed and generally completed in the experimental system. This mainly includes noise reduction and baseline correction, peak detection and deconvolution, normalization, and data standardization [92,93]. The classical analysis method is to use a single variable, i.e., parameter by parameter, or to use multivariable techniques to evaluate group differences. Although the univariate analysis method is simple and convenient, it cannot accurately distinguish the groups when the difference is small. Multivariate analysis can be used to analyze the changes in single metabolites between different groups and the dependent structure of individual molecules [94]. Multivariate analysis can be divided into two categories; one is the unsupervised learning method, which classifies the original data directly, including principal component analysis (PCA), hierarchical clustering analysis (HCA), and self-organizing maps (SOMs). The other is the supervised learning method, i.e., learning the training samples with a given sample label, such as partial least squares discriminant analysis (PLS-DA), partial least squares discriminant analysis based on orthogonal signal correction (OPLS-DA), artificial neural network (ANN), and support vector machine (SVM). Among them, PCA, PLS-DA, and OPLS-DA are the most frequently used multivariate statistical analysis methods in the field of metabolomics [95–98]. The generally used analysis software includes MetAlign [99], MZmine [100], XCMS [101], Metabolomic Analysis and Visualization Engine (MAVEN), Metabolite Biological Role (MBRole), MetaCoreTM, MetaboAnalyst, InCroMA, and 3Omics [102–104].

3. Application of Metabolomics in the Field of Fungal Research

3.1. Application of Metabolomics in Classification and Identification of Fungal Research

Morphological methods, as the traditional fungal classification method [105], have some limitations, i.e., the classification from appearance characteristics is affected by the fungal growth environment, similar morphology, and other factors, thus affecting the accuracy of classification. Genomic DNA/DNA hybridization [106,107], ribosomal typing [108,109], multilocus sequence typing [110,111], ITS rDNA sequences [112], and lipid profile analysis [113] are popular methods in fungal taxonomic identification. Chemical taxonomy was initially considered complementary to morphological methods based on primary and, more often, secondary metabolites. However, with advances in HPLC and mass spectrometry, the application of metabolomic chemotaxonomic in fungal taxonomic identification has progressed considerably (Table 2). Kang et al. analyzed the secondary metabolites of seven species of *Trichoderma* (33 strains) using its sequence and metabolome-based chemotaxonomic comparison. They found that the chemical taxonomy based on secondary metabolites was more accurate than its sequence and identified an unknown group of *Trichoderma* [114]. Chen used HPLC fingerprinting combined with stoichiometric analysis of *Ganoderma lucidum* fruiting bodies and screened four marker components as discriminative variables to distinguish *Ganoderma lucidum* [115]. However, Wen et al. believed that the identification of distinguishing *Ganoderma lucidum* with the NMR metabolomics method was less time-consuming and faster, which was in line with the quality control of large-scale production. Through NMR metabolomics, labeling choline and propionic amino acids as discriminating variables not only successfully differentiated between Chinese and Korean *Ganoderma lucidum* but also differentiated the cultivated origin of Chinese *Ganoderma lucidum* [116]. This method of taxonomic identification based on the specific metabolites of fungal species effectively avoids the limitation of low accuracy of traditional methods and identifying fungi from different regions or even different growth stages.

Table 2. Metabolomics for taxonomic identification of fungi.

Fungal Species	Analysis Platform	Extraction Method	Data Processing	Achievement	Ref.
T. harzianum *T. aggressivum* *T. virens* *T. longibrachiatum* *T. hamatum* *T. koningii* *T. atroviride*	LC–ESI-MS-MS	The concentrate was pooled into 100 μL of methanol and filtered through a 0.45 umptf filter	Varian MS Workstation 6.9, Vx Capture 2.1, MetAlign, SIMCA-P+ 12.0, Statistica 7	Chemical taxonomy based on secondary metabolite profiling was found to be advantageous over other classification methods	[114]
Ganoderma lucidum	NMR spectroscopy	CD3OD and D2O (*v*/*v*, 1:1), 10 mM sodium phosphate, and 0.025% TMSP were mixed and extracted, followed by centrifugation	Matlab, SIMCA-P version 11.0, Chenomx, and Excel	Development of a method to effectively distinguish between national and even regional sources of *G. lucidum* cultivation	[116]
Rhizoctonia solani	GC/MS	Derivatization in autosampler vials, upon addition of 80 μL of methoxyamine hydrochloride solution (30 °C, 120 min) and 80 μL of MSTFA (37 °C, 90 min)	ACD/Spec Manager v.12.00, mass spectra matching the National Institute of standards and Technology Library, SIMCA-P 12.0	Characterization and identification of an isolate of *Rhizoctonia solani*	[117]
Aspergillus	MALDI-TOF-MS	Bead disruption sample pretreatment followed by centrifugation	BioRad data processing suite	Can be used to unambiguously identify members of the genus *Aspergillus* at the species and strain level	[118]

Table 2. *Cont.*

Fungal Species	Analysis Platform	Extraction Method	Data Processing	Achievement	Ref.
Candida species, Aspergillus species, and other yeast genera	MALDI-TOF-MS	Washed yeast cells were fixed by suspension in 50% methanol/water (v/v) or stored at 4–6 °C for 45 days for subsequent comparative analysis	External alignment was performed using cytosolic picolinic acid A, etc.; MALDI mass spectra were processed using "Data Explorer" (Applied Biosystems), and data were processed in MATLAB	Was used to identify yeast and group strains, as well as follow morphogenesis of *C. albicans*	[119]
Epichloë festucae	LC–HR-MS/MS	MTBE, methanol, and water were extracted in two phases, dissolved in 60 µL of methanol/acetonitrile/water ($v/v/v$, 1:1:12), and centrifuged	The datasets were processed with markerlynx XS for maslynx v.4.1, and the software suite Marvis did the subsequent processing	A genetic approach combined with tandem mass spectrometry was used to identify novel products of secondary metabolite gene clusters and to discover novel Leu/Ile glycoside metabolites	[120]
Wide edible mushrooms	UHPLC–QE Orbitrap/MS/MS	Chloroform and methanol are mixed (2:1 v/v), then centrifuged	SPSS 16.0 statistical analysis, xcalibur 4.0, ms-dial 4.36, and lipidmaps for identification and quantification of lipids, SIMCA 14.1, metaboanalyst 4.0 follow-up analysis	It is helpful for improving the sensitivity, reproducibility, and accuracy of trace-level analysis of triterpenoids in complex biological samples	[121]
Ganoderma lucidum mycelium	UPLC–ESI-HR-QTOF-MRM	Methanol post-extraction filtration	Masslynx 4.1 performed data acquisition, targetlynx quantification, and SPSS 17.0	Highly precise identification and quantification of triterpenoids present in trace amounts in mycelia of *G. lucidum*	[122]

Nevertheless, metabolome-based chemotaxonomy also has drawbacks. First, how to overcome the problem of finding specific markers from a plethora of metabolites for optimal biological interpretation is still unanswered. Second, changes in various environmentally relevant regulatory genes may not affect the expression of taxonomic-related genes. Lastly, while it is desirable to analyze metabolites of specific organelles, how organelles can be isolated while maintaining a structural, metabolic state remains unattainable.

3.2. *Application of Metabolomics in the Study of Fungal Response to Stress*

Fungi can survive only under various stress reactions such as ionizing radiation, hydration activity, acid–base environment, hypoxic stress, solar ultraviolet radiation, agricultural and industrial pollutants, biological stress, nutrient stress, oxidative stress, heat stress, and cold stress to survive. Ecological metabolomics studies changes in endogenous metabolites produced by biological systems that are affected by environmental factors [123]. Using metabolomics to study fungi, we can understand how fungi respond to stress when they grow or infect their hosts. It is helpful to optimize the application of fungi in biotechnology, improve the environment, and even prevent fungal diseases.

Environmental stress, i.e., abiotic stress, is usually the most severe situation faced by fungi. Oliveira et al. found that the nutrient content of low-molecular-weight metabolites of wild mushrooms is higher than that of indoor cultivated ones [19], perhaps because the functions of the low-molecular-weight metabolite gene clusters of these mushrooms and their relative expression have differences in the two environments. Metabolomic analysis of the white rot basidomycete *Phanerochaete chrysosporium* under air and 100% oxygen revealed that three metabolites associated with the oxygen stress response were veratryl alcohol (VA), threonate, and erythronate. High concentrations of ROS can directly activate the VA synthesis pathway, and the intracellular oxygen concentration is significantly elevated. In contrast, threonate and erythronate resistance to hyperoxia is a process of progressive gradient accumulation [20]. When the fungal *Alternaria* sp. MG1 grown on grape interiors was subjected to starvation treatment, the shikimate pathway and the phenylpropanoid (PPPN) pathway were strongly activated, and relevant metabolites such as resveratrol were significantly upregulated [21]. The metabolic profiles of *Cryptococcus neoformans* changed under Cu stress, and the differential metabolites were mainly related to the metabolism of amino acids and carbohydrates. Replacing the carbon source with glycerol and ethanol can counteract the toxic effect of copper on *Cryptococcus neoformans* and improve urea clearance [22]. Yan et al. analyzed the metabolites of *Pleurotus ostreatus* under different heat stress times (6, 12, 24, and 48 h) for dynamic metabolite changes. They found that the contents of metabolites such as amino acids, nucleotides, and lipids showed an increasing trend with increased heat stress time [23]. Zhao et al. found that *Volvariella volvacea* showed little resistance to low temperatures. Nevertheless, under chilling stress, the relative levels of compounds such as amino acids and organic acids inside *Volvariella volvacea* increase significantly, and soluble sugars such as sorbitol are induced to be produced, improving its osmoregulatory capacity [24]. Interestingly, *Aspergillus aculeatus*, under drought and heat stress, increased the accumulation of amino acids and sugars and enhanced the total photosynthesis of tall fescue, resulting in a vastly improved ability of tall fescue inoculated with *Aspergillus aculeatus* to resist cold and heat stress [25].

Aspergillus flavus showed significant changes in carbohydrates, sulfur-containing amino acids and their derivatives, fatty acids, etc. under drought stress [124]. 1-Nonanol can destroy the integrity of the cell membrane in *Aspergillus flavus* and affect mitochondrial function, which induces apoptosis in *Aspergillus flavus* [26]. *Aspergillus niger* resisted copper stress by converting sorbitol from glucose to produce a large amount of mannitol [27]. When *Aspergillus niger* was exposed to 5% ethanol stress, its growth amount was about 70% less than that under normal growth conditions [28]. Using untargeted metabolomics to study its reaction mechanism, it was found that TAG, DAG, and hTAGs significantly accumulated. These neutral glycerolipids were previously believed to be associated with the fungi's exposure to abiotic stress factors [125,126]. Whether glycerides in the response of *Aspergillus niger* strain Es4 to ethanol stress can be used as a new response of the fungus to ethanol stress still needs further confirmation. Hammerl et al. further established a differential offline LC–NMR (DOLC–NMR) method to qualitatively and quantitatively analyze metabolic changes in *Penicillium roqueforti* when L-tyrosine levels are perturbed. Twenty-three metabolites were affected by the amino-acid perturbation method, among which the amino-acid degradation products 2-(4-hydroxyphenyl) acetic acid and 2-(3,4-dihydroxyphenyl) acetic acid were significantly upregulated [127]. Jiang et al. found that treatment of *Ganoderma lucidum* with methyl jasmonate (MeJA) for 24 h was the optimal condition to induce the biosynthesis of *Ganoderma lucidum*. MeJA induction can lead to metabolic rearrangements in *Ganoderma lucidum*, inhibit its normal glucose metabolism, energy supply, and protein synthesis, and promote cellular secondary metabolic production [29]. After treatment with tributyltin (TBT), the mycelial morphology of *Cunninghamella echinulata* changed, the metabolic activity was inhibited, and glycolysis and the TCA cycle were dysregulated. This fungus can eliminate the hazard of tributyltin compounds to the organisms by accumulating amino acids with antioxidant functions, such as betaine, proline, and GABA, to recover from the toxic TBT environment [30].

In contrast, benzoic acid derivatives such as sodium benzoate, p-aminobenzoic acid, and p-methylbenzoic acid all promote lipid synthesis in *Penicillium sr21*. Furthermore, 200 mg/L p-aminobenzoic acid even promoted glucose catabolism during glycolysis, increased the mevalonate pathway, weakened the tricarboxylic acid cycle, and promoted the production of tetrahydrofolate and NADPH [31]. The antibacterial polycationic peptide ε-poly-l-lysine (ε-PL) enhances the freeze–thaw tolerance of industrial *yeast* by promoting cell membrane-associated fatty-acid synthesis before freeze–thaw and promoting alglucan biosynthesis and glycerophospholipid metabolism after freeze–thaw [32]. From the above studies, it is easy to see that both biotic and abiotic stresses involve a variety of metabolic pathways in fungi, the most important of which are the metabolism of amino acids and their derivatives, glycolytic pathways, etc. Significant marginal changes in the levels of metabolites such as sugars, nucleotides, and lipids are the main mechanisms of their dynamic regulation.

3.3. *Application of Metabolomics in the Discovery of Fungal Metabolites*

A large number of metabolites such as primary and secondary metabolites exist in fungi. Primary metabolites are monomers synthesized by primary metabolisms, such as monosaccharides or monosaccharide derivatives, nucleotides, vitamins, amino acids, fatty acids, and various macromolecular polymers composed of them, including proteins, nucleic acids, polysaccharides, lipids, and other essential substances. Secondary metabolites refer to substances synthesized by fungi in which primary metabolites serve as precursors with no clear function in their life activities, such as gibberellins, penicillins, aflatoxins, and cordycepin [128]. Fungi can provide diverse and unique secondary metabolites, making them potential drug sources. Traditional assays can easily lead to the rediscovery of known compounds. With the advancement of analytical technology platforms, MS-based metabolomics workflows are mainly suitable for screening hundreds of natural products simultaneously for dereplication studies and extractions of bioactive compounds, which is of great benefit for the comprehensive exploration of potentially useful secondary metabolites. In addition to drug discovery, screening of bioactive compounds or discovery of unknown fungal metabolites that play critical roles in host fungal interactions are also required to identify fungal secondary metabolites. A metabolomics approach can aid in discovering and detecting novel metabolites in fungi (Table 3).

Table 3. Novel metabolites discovered and detected using metabolomics.

Structure	Molecular Weight	Molecular Formula	Compound Name	Reference
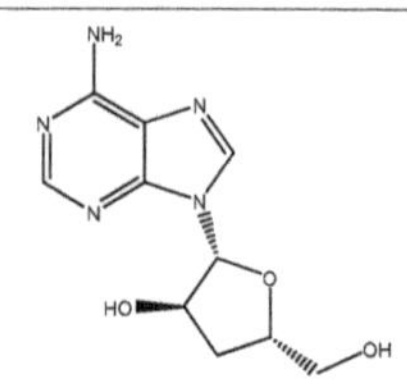	251.2460	$C_{10}H_{13}N_5O_3$	Cordycepin	[34]
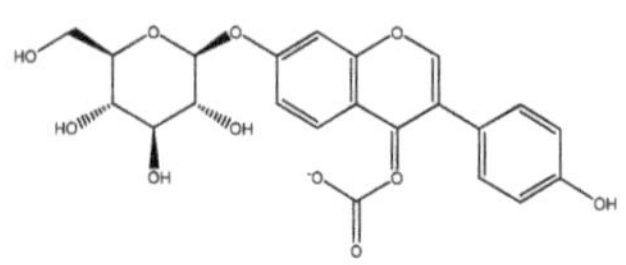	460.3915	$C_{22}H_{22}O_9$	Daidzein 7-O-beta-D-glucoside 4-O-methylate	[38]
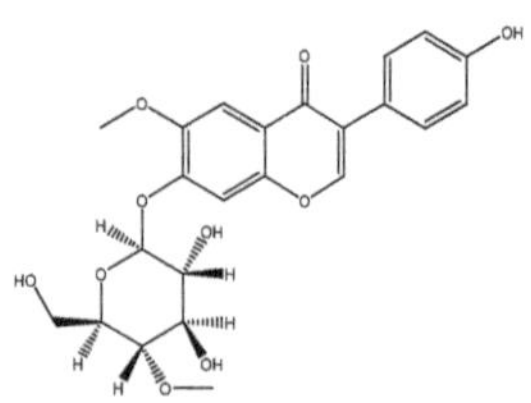	460.4350	$C_{23}H_{24}O_{10}$	Glycitein 7-O-beta-d-glucoside 4″-O-methylate	[38]

Table 3. *Cont.*

Structure	Molecular Weight	Molecular Formula	Compound Name	Reference
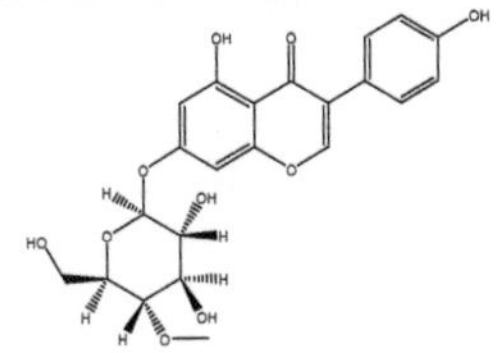	446.4080	$C_{22}H_{22}O_{10}$	Genistein 7-O-beta-d-glucoside 4″-O-methylate	[38]
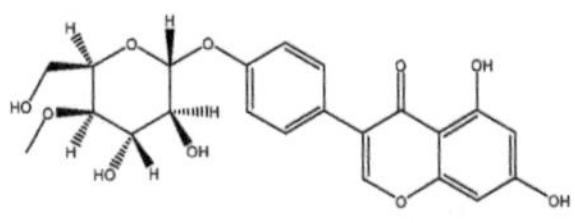	446.4080	$C_{22}H_{22}O_{10}$	Genistein 4′-O-beta-d-glucoside 4″-O-methylate	[38]
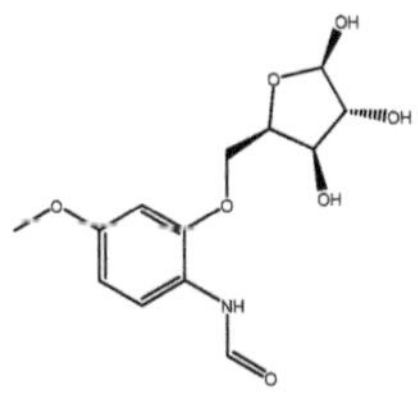	299.2970	$C_{13}H_{17}NO_7$	N-(4-Methoxyphenyl)formamide 2-O-beta-D-xyloside	[39]
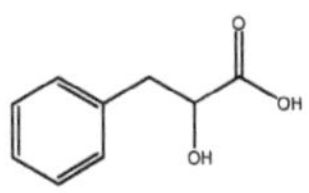	166.1760	$C_9H_{10}O_3$	3-Phenyllactic acid	[39]
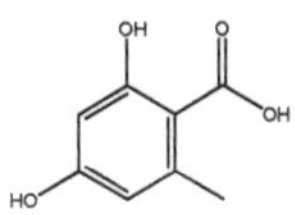	168.1480	$C_8H_8O_4$	Orsellinic acid	[39]
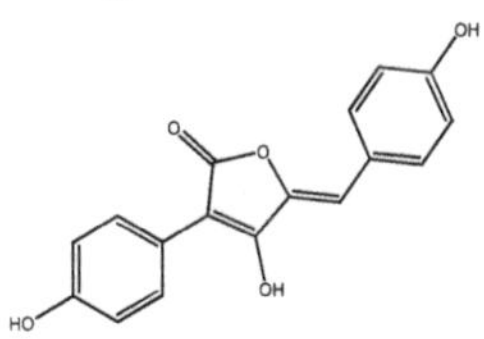	296.2780	$C_{17}H_{12}O_5$	Aspergillide B1	[43]
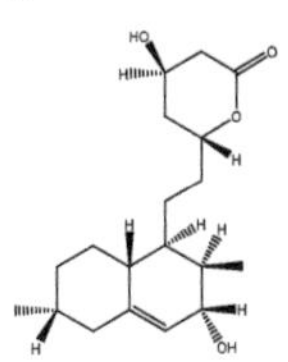	322.4450	$C_{19}H_{30}O_4$	3a-Hydroxy-3, 5-dihydromonacolin L	[43]
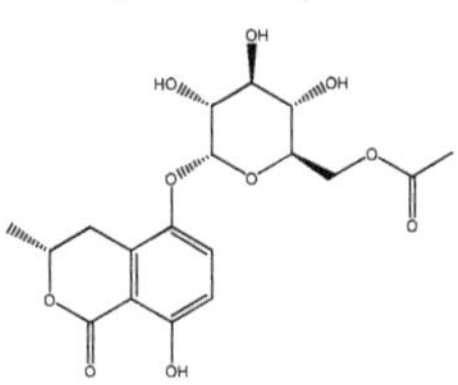	398.3640	$C_{18}H_{22}O_{10}$	5-Hydroxymellein	[129]
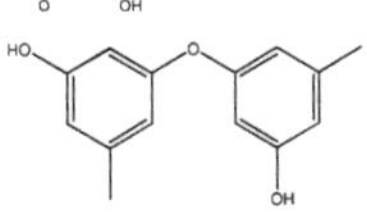	230.2630	$C_{14}H_{14}O_3$	Diorcinol	[129]

Table 3. *Cont.*

Structure	Molecular Weight	Molecular Formula	Compound Name	Reference
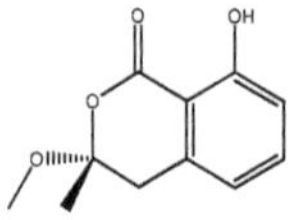	208.2130	$C_{11}H_{12}O_4$	Botryoisocoumarin A	[129]
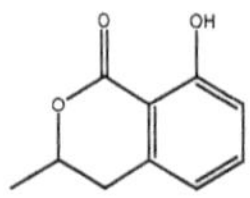	178.1870	$C_{10}H_{10}O_3$	Mellein	[129]
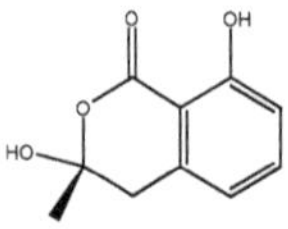	194.1860	$C_{10}H_{10}O_4$	3-Hydroxymellein	[129]
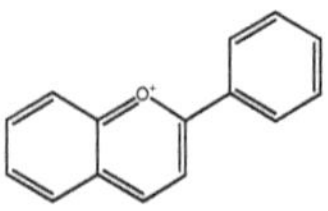	207.2515	$C_{15}H_{11}O^+$	Anthocyanins	[130]
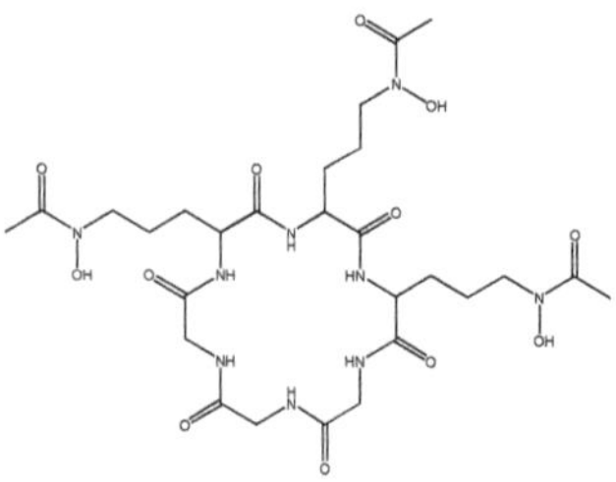	687.7080	$C_{27}H_{45}N_9O_{12}$	Desferriferrichrome	[44]
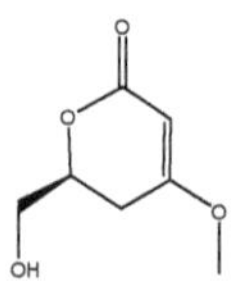	158.1530	$C_7H_{10}O_4$	5,6-Dihydro-6s-hydroxymethyl-4-methoxy-2h-pyrene-2-one	[45]
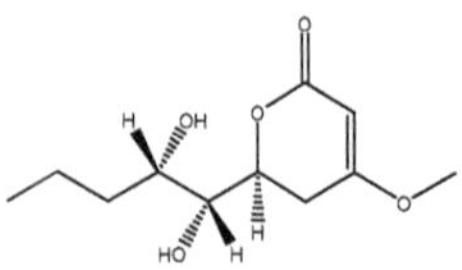	230.2600	$C_{11}H_{18}O_5$	(6S, 1′r, 2′s)—ll-p880 β	[45]
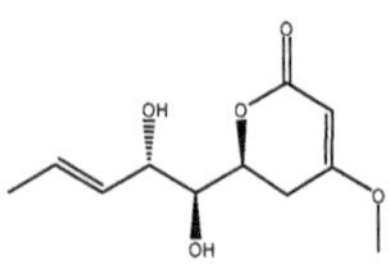	228.2240	$C_{11}H_{16}O_5$	5,6-Dihydro-4-methoxy-6S-(1′S, 2′S-dihydroxy pent-3′ (E)-enyl)-2H-pyran-2-one	[45]
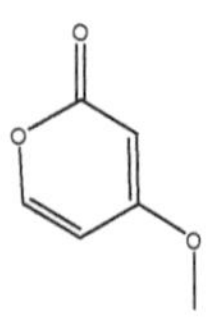	126.1110	$C_6H_6O_3$	4-Methoxy-6-(1′R, 2′S-dihydroxy pent-3′ (E)-enyl)-2H-pyran-2-one	[45]
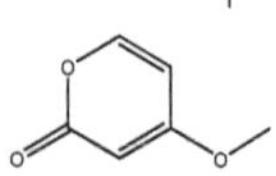	126.1110	$C_6H_60_3$	4-Methoxy-2H-pyran-2-one	[45]

Cordycepin is considered to be an important marker for the identification of *Cordyceps sinensis*. While nucleosides such as thymine, uracil, adenine, and guanosine are also the main substances used by researchers to identify and analyze *Cordyceps sinensis*. Mishra reidentified the nucleobases in samples of *Ganoderma lucidum* and *Cordyceps sinensis* by HPLC–MS and found that both had abundant nucleosides. Furthermore, the water extract of *Ganoderma lucidum* and the ethanolic extract of *Cordyceps sinensis* had the highest nucleobase content [33]. Joshi et al., for the first time, identified the presence of cordycepin using ion mobility mass spectrometry (IMMS), which provided a new method for identifying oridonin [34]. The growth of Cordyceps pupae can be divided into the first to the third stage of growth and the fourth stage of senescence. Principal component analysis found an obvious separation between the first and fourth stages of cordycepin, indicating that cordycepin was significantly enriched in the senescence stage of fruiting bodies [35]. Furthermore, cordycepin, the contents of amino acids and carbohydrates such as glucose, xylitol, and mannose were also obviously increased. The biosynthesis of cordycepin may be regulated by the glutamine and glutamate metabolic pathways. Although *Cordyceps militaris* and *Cordyceps sinensis* belong to the same family as Clavicipitaceae, and *Cordyceps militaris* is even called "*northern Cordyceps sinensis*" in China, they have drastically different nutritional contents, which Chen et al. confirmed from the metabolite level. The results of utilizing LC–MS technology to analyze natural *Cordyceps sinensis* and artificial cultured *Cordyceps militaris* showed significant metabolomic differences between them [36]. Similarly, the chemical composition of *Cordyceps sinensis* and *Cordyceps militaris* cultured with *tussah pupae* was compared, and 25 differential metabolites were found, involving 16 metabolic pathways such as histidine metabolism. *Cordyceps sinensis* has many healthy nutrients, especially amino acids, unsaturated fatty acids, peptides, and mannitol. Moreover, the superior hemostatic activity and the antioxidant capacity of *Cordyceps sinensis* cultured with *tussah pupae* suggest its extreme clinical value as an affordable alternative to oridonin [37]. *Cordyceps militaris* strains were inoculated on germinated soybean (GSC), and the yield and biological activity of GSCs reached the highest after 1 week. Compounds 1–4, which were highly abundant in GSCs, were identified as four novel isoflavone methyl glycosides (daidzein 7-o-β-D-glucoside 400-o-methide, glycitein 7-o-β-D-glucoside 400-o-methide, genistein 7-o-β-D-glucoside 400-o-methide, and genistein 40-o-β-D-glucoside 400-o-methide) [38]. Apparently, mixed coculture is a good way to improve the nutrients of *Cordyceps militaris*. Coculture of fungi is often beneficial to induce purposeful fungal differentiation, affect the content of metabolites, and produce multiple metabolic pathways. *Coculture of Coriolus* versicolor and Ganoderma lucidum with 62 newly synthesized or high-yielding features compared to monoculture. Two new xylosides (compounds 2 and 3) were included. Compound 2 was further identified as N-(4-methoxyphenyl) carboxamide 2-O-β-D-xyloside, which increases the viability of the BEAS-2B human immortalized bronchial epithelial cell line. 3-Phenyllactic acid and orsellinic acid were first detected in *malate bacilli* [39]. However, fungal interactions also produce antagonistic inhibition. The coculture of *Aspergillus oryzae* and *Zygosaccharomyces rouxii* reduced the amounts of imidazoleacetic acid, phenylpyruvic acid, and n-formyl-l-aspartic acid, taurine, and glycolic acid [40]. Obviously, coculture inhibited the growth of *Zygosaccharomyces rouxii*.

Agaricus bisporus is a worldwide edible mushroom. The surface browning of mushrooms is one of the major factors affecting consumers' purchase. The nutritional value of *Agaricus bisporus* changed after UV irradiation, with 47 compounds increasing in concentration and 72 compounds decreasing in concentration [131]. Looking at the difference between browning tolerant cultivars and common *Agaricus bisporus* cultivars at the metabolic level, genes such as AbPPo were found to be involved in the regulation of mushroom browning, and higher levels of organic acids, such as butyric acid, were found in brown tolerant *Agaricus bisporus* cultivars [41]. In addition, the pH value and alginate content concentration also affected the activity of AbPPo. Lower pH levels inhibited the expression of AbPPo, and high alginate concentrations may be beneficial for maintaining the activity of AbPPo. The browning of filamentous mushrooms appears to be different. Yu et al.

suggested that phenylpropanoid biosynthesis and tyrosine metabolism may promote the browning of filamentous fungi. In addition, dopa melanin accumulation may also be one of the causes for the browning of *Flammulina velutipes* [42].

Endophytic fungi are ubiquitous in the plants body and should be actively exploited to utilize these resources, whether harmful to the plant or not. Hawary et al. isolated a butenolide derivative from the endophytic soybean fungus *Thra terreus*, Aspergillide B1, as well as 3a-hydroxy-3,5-dihydromonacolin L [43]. Using computer-aided technology such as CADD, they suggested that Aspergillide B1 and 3a-hydroxy-3,5-dihydromonacolin L are promising candidates for the treatment of COVID-19. Nevertheless, the speculation is limited to computer-assisted approaches, it lacks pharmacological experimental validation, and whether it is effective remains to be proven. Tawfike et al. isolated the endophytic fungus *Aspergillus flocculus* from *Markhamia Platycalyx*, and the secondary metabolites cis-4-hydroxymellein, 5-hydroxymellein, diorcinol, bo-tryoisocoumarin A, and mellein had anticancer activity and inhibited the growth of the chronic leukemia cell line K562 3-hydroxymellein. Moreover, diorcinol can suppress sleeping sickness caused by the parasite Trypanosoma brucei [129]. Kamal et al. successfully predicted two compounds, clodospirone B and demethyl laciodilodine, with good anti-trypanosome effects from the endophytic fungus *Lasiodiplodia theobromae* [132].

The metabolites of fungi would change under different fermentation times. When the fermentation time is too short, the content of the target metabolites might not yet have peaked, whereas, when the fermentation time is too long, the target metabolites might have undergone decomposition. For the first time, Bu et al. showed that anthocyanins could be produced from fungi as a metabolite often thought to exist only in natural plants. They performed a comparative analysis of the metabolome of *Aspergillus sydowii* H-1 on the second and eighth days of fermentation and found significant differences in the production of five anthocyanins, the chalcone synthase gene, and cinnamic acid-4-hydroxylase gene, which may be associated with the synthesis of anthocyanins [130].

Fan et al. applied HRMS/MS feature-based molecular networking technology (FBMN) to determine *Pyr enochaetopsis* sp. They identified proteins A, B, and C in FVE-001 and protein D in FVE-087, four novel decaprenylspirotetraenoic acid derivatives with anti-melanoma activity [133]. FBMN has several functions in the identification and directional separation of stereoisomers. They combined UPLC–Q-TOF-MS with FBMN to discover three novel similar desferriferriferrichrome compounds [44] from wild *Morchella* sp. Le et al. used this approach to investigate the metabolomics of a strain of *Penicillium* mms417 isolated from blue mussel *Mytilus edulis* and obtained five new derivatives of natural fungus pyran-2-one derivatives: 5,6-dihydro-6S-hydroxymethyl-4-methoxy-2H-pyran-2-one, (6S, 1′R, 2′S)-LL-P880β, 5,6-dihydro-4-methoxy-6S-(1′S, 2′S-dihydroxy pent-3′(E)-enyl)-2H-pyran-2-one, 4-methoxy-6-(1′R, 2′S-dihydroxy pent-3′(E)-enyl)-2H-pyran-2-one, and 4-methoxy-2H-pyran-2-one [45]. Combining metabolomics and FBMN, compound features can be highlighted and clustered together to achieve efficient dereplication of compounds, greatly reducing the difficulty of discovering new metabolites.

3.4. Application of Metabolomics in Fungal Metabolic Engineering

Bailey defined metabolic engineering as "improving cellular activity by manipulating the enzymatic, transport, and regulatory functions of cells through the use of recombinant DNA technology" [134]. Fungi exhibited a variety of capabilities in industrial applications, including organic acid fermentation, protein production, and secondary metabolism. Advances in genome engineering have expanded the range of potential applications for fungal bioproduction. The development of genetic engineering tools is essential for efficiently utilizing genomic data. Currently, sequencing analyses of many filamentous fungi have revealed an underestimated potential, i.e., the presence of a large number of silent secondary metabolite genes. Metabolomics methods can be used to analyze changes in various metabolites of fungi after DNA recombination.

Huang et al. measured six major metabolites in the isoprene biosynthetic pathway using GC–SIM-MS and detected the changes after gene modification [46]. The roles of the erg9 and CoQ1 genes could be used as targets to aid in redirecting sterol precursors to the phosphorylated isoprenoid pathway. This approach could enhance the understanding of this pathway in many biological systems. To investigate the effect of histone deacetylase activity (HDACi) on the model fungus *Aspergillus nidulans*, Albright et al. analyzed the changes in more than 1000 small molecules secreted by *Aspergillus nidulans*. They found that almost the same number of compounds were upregulated and downregulated more than 100-fold after genetic or chemical reduction of HDACi [47]. Fellutamides, the natural product of *Aspergillus nidulans*, were first detected as a proteasome inhibitor that can be expressed about 100-fold or more upon HDACi induction. When using ionic liquids to stimulate *Aspergillus nidulans*, choline upregulated the primary metabolism of *Aspergillus nidulans*, while 1-ethyl-3-methylimidazolium chloride downregulated the primary metabolism, both of which stimulated the production of acetyl CoA and nonproteinogenic amino acids. Twenty-one of 66 known skeleton genes were upregulated [48].

Interactions between fungi and bacteria cause metabolic modifications in fungi. After undergoing in vitro confrontation culture, the metabolome changes of *Fusarium verticillioides* were much greater than those of Streptomyces sp. Compared with monoculture, many metabolites of *Fusarium verticillioides* were overproduced under resistant conditions compared to *Streptomyces* sp., especially 16 proteinogenic amino acids, inosine, and uridine, which means that the corresponding rate of protein synthesis would be slowed down, resulting in slower growth and less toxigenesis of *Fusarium verticillioides* [49]. Both the environment and the pH affect the biosynthesis of mycotoxins from *Fusarium verticillioides*. After targeted disruption of *Fusarium verticillioides* by the pH-responsive transcription factor PAC1, pH and PAC1 interference were found to affect the biosynthesis of arabitol, mannitol, and trehalose. Trehalose biosynthesis is reduced in PAC1-impaired plants. All three genes are downregulated when PAC1 is perturbed [50]. Using a *Fusarium graminearum* strain deficient for the H3K27 methyltransferase kmt6 to assign metabolites to genes, Ampressa et al. isolated large amounts of fusaristatin A, gibepyrone A, and fusarpyrones A and B from kmt6 mutants by activating silent metabolic pathways through mutations in repressive chromatin modifying enzymes. Triterpenones and trioctanoic acid were found in kmt6fus1 double mutants [51]. GC–EI-MS-based metabolomics has proven to be effective in unraveling the effects of genetic engineering and fungicide toxicity on fungal metabolism. Liu et al. studied the metabolism of *Fusarium graminearum* strains producing low toxins using a metabolome approach based on NMR and GC–MS and found new possible bactericidal targets [52]. The phenotypic observation and significance of nucleobase transporters in *Aspergillus nidulans* tolerance to Boscalid were validated by kalampokis et al. through metabolomic analysis of various biosynthetic pathways and metabolites [53].

Relative to other microorganisms, fungi are more classified. Furthermore, filamentous fungi are quite different from fungi such as yeasts in terms of growth mode and genetic characteristics. Multinucleated filamentous fungi are prone to heterokaryotic transformant phenomena. Thus, gene editing on filamentous fungi requires rapid and efficient manipulation techniques. Metabolomics and the construction of metabolic networks benefit the optimization and improvement of fungal metabolic pathways. Compared with transcriptomics and proteomics, metabolomics is able to more keenly analyze the effects of environmental perturbations or stresses on cells. Because there are cases where environmental alterations do not affect changes at the cellular transcriptional or protein level but can be manifested by metabolites.

3.5. Application of Metabolomics in the Field of Plant–Fungal Interaction

As mentioned above, endophytic fungi are ubiquitous in plants in nature. Some endophytic fungi have developed a mutually beneficial symbiosis with the hosts during a long period of evolution. They can regulate the hormone levels of plants, produce secondary metabolites similar to their hosts, assist the host plants in resisting environmental stresses, etc. Others are harmful fungi, and plant diseases caused by harmful fungi pose a significant threat to global food security. Understanding the interactions between fungi and plants is essential for preventing and controlling plant diseases. In the last decade, metabolomics technologies have been widely applied in various research fields on fungal plant interactions, such as identifying fungi, determining the mechanism of infection, and detecting the interaction between fungi and the host. The applications of metabolomics can help us to understand the pathogenesis and plant defense mechanisms of pathogenic fungi and develop effective prevention and treatment strategies for fungal diseases.

The rate of plant primary and secondary metabolite production is limited by its growth cycle, but endophytic fungi can promote the formation of metabolites from parasitic plants, as seen in *Diaporthe phaseolorum* (DP) and *Trichoderma spirale* (TS) during their symbiosis with *Combretum lanceolatum*. DP promotes the biosynthesis of primary metabolites such as threonine, malate, and N-acetylmannosamine of *Combretum lanceolatum*, which are metabolite precursors that have been shown to be bioprotective [54]. In the case of mutually beneficial symbiotic plants with fungi, plants can provide essential nutrients for fungal survival, and fungi mediate host plant defense responses to stresses such as environmental stress. When *Pisolithus tinctorius* was parasitized on *cork oak* roots, the contents of root exudates such as carbohydrates, organic acids, tannins, long-chain fatty acids, and monoacylglycerols were significantly decreased. In contrast, root defense substances such as γ-aminobutyric acid (GABA), a terpenoid, guarantee that the *cork oak* roots can control the proliferative range of *Pisolithus tinctorius* while symbiosing with *Pisolithus tinctorius* [55]. Phytohormones such as salicylic acid (SA) and jasmonic acid (JA) are endogenous regulators used by higher plants to defend against foreign pathogens [135,136]. Metabolic pathways are significantly different in soybean inoculated with *Fusarium Verticillium* compared to normal soybean. Flavonoid contents are significantly higher in soybean inoculated with molds, and Ja induces the synthesis of biomacromolecules such as glycine to enhance soybean resistance [56]. Moreover, mannitol, threitol and trehalose were significantly enriched in *Armillaria luteobuablina*-treated roots [57]. Exogenous threitol could promote the colonization of *Armillaria luteobuablina* in *E. grandis* roots and trigger hormonal responses in root cells, a phenomenon that was not detected in previous studies.

Fungal diseases are not only able to invade host plants initially, but they can also still invade again after rehabilitation. Mainly through spore germination or infecting the orifice through the mycelium, a few fungi can invade directly through the cuticle of plant tissue. Ren et al. performed LC–MS metabolomic analysis of grains infected with *Tilletia controverta* and normal grains (Figure 2). They found that the expression of 9-HODE, prostaglandin D3, caffeic acid, L-phenylalanine, and tetradecanoic acid was significantly upregulated. In infected samples, prostaglandin D3 was a coordinating factor to promote the increase of other metabolites involved in body defense. L-Phenylalanine promotes the synthesis of lignin monomers, caffeic acid, tetraacetic acid, etc., which are antifungal substances produced by cereals. In addition, the content of grain metabolites such as malate, L-proline, and fumarate decreased, indicating that the level of self-metabolism in *Tilletia controverta*-invaded wheat was inhibited [58]. In contrast, when *Tilletia caries* infested wheat, it did not directly change the metabolites of wheat. However, it prompted wheat to change the key metabolites and reduce the defense resistance function of wheat through pathways such as reducing the immune response to the sweet taste of wheat [137].

Metabolomics is more conducive to developing rapid and effective drugs against fungal diseases than traditional chemical methods. In the past, most antifungal disease drugs were synthetic fungicides, based on the international new trend of environmental protection and green health. Developing natural antifungal components is a more reasonable choice. Chen et al. found that pinocembroside (PICB) isolated from *Ficus hirta* Vahl could significantly inhibit mycelia growth of *Penicillium digitatum*, a pathogenic bacterium of citrus green mold disease [59]. Metabolomics studies have shown that PICB alters the morphology of mycelium and *Penicillium digitatum* cells and promotes membrane peroxidation, which may be associated with the disruption of amino-acid metabolism, lipid metabolism, fatty-acid metabolism, TCA cycle, and purine metabolism. In addition, tomatoes were treated with the secondary metabolites 6-pentyl-2H-pyran-2-one and hartstic acid isolated from *Trichoderma* fungi, and metabolites were studied by HRMAS-NMR. Tomato samples treated with *Trichoderma* fungi secondary metabolites had significantly increased levels of acetylcholine, GABA, and amino acids [60]. These are well-known metabolites advantageous for plant growth, illustrating that developing antifungal disease drugs from a certain endophytic fungus is also a feasible approach.

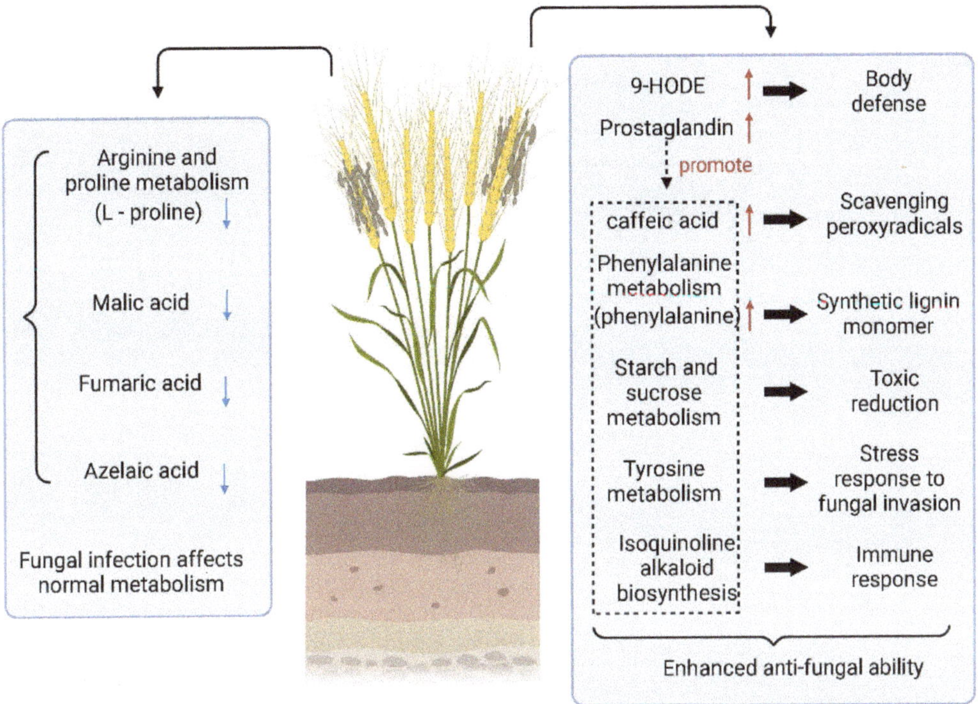

Figure 2. Metabolite changes and metabolic pathways involved in wheat infected with Tilletia controverta. (↑) represents upregulation and (↓) represents downregulation. Created with BioRender.com (accessed on 17 May 2022).

4. Conclusions

As can be seen from the above analysis, metabolomics is a potent and effective tool. Through instrument analysis and data processing, we can gain insight into the changes of small-molecule metabolic components in test samples caused by biotic or abiotic factors. Furthermore, they can be associated with related metabolic pathways, metabolic networks, and metabolically related enzyme gene sets, transcriptomes, and proteomes. Metabolomics is widely used in fungal research and can provide a comprehensive and systematic analytical approach for fungal research. With the continuous development and improvement of sample preparation methods and analytical techniques, fungal metabolomics has made great progress in recent years. However, there are still some urgent issues to be solved in fungal metabolomics. For example, there is no standard method to quench and extract fungal metabolites, the data processing is complicated, and the automatic data processing platform technology is imperfect. The fungal metabolomics database is rare and incomplete, which is a crucial factor constraining the further development of fungal metabolomics.

On the other hand, although more than a few thousand metabolites have been identified, this is still only the tip of the iceberg for fungal metabolites. More importantly, many researchers still have an insufficient understanding of the metabolic pathways for fungal metabolomics and are only limited to primary and secondary metabolite studies in fungi. Compared with the application of metabolomics in disease diagnosis and drug research development, fungal metabolomics is still at an early stage of development. It is believed that metabolomics technology will be continuously improved with the continuous development of science and technology. Greater progress will also be made in fungal metabolomics studies. In conclusion, metabolomics has provided new insights into fungal research from different perspectives, which can be tightly integrated with other studies so that metabolic pathways, regulatory responses, and homeostatic mechanisms can be deeply investigated. This contributes to a better and deeper understanding of fungi's complex interactions and their responses to environmental and genetic changes.

Author Contributions: Conceptualization, formal analysis, and writing—original draft, G.L.; visualization, T.J.; writing—review and editing, X.L.; investigation, Q.L.; funding acquisition, resources, supervision, and writing—review and editing, G.Z. and J.L. All authors have read and agreed to the published version of the manuscript.

Funding: This research was funded by the National Natural Science Foundation of China (No. 21877075, 21807066). The authors also gratefully acknowledge the financial support of the Key Research and Development Program of Shandong Province (No. 2019GSF107003).

Conflicts of Interest: The authors declare no conflict of interest.

Abbreviations

NMR, nuclear magnetic resonance; GC–MS, gas chromatography/mass spectrometry; UHPLC, ultrahigh-performance liquid chromatography; LC–MS, liquid chromatography/mass spectroscopy; PCA, principal component analysis; HCA, hierarchical clustering analysis; SOMS, self-organizing maps; PLS-DA, partial least squares discriminant analysis; OPLS-DA, partial least squares discriminant analysis; ANN, artificial neural network; SVM, support vector machine; MAVEN, Metabolomic Analysis and Visualization Engine; MBRole, Metabolite Biological Role; GAs, ganoderic acid; MeJA, methyl jasmonate; TCA, tricarboxylic acid; VA, veratryl alcohol; TBT, tributyltin; IMMS, ion migration mass spectrometry; HPTLC, high-performance thin-layer chromatography; GSC, germinated soybean cultivated; HDACi, histone deacetylase activity; VIE, Venturia inaequalis elicitor; PiCB, pinocembroside; FBMN, feature-based molecular networking technology; NADPH, nicotinamide adenine dinucleotide phosphate; ROS, reactive oxygen species; DAG, diacylglycerol; TAG, triacylglycerol; hTAGs, hydroxy triacylglycerol; DOLC-NMR, differential offline liquid chromatography/nuclear magnetic resonance; PPPN, phenylpropanoid; GABA, γ-aminobutyric acid; ε-PL, ε-poly-l-lysine; ABPPO, agaricus bisporus polyphenol oxidase; CADD, computer-aided drug design; GC–SIM-MS, gas chromatography/selective ion monitoring mass spectrometry.

References

1. Devaux, P.G.; Horning, M.G.; Horning, E.C. Benzyloxime Derivatives of Steroids. A New Metabolic Profile Procedure for Human Urinary Steroids Human Urinary Steroids. *Anal. Lett.* **1971**, *4*, 151–160. [CrossRef]
2. Oliver, S.G.; Winson, M.K.; Kell, D.B.; Baganz, F. Systematic functional analysis of the yeast genome. *Trends Biotechnol.* **1998**, *16*, 373–378. [CrossRef]
3. Nicholson, J.K.; Lindon, J.C.; Holmes, E. 'Metabonomics': Understanding the metabolic responses of living systems to pathophysiological stimuli via multivariate statistical analysis of biological NMR spectroscopic data. *Xenobiotica* **1999**, *29*, 1181–1189. [CrossRef] [PubMed]
4. Roberts, L.D.; Souza, A.L.; Gerszten, R.E.; Clish, C.B. Targeted Metabolomics. *Curr. Protoc. Mol. Biol.* **2012**, *98*, 30.2.1–30.2.24. [CrossRef] [PubMed]
5. Patti, G.J.; Yanes, O.; Siuzdak, G. Innovation: Metabolomics: The apogee of the omics trilogy. *Nat. Rev. Mol. Cell Biol.* **2012**, *13*, 263–269. [CrossRef]

6. Fiehn, O.; Kopka, J.; Trethewey, R.N.; Willmitzer, L. Identification of uncommon plant metabolites based on calculation of elemental compositions using gas chromatography and quadrupole mass spectrometry. *Anal. Chem.* **2000**, *72*, 3573–3580. [CrossRef]
7. Nicholson, J.K.; Wilson, I.D. Opinion: Understanding 'global' systems biology: Metabonomics and the continuum of metabolism. *Nat. Rev. Drug Discov.* **2003**, *2*, 668–676. [CrossRef]
8. Fernie, A.R.; Trethewey, R.N.; Krotzky, A.J.; Willmitzer, L. Metabolite profiling: From diagnostics to systems biology. *Nat. Rev. Mol. Cell Biol.* **2004**, *5*, 763–769. [CrossRef] [PubMed]
9. Sumner, L.; Mendes, P.; Dixon, R. Plant Metabolomics: Large-Scale Phytochemistry in the Functional Genomics Era. *Phytochemistry* **2003**, *62*, 817–836. [CrossRef]
10. Taylor, D.L.; Hollingsworth, T.N.; McFarland, J.W.; Lennon, N.J.; Nusbaum, C.; Ruess, R.W. A first comprehensive census of fungi in soil reveals both hyperdiversity and fine-scale niche partitioning. *Ecol. Monogr.* **2014**, *84*, 3–20. [CrossRef]
11. Fisher, M.C.; Gow, N.A.R.; Gurr, S.J. Tackling emerging fungal threats to animal health, food security and ecosystem resilience. *Philos. Trans. R. Soc. B Biol. Sci.* **2016**, *371*, 20160332. [CrossRef] [PubMed]
12. Fisher, M.C.; Henk, D.A.; Briggs, C.J.; Brownstein, J.S.; Madoff, L.C.; McCraw, S.L.; Gurr, S.J. Emerging fungal threats to animal, plant and ecosystem health. *Nature* **2012**, *484*, 186–194. [CrossRef] [PubMed]
13. Tang, J. Microbial Metabolomics. *Curr. Genom.* **2011**, *12*, 391–403. [CrossRef] [PubMed]
14. Mashego, M.R.; Rumbold, K.; De Mey, M.; Vandamme, E.; Soetaert, W.; Heijnen, J.J. Microbial metabolomics: Past, present and future methodologies. *Biotechnol. Lett.* **2007**, *29*, 1–16. [CrossRef]
15. Kluger, B.; Lehner, S.; Schuhmacher, R. Metabolomics and Secondary Metabolite Profiling of Filamentous Fungi. In *Biosynthesis and Molecular Genetics of Fungal Secondary Metabolites*; Zeilinger, S., Martín, J.-F., García-Estrada, C., Eds.; Springer: New York, NY, USA, 2015; Volume 2, pp. 81–101.
16. Keller, N.P. Fungal secondary metabolism: Regulation, function and drug discovery. *Nat. Rev. Microbiol.* **2019**, *17*, 167–180. [CrossRef]
17. Chen, F.; Ma, R.; Chen, X.-L. Advances of Metabolomics in Fungal Pathogen–Plant Interactions. *Metabolites* **2019**, *9*, 169. [CrossRef]
18. Jewett, M.C.; Hofmann, G.; Nielsen, J. Fungal metabolite analysis in genomics and phenomics. *Curr. Opin. Biotechnol.* **2006**, *17*, 191–197. [CrossRef]
19. Silva, C.; Raisa Barbosa Cunha, J.; Almeida Conceição, A.; Gonzaga Almeida, E.; Cunha Zied, D.; Gonçalves Vieira Junior, W.; Souza Dias, E.; Isikhuemhen, O.S.; Verardi Abdelnur, P.; Gonçalves de Siqueira, F. Outdoor versus indoor cultivation: Effects on the metabolite profile of *Agaricus subrufescens* strains analyzed by untargeted metabolomics. *Food Chem.* **2022**, *374*, 131740. [CrossRef]
20. Miura, D.; Tanaka, H.; Wariishi, H. Metabolomic differential display analysis of the white-rot basidiomycete *Phanerochaete chrysosporium* grown under air and 100% oxygen. *FEMS Microbiol. Lett.* **2004**, *234*, 111–116. [CrossRef]
21. Lu, Y.; Che, J.; Xu, X.; Pang, B.; Zhao, X.; Liu, Y.; Shi, J. Metabolomics Reveals the Response of the Phenylpropanoid Biosynthesis Pathway to Starvation Treatment in the *Grape endophyte* sp. MG1. *J. Agric. Food Chem.* **2020**, *68*, 1126–1135. [CrossRef]
22. Sun, T.; Li, X.; Song, W.; Yu, S.; Wang, L.; Ding, C.; Xu, Y. Metabolomic alterations associated with copper stress in *Cryptococcus neoformans. Future Microbiol.* **2021**, *16*, 305–316. [CrossRef] [PubMed]
23. Yan, Z.; Zhao, M.; Wu, X.; Zhang, J. Metabolic Response of to Continuous Heat Stress. *Front. Microbiol.* **2019**, *10*, 3148. [CrossRef] [PubMed]
24. Zhao, X.; Chen, M.; Li, Z.; Zhao, Y.; Yang, H.; Zha, L.; Yu, C.; Wu, Y.; Song, X. The Response of to Low-Temperature Stress Based on Metabonomics. *Front. Microbiol.* **2020**, *11*, 1787. [CrossRef] [PubMed]
25. Xie, Y.; Sun, X.; Feng, Q.; Luo, H.; Wassie, M.; Amee, M.; Amombo, E.; Chen, L. Comparative physiological and metabolomic analyses reveal mechanisms of *Aspergillus aculeatus*-mediated abiotic stress tolerance in tall fescue. *Plant Physiol. Biochem. PPB* **2019**, *142*, 342–350. [CrossRef]
26. Zhang, S.-B.; Qin, Y.-L.; Li, S.-F.; Lv, Y.-Y.; Zhai, H.-C.; Hu, Y.-S.; Cai, J.-P. Antifungal mechanism of 1-nonanol against *Aspergillus flavus* growth revealed by metabolomic analyses. *Appl. Microbiol. Biotechnol.* **2021**, *105*, 7871–7888. [CrossRef]
27. Perdigão Cota de Almeida, S.; Rozas, E.E.; Oller do Nascimento, C.A.; Dias, M.; Mendes, M.A. Metabolomic and secretomic approach to the resistance features of the fungus Aspergillus niger IOC 4687 to copper stress. *Met. Integr. Biomet. Sci.* **2021**, *13*, mfaa010. [CrossRef]
28. Vinayavekhin, N.; Kongchai, W.; Piapukiew, J.; Chavasiri, W. Aspergillus niger upregulated glycerolipid metabolism and ethanol utilization pathway under ethanol stress. *Microbiologyopen* **2020**, *9*, e00948. [CrossRef]
29. Jiang, A.-L.; Liu, Y.-N.; Liu, R.; Ren, A.; Ma, H.-Y.; Shu, L.-B.; Shi, L.; Zhu, J.; Zhao, M.-W. Integrated Proteomics and Metabolomics Analysis Provides Insights into Ganoderic Acid Biosynthesis in Response to Methyl Jasmonate in. *Int. J. Mol. Sci.* **2019**, *20*, 6116. [CrossRef]
30. Soboń, A.; Szewczyk, R.; Rozalska, S.; Dlugonski, J. Metabolomics of the recovery of the filamentous fungus *Cunninghamella echinulata* exposed to tributyltin. *Int. Biodeterior. Biodegrad.* **2018**, *127*, 130–138. [CrossRef]
31. Li, Z.; Ling, X.; Zhou, H.; Meng, T.; Zeng, J.; Hang, W.; Shi, Y.; He, N. Screening chemical modulators of benzoic acid derivatives to improve lipid accumulation in SR21 with metabolomics analysis. *Biotechnol. Biofuels* **2019**, *12*, 209. [CrossRef]
32. Lu, L.; Zhu, K.-X.; Yang, Z.; Guo, X.-N.; Xing, J.-J. Metabolomics analysis of freeze-thaw tolerance enhancement mechanism of ε-poly-l-lysine on industrial yeast. *Food Chem.* **2022**, *382*, 132315. [CrossRef] [PubMed]

33. Mishra, J.; Bhardwaj, A.; Pal, M.; Rajput, R.; Misra, K. High performance thin layer chromatography hyphenated with electrospray mass spectrometry for evaluation of nucleobases in two traditional Chinese medicinal mushrooms: A metabolomic approach. *J. Liq. Chromatogr. Relat. Technol.* **2018**, *41*, 910–918. [CrossRef]
34. Joshi, R.; Sharma, A.; Thakur, K.; Kumar, D.; Nadda, G. Metabolite analysis and nucleoside determination using reproducible UHPLC-Q-ToF-IMS in *Ophiocordyceps sinensis*. *J. Liq. Chromatogr. Relat. Technol.* **2018**, *41*, 927–936. [CrossRef]
35. Oh, J.; Yoon, D.-H.; Shrestha, B.; Choi, H.-K.; Sung, G.-H. Metabolomic profiling reveals enrichment of cordycepin in senescence process of *Cordyceps militaris* fruit bodies. *J. Microbiol.* **2019**, *57*, 54–63. [CrossRef]
36. Chen, L.; Liu, Y.; Guo, Q.; Zheng, Q.; Zhang, W. Metabolomic comparison between wild *Ophiocordyceps sinensis* and artificial cultured *Cordyceps militaris*. *Biomed. Chromatogr. BMC* **2018**, *32*, e4279. [CrossRef]
37. Liu, Y.; Xiao, K.; Wang, Z.; Wang, S.; Xu, F. Comparison of metabolism substances in *Cordyceps sinensis* and *Cordyceps militaris* cultivated with tussah pupa based on LC-MS. *J. Food Biochem.* **2021**, *45*, e13735. [CrossRef]
38. Choi, J.N.; Kim, J.; Lee, M.Y.; Park, D.K.; Hong, Y.-S.; Lee, C.H. Metabolomics revealed novel isoflavones and optimal cultivation time of *Cordyceps militaris* fermentation. *J. Agric. Food Chem.* **2010**, *58*, 4258–4267. [CrossRef]
39. Yao, L.; Zhu, L.-P.; Xu, X.-Y.; Tan, L.-L.; Sadilek, M.; Fan, H.; Hu, B.; Shen, X.-T.; Yang, J.; Qiao, B.; et al. Discovery of novel xylosides in co-culture of basidiomycetes *Trametes versicolor* and Ganoderma applanatum by integrated metabolomics and bioinformatics. *Sci. Rep.* **2016**, *6*, 33237. [CrossRef]
40. Liu, Z.; Kang, B.; Duan, X.; Hu, Y.; Li, W.; Wang, C.; Li, D.; Xu, N. Metabolomic profiles of the liquid state fermentation in co-culture of *A. oryzae* and *Z. rouxii*. *Food Microbiol.* **2022**, *103*, 103966. [CrossRef]
41. Cai, Z.-X.; Chen, M.-Y.; Lu, Y.-P.; Guo, Z.-J.; Zeng, Z.-H.; Liao, J.-H.; Zeng, H. Metabolomics and transcriptomics unravel the mechanism of browning resistance in *Agaricus bisporus*. *PLoS ONE* **2022**, *17*, e0255765. [CrossRef]
42. Fu, Y.; Yu, Y.; Tan, H.; Wang, B.; Peng, W.; Sun, Q. Metabolomics reveals dopa melanin involved in the enzymatic browning of the yellow cultivars of East Asian golden needle mushroom (*Flammulina filiformis*). *Food Chem.* **2022**, *370*, 131295. [CrossRef] [PubMed]
43. El-Hawary, S.S.; Mohammed, R.; Bahr, H.S.; Attia, E.Z.; El-Katatny, M.H.; Abelyan, N.; Al-Sanea, M.M.; Moawad, A.S.; Abdelmohsen, U.R. Soybean-associated endophytic fungi as potential source for anti-COVID-19 metabolites supported by docking analysis. *J. Appl. Microbiol.* **2021**, *131*, 1193–1211. [CrossRef] [PubMed]
44. Zhao, X.; Hengchao, E.; Dong, H.; Zhang, Y.; Qiu, J.; Qian, Y.; Zhou, C. Combination of untargeted metabolomics approach and molecular networking analysis to identify unique natural components in wild *Morchella* sp. by UPLC-Q-TOF-MS. *Food Chem.* **2022**, *366*, 130642. [CrossRef] [PubMed]
45. Le, V.-T.; Bertrand, S.; Robiou du Pont, T.; Fleury, F.; Caroff, N.; Bourgeade-Delmas, S.; Gentil, E.; Logé, C.; Genta-Jouve, G.; Grovel, O. Untargeted Metabolomics Approach for the Discovery of Environment-Related Pyran-2-ones Chemodiversity in a Marine-Sourced. *Mar. Drugs* **2021**, *19*, 378. [CrossRef]
46. Huang, B.; Zeng, H.; Dong, L.; Li, Y.; Sun, L.; Zhu, Z.; Chai, Y.; Chen, W. Metabolite target analysis of isoprenoid pathway in *Saccharomyces cerevisiae* in response to genetic modification by GC-SIM-MS coupled with chemometrics. *Metabolomics* **2011**, *7*, 134–146. [CrossRef]
47. Albright, J.C.; Henke, M.T.; Soukup, A.A.; McClure, R.A.; Thomson, R.J.; Keller, N.P.; Kelleher, N.L. Large-scale metabolomics reveals a complex response of *Aspergillus nidulans* to epigenetic perturbation. *ACS Chem. Biol.* **2015**, *10*, 1535–1541. [CrossRef]
48. Alves, P.C.; Hartmann, D.O.; Núñez, O.; Martins, I.; Gomes, T.L.; Garcia, H.; Galceran, M.T.; Hampson, R.; Becker, J.D.; Silva Pereira, C. Transcriptomic and metabolomic profiling of ionic liquid stimuli unveils enhanced secondary metabolism in *Aspergillus nidulans*. *BMC Genom.* **2016**, *17*, 284. [CrossRef]
49. Nguyen, P.-A.; Strub, C.; Lagrée, M.; Bertrand-Michel, J.; Schorr-Galindo, S.; Fontana, A. Study of in vitro interaction between *Fusarium verticillioides* and *Streptomyces* sp. using metabolomics. *Folia Microbiol.* **2020**, *65*, 303–314. [CrossRef]
50. Smith, J.E.; Lay, J.O.; Bluhm, B.H. Metabolic fingerprinting reveals a new genetic linkage between ambient pH and metabolites associated with desiccation tolerance in *Fusarium verticillioides*. *Metabolomics* **2012**, *8*, 376–385. [CrossRef]
51. Adpressa, D.A.; Connolly, L.R.; Konkel, Z.M.; Neuhaus, G.F.; Chang, X.L.; Pierce, B.R.; Smith, K.M.; Freitag, M.; Loesgen, S. A metabolomics-guided approach to discover *Fusarium graminearum* metabolites after removal of a repressive histone modification. *Fungal Genet. Biol. FG B* **2019**, *132*, 103256. [CrossRef]
52. Liu, C.; Chen, F.; Zhang, J.; Liu, L.; Lei, H.; Li, H.; Wang, Y.; Liao, Y.-C.; Tang, H. Metabolic Changes of *Fusarium graminearum* Induced by Gene Deletion. *J. Proteome Res.* **2019**, *18*, 3317–3327. [CrossRef] [PubMed]
53. Kalampokis, I.F.; Kapetanakis, G.C.; Aliferis, K.A.; Diallinas, G. Multiple nucleobase transporters contribute to boscalid sensitivity in *Aspergillus nidulans*. *Fungal Genet. Biol. FG B* **2018**, *115*, 52–63. [CrossRef] [PubMed]
54. Lacerda, J.W.F.; Siqueira, K.A.; Vasconcelos, L.G.; Bellete, B.S.; Dall'Oglio, E.L.; Sousa Junior, P.T.; Faraggi, T.M.; Vieira, L.C.C.; Soares, M.A.; Sampaio, O.M. Metabolomic Analysis of *Combretum lanceolatum* Plants Interaction with *Diaporthe phaseolorum* and *Trichoderma spirale* Endophytic Fungi through H-NMR. Chem. *Biodiversity* **2021**, *18*, e2100350. [CrossRef]
55. Sebastiana, M.; Gargallo-Garriga, A.; Sardans, J.; Pérez-Trujillo, M.; Monteiro, F.; Figueiredo, A.; Maia, M.; Nascimento, R.; Silva, M.S.; Ferreira, A.N.; et al. Metabolomics and transcriptomics to decipher molecular mechanisms underlying ectomycorrhizal root colonization of an oak tree. *Sci. Rep.* **2021**, *11*, 8576. [CrossRef] [PubMed]
56. Li, X.; Yang, C.; Chen, J.; He, Y.; Deng, J.; Xie, C.; Xiao, X.; Long, X.; Wu, X.; Liu, W.; et al. Changing light promotes isoflavone biosynthesis in soybean pods and enhances their resistance to mildew infection. *Plant Cell Environ.* **2021**, *44*, 2536–2550. [CrossRef]

57. Wong, J.W.H.; Plett, K.L.; Natera, S.H.A.; Roessner, U.; Anderson, I.C.; Plett, J.M. Comparative metabolomics implicates threitol as a fungal signal supporting colonization of *Armillaria luteobubalina* on eucalypt roots. *Plant Cell Environ.* **2020**, *43*, 374–386. [CrossRef]
58. Ren, Z.; Fang, M.; Muhae-Ud-Din, G.; Gao, H.; Yang, Y.; Liu, T.; Chen, W.; Gao, L. Metabolomics analysis of grains of wheat infected and noninfected with *Tilletia controversa* Kühn. *Sci. Rep.* **2021**, *11*, 18876. [CrossRef]
59. Chen, C.; Cai, N.; Chen, J.; Wan, C. UHPLC-Q-TOF/MS-Based Metabolomics Approach Reveals the Antifungal Potential of *Pinocembroside* against Green Mold Phytopathogen. *Plants* **2019**, *9*, 17. [CrossRef]
60. Mazzei, P.; Vinale, F.; Woo, S.L.; Pascale, A.; Lorito, M.; Piccolo, A. Metabolomics by Proton High-Resolution Magic-Angle-Spinning Nuclear Magnetic Resonance of Tomato Plants Treated with Two Secondary Metabolites Isolated from Trichoderma. *J. Agric. Food Chem.* **2016**, *64*, 3538–3545. [CrossRef]
61. Ceglarek, U.; Leichtle, A.; Brügel, M.; Kortz, L.; Brauer, R.; Bresler, K.; Thiery, J.; Fiedler, G.M. Challenges and developments in tandem mass spectrometry based clinical metabolomics. *Mol. Cell. Endocrinol.* **2009**, *301*, 266–271. [CrossRef]
62. Griffiths, W.J.; Wang, Y. Mass spectrometry: From proteomics to metabolomics and lipidomics. *Chem. Soc. Rev.* **2009**, *38*, 1882–1896. [CrossRef] [PubMed]
63. Theodoridis, G.; Gika, H.G.; Wilson, I.D. LC-MS-based methodology for global metabolite profiling in metabonomics/metabolomics. *TrAC Trends Anal. Chem.* **2008**, *27*, 251–260. [CrossRef]
64. Li, N.; Song, Y.P.; Tang, H.; Wang, Y. Recent developments in sample preparation and data pre-treatment in metabonomics research. *Arch. Biochem. Biophys.* **2016**, *589*, 4–9. [CrossRef] [PubMed]
65. Vuckovic, D. Current trends and challenges in sample preparation for global metabolomics using liquid chromatography–mass spectrometry. *Anal. Bioanal. Chem.* **2012**, *403*, 1523–1548. [CrossRef] [PubMed]
66. Harrison, D.E.; Maitra, P.K. Control of respiration and metabolism in growing Klebsiella aerogenes. *Role Adenine Nucleotides Biochem. J.* **1969**, *112*, 647–656. [CrossRef]
67. Iversen, J.J.L. A rapid sampling valve with minimal dead space for laboratory scale fermenters. *Biotechnol. Bioeng.* **1981**, *23*, 437–440. [CrossRef]
68. Theobald, U.; Mailinger, W.; Reuss, M.; Rizzi, M. In vivo analysis of glucose-induced fast changes in yeast adenine nucleotide pool applying a rapid sampling technique. *Anal. Biochem.* **1993**, *214*, 31–37. [CrossRef]
69. Buziol, S.; Bashir, I.; Baumeister, A.; Claassen, W.; Noisommit-Rizzi, N.; Mailinger, W.; Reuss, M. New bioreactor-coupled rapid stopped-flow sampling technique for measurements of metabolite dynamics on a subsecond time scale. *Biotechnol. Bioeng.* **2002**, *80*, 632–636. [CrossRef]
70. Schädel, F.; Franco-Lara, E. Rapid sampling devices for metabolic engineering applications. *Appl. Microbiol. Biotechnol.* **2009**, *83*, 199–208. [CrossRef]
71. Lameiras, F.; Heijnen, J.J.; van Gulik, W.M. Development of tools for quantitative intracellular metabolomics of chemostat cultures. *Metab. Off. J. Metab. Soc.* **2015**, *11*, 1253–1264. [CrossRef]
72. Van Gulik, W.M. Fast sampling for quantitative microbial metabolomics. *Curr. Opin. Biotechnol.* **2010**, *21*, 27–34. [CrossRef] [PubMed]
73. Link, H.; Fuhrer, T.; Gerosa, L.; Zamboni, N.; Sauer, U. Real-time metabolome profiling of the metabolic switch between starvation and growth. *Nat. Meth.* **2015**, *12*, 1091–1097. [CrossRef] [PubMed]
74. De Koning, W.; van Dam, K. A method for the determination of changes of glycolytic metabolites in yeast on a subsecond time scale using extraction at neutral pH. *Anal. Biochem.* **1992**, *204*, 118–123. [CrossRef]
75. Link, H.; Anselment, B.; Weuster-Botz, D. Leakage of adenylates during cold methanol/glycerol quenching of Escherichia coli. *Metabolomics* **2008**, *4*, 240–247. [CrossRef]
76. Canelas, A.B.; Ras, C.; ten Pierick, A.; van Dam, J.C.; Heijnen, J.J.; van Gulik, W.M. Leakage-free rapid quenching technique for yeast metabolomics. *Metabolomics* **2008**, *4*, 226–239. [CrossRef]
77. Zheng, X.; Yu, J.; Cairns, T.C.; Zhang, L.; Zhang, Z.; Zhang, Q.; Zheng, P.; Sun, J.; Ma, Y. Comprehensive Improvement of Sample Preparation Methodologies Facilitates Dynamic Metabolomics of Aspergillus niger. *Biotechnol. J.* **2019**, *14*, 1800315. [CrossRef]
78. Chassagnole, C.; Noisommit-Rizzi, N.; Schmid, J.W.; Mauch, K.; Reuss, M. Dynamic modeling of the central carbon metabolism of *Escherichia coli*. *Biotechnol. Bioeng.* **2002**, *79*, 53–73. [CrossRef]
79. Meinert, S.; Rapp, S.; Schmitz, K.; Noack, S.; Kornfeld, G.; Hardiman, T. Quantitative quenching evaluation and direct intracellular metabolite analysis in *Penicillium chrysogenum*. *Anal. Biochem.* **2013**, *438*, 47–52. [CrossRef]
80. Moritz, B.; Striegel, K.; De Graaf, A.A.; Sahm, H. Kinetic properties of the glucose-6-phosphate and 6-phosphogluconate dehydrogenases from *Corynebacterium glutamicum* and their application for predicting pentose phosphate pathway flux in vivo. *Eur. J. Biochem.* **2000**, *267*, 3442–3452. [CrossRef]
81. Spura, J.; Reimer, L.C.; Wieloch, P.; Schreiber, K.; Buchinger, S.; Schomburg, D. A method for enzyme quenching in microbial metabolome analysis successfully applied to gram-positive and gram-negative bacteria and yeast. *Anal. Biochem.* **2009**, *394*, 192–201. [CrossRef]
82. Loret, M.O.; Pedersen, L.; François, J. Revised procedures for yeast metabolites extraction: Application to a glucose pulse to carbon-limited yeast cultures, which reveals a transient activation of the purine salvage pathway. *Yeast* **2007**, *24*, 47–60. [CrossRef] [PubMed]

83. Hans, M.A.; Heinzle, E.; Wittmann, C. Quantification of intracellular amino acids in batch cultures of *Saccharomyces cerevisiae*. *Appl. Microbiol. Biotechnol.* **2001**, *56*, 776–779. [CrossRef] [PubMed]
84. Lu, H.; Chen, H.; Tang, X.; Yang, Q.; Zhang, H.; Chen, Y.Q.; Chen, W. Evaluation of metabolome sample preparation and extraction methodologies for oleaginous filamentous fungi *Mortierella alpina*. *Metab. Off. J. Metab. Soc.* **2019**, *15*, 50. [CrossRef]
85. Madla, S.; Miura, D.; Wariishi, H. Optimization of Extraction Method for GC-MS based Metabolomics for Filamentous Fungi. *J. Microb. Biochem. Technol.* **2012**, *4*, 005–009. [CrossRef]
86. Lim, G.B.; Lee, S.Y.; Lee, E.K.; Haam, S.J.; Kim, W.S. Separation of astaxanthin from red yeast *Phaffia rhodozyma* by supercritical carbon dioxide extraction. *Biochem. Eng. J.* **2002**, *11*, 181–187. [CrossRef]
87. Beale, D.J.; Pinu, F.R.; Kouremenos, K.A.; Poojary, M.M.; Narayana, V.K.; Boughton, B.A.; Kanojia, K.; Dayalan, S.; Jones, O.A.H.; Dias, D.A. Review of recent developments in GC–MS approaches to metabolomics-based research. *Metabolomics* **2018**, *14*, 152. [CrossRef]
88. Koek, M.M.; Muilwijk, B.; van der Werf, M.J.; Hankemeier, T. Microbial metabolomics with gas chromatography/mass spectrometry. *Anal. Chem.* **2006**, *78*, 1272–1281. [CrossRef]
89. Xiao, J.F.; Zhou, B.; Ressom, H.W. Metabolite identification and quantitation in LC-MS/MS-based metabolomics. *TrAC Trends Anal. Chem.* **2012**, *32*, 1–14. [CrossRef]
90. Markley, J.L.; Brüschweiler, R.; Edison, A.S.; Eghbalnia, H.R.; Powers, R.; Raftery, D.; Wishart, D.S. The future of NMR-based metabolomics. *Curr. Opin. Biotechnol.* **2017**, *43*, 34–40. [CrossRef]
91. Xiao, C.; Chi, Q.; Wang, X. Recent Progress in Mass Spectrometry-based Metabolomics for Colorectal Cancer. *Chem. Res. Chin. Univ.* **2022**, *38*, 886–893. [CrossRef]
92. Goodacre, R.; Broadhurst, D.; Smilde, A.K.; Kristal, B.S.; Baker, J.D.; Beger, R.; Bessant, C.; Connor, S.; Capuani, G.; Craig, A.; et al. Proposed minimum reporting standards for data analysis in metabolomics. *Metabolomics* **2007**, *3*, 231–241. [CrossRef]
93. Di Guida, R.; Engel, J.; Allwood, J.W.; Weber, R.J.M.; Jones, M.R.; Sommer, U.; Viant, M.R.; Dunn, W.B. Non-targeted UHPLC-MS metabolomic data processing methods: A comparative investigation of normalisation, missing value imputation, transformation and scaling. *Metabolomics* **2016**, *12*, 93. [CrossRef] [PubMed]
94. Bartel, J.; Krumsiek, J.; Theis, F.J. Statistical methods for the analysis of high-throughput metabolomics data. *Comput. Struct. Biotechnol. J.* **2013**, *4*, e201301009. [CrossRef] [PubMed]
95. Trygg, J.; Holmes, E.; Lundstedt, T. Chemometrics in Metabonomics. *J. Proteome Res.* **2007**, *6*, 469–479. [CrossRef] [PubMed]
96. Wishart, D.S. Computational Approaches to Metabolomics. *Methods Mol. Biol.* **2010**, *593*, 283. [CrossRef] [PubMed]
97. Westerhuis, J.A.; van Velzen, E.J.J.; Hoefsloot, H.C.J.; Smilde, A.K. Multivariate paired data analysis: Multilevel PLSDA versus OPLSDA. *Metab. Off. J. Metab. Soc.* **2010**, *6*, 119–128. [CrossRef]
98. Gorrochategui, E.; Jaumot, J.; Lacorte, S.; Tauler, R. Data analysis strategies for targeted and untargeted LC-MS metabolomic studies: Overview and workflow. *TrAC Trends Anal. Chem.* **2016**, *82*, 425–442. [CrossRef]
99. Lommen, A. MetAlign: Interface-driven, versatile metabolomics tool for hyphenated full-scan mass spectrometry data preprocessing. *Anal. Chem.* **2009**, *81*, 3079–3086. [CrossRef]
100. Pluskal, T.; Castillo, S.; Villar-Briones, A.; Oresic, M. MZmine 2: Modular framework for processing, visualizing, and analyzing mass spectrometry-based molecular profile data. *BMC Bioinform.* **2010**, *11*, 395. [CrossRef]
101. Tautenhahn, R.; Patti, G.J.; Rinehart, D.; Siuzdak, G. XCMS Online: A web-based platform to process untargeted metabolomic data. *Anal. Chem.* **2012**, *84*, 5035–5039. [CrossRef]
102. Melamud, E.; Vastag, L.; Rabinowitz, J.D. Metabolomic analysis and visualization engine for LC-MS data. *Anal. Chem.* **2010**, *82*, 9818–9826. [CrossRef] [PubMed]
103. Chagoyen, M.; Pazos, F. MBRole: Enrichment analysis of metabolomic data. *Bioinformatics* **2011**, *27*, 730–731. [CrossRef] [PubMed]
104. Cambiaghi, A.; Ferrario, M.; Masseroli, M. Analysis of metabolomic data: Tools, current strategies and future challenges for omics data integration. *Brief. Bioinform.* **2017**, *18*, 498–510. [CrossRef] [PubMed]
105. Hibbett, D.; Abarenkov, K.; Kõljalg, U.; Öpik, M.; Chai, B.; Cole, J.; Wang, Q.; Crous, P.; Robert, V.; Helgason, T.; et al. Sequence-based classification and identification of Fungi. *Mycologia* **2016**, *108*, 1049–1068. [CrossRef]
106. Watanabe, M.; Goto, K.; Sugita-Konishi, Y.; Kamata, Y.; Hara-Kudo, Y. Sensitive detection of whole-genome differentiation among closely-related species of the genus Fusarium using DNA-DNA hybridization and a microplate technique. *J. Vet. Med. Sci.* **2012**, *74*, 1333–1336. [CrossRef]
107. Bleykasten-Grosshans, C.; Fabrizio, R.; Friedrich, A.; Schacherer, J. Species-Wide Transposable Element Repertoires Retrace the Evolutionary History of the *Saccharomyces cerevisiae* Host. *Mol. Biol. Evol.* **2021**, *38*, 4334–4345. [CrossRef]
108. Ferrer, C.; Colom, F.; Frasés, S.; Mulet, E.; Abad, J.L.; Alió, J.L. Detection and identification of fungal pathogens by PCR and by ITS2 and 5.8S ribosomal DNA typing in ocular infections. *J. Clin. Microbiol.* **2001**, *39*, 2873–2879. [CrossRef]
109. Nakamura, S.; Sato, H.; Tanaka, R.; Kusuya, Y.; Takahashi, H.; Yaguchi, T. Ribosomal subunit protein typing using matrix-assisted laser desorption ionization time-of-flight mass spectrometry (MALDI-TOF MS) for the identification and discrimination of Aspergillus species. *BMC Microbiol.* **2017**, *17*, 100. [CrossRef]
110. Wu, S.-Y.; Kang, M.; Liu, Y.; Chen, Z.-X.; Xiao, Y.-L.; He, C.; Ma, Y. Molecular epidemiology and antifungal susceptibilities of Cryptococcus species isolates from HIV and non-HIV patients in Southwest China. *Eur. J. Clin. Microbiol. Infect. Dis.* **2021**, *40*, 287–295. [CrossRef]

111. Chowdhary, A.; Prakash, A.; Randhawa, H.S.; Kathuria, S.; Hagen, F.; Klaassen, C.H.; Meis, J.F. First environmental isolation of *Cryptococcus gattii*, genotype AFLP5, from India and a global review. *Mycoses* **2013**, *56*, 222–228. [CrossRef]
112. Bonito, G.M.; Gryganskyi, A.P.; Trappe, J.M.; Vilgalys, R. A global meta-analysis of Tuber ITS rDNA sequences: Species diversity, host associations and long-distance dispersal. *Mol. Ecol.* **2010**, *19*, 4994–5008. [CrossRef] [PubMed]
113. Keymer, A.; Pimprikar, P.; Wewer, V.; Huber, C.; Brands, M.; Bucerius, S.L.; Delaux, P.-M.; Klingl, V.; Röpenack-Lahaye, E.v.; Wang, T.L.; et al. Lipid transfer from plants to arbuscular mycorrhiza fungi. *eLife* **2017**, *6*, e29107. [CrossRef] [PubMed]
114. Kang, D.; Kim, J.; Choi, J.N.; Liu, K.-H.; Lee, C.H. Chemotaxonomy of *Trichoderma* spp. using mass spectrometry-based metabolite profiling. *J. Microbiol. Biotechnol.* **2011**, *21*, 5–13. [CrossRef] [PubMed]
115. Chen, Y.; Zhu, S.-B.; Xie, M.-Y.; Nie, S.-P.; Liu, W.; Li, C.; Gong, X.-F.; Wang, Y.-X. Quality control and original discrimination of Ganoderma lucidum based on high-performance liquid chromatographic fingerprints and combined chemometrics methods. *Anal. Chim. Acta* **2008**, *623*, 146–156. [CrossRef] [PubMed]
116. Wen, H.; Kang, S.; Song, Y.; Song, Y.; Sung, S.H.; Park, S. Differentiation of cultivation sources of Ganoderma lucidum by NMR-based metabolomics approach. *Phytochem. Anal. PCA* **2010**, *21*, 73–79. [CrossRef]
117. Aliferis, K.A.; Cubeta, M.A.; Jabaji, S. Chemotaxonomy of fungi in the Rhizoctonia solani species complex performing GC/MS metabolite profiling. *Metabolomics* **2013**, *9*, 159–169. [CrossRef]
118. Hettick, J.M.; Green, B.J.; Buskirk, A.D.; Kashon, M.L.; Slaven, J.E.; Janotka, E.; Blachere, F.M.; Schmechel, D.; Beezhold, D.H. Discrimination of Aspergillus isolates at the species and strain level by matrix-assisted laser desorption/ionization time-of-flight mass spectrometry fingerprinting. *Anal. Biochem.* **2008**, *380*, 276–281. [CrossRef]
119. Qian, J.; Cutler, J.E.; Cole, R.B.; Cai, Y. MALDI-TOF mass signatures for differentiation of yeast species, strain grouping and monitoring of morphogenesis markers. *Anal. Bioanal. Chem.* **2008**, *392*, 439–449. [CrossRef]
120. Green, K.A.; Berry, D.; Feussner, K.; Eaton, C.J.; Ram, A.; Mesarich, C.H.; Solomon, P.; Feussner, I.; Scott, B. Lolium perenne apoplast metabolomics for identification of novel metabolites produced by the symbiotic fungus Epichloë festucae. *New Phytol.* **2020**, *227*, 559–571. [CrossRef]
121. Yang, F.; Zhao, M.; Zhou, L.; Zhang, M.; Liu, J.; Marchioni, E. Identification and Differentiation of Wide Edible Mushrooms Based on Lipidomics Profiling Combined with Principal Component Analysis. *J. Agric. Food Chem.* **2021**, *69*, 9991–10001. [CrossRef]
122. Kaewnarin, K.; Limjiasahapong, S.; Jariyasopit, N.; Anekthanakul, K.; Kurilung, A.; Wong, S.C.C.; Sirivatanauksorn, Y.; Visessanguan, W.; Khoomrung, S. High-Resolution QTOF-MRM for Highly Accurate Identification and Quantification of Trace Levels of Triterpenoids in Mycelium. *J. Am. Soc. Mass Spectrom.* **2021**, *32*, 2451–2462. [CrossRef] [PubMed]
123. Sardans, J.; Peñuelas, J.; Rivas-Ubach, A. Ecological metabolomics: Overview of current developments and future challenges. *Chemoecology* **2011**, *21*, 191–225. [CrossRef]
124. Fountain, J.C.; Yang, L.; Pandey, M.K.; Bajaj, P.; Alexander, D.; Chen, S.; Kemerait, R.C.; Varshney, R.K.; Guo, B. Carbohydrate, glutathione, and polyamine metabolism are central to *Aspergillus flavus* oxidative stress responses over time. *BMC Microbiol.* **2019**, *19*, 209. [CrossRef] [PubMed]
125. Tan, W.-J.; Yang, Y.-C.; Zhou, Y.; Huang, L.-P.; Xu, L.; Chen, Q.-F.; Yu, L.-J.; Xiao, S. Diacylglycerol Acyltransferase and Diacylglycerol Kinase Modulate Triacylglycerol and Phosphatidic Acid Production in the Plant Response to Freezing Stress. *Plant Physiol.* **2018**, *177*, 1303–1318. [CrossRef]
126. Fan, J.; Yu, L.; Xu, C. A Central Role for Triacylglycerol in Membrane Lipid Breakdown, Fatty Acid-Oxidation, and Plant Survival under Extended Darkness. *Plant Physiol.* **2017**, *174*, 1517–1530. [CrossRef]
127. Hammerl, R.; Frank, O.; Dietz, M.; Hirschmann, J.; Hofmann, T. Tyrosine Induced Metabolome Alterations of and Quantitation of Secondary Key Metabolites in Blue-Mold Cheese. *J. Agric. Food Chem.* **2019**, *67*, 8500–8509. [CrossRef]
128. Keller, N.P.; Turner, G.; Bennett, J.W. Fungal secondary metabolism—From biochemistry to genomics. *Nat. Rev. Microbiol.* **2005**, *3*, 937–947. [CrossRef]
129. Tawfike, A.F.; Romli, M.; Clements, C.; Abbott, G.; Young, L.; Schumacher, M.; Diederich, M.; Farag, M.; Edrada-Ebel, R. Isolation of anticancer and anti-trypanosome secondary metabolites from the endophytic fungus *Aspergillus flocculus* via bioactivity guided isolation and MS based metabolomics. *J. Chromatogr. B Analyt. Technol. Biomed. Life. Sci.* **2019**, *1106–1107*, 71–83. [CrossRef]
130. Bu, C.; Zhang, Q.; Zeng, J.; Cao, X.; Hao, Z.; Qiao, D.; Cao, Y.; Xu, H. Identification of a novel anthocyanin synthesis pathway in the fungus *Aspergillus sydowii* H-1. *BMC Genom.* **2020**, *21*, 29. [CrossRef]
131. Pandohee, J.; Stevenson, P.G.; Conlan, X.A.; Zhou, X.-R.; Jones, O.A.H. Off-line two-dimensional liquid chromatography for metabolomics: An example using *Agaricus bisporus* mushrooms exposed to UV irradiation. *Metabolomics* **2015**, *11*, 939–951. [CrossRef]
132. Kamal, N.; Viegelmann, C.V.; Clements, C.J.; Edrada-Ebel, R. Metabolomics-Guided Isolation of Anti-trypanosomal Metabolites from the Endophytic Fungus *Lasiodiplodia theobromae*. *Planta Med.* **2017**, *83*, 565–573. [CrossRef] [PubMed]
133. Fan, B.; Grauso, L.; Li, F.; Scarpato, S.; Mangoni, A.; Tasdemir, D. Application of Feature-Based Molecular Networking for Comparative Metabolomics and Targeted Isolation of Stereoisomers from *Algicolous fungi*. *Mar. Drugs* **2022**, *20*, 210. [CrossRef] [PubMed]
134. Bailey, J.E. Toward a science of metabolic engineering. *Science* **1991**, *252*, 1668–1675. [CrossRef] [PubMed]
135. Mishra, A.; Singh, S.P.; Mahfooz, S.; Singh, S.P.; Bhattacharya, A.; Mishra, N.; Nautiyal, C.S. Endophyte-Mediated Modulation of Defense-Related Genes and Systemic Resistance in *Withania somnifera* (L.) Dunal under *Alternaria alternata* Stress. *Appl. Environ. Microbiol.* **2018**, *84*, e02845-17. [CrossRef] [PubMed]

136. Pappas, M.L.; Liapoura, M.; Papantoniou, D.; Avramidou, M.; Kavroulakis, N.; Weinhold, A.; Broufas, G.D.; Papadopoulou, K.K. The Beneficial Endophytic Fungus Strain K Alters Tomato Responses against Spider Mites to the Benefit of the Plant. *Front. Plant Sci.* **2018**, *9*, 1603. [CrossRef]
137. Weed, R.A.; Savchenko, K.G.; Lessin, L.M.; Carris, L.M.; Gang, D.R. Untargeted Metabolomic Investigation of Wheat Infected with Stinking Smut. *Phytopathology* **2021**, *111*, 2343–2354. [CrossRef]

molecules

MDPI

Article

Fatty-Acid-Rich *Agave angustifolia* Fraction Shows Antiarthritic and Immunomodulatory Effect

Enrique Jiménez-Ferrer [1], Gabriela Vargas-Villa [1], Gabriela Belen Martínez-Hernández [1], Manases González-Cortazar [1], Alejandro Zamilpa [1], Maribel Patricia García-Aguilar [1,2], Martha Lucía Arenas-Ocampo [2] and Maribel Herrera-Ruiz [1,*]

1 Centro de Investigación Biomédica del Sur, Instituto Mexicano del Seguro Social, Argentina No. 1, Col. Centro, Xochitepec 62790, Morelos, Mexico
2 Centro de Desarrollo de Productos Bióticos, Instituto Politécnico Nacional, Col. San Isidro, Yautepec 62739, Morelos, Mexico
* Correspondence: cibis_herj@yahoo.com.mx

Abstract: *Agave angustifolia* is a xerophytic species widely used in Mexico as an ingredient in sweet food and fermented beverages; it is also used in traditional medicine to treat wound pain and rheumatic damage, and as a remedy for psoriasis. Among the various *A. angustifolia* extracts and extract fractions that have been evaluated for their anti-inflammatory effects, the acetonic extract (AaAc) and its acetonic (F-Ac) and methanolic (F-MeOH) fractions were the most active in a xylene-induced ear edema model in mice, when orally administered. Four fractions resulting from chemically resolving F-Ac (F1–F4) were locally applied to mice with phorbol 12-myristate 13-acetate (TPA)-induced ear inflammation; F1 inhibited inflammation by 70% and was further evaluated in a carrageenan-induced mono-arthritis model. When administered at doses of 12.5, 25, and 50 mg/kg, F1 reduced articular edema and the spleen index. In addition, it modulated spleen and joint cytokine levels and decreased pain. According to a GC–MS analysis, the main components of F1 are fatty-acid derivatives: palmitic acid methyl ester, palmitic acid ethyl ester, octadecenoic acid methyl ester, linoleic acid ethyl ester, and oleic acid ethyl ester.

Keywords: *Agave angustifolia*; edible plant; fatty acids; rheumatoid arthritis; pain; inflammation; cytokines

Citation: Jiménez-Ferrer, E.; Vargas-Villa, G.; Martínez-Hernández, G.B.; González-Cortazar, M.; Zamilpa, A.; García-Aguilar, M.P.; Arenas-Ocampo, M.L.; Herrera-Ruiz, M. Fatty-Acid-Rich *Agave angustifolia* Fraction Shows Antiarthritic and Immunomodulatory Effect. *Molecules* **2022**, *27*, 7204. https://doi.org/10.3390/molecules27217204

Academic Editors: Arunaksharan Narayanankutty, Ademola C. Famurewa and Eliza Oprea

Received: 28 September 2022
Accepted: 19 October 2022
Published: 24 October 2022

Publisher's Note: MDPI stays neutral with regard to jurisdictional claims in published maps and institutional affiliations.

Copyright: © 2022 by the authors. Licensee MDPI, Basel, Switzerland. This article is an open access article distributed under the terms and conditions of the Creative Commons Attribution (CC BY) license (https://creativecommons.org/licenses/by/4.0/).

1. Introduction

Agave, a Latin term that means "something great, illustrious, dignified", is the name Carl Linnaeus assigned to various species [1]. *Agave*, a specific genus in the Agavaceae family, has been used historically as a source of fiber, as an ingredient in beverages and food, to produce clothing, as fodder, as an ornament, as a building material, and as a remedy. The genus has approximately two hundred species, 75% of which are distributed in Mexico, and this country is regarded as the center of origin of these plants, since about 50% of the species are endemic [2,3]. Agaves were of great economic importance to Mesoamerican cultures for nine thousand years [4], and some of their pre-Hispanic uses are still alive today, e.g., as a source of fiber; for the consumption of their sap (aguamiel); for alcoholic fermentation; and the use of the plant's core (piña) and stems as food, fodder, and for medicinal purposes [1]. *Agave angustifolia* Haw var. *angustifolia*, an *Agave* subgenus in the Rigidae group, commonly called "maguey espadín" or "maguey de monte" [5], is widely distributed in Mexico, from Sonora and Chihuahua to the south of the country, and as far as Nicaragua. It grows in a wide range of ecosystems: coastal dunes, low deciduous forest, the margins of medium sub-deciduous forests, some types of xerophytic scrub, palm groves, and oak–pine forests. This species lives both at sea level and above an altitude of 2000 m. It has fibrous, thin, narrow leaves, with small teeth on the margins and a dark-colored terminal spine; its color varies from gray to olive green; its inflorescence measures 2–4 m, with open branches, yellow-green flowers, and abundant nectar; and once pollinated, it produces a fruit with black seeds [3,5]. Its fibrous stems are used to make rope, twine, and bags; both its flowers and stems are edible and are often used to make traditional foods and sweets, and its sap is consumed either raw or fermented [3]. The uses of *A. angustifolia* in traditional medicine are diverse. The fiber is used to treat urticaria; the sap, cooked leaves, and root infusions have been used for "internal injuries"; a concoction of the root has been used against dysentery; fresh leaves are useful to stop bleeding and alleviate wound pain, to treat skin pimples, and for coughs; and roasted leaves are used to relieve rheumatic pain and as a remedy for sprains or broken bones. Interestingly, the plant has also been used to treat psoriasis, an autoimmune disease that causes joint pain and inflammation [6,7]. Several studies on the pharmacologic activity of *A. angustifolia* have been conducted. Ethanolic extract from the leaves of this agave allowed the identification of its antioxidant capacity in vitro using DPPH, ABTS, and FRAP methods; the same treatment also demonstrated antimicrobial activity against *Staphylococcus epidermidis* and *Escherichia coli* [8]. The administration o.p. of a hydroethanolic extract of the leaf of *A. angustifolia* to Wistar rats decreased the presence of gastric ulcers induced with absolute ethanol by 90% [9]. It has been shown that *A. angustifolia* has effects on metabolism; for example, the administration of fructans of this plant to a group of healthy rats and another group of diabetic animals showed a decrease in the blood concentration of cholesterol and LDL proteins, and hepatic steatosis was observed in the latter group; in addition, the fecal *Lactobacillus* spp. and *Bifidobacterium* spp. counts were greater than in those who did not receive fructans [10]. Furthermore, the incorporation of agavinas, which are also considered a source of dietary fiber, into the diet of mice caused beneficial effects by increasing the concentration of glucagon-like peptide-1 (GLP-1) and decreasing ghrelin (involved in the regulation of eating behavior) [11].

The medicinal use of and biological studies on *A. angustifolia* led to research into its effect on rheumatoid arthritis (RA), which is an autoimmune inflammatory disease characterized by aggressive synovial hyperplasia, resulting in joint destruction manifested by pain, swelling, and stiffness, along with bone erosion and rheumatoid nodules under the skin. Fatigue; weight loss; and extra-articular disorders involving the eyes, oral cavity, blood, lungs, skin, heart, kidneys, nerves, and lymph nodes may also be observed [12]. The World Health Organization (WHO) has reported a prevalence of 0.3–1% for RA; this condition, which is more common in women and in developed countries, contributes 14 million patients to the overall burden of musculoskeletal conditions worldwide [13]. In Mexico, a prevalence of 1.6% was determined for RA in the general population. The disease has a significant impact on the most productive age group, resulting in high rates of work incapacity that affect the economy and quality of life of patients [14,15]. In 2018, in México, RA was a significant health problem in the population aged 50 years and over, with a female/male ratio of 3:1 [16].

The joints of patients with RA suffer from the constant and sustained entry of immune cells, which form a vicious circuit in the promotion and release of proinflammatory mediators, such as cytokines, which trigger the activation of fibroblast-like synoviocytes, contributing to the bone and cartilage damage. Then, the deregulation in the release of cytokines plays an essential role in the pathology of RA; one of the primary sources of these molecules is macrophages. In mice, two types of these synovial cells are recognized, intrinsic macrophages present in the joint synovium from birth and extrinsic bone-marrow-derived macrophages. The first express the anti-inflammatory cytokines IL-10 and IL-4, the latter the proinflammatory cytokines, such as tumor necrosis factor-alpha (TNF-α) and interleukin (IL)-1β [17]. TNF is a crucial cytokine acting as a regulator in the pathogenesis of this disease, and its expression is increased in patients with RA. It activates endothelial cells and recruits synovial fibroblasts and macrophages to release IL-1β, IL-6, and TNF [18,19]. IL-17 is a pro-inflammatory cytokine that controls osteoclast differentiation and antibody increase, together with TNF-a and IL-6. IL-17 participates in the initial stages, and when the disease is already established, its overexpression correlates with clinical parameters of RA; this cytokine promotes the activation of fibroblast-like synoviocytes and the recruitment of neutrophils, macrophages, and B cells [20,21]. Another inflammatory marker of the cytokine family is IL-1β, produced mainly by monocytes and macrophages, but also by other types of cells, such as synovial cells. In turn, this molecule activates monocytes/macrophages and increases the proliferation of fibroblasts, prolonging and increasing the inflammation of the synovial membrane. In addition, it induces the activation of chondrocytes and osteoclasts, which causes cartilage damage and bone resorption [22]. In this context of joint damage potentiated by the action of proinflammatory cytokines, IL-10 is described as a potent immunomodulator that inhibits neutrophil infiltration and synovial tissue activation. Immune modulation is fundamental, and cytokine IL-10 actively participates in this line of action, due to its potent ability to regulate the immune response by inhibiting the infiltration of neutrophils and the activation of synovial tissue. In addition, this molecule promotes the polarization of macrophages towards an M2 phenotype and inhibits the expression of TNF [23,24].

RA treatment usually starts with non-steroidal anti-inflammatory drugs (NSAIDs) such as paracetamol, celecoxib, indomethacin, naproxen, and diclofenac to alleviate pain and inflammation. Since NSAIDs do not address the cause of the disease, they are accompanied by disease-modifying drugs (DMARDs) such as methotrexate (MTX), sulfasalazine, and azathioprine; in a more recent approach, biologics such as anti-TNF, abatacept, and rituximab have been used [25].

In searching for valuable therapies to improve patients' quality of life with fewer adverse effects, the use of medicinal plants is an attractive approach. Considering the wide use of *A. angustifolia* in Mexico and its medicinal properties, including its reported efficacy in inflammatory and autoimmune disorders, this work aimed to evaluate the anti-inflammatory effects of extracts and fractions from this plant in murine models of xylene and ear edema, as well as its immunomodulatory activity in carrageenan/kaolin (CK)-induced mono-arthritis through the quantification of the relevant cytokines, such as TNF-α, IL-1α, IL-17, and IL-10.

2. Results

2.1. Chemical Analysis

One kilogram of lyophilized product was obtained from fresh *A. angustifolia* Haw leaves (4.3 kg). After a first extraction of the leaves with acetone, 35 g of product (AaAc) was obtained, providing a yield of 3.5%. Then, 4.9 g of AaAc was fractioned, with yields of 6.4% (0.31 g) for the n-hexane fraction (F-Hex), 68.12% (3.33 g) for the ethyl acetate fraction (F-AcOEt), 20.87% (1.02 g) for the acetonic fraction (F-Ac), and 4.61% (0.22 g) for the methanolic fraction (F-MeOH). F-Ac was sub-fractioned, with yields of 8.6% (88 mg) for F1, 5.6% (58 mg) for F2, 3.13% (32 mg) for F3, and 2.54% (25 mg) for F4.

2.2. Anti-Inflammatory Activity of *A. angustifolia* Extracts

Local xylene application to the ears of mice induced edema, resulting in a weight difference of 8.87 mg in the vehicle-administered group (VEH). Edema weight was reduced by 72% after the oral administration of indomethacin (INDO). Treating mice with AaAc at a dose of 200 mg/kg also inhibited ear edema with respect to the VEH group (* $p < 0.05$). While all extract fractions caused a significant reduction in edema weight with respect to the VEH group, F-MeOH and F-Ac showed the highest inhibitory effect (Table 1).

To evaluate the effect of the F-MeOH-derived fractions, the TPA (phorbol 12-myristate 13-acetate)-induced edema assay was used. As shown in Table 2, TPA-induced edema resulted in a weight difference of 10.24 mg in vehicle-treated mice. This effect was inhibited by INDO and F1 at a dose of 1.0 mg/ear. While F2, F3, and F4 also induced a statistically significant effect, the percentage of inhibition was less than 40%.

Table 1. Effect of the extract and fractions from *A. angustifolia* on xylene-induced ear edema in mice.

Treatment (mg/kg)	Edema (mg)	Inflammation Inhibition (%)
VEH	8.87 ± 2.8	-
IND (5.0)	2.40 ± 1.8 *	72.9
AaAc (200.0)	1.52 ± 0.8 *	82.9
Fractions		
F-Hex (50.0)	5.36 ± 0.6 *	39.6
F-AcOEt (50.0)	5.10 ± 2.2 *	42.6
F-Ac (50.0)	4.75 ± 1.6 *	46.4
F-MeOH (50.0)	4.50 ± 2.1 *	49.3

Data are reported as mean ± standard deviation (SD) and were analyzed by analysis of variance (ANOVA) with post hoc Bonferroni test ($n = 7$). * $p < 0.05$ with respect to the negative control group. IND = indomethacin.

Table 2. Effect of the extract and fractions from *A. angustifolia* on TPA-induced ear edema.

Treatment (mg/ear)	Edema (mg)	Inflammation Inhibition (%)
VEH	10.24 ± 2.10	-
IND (1.0)	1.65 ± 0.21 *	83.8
F1 (1.0)	3.06 ± 2.00 *	70.1
F2 (1.0)	7.87 ± 0.23 *	23.1
F3 (1.0)	6.58 ± 0.51 *	35.7
F4 (1.0)	7.08 ± 1.20 *	30.8

Data are reported as mean ± SD and were analyzed by ANOVA with post hoc Bonferroni test ($n = 7$). * $p < 0.05$ with respect to the negative control group. IND = indomethacin.

2.3. *Effect of A. angustifolia Extract on Mono-Arthritis Induced by K/C*

Based on the results of the inhibition of TPA-induced edema, the anti-inflammatory activity of F1 was assessed at three different doses in a mono-arthritis model. The area under the curve for joint inflammation (mm) 24 h after the administration of either K/C or SSF (basal) is shown in Figure 1. The positive control drug, methotrexate, reduced inflammation levels for the duration of the experiment. At the same time, F1 failed to modify K/C-induced inflammation at the lowest dose, but it reduced articular edema at doses of 50 mg/kg (* $p < 0.05$).

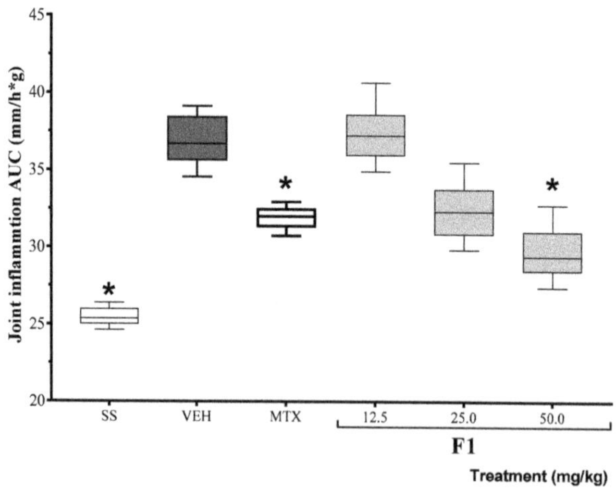

Figure 1. Effect of the F1 fraction of *A. angustifolia* extract at different doses on the AUC of the time course plot of joint inflammation in a murine model of K/C-induced mono-arthritis. The plots show the mean and SD for each group ($n = 8$). Asterisks indicate significant differences with respect to the vehicle group, administered with 1% Tween-20 (VEH), * $p \leq 0.05$, by ANOVA with post hoc Bonferroni test. SS: saline solution; MTX: methotrexate.

As shown in Figure 2, the time elapsed before mouse response on the hot plate test was significantly shorter in K/C-treated than in SS-treated animals (* $p < 0.05$). This effect was offset by the administration of F1 at doses of 25 and 50 mg/kg (* $p < 0.05$).

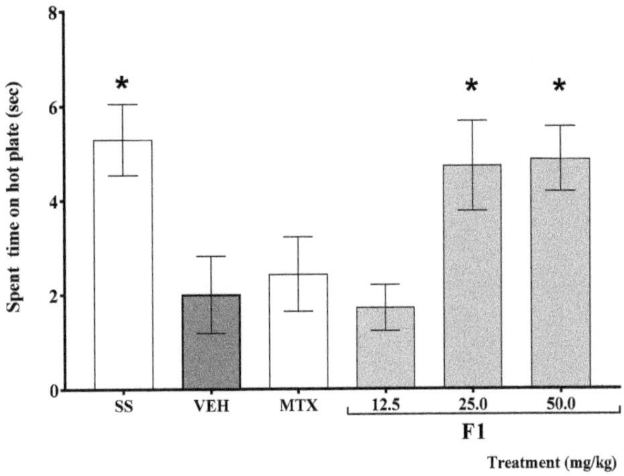

Figure 2. Effect of the F1 fraction of *A. angustifolia* extract at different doses on thermal hyperalgesia in inflamed joint in a murine K/C-induced mono-arthritis model. Data are reported as the mean ± SD for each group (n = 12) and were analyzed by ANOVA with post hoc Bonferroni test. * $p \leq 0.05$ with respect to the vehicle-treated group (Tween-20, VEH). SS: saline solution; MTX: methotrexate.

Spleen size, expressed as a percentage of total body weight, increased in animals treated with K/C with respect to SS-treated mice. Both methotrexate and F1 at doses of 12.65 and 50 mg/kg significantly reduced this effect with respect to the VEH group (* $p < 0.05$, Table 3). Interestingly, an intermediate dose of F1 failed to revert spleen growth.

Table 3. Effect of the F1 fraction of *A. angustifolia* extract on spleen index (percentage of total body weight).

Treatment (mg/kg)	Relative Weight of Spleen (%)
SS	0.552 ± 0.045 *
VEH	0.638 ± 0.035
MTX (1.0)	0.582 ± 0.050 *
F1 (12.5)	0.573 ± 0.043 *
F1 (25.0)	0.632 ± 0.030
F1 (50.0)	0.587 ± 0.044 *

Data are reported as mean ± SD and were analyzed by ANOVA with post hoc Bonferroni test (n = 7). * $p < 0.05$ with respect to the negative control group. SS: saline solution; VEH: Tween-20; MTX: methotrexate.

2.3.1. Effect of *A. angustifolia* Extract on Cytokine Concentration in Organs from K/C-Treated Mice

Cytokine levels were measured in the spleen and left knee joint of mice. As shown in Table 4, K/C-treated animals showed increased levels of IL-1, TNF-α, and IL-17 in the spleen, whereas IL-10 levels were significantly decreased with respect to SS-treated control animals (* $p < 0.05$). A significant decrease in TNF-α levels and an increase in IL-10 levels were observed in all treatment groups; however, while a decrease in IL-1 levels was observed in all groups with respect to the VEH group, it was significant only for mice treated with F1 at doses of 12.5 and 25 mg/kg (* $p < 0.05$). Finally, no significant changes in spleen IL-17 levels were observed under any treatment.

Table 4. Effect of the extract and fractions from *A. angustifolia* on cytokine levels in spleen and left knee joint in a K/C-induced mono-arthritis model.

Treatment (mg/kg)	Cytokine Levels (pg/mg Protein)			
	IL-1β	TNF-α	IL-17	IL-10
		Spleen		
SS	15,589.6 ± 3110.3 *	10,339.97 ± 2379.4 *	4588.2 ± 253.6 *	17,474.8 ± 1768.0 *
VEH	20,661.5 ± 1302.9	23,015.1 ± 4134.2	5917.7 ± 997.5	7742.0 ± 3688.6
MTX (1.0)	18,360.4 ± 1396.1	11,751.53 ± 3765.9 *	5090.7 ± 767.7	19,709.9 ± 3429.0 *
F1 (12.5)	18,070.5 ± 596.1 *	11,547.29 ± 1792.1 *	5576.8 ± 916.1	17,271.0 ± 625.0 *
F1 (25.0)	17,639.1 ± 409.7 *	12,296.05 ± 1541.5 *	6291.1 ± 669.0	16,670.3 ± 370.5 *
F1 (50.0)	18,481.0 ± 1381.8	11,844.34 ± 1669.6 *	5212.9 ± 738.2	17,142.5 ± 1728.8 *
		Left knee joint		
SS	13,883.5 ± 2957.9	6808.08 ± 2761.4 *	1258.7 ± 261.5 *	15,938.4 ± 1855.6 *
VEH	18,382.1 ± 3616.2	10,762.80 ± 564.4	2188.5 ± 249.1	9832.6 ± 1514.5
MTX (1.0)	12,242.5 ± 2605.4	8080.48 ± 1703.3 *	1561.2 ± 324.1 *	13,501.1 ± 1382.2 *
F1 (12.5)	18,807.9 ± 1401.3	9569.87 ± 1038.7	2337.4 ± 168.6	15,803.1 ± 1362.1 *
F1 (25.0)	17,465.7 ± 1398.0	9145.84 ± 588.6 *	1556.8 ± 196.8 *	15,506.8 ± 1075.4 *
F1 (50.0)	15,112.1 ± 780.9	9064.67 ± 313.6 *	1713.1 ± 232.5	14,851.6 ± 989.2 *

Data are reported as mean ± SD and were analyzed by ANOVA with post hoc Bonferroni test ($n = 7$). * $p < 0.05$ with respect to the negative control group. SS: saline solution; VEH: Tween-20; MTX: methotrexate.

On the other hand, TNF-α and IL-17 levels in the joint were increased in mice in the VEH group, while the IL-10 concentration was much lower with respect to SS-treated animals (* $p < 0.05$). Both methotrexate and F1 significantly reduced these effects, and the reduction was greater when F1 was administered at a dose of 25 mg/kg (* $p < 0.05$). No significant differences in IL-1 levels were observed in either group.

2.3.2. Composition of F1

A GC–MS analysis identified five fatty acids and derivatives, shown in Table 5, as the main constituents of the active fraction (F1).

Table 5. Chemical composition of F1, according to GC–MS analysis.

RT (min)	Fatty Acid Name	Composition (%)	General Data
18.71	Palmitic acid, methyl ester (methyl palmitate)	20	A C16 saturated fatty acid, a natural product with a vasodilator effect on rat/rabbit thoracic aorta [26]; anti-inflammatory activity in different assays [27,28]
19.32	Palmitic acid, ethyl ester (ethyl palmitate)	10	Ethyl hexadecanoate is a long-chain fatty acid ethyl ester resulting from the formal condensation of the carboxy group of palmitic acid with the hydroxy group of ethanol; it has a role as a plant metabolite; anti-inflammatory activity in different assays [27–29]
20.43	Octadecenoic acid, methyl ester (methyl octadecenoate or ethyl stearate)	35	This aliphatic-chain fatty acid had a neuroprotective effect on dopaminergic cells in a model of Parkinson's disease [30]
20.93	Linoleic acid, ethyl ester (ethyl linoleate)	13	An essential fatty acid, which among other things reduces the levels of nitric oxide and PGE2 by downregulating the enzymes involved in their production (iNOS and COX-2) [31]
20.98	Oleic acid, ethyl ester (ethyl oleate)	8	

3. Discussion

A. angustifolia is widely used in Mexico for its nutritional and medicinal properties. Some of the many beneficial effects traditionally attributed to this plant species have been confirmed, especially in conditions with an inflammatory background, such as sprains and psoriasis. An acetonic extract of this plant was reported to show anti-inflammatory activity in a TPA-induced ear edema model [32], with 3-O-((6′-O-palmitoyl)-β-D-glucopyranosyl) sitosterol being identified as the active constituent. This compound induced a dose-dependent effect and was able to modulate local cytokine levels in the ear of mice exposed to the stimulus [33]. Additionally, the oral administration of 200 mg/kg of a methanolic *A. angustifolia* extract inhibited carrageenan-induced plantar edema by 55% in mice [9].

The anti-inflammatory activity of an orally administered acetone extract of *A. angustifolia* (AaAc) and its fractions was herein assessed in a model of ear edema induced by xylene, a volatile and toxic substance that causes skin irritation [34]. Xylene application to mouse ears results in robust edema and long-lasting erythema. Local xylene application has been shown to promote vascular reactions mediated by the activation of the TRPA1 channel [35], which is expressed in mast cells, keratinocytes, melanocytes, and sensory nerves, among other cell types [36,37]. Xylene application causes severe vasodilatation and inflammatory cell infiltration [38], with substance P also reported to mediate its effects [39].

AaAc showed the highest anti-inflammatory activity, followed by F-MeOH and F-Ac, indicating that these products might compensate for some of the mechanisms underlying the irritant activity of xylene. F-Ac resolution yielded four fractions with different chemical compositions. Among them, F1 showed the highest anti-inflammatory effect in the TPA assay; therefore, its capacity was evaluated at different doses in a K/C-induced monoarthritis model.

GC–MS analysis showed that F1 is composed of a mixture of fatty acids and their esters, including methyl palmitate, ethyl palmitate, ethyl octadecenoate (ethyl stearate), ethyl linoleate, and ethyl oleate. These compounds have shown anti-inflammatory activity in several biological models, including some experimental arthritis models. For instance, methyl palmitate (the methyl ester of palmitic acid) inhibited phagocyte function in a culture of RAW cells when stimulated with LPS; it also decreased TNF-α and increased IL-10 levels, while decreasing the phosphorylation of the inhibitory protein kappa B (IκBα) [27,31]. Both methyl palmitate and ethyl palmitate were found to reduce carrageenan-induced plantar edema in rats, in addition to decreasing PGE2 levels in inflammatory exudates. In another study, methyl palmitate and ethyl palmitate inhibited carrageenan-induced plantar inflammation by decreasing prostaglandin E2 (PGE2) levels in the exudate. Furthermore, in a rat model of endotoxemia, both compounds reduced TNF-α and IL-6 levels by decreasing NF-κB expression in the lung and liver, in addition to counteracting histopathological changes caused by LPS. Both fatty acid esters reduced croton oil-induced ear edema, decreasing neutrophil infiltration by inhibiting myeloperoxidase activity [28]. Moreover, some fractions of a *Jatropha curcas* extract that showed anti-inflammatory activity in RAW264.7 cells contain fatty acids such as octadecanoic acid methyl ester and free octadecanoic acid [40]. Furthermore, when added to the diet of mice with collagen-induced experimental arthritis, linoleic acid was able to offset the damage, reducing joint edema, blood IL-1β levels [41], and hepatic arachidonic acid levels [42].

As mentioned above, the major components of F1 have been proved to modulate the immune response mediated by the proinflammatory cytokines TNF-α, IL-1β, and IL-17, and the regulatory molecule IL-10. On the other hand, it is an established fact that none of these proteins is totally pro- or anti-inflammatory. In this case, IL-10 is a pleiotropic cytokine that exerts regulatory effects through various mechanisms; it is capable, for example, of suppressing the activation and function of immune cells, inhibiting edema and the production of TNF-α, and downregulating the expression of cyclooxygenase-2 (COX-2) and PGE2, among other mediators. IL-10 blocks neutrophil infiltration and synovial tissue activation, prevents macrophage polarization towards the M2 phenotype, and selectively inhibits the expression of pro-inflammatory cytokines, especially IL-1β and TNF-α [24]; both molecules are known to promote the progression of rheumatoid arthritis.

In the carrageenan-induced intraarticular inflammation model studied in this work, animals treated with F1 at all doses consistently showed increased levels of splenic and joint IL-10 with respect to the VEH group. F1, whose main components are fatty acids, could modulate IL-10 release; in turn, this could partially inhibit IL-1β expression. Thus, F1 showed no effect on the levels of that cytokine in the joint, although a slight but significant decrease in IL-1β levels in the spleen was observed at low doses of F1. This modulation could be evidenced by a significant decrease in TNF-α levels in both tissues.

Finally, at doses of 25 and 50 mg/kg, F1 increased the response time in the hot plate assay, suggesting an analgesic action. The imbalance in the central and peripheral mechanisms of pain includes the direct activation or the increment in the sensitivity of the nociceptors to different stimuli, such as joint inflammation, through the elevation of pro-inflammatory cytokines released by immune cells in the synovium, including TNF-α, IL-1β, IL-6, and IL-17, which directly alter the responses of nociceptive neurons [43]. Therefore, the analgesic effect of F1 could be related to a decrease in proinflammatory molecules and, with them, the inflammation and, consequently, the overactivation of nociceptors. In addition, the increase in the concentration of IL-10 in joints, a cytokine that decreases inflammation, reduces fever, acute-phase protein release, and vascular permeability and is an inhibitor of hyperalgesia [44].

F1 is an active mixture of aliphatic-chain fatty acids, capable of reducing local and joint inflammation, which modulates the immune response by blocking the increase in the concentration of pro-inflammatory cytokines and raising IL-10 levels. Therefore, it is necessary to continue with the pharmacological investigation of this mixture of compounds in more complex models of RA, which could lead to the identification of a mode of action for a possible antiarthritic treatment.

4. Material and Methods

4.1. Plant Material, Extract, and Fraction Preparation

A. angustifolia leaves were obtained from a 5-year-old specimen collected in a controlled cultivation field located in Tlalquitenango (Morelos State), at 18°37′ N latitude, 99°10′ W longitude, and an elevation of 911 m above sea level. This culture belonged to the company "Yautli", which houses the identity register depositaries of the specimens of this species. An herbarium specimen was sent to the herbarium of the Institute of Biology of the National Autonomous University of Mexico and was identified by Dr. Abisaí Mendoza with the number 1,419,710.

Stalks were cut 5 cm from the core (piña). The material was cut into small pieces, frozen, and freeze-dried; the dried material was ground and macerated with acetone for 24 h. The resulting liquid was filtered and concentrated to dryness on a rotary evaporator (R-114, Buchi, Flawil, Switzerland). This procedure was repeated three times to yield the acetone extract (AaAc). The extract was kept at 4 °C until it was used for chemical analysis and in biological tests.

4.2. Chemical Analysis

4.2.1. Fractioning the Acetone Extract (AaAc)

Based on previous studies, the acetone extract [33], AaAc (28 g), was absorbed on silica gel in a 2:1 ratio to be placed in a glass column (100 × 400 mm^2) packed with normal-phase silica gel (200 g, 70–230 mesh, Merck, Kenilworth, NJ, USA). After elution with 1500 mL of hexane as the mobile phase, the solvent was concentrated on a rotary evaporator to yield the hexane fraction (F-Hex). Then, 1500 mL of ethyl acetate was added to the column, which was concentrated to obtain the ethyl acetate fraction (F-AcOEt), and the process was repeated with acetone and methanol as mobile phases to obtain the acetone (F-Ac) and methanolic (F-MeOH) fractions, respectively. These fractions were evaluated in a murine model of xylene-induced local edema [38].

4.2.2. Separation of Acetone Fraction (F-Ac)

F-Ac (1.2 g) was selected for chromatographic separation, given its higher activity. The fraction was absorbed on silica gel (10 g, 70–230 mesh, Merck, Kenilworth, NJ, USA), placed in a glass column (50 × 300 mm^2) packed with silica gel (80 g), and eluted with a dichloromethane:methanol gradient system as the mobile phase with a 10% increase in polarity, collecting fractional volumes of 100 mL. In total, 30 fractions were obtained, which were then pooled according to similarity via thin-layer chromatography (TLC) into four fractions (F1, F2, F3, and F4).

Due to the low availability of xylene and the yield of fractions, a TPA-induced local edema assay was used to evaluate the anti-inflammatory activity of F1–F4. TPA is a phorbol ester, widely used to study local inflammation by activating PKC with leukocyte infiltration and promoting eicosanoid release [45].

4.2.3. Gas Chromatography–Mass Spectrometry Analysis

The chemical composition of the most active fraction (F1) was determined on a gas chromatograph (GC, Agilent Technology 6890 masses coupled 5973N, Santa Clara, CA, USA) fitted with a quadrupole mass detector in electron impact mode at 70 eV. Volatile compounds were resolved on an HP 5MS capillary column (25 m length, 0.2 mm inner diameter, 0.3 μm film thickness). The oven temperature was set at 40 °C for 2 min, then increased from 40 to 260 °C at a rate of 10 °C/min and maintained at 260 °C for 20 min. The mass detector interface temperature was set to 200 °C and the mass acquisition range was 20–550. Injector and detector temperatures were set at 250 and 280 °C, respectively. Helium was used as the carrier gas, with a flow rate of 1 mL/min. The major compounds in the mixture were identified by comparing their mass spectra with those of the National Institute of Standards and Technology (NIST) Library 1.7 [46].

4.3. Biological Tests

4.3.1. Animals

Female BALB/c mice (20–25 g) were acquired from the vivarium of the Centro Médico Nacional Siglo XXI-IMSS. Mice were randomly distributed into groups of 7 animals and were maintained under controlled conditions for at least 3 weeks before the experiments, with a 12 h light/dark cycle at 20 ± 1 °C; food (Labdiet 5008, Brentwood, MO, USA) and water were provided ad libitum. Behavioral tests were performed in an isolated, well-lit room, with a video recording system. Mice were handled in accordance with the Mexican Official Standard for the Care and Handling of Animals (NOM-062-ZOO-1999) and in compliance with all international regulations. The IMSS Research Committee approved the experimental protocol (permit number R-2015-1702-4).

4.3.2. Xylene-Induced Mouse Ear Edema

Each treatment was administered orally (p.o.) to groups of 7 mice 1 h before the application of xylene (irritant) to the pinna. Different groups received AaAc 200 mg/kg, F-Hex, F-AcOEt, F-Ac, or F-MeOH at a dose of 50 mg/kg each; a positive control group received indomethacin (INDO, ≥99% purity by TLC, Sigma Chemical Co., Saint Louis, MO, USA) at a dose of 5.0 mg/kg; and a negative control group was administered with 1% Tween-20 solution (VEH, 100 μL/10 g body weight). The assay was performed as described in [29]. Briefly, 40 μL of xylene was applied to the right pinna, 20 μL to the inner side and 20 μL to the outer side; for reference, 20 μL of 70% ethanol was applied to the inner and outer sides of the left ear. One hour after the irritant stimulus, the animals were sacrificed by cervical dislocation, and 6 mm diameter circular sections were obtained from the right and left ears. Both treated (*t*) and untreated (*nt*) ears were weighed to measure the level of inflammation. The percentage of edema inhibition was calculated as follows:

$$\text{Inhibition}\ (\%) = \left|\frac{\Delta\text{veh} - \Delta_{\text{tr}}}{\Delta\text{veh}}\right| \times 100$$

Δveh = ear weight of VEH group
Δtr = ear weight of treatments

4.3.3. TPA-Induced Mouse Ear Edema

Edema was induced in the ear of mice following a previously described method [46]. A negative control group administered with vehicle (VEH) and a positive group receiving indomethacin (INDO, 1.0 mg/ear dissolved in acetone) as described above were included. Fractions F1–F4 were administered topically at a dose of 1.0 mg/ear dissolved in acetone, 10 μL on the outer and 10 μL on the inner side of the left ear of each mouse, leaving the right ear untreated as a control. Ten minutes later, 2.5 μg/ear of 12-O-tretradecanoylphorbol-13-acetate (TPA, ≥99% purity by HPLC) was administered. Four hours later, the mice were sacrificed by cervical dislocation and 6 mm diameter circular sections were obtained from each ear. Both treated (*t*) and untreated (*nt*) ears were weighed to measure the level of inflammation. The percentage of edema inhibition was calculated as follows:

$$\text{Inhibition }(\%) = \left|\frac{\Delta \text{veh} - \Delta_{\text{tr}}}{\Delta \text{veh}}\right| \times 100$$

Δveh = ear weight VEH group
Δtr = ear weight of treatment groups

4.3.4. Kaolin/Carrageenan (K/C)-Induced Mono-Arthritis

Mice were distributed into groups of 8 animals. One arthritis-free group was administered with sterile saline (SS), and the other groups were given an intraarticular kaolin/carrageenan (K/C) challenge on the first day of the experiment, as described below. Based on the previous study [47], F1 was administered p.o. at a dose of 12.5, 25, or 50 mg/kg to the different groups, starting on day 2 and continuing until day 10. A negative control group (VEH) and a positive control group, treated with methotrexate (MTX), were also included.

For the induction of mono-arthritis [48], the base of the left knee joint was measured preoperatively in animals of all groups as a reference, using a digital micrometer (Mod. MDC-1" SB, Mitutoyo Products, Kanagawa, Japan). The animals were then anesthetized with 55 mg/kg sodium pentobarbital intraperitoneally, and 40 μL of 40% kaolin solution was injected into the joint cavity of the left knee; immediately thereafter, a series of flexions and extensions were performed for 15 min. Then, 40 μL of 2% carrageenan solution was injected into the joint cavity, and flexions and extensions were performed again for 5 min. Mice in an untreated group were handled similarly, but SS was injected instead of K/C. Joint edema was measured once daily as described above. Subsequent treatments were administered every 24 h after the initial K/C administration. Mice were sacrificed on day 10; the left knee and spleen were dissected out and stored at −70 °C until use.

4.3.5. Thermal Hyperalgesia Assay

Thermal hyperalgesia was assessed in a hot plate assay. The animals were placed one at a time on a surface at 50 ± 2 °C. The latency time to respond to the stimulus by licking and/or shaking the hind paw or jumping was measured on the last day of the experiment with a hand-held stopwatch, with a maximum time allowed of 20 s [49,50].

4.3.6. Spleen Index

The spleen of each animal was weighed, and the spleen index was calculated as a percentage with respect to the total body weight of the animal [51].

4.3.7. Homogenization of Spleen and Joint Tissues

Joint and spleen samples were triturated with dry ice in a mortar until the dry ice evaporated. Joint and spleen tissues were placed in a vial with 2 mL of PBS, pH 7.4, and

0.01% PMFS in isopropyl alcohol. The tissues were homogenized (Mod. D-500 Pack 1, Dragon Lab, Meriden, CT, USA) for 10–15 s and centrifuged at 12,000 RPM for 5 min. Five 300 μL aliquots of supernatant were stored at −70 °C until cytokine quantification.

4.3.8. Quantification of Anti-Inflammatory and Pro-Inflammatory Cytokines

Cytokines were quantified by ELISA using a commercial kit (OptEIATM, BD Biosciences, Franklin Lakes, NJ, USA) following the manufacturer's instructions. Briefly, 100 μL/well of the capture antibody were placed in 96-well plates and incubated for 12 h at 4 °C. Then, the plates were washed three times with 300 μL/well of 0.05% Tween-20 in PBS. The plates were supplemented with 100 μL of PBS containing 10% fetal bovine serum (FBS), pH 7.0, and left to stand for 1 h at room temperature. The contents were discarded, and the plate was washed three times with 300 μL/well of 0.05% Tween-20 in PBS. Then, 100 μL of standard, target (PBS with FBS), and sample was added. The plates were incubated for 2 h at room temperature. The contents were discarded, and the plates were washed five times with 300 μL/well of 0.05% Tween-20 in PBS. For TNF-α, IL-6, IL-4, and IL-10 quantification, 100 μL/well of streptavidin-HRP-conjugated detection antibody was added. The plates were incubated for 1 h and washed seven times with 300 μL/well of 0.05% Tween-20 in PBS.

For IL-1β quantification, 100 μL/well of antibody detection was added, incubated for 1 h at room temperature and washed five times with 300 μL/well of 0.05% Tween-20 in PBS, followed by 100 μL/well of streptavidin-HRP. The plates were incubated at room temperature for 1 h and washed seven times with 300 μL/well of 0.05% Tween-20 in PBS.

Then, 100 μL/well of o-phenylenediamine substrate was added, and the plates were incubated for 30 min at room temperature in the dark. The reaction was stopped with 2N H_2SO_4, and the plates were read in a Stat Fax 2100 spectrophotometer (Awareness Technologies, Bellport, NY, USA) at 450 nm and 37 °C.

For IL-17 quantification, 50 μL/well of RD1–38 or a standard was added to the plates. The plates were gently shaken for 1 min and incubated at room temperature for 2 h. The contents were discarded, and the plates were washed five times with PBS. Then, 100 μL of HBR-conjugated antibody was added, incubated for 2 h at room temperature, and washed five times. Subsequently, 100 μL of substrate solution was added, and the plates were incubated for 30 min in the dark. The reaction was stopped with 2N H_2SO_4, and the plates were read at 450 nm and 37 °C.

4.4. Statistical Analysis

Data were analyzed with SPSS v.22.0 software, using one-way ANOVA followed by a Bonferroni test. Differences with respect to the negative control were regarded as significant for $p \leq 0.05$. Data are reported as mean ± SD.

5. Conclusions

Agave angustifolia is an edible plant to which various medicinal properties have been attributed. Its value as an anti-inflammatory and immunomodulatory remedy could be due to its content of free fatty acids and their esters. The extract herein evaluated showed a clear anti-inflammatory effect in TPA- and carrageenan-induced arthritis models. Interestingly, the extract also showed an analgesic effect, which should be further studied in other models. It is necessary to continue with the pharmacological study of F1 and the fatty acids individually, through histology and in vitro studies (culture cells). This information will allow future researchers to describe the mode of action of the antiarthritic effect (anti-inflammatory and immunomodulatory) and propose this plant as a possible antiarthritic drug.

Author Contributions: M.H.-R. Designed the experiments, analyzed data, edited and wrote the paper. E.J.-F. Designed the experiments and analyzed the data. M.L.A.-O. Revised the manuscript, collected the plant, and analyzed data. G.B.M.-H. Performed the experiments. M.P.G.-A. Performed the experiments. G.V.-V. Performed the experiments and analyzed the data. M.G.-C. and A.Z.

conducted the chemical analysis. All authors have read and agreed to the published version of the manuscript.

Funding: This research was funded by Fondo de Investigación en Salud, with the financing number FIS/IMSS/PROT/G18/1820.

Institutional Review Board Statement: This work was approved by the Institutional Ethics Committee, Instituto Mexicano del Seguro Social, which after review assigned to it the registration number R-2015-1702-04 on 15 May 2015, and it was submitted for reapproval to the same committee on 30 March 2022.

Data Availability Statement: Not applicable.

Acknowledgments: We would like to thank the Mexican Social Security Institute for its support in carrying out this project.

Conflicts of Interest: The authors declare no conflict of interest.

Sample Availability: Samples of the compounds are available from the authors.

References

1. Jaques-Hernández, C.; Salazar-Bravo, A. Caracterización y usos de las especies de agave en el Estado de Tamaulipas. *Cienc. Conoc. Tecnol.* **2009**, *89*, 91–100.
2. García-Mendoza, A. Los Agaves de México. *Ciencias* **2007**, *8*, 14–23.
3. García-Mendoza, A.; Galván, V. Riqueza de las familias Agavaceae y Nolinaceae en México. *Biol. Soc. Bot. Mex.* **1995**, *56*, 7–24. [CrossRef]
4. Domínguez, M.; González, M.; Rosales, C.; Quiñones, C.; Delgadillo, S.; Mireles, J.; Pérez, E. El cultivo in vitro como herramienta para el aprovechamiento, mejoramiento y conservación de especies del género Agave. Universidad Autónoma de Aguascalientes, Aguascalientes, Mexico. *Investig. Cienc.* **2008**, *16*, 53–62.
5. Herrera-Ruiz, M.; Jiménez-Ferrer, E.; González-Cortazar, M.; Zamilpa, A.; Cardoso-Taketa, A.; Arenas-Ocampo, M.L.; Jiménez-Aparicio, A.R.; Monterrosas-Brisson, N. Potential Use of Agave Genus in Neuroinflammation Management. *Plants* **2022**, *11*, 2208. [CrossRef]
6. Cano, L. Flora Medicinal de Veracruz. In *Inventario Etnobotánico*; Universidad Veracruzana: Xalapa, Mexico, 1998; Volume 1, pp. 48–49.
7. Monroy-Ortiz, C.; Castillo-España, P. Agave Angustifolia. In *Plantas Medicinales Utilizadas en el estado de Morelos*, 1st ed.; Universidad Autónoma del Estado de Morelos: Cuernavaca, Mexico, 2007; Volume 1, pp. 265–266.
8. López-Romero, J.C.; Ayala-Zavala, J.F.; Peña-Ramos, E.A.; Hernández, J.; González-Ríos, H. Antioxidant and antimicrobial activity of Agave angustifolia extract on overall quality and shelf life of pork patties stored under refrigeration. *J. Food Sci. Technol.* **2018**, *55*, 4413–4423. [CrossRef] [PubMed]
9. El-Hawary, S.; El-Kammar, H.A.; Farag, M.A.; Saleh, D.O.; El Dine, R.H. Metabolomic profiling of five Agave leaf taxa via UHPLC/PDA/ESI-MS inrelation to their anti-inflammatory, immunomodulatory and ulcero-protective activities. *Steroids* **2020**, *160*, 108648. [CrossRef]
10. Rendón-Huerta, J.A.; Juárez-Flores, B.; Pinos-Rodríguez, J.M.; Aguirre-Rivera, J.R.; Delgado-Portales, R.E. Effects of different sources of fructans on body weight, blood metabolites and fecal bacteria in normal and obese non-diabetic and diabetic rats. *Plant Foods Hum. Nutr.* **2012**, *67*, 64–70. [CrossRef] [PubMed]
11. Santiago-García, P.A.; López, M.G. Agavins from Agave angustifolia and Agave potatorum affect food intake, body weight gain and satiety-related hormones (GLP-1 and ghrelin) in mice. *Food Funct.* **2014**, *5*, 3311–3319. [CrossRef] [PubMed]
12. Tanaka, Y. Rheumatoid Arthritis. *Inflamm. Regen.* **2020**, *40*, 20. [CrossRef]
13. WHO. Musculoskeletal Conditions. Key Facts. 2021. Available online: https://www.who.int/news-room/fact-sheets/detail/musculoskeletal-conditions (accessed on 3 June 2022).
14. Cardiel, M.H.; Rojas-Serrano, J. Community based study to estimate prevalence, burden of illness and help seeking behaviour in rheumatic diseases in Mexico City. A COPCORD study. *Clin. Exp. Rheumatol.* **2002**, *20*, 617–624.
15. Peláez-Ballestas, I.; Sanin, L.H.; Moreno-Montoya, J.; Alvarez-Nemegyei, J.; Burgos-Vargas, R.; Garza-Elizondo, M. Epidemiology of the rheumatic diseases in Mexico. A study of 5 regions based on the COPCORD methodology. *J. Rheumatol.* **2011**, *86*, 3–8. [CrossRef]
16. INEGI. Comunicado de Prensa Núm. 450/20, del 1 Oct 2020, Pages 1/3. Available online: https://www.inegi.org.mx/contenidos/saladeprensa/boletines/2020/ENASEM/Enasem_Nal20.pdf (accessed on 4 April 2022).
17. Tu, J.; Hong, W.; Guo, Y.; Zhang, P.; Fang, Y.; Wang, X.; Chen, X.; Lu, S.; Wei, W. Ontogeny of synovial macrophages and the roles of synovial macrophages from different origins in arthritis. *Front. Immunol.* **2019**, *10*, 1146. [CrossRef]
18. Kalliolias, G.D.; Ivashkiv, L.B. TNF biology, pathogenic mechanisms and emerging therapeutic strategies. *Nat. Rev. Rheumatol.* **2016**, *12*, 49–62. [CrossRef] [PubMed]
19. Okamoto, K.; Takayanagi, H. Osteoimmunology. *Cold Spring Harb. Perspect. Med.* **2019**, *9*, a031245. [CrossRef]

20. Lubberts, E. The IL-23-IL-17 axis in inflammatory arthritis. *Nat. Rev. Rheumatol.* **2015**, *11*, 415–429. [CrossRef]
21. Burska, A.N.; Neilan, J.; Chisman, R.E.; Pitaksalee, R.; Hodgett, R.; Marzo-Ortega, H.; Conaghan, P.G.; West, R.; Emery, P.; Ponchel, F. Serum IL-7 as diagnostic biomarker for rheumatoid arthritis, validation with EULAR 2010 classification criteria. *Clin. Exp. Rheumatol.* **2018**, *36*, 115–120. [PubMed]
22. Lopez-Castejon, G.; Brough, D. Understanding the mechanism of IL-1β secretion. *Cytokine Growth Factor Rev.* **2011**, *22*, 189–195. [CrossRef]
23. Bober, L.A.; Rojas-Riana, A.; Jackson, J.V.; Leach, M.W.; Manfra, D.; Narula, S.K.; Grace, M.J. Regulatory effects of interleukin-4 and interleukin-10 on human neutrophil function ex vivo and on neutrophil influx in a rat model of arthritis. *Arthritis Rheum.* **2000**, *43*, 2660–2667. [CrossRef]
24. Smallie, T.; Ricchetti, G.; Horwood, N.J.; Feldmann, M.; Clark, A.R.; Williams, L.M. IL-10 inhibits transcription elongation of the human TNF gene in primary macrophages. *J. Exp. Med.* **2010**, *207*, 2081–2088. [CrossRef]
25. Cardiel, M.H.; Díaz-Borjón, A.; Espinosa, M.V.D.M. Update of the Mexican College of Rheumatology guidelines for the pharmacologic treatment of rheumatoid arthritis. *Reumatol. Clin.* **2014**, *10*, 227–240. [CrossRef] [PubMed]
26. Lin, H.W.; Liu, C.Z.; Cao, D.; Chen, P.Y.; Chen, M.-F.; Lin, S.Z.; Mozayan, M.; Chen, A.F.; Premkumar, L.S.; Torry, D.S.; et al. Endogenous methyl palmitate modulates nicotinic receptor-mediated transmission in the superior cervical ganglion. *Proc. Natl. Acad. Sci. USA* **2008**, *105*, 19526–19531. [CrossRef] [PubMed]
27. El-Demerdash, E. Anti-inflammatory and antifibrotic effects of methyl palmitate. *Toxicol. Appl. Pharmacol.* **2011**, *254*, 238–244. [CrossRef]
28. Saeed, N.M.; El-Demerdash, E.; Abdel-Rahman, H.M.; Algandaby, M.M.; Al-Abbasi, F.A.; Abdel-Naim, A.B. Anti-inflammatory activity of methyl palmitate and ethyl palmitate in different experimental rat models. *Toxicol. Appl. Pharmacol.* **2012**, *264*, 84–93. [CrossRef]
29. Ethyl Palmitate. Available online: https://pubchem.ncbi.nlm.nih.gov/compound/Ethyl-palmitate (accessed on 30 March 2022).
30. Ye, S.; Zhong, J.; Huang, J.; Chen, L.; Yi, L.; Li, X.; Lv, J.; Miao, J.; Li, H.; Chen, D.; et al. Protective effect of plastrum testudinis extract on dopaminergic neurons in a Parkinson's disease model through DNMT1 nuclear translocation and SNCA's methylation. *Biomed. Pharmacother.* **2021**, *141*, 111832. [CrossRef] [PubMed]
31. Park, S.Y.; Seetharaman, R.; JungKo, M.; Kim, D.Y.; Kim, T.H.; Yoon, M.K.; Kwak, J.H.; Lee, S.J.; Bae, Y.S.; Choi, Y.W. Ethyl linoleate from garlic attenuates lipopolysaccharide-induced pro-inflammatory cytokine production by inducing heme oxygenase-1 in RAW264.7 cells. *Int. Immunopharmacol.* **2014**, *19*, 253–261. [CrossRef] [PubMed]
32. Monterrosas-Brisson, N.; Ocampo, M.L.; Jiménez-Ferrer, E.; Jiménez-Aparicio, A.R.; Zamilpa, A.; Gonzalez-Cortazar, M.; Tortoriello, J.; Herrera-Ruiz, M. Anti-inflammatory activity of different agave plants and the compound cantalasaponin-1. *Molecules* **2013**, *18*, 8136–8146. [CrossRef]
33. Hernández-Valle, E.; Herrera-Ruiz, M.; Salgado, G.R.; Zamilpa, A.; Ocampo, M.L.; Aparicio, A.J.; Tortoriello, J.; Jiménez-Ferrer, E. Anti-inflammatory effect of 3-O-[(6′-O-palmitoyl)-beta-D-glucopyranosyl sitosterol] from Agave angustifolia on ear e in mice. *Molecules* **2014**, *19*, 15624–15637. [CrossRef]
34. Bautista, D.M.; Jordt, S.E.; Nikai, T.; Tsuruda, P.R.; Read, A.J.; Poblete, J.; Yamoah, E.N.; Basbaum, A.I.; Julius, D. TRPA1 Mediates the Inflammatory Actions of Environmental Irritants and Proalgesic Agents. *Cell* **2006**, *124*, 1269–1282. [CrossRef]
35. Norões, M.M.; Santos, L.G.; Gavioli, E.C.; Rachetti, V.D.P.S.; Otuki, M.F.; de Almeida Cabrini, D.; da Silveira Prudente, A.; Oliveira, J.R.J.M.; de Carvalho Gonçalves, M.; Ferreira, J.; et al. Role of TRPA1 receptors in skin inflammation induced by volatile chemical irritants in mice. *Eur. J. Pharmacol.* **2019**, *858*, 172460. [CrossRef]
36. Bellono, N.W.; Kammel, L.G.; Zimmerman, A.L.; Oancea, E. UV light phototransduction activates transient receptor potential A1 ion channels in human melanocytes. *Proc. Natl. Acad. Sci. USA* **2013**, *110*, 2383–2388. [CrossRef] [PubMed]
37. Bíró, T.; Kovács, L. An ice-Cold TR(i)P to skin biology: The role of TRPA1 in human epidermal keratinocytes. *J. Investig. Dermatol.* **2009**, *129*, 2096–2209. [CrossRef] [PubMed]
38. Hyeong-Dong, K.; Hyung-Rae, C.; Seung-Bae, M.; Hyun-Dong, S.; Kun-Ju, Y.; Bok-Ryeon, P.; Hee-Jeong, J.; Lin-Su, K.; Hyeung-Sik, L.; Sae-Kwang, K. Effects of beta-glucan from Aureobasidium pullulans on acute inflammation in mice. *Arch. Pharm. Res.* **2007**, *30*, 323–328.
39. Eidi, A.; Oryan, S.; Zaringhalam, J.; Rad, M. Antinociceptive and anti-inflammatory effects of the aerial parts of Artemisia dracunculus in mice. *Pharm. Biol.* **2016**, *54*, 549–554. [CrossRef]
40. Othman, A.R.; Abdullah, N.; Ahmad, S.; Ismail, I.S.; Zakaria, M.P. Elucidation of in-vitro anti-inflammatory bioactive compounds isolated from Jatropha curcas L. plant root. *BMC Complement. Altern. Med.* **2015**, *15*, 11–21. [CrossRef]
41. Huebner, S.M.; Campbell, J.P.; Butz, D.E.; Fulmer, T.G.; Gendron-Fitzpatrick, A.; Cook, M.E. Individual isomers of conjugated linoleic acid reduce inflammation associated with established collagen-induced arthritis in DBA/1 mice. *J. Nutr.* **2010**, *140*, 1454–1461. [CrossRef]
42. Muhlenbeck, J.A.; Olson, J.M.; Hughes, A.B.; Cook, M.E. Conjugated Linoleic Acid Isomers Trans-10, Cis-12 and Cis-9, Trans-11 Prevent Collagen-Induced Arthritis in a Direct Comparison. *Lipids* **2018**, *53*, 689–698. [CrossRef]
43. Schaible, H.G. Nociceptive neurons detect cytokines in arthritis. *Arthritis Res. Ther.* **2014**, *16*, 470. [CrossRef]
44. Da Silva, M.D.; Bobinski, F.; Sato, K.L.; Kolker, S.J.; Sluka, K.A.; Santos, A.R. IL-10 cytokine released from M2 macrophages is crucial for analgesic and anti-inflammatory effects of acupuncture in a model of inflammatory muscle pain. *Mol. Neurobiol.* **2015**, *51*, 19–31. [CrossRef]

45. Gárbor, M. 12-O-tetradecanoylphorbol-13-acetate ear oedema (TPA). In *Mouse Ear Inflammation Models and Their Pharmacological Applications*; Gárbor, M., Ed.; Hungarian Academy of Sciences: Budapest, Hungary, 2000; pp. 28–33.
46. Gómez-Rivera, A.; González-Cortazar, M.; Gallegos-García, A.J.; Escobar-Ramos, A.; Flores-Franco, G.; Lobato-García, C. Spasmolytic, anti-inflammatory, and antioxidant activities of Salvia gesneriflora Lindley. *Afr. J. Tradit. Complement. Altern. Med.* **2018**, *15*, 78–82. [CrossRef]
47. Gutiérrez-Román, A.S.; Trejo-Tapia, G.; González-Cortazar, M.; Jiménez-Ferrer, E.; Trejo-Espino, J.L.; Zamilpa, A.; Ble-González, E.; Camacho-Díaz, B.H.; Herrera-Ruiz, M. Anti-arthritic and anti-inflammatory effects of Baccharis conferta Kunth in a kaolin/carrageenan-induced monoarthritis model. *J. Ethnopharmacol.* **2022**, *288*, 114996. [CrossRef] [PubMed]
48. Neugebauer, V.; Han, J.S.; Adwanikar, H.; Fu, Y.; Ji, G. Techniques for assessing knee joint pain in arthritis. *Mol. Pain* **2007**. [CrossRef] [PubMed]
49. Sluka, K.A.; Bailey, K.; Bogush, J.; Olson, R.; Ricketts, A. Treatment with either high or low frequency TENS reduces the secondary hyperalgesia observed after injection of kaolin and carrageenan into the knee joint. *Pain* **1998**, *77*, 97–102. [CrossRef]
50. Langford, D.J.; Mogil, J.S. Pain testing in the laboratory mouse. In *Anesthesia and Analgesia in Laboratory Animals*; Fish, R.E., Brown, M.J., Danneman, P.J., Karas, A.Z., Eds.; Academic Press: Cambridge, MA, USA, 2008; pp. 549–560.
51. Martínez-Hernández, G.B.; Vargas-Villa, G.; Jiménez-Ferrer, E.; García-Aguilar, M.P.; Zamilpa, A.; González-Cortazar, M.; Avilés-Flores, M.; Herrera-Ruiz, M. Anti-arthritic and anti- inflammatory effects of extract and fractions of Malva parviflora in a mono-arthritis model induced with kaolin. *Naunyn-Schmiedeberg's Arch. Pharmacol.* **2020**, *393*, 1281–1291. [CrossRef] [PubMed]

molecules

MDPI

Article

Phytochemical Composition and In Vitro Antioxidant, Anti-Inflammatory, Anticancer, and Enzyme-Inhibitory Activities of *Artemisia nilagirica* (C.B. Clarke) Pamp

Jawaher J. Albaqami [1,†], Tancia P. Benny [2,†], Hamida Hamdi [1,3], Ammar B. Altemimi [4,5], Aswathi Moothakoottil Kuttithodi [2], Joice Tom Job [2,*], Anju Sasidharan [2] and Arunaksharan Narayanankutty [2,*]

1 Department of Biology, College of Science, Taif University, P.O. Box 11099, Taif 21944, Saudi Arabia
2 Division of Cell and Molecular Biology, PG and Research Department of Zoology, St. Joseph's College (Autonomous), Devagiri, Calicut 673 008, Kerala, India
3 Zoology Department, Faculty of Science, Cairo University, Giza 12613, Egypt
4 Department of Food Science, College of Agriculture, University of Basrah, Basrah 61004, Iraq
5 College of Medicine, University of Warith Al-Anbiyaa, Karbala 56001, Iraq
* Correspondence: joicetomj@devagiricollege.org (J.T.J.); arunaksharann@devagiricollege.org (A.N.)
† These authors contributed equally to this work.

Abstract: Plants have been employed in therapeutic applications against various infectious and chronic diseases from ancient times. Various traditional medicines and folk systems have utilized numerous plants and plant products, which act as sources of drug candidates for modern medicine. Artemisia is a genus of the *Asteraceae* family with more than 500 species; however, many of these species are less explored for their biological efficacy, and several others are lacking scientific explanations for their uses. *Artemisia nilagirica* is a plant that is widely found in the Western Ghats, Kerala, India and is a prominent member of the genus. In the current study, the phytochemical composition and the antioxidant, enzyme-inhibitory, anti-inflammatory, and anticancer activities were examined. The results indicated that the ethanol extract of *A. nilagirica* indicated in vitro DPPH scavenging (23.12 ± 1.28 μg/mL), ABTS scavenging (27.44 ± 1.88 μg/mL), H_2O_2 scavenging (12.92 ± 1.05 μg/mL), and FRAP (5.42 ± 0.19 μg/mL). The anti-inflammatory effect was also noticed in the Raw 264.7 macrophages, where pretreatment with the extract reduced the LPS-stimulated production of cytokines ($p < 0.05$). *A. nilagirica* was also efficient in inhibiting the activities of α-amylase (38.42 ± 2.71 μg/mL), α-glucosidase (55.31 ± 2.16 μg/mL), aldose reductase (17.42 ± 0.87 μg/mL), and sorbitol dehydrogenase (29.57 ± 1.46 μg/mL). It also induced significant inhibition of proliferation in breast (MCF7 IC_{50} = 41.79 ± 1.07, MDAMB231 IC_{50} = 55.37 ± 2.11μg/mL) and colon (49.57 ± 1.46 μg/mL) cancer cells. The results of the phytochemical screening indicated a higher level of polyphenols and flavonoids in the extract and the LCMS analysis revealed the presence of various bioactive constituents including artemisinin.

Keywords: phytochemistry; *Artemisia nilagirica*; *Asteraceae*; antioxidant; anti-inflammatory activity; anticancer activity

Citation: Albaqami, J.J.; Benny, T.P.; Hamdi, H.; Altemimi, A.B.; Kuttithodi, A.M.; Job, J.T.; Sasidharan, A.; Narayanankutty, A. Phytochemical Composition and In Vitro Antioxidant, Anti-Inflammatory, Anticancer, and Enzyme-Inhibitory Activities of *Artemisia nilagirica* (C.B. Clarke) Pamp. *Molecules* **2022**, 27, 7119. https://doi.org/10.3390/molecules27207119

Academic Editor: Lucia Panzella

Received: 27 September 2022
Accepted: 18 October 2022
Published: 21 October 2022

Publisher's Note: MDPI stays neutral with regard to jurisdictional claims in published maps and institutional affiliations.

Copyright: © 2022 by the authors. Licensee MDPI, Basel, Switzerland. This article is an open access article distributed under the terms and conditions of the Creative Commons Attribution (CC BY) license (https://creativecommons.org/licenses/by/4.0/).

1. Introduction

Medicinal plants are important sources of various biologically and pharmacologically active compounds [1,2]. Several traditional medicinal plants have been shown to have strong pharmacological properties, such as radical neutralizing, inflammation-preventing, antiproliferative, hypolipidemic, hepatoprotective, neuroprotective, antithrombotic, and immunomodulatory activities [3,4]. Among the various plant families, *Asteraceae* is one of the most widely utilized ones, and it is also equipped with numerous biological and pharmacological activities. Among the various genera, the Artemesia genus is well-known [5–7].

The Artemisia genus and the member species are well-studied for their various biological activities [8–13]. *Artemisia annua* L. has demonstrated significant medicinal benefits because of the presence of artemisinin [14]. *Artemisia mongolica* is another important member of the genus, which is rich in lactone derivatives of Sesquiterpene and a wide range of pharmacological activities [15]. The different species of the genus were found to have strong antibacterial and antifungal properties against pathogenic organisms in humans, livestock, and plants [16–20]. Antiproliferative and apoptotic effects are attributed to the bioactive compounds and extracts of various species of *Artemisia* [21–24].

Artemisia nilagirica is distributed throughout the Western Ghats, India; it has been traditionally applied by various tribal healers in the area for the treatment of infectious diseases and toxicity prevention. The plant has been shown to have significant biological and pharmacological activities based on various in vitro and in vivo studies. The initial studies by Ahameethunisa and Hopper [25] identified the antibacterial potential of the methanol extract of *A. nilagirica* against 15 bacterial strains. Further, the extract was found to be effective against *Mycobacterium smegmatis* and *M. bovis* [26]. The extract was also found to be effective against the malarial parasite *Plasmodium falciparum* [27].

The anticancer activities of the methanol and ethyl acetate extracts were also elucidated against the human monocytic leukemia cell (THP-1) [28]. Later, studies by Sahu, Meena, Shukla, Chaturvedi, Kumar, Datta, and Arya [24] also supported these results in colorectal cancer cell models. Studies by Raju et al. [29] indicated that the anticancer activity was mediated through the inhibition of TGF-beta signaling. The plant extract was also found to inhibit inflammatory insults in human red blood cell models [30]. The fruit of *A. nilagirica* was found to have significant antiradical activity via scavenging DPPH and nitric oxide radicals [31]. The essential oil extracted from *A. nilagirica* was a rich source of monoterpenoid compounds such as thujone, and by virtue of these compounds, the essential oil inhibited the growth of various fungal pathogens [32]. The essential oil was also effective against the phytopathogenic fungal groups of table grapes [33]. Additionally, the essential oil was also effective against various bacterial populations and capable of repelling mosquitoes [34].

Although several studies have reported the preliminary pharmacological activity of the plant, there is no clear-cut information on its quantitative chemical profile and nutritional value. Additionally, the anti-inflammatory properties are yet to be discovered in cell line models, and its mechanism of action is also not specified. Therefore, the present study aimed to analyze the chemical composition of the ethanol extract of *Artemisia nilagirica* leaves in terms of the bioactive compounds and proximate composition, as well as their antioxidant potential. Further, this study for the first time attempted to analyze the enzyme-inhibitory and anti-inflammatory activities of the extract in Raw 264.7 cells stimulated by lipopolysaccharides.

2. Results and Discussion

2.1. Determination of Proximate Composition of A. nilagirica

The Artemisia species, which includes 200–400 identified plants, are extensively spread in tropical and temperate areas [6]. The importance of the artemisia species in traditional medicine is well established [5]. The plant's antiviral, antifungal, antibiotic, insecticidal, hepatoprotective, and neuroprotective qualities make it useful in both Chinese and Ayurvedic medical systems [35]. The current study examined a specific member *Artemisia nilagirica*, its phytochemical makeup, and its pharmacological effects.

The physicochemical parameters of the *A. nilagirica* leaf powder are shown in Table 1. The predominant compounds were carbohydrate, protein, fat, and ash contents. The moisture content was estimated to be 87.4 $\pm$ 2.12%.

2.2. Quantitative and Qualitative Estimation of Phytochemicals in A. nilagirica

The qualitative phytochemical screening identified the presence of compounds such as alkaloids, flavonoids, glycosides, sterols, and triterpenes (Table 2). The LCMS analysis of the *A. nilagirica* ethanol extract indicated the presence of various phytocompounds, including artemisinin, quercetin, apigenin, B-caryophyllene, luteolin, and simple phenolic acids (Figure 1 and Table 3). Previous reports have also confirmed that numerous kinds of bioactive substances are found in *A. vulgaris*, *A. annua*, and other species, including flavonoids, sesquiterpenoids, essential oils, tannins, phenols, and saponins [15,36]. The total polyphenol content of *A. nilagirica* was estimated to be 89.51 ± 2.5 mg gallic acid equivalent/g of extract. The total flavonoid content was 14.35 ± 0.9 mg quercetin equivalent/g of extract (Table 4). Further, the HPLC quantification indicated higher levels of quercetin (240.39 ± 4.87 µg/g extract), luteolin (146.87 ± 5.29 µg/g extract), and apigenin (103.41 ± 3.35 µg/g extract) in the *A. nilagirica* extract (Table 5). These compounds are known to possess strong anti-inflammatory, antiproliferative and antidiabetic activities [37–40].

Table 1. Physicochemical parameters of *A. nilagirica* leaf powder.

Physicochemical Parameters	Result
Moisture content (%)	87.4 ± 2.12
Carbohydrate (%)	55.80 ± 4.1
Protein (%)	3.90 ± 0.16
Crude fat (%)	2.12 ± 0.18
Ash content (%)	0.74 ± 0.04

Table 2. Phytochemical constituents in the ethanol extract of *A. nilagirica*.

Test	Reaction
Alkaloids	
Marqui's test	++
Wagner's test	++
Mayer's test	+++
Hager's test	+
Froehde's test	++
Dragendorff test	++
Glycosides	
Legal's test	+
Keller-Kiliani test	+
Flavonoids	
Alkaline reagent test	++
Lead acetate test	++
Shinoda's test	+++
Tannins	
Ferric Chloride test	++
Gelatin test	++
Phytosterols	
Salkowski's test	++
Liebermann-Burchard test	+++
Saponins	
Froth test	+
Foam test	+
Carbohydrates	
Fehling test	++
Molish test	++
Benedict's test	++
Phenols	
Folin-Ciocalteu test	+++
Resin	
Acetone-water test	+

Table 2. *Cont.*

Test	Reaction
Alkaloids	
Fixed oils and fats	
Stain test	-
Triterpenes	
Liebermann-Burchardt's test	+++

Note: +++ high level, ++ moderate level, and + low-level presence of the compound.

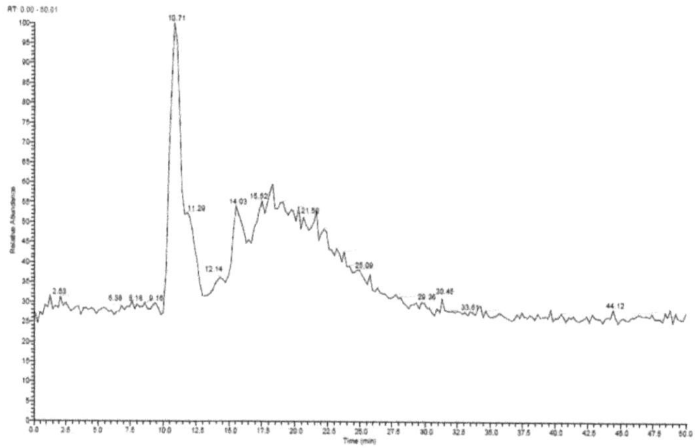

Figure 1. The LC-MS total ion chromatogram of the *A. nilagirica* extract.

Table 3. LCMS profiling of *A. nilagirica* with the retention time (RT), molecular mass, and chemical formula.

Sl. No.	RT (mins)	Compound Name	Formula	Mass
1	2.53	Ferulic acid	$C_{10}H_{10}O_4$	194.00
2	6.38	Eugenol	$C_{10}H_{12}O_2$	164.08
3	8.18	B-caryophyllene	$C_{21}H_{20}O_{11}$	448.40
4	9.06	Luteolin	$C_{15}H_{10}O_6$	286.00
5	10.71	caffeic acid	$C_9H_8O_4$	180.16
6	11.29	Quercetin	$C_{15}H_{10}O_7$	302.00
7	12.14	Myricetin	$C_{15}H_{10}O_8$	318.00
8	12.89	Apigenin	$C_{15}H_{10}O_5$	270.05
9	14.03	Luteolin 5-0-beta-d-glucopyranoside	$C_{21}H_{20}O_{11}$	448.13
10	15.52	Kaempferol	$C_{15}H_{10}O_6$	286.23
11	21.56	Carnosic acid	$C_{20}H_{28}O_4$	332.19
12	25.09	Artemisinin	$C_{20}H_{20}O_8$	388.11
13	29.36	2alpha, 3 beta-Dihydroxyolean-12en-28-oic acid	$C_{30}H_{48}O_4$	472.35
14	30.45	Menthyl acetate	$C_{12}H_{22}O_2$	198.16
15	33.61	Oleanolic acid	$C_{30}H_{48}O_3$	456.36
16	44.12	Basilimoside	$C_{36}H_{60}O_6$	588.47

Table 4. The total polyphenol and flavonoid contents of *A. nilagirica* ethanol extract.

Assay	mg Equivalent/g
Total phenolic content	89.51 ± 2.5
Total flavonoid content	14.35 ± 0.9

Table 5. The quantification of selected compounds in the extract via HPLC.

RT (mins)	Compound Name	Quantity (μg/g Extract)
2.50	Ferulic acid	18.51 ± 1.82
9.05	Luteolin	146.87 ± 5.29
10.70	caffeic acid	88.62 ± 1.30
11.30	Quercetin	240.39 ± 4.87
12.87	Apigenin	103.41 ± 3.35

2.3. *In Vitro Antioxidant Activities of A. nilagirica Extract*

The Artemisia genus members frequently display antioxidant activity [41]; our study also confirmed the antioxidant activity of *A. nilagirica* for the first time in terms of the radical generation inhibition and reducing potentials. The IC_{50} value of the *A. nilagirica* extract in the anti-DPPH radical assay was estimated to be 23.12 ± 1.28 μg/mL. Likewise, Table 6 shows the other antioxidant activities in terms of the ABTS radical scavenging activity, hydrogen peroxide scavenging potential, and ferric-reducing antioxidant power; the respective IC_{50} values were found to be 27.44 ± 1.88, 12.92 ± 1.05, and 5.42 ± 0.19 μg/mL. On the contrary, the level of inhibition of nitric oxide radical generation (IC_{50}) was determined to be 367.09 ± 12.05 μg/mL for the extract. However, in comparison with the standard antioxidant ascorbic acid (Table 6), the activity was much lower in the *A. nilagirica* extract; further purification of the extract may yield more active antioxidant compounds. The antioxidant properties are attributed to the bioactive compounds identified in the plant via LC-MS. Oxidative stress is the central independent factor that drives many chronic diseases, including cancers [42,43]; hence, the antioxidant properties of the plant may be useful in the management of diseases associated with oxidative stress.

Table 6. In vitro antioxidant activities of *A. nilagirica* extract (AN) expressed as IC_{50} values (μg/mL).

Antioxidant Activity	IC_{50} Value (μg/mL)	
	AN	Ascorbic Acid
DPPH scavenging	23.12 ± 1.28	9.64 ± 0.89
ABTS scavenging	27.44 ± 1.88	35.19 ± 1.47
H_2O_2 scavenging	12.92 ± 1.05	19.08 ± 1.65
FRAP value (EC_{50})	5.42 ± 0.19	3.22 ± 0.15
Nitric oxide scavenging	367.09 ± 12.05	68.10 ± 2.11

2.4. *Enzyme-Inhibitory Activities of A. nilagirica Ethanol Extract*

The enzyme-inhibitory properties of the extract were analyzed against four enzymes involved in type 2 diabetes mellitus, including α-amylase, α-glucosidase, aldose reductase, and sorbitol dehydrogenase (Table 7). The IC_{50} values for these enzymes were 38.42 ± 2.71, 55.31 ± 2.16, 17.42 ± 0.87, and 29.57 ± 1.46 μg/mL, respectively. Furthermore, α-amylase and α-glucosidase are enzymes involved in carbohydrate metabolism and are common targets of antidiabetic drugs [44]. Similarly, the polyol pathway enzymes, including aldose reductase and sorbitol dehydrogenase, are involved in diabetic complications [45,46]. Hence, the inhibition of these enzymes could result in strong antidiabetic activity for the *A. nilagirica* extract.

Table 7. In vitro enzyme-inhibitory properties of *A. nilagirica* expressed as IC_{50} values (μg/mL).

Enzyme	IC_{50} Value (μg/mL)
α-Amylase	38.42 ± 2.71
α-Glucosidase	55.31 ± 2.16
Aldose reductase	17.42 ± 0.87
Sorbitol dehydrogenase	29.57 ± 1.46

2.5. Antiproliferative Activity of the A. nilagirica

Additionally, the results showed the anticancer properties of *A. nilagirica* in human breast and colon cancer cells. The anticancer activity was analyzed in three cancer cell lines, including MCF-7, MDA-MB-231, and HCT-15. We observed dose-dependent cytotoxicity in these three cell lines (Figure 2). The IC_{50} values against the three cells were estimated to be 41.79 ± 1.07, 55.37 ± 2.11, and 49.57 ± 1.46 µg/mL, respectively. In comparison, the standard cyclophosphamide was more toxic to these cells, with respective IC_{50} values of 3.12 ± 0.13, 5.74 ± 0.20, and 6.04 ± 0.21 µg/mL. Previous studies have also shown different species of *Artemisia* in various cancer cells [47–50]. In addition, the green synthesized nanoparticles from different *Artemisia* species are also reported to exert antiproliferative effects on cancer cells mediated through apoptotic cell death [23,51,52]. A study by Sahu, Meena, Shukla, Chaturvedi, Kumar, Datta, and Arya [24] indicated that ethyl acetate and hexane fractions of *A. nilagirica* induced cell death in colon, lung, and breast cancer cells. In addition, the bioactive compounds, including quercetin, apigenin, and eugenol, have also been shown to have significant antiproliferative effects by modulating different signaling pathways [53,54].

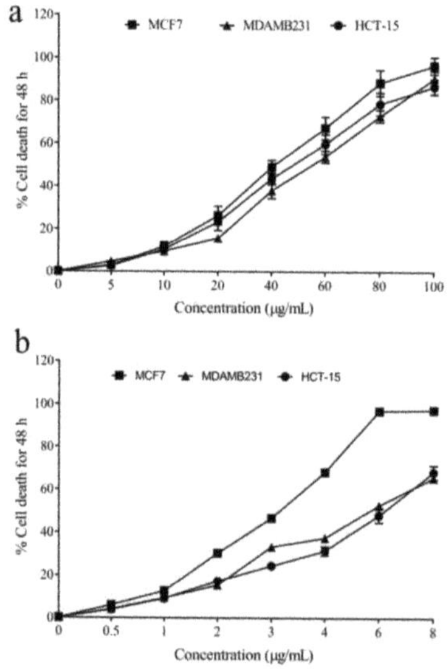

Figure 2. The anticancer potentials of the leaf extract of *A. nilagirica* (**a**) and cyclophosphamide (**b**).

2.6. Anti-Inflammatory Activity of A. nilagirica

The *Artemisia nilagirica* extract was shown to inhibit the production of nitric oxide radicals in vitro. Further, the pretreatment of the extract also inhibited cytokine production and inflammatory insults in lipopolysaccharide-stimulated macrophages. The LPS is a microbial component that is known to stimulate inflammatory insults [55,56]. The Artemisia *nilagirica* leaf ethanol extract (AN) was found to inhibit the lipopolysaccharide-induced activation of macrophages and the subsequent cytokine release. The level of IL-1β was found to be significantly increased after LPS stimulation in macrophages; however, the pretreatment with AN at different doses significantly brought down the IL-1β levels in the macrophages (Table 8). Likewise, the levels of IL-6 and TNF-α also showed a similar increase during LPS exposure, which were successfully brought down by the treatment with different concentrations of *A. nilagirica*. The level of nitric oxide was determined biochemically and was also significantly elevated in LPS control cells. Pretreatment with 2.5, 5.0, and 7.5 µg/mL of AN successfully brought down the levels to 40.7 ± 1.6 ($p < 0.05$), 32.2 ± 2.4 ($p < 0.05$), and 25.7 ± 2.1 ($p < 0.01$). In addition, it is noted that the high dose of the extract resulted in stronger anti-inflammatory molecules compared to quercetin, which is a well-known anti-inflammatory molecule [57,58]. The LPS is known to stimulate cytokine production in macrophages by upregulating the NF-κB translocation to the nuclear compartment [59,60]. It is, therefore, possible that the *A. nilagirica* extract may also influence the LPS-induced activation of intracellular NF-KB signaling.

Table 8. Effect of the *Artemisia nilagirica* leaf ethanol extract (AN) against lipopolysaccharide-induced macrophage (Raw 264.7) activation, cytokine release (in pg/mg protein), and nitric oxide production (µM/mg protein).

Nature		Tumor Necrosis Factor α	Interleukin 6	Interleukin 1β	NO
Untreated		97.6 ± 2.8	76.4 ± 3.1	67.8 ± 2.8	7.4 ± 0.57
Negative Control (LPS alone)		420.8 ± 10.6	795.2 ± 11.7	628.9 ± 14.2	52.1 ± 2.0
Quercetin (4.5 µg/mL)		279.1 ± 11.3 **	414.2 ± 10.7 ***	334.8 ± 11.7 **	30.7 ± 1.2 *
Artemisia nilagirica extract	2.5 µg/mL	314.1 ± 14.5 *	698.0 ± 17.3 **	477.6 ± 11.8 **	40.7 ± 1.6 *
	5.0 µg/mL	265.7 ± 10.7 **	524.3 ± 15.6 **	389.5 ± 14.6 **	32.2 ± 2.4 *
	7.5 µg/mL	190.9 ± 14.8 ***	388.2 ± 15.8 ***	298.7 ± 15.2 ***	25.7 ± 2.1 **

Artemisia nilagirica leaf ethanol extract (AN), lipopolysaccharide (LPS), nitric oxide (NO). The significance is indicated as * ($p < 0.05$), ** ($p < 0.01$), *** ($p < 0.001$).

Thus, the study concludes that the *Artemisia nilagirica* ethanol extract exhibits antioxidant and anti-inflammatory properties in vitro and cultured cells. Further, the extract is also capable of inhibiting the proliferation of various cancer cells. The inhibition of enzymes associated with type 2 diabetes mellitus is also indicative of its anti-diabetic properties. The biological properties of the plant are expected to be due to the bioactive compounds identified in the *A. nilagirica* extract.

3. Materials and Methods

3.1. Artemisia Nilagirica (C.B.Clarke) Pamp. Collection and Extraction Using 100% Ethanol

The *Artemisia nilagirica* plant samples were collected from the Wayanad District, Kerala (11.7917° N, 76.1716° E). The mature leaves were carefully cleaned of all kinds of dust via washing. These leaves were dried under shade for 2 weeks and powdered using a mixer grinder; the powder was extracted with 100% ethanol using the Soxhlet method. Briefly, 100 g of the powder was extracted with ethanol at 80 °C for 8 h and the extract was collected, filtered, and concentrated before storage.

3.2. Phytochemical Analysis of Artemesia nilagirica

The leaf powder of *A. nilagirica* was analyzed for the proximate composition according to the methods used by Shukla et al. [61]. The qualitative phytochemical screening was carried out for the detection of alkaloids, flavonoids, glycosides, sterols, and triterpenes by referring to standard protocols [62,63]. The LC-MS analysis (Shimadzu LC- 8045, Kyoto, Japan) was used for phytochemical screening [64]; briefly, the C18 column measuring 4.6 × 150 mm and 5 µm in size was used for the study, with methanol (A) and water with 0.1% formic acid (B) as the mobile phase (gradient elution mode). The gradient was set as 95% solution A (0–5 min), 70% solution A (5 to 10 min), 65% solution A (10 to 20 min), 50% solution A (20 to 30 min), and 90% of solution B (until 50 min), with a flow rate of 1.0 mL/min.

The quantitative profiling was estimated in terms of the total polyphenols [65] and total flavonoids [66], and the concentrations of ferulic acid, luteolin, caffeic acid, quercetin, and apigenin were determined using an HPLC analysis according to the same LC-MS conditions mentioned above.

3.3. *Analysis of the Antioxidant Activity of A. nilagirica Ethanol Extract*

The antioxidant activities were determined as the scavenging potentials of different radicals, including diphenyl picryl hydrazyl (DPPH), ABTS [67], and hydrogen peroxide [68]; the reducing potential on ferric ions was also estimated using the procedures described in [69]. The nitric oxide radical removal rate was used as an indicator of the inflammatory process inhibition model [70]. The DPPH was dissolved in methanol (0.1 mM) and varying concentrations of the extract were mixed with it. The solution was incubated for 20 min in the dark at 30 °C and the change in absorbance was used to estimate the percentage inhibition. Likewise, the ABTS radical generated was mixed with different doses of the *A. nilagirica* extract and the % inhibition was calculated spectrophotometrically. The nitric oxide scavenging was determined using sodium nitroprusside (8 mM) as the radical source; the Griess reagent was used to estimate the nitrite remaining in the treated samples using spectrophotometry at 596 nm.

Ascorbic acid was used as a positive control and standard for the antioxidant assays. The percentage inhibition was determined using the formula

$$Percentage\ inhibition = \frac{Absorbance\ of\ Control - Absorbance\ of\ Sample}{Absorbance\ of\ Control} \times 100$$

3.4. *Efficacy of A. nilagirica Ethanol Extract on Activities of Enzymes*

The enzyme-inhibitory properties were analyzed against the selected enzymes involved in diabetes and secondary diabetic complications. The inhibitory effect on α-amylase [71], α-glucosidase [72], aldose reductase [73], and sorbitol dehydrogenase [46] was assessed according to the standard methods.

3.5. *Effect of A. nilagirica Ethanol Extract on Cancer Cell Proliferation*

The human breast cancer cell lines MCF7 and MDA-MB-231 and a colon cancer cell line (HCT-15) were collected from NCCS, Pune, India. These cells were maintained in complete MEM, Leibovitz's L-15, and RPMI-1640 media. The cells were selected as they are widely used in the anticancer screening of phytochemicals.

The inhibitory potential of the extract on human cancer cell proliferation (MCF7, MDA-MB-231, and HCT-15) was assessed using the MTT assay [74]. The IC_{50} value was determined using probit analysis.

3.6. *Effect of A. nilagirica Extract on Lipopolysaccharide-Induced Cytokine Production in Macrophages*

The murine Raw 264.7 cells were allowed to attach (1×10^7 cells/mL) in a 24-well plate in complete growth media. The RPMI-1640 media was used to dilute the different concentrations of *A. nilagirica* (AN) (2.5, 5.0, and 7.5 μg/mL). Next, the cells were exposed to 1 μg/mL lipopolysaccharide for another 24 h. The protein expression of cytokines such as interleukin-1β and interleukin-6 and the tumor necrosis factor-α release were determined using PeproTech ELISA kits (Rocky Hill, CT, USA), as per the commercially prescribed methods. The nitric oxide release was quantified using the Griess reaction method [64]. Quercetin was used as a standard anti-inflammatory compound in the study.

3.7. *Presentation of the Data, Software Used, and Statistical Analysis*

The accuracy of the results obtained was ensured by conducting three independent assignments, with each having four replicates. Microsoft Excel 2010 was used for data consolidation and verification. The processed data are presented as means ± standard deviations; the IC_{50} values were estimated using probit analysis (GraphPad Prism 7.0, San Diego, CA, USA).

4. Conclusions

Artemisia nilagirica is an ethnomedicinal plant in India. In our study, the ethanol extract of *A. nilagirica* leaves showed significant antiradical and reducing potentials, which are indicative of its antioxidant potential. The IC_{50} values were lower but comparable with those of the standard ascorbic acid. The extract also inhibited enzymes associated with diabetes mellitus, including alpha-amylase and α-glucosidase. Additionally, the extract treatment significantly reduced the proliferative potential of breast and colon cancer cells. In Raw 264.7 macrophages, the pretreatment with the extract inhibited the LPS-stimulated production of cytokines and proved itself to be anti-inflammatory. Most importantly, the higher dose of the extract caused significantly higher activity than the standard quercetin used. Hence, we conclude that the ethanol extract of *A. nilagirica* leaves has antioxidant, anti-inflammatory, and anticancer properties; further studies on animal models and with bioassay-guided purification are necessary to identify the bioactive components.

Author Contributions: A.N.: Study design, methodology, analysis, manuscript editing. J.J.A. and H.H.: Study design, methodology, experimentation, analysis, funding acquisition, manuscript preparation, manuscript editing. A.M.K., A.S., T.P.B. and J.T.J.: Experimentation, analysis; manuscript draft preparation. A.B.A.: Analysis, manuscript preparation, manuscript editing. All authors have read and agreed to the published version of the manuscript.

Funding: This study received no funding support.

Institutional Review Board Statement: Not applicable.

Informed Consent Statement: Not applicable.

Data Availability Statement: Data is contained within the article.

Acknowledgments: ARN and JTJ acknowledge the DBT star scheme for the infrastructural developments in the college. ARN and JTJ is thankful to St. Joseph's College (Autonomous), Devagiri for the research promotion (seed grant) scheme RPSG-2022-23.

Conflicts of Interest: The authors declare no conflict of interest.

Sample Availability: Samples of the compounds are available from the corresponding author.

References

1. Behera, N.K.; Mahalakshmi, G.S. A cloud based knowledge discovery framework, for medicinal plants from PubMed literature. *Inform. Med. Unlocked* **2019**, *16*, 100105. [CrossRef]
2. Karpavičienė, B. Traditional Uses of Medicinal Plants in South-Western Part of Lithuania. *Plants* **2022**, *11*, 2093. [CrossRef] [PubMed]
3. Noor, F.; Tahir ul Qamar, M.; Ashfaq, U.A.; Albutti, A.; Alwashmi, A.S.S.; Aljasir, M.A. Network Pharmacology Approach for Medicinal Plants: Review and Assessment. *Pharmaceuticals* **2022**, *15*, 572. [CrossRef] [PubMed]
4. Adeleke, B.S.; Babalola, O.O. Pharmacological Potential of Fungal Endophytes Associated with Medicinal Plants: A Review. *J. Fungi* **2021**, *7*, 147. [CrossRef] [PubMed]
5. Septembre-Malaterre, A.; Lalarizo Rakoto, M.; Marodon, C.; Bedoui, Y.; Nakab, J.; Simon, E.; Hoarau, L.; Savriama, S.; Strasberg, D.; Guiraud, P.; et al. *Artemisia annua*, a Traditional Plant Brought to Light. *Int. J. Mol. Sci.* **2020**, *21*, 4986. [CrossRef] [PubMed]
6. Bora, K.S.; Sharma, A. The genus Artemisia: A comprehensive review. *Pharm Biol.* **2011**, *49*, 101–109. [CrossRef] [PubMed]
7. Bisht, D.; Kumar, D.; Kumar, D.; Dua, K.; Chellappan, D.K. Phytochemistry and pharmacological activity of the genus artemisia. *Arch. Pharmacal Res.* **2021**, *44*, 439–474. [CrossRef]
8. Salaroli, R.; Andreani, G.; Bernardini, C.; Zannoni, A.; La Mantia, D.; Protti, M.; Forni, M.; Mercolini, L.; Isani, G. Anticancer activity of an *Artemisia annua* L. hydroalcoholic extract on canine osteosarcoma cell lines. *Res. Vet. Sci.* **2022**, *152*, 476–484. [CrossRef]
9. Baies, M.H.; Gherman, C.; Boros, Z.; Olah, D.; Vlase, A.M.; Cozma-Petrut, A.; Gyorke, A.; Miere, D.; Vlase, L.; Crisan, G.; et al. The Effects of *Allium sativum* L., *Artemisia absinthium* L., *Cucurbita pepo* L., *Coriandrum sativum* L., *Satureja hortensis* L. and *Calendula officinalis* L. on the Embryogenesis of Ascaris suum Eggs during an In Vitro Experimental Study. *Pathogens* **2022**, *11*, 1065. [CrossRef]
10. Suroowan, S.; Llorent-Martinez, E.J.; Zengin, G.; Dall'Acqua, S.; Sut, S.; Buskaran, K.; Fakurazi, S.; Mahomoodally, M.F. Phytochemical Characterization, Anti-Oxidant, Anti-Enzymatic and Cytotoxic Effects of Artemisia verlotiorum Lamotte Extracts: A New Source of Bioactive Agents. *Molecules* **2022**, *27*, 5886. [CrossRef]
11. Chen, X.Y.; Liu, T.; Hu, Y.Z.; Qiao, T.T.; Wu, X.J.; Sun, P.H.; Qian, C.W.; Ren, Z.; Zheng, J.X.; Wang, Y.F. Sesquiterpene lactones from Artemisia vulgaris L. as potential NO inhibitors in LPS-induced RAW264.7 macrophage cells. *Front. Chem.* **2022**, *10*, 948714. [CrossRef]

12. Chen, J.; Chen, F.; Peng, S.; Ou, Y.; He, B.; Li, Y.; Lin, Q. Effects of Artemisia argyi Powder on Egg Quality, Antioxidant Capacity, and Intestinal Development of Roman Laying Hens. *Front. Physiol.* **2022**, *13*, 902568. [CrossRef]
13. Su, S.H.; Sundhar, N.; Kuo, W.W.; Lai, S.C.; Kuo, C.H.; Ho, T.J.; Lin, P.Y.; Lin, S.Z.; Shih, C.Y.; Lin, Y.J.; et al. Artemisia argyi extract induces apoptosis in human gemcitabine-resistant lung cancer cells via the PI3K/MAPK signaling pathway. *J. Ethnopharmacol.* **2022**, *299*, 115658. [CrossRef]
14. Agrawal, P.K.; Agrawal, C.; Blunden, G. Artemisia Extracts and Artemisinin-Based Antimalarials for COVID-19 Management: Could These Be Effective Antivirals for COVID-19 Treatment? *Molecules* **2022**, *27*, 3828. [CrossRef]
15. Zhu, Z.; Turak, A.; Aisa, H.A. Sesquiterpene lactones from Artemisia mongolica. *Phytochemistry* **2022**, *199*, 113158. [CrossRef]
16. Zhang, J.J.; Qu, L.B.; Bi, Y.F.; Pan, C.X.; Yang, R.; Zeng, H.J. Antibacterial activity and mechanism of chloroform fraction from aqueous extract of mugwort leaves (*Artemisia argyi* L.) against Staphylococcus aureus. *Lett. Appl. Microbiol.* **2022**, *74*, 893–900. [CrossRef]
17. Wang, Y.; Li, J.; Chen, Q.; Zhou, J.; Xu, J.; Zhao, T.; Huang, B.; Miao, Y.; Liu, D. The role of antifungal activity of ethyl acetate extract from Artemisia argyi on Verticillium dahliae. *J. Appl. Microbiol.* **2022**, *132*, 1343–1356. [CrossRef]
18. Tao, A.; Feng, X.; Sheng, Y.; Song, Z. Optimization of the Artemisia Polysaccharide Fermentation Process by Aspergillus niger. *Front. Nutr.* **2022**, *9*, 842766. [CrossRef]
19. Suvaithenamudhan, S.; Ananth, S.; Mariappan, V.; Dhayabaran, V.V.; Parthasarathy, S.; Ganesh, P.S.; Shankar, E.M. In Silico Evaluation of Bioactive Compounds of Artemisia pallens Targeting the Efflux Protein of Multidrug-Resistant Acinetobacter baumannii (LAC-4 Strain). *Molecules* **2022**, *27*, 5188. [CrossRef]
20. Son, S.R.; Ju, I.G.; Kim, J.; Park, K.T.; Oh, M.S.; Jang, D.S. Chemical Constituents from the Aerial Parts of Artemisia iwayomogi and Their Anti-Neuroinflammatory Activities. *Plants* **2022**, *11*, 1954. [CrossRef]
21. Kolesar, J.M.; Seeberger, P.H. Editorial: Anticancer Potential of *Artemisia annua*. *Front. Oncol.* **2022**, *12*, 853406. [CrossRef]
22. Jung, E.J.; Paramanantham, A.; Kim, H.J.; Shin, S.C.; Kim, G.S.; Jung, J.M.; Hong, S.C.; Chung, K.H.; Kim, C.W.; Lee, W.S. Identification of Growth Factors, Cytokines and Mediators Regulated by *Artemisia annua* L. Polyphenols (pKAL) in HCT116 Colorectal Cancer Cells: TGF-beta1 and NGF-beta Attenuate pKAL-Induced Anticancer Effects via NF-kappaB p65 Upregulation. *Int. J. Mol. Sci.* **2022**, *23*, 1598. [CrossRef]
23. Bordoni, V.; Sanna, L.; Lyu, W.; Avitabile, E.; Zoroddu, S.; Medici, S.; Kelvin, D.J.; Bagella, L. Silver Nanoparticles Derived by Artemisia arborescens Reveal Anticancer and Apoptosis-Inducing Effects. *Int. J. Mol. Sci.* **2021**, *22*, 8621. [CrossRef]
24. Sahu, N.; Meena, S.; Shukla, V.; Chaturvedi, P.; Kumar, B.; Datta, D.; Arya, K.R. Extraction, fractionation and re-fractionation of Artemisia nilagirica for anticancer activity and HPLC-ESI-QTOF-MS/MS determination. *J. Ethnopharmacol.* **2018**, *213*, 72–80. [CrossRef]
25. Ahameethunisa, A.R.; Hopper, W. Antibacterial activity of Artemisia nilagirica leaf extracts against clinical and phytopathogenic bacteria. *BMC Complement. Altern. Med.* **2010**, *10*, 6. [CrossRef]
26. Naik, S.K.; Mohanty, S.; Padhi, A.; Pati, R.; Sonawane, A. Evaluation of antibacterial and cytotoxic activity of Artemisia nilagirica and Murraya koenigii leaf extracts against mycobacteria and macrophages. *BMC Complement. Altern. Med.* **2014**, *14*, 87. [CrossRef]
27. Panda, S.; Rout, J.R.; Pati, P.; Ranjit, M.; Sahoo, S.L. Antimalarial activity of Artemisia nilagirica against Plasmodium falciparum. *J. Parasit. Dis.* **2018**, *42*, 22–27. [CrossRef]
28. Gul, M.Z.; Chandrasekaran, S.; Manjulatha, K.; Bhat, M.Y.; Maurya, R.; Qureshi, I.A.; Ghazi, I.A. Bioassay-Guided Fractionation and In Vitro Antiproliferative Effects of Fractions of Artemisia nilagirica on THP-1 cell line. *Nutr. Cancer* **2016**, *68*, 1210–1224. [CrossRef]
29. Raju, S.R.; Balakrishnan, S.; Kollimada, S.; Chandrashekara, K.N.; Jampani, A. Anti-tumor effects of Artemisia nilagirica extract on MDA-MB-231 breast cancer cells: Deciphering the biochemical and biomechanical properties via TGF-beta upregulation. *Heliyon* **2020**, *6*, e05088. [CrossRef]
30. Parameswari, P.; Devika, R.; Vijayaraghavan, P. In vitro anti-inflammatory and antimicrobial potential of leaf extract from Artemisia nilagirica (Clarke) Pamp. *Saudi J. Biol. Sci.* **2019**, *26*, 460–463. [CrossRef]
31. Suseela, V.; Gopalakrishnan, V.K.; Varghese, S. In vitro Antioxidant Studies of Fruits of Artemisia nilagirica (Clarke) Pamp. *Indian J. Pharm. Sci.* **2010**, *72*, 644–649. [CrossRef] [PubMed]
32. Sati, S.C.; Sati, N.; Ahluwalia, V.; Walia, S.; Sati, O.P. Chemical composition and antifungal activity of Artemisia nilagirica essential oil growing in northern hilly areas of India. *Nat. Prod. Res.* **2013**, *27*, 45–48. [CrossRef] [PubMed]
33. Sonker, N.; Pandey, A.K.; Singh, P. Efficiency of Artemisia nilagirica (Clarke) Pamp. essential oil as a mycotoxicant against postharvest mycobiota of table grapes. *J. Sci. Food Agric.* **2015**, *95*, 1932–1939. [CrossRef] [PubMed]
34. Stappen, I.; Wanner, J.; Tabanca, N.; Wedge, D.E.; Ali, A.; Khan, I.A.; Kaul, V.K.; Lal, B.; Jaitak, V.; Gochev, V.; et al. Chemical composition and biological effects of Artemisia maritima and Artemisia nilagirica essential oils from wild plants of western Himalaya. *Planta Med.* **2014**, *80*, 1079–1087. [CrossRef]
35. Ekiert, H.; Świątkowska, J.; Klin, P.; Rzepiela, A.; Szopa, A. *Artemisia annua*—Importance in Traditional Medicine and Current State of Knowledge on the Chemistry, Biological Activity and Possible Applications. *Planta Med.* **2021**, *87*, 584–599. [CrossRef]
36. Matvieieva, N.; Drobot, K.; Duplij, V.; Ratushniak, Y.; Shakhovsky, A.; Kyrpa-Nesmiian, T.; Mickevičius, S.; Brindza, J. Flavonoid content and antioxidant activity of *Artemisia vulgaris* L. "hairy" roots. *Prep. Biochem. Biotechnol.* **2019**, *49*, 82–87. [CrossRef]
37. Tavsan, Z.; Kayali, H.A. Flavonoids showed anticancer effects on the ovarian cancer cells: Involvement of reactive oxygen species, apoptosis, cell cycle and invasion. *Biomed. Pharmacother.* **2019**, *116*, 109004. [CrossRef]

38. Al-Ishaq, R.K.; Abotaleb, M.; Kubatka, P.; Kajo, K.; Büsselberg, D. Flavonoids and Their Anti-Diabetic Effects: Cellular Mechanisms and Effects to Improve Blood Sugar Levels. *Biomolecules* **2019**, *9*, 430. [CrossRef]
39. Tian, C.; Liu, X.; Chang, Y.; Wang, R.; Lv, T.; Cui, C.; Liu, M. Investigation of the anti-inflammatory and antioxidant activities of luteolin, kaempferol, apigenin and quercetin. *S. Afr. J. Bot.* **2021**, *137*, 257–264. [CrossRef]
40. Crascì, L.; Cardile, V.; Longhitano, G.; Nanfitò, F.; Panico, A. Anti-degenerative effect of Apigenin, Luteolin and Quercetin on human keratinocyte and chondrocyte cultures: SAR evaluation. *Drug Res.* **2018**, *68*, 132–138. [CrossRef]
41. Skowyra, M.; Gallego, M.G.; Segovia, F.; Almajano, M.P. Antioxidant Properties of *Artemisia annua* Extracts in Model Food Emulsions. *Antioxidants* **2014**, *3*, 116–128. [CrossRef]
42. Perillo, B.; Di Donato, M.; Pezone, A.; Di Zazzo, E.; Giovannelli, P.; Galasso, G.; Castoria, G.; Migliaccio, A. ROS in cancer therapy: The bright side of the moon. *Exp. Mol. Med.* **2020**, *52*, 192–203. [CrossRef]
43. Hayes, J.D.; Dinkova-Kostova, A.T.; Tew, K.D. Oxidative Stress in Cancer. *Cancer Cell* **2020**, *38*, 167–197. [CrossRef]
44. Alqahtani, A.S.; Hidayathulla, S.; Rehman, M.T.; ElGamal, A.A.; Al-Massarani, S.; Razmovski-Naumovski, V.; Alqahtani, M.S.; El Dib, R.A.; AlAjmi, M.F. Alpha-Amylase and Alpha-Glucosidase Enzyme Inhibition and Antioxidant Potential of 3-Oxolupenal and Katononic Acid Isolated from Nuxia oppositifolia. *Biomolecules* **2020**, *10*, 61. [CrossRef]
45. Jannapureddy, S.; Sharma, M.; Yepuri, G.; Schmidt, A.M.; Ramasamy, R. Aldose Reductase: An Emerging Target for Development of Interventions for Diabetic Cardiovascular Complications. *Front. Endocrinol.* **2021**, *12*, 636267. [CrossRef]
46. Kazeem, M.I.; Adeyemi, A.A.; Adenowo, A.F.; Akinsanya, M.A. Carica papaya Linn. fruit extract inhibited the activities of aldose reductase and sorbitol dehydrogenase: Possible mechanism for amelioration of diabetic complications. *Future J. Pharm. Sci.* **2020**, *6*, 96. [CrossRef]
47. Ali, A.N.M.; Saeed, N.; Omear, H.A. The Anticancer Properties of Artemisia aucheri Boiss Extract on HT29 Colon Cancer Cells. *J. Gastrointest. Cancer* **2021**, *52*, 113–119. [CrossRef]
48. Mashati, P.; Esmaeili, S.; Dehghan-Nayeri, N.; Bashash, D.; Darvishi, M.; Gharehbaghian, A. Methanolic Extract from Aerial Parts of *Artemisia annua* L. Induces Cytotoxicity and Enhances Vincristine-Induced Anticancer Effect in Pre-B Acute Lymphoblastic Leukemia Cells. *Int. J. Hematol. Oncol. Stem Cell Res.* **2019**, *13*, 132–139. [CrossRef]
49. Choi, E.J.; Kim, G.H. Antioxidant and anticancer activity of Artemisia princeps var. orientalis extract in HepG2 and Hep3B hepatocellular carcinoma cells. *Chin. J. Cancer Res. Chung-Kuo Yen Cheng Yen Chiu* **2013**, *25*, 536–543. [CrossRef]
50. Choi, E.; Park, H.; Lee, J.; Kim, G. Anticancer, antiobesity, and anti-inflammatory activity of Artemisia species in vitro. *J. Tradit. Chin. Med. Chung I Tsa Chih Ying Wen Pan* **2013**, *33*, 92–97. [CrossRef]
51. Ghanbar, F.; Mirzaie, A.; Ashrafi, F.; Noorbazargan, H.; Dalirsaber Jalali, M.; Salehi, S.; Sadat Shandiz, S.A. Antioxidant, antibacterial and anticancer properties of phyto-synthesised Artemisia quttensis Podlech extract mediated AgNPs. *IET Nanobiotechnol.* **2017**, *11*, 485–492. [CrossRef] [PubMed]
52. Salehi, S.; Shandiz, S.A.; Ghanbar, F.; Darvish, M.R.; Ardestani, M.S.; Mirzaie, A.; Jafari, M. Phytosynthesis of silver nanoparticles using Artemisia marschalliana Sprengel aerial part extract and assessment of their antioxidant, anticancer, and antibacterial properties. *Int. J. Nanomed.* **2016**, *11*, 1835–1846. [CrossRef]
53. Briguglio, G.; Costa, C.; Pollicino, M.; Giambò, F.; Catania, S.; Fenga, C. Polyphenols in cancer prevention: New insights (Review). *Int. J. Funct. Nutr.* **2020**, *1*, 9. [CrossRef]
54. Kopustinskiene, D.M.; Jakstas, V.; Savickas, A.; Bernatoniene, J. Flavonoids as Anticancer Agents. *Nutrients* **2020**, *12*, 457. [CrossRef]
55. Tong, W.; Chen, X.; Song, X.; Chen, Y.; Jia, R.; Zou, Y.; Li, L.; Yin, L.; He, C.; Liang, X.; et al. Resveratrol inhibits LPS-induced inflammation through suppressing the signaling cascades of TLR4-NF-κB/MAPKs/IRF3. *Exp. Ther. Med.* **2020**, *19*, 1824–1834. [CrossRef]
56. Yücel, G.; Zhao, Z.; El-Battrawy, I.; Lan, H.; Lang, S.; Li, X.; Buljubasic, F.; Zimmermann, W.-H.; Cyganek, L.; Utikal, J.; et al. Lipopolysaccharides induced inflammatory responses and electrophysiological dysfunctions in human-induced pluripotent stem cell derived cardiomyocytes. *Sci. Rep.* **2017**, *7*, 2935. [CrossRef]
57. Tang, J.; Diao, P.; Shu, X.; Li, L.; Xiong, L. Quercetin and Quercitrin Attenuates the Inflammatory Response and Oxidative Stress in LPS-Induced RAW264.7 Cells: In Vitro Assessment and a Theoretical Model. *Biomed. Res. Int.* **2019**, *2019*, 7039802. [CrossRef]
58. Kim, Y.J.; Park, W. Anti-Inflammatory Effect of Quercetin on RAW 264.7 Mouse Macrophages Induced with Polyinosinic-Polycytidylic Acid. *Molecules* **2016**, *21*, 450. [CrossRef]
59. Lee, A.J.; Cho, K.-J.; Kim, J.-H. MyD88–BLT2-dependent cascade contributes to LPS-induced interleukin-6 production in mouse macrophage. *Exp. Mol. Med.* **2015**, *47*, e156. [CrossRef]
60. Sakai, J.; Cammarota, E.; Wright, J.A.; Cicuta, P.; Gottschalk, R.A.; Li, N.; Fraser, I.D.C.; Bryant, C.E. Lipopolysaccharide-induced NF-κB nuclear translocation is primarily dependent on MyD88, but TNFα expression requires TRIF and MyD88. *Sci. Rep.* **2017**, *7*, 1428. [CrossRef]
61. Shukla, A.; Vats, S.; Shukla, R.K. Phytochemical Screening, Proximate Analysis and Antioxidant Activity of Dracaena reflexa Lam. Leaves. *Indian J. Pharm. Sci.* **2015**, *77*, 640–644. [CrossRef]
62. Agidew, M.G. Phytochemical analysis of some selected traditional medicinal plants in Ethiopia. *Bull. Natl. Res. Cent.* **2022**, *46*, 87. [CrossRef]
63. Owolabi, O.O.; James, D.B.; Sani, I.; Andongma, B.T.; Fasanya, O.O.; Kure, B. Phytochemical analysis, antioxidant and anti-inflammatory potential of FERETIA APODANTHERA root bark extracts. *BMC Complement. Altern. Med.* **2018**, *18*, 12. [CrossRef]

64. House, N.C.; Puthenparampil, D.; Malayil, D.; Narayanankutty, A. Variation in the polyphenol composition, antioxidant, and anticancer activity among different Amaranthus species. *S. Afr. J. Bot.* **2020**, *135*, 408–412. [CrossRef]
65. Ortiz-Cruz, R.A.; Ramírez-Wong, B.; Ledesma-Osuna, A.I.; Torres-Chávez, P.I.; Sánchez-Machado, D.I.; Montaño-Leyva, B.; López-Cervantes, J.; Gutiérrez-Dorado, R. Effect of Extrusion Processing Conditions on the Phenolic Compound Content and Antioxidant Capacity of Sorghum (*Sorghum bicolor* (L.) Moench) Bran. *Plant Foods Hum. Nutr.* **2020**, *75*, 252–257. [CrossRef]
66. Wang, B.; Liu, L.; Huang, Q.; Luo, Y. Quantitative Assessment of Phenolic Acids, Flavonoids and Antioxidant Activities of Sixteen Jujube Cultivars from China. *Plant Foods Hum. Nutr.* **2020**, *75*, 154–160. [CrossRef]
67. Liu, D.; Guo, Y.; Wu, P.; Wang, Y.; Kwaku Golly, M.; Ma, H. The necessity of walnut proteolysis based on evaluation after in vitro simulated digestion: ACE inhibition and DPPH radical-scavenging activities. *Food Chem.* **2020**, *311*, 125960. [CrossRef]
68. Bi, X.; Zhang, J.; Chen, C.; Zhang, D.; Li, P.; Ma, F. Anthocyanin contributes more to hydrogen peroxide scavenging than other phenolics in apple peel. *Food Chem.* **2014**, *152*, 205–209. [CrossRef]
69. Dutta, S.; Ray, S. Comparative assessment of total phenolic content and in vitro antioxidant activities of bark and leaf methanolic extracts of Manilkara hexandra (Roxb.) Dubard. *J. King Saud Univ. Sci.* **2020**, *32*, 643–647. [CrossRef]
70. Tonisi, S.; Okaiyeto, K.; Mabinya, L.V.; Okoh, A.I. Evaluation of bioactive compounds, free radical scavenging and anticancer activities of bulb extracts of Boophone disticha from Eastern Cape Province, South Africa. *Saudi J. Biol. Sci.* **2020**, *27*, 3559–3569. [CrossRef]
71. Mechchate, H.; Es-safi, I.; Louba, A.; Alqahtani, A.S.; Nasr, F.A.; Noman, O.M.; Farooq, M.; Alharbi, M.S.; Alqahtani, A.; Bari, A.; et al. In Vitro Alpha-Amylase and Alpha-Glucosidase Inhibitory Activity and In Vivo Antidiabetic Activity of *Withania frutescens* L. Foliar Extract. *Molecules* **2021**, *26*, 293. [CrossRef]
72. Karakaya, S.; Gözcü, S.; Güvenalp, Z.; Özbek, H.; Yuca, H.; Dursunoğlu, B.; Kazaz, C.; Kılıç, C.S. The α-amylase and α-glucosidase inhibitory activities of the dichloromethane extracts and constituents of Ferulago bracteata roots. *Pharm. Biol.* **2018**, *56*, 18–24. [CrossRef]
73. Kim, T.H.; Kim, J.K.; Kang, Y.-H.; Lee, J.-Y.; Kang, I.J.; Lim, S.S. Aldose Reductase Inhibitory Activity of Compounds from *Zea mays* L. *Biomed. Res. Int.* **2013**, *2013*, 727143. [CrossRef]
74. Al-Yousef, H.M.; Fantoukh, O.I.; El-Sayed, M.A.; Amina, M.; Adel, R.; Hassan, W.H.B.; Abdelaziz, S. Metabolic profiling and biological activities of the aerial parts of Micromeria imbricata Forssk. growing in Saudi Arabia. *Saudi J. Biol. Sci.* **2021**, *28*, 5609–5616. [CrossRef]

molecules

MDPI

Article

Evaluation of the Anti-Cancer Potential of *Rosa damascena* Mill. Callus Extracts against the Human Colorectal Adenocarcinoma Cell Line

Hadeer Darwish [1,*], Sarah Alharthi [2], Radwa A. Mehanna [3,4], Samar S. Ibrahim [4], Mustafa A. Fawzy [5], Saqer S. Alotaibi [1], Sarah M. Albogami [1], Bander Albogami [5], Sedky H. A. Hassan [6] and Ahmed Noureldeen [5]

1 Department of Biotechnology, College of Science, Taif University, P.O. Box 11099, Taif 21944, Saudi Arabia
2 Department of Chemistry, College of Science, Taif University, P.O. Box 11099, Taif 21944, Saudi Arabia
3 Department of Physiology, Faculty of Medicine, Alexandria University, Alexandria P.O. 21544, Egypt
4 Center of Excellence for Research in Regenerative Medicine and Applications (CERRMA), Faculty of Medicine, Alexandria University, Alexandria P.O. 21544, Egypt
5 Department of Biology, College of Science, Taif University, P.O. Box 11099, Taif 21944, Saudi Arabia
6 Department of Biology, College of Science, Sultan Qaboos University, P.O. Box 36, Muscat 123, Oman
* Correspondence: hadeer@tu.edu.sa; Tel.: +966-55-621-4209

Abstract: Chemotherapy is an aggressive form of chemical drug therapy aiming to destroy cancer cells. Adjuvant therapy may reduce hazards of chemotherapy and help in destroying these cells when obtained from natural products, such as medical plants. In this study, the potential therapeutic effect of *Rosa damascena* callus crude extract produced in vitamin-enhanced media is investigated on colorectal cancer cell line Caco-2. Two elicitors, i.e., L-ascorbic acid and citric acid at a concentration of 0.5 g/L were added to the callus induction medium. Callus extraction and the GC–MS analysis of methanolic crude extracts were also determined. Cytotoxicity, clonogenicity, proliferation and migration of Caco-2 colorectal cancer cells were investigated using MTT cytotoxicity, colony-forming, Ki-67 flow cytometry proliferation and Migration Scratch assays, respectively. Our results indicated that L-ascorbic acid treatment enhanced callus growth parameters and improved secondary metabolite contents. It showed the least IC50 value of 137 ug/mL compared to 237 ug/mL and 180 ug/mL in the citric acid-treated and control group. We can conclude that *R. damascena* callus elicited by L-ascorbic acid improved growth and secondary metabolite contents as well as having an efficient antiproliferative, anti-clonogenic and anti-migratory effect on Caco-2 cancer cells, thus, can be used as an adjuvant anti-cancer therapy.

Keywords: *Rosa damascena*; callus induction; bio-elicitors; colorectal cancer cell line; anti-cancer activity

Citation: Darwish, H.; Alharthi, S.; Mehanna, R.A.; Ibrahim, S.S.; Fawzy, M.A.; Alotaibi, S.S.; Albogami, S.M.; Albogami, B.; Hassan, S.H.A.; Noureldeen, A. Evaluation of the Anti-Cancer Potential of *Rosa damascena* Mill. Callus Extracts against the Human Colorectal Adenocarcinoma Cell Line. *Molecules* **2022**, *27*, 6241. https://doi.org/10.3390/molecules27196241

Academic Editors: Arunaksharan Narayanankutty, Ademola C. Famurewa and Eliza Oprea

Received: 19 August 2022
Accepted: 7 September 2022
Published: 22 September 2022

Publisher's Note: MDPI stays neutral with regard to jurisdictional claims in published maps and institutional affiliations.

Copyright: © 2022 by the authors. Licensee MDPI, Basel, Switzerland. This article is an open access article distributed under the terms and conditions of the Creative Commons Attribution (CC BY) license (https://creativecommons.org/licenses/by/4.0/).

1. Introduction

Colorectal cancer is the fourth frequently diagnosed cancer after lung, prostate and breast cancers [1]. In 2020, 104,610 new cases of colon cancer and 43,340 cases of rectal cancer were estimated to occur. During the same year, 53,200 people were estimated to die of colon and rectal cancer combined [2]. In the past 30 years, the mortality from colon cancer has reduced slightly due to early diagnosis [3]. However, the incidence of colon cancer has increased in people younger than 65 years old, with a 1% yearly rise in those aged between 50 to 64 years and 2% for those younger than 50 years old [1]. The risk of increasing colon cancer is influenced by genetic and acquired risk factors. Acquired risk factors associated with this disease include (i) dietary factors, for example low intake of fruit, fiber or vegetables, and high intake of red meat, caffeine, saturated fat and alcohol; (ii) lifestyle factors such as smoking and absence of exercise; (iii) side effects of some medical or surgical procedures, such as pelvic irradiation, cholecystectomy and ureterocolic anastomosis and (iv) co-morbid medical conditions such as human immunodeficiency virus infection, diabetes mellitus and inflammatory bowel disease [4]. Surgical resection is the best treatment for colon cancer but chemotherapy, radiation therapy and immunotherapy also play an important role in inhibiting recurrence and metastasis. Chemotherapy is the main therapeutic strategy in many incidences. It uses various drugs or combinations of drugs to diminish the cancer cell division. The classical route for delivering chemotherapy for colon cancer comprises delivering drugs to non-target positions; thus, patients suffer from side effects such as gastrointestinal toxicity, anemia, diarrhea, neutropenia, vomiting, mucositis, liver toxicity, hematologic disorders, damage to the nervous system and memory problems [5]. To date, the results of used treatments have not reached an acceptable outcome due to the risk of side effects and resistance to chemotherapy.

The use of medicinal plants as a natural source of pharmaceuticals has become increasingly popular in recent years. As a result of urbanization, overgrazing, pollution and the growth of agricultural regions, medicinal plants have been targeted for uncontrolled gathering and destruction. Secondary metabolism in plants produces a diverse range of chemically complicated chemicals, many of which are commercially valuable. Secondary metabolites are plant products of great pharmacological value, such as phenolics, terpenoids, glycosides, alkaloids and other compounds [6]. Secondary metabolites are often extracted from intact plants for commercial use. The three main chemical categories of secondary metabolites in plants are nitrogen compounds, terpenes and phenolic compounds [7]. Secondary metabolites that contain nitrogen have a basic nature and are identified by the presence of nitrogen in their fundamental structure. Alkaloids, glycosides and nonprotein amino acids are among plants' most prevalent nitrogen-containing substances [8,9]. These metabolites are crucial for plants' defense against insects and animals. The majority of alkaloids are extremely dangerous to humans, although small dosages of these substances may have medicinal benefits. Alkaloids or extracts containing alkaloids have been used as muscle relaxants, analgesics and tranquilizers from ancient times to the present [10]. The building block for terpenes, which are organic compounds, is an isoprene compound with five carbon atoms. Monoterpenes (C10), sesquiterpenes (C15), diterpenes (C20), triterpenes (C30), tetraterpenes (C40) and polyterpenes, with more than 40 carbons, are the different subgroups of terpenes [8,11]. Sesquiterpenes and volatile monoterpenes make up the majority of an essential oil's chemical makeup [12,13]. Triterpenes, tetraterpenes and polyterpenes are the most prevalent terpenes [8]. Some terpenes, such as gibberellins, carotenoids and brassinosteroids, are also crucial for plant development in addition to their roles as anti-herbivore defensive chemicals in plants [9,14,15]. Phenolic chemicals are aromatic molecules made up of two groups: a hydroxyl (OH) group and a phenyl (C6) group. Simple molecules (such as phenolic acids) and highly polymerized molecules have different types of structures in these molecules (condensed tannins). Lignin, flavonoids, tannins, phenolic acids and coumarins are the five subgroups that make up phenolic chemicals [16]. The development of plant structure is affected by phenolic compounds, and they are also linked to a number of physiological functions, such as the defense against infections, insects and animals [17]. These substances are also recognized for their cardiovascular, gastrointestinal, antibacterial, antiviral, anti-cancer, anti-inflammatory, antiatherogenic, antithrombotic, analgesic and antithrombotic properties [18–23].

Plant cell culture is a renewable and environmentally acceptable source of secondary metabolites. Several findings show that plants can create a wide range of secondary metabolites, with some of them being commercially produced. Tissue culture methods have been created for a number of plants, but there are many others that are being over-exploited in the pharmaceutical industry and other disciplines, and need to be protected [24]. For generations, roses have been one of the most popular ornamental plants in the world. *Rosa damascena* Mill. (Damask rose) is one of the oldest and most valuable variants among the Rosaceae family. This rose is also used to make products with a variety of uses, including aromatherapy, antiseptic, antispasmodic, astringent agent, sedative, blood cholesterol altering, antibacterial, antimicrobial [25,26], anti-oxidizing [27] and anti-HIV effects [28]. *R.*

damascena also has a variety of uses in the perfume, cosmetic and food sectors, such as the creation of rosewater, jam and dried flowers [29].

Vitamins may be conceived as bioregulator or hormone precursor chemicals that have a beneficial effect on plant growth and development when present in minute amounts. Therefore, these compounds might have an impact on the metabolic process for energy [30,31]. Ascorbic acid, a form of vitamin C, is more or less necessary for many vital physiological functions, including cell division, nutrition and water absorption, photosynthesis and the manufacture of enzymes and secondary metabolites. It serves as an enzyme cofactor, an antioxidant and a precursor for the production of oxalate and tartrate. Ascorbic acid is connected to chloroplasts that reduce the impact of oxidative stress during photosynthesis. Additionally, it prevents cell division from altering and functions as the main substrate in the cyclic pathway of hydrogen peroxide enzymatic detoxification [32].

The effect of these metabolites on cancer cell growth and behavior needs to be explored. Thus, the aim of this study is to investigate the potential effect of *R. damascena* callus elicited by two vitamins, ascorbic and citric acids, on cancer cells using colorectal cancer cell line Caco-2 as an in vitro model.

2. Results

2.1. *R. damascena Callus Fresh, Dry and Crude Weight (g)*

We looked for the suitable bio-elicitor to promote the productivity of phenolic content obtained from *R. damascena* callus' fresh or dry weight in a series of early studies. This experiment's outcomes in Table 1 and Figure 1A show that the ascorbic acid concentration of 0.5 g/L resulted in the highest fresh weight (2.878 g) and the maximum dry weight (0.306 g). The increase in fresh weight was not significant when compared to the control, but it was substantial in the callus' dry weight when compared to the control and citric acid treatment. Table 1 also presents that the control treatment showed the biggest increase in crude weight (0.038 g). The presence of ascorbic acid or citric acid in the media resulted in a lower weight of crude output than the control treatment, with values of 0.032 g and 0.022 g, respectively, without significant differences between them.

Table 1. The effect of bio-elicitors on *R. damascena* callus development and total crude weight.

Treatments	[1] Fresh Weight (g)	Dry Weight (g)	Crude Weight (g)
Control	[2] 2.166 ab	0.214 b	0.038 a
Citric acid 0.5 g/L	1.267 b	0.168 b	0.022 a
L-ascorbic acid 0.5 g/L	2.878 a	0.306 a	0.032 a
L.S.D at 5%	1.019	0.075	2922.8

[1] Each treatment was represented by ten replicates, each with three explants. [2] Means with different letters within the same column or row differ significantly ($p < 0.05$).

2.2. *GC–MS Analysis of R. damascena Callus*

2.2.1. Control Treatment

The crude yield of *R. damascena* callus in the control treatment was 0.038 g. Table 2 and Figure 1B represent the GC–MS analysis for *R. damascena* callus in the control treatment. The obtained result shows the presence of eight compounds in the chromatogram. The major compounds were 13.75% of Octadecanoic acid, methyl ester (CAS)—which was reported to be the highest composition of the compounds—followed by benzonitrile (CAS) with 9.15% and 1,2-Benzenedicarboxylic acid, bis(2-ethylhexyl) ester (CAS) with 3.05%. The 4-Octanol, propanoate, 1-Pentanol, 2,2-dimethyl-(CAS) and 2,6-Nonadien-1-ol had values of 1.55%, 1.60% and 1.74%, respectively, whereas (5,10,15,20-Tetraphenyl [2-(2)H1]prophyrinato)zinx (II) achieved the lowest amount (0.96%). The components of *R. damascena* crude differ according to the bio-elicitor callus they were exposed to.

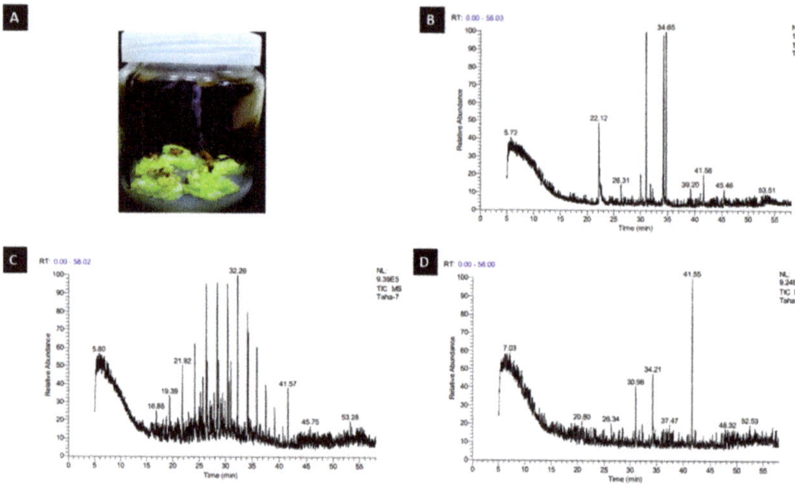

Figure 1. Callus adequate mass (**A**); GC–MS chromatograms of callus elicited by control (**B**); citric acid (**C**); L–ascorbic acid (**D**).

Table 2. GC–MS analysis of *R. damascena* callus in the control treatment.

Peak	R. Time	Area	Area %	Name
1	5.72	2561	1.04	(2,3-Dihydro-5,10,15,20- tetraphenyl [2-(2)H1]prop hyinato) copper(II)
2	22.12	2251	9.15	Benzonitrile (CAS)
3	26.31	3938	1.60	1-Pentanol, 2,2-dimethyl-(CAS)
4	34.65	3385	13.75	Octadecanoic acid, methyl ester (CAS)
5	39.20	3807	1.55	4-Octanol, propanoate
6	41.56	7511	3.05	1,2-Benzenedicarboxylic acid, bis(2-ethylhexyl) ester (CAS)
7	45.46	4280	1.74	2,6-Nonadien-1-ol
8	53.51	2360	0.96	(5,10,15,20-Tetraphenyl [2-(2)H1]prophyrinato)zinx(II)

2.2.2. Citric Acid Treatment

Table 3 and Figure 1C represent the GC–MS analysis for *R. damascena* callus in the citric acid treatment. The obtained result shows the presence of eight compounds in the chromatogram. The major compounds included Octadecane, 2-methyl—which is reported to be highest composition of the compounds (5.82%)—followed by Tetratetracontane (CAS) with 3.58% and 1,2-Benzenedicarboxylic acid, diisooctyl ester (CAS) with 2.69%. However, Tetradecane (CAS) and Dodecane 5,8-diethyl-(CAS) recorded 2.00% and 1.16%, respectively. Dichloroacetaldehyde valued 0.95%, whereas 3,3′,5,5′-Tetrabromo-2-nitro-2′-propylsulfonylbiphenyl and 5á-Pregnan-20-one, 3à,11á,17,21-tetrakis(trim ethylsiloxy)-, O-methyloxime revealed the lowest amounts—0.76% and 0.73%, respectively.

2.2.3. L-Ascorbic Acid Treatment

Table 4 and Figure 1D represent the GC–MS analysis for *R. damascena* callus in the ascorbic acid treatment. The obtained result shows the presence of eight compounds in the chromatogram. The most prevalent compounds were 1,2-Benzenedicarboxylic acid, mono(2-ethylhexyl) ester (17.75%), 7-Octadecenoic acid, methyl ester (CAS) with 8.57% and Hexadecanoic acid, methyl ester (CAS) with 5.82%. Nonane, 1-chloro-(CAS) and11-[(t-Butyldimethylsilyl)oxy]-6,9a-dimethyl-6-(methoxycarbonyl)-(perhydro)napthaleno[a]benzofulvene valued 1.98% and 1.79%, respectively, whereas (2,3-Dihydro-5,10,15,20-tetraphenyl [2-(2)H1]prop hyrinato)copper(II) achieved the lowest amount (1.09%).

Table 3. GC–MS analysis of *R. damascena* callus in the citric acid treatment.

Peak	R. Time	Area	Area %	Name
1	5.80	3614	0.95	Dichloroacetaldehyde
2	16.85	4391	1.16	Dodecane, 5,8-diethyl-(CAS)
3	19.39	7554	2.00	Tetradecane (CAS)
4	21.82	1355	3.58	Tetratetracontane (CAS)
5	32.26	2202	5.82	Octadecane, 2-methyl-
6	41.57	1018	2.69	1,2-Benzenedicarboxylic acid, diisooctyl ester (CAS)
7	45.75	2866	0.76	3,3′,5,5′-Tetrabromo-2-nitro-2′-propylsulfonylbiphenyl
8	53.28	2764	0.73	5á-Pregnan-20-one, 3à,11á,17,21-tetrakis(trim ethylsiloxy)-, O-methyloxime

Table 4. GC–MS analysis of *R. damascena* callus in the L-ascorbic acid treatment.

Peak	R. Time	Area	Area %	Name
1	7.03	2904	1.79	11-[(t-Butyldimethylsilyl)oxy]-6,9a-dimethyl-6-(m ethoxycarbonyl)-(perhydro)napthaleno[a]benzofulvene
2	20.80	1980	1.22	Mixture of: 5,6-Dihydro-6-methy l-2H-pyran-2-one and 5-methoxy-3-pente ne-2-ol
3	26.34	3210	1.98	Nonane, 1-chloro-(CAS)
4	30.98	9453	5.82	Hexadecanoic acid, methyl ester (CAS)
5	34.21	1391	8.57	7-Octadecenoic acid, methyl ester (CAS)
6	37.47	1777	1.09	(2,3-Dihydro-5,10,15,20-tetraphenyl [2-(2)H1]prophyrinato)copper(II)
7	41.55	2884	17.75	1,2-Benzenedicarboxylic acid, mono(2-ethylhexyl) ester
8	48.32	1799	1.11	6-[2′-(4″-Phenyl)ethyl]-1,2,3-triphenyl-9H-tribenzo[a,c,e]cycloheptatriene

2.3. *In Vitro Anti-Cancer Study (Cell Line Studies)*

2.3.1. Cellular Cytotoxicity Assay

Cytotoxicity of control and treated callus were investigated by an MTT assay. A concentration that kills 50% of the cells (IC50) was determined and used to compare the activity of different preparations. The IC50 values were 180.6 ug/mL for the control group, and 137.8 ug/mL and 237 ug/mL for the LAA- and CA-treated groups, respectively. The results prove that LAA treatment is the most potent cytotoxic one (Figure 2A–D).

2.3.2. Clonogenic Assay

Digital images of the colonies were obtained using a camera, and colonies were counted for the calculation of plating efficiency for three culture dishes in each group in three separate experiments. The results revealed a significant decrease in the LAA-treated group compared to the untreated, control- and CA-treated groups. There was no significant difference between the control- and CA-treated groups, yet they were both significantly lower than the untreated one (Figure 2D–H).

2.3.3. Ki-67 Flow Cytometry Proliferation Assay

A negative control of untreated Caco-2 cells was used to gate the Ki-67 negative population. The Ki-67 expression on untreated cells showed that 69.3% of cells were undergoing proliferation. Proliferation was significantly decreased in cells with the control-, LAA- and CA-treated callus, showing 38.4%, 31.7% and 40.4%, respectively. The control- and CA-treated cells showed no significant difference compared to each other. On the contrary, the LAA-treated cells group showed a significant reduction in cell proliferation in comparison to the CA and control groups (Figure 3).

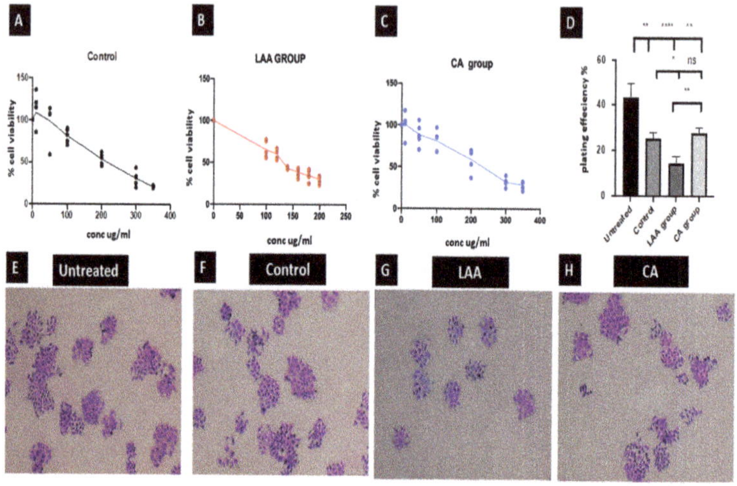

Figure 2. Cellular cytotoxicity (**A–C**); clonogenic assay statistical analysis of plating efficiency (PE) of tested groups blotted at the X axis, PE % = the number of colonies formed/number of cells plated) × 100 blotted at the Y axis, the untreated group showed the highest PE while the LAA group showed the least PE, being significantly lower than the CA, control and untreated groups. * Significance at $p < 0.05$, ** Significance at $p < 0.001$, **** Significance at $p < 0.0001$, ns No significant difference. (**D**) Clonogenic assay images showing cell colonies stained with crystal violet imaged using inverted light microscope X40 (Olympus CKX41SF, Tokyo, Japan) (**E–H**). Untreated (**E**); control (**A**,**F**); L-ascorbic acid (**B**,**G**); citric acid (**C**,**H**).

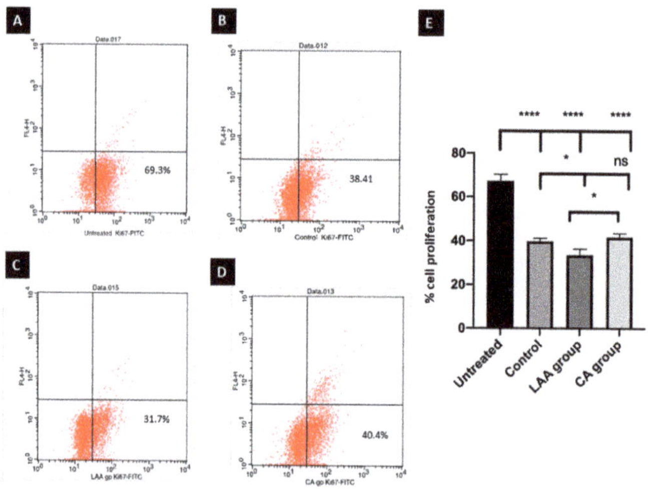

Figure 3. Ki-67 flow cytometry proliferation assay. Untreated (**A**); control (**B**); L-ascorbic acid (**C**); citric acid (**D**). Statistical analysis for the cell proliferation of different groups blotted on X axis and % of proliferation on Y axis, where proliferation of cells in all groups was significantly lower than the untreated group. **** Significance at $p < 0.0001$, LAA group showed the lowest proliferation compared to the crude callus-treated (control) and CA-treated groups. * Significance at $p < 0.05$; there was no significant difference between the control and CA groups (ns) (**E**).

2.3.4. Migration Assay (Scratch Assay)

A comparison between untreated cells and the control-, LAA- and CA-treated groups was based on measuring the percentage of wound closure—a decrease in this percentage was taken as an indication of the decrease in cells' migratory ability. Areas were quantified using Image J software (1.52p software 32, NIH, USA). The results showed that the least average wound closure percentage was for the LAA-treated group, with 17% after 24 h and 33% after 48 h. This was significantly less than that of the control- and CA-treated groups, showing 26% and 34% after 24 h, and 44% and 56% after 48 h, respectively (Figure 4).

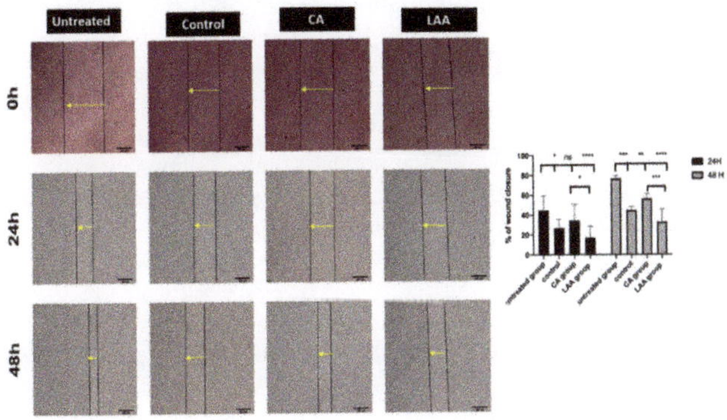

Figure 4. Migration assay at 0, 24 and 48 h. Statistical analysis for the % of wound closure of different groups at 24 and 48 h. * Significance at $p < 0.05$, ** significance at $p < 0.001$, *** significance at $p = 0.0001$, **** significance at $p < 0.0001$.

3. Discussion

The components of *R. damascena* crude differ according to regions and the elicitors used in the experiments. Obviously, the results showed that ascorbic acid treatment recorded the highest values of fresh and dry weights of rose callus with a reliable value of crude weight when compared with the control treatment. Ascorbic acid is the major compound functioning in plant antioxidant systems [33]. It is implicated in cell division, cell elongation and synthesis of phytohormones [34]. Furthermore, it plays a key role in protecting plants from ROS through the ascorbate–glutathione cycle, as well as acting as a cofactor to violaxanthin de-epoxidase, which is considered to be an important enzyme in the photoprotective xanthophyll cycle during photosynthesis [35]. All of the previously mentioned positive effects of ascorbic acid can explain its significant improvement in the concentration of fresh, dry and crude weight under the circumstances of this study. This study evaluated the compounds obtained from the GC–MS analysis of *R. damascena* callus methanol extracts. Octadecanoic acid, methyl ester (CAS) is reported to be the major compound obtained from the GC–MS analysis of *R. damascena* callus in the control treatment, whereas (5,10,15,20-tetraphenyl [2-(2)H1]prophyrinato)zinx (II) recorded the lowest value. Our investigation was in line with those recorded by [36], who mentioned a significant presence of βcitronellol (30.24–31.15%); trans-geraniol (20.62–21.24%), n-heneicosane (8.79–9.05%), n-nonadecane (8.51–8.77%), nonadecene (4.42–4.55%) and phenylethyl alcohol (4.04–4.16%) was detected from the GC–MS chromatograph profile of *R. damascena* essential oil. GC–MS analysis for the methanol extract of *R. damascena* callus exposed to citric and ascorbic acids resulted in 16 compounds in the chromatogram, which confirmed that biotic elicitors increased the secondary products content. Obtained results of our study agreed with [37], who demonstrated that the exogenous application of ascorbic acid could enhance foliar growth, which may contribute to increased plant biomass and the yield of secondary products. Ascorbic acid was considered as an antioxidant involved in cell division and elongation [38]. Moreover, it also enhanced shoot formation in both young and old tobacco callus [39]. Furthermore, the application of ascorbic acid significantly improved photosynthetic pigments, gas exchange and nutrient content. This enhancement could be due to the functions of ascorbic acid as an essential cofactor of various enzymes or protein complexes [40,41]. This finding is in harmony with [42,43], who found that ascorbic acid foliar spray had a positive impact on total phenols, total chlorophyll, total carbs and percentages of N, P and K in lettuce. Ascorbic acid supplementation, in accordance with [44], enhanced N, P and K accumulation, and had a significant effect on nutrient uptake. Additionally, ascorbic acid enhanced the osmoprotectant proline and soluble carbohydrates, as well as the antioxidant enzymes CAT, POD, APX and SOD. This impact might be brought on by ascorbic acid, a water-soluble antioxidant that modulates a wide range of biological processes that influence plant development [45,46]. It is also involved in a variety of metabolic processes and has a dynamic connection with reactive oxygen species (ROS) [47–49]. In this regard, Sajid and Aftab [50] discovered that a foliar application of ascorbic acid significantly boosted the antioxidant enzyme activity of SOD, POD, CAT and APX in potatoes under salinity stress conditions. The variation in chemical composition between our research (greater 7-Octadecenoic acid, methyl ester (CAS) content and lower n-heptadecane content) and the presented data could be attributable to the bio-elicitors to which the plant was subjected.

Investigating the potential effect of the crude, ascorbic acid and citric acid-treated callus as an anti-cancerous adjuvant, human colorectal carcinoma cell line Caco-2 was employed in this study. The LAA-treated group was the most potent regarding cytotoxicity, where the IC50 was 137 ug/mL compared to 237 ug/mL and 180 ug/mL in the CA and control groups, respectively. This was further confirmed by the results of the clonogenic assay, where the colonies formed were the lowest in cells treated with LAA, denoting the lowest reproductive viability. The control- and CA-treated groups' ability to produce progeny was also decreased in comparison to the untreated cells, but with no significance between them. Moreover, a negative control of untreated Caco-2 cells was used to gate the Ki-67 negative population. The Ki-67 status of Caco-2 cells treated with LAA showed the least proliferation capacity. Proliferation was also decreased in the CA-treated and control groups rather than the untreated group, suggesting that the anti-cancerous potential of *R. damascena* metabolites may be due to inhibiting the proliferation of cancer cells. In this study, a wound-healing scratch test was employed to reflect the migration potential of Caco-2 cells, and thus, their metastasizing ability. The results indicated that the ability of the LAA-treated group would significantly decrease the migration of Caco-2 cancer cells more than the CA-treated and control crude groups, which would result in better therapeutic outcomes.

4. Materials and Methods

This research was carried out in the Tissue Culture Laboratory of Taif University—where tissue culture techniques were used to increase the production of secondary metabolites in *R. damascena* (Family: Rosaceae)—and in the Center of Excellence for Research in Regenerative Medicine and its Applications (CERRMA) in Faculty of Medicine, Alexandria University—for the application of *R. damascena* callus crude extracts on cancer cell line Caco-2 to evaluate their potential anti-cancer effect. The current study was carried in three parts: the first was the induction of *R. damascena* callus; the second included an increment of secondary metabolites utilizing various elicitors, extraction, and GC–MS measurement; the third applied these metabolites on a cultured cancer colon cell line Caco-2 to evaluate their effect on cell viability, proliferation, clonogenicity and migration.

4.1. Callus Initiation of R. damascena

The goal of this section was to receive adequate quantities of callus. *R. damascena* mature closed flowering buds (1–1.5 cm long) were collected from the Al-hada district of Taif governorate for this investigation.

4.1.1. Explant Sterilization

Closed mature flowering buds were washed in tap water with soap and a few drops of Tween-20 (a wetting agent that lowers the surface tension, enabling better surface contact), then submerged for 1 min in 70% ethanol, washed with sterile distilled water, and immersed for 5 min in a 5% marketing Clorox solution. To eradicate residuals, explants were disinfected three times with sterile distilled water in a laminar air flow hood.

4.1.2. *R. damascena* Callus Generation in MS Medium Culture

For callus initiation, sterilized explants were grown in MS medium supplemented with 0.1 mg/L kinetin (Kin) and 1.0 mg/L naphthaleneacetic acid (NAA). Callus was obtained after 6 weeks of culture on medium.

4.2. R. damascena Elicitation Secondary Metabolites

Flowering callus buds were cultivated on Murashige and Skoog (MS) medium with citric acid at 0.5 g/L, L-ascorbic acid at 0.5 g/L, sucrose at 30 g/L, agar at 7 g/L, and produced from MS media supported with 0.1 mg/L kinetin (Kin) and 1.0 mg/L naphthaleneacetic acid (NAA).

4.2.1. Experiment Design and Setting

The three following treatments were performed: 1-0.1 mg/L KIN and 1.0 mg/L NAA (control group), 2-L-ascorbic acid at 0.5 g/L (LAA group) and 3-citric acid at 0.5 g/L (CA group). Every three explants received ten jars in each treatment. Before autoclaving, the pH of the media was adjusted to 5.7–5.8 using a pH meter and an appropriate amount of 0.1 N HCl or 0.1 NaOH. Clean jars were used to deliver the material. Each one had 30 mL of nutritional media in it. After that, the jars were autoclaved for 15 min at 121 °C, 1.5 kg/cm^3 at 1.5 kg/cm^3. All treatments were incubated at 26 ± 2 °C and exposed to a 16 h light/day photoperiod under constant fluorescent light of 1500 Lux in the growth chamber. After 4 weeks of culturing on media, the following data were recorded: 1—Average callus fresh weight (g), 2—Average callus dry weight (g).

4.2.2. Callus Crude Extraction

The obtained callus was dried, and ground to fine powder by a mortar and pestle. Five grams of dried powdered callus of each treatment was extracted for roughly 6 h at 60 °C using a Soxhlet apparatus with 50 mL of 100% methanol. The solution was then evaporated to dryness at 40 °C using a rotary evaporator [51]. Next, the crudes were stored in glass bottles at −20 °C for further bioassays.

4.2.3. The GC–MS Analysis

The GC–MS analysis of methanolic crude extracts was conducted for the identification and characterization of various chemical components, as well as the presentation of the total extract from samples. A Thermo Scientific Trace GC Ultra/ISQ Single Quadrupole MS (Thermo Scientific, Waltham, MA, USA) TG-5MS-fused silica capillary column was used for the GC–MS study (30 m, 0.251 mm, 0.1 mm film thickness). An electron ionization system with 70 eV ionization energy was employed for GC–MS detection, with Helium as the carrier gas at a constant flow rate of 1 mL/min. The temperature of the injector and MS transfer line was fixed to 280 °C. The oven temperature was set to rise from 50 °C (hold for 2 min) to 150° C at a pace of 7 °C per min, then to 270 °C at a pace of 5 °C per min (hold for 2 min), then to 310 °C at a rate of 3.5 °C per min as a final temperature (hold 10 min). A percent relative peak area was used to evaluate the quantification of all the discovered components. The chemicals were tentatively identified by comparing their respective retention times and mass spectra to those of the NIST, WILLY library data from the GC–MS instrument.

4.3. *In Vitro Anti-Cancer Study (Cell Line Studies)*

Following the method of Etman et al., 2020 [52], human colorectal adenocarcinoma cell line (Caco-2) (ATCC® HTB-37™) was employed in this study and all experiments were carried out in CERRMA (Center of Excellence for Research in Regenerative Medicine and its Applications), Faculty of Medicine, Alexandria University. Cells were cultured in Dulbecco's modified eagle medium (DMEM)-high glucose enriched with (10% *v/v*) fetal bovine serum (FBS) and antibiotics (100 U/mL penicillin, 100 µg/mL streptomycin). Cells were incubated in 5% CO_2 at 37 °C for maintenance and media was changed every 2 days. Cells were passaged on reaching 80–90% confluence using 0.25% (*w/v*) trypsin/ethylene diamine tetra acetic acid (EDTA), then plated in T75 cm^2 flasks or in other culture vessels (6 or 96 well plate) according to the experiment conducted.

4.3.1. Cellular Cytotoxicity Assay

Different callus extracts were dissolved in 1 mL dimethyl sulfoxide (DMSO) to assess their cytotoxicity through applying different concentrations, in which the highest concentration of DMSO would never exceed 0.1% DMSO to avoid its cytotoxic effect. Cellular cytotoxicity was assessed using MTT assay as described by El-Habashy et al., 2020 [53]. Caco-2 cells were seeded at a density of (5×10^3) in 96-well plate. Each well contained 100 µL of culture media. Cells were allowed to adhere for 24 h. Then, they were treated with different concentrations ranging from 10 to 200 ug/mL or 100 to 350 ug/mL of the LAA-, CA- and control-treated groups, respectively, then incubated for 48 h. After incubation, 100 µL of fresh media containing a 10 µL MTT solution (5 mg/mL) was added and incubated for another 4 h in a CO_2 incubator. Finally, 100 µL DMSO was added to dissolve the produced formazan crystals. Absorbance was measured at 570 nm using a microplate reader. The viability of cells was determined according to the following equation: % viability = Absorbance of sample at 570 nm/Absorbance of untreated at 570 nm $\times$ 100, where untreated cells were treated with culture media only. The effect of different concentrations on cell viability was expressed as % inhibition against concentration and used to calculate IC50 (concentration required to kill 50% of the cells). Results were expressed as mean ± SD ($n = 8$).

4.3.2. Colony-Forming Assay

Clonogenic assay is an in vitro cell-survival assay based on the ability of a single cell to grow into a colony. The assay tests the cell for its ability to undergo division with and without treatment, thus, used to determine the effectiveness of cytotoxicity. The assay was performed by plating 10^3 cells on a 60 mm culture dish in CCM; after 24 h, the media -was discarded from the wells and replaced with culture media only or media with added treatment of half-calculated IC50 of the LAA, CA and control for 14 days. The medium was changed every 2–3 days; then, after 14 days, cells were washed with PBS, fixed and stained using Crystal Violet (Sigma-Aldrich, Burlington, MA, USA) at 3% (*w/v*). The number of visible colonies was counted and plating efficiency, (the number of colonies formed/number of cells plated) $\times$ 100 was calculated for comparison.

4.3.3. Ki-67 Flow Cytometry Proliferation Assay

Caco-2 cells were plated in 6-well plates at a density of 2×10^4 cells per well for the cell proliferation assay. After incubation for 24 h for adherence, the media was discarded from the wells and replaced with culture media only or media with added treatment of half-calculated IC50 of LAA, CA and control groups for 48 h. The culture medium was then removed from the wells, cells were washed twice with sterile PBS, and 0.25% of the trypsin-EDTA solution was added to detach the cells. After 5 min of incubation at 37 °C and 5% CO_2, fresh culture medium was added to inactivate trypsin, and cells were collected in flow cytometry tubes. Cells were then labelled using the Ki67 Proliferation Kit (D3B5, Rabbit mAb Alexa Fluor® 488 Conjugate, Cell signaling technology) according to the manufacturer's instructions (cell signaling technology; flow cytometry, methanol permeabilization protocol) and analyzed by using BD FACS Calibur flow cytometry. Unstained control cells were used for gating to determine the percentage of proliferation of the Ki67-positive cells in the samples. Percentages of proliferating (Ki67 positive) cells were used to calculate the means ± SD for each group in triplicate [54].

4.3.4. Migration Assay (Scratch Assay)

The scratch assay was carried out to test the ability of the LAA, CA and control groups to attenuate the migration of cancer cells [55]. The cells were grown until 70–80% confluency in complete media. Then, they were incubated for 24 h in a 6-well plate to allow cellular adhesion. Next, media was removed and replaced with starvation (serum-free) medium for another 24 h. Then, a scratch was carried out using a sterile 200 μL pipette tip and cells were washed with PBS twice to remove any detached cells; then, the cells were treated with half-calculated IC50 in the LAA, CA and control groups in starvation media, or just starvation medium in the control–untreated group, for 48 h. Images of the scratch were taken using an inverted phase contrast microscope (Olympus, Waltham, MA, USA) after scratching and this was marked as zero time. Images were then taken after 24 and 48 h. Representative images were taken, and the area of wound-healing was calculated using Image J software (Version 1.52p software 32, NIH, USA).

4.4. Statistical Analysis

GraphPad Prism 8 was used for statistical analysis (GraphPad Software, La Jolla, CA, USA). The data were analyzed using one-way analysis of variance (ANOVA). Significant differences were determined to have *p*-values less than 0.05.

5. Conclusions

In the current study, we found that adding L-ascorbic acid to media enhanced the secondary metabolite production in *R. damascena* callus and it has an efficient antiproliferative, anti-clonogenic and anti-migratory effect on Caco-2 cancer cells, thus, can be used as an adjuvant anti-cancer therapy.

Author Contributions: Conceptualization: H.D., R.A.M. and A.N.; Data curation: H.D., R.A.M., S.S.I. and S.A.; Formal analysis: H.D., A.N., S.S.A. and M.A.F.; Investigation: H.D., R.A.M., S.A. and S.S.A.; Methodology: H.D., R.A.M., S.S.I., A.N., S.A., S.M.A., S.S.A. and M.A.F.; Project administration: A.N. and M.A.F.; Resources: H.D., R.A.M., S.S.A., S.A. and B.A.; Validation: H.D., S.H.A.H. and A.N.; Visualization: H.D., R.A.M., S.S.I. and S.M.A.; Writing—original draft: H.D., R.A.M., A.N. and S.A.; Writing—review and editing: S.S.A., S.M.A., B.A. and S.H.A.H. All authors have read and agreed to the published version of the manuscript.

Funding: This research was funded by the Deputyship for Research & Innovation, Ministry of Education in Saudi Arabia through the project number 1-441-123.

Institutional Review Board Statement: Not applicable.

Informed Consent Statement: Not applicable.

Data Availability Statement: Not applicable.

Acknowledgments: The authors extend their appreciation to the Deputyship for Research & Innovation, Ministry of Education in Saudi Arabia for funding this research work through the project number 1-441-123.

Conflicts of Interest: The authors declare no conflict of interest.

References

1. Labianca, R.; Beretta, G.D.; Kildani, B.; Milesi, L.; Merlin, F.; Mosconi, S.; Pessi, M.A.; Prochilo, T.; Quadri, A.; Gatta, G.; et al. Colon cancer. *Crit. Rev. Oncol. Hematol.* **2010**, *74*, 106–133. [CrossRef]
2. Siegel, R.L.; Miller, K.D.; Jemal, A. Cancer statistics. *CA Cancer J. Clin.* **2020**, *70*, 7–30. [CrossRef] [PubMed]

3. Benson, A.B.; Venook, A.P.; Al-Hawary, M.M.; Arain, M.A.; Chen, Y.J.; Ciombor, K.K.; Cohen, S.; Cooper, H.S.; Deming, D.; Farkas, L.; et al. Colon cancer, version 2.2021, NCCN clinical practice guidelines in oncology. *J. Natl. Compr. Cancer Netw.* **2021**, *19*, 329–359. [CrossRef] [PubMed]
4. Yuhara, H.; Steinmaus, C.; Cohen, S.E.; Corley, D.A.; Tei, Y.; Buffler, P.A. Is diabetes mellitus an independent risk factor for colon cancer and rectal cancer? *Am. J. Gastroenterol.* **2011**, *106*, 1911. [CrossRef]
5. Banerjee, A.; Pathak, S.; Subramanium, V.D.; Dharanivasan, G.; Murugesan, R.; Verma, R.S. Strategies for targeted drug delivery in treatment of colon cancer: Current trends and future perspectives. *Drug Discov. Today* **2017**, *22*, 1224–1232. [CrossRef]
6. Shan, B.; Cai, Y.Z.; Sun, M.; Corke, H. Antioxidant capacity of 26 spice extracts and characterization of their phenolic constituents. *J. Agric. Food Chem.* **2005**, *53*, 7749–7759. [CrossRef]
7. Michalak, A. Phenolic compounds and their antioxidant activity in plants growing under heavy metal stress. *Pol. J. Environ. Stud.* **2006**, *4*, 523–530.
8. Taiz, L.; Zeiger, E. *Plant Physiology*, 5th ed.; Sinauer Associates Inc.: Sunderland, MA, USA, 2010.
9. Olivoto, T.; Nardino, M.; Carvalho, I.R.; Follmann, D.N.; Szareski, V.J.; Ferrari, M.; de Pelegrin, A.J.; de Souza, V.Q. Plant secondary metabolites and its dynamical systems of induction in response to environmental factors: A review. *Afr. J. Agr. Res.* **2017**, *12*, 71–84.
10. Kittakoop, P.; Mahidol, C.; Ruchirawat, S. Alkaloids as important scaffolds in therapeutic drugs for the treatments of cancer, tuberculosis, and smoking cessation. *Curr. Top. Med. Chem.* **2014**, *14*, 239–252. [CrossRef]
11. Takemura, M.; Tanaka, R.; Misawa, N. Pathway engineering for the production of β-amyrin and cycloartenol in *Escherichia coli*, a method to biosynthesize plant-derived triterpene skeletons in *E. coli*. *Appl. Microbiol. Biotechnol.* **2017**, *101*, 6615–6625. [CrossRef]
12. Lamarti, A.; Badoc, A.; Deffieux, G.; Carde, J.P. Biogenesis of monoterpenes: The isoprenic chain. *Bull. Soc. Pharm. Bordx.* **1994**, *133*, 79–99.
13. Bakkali, F.; Averbeck, S.; Averbeck, D.; Idaomar, M. Biological effects of essential oils—A review. *Food Chem. Toxicol.* **2008**, *46*, 446–475. [CrossRef] [PubMed]
14. Soriano, I.R.; Riley, I.T.; Potter, M.J.; Bowers, W.S. Phytoecdysteroids: A novel defense against plant-parasitic nematodes. *J. Chem. Ecol.* **2004**, *30*, 1885–1899. [CrossRef] [PubMed]
15. Veitch, G.E.; Boyer, A.; Ley, S.V. The azadirachtin story. *Angew. Chem. Int. Ed. Engl.* **2008**, *47*, 9402–9429. [CrossRef] [PubMed]
16. Agrawal, A.A.; Weber, M.G. On the study of plant defence and herbivory using comparative approaches: How important are secondary plant compounds. *Ecol. Lett.* **2015**, *18*, 985–991. [CrossRef] [PubMed]
17. Bartwal, A.; Mall, R.; Lohani, P.; Guru, S.K.; Arora, S. Role of secondary metabolites and brassinosteroids in plant defense against environmental stresses. *J. Plant Growth Regul.* **2012**, *32*, 216–232. [CrossRef]
18. Gómez-Caravaca, A.M.; Gómez-Romero, M.; Arráez-Román, D.; Segura-Carretero, A.; Fernández-Gutiérrez, A. Advances in the analysis of phenolic compounds in products derived from bees. *J. Pharm. Biomed. Anal.* **2006**, *41*, 1220–1234. [CrossRef]
19. Ali, M.B.; Hahn, E.J.; Paek, K.Y. Methyl jasmonate and salicylic acid induced oxidative stress and accumulation of phenolics in *Panax ginseng* bioreactor root suspension cultures. *Molecules* **2007**, *12*, 607–621. [CrossRef]
20. Homoki, J.R.; Nemes, A.; Fazekas, E.; Gyémánt, G.; Balogh, P.; Gál, F.; Al-Asri, J.; Mortier, J.; Wolber, G.; Babinszky, L.; et al. Anthocyanin composition, antioxidant efficiency, and α-amylase inhibitor activity of different hungarian sour cherry varieties (*Prunus cerasus* L.). *Food Chem.* **2016**, *194*, 222–229. [CrossRef]
21. Salawu, S.O.; Alao, O.F.; Faloye, O.F.; Akindahunsi, A.A.; Boligon, A.A.; Athayde, M.L. Antioxidant potential of phenolic-rich two varieties of Nigerian local rice and their anti-cholinesterase activities after in vitro digestion. *Nutr. Food Sci.* **2016**, *46*, 171–189. [CrossRef]
22. Ben Mrid, R.; Bouargalne, Y.; El Omari, R.; Nhiri, M. New insights into the therapeutic effects of phenolic acids from sorghum seeds. *J. Rep. Pharma. Sci.* **2019**, *8*, 91–101.
23. Darwish, H.; Abdelmigid, H.; Albogami, S.; Alotaibi, S.; Noureldeen, A.; Alnefaie, A. Induction of biosynthetic genes related to rosmarinic acid in plant callus culture and antiproliferative activity against breast cancer cell line. *Pak. J. Biol. Sci.* **2020**, *23*, 1025–1036.
24. Veeresham, C.; Chitti, P. Therapeutic Agents from Tissue Cultures of Medicinal Plants. College of Pharmaceutical Sciences, Kakatiya University, Warangal, Andhra Pradesh, India Natural Products Chemistry and Research. *Nat. Prod. Chem. Res.* **2013**, *1*, 4.
25. Basim, E.; Basim, H. Antibacterial activity of *Rosa damascene* essential oil. *Fitoterapia* **2003**, *74*, 394–396. [CrossRef]
26. Ozkan, G.; Sagdic, O.; Baydar, N.G.; Baydar, H. Antioxidant and antibacterial activities of *Rosa damascena* flower extracts. *Food Sci. Technol. Int.* **2004**, *10*, 277–281. [CrossRef]
27. Achuthan, C.R.; Babu, B.H.; Padikkala, J. Antioxidant and hepatoprotective effects of *Rosa damascene*. *J. Padikkala Pharm. Biol.* **2003**, *41*, 357–361. [CrossRef]
28. Mahmood, N.; Piacente, S.; Pizza, C.; Burke, A.; Khan, A.L.; Hay, A.J. The anti-HIV activity and mechanisms of action of pure com-pounds isolated from *Rosa damascena*. *Biochem. Biophys. Res. Commun.* **1996**, *229*, 73–79. [CrossRef]
29. Jabbarzadeh, Z.; Khosh-Khui, M. Factors affecting tissue culture of Damask rose (*Rosa damascena* Mill.). *Sci. Hortic.* **2005**, *105*, 475–482. [CrossRef]
30. Rady, M.M.; Sadak, M.S.; El-Lethy, S.R.; Abdelhamid, E.M.; Abdelhamid, M.T. Exogenous α-tocopherolhasa beneficial effect on *Glycine max* (L.) plants irrigated with diluted sea water. *J. Hortic. Sci. Biotechnol.* **2015**, *90*, 195–202.

31. Orabi, S.A.; Abdelhamid, M.T. Protective role of α-tocopherol on two *Vicia faba* cultivars against seawater-induced lipid peroxidation by enhancing capacity of anti-oxidative system. *J. Saudi Soc. Agric. Sci.* **2016**, *15*, 145–154. [CrossRef]
32. Beltaji, M.S. Exogenous ascorbic acid (vitamin C) induced anabolic changes for salt tolerance in chickpea (*Cicer arietinum* L.) plants. *Afr. J. Plant Sci.* **2008**, *2*, 118–123.
33. Zechmann, B. Subcellular distribution of ascorbate in plants. *Plant Signal. Behav.* **2011**, *6*, 360–363. [CrossRef] [PubMed]
34. Gallie, D.R. L-Ascorbic acid, a multifunctional molecule supporting plant growth and development. *Scientifica* **2013**, 795964. [CrossRef]
35. Smirnoff, N.; Wheeler, G.L. Ascorbic acid in plants: Biosynthesis and function. *Crit. Rev. Biochem. Mol. Biol.* **2000**, *35*, 291–314. [CrossRef] [PubMed]
36. Atanasova, T.; Kakalova, M.; Stefanof, L.; Petkova, M.; Stoyanova, A.; Damyanova, S.; Desyk, M. Chemical composition of essential oil from *Rosa damascena* mill., growing in new region of Bulgaria. *Ukr. Food J.* **2016**, *5*, 492–498. [CrossRef]
37. Hussein, M.M.; Alva, A.K. Effects of Zinc and Ascorbic Acid Application on the Growth and Photosynthetic Pigments of Millet Plants Grown under Different Salinity. *Agric. Sci.* **2014**, *5*, 1253–1260. [CrossRef]
38. De Pinto, M.C.; Francis, D.; De Gara, D. The redox state of the ascorbate-dehydroascrobate pair as a specific sensor of cell division in tobacco BY-2 cells. *Protoplasma* **1999**, *209*, 90–97. [CrossRef]
39. Joy, R.W.; Patel, K.R.; Thorpe, T.A. Ascorbic acid enhancement of organogenesis in tobacco. *Plant Cell Tissue Organ Cult.* **1988**, *13*, 219–228. [CrossRef]
40. El-Sayed, O.M.; El-Gammal, O.; Salama, A. Effect of ascorbic acid, proline and jasmonic acid foliar spraying on fruit set and yield of Manzanillo olive trees under salt stress. *Sci. Hortic.* **2014**, *176*, 32–37. [CrossRef]
41. Chen, Z.; Gallie, D.R. The ascorbic acid redox state controls guard cell signaling and stomatal movement. *Plant Cell* **2004**, *16*, 1143–1162. [CrossRef]
42. El-Hifny, I.M.; El-Sayed, M. Response of sweet pepper plant growth and productivity to application of ascorbic acid and biofertilizers under saline conditions. *Aust. J. Basic. Appl. Sci.* **2011**, *5*, 1273–1283.
43. Midan, S.A.; Sorial, M.E. Some antioxidants application in relation to lettuce growth, chemical constituents and yield. *Aust. J. Basic Appl. Sci.* **2011**, *5*, 127.
44. Shalaby, T.A.; El-Ramady, H. Effect of foliar application of bio-stimulants on growth, yield, components, and storability of garlic (*Allium sativum* L.). *Aust. J. Crop Sci.* **2014**, *8*, 271–275.
45. Fatima, A.; Singh, A.A.; Mukherjee, A.; Agrawal, M.; Agrawal, S.B. Ascorbic acid and thiols as potential biomarkers of ozone tolerance in tropical wheat cultivars. *Ecotoxicol. Environ. Saf.* **2019**, *171*, 701–708. [CrossRef] [PubMed]
46. Desoky, E.M.; Mansour, E.; Yasin, M.A.; El Sobky, E.-S.E.; Rady, M.M. Improvement of drought tolerance in five different cultivars of *Vicia faba* with foliar application of ascorbic acid or silicon. *Span. J. Agric. Res.* **2020**, *18*, 16. [CrossRef]
47. Mazher, A.A.; Zaghloul, S.M.; Mahmoud, S.A.; Siam, H.S. Stimulatory effect of kinetin, ascorbic acid and glutamic acid ongrowth and chemical constituents of *Codiaeum variegatum* L. plants. *Am. Eurasian. J. Agric. Environ. Sci.* **2011**, *10*, 318–323.
48. El-Sanatawy, A.M.; Ash-Shormillesy, S.M.; Qabil, N.; Awad, M.F.; Mansour, E. Seed halo-priming improves seedling vigor, grain yield, and water use efficiency of maize under varying irrigation regimes. *Water* **2021**, *13*, 2115. [CrossRef]
49. Rafique, N.; Raza, S.H.; Qasim, M.; Iqbal, N. Pre-sowing application of ascorbic acid and salicylic acid to seed of pumpkin and seedling response to salt. *Pak. J. Bot.* **2011**, *43*, 2677–2682.
50. Sajid, Z.A.; Aftab, F. Amelioration of salinity tolerance in *Solanum tuberosum* L. by exogenous application of ascorbic acid. *In Vitro Cell. Dev. Biol. Plant.* **2009**, *45*, 540–549. [CrossRef]
51. Albogami, S.; Darwish, H. A novel therapeutic effect Callus from *Rosa damascena* that suppresses Human T-cell Activation: Medicine from Cultured Cells. *Res. J. Biotechnol.* **2017**, *12*, 20–24.
52. Etman, S.M.; Abdallah, O.Y.; Mehanna, R.A.; Elnaggar, Y.R. Lactoferrin/Hyaluronic acid double-coated lignosulfonate nanoparticles of quinacrine as a controlled release biodegradable nanomedicine targeting pancreatic cancer. *Int. J. Pharm.* **2020**, *30*, 119097. [CrossRef] [PubMed]
53. El-Habashy, S.E.; Eltaher, H.M.; Gaballah, A.; Eiman, I.Z.; Mehanna, R.A.; El-Kamel, A.H. Hybrid bioactive hydroxyapatite/polycaprolactone nanoparticles for enhanced osteogenesis. *Mater. Sci. Eng. C* **2021**, *119*, 111599. [CrossRef] [PubMed]
54. Elsehly, W.M.; Mourad, G.M.; Mehanna, R.A.; Kholief, M.A.; El-Nikhely, N.A.; Awaad, A.K.; Attia, M.H. The potential implications of estrogenic and antioxidant-dependent activities of high doses of methyl paraben on MCF7 breast cancer cells. *J. Biochem. Mol. Toxicol.* **2022**, *36*, e23012. [CrossRef] [PubMed]
55. Cory, G. Scratch-Wound Assay. In *Cell Migration. Methods in Molecular Biology*; Wells, C., Parsons, M., Eds.; Humana Press: London, UK, 2011; Volume 769.

molecules

MDPI

Article

Chemical Composition and Antioxidant, Antiviral, Antifungal, Antibacterial and Anticancer Potentials of *Opuntia ficus-indica* Seed Oil

Abeer S. Alqurashi [1], Luluah M. Al Masoudi [1], Hamida Hamdi [1,2,*] and Abeer Abu Zaid [3]

1 Department of Biology, Faculty of Sciences, Taif University, P.O. Box 11099, Taif 21944, Saudi Arabia
2 Zoology Department, Faculty of Science, Cairo University, Giza 12613, Egypt
3 Department of Food Science and Nutrition, Alkhurmah University College, Taif University, Taif 21974, Saudi Arabia
* Correspondence: shimaa76sl@tu.edu.sa

Abstract: *Opuntia ficus-indica* (OFI) is a cactus that is widely cultivated in the Kingdom of Saudi Arabia especially in the Taif region due to its favorable weather for growing, and it has benefits as a food and traditional medicine. The aim of the current study was to chemically characterize *Opuntia ficus-indica* seed oil from Taif, Kingdom of Saudi Arabia, using GC-MS and HPLC analysis and evaluate its antioxidant, antiviral, antifungal, antibacterial and anticancer activities. Linolenic acid was the dominating fatty acid in OFI oil, followed by oleic acid, linoleic acid, palmitic acid and stearic acid. Total tocopherol (α-, β-, γ-tocopherol) was found to be 24.02 µg/mL. Campesterol was the main phytosterol, followed by γ- & β -sitosterol, and Stigmasterol. The phenolic components scored 30.5 mg gallic acid equivalent per ml of oil with 89.2% antioxidant activity (% DPPH radical inhibition) at 200 µL/mL of OFI oil. OFI oil showed an inhibition efficacy against microbial strains especially *Saccharomyces cervisiae* with a diameter (28.3 $\pm$ 0.4), MBC (15 µg/mL) and MIC bacteriostatic (10 µg/mL). While OFI oil had no effect against *Aspergillus niger*, OFI oil showed weak inhibitory activity against A-2780 (Ovarian carcinoma) cell line, although it showed significant inhibitory activity against PC-3 (Prostate carcinoma) cell line. OFI oil exhibited an antiviral effect (22.67 $\pm$ 2.79%) at 300 µg/mL of Oil against herpes simplex type 2 (HSV-2) virus. The bioactive compounds of OFI oil, as well as its main biological activities, make it a promising candidate for the non-communicable disease management.

Keywords: *Opuntia ficus-indica*; fixed oils; antioxidant; antifungal; antibacterial; antiviral; cytotoxicity

Citation: Alqurashi, A.S.; Al Masoudi, L.M.; Hamdi, H.; Abu Zaid, A. Chemical Composition and Antioxidant, Antiviral, Antifungal, Antibacterial and Anticancer Potentials of *Opuntia ficus-indica* Seed Oil. *Molecules* **2022**, *27*, 5453. https://doi.org/10.3390/molecules27175453

Academic Editors: Arunaksharan Narayanankutty, Ademola C. Famurewa and Eliza Oprea

Received: 27 July 2022
Accepted: 21 August 2022
Published: 25 August 2022

Publisher's Note: MDPI stays neutral with regard to jurisdictional claims in published maps and institutional affiliations.

Copyright: © 2022 by the authors. Licensee MDPI, Basel, Switzerland. This article is an open access article distributed under the terms and conditions of the Creative Commons Attribution (CC BY) license (https://creativecommons.org/licenses/by/4.0/).

1. Introduction

Botanical medicines are widely used due to their reliable efficacy, reduced side effects and relative economic cost [1]. Nowadays, the demand for nutrients, natural components and health-boosting foods is permanently increasing [2,3]. There are about 2,253 medicinal plants in various regions of the Kingdom of Saudi Arabia [4,5]. *Opuntia ficus-indica* (OFI) of the family Cactaceae, known as prickly pear, comprises about 1500 species [6]. It was originally grown in different regions of Saudi Arabia especially in the City of Taif for the edible prickly pear fruit and consumed by local populations as an important food source [7]. Moreover, the OFI plant is widely spread in South America, Australia, South Africa and the Mediterranean area [8]. It is a tropical or subtropical plant up to five meters high with a thick, woody stem [9].

Various studies have shown that prickly pear seed oil is edible, and with potential importance to the agriculture industry [10]. Many researchers have been interested in studying the phytochemical profile of the seed oils of two Opuntia species *O. ficus-indica* and *O.dillenii* and have found that the two seed oils are rich in very active molecules, such as unsaturated fatty acids, sterols, tocopherols and polyphenols [3,11].

The highest benefit of this oil due to its high amount of polyunsaturated fatty acids, especially linoleic and linolenic acids, which have potential health avails due to their roles as the eicosanoids biosynthesis precursors [3,12]. Oil extraction by cold screw pressing is an alternative method and has been found to be a substitute to extraction of solvent [13]. This process has the advantage of being less oil producing than others, but is safer, simpler, less expensive, hygienic, no chemical residue and ecologically friendly [14]. Cold pressed oils improve the quality of oil and rich with bioactive components such as essential fatty acids, sterols, tocopherol, and phenolics [15].

To our knowledge, there are few studies about *Opuntia ficus-indica* seed oil growing in Saudi Arabia. Therefore, the aim of current study was to chemically characterize *Opuntia ficus-indica* seed oil growing in Saudi Arabia using GC-MS and HPLC analysis and evaluate its antioxidant, antiviral, antifungal, antibacterial and anticancer activities.

2. Materials and Methods

2.1. Oil Extraction

Seeds of *Opuntia ficus-indica* were collected from Taif City. Natural oil 100% extracted by cold pressing using a screw extractor in a local maesarat, Taif, Saudi Arabia. Finally, the oil was stored at 20 °C until analysis.

2.2. Identification of Opuntia ficus-indica *Seed Oil*

2.2.1. Fatty Acid Composition

Fatty acid analysis was carried out in triplicate, consisting of two successive steps, fatty acid methyl ester (FAME) preparation and chromatographic analysis. Lipids-extract esterification was performed according to the method of [16]. Determination of fatty acid methyl esters was performed by comparing their retention times with pure standards. Their quantification according to their percentage are taken out by the peak integration. Data were expressed as individual fatty acids percentages in the lipid fraction.

2.2.2. Sterols and the Various Components

Gas chromatography/mass spectrophotometer was used for the identification and quantification of sterols and the various components of *Opuntia ficus-indica* oil. Sterols were converted to trimethylsilyl (TMS) ether derivatives prior to analysis by gas chromatography [17]. Sterols were analyzed as their TMS ethers by capillary gas chromatography with flame ionization detection. The GC parameters were as previously described [17]. Identification of sterols and various components was based on relative retention times of commercially-available compounds, comparison with literature data [18,19] and mass spectral analyses (NIST/EPA/NIH 1999). Quantitative data were calculated by comparing the average peak area of the component to the total areas.

2.2.3. Tocopherols

Tocopherol Analysis was performed by HPLC-(Agilent 1100), consisting of two LC-pumps and a UV/Vis detector with a C18 column (125 mm × 4.60 mm, 5 µm particle size). Agilent ChemStation is used to analyze the obtained Chromatograms. Conditions of Chromatography were as previously described [20].

2.2.4. Determination of Total Phenolic Compounds

Total phenolic compounds (TPC) in *Opuntia ficus-indica* seed oil were determined spectrophotometrically according to the colorimetric method of Folin–Ciocalteu [21]. Data expressed as mg gallic acid equivalent (GAE) per ml of oil.

2.3. Antioxidant Activity by Free Radical Scavenging Assay:

The free radical scavenging activity was estimated using 1,1-diphenyl-2-picrylhydrazyl (DPPH) radical as illustrated by [22]. The positive control was BHT. Results expressed as % of inhibition of the DPPH radical (Equation (1)). The IC_{50} is equivalent to 50% of DPPH inhibition.

$$\% \, of \, inhibition \, of \, the \, DPPH \, radical = \left[\frac{Abs_{control} - \left(Abs_{sample} - Abs_{blank} \right)}{Abs_{control}} \right] \times 100 \quad (1)$$

$Abs_{control}$ = The DPPH absorbance
Abs_{sample} = The sample absorbance
Abs_{blank} = The ethanol negative control absorbance

2.4. Antimicrobial Activity of Opuntia ficus-indica Seed Oil

2.4.1. Microbial Strains

The OFI oil's antimicrobial activity was determined versus seven pathogenic microorganisms mentioned in the following: Gram-positive bacterial strains (*Staphylococcus aureus*, *Bacillus subtilis*); Gram-negative strains (*Escherichia coli*, and *Klebsiella pneumoniae*); a strain of yeast (*Saccharomyces cerevisiae*) and fungi (*Aspergillus niger*, *Penicillium digitatum*). All pathogenic isolates were obtained from department of microbiological laboratories, Faculty of science, Cairo University, after its isolation and identification. Oxytetracycline (OT30) and penicillin (P10) were used as positive control [23,24].

2.4.2. Antimicrobial Activity

Disk diffusion agar method was used to determine the antibacterial and antifungal activities of OFI oil [25]. Microdilution assay for bacterial strains by using sterile Mueller–Hinton media and for antifungal tests potato dextrose agar (Scharlab, S.L, Barcelona, Spain) was performed. Cell suspensions (0.1 mL) of bacterial strains were adjusted to 10^8 CFU/mL Cell Forming Units and 10^5 spores/mL for fungus by MacFarland, and then inoculated onto the surface of agar plates. Then, sterile discs were made (3 mm in diameter) into inoculated plates, and 25 µL of oil filled into each disc. Dishes were placed for 2 h to allow the oil to diffuse and incubated at 37 °C for 48 h for yeast, 24 h for bacterial strains, and 3–4 days for fungi. The negative control was carried out without oils. Antimicrobial activity was calculated by measuring the area of inhibition zone around the discs. Strains tests were replicated three times.

2.4.3. Evaluation of MIC and MBC

The broth dilution method was used to determine Minimum Inhibitory Concentration (MIC) of OFI oil against microorganisms. Pre-modified 0.01mL microbial strains were inoculated into tubes containing. 50.0, 45.0, 40.0, 35.0, 30.0, 25.0, 20.0, 15.0, 10.0, 5.0 µL/mL OFI oil and incubated 24 h at 37 °C. The results were evaluated by showing visible growth inhibition of microbial tubes (no turbidity). The Minimum Bactericidal Concentration (MBC) was introduced by subculture, in which partition of the ~10 µL of each tube invisible growth used in MIC onto Mueller–Hinton agar medium and at 37 °C, 24 h. Colony growth was examined, and all tests were repeated three times.

2.5. Cytotoxic Assay of Opuntia ficus-indica Seed Oil

2.5.1. Mammalian Cell Line

Vero cell (derived from the kidney of African green monkey), PC-3 cell line (Prostate carcinoma), and A2780 cell line (Ovarian carcinoma) were purchased from the American Type Culture Collection (ATCC, Manassas, VA, USA). The cells were cultured in Dulbecco's modified Eagle's medium (DMEM) boosted with 10% heat-inactivated fetal bovine serum, 1% L-glutamine, buffer of HEPES and 50 µg/mL gentamycin. The cells were maintained at 37 °C in a humidified 5% CO_2 atmosphere and were subcultured twice a week [26].

2.5.2. Cytotoxicity Evaluation

The cytotoxicity of fixed oil against each cell lines (Vero, PC-3 and A-2780) were determined through MTT colorimetric method by [27]. The 50% inhibitory concentration (IC_{50}), the concentration demanded to produce toxic effects in 50% of healthy cells, was determined.

2.5.3. Antiviral Evaluation

The cytopathogenic herpes simplex type (2HSV-2) virus was propagated and assayed in confluent Vero cells [28]. Spearman–Karber method was used to enumerate infectious viruses by determining the tissue culture infectious dose 50% ($TCID_{50}$) with eight wells per dilution and 20 µL of inoculum per well [29].

Antiviral Activity

The cytopathic effect inhibition assay was used to estimate the antiviral screening. This assay will be chosen to demonstrate specific inhibition of a biological function, and the MTT method was used to measure a cytopathic effect in sensitive mammalian cells [30,31]. Acyclovir was used as positive control in this assay.

2.6. Statistical Analysis

Statistical analysis of data was carried out using GraphPad Prism 5. The data were analyzed for statistical significance by the one-way analysis of variance, followed by Tukey's multiple comparison tests. The data represented as mean ± standard error (*SE*). Data at $p < 0.05$ were considered significant.

3. Results

3.1. Fatty Acid Composition

Table 1 summarizes the data of fatty acid composition, total saturated fatty acids (SFA), monounsaturated (MUFA) and polyunsaturated fatty acids (PUFA) of *Opuntia ficus-indica* seed oil. The five major fatty acids were linolenic acid (C18:3), oleic acid (C18:1) linoleic acid (ω 6; C18:2) followed by palmitic acid (C16:0) and stearic acid (C18:0), representing, respectively 50.69, 21.10, 14.00, 6.73 and 5.74%. Minimal quantities of myristic (C14:0), margaric (C17:0), palmitoleic (C16:1), Arachidic (C20:0), Eicosanoic (Gondoic) (C20:1), Behenic (C22:0), and Erucic (22:1) fatty acids were also identified and quantified. Polyunsaturated fatty acids were the major group of fatty acids, representing 64.69, followed by monounsaturated fatty acids 22.30% and saturated fatty acids 12.98%. The ratio of saturated/unsaturated acid of *Opuntia ficus-indica* seed oil was 0.1, which is low due to the high quantity of unsaturated fatty acid such as C18:3n9, C18:1n9 and C18:2n9.

Table 1. Fatty acids percentage of *Opuntia ficus-indica* seed oil.

Fatty Acids	%
Saturated fatty acids	
Myristic acid C14:0	0.05
Palmitic acid C16:0	6.73
Margaric acid C17:0	0.06
Stearic acid C18:0	5.74
Arachidic acid C20:0	0.23
Behenic acid C22:0	0.17

Table 1. *Cont.*

Fatty Acids	%
Monounsaturated fatty acids	
Palmitoleic acid C16:1	0.10
Oleic acid C18:1	21.10
Eicosanoic acid (Gondoic acid) C20:1	0.34
Erucic acid C22:1	0.76
Polyunsaturated fatty acids	
Linoleic acid C18:2	14.00
Linolenic acid C18:3	50.69
Σ SFA	12.98
Σ MUFA	22.30
Σ PUFA	64.69

3.2. Tocopherol Content

Table 2 revealed that *Opuntia ficus-indica* seed oil has a high tocopherol profile, which consists of α-, γ- and β-tocopherols. These results showed that total tocopherol was found to be 24.02 μg/mL, where β-tocopherol was found to be the main form of tocopherols in OFI oil scored 42.21%, γ-tocopherol scored 41.13%, and α-tocopherol scored 16.65%.

Table 2. Tocopherol content of *Opuntia ficus-indica* seed oil.

Compound	RT	Concentration μg/mL
A-Tocopherol	5.0	4.00
γ-Tocopherol	7.0	9.88
Β-Tocopherol	11.0	10.14
Total tocopherols	-	24.02

3.3. GC-MS Analysis of Opuntia ficus-indica Seed Oil

Totally 21 components were identified through GC-MS according to their peak area and retention time as shown in Table 3 and Figure 1. The studied oil contain different percentages of the major phytosterols, β & γ -sitosterol, Stigmasterol, Campesterol, Stigmast-5-en-3 ol, (3á,24S), which represents (1.67, 2.05,1.36, 4.15, 2.34%) for OFI oil. Furthermore, other constituents detected in the OFI oil in different percentages, such as alcohol triterpenic (9,19-Cyclolanost-24-en-3-ol,acetate, (3á)) or Cycloartenol (3.24%). Esters of fatty acid were found, such as Hexadecanoic acid, methyl ester (6.95%), 9,12,15-Octadecatrienoic acid, methyl ester, (Z,Z,Z)- (5.52%), 7,10,13-Eicosatrienoic acid, methyl ester (7.52%), 9-Octadecenoic acid, 12-hydroxy-, methyl ester, [R-(Z)]- (6.88%), 6,9,12-Octadecatrienoic acid, methyl ester (0.88%), 13-Docosenoic acid, methyl ester (15.10%), Docosanoic acid, methyl ester (0.89%), Eicosanoic acid, methyl ester (4.00%), Tetracosanoic acid, methyl ester (2.19%), Hexacosanoic acid, methyl ester (1.58%). Aromatic compounds (phenolics) were found, such as 1H-Purin-6-amine, (2fluorophenyl)methyl)- (4.75%), (flavonoids) such as 3′,4′,7-Trimethylquercetin (1.77%), 6,8-di-c-á-glucosylluteolin (1.37%).

Table 3. GC-MS analysis of *Opuntia ficus-indica* seed oil.

Compound Name	RT	Area %	Molecular Formula	Molecular Weight
Hexadecanoic acid, methyl ester (Palmitic acid, methyl ester)	7.59	6.95	C17H34O2	270
9,12,15-Octadecatrienoic acid, methyl ester, (Z,Z,Z)- (Linolenic acid, methyl ester)	9.13 9.65	5.52 8.97	C19H32O2	292
7,10,13-Eicosatrienoic acid, methyl ester (Methyl eicosa-7,10,13-trienoate)	11.55	7.52	C21H36O2	320
9-Octadecenoic acid, 12-hydroxy-, methyl ester, [R-(Z)]- (Methyl ricinoleate)	11.81	6.88	C19H36O3	312
Eicosanoic acid, methyl ester (Arachidic acid—methyl ester)	11.87	4.00	C21H42O2	326
Stigmast-5-en-3 ol,(3á,24S)-	12.01	2.34	C29H50O	414
6,9,12-Octadecatrienoic acid, methyl ester	12.31	0.88	C19H32O2	292
13-Docosenoic acid, methyl ester	12.74	15.10	C23H44O2	352
Tetracosanoic acid, methyl ester	14.25	2.19	C25H50O2	382
Docosanoic acid, methyl ester (Behenic acid, methyl ester)	15.53	0.89	C23H46O2	354
Hexacosanoic acid, methyl ester	15.59	1.58	C27H54O2	410
6,8-di-c-á-glucosylluteolin	16.45	1.37	C27H30O16	610
á-Sitosterol	16.76	1.67	C29H50O	414
1H-Purin-6-amine, (2fluorophenyl)methyl)-	17.49	4.75	C12H10FN5	243
Campesterol	18.04	4.15	C28H48O	400
Stigmasterol	18.30	1.36	C29H48O	412
ç-Sitosterol	19.07	2.05	C29H50O	414
9,19-Cyclolanost-24-en-3-ol, (3á)-(Cycloartenol)	19.98	3.24	C30H50O	426
3′,4′,7Trimethylquercetin	22.80	1.77	C18H16O7	344

3.4. Total Phenolic Contents

Our study revealed that OFI oil has a total phenolic content of 30.5 mg Gallic acid equivalents (GAE)/mL oil.

3.5. Scavenging Activity against DPPH

The scavenging activity of OFI oil versus free radical DPPH recorded a high inhibition rate of 89.2 % at 200 μL/mL of oil (IC_{50} value: 42.12 μg/mL) as compared to beta hydroxy butyrate BHT (IC_{50} = 57.10 μg/mL) as reference scored 80.40%.

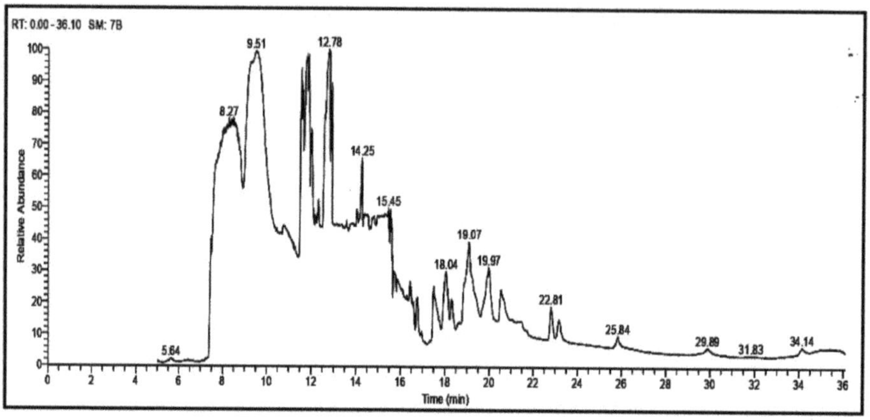

Figure 1. GC-MS analysis of *Opuntia ficus-indica* seed oil.

3.6. Antimicrobial Activity

Table 4 and Figure 2 summarize the antimicrobial potency of the OFI oil using the disc diffusion method. OFI oil was able to inhibit all tested strains with varying diameter from 9.4 to 28.4 mm, except *Aspergillus niger* which showed greater resistance to OFI oil without zone of inhibition (Table 4). The least efficacy of OFI oil against *Pen.digitatum* (fungus) was recorded by region (9.4 ± 0.5), while the most sensitive strain was *Saccharomyces cerevisiae* with a diameter (28.3 ± 0.4 mm) compared to (9.2 ± 0.6 mm) of Oxytetracycline as a positive control. Diversity of antibacterial power in OFI oil; *E. coli* as Gram-negative bacteria was more sensitive than the *S. aureus* as Gram-positive bacteria by area of inhibition (21.2 ± 0.2 mm) and (17.3 ± 0.4 mm) respectively. MIC and MBC value were determined to generate the specific dose and nature activity of the OFI oil for use as bacteriostatic or bactericidal. MIC and MBC values in the OFI oil differed according to resistance of microbial strains examined. The lowest MIC and MBC were recorded at 10, 15 μg/mL and 15, 20 μg/mL OFI oil against *Saccharomyces cerevisiae* and *E.coli* respectively.

Table 4. Measurement of inhibition zone diameter, Minimum Inhibition Concentration MIC, and Minimum Bactericidal Concentration of *Opuntia ficus-indica* seed oil.

Microbial Species	Zone of Inhibition (mm, Mean ± SEM)	MIC (μg/mL)	MBC (μg/mL)	* Oxytetracycline 30 mg	* Penicillin 10 mg
S. aureus	17.3± 0.4	20	25	ND	____
B. subtilis	14.4 ± 0.5	25	35	5.1 ± 0.5	____
E. coli	21.2 ± 0.2	15	20	17.0 ± 0.3	____
K.pneumoniae	18.3 ± 0.4	20	25	ND	____
S. cerevisiae	28.3 ± 0.4	10	15	9.2 ± 0.6	____
Asp.niger	ND	ND	ND	ND	ND
Pen.digitatum	9.4 ± 0.5	35	40	____	4.4 ± 0.5

* (OT30) and (P10) were used as positive controls: ND; Not detected.

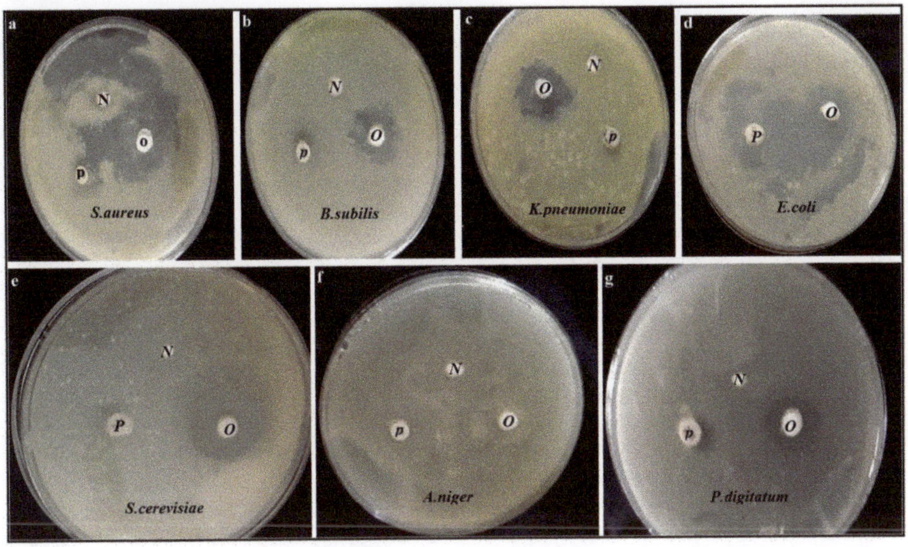

Figure 2. The antimicrobial potency of the OFI oil against tested microbial strains. OFI oil (**O**), negative control (**N**), positive control (**P**).

3.7. Cytotoxic Activity

Results revealed that OFI oil exhibited highest cell viability (99.63 ± 0.45%) at 250 µg/mL of oil against Mammalian cells from African green monkey kidney (Vero) cells. In case of A-2780 (Ovarian carcinoma) cell line, OFI oil showed weak inhibitory activity (0.72 ± 0.64%, 15.31 ± 1.25%) at 250,500 µg/mL of Oil respectively. Although in the case of PC-3 (Prostate carcinoma) cell line, OFI oil showed significant inhibitory activity (69.33 ± 2.19%, 83.06 ± 1.78%) at 250,500 µg/mL of oil respectively with IC_{50} = 110.28 ± 2.16 µg/mL. Figure 3 represented the microscopic observation of the prostate carcinoma cells (PC3) treated with different concentrations of *Opuntia ficus-indica* seed oil.

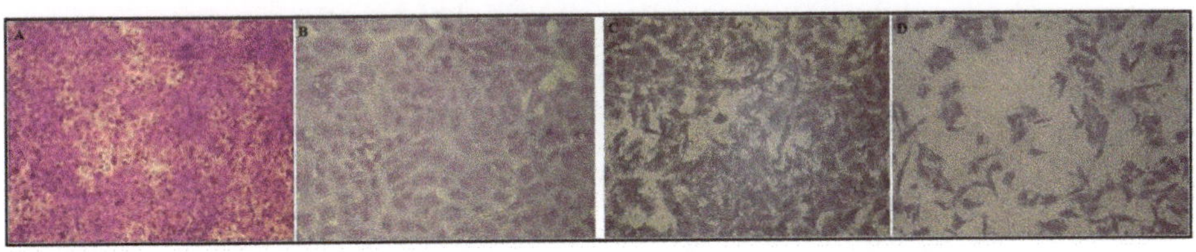

Figure 3. Microscopic observation of the cytopathic effects (morphological alterations) of the prostate carcinoma cells (PC3) treated with *Opuntia ficus-indica* seed oil concentrations (**A**) Control; (**B**) 20 µg/mL; (**C**) 100 µg/mL; (**D**) 500 µg/mL. Figure is at 100× magnification.

3.8. Antiviral Activity

Results revealed that OFI oil exhibited low viral inhibition rate (Antiviral effect %) (22.67 ± 2.79%) at 300 µg/mL of oil against herpes simplex type 2 (HSV-2) virus as compared with Acyclovir (93.70 ± 1.19%) as a positive control at 20 µg/mL of it.

4. Discussion

The fatty acid composition study of the seed oil of *O. ficus-indica* has shown that this oil belongs to the class of "polyunsaturated" oils [12]. Our results are in agreement with those published by [32] stated that linolenic acid represents the major component of fatty acids in the oil of total lipids of *O. dillenii*, and also the study by [33] where they reported that

O. ficus-indica oil contains 20.19% of oleic acid, 12.24% of palmitic acid and 3.69% of stearic acid. Also, other fatty acids were identified in minimal quantities in this oil: palmitoleic, myristic, arachidic and behenic. On the contrary, [33] reported that linoleic acid represents the main fatty acid. Furthermore; the fatty acid composition of cactus grown in various regions is significantly different. It is known that this composition is strongly affected by the climatic factors and the type of soil in which it was grown [34].

The high level of total tocopherols is the peculiarity of cactus seed oils [35–37]. Our data are concordant with that previously published by [35,37,38] reported that *O. ficus-indica* oil is very rich in tocopherols, generally β-tocopherol, γ-tocopherol, α-tocopherol. Tocopherols, also called Vitamin E, are an important family of lipophilic compounds which have antioxidant activity where the interest is determining the tocopherols composition in *O. ficus-indica* seed oil. Our results concluded that the O. ficus-indica oil is rich in tocopherols (24.02 μg/mL). Thus, high content of vitamin E, seen in oils, may contribute to significant oxidative stability [39].

Our study revealed that the *O. ficus-indica* seed oil contains phytosterols, esterified fatty acids and organic acids with varied percentage. Our data are concordant with that previously published by [3,11,37–40], which stated that *O. ficus-indica* seed oil is the richest in sterol constituent, compared to oils from other Opuntia species. The next main sterol component of Opuntia oil was campesterol, which is effective for the inhibition of proinflammatory cytokines [41] and inducing cell cycle arrest and prostaglandin release in response to the increased ROS level [42].

The radical scavenging activity of DPPH is one of the accreditation methods for investigating the antioxidant activity of plant extracts [43]. The scavenging activity against free radical DPPH of OFI oil scored a high inhibition rate of 89.2 % at 200 μL/mL of oil as compared to butylated hydroxytoluene (BHT) as reference scored 80.40%. While the total phenolic content was 30.5 mg gallic acid equivalents (GAE)/mL oil, this was in the same line with [44]. The scavenging activity and phenolic components increased by increasing the concentration of OFI oil concentration. The higher antioxidant activity observed in OFI may be relative to higher levels of phenolic compounds and other tocopherols, and sterols present in it. The effect of relationship of phenolic composition in the antioxidant capacity is a well-known fact [45]. The antioxidant potential of OFI oil attributed to its bioactive compounds such as flavonoids, polyphenols, chlorophylls, carotenoids, and tocopherols against the harmful effects of free radicals that cause pathophysiological condition such as diabetes, cardiovascular diseases, and degenerative disorders such as dementia and Parkinson's disease [46–48]. The study of [49] showed that phenolic compounds play a role in extending the food's shelf-life and act as antioxidants in many biological systems. Ref [50] reported that, in vitro, the inhibition of lipid peroxidation attributed to their ability to isolate free radicals and act as metal chelators, which increased by increasing concentration of OFI oil. In this aspect, [51] concluded that there was a significant relationship between phenolic content and DPPH root scavenging in all examined leafy vegetables ($r = 0.993$, $p < 0.5$). They have shown high efficacy in free radical scavengers due to their redox properties, which can play an essential role in the uptake and neutralization of free radicals, and the quenching of single and triple oxygen or decomposition peroxides.

Phytosterols play key roles in many areas such as nutrition (anticancer properties), medicine (therapeutic production steroids), and cosmetics (creams, lipstick). Furthermore, they have been suggested to have anti-inflammatory, antibacterial, antifungal, antioxidant, anti-ulcerative and antitumor activities [52–54]. Moreover, our study revealed that the OFI oil contain esterified fatty acids, alcohol triterpenic (Cycloartenol), Propanoic acid, 2-(3-acetoxy-4,4,14-trimethylandrost-8-en-17-yl),6,8-di-c-á-glucosylluteolin, 3′,4′,7Trimethylquercetin, Hexadecanoic acid, methyl ester, 9,12,15-Octadecatrienoic acid, methyl ester, (Z,Z,Z)-which have antioxidant, anti-inflammatory, antibacterial and antifungal, anticancer, hemolytic and 5-alphareductase inhibitor cancer enzyme inhibitors in pharmaceutical, cosmetics, and food industry actions as reported by other studies [55–60].

From the result of this paper, *Saccharomyces cerevisiae* showed a high sensitivity to OFI oil. These data are similar to the results observed by [61] that *S.cerevisiae* had area of inhibition (38–40 mm), while *Candida albicans (yeast)* exhibited smaller area of inhibition against oil red: *Opuntia ficus- indica.* Cactus pear seed oil is rich in compounds that lead to have antimicrobial activity; many researchers have come to the similar results [62]. The antimicrobial effect of OFI oil was varied may be due to a variable chemical component of the oil [63]. The present data revealed OFI oil had antibacterial activity against gram negative bacteria that is inconsistent with [61], who demonstrated that antimicrobial effect of OFI oil against Salmonella Typhi as a Gram-negative bacterium was not detected. This paper showed that OFI oil is more effective against Gram-negative bacteria than Gram-positive bacteria, and is incompatible with [64], and the opposite result may be related to less permeable outer membrane [65]. These results indicated that the effect of OFI oil on *Escherichia coli* was in the same line with [64]. They concluded that *Pseudomonas aeruginosa* was more resistant than *Escherichia coli*, due to its outer membrane rich with lipopolysaccharides, which makes it less permeable [66]. Another reason for resistance exclusion systems, is the pumps that extrude antimicrobial compounds from inside the cell before they cause infection [67]. The present study showed that OFI oil had the lowest bacteriostatic concentration (10 μg/mL (MIC)) and bactericidal concentration (15 μg/mL (MBC) on S. cerevisiae). These data are similar to the results observed by [68] that the OFI oil exerts both a bactericidal and bacteriostatic effects against Enterobacter cloacae.

Our study revealed that OFI oil showed weak inhibitory activity against A-2780 (Ovarian carcinoma) cell line, although, in the case of PC-3 (Prostate carcinoma) cell line, OFI oil showed significant inhibitory activity. These data are similar to the results of [69], who investigated the in vitro chemoprevention effect of prickly pear seed oil at various concentrations (0.01, 0.1, 1, 10, 100 M) versus the growth of HepG2 and Colo-205 cells. On contrary, ref [70] stated that the prickly pear seed extracts taken from different cultivars of prickly pear showed no toxicity to colon, prostate or breast cancer cells in the concentration range of 0.2–0.16 g/mL estimated by the MTT assay. Many in vitro studies have shown that PUFAs have growth-inhibitory and pro-apoptotic effects on various kinds of cancer cell lines [71]. For this reason, the inhibitory effect of OFI oil is due to the high content of PUFAs (linolenic acid and linoleic acid), which are compounds known for their anticancer effect in cancer cells. The α-tocopherol is a predominant component of *Opuntia ficus-indica* oils accounting for 56 mg/kg in oil. The potential health effects of α-tocopherol are powerful as antioxidant effects and the active form of vitamin E that protects the body from cardiovascular and cancer disease. As γ-tocopherol is more powerful than α-tocopherol in preventing prostate cancer cell growth, reducing oxidative DNA damage, scavenging complex and mutagenic and nitrifying oxidative stress [34,72–74]. These data are similar to the results of [75], who concluded that OFI oil may have anti-cancer therapeutic effects against colon cancer and adenocarcinoma cell lines. This effect could be elucidated by inducing programmed cell death (apoptosis). OFI seed oil is rich with unsaturated fatty acids (USFA) such as oleic acid (omega-9) plus to β-sitosterol, which led to reducing prostaglandin concentrations (PGE2), and myeloperoxidase activity (MPO) in the inflamed tissues. It has an anti-inflammatory effect [76]. The interest in plant materials containing phenolic compounds is increasing due to their high antioxidant efficacy, which may provide protection against cancer by inhibiting oxidative damage, known to be a possible cause of mutation [77].

Our results revealed that OFI oil exhibited low viral inhibition rate against herpes simplex type 2 (HSV-2) virus in the same line of the only report on antiviral activity of the Opuntia genus is by [78], who found an antiviral effect of the crude extract of *Opuntia streptacantha Lem.* against some viruses of human, horses and mice in cell culture.

5. Conclusions

Our data concluded that the cold pressing oil of *Opuntia ficus-indica*, produces interesting bioactive compounds such as fatty acids, tocopherols, sterols, flavonoids and polyphenols, as well as its main biological potentials such as, antioxidant, antiviral, antifungal, antibacterial and anticancer potentials, making it a promising candidate for the application in pharmacology and cosmetics industry.

Author Contributions: Conceptualization, H.H.; methodology, H.H., L.M.A.M., A.A.Z. and A.S.A.; validation, L.M.A.M., H.H., A.S.A. and A.A.Z.; formal analysis, H.H. and A.A.Z.; investigation, H.H., L.M.A.M., A.A.Z. and A.S.A.; resources, H.H. and L.M.A.M.; data curation, A.A.Z. and A.S.A.; writing—original draft preparation, H.H., L.M.A.M., A.A.Z. and A.S.A.; writing—review and editing, H.H. and A.A.Z. All authors have read and agreed to the published version of the manuscript.

Funding: This research received no external funding.

Institutional Review Board Statement: Not applicable.

Informed Consent Statement: Not applicable.

Data Availability Statement: Not applicable.

Conflicts of Interest: The authors declare no conflict of interest.

Sample Availability: Samples of the compounds are available from the authors.

References

1. Abdallah, H.M.; Almowallad, F.M.; Esmat, A.; Shehata, I.A.; Abdel-Sattar, E.A. Anti-inflammatory activity of flavonoids from *Chrozophora tinctoria*. *Phytochem. Lett.* **2015**, *13*, 74–80. [CrossRef]
2. Trombetta, D.; Puglia, C.; Perri, D.; Licata, A.; Pergolizzi, S.; Lauriano, E.R.; De Pasquale, A.; Saija, A.; Bonina, F.P. Effect of polysaccharides from *Opuntia ficus-indica* (L.) cladodes on the healing of dermal wounds in the rat. *Phytomedicine* **2006**, *13*, 352–357. [CrossRef]
3. El Mannoubi, I.; Barrek, S.; Skanji, T.; Casabianca, H.; Zarrouk, H. Characterization of *Opuntia ficus indica* seed oil from Tunisia. *Chem. Nat. Compd.* **2009**, *45*, 616–620. [CrossRef]
4. Collenette, S. *Checklist of Botanical Species in Saudi Arabia*; International Asclepiad Society: West Sussex, UK, 1998; p. 78.
5. Aati, H.; El-Gamal, A.; Shaheen, H.; Kayser, O. Traditional use of ethnomedicinal native plants in the Kingdom of Saudi Arabia. *J. Ethnobiol. Ethnomed.* **2019**, *15*, 2. [CrossRef] [PubMed]
6. Mulas, M.; Mulas, G. *Potentialités d'Utilisation Stratégique des Plantes des Genres Atriplex et Opuntia dans la Lutte Contre la Désertification*; Université des Études de Sassari: Sassari, Italia, 2004.
7. Abdel-Hameed, E.S.S.; Nagaty, M.A.; Salman, M.S.; Bazaid, S.A. Phytochemicals, nutritional's and antioxidant properties of two prickly pear cactus cultivars (*Opuntia ficus indica* Mill.) growing in Taif, KSA. *Food Chem.* **2014**, *160*, 31–38. [CrossRef] [PubMed]
8. Leo, M.; Abreu, M.B.; Pawlowska, A.M.; Cioni, P.L.; Braca, A. Profiling the chemical content of *Opuntia ficus-indica* flowers by HPLC–PDAESI-MS and GC/EIMS analyses. *Phytochem. Lett.* **2010**, *3*, 48–52. [CrossRef]
9. Habibi, Y. Contribution à l'Étude Morphologique, Ultrastructurale et Chimique de la Figue de Barbarie. Les Polysaccharides Pariétaux: Charactérisation et Modification Chimique. Ph.D. Thesis, Autre. Université Joseph-Fourier-Grenoble I, Saint-Martin-d'Hères, France, 2004.
10. Inglese, P. *Ecologie, Culture Etutilisations du Figuier De Barbarie*; Food and Agriculture Organization of the United Nations (FAO): Rome, Italy, 2019.
11. Ghazi, Z.; Ramdani, M.; Fauconnier, M.L.; El Mahi, B.; Cheikh, R. Fatty acids sterols and vitamin E composition of seed oil of *Opuntia ficus-indica* and *Opuntia dillenii* from Morocco. *J. Mater. Environ. Sci.* **2013**, *4*, 967–972.
12. Simopoulos, A.P. The importance of the ratio of omega-6/omega-3 essential fatty acids. *Biomed. Pharmacother.* **2002**, *56*, 365–379. [CrossRef]
13. Martínez, M.; Penci, M.C.; Marin, M.A.; Ribotta, P.D.; Maestri, D.M. Screw press extraction of almond (*Prunus dulcis* (Miller) D.A. Webb): Oil recovery and oxidative stability. *J. Food Eng.* **2013**, *119*, 40–45. [CrossRef]
14. Thanonkaew, A.; Wongyai, S.; McClements, D.; Decker, E.A. Effect of stabilization of rice bran by domestic heating on mechanical extraction yield, quality, and antioxidant properties of cold-pressed rice bran oil (*Oryza saltiva* L.). *LWT Food Sci. Technol.* **2012**, *48*, 231–236. [CrossRef]
15. Teh, S.-S.; Birch, J. Physico-chemical and quality characteristics of cold-pressed hemp 2012, flax and canola seed oils. *J. Food Compos. Anal.* **2013**, *30*, 26–31. [CrossRef]
16. AOAC. *Official Methods of Analysis*, 15th ed.; Association of Official Analytical Chemists: Washington, DC, USA, 1997; p. 684.
17. Piironen, V.; Toivo, J.; Lampi, A.-M. Plant sterols in cereals and cereal products. *Cereal Chem.* **2002**, *79*, 148–154. [CrossRef]

18. Mattila, P.; Lampi, A.-M.; Ronkainen, R.; Toivo, J.; Piironen, V. Sterol and vitamin D2 contents in some wild and cultivated mushrooms. *Food Chem.* **2002**, *76*, 293–298. [CrossRef]
19. Kamal-Eldin, A.; Appelqvist, L.; Yosif, G.; Iskander, G.M. Seed lipids of *Sesamum indicum* and related wild species in Sudan. The sterols. *J. Sci. Food Agric.* **1992**, *59*, 327–334. [CrossRef]
20. Liu, Q.; Scheller, K.K.; Schaefer, D.M. Technical note. A simplified procedure for vitamin E determination in beef muscle. *J. Anim. Sci.* **1996**, *74*, 2406–2410. [CrossRef]
21. Singleton, V.L.; Orthofer, R.; Lamuela-Raventos, R.M. Analysis of total phenols and other oxidation substrates and antioxidants by means of Folin–Ciocalteau reagent. *Method. Enzymol.* **1999**, *299*, 152–178.
22. Abu-Zaid, A.A.; Al-Barty, A.; Morsy, K.; Hamdi, H. In vitro study of antimicrobial activity of some plant seeds against bacterial strains causing food poisoning diseases. *Braz. J. Biol.* **2021**, *82*, e256409. [CrossRef]
23. Somchit, M.N.; Reezal, I.; Nur, I.E.; Mutalib, A.R. In vitro antimicrobial activity of ethanol and water extracts of *Cassia alata*. *J. Ethnopharmacol.* **2003**, *84*, 1–4. [CrossRef]
24. Morales, F.J.; Jiménez-Pérez, S. Free radical scavenging capacity of Maillard reaction products as related to colour and fluorescence. *Food Chem.* **2001**, *72*, 119–125. [CrossRef]
25. Tagg, J.R.; McGiven, A.R. Assay system for bacteriocins. *Appl. Microbiol.* **1971**, *21*, 943. [CrossRef]
26. Vijayan, P.; Raghu, C.; Ashok, G.; Dhanaraj, S.A.; Suresh, B. Antiviral activity of medicinal plants of Nilgiris. *Indian J. Med. Res.* **2004**, *120*, 24–29. [PubMed]
27. Mosmann, T. Rapid colorimetric assay for cellular growth and survival: Application to proliferation and cytotoxicity assays. *J. Immunol. Methods* **1983**, *65*, 55–63. [CrossRef]
28. Randazzo, W.; Piqueras, J.; Rodrıguez-Diaz, J.; Aznar, R.; Sanchez, G. Improving efficiency of viability-qPCR for selective detection of infectious HAV in food and water samples. *J. Appl. Microbiol.* **2017**, *124*, 958–964. [CrossRef] [PubMed]
29. Pinto, R.M.; Diez, J.M.; Bosch, A. Use of the colonic carcinoma cell line CaCo-2 for in vivo amplification and detection of enteric viruses. *J. Med. Virol.* **1994**, *44*, 310–315. [CrossRef] [PubMed]
30. Hu, J.M.; Hsiung, G.D. Evaluation of new antiviral agents I: In vitro prospective. *Antivir. Res.* **1989**, *11*, 217–232. [CrossRef]
31. Al-Salahi, R.; Alswaidan, I.; Ghabbour, H.A.; Ezzeldin, E.; Elaasser, M.M.; Marzouk, M. Docking and antiherpetic activity of 2-aminobenzo[*de*]-isoquinoline-1,3-diones. *Molecules* **2015**, *20*, 5099–5111. [CrossRef] [PubMed]
32. Liu, W.; Fu, Y.J.; Zu, Y.G.; Tong, M.H.; Wu, N.; Liu, X.L.; Zhang, S. Supercritical carbon dioxide extraction of seed oil from *Opuntia dillenii* Haw. and its antioxidant activity. *Food Chem.* **2009**, *114*, 334–339. [CrossRef]
33. Tlili, N.; Bargougui, A.; Elfalleh, W.; Triki, S.; Nasri, N. Phenolic compounds, protein, lipid content and fatty acids compositions of cactus seeds. *J. Med. Plants Res.* **2011**, *5*, 4519.
34. Ramadan, M.F.; Mörsel, J.-T. Recovered lipids from prickly pear [*Opuntia ficus-indica* (L.) Mill] peel: A good source of polyunsaturated fatty acids 4524, natural antioxidant vitamins and sterols. *Food Chem.* **2003**, *83*, 447–456. [CrossRef]
35. Matthäus, B.; Özcan, M.M. Habitat effects on yield, fatty acid composition and tocopherol contents of prickly pear (*Opuntia ficus-indica* L.) seed oils. *Sci. Hortic.* **2011**, *131*, 95–98. [CrossRef]
36. Zine, S.; Gharby, S.; El Hadek, M. Physicochemical characterization of *Opuntia ficus-indica* seed oil from Morocco. *Biosci. Biotechnol. Res. Asia* **2013**, *10*, 1–7. [CrossRef]
37. Ramadan, M.F.; Mörsel, J.-T. Oil cactus pear (*Opuntia ficus-indica* L.). *Food Chem.* **2003**, *82*, 339–345. [CrossRef]
38. Gharby, S.; Harhar, H.; Charrouf, Z.; Bouzobaa, Z.; Boujghagh, M.; Zine, S. Physicochemical composition and oxidative stability of *Opuntia ficus-indica* seed oil from Morocco. *Acta Hortic.* **2015**, *1067*, 83–88. [CrossRef]
39. Pope, L.E.; Marcelletti, J.F.; Katz, L.R.; Lin, J.Y.; Katz, D.H.; Parish, M.L.; Spear, P.G. The anti-herpes simplex virus activity of n-docosanol includes inhibition of the viral entry process. *Antivir. Res.* **1998**, *40*, 85–94. [CrossRef]
40. El Adib, S.; Slim, S.; Hamdeni, E. Characterization and comparison of the three cultivars seed oil of *Opuntia ficus-indica* in Tunisia. *Indo Am. J. Pharm. Sci.* **2015**, *2*, 967–973.
41. Nashed, B.; Yeganeh, B.; HayGlass, K.T.; Moghadasian, M.H. Antiatherogenic effects of dietary plant sterols are associated with inhibition of proinflammatory cytokine production in Apo E-KO mice. *J. Nutr.* **2005**, *135*, 2438–2444. [CrossRef]
42. Awad, A.B.; Burr, A.T.; Fink, C. Effect of resveratrol and β-sitosterol in combination on reactive oxygen species and prostaglandin release by PC-3 cells. *Prostaglandins Leucot. Essent. Fat. Acids.* **2005**, *72*, 219–226. [CrossRef]
43. Brand-William, W.; Cuvelier, M.E.; Berset, C. Use of a free radical method to evaluate antioxidant activity. *LWT* **1995**, *28*, 25–30. [CrossRef]
44. Khémiri, I.; Bitri, L. Effectiveness of *Opuntia ficus indica* L. *inermis seed oil in the protection and the healing of experimentally induced gastric mucosa ulcer. Oxid. Med. Cell Longev.* **2019**, *2019*, 1568720. [CrossRef]
45. Lien, E.J.; Ren, S.; Bui, H.-H.B.; Wang, R. Quantitative structure-activity relationship analysis of phenolic antioxidants. *Free Radic. Biol. Med.* **1999**, *26*, 285–294. [CrossRef]
46. Sindhu, E.R.; Kuttan, R. Carotenoid lutein protects rats from gastric ulcer induced by ethanol. *J. Basic Clin. Physiol. Pharmacol.* **2012**, *23*, 33–37. [CrossRef] [PubMed]
47. Stevenson, D.E. Polyphenols as adaptogens—The real mechanism of the antioxidant effect. In *Bioactive Compounds in Phytomedicine*; InTech: Rijeka, Croatia, 2012; pp. 143–162.

48. Atwood, C.S.; Huang, X.; Moir, R.D.; Tanzi, R.E.; Bush, A.I. Role of free radicals and metal ions in the pathogenesis of Alzheimer's disease. In *Metal Ions in Biological Systems: Interrelations Between Free Radicals and Metal Ions in Life Processes*; Sigel, A., Ed.; Routledge: London, UK, 1999; Volume 36, pp. 309–364.
49. Colla, L.M.; Furlong, E.B.; Costa, J.A.V. Antioxidant properties of Spirulina (Arthospira) platensis cultivated under different temperatures and nitrogen regimes. *Braz. Arch. Biol. Technol.* **2007**, *50*, 161–167. [CrossRef]
50. Gershwin, M.E.; Belay, A. *Spirulina in Human Nutrition and Health*; CRC Press: Boca Raton, FL, USA, 2007; 328p, ISBN 9781420052572.
51. Sahu, R.K.; Kar, M.; Routray, R. DPPH free radical scavenging activity of some leafy vegetables used by tribals of Odisha, India. *J. Med. Plants Stud.* **2013**, *1*, 21–27.
52. Awad, A.B.; Downie, A.; Fink, C.S.; Kim, U. Dietary phytosterols inhibits the growth and metastasis of MDA-MB-231 human breast cancer cells grown in SCID mice. *Anticancer Res.* **2000**, *20*, 821–824. [PubMed]
53. Sudha, T.; Chidambarampillai, S.; Mohan, V.R. GC-MS analysis of bioactive components of aerial parts of *Fluggea leucopyrus* Willd. (*Euphorbiaceae*). *J. Appl. Pharm. Sci.* **2013**, *3*, 126–130.
54. Stuchlı'k, M.; Zak, S. Vegetable lipids as components of functional foods. *Biomed. Papers* **2002**, *146*, 3–10. [CrossRef]
55. Akihisa, T.; Yasukawa, K.; Yamaura, M.; Ukiya, M.; Kimura, Y.; Shimizu, N.; Arai, K. Triterpene alcohol and sterol ferulates from rice bran and their anti-inflammatory effects. *J. Agric. Food Chem.* **2000**, *48*, 2313–2319. [CrossRef]
56. Ouyang, H.; Kong, X.; He, W.; Qin, N.; He, Q.; Wang, Y.; Wang, R.; Xu, F. Effects of five heavy metals at sub-lethal concentrations on the growth and photosynthesis of *Chlorella vulgaris*. *Chin. Sci. Bull.* **2012**, *57*, 3363–3370. [CrossRef]
57. Mohamed, A.A.; Ali, S.I.; Darwesh, O.M.; El-Hallouty, S.M.; Sameeh, M.Y. Chemical compositions, potential cytotoxic and antimicrobial activities of *Nitraria retusa* methanolic extract sub-fractions. *Int. J. Toxicol. Pharmacol. Res.* **2015**, *7*, 204–212.
58. Sayik, A.; Yusufoglu, A.S.; Leyla, A.Ç.; Turer, G.; Aydin, B.; Arslan, L. DNA binding, biological activities and chemical composition of wild growing *Epilobium angustifolium* L. extracts from Canakkale, Turkey. *J. Turk. Chem. Soc. Sect.* **2017**, *4*, 811–840. [CrossRef]
59. Sosa, A.A.; Suhaila Husaein Bagi, S.H.; Hameed, I.H. Analysis of bioactive chemical compounds of *Euphorbia lathyrus* using gas chromatography-mass spectrometry and Fourier-transform infrared spectroscopy. *J. Pharmacogn. Phytother.* **2016**, *8*, 109–126.
60. Zhang, Z.L.; Luo, Z.H.; Shi, H.W.; Zhang, L.X.; Ma, X.J. Research advance of functional plant pharmaceutical cycloartenol about pharmacological and physiological activity. *China J. Chin. Mater. Med.* **2017**, *42*, 433–437.
61. Ramírez-Moreno, E.; Cariño-Cortés, R.; Cruz-Cansino, N.; Delgado-Olivares, L.; Ariza-Ortega, J.A.; Montañez-Izquierdo, V.Y.; Hernández-Herrero, M.M.; Filardo-Kerstupp, T. Antioxidant and antimicrobial properties of cactus pear (*Opuntia*) seed oils. *J. Food Qual.* **2017**, *2017*, 3075907. [CrossRef]
62. Moosazadeh, E.; Akhgar, M.R.; Kariminik, A. Chemical composition and antimicrobial activity of *Opuntia stricta* F. essential oil. *J. Biodivers. Environ. Sci.* **2014**, *4*, 94–101.
63. Burt, S. Essential oils: Their antibacterial properties and potential applications in foods—A review. *Int. J. Food Microbiol.* **2004**, *94*, 223–253. [CrossRef]
64. Ortega, M.D.L.A.O.; Cruz-Cansino, N.D.S.; Alanís-García, E.; Delgado-Olivares, L.; Ariza-Ortega, J.A.; Ramírez-Moreno, E.; Manríquez-Torres, J.J. Optimization of ultrasound extraction of cactus pear (*Opuntia ficus indica*) seed oil based on antioxidant activity and evaluation of its antimicrobial activity. *J. Food Qual.* **2017**, *2017*, 9315360. [CrossRef]
65. Zekaria, D. Los Aceites Esenciales Una Alternativa a los Antimicrobianos. Laboratorios Calier. 2007. Available online: https://www.calier.es/pdf/Microsoft_Word__Aceites_esen_como_promotores.pdf (accessed on 20 August 2022).
66. Viuda-Martos, M.; Mohamady, M.A.; Fernández-López, J.; Abd ElRazik, K.A.; Omer, E.A.; Pérez-Alvarez, J.A.; Sendra, E. In vitro antioxidant and antibacterial activities of essentials oils obtained from Egyptian aromatic plants. *Food Control* **2011**, *22*, 1715–1722. [CrossRef]
67. Gomez-Alvarez, C.A.; Leal-Castro, A.L.; Perez de Gonzalez, P.; Navarrete-Jiménez, M. L Mecanismos de resistenciaen *Pseudomonas aeruginosa*: Entendiendo a un peligroso enemigo. *Rev. Fac. Med. Univ. Nac. Colomb.* **2005**, *53*, 27–34.
68. Khémiri, I.; Hédi, B.E.; Zouaoui, N.S.; Gdara, N.B.; Bitri, L. The antimicrobial and wound healing potential of *Opuntia ficus- indica* L. *inermis* extracted oil from Tunisia. *Hindawi Evid. Based Complement. Altern. Med.* **2019**, *2019*, 9148782. [CrossRef]
69. Abdel Fattah, M.S.; Badr, S.E.A.; Khalil, E.M.; Elsaid, A.S. Feed efficiency, some blood parameters and In-vitro chemoprevention of prickly pear (*Opuntia ficus indica* L.) seeds oil growing in Egypt. *IBSPR* **2020**, *8*, 20–28.
70. Camarena-Ordonez, D.R.; Gutierrez-Uribe, J.; Hernandez-Brenes, C.; Mertens-Talcott, S. Prickly pear seed extracts with antioxidant and cancer-preventing capacity. *FASEB J.* **2010**, *24*, 928.17. [CrossRef]
71. Zhang, C.; Yu, H.; Shen, Y.; Ni, X.; Shen, S.; Das, U.N. Polyunsaturated fatty acids trigger apoptosis of colon cancer cells through a mitochondrial pathway. *Arch. Med. Sci.* **2015**, *11*, 1081–1094.
72. Hasani, N.A.; Yussof, P.A.; Khalid, B.A.K.; Ghapor, M.T.A.; Ngah, W.Z.W. The possible mechanism of action of palm oil gammatocotrienol and alpha-tocopherol on the cervical carcinoma caski cell apoptosis. *Biomed. Res. Int.* **2008**, *19*, 194–200.
73. Zingg, J.M. Modulation of signal transduction by vitamin E. *Mol. Asp. Med.* **2007**, *28*, 481–506. [CrossRef]
74. Christen, S.; Woodall, A.A.; Shigenaga, M.K.; Southwell-Keely, P.T.; Duncan, M.W.; Ames, B.N. γ-tocopherol traps mutagenic electrophiles such as NO_X and complements α-tocopherol: Physiological implications. *Proc. Natl. Acad. Sci. USA* **1997**, *94*, 3217–3222. [CrossRef]
75. Becer, E.; Kabadayı, H.; Meriçli, F.; Meriçli, A.H.; Kıvançlı, B.; Vatansever, S. Apoptotic effects of *Opuntia ficus indica* L. seed oils on colon adenocarcinoma cell lines. *Proceedings* **2018**, *2*, 1566.

76. Koshak, A.E.; Abdallah, H.M.; Esmat, A.; Rateb, M. Anti-inflammatory activity and chemical characterization of *Opuntia ficus-indica* seed oil cultivated in Saudi Arabia. *Arab. J. Sci. Eng.* **2020**, *45*, 4571–4578. [CrossRef]
77. Khomdram, S.D.; Singh, P.K. Polyphenolic compounds and free radical scavenging activity in eight *Lamiaceae* herbs of Manipur. *Not. Sci. Biol.* **2011**, *3*, 108–113. [CrossRef]
78. Ahmad, A.; Davies, J.; Randall, S.; Skinner, G.R. Antiviral properties of extract of *Opuntia streptacantha*. *Antivir. Res.* **1996**, *30*, 75–85. [CrossRef]

molecules

MDPI

Article

Assessment of Cosmetic Properties and Safety of Use of Model Washing Gels with Reishi, Maitake and Lion's Mane Extracts

Aleksandra Ziemlewska [1,*], Magdalena Wójciak [2], Kamila Mroziak-Lal [1], Martyna Zagórska-Dziok [1], Tomasz Bujak [1], Zofia Nizioł-Łukaszewska [1], Dariusz Szczepanek [3] and Ireneusz Sowa [2]

[1] Department of Technology of Cosmetic and Pharmaceutical Products, Medical College, University of Information Technology and Management in Rzeszow, Sucharskiego 2, 35-225 Rzeszow, Poland
[2] Department of Analytical Chemistry, Medical University of Lublin, AlejeRaclawickie 1, 20-059 Lublin, Poland
[3] Chair and Department of Neurosurgery and Paediatric Neurosurgery, Medical University of Lublin, 20-090 Lublin, Poland
* Correspondence: aziemlewska@wsiz.edu.pl

Abstract: Natural cosmetics are becoming more and more popular every day. For this reason, this work investigates the properties of mushroom extracts, which are not as widely used in the cosmetics industry as plant ingredients. Water extracts of *Grifolafrondosa* (Maitake), *Hericiumerinaceus* (Lion's Mane) and *Ganoderma lucidum* (Reishi) were tested for their antioxidant properties, bioactive substances content, skin cell toxicity, ability to limit TEWL, effect on skin hydration and pH, and skin irritation. Our research showed that Maitake extract contained the highest amount of flavonoids and phenols, and also showed the most effective scavenging of DPPH and ABTS radicals as well as Chelation of Fe^{2+} and FRAP radicals, which were 39.84% and 82.12% in a concentration of 1000 µg/mL, respectively. All tested extracts did not increase the amount of ROS in fibroblasts and keratinocytes. The addition of mushroom extracts to washing gels reduced the irritating effect on skin, and reduced the intracellular production of free radicals, compared with the cosmetic base. Moreover, it was shown that the analyzedcosmetics had a positive effect on the pH and hydration of the skin, and reduced TEWL.

Keywords: mushroom extracts; washing gel; bioactive compounds; skin cells; irritant potential

Citation: Ziemlewska, A.; Wójciak, M.; Mroziak-Lal, K.; Zagórska-Dziok, M.; Bujak, T.; Nizioł-Łukaszewska, Z.; Szczepanek, D.; Sowa, I. Assessment of Cosmetic Properties and Safety of Use of Model Washing Gels with Reishi, Maitake and Lion's Mane Extracts. *Molecules* **2022**, *27*, 5090. https://doi.org/10.3390/molecules27165090

Academic Editors: Arunaksharan Narayanankutty, Ademola C. Famurewa and Eliza Oprea

Received: 21 June 2022
Accepted: 5 August 2022
Published: 10 August 2022

Publisher's Note: MDPI stays neutral with regard to jurisdictional claims in published maps and institutional affiliations.

Copyright: © 2022 by the authors. Licensee MDPI, Basel, Switzerland. This article is an open access article distributed under the terms and conditions of the Creative Commons Attribution (CC BY) license (https://creativecommons.org/licenses/by/4.0/).

1. Introduction

Currently, natural cosmetics are becoming more and more popular. They contain compounds mainly derived from plants, instead of synthetic substances which are often harmful [1,2]. The adverse health effects of these substances are primarily associated with their allergenic and irritating effects, as well as the harmful effects of heavy metals on the entire body. Their presence in cosmetic preparations often leads to atopic dermatitis, erythema, rashes and other allergic reactions, which is often the result of the additive effect associated with the presence of the same ingredient in many cosmetic products [3–5]. Currently, there are many studies describing various plant species that have a beneficial effect on skin and hair, and, therefore, are used in the production of cosmetics. However, it is not only plants that have this effect. Mushrooms also contain many bioactive compounds such as flavonoids, phenolic compounds, terpenes, polysaccharides, and fatty acids [6,7]. Thanks to their presence, mushroom extracts show anti-cancer, anti-inflammatory, antioxidant and antimicrobial properties [6,8–13]. Both fruiting bodies and mycelia have such features, and due to them, are used in cosmetology [6,14,15]. Due to their antioxidant properties and ability to absorb UV radiation, they are added to anti-aging cosmetics [15]. Mushroom extracts show a moisturizing and lightening effect on the skin, which also makes them useful in the cosmetics industry [15]. Mushrooms withanti-inflammatory and antimicrobial effects can be found in preparations for sensitive and problematic skin [14,15], however, cosmetics based on mushroom extracts are still not as popular as, for example, those containing compounds of plant origin. This is the reason for further research in this direction.

In these studies, we tested water extracts of three mushroom species: *Grifolafrondosa*, *Hericiumerinaceus* and *Ganoderma lucidum*. *G. frondosa*, also known as Maitake, is commonly used in Japan and China in herbal medicine [16]. It belongs to the Basidiomycota division and Polyporales family. The fruiting bodies of these fungi are bush-branched, composed of many wavy caps growing from a common trunk. The edges of the caps are uneven, cut, and wrinkled [17]. It has been proven that Maitake mushrooms have antitumor, antioxidant, and anti-diabetic activities. These properties are due to the high content of polysaccharides, as well as the presence of other compounds such as phenolic compounds [16]. *Hericiumerinaceus*, also known as Lion's Mane, is a commonly known mushroom in Japan and China, used in alternative medicines and as a food supplement. It be-longs to the family Hericiaceae, order Russulales, and class Agaricomycetes. Due to the presence of polysaccharides, phenolic compounds and other biologically active substances, *H. erinaceus* shows antimicrobial, antioxidant, and anti-aging activity. Additionally, it has been shown that the water extract of *H. erinaceus* promotes wound healing [18,19]. *Ganoderma lucidum*, also called Reishi, is a species belonging to the class Agaricomycetes and Ganodermataceae family. In Korea, China, and Japan, it is a well-known medicinal mushroom which exhibits antioxidant, antiperoxidative, antibacterial, antitumor, antimutagenic and anti-inflammatory activity. For this reason, it is used as a food additive, in pharmaceutical products, and also as an ingredient in cosmetics [20–22].

Due to the high concentration of polysaccharides in the dry matter of mushrooms, mushroom extracts may be a valuable raw material for cleansing cosmetics. These groups of cosmetic products are used by consumers in high amounts, many times a day. The biggest disadvantage of cleansing cosmetics is the potential to cause skin irritation, which is induced by surfactants—the main ingredients of this group of cosmetics [23–25]. Surfactants have the ability to interact with skin proteins, leading to their denaturation and washing out of the skin. They are the main cause of skin irritations appearing after using cosmetics such as shower gels, shampoos, and hand soap liquids [23,24,26–30]. The force of irritant potential depends on the type of surfactants used in cosmetic formulations. Anionic surfactants have the strongest ability to induce an irritant effect because of their strong, electrostatic interaction with skin proteins. The irritant effect also depends on the form of surfactant molecules in the bulk phase of the products. The strongest irritant properties have been proven for surfactant monomers that have been identified, in the highest concentration (below critical micelle concentration (CMC)). Irritant potential is lower above CMC because of the formation of micelle aggregates. Monomers, as individual molecules of surfactants, are characterized by a smaller size than micelles, may penetrate deeper into the skin, and the probability of their interaction with skin proteins is higher. After micelles' formation, the concentration of monomers in the solutions of surfactants is very low, and they arise as the result of the disintegration of thermodynamically unstable micelles [23,25,30,31]. There are many methods to decrease the irritant potential of surfactants. One of the most popular methods, and used in industrial practice, is the use of mixtures of various types of surfactants (for example, anionic, nonionic and amphoteric), which increases the stability of micelles and reduces the amount of monomers released from micelles to the bulk phase of a surfactant's solution. The second methodis the use of high molecular weight substances such as natural and synthetically derived polymers. Such substances create so-called "necklace structures" with micelles (micelles connected with the active centers of polymer chains) which leads to anincrease in the micellestability [31–34]. Mushroom polysaccharides, proteins and carbohydrates, as natural polymers, may interact with surfactant micelles and cause an increase in their stability and, in fact, lead to a reduction in irritant potential.

In order to investigate the properties of the water extracts of mushrooms, a cytotoxicity test was performed on skin cells. Furthermore, the content of flavonoids and phenolic compounds was determined, and the antioxidant properties of the examined extracts were examined. Obtained extracts were used as the active ingredient of model body wash gels, in order to increase their safety of use. For the obtained cosmetics containing the analyzed extracts and the reference sample (without the addition of the extracts), their irritant potential, effect on skin hydration and pH, as well as transepidermal water loss from the epidermis (TEWL), were determined.

2. Results and Discussion

2.1. Determination of Biologically Active Compounds

2.1.1. Determination of Bioactive Compounds by HPLC-ESI/TOF

The HPLC-ESI/TOF method was employed to investigate the chemical profiles of the Lion's Mane, Reishi, and Maitake extracts. As can be seen in Figure 1, the extracts showed diverse phytochemical characteristics. The chromatogram of the Lion's Mane extract shows many peaks at retention times from 0 to 4 min, which correspond to highly polar components; moreover, there are many peaks in the range of 53–70 min with one characteristic peak at 53.4 min (not identified, m/z 526.21). In the case of the Reishi extract, many peaks can be observed at retention times from 54 to 72 min. In turn, in the case of the Maitake extract, the number of registered peaks aresignificantly lower; the majority of peaks can beobserved in the region 0–4 min and some minor peaks are visible at 61–65 min. These observations demonstrate considerable differences in the composition of each extract.

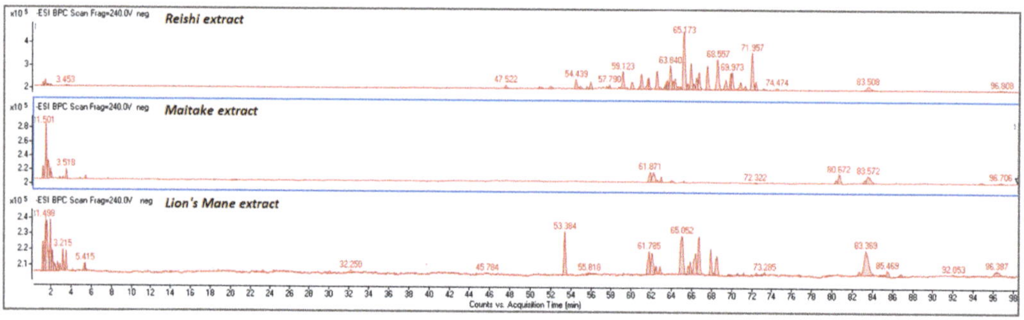

Figure 1. Representative BIC chromatogram for Reishi, Maitake and Lion's Mane extracts.

Molecular weight, elution order and mass spectra were taken from literature data. Detailed results are shown in Tables 1–3. In this study, chemical compounds were identified in Lion's Mane extract, including organic acids, aromatic compounds (hericerins) and diterpenoids (erinacines) (Table 1). Most of the compounds in Lion's Mane exhibit good biological activity. Erinacines are able topromote the synthesis of nerve growth factors, thus have neuroprotective properties [35,36]. Phenolics compounds, as well as terpenoid lactones such as hericenone C, hericenone D, hericenone E, and hericenone H, can also promote the synthesis of nerve growth factors [37–39]. The anticancer properties of some of these compounds are also mentioned [40]. Table 2 shows the structure of the triterpenoids identified in the Reishi extract, most of which were ganoderic and lucidenic acids. The triterpenoids were extracted for HPLC-MS analysis at negative mode, which gave [M-H]- and [2 M-H]-ions. Furthermore, all tested mushroom extracts contained organic acids such as malic, fumaric, citric, pyroglutamic and (15z)-9,12,13-Trihydroxy-15-octadecenoic acid.

Table 1. Results of compound identification in Lion's Mane extract using HPLC-ESI/TOF.

Rt (min.)	*m/z*-H	Error (ppm)	Molecular Formula	Compound	Identification
1.90	133.0142	−0.35	$C_4H_6O_5$	Malic acid	[41]
1.96	479.1275	3.93	$C_{15}H_{28}O_{17}$	Unknown	
2.09	115.0042	4.46	$C_4H_4O_4$	Fumaric acid	[41]
2.15	523.0952	2.15	$C_{19}H_{24}O_{17}$	Unknown	
3.32	128.0350	−2.45	$C_5H_7NO_3$	Pyroglutamic acid	[41]
5.41	474.1567	−4.88	$C_{17}H_{31}O_{15}$	Unknown	
32.25	209.0445	−4.98	$C_{10}H_{10}O_5$	Unknown	
53.38	526.1752	0.26	$C_{17}H_{35}O_{18}$	Unknown	
61.78	329.2340	1.98	$C_{18}H_{34}O_5$	(15z)-9,12,13-Trihydroxy-15-octadecenoic acid	[41]
62.05	329.1389	−1.66	$C_{19}H_{22}O_5$	Hericenone A	[42]
62.50	329.2331	−0.75	$C_{18}H_{34}O_5$	Octadecenoic acid derivatives	[41]
62.90	329.2342	2.58	$C_{18}H_{34}O_5$	Octadecenoic acid derivatives	[41]
65.05	431.2445	1.36	$C_{25}H_{36}O_6$	Erinacine A	[43]
67.70	473.1812	−1.07	$C_{25}H_{30}O_9$	Unknown	
66.39	473.1803	−2.97	$C_{25}H_{30}O_9$	Unknown	
66.77	477.2868	2.14	$C_{27}H_{42}O_7$	Erinacine D	[42]
67.65	432.2188	1.77	$C_{27}H_{31}NO_4$	Hericenone B	[42]
68.57	473.1821	0.83	$C_{25}H_{30}O_9$	Unknown	
83.37	265.1471	−3.02	$C_{12}H_{26}O_4S$	Dodecyl sulfate	[41]
96.725	309.2790	−2.91	$C_{20}H_{38}O_2$	Ethyl oleate	[41]

Table 2. Results of compound identification in Reishi extract using HPLC-ESI/TOF.

Rt (min.)	*m/z*-H	Error (ppm)	Molecular Formula	Compound	Identification
1.95	133.0148	4.13	$C_4H_6O_5$	Malic acid	[44]
3.17	128.0355	1.42	$C_5H_7NO_3$	Pyroglutamic acid	[44]
47.52	533.3141	3.95	$C_{30}H_{46}O_8$	Unknown	
54.39	531.2972	1.63	$C_{30}H_{44}O_8$	Ganoderic acid G	[45]
55.92	531.2969	1.05	$C_{30}H_{44}O_8$	Ganoderic acid G	[45]
57.240	529.2821	2.66	$C_{30}H_{42}O_8$	12-Hydroxy-Ganoderic acid D	[45]
57.906	515.3029	2.85	$C_{30}H_{44}O_7$	Ganoderic acid A/B	[45]
59.223	517.2815	1.56	$C_{29}H_{42}O_8$	Lucidenic acid P	[46]
60.96	513.2856	−0.34	$C_{30}H_{42}O_7$	Unknown	
61.69	513.2867	1.79	$C_{30}H_{42}O_7$	Unknown	
62.49	515.3030	3.05	$C_{30}H_{44}O_7$	Ganoderic acid A/B	[45]
63.84	513.2862	0.82	$C_{30}H_{42}O_7$	Unknown	
65.140	515.3006	−1.6	$C_{30}H_{44}O_7$	Ganoderic acid A/B	[45]
65.87	571.2921	1.47	$C_{32}H_{44}O_9$	Ganoderic acid K	[45]
66.72	457.2654	−0.08	$C_{20}H_{42}O_{11}$	Unknown	
67.657	511.2715	2.68	$C_{30}H_{40}O_7$	Ganoderic acid D	[45]
68.56	513.2868	1.99	$C_{30}H_{42}O_7$	Ganoderic acid AM1	[45]
69.97	511.2720	3.66	$C_{30}H_{40}O_7$	Ganoderic acid D	[45]
71.967	569.2768	2.09	$C_{32}H_{42}O_9$	12-Acetoxy-ganoderic acid F	[45]
83.647	265.1483	1.49	$C_{12}H_{26}O_4S$	Dodecyl sulfate	[41]
96.725	309.2810	3.53	$C_{20}H_{38}O_2$	Ethyl oleate	[41]

Table 3. Results of compound identification in Maitake extract using HPLC-ESI/TOF.

Rt (min.)	*m/z*-H	Error (ppm)	Molecular Formula	Compound	Identification
1.84	133.0144	1.14	$C_4H_6O_5$	Malic acid	[41]
2.11	115.0039	1.88	$C_4H_4O_4$	Fumaric acid	[41]
2.34	191.0194	−1.7	$C_6H_8O_7$	Citric acid	[41]
3.518	382.128	2.79	$C_{18}H_{23}O_9$	Unknown	
61.870	329.2349	4.70	$C_{18}H_{34}O_5$	(15z)-9,12,13-Trihydroxy-15-octadecenoic acid	[41]
62.17	329.2340	1.98	$C_{18}H_{34}O_5$	Octadecenoic acid derivatives	[41]
80.67	313.2400	4.99	$C_{18}H_{34}O_4$	Unknown	
83.647	265.1466	−4.9	$C_{12}H_{26}O_4S$	Dodecyl sulfate	[41]
96.725	309.2792	−2.27	$C_{20}H_{38}O_2$	Ethyl oleate	[41]

2.1.2. Total Phenolic Compounds and Flavonoids Content

Phenolic compounds and flavonoids are bioactive substances known for their antioxidant properties. In order to determine total phenolic content in water extracts of the mushrooms, the spectrophotometric method with the use of the Folin&Ciocalteu reagent was used. In order to determine the total flavonoids content, a spectrophotometric method with the use of the aluminum nitrate nonahydrate, was carried out.

The obtained results are presented in Table 4. For all investigated mushroom extracts, the total content of phenolic compounds was higher than the total content of flavonoids. The highest content of phenolic compounds was found in Maitake (183.75 ± 0.21 µg GAE/g DW) which was about 13 times higher than the lowest content, which was obtained by Reishi (13.23 ± 0.07 µg GAE/g DW). The situation was similar in the case of the content of flavonoids. The highest value determined was for Maitake extract (38.38 ± 0.07 µg QE/g DW) and the lowest for Reishiextract (3.43 ± 0.03 µg QE/g DW). Although no polyphenolic and flavonoid compounds were found in the chromatographic analysis, the obtained results can be explained by the presence of non-phenolic components able to react with the aforementioned reagents. Reducing sugars are an example of such constituents, so total carbohydrates were determined in our study. The determined carbohydrate contents were 0.121 ± 0.011 mg/mL, 0.054 ± 0.003 mg/mL, and 0.015 ± 0.002 mg/mL for Maitake, Lion's Mane, and Reishi, respectively. It can, therefore, be assumed that carbohydrates wereresponsible for the positive reaction with the Folin&Ciocalteu reagent.

Table 4. Total phenolic and flavonoid content of Maitake, Lion's Mane and Reishi water extracts. Values are the mean of three replicate determinations ± SD.

Maitake	Lion's Mane	Reishi
	Total phenolic compounds content [µg GAE/g DW]	
183.75 ± 0.21	59.70 ± 0.14	13.23 ± 0.07
	Total flavonoids content [µg QE/g DW]	
38.38 ± 0.07	13.68 ± 0.21	3.43 ± 0.03

2.2. Determination of Antioxidant Properties

In order to determine the antioxidant properties of the tested extracts, a DPPH radical scavenging assay was performed. The obtained results, for the samples in a concentration of 1000 µg/mL, are presented in Figure 2. The most effective scavenging of the DPPH radical was noted for the Maitake extract, which reached the highest value after 20 min of incubation (39.84%). The Reishiextract obtained the least efficient scavenging of the DPPH radical and its value did not change during the 20 min of the measurement. At the beginning of the measurement, it was 6.47%, while after 20 min it was only 9.29%. The situation wassimilar in the case of the Lion's Mane extract. At the beginning of the measurement, it was 7.11%, and after 20 min it was 12.29%.

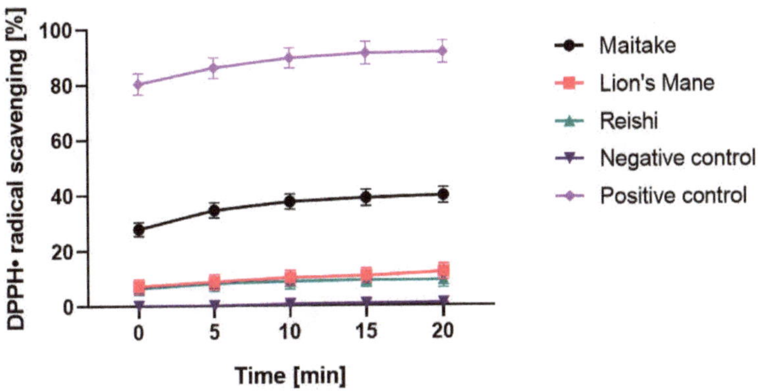

Figure 2. Kinetics of the absorbance changes in DPPH• solutions in the presence of water extract of Maitake, Lion's Mane and Reishi. Values are the mean of three replicate determinations Negative control: ethanolsolution (1000 µg/mL). Positive control: ascorbic acid solution (1000 µg/mL).

The second test used to determine the antioxidant properties of the examined extracts was ABTS• scavengingassay. The obtained results, for the samples in concentrations of 500 and 1000 µg/mL, are presented on the Figure 3. For all extracts, the highest values of ABTS• radical scavenging wereobtained in the concentration of 1000 µg/mL. The most effective scavenging was noted for Maitake (82.12%), and the least for Reishi (38.50%). The situation wassimilar in the case of the 500 µg/mLconcentration. The highest value was obtained for Maitake (64.19%) and the lowest for Reishi (16.22%). Due to the fact that Fe^{2+} ions are involved in the formation of reactive oxygen species, the next test evaluating the antioxidant properties was the test assessing the ability to chelate these ions (Figure 4). Results were obtained for extract samples also at concentrations of 500 and 1000 µg/mL. As in previous tests, the highest chelating capacity for Fe^{2+} ions was recorded for the Maitake extract (31.67%) in the concentration of 1000 µg/mL. The reducing activity of the tested Maitake, Lion's Mane and Reishi extracts was also measured by the FRAP method (Table 5). The results of the experiments were expressed as µmol of Trolox equivalent/g of dry weight of individual mushrooms. As for the other tests assessing antioxidant activity, Maitake extract also showed the most favorable properties obtaining 21.1 ± 3.2 µmol Trolox/g dry weight.

Table 5. Reducing activity of Maitake, Lion's Mane and Reishi extracts using FRAP assay. Data are the mean ± SD of three independent experiments conducted in triplicate for each sample.

Mushroom Extract Type	µmol Trolox/g Dry Weight
Maitake	21.1 ± 3.2
Lion's Mane	7.6 ± 1.1
Reishi	9.8 ± 2.4

Both for the extracts and for the prepared cosmetics containing the analyzed mushrooms, a test was carried out to detect the ability of the compounds to generate intracellular production of reactive oxygen species. The test was performed using fluorogenic H_2DCFDA dye, which is oxidized inside cells into highly fluorescent 20,70-dichlorofluorescein in the presence of ROS.

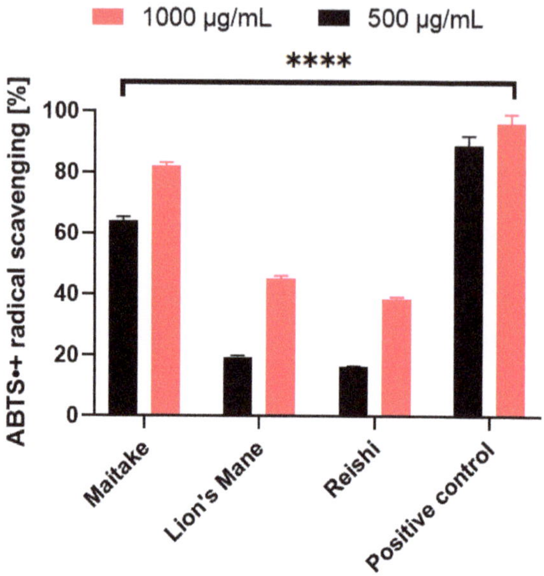

Figure 3. ABTS scavenging by extracts of Maitake, Lion's Mane and Reishi. Data are the mean ± SD of three independent experiments, each of which consisted of three replicates per treatment group, **** $p < 0.0001$ versus the control. Negative control: ethanol solution (1000 µg/mL) (not shown). Positive control: ascorbic acid solution (500 and 1000 µg/mL).

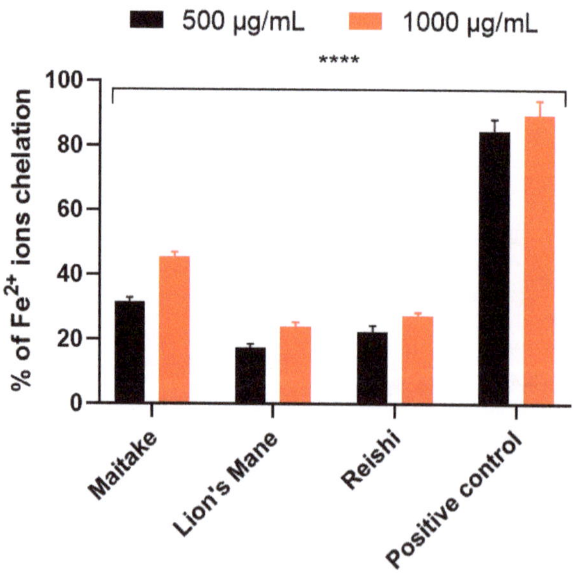

Figure 4. Chelation of Fe^{2+} by extracts of Maitake, Lion's Mane and Reishi. Data are the mean ± SD of three independent experiments, each of which consisted of three replicates per treatment group, **** $p < 0.0001$ versus the control. Negative control: ethanol solution (1000 µg/mL) (not shown). Positive control: ascorbic acid solution (500 and 1000 µg/mL).

As shown in Figure 5, all tested extracts did not increase the amount of ROS in fibroblasts. Reishi showed values similar to the control (cells not treated with the extracts), while

Maitake and Lion's Mane achieved values lower than the control, which means that the presence of these extracts lowers ROS levels in the cells. In this respect, the Maitake extract wasthe most effective, compared with the Reishi and Lion's Mane extracts. The situation wassimilar with keratinocytes; only Reishi extract obtained a higher fluorescencevalue than the control. The rest of the extracts did not increase the intracellular production of ROS.

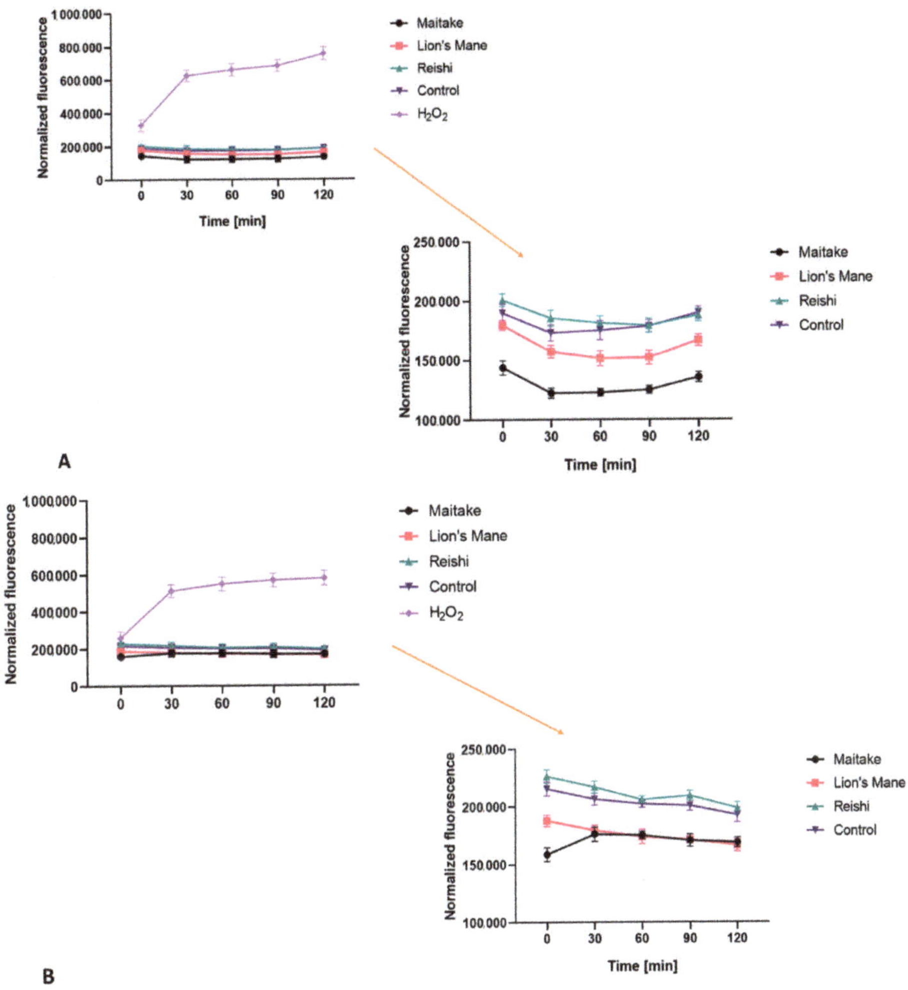

Figure 5. The effect of mushroom extracts at a concentration of 500 μg/mL on the 20,70-dichlorofluorescein (DCF) fluorescence in BJ (**A**), and HaCaT (**B**), cells. Data are the mean ± SD of three independent experiments.

A test was also carried out on washing gels containing extracts from the studied mushrooms, the results of which are presented in Figure 6. In the case of fibroblasts, all washing gels containing mushrooms extracts achieved higher fluorescence values than the control, but lower than cells treated with only base. In the case of keratinocytes, the situation wassimilar. The Maitake, Reishi and Lion's Mane washing gels obtained a fluorescence value higher than the control, but lower than the base, which may suggest that the presence of the mushroom extract reduces the harmful effects of the compounds present in the cosmetic base, both on fibroblasts and keratinocytes.

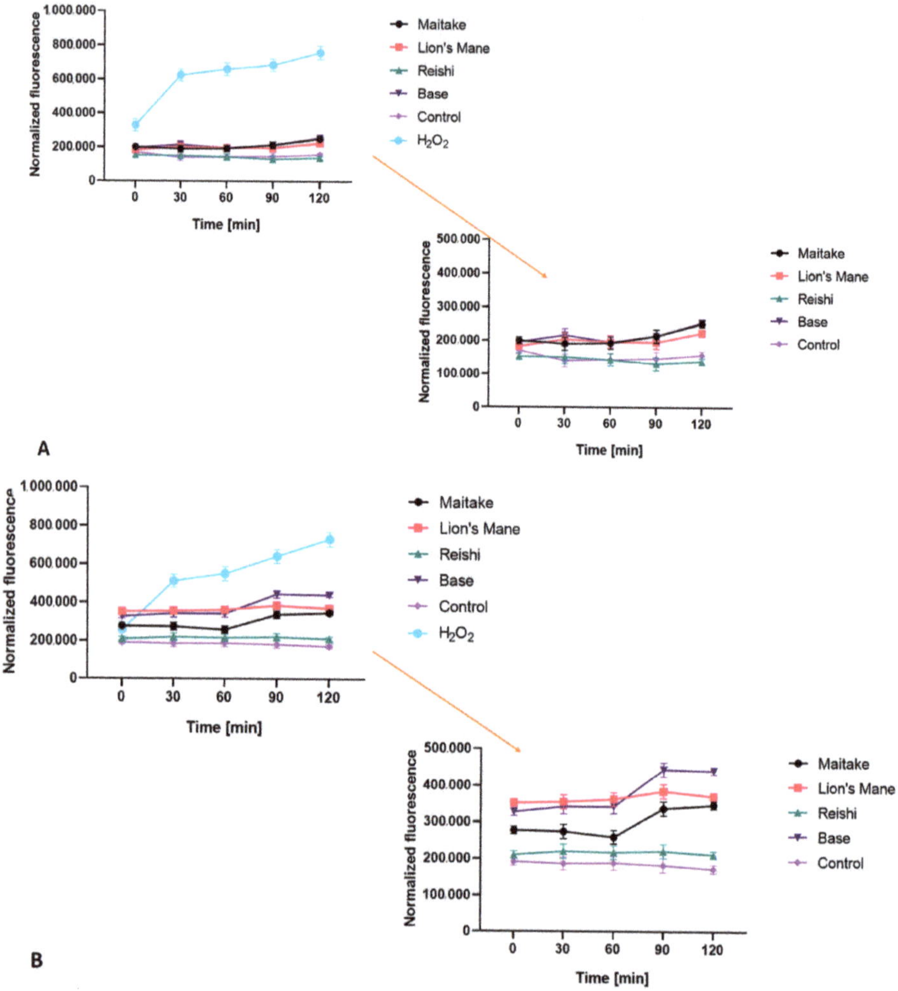

Figure 6. The effect of washing gels at a concentration of 0.01% on the 20,70-dichlorofluorescein (DCF) fluorescence in BJ (**A**), and HaCaT (**B**), cells. Data are the mean ± SD of three independent experiments.

When analyzing the obtained results, it can be concluded that the tested mushrooms showed antioxidant activity. This is caused by the presence of compounds such as malic acid, fumaric acid, citric acid, hericenone, erinacine, ganoderic acid and lucidenic acid, which have the ability to protect cells against the harmful effects of free radicals [47–56]. Maitake extract showed the best antioxidant properties.

Wang and Xu investigated the antioxidant properties of various mushroom species, including Lion's Mane and Maitake. They compared, among other, total phenolic content in acetone, ethanol, water, and hot water extracts. For both Lion's Mane and Maitake, the highest phenolics were found in the aqueous extracts, namely, 3.08 mg GAE/g (Lion's Mane) and 3.78 mg GAE/g (Maitake). Moreover, DPPH free radical scavenging capacities were determined, which proved to be better for Lion's Mane (2.85 µmole TE/g) than Maitake (1.75 µmole TE/g) [57]. Yeh et al. tested two Maitake strains, for which the phenol content in the water extracts was 39.78 mg/g (T1 strain) and 38.96 m/g (T2 strain). The researchers also determined the content of flavonoids, which was lower: 1.09 mg/g for the T1 strain, and 0.52 mg/g for the T2 strain. The presence of ascorbic acid, which also exhibits antioxidant properties, was also confirmed in these extracts. The ability of the extracts to scavenge DPPH radicals was also tested; at a concentration of 20 mg/mL for cold water extracts of Maitake T1 and T2, it was 50.62% and 59.58%, respectively. Much higher results were recorded for ethanolic extracts: 99.19% (T1) and 84.36% (T2) at 20 mg/mL [58]. On the other hand, in the work of Rahman et al., the antioxidant properties of two Reishi strains were compared. Total polyphenol content for Ganoderma lucidum-5 was 33.30 mg/100g, and for Ganoderma lucidum-7 was 43.49 mg/100g. Total flavonoid content was 34.09 mg/100g (Ganoderma lucidum-5) and 38.08 mg/100g (Ganoderma lucidum-7). The ability to scavenge the DPPH radicals by strains Ganoderma lucidum-5 and Ganoderma lucidum-7, was 24.27% and 23.66%, respectively [59].

2.3. *Cytotoxicity Assessment*

In assessing the potential use of extracts from various natural raw materials, including fungi, it is very important to assess their cytotoxicity to skin cells. Hence, as part of this study, the impact of the three prepared extracts, and gels containing these extracts, on the viability of fibroblasts and keratinocytes was assessed. The first of the tests to assess the metabolic activity of the studied cells was alamarBlueassay (AB). As shown in Figure 6, Lion's Mane and Reishi extracts at all tested concentrations showed no toxic effects on keratinocytes (HaCaT) and fibroblasts (BJ cells). Maitake extract at the concentration of 100 μg/mL was not cytotoxic for BJ and HaCaT cells, but with the increase in concentration the cytotoxicity increased, and at the highest concentration (1000 μg/mL) the viability of these cells decreased below 60%. The developed gels in a concentration of 0.01%, containing extracts from the tested fungi, did not show any cytotoxicity. However, in the case of their higher concentration (0.1%), they showed cytotoxicity towards keratinocytes, depending on the type of fungus used. It should be noted that compared with the gels base, the addition of the extract significantly reduced their cytotoxicity, which indicates an increase in their safety of application to skin. The positive effect was observed to the greatest extent in the case of Lion's Mane, for which at a concentration of 0.1%, an increase in keratinocyte viability by 70% compared with the gel base was observed. The cytotoxic effect of the gel base itself is probably related to the activity of surfactants present in the cosmetic base, the cytotoxic effect of which has been proven in numerous studies [60,61]. Based on the results presented in Figure 7C, it can be concluded that the presence of fungal extracts in the developed washing preparations minimizes the unfavorable effect of surfactants, and thus increases the proliferation and viability of cells.

The evaluation of cytotoxicity was also carried out using the Neutral Red uptake assay (NR), and the results are shown in Figure 8. The obtained results indicate that Lion's Mane and Reishi extracts at all tested concentrations showed no cytotoxicity to both BJ and HaCaT cells. Moreover, treating keratinocytes with Reishi extracts increased the viability of these cells. On the other hand, Maitake extract showed cytotoxicity to BJ cells at all tested concentrations, and this effect was stronger with increasing concentration. In the case of HaCaT cells, this extract showed no cytotoxicity at the concentrations of 100 μg/mL and 500 μg/mL, but it inhibited the viability of these cells at the concentration of 1000 μg/mL. In the case of the developed gels containing extracts of the tested mushrooms, none showed cytotoxicity. At a concentration of 0.01%, all showed the ability to increase the proliferation of HaCaT cells by about 20% compared with the control (cells not treated with gels). Similar to the AB test, the studies showed that the gel base at a concentration of 0.01% did not show cytotoxicity, while when 0.1% concentration wasused, the cytotoxicity wassignificant and reduced the keratinocyte viability to just over 20%. On the other hand, the addition of extracts from all three tested fungi to the gel formula eliminated the cytotoxicity of the tested washing preparations, which indicates the legitimacy of including these extracts in the formulathedeveloped gels.

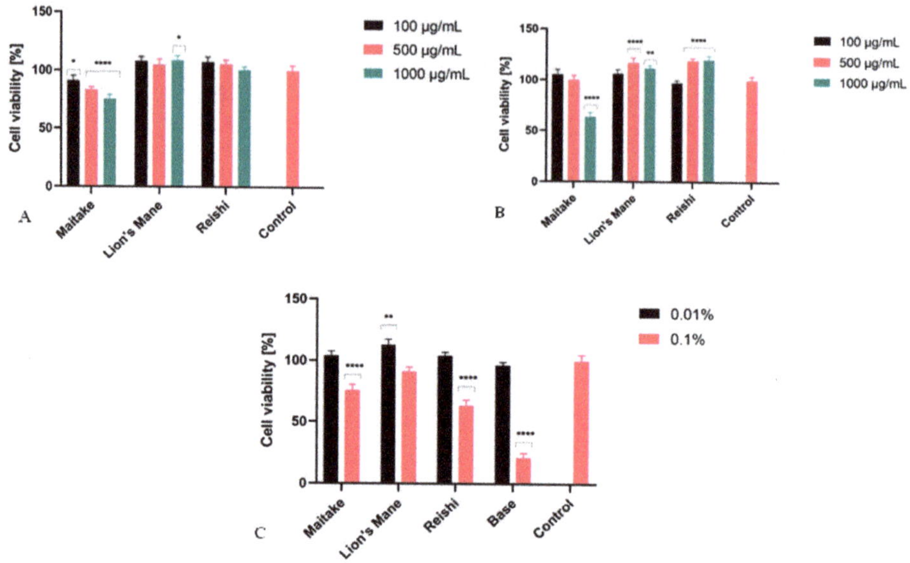

Figure 7. The reduction in resazurin after 24 h (for extract) and 1 h (for washing gels) exposure to the Maitake, Lion's Mane and Reishi in cultured (**A**) fibroblasts (treated with extracts), (**B**) keratinocytes (treated with extracts), and (**C**) keratinocytes (treated with washing gels). Data are the mean ± SD of three independent experiments, each of which consisted of three replicates per treatment group, **** $p < 0.0001$, ** $p < 0.005$, * $p < 0.8$ versus the control (100%).

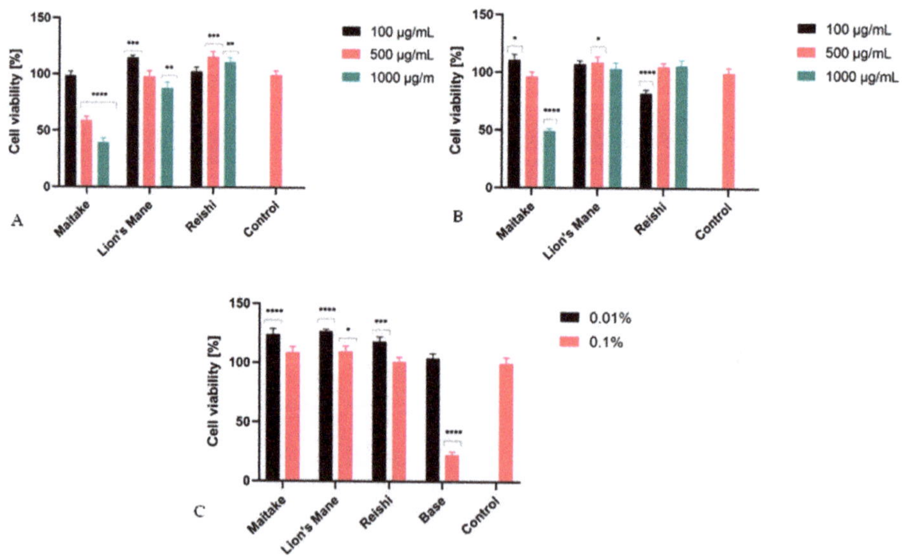

Figure 8. The effect of Maitake, Lion's Mane and Reishi on Neutral Red dye uptake in cultured (**A**) fibroblasts (treated with extracts), (**B**) keratinocytes (treated with extracts), and (**C**) keratinocytes (treated with washing gels). The exposure time of cells to individual extracts was 24 h, and to washing gels, 1 h. Data are the mean ± SD of three independent experiments, each of which consisted of three replicates per treatment group, **** $p < 0.0001$, *** $p < 0.0005$, ** $p < 0.005$, * $p < 0.05$ versus the control (100%).

Research on skin cell lines is an important element that allows assessment of the safety of the obtained extracts, and to predict their possible effects in the next stages of research, such as in vivo tests or clinical trials. The results obtained in this study, indicating the lack of cytotoxicity of the tested extracts used in a given concentration range and the possibility of reducing the negative effects of various cosmetic ingredients, allow the estimate that these raw materials can be more and more willingly included in the formulations of a wide range of cosmetic preparations. In addition to the lack of a negative impact on the viability and proliferation of keratinocytes and fibroblasts, scientific works indicate that these extracts also exhibit multidirectional activity in in vitro conditions.

The available literature data do not contain any reports on the cytotoxicity of Maitake extracts to skin cells. There are reports mainly describing antitumor, immunomodulating and antioxidant properties of this fungus, but the influence of this type of extract on the viability of keratinocytes and fibroblasts has not been assessed [62–64]. Contrary to the extracts of Lion's Mane and Reishi, the analyses carried out in this study showed that these extracts, in the higher concentrations used, exert a cytotoxic effect on skin cells in vitro, mainly fibroblasts. This may be the result of the action of compounds contained in these extracts, including alpha-hydroxy acids, which, despite their antioxidant activity, may exert an antiproliferative effect on HaCaT cells [65]. Scientific research shows that malic acid inhibits the proliferation of keratinocytes by inhibiting the progression of the cell cycle in the G0/G1 phase. Additionally, this acid can induce the expression of endoplasmic reticulum stress-related proteins such as GRP78, GADD153 and ATF6α [65]. Other studies carried out on human dermal fibroblasts (HDF) exposed to ultraviolet A radiation, however, indicate that exopolysaccharide isolated from this fungus has photoprotective potential. This polysaccharide has an inhibitory effect on the expression of human interstitial collagenase (matrix metalloproteinase, MMP-1), which may reduce skin photoaging by reducing the matrix degradation system associated with MMP-1 [66]. Thus, in order to use the potential of this fungus, it is necessary to select theconcentration of the extract that will show the desired biological activity with the simultaneous lack of cytotoxicity. This is important because, as shown in this work, the addition of Maitake extract may reduce the negative effect of cosmetic cleansing preparations on skin cells.

The effect of Lion's Mane extract on keratinocytes and fibroblasts in vitro has not been described to date. Thus, in this study, for the first time, we demonstrated the lack of cytotoxic effect of the studied extract on these skin cells, and the possibility of stimulating their viability and proliferation. This may be the result of reducing the level of free radicals in these cells, which was also shown in this work. The inhibitory effect on the production of ROS was also indicated by Chang et al., who demonstrated that the extracts of this fungus trigger the expression of the antioxidant genes of heme oxidase-1 (HO-1), γ-glutamylcysteine synthetase (γ-GCLC), and affect the level of glutathione. The antioxidant activity demonstrated by these authors on human endothelial cells (EA.hy926) is associated with increased nuclear translocation and transcriptional activation of NF-E2 associated factor 2 (Nrf2) [67].

Due to the fact that Reishi is considered to be one of the strongest adaptogens found in nature and exhibits not only antioxidant but also anti-inflammatory, immunomodulating and anti-cancer properties, interest in it in the context of skin care and treatment of skin diseases is growing [68]. Abate et al. showed that Reishiextract inducesthe proliferation of keratinocytes and increases the expression of cyclic protein kinases, such as CDK2 and CDK6. Additionally, these authors indicated that keratinocytes treated with Reishiextract show an increased migration rate and an increase in activation of tissue remodeling factors such as matrix metalloproteinases 2 (MMP2) and 9 (and MMP-9). Moreover, this extract, through its antioxidant activity, protects keratinocytes against H_2O_2-induced cytotoxicity. These studies indicate the legitimacy of the cosmetic use of this fungus due to the possibility of accelerating wound healing processes, protecting cells against oxidative stress and intensified re-epithelialization [69]. Kim et al. indicated the inhibitory effect of Reishi extract on the activity of tyrosinase and melanin biosynthesis in B16F10 melanoma cells.

They also demonstrated the possibility of inhibiting the expression of tyrosinase-related protein 1 (TRP-1), TRP-2, as well as microphthalmia-related transcription factor (MITF), thus reducing the production of melanin. This extract also influences the mitogen-activated kinase (MAPK) cascade and cyclic adenosine monophosphate (cAMP)-dependent signaling pathway, which has a significant effect on the melanogenesis of B16F10 melanoma cells [20]. Hu et al., on the other hand, indicated that the polysaccharides contained in Reishi increase the viability of fibroblasts and the ability to migrate these cells. Moreover, these polysaccharides increase the expression of β-catenin, CICP and TGF-β1 in fibroblasts in vitro. Additionally, in vivo studies in mice indicated that these compounds significantly improved healing rates and wound healing time. This is likely the result of activation of the Wnt/β-catenin signaling pathway and elevated levels of TGF-β1 [1].

To sum up, the lack of cytotoxic effect of Lion's Mane and Reishi extracts on skin cells, and a reduction in the cytotoxicity of the base of cleansing preparations by extracts from all three tested mushrooms, indicate that they can be perceived as valuable cosmetic raw materials with a broad spectrum of activity.

2.4. Transepidermal Water Loss (TEWL), Skin Hydration and Skin pH Measurements

Due to the content of hydroxyl groups in molecules, bioactive ingredients of the studied extracts have a positive effect on the condition of our skin. They mainly affect the retention of water in the skin, but also reduce the evaporation of water from the upper layer of the epidermis [70–72]. In addition, these extracts are characterized by several health-promoting properties and therefore can be successfully used as active ingredients in cosmetic products. The conducted research assessed the effect of preparations containing extracts on basic skin parameters, such as skin hydration, TEWL and pH. The tested samples were products containing 1% mushroom extract (Reishi, Lion's Mane, Maitake) and the base sample, for conductinga comparison. The analyses were carried out at twotime intervals, 1 h and 3 h, after the application of the product. The analysis of the obtained results showed the positive effect of the contained extract on the condition of the skin. The results are presented in the Figures 8–10.

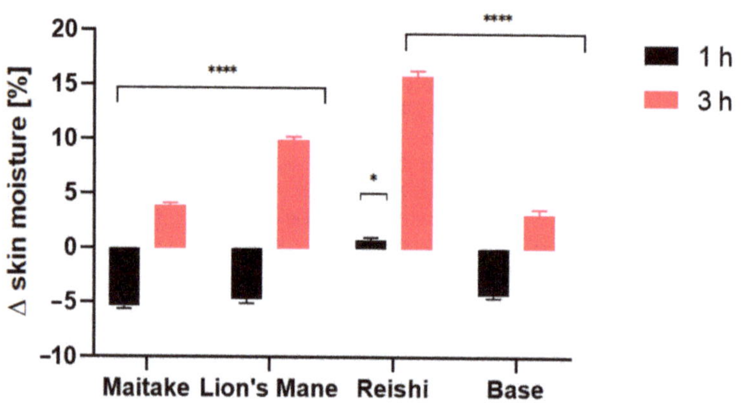

Figure 9. The influence of model gels with Maitake, Lion's Mane and Reishi extracts, on skin hydration. Data are the mean ± SD of three independent measurements. **** $p < 0.0001$, * $p < 0.015$ versus the control.

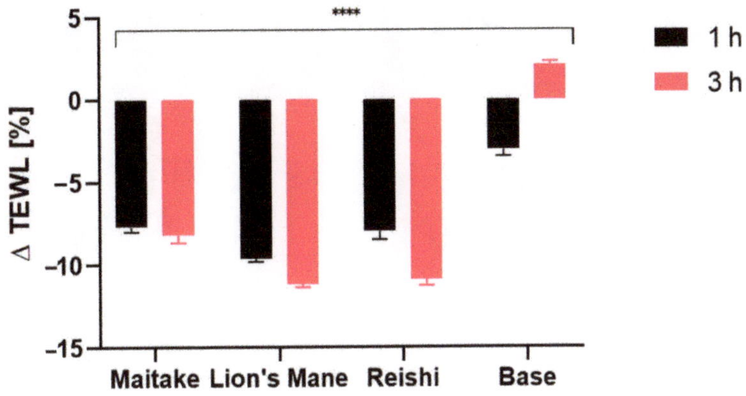

Figure 10. The influence of model gels with Maitake, Lion's Mane and Reishi extracts, on transepidermal water loss (TEWL). Data are the mean ± SD of three independent measurements. **** $p < 0.0001$ versus the control.

Analysis of skin moisture showed differences between examined model gels with extracts, which will be discussed one by one. For the gel containing Maitake extract, aminimal influence was observed. After 1 h, the effect was similar to the base, for which decreased moisture was recorded, but registered values were around 20% lower when compared with the base. After 3 h it turned into a 20% relative increase with reference to the base level. In the case of Lion's Mane, the initial effect measured after 1 h was also negative, showing around 10% decrease in moisture compared with the base, but after 3 h the result turned into a significant 200% increase in moisture. The best results were noted for the Reishi extract, where, even after the first 1 h, a slight increase in moisture was observed, bearing in mind that for the base a noticeable decrease was observed. After 3 h, high values were observed, as the moisture was 350% higher when compared with the base. From that perspective, Reishi showed the most preferable properties, with better results, compared with the rest of the examined extracts.

Analysis of the TEWL showed a positive effect of each of the model gels with mushroom extracts, however there were still some differences between them which are worth discussing. First, values observed after 1 h were all preferable, showing values around −7.5% and −8% for Maitake and Reishi, respectively, and even better valuearound −9.5% for Lion's Mane. Values measured after 3 h showed, in case of Maitake extract, the TEWL value stayed at almost the same level of −8%, but in the case of Lion's Mane and Reishi extracts even better values were observed, equaling −11% for both. In conclusion, the Lion's Mane extract exhibited the most preferable properties, slightly ahead of the Reishi extract.

Analysis of the pH showed noticeable differences between model gels with mushroom extracts (Figure 11). Zero level means no change in the skin pH in relation to the control field, i.e., the physiological pH of the test volunteers, therefore, the most favorable values would beclose to zero. First, values observed for the base werethe most distant from zero level, being the least preferable. Values observed for the Maitake extract appeared to be the best among those compared, showing differences of −0.5 and −2, respectively, after 1 h and 3 h. It is worth emphasizing that the deviation observed after 1h wasvery close to the perfect natural skin pH.NextwastheLion's Mane extract, for which differences of −2 and −3.5,after 1 h and 3 h, respectively, wereobserved. The worst results were measured for the Reishi extract, where the deviation from the optimal pH was close to −4 after both 1 h and 3 h. From that perspective, the Maitake extract results werethe most preferable.

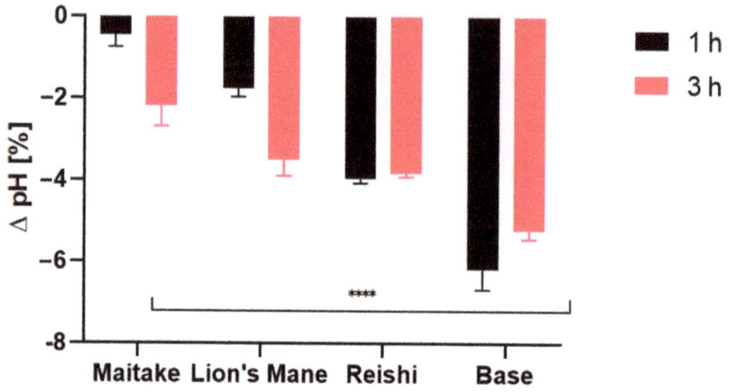

Figure 11. The influence of model gels with Maitake, Lion's Mane and Reishi extracts on skin pH. Data are the mean ± SD of three independent measurements. **** $p < 0.0001$ versus the control.

2.5. Irritant Potential of Model Body Wash Gels

The lack of adverse effects of washing cosmetics on skin is one of the most important requirements for this group of products. The measurement results of the irritating potential of the analyzed washing gels are presented in Figure 12.

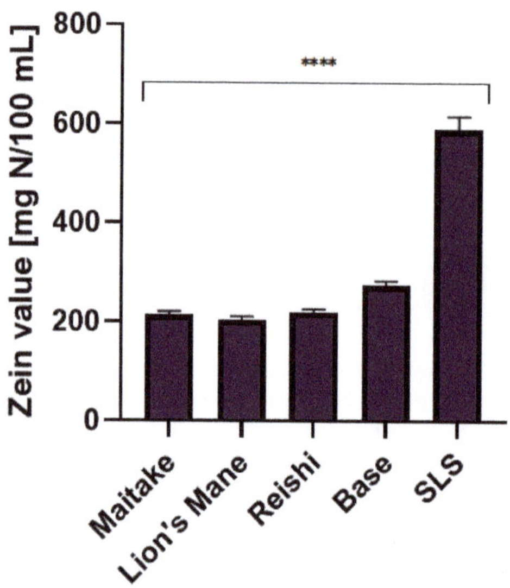

Figure 12. Irritant potential of model washing gels with Maitake, Lion's Mane and Reishi extracts. Data are the mean ± SD of three independent measurements. **** $p < 0.0001$ versus the control. Positive control: sodium lauryl sulfate (SLS) solution (1%).

The method used to determine the irritation potential of the model washing gels was the determination of the zein number. Due to the fact that the irritant potential of washing cosmetics is caused by interactions of surfactants with proteins that build the stratum corneum layer, this method closely imitates the strength of the interactions of surfactants used in the formulation of the model washing gels, with skin proteins. The irritating potential generated by surfactants results from the ability of these substances to denature epidermal proteins and then wash them out of the skin. Contact of the analyzed samples with zein, a protein with a structure similar to the main proteins in skin, may indicate their protein-denaturing capacity, which is a measure of irritating potential.

Obtained results showed that the addition of analyzed mushroom extracts to the base model washing gel reduced the irritant potential (zein number). The base sample (without the addition of the extracts) was characterized by a zein number at the level of 270 mg N/100 mL and, according to the accepted classification, was classified as moderately irritant for the skin. Samples containing Maitake and Reishi extracts had lower irritant potential and their zein valueswere about 20% lower in relation to the base sample (about 220 mg N/100 mL). The lowest irritant potential was observed for the sample that contained Lion's Mane extract. The zein value of this sample was about 25% lower than the base sample. The obtained results indicate that samples with the addition of the analyzed extracts can be classified on the border of non-irritating and slightly irritating.

The anionic surfactant, sodium coco sulfate, was used in the formulation of the analyzed model washing gels as the main washing agent. As mentioned in the introduction, anionic surfactants have the strongest irritant potential due to the fact that they interact with skin proteins through strong ionic bonds, and their denaturing potential is significantly higher than for nonionic and amphoteric surfactants. The lower irritant potential observed for the samples containing mushroom extracts wasmost probably caused by carbohydrates and proteins, that are the main ingredients of mushrooms' dry weight, and are extracted, in the extraction process, by polar solvents. These substances may be incorporated into surfactant's micelles, causing anincrease in their stability and size. Formation of mixed micelles that contain both surfactants and high molecular weight carbohydrates and proteins limits the number of free surfactants in the form of monomers that are released to the bulk phase, and leads to a lowering of the irritant potential of the product [23,25–27].

3. Materials and Methods

3.1. Chemicals

2′,7′-dichlorodihydrofluorescein diacetate (H_2DCFDA, Sigma-Aldrich, Poznan, Poland), acetic acid (CH_3COOH, ≥99%, Sigma-Aldrich, Poznan, Poland), acetonitrile (CH_3CN, ≥99.9%, Sigma-Aldrich, Poznan, Poland), aluminum nitrate nonahydrate ($Al(NO_3)_3 \cdot 9H_2O$, Sigma-Aldrich, Poznan, Poland), antibiotics (Penicillin–Streptomycin, Life Technologies, Bleiswijk, The Netherlands), DMEM (Dulbecco's modification of Eagle's medium, Biological Industries, Beit Haemek, Israel), ethyl alcohol (ethanol, C_2H_5OH, 96%, Sigma-Aldrich, Poznan, Poland), FBS (fetal bovine serum, Biological Industries, Genos, Lodz, Poland), Folin &Ciocalteu's phenol reagent (Sigma-Aldrich, Poznan, Poland), formic acid (HCOOH, Merck Life Science, Poznan, Poland), gallic acid ($C_7H_6O_5$, Sigma-Aldrich, Poznan, Poland), methanol (CH_3OH, ≥99.9%, Sigma-Aldrich, Poznan, Poland), Neutral Red solution (NR, 0.33%, Sigma-Aldrich, Poznan, Poland), phosphate buffered saline (PBS, pH 7.00 ± 0.05, ChemPur, Piekary Slaskie, Poland), potassium persulfate (99.99%, $K_2S_2O_8$, Sigma-Aldrich, Poznan, Poland), resazurin sodium salt (RES, Sigma-Aldrich, Poznan, Poland), sodium carbonate (Na_2CO_3, ≥99.5%, Sigma-Aldrich, Poznan, Poland), Sodium Pyruvate Solution (100 mM, Genos, Lodz, Poland), trypsin-EDTA solution (Sigma-Aldrich, Poznan, Poland), Quercetin hydrate (≥95%, Sigma-Aldrich, Poznan, Poland), were used as received.

3.2. Preparation of Extracts

The material used in the research was dry, powdered mushrooms: Maitake, Lion's Mane and Reishi, purchased from the Polish certified dealer, MagicznyOgród. To prepare the extracts, 5 g of powdered mushroom was weighed into a beaker. and 100 mL of water was added. Afterwards, the beakers were placed in an ultrasonic bath (Digital Ultrasonic Cleaner) for 5 min. After this time, the obtained extracts were filtered through Whatman No. 1 filters. After filtration, the extracts were evaporated in a concentrator at a temperature of 40 °C under reduced pressure. From the obtained dry extracts, stocks with a concentration of 100 mg/mL were prepared and stored at 4 °C for further analysis.

3.3. Determination of Biologically Active Compounds

3.3.1. Determination of Bioactive Compounds by HPLC-ESI/TOF

Extracts were analyzed using ultra-high performance liquid chromatograph (UHPLC) with an ESI/TOF detector (Agilent Technologies, Santa Clara, CA, USA). The separation was carried out on an RP18 reversed-phase column Titan (10 cm × 2.1 mm i.d., 1.9 µm particle size) (Supelco, Sigma-Aldrich, Burlington, MA, USA) using a mixture of water with 0.05% of formic acid (solvent A) and acetonitrile with 0.05% of formic acid (solvent B) at a flow rate of 0.2 mL/min according to gradient as follows: 0–5 min from 100% A to 98% A (from 0% to 2% B), 5–50 min from 98% A to 75% A (from 2% to 25% B), 50–70 min from 75% A to 60% A (from 25% to 40% B), and 70–100 min from 60% A to 40% A (from 40% B to 60% B). Thermostat temperature was 30 °C. The ion source operating parameters were as follows: drying gas temperature 325 °C, drying gas flow 5 L min^{-1}, nebulizer pressure 30 psi, capillary voltage 3500 V, fragmentator 240 V, and skimmer 65 V. Ions were acquired in the range of 100 to 1000 m/z.

3.3.2. Determination of the Total Phenolic Content (TPC)

In order to determine the TPC of analyzed extracts, the spectrophotometric method involving the use of the Folin&Ciocalteu reagent, was carried out [70]. Gallic acid solution in a concentration range of 10–100 mg/mL was used as astandard. First, 1500 µL of 1:10 diluted Folin&Ciocalteu reagent was added to the test tubes containing 300 µL of the analyzed samples (in concentrations of 100, 500 and 1000 µg/mL) and incubated in adark room for 6 min. Then, 1200 µL of 7.5% sodium carbonate solution was added to the test tubes. Samples were mixed and incubated in the dark room for 1.5 h. After this time, the absorbance was measured at wavelength of λ = 740 nm. To calculate the total concentration of phenols in the analyzed samples, a gallic acid (GA) calibration curve (in the 10–100 mg/mL concentration range) was used. The measurements were made in triplicate and the results obtained were averaged. The TPC results are presented as µg of GA equivalents (GAE) per g of dry weight.

3.3.3. Determination of the Total Flavonoids Content (TFC)

In order to determine the TFC of analyzed extracts, the spectrophotometric method involving the use of aluminum nitrate nonahydrate, was carried out [71]. Quercetin hydrate solution in a concentration range of 10–100 mg/mL was used as atandard. In the first step, 1200 µL of reaction mixture containing 80% C_2H_5OH, 10% $Al(NO_3)_3 \cdot 9H_2O$, and 1M $C_2H_3KO_2$, was added to 300 µL of the analyzed samples (in concentrations of 100, 500 and 1000 µg/mL). Then, samples were mixed and incubated in a dark room for 40 min. After this time, the absorbance was measured at wavelength of λ = 415 nm. To calculate the total concentration of flavonoids in the analyzed samples, a quercetin (Q) calibration curve was used. The measurements were made in triplicate and the results obtained were averaged. The TFC results are presented as µg of Q equivalents (QE) per g of dry weight.

3.3.4. Determination of Total Carbohydrate

The total carbohydrate content in the extracts was estimated using the phenol sulfuric acid method. An amount of 0.25 mL of sample was mixed with 1.25 mL of concentrated sulfuric acid (95% v/v) and 0.25 mL of phenol (5% v/v). The mixture was heated at 100 °C for 5 min, cooled to room temperature and absorbance was measured at 490 nm. The results were calculated using the glucose calibration curve [72].

3.4. Determination of Antioxidant Properties

3.4.1. DPPH Radical Scavenging Assay

The ability to scavenge free radicals was determined using the method described by Brand-Williams et al. [73], which is based on using the 1,1-diphenyl-2-picrylhydrazyl (DPPH) radical. In the first step, 100 µL of water solutions of analyzed extracts at concentration of 1000 µg/mL were transferred to a 96-well plate. Then, 100 µL of ethanol solution of DPPH was added to the samples and mixed. An ethanol solution was used as a negative control, and an ascorbic acid solution as a positive control, at a concentration of 1000 μg/mL. The absorbance was measured at wavelength of 517 nm, every 5 min for 20 min, using a UV-VIS Filter Max spectrophotometer (Thermo Fisher Scientific, Waltham, MA, USA). Measurements were carried out in triplicate for each extract sample. The antioxidant capacity was expressed as a percentage of DPPH inhibition using Equation (1):

$$\%\text{DPPH scavenging} = \frac{\text{Abs control} - \text{Abs sample}}{\text{Abs control}} \times 100 \quad (1)$$

where: Abs sample—absorbance of the sample; Abs control—absorbance of the control sample.

3.4.2. ABTS• Scavenging Assay

The second method for determining the antioxidant properties of mushroom water extracts was by ABTS• scavenging assay, described by Gaweł-Beben et al. [74]. First, a 7 mM aqueous ABTS solution and 2.4 mM potassium persulfate were prepared. Then, ABTS and potassium persulfate were mixed in equal proportions and left at room temperature, in darkness for 14 h. After this time, the solution was diluted in methanol to an absorbance at the level of about 1.0 (λ = 734 nm). In the next step, 1 mL of analyzed extract in a concentration of 1000 μg/mL was mixed with 1 mL of ABTS. An ethanol solution was used as a negative control, and an ascorbic acid solution as a positive control, at a concentration of 500 and 1000 μg/mL. Distilled water was added to the blank instead of the extract. The absorbance was measured at wavelength of λ = 734 nm using a UV/VIS spectrophotometer Aquamate Helion (Thermo Fisher Scientific, Waltham, MA, USA). The ABTS• scavenging was calculated from Equation (2):

$$\% \text{ of ABTS}^{\bullet} + \text{scavenging} = 1 - \frac{\text{As}}{\text{Ac}} \times 100 \quad (2)$$

where: As—absorbance of the sample; Ac—absorbance of the control sample.

3.4.3. Fe^{2+} Chelation Assay

Chelation of iron (II) ions by Maitake, Lion's Mane and Reishi extracts was measured according to the methodology described by Gaweł-Bęben et al. [74] with slight modifications. In the first step, 0.5 mL of each extract (with a concentration of 500 and 1000 μg/mL) was mixed with 3.7 mL of distilled water, 0.1 mL of $FeCl_2$ (1 mM) and 0.2 mL of ferrosine (5 mM). An ethanol solution was used as a negative control, and an ascorbic acid solution as a positive control, at a concentration of 500 and 1000 μg/mL.Then, the reaction mixture was thoroughly mixed and incubated at room temperature for 10 min. The absorbance of the tested samples was measured at wavelength of λ = 562 nm using a UV/VIS spectrophotometer Aquamate Helion (Thermo Fisher Scientific, Waltham, MA, USA). Each sample was analyzed in triplicate. The chelating activity of the analyzed mushroom extracts was calculated as the percentage inhibition of the ferrosine-Fe^{2+} complex formation from the formula:

$$\% \ Fe^{2+} \text{ chelating activity} = [1 - (\text{As}/\text{Ac})] \times 100\%$$

where: As—sample absorbance; Ac—absorbance of the control sample.

3.4.4. Ferric Reducing Antioxidant Power (FRAP) Assay

The reducing activity of the studied Maitake, Lion's Mane and Reishi extracts was also measured on the basis of the FRAP method procedure described by Benzie and Strain, with minor modifications [75]. In the first step, fresh FRAP reagent was prepared. For this purpose, 25 mL of acetate buffer (CH_3COOH:CH_3COONa, 0.3 M, pH = 3.6) wasmixed with 25 mL of methanol (Alchem, Toruń, Poland). An amount of 5 mL of 2,4,6-tris (2-pyridyl)-s-triazine (10 mM) dissolved in HCl (0.04 M) and 5 mL of $FeCl_3 \cdot 6H_2O$ (0.02 M) were then added to the solution. In the next step, 225 μL of 50% methanol solution and

75 μL of the tested 1000 μg/mL mushroom extracts were added to 2.25 mL of FRAP reagent and mixed thoroughly. The mixture was incubated in a 37 °C water bath for 30 min. The absorbance was then measured at λ = 593 nm in a glass cuvette using an Aquamate Helion spectrophotometer (Thermo Fisher Scientific, Waltham, MA, USA). A standard curve was prepared using different Trolox concentrations. For each extract, measurements were made in triplicate. The results of the experiments were expressed as μmol of Trolox equivalent/g of dry weight of individual mushrooms.

3.4.5. Detection of Intracellular Levels of Reactive Oxygen Species (ROS)

In order to determine the ability of the analyzed extracts and washing gels to generate the intracellular production of reactive oxygen species in skin cells, a fluorogenic H_2DCFDA dye was used. This dye has the ability to penetrate inside the cell, where it is transformed into a non-fluorescent compound. If reactive oxygen species are present in the cell, this compound is then transformed into highly fluorescent DCF. To determine the intracellular level of ROS in HaCaTs and BJ, cells were seeded in 96-well plates and cultured in an incubator for 24 h. Next, DMEM medium was removed and replaced with 10 μM H_2DCFDA (Sigma Aldrich, St. Louis, MO, USA) dissolved in serum-free DMEM medium. Cells were incubated for 45 min and then incubated with the extracts in concentrations of 100, 500, and 1000 μg/mL, and with samples of washing gels in concentrations of 0.1% and 0.01%. Cells treated with 1 mM hydrogen peroxide (H_2O_2) were used as positive controls. The control samples were cells untreated with the tested extracts. The fluorescence of DCF was measured every 30 min for 120 min using a FilterMax F5 microplate reader (Thermo Fisher Scientific) at a maximum excitation of 485 nm and emission spectra of 530 nm [76].

3.5. *Cytotoxicity Analysis*

3.5.1. Cell Culture

In this research, two types of skin cells were used: fibroblasts (American Type Culture Collection Manassas, VA, USA) and keratinocytes (CLS Cell Lines Service GmbH, Eppelheim, Germany). Cells were grown in Dulbecco's Modification of Eagle's Medium (DMEM, Biological Industries, Cromwell, CO, USA) high glucose content (4.5 g/L), enriched with sodium pyruvate, L-glutamine, 10% fetal bovine serum (Gibco, Waltham, MA, USA) and 1% antibiotics (100 U/mL penicillin and 1000 μg/mL streptomycin, Gibco). Cells were grown in an incubator in a humidified atmosphere of 95% air and 5% carbon dioxide at 37 °C. After obtaining the required confluence, the medium was removed. Cells were washed with sterile phosphate buffered saline and then were detached from the bottom of the culture flasks with trypsin. Next, cells were placed in fresh medium, plated in 96-well plates and incubated for 24 h. After this time, cells were treated with mushroom extracts in concentrations of 100, 500 and 1000 μg/mL, and washing gels in concentrations of 0.01% and 0.1% containing mushrooms extracts, and incubated for another 24 h.

3.5.2. AlamarBlue Assay

In order to evaluate the viability of BJ and HaCaT cells treated with analyzed extracts, the alamarBlueassay was performed. After incubation, analyzed samples were removed from the wells and then Resazurin solution (60 μM) was added. Plates were placed in an incubator for 2 h at 37 °C. Then, fluorescence was measured at wavelength of λ = 570 nm. Each sample was performed in three replications.

3.5.3. Neutral Red Uptake Assay

In addition, the Neutral Red uptake assay was performed. After incubation, analyzed samples were removed from the wells and then Neutral Red dye (40 μg/mL) was added to the wells. Plates were placed in an incubator for 2 h at 37 °C, then the Neutral Red dye was removed, and the cells were washed with phosphate buffered saline. After this, phosphate buffered saline was removed and 150 μL of decolorizing buffer was added. The absorbance measurements were performed at wavelength of λ = 540 nm. Each sample was performed in three replications.

3.6. Transepidermal Water Loss (TEWL), Skin Hydration and Skin pH Measurements

TEWL, skin pH and skin hydration measurements were conducted using a Tewameter TM 300 probe, Skin pH Meter PH950 and a Corneometer CM825 probe connected to a MPA adapter (Courage + Khazaka Electronic, Köln, Germany). The study was conducted on 10 volunteers, according to the procedure described by Nizioł-Łukaszewska et al. [77]. Five areas (2 cm × 2 cm in size) were marked on the forearm skin of volunteers. Then, 0.2 mL of 100 μg/mL solution of obtained gelswas applied to four fields (50 μL/cm^2). One field (control field) was not treated with any sample. Sample solutions were gently spread over every part of the four fragments of the skin in the marked area, and after 20 min, dried with a paper towel. After 60 and 180 min, the hydration and TEWL measurements were taken. The final result was the arithmetic mean (from each volunteer) of five independent measurements (skin hydration and pH) and 20 measurements (TEWL). The final result was expressed as the change in skin hydration, TEWL and pH after the application of the analyzed gels relative to the control field to which the product was not applied.

3.7. Determination of Irritant Potential–Zein Value

Irritant potential of the products was measured using the zein test [18,78]. In a solution of surfactants, the zein protein denatures and then dissolves in the solution. This process simulates the behavior of surfactants towards skin proteins. Zein from corn, in the amount of 2 ± 0.05 g, was added to 40 mL of each of the samples. Sodium lauryl sulfate (1%) was used as a positive control. The solutions with zein were shaken in a shaker in a water bath (60 min at 35 °C). The solutions were filtered using Whatman No. 1 filters and then centrifuged at 5000 rpm for 10 min. The nitrogen content in the solutions was determined by Kjeldahl method. One milliliter of the filtrate was mineralized in sulfuric acid (98%) containing copper sulphate pentahydrate and potassium sulphate. After mineralization, the solution was transferred (with 50 mL of Milli-Q water) into the flask of the Wagner–Parnas apparatus. In the next step, 20 mL of sodium hydroxide (25 wt%) was added. The released ammonia was distilled with steam. Ammonia was bound by sulfuric acid (5 mL of 0.1 N H_2SO_4) in the receiver of the Wagner–Parnas apparatus. The unbound sulfuric acid was titrated with 0.1 N sodium hydroxide. Tashiro solution was used as an indicator. The zein number (ZN) was calculated using Equation (3):

$$ZN = (10 - V1) \times 100 \times 0.7\ [\text{mg N}/100\ \text{mL}] \quad (3)$$

where V1 is the volume (mL) of sodium hydroxide used for titration of the sample. The final result was the arithmetic mean of five independent measurements.

3.8. Preparation of the Model Washing Gels

The final formulation of the analyzed model washing gels is presented in Table 6. Raw materials widely used in the cosmetics industry were used to prepare the samples.

We prepared 600 g of the washing gel according to the following procedure: water was weighed into the glass beaker and then heated to 45 °C. Raw materials, in the order given in the formulation, were added to the water and mixed (mechanic stirrer Chemland O20) until fully dissolved. Then, the obtained gel was cooled to room temperature and divided into 4 equal parts. One was the base model washing gel without the addition of the extract. Maitake, Reishi and Lion's Mane extracts were added to the three remaining samples at a concentration of 1 wt% in each sample.

Table 6. Formulation of the analyzed model washing gels.

INCI Name	Concentration [wt%]
Aqua	84.85
Sodium coco sulfate	7.00
Cocamidopropyl betaine	1.50
Coco glucoside	1.50
Glycerin	5.00
Sodium benzoate	0.50
Potassium sorbate	0.20
Sodium chloride	1.00
Lactic acid	0.45
Extract	1.00

3.9. Statistical Analysis

Values of different parameters were expressed as the mean ±standard deviation (SD). Two-way analysis of variance (ANOVA) was performed at the *p*-value level of <0.05 to evaluate the significance of differences between values. Statistical analysis was performed using GraphPad Prism 8.0.1 (GraphPad Software, Inc., San Diego, CA, USA).

4. Conclusions

The paper presents the potential use of mushroom extracts in cosmetics. Three types of mushrooms were tested—*Grifolafrondosa* (Maitake), *Hericiumerinaceus* (Lion's Mane) and *Ganoderma Lucidum* (Reishi)—as potential ingredients in cleansing cosmetics. The obtained extracts were characterized by moderate antioxidant properties, and the strongest properties were demonstrated by the Maitake extract. Reishi and Lion's Mane extracts did not show cytotoxic activity, while Maitake extract in concentrations above 500 µg/mL may be toxic to skin cells, keratinocytes and fibroblasts. The introduction of extracts from the analyzed mushrooms to the recipes of model washing gels had a positive effect on their safety. It was shown that samples with the addition of the extract at the level of 0.1% were characterized by significantly lower toxicity to skin cells, compared with the sample without the addition of mushroom extract. Furthermore, addition of the analyzed extracts to the formulas of shower gels had a positive effect on skin hydration, TEWL and skin pH after their application. Compared with the base washing gel (without the addition of an extract), samples containing extracts in the recipe moisturized the skin more strongly and reduced TEWL. The most favorable properties were observed for a gel sample with the addition of Reishi extract. This study found that a very important property of mushroom extracts was their ability to reduce the irritating effect of washing gels; the analyzed extracts used in a concentration of 1% reduced the irritating effect of the product by up to 20%. The conducted research confirmed that the analyzed mushroom extracts contain a number of bioactive substances that have a positive effect on skin and may be a valuable active ingredient in cosmetic products.

Author Contributions: Conceptualization, A.Z., K.M.-L., M.Z.-D., T.B. and Z.N.-Ł.; methodology, A.Z., M.W., K.M.-L., M.Z.-D., T.B., Z.N.-Ł., D.S. and I.S.; formal analysis, A.Z., M.W., K.M.-L., M.Z.-D., T.B., Z.N.-Ł., D.S. and I.S.; investigation, A.Z., M.W., K.M.-L., M.Z.-D., T.B., Z.N.-Ł., D.S. and I.S.; writing—original draft preparation, A.Z., M.W., K.M.-L., M.Z.-D., T.B., Z.N.-Ł., D.S. and I.S.; writing—review and editing, A.Z., M.W., K.M.-L., M.Z.-D., T.B., Z.N.-Ł., D.S. and I.S.; visualization, A.Z., K.M.-L., M.Z.-D., T.B. and Z.N.-Ł.; supervision, A.Z., M.W., K.M.-L., M.Z.-D., T.B., Z.N.-Ł., D.S. and I.S. All authors have read and agreed to the published version of the manuscript.

Funding: This research received no external funding.

Institutional Review Board Statement: The analyzed extracts are well-known raw materials with a long history of use in the food and pharmaceutical industries. The toxicological profile does not indicate any risks and the test was carried out by topical application of analyzed extracts to the skin in a low concentration, without disturbing the continuity of the epidermis. The substances have a CAS number and are not subject to any restrictions in the cosmetics industry.

Informed Consent Statement: Informed consent was obtained from all subjects involved in the study.

Data Availability Statement: Data are contained within the article.

Conflicts of Interest: The authors declare no conflict of interest.

Sample Availability: Not available.

References

1. Dini, I.; Laneri, S. The new challenge of green cosmetics: Natural food ingredients for cosmetic formulations. *Molecules* **2021**, *26*, 3921. [CrossRef] [PubMed]
2. Nadim, S.D.K.; Jani, J.M. Millenial'sbehaviour and attitude towards natural cosmetics: A case study in university Malysia Terengganu. *UMT J. Undergrad. Res.* **2021**, *3*, 63–74. [CrossRef]
3. Bom, S.; Fitas, M.; Martins, A.M.; Pinto, P.; Ribeiro, H.M.; Marto, J. Replacing Synthetic Ingredients by Sustainable Natural Alternatives: A Case Study Using Topical O/W Emulsions. *Molecules* **2020**, *25*, 4887. [CrossRef] [PubMed]
4. Kaličanin, B.; Velimirović, D. A Study of the Possible Harmful Effects of Cosmetic Beauty Products on Human Health. *Biol. Trace Elem. Res.* **2016**, *170*, 476–484. [CrossRef]
5. Panico, A.; Serio, F.; Bagordo, F.; Grassi, T.; Idolo, A.; Giorgi, M.; Guido, M.; Congedo, M.; Donno, A. Skin safety and health prevention: An overview of chemicals in cosmetic products. *J. Prev. Med. Hyg.* **2019**, *60*, E50–E57. [CrossRef]
6. Sułkowska-Ziaja, K.; Grabowska, K.; Apola, A.; Kryczyk-Poprawa, A.; Muszyńska, B. Mycelial culture extracts of selected wood-decay mushrooms as a source of skin-protecting factors. *Biotechnol. Lett.* **2021**, *43*, 1051–1061. [CrossRef]
7. Rijal, R.; Rana, M.; Srivastava, S. Biochemical Composition, Nutritional Values and Medicinal Properties of Some Edible Mushrooms: A Review. *Environ. Ecol.* **2021**, *39*, 358–369.
8. Patel, S.; Goyal, A. Recent developments in mushrooms as anti-cancer therapeutics: A review. *3 Biotech* **2012**, *2*, 1–15. [CrossRef]
9. Nowakowski, P.; Markiewicz-Żukowska, R.; Bielecka, J.; Mielcarek, K.; Grabia, M.; Socha, K. Treasures from the forest: Evaluation of mushroom extracts as anti-cancer agents. *Biomed. Pharmacother.* **2021**, *143*, 112106. [CrossRef]
10. Muszyńska, B.; Grzywacz-Kisielewska, A.; Kała, K.; Gdula-Argasińska, J. Anti-inflammatory properties of edible mushrooms: A review. *Food Chem.* **2018**, *15*, 373–381. [CrossRef]
11. Elsayed, E.A.; ElEnshasy, H.; Wadaan, M.A.; Aziz, R. Mushrooms: A potential natural source of anti-inflammatory compounds for medical applications. *Mediat. Inflamm.* **2014**, *2014*, 805841. [CrossRef]
12. Sánchez, C. Reactive oxygen species and antioxidant properties from mushrooms. *Synth. Syst. Biotechnol.* **2016**, *24*, 13–22. [CrossRef]
13. Kosanić, M.; Ranković, B.; Dašić, M. Mushrooms as possible antioxidant and antimicrobial agents. *Iran. J. Pharm. Res.* **2012**, *11*, 1095–1102.
14. Mwangi, R.W.; Macharia, J.M.; Wagara, I.N.; Bence, R.L. The antioxidant potential of different edible and medicinal mushrooms. *Biomed. Pharmacother.* **2022**, *147*, 112621. [CrossRef]
15. Hyde, K.D.; Bahkali, A.H.; Moslem, M.A. Fungi—An unusual source for cosmetics. *Fungal Divers.* **2010**, *43*, 1–9. [CrossRef]
16. Shin, Y.J.; Lee, S.C. Antioxidant activity and β-glucan contents of hydrothermal extracts from maitake (*Grifolafrondosa*). *Food Sci. Biotechnol.* **2014**, *23*, 277–282. [CrossRef]
17. He, X.; Wang, X.; Fang, J.; Chang, Y.; Ning, N.; Guo, H.; Huang, L.; Huang, X.; Zhao, Z. Polysaccharides in *Grifolafrondosa* mushroom and their health promoting properties: A review. *Int. J. Biol. Macromol.* **2017**, *101*, 910–921. [CrossRef]
18. Thongbai, B.; Rapior, S.; Hyde, K.D.; Wittstein, K.; Stadler, M. Hericiumerinaceus, an amazing medicinal mushroom. *Mycol. Prog.* **2015**, *14*, 91. [CrossRef]
19. Elkhateeb, W.A.; Elnahas, M.O.; Thomas, P.W.; Daba, G.M. To heal or not to heal? Medicinal mushrooms wound healing capacities. *ARC J. Pharm. Sci.* **2019**, *5*, 28–35.
20. Kim, J.W.; Kim, H.I.; Kim, J.H.; Kwon, O.C.; Son, E.S.; Lee, C.S.; Park, Y.J. Effects of ganodermanondiol, a new melanogenesis inhibitor from the medicinal mushroom *Ganoderma lucidum*. *Int. J. Mol. Sci.* **2016**, *17*, 1798. [CrossRef]
21. Shi, M.; Zhang, Z.; Yang, Y. Antioxidant and immunoregulatory activity of *Ganoderma lucidum* polysaccharide (GLP). *Carbohydr. Polym.* **2013**, *95*, 200–206. [CrossRef]
22. Li, L.D.; Mao, P.W.; Shao, K.D.; Bai, X.H.; Zhou, X.W. Ganoderma proteins and their potential applications in cosmetics. *Appl. Microbiol. Biotechnol.* **2019**, *103*, 9239–9250. [CrossRef]
23. Seweryn, A. Interactions between surfactants and the skin—Theory and practice. *Adv. Coll. Int. Sci.* **2018**, *256*, 242–255. [CrossRef]
24. Pezron, I.; Galet, L.; Clausse, D. Surface interaction between a protein monolayer and surfactants and its correlation with skin irritation by surfactants. *J. Colloid Interface Sci.* **1996**, *180*, 285–289. [CrossRef]
25. Rosen, M.J. *Surfactants and Interfacial Phenomena*, 3rd ed.; John Wiley & Sons Inc.: New York, NY, USA, 2006.

26. Moore, P.N.; Puvvada, S.; Blankschtein, D. Role of the Surfactant Polar Head Structure in Protein-Surfactant Complexation: Zein Protein Solubilization by SDS and by SDS/C12En Surfactant Solutions. *Langmuir* **2003**, *19*, 1009–1016. [CrossRef]
27. Dasilva, S.C.; Sahu, R.P.; Konger, R.L.; Perkins, S.M.; Kaplan, M.H.; Travers, J.B. Increased skin barrier disruption by sodium lauryl sulfate in mice expressing a constitutively active STAT6 in T cells. *Arch. Dermatol. Res.* **2012**, *304*, 65–71. [CrossRef]
28. Moore, P.N.; Puvvada, S.; Blankschtein, D. Challenging the surfactant monomer skin penetration model: Penetration of sodium dodecyl sulfate micelles into the epidermis. *J. Cosmet. Sci.* **2003**, *54*, 29–46. [PubMed]
29. Nielsen, G.D.; Nielsen, J.B.; Andersen, K.E.; Grandjean, P. Effects of industrial detergents on the barrier function of human skin. *Int. J. Occup. Environ. Health* **2000**, *6*, 138–142. [CrossRef] [PubMed]
30. Faucher, J.A.; Goddard, E.D. Interaction of keratinous substrates with sodium lauryl sulfate: I. Sorption. *J. Soc. Cosmet. Chem.* **1978**, *29*, 323–337.
31. Hall-Manning, T.J.; Holland, G.H.; Rennie, G.; Revell, P.; Hines, J.; Barratt, M.D.; Basketter, D.A. Skin irritation potential of mixed surfactant systems. *Food Chem. Toxicol.* **1998**, *36*, 233–238. [CrossRef]
32. Bujak, T.; Wasilewski, T.; Nizioł-Łukaszewska, Z. Role of macromolecules in the safety of use of body wash cosmetics. *Colloids Surf. B* **2015**, *135*, 497–503. [CrossRef]
33. Bujak, T.; Wasilewski, T.; Nizioł-Łukaszewska, Z. Effect of molecular weight of polyvinylpyrrolidone on the skin irritation potential and properties of body wash cosmetics in the coacervate form. *Pure Appl. Chem.* **2019**, *91*, 1521–1532. [CrossRef]
34. Bujak, T.; Nizioł-Łukaszewska, Z.; Ziemlewska, A. Amphiphilic cationic polymers as effective substances improving the safety of use of body wash gels. *Int. J. Biol. Macromol.* **2020**, *15*, 973–979. [CrossRef]
35. Tang, H.Y.; Yin, X.; Zhang, C.C.; Jia, Q.; Gao, J.M. Structure diversity, synthesis, and biological activity of cyathane diterpenoids in higher fungi. *Curr. Med. Chem.* **2015**, *22*, 2375–2391. [CrossRef]
36. Kawagishi, H.; Shimada, A.; Hosokawa, S.; Mori, H.; Sakamoto, H.; Ishiguro, Y.; Sakemi, S.; Bordner, J.; Kojima, N.; Furukawa, S. Erinacines E, F, and G, stimulators of nerve growth factor (NGF)-synthesis, from the mycelia of Hericiumerinaceum. *Tetrahedron Lett.* **1996**, *37*, 7399–7402. [CrossRef]
37. Kawagishi, H.; Ando, M.; Shinba, K.; Sakamoto, H.; Yoshida, S.; Ojima, F.; Sakemi, S.; Bordner, J.; Kojima, N.; Furukawa, S.; et al. Chromans, hericenones F, G and H from the mushroom *Hericium erinaceum*. *Phytochemistry* **1992**, *32*, 175–178. [CrossRef]
38. Kawagishi, H.; Ando, M.; Mizuno, T. Hericenone A and B as cytotoxic principles from the mushroom Hericiumerinaceum. *Tetrahedron Lett.* **1990**, *31*, 373–376. [CrossRef]
39. Mori, K.; Kikuchi, H.; Obara, Y.; Iwashita, M.; Azumi, Y.; Kinugasa, S.; Inatomi, S.; Oshima, Y.; Nakahata, N. Inhibitory effect of hericenone B from Hericiumerinaceus on collagen-induced platelet aggregation. *Phytomedicine* **2010**, *17*, 1082–1085. [CrossRef]
40. Kobayashi, S.; Tamanoi, H.; Hasegawa, Y.; Segawa, Y.; Masuyama, A. Divergent synthesis of bioactive resorcinols isolated from the fruiting bodies of Hericiumerinaceum: Total syntheses of hericenones A, B, and I, hericenols B–D, and erinacerins A and B. *J. Org. Chem.* **2014**, *79*, 5227–5238. [CrossRef]
41. Yang, F.; Wang, H.; Feng, G.; Zhang, S.; Wang, J.; Cui, L. Rapid identification of chemical constituents in Hericiumerinaceus based on LC-MS/MS metabolomics. *J. Food Qual.* **2021**, *2021*, 5560626. [CrossRef]
42. Friedman, M. Chemistry, nutrition, and health-promoting properties of *Hericiumerinaceus* (Lion's Mane) mushroom fruiting bodies and mycelia and their bioactive compounds. *J. Agric. Food Chem.* **2015**, *63*, 7108–7123. [CrossRef]
43. Kuo, Y.H.; Lin, T.W.; Lin, J.Y.; Chen, Y.W.; Li, T.J.; Chen, C.C. Identification of Common Liver Metabolites of the Natural Bioactive Compound Erinacine A, Purified from *Hericiumerinaceus* Mycelium. *Appl. Sci.* **2022**, *12*, 1201. [CrossRef]
44. Stojković, D.S.; Barros, L.; Calhelha, R.C.; Glamočlija, J.; Ćirić, A.; van Griensven, L.J.L.D.; Soković, M.; Ferreira, I.C.F.R. A detailed comparative study between chemical and bioactive properties of *Ganodermalucidum* from different origins. *Int. J. Food Sci. Nutr.* **2014**, *65*, 42–47. [CrossRef]
45. Hennicke, F.; Cheikh-Ali, Z.; Liebisch, T.; Maciá-Vicente, J.G.; Bode, H.B.; Piepenbring, M. Distinguishing commercially grown *Ganoderma lucidum* from Ganoderma lingzhi from Europe and East Asia on the basis of morphology, molecular phylogeny, and triterpenic acid profiles. *Phytochemistry* **2016**, *127*, 29–37. [CrossRef]
46. Yang, M.; Wang, X.; Guan, S.; Xia, J.; Sun, J.; Guo, H.; Guo, D.A. Analysis of triterpenoids in *Ganoderma lucidum* using liquid chromatography coupled with electrospray ionization mass spectrometry. *J. Am. Soc. Mass Spectrom.* **2007**, *18*, 927–939. [CrossRef]
47. Hanachi, P.; Golkho, S.H. Using HPLC to determination the composition and antioxidant activity of Berberis vulgaris. *Eur. J. Sci. Res.* **2009**, *29*, 47–54.
48. Khan, I.; Ullah, S.; Oh, D.H. Chitosan grafted monomethyl fumaric acid as a potential food preservative. *Carbohydr. Polym.* **2016**, *152*, 87–96. [CrossRef]
49. Özer, Z.; Kılıç, T.; Çarıkçı, S.; Yılmaz, H. Investigation of phenolic compounds and antioxidant activity of *Teucrium polium* L. decoction and infusion. *Balıkesir Üniversitesi Fen Bilimleri Enstitüsü Dergisi* **2018**, *20*, 212–218. [CrossRef]
50. Hraš, A.R.; Hadolin, M.; Knez, Ž.; Bauman, D. Comparison of antioxidative and synergistic effects of rosemary extract with α-tocopherol, ascorbyl palmitate and citric acid in sunflower oil. *Food Chem.* **2000**, *71*, 229–233. [CrossRef]
51. Allahveran, A.; Farokhzad, A.; Asghari, M.; Sarkhosh, A. Foliar application of ascorbic and citric acids enhanced 'Red Spur'apple fruit quality, bioactive compounds and antioxidant activity. *Physiol. Mol. Biol. Plants* **2018**, *24*, 433–440. [CrossRef]
52. Valu, M.-V.; Soare, L.C.; Sutan, N.A.; Ducu, C.; Moga, S.; Hritcu, L.; Boiangiu, R.S.; Carradori, S. Optimization of Ultrasonic Extraction to Obtain Erinacine A and Polyphenols with Antioxidant Activity from the Fungal Biomass of *Hericiumerinaceus*. *Foods* **2020**, *9*, 1889. [CrossRef] [PubMed]

53. Liang, C.; Tian, D.; Liu, Y.; Li, H.; Zhu, J.; Li, M.; Xin, M.; Xia, J. Review of the molecular mechanisms of *Ganoderma lucidum* triterpenoids: Ganoderic acids A, C2, D, F, DM, X and Y. *Eur. J. Med. Chem.* **2019**, *174*, 130–141. [CrossRef] [PubMed]
54. Shi, L.; Ren, A.; Mu, D.; Zhao, M. Current progress in the study on biosynthesis and regulation of ganoderic acids. *Appl. Microbiol. Biotechnol.* **2010**, *88*, 1243–1251. [CrossRef] [PubMed]
55. Singh, C.; Pathak, P.; Chaudhary, N.; Rathi, A.; Vyas, D. Ganoderma lucidum: Cultivation and Production of Ganoderic and Lucidenic Acid. In *Recent Trends in Mushroom Biology*; Global books Organisation: Delhi, India, 2021.
56. Cör, D.; Knez, Ž.; Knez Hrnčič, M. Antitumour, antimicrobial, antioxidant and antiacetylcholinesterase effect of *Ganoderma lucidum* terpenoids and polysaccharides: A review. *Molecules* **2018**, *23*, 649. [CrossRef]
57. Wang, Y.; Xu, B. Distribution of antioxidant activities and total phenolic contents in acetone, ethanol, water and hot water extracts from 20 edible mushrooms via sequential extraction. *Austin J. Nutr. Food Sci.* **2014**, *2*, 5.
58. Yeh, J.Y.; Hsieh, L.H.; Wu, K.T.; Tsai, C.F. Antioxidant properties and antioxidant compounds of various extracts from the edible basidiomycete *Grifolafrondosa* (Maitake). *Molecules* **2011**, *16*, 3197–3211. [CrossRef]
59. Rahman, M.A.; Al Masud, A.; Lira, N.Y.; Shakil, S. Proximate analysis, phtochemical screening and antioxidant activity of different strains of *Ganoderma lucidum* (Reishi mushroom). *Open J. Biol. Sci.* **2020**, *5*, 024–027. [CrossRef]
60. Dickson, F.M.; Lawrence, J.N.; Benford, D.J. Surfactant-induced cytotoxicity in cultures of human keratinocytes and a commercially available cell line (3T3). *Toxicol. Vitr.* **1993**, *7*, 381–384. [CrossRef]
61. Bujak, T.; Zagórska-Dziok, M.; Nizioł-Łukaszewska, Z. Complexes of ectoine with the anionic surfactants as active ingredients of cleansing cosmetics with reduced irritating potential. *Molecules* **2020**, *25*, 1433. [CrossRef]
62. Wu, J.Y.; Siu, K.C.; Geng, P. Bioactive ingredients and medicinal values of *Grifolafrondosa* (Maitake). *Foods* **2021**, *10*, 95. [CrossRef]
63. Kodama, N.; Komuta, K.; Nanba, H. Effect of Maitake (Grifolafrondosa) D-Fraction on the activation of NK cells in cancer patients. *J. Med. Food* **2003**, *6*, 371–377. [CrossRef]
64. Hong, L.; Weiyu, W.; Qin, W.; Shuzhen, G.; Lebin, W. Antioxidant and immunomodulatory effects of a α-glucan from fruit body of maitake (*Grifolafrondosa*). *Food Agric. Immunol.* **2013**, *24*, 409–418. [CrossRef]
65. Tang, S.C.; Yang, J.H. Dual effects of alpha-hydroxy acids on the skin. *Molecules* **2018**, *23*, 863. [CrossRef]
66. Bae, J.T.; Sim, G.S.; Lee, D.H.; Lee, B.C.; Pyo, H.B.; Choe, T.B.; Yun, J.W. Production of exopolysaccharide from mycelial culture of Grifolafrondosa and its inhibitory effect on matrix metalloproteinase-1 expression in UV-irradiated human dermal fibroblasts. *FEMS Microbiol. Lett.* **2005**, *251*, 347–354. [CrossRef]
67. Chang, H.C.; Yang, H.-L.; Pan, J.-H.; Korivi, M.; Pan, J.-Y.; Hsieh, M.-C.; Chao, P.-M.; Huang, P.-J.; Tsai, C.-T.; Hseu, Y.-C. *Hericium erinaceus* inhibits TNF-α-induced angiogenesis and ROS generation through suppression of MMP-9/NF-κB signaling and activation of Nrf2-mediated antioxidant genes in human EA. hy926 endothelial cells. *Oxidative Med. Cell. Longev.* **2015**, *2016*, 8257238. [CrossRef]
68. Wachtel-Galor, S.; Yuen, J.; Buswell, J.A.; Benzie, I.F. Ganoderma lucidum (Lingzhi or Reishi). In *Herbal Medicine: Biomolecular and Clinical Aspects*, 2nd ed.; CRC Press/Taylor & Francis: Boca Raton, FL, USA, 2011.
69. Abate, M.; Pepe, G.; Randino, R.; Pisanti, S.; Basilicata, M.G.; Covelli, V.; Bifulco, M.; Cabri, W.; D'Ursi, A.M.; Campiglia, P.; et al. *Ganoderma lucidum* ethanol extracts enhance re-epithelialization and prevent keratinocytes from free-radical injury. *Pharmaceuticals* **2020**, *13*, 224. [CrossRef]
70. Singleton, V.L.; Orthofer, R.; Lamuela-Raventos, R.M. Analysis of total phenols and other oxidation substrates and antioxidants by means of Folin–Ciocalteu reagent. *Meth. Enzymol.* **1999**, *299*, 152–178. [CrossRef]
71. Matejić, J.; Džamić, A.; Mihajilov-Krstev, T.; Ranđelović, V.; Krivošej, Z.; Marin, P. Total phenolic content, flavonoid concentration, antioxidant and antimicrobial activity of methanol extracts from three Seseli L. taxa. *Open Life Sci.* **2012**, *7*, 1116–1122. [CrossRef]
72. Sawangwan, T.; Wansanit, W.; Pattani, L.; Noysang, C. Study of prebiotic properties from edible mushroom extraction. *Agric. Nat. Resour.* **2018**, *52*, 519–524. [CrossRef]
73. Brand-Williams, W.; Cuvelier, M.E.; Berset, C.L.W.T. Use of a free radical method to evaluate antioxidant activity. *LWT-Food Sci. Technol.* **1995**, *28*, 25–30. [CrossRef]
74. Gaweł-Bęben, K.; Tomasz, B.; Nizioł-Łukaszewska, Z.; Antosiewicz, B.; Jakubczyk, A.; Karaś, M.; Rybczyńska, K. *Stevia Rebaudiana* Bert. leaf extracts as a multifunctional source of natural antioxidants. *Molecules* **2015**, *20*, 5468–5486. [CrossRef]
75. Benzie, I.F.F.; Strain, J.J. Ferric reducing/antioxidant power assay: Direct measure of total antioxidant activity of biological fluids and modified version for simultaneous measurement of total antioxidant power and ascorbic acid concentration. *Methods Enzymol.* **1999**, *299*, 15–27. [CrossRef]
76. Evangelista-Vargas, S.; Santiani, A. Detection of intracellular reactive oxygen species (superoxide anion and hydrogen peroxide) and lipid peroxidation during cryopreservation of alpaca spermatozoa. *Reprod. Domest. Anim.* **2017**, *52*, 819–824. [CrossRef]
77. Nizioł-Łukaszewska, Z.; Ziemlewska, A.; Bujak, T.; Zagórska-Dziok, M.; Zarębska, M.; Hordyjewicz-Baran, Z.; Wasilewski, T. Effect of fermentation time on antioxidant and anti-ageing properties of green coffee Kombucha ferments. *Molecules* **2020**, *25*, 5394. [CrossRef]
78. Deo, N.; Jockusch, S.; Turro, N.J.; Somasundaran, P. Surfactant interactions with zein protein. *Langmuir* **2003**, *19*, 5083–5088. [CrossRef]

molecules

MDPI

Article

Improvement in Solubility–Permeability Interplay of Psoralens from *Brosimum gaudichaudii* Plant Extract upon Complexation with Hydroxypropyl-β-cyclodextrin

Rúbia Darc Machado [1], Júlio C. G. Silva [1], Luís A. D. Silva [2], Gerlon de A. R. Oliveira [3], Luciano M. Lião [3], Eliana M. Lima [2], Mariana C. de Morais [4], Edemilson C. da Conceição [4] and Kênnia R. Rezende [1,*]

[1] Laboratório de Biofarmácia e Farmacocinética (BioPk), Faculdade de Farmácia, Universidade Federal de Goiás, Goiânia 74605-170, GO, Brazil; drubia@egresso.ufg.br (R.D.M.); jcesagonzaga@gmail.com (J.C.G.S.)

[2] Laboratório de Nanotecnologia Farmacêutica e Sistemas de Liberação de Fármacos (FarmaTec), Faculdade de Farmácia, Universidade Federal de Goiás, Goiânia 74605-170, GO, Brazil; luis.dantas@ufg.br (L.A.D.S.); emlima@ufg.br (E.M.L.)

[3] Laboratório de Ressonância Magnética Nuclear (LabRMN), Instituto de Química, Universidade Federal de Goiás, Goiânia 74605-170, GO, Brazil; gerlonribeiro@gmail.com (G.d.A.R.O.); lucianoliao@ufg.br (L.M.L.)

[4] Laboratório de PD&I de Bioprodutos, Faculdade de Farmácia, Universidade Federal de Goiás, Goiânia 74605-170, GO, Brazil; marianacmfarma@gmail.com (M.C.d.M.); edemilson_conceicao@ufg.br (E.C.d.C.)

* Correspondence: kennia@ufg.br; Tel.: +55-(62)3209-6470

Abstract: Psoralen (PSO) and 5-methoxypsoralen (5-MOP) are widely used drugs in oral photochemotherapy against vitiligo and major bioactive components of root bark extract of *Brosimum gaudichaudii* Trécul (EBGT), previously standardized by LC-MS. However, the exceptionally low water solubility of these psoralens can cause incomplete and variable bioavailability limiting their applications and patient adherence to treatment. Therefore, the purpose of this work was to investigate the effects of 2-hydroxypropyl-β-cyclodextrin (HP-β-CD) inclusion complex on the solubility and jejunal permeability of PSO and 5-MOP from EBGT. Characterization of inclusion complexes were evaluated by current methods in nuclear magnetic resonance studies on aqueous solution, Fourier transform infrared spectroscopy, thermal analysis, and scanning electron microscopy in solid state. Ex vivo rat jejunal permeability was also investigated and compared for both pure psoralens and plant extract formulation over a wide HP-β-CD concentration range (2.5 to 70 mM). Phase solubility studies of the PSO- and 5-MOP-HP-β-CD inclusion complex showed 1:1 inclusion complex formation with small stability constants ($K_c < 500\ M^{-1}$). PSO and 5-MOP permeability rate decreased after adding HP-β-CD by 6- and 4-fold for pure standards and EBGT markers, respectively. Nevertheless, the complexation with HP-β-CD significantly improved solubility of PSO (until 10-fold) and 5-MOP (until 31-fold). As a result, the permeability drop could be overcome by solubility augmentation, implying that the HP-β-CD inclusion complexes with PSO, 5-MOP, or EBGT can be a valuable tool for designing and developing novel oral drug product formulation containing these psoralens for the treatment of vitiligo.

Keywords: cyclodextrin; HP-β-CD; inclusion complexes; physical–chemical characterization; host–guest interaction; solubilization; ex vivo permeability; intestinal absorption; in vitro studies; furanocoumarins

Citation: Machado, R.D.; Silva, J.C.G.; Silva, L.A.D.; Oliveira, G.d.A.R.; Lião, L.M.; Lima, E.M.; de Morais, M.C.; da Conceição, E.C.; Rezende, K.R. Improvement in Solubility–Permeability Interplay of Psoralens from *Brosimum gaudichaudii* Plant Extract upon Complexation with Hydroxypropyl-β-cyclodextrin. *Molecules* **2022**, *27*, 4580. https://doi.org/10.3390/molecules27144580

Academic Editor: Arunaksharan Narayanankutty

Received: 9 June 2022
Accepted: 13 July 2022
Published: 19 July 2022

Publisher's Note: MDPI stays neutral with regard to jurisdictional claims in published maps and institutional affiliations.

Copyright: © 2022 by the authors. Licensee MDPI, Basel, Switzerland. This article is an open access article distributed under the terms and conditions of the Creative Commons Attribution (CC BY) license (https://creativecommons.org/licenses/by/4.0/).

1. Introduction

Psoralen (PSO) and 5-methoxypsoralen (5-MOP) are linear furocoumarins, also known psoralens, commonly employed in oral photochemotherapy against different skin disorders, such as psoriasis and vitiligo. Mostly, 5-MOP shows less acute side effects and a slightly higher tolerance in patients in comparison to other photosensitizers [1]. In this context, *Brosimum gaudichaudii* Trécul (Moraceae) has attracted a great deal of attention due to large accumulation of these psoralens in its roots. Other chemically related coumarins previously described for *B. gaudichaudii* include gaudichaudine, xanthyletin, luvangetin, (+)-(2′S,3′R)-1′-hydroxy-marmesin, marmesin, 1′,2′-dehydromarmesin, 8-methoxymarmesin, and marmesinin [2–4]. Furthermore, this species has long been used as folk medicine in vitiligo's treatment [5–7]. Recently, Quintão et al. (2019) brought additional evidence on eliciting properties of *B gaudichaudii* inducing melanocyte migration and skin pigmentation in vitro [8]. However, very low water solubility of psoralens can cause incomplete and variable bioavailability, limiting oral administration [9–14].

During both drug development and quality control process, complexation of poorly water-soluble phytochemicals with parent-cyclodextrin (CD) and their derivatives have emerged as a valuable tool to optimize pharmaceutical properties, such as solubility/permeability, stability, and toxicity and thus enhance drug delivery systems and quality control issues [15]. Thus, improving physicochemical properties of phytochemicals can enhance their biological activities, sensory properties, and effectively improve patient adherence to treatment. Particularly, CD-inclusion complexes with phenolic plant compounds can enhance the water solubility and mask the bitterness of catechins [16].

Currently, there are different techniques to obtain and characterize cyclodextrin inclusion complexes. The most used in the last few years include co-precipitation, kneading, super critical carbon dioxide, spray drying, and freeze drying, among others, with their pros and cons, as elsewhere discussed [17]. Characterization methods involve the use of several analytical techniques, mostly depending on the physical nature of the complex (liquid or solid state). Typically, data-based evidence is shown by physical or chemical property variations of the guest molecule due to the host-guest interactions. For solid state, the major analytical techniques are thermogravimetric analysis (TGA) and differential scanning calorimetry (DSC), together with scanning electron microscopy (SEM), Fourier-transform infrared spectroscopy (FTIR), powder X-ray diffraction (PXRD), and Raman spectroscopy [18]. Some of them, such as powder X-ray diffraction (PXRD), are usually not suitable for providing structural information when putative complexes are as an amorphous powder [17].

Psoralen inclusion complexes with heptakis-(2,3,6-tri-O-methyl)-β-cyclodextrin (TRIMEB), β-CD, heptakis-(2,6-di-O-methyl)-β-cyclodextrin (DM-β-CD), and hepatakis-(2,3,6-tri-O-methyl)-β-cyclodextrin (TM-β-CD) in solid state have been investigated and confirmed by spectroscopic and/or thermal analyses [19–21]. At present, oral administration of methylated β-cyclodextrin is limited by its potential cytotoxic properties [22–24]. On the other hand, drug complex formation with HP-β-CD is most favorable both when increasing solubility and decreasing toxicity. Thus, HP-β-CD has often been used as an alternative to parent β-CD, and its methylated derivatives, except the HP-β-CD impact on intestinal permeability, is still poorly understood [25].

In this sense, drug/CD complexes formation can increase drug solubility, but only the free guest molecules in the equilibrium with the complex itself are able to permeate biological membranes. It is well-known that cyclodextrins can modulate, improve, or hamper the permeability through biological barriers. Improved permeability can be seen for hydrophobic drugs (BCS class II) as the water-soluble inclusion complex may eliminate the unstirred water layer (UWL), facilitating drug transport to membrane surface. Instead, permeability decreases in an excess amount of cyclodextrin in oral dosage forms [23,26–28]. Therefore, to maximize drug absorption, the CD amount added into pharmaceutical preparation needs to be carefully evaluated.

From our previous studies [2], PSO and 5-MOP showed limited aqueous solubility, both being classified as class II drugs, according to the Biopharmaceutical Classification System (BCS), suggesting that their absorption is markedly affected by its aqueous solubility. Therefore, these psoralens are very promising drugs for cyclodextrin complexation. The purpose of this work was to evaluate and optimize their solubility/permeability by inclusion with HP-β-cyclodextrin, characterizing host-guest interactions between HP-β-CD, PSO, and 5-MOP from *B. gaudichaudii*.

2. Results and Discussion

2.1. Phase Solubility Studies

The phase solubility diagram obtained for the PSO- and 5-MOP-HP-β-CD inclusion complex is shown in Figure 1A. The intrinsic solubility (S_0) was found to be 0.273 ± 0.007 mM (n = 3) and 0.032 ± 0.001 mM (n = 3), respectively. The solubility of these psoralens linearly increased (r > 0.999) as a function of HP-β-CD concentration, and slopes were less than one, so the profile of the diagram corresponds to A_L type, indicating 1:1 molecular complex formation. The K_c calculated for PSO- and 5-MOP-HP-β-CD were 259.2 ± 29.1 M^{-1} (n = 3) and 449.9 ± 16.6 M^{-1} (n = 3), respectively, exhibiting a very weak interaction profile and values within the range previously stated (50–5000 M^{-1}) [29].

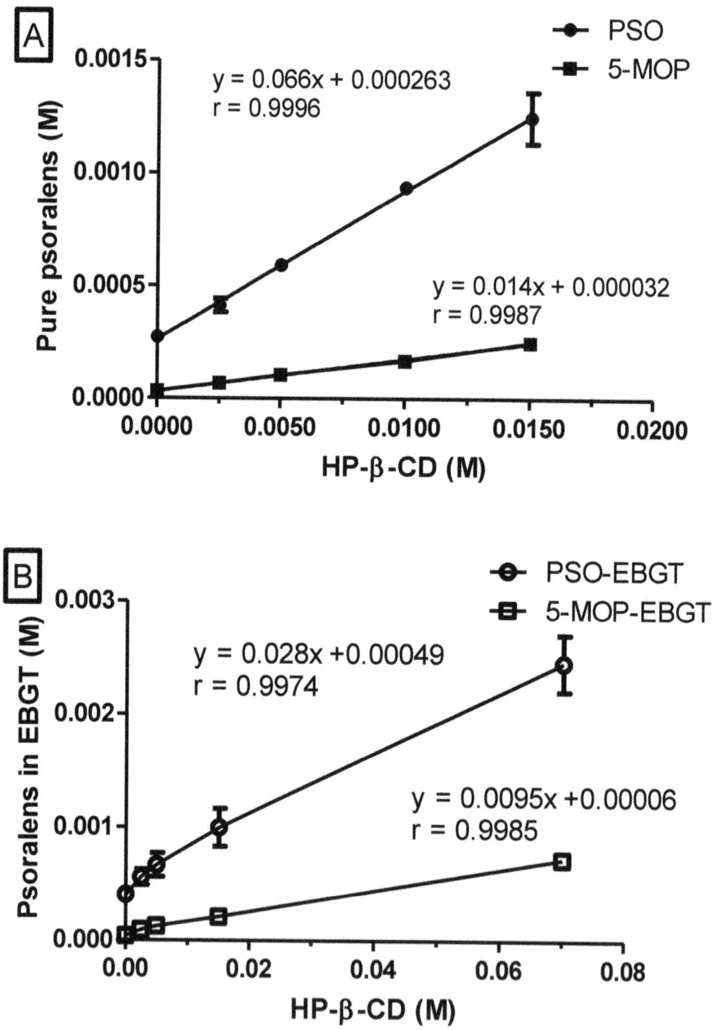

Figure 1. Solubility of psoralens standards (**A**) and psoralens in EBGT (**B**) as a function of HP-β-CD concentration in KRB buffer solutions (pH 7.4; 37 °C). Data shown as mean ± SD (error bars smaller than symbols); n = 3.

For the EBGT markers, the phase solubility diagram obtained for PSO- and 5-MOP-HP-β-CD inclusion complex was also A_L type, indicating 1:1 molecular complex formation, according to the Higuchi-Connor classification system (Figure 1B). However, The K_c was about 3.6- and 1.8-fold lower when compared to the respective pure compounds. The K_c values for PSO and 5-MOP in EBGT were 72.5 ± 0.9 M^{-1} (n = 3) and 251.2 ± 1.4 M^{-1} (n = 3), respectively. Such values indicate that, under in vivo conditions, these complexes will readily dissociate into separate components [25,29].

The intrinsic solubility (S_0) of PSO and 5-MOP in EBGT (9.0 mg/mL) was found to be 0.40 ± 0.02 mM (n = 3) and 0.038 ± 0.002 mM (n = 3), respectively. Thus, the experimental value of intrinsic psoralens solubility (S_0) was also affected by the plant matrix and demonstrates positive deviation of the extrapolated value from the phase-solubility line. As demonstrated by Kurkov, Ukhatskaya, and Loftsson (2011) [30], this behavior is common among poorly soluble drugs.

Such intrinsic solubility uncertainty can yield great fluctuations in the stability constant values obtained from phase-solubility diagrams. Thus, complexation efficiency (CE) is taken as a better parameter for comparing solubilization effects of HP-β-CD over psoralens at different environments, since it is independent of intrinsic solubility and the intercept [31].

CE values of PSO/ and 5-MOP/HP-β-CD complexes decreased from 0.071 ± 0.008 (n = 3) to 0.029 ± 0.000 (n = 3) and from 0.014 ± 0.001 (n = 3) to 0.010 ± 0.001 (n = 6), respectively, in EBGT. This means that around 7.1% of the HP-β-CD molecules in the solution are forming water soluble complexes with pure PSO, while only about 2.9% of cyclodextrin is forming complexes with PSO in EBGT [25,31,32]. For 5-MOP, there was no important change in the complexation efficiencies as a function of plant matrix. Taken together, results suggest a competitive role played by other phytochemicals in plant matrix, mainly with PSO, by cyclodextrin cavity.

To our knowledge, there is no previous report regarding the encapsulation of 5-MOP and *B. gaudichaudii* extract in HP-β-CD. Regarding the EBGT, it is worthwhile to consider that the herbal extracts are multicomponent mixtures where hundreds of compounds can coexist, albeit coming from a single plant. Thus, providing definitive experimental evidence of the inclusion complex formation of one specific phytochemical in their plant matrix with cyclodextrins is a rather complicated task, especially in solid state. Although the determination of phase-solubility profiles does not indicate whether EBGT markers form an inclusion complex with HP-β-CD, it can be a source of valuable information on how the HP-β-CD influences the psoralens solubility in plant matrix.

Additionally, phase-solubility profile indicated that the highest enhancement in the solubility was seen for 5-MOP-HP-β-CD inclusion complexes, followed by 5-MOP/HP-β-CD in plant extract. Next, it was seen for PSO complex as a pure compound and then for PSO-HP-β-CD in plant extract.

2.2. *Nuclear Magnetic Resonance Characterization*

Most CD applications take place in solution, so NMR spectra are especially important once allowing the characterization of complex formation to be in this state. The furanocoumarins signals observed in the ^{1}H NMR spectra are shown in Figure 2. Their signals were less intense than those from HP-β-CD (δ 3.0 to 5.5) due to the lower concentration of the coumarins.

Selective irradiation of H-10 hydrogen in 5-MOP and PSO (Figure 3) clearly shows an increase of the HP-β-CD (δ 3.0 to 4.0) signals due to the polarization transfer. This kind of interaction, typical of nuclei with spatial proximity, is strong evidence of complexation between furanocoumarins and HP-β-CD.

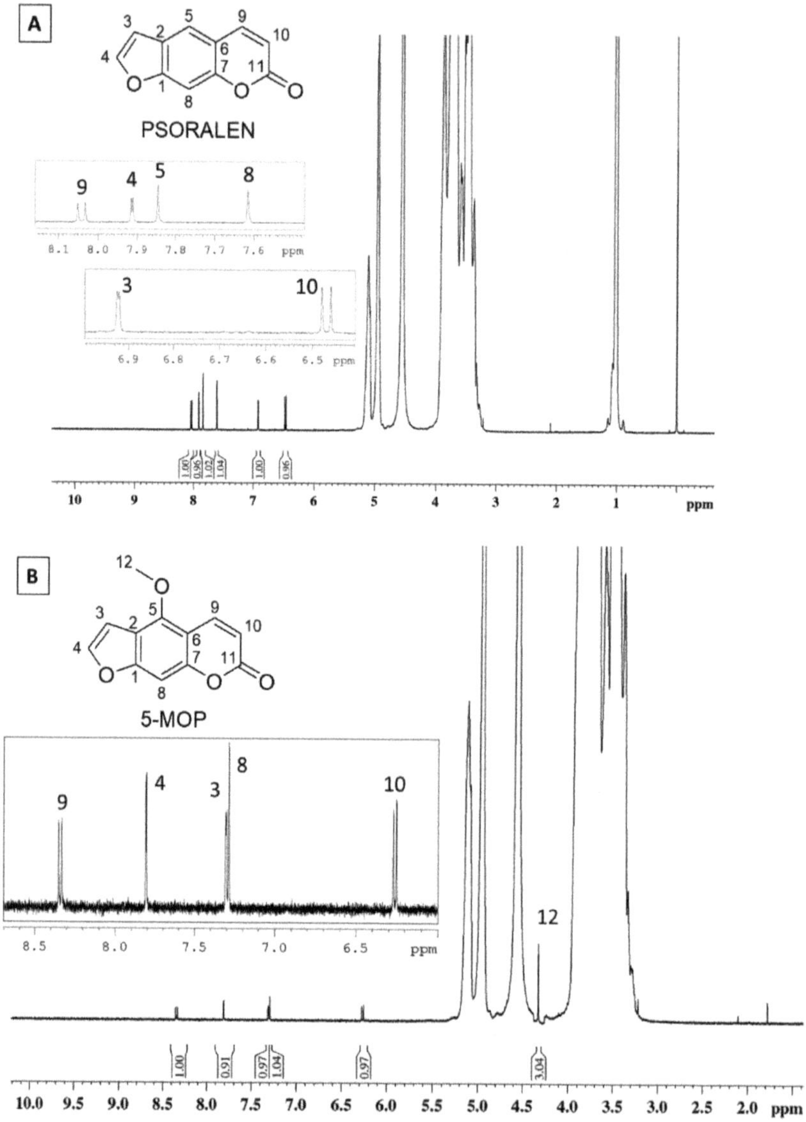

Figure 2. ^{1}H NMR spectrum of the (**A**) PSO- and (**B**) 5-MOP-HP-β-CD inclusion complex (HP-β-CD 15 Mm, D_2O, 500 MHz).

The complexation between the furanocoumarins and HP-β-CD was also corroborated by Diffusion Ordered Spectroscopy (DOSY). In this experiment, chemical shifts and diffusion coefficients were presented in two orthogonal directions that effectively separate out the NMR signals according to the diffusion coefficients of the components [33]. Furanocoumarins and HP-β-CD have a large difference in molecular weight, so they have different diffusion coefficients. However, in DOSY experiments (Figure 4), only one component signal was observed (0.45×10^{-10} m^2 s^{-1} of diffusion), in addition to the D_2O signal, demonstrating the host-guest complex between the furanocoumarins and HP-β-CD. In order to confirm the joint diffusion demonstrating complexation of these compounds, the values of the total diffusion-encoding pulse duration and the diffusion delays were varied (p30 = 1 to 2 ms; d20 = 50 to 100 ms), and no separation between furanocoumarins and HP-β-CD was observed.

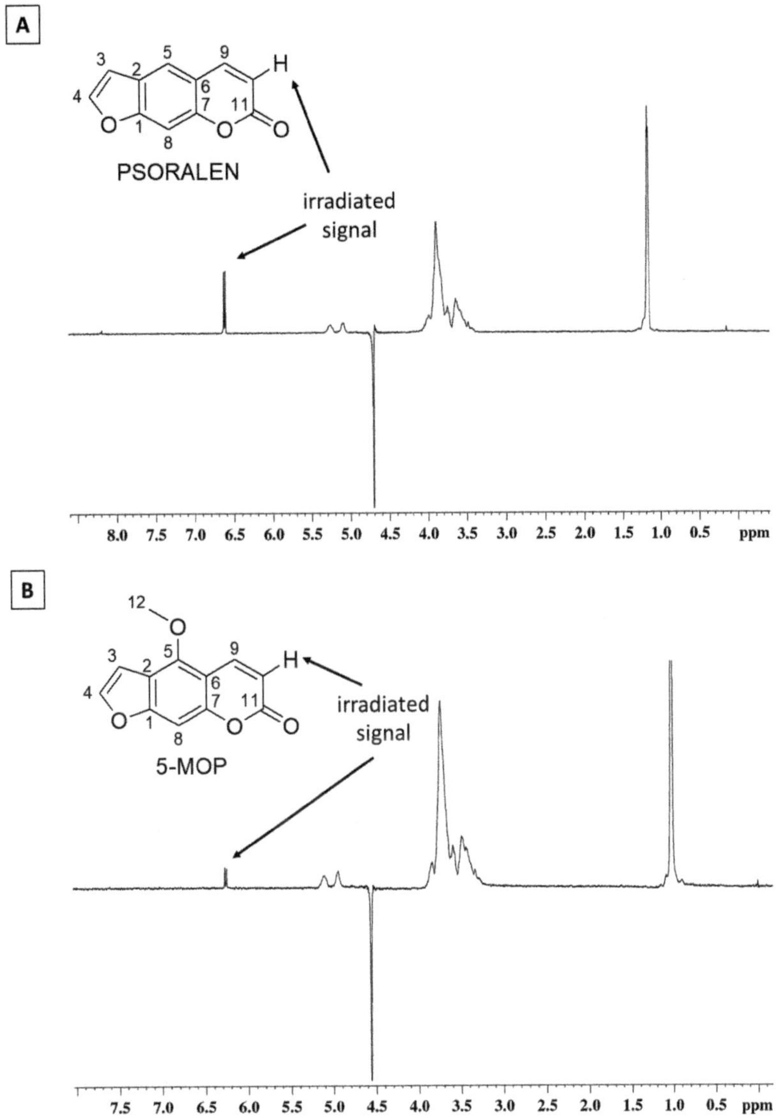

Figure 3. NOESY experiments for PSO (**A**) and 5-MOP (**B**) solutions. H-10 signals were irradiated, and polarization transfers to HP-β-CD were observed.

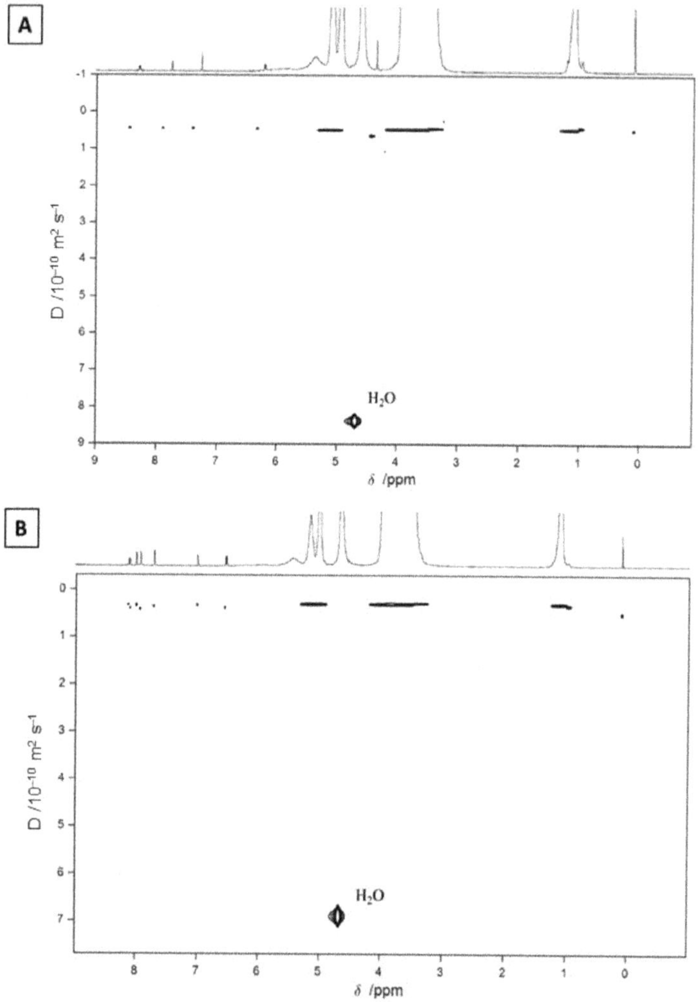

Figure 4. [1]H DOSY experiments for 5-MOP/HP-β-CD (**A**) and PSO/HP-β-CD (**B**) (500 MHz, D_2O).

2.3. Characterization of the PSO- and 5-MOP-HP-β-CD Inclusion Complex in Solid State

2.3.1. Thermogravimetry (TG)

The TGA curves of PSO, 5-MOP, HP-β-CD, PSO-HP-β-CD, or 5-MOP-HP-β-CD inclusion complex (1:1) and physical mixtures are shown in Figure 5. The TGA curves of HP-β-CD show a typical decomposition temperature at 343.0 °C (onset temperature = 331.7 °C), with a mass change of 88.95%. The overall mass change was 93.6%. Dehydration of HP-β-CD (loss of water molecules from the cavity) was observed in a range of 36.5 to 65.0 °C.

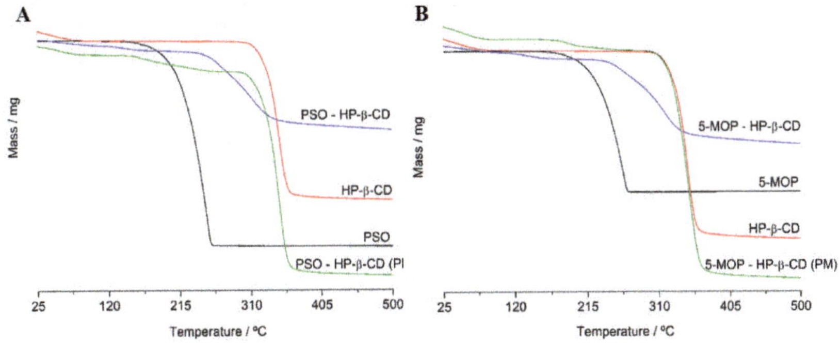

Figure 5. (**A**) TGA curves of PSO, HP-β-CD, physical mixture (PM), and PSO-HP-β-CD inclusion complex (1:1). (**B**) TGA curves of 5-MOP, HP-β-CD, physical mixture (PM), and 5-MOP-HP-β-CD inclusion complex (1:1).

The TGA curve of free PSO revealed a unique representative decomposition peak at the onset temperature of 223.2 °C (weight loss of 99.5%), while the 5-MOP decomposition peak occurred at the onset temperature of 231.9 °C (weight loss of 98.0%).

The physical mixture exhibited a similar three-step weight loss for PSO (30–65 °C, 223.2 °C, and 328.6 °C) and 5-MOP (30–65 °C, 231.9 °C, and 330.6 °C), corresponding to evaporation of water from HP-β-CD, drug decomposition, and HP-β-CD decomposition, respectively. This indicated that a physical mixture was not able to incorporate psoralens inside the nonpolar cavity of HP-β-CD.

On the other hand, the TGA curves of the PSO and 5-MOP in inclusion complex showed a higher degradation temperature compared to single components: shift from 223.18 to 271.50 °C for PSO-HP-β-CD and from 231.89 to 275.78 °C for 5-MOP-HP-β-CD inclusion complex. In general, the degradation of the guest molecule takes place at higher temperatures when complexation occurs, since the drug is protected by cyclodextrin [34].

2.3.2. Differential Scanning Calorimetry (DSC)

In particular, DSC is the most widely used technique for the evaluation of inclusion compound formation in solid-state [17]. Figure 6 shows DSC thermal curves of both pure psoralens (PSO, 5-MOP), guest molecule (HP-β-CD), the putative (1:1) inclusion complexes (PSO-HP-β-CD or 5-MOP-HP-β-CD), and their physical mixtures.

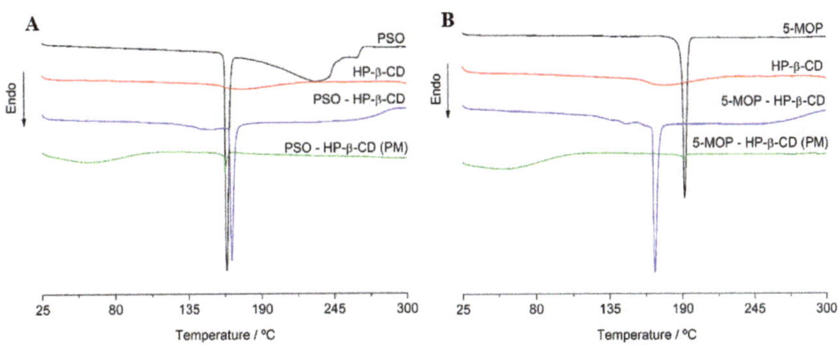

Figure 6. (**A**) Differential scanning calorimetry thermograms of PSO, HP-β-CD, physical mixture, and PSO-HP-β-CD inclusion complex (1:1). (**B**) Differential scanning calorimetry thermograms of 5-MOP, HP-β-CD, physical mixture (PM), and 5-MOP-HP-β-CD inclusion complex (1:1).

For the guest molecule spectra (HP-β-CD), a small endothermic state transition effect appears at 176.0 °C with an exceptionally low mass loss. The typical broad endothermic peak of HP-β-CD (around 341 °C), which is attributed to the decomposition of the HP-β-CD, could not be observed in the temperature range herein evaluated.

Free PSO and 5-MOP showed a well-defined endothermic peak in 163.8 °C and 190 °C, respectively, which corresponds to their melting points of crystal forms.

DSC thermal curve of the putative inclusion complexes (PSO-HP-β-CD or 5-MOP-HP-β-CD) showed temperature profiles differing from both host molecules, guest molecules, and their physical mixture, which still exhibited the endothermic peak of the drugs at the same melting points of crystal forms.

Notably, the spectra of PSO-HP-β-CD and 5-MOP-HP-β-CD did not show PSO or 5-MOP peaks correspondent to pure compounds, respectively. Instead, it showed a shift of melting peak of PSO (163.81 to 168.28 °C) and 5-MOP (190.02 to 168.11 °C), which is generally taken as evidence of complexation and/or amorphization of the inclusion complex in the solid state.

The disappearance of the melting peak of the drug or the shift and/or the flattening of the DSC profile in the melting point region of the crystalline drug is often taken as robust evidence of inclusion complex formation [35]. Additionally, it can also be seen as indicative of a residual presence of the drug with a partial reduction in the crystallinity, suggesting amorphization and/or incomplete inclusion in the guest molecule. Both findings are well documented by an overwhelming number of publications covered by a recent review article [17].

During drug development, the inclusion complex encapsulation ratio can be measured by a joint analytical approach, together with further studies, as reported for apigenin-HP-β-CD with -chitosan ternary complex formulation [36], including additional analysis, such as XRD evaluation and formulation dissolution. Here, the formation of the inclusion complex of psoralens and HP-β-CD were investigated by four different methods in solid state: Differential Scanning Calorimetry (DSC); Thermogravimetry (TG); Fourier transform infrared spectroscopy (FTIR), and Scanning electron microscopy (SEM), showing findings corroborating with the evidence of inclusion complex formation (see details on the following sections). Although evidenced in our earlier study [2], the EBGT showed amorphous properties, seen by the absence of peaks in X-ray diffraction spectra after the freeze-drying process, so that XRD would not be very suitable to provide useful structural information. It is worth remarking that the primary aim of our work was to investigate HP-β-CD complex formation with PSO and 5-MOP from *B. gaudichaudii* impacting on solubility and permeability through GT mucosa. When practical, such remarkable analytical progress may allow a more comprehensive evaluation of the prepared inclusion complex in order to prevent loss of formulation quality control, e.g., drug recrystallization.

2.3.3. Fourier Transform Infrared Spectroscopy (FTIR)

The FTIR spectrum of PSO, 5-MOP, HP-β-CD, PSO-HP-β-CD, or 5-MOP-HP-β-CD inclusion complex (1:1) and physical mixtures are shown in Figure 7. The FTIR spectral analysis of pure PSO showed principal peaks at wavenumbers 3155, 3120, and 3061 (aromatic C–H stretch); 1711 (C=O stretching vibration of unsaturated lactones); 1448; 1573 (C=C of ring); 1226 (C–O–C ether group); 1092–1130 (C–O stretching vibration); 1020 (O–C–C band); and 748 (out of plane C–H bend).

For 5-MOP, the values of absorption band integration were noted at wavenumbers 3074 and 3086 (aromatic C–H stretch); 2847 and 2958 (C–H stretch from methyl group); 1726 (C=O stretching vibration of unsaturated lactones); 1605; 1468 (C=C of ring); 1257 (asymmetric C–O–C stretch of aryl ethers); 1030–1121 (C–O stretching vibration); 1153 (O–C–C band); and 760 (out of plane C–H bend).

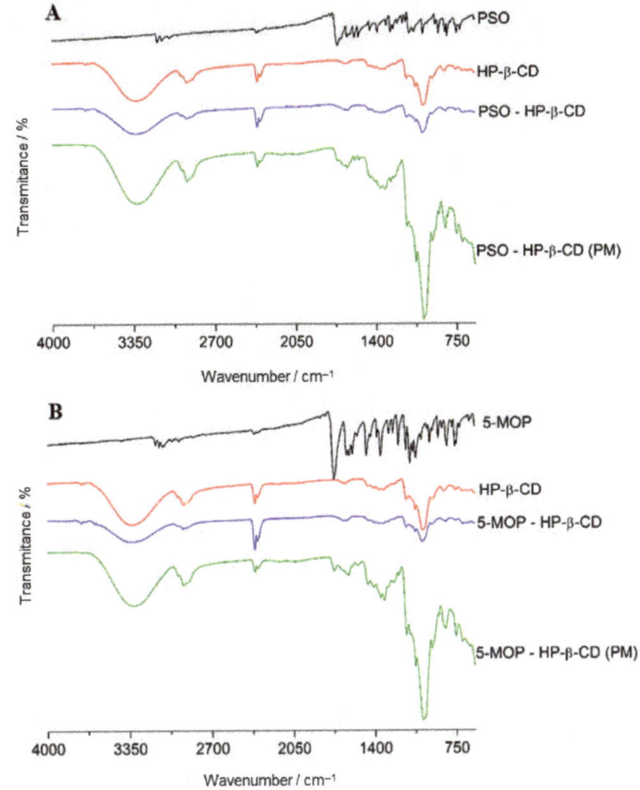

Figure 7. (**A**) ATR-FTIR spectra of PSO, HP-β-CD, physical mixture (PM), and PSO-HP-β-CD inclusion complex (1:1). (**B**) ATR-FTIR spectra of 5-MOP, HP-β-CD, physical mixture (PM), and 5-MOP-HP-β-CD inclusion complex (1:1).

The FTIR spectra of HP-β-CD exhibited intense absorption peak at 3326.6 cm^{-1}, related to the O–H stretching vibration. Other HP-β-CD characteristic bands were shown in the regions of 2925.5 cm^{-1} (C–H bond stretching vibrations), 1151 cm^{-1} (C−H bond stretching vibrations), and 1080 cm^{-1} (C–O bond stretching vibrations), as reported by Medarević et al. [37].

The spectrum of physical mixtures of these psoralens with HP-β-CD was almost a simple overlapping of pure components spectra. On the other hand, the spectrum of the inclusion complex of both PSO and 5-MOP was identical to that of pure HP-β-CD, suggesting that the drugs are completely inside the truncated-cone formed by cyclodextrin.

2.3.4. Scanning Electron Microscopy (SEM)

Representative SEM images of PSO, 5-MOP, HP-β-CD, PSO-, and 5-MOP-HP-β-CD inclusion complex (1:1) are shown in Figure 8. SEM analysis of HP-β-CD shows smooth and spherical shaped particles of different sizes. In contrast, pure psoralen was characterized by the presence of uneven and broken particles, while pure 5-MOP was characterized by the presence of crystals, showing a rectangular shape. The typical morphological characteristics of pure psoralens were changed in the inclusion complex that shows different sizes, with a predominantly smooth surface and with a porous inside for both PSO- and 5-MOP-HP-β-CD inclusion complexes. These results suggested the formation of complexes of psoralens with HP-β-CD in the solid-state. Our findings are in agreement with observations from formulation of genistein-HP-β-C-Poloxamer 188 ternary inclusion complex [38].

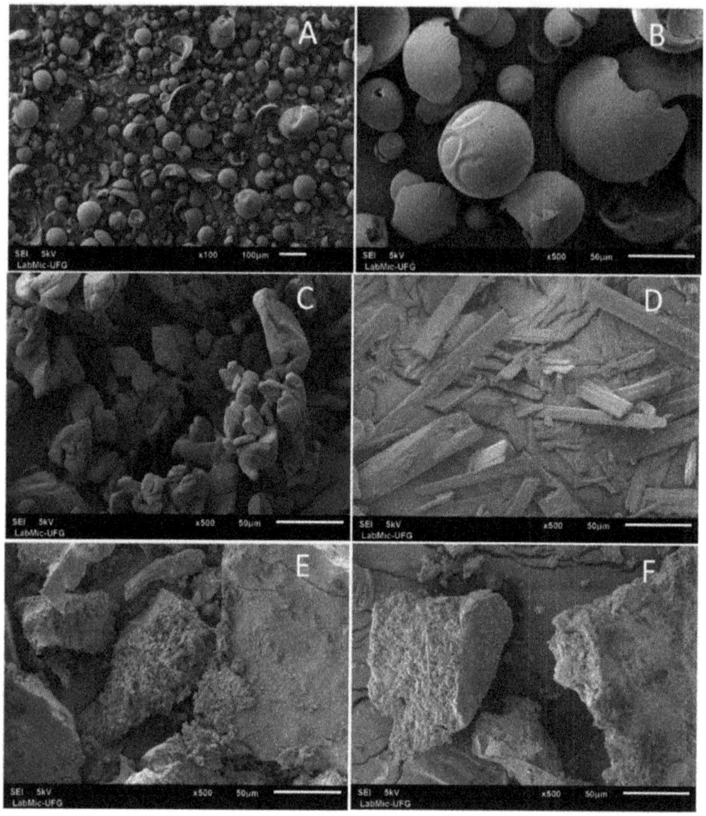

Figure 8. Scanning electron photomicrographs of (**A**) HP-β-CD; (**B**) HP-β-CD; (**C**) PSO; (**D**) 5-MOP; (**E**) PSO-HP-β-CD inclusion complex (1:1); (**F**) 5-MOP-HP-β-CD inclusion complex (1:1). 100× magnification (**A**) and 500× magnification (**B–F**).

2.4. Ex Vivo Oral Permeability Assessment

Permeability of psoralens was evaluated by using increasing concentrations of HP-β-CD by using rat jejunal ex vivo assay, previously validated showing excellent correlation to intestinal absorption in humans [39].

The effects of HP-β-CD on the viability of intestinal tissue segments mounted in the MTS-Snapwell system was evaluated during incubation time (120 min) at 37 °C. For this purpose, TEER values in the presence and absence of 20% (*w*/*w*) HP-β-CD were compared. Values of the TEER, after incubation, containing this cyclodextrin derivative ($29 \pm 7\ \Omega \cdot cm^2$, $n = 16$), did not differ ($p = 0.99$) from the values for control KRB buffer solution ($32 \pm 9\ \Omega \cdot cm^2$, $n = 16$). These results clearly suggest that the HP-β-CD, even at high concentration, does not affect the viability of jejunal tissue.

To permeate the enterocytes cells, most lipophilic drugs need to overcome the unstirred water layer (UWL) adjacent to mucosal epithelium. Depending on thickness, UWL becomes the main barrier, i.e., permeation becomes diffusion controlled. According to Loftsson [40], under unstirred in vitro condition, the UWL can be significantly thicker (until 1000 μm or more) than in a typical in vivo environment (up to about 100 μm). Therefore, it is important to monitor the relative resistance of the drug flux through UWL. This resistance can be examined through a comparison of the permeation on stagnated and stirring conditions [28].

In order to evaluate UWL contribution on permeability of psoralens, the assay was carried out on MTS-Snapwell plate with (60 rpm) and without stirring (0 rpm) during 120 min. Results showed no statistical difference on P_{app} of PSO (21.38 ± 3.81 vs. 20.49 ± 2.82 × 10^{-6} cm/s; $n = 6$; $p = 0.63$) or 5-MOP (11.16 ± 3.28 vs. 7.80 ± 1.91 × 10^{-6} cm/s; $n = 6$; $p = 0.052$), respectively, suggesting that membrane plays the main role in the permeation of psoralens, i.e., the UWL barrier is negligible. In general, hydrophilic CDs, such as HP-β-CD, do not enhance drug delivery through membranes if no UWL is present [41].

Regarding solubility of psoralens, it was significantly improved by HP-β-CD. By adding 70 mM HP-β-CD (*w*/*w*), PSO aqueous solubility, as pure compound or in plant extract, was improved from 0.27 ± 0.007 to 2.72 ± 0.03 mM and 0.40 ± 0.02 to 2.46 ± 0.03 mM, respectively, corresponding to a solubility increase of 10- and 6-fold (Figure 9A,B). Nevertheless, it is worth mentioning that, for the pure compound, a plateau was achieved upon completion of existing PSO (2.5 mg) in the HP-β-CD complexation media (5.0 mL) at 70 mM (Figure 9A). It is assumed that if more PSO is added by extrapolation, the expected value should be approximately 4.88 mM.

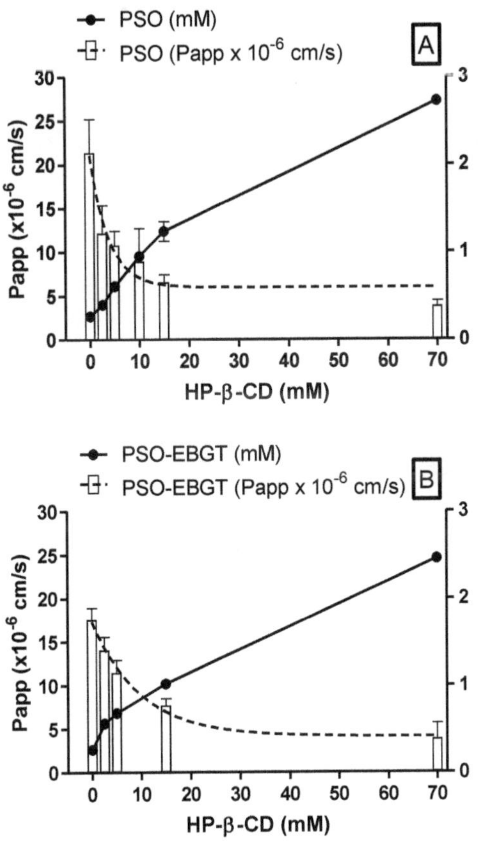

Figure 9. Ex vivo P_{app} and solubility of PSO standard (**A**) and PSO EBGT marker (**B**) as a function of increasing HP-β-CD concentration. P_{app} shown as mean ± SD ($n = 6$) and expressed as ×10^{-6} cm/s; solubility shown as mean ± SD ($n = 3$) and expressed as mM.

In contrast, PSO permeability as a pure compound and in plant extract decreased exponentially from 21.38 ± 3.81 and 17.56 ± 1.35 × 10^{-6} cm/s in KRB solution to 3.82 ± 0.62 and 3.86 ± 1.81 × 10^{-6} cm/s in HP-β-CD 70 mM, respectively, i.e., about 5.5-fold of the permeability rate (Figure 9A,B).

Aqueous solubility of 5-MOP, which was nearly insoluble as a free compound (uncomplexed), had drastically increased after its inclusion on HP-β-CD cavity. Aqueous solubility of 5-MOP as a pure compound and in plant extract was improved from 0.03 ± 0.001 to 0.93 ± 0.007 mM and 0.04 ± 0.002 to 0.72 ± 0.04 mM, respectively, by adding 70 mM HP-β-CD (*w*/*w*), corresponding to a solubility increase of 31- and 18-fold (Figure 10A,B).

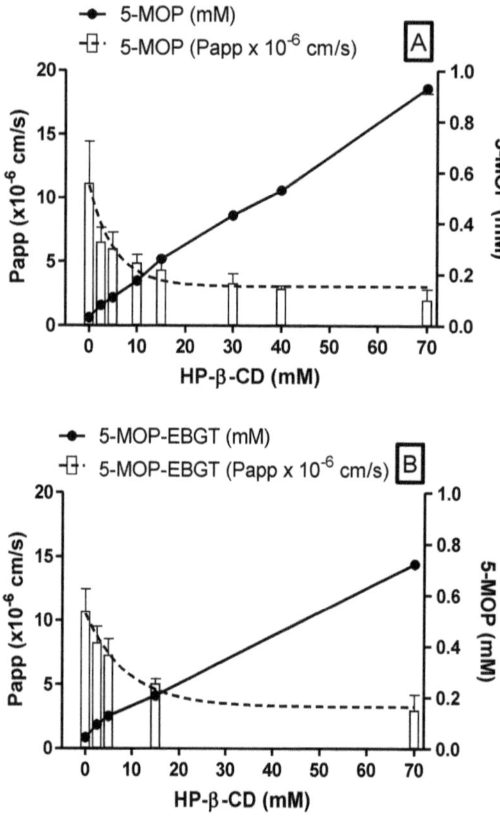

Figure 10. Ex vivo P_{app} and solubility of 5-MOP standard (**A**) and 5-MOP EBGT marker (**B**) as a function of increasing HP-β-CD concentration. P_{app} shown as mean ± SD (n = 6) and expressed as $\times 10^{-6}$ cm/s; solubility shown as mean ± SD (n = 3) and expressed as mM.

Ex vivo permeability studies of 5-MOP also showed an exponential decrease when increasing HP-β-CD concentration. Permeability of 5-MOP-HP-β-CD inclusion complex, both as a pure compound or as plant extract decreased from 11.16 ± 3.28 and 10.73 ± 1.73 × 10^{-6} cm/s in KRB buffer to 2.05 ± 0.81 and 3.04 ± 1.19 × 10^{-6} cm/s in HP-β-CD 70 mM solution, respectively (Figure 10A,B).

These observations are in agreement with findings from permeability studies of progesterone (BCS class II drug), where cyclodextrins may exponentially reduce permeability even for lipophilic compounds [28], indicating the existence of solubility-permeability interplay/tradeoff phenomenon. The exact mechanism by which this phenomenon occurs is not fully understood. For EBGT markers, P_{app} decreases slower than pure compounds. A higher permeability rate for EBGT was probably due to a higher free drug amount available for membrane permeation as a result of its higher solubility and lower affinity for HP-β-CD, when compared with pure compounds.

Introducing pharmaceutical formulations into the field of natural medicine is useful and promising once it offers effective and reliable delivery of medicinal phytochemicals [42]. Although various solubility-enabling formulations can be applied to the development of oral dosage forms, most of the pharmaceutical excipients (e.g., cosolvents, cyclodextrins, surfactants, hydrotropes, etc.) may affect the permeability in opposite directions [43,44]. However, its drug permeability impact depends on the amount of excipient used. Then, to maximize psoralens absorption, the CD amount added into pharmaceutical preparation needs to be carefully evaluated, aiming at the desired drug bioavailability intended for therapeutic benefits.

The solubility-permeability trade-off phenomenon, as well as defining their interplay, is an emerging challenge in today's drug development. The primary approach to overcome this issue is using the minimal excipient amounts sufficient to dissolve the drug dose throughout the GIT. In other words, when the intrinsic permeability of the drug is very high, it may be useful to waste some permeability to gain solubility [43,45].

In this context, permeability drawback of the obtained psoralens complexes could be overcome by their improved solubility, meaning that comparable plasma concentrations can be reached at a lower dose than uncomplexed ones, as also evidenced by the closely resembling AUC of 5-MOP (30,977 $\pm$ 4742 vs. 22,645 $\pm$ 8630 ng/mL·min, n = 3) herein obtained in preliminary in vivo studies in rats at about a 10-fold lower dose of EBGT in HP-β-CD 70 mM solution (1.5 g/kg vs. 0.12 mg/kg) (*data not shown*). For comparison purposes, P_{app} found for metoprolol (0.58 mM), a BCS-marker used as low/high permeability class boundary drug, was 6.21 $\pm$ 2.63 $\times$ 10^{-6} cm/s (n = 41) [2]. Hence, psoralens complexed with HP-β-CD in the concentration range of 2.5 to 15.0 mM, in either the EBGT's plant matrix or as pure compound, could be considered as high to moderate permeability compounds.

3. Materials and Methods

3.1. Solvents and Chemicals

Analytical grade 5-methoxypsoralen (≥99%) and psoralen (≥99%) were purchased from Sigma-Aldrich (St. Louis, MO, USA). Chemicals used for the Krebs-Ringer Bicarbonate buffers (KRB), NaCl, KCl, $CaCl_2.2H_2O$, NaH_2PO_4, $MgSO_4.7H_2O$, and $NaHCO_3$, were purchased from Sigma-Aldrich (St. Louis, MO, USA) as well. HPLC-grade methanol, acetonitrile, and trifluoroacetic acid were obtained from Merck (Darmstadt, Germany). HP-β-CD (Cavitron W7 HP5 Pharma HPB®) was kindly provided by Ashland.

3.2. Plant Material

Root bark extract of *B. gaudichaudii* Trécul (EBGT) was provided by Dr. Edemilson Cardoso da Conceição (Laboratório de PD&I de Bioprodutos, Federal University of Goiás, Goiânia, Brazil). Briefly, *B. gaudichaudii* root bark (voucher specimen number 45517) was cleaned, dried, crushed, ground in a mill, and percolated in 55% ethanol. The solution was evaporated to yield a solid content higher than 75% (*w*/*w*) [46], which was freeze-dried (−20 °C, 905 mTOR; Genesis SQ Super X-70, SP Scientifc®, Warminster, PA, USA). Plant extract was frozen just once and immediately before freeze-drying in the absence of cryoprotectant, as HP-β-CD itself is considered to have such properties. The PSO and 5-MOP content (%, *w*/*w*) in EBGT was determined by HPLC-DAD following a validated method described in the *HPLC-DAD method* section.

3.3. Preparation of PSO- and 5-MOP-HP-β-CD Inclusion Complex

The inclusion complexes of PSO, 5-MOP, and HP-β-CD were prepared by the freeze-drying method at a 1:1 molar ratio based on the results of phase solubility studies. PSO and 5-MOP were dissolved in 10 mL of HP-β-CD aqueous solution and individually shaken in borosilicate tubes (150 rpm, 37 °C, 7 d). Samples were centrifuged (4000 rpm, for 5 min) and filtered (Nylon filter; 0.45 μm) to remove any free drug. The solutions were then frozen and lyophilized (−20 °C, 905 mTOR) to obtain the inclusion complexes.

3.4. Physical Mixture (PM)

A physical mixture of PSO or 5-MOP and HP-β-CD (molar ratio of 1:1) was formed by mixing both pulverized powder components in a suitable flask, for about 10 min, until a uniform mixture was obtained and then passed through sieve (#100). The obtained products were stored in a desiccator at room temperature.

3.5. Characterization of the PSO-and 5-MOP-HP-β-CD Inclusion Complex in Solid State

Inclusion complex of PSO- and 5-MOP-HP-β-CD was investigated in solid state by means of Fourier transform infrared spectroscopy (FTIR) with an attenuated total internal reflection (ATR), differential scanning calorimetry (DSC), thermogravimetric analysis (TGA), and scanning electron microscopy (SEM).

3.5.1. Differential Scanning Calorimetry (DSC)

The DSC curves of PSO, 5-MOP, HP-β-CD, PSO-HP-β-CD, or 5-MOP-HP-β-CD inclusion complex (1:1) and physical mixtures were obtained using a DSC-60® calorimeter (Shimadzu, Kyoto, Japan) operated with a nitrogen atmosphere at 50 mL.min^{-1}, heating rate of 10 °C·min^{-1}, at the temperature range of 25–300 °C. Samples of about 1.0 ± 0.5 mg were sealed into closed aluminum crucible for the analyses.

3.5.2. Thermogravimetry (TG)

TG analyses were performed on a TGA/DTA-60® thermobalance (Shimadzu, Kyoto, Japan) under nitrogen atmosphere (flow rate of 50 mL min^{-1}). Samples (5.0 ± 1.0 mg) were placed in platinum crucibles and heated at a rate of 10 °C min^{-1}, from 25 to 500 °C. All analyses were performed on PSO, 5-MOP, HP-β-CD, PSO-HP-β-CD, or 5-MOP-HP-β-CD inclusion complex (1:1) and physical mixtures.

3.5.3. Fourier Transform Infrared Spectroscopy (FTIR)

FTIR spectra were obtained with a Varian 640-IR FTIR® spectrometer (Varian Inc., Palo Alto, CA, USA), equipped with a universal ATR sampling accessory. A small quantity of the individual compounds (PSO, 5-MOP, and HP-β-CD) or PSO-HP-β-CD and 5-MOP-HP-β-CD inclusion complex (1:1) and physical mixtures were placed on the diamond ATR crystal. Analysis was performed over a wavelength ranging from 4000 to 400 cm^{-1} with a resolution of 4 cm^{-1}.

3.5.4. Scanning Electron Microscopy (SEM)

Morphology of PSO, 5-MOP, HP-β-CD, PSO-, and 5-MOP-HP-β-CD inclusion complex were determined using scanning electron microscopy (SEM). All samples were fixed in stubs using double-sided carbon tape and coated with a thin layer of gold by a sputter-coated unit (Denton Vacuum Desk V, Moorestown, NJ, USA) to make them electrically conductive prior to imaging. In this experiment, a JEOL JSM-6610 (Thermo Scientific, Madison, WI, USA) scanning electron microscope was used. The surface topography was analyzed with a scanning electron microscope operated at an acceleration voltage of 5 kV, and obtained micrographs were examined at ×100, ×500, and ×1000.

3.6. Phase Solubility of the PSO- and 5-MOP-HP-β-CD Inclusion Complex

Phase solubility studies were performed in KRB buffer solutions (pH 7.4), as described by Higuchi and Connors [47]. Excess amounts of PSO (2.5 mg), 5-MOP (2.5 mg), and EBGT (45.0 mg) were individually shaken in borosilicate tubes (150 rpm, 37 °C, 7 d), containing HP-β-CD of different concentrations (2.5–70.0 mM; 5 mL). Samples were centrifuged (4000 rpm, 5 min) and filtered (Nylon filter; 0.45 μm) prior to the psoralens content analysis by the HPLC-DAD validated method. The K_c (stability constant) was calculated based on the phase solubility diagram according to equation:

$$K_c = slope/S_0\ (1 - slope) \quad (1)$$

where S_0 is the intrinsic solubility of PSO and 5-MOP (in the absence of HP-β-CD).

The complexation efficiency (CE) was determined by using the slope of the linear phase solubility diagram as follows:

$$CE = slope/(1 - slope) \quad (2)$$

3.7. Nuclear Magnetic Resonance Characterization (NMR)

NMR experiments were performed at 25 °C on a Bruker Avance III 500 spectrometer, operating at 11.75 T, observing ^1H at 500.13 MHz, and using D_2O as solvent. The spectrometer was equipped with a 5 mm triple inverse detection three-channel (^1H, ^2H, and X-nucleus—BBI) probe, tuned and matched for each sample analysis. ^1H NMR experiments were acquired using a single excitation pulse sequence (*zg* Bruker), 8 scans with an acquisition time of 2.62 s, 64 k time domain points distributed in a spectral width of 25 ppm, and recycle delay of 1 s. For each sample analyzed, shimming was adjusted for a 90° calibrated pulse length. The results were analyzed using TopSpin and GNAT software. ^1H and ^{13}C chemical shifts are given in δ (ppm), related to a sodium 3-trimethylsilyl-propionate-2,2,3,3-d_4 (TMSP) signal at δ 0.00 as an internal reference.

^1H NMR and Nuclear Overhauser Effect (NOESY) experiments were carried out with the same solutions from PSO and 5-MOP in HP-β-CD (15 mM) used in the phase solubility studies, then dissolved in D_2O. Samples were transferred to 5 mm NMR tubes, and each signal of the furanocoumarins observed in ^1H NMR spectra was irradiated in a NOE experiment (*selnogp* pulse sequence, Bruker, Billerica, MA, USA). Selective refocusing with a shaped pulse was used in the NOE experiments [48], with a mixing time of 500 ms and a relaxation delay of 2 s.

^1H diffusion measurements (DOSY—Diffusion Ordered Spectroscopy) were carried out, applying a stimulated-echo NMR pulse sequence (*stebpgp1s*, Bruker) and delay of 0.2 and 0.6 ms, respectively, for gradient recovery (d16) and duration of the gradient purge pulse (p19). The total diffusion-encoding pulse duration (p30) and diffusion delay Δ (d20) were, respectively, 2 ms and 100 ms. These values provided an attenuation of ~99% in the water signal with the maximum gradient strength. A total of 32 nominal gradient amplitudes ranging from 5 to 95% of the gradient magnitude were chosen to give equal steps in the gradient squared. The ^1H DOSY spectra were constructed with 32 k data points in the GNAT processing with one zero filling and line broadening factor of 1 Hz.

All prepared samples (PSO and 5-MOP in HP-β-CD 15 mM) had a high amount of KRB buffer and lower concentration of psoralens, resulting in low-intensity ^1H signals and difficulties to tune the probe. Thus, new samples without KRB were prepared to assess the diffusion of the furanocoumarins by DOSY as follows: 2.50 mg of PSO + 303 mg de HP-β-CD + 550 μL de D_2O; 2.62 mg de 5-MOP + 238 mg de HP-β-CD + 500 μL de D_2O; 210 mg of HP-β-CD + 450 μL de D_2O (blank).

3.8. Ex Vivo Permeability

Study protocol for female Wistar rats (12–14 weeks old) was approved by the Ethics Committee on Animal Use (#066/15) at the Federal University of Goiás, Goiás, Brazil. One week before starting the experiment, all animals were acclimatized at controlled conditions with regulated access to a standard food (22 ± 2 °C, 12-h light/dark cycle). After this, they fasted for 12 h with free access to water. All animals (body weight = 215 ± 12 g) were then anesthetized using a mixture of ketamine/xylazine (90.0/7.5 mg/kg). Rat jejunum was removed and immediately placed into an ice-cold KRB solution. The jejunum (20 mm) seromuscular layer was carefully stripped off from the membrane and mounted in the MTS-Snapwell system [39]. Next, unidirectional drug transport was measured from the apical to basolateral side (A–B).

Solutions from PSO, 5-MOP, and EBGT in HP-β-CD (2.5–70.0 mM) were prepared in the physiological media, i.e., KRB (pH 7.4), as described in phase solubility studies. Sample aliquots (400 μL) were placed in the donor compartment (apical side).

Drug transport was measured by collecting samples (500 μL) from the receiver chamber (basolateral side) at 0, 20, 30, 40, 60, 90, and 120 min. Receiver samples were replaced with equal volume of fresh KRB solution (37 °C). The system was maintained on a plate shaker (60 rpm) inside a humidified incubator (37 ± 0.5 °C) in a controlled atmosphere (95% O_2, and 5% CO_2), before and during the transport experiment. Samples were analyzed by the HPLC-DAD validated method.

Transport experiments were carried out under sink conditions, i.e., the concentration in the receiver chamber did not exceed 10% of the concentration in the donor chamber [49]. The change in donor and receiver concentrations was taken into consideration when the apparent permeability coefficient (P_{app}) was calculated using to the following equation:

$$P_{app} = dQ/dt(1/C_D\ A) \tag{3}$$

where dQ/dt is the appearance rate of the drug in the receiver compartment, C_D is the concentration of the drug in the donor chamber at each sample interval, and A is the surface area (0.1 cm^2) of the tissue segment. The initial concentration of the drug in the donor compartment was determined experimentally, and donor concentrations at each time point were obtained by the calculation of the mass balance [49,50].

Integrity of intestinal tissue segments mounted in the MTS-Snapwell system was assessed by measuring the transepithelial electrical resistance (TEER) using the KRB solution at 37 °C as the conducting medium. Viable intestinal segments (>30 $\Omega \cdot cm^2$) were mounted on the permeation apparatus and subjected to a preincubation period (30 min) under a controlled atmosphere (95% O_2 and 5% CO_2) before starting the experiments.

3.9. HPLC-DAD Method

Analysis of EBGT markers was carried out according to a previously validated method [2]. Briefly, chromatographic analysis was performed using an HPLC Agilent 1260 system (Agilent Technologies, Santa Clara, CA, USA) equipped with a photodiode array (DAD) and a fluorescence detector (FLD). The mobile phase consisted of acetonitrile as solvent A and 0.04% trifluoroacetic acid as solvent B. PSO and 5-MOP were assayed by the HPLC-DAD (Infinity 1260, Agilent Technologies) on a Phenyl-hexyl column (150 × 4.6 mm, 5 μm, Phenomenex) using 40% A for 7.0 min at a flow rate of 1.0 mL min^{-1} and detection at 245 nm. Sample injection volume was 25 μL.

3.10. Statistical Analysis

GraphPad Prism 5.0 and Statistica 7 softwares were used for statistical analysis. Results are expressed as means ± standard deviation (SD). Phase solubility studies were run in triplicate (n = 3), and P_{app} experiments were run in sextuplicate (n = 6). Data analysis was evaluated using a Student's t-test or a one-way analysis of variance (ANOVA), followed by Tukey's multiple comparison tests ($\alpha = 0.05$). Differences were scored as significant if $p < 0.05$. Thermal analyses and Fourier transform infrared spectroscopy FTIR figures were obtained by means of Origin 8 software (OriginLab Corporation, Northampton, MA, USA).

4. Conclusions

The complex formation of PSO and 5-MOP with HP-B-CD was shown by FTIR-ATR, DSC, TGA, NMR, and SEM analysis in both aqueous solution and solid state. Moreover, phase solubility studies indicated 1:1 inclusion complex formation with small stability constants, particularly for EBGT markers, which could explain the fact that P_{app} was less affected in plant extract than in pure compounds.

HP-β-CD was shown to be a potent aqueous solubility enhancer for PSO and 5-MOP, exerting only a mild negative effect on ex vivo jejunal permeability. PSO and 5-MOP solubility were improved (up to 10- and 31-fold, respectively) upon the addition of 70 mM

HP-β-CD to aqueous media while reducing permeability for psoralens and EBGT markers around to 6- to 4-fold, respectively, so that there remained at least a 2-fold improvement in the potential absorption. Therefore, using HP-β-CD as a host agent for PSO, 5-MOP, or EBGT inclusion complexes may result in similar drug plasma profiles at lower doses, decreasing possible collateral effects and toxicity. Thus, it could be a useful approach when designing future pharmaceutical formulation for the treatment of vitiligo. However, HP-β-CD amount should be carefully selected to ensure an optimum solubility and permeability balance, aiming at optimal drug bioavailability and intended therapeutic activity.

Author Contributions: Conceptualization, R.D.M. and K.R.R.; methodology, R.D.M., J.C.G.S., L.A.D.S., G.d.A.R.O. and M.C.d.M.; formal analysis, R.D.M., J.C.G.S., L.A.D.S., G.d.A.R.O., L.M.L., E.M.L., M.C.d.M., E.C.d.C. and K.R.R.; investigation, R.D.M. and K.R.R.; resources, K.R.R., L.M.L., E.M.L. and E.C.d.C.; data curation, K.R.R.; writing—original draft preparation, R.D.M.; writing—review and editing, K.R.R.; supervision, K.R.R.; project administration, K.R.R.; funding acquisition, K.R.R. All authors have read and agreed to the published version of the manuscript.

Funding: This research was funded by the Conselho Nacional de Desenvolvimento Científico e Tecnológico (CNPq) (grant number 407985/2013-1); Coordenação de Aperfeiçoamento de Pessoal de Nível Superior (CAPES), and Fundação de Amparo à Pesquisa do Estado de Goiás (FAPEG).

Institutional Review Board Statement: The study was conducted according to the Brazilian animal use guidelines and approved by the Institutional Ethics Committee on Animal Use (protocol # 066/15) at the Federal University of Goiás, Brazil.

Informed Consent Statement: Not applicable.

Data Availability Statement: Not applicable.

Acknowledgments: The authors would like to thank the Laboratório Multiusuário de Microscopia de Alta Resolução (LabMic), Federal University of Goiás, Brazil, for providing analytical facilities to perform SEM studies. The gift sample of HP-β-CD, provided by Ashland, for research work is gratefully acknowledged.

Conflicts of Interest: The authors declare no conflict of interest.

Sample Availability: Samples of the compounds are not available from the authors.

Abbreviations

EBGT, extract of *Brosimum gaudichaudii* Trécul; PSO, psoralen; 5-MOP, 5-methoxypsoralen; BCS, Biopharmaceutics Classification System; P_{app}, apparent permeability coefficient; HPLC, high performance liquid chromatography; DAD, diode-array detection; KRB, Krebs-Ringer Bicarbonate buffers; TEER, transepithelial electrical resistance; CD, cyclodextrin;HP-β-CD, hydroxypropyl-β-cyclodextrin; PM, physical mixture; FTIR, Fourier transform infrared spectroscopy; DSC, differential scanning calorimetry; TGA, thermogravimetric analysis; SEM, scanning electron microscopy; NMR, nuclear magnetic resonance; KC, stability constant; CE, complexation efficiency; UWL, Unstirred water layer; AUC, Area under curve; GIT, Gastrointestinal tract.

References

1. Pacifico, A.; Leone, G. Photo(Chemo)Therapy for Vitiligo. *Photodermatol. Photoimmunol. Photomed.* **2011**, *27*, 261–277. [CrossRef] [PubMed]
2. Machado, R.D.; de Morais, M.C.; da Conceição, E.C.; Vaz, B.G.; Chaves, A.R.; Rezende, K.R. Crude Plant Extract versus Single Compounds for Vitiligo Treatment: Ex Vivo Intestinal Permeability Assessment on *Brosimum gaudichaudii* Trécul. *J. Pharm. Biomed. Anal.* **2020**, *191*, 113593. [CrossRef] [PubMed]
3. Vilegas, W.; Pozetti, G.L.; Vilegas, J.H.Y. Coumarins from *Brosimum gaudichaudii*. *J. Nat. Prod.* **1993**, *56*, 416–417. [CrossRef]
4. De Monteiro, V.F.F.; Mathias, L.; Vieira, I.J.C.; Schripsema, J.; Braz-Filho, R. Prenylated Coumarins, Chalcone and New Cinnamic Acid and Dihydrocinnamic Acid Derivatives from *Brosimum gaudichaudii*. *J. Braz. Chem. Soc.* **2002**, *13*, 281–287. [CrossRef]
5. Jacomassi, E.; Moscheta, I.S.; Machado, S.R. Morfoanatomia e Histoquímica de *Brosimum gaudichaudii* Trécul (Moraceae). *Acta Bot. Bras.* **2007**, *21*, 575–597. [CrossRef]

6. Palhares, D.; de Paula, J.E.; dos Santos Silveira, C.E. Morphology of Stem and Subterranean System of *Brosimum gaudichaudii* (Moraceae). *Acta Bot. Hung.* **2006**, *48*, 89–101. [CrossRef]
7. Ribeiro, R.V.; Bieski, I.G.C.; Balogun, S.O.; Martins, D.T.d.O. Ethnobotanical Study of Medicinal Plants Used by Ribeirinhos in the North Araguaia Microregion, Mato Grosso, Brazil. *J. Ethnopharmacol.* **2017**, *205*, 69–102. [CrossRef]
8. Quintão, W.d.S.C.; Alencar-Silva, T.; Borin, M.d.F.; Rezende, K.R.; Albernaz, L.C.; Cunha-Filho, M.; Gratieri, T.; de Carvalho, J.L.; Sá-Barreto, L.C.L.; Gelfuso, G.M. Microemulsions Incorporating *Brosimum gaudichaudii* Extracts as a Topical Treatment for Vitiligo: In Vitro Stimulation of Melanocyte Migration and Pigmentation. *J. Mol. Liq.* **2019**, *294*, 111685. [CrossRef]
9. Feng, L.; Wang, L.; Jiang, X. Pharmacokinetics, Tissue Distribution and Excretion of Coumarin Components from *Psoralea corylifolia* L. in Rats. *Arch. Pharm. Res.* **2010**, *33*, 225–230. [CrossRef]
10. Balogh, A.; Merkel, U.; Looks, A.; Vollandt, R.; Wollina, U. Drug Monitoring of Orally Administered 8-Methoxypsoralen in Patients Treated with Extracorporeal Photopheresis. *Skin Pharmacol. Physiol.* **1998**, *11*, 258–265. [CrossRef]
11. Bräutigam, L.; Seegel, M.; Tegeder, I.; Schmidt, H.; Meier, S.; Podda, M.; Kaufmann, R.; Grundmann-Kollmann, M.; Geisslinger, G. Determination of 8-Methoxypsoralen in Human Plasma, and Microdialysates Using Liquid Chromatography–Tandem Mass Spectrometry. *J. Chromatogr. B* **2003**, *798*, 223–229. [CrossRef] [PubMed]
12. Kobyletzki, G.; Hoffmann, K.; Kerscher, M.; Altmeyer, P. Plasma Levels of 8-Methoxypsoralen Following PUVA-Bath Photochemotherapy. *Photodermatol. Photoimmunol. Photomed.* **1998**, *14*, 136–138. [CrossRef]
13. Shephard, S.E.; Nestle, F.O.; Panizzon, R. Pharmacokinetics of 8-Methoxypsoralen during Extracorporeal Photopheresis. *Photodermatol. Photoimmunol. Photomed.* **1999**, *15*, 64–74. [CrossRef]
14. Thomas, S.E.; O'Sullivan, J.; Balac, N. Plasma Levels of 8-Methoxypsoralen Following Oral or Bath-Water Treatment. *Br. J. Dermatol.* **1991**, *125*, 56–58. [CrossRef] [PubMed]
15. Suvarna, V.; Gujar, P.; Murahari, M. Complexation of Phytochemicals with Cyclodextrin Derivatives—An Insight. *Biomed. Pharmacother.* **2017**, *88*, 1122–1144. [CrossRef] [PubMed]
16. Astray, G.; Mejuto, J.C.; Simal-Gandara, J. Latest Developments in the Application of Cyclodextrin Host-Guest Complexes in Beverage Technology Processes. *Food Hydrocoll.* **2020**, *106*, 105882. [CrossRef]
17. Mura, P. Analytical Techniques for Characterization of Cyclodextrin Complexes in the Solid State: A Review. *J. Pharm. Biomed. Anal.* **2015**, *113*, 226–238. [CrossRef] [PubMed]
18. Cid-Samamed, A.; Rakmai, J.; Mejuto, J.C.; Simal-Gandara, J.; Astray, G. Cyclodextrins Inclusion Complex: Preparation Methods, Analytical Techniques and Food Industry Applications. *Food Chem.* **2022**, *384*, 132467. [CrossRef] [PubMed]
19. Caira, M.R.; Giordano, F.; Vilakazi, S.L. X-Ray Structure and Thermal Properties of a 1:1 Inclusion Complex Between Permethylated β-Cyclodextrin and Psoralen. *Supramol. Chem.* **2004**, *16*, 389–393. [CrossRef]
20. Pathak, K.; Kumari, S. Cavamax W7 Composite Psoralen Ethosomal Gel versus Cavamax W7 Psoralen Solid Complex Gel for Topical Delivery: A Comparative Evaluation. *Int. J. Pharm. Investig.* **2013**, *3*, 171. [CrossRef]
21. Vincieri, F.F.; Mazzi, G.; Mulinacci, N.; Bambagiotti-Alberti, M.; Dall' Acqua, F.; Vedaldi, D. Improvement of Dissolution Characteristics of Psoralen by Cyclodextrins Complexation. *Farmaco* **1995**, *50*, 543–547.
22. Loftsson, T.; Brewster, M.E.; Másson, M. Role of Cyclodextrins in Improving Oral Drug Delivery. *Am. J. Drug Deliv.* **2004**, *2*, 261–275. [CrossRef]
23. Jansook, P.; Ogawa, N.; Loftsson, T. Cyclodextrins: Structure, Physicochemical Properties and Pharmaceutical Applications. *Int. J. Pharm.* **2018**, *535*, 272–284. [CrossRef] [PubMed]
24. Kiss, T.; Fenyvesi, F.; Bácskay, I.; Váradi, J.; Fenyvesi, É.; Iványi, R.; Szente, L.; Tósaki, Á.; Vecsernyés, M. Evaluation of the Cytotoxicity of β-Cyclodextrin Derivatives: Evidence for the Role of Cholesterol Extraction. *Eur. J. Pharm. Sci.* **2010**, *40*, 376–380. [CrossRef] [PubMed]
25. Jambhekar, S.S.; Breen, P. Cyclodextrins in Pharmaceutical Formulations II: Solubilization, Binding Constant, and Complexation Efficiency. *Drug Discov. Today* **2016**, *21*, 363–368. [CrossRef] [PubMed]
26. Brewster, M.E.; Noppe, M.; Peeters, J.; Loftsson, T. Effect of the Unstirred Water Layer on Permeability Enhancement by Hydrophilic Cyclodextrins. *Int. J. Pharm.* **2007**, *342*, 250–253. [CrossRef]
27. Loftsson, T.; Jarho, P.; Másson, M.; Järvinen, T. Cyclodextrins in Drug Delivery. *Expert Opin. Drug Deliv.* **2005**, *2*, 335–351. [CrossRef]
28. Dahan, A.; Miller, J.M.; Hoffman, A.; Amidon, G.E.; Amidon, G.L. The Solubility-Permeability Interplay in Using Cyclodextrins as Pharmaceutical Solubilizers: Mechanistic Modeling and Application to Progesterone. *J. Pharm. Sci.* **2010**, *99*, 2739–2749. [CrossRef]
29. Takahashi, A.I.; Veiga, F.J.B.; Ferraz, H.G. Literature Review of Cyclodextrins Inclusion Complexes Characterization—Part I: Phase Solubility Diagram, Dissolution and Scanning Electron Microscopy. *Int. J. Pharm. Sci. Rev. Res.* **2012**, *12*, 531–539.
30. Kurkov, S.V.; Ukhatskaya, E.V.; Loftsson, T. Drug/Cyclodextrin: Beyond Inclusion Complexation. *J. Incl. Phenom. Macrocycl. Chem.* **2011**, *69*, 297–301. [CrossRef]
31. Loftsson, T.; Hreinsdóttir, D.; Másson, M. Evaluation of Cyclodextrin Solubilization of Drugs. *Int. J. Pharm.* **2005**, *302*, 18–28. [CrossRef] [PubMed]
32. Jansook, P.; Kurkov, S.V.; Loftsson, T. Cyclodextrins as Solubilizers: Formation of Complex Aggregates. *J. Pharm. Sci.* **2010**, *99*, 719–729. [CrossRef] [PubMed]

33. Jaski, J.M.; Barão, C.E.; Morais Lião, L.; Silva Pinto, V.; Zanoelo, E.F.; Cardozo-Filho, L. β-Cyclodextrin Complexation of Extracts of Olive Leaves Obtained by Pressurized Liquid Extraction. *Ind. Crop. Prod.* **2019**, *129*, 662–672. [CrossRef]
34. Takahashi, A.I.; Veiga, F.J.B.; Ferraz, H.G. A Literature Review of Cyclodextrin Inclusion Complexes Characterization—Part III: Differential Scanning Calorimetry and Thermogravimetry. *Int. J. Pharm. Sci. Rev. Res.* **2012**, *12*, 16–20.
35. Giordano, F.; Novak, C.; Moyano, J.R. Thermal Analysis of Cyclodextrins and Their Inclusion Compounds. *Thermochim. Acta* **2001**, *380*, 123–151. [CrossRef]
36. Jafar, M.; Khalid, M.S.; Alghamdi, H.; Amir, M.; Al Makki, S.A.; Alotaibi, O.S.; Al Rmais, A.A.; Imam, S.S.; Alshehri, S.; Gilani, S.J. Formulation of Apigenin-Cyclodextrin-Chitosan Ternary Complex: Physicochemical Characterization, In Vitro and In Vivo Studies. *AAPS PharmSciTech* **2022**, *23*, 1–11. [CrossRef] [PubMed]
37. Medarević, D.; Kachrimanis, K.; Djurić, Z.; Ibrić, S. Influence of Hydrophilic Polymers on the Complexation of Carbamazepine with Hydroxypropyl-β-Cyclodextrin. *Eur. J. Pharm. Sci.* **2015**, *78*, 273–285. [CrossRef] [PubMed]
38. Zafar, A.; Alruwaili, N.K.; Imam, S.S.; Alsaidan, O.A.; Alkholifi, F.K.; Alharbi, K.S.; Mostafa, E.M.; Alanazi, A.S.; Gilani, S.J.; Musa, A.; et al. Formulation of Genistein-HP β Cyclodextrin-Poloxamer 188 Ternary Inclusion Complex: Solubility to Cytotoxicity Assessment. *Pharmaceutics* **2021**, *13*, 1997. [CrossRef]
39. Da Silva, L.C.; Da Silva, T.L.; Antunes, A.H.; Rezende, K.R. A Sensitive Medium-Throughput Method to Predict Intestinal Absorption in Humans Using Rat Intestinal Tissue Segments. *J. Pharm. Sci.* **2015**, *104*, 2807–2812. [CrossRef]
40. Loftsson, T.; Duchene, D. Cyclodextrins and Their Pharmaceutical Applications. *Int. J. Pharm.* **2007**, *329*, 1–11. [CrossRef]
41. Loftsson, T.; Saokham, P.; Sá Couto, A.R. Self-Association of Cyclodextrins and Cyclodextrin Complexes in Aqueous Solutions. *Int. J. Pharm.* **2019**, *560*, 228–234. [CrossRef] [PubMed]
42. Ben-Shabat, S.; Yarmolinsky, L.; Porat, D.; Dahan, A. Antiviral Effect of Phytochemicals from Medicinal Plants: Applications and Drug Delivery Strategies. *Drug Deliv. Transl. Res.* **2020**, *10*, 354–367. [CrossRef] [PubMed]
43. Pawar, B.M.; Rahman, S.N.R.; Pawde, D.M.; Goswami, A.; Shunmugaperumal, T. Orally Administered Drug Solubility-Enhancing Formulations: Lesson Learnt from Optimum Solubility-Permeability Balance. *AAPS PharmSciTech* **2021**, *22*, 1–14. [CrossRef] [PubMed]
44. Porat, D.; Dahan, A. Active Intestinal Drug Absorption and the Solubility-Permeability Interplay. *Int. J. Pharm.* **2018**, *537*, 84–93. [CrossRef] [PubMed]
45. Fine-Shamir, N.; Dahan, A. Methacrylate-Copolymer Eudragit EPO as a Solubility-Enabling Excipient for Anionic Drugs: Investigation of Drug Solubility, Intestinal Permeability, and Their Interplay. *Mol. Pharm.* **2019**, *16*, 2884–2891. [CrossRef] [PubMed]
46. De Morais, M.C.; de Almeida, P.H.G.; Ferreira, N.L.O.; Arruda, R.L.; Borges, L.L.; de Freitas, O.; da Conceição, E.C. Validation of a Photostability Indicating Method for Quantification of Furanocoumarins from *Brosimum gaudichaudii* Soft Extract. *Rev. Bras. Farmacogn.* **2018**, *28*, 118–123. [CrossRef]
47. Higuchi, T.; Connors, K.A. Phase Solubility Techniques. *Adv. Anal. Chem. Instrum.* **1965**, *4*, 117–212.
48. Stott, K.; Stonehouse, J.; Keeler, J.; Hwang, T.-L.; Shaka, A.J. Excitation Sculpting in High-Resolution Nuclear Magnetic Resonance Spectroscopy: Application to Selective NOE Experiments. *J. Am. Chem. Soc.* **1995**, *117*, 4199–4200. [CrossRef]
49. Tavelin, S.; Gråsjö, J.; Taipalensuu, J.; Ocklind, G.; Artursson, P. Applications of Epithelial Cell Culture in Studies of Drug Transport. In *Epithelial Cell Culture Protocols*; Humana Press: Totowa, NJ, USA, 2002; Volume 188, pp. 233–272.
50. Oltra-Noguera, D.; Mangas-Sanjuan, V.; Centelles-Sangüesa, A.; Gonzalez-Garcia, I.; Sanchez-Castaño, G.; Gonzalez-Alvarez, M.; Casabo, V.-G.; Merino, V.; Gonzalez-Alvarez, I.; Bermejo, M. Variability of Permeability Estimation from Different Protocols of Subculture and Transport Experiments in Cell Monolayers. *J. Pharmacol. Toxicol. Methods* **2015**, *71*, 21–32. [CrossRef]

MDPI AG
Grosspeteranlage 5
4052 Basel
Switzerland
Tel.: +41 61 683 77 34

Molecules Editorial Office
E-mail: molecules@mdpi.com
www.mdpi.com/journal/molecules

Disclaimer/Publisher's Note: The statements, opinions and data contained in all publications are solely those of the individual author(s) and contributor(s) and not of MDPI and/or the editor(s). MDPI and/or the editor(s) disclaim responsibility for any injury to people or property resulting from any ideas, methods, instructions or products referred to in the content.

www.ingramcontent.com/pod-product-compliance
Lightning Source LLC
Chambersburg PA
CBHW061450220926
51850CB00004B/59
* 9 7 8 3 7 2 5 8 1 6 6 4 4 *